MORE TOOLS FOR YOUR MATHEMATICS SUCCESS

Student Study Pack

Get math help when YOU need it! The Student Study Pack provides you with the ultimate set of support resources to go along with your text. The Student Study Pack can be packaged with your textbook, and contains these invaluable tools:

- ☑ ***Student Solutions Manual***

  A printed manual containing full solutions to odd-numbered

- ☑ **Prentice Hall Math Tutor Center**

  Tutors provide one-on-one tutoring for any problem with an answer at the back of the book. You can contact the Tutor Center via a toll-free phone number, fax, or email.

- ☑ ***CD-ROM Lecture Series***

  A comprehensive set of textbook-specific CD-ROMs containing short video clips of the textbook objectives being reviewed and key examples being solved.

Tutorial and Homework Options

MYMATHLAB

MyMathLab can be packaged with your textbook. It is a complete online multimedia resource to help you succeed in learning. MyMathLab features:

- ☑ *The entire textbook online.*
- ☑ *Problem-solving video clips and practice exercises, correlated to the examples and exercises in the text.*
- ☑ *Online tutorial exercises with guided examples.*
- ☑ *Online homework and tests.*
- ☑ *Generates a personalized study plan based on your test results.*
- ☑ *Tracks all online homework, tests, and tutorial work you do in MyMathLab gradebook.*

Extensive and Varied Exercise Sets	*An abundant collection of exercises is included in an exercise set at the end of each section. Exercises are organized within categories. Your instructor will usually provide guidance on which exercises to work. The exercises in the first category, Practice Exercises, follow the same order as the section's worked examples.*	*The parallel order of the Practice Exercises lets you refer to the worked examples and use them as models for solving these problems.*	408
Practice Plus Problems	*This category of exercises contains more challenging problems that often require you to combine several skills or concepts.*	*It is important to dig in and develop your problem-solving skills. Practice Plus Exercises provide you with ample opportunity to do so.*	409

3 Review for Quizzes and Tests

Feature	***Description***	***Benefit***	***Page***
Mid-Chapter Check Points	*At approximately the midway point in the chapter, an integrated set of review exercises allows you to review the skills and concepts you learned separately over several sections.*	*By combining exercises from the first half of the chapter, the Mid-Chapter Check Points give a comprehensive review before you move on to the material in the remainder of the chapter.*	433
Chapter Review Grids	*Each chapter contains a review chart that summarizes the definitions and concepts in every section of the chapter. Examples that illustrate these key concepts are also included in the chart.*	*Review this chart and you'll know the most important material in the chapter!*	459-461
Chapter Review Exercises	*A comprehensive collection of review exercises for each of the chapter's sections follows the review grid.*	*Practice makes perfect. These exercises contain the most significant problems for each of the chapter's sections.*	461-464
Chapter Tests	*Each chapter contains a practice test with approximately 25 problems that cover the important concepts in the chapter. Take the practice test, check your answers, and then watch the Chapter Test Prep Video CD to see worked-out solutions for any exercises you miss.*	*You can use the chapter test to determine whether you have mastered the material covered in the chapter.*	465
Chapter Test Prep Video CDs	*These video CDs found at the back of your text contain worked-out solutions to every exercise in each chapter test.*	*The videos let you review any exercises you miss on the chapter test.*	465
Cumulative Review Exercises	*Beginning with Chapter 2, each chapter concludes with a comprehensive collection of mixed cumulative review exercises. These exercises combine problems from previous chapters and the present chapter, providing an ongoing cumulative review.*	*Ever forget what you've learned? These exercises ensure that you are not forgetting anything as you move forward.*	466

College Algebra Essentials

College Algebra Essentials

SECOND EDITION

Robert Blitzer
Miami Dade College

PEARSON
Prentice Hall

Upper Saddle River, NJ 07458

Library of Congress Cataloging-in-Publication Data
Blitzer, Robert.
College algebra essentials / Robert Blitzer.—2nd ed.
p. cm.
Includes index.
ISBN 0-13-220313-8
1. Algebra—Textbooks. I. Title.

QA152.3.B645 2007
512.9—dc22 2005057989

Acquisitions Editor: *Adam Jaworski*
Editor-in-Chief: *Sally Yagan*
Project Manager: *Dawn Murrin*
Production Editor: *Prepare, Inc.*
Assistant Managing Editor: *Bayani Mendoza de Leon*
Senior Managing Editor: *Linda Mihatov Behrens*
Executive Managing Editor: *Kathleen Schiaparelli*
Manufacturing Manager: *Alexis Heydt-Long*
Manufacturing Buyer: *Maura Zaldivar*
Director of Marketing: *Patrice Jones*
Marketing Manager: *Halee Dinsey*
Marketing Assistant: *Joon Won Moon*
Managing Editor, Digital Supplements: *Nicole M. Jackson*
Media Production Editor: *Tyler Suydam*
Director of Creative Services: *Paul Belfanti*
Creative Director: *Juan R. López*
Art Director: *Kenny Beck*
Assistant to the Art Director: *Dina Curro*
Art Editor: *Thomas Benfatti*
Interior Design: *ESM Design*
Manager, Cover Visual Research & Permissions: *Karen Sanatar*
Director, Image Resource Center: *Melinda Reo*
Manager, Rights and Permissions: *Zina Arabia*
Manager, Visual Research: *Beth Brenzel*
Image Permission Coordinator: *Debbie Hewitson*
Photo Researcher: *Melinda Alexander*
Composition: *Prepare, Inc.*
Art Studio: *Scientific Illustrators/Laserwords*
Cover Designers: Cowboy boot pepper—*Suzanne Behnke*; Tie dye pepper—*Dianne Densberger*; Pink pop pepper, Tattoo pepper—*Stacey Abraham*; Basketball shoe pepper: *Geoffrey Cassar*; Literary pepper—*Kristine Carney*
Cover Photos: Jalapeño pepper, *photographed by John E. Kelly*, © Getty images. Cowboy boot pepper—Cowboy boots, *photographed by Jules Frazier*, © Getty Images; Old wood, *photographed by Siede Preis*, © Getty images; Basketball shoe pepper—Basketball gym floor, *photographed by John Giustina*, © Getty images; Basketball shoe, *photographed by David Mager*, © Pearson Education.

Pearson Prentice Hall
Pearson Education, Inc.
Upper Saddle River, New Jersey 07458

Printed in the United States of America

10 9 8 7 6 5 4 3

ISBN 0-13-220313-8

Pearson Education LTD., *London*
Pearson Education Australia PTY, Limited, *Sydney*
Pearson Education Singapore, Pte. Ltd
Pearson Education North Asia Ltd, *Hong Kong*
Pearson Education Canada, Ltd., *Toronto*
Pearson Educación de Mexico, S.A. de C.V.
Pearson Education—Japan, *Tokyo*
Pearson Education Malaysia, Pte. Ltd

Contents

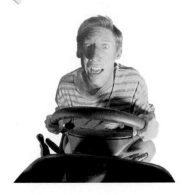

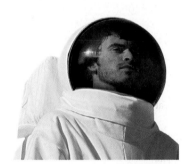

Chapter 2

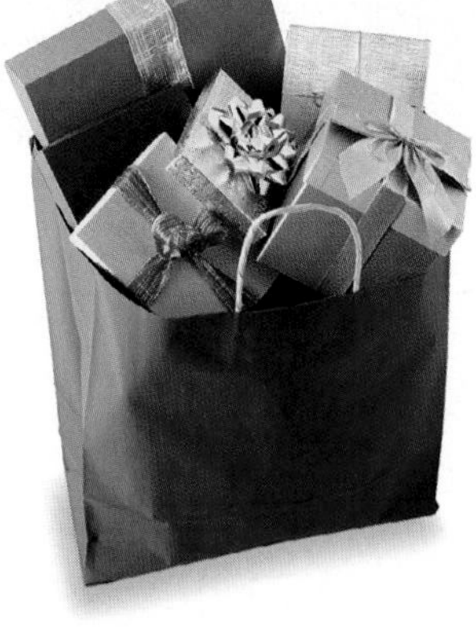

Chapter 3

Chapter 4

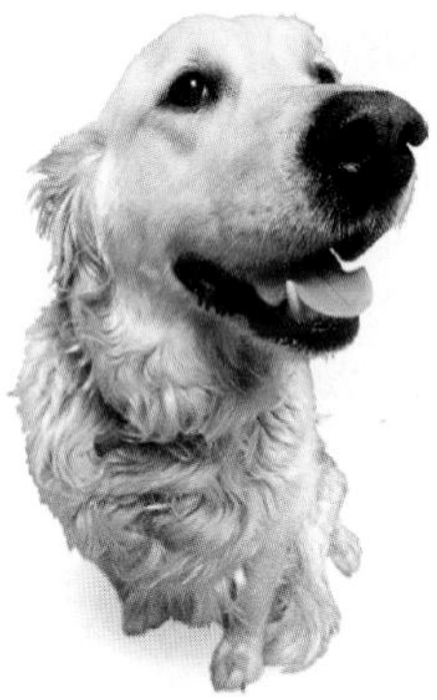

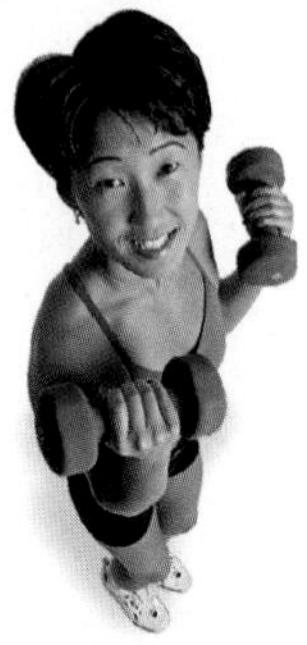

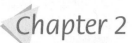

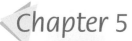

Chapter 5

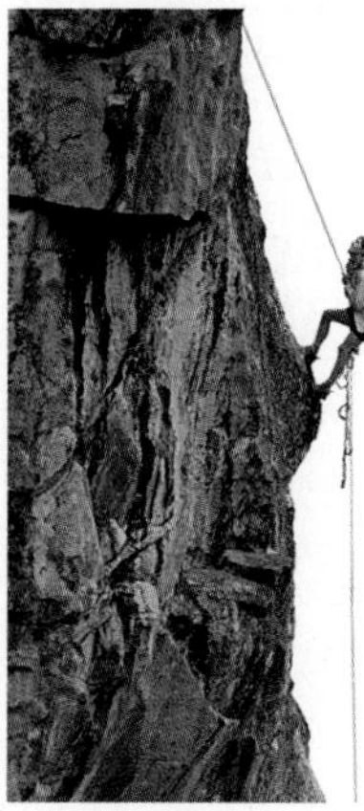

Systems of Equations and Inequalities 467

Shaded chapters available in Blitzer, *College Algebra, Fourth Edition*

Chapter 6

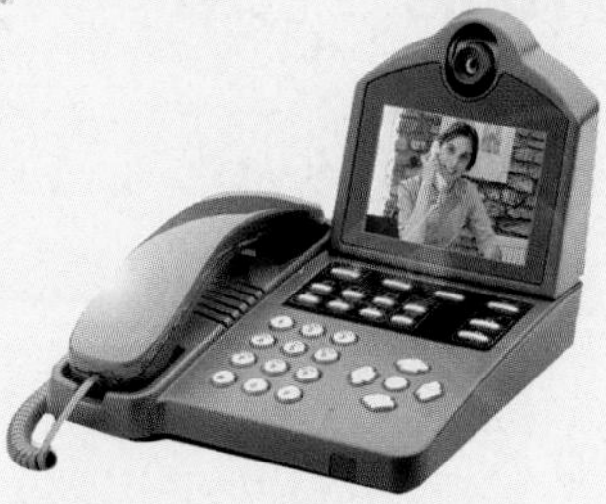

Matrices and Determinants 535

Chapter 7

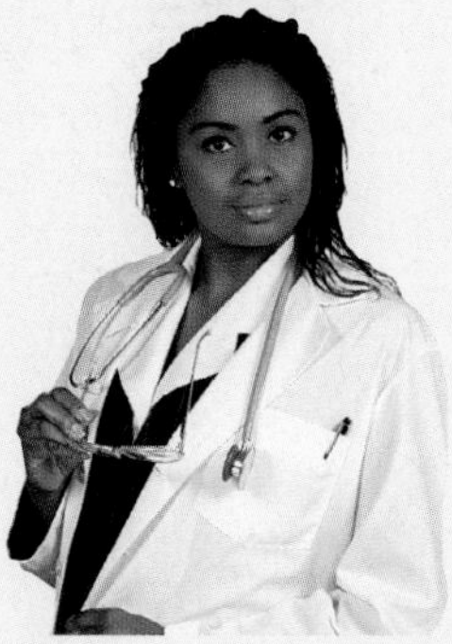

Conic Sections 603

Chapter 8

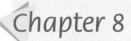

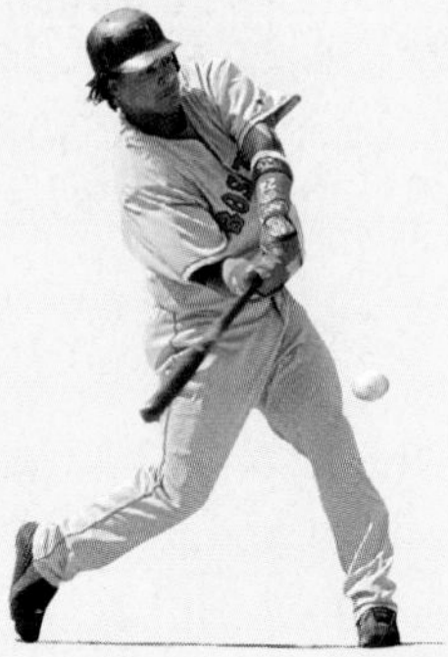

Appendix

Preface

I've written **College Algebra Essentials, Second Edition** to help diverse students, with different backgrounds and future goals, to succeed. The book has three fundamental goals:

1. To help students acquire a solid foundation in algebra, preparing them for other courses such as calculus, business calculus, and finite mathematics.
2. To show students how algebra can model and solve authentic real-world problems.
3. To enable students to develop problem-solving skills, while fostering critical thinking, within an interesting setting.

One major obstacle in the way of achieving these goals is the fact that very few students actually read their textbook. This has been a regular source of frustration for me and for my colleagues in the classroom. Anectodal evidence gathered over years highlights two basic reasons that students do not take advantage of their textbook:

- "I'll never use this information."
- "I can't follow the explanations."

I've written every page of the Second Edition with the intent of eliminating these two objections. The ideas and tools I've used to do so are described for the student in "A Brief Guide to Getting the Most from This Book" which appears inside the front cover.

A Note on the Essentials Version of College Algebra

College Algebra Essentials, Second Edition is a concise version of the **Fourth Edition** of **College Algebra**. The essentials version differs from the Fourth Edition only in terms of length. Chapter 6 (Matrices and Determinants), Chapter 7 (Conic Sections), and Chapter 8 (Sequences, Induction, and Probability) have been eliminated. The omitted chapters and their sections are highlighted for your reference in the Table of Contents. The essentials version provides a lighter, less expensive alternative to the Fourth Edition for instructors who do not cover the topics in Chapters 6, 7, and 8.

What's New in the Second Edition?

- **Practice Plus Exercises.** More challenging practice exercises that often require students to combine several skills or concepts have been added to the exercise sets. The 336 Practice Plus Exercises in the Second Edition, averaging between 8 and 10 of these exercises per exercise set, provide instructors with the option of creating assignments that take practice exercises to a more challenging level than in the previous edition.
- **Mid-Chapter Check Points.** At approximately the midway point in each chapter, an integrated set of review exercises allows students to review and assimilate the skills and concepts they learned separately over several sections. The 200 exercises that make up the Mid-Chapter Check Points, averaging 30 exercises per check point, are of a mixed nature, requiring students to discriminate which concepts or skills to apply. The Mid-Chapter Check Points should help students bring together the different objectives covered in the first half of the chapter before they move on to the material in the remainder of the chapter.

- **New Applications and Real-World Data.** I researched hundreds of books, magazines, almanacs, and online data sites to prepare the Second Edition. The Second Edition contains more new, innovative, applications, supported by data that extend as far up to the present as possible, than any previous revisions of the complete version of **College Algebra**.
- **More Than 1100 New Examples and Exercises.** In addition to the 336 Practice Plus Exercises and the 200 Mid-Chapter Check Points, the Second Edition contains more than 600 new exercises that appear primarily in the practice and application categories of the exercise sets.
- **Integration of Technology Using Graphical and Numerical Approaches to Problems.** New side-by-side features in the technology boxes connect algebraic solutions to graphical and numerical approaches to problems. Although the use of graphing utilites is optional, students can use the explanatory voice balloons to understand different approaches to problems even if they are not using a graphing utility in the course.
- **Increased Study Tip Boxes.** The book's Study Tip boxes offer suggestions for problem solving, point out common errors to avoid, and provide informal hints and suggestions. These invaluable hints appear in greater abundance in the Second Edition.
- **Chapter Test Prep Video CD.** Packaged at the back of the text, this video CD provides students with step-by-step solutions for each of the exercises in the book's chapter tests.

What Content and Organizational Changes Have Been Made to the Second Edition?

- **Section P.1 (Algebraic Expressions and Real Numbers)** now includes an early discussion of mathematical modeling and mathematical models, central themes of the book. Mathematical models now appear throughout Chapter P. A new discussion of intersection and union of sets paves the way for this notation to be used throughout the book.
- **Section P.2 (Exponents and Scientific Notation)** includes negative numbers in scientific notation, as well as an expanded discussion of converting from decimal to scientific notation.
- **Section P.6 (Rational Expressions)** takes the discussion of simplifying complex rational expressions to a slightly higher level, including expressions such as

$$\frac{\frac{1}{x+h}-\frac{1}{x}}{h}.$$

- **Section 1.2 (Linear Equations and Rational Equations)** has more to say on solving rational equations, including examples that illustrate the solutions of equations such as

$$\frac{x+2}{4}-\frac{x-1}{3}=2 \quad \text{and} \quad \frac{3}{x+6}+\frac{1}{x-2}=\frac{4}{x^2+4x-12}.$$

- **Section 1.6 (Other Types of Equations) and Section 1.7 (Linear Inequalities and Absolute Value Inequalities)** take the discussion of absolute value to a slightly higher level, including solutions of equations and inequalities such as

$$5|1-4x|-15=0 \quad \text{and} \quad -2|3x+5|+7>-13.$$

  Section 1.7 contains a new discussion of intersections and unions of intervals.
- **Chapter 2 (Functions and Graphs)** has been reorganized around the book's central idea: functions. Functions are introduced in the first section. All subsequent topics are viewed from the perspective of functions and relations. New multipart exercises that require students to bring together their knowledge of functions appear throughout the chapter.

- **Section 2.1 (Basics of Functions and Their Graphs)** contains a more detailed discussion, including new graphics, of identifying domain and range from a function's graph.
- **Section 2.3 (Linear Functions and Slope) and Section 2.4 (More on Slope)** develop lines and slope from the perspective of functions. Section 2.3 contains a new example on using intercepts to graph the general form of a line's equation. Section 2.4 contains a more thoroughly developed example on writing equations of a line perpendicular to a given line.
- **Section 2.5 (Transformations of Functions)** adds the graph of the cube root function, $f(x) = \sqrt[3]{x}$, to the table of common graphs, using this graph, as well as the other six graphs in the table, in the discussion of transformations. New graphics with clarifying voice balloons illustrate transformations. A new discussion of horizontally stretching and shrinking a graph is included among the transformations.
- **Section 2.7 (Inverse Functions)** takes finding the inverse of a function to a slightly higher level, with examples showing how to find f^{-1} for

  $$f(x) = \frac{5}{x} + 4 \quad \text{and} \quad f(x) = x^2 - 1, x \geq 0.$$

  In the latter case, both f and f^{-1} are graphed on the same axes, using transformations to obtain the graph of f^{-1}.
- **Section 3.1 (Quadratic Functions)** contains a new discussion on modeling quadratic functions from verbal condtions and solving problems that involve maximizing or minimizing these functions.
- **Section 3.2 (Polynomial Functions and Their Graphs)** now includes the Intermediate Value Theorem.
- **Section 3.4 (Zeros of Polynomial Functions)** now covers this topic in one section, rather than two. Although there is a new discussion on the various kinds of zeros and a new example on finding these zeros, finding bounds for the roots of a polynomial equation has been omitted.
- **Section 3.5 (Rational Functions and Their Graphs)** includes a new discussion on using transformations of $f(x) = \frac{1}{x}$ and $f(x) = \frac{1}{x^2}$ to graph rational functions.
- **Section 3.6 (Polynomial and Rational Inequalities)** is a reworked and expanded discussion of quadratic and rational inequalities that appeared in Chapter 1 of the previous edition. Solution procedures are now developed and organized around polynomial functions, rational functions, and their graphs.
- **Section 4.1 (Exponential Functions)** now defines e as the value that $\left(1 + \frac{1}{n}\right)^n$ approaches as $n \to \infty$, using this definition to develop the formula for compound interest subject to continuous compounding. New graphics illustrate transformations of exponential functions.
- **Section 4.4 (Exponential and Logarithmic Equations)** has been reorganized into four categories:
  - ➜ Solving exponential equations using like bases
  - ➜ Solving exponential equations using logarithms and logarithmic properties
  - ➜ Solving logarithmic equations using the definition of a logarithm
  - ➜ Solving logarithmic equations using the one-to-one property of logarithms.

  New examples appear throughout the section to ensure adequate coverage of each category.
- **Section 4.5 (Exponential Growth and Decay; Modeling Data)** contains new examples involving choosing models for data before technology is used to obtain these models.

I hope that my love for learning, as well as my respect for the diversity of students I have taught and learned from over the years, is apparent throughout this new edition. By connecting algebra to the whole spectrum of learning, it is my intent to show students that their world is profoundly mathematical, and indeed, π is in the sky.

Robert Blitzer

STUDENT RESOURCES

Student Study Pack

Everything a student needs to succeed in one place. It is available, packaged with the book, or it is available for purchase stand-alone. Study Pack contains:

- ***Student Solutions Manual***

  Fully worked solutions to odd-numbered exercises.
- ***Pearson Tutor Center***

  Tutors provide one-on-one tutoring for any problem with an answer at the back of the book. Students access the Tutor Center via toll-free phone, fax, or email. Available only to college students in the U.S. and Canada.
- ***CD Lecture Series***

  A comprehensive set of CD-ROMs, tied to the textbook, containing short video clips of an instructor working key book examples.

INSTRUCTOR RESOURCES

Instructor Resource Distribution

All instructor resources can be downloaded from the web site, www.prenhall.com. Select "Browse our catalog," then, click on "Mathematics"; select your course and choose your text. Under "Resources," on the left side, select "instructor" and choose the supplement you need to download. You will be required to run through a one time registration before you can complete this process.

- ***TestGen***

  Easily create tests from textbook section objectives. Questions are algorithmically generated allowing for unlimited versions. Edit problems or create your own.
- ***Test Item File***

  A printed test bank derived from TestGen.
- ***PowerPoint Lecture Slides***

  Fully editable slides that follow the textbook. Project in class or post to a website in an online course.
- ***Instructor Solutions Manual***

  Fully worked solutions to all textbook exercises and chapter projects.

Instructor's Edition

Provides answers to *all* exercises in the back of the text.

MathXL®

MathXL® is a powerful online homework, tutorial, and assessment system that accompanies your textbook. Instructors can create edit, and assign online homework and tests using algorithmically generated exercises correlated at the objective level to the textbook. Student work is tracked in an online gradebook. Students can take chapter tests and receive personalized study plans based on their results. The study plan diagnoses weaknesses and links students to tutorial exercises for objectives they need to study. Students can also access video clips from selected exercises. MathXL® is available to qualified adopters. For more information, visit our website at *www.mathxl.con*, or contact your Prentice Hall sales representative for a demonstration.

MyMathLab

MyMathLab is a text-specific, customizable online course for your textbooks, MyMathLab is powered by CourseCompass™—Pearson Education's online teaching and learning environment—and by MathXL®—our online homework, tutorial, and assessment system. MyMathLab gives you the tools you need to deliver all or a portion of your course online, whether your students are in a lab setting or working from home.

MyMathLab provides a rich and flexible set of course materials, featuring free-response exercises that are algorithmically generated for unlimited practice. Students can use online tools such as video lectures and a multimedia textbook to improve their performance. Instructors can use MyMathLab's homework and test managers to select and assign online exercises correlated to the textbook, and can import TestGen tests for added flexibility. The only gradebook—designed specifically for mathematics—automatically tracks students' homework and test results and gives the instructor control over how to calculate final grades. MyMathLab is available to qualified adopters. For more information, visit our website at *www.mymathlab.com* or contact your Prentice Hall sales representative for a product demonstration.

Acknowledgments

I wish to express my appreciation to all the reviewers of my precalculus series for their helpful feedback, frequently transmitted with wit, humor, and intelligence. Every change to this edition is the result of their thoughtful comments and suggestions. In particular, I would like to thank the following people for reviewing **College Algebra**, **Algebra and Trigonometry**, and **Precalculus**.

Barnhill, Kayoko Yates, *Clark College*
Beaver, Timothy, *Isothermal Community College*
Best, Lloyd, *Pacific Union College*
Burgin, Bill, *Gaston College*
Chang, Jimmy, *St. Petersburg College*
Colt, Diana, *University of Minnesota-Duluth*
Densmore, Donna, *Bossier Parish Community College*
Enegren, Disa, *Rose State College*
Fisher, Nancy, *University of Alabama*
Glickman, Cynthia, *Community College of Southern Nevada*
Goel, Sudhir Kumar, *Valdosta State University*
Gordon, Donald, *Manatee Community College*
Gross, David L., *University of Connecticut*
Haack, Joelx K., *University of Northern Iowa*
Haefner, Jeremy, *University of Colorado*
Hague, Joyce, *University of Wisconsin at River Falls*
Hall, Mike, *Univeristy of Mississippi*
Hay-Jahans, Christopher N., *University of South Dakota*
Hernandez, Celeste, *Richland College*
Ihlow, Winfield A., *SUNY College at Oswego*
Johnson, Nancy Raye, *Manatee Community College*
Leesburg, Mary, *Manatee Community College*
Lehmann, Christine Heinecke, *Purdue University North Central*
Levichev, Alexander, *Boston University*
Lin, Zongzhu, *Kansas State University*
Marlin, Benjamin, *Northwestern Oklahoma State University*
Massey, Marilyn, *Collin County Community College*
McCarthy-Germain, Yvelyne, *University of New Orleans*
Miller, James, *West Virginia University*
Pharo, Debra A., *Northwestern Michigan College*
Phoenix, Gloria, *North Carolina Agricultural and Technical State University*
Platt, David, *Front Range Community College*
Pohjanpelto, Juha, *Oregon State University*
Rech, Janice, *University of Nebraska at Omaha*
Salmon, Judith, *Fitchburg State College*
Schultz, Cynthia, *Illinois Valley Community College*
Stump, Chris, *Bethel College*
Trim, Pamela, *Southwest Tennessee Community College*
Turner, Chris, *Arkansas State University*
Van Lommel, Richard E., *California State University-Sacramento*
Van Peursem, Dan, *University of South Dakota*
Van Veldhuizen, Philip, *University of Nevada at Reno*
White, David, *The Victoria College*
Wienckowski, Tracy, *Univesity of Buffalo*

Special thanks to Professor Phoebe Rousse at Louisiana State University for your detailed notebooks of suggestions, including examples and exercise descriptions, as well as actual problems, that helped increase the rigor of the text without affecting its tone or diversity of applications.

Additional acknowledgments are extended to Dan Miller, for the Herculean task of preparing the solutions manuals, Brad Davis, for preparing the answer section and serving as accuracy checker, the Preparè Inc. formatting and production team, including Frank Weihenig and Linda Martino, for the book's brilliant paging, as well as keeping this complex project moving through its many stages, Aaron Darnall at Scientific Illustrators, for superbly illustrating the book, Melinda Alexander, photo researcher, for obtaining the book's new photographs, Kirk Trigsted and Scott Satake, for preparing the videotape series, including the chapter test prep video CDs, and Bayani Mendoza de Leon, assistant managing editor, for orchestrating the entire production process.

I would like to thank my editor at Prentice Hall, Adam Jaworski, and Project Manager, Dawn Murrin, who guided and coordinated the book from manuscript through production. Thanks to Kenny Beck for the beautiful covers and interior design. Finally, thanks to Halee Dinsey and Patrice Jones, for your innovative marketing efforts, to Sally Yagan, for your continuing support, and to the entire Prentice Hall sales force, for your confidence and enthusiasm about the book.

To the Student

I've written this book so that you can learn about the power of algebra and how it relates directly to your life outside the classroom. All concepts are carefully explained, important definitions and procedures are set off in boxes, and worked-out examples that present solutions in a step-by-step manner appear in every section. Each example is followed by a similar matched problem, called a Check Point, for you to try so that you can actively participate in the learning process as you read the book. (Answers to all Check Points appear in the back of the book.) Study Tips offer hints and suggestions and often point out common errors to avoid. A great deal of attention has been given to applying algebra to your life to make your learning experience both interesting and relevant.

As you begin your studies, I would like to offer some specific suggestions for using this book and for being successful in this course:

1. **Read the book.** Read each section with pen (or pencil) in hand. Move through the worked-out examples with great care. These examples provide a model for doing exercises in the exercise sets. As you proceed through the reading, do not give up if you do not understand every single word. Things will become clearer as you read on and see how various procedures are applied to specific worked-out examples.
2. **Work problems every day and check your answers.** The way to learn mathematics is by doing mathematics, which means working the Check Points and assigned exercises in the exercise sets. The more exercises you work, the better you will understand the material.
3. **Review for quizzes and tests.** After completing a chapter, study the chapter summary, work the exercises in the Chapter Review, and work the exercises in the Chapter Test. Answers to all these exercises are given in the back of the book.

The methods that I've used to help you read the book, work the problems, and review for tests are described in "A Brief Guide to Getting the Most from This Book," which appears inside the front cover. Spend a few minutes reviewing the guide to familiarize yourself with the book's features and their benefits.

4. **Use the resources available with this book.** Additional resources to aid your study are described following the guide to getting the most from your book. These resources include a Solutions Manual, a Chapter Test Prep Video CD, MyMathLab, an online version of the book with links to multimedia resources, MathXL®, an online homework, tutorial, and assessment system of the text, and tutorial support at no charge at the PH Tutor Center.
5. **Attend all lectures.** No book is intended to be a substitute for valuable insights and interactions that occur in the classroom. In addition to arriving for lecture on time and being prepared, you will find it useful to read a section before it is covered in lecture. This will give you a clear idea of the new material that will be discussed.

I wrote this book in Point Reyes National Seashore, 40 miles north of San Francisco. The park consists of 75,000 acres with miles of pristine surf-washed beaches, forested ridges, and bays bordered by white cliffs. It was my hope to convey the beauty and excitement of mathematics using nature's unspoiled beauty as a source of inspiration and creativity. Enjoy the pages that follow as you empower yourself with the algebra needed to succeed in college, your career, and in your life.

Regards,

Bob

Robert Blitzer

About the Author

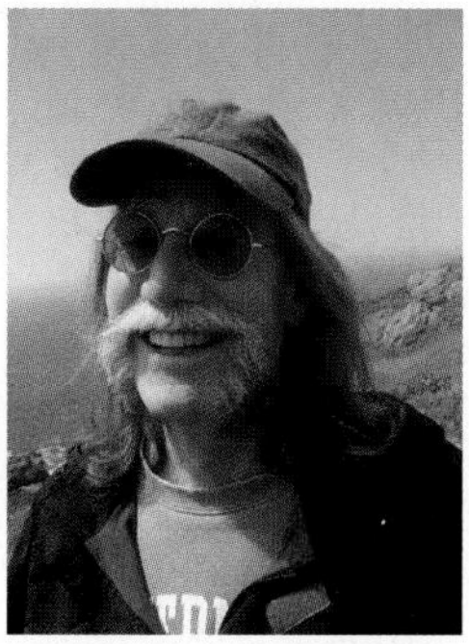

Bob Blitzer is a native of Manhattan and received a Bachelor of Arts degree with dual majors in mathematics and psychology (minor: English literature) from the City College of New York. His unusual combination of academic interests led him toward a Master of Arts in mathematics from the University of Miami and a doctorate in behavioral sciences from Nova University. Bob is most energized by teaching mathematics and has taught a variety of mathematics courses at Miami Dade College for nearly 30 years. He has received numerous teaching awards, including Innovator of the Year from the League for Innovations in the Community College, and was among the first group of recipients at Miami Dade College for an endowed chair based on excellence in the classroom. In addition to *College Algebra Essentials*, Bob has written textbooks covering introductory algebra, intermediate algebra, algebra and trigonometry, precalculus, and liberal arts mathematics, all published by Prentice Hall.

Applications Index

D

E

F

College Algebra Essentials

CHAPTER P

Prerequisites: Fundamental Concepts of Algebra

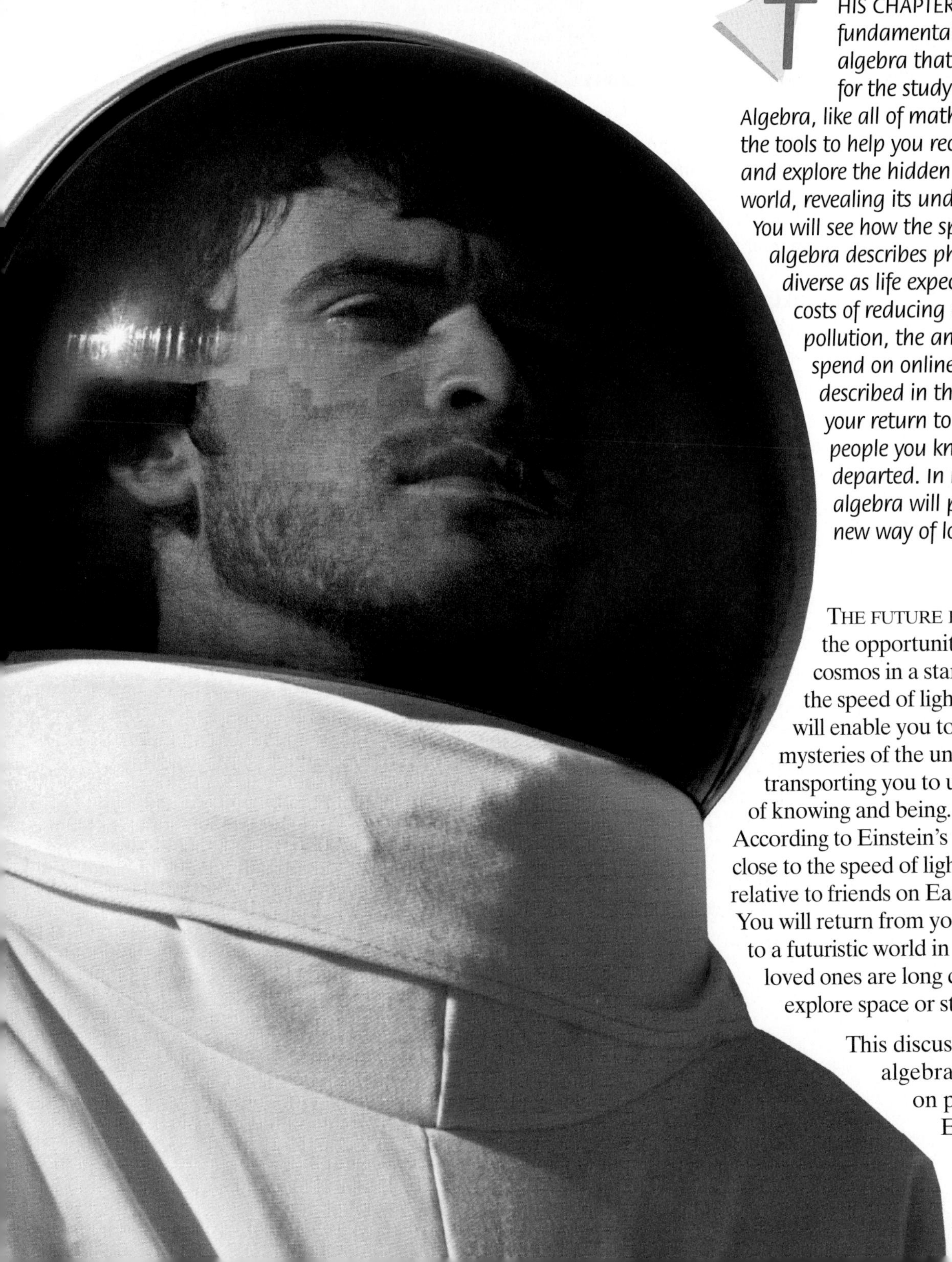

THIS CHAPTER REVIEWS fundamental concepts of algebra that are prerequisites for the study of college algebra. Algebra, like all of mathematics, provides the tools to help you recognize, classify, and explore the hidden patterns of your world, revealing its underlying structure. You will see how the special language of algebra describes phenomena as diverse as life expectancy, windchill, costs of reducing environmental pollution, the amount Americans spend on online dating, and, as described in the photo caption, your return to a world where the people you knew have long since departed. In many ways, algebra will provide you with a new way of looking at our world.

THE FUTURE IS NOW: YOU HAVE the opportunity to explore the cosmos in a starship traveling near the speed of light. The experience will enable you to understand the mysteries of the universe first hand, transporting you to unimagined levels of knowing and being. The down side: According to Einstein's theory of relativity, close to the speed of light, your aging rate relative to friends on Earth is nearly zero. You will return from your two-year journey to a futuristic world in which friends and loved ones are long dead. Do you explore space or stay here on Earth?

This discussion is developed algebraically in the essay on page 35 and in Exercise 120 in Exercise Set P. 3.

SECTION P.1 Algebraic Expressions and Real Numbers

Objectives

1. Evaluate algebraic expressions.
2. Use mathematical models.
3. Find the intersection of two sets.
4. Find the union of two sets.
5. Recognize subsets of the real numbers.
6. Use inequality symbols.
7. Evaluate absolute value.
8. Use absolute value to express distance.
9. Identify properties of the real numbers.
10. Simplify algebraic expressions.

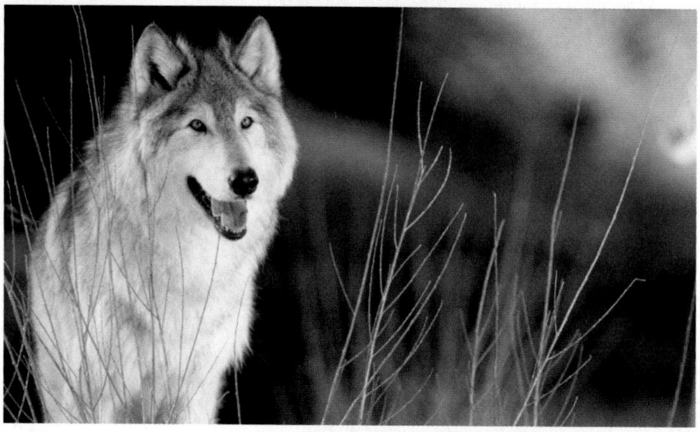

Insatiable killer. That's the reputation the gray wolf acquired in the United States in the nineteenth and early twentieth centuries. Although the label was undeserved, an estimated 2 million wolves were shot, trapped, or poisoned. By 1960, the population was reduced to 800 wolves. In this section, you will learn how the special language of algebra describes your world, including the increasing wolf population in the continental United States following the Endangered Species Act of 1973.

Algebraic Expressions

Algebra uses letters, such as x and y, to represent numbers. If a letter is used to represent various numbers, it is called a **variable**. For example, imagine that you are basking in the sun on the beach. We can let x represent the number of minutes that you can stay in the sun without burning with no sunscreen. With a number 6 sunscreen, exposure time without burning is six times as long, or 6 times x. This can be written $6 \cdot x$, but it is usually expressed as $6x$. Placing a number and a letter next to one another indicates multiplication.

Notice that $6x$ combines the number 6 and the variable x using the operation of multiplication. A combination of variables and numbers using the operations of addition, subtraction, multiplication, or division, as well as powers or roots, is called an **algebraic expression**. Here are some examples of algebraic expressions:

$$x + 6, \quad x - 6, \quad 6x, \quad \frac{x}{6}, \quad 3x + 5, \quad x^2 - 3, \quad \sqrt{x} + 7.$$

Many algebraic expressions involve *exponents*. For example, the algebraic expression

$$0.72x^2 + 9.4x + 783$$

approximates the gray wolf population in the United States x years after 1960. The expression x^2 means $x \cdot x$, and is read "x to the second power" or "x squared." The exponent, 2, indicates that the base, x, appears as a factor two times.

Exponential Notation

If n is a counting number (1, 2, 3, and so on),

$$b^n = \underbrace{b \cdot b \cdot b \cdot \cdots \cdot b}_{b \text{ appears as a factor } n \text{ times.}}$$

(Exponent or Power: n; Base: b)

b^n is read "the nth power of b" or "b to the nth power." Thus, the nth power of b is defined as the product of n factors of b. The expression b^n is called an **exponential expression**. Furthermore, $b^1 = b$.

For example,

$$8^2 = 8 \cdot 8 = 64, \quad 5^3 = 5 \cdot 5 \cdot 5 = 125, \quad \text{and} \quad 2^4 = 2 \cdot 2 \cdot 2 \cdot 2 = 16.$$

1 Evaluate algebraic expressions.

Evaluating Algebraic Expressions

Evaluating an algebraic expression means to find the value of the expression for a given value of the variable. For example, we can evaluate $6x$ (from the sunscreen example) when $x = 15$. We substitute 15 for x. We obtain $6 \cdot 15$, or 90. This means that if you can stay in the sun for 15 minutes without burning when you don't put on any lotion, then with a number 6 lotion, you can "cook" for 90 minutes without burning.

Many algebraic expressions involve more than one operation. Evaluating an algebraic expression without a calculator involves carefully applying the following order of operations agreement:

The Order of Operations Agreement

1. Perform operations within the innermost parentheses and work outward. If the algebraic expression involves a fraction, treat the numerator and the denominator as if they were each enclosed in parentheses.
2. Evaluate all exponential expressions.
3. Perform multiplications and divisions as they occur, working from left to right.
4. Perform additions and subtractions as they occur, working from left to right.

EXAMPLE 1 Evaluating an Algebraic Expression

Evaluate $7 + 5(x - 4)^3$ for $x = 6$.

Solution

$7 + 5(x - 4)^3 = 7 + 5(6 - 4)^3$ Replace x with 6.

$= 7 + 5(2)^3$ First work inside parentheses: $6 - 4 = 2$.

$= 7 + 5(8)$ Evaluate the exponential expression: $2^3 = 2 \cdot 2 \cdot 2 = 8$.

$= 7 + 40$ Multiply: $5(8) = 40$.

$= 47$ Add.

Check Point 1 Evaluate $8 + 6(x - 3)^2$ for $x = 13$.

2 Use mathematical models.

Formulas and Mathematical Models

An **equation** is formed when an equal sign is placed between two algebraic expressions. One aim of algebra is to provide a compact, symbolic description of the world. These descriptions involve the use of *formulas*. A **formula** is an equation that uses letters to express a relationship between two or more variables. Here is an example of a formula:

$$C = \frac{5}{9}(F - 32).$$

Celsius temperature | is | $\frac{5}{9}$ of | the difference between Fahrenheit temperature and 32°.

The process of finding formulas to describe real-world phenomena is called **mathematical modeling**. Such formulas, together with the meaning assigned to the variables, are called **mathematical models**. We often say that these formulas model, or describe, the relationships among the variables.

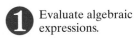

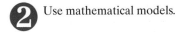

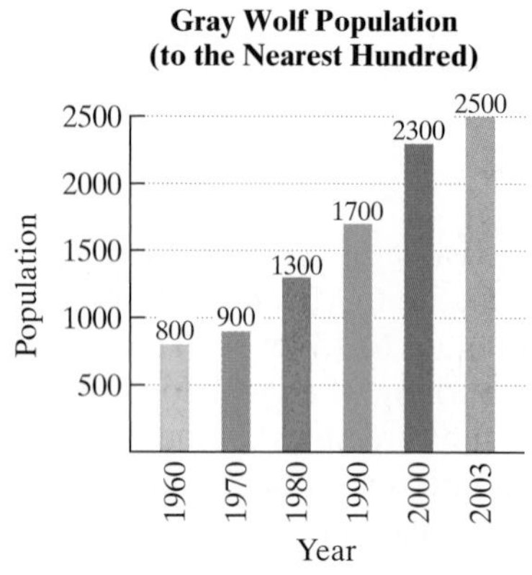

Figure P.1

Source: U.S. Department of the Interior

EXAMPLE 2 Modeling the Gray Wolf Population

The formula

$$P = 0.72x^2 + 9.4x + 783$$

models the gray wolf population, P, in the United States, x years after 1960. Use the formula to find the population in 1990. How well does the formula model the actual data shown in the bar graph in Figure P.1?

Solution Because 1990 is 30 years after 1960, we substitute 30 for x in the given formula. Then we use the order of operations to find P, the gray wolf population in 1990.

$P = 0.72x^2 + 9.4x + 783$ — This is the given mathematical model.

$P = 0.72(30)^2 + 9.4(30) + 783$ — Replace each occurrence of x with 30.

$P = 0.72(900) + 9.4(30) + 783$ — Evaluate the exponential expression: $30^2 = 30 \cdot 30 = 900$.

$P = 648 + 282 + 783$ — Multiply from left to right: $0.72(900) = 648$ and $9.4(30) = 282$.

$P = 1713$ — Add.

The formula indicates that in 1990, the gray wolf population in the United States was 1713. The number given in Figure P.1 is 1700, so the formula models the data quite well.

Check Point 2 Use the formula in Example 2 to find the gray wolf population in 2000. How well does the formula model the data in Figure P.1?

Sometimes a mathematical model gives an estimate that is not a good approximation or is extended to include values of the variable that do not make sense. In these cases, we say that **model breakdown** has occurred.

Sets

Before we describe the set of real numbers, let's be sure you are familiar with some basic ideas about sets. A **set** is a collection of objects whose contents can be clearly determined. The objects in a set are called the **elements** of the set. For example, the set of numbers used for counting can be represented by

$$\{1, 2, 3, 4, 5, \dots\}.$$

The braces, { }, indicate that we are representing a set. This form of representation, called the **roster method**, uses commas to separate the elements of the set. The three dots after the 5, called an *ellipsis*, indicate that there is no final element and that the listing goes on forever.

A set can also be written in **set-builder notation**. In this notation, the elements of the set are described, but not listed. Here is an example:

$$\{x | x \text{ is a counting number less than } 6\}.$$

The set of all x — such that — x is a counting number less than 6.

The same set written using the roster method is

$$\{1, 2, 3, 4, 5\}.$$

Study Tip

Grouping symbols such as parentheses, ( ), and square brackets, [ ], are not used to represent sets. Only commas are used to separate the elements of a set. Separators such as colons or semicolons are not used.

Find the intersection
of two sets.

$A \cap B$

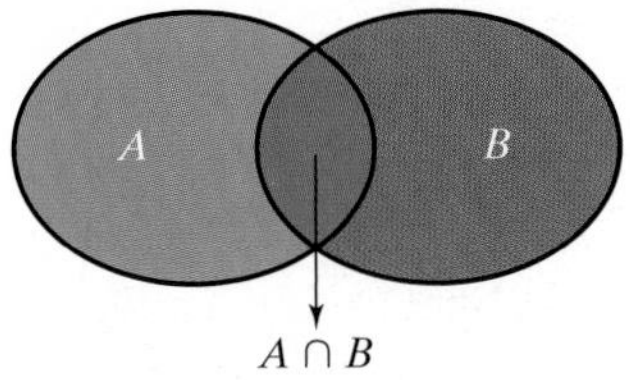

Figure P.2 Picturing the intersection of two sets

If A and B are sets, we can form a new set consisting of all elements that are in both A and B. This set is called the *intersection* of the two sets.

Definition of the Intersection of Sets

The **intersection** of sets A and B, written $A \cap B$, is the set of elements common to both set A **and** set B. This definition can be expressed in set-builder notation as follows:

$$A \cap B = \{x \mid x \text{ is an element of } A \text{ AND } x \text{ is an element of } B\}.$$

Figure P.2 shows a useful way of picturing the intersection of sets A and B. The figure indicates that $A \cap B$ contains those elements that belong to both A and B at the same time.

EXAMPLE 3   Finding the Intersection of Two Sets

Find the intersection: $\{7, 8, 9, 10, 11\} \cap \{6, 8, 10, 12\}$.

Solution   The elements common to $\{7, 8, 9, 10, 11\}$ and $\{6, 8, 10, 12\}$ are 8 and 10. Thus,

$$\{7, 8, 9, 10, 11\} \cap \{6, 8, 10, 12\} = \{8, 10\}.$$

Check Point 3   Find the intersection: $\{3, 4, 5, 6, 7\} \cap \{3, 7, 8, 9\}$.

If a set has no elements, it is called the **empty set**, or the **null set**, and is represented by the symbol ∅ (the Greek letter phi). Here is an example that shows how the empty set can result when finding the intersection of two sets:

$$\{2, 4, 6\} \cap \{3, 5, 7\} = \emptyset.$$

These sets have no common elements.   Their intersection has no elements and is the empty set.

Find the union of two sets.

$A \cup B$

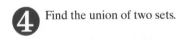

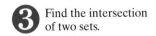

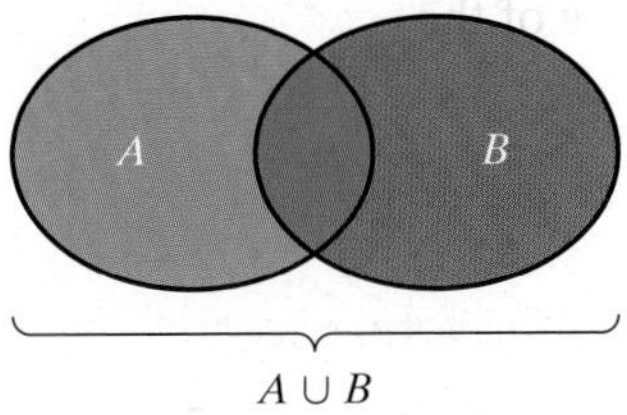

Figure P.3 Picturing the union of two sets

Another set that we can form from sets A and B consists of elements that are in A or B or in both sets. This set is called the *union* of the two sets.

Definition of the Union of Sets

The **union** of sets A and B, written $A \cup B$, is the set of elements that are members of set A **or** of set B or of both sets. This definition can be expressed in set-builder notation as follows:

$$A \cup B = \{x \mid x \text{ is an element of } A \text{ OR } x \text{ is an element of } B\}.$$

Figure P.3 shows a useful way of picturing the union of sets A and B. The figure indicates that $A \cup B$ is formed by joining the sets together.

We can find the union of set A and set B by listing the elements of set A. Then, we include any elements of set B that have not already been listed. Enclose all elements that are listed with braces. This shows that the union of two sets is also a set.

Study Tip

When finding the union of two sets, do not list twice any elements that appear in both sets.

EXAMPLE 4   Finding the Union of Two Sets

Find the union: $\{7, 8, 9, 10, 11\} \cup \{6, 8, 10, 12\}$.

Solution   To find $\{7, 8, 9, 10, 11\} \cup \{6, 8, 10, 12\}$, start by listing all the elements from the first set, namely 7, 8, 9, 10, and 11. Now list all the elements from the second set that are not in the first set, namely 6 and 12. The union is the set consisting of all these elements. Thus,

$$\{7, 8, 9, 10, 11\} \cup \{6, 8, 10, 12\} = \{6, 7, 8, 9, 10, 11, 12\}.$$

Check Point 4 Find the union: $\{3, 4, 5, 6, 7\} \cup \{3, 7, 8, 9\}$.

5 Recognize subsets of the real numbers.

The Set of Real Numbers

The sets that make up the real numbers are summarized in Table P.1. We refer to these sets as **subsets** of the real numbers, meaning that all elements in each subset are also elements in the set of real numbers.

Notice the use of the symbol $\approx$ in the examples of irrational numbers. The symbol means "is approximately equal to." Thus,

$$\sqrt{2} \approx 1.414214.$$

We can verify that this is only an approximation by multiplying 1.414214 by itself. The product is very close to, but not exactly, 2:

$$1.414214 \times 1.414214 = 2.000001237796.$$

Technology

A calculator with a square root key gives a decimal approximation for $\sqrt{2}$, not the exact value.

Table P.1 Important Subsets of the Real Numbers

Name	Description	Examples
Natural numbers $\mathbb{N}$	$\{1, 2, 3, 4, 5, \dots\}$ These are the numbers that we use for counting.	2, 3, 5, 17
Whole numbers $\mathbb{W}$	$\{0, 1, 2, 3, 4, 5, \dots\}$ The set of whole numbers includes 0 and the natural numbers.	0, 2, 3, 5, 17
Integers $\mathbb{Z}$	$\{\dots, -5, -4, -3, -2, -1, 0, 1, 2, 3, 4, 5, \dots\}$ The set of integers includes the negatives of the natural numbers and the whole numbers.	−17, −5, −3, −2, 0, 2, 3, 5, 17
Rational numbers $\mathbb{Q}$	$\left\{\frac{a}{b} \mid a \text{ and } b \text{ are integers and } b \neq 0\right\}$ This means that b is not equal to zero. The set of rational numbers is the set of all numbers that can be expressed as a quotient of two integers, with the denominator not 0. Rational numbers can be expressed as terminating or repeating decimals.	$-17 = \frac{-17}{1}$, $-5 = \frac{-5}{1}$, −3, −2, 0, 2, 3, 5, 17, $\frac{2}{5} = 0.4$, $\frac{-2}{3} = -0.6666\dots = -0.\overline{6}$
Irrational numbers $\mathbb{I}$	The set of irrational numbers is the set of all numbers whose decimal representations are neither terminating nor repeating. Irrational numbers cannot be expressed as a quotient of integers.	$\sqrt{2} \approx 1.414214$, $-\sqrt{3} \approx -1.73205$, $\pi \approx 3.142$, $-\frac{\pi}{2} \approx -1.571$

Not all square roots are irrational. For example, $\sqrt{25} = 5$ because $5^2 = 5 \cdot 5 = 25$. Thus, $\sqrt{25}$ is a natural number, a whole number, an integer, and a rational number $\left(\sqrt{25} = \frac{5}{1}\right)$.

The set of *real numbers* is formed by taking the union of the sets of rational numbers and irrational numbers. Thus, every real number is either rational or irrational, as shown in Figure P.4.

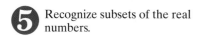

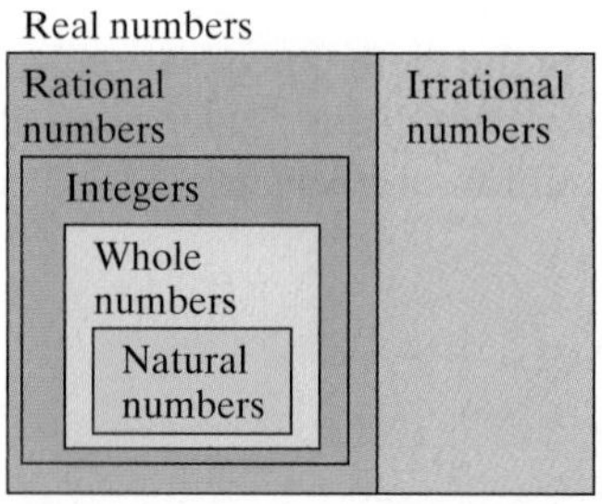

Figure P.4 Every real number is either rational or irrational.

Real Numbers

The set of **real numbers** is the set of numbers that are either rational or irrational:

$$\{x|x \text{ is rational or } x \text{ is irrational}\}.$$

The symbol $\mathbb{R}$ is used to represent the set of real numbers. Thus,

$$\mathbb{R} = \{x|x \text{ is rational}\} \cup \{x|x \text{ is irrational}\}.$$

EXAMPLE 5 Recognizing Subsets of the Real Numbers

Consider the following set of numbers:

$$\left\{-7, -\frac{3}{4}, 0, 0.\overline{6}, \sqrt{5}, \pi, 7.3, \sqrt{81}\right\}.$$

List the numbers in the set that are

a. natural numbers. **b.** whole numbers. **c.** integers.
d. rational numbers. **e.** irrational numbers. **f.** real numbers.

Solution

a. Natural numbers: The natural numbers are the numbers used for counting. The only natural number in the set is $\sqrt{81}$ because $\sqrt{81} = 9$. (9 multiplied by itself, or 9^2, is 81.)

b. Whole numbers: The whole numbers consist of the natural numbers and 0. The elements of the set that are whole numbers are 0 and $\sqrt{81}$.

c. Integers: The integers consist of the natural numbers, 0, and the negatives of the natural numbers. The elements of the set that are integers are $\sqrt{81}$, 0, and -7.

d. Rational numbers: All numbers in the set that can be expressed as the quotient of integers are rational numbers. These include $-7\left(-7 = \frac{-7}{1}\right)$, $-\frac{3}{4}$, $0\left(0 = \frac{0}{1}\right)$, and $\sqrt{81}\left(\sqrt{81} = \frac{9}{1}\right)$. Furthermore, all numbers in the set that are terminating or repeating decimals are also rational numbers. These include $0.\overline{6}$ and 7.3.

e. Irrational numbers: The irrational numbers in the set are $\sqrt{5}\left(\sqrt{5} \approx 2.236\right)$ and $\pi(\pi \approx 3.14)$. Both $\sqrt{5}$ and π are only approximately equal to 2.236 and 3.14, respectively. In decimal form, $\sqrt{5}$ and π neither terminate nor have blocks of repeating digits.

f. Real numbers: All the numbers in the given set are real numbers.

Check Point 5 Consider the following set of numbers:

$$\left\{-9, -1.3, 0, 0.\overline{3}, \frac{\pi}{2}, \sqrt{9}, \sqrt{10}\right\}.$$

List the numbers in the set that are

a. natural numbers. **b.** whole numbers.
c. integers. **d.** rational numbers.
e. irrational numbers. **f.** real numbers.

The Real Number Line

The **real number line** is a graph used to represent the set of real numbers. An arbitrary point, called the **origin**, is labeled 0. Select a point to the right of 0 and label it 1. The distance from 0 to 1 is called the **unit distance**. Numbers to the right of the origin are **positive** and numbers to the left of the origin are **negative**. The real number line is shown in Figure P.5.

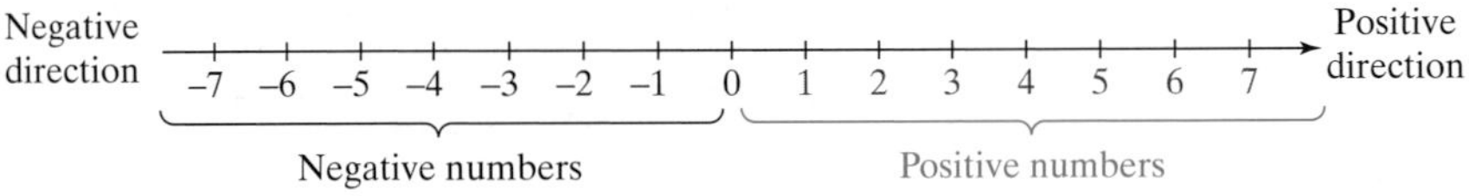

Figure P.5 The real number line

Real numbers are **graphed** on a number line by placing a dot at the correct location for each number. The integers are easiest to locate. In Figure P.6, we've graphed the integers -3, 0, and 4.

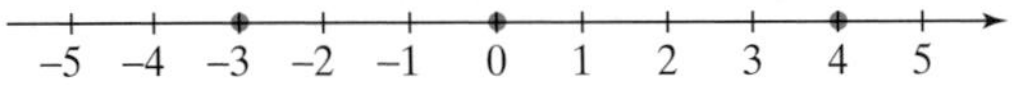

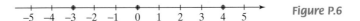

Figure P.6

Every real number corresponds to a point on the number line and every point on the number line corresponds to a real number. We say that there is a **one-to-one correspondence** between all the real numbers and all points on a real number line.

6 Use inequality symbols.

Ordering the Real Numbers

On the real number line, the real numbers increase from left to right. The lesser of two real numbers is the one farther to the left on a number line. The greater of two real numbers is the one farther to the right on a number line.

Look at the number line in Figure P.7. The integers -4 and -1 are graphed.

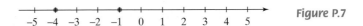

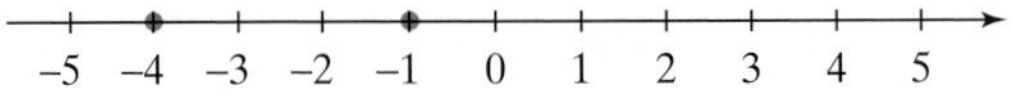

Figure P.7

Observe that -4 is to the left of -1 on the number line. This means that -4 is less than -1.

$$-4 < -1$$

-4 is less than -1 because -4 is to the **left** of -1 on the number line.

In Figure P.7, we can also observe that -1 is to the right of -4 on the number line. This means that -1 is greater than -4.

$$-1 > -4$$

-1 is greater than -4 because -1 is to the **right** of -4 on the number line.

The symbols $<$ and $>$ are called **inequality symbols**. These symbols always point to the lesser of the two real numbers when the inequality statement is true.

-4 is less than -1. $\quad -4 < -1 \quad$ *The symbol points to -4, the lesser number.*

-1 is greater than -4. $\quad -1 > -4 \quad$ *The symbol still points to -4, the lesser number.*

The symbols $<$ and $>$ may be combined with an equal sign, as shown in the following table:

Symbols	Meaning	Examples	Explanation
$a \le b$	a is less than or equal to b.	$2 \le 9$ $9 \le 9$	Because $2 < 9$ Because $9 = 9$
$b \ge a$	b is greater than or equal to a.	$9 \ge 2$ $2 \ge 2$	Because $9 > 2$ Because $2 = 2$

This inequality is true if either the $<$ part or the $=$ part is true.

This inequality is true if either the $>$ part or the $=$ part is true.

7 Evaluate absolute value.

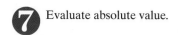

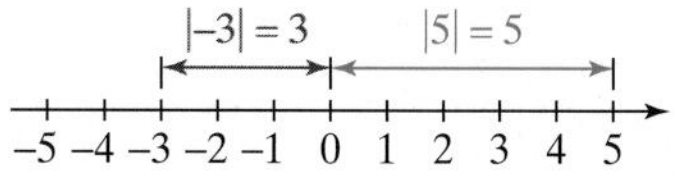

Figure P.8 Absolute value as the distance from 0

Absolute Value

The **absolute value** of a real number a, denoted by $|a|$, is the distance from 0 to a on the number line. This distance is always taken to be nonnegative. For example, the real number line in Figure P.8 shows that

$$|-3| = 3 \quad \text{and} \quad |5| = 5.$$

The absolute value of -3 is 3 because -3 is 3 units from 0 on the number line. The absolute value of 5 is 5 because 5 is 5 units from 0 on the number line. The absolute value of a positive real number or 0 is the number itself. The absolute value of a negative real number, such as -3, is the number without the negative sign.

We can define the absolute value of the real number x without referring to a number line. The algebraic definition of the absolute value of x is given as follows:

Definition of Absolute Value

$$|x| = \begin{cases} x & \text{if } x \geq 0 \\ -x & \text{if } x < 0 \end{cases}$$

If x is nonnegative (that is, $x \geq 0$), the absolute value of x is the number itself. For example,

$$|5| = 5 \qquad |\pi| = \pi \qquad \left|\frac{1}{3}\right| = \frac{1}{3} \qquad |0| = 0.$$

Zero is the only number whose absolute value is 0.

If x is a negative number (that is, $x < 0$), the absolute value of x is the opposite of x. This makes the absolute value positive. For example,

$$|-3| = -(-3) = 3 \qquad |-\pi| = -(-\pi) = \pi \qquad \left|-\frac{1}{3}\right| = -\left(-\frac{1}{3}\right) = \frac{1}{3}.$$

This middle step is usually omitted.

EXAMPLE 6 Evaluating Absolute Value

Rewrite each expression without absolute value bars:

a. $|\sqrt{3} - 1|$ **b.** $|2 - \pi|$ **c.** $\dfrac{|x|}{x}$ if $x < 0$.

Solution

a. Because $\sqrt{3} \approx 1.7$, the number inside the absolute value bars, $\sqrt{3} - 1$, is positive. The absolute value of a positive number is the number itself. Thus,

$$|\sqrt{3} - 1| = \sqrt{3} - 1.$$

b. Because $\pi \approx 3.14$, the number inside the absolute value bars, $2 - \pi$, is negative. The absolute value of x when $x < 0$ is $-x$. Thus,

$$|2 - \pi| = -(2 - \pi) = \pi - 2.$$

c. If $x < 0$, then $|x| = -x$. Thus,

$$\frac{|x|}{x} = \frac{-x}{x} = -1.$$

Check Point 6 Rewrite each expression without absolute value bars:

a. $|1 - \sqrt{2}|$ **b.** $|\pi - 3|$ **c.** $\dfrac{|x|}{x}$ if $x > 0$.

Listed below are several basic properties of absolute value. Each of these properties can be derived from the definition of absolute value.

Properties of Absolute Value

For all real numbers a and b,

1. $|a| \geq 0$ **2.** $|-a| = |a|$ **3.** $a \leq |a|$

4. $|ab| = |a||b|$ **5.** $\left|\dfrac{a}{b}\right| = \dfrac{|a|}{|b|}, \quad b \neq 0$

6. $|a + b| \leq |a| + |b|$ (called the triangle inequality)

 Use absolute value to express distance.

Distance between Points on a Real Number Line

Absolute value is used to find the distance between two points on a real number line. If a and b are any real numbers, the **distance between a and b** is the absolute value of their difference. For example, the distance between 4 and 10 is 6. Using absolute value, we find this distance in one of two ways:

$$|10 - 4| = |6| = 6 \quad \text{or} \quad |4 - 10| = |-6| = 6.$$

The distance between 4 and 10 on the real number line is 6.

Notice that we obtain the same distance regardless of the order in which we subtract.

Distance between Two Points on the Real Number Line

If a and b are any two points on a real number line, then the distance between a and b is given by

$$|a - b| \quad \text{or} \quad |b - a|.$$

EXAMPLE 7   Distance between Two Points on a Number Line

Find the distance between -5 and 3 on the real number line.

Solution   Because the distance between a and b is given by $|a - b|$, the distance between -5 and 3 is

$$|-5 - 3| = |-8| = 8.$$

$a = -5$ $b = 3$

Figure P.9 verifies that there are 8 units between -5 and 3 on the real number line. We obtain the same distance if we reverse the order of the subtraction:

$$|3 - (-5)| = |8| = 8.$$

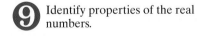

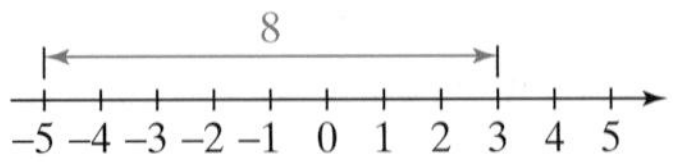

Figure P.9 The distance between -5 and 3 is 8.

Check Point 7   Find the distance between -4 and 5 on the real number line.

 Identify properties of the real numbers.

Properties of Real Numbers and Algebraic Expressions

When you use your calculator to add two real numbers, you can enter them in any order. The fact that two real numbers can be added in any order is called the **commutative property of addition**. You probably use this property, as well as other properties of real numbers listed in Table P.2, without giving it much thought. The properties of the real numbers are especially useful when working with algebraic expressions. For each property listed in Table P.2, *a*, *b*, and *c* represent real numbers, variables, or algebraic expressions.

Table P.2 Properties of the Real Numbers

Name	Meaning	Examples
Commutative Property of Addition	Changing order when adding does not affect the sum. $a + b = b + a$	• $13 + 7 = 7 + 13$ • $13x + 7 = 7 + 13x$
Commutative Property of Multiplication	Changing order when multiplying does not affect the product. $ab = ba$	• $\sqrt{2} \cdot \sqrt{5} = \sqrt{5} \cdot \sqrt{2}$ • $x \cdot 6 = 6x$
Associative Property of Addition	Changing grouping when adding does not affect the sum. $(a + b) + c = a + (b + c)$	• $3 + (8 + x) = (3 + 8) + x = 11 + x$
Associative Property of Multiplication	Changing grouping when multiplying does not affect the product. $(ab)c = a(bc)$	• $-2(3x) = (-2 \cdot 3)x = -6x$
Distributive Property of Multiplication over Addition	Multiplication distributes over addition. $a \cdot (b + c) = a \cdot b + a \cdot c$	• $7(4 + \sqrt{3}) = 7 \cdot 4 + 7 \cdot \sqrt{3} = 28 + 7\sqrt{3}$ • $5(3x + 7) = 5 \cdot 3x + 5 \cdot 7 = 15x + 35$
Identity Property of Addition	Zero can be deleted from a sum. $a + 0 = a$ $0 + a = a$	• $\sqrt{3} + 0 = \sqrt{3}$ • $0 + 6x = 6x$
Identity Property of Multiplication	One can be deleted from a product. $a \cdot 1 = a$ $1 \cdot a = a$	• $1 \cdot \pi = \pi$ • $13x \cdot 1 = 13x$
Inverse Property of Addition	The sum of a real number and its additive inverse gives 0, the additive identity. $a + (-a) = 0$ $(-a) + a = 0$	• $\sqrt{5} + (-\sqrt{5}) = 0$ • $-\pi + \pi = 0$ • $6x + (-6x) = 0$ • $(-4y) + 4y = 0$
Inverse Property of Multiplication	The product of a nonzero real number and its multiplicative inverse gives 1, the multiplicative identity. $a \cdot \frac{1}{a} = 1, \quad a \neq 0$ $\frac{1}{a} \cdot a = 1, \quad a \neq 0$	• $7 \cdot \frac{1}{7} = 1$ • $\left(\frac{1}{x-3}\right)(x - 3) = 1, \quad x \neq 3$

The Associative Property and the English Language

In the English language, phrases can take on different meanings depending on the way the words are associated with commas. Here are three examples.

- Woman, without her man, is nothing.
  Woman, without her, man is nothing.
- What's the latest dope?
  What's the latest, dope?
- Population of Amsterdam broken down by age and sex
  Population of Amsterdam, broken down by age and sex

Commutative Words and Sentences

The commutative property states that a change in order produces no change in the answer. The words and sentences listed here suggest a characteristic of the commutative property; they read the same from left to right and from right to left!

- dad
- repaper
- never odd or even
- Go deliver a dare, vile dog!
- May a moody baby doom a yam?
- Madam, in Eden I'm Adam.
- Ma is a nun, as I am.
- A man, a plan, a canal: Panama
- Are we not drawn onward, we few, drawn onward to new era?

The properties of the real numbers in Table P.2 on page 11 apply to the operations of addition and multiplication. Subtraction and division are defined in terms of addition and multiplication.

Definitions of Subtraction and Division

Let a and b represent real numbers.

Subtraction: $a - b = a + (-b)$
We call $-b$ the **additive inverse** or **opposite** of b.

Division: $a \div b = a \cdot \frac{1}{b}$, where $b \neq 0$
We call $\frac{1}{b}$ the **multiplicative inverse** or **reciprocal** of b. The quotient of a and b, $a \div b$, can be written in the form $\frac{a}{b}$, where a is the **numerator** and b the **denominator** of the fraction.

Because subtraction is defined in terms of adding an inverse, the distributive property can be applied to subtraction:

$$a(b - c) = ab - ac$$

$$(b - c)a = ba - ca.$$

For example,

$$4(2x - 5) = 4 \cdot 2x - 4 \cdot 5 = 8x - 20.$$

10 Simplify algebraic expressions.

Simplifying Algebraic Expressions

The **terms** of an algebraic expression are those parts that are separated by addition. For example, consider the algebraic expression

$$7x - 9y + z - 3,$$

which can be expressed as

$$7x + (-9y) + z + (-3).$$

This expression contains four terms, namely $7x$, $-9y$, z, and -3.

The numerical part of a term is called its **coefficient**. In the term $7x$, the 7 is the coefficient. If a term containing one or more variables is written without a coefficient, the coefficient is understood to be 1. Thus, z means $1z$. If a term is a constant, its coefficient is that constant. Thus, the coefficient of the constant term -3 is -3.

$$7x + (-9y) + z + (-3)$$

Coefficient is 7. | Coefficient is −9. | Coefficient is 1; z means $1z$. | Coefficient is −3.

The parts of each term that are multiplied are called the **factors** of the term. The factors of the term $7x$ are 7 and x.

Like terms are terms that have exactly the same variable factors. For example, $3x$ and $7x$ are like terms. The distributive property in the form

$$ba + ca = (b + c)a$$

enables us to add or subtract like terms. For example,

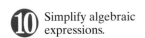

Study Tip

To combine like terms mentally, add or subtract the coefficients of the terms. Use this result as the coefficient of the terms' variable factor(s).

$$3x + 7x = (3 + 7)x = 10x$$
$$7y^2 - y^2 = 7y^2 - 1y^2 = (7 - 1)y^2 = 6y^2.$$

This process is called **combining like terms**.

An algebraic expression is **simplified** when parentheses have been removed and like terms have been combined.

EXAMPLE 8 Simplifying an Algebraic Expression

Simplify: $6(2x^2 + 4x) + 10(4x^2 + 3x)$.

Solution

$$6(2x^2 + 4x) + 10(4x^2 + 3x)$$
$$= 6 \cdot 2x^2 + 6 \cdot 4x + 10 \cdot 4x^2 + 10 \cdot 3x \quad \text{Use the distributive property to remove the parentheses.}$$
$$= 12x^2 + 24x + 40x^2 + 30x \quad \text{Multiply.}$$
$$= (12x^2 + 40x^2) + (24x + 30x) \quad \text{Group like terms.}$$
$$= 52x^2 + 54x \quad \text{Combine like terms.}$$

$52x^2$ and $54x$ are not like terms. They contain different variable factors, x^2 and x, and cannot be combined.

Check Point 8 Simplify: $7(4x^2 + 3x) + 2(5x^2 + x)$.

Properties of Negatives

The distributive property can be extended to cover more than two terms within parentheses. For example,

$$-3(4x - 2y + 6) = -3 \cdot 4x - (-3) \cdot 2y - 3 \cdot 6$$
$$= -12x - (-6y) - 18$$
$$= -12x + 6y - 18.$$

This sign represents subtraction. This sign tells us that the number is negative.

The voice balloons illustrate that negative signs can appear side by side. They can represent the operation of subtraction or the fact that a real number is negative. Here is a list of properties of negatives and how they are applied to algebraic expressions:

Properties of Negatives

Let a and b represent real numbers, variables, or algebraic expressions.

Property	Examples
1. $(-1)a = -a$	$(-1)4xy = -4xy$
2. $-(-a) = a$	$-(-6y) = 6y$
3. $(-a)b = -ab$	$(-7)4xy = -7 \cdot 4xy = -28xy$
4. $a(-b) = -ab$	$5x(-3y) = -5x \cdot 3y = -15xy$
5. $-(a + b) = -a - b$	$-(7x + 6y) = -7x - 6y$
6. $-(a - b) = -a + b$	$-(3x - 7y) = -3x + 7y$
$= b - a$	$= 7y - 3x$

It is not uncommon to see algebraic expressions with parentheses preceded by a negative sign or subtraction. Properties 5 and 6 in the box on the previous page, $-(a + b) = -a - b$ and $-(a - b) = -a + b$, are related to this situation. An expression of the form $-(a + b)$ can be simplified as follows:

$$-(a + b) = -1(a + b) = (-1)a + (-1)b = -a + (-b) = -a - b.$$

Do you see a fast way to obtain the simplified expression on the right? **If a negative sign or a subtraction symbol appears outside parentheses, drop the parentheses and change the sign of every term within the parentheses.** For example,

$$-(3x^2 - 7x - 4) = -3x^2 + 7x + 4.$$

EXAMPLE 9 Simplifying an Algebraic Expression

Simplify: $8x + 2[5 - (x - 3)]$.

Solution

$8x + 2[5 - (x - 3)]$

$= 8x + 2[5 - x + 3]$ — Drop parentheses and change the sign of each term in parentheses: $-(x - 3) = -x + 3$.

$= 8x + 2[8 - x]$ — Simplify inside brackets: $5 + 3 = 8$.

$= 8x + 16 - 2x$ — Apply the distributive property: $2[8 - x] = 2 \cdot 8 - 2x = 16 - 2x$.

$= (8x - 2x) + 16$ — Group like terms.

$= (8 - 2)x + 16$ — Apply the distributive property.

$= 6x + 16$ — Simplify.

Check Point 9 Simplify: $6 + 4[7 - (x - 2)]$.

EXERCISE SET P.1

Practice Exercises

In Exercises 1–16, evaluate each algebraic expression for the given value or values of the variable(s).

1. $7 + 5x$, for $x = 10$
2. $8 + 6x$, for $x = 5$
3. $6x - y$, for $x = 3$ and $y = 8$
4. $8x - y$, for $x = 3$ and $y = 4$
5. $x^2 + 3x$, for $x = 8$
6. $x^2 + 5x$, for $x = 6$
7. $x^2 - 6x + 3$, for $x = 7$
8. $x^2 - 7x + 4$, for $x = 8$
9. $4 + 5(x - 7)^3$, for $x = 9$
10. $6 + 5(x - 6)^3$, for $x = 8$
11. $x^2 - 3(x - y)$, for $x = 8$ and $y = 2$
12. $x^2 - 4(x - y)$, for $x = 8$ and $y = 3$
13. $\frac{5(x + 2)}{2x - 14}$, for $x = 10$
14. $\frac{7(x - 3)}{2x - 16}$, for $x = 9$
15. $\frac{2x + 3y}{x + 1}$, for $x = -2$ and $y = 4$
16. $\frac{2x + y}{xy - 2x}$, for $x = -2$ and $y = 4$

The formula

$$C = \frac{5}{9}(F - 32)$$

expresses the relationship between Fahrenheit temperature, F, and Celsius temperature, C. In Exercises 17–18, use the formula to convert the given Fahrenheit temperature to its equivalent temperature on the Celsius scale.

17. 50°F
18. 86°F

A football was kicked vertically upward from a height of 4 feet with an initial speed of 60 feet per second. The formula

$$h = 4 + 60t - 16t^2$$

describes the ball's height above the ground, h, in feet, t seconds after it was kicked. Use this formula to solve Exercises 19–20.

19. What was the ball's height 2 seconds after it was kicked?
20. What was the ball's height 3 seconds after it was kicked?

In Exercises 21–28, find the intersection of the sets.

21. $\{1, 2, 3, 4\} \cap \{2, 4, 5\}$
22. $\{1, 3, 7\} \cap \{2, 3, 8\}$
23. $\{s, e, t\} \cap \{t, e, s\}$
24. $\{r, e, a, l\} \cap \{l, e, a, r\}$

25. $\{1, 3, 5, 7\} \cap \{2, 4, 6, 8, 10\}$
26. $\{0, 1, 3, 5\} \cap \{-5, -3, -1\}$
27. $\{a, b, c, d\} \cap \emptyset$
28. $\{w, y, z\} \cap \emptyset$

In Exercises 29–34, find the union of the sets.

29. $\{1, 2, 3, 4\} \cup \{2, 4, 5\}$
30. $\{1, 3, 7, 8\} \cup \{2, 3, 8\}$
31. $\{1, 3, 5, 7\} \cup \{2, 4, 6, 8, 10\}$
32. $\{0, 1, 3, 5\} \cup \{2, 4, 6\}$
33. $\{a, e, i, o, u\} \cup \emptyset$
34. $\{e, m, p, t, y\} \cup \emptyset$

In Exercises 35–38, list all numbers from the given set that are **a.** *natural numbers,* **b.** *whole numbers,* **c.** *integers,* **d.** *rational numbers,* **e.** *irrational numbers,* **f.** *real numbers.*

35. $\left\{-9, -\frac{4}{5}, 0, 0.25, \sqrt{3}, 9.2, \sqrt{100}\right\}$
36. $\left\{-7, -0.\overline{6}, 0, \sqrt{49}, \sqrt{50}\right\}$
37. $\left\{-11, -\frac{5}{6}, 0, 0.75, \sqrt{5}, \pi, \sqrt{64}\right\}$
38. $\left\{-5, -0.\overline{3}, 0, \sqrt{2}, \sqrt{4}\right\}$
39. Give an example of a whole number that is not a natural number.
40. Give an example of a rational number that is not an integer.
41. Give an example of a number that is an integer, a whole number, and a natural number.
42. Give an example of a number that is a rational number, an integer, and a real number.

Determine whether each statement in Exercises 43–50 is true or false.

43. $-13 \leq -2$
44. $-6 > 2$
45. $4 \geq -7$
46. $-13 < -5$
47. $-\pi \geq -\pi$
48. $-3 > -13$
49. $0 \geq -6$
50. $0 \geq -13$

In Exercises 51–60, rewrite each expression without absolute value bars.

51. $|300|$
52. $|-203|$
53. $|12 - \pi|$
54. $|7 - \pi|$
55. $|\sqrt{2} - 5|$
56. $|\sqrt{5} - 13|$
57. $\dfrac{-3}{|-3|}$
58. $\dfrac{-7}{|-7|}$
59. $||-3| - |-7||$
60. $||-5| - |-13||$

In Exercises 61–66, evaluate each algebraic expression for $x = 2$ and $y = -5$.

61. $|x + y|$
62. $|x - y|$
63. $|x| + |y|$
64. $|x| - |y|$
65. $\dfrac{y}{|y|}$
66. $\dfrac{|x|}{x} + \dfrac{|y|}{y}$

In Exercises 67–74, express the distance between the given numbers using absolute value. Then find the distance by evaluating the absolute value expression.

67. 2 and 17
68. 4 and 15
69. -2 and 5
70. -6 and 8
71. -19 and -4
72. -26 and -3
73. -3.6 and -1.4
74. -5.4 and -1.2

In Exercises 75–84, state the name of the property illustrated.

75. $6 + (-4) = (-4) + 6$
76. $11 \cdot (7 + 4) = 11 \cdot 7 + 11 \cdot 4$
77. $6 + (2 + 7) = (6 + 2) + 7$
78. $6 \cdot (2 \cdot 3) = 6 \cdot (3 \cdot 2)$
79. $(2 + 3) + (4 + 5) = (4 + 5) + (2 + 3)$
80. $7 \cdot (11 \cdot 8) = (11 \cdot 8) \cdot 7$
81. $2(-8 + 6) = -16 + 12$
82. $-8(3 + 11) = -24 + (-88)$
83. $\dfrac{1}{(x + 3)}(x + 3) = 1, x \neq -3$
84. $(x + 4) + [-(x + 4)] = 0$

In Exercises 85–96, simplify each algebraic expression.

85. $5(3x + 4) - 4$
86. $2(5x + 4) - 3$
87. $5(3x - 2) + 12x$
88. $2(5x - 1) + 14x$
89. $7(3y - 5) + 2(4y + 3)$
90. $4(2y - 6) + 3(5y + 10)$
91. $5(3y - 2) - (7y + 2)$
92. $4(5y - 3) - (6y + 3)$
93. $7 - 4[3 - (4y - 5)]$
94. $6 - 5[8 - (2y - 4)]$
95. $18x^2 + 4 - [6(x^2 - 2) + 5]$
96. $14x^2 + 5 - [7(x^2 - 2) + 4]$

In Exercises 97–102, write each algebraic expression without parentheses.

97. $-(-14x)$
98. $-(-17y)$
99. $-(2x - 3y - 6)$
100. $-(5x - 13y - 1)$
101. $\frac{1}{3}(3x) + [(4y) + (-4y)]$
102. $\frac{1}{2}(2y) + [(-7x) + 7x]$

Practice Plus

In Exercises 103–110, insert either <, >, or = in the shaded area to make a true statement.

103. $|-6| \;\square\; |-3|$
104. $|-20| \;\square\; |-50|$
105. $\left|\dfrac{3}{5}\right| \;\square\; |-0.6|$
106. $\left|\dfrac{5}{2}\right| \;\square\; |-2.5|$
107. $\dfrac{30}{40} - \dfrac{3}{4} \;\square\; \dfrac{14}{15} \cdot \dfrac{15}{14}$
108. $\dfrac{17}{18} \cdot \dfrac{18}{17} \;\square\; \dfrac{50}{60} - \dfrac{5}{6}$
109. $\dfrac{8}{13} \div \dfrac{8}{13} \;\square\; |-1|$
110. $|-2| \;\square\; \dfrac{4}{17} \div \dfrac{4}{17}$

In Exercises 111–118, write each English phrase as an algebraic expression. Then simplify the expression. Let x represent the number.

111. A number decreased by the sum of the number and four
112. A number decreased by the difference between eight and the number
113. Six times the product of negative five and a number
114. Ten times the product of negative four and a number
115. The difference between the product of five and a number and twice the number
116. The difference between the product of six and a number and negative two times the number
117. The difference between eight times a number and six more than three times the number
118. Eight decreased by three times the sum of a number and six

Application Exercises

The bar graph shows the number of billionaires in the United States from 2000 through 2004.

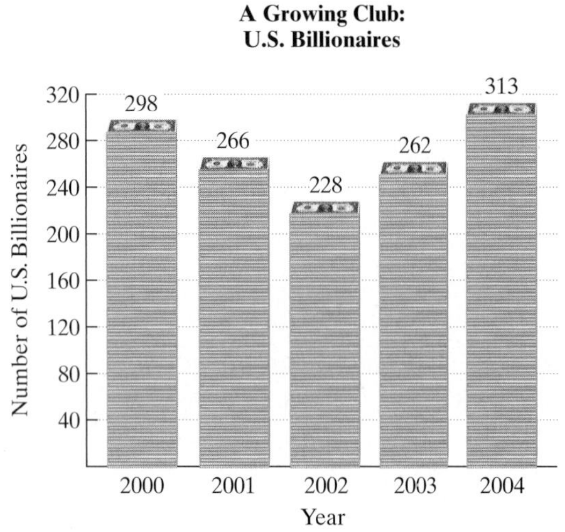

Source: Forbes magazine

The formula

$$N = 17x^2 - 65.4x + 302.2$$

models the number of billionaires, N, in the United States, x years after 2000. Use the formula to solve Exercises 119–122.

119. According to the formula, how many U.S. billionaires, to the nearest whole number, were there in 2004? How well does the formula model the actual data shown in the bar graph?

120. According to the formula, how many U.S. billionaires, to the nearest whole number, were there in 2003? How well does the formula model the actual data shown in the bar graph?

121. According to the formula, how many U.S. billionaires, to the nearest whole number, will there be in 2006?

122. According to the formula, how many U.S. billionaires, to the nearest whole number, will there be in 2007?

The bar graph shows that since her blockbuster debut, Britney Spears's album sales have slid downward.

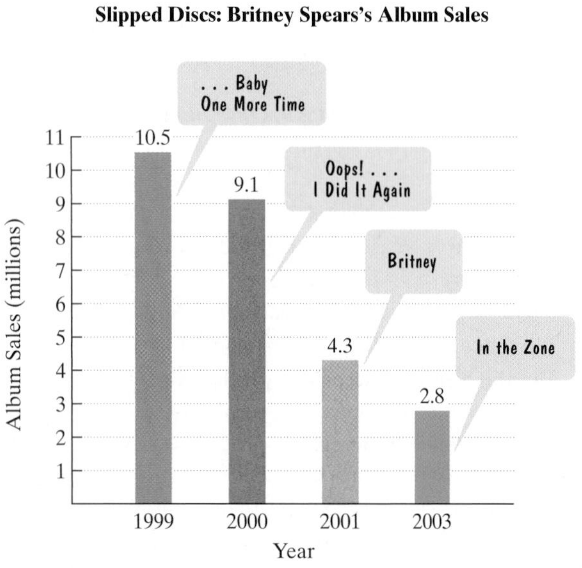

Source: Entertainment Weekly

Here are three mathematical models for the data shown in the graph. In each formula, N represents Britney Spears's album sales, in millions, x years after 1999.

Model 1 $N = -2.04x + 10.24$

Model 2 $N = 0.04x^2 - 3.6x + 11$

Model 3 $N = 0.76x^3 - 4x^2 + 1.8x + 10.5$

Use these models to solve Exercises 123–126.

123. Which of the three models best describes the actual number of album sales in 1999?

124. Which of the three models best describes the actual number of album sales in 2000?

125. Which of the three models best describes the actual number of album sales in 2003?

126. For which formula does model breakdown occur when describing the number of album sales in 2003?

127. You had \$10,000 to invest. You put x dollars in a safe, government-insured certificate of deposit paying 5% per year. You invested the remainder of the money in noninsured corporate bonds paying 12% per year. Your total interest earned at the end of the year is given by the algebraic expression

$$0.05x + 0.12(10{,}000 - x).$$

a. Simplify the algebraic expression.

b. Use each form of the algebraic expression to determine your total interest earned at the end of the year if you invested \$6000 in the safe, government-insured certificate of deposit.

128. It takes you 50 minutes to get to campus. You spend t minutes walking to the bus stop and the rest of the time riding the bus. Your walking rate is 0.06 miles per minute and the bus travels at a rate of 0.5 miles per minute. The total distance walking and traveling by bus is given by the algebraic expression

$$0.06t + 0.5(50 - t).$$

a. Simplify the algebraic expression.

b. Use each form of the algebraic expression to determine the total distance that you travel if you spend 20 minutes walking to the bus stop.

Writing in Mathematics

Writing about mathematics will help you learn mathematics. For all writing exercises in this book, use complete sentences to respond to the question. Some writing exercises can be answered in a sentence; others require a paragraph or two. You can decide how much you need to write as long as your writing clearly and directly answers the question in the exercise. Standard references such as a dictionary and a thesaurus should be helpful.

129. What is an algebraic expression? Give an example with your explanation.

130. If n is a natural number, what does b^n mean? Give an example with your explanation.

131. What does it mean when we say that a formula models real-world phenomena?

132. What is the intersection of sets A and B?

133. What is the union of sets A and B?

134. How do the whole numbers differ from the natural numbers?

135. Can a real number be both rational and irrational? Explain your answer.

136. If you are given two real numbers, explain how to determine which is the lesser.

137. How can $\frac{|x|}{x}$ be equal to 1 or -1?

138. Describe the difference between the commutative and the associative properties of addition.

139. Why is $3(x + 7) - 4x$ not simplified? What must be done to simplify the expression?

Critical Thinking Exercises

140. Which one of the following statements is true?

a. Every rational number is an integer.
b. Some whole numbers are not integers.
c. Some rational numbers are not positive.
d. Irrational numbers cannot be negative.

141. Which of the following is true?

a. The term x has no coefficient.
b. $5 + 3(x - 4) = 8(x - 4) = 8x - 32$
c. $-x - x = -x + (-x) = 0$
d. $x - 0.02(x + 200) = 0.98x - 4$

In Exercises 142–144, insert either < or > in the shaded area between the numbers to make the statement true.

142. $\sqrt{2}$ ▒ 1.5 **143.** $-\pi$ ▒ -3.5

144. $-\frac{3.14}{2}$ ▒ $-\frac{\pi}{2}$

145. A business that manufactures small alarm clocks has a weekly fixed cost of \$5000. The average cost per clock for the business to manufacture x clocks is described by

$$\frac{0.5x + 5000}{x}.$$

a. Find the average cost when $x = 100$, 1000, and 10,000.

b. Like all other businesses, the alarm clock manufacturer must make a profit. To do this, each clock must be sold for at least 50¢ more than what it costs to manufacture. Due to competition from a larger company, the clocks can be sold for \$1.50 each and no more. Our small manufacturer can only produce 2000 clocks weekly. Does this business have much of a future? Explain.

SECTION P.2 Exponents and Scientific Notation

Objectives

1. Use the product rule.
2. Use the quotient rule.
3. Use the zero-exponent rule.
4. Use the negative-exponent rule.
5. Use the power rule.
6. Find the power of a product.
7. Find the power of a quotient.
8. Simplify exponential expressions.
9. Use scientific notation.

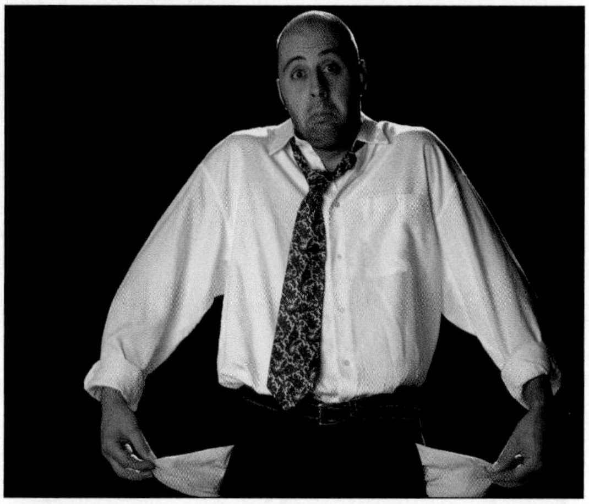

People who complain about paying their income tax can be divided into two types: men and women. Perhaps we can quantify the complaining by examining the data in Figure P.10. The bar graphs show the U.S. population, in millions, and the total amount we paid in federal taxes, in trillions of dollars, for six selected years.

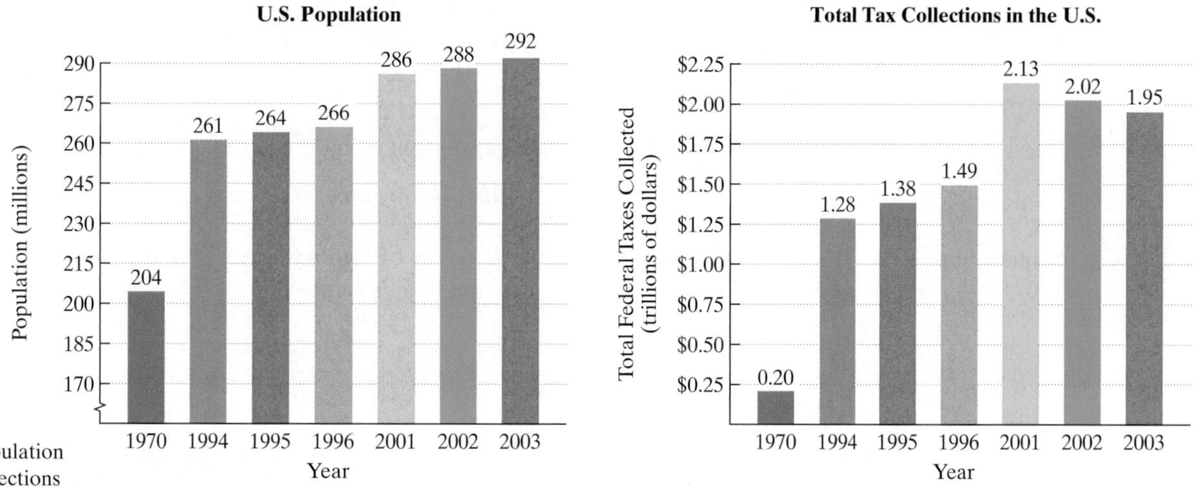

Source: U.S. Census Bureau

Source: Internal Revenue Service

Figure P.10 Population and total tax collections in the U.S.

The bar graph in Figure P.10 shows that in 2003, total tax collections were \$1.95 trillion. How can we place this amount in the proper perspective? If the total tax collections were evenly divided among all Americans, how much would each citizen pay in taxes?

In this section, you will learn to use exponents to provide a way of putting large and small numbers in perspective. Using this skill, we will explore the per capita tax for some of the years shown in Figure P.10 on the previous page.

1 Use the product rule.

The Product and Quotient Rules

We have seen that exponents are used to indicate repeated multiplication. Now consider the multiplication of two exponential expressions, such as $b^4 \cdot b^3$. We are multiplying 4 factors of b and 3 factors of b. We have a total of 7 factors of b:

4 factors of b | 3 factors of b

$$b^4 \cdot b^3 = (b \cdot b \cdot b \cdot b)(b \cdot b \cdot b) = b^7.$$

Total: 7 factors of b

The product is exactly the same if we add the exponents:

$$b^4 \cdot b^3 = b^{4+3} = b^7.$$

This suggests the following rule:

The Product Rule

$$b^m \cdot b^n = b^{m+n}$$

When multiplying exponential expressions with the same base, add the exponents. Use this sum as the exponent of the common base.

EXAMPLE 1 Using the Product Rule

Multiply each expression using the product rule:

a. $2^2 \cdot 2^3$ **b.** $(6x^4y^3)(5x^2y^7)$.

Solution

a. $2^2 \cdot 2^3 = 2^{2+3} = 2^5$ or 32

$2^5 = 2 \cdot 2 \cdot 2 \cdot 2 \cdot 2 = 32$

b. $(6x^4y^3)(5x^2y^7)$

$= 6 \cdot 5 \cdot x^4 \cdot x^2 \cdot y^3 \cdot y^7$ Use the associative and commutative properties. This step can be done mentally.

$= 30x^{4+2}y^{3+7}$

$= 30x^6y^{10}$

Check Point 1 Multiply each expression using the product rule:

a. $3^3 \cdot 3^2$ **b.** $(4x^3y^4)(10x^2y^6)$.

2 Use the quotient rule.

Now, consider the division of two exponential expressions, such as the quotient of b^7 and b^3. We are dividing 7 factors of b by 3 factors of b.

$$\frac{b^7}{b^3} = \frac{b \cdot b \cdot b \cdot b \cdot b \cdot b \cdot b}{b \cdot b \cdot b} = \boxed{\frac{b \cdot b \cdot b}{b \cdot b \cdot b}} \cdot b \cdot b \cdot b \cdot b = 1 \cdot b \cdot b \cdot b \cdot b = b^4$$

This factor is equal to 1.

The quotient is exactly the same if we subtract the exponents:

$$\frac{b^7}{b^3} = b^{7-3} = b^4.$$

This suggests the following rule:

The Quotient Rule

$$\frac{b^m}{b^n} = b^{m-n}, \quad b \neq 0$$

When dividing exponential expressions with the same nonzero base, subtract the exponent in the denominator from the exponent in the numerator. Use this difference as the exponent of the common base.

EXAMPLE 2

Divide each expression using the quotient rule:

a. $\frac{(-2)^7}{(-2)^4}$ **b.** $\frac{30x^{12}y^9}{5x^3y^7}$.

Solution

a. $\frac{(-2)^7}{(-2)^4} = (-2)^{7-4} = (-2)^3$ or -8

$(-2)^3 = (-2)(-2)(-2) = -8$

b. $\frac{30x^{12}y^9}{5x^3y^7} = \frac{30}{5} \cdot \frac{x^{12}}{x^3} \cdot \frac{y^9}{y^7} = 6x^{12-3}y^{9-7} = 6x^9y^2$

Check Point 2 Divide each expression using the quotient rule:

a. $\frac{(-3)^6}{(-3)^3}$ **b.** $\frac{27x^{14}y^8}{3x^3y^5}$.

3 Use the zero-exponent rule.

Zero as an Exponent

A nonzero base can be raised to the 0 power. The quotient rule can be used to help determine what zero as an exponent should mean. Consider the quotient of b^4 and b^4, where b is not zero. We can determine this quotient in two ways.

$$\frac{b^4}{b^4} = 1 \qquad \frac{b^4}{b^4} = b^{4-4} = b^0$$

Any nonzero expression divided by itself is 1.

Use the quotient rule and subtract exponents.

This means that b^0 must equal 1.

The Zero-Exponent Rule

If b is any real number other than 0,

$$b^0 = 1.$$

Here are examples involving simplification using the zero-exponent rule:

$$8^0 = 1, \quad (-6)^0 = 1, \quad -6^0 = -1, \quad (5x)^0 = 1, \quad 5x^0 = 5.$$

Only 6 is raised to the 0 power: $-6^0 = -(6^0) = -1$.

Only x is raised to the 0 power: $5x^0 = 5 \cdot 1 = 5$.

4 Use the negative-exponent rule.

Negative Integers as Exponents

A nonzero base can be raised to a negative power. The quotient rule can be used to help determine what a negative integer as an exponent should mean. Consider the quotient of b^3 and b^5, where b is not zero. We can determine this quotient in two ways.

$$\frac{b^3}{b^5} = \frac{\cancel{b}\cdot\cancel{b}\cdot\cancel{b}}{\cancel{b}\cdot\cancel{b}\cdot\cancel{b}\cdot b\cdot b} = \frac{1}{b^2} \qquad \frac{b^3}{b^5} = b^{3-5} = b^{-2}$$

After dividing common factors, we have two factors of b in the denominator.

Use the quotient rule and subtract exponents.

Notice that $\frac{b^3}{b^5}$ equals both b^{-2} and $\frac{1}{b^2}$. This means that b^{-2} must equal $\frac{1}{b^2}$. This example is a special case of the **negative-exponent rule.**

The Negative-Exponent Rule

If b is any real number other than 0 and n is a natural number, then

$$b^{-n} = \frac{1}{b^n}.$$

EXAMPLE 3 Using the Negative-Exponent Rule

Use the negative-exponent rule to write each expression with a positive exponent. Simplify, if possible:

a. 9^{-2} **b.** $(-2)^{-5}$ **c.** $\frac{1}{6^{-2}}$ **d.** $7x^{-5}y^2$.

Solution

a. $9^{-2} = \frac{1}{9^2} = \frac{1}{81}$

b. $(-2)^{-5} = \frac{1}{(-2)^5} = \frac{1}{(-2)(-2)(-2)(-2)(-2)} = \frac{1}{-32} = -\frac{1}{32}$

Only the sign of the exponent, −5, changes. The base, −2, does not change sign.

c. $\frac{1}{6^{-2}} = \frac{1}{\frac{1}{6^2}} = 1\cdot\frac{6^2}{1} = 6^2 = 36$

d. $7x^{-5}y^2 = 7\cdot\frac{1}{x^5}\cdot y^2 = \frac{7y^2}{x^5}$

Check Point 3 Use the negative-exponent rule to write each expression with a positive exponent. Simplify, if possible:

a. 5^{-2} **b.** $(-3)^{-3}$ **c.** $\frac{1}{4^{-2}}$ **d.** $3x^{-6}y^4$.

In Example 3 and Check Point 3, did you notice that

$$\frac{1}{6^{-2}} = 6^2 \quad \text{and} \quad \frac{1}{4^{-2}} = 4^2?$$

In general, if a negative exponent appears in a denominator, an expression can be written with a positive exponent using

$$\frac{1}{b^{-n}} = b^n.$$

Negative Exponents in Numerators and Denominators

If b is any real number other than 0 and n is a natural number, then

$$b^{-n} = \frac{1}{b^n} \quad \text{and} \quad \frac{1}{b^{-n}} = b^n.$$

When a negative number appears as an exponent, switch the position of the base (from numerator to denominator or from denominator to numerator) and make the exponent positive. The sign of the base does not change.

5 Use the power rule.

The Power Rule for Exponents (Powers to Powers)

The next property of exponents applies when an exponential expression is raised to a power. Here is an example:

$$(b^2)^4.$$

The exponential expression b^2 is raised to the fourth power.

There are 4 factors of b^2. Thus,

$$(b^2)^4 = b^2 \cdot b^2 \cdot b^2 \cdot b^2 = b^{2+2+2+2} = b^8.$$

Add exponents when multiplying with the same base.

We can obtain the answer, b^8, by multiplying the exponents:

$$(b^2)^4 = b^{2 \cdot 4} = b^8.$$

This suggests the following rule:

The Power Rule (Powers to Powers)

$$(b^m)^n = b^{mn}$$

When an exponential expression is raised to a power, multiply the exponents. Place the product of the exponents on the base and remove the parentheses.

EXAMPLE 4 Using the Power Rule (Powers to Powers)

Simplify each expression using the power rule:

a. $(2^2)^3$ **b.** $(y^5)^{-3}$ **c.** $(b^{-4})^{-2}$.

Solution

a. $(2^2)^3 = 2^{2 \cdot 3} = 2^6$ or 64

b. $(y^5)^{-3} = y^{5(-3)} = y^{-15} = \dfrac{1}{y^{15}}$

c. $(b^{-4})^{-2} = b^{(-4)(-2)} = b^8$

Check Point 4 Simplify each expression using the power rule:

a. $(3^3)^2$ **b.** $(y^7)^{-2}$ **c.** $(b^{-3})^{-4}$.

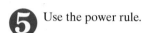

6 Find the power of a product.

The Products-to-Powers Rule for Exponents

The next property of exponents applies when we are raising a product to a power. Here is an example:

$$(2x)^4.$$

The product $2x$ is raised to the fourth power.

There are four factors of $2x$. Thus,

$$(2x)^4 = 2x \cdot 2x \cdot 2x \cdot 2x = 2 \cdot 2 \cdot 2 \cdot 2 \cdot x \cdot x \cdot x \cdot x = 2^4x^4.$$

We can obtain the answer, 2^4x^4, by raising each factor within the parentheses to the fourth power:

$$(2x)^4 = 2^4x^4.$$

This suggests the following rule:

Products to Powers

$$(ab)^n = a^nb^n$$

When a product is raised to a power, raise each factor to that power.

EXAMPLE 5 Raising a Product to a Power

Simplify: $(-2y^2)^4$.

Solution

$(-2y^2)^4 = (-2)^4(y^2)^4$ Raise each factor to the fourth power.

$= (-2)^4y^{2\cdot4}$ To raise an exponential expression to a power, multiply exponents: $(b^m)^n = b^{mn}$.

$= 16y^8$ Simplify: $(-2)^4 = (-2)(-2)(-2)(-2) = 16$.

Check Point 5 Simplify: $(-4x)^3$.

The rule for raising a product to a power can be extended to cover three or more factors. For example,

$$(-2xy)^3 = (-2)^3x^3y^3 = -8x^3y^3.$$

7 Find the power of a quotient.

The Quotients-to-Powers Rule for Exponents

The following rule is used to raise a quotient to a power:

Quotients to Powers

If b is a nonzero real number, then

$$\left(\frac{a}{b}\right)^n = \frac{a^n}{b^n}.$$

When a quotient is raised to a power, raise the numerator to that power and divide by the denominator to that power.

EXAMPLE 6 Raising Quotients to Powers

Simplify by raising each quotient to the given power:

a. $\left(-\frac{3}{x}\right)^4$ **b.** $\left(\frac{x^2}{4}\right)^3$.

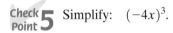

Solution

a. $\left(-\frac{3}{x}\right)^4 = \frac{(-3)^4}{x^4} = \frac{(-3)(-3)(-3)(-3)}{x^4} = \frac{81}{x^4}$

b. $\left(\frac{x^2}{4}\right)^3 = \frac{(x^2)^3}{4^3} = \frac{x^{2\cdot 3}}{4\cdot 4\cdot 4} = \frac{x^6}{64}$

Check Point 6 Simplify:

a. $\left(-\frac{2}{y}\right)^5$ **b.** $\left(\frac{x^5}{3}\right)^3$.

8 Simplify exponential expressions.

Simplifying Exponential Expressions

Properties of exponents are used to simplify exponential expressions. An exponential expression is **simplified** when

- No parentheses appear.
- No powers are raised to powers.
- Each base occurs only once.
- No negative or zero exponents appear.

Simplifying Exponential Expressions

	Example
1. If necessary, remove parentheses by using $(ab)^n = a^n b^n$ or $\left(\frac{a}{b}\right)^n = \frac{a^n}{b^n}$.	$(xy)^3 = x^3y^3$
2. If necessary, simplify powers to powers by using $(b^m)^n = b^{mn}$.	$(x^4)^3 = x^{4\cdot 3} = x^{12}$
3. If necessary, be sure that each base appears only once by using $b^m \cdot b^n = b^{m+n}$ or $\frac{b^m}{b^n} = b^{m-n}$.	$x^4 \cdot x^3 = x^{4+3} = x^7$
4. If necessary, rewrite exponential expressions with zero powers as 1 ($b^0 = 1$). Furthermore, write the answer with positive exponents by using $b^{-n} = \frac{1}{b^n}$ or $\frac{1}{b^{-n}} = b^n$.	$\frac{x^5}{x^8} = x^{5-8} = x^{-3} = \frac{1}{x^3}$

The following example shows how to simplify exponential expressions. Throughout the example, assume that no variable in a denominator is equal to zero.

EXAMPLE 7 Simplifying Exponential Expressions

Simplify:

a. $(-3x^4y^5)^3$ **b.** $(-7xy^4)(-2x^5y^6)$ **c.** $\frac{-35x^2y^4}{5x^6y^{-8}}$ **d.** $\left(\frac{4x^2}{y}\right)^{-3}$.

Solution

a. $(-3x^4y^5)^3 = (-3)^3(x^4)^3(y^5)^3$ Raise each factor inside the parentheses to the third power.

$= (-3)^3x^{4\cdot 3}y^{5\cdot 3}$ Multiply the exponents when raising powers to powers.

$= -27x^{12}y^{15}$ $(-3)^3 = (-3)(-3)(-3) = -27$

b. $(-7xy^4)(-2x^5y^6) = (-7)(-2)xx^5y^4y^6$ Group factors with the same base.

$= 14x^{1+5}y^{4+6}$ When multiplying expressions with the same base, add the exponents.

$= 14x^6y^{10}$ Simplify.

c. $\dfrac{-35x^2y^4}{5x^6y^{-8}} = \left(\dfrac{-35}{5}\right)\left(\dfrac{x^2}{x^6}\right)\left(\dfrac{y^4}{y^{-8}}\right)$ Group factors with the same base.

$= -7x^{2-6}y^{4-(-8)}$ When dividing expressions with the same base, subtract the exponents.

$= -7x^{-4}y^{12}$ Simplify. Notice that $4 - (-8) = 4 + 8 = 12$.

$= \dfrac{-7y^{12}}{x^4}$ Move the base with the negative exponent, x^{-4}, to the other side of the fraction bar and make the negative exponent positive.

d. $\left(\dfrac{4x^2}{y}\right)^{-3} = \dfrac{(4x^2)^{-3}}{y^{-3}}$ Raise the numerator and the denominator to the -3 power.

$= \dfrac{4^{-3}(x^2)^{-3}}{y^{-3}}$ Raise each factor in the numerator to the -3 power.

$= \dfrac{4^{-3}x^{-6}}{y^{-3}}$ Multiply the exponents when raising a power to a power: $(x^2)^{-3} = x^{2(-3)} = x^{-6}$.

$= \dfrac{y^3}{4^3x^6}$ Move each base with a negative exponent to the other side of the fraction bar and make each negative exponent positive.

$= \dfrac{y^3}{64x^6}$ $4^3 = 4 \cdot 4 \cdot 4 = 64$

Check Point 7 Simplify:

a. $(2x^3y^6)^4$ **b.** $(-6x^2y^5)(3xy^3)$ **c.** $\dfrac{100x^{12}y^2}{20x^{16}y^{-4}}$ **d.** $\left(\dfrac{5x}{y^4}\right)^{-2}$.

Study Tip

Try to avoid the following common errors that can occur when simplifying exponential expressions.

Correct	Incorrect	Description of Error
$b^3 \cdot b^4 = b^7$	~~$b^3 \cdot b^4 = b^{12}$~~	The exponents should be added, not multiplied.
$3^2 \cdot 3^4 = 3^6$	~~$3^2 \cdot 3^4 = 9^6$~~	The common base should be retained, not multiplied.
$\dfrac{5^{16}}{5^4} = 5^{12}$	~~$\dfrac{5^{16}}{5^4} = 5^4$~~	The exponents should be subtracted, not divided.
$(4a)^3 = 64a^3$	~~$(4a)^3 = 4a^3$~~	Both factors should be cubed.
$b^{-n} = \dfrac{1}{b^n}$	~~$b^{-n} = -\dfrac{1}{b^n}$~~	Only the exponent should change sign.
$(a+b)^{-1} = \dfrac{1}{a+b}$	~~$(a+b)^{-1} = \dfrac{1}{a} + \dfrac{1}{b}$~~	The exponent applies to the entire expression $a + b$.

9 Use scientific notation.

Scientific Notation

We have seen that in 2003, total tax collections were \$1.95 trillion. Because a trillion is 10^{12} (see Table P.3), this amount can be expressed as

$$1.95 \times 10^{12}.$$

The number 1.95×10^{12} is written in a form called *scientific notation.*

Table P.3 Names of Large Numbers

10^2	hundred
10^3	thousand
10^6	million
10^9	billion
10^{12}	trillion
10^{15}	quadrillion
10^{18}	quintillion
10^{21}	sextillion
10^{24}	septillion
10^{27}	octillion
10^{30}	nonillion
10^{100}	googol

Scientific Notation

A number is written in **scientific notation** when it is expressed in the form

$$a \times 10^n,$$

where the absolute value of a is greater than or equal to 1 and less than 10 $(1 \le |a| < 10)$, and n is an integer.

It is customary to use the multiplication symbol, $\times$, rather than a dot, when writing a number in scientific notation.

Converting from Scientific to Decimal Notation

Here are two examples of numbers in scientific notation:

$$6.4 \times 10^5 \quad \text{means} \quad 640{,}000.$$
$$2.17 \times 10^{-3} \quad \text{means} \quad 0.00217.$$

Do you see that the number with the positive exponent is relatively large and the number with the negative exponent is relatively small?

We can use n, the exponent on the 10 in $a \times 10^n$, to change a number in scientific notation to decimal notation. If n is **positive**, move the decimal point in a to the **right** n places. If n is **negative**, move the decimal point in a to the **left** $|n|$ places.

EXAMPLE 8 Converting from Scientific to Decimal Notation

Write each number in decimal notation:

a. 6.2×10^7 **b.** -6.2×10^7 **c.** 2.019×10^{-3} **d.** -2.019×10^{-3}.

Solution In each case, we use the exponent on the 10 to move the decimal point. In parts (a) and (b), the exponent is positive, so we move the decimal point to the right. In parts (c) and (d), the exponent is negative, so we move the decimal point to the left.

a. $6.2 \times 10^7 = 62{,}000{,}000$

$n = 7$ — Move the decimal point 7 places to the right.

b. $-6.2 \times 10^7 = -62{,}000{,}000$

$n = 7$ — Move the decimal point 7 places to the right.

c. $2.019 \times 10^{-3} = 0.002019$

$n = -3$ — Move the decimal point $|-3|$ places, or 3 places, to the left.

d. $-2.019 \times 10^{-3} = -0.002019$

$n = -3$ — Move the decimal point $|-3|$ places, or 3 places, to the left.

Check Point 8 Write each number in decimal notation:

a. -2.6×10^9 **b.** 3.017×10^{-6}.

Converting from Decimal to Scientific Notation

To convert from decimal notation to scientific notation, we reverse the procedure of Example 8.

Converting from Decimal to Scientific Notation

Write the number in the form $a \times 10^n$.

- Determine a, the numerical factor. Move the decimal point in the given number to obtain a number whose absolute value is between 1 and 10, including 1.
- Determine n, the exponent on 10^n. The absolute value of n is the number of places the decimal point was moved. The exponent n is positive if the decimal point was moved to the left, negative if the decimal point was moved to the right, and 0 if the decimal point was not moved.

EXAMPLE 9 Converting from Decimal Notation to Scientific Notation

Write each number in scientific notation:

a. 34,970,000,000,000 **b.** −34,970,000,000,000

b. 0.0000000000802 **d.** −0.0000000000802.

Solution

a. $34{,}970{,}000{,}000{,}000 = 3.497 \times 10^{13}$

Move the decimal point to get a number whose absolute value is between 1 and 10.

The decimal point was moved 13 places to the left, so $n = 13$.

b. $-34{,}970{,}000{,}000{,}000 = -3.497 \times 10^{13}$

c. $0.0000000000802 = 8.02 \times 10^{-11}$

Move the decimal point to get a number whose absolute value is between 1 and 10.

The decimal point was moved 11 places to the right, so $n = -11$.

d. $-0.0000000000802 = -8.02 \times 10^{-11}$

Technology

You can use your calculator's EE (enter exponent) or EXP key to convert from decimal to scientific notation. Here is how it's done for 0.0000000000802.

Many Scientific Calculators

Keystrokes

.0000000000802 EE =

Display

8.02 − 11

Many Graphing Calculators

Use the mode setting for scientific notation.

Keystrokes

.0000000000802 ENTER

Display

8.02E −11

Study Tip

If the absolute value of a number is greater than 10, it will have a positive exponent in scientific notation. If the absolute value of a number is less than 1, it will have a negative exponent in scientific notation.

Check Point 9 Write each number in scientific notation:

a. 5,210,000,000 **b.** −0.00000006893.

EXAMPLE 10 Expressing the U.S. Population in Scientific Notation

In 2003, the population of the United States was approximately 292 million. Express the population in scientific notation.

Solution Because one million is 10^6 (see Table P.3 on page 25), the 2003 population can be expressed as

$$292 \times 10^6.$$

This factor is not between 1 and 10, so the number is not in scientific notation.

The voice balloon indicates that we need to convert 292 to scientific notation.

$$292 \times 10^6 = (2.92 \times 10^2) \times 10^6 = 2.92 \times 10^{2+6} = 2.92 \times 10^8$$

$292 = 2.92 \times 10^2$

In scientific notation, the population is 2.92×10^8.

Check Point 10 Express 410×10^7 in scientific notation.

Computations with Scientific Notation

Properties of exponents are used to perform computations with numbers that are expressed in scientific notation.

Technology

$(6.1 \times 10^5)(4 \times 10^{-9})$
On a Calculator:

Many Scientific Calculators

6.1 [EE] 5 [×] 4 [EE] 9 [+/−] [=]

Display

2.44 − 03

Many Graphing Calculators

6.1 [EE] 5 [×] 4 [EE] [(−)] 9 [ENTER]

Display (in scientific notation mode)

2.44E − 3

EXAMPLE 11 Computations with Scientific Notation

Perform the indicated computations, writing the answers in scientific notation:

a. $(6.1 \times 10^5)(4 \times 10^{-9})$ **b.** $\dfrac{1.8 \times 10^4}{3 \times 10^{-2}}$.

Solution

a. $(6.1 \times 10^5)(4 \times 10^{-9})$

$= (6.1 \times 4) \times (10^5 \times 10^{-9})$ Regroup factors.

$= 24.4 \times 10^{5+(-9)}$ Add the exponents on 10 and multiply the other parts.

$= 24.4 \times 10^{-4}$ Simplify.

$= (2.44 \times 10^1) \times 10^{-4}$ Convert 24.4 to scientific notation: $24.4 = 2.44 \times 10^1$.

$= 2.44 \times 10^{-3}$ $10^1 \times 10^{-4} = 10^{1+(-4)} = 10^{-3}$

b. $\dfrac{1.8 \times 10^4}{3 \times 10^{-2}} = \left(\dfrac{1.8}{3}\right) \times \left(\dfrac{10^4}{10^{-2}}\right)$ Regroup factors.

$= 0.6 \times 10^{4-(-2)}$ Subtract the exponents on 10 and divide the other parts.

$= 0.6 \times 10^6$ Simplify: $4 - (-2) = 4 + 2 = 6$.

$= (6 \times 10^{-1}) \times 10^6$ Convert 0.6 to scientific notation: $0.6 = 6 \times 10^{-1}$.

$= 6 \times 10^5$ $10^{-1} \times 10^6 = 10^{-1+6} = 10^5$

Check Point 11 Perform the indicated computations, writing the answers in scientific notation:

a. $(7.1 \times 10^5)(5 \times 10^{-7})$ **b.** $\dfrac{1.2 \times 10^6}{3 \times 10^{-3}}$.

Applications: Putting Numbers in Perspective

We have seen that in 2003, the U.S. government collected \$1.95 trillion in taxes. Example 12 shows how we can use scientific notation to comprehend the meaning of a number such as 1.95 trillion.

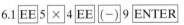

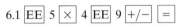

EXAMPLE 12 Tax per Capita

In 2003, the U.S. government collected 1.95×10^{12} dollars in taxes. At that time, the U.S. population was approximately 292 million, or 2.92×10^8. If the total tax collections were evenly divided among all Americans, how much would each citizen pay? Express the answer in decimal notation, rounded to the nearest dollar.

Solution The amount that we would each pay, or the tax per capita, is the total amount collected, 1.95×10^{12}, divided by the number of Americans, 2.92×10^8.

$$\frac{1.95 \times 10^{12}}{2.92 \times 10^8} = \left(\frac{1.95}{2.92}\right) \times \left(\frac{10^{12}}{10^8}\right) \approx 0.6678 \times 10^{12-8} = 0.6678 \times 10^4 = 6678$$

To obtain an answer in decimal notation, it is not necessary to express this number in scientific notation.

Move the decimal point 4 places to the right.

If total tax collections were evenly divided, we would each pay approximately $6678 in taxes.

Check Point 12 In 2002, the U.S. government collected 2.02×10^{12} dollars in taxes. At that time, the U.S. population was approximately 288 million, or 2.88×10^8. Find the per capita tax, rounded to the nearest dollar, in 2002.

EXERCISE SET P.2

Practice Exercises

Evaluate each exponential expression in Exercises 1–22.

1. $5^2 \cdot 2$
2. $6^2 \cdot 2$
3. $(-2)^6$
4. $(-2)^4$
5. -2^6
6. -2^4
7. $(-3)^0$
8. $(-9)^0$
9. -3^0
10. -9^0
11. 4^{-3}
12. 2^{-6}
13. $2^2 \cdot 2^3$
14. $3^3 \cdot 3^2$
15. $(2^2)^3$
16. $(3^3)^2$
17. $\frac{2^8}{2^4}$
18. $\frac{3^8}{3^4}$
19. $3^{-3} \cdot 3$
20. $2^{-3} \cdot 2$
21. $\frac{2^3}{2^7}$
22. $\frac{3^4}{3^7}$

Simplify each exponential expression in Exercises 23–64.

23. $x^{-2}y$
24. xy^{-3}
25. x^0y^5
26. x^7y^0
27. $x^3 \cdot x^7$
28. $x^{11} \cdot x^5$
29. $x^{-5} \cdot x^{10}$
30. $x^{-6} \cdot x^{12}$
31. $(x^3)^7$
32. $(x^{11})^5$
33. $(x^{-5})^3$
34. $(x^{-6})^4$
35. $\frac{x^{14}}{x^7}$
36. $\frac{x^{30}}{x^{10}}$
37. $\frac{x^{14}}{x^{-7}}$
38. $\frac{x^{30}}{x^{-10}}$
39. $(8x^3)^2$
40. $(6x^4)^2$
41. $\left(-\frac{4}{x}\right)^3$
42. $\left(-\frac{6}{y}\right)^3$
43. $(-3x^2y^5)^2$
44. $(-3x^4y^6)^3$
45. $(3x^4)(2x^7)$
46. $(11x^5)(9x^{12})$
47. $(-9x^3y)(-2x^6y^4)$
48. $(-5x^4y)(-6x^7y^{11})$
49. $\frac{8x^{20}}{2x^4}$
50. $\frac{20x^{24}}{10x^6}$
51. $\frac{25a^{13}b^4}{-5a^2b^3}$
52. $\frac{35a^{14}b^6}{-7a^7b^3}$
53. $\frac{14b^7}{7b^{14}}$
54. $\frac{20b^{10}}{10b^{20}}$
55. $(4x^3)^{-2}$
56. $(10x^2)^{-3}$
57. $\frac{24x^3y^5}{32x^7y^{-9}}$
58. $\frac{10x^4y^9}{30x^{12}y^{-3}}$
59. $\left(\frac{5x^3}{y}\right)^{-2}$
60. $\left(\frac{3x^4}{y}\right)^{-3}$
61. $\left(\frac{-15a^4b^2}{5a^{10}b^{-3}}\right)^3$
62. $\left(\frac{-30a^{14}b^8}{10a^{17}b^{-2}}\right)^3$
63. $\left(\frac{3a^{-5}b^2}{12a^3b^{-4}}\right)^0$
64. $\left(\frac{4a^{-5}b^3}{12a^3b^{-5}}\right)^0$

In Exercises 65–76, write each number in decimal notation without the use of exponents.

65. 3.8×10^2 **66.** 9.2×10^2

67. 6×10^{-4} **68.** 7×10^{-5}

69. -7.16×10^6 **70.** -8.17×10^6

71. 7.9×10^{-1} **72.** 6.8×10^{-1}

73. -4.15×10^{-3} **74.** -3.14×10^{-3}

75. -6.00001×10^{10} **76.** -7.00001×10^{10}

In Exercises 77–86, write each number in scientific notation.

77. 32,000 **78.** 64,000

79. 638,000,000,000,000,000 **80.** 579,000,000,000,000,000

81. −5716 **82.** −3829

83. 0.0027 **84.** 0.0083

85. −0.00000000504 **86.** −0.00000000405

In Exercises 87–106, perform the indicated computations. Write the answers in scientific notation. If necessary, round the decimal factor in your scientific notation answer to two decimal places.

87. $(3 \times 10^4)(2.1 \times 10^3)$ **88.** $(2 \times 10^4)(4.1 \times 10^3)$

89. $(1.6 \times 10^{15})(4 \times 10^{-11})$ **90.** $(1.4 \times 10^{15})(3 \times 10^{-11})$

91. $(6.1 \times 10^{-8})(2 \times 10^{-4})$ **92.** $(5.1 \times 10^{-8})(3 \times 10^{-4})$

93. $(4.3 \times 10^8)(6.2 \times 10^4)$ **94.** $(8.2 \times 10^8)(4.6 \times 10^4)$

95. $\dfrac{8.4 \times 10^8}{4 \times 10^5}$ **96.** $\dfrac{6.9 \times 10^8}{3 \times 10^5}$

97. $\dfrac{3.6 \times 10^4}{9 \times 10^{-2}}$ **98.** $\dfrac{1.2 \times 10^4}{2 \times 10^{-2}}$

99. $\dfrac{4.8 \times 10^{-2}}{2.4 \times 10^6}$ **100.** $\dfrac{7.5 \times 10^{-2}}{2.5 \times 10^6}$

101. $\dfrac{2.4 \times 10^{-2}}{4.8 \times 10^{-6}}$ **102.** $\dfrac{1.5 \times 10^{-2}}{3 \times 10^{-6}}$

103. $\dfrac{480{,}000{,}000{,}000}{0.00012}$ **104.** $\dfrac{282{,}000{,}000{,}000}{0.00141}$

105. $\dfrac{0.00072 \times 0.003}{0.00024}$ **106.** $\dfrac{66{,}000 \times 0.001}{0.003 \times 0.002}$

Practice Plus

In Exercises 107–114, simplify each exponential expression. Assume that variables represent nonzero real numbers.

107. $\dfrac{(x^{-2}y)^{-3}}{(x^2y^{-1})^3}$ **108.** $\dfrac{(xy^{-2})^{-2}}{(x^{-2}y)^{-3}}$

109. $(2x^{-3}yz^{-6})(2x)^{-5}$ **110.** $(3x^{-4}yz^{-7})(3x)^{-3}$

111. $\left(\dfrac{x^3y^4z^5}{x^{-3}y^{-4}z^{-5}}\right)^{-2}$ **112.** $\left(\dfrac{x^4y^5z^6}{x^{-4}y^{-5}z^{-6}}\right)^{-4}$

113. $\dfrac{(2^{-1}x^{-2}y^{-1})^{-2}(2x^{-4}y^3)^{-2}(16x^{-3}y^3)^0}{(2x^{-3}y^{-5})^2}$

114. $\dfrac{(2^{-1}x^{-3}y^{-1})^{-2}(2x^{-6}y^4)^{-2}(9x^3y^{-3})^0}{(2x^{-4}y^{-6})^2}$

Application Exercises

The graph shows the number of people in the United States ages 65 and over for the year 2000 and projections beyond. Use 10^6 for one million and the figures shown to solve Exercises 115–118. Express all answers in scientific notation.

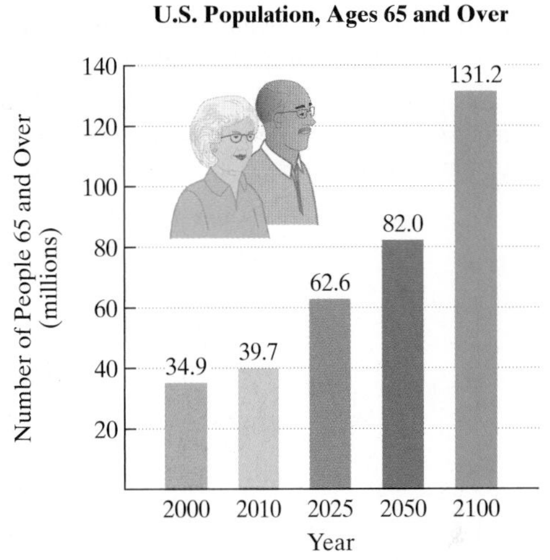

Source: U.S. Bureau of the Census

115. How many people 65 and over will there be in 2025?

116. How many people 65 and over will there be in 2050?

117. How many more people 65 and over will there be in 2100 than in 2000?

118. How many more people 65 and over will there be in 2050 than in 2000?

Our ancient ancestors hunted for their meat and expended a great deal of energy chasing it down. Today, our animal protein is raised in cages and on feedlots, delivered in great abundance nearly to our door. Use the numbers shown below to solve Exercises 119–122. Use 10^6 for one million and 10^9 for one billion.

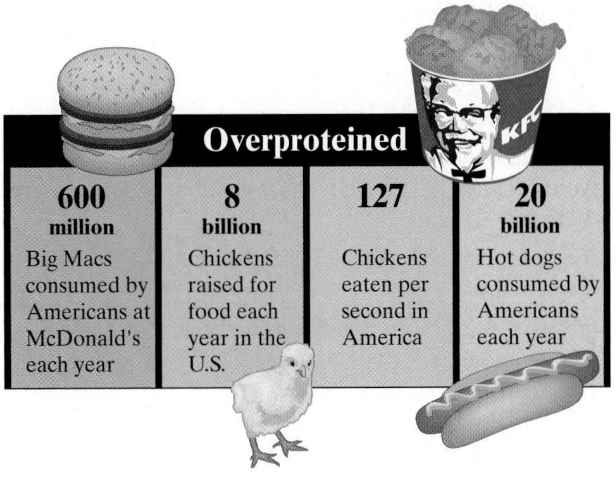

Source: Time, October 20, 2003

In Exercises 119–120, use 292 million, or 2.92×10^8, for the U.S. population. Express answers in decimal notation, rounded to the nearest whole number.

119. Find the number of hot dogs consumed by each American in a year.

120. If the consumption of Big Macs was divided evenly among all Americans, how many Big Macs would we each consume in a year?

In Exercises 121–122, use the Overproteined table on the previous page and the fact that there are approximately 3.2×10^7 *seconds in a year.*

121. How many chickens are raised for food each second in the United States? Express the answer in scientific and decimal notations.

122. How many chickens are eaten per year in the United States? Express the answer in scientific notation.

123. Due to tax cuts and spending increases, the United States began accumulating large deficits in the 1980s. The graph shows the national debt increasing over time.

The National Debt

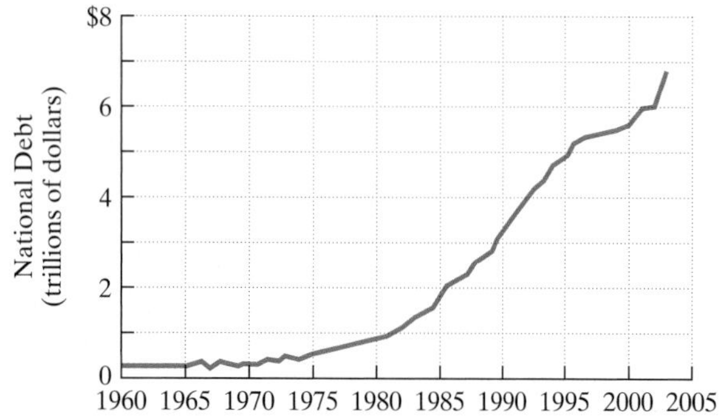

As of November 2003, to finance the deficit, the government had borrowed \$6.8 trillion and the national debt was \$6.8 trillion, or 6.8×10^{12} dollars. At that time, the U.S. population was approximately 290,000,000 (290 million), or 2.9×10^8. If the national debt was evenly divided among every individual in the United States, how much would each citizen have to pay?

124. In Exercises 121–122, we used 3.2×10^7 as an approximation for the number of seconds in a year. Convert 365 days (one year) to hours, to minutes, and, finally, to seconds, to determine precisely how many seconds there are in a year. Express the answer in scientific notation.

Writing in Mathematics

125. Describe what it means to raise a number to a power. In your description, include a discussion of the difference between -5^2 and $(-5)^2$.

126. Explain the product rule for exponents. Use $2^3 \cdot 2^5$ in your explanation.

127. Explain the power rule for exponents. Use $(3^2)^4$ in your explanation.

128. Explain the quotient rule for exponents. Use $\frac{5^8}{5^2}$ in your explanation.

129. Why is $(-3x^2)(2x^{-5})$ not simplified? What must be done to simplify the expression?

130. How do you know if a number is written in scientific notation?

131. Explain how to convert from scientific to decimal notation and give an example.

132. Explain how to convert from decimal to scientific notation and give an example.

Critical Thinking Exercises

133. Which one of the following is true?

a. $4^{-2} < 4^{-3}$ **b.** $5^{-2} > 2^{-5}$

c. $(-2)^4 = 2^{-4}$ **d.** $5^2 \cdot 5^{-2} > 2^5 \cdot 2^{-5}$

134. The mad Dr. Frankenstein has gathered enough bits and pieces (so to speak) for $2^{-1} + 2^{-2}$ of his creature-to-be. Write a fraction that represents the amount of his creature that must still be obtained.

135. If $b^A = MN$, $b^C = M$, and $b^D = N$, what is the relationship among A, C, and D?

136. Our hearts beat approximately 70 times per minute. Express in scientific notation how many times the heart beats over a lifetime of 80 years. Round the decimal factor in your scientific notation answer to two decimal places.

Group Exercise

137. Putting Numbers into Perspective. A large number can be put into perspective by comparing it with another number. For example, we put the \$1.95 trillion the government collected in taxes (Example 12) and the \$6.8 trillion national debt (Exercise 123) by comparing these numbers to the number of U.S. citizens.

For this project, each group member should consult an almanac, a newspaper, or the World Wide Web to find a number greater than one million. Explain to other members of the group the context in which the large number is used. Express the number in scientific notation. Then put the number into perspective by comparing it with another number.

SECTION P.3 Radicals and Rational Exponents

Objectives

1. Evaluate square roots.
2. Simplify expressions of the form $\sqrt{a^2}$.
3. Use the product rule to simplify square roots.
4. Use the quotient rule to simplify square roots.
5. Add and subtract square roots.
6. Rationalize denominators.
7. Evaluate and perform operations with higher roots.
8. Understand and use rational exponents.

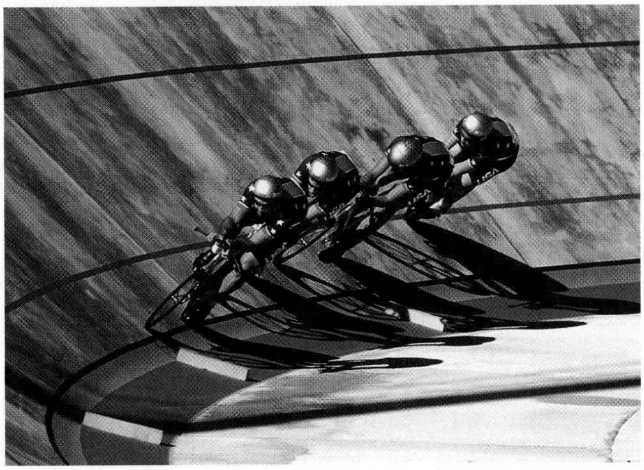

What is the maximum speed at which a racing cyclist can turn a corner without tipping over? The answer, in miles per hour, is given by the algebraic expression $4\sqrt{x}$, where x is the radius of the corner, in feet. Algebraic expressions containing roots describe phenomena as diverse as the distance we can see to the horizon, how we perceive the temperature on a cold day, and Albert Einstein's bizarre concept of how an astronaut moving close to the speed of light would barely age relative to friends watching from Earth. No description of your world can be complete without roots and radicals. In this section, we review the basics of radical expressions and the use of rational exponents to indicate radicals.

1 Evaluate square roots.

Square Roots

From our earlier work with exponents, we are aware that the square of both 5 and −5 is 25:

$$5^2 = 25 \quad \text{and} \quad (-5)^2 = 25.$$

The reverse operation of squaring a number is finding the *square root* of the number. For example,

- One square root of 25 is 5 because $5^2 = 25$.
- Another square root of 25 is −5 because $(-5)^2 = 25$.

In general, **if $b^2 = a$, then b is a square root of a.**

The symbol $\sqrt{}$ is used to denote the *positive* or *principal square root* of a number. For example,

- $\sqrt{25} = 5$ because $5^2 = 25$ and 5 is positive.
- $\sqrt{100} = 10$ because $10^2 = 100$ and 10 is positive.

The symbol $\sqrt{}$ that we use to denote the principal square root is called a **radical sign**. The number under the radical sign is called the **radicand**. Together we refer to the radical sign and its radicand as a **radical expression**.

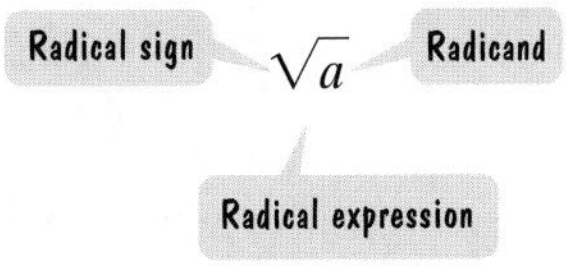

Definition of the Principal Square Root

If a is a nonnegative real number, the nonnegative number b such that $b^2 = a$, denoted by $b = \sqrt{a}$, is the **principal square root** of a.

The symbol $-\sqrt{\ }$ is used to denote the negative square root of a number. For example,

- $-\sqrt{25} = -5$ because $(-5)^2 = 25$ and -5 is negative.
- $-\sqrt{100} = -10$ because $(-10)^2 = 100$ and -10 is negative.

EXAMPLE 1 Evaluating Square Roots

Evaluate:

a. $\sqrt{64}$ **b.** $-\sqrt{49}$ **c.** $\sqrt{\frac{1}{4}}$ **d.** $\sqrt{9+16}$ **e.** $\sqrt{9}+\sqrt{16}$.

Solution

a. $\sqrt{64} = 8$ The principal square root of 64 is 8. Check: $8^2 = 64$.

b. $-\sqrt{49} = -7$ The negative square root of 49 is -7. Check: $(-7)^2 = 49$.

c. $\sqrt{\frac{1}{4}} = \frac{1}{2}$ The principal square root of $\frac{1}{4}$ is $\frac{1}{2}$. Check: $\left(\frac{1}{2}\right)^2 = \frac{1}{4}$.

d. $\sqrt{9+16} = \sqrt{25}$ First simplify the expression under the radical sign.

$= 5$ Then take the principal square root of 25, which is 5.

e. $\sqrt{9}+\sqrt{16} = 3 + 4$ $\sqrt{9} = 3$ because $3^2 = 9$. $\sqrt{16} = 4$ because $4^2 = 16$.

$= 7$

Study Tip

In Example 1, parts (d) and (e), observe that $\sqrt{9+16}$ is not equal to $\sqrt{9}+\sqrt{16}$. In general,

$$\sqrt{a+b} \neq \sqrt{a}+\sqrt{b}$$

and

$$\sqrt{a-b} \neq \sqrt{a}-\sqrt{b}.$$

Check Point 1 Evaluate:

a. $\sqrt{81}$ **b.** $-\sqrt{9}$ **c.** $\sqrt{\frac{1}{25}}$

d. $\sqrt{36+64}$ **e.** $\sqrt{36}+\sqrt{64}$.

A number that is the square of a rational number is called a **perfect square**. All the radicands in Example 1 and Check Point 1 are perfect squares.

- 64 is a perfect square because $64 = 8^2$. Thus, $\sqrt{64} = 8$.
- $\frac{1}{4}$ is a perfect square because $\frac{1}{4} = \left(\frac{1}{2}\right)^2$. Thus, $\sqrt{\frac{1}{4}} = \frac{1}{2}$.

Let's see what happens to the radical expression $\sqrt{x}$ if x is a negative number. Is the square root of a negative number a real number? For example, consider $\sqrt{-25}$. Is there a real number whose square is -25? No. Thus, $\sqrt{-25}$ is not a real number. In general, **a square root of a negative number is not a real number.**

If a number is nonnegative $(a \geq 0)$, then $(\sqrt{a})^2 = a$. For example,

$$(\sqrt{2})^2 = 2, \quad (\sqrt{3})^2 = 3, \quad (\sqrt{4})^2 = 4, \quad \text{and} \quad (\sqrt{5})^2 = 5.$$

2 Simplify expressions of the form $\sqrt{a^2}$.

Simplifying Expressions of the Form $\sqrt{a^2}$

You may think that $\sqrt{a^2} = a$. However, this is not necessarily true. Consider the following examples:

$$\sqrt{4^2} = \sqrt{16} = 4$$
$$\sqrt{(-4)^2} = \sqrt{16} = 4.$$

The result is not -4, but rather the absolute value of -4, or 4.

Here is a rule for simplifying expressions of the form $\sqrt{a^2}$:

Simplifying $\sqrt{a^2}$

For any real number a,

$$\sqrt{a^2} = |a|.$$

In words, the principal square root of a^2 is the absolute value of a.

For example, $\sqrt{6^2} = |6| = 6$ and $\sqrt{(-6)^2} = |-6| = 6$.

3 Use the product rule to simplify square roots.

The Product Rule for Square Roots

A rule for multiplying square roots can be generalized by comparing $\sqrt{25} \cdot \sqrt{4}$ and $\sqrt{25 \cdot 4}$. Notice that

$$\sqrt{25} \cdot \sqrt{4} = 5 \cdot 2 = 10 \quad \text{and} \quad \sqrt{25 \cdot 4} = \sqrt{100} = 10.$$

Because we obtain 10 in both situations, the original radical expressions must be equal. That is,

$$\sqrt{25} \cdot \sqrt{4} = \sqrt{25 \cdot 4}.$$

This result is a special case of the **product rule for square roots** that can be generalized as follows:

The Product Rule for Square Roots

If a and b represent nonnegative real numbers, then

$$\sqrt{ab} = \sqrt{a} \cdot \sqrt{b} \quad \text{and} \quad \sqrt{a} \cdot \sqrt{b} = \sqrt{ab}.$$

The square root of a product is the product of the square roots.

A square root is **simplified** when its radicand has no factors other than 1 that are perfect squares. For example, $\sqrt{500}$ is not simplified because it can be expressed as $\sqrt{100 \cdot 5}$ and 100 is a perfect square. Example 2 shows how the product rule is used to remove from the square root any perfect squares that occur as factors.

EXAMPLE 2 Using the Product Rule to Simplify Square Roots

Simplify: **a.** $\sqrt{500}$ **b.** $\sqrt{6x} \cdot \sqrt{3x}$.

Solution

a.

$\sqrt{500} = \sqrt{100 \cdot 5}$ Factor 500. 100 is the greatest perfect square factor.

$= \sqrt{100}\sqrt{5}$ Use the product rule: $\sqrt{ab} = \sqrt{a}\sqrt{b}$.

$= 10\sqrt{5}$ Write $\sqrt{100}$ as 10. We read $10\sqrt{5}$ as "ten times the square root of 5."

b. We can simplify $\sqrt{6x} \cdot \sqrt{3x}$ using the product rule only if $6x$ and $3x$ represent nonnegative real numbers. Thus, $x \geq 0$.

$\sqrt{6x} \cdot \sqrt{3x} = \sqrt{6x \cdot 3x}$ Use the product rule: $\sqrt{a}\sqrt{b} = \sqrt{ab}$.

$= \sqrt{18x^2}$ Multiply in the radicand.

$= \sqrt{9x^2 \cdot 2}$ Factor 18. 9 is the greatest perfect square factor.

$= \sqrt{9x^2}\sqrt{2}$ Use the product rule: $\sqrt{ab} = \sqrt{a}\sqrt{b}$.

$= \sqrt{9}\sqrt{x^2}\sqrt{2}$ Use the product rule to write $\sqrt{9x^2}$ as the product of two square roots.

$= 3x\sqrt{2}$ $\sqrt{x^2} = |x| = x$ because $x \geq 0$.

Study Tip

When simplifying square roots, always look for the *greatest* perfect square factor possible. The following factorization will lead to further simplification:

$$\sqrt{500} = \sqrt{25 \cdot 20} = \sqrt{25}\sqrt{20} = 5\sqrt{20}.$$

25 is a perfect square factor of 500, but not the greatest perfect square factor.

Because 20 contains a perfect square factor, 4, the simplification is not complete.

$$5\sqrt{20} = 5\sqrt{4 \cdot 5} = 5\sqrt{4}\,\sqrt{5} = 5 \cdot 2\sqrt{5} = 10\sqrt{5}$$

Although the result checks with our simplification using $\sqrt{500} = \sqrt{100 \cdot 5}$, more work is required when the greatest perfect square factor is not used.

Check Point 2 Simplify:

a. $\sqrt{75}$ **b.** $\sqrt{5x} \cdot \sqrt{10x}$.

4 Use the quotient rule to simplify square roots.

The Quotient Rule for Square Roots

Another property for square roots involves division.

The Quotient Rule for Square Roots

If a and b represent nonnegative real numbers and $b \neq 0$, then

$$\sqrt{\frac{a}{b}} = \frac{\sqrt{a}}{\sqrt{b}} \quad \text{and} \quad \frac{\sqrt{a}}{\sqrt{b}} = \sqrt{\frac{a}{b}}.$$

The square root of a quotient is the quotient of the square roots.

EXAMPLE 3 Using the Quotient Rule to Simplify Square Roots

Simplify: **a.** $\sqrt{\frac{100}{9}}$ **b.** $\frac{\sqrt{48x^3}}{\sqrt{6x}}$.

Solution

a. $\sqrt{\frac{100}{9}} = \frac{\sqrt{100}}{\sqrt{9}} = \frac{10}{3}$

b. We can simplify the quotient of $\sqrt{48x^3}$ and $\sqrt{6x}$ using the quotient rule only if $48x^3$ and $6x$ represent nonnegative real numbers and $6x \neq 0$. Thus, $x > 0$.

$$\frac{\sqrt{48x^3}}{\sqrt{6x}} = \sqrt{\frac{48x^3}{6x}} = \sqrt{8x^2} = \sqrt{4x^2}\sqrt{2} = \sqrt{4}\sqrt{x^2}\sqrt{2} = 2x\sqrt{2}$$

$\sqrt{x^2} = |x| = x$ because $x > 0$.

Check Point 3 Simplify: **a.** $\sqrt{\frac{25}{16}}$ **b.** $\frac{\sqrt{150x^3}}{\sqrt{2x}}$.

5 Add and subtract square roots.

Adding and Subtracting Square Roots

Two or more square roots can be combined using the distributive property provided that they have the same radicand. Such radicals are called **like radicals**. For example,

$$7\sqrt{11} + 6\sqrt{11} = (7 + 6)\sqrt{11} = 13\sqrt{11}.$$

7 square roots of 11 plus 6 square roots of 11 result in 13 square roots of 11.

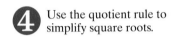

A Radical Idea: Time Is Relative

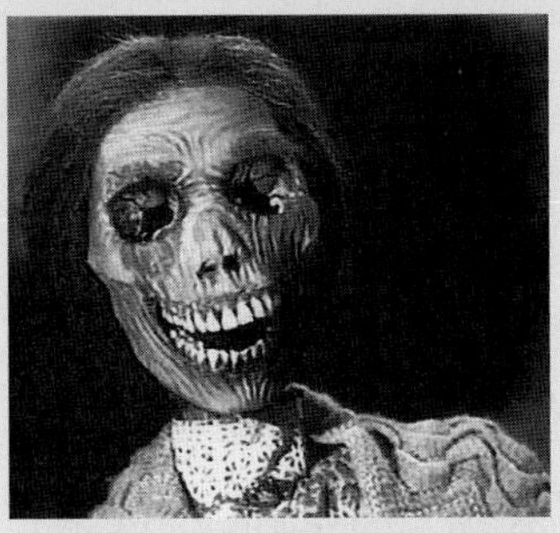

What does travel in space have to do with radicals? Imagine that in the future we will be able to travel at velocities approaching the speed of light (approximately 186,000 miles per second). According to Einstein's theory of relativity, time would pass more quickly on Earth than it would in the moving spaceship. The radical expression

$$R_f\sqrt{1-\left(\frac{v}{c}\right)^2}$$

gives the aging rate of an astronaut relative to the aging rate of a friend on Earth, R_f. In the expression, v is the astronaut's speed and c is the speed of light. As the astronaut's speed approaches the speed of light, we can substitute c for v:

$$R_f\sqrt{1-\left(\frac{v}{c}\right)^2} \quad \text{Let } v = c.$$
$$= R_f\sqrt{1-\left(\frac{c}{c}\right)^2}$$
$$= R_f\sqrt{1-1^2}$$
$$= R_f\sqrt{0} = 0$$

Close to the speed of light, the astronaut's aging rate relative to a friend on Earth is nearly 0. What does this mean? As we age here on Earth, the space traveler would barely get older. The space traveler would return to a futuristic world in which friends and loved ones would be long dead.

EXAMPLE 4 Adding and Subtracting Like Radicals

Add or subtract as indicated:

a. $7\sqrt{2} + 5\sqrt{2}$ **b.** $\sqrt{5x} - 7\sqrt{5x}$.

Solution

a. $7\sqrt{2} + 5\sqrt{2} = (7+5)\sqrt{2}$ Apply the distributive property.

$= 12\sqrt{2}$ Simplify.

b. $\sqrt{5x} - 7\sqrt{5x} = 1\sqrt{5x} - 7\sqrt{5x}$ Write $\sqrt{5x}$ as $1\sqrt{5x}$.

$= (1-7)\sqrt{5x}$ Apply the distributive property.

$= -6\sqrt{5x}$ Simplify.

Check Point 4 Add or subtract as indicated:

a. $8\sqrt{13} + 9\sqrt{13}$ **b.** $\sqrt{17x} - 20\sqrt{17x}$.

In some cases, radicals can be combined once they have been simplified. For example, to add $\sqrt{2}$ and $\sqrt{8}$, we can write $\sqrt{8}$ as $\sqrt{4\cdot 2}$ because 4 is a perfect square factor of 8.

$$\sqrt{2} + \sqrt{8} = \sqrt{2} + \sqrt{4\cdot 2} = 1\sqrt{2} + 2\sqrt{2} = (1+2)\sqrt{2} = 3\sqrt{2}$$

EXAMPLE 5 Combining Radicals That First Require Simplification

Add or subtract as indicated:

a. $7\sqrt{3} + \sqrt{12}$ **b.** $4\sqrt{50x} - 6\sqrt{32x}$.

Solution

a. $7\sqrt{3} + \sqrt{12}$

$= 7\sqrt{3} + \sqrt{4\cdot 3}$ Split 12 into two factors such that one is a perfect square.

$= 7\sqrt{3} + 2\sqrt{3}$ $\sqrt{4\cdot 3} = \sqrt{4}\sqrt{3} = 2\sqrt{3}$

$= (7+2)\sqrt{3}$ Apply the distributive property. You will find that this step is usually done mentally.

$= 9\sqrt{3}$ Simplify.

b. $4\sqrt{50x} - 6\sqrt{32x}$

$= 4\sqrt{25\cdot 2x} - 6\sqrt{16\cdot 2x}$ 25 is the greatest perfect square factor of 50x and 16 is the greatest perfect square factor of 32x.

$= 4\cdot 5\sqrt{2x} - 6\cdot 4\sqrt{2x}$ $\sqrt{25\cdot 2x} = \sqrt{25}\sqrt{2x} = 5\sqrt{2x}$ and $\sqrt{16\cdot 2x} = \sqrt{16}\sqrt{2x} = 4\sqrt{2x}$.

$= 20\sqrt{2x} - 24\sqrt{2x}$ Multiply: $4\cdot 5 = 20$ and $6\cdot 4 = 24$.

$= (20-24)\sqrt{2x}$ Apply the distributive property.

$= -4\sqrt{2x}$ Simplify.

Check Point 5 Add or subtract as indicated:

a. $5\sqrt{27} + \sqrt{12}$ **b.** $6\sqrt{18x} - 4\sqrt{8x}$.

6 Rationalize denominators.

Rationalizing Denominators

You can use a calculator to compare the approximate values for $\frac{1}{\sqrt{3}}$ and $\frac{\sqrt{3}}{3}$. The two approximations are the same. This is not a coincidence:

$$\frac{1}{\sqrt{3}} = \frac{1}{\sqrt{3}} \cdot \boxed{\frac{\sqrt{3}}{\sqrt{3}}} = \frac{\sqrt{3}}{\sqrt{9}} = \frac{\sqrt{3}}{3}.$$

Any number divided by itself is 1. Multiplication by 1 does not change the value of $\frac{1}{\sqrt{3}}$.

This process involves rewriting a radical expression as an equivalent expression in which the denominator no longer contains any radicals. The process is called **rationalizing the denominator**. If the denominator contains the square root of a natural number that is not a perfect square, **multiply the numerator and the denominator by the smallest number that produces the square root of a perfect square in the denominator**.

EXAMPLE 6 Rationalizing Denominators

Rationalize the denominator: **a.** $\frac{15}{\sqrt{6}}$ **b.** $\frac{12}{\sqrt{8}}$.

Solution

a. If we multiply the numerator and the denominator of $\frac{15}{\sqrt{6}}$ by $\sqrt{6}$, the denominator becomes $\sqrt{6} \cdot \sqrt{6} = \sqrt{36} = 6$. Therefore, we multiply by 1, choosing $\frac{\sqrt{6}}{\sqrt{6}}$ for 1.

$$\frac{15}{\sqrt{6}} = \frac{15}{\sqrt{6}} \cdot \frac{\sqrt{6}}{\sqrt{6}} = \frac{15\sqrt{6}}{\sqrt{36}} = \frac{15\sqrt{6}}{6} = \frac{5\sqrt{6}}{2}$$

Multiply by 1.

Simplify: $\frac{15}{6} = \frac{15 \div 3}{6 \div 3} = \frac{5}{2}$.

b. The *smallest* number that will produce a perfect square in the denominator of $\frac{12}{\sqrt{8}}$ is $\sqrt{2}$, because $\sqrt{8} \cdot \sqrt{2} = \sqrt{16} = 4$. We multiply by 1, choosing $\frac{\sqrt{2}}{\sqrt{2}}$ for 1.

$$\frac{12}{\sqrt{8}} = \frac{12}{\sqrt{8}} \cdot \frac{\sqrt{2}}{\sqrt{2}} = \frac{12\sqrt{2}}{\sqrt{16}} = \frac{12\sqrt{2}}{4} = 3\sqrt{2}$$

Check Point 6 Rationalize the denominator: **a.** $\frac{5}{\sqrt{3}}$ **b.** $\frac{6}{\sqrt{12}}$.

Radical expressions that involve the sum and difference of the same two terms are called **conjugates**. Thus,

$$\sqrt{a} + \sqrt{b} \quad \text{and} \quad \sqrt{a} - \sqrt{b}$$

are conjugates. Conjugates are used to rationalize denominators because the product of such pairs contains no radicals:

Multiply each term of $\sqrt{a}-\sqrt{b}$ by each term of $\sqrt{a}+\sqrt{b}$.

$$(\sqrt{a}+\sqrt{b})(\sqrt{a}-\sqrt{b})$$
$$=\sqrt{a}(\sqrt{a}-\sqrt{b})+\sqrt{b}(\sqrt{a}-\sqrt{b})$$

Distribute $\sqrt{a}$ over $\sqrt{a}-\sqrt{b}$. Distribute $\sqrt{b}$ over $\sqrt{a}-\sqrt{b}$.

$$=\sqrt{a}\cdot\sqrt{a}-\sqrt{a}\cdot\sqrt{b}+\sqrt{b}\cdot\sqrt{a}-\sqrt{b}\cdot\sqrt{b}$$
$$=(\sqrt{a})^2-\sqrt{ab}+\sqrt{ab}-(\sqrt{b})^2$$

$-\sqrt{ab}+\sqrt{ab}=0$

$$=(\sqrt{a})^2-(\sqrt{b})^2$$
$$=a-b.$$

Multiplying Conjugates

$$(\sqrt{a}+\sqrt{b})(\sqrt{a}-\sqrt{b})=(\sqrt{a})^2-(\sqrt{b})^2=a-b$$

How can we rationalize a denominator if the denominator contains two terms with one or more square roots? **Multiply the numerator and the denominator by the conjugate of the denominator.** Here are three examples of such expressions:

- $\dfrac{7}{5+\sqrt{3}}$ — The conjugate of the denominator is $5-\sqrt{3}$.
- $\dfrac{8}{3\sqrt{2}-4}$ — The conjugate of the denominator is $3\sqrt{2}+4$.
- $\dfrac{h}{\sqrt{x+h}-\sqrt{x}}$ — The conjugate of the denominator is $\sqrt{x+h}+\sqrt{x}$.

The product of the denominator and its conjugate is found using the formula

$$(\sqrt{a}+\sqrt{b})(\sqrt{a}-\sqrt{b})=(\sqrt{a})^2-(\sqrt{b})^2=a-b.$$

The simplified product will not contain a radical.

EXAMPLE 7 Rationalizing a Denominator Containing Two Terms

Rationalize the denominator: $\dfrac{7}{5+\sqrt{3}}$.

Solution The conjugate of the denominator is $5-\sqrt{3}$. If we multiply the numerator and denominator by $5-\sqrt{3}$, the simplified denominator will not contain a radical. Therefore, we multiply by 1, choosing $\dfrac{5-\sqrt{3}}{5-\sqrt{3}}$ for 1.

$$\frac{7}{5+\sqrt{3}}=\frac{7}{5+\sqrt{3}}\cdot\frac{5-\sqrt{3}}{5-\sqrt{3}}=\frac{7(5-\sqrt{3})}{5^2-(\sqrt{3})^2}=\frac{7(5-\sqrt{3})}{25-3}$$

Multiply by 1. $(\sqrt{a}+\sqrt{b})(\sqrt{a}-\sqrt{b})=(\sqrt{a})^2-(\sqrt{b})^2$

$$=\frac{7(5-\sqrt{3})}{22}\text{ or }\frac{35-7\sqrt{3}}{22}$$

In either form of the answer, there is no radical in the denominator.

Check Point 7 Rationalize the denominator: $\dfrac{8}{4 + \sqrt{5}}$.

⑦ Evaluate and perform operations with higher roots.

Study Tip

Some higher even and odd roots occur so frequently that you might want to memorize them.

Cube Roots

$\sqrt[3]{1} = 1$	$\sqrt[3]{125} = 5$
$\sqrt[3]{8} = 2$	$\sqrt[3]{216} = 6$
$\sqrt[3]{27} = 3$	$\sqrt[3]{1000} = 10$
$\sqrt[3]{64} = 4$	

Fourth Roots

$\sqrt[4]{1} = 1$
$\sqrt[4]{16} = 2$
$\sqrt[4]{81} = 3$
$\sqrt[4]{256} = 4$
$\sqrt[4]{625} = 5$

Fifth Roots

$\sqrt[5]{1} = 1$
$\sqrt[5]{32} = 2$
$\sqrt[5]{243} = 3$

Other Kinds of Roots

We define the **principal nth root** of a real number a, symbolized by $\sqrt[n]{a}$, as follows:

> ### Definition of the Principal nth Root of a Real Number
>
> $$\sqrt[n]{a} = b \text{ means that } b^n = a.$$
>
> If n, the **index**, is even, then a is nonnegative ($a \geq 0$) and b is also nonnegative ($b \geq 0$). If n is odd, a and b can be any real numbers.

For example,

$$\sqrt[3]{64} = 4 \text{ because } 4^3 = 64 \quad \text{and} \quad \sqrt[5]{-32} = -2 \text{ because } (-2)^5 = -32.$$

The same vocabulary that we learned for square roots applies to nth roots. The symbol $\sqrt[n]{a}$ is called a **radical** and a is called the **radicand**.

A number that is the nth power of a rational number is called a **perfect nth power**. For example, 8 is a perfect third power, or perfect cube, because $8 = 2^3$. Thus, $\sqrt[3]{8} = \sqrt[3]{2^3} = 2$. In general, one of the following rules can be used to find nth roots of perfect nth powers:

> ### Finding nth Roots of Perfect nth Powers
>
> If n is odd, $\sqrt[n]{a^n} = a$.
> If n is even, $\sqrt[n]{a^n} = |a|$.

For example,

$$\sqrt[3]{(-2)^3} = -2 \quad \text{and} \quad \sqrt[4]{(-2)^4} = |-2| = 2.$$

Absolute value is not needed with odd roots, but is necessary with even roots.

The Product and Quotient Rules for Other Roots

The product and quotient rules apply to cube roots, fourth roots, and all higher roots.

> ### The Product and Quotient Rules for nth Roots
>
> For all real numbers, where the indicated roots represent real numbers,
>
> $$\sqrt[n]{ab} = \sqrt[n]{a} \cdot \sqrt[n]{b} \quad \text{and} \quad \sqrt[n]{\frac{a}{b}} = \frac{\sqrt[n]{a}}{\sqrt[n]{b}}, \quad b \neq 0.$$

EXAMPLE 8 Simplifying, Multiplying, and Dividing Higher Roots

Simplify: **a.** $\sqrt[3]{24}$ **b.** $\sqrt[4]{8} \cdot \sqrt[4]{4}$ **c.** $\sqrt[4]{\dfrac{81}{16}}$.

Solution

a. $\sqrt[3]{24} = \sqrt[3]{8 \cdot 3}$ Find the greatest perfect cube that is a factor of 24. $2^3 = 8$, so 8 is a perfect cube and is the greatest perfect cube factor of 24.

$= \sqrt[3]{8} \cdot \sqrt[3]{3}$ $\sqrt[n]{ab} = \sqrt[n]{a} \ \sqrt[n]{b}$

$= 2\sqrt[3]{3}$ $\sqrt[3]{8} = 2$

b. $\sqrt[4]{8}\cdot\sqrt[4]{4} = \sqrt[4]{8\cdot 4}$ $\quad$ $\sqrt[n]{a}\cdot\sqrt[n]{b} = \sqrt[n]{ab}$

$= \sqrt[4]{32}$ $\quad$ Find the greatest perfect fourth power that is a factor of 32.

$= \sqrt[4]{16\cdot 2}$ $\quad$ $2^4 = 16$, so 16 is a perfect fourth power and is the greatest perfect fourth power that is a factor of 32.

$= \sqrt[4]{16}\cdot\sqrt[4]{2}$ $\quad$ $\sqrt[n]{ab} = \sqrt[n]{a}\cdot\sqrt[n]{b}$

$= 2\sqrt[4]{2}$ $\quad$ $\sqrt[4]{16} = 2$

c. $\sqrt[4]{\frac{81}{16}} = \frac{\sqrt[4]{81}}{\sqrt[4]{16}}$ $\quad$ $\sqrt[n]{\frac{a}{b}} = \frac{\sqrt[n]{a}}{\sqrt[n]{b}}$

$= \frac{3}{2}$ $\quad$ $\sqrt[4]{81} = 3$ because $3^4 = 81$ and $\sqrt[4]{16} = 2$ because $2^4 = 16$.

Check Point 8 Simplify: **a.** $\sqrt[3]{40}$ **b.** $\sqrt[5]{8}\cdot\sqrt[5]{8}$ **c.** $\sqrt[3]{\frac{125}{27}}$.

We have seen that adding and subtracting square roots often involves simplifying terms. The same idea applies to adding and subtracting *n*th roots.

EXAMPLE 9 Combining Cube Roots

Subtract: $5\sqrt[3]{16} - 11\sqrt[3]{2}$.

Solution

$5\sqrt[3]{16} - 11\sqrt[3]{2}$

$= 5\sqrt[3]{8\cdot 2} - 11\sqrt[3]{2}$ $\quad$ Factor 16. 8 is the greatest perfect cube factor: $2^3 = 8$ and $\sqrt[3]{8} = 2$.

$= 5\cdot 2\sqrt[3]{2} - 11\sqrt[3]{2}$ $\quad$ $\sqrt[3]{8\cdot 2} = \sqrt[3]{8}\sqrt[3]{2} = 2\sqrt[3]{2}$

$= 10\sqrt[3]{2} - 11\sqrt[3]{2}$ $\quad$ Multiply: $5\cdot 2 = 10$.

$= (10 - 11)\sqrt[3]{2}$ $\quad$ Apply the distributive property.

$= -1\sqrt[3]{2}$ or $-\sqrt[3]{2}$ $\quad$ Simplify.

Check Point 9 Subtract: $3\sqrt[3]{81} - 4\sqrt[3]{3}$.

8 Understand and use rational exponents.

Rational Exponents

We define rational exponents so that their properties are the same as the properties for integer exponents. For example, we know that exponents are multiplied when an exponential expression is raised to a power. For this to be true,

$$\left(7^{\frac{1}{2}}\right)^2 = 7^{\frac{1}{2}\cdot 2} = 7^1 = 7.$$

We also know that

$$\left(\sqrt{7}\right)^2 = \sqrt{7}\cdot\sqrt{7} = \sqrt{49} = 7.$$

Can you see that the square of both $7^{\frac{1}{2}}$ and $\sqrt{7}$ is 7? It is reasonable to conclude that

$$7^{\frac{1}{2}} \text{ means } \sqrt{7}.$$

We can generalize the fact that $7^{\frac{1}{2}}$ means $\sqrt{7}$ with the following definition:

The Definition of $a^{\frac{1}{n}}$

If $\sqrt[n]{a}$ represents a real number and $n \geq 2$ is an integer, then

$$a^{\frac{1}{n}} = \sqrt[n]{a}.$$

The denominator of the rational exponent is the radical's index.

Furthermore,

$$a^{-\frac{1}{n}} = \frac{1}{a^{\frac{1}{n}}} = \frac{1}{\sqrt[n]{a}}, \quad a \neq 0.$$

EXAMPLE 10 Using the Definition of $a^{\frac{1}{n}}$

Simplify:

a. $64^{\frac{1}{2}}$ **b.** $125^{\frac{1}{3}}$ **c.** $-16^{\frac{1}{4}}$ **d.** $(-27)^{\frac{1}{3}}$ **e.** $64^{-\frac{1}{3}}$.

Solution

a. $64^{\frac{1}{2}} = \sqrt{64} = 8$

b. $125^{\frac{1}{3}} = \sqrt[3]{125} = 5$

The denominator is the index.

c. $-16^{\frac{1}{4}} = -(\sqrt[4]{16}) = -2$

The base is 16 and the negative sign is not affected by the exponent.

d. $(-27)^{\frac{1}{3}} = \sqrt[3]{-27} = -3$

Parentheses show that the base is −27 and that the negative sign is affected by the exponent.

e. $64^{-\frac{1}{3}} = \frac{1}{64^{\frac{1}{3}}} = \frac{1}{\sqrt[3]{64}} = \frac{1}{4}$

Check Point 10 Simplify:

a. $25^{\frac{1}{2}}$ **b.** $8^{\frac{1}{3}}$ **c.** $-81^{\frac{1}{4}}$ **d.** $(-8)^{\frac{1}{3}}$ **e.** $27^{-\frac{1}{3}}$

In Example 10 and Check Point 10, each rational exponent had a numerator of 1. If the numerator is some other integer, we still want to multiply exponents when raising a power to a power. For this reason,

$$a^{\frac{2}{3}} = (a^{\frac{1}{3}})^2 \text{ and } a^{\frac{2}{3}} = (a^2)^{\frac{1}{3}}.$$

This means $(\sqrt[3]{a})^2$. This means $\sqrt[3]{a^2}$.

Thus,

$$a^{\frac{2}{3}} = (\sqrt[3]{a})^2 = \sqrt[3]{a^2}.$$

Do you see that the denominator, 3, of the rational exponent is the same as the index of the radical? The numerator, 2, of the rational exponent serves as an exponent in each of the two radical forms. We generalize these ideas with the following definition:

The Definition of $a^{\frac{m}{n}}$

If $\sqrt[n]{a}$ represents a real number and $\frac{m}{n}$ is a positive rational number, $n \geq 2$, then

$$a^{\frac{m}{n}} = (\sqrt[n]{a})^m.$$

Also,

$$a^{\frac{m}{n}} = \sqrt[n]{a^m}.$$

Furthermore, if $a^{-\frac{m}{n}}$ is a nonzero real number, then

$$a^{-\frac{m}{n}} = \frac{1}{a^{\frac{m}{n}}}.$$

The first form of the definition of $a^{\frac{m}{n}}$, shown again below, involves taking the root first. This form is often preferable because smaller numbers are involved. Notice that the rational exponent consists of two parts, indicated by the following voice balloons:

The numerator is the exponent.

$$a^{\frac{m}{n}} = (\sqrt[n]{a})^m.$$

The denominator is the radical's index.

EXAMPLE 11 Using the Definition of $a^{\frac{m}{n}}$

Simplify:

a. $27^{\frac{2}{3}}$ **b.** $9^{\frac{3}{2}}$ **c.** $81^{-\frac{3}{4}}$.

Solution

a. $27^{\frac{2}{3}} = (\sqrt[3]{27})^2 = 3^2 = 9$

b. $9^{\frac{3}{2}} = (\sqrt{9})^3 = 3^3 = 27$

c. $81^{-\frac{3}{4}} = \frac{1}{81^{\frac{3}{4}}} = \frac{1}{(\sqrt[4]{81})^3} = \frac{1}{3^3} = \frac{1}{27}$

Technology

Here are the calculator keystroke sequences for $81^{-\frac{3}{4}}$:

Many Scientific Calculators

81 [y^x] [(] 3 [+/−] [÷] 4 [)] [=]

Many Graphing Calculators

81 [∧] [(] [(−)] 3 [÷] 4 [)] [ENTER].

Check Point 11 Simplify: **a.** $27^{\frac{4}{3}}$ **b.** $4^{\frac{3}{2}}$ **c.** $32^{-\frac{2}{5}}$

Properties of exponents can be applied to expressions containing rational exponents.

EXAMPLE 12 Simplifying Expressions with Rational Exponents

Simplify using properties of exponents:

a. $\left(5x^{\frac{1}{2}}\right)\left(7x^{\frac{3}{4}}\right)$ **b.** $\frac{32x^{\frac{5}{3}}}{16x^{\frac{3}{4}}}$.

Solution

a. $\left(5x^{\frac{1}{2}}\right)\left(7x^{\frac{3}{4}}\right) = 5 \cdot 7x^{\frac{1}{2}} \cdot x^{\frac{3}{4}}$ Group factors with the same base.

$= 35x^{\frac{1}{2}+\frac{3}{4}}$ When multiplying expressions with the same base, add the exponents.

$= 35x^{\frac{5}{4}}$ $\frac{1}{2} + \frac{3}{4} = \frac{2}{4} + \frac{3}{4} = \frac{5}{4}$

b. $\dfrac{32x^{\frac{5}{3}}}{16x^{\frac{3}{4}}} = \left(\dfrac{32}{16}\right)\left(\dfrac{x^{\frac{5}{3}}}{x^{\frac{3}{4}}}\right)$ Group factors with the same base.

$= 2x^{\frac{5}{3}-\frac{3}{4}}$ When dividing expressions with the same base, subtract the exponents.

$= 2x^{\frac{11}{12}}$ $\frac{5}{3} - \frac{3}{4} = \frac{20}{12} - \frac{9}{12} = \frac{11}{12}$

Check Point 12 Simplify: **a.** $\left(2x^{\frac{4}{3}}\right)\left(5x^{\frac{8}{3}}\right)$ **b.** $\dfrac{20x^4}{5x^{\frac{3}{2}}}$.

Rational exponents are sometimes useful for simplifying radicals by reducing their index.

EXAMPLE 13 Reducing the Index of a Radical

Simplify: $\sqrt[9]{x^3}$.

Solution $\sqrt[9]{x^3} = x^{\frac{3}{9}} = x^{\frac{1}{3}} = \sqrt[3]{x}$

Check Point 13 Simplify: $\sqrt[6]{x^3}$.

EXERCISE SET P.3

Practice Exercises

Evaluate each expression in Exercises 1–12, or indicate that the root is not a real number.

1. $\sqrt{36}$
2. $\sqrt{25}$
3. $-\sqrt{36}$
4. $-\sqrt{25}$
5. $\sqrt{-36}$
6. $\sqrt{-25}$
7. $\sqrt{25 - 16}$
8. $\sqrt{144 + 25}$
9. $\sqrt{25} - \sqrt{16}$
10. $\sqrt{144} + \sqrt{25}$
11. $\sqrt{(-13)^2}$
12. $\sqrt{(-17)^2}$

Use the product rule to simplify the expressions in Exercises 13–22. In Exercises 17–22, assume that variables represent nonnegative real numbers.

13. $\sqrt{50}$
14. $\sqrt{27}$
15. $\sqrt{45x^2}$
16. $\sqrt{125x^2}$
17. $\sqrt{2x} \cdot \sqrt{6x}$
18. $\sqrt{10x} \cdot \sqrt{8x}$
19. $\sqrt{x^3}$
20. $\sqrt{y^3}$
21. $\sqrt{2x^2} \cdot \sqrt{6x}$
22. $\sqrt{6x} \cdot \sqrt{3x^2}$

Use the quotient rule to simplify the expressions in Exercises 23–32. Assume that $x > 0$.

23. $\sqrt{\dfrac{1}{81}}$
24. $\sqrt{\dfrac{1}{49}}$
25. $\sqrt{\dfrac{49}{16}}$
26. $\sqrt{\dfrac{121}{9}}$
27. $\dfrac{\sqrt{48x^3}}{\sqrt{3x}}$
28. $\dfrac{\sqrt{72x^3}}{\sqrt{8x}}$
29. $\dfrac{\sqrt{150x^4}}{\sqrt{3x}}$
30. $\dfrac{\sqrt{24x^4}}{\sqrt{3x}}$
31. $\dfrac{\sqrt{200x^3}}{\sqrt{10x^{-1}}}$
32. $\dfrac{\sqrt{500x^3}}{\sqrt{10x^{-1}}}$

In Exercises 33–44, add or subtract terms whenever possible.

33. $7\sqrt{3} + 6\sqrt{3}$
34. $8\sqrt{5} + 11\sqrt{5}$
35. $6\sqrt{17x} - 8\sqrt{17x}$
36. $4\sqrt{13x} - 6\sqrt{13x}$
37. $\sqrt{8} + 3\sqrt{2}$
38. $\sqrt{20} + 6\sqrt{5}$
39. $\sqrt{50x} - \sqrt{8x}$
40. $\sqrt{63x} - \sqrt{28x}$
41. $3\sqrt{18} + 5\sqrt{50}$
42. $4\sqrt{12} - 2\sqrt{75}$
43. $3\sqrt{8} - \sqrt{32} + 3\sqrt{72} - \sqrt{75}$
44. $3\sqrt{54} - 2\sqrt{24} - \sqrt{96} + 4\sqrt{63}$

In Exercises 45–54, rationalize the denominator.

45. $\dfrac{1}{\sqrt{7}}$
46. $\dfrac{2}{\sqrt{10}}$
47. $\dfrac{\sqrt{2}}{\sqrt{5}}$
48. $\dfrac{\sqrt{7}}{\sqrt{3}}$
49. $\dfrac{13}{3 + \sqrt{11}}$
50. $\dfrac{3}{3 + \sqrt{7}}$
51. $\dfrac{7}{\sqrt{5} - 2}$
52. $\dfrac{5}{\sqrt{3} - 1}$
53. $\dfrac{6}{\sqrt{5} + \sqrt{3}}$
54. $\dfrac{11}{\sqrt{7} - \sqrt{3}}$

Evaluate each expression in Exercises 55–66, or indicate that the root is not a real number.

55. $\sqrt[3]{125}$
56. $\sqrt[3]{8}$
57. $\sqrt[3]{-8}$
58. $\sqrt[3]{-125}$

59. $\sqrt[4]{-16}$ **60.** $\sqrt[4]{-81}$

61. $\sqrt[4]{(-3)^4}$ **62.** $\sqrt[4]{(-2)^4}$

63. $\sqrt[5]{(-3)^5}$ **64.** $\sqrt[5]{(-2)^5}$

65. $\sqrt[5]{-\frac{1}{32}}$ **66.** $\sqrt[6]{\frac{1}{64}}$

Simplify the radical expressions in Exercises 67–74.

67. $\sqrt[3]{32}$ **68.** $\sqrt[3]{150}$

69. $\sqrt[3]{x^4}$ **70.** $\sqrt[3]{x^5}$

71. $\sqrt[3]{9}\cdot\sqrt[3]{6}$ **72.** $\sqrt[3]{12}\cdot\sqrt[3]{4}$

73. $\dfrac{\sqrt[5]{64x^6}}{\sqrt[5]{2x}}$ **74.** $\dfrac{\sqrt[4]{162x^5}}{\sqrt[4]{2x}}$

In Exercises 75–82, add or subtract terms whenever possible.

75. $4\sqrt[5]{2} + 3\sqrt[5]{2}$ **76.** $6\sqrt[5]{3} + 2\sqrt[5]{3}$

77. $5\sqrt[3]{16} + \sqrt[3]{54}$ **78.** $3\sqrt[3]{24} + \sqrt[3]{81}$

79. $\sqrt[3]{54xy^3} - y\sqrt[3]{128x}$ **80.** $\sqrt[3]{24xy^3} - y\sqrt[3]{81x}$

81. $\sqrt{2} + \sqrt[3]{8}$ **82.** $\sqrt{3} + \sqrt[3]{15}$

In Exercises 83–90, evaluate each expression without using a calculator.

83. $36^{\frac{1}{2}}$ **84.** $121^{\frac{1}{2}}$

85. $8^{\frac{1}{3}}$ **86.** $27^{\frac{1}{3}}$

87. $125^{\frac{2}{3}}$ **88.** $8^{\frac{2}{3}}$

89. $32^{-\frac{4}{5}}$ **90.** $16^{-\frac{5}{2}}$

In Exercises 91–100, simplify using properties of exponents.

91. $\left(7x^{\frac{1}{3}}\right)\left(2x^{\frac{1}{4}}\right)$ **92.** $\left(3x^{\frac{2}{3}}\right)\left(4x^{\frac{3}{4}}\right)$

93. $\dfrac{20x^{\frac{1}{2}}}{5x^{\frac{1}{4}}}$ **94.** $\dfrac{72x^{\frac{3}{4}}}{9x^{\frac{1}{3}}}$

95. $\left(x^{\frac{2}{3}}\right)^3$ **96.** $\left(x^{\frac{4}{5}}\right)^5$

97. $(25x^4y^6)^{\frac{1}{2}}$ **98.** $(125x^9y^6)^{\frac{1}{3}}$

99. $\dfrac{\left(3y^{\frac{1}{4}}\right)^3}{y^{\frac{1}{12}}}$ **100.** $\dfrac{\left(2y^{\frac{1}{5}}\right)^4}{y^{\frac{3}{10}}}$

In Exercises 101–108, simplify by reducing the index of the radical.

101. $\sqrt[4]{5^2}$ **102.** $\sqrt[4]{7^2}$

103. $\sqrt[3]{x^6}$ **104.** $\sqrt[4]{x^{12}}$

105. $\sqrt[6]{x^4}$ **106.** $\sqrt[9]{x^6}$

107. $\sqrt[9]{x^6y^3}$ **108.** $\sqrt[12]{x^4y^8}$

Practice Plus

In Exercises 109–110, evaluate each expression.

109. $\sqrt[3]{\sqrt[4]{16} + \sqrt{625}}$

110. $\sqrt[3]{\sqrt{\sqrt{169} + \sqrt{9}} + \sqrt{\sqrt[3]{1000} + \sqrt[3]{216}}}$

In Exercises 111–114, simplify each expression. Assume that all variables represent positive numbers.

111. $(49x^{-2}y^4)^{-\frac{1}{2}}\left(xy^{\frac{1}{2}}\right)$ **112.** $(8x^{-6}y^3)^{\frac{1}{3}}\left(x^{\frac{5}{6}}y^{-\frac{1}{3}}\right)^6$

113. $\left(\dfrac{x^{-\frac{5}{4}}y^{\frac{1}{3}}}{x^{-\frac{3}{4}}}\right)^{-6}$ **114.** $\left(\dfrac{x^{\frac{1}{2}}y^{-\frac{7}{4}}}{y^{-\frac{5}{4}}}\right)^{-4}$

Application Exercises

The formula

$$d = \sqrt{\frac{3h}{2}}$$

models the distance, d, in miles, that a person h feet high can see to the horizon. Use this formula to solve Exercises 115–116.

115. The pool deck on a cruise ship is 72 feet above the water. How far can passengers on the pool deck see? Write the answer in simplified radical form. Then use the simplified radical form and a calculator to express the answer to the nearest tenth of a mile.

116. The captain of a cruise ship is on the star deck, which is 120 feet above the water. How far can the captain see? Write the answer in simplified radical form. Then use the simplified radical form and a calculator to express the answer to the nearest tenth of a mile.

Police use the formula $v = 2\sqrt{5L}$ to estimate the speed of a car, v, in miles per hour, based on the length, L, in feet, of its skid marks upon sudden braking on a dry asphalt road. Use the formula to solve Exercises 117–118.

117. A motorist is involved in an accident. A police officer measures the car's skid marks to be 245 feet long. Estimate the speed at which the motorist was traveling before braking. If the posted speed limit is 50 miles per hour and the motorist tells the officer he was not speeding, should the officer believe him? Explain.

118. A motorist is involved in an accident. A police officer measures the car's skid marks to be 45 feet long. Estimate the speed at which the motorist was traveling before braking. If the posted speed limit is 35 miles per hour and the motorist tells the officer she was not speeding, should the officer believe her? Explain.

119. In the Peanuts cartoon shown below, Woodstock appears to be working steps mentally. Fill in the missing steps that show how to go from $\dfrac{7\sqrt{2\cdot2\cdot3}}{6}$ to $\dfrac{7}{3}\sqrt{3}$.

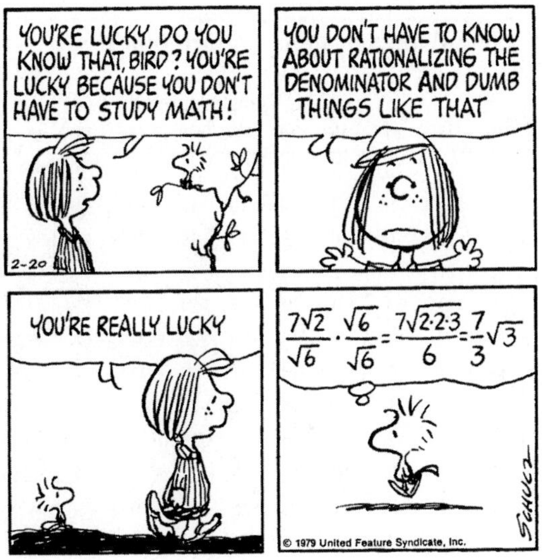

PEANUTS reprinted by permission of United Feature Syndicate, Inc.

120. According to Einstein's theory of relativity, traveling in starships at velocities approaching the speed of light (approximately 186,000 miles per second), time would pass more quickly on Earth than it would in the moving starship. The radical expression

$$R_f \frac{\sqrt{c^2 - v^2}}{\sqrt{c^2}}$$

gives the aging rate of an astronaut relative to the aging rate of a friend, R_f, on Earth. In the expression, v is the astronaut's velocity and c is the speed of light. Use the expression to solve this exercise. Imagine that you are the astronaut on the starship.

a. Use the quotient rule and simplify the expression that shows your aging rate relative to a friend on Earth. Working step-by-step, express your aging rate as

$$R_f \sqrt{1 - \left(\frac{v}{c}\right)^2}.$$

b. You are moving at 90% of the speed of light. Substitute $0.9c$ for v, your velocity, in the simplified expression from part (a). What is your aging rate, correct to two decimal places, relative to a friend on Earth? If you are gone for 44 weeks, approximately how many weeks have passed for your friend?

The way that we perceive the temperature on a cold day depends on both air temperature and wind speed. The windchill is what the air temperature would have to be with no wind to achieve the same chilling effect on the skin. In 2002, the National Weather Service issued new windchill temperatures, shown in the table below. (One reason for this new windchill index is that the wind speed is now calculated at 5 feet, the average height of the human body's face, rather than 33 feet, the height of the standard anemometer, an instrument that calculates wind speed.)

New Windchill Temperature Index

Wind Speed (miles per hour)	Air Temperature (°F) 30	25	20	15	10	5	0	−5	−10	−15	−20	−25
5	25	19	13	7	1	−5	−11	−16	−22	−28	−34	−40
10	21	15	9	3	−4	−10	−16	−22	−28	−35	−41	−47
15	19	13	6	0	−7	−13	−19	−26	−32	−39	−45	−51
20	17	11	4	−2	−9	−15	−22	−29	−35	−42	−48	−55
25	16	9	3	−4	−11	−17	−24	−31	−37	−44	−51	−58
30	15	8	1	−5	−12	−19	−26	−33	−39	−46	−53	−60
35	14	7	0	−7	−14	−21	−27	−34	−41	−48	−55	−62
40	13	6	−1	−8	−15	−22	−29	−36	−43	−50	−57	−64
45	12	5	−2	−9	−16	−23	−30	−37	−44	−51	−58	−65
50	12	4	−3	−10	−17	−24	−31	−38	−45	−52	−60	−67
55	11	4	−3	−11	−18	−25	−32	−39	−46	−54	−61	−68
60	10	3	−4	−11	−19	−26	−33	−40	−48	−55	−62	−69

Frostbite occurs in 15 minutes or less.

Source: National Weather Service

The windchill temperatures shown in the table can be calculated using

$$C = 35.74 + 0.6215t - 35.74\sqrt[25]{v^4} + 0.4275t\sqrt[25]{v^4},$$

in which C is the windchill, in degrees Fahrenheit, t is the air temperature, in degrees Fahrenheit, and v is the wind speed, in miles per hour. Use the formula to solve Exercises 121–122.

121. a. Rewrite the equation for calculating windchill temperatures using rational exponents.

b. Use the form of the equation in part (a) and a calculator to find the windchill temperature, to the nearest degree, when the air temperature is 25°F and the wind speed is 30 miles per hour.

122. a. Rewrite the equation for calculating windchill temperatures using rational exponents.

b. Use the form of the equation in part (a) and a calculator to find the windchill temperature, to the nearest degree, when the air temperature is 35°F and the wind speed is 15 miles per hour.

In Exercises 123–124, find the perimeter and area of each rectangle. Express answers in simplified radical form.

123.

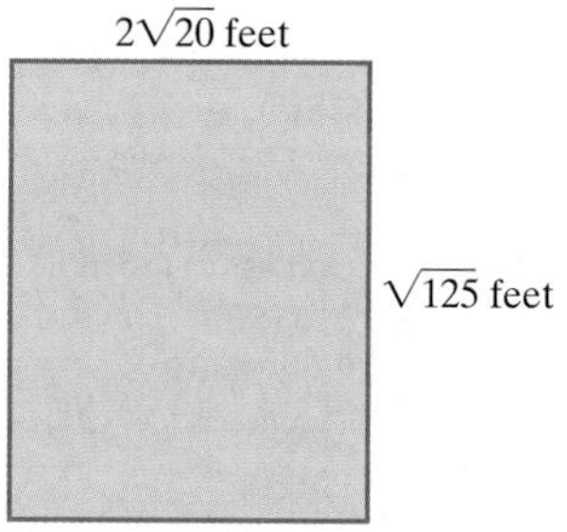

124.

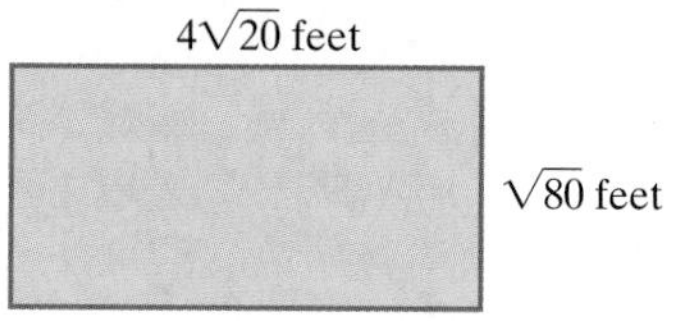

Writing in Mathematics

125. Explain how to simplify $\sqrt{10} \cdot \sqrt{5}$.

126. Explain how to add $\sqrt{3} + \sqrt{12}$.

127. Describe what it means to rationalize a denominator. Use both $\frac{1}{\sqrt{5}}$ and $\frac{1}{5 + \sqrt{5}}$ in your explanation.

128. What difference is there in simplifying $\sqrt[3]{(-5)^3}$ and $\sqrt[4]{(-5)^4}$?

129. What does $a^{\frac{m}{n}}$ mean?

130. Describe the kinds of numbers that have rational fifth roots.

131. Why must a and b represent nonnegative numbers when we write $\sqrt{a} \cdot \sqrt{b} = \sqrt{ab}$? Is it necessary to use this restriction in the case of $\sqrt[3]{a} \cdot \sqrt[3]{b} = \sqrt[3]{ab}$? Explain.

132. Answer the question posed in the chapter opener on page 1. What will you do: explore space or stay here on Earth? What are the reasons for your choice?

Critical Thinking Exercises

133. Which one of the following is true?

  a. Neither $(-8)^{\frac{1}{2}}$ nor $(-8)^{\frac{1}{3}}$ represents real numbers.

  b. $\sqrt{x^2 + y^2} = x + y$

  c. $8^{-\frac{1}{3}} = -2$

  d. $2^{\frac{1}{2}} \cdot 2^{\frac{1}{2}} = 2$

In Exercises 134–135, fill in each box to make the statement true.

134. $\left(5 + \sqrt{\Box}\right)\left(5 - \sqrt{\Box}\right) = 22$

135. $\sqrt{\Box x^{\Box}} = 5x^7$

136. Find exact value of $\sqrt{13 + \sqrt{2} + \dfrac{7}{3 + \sqrt{2}}}$ without the use of a calculator.

137. Place the correct symbol, $>$ or $<$, in the shaded area between each of the given numbers. *Do not use a calculator.* Then check your result with a calculator.

  a. $3^{\frac{1}{2}} \;\rule{1cm}{0.3mm}\; 3^{\frac{1}{3}}$ **b.** $\sqrt{7} + \sqrt{18} \;\rule{1cm}{0.3mm}\; \sqrt{7 + 18}$

138. a. A mathematics professor recently purchased a birthday cake for her son with the inscription

$$\text{Happy } \left(2^{\frac{5}{2}} \cdot 2^{\frac{3}{4}} \div 2^{\frac{1}{4}}\right)\text{th Birthday.}$$

  How old is the son?

  b. The birthday boy, excited by the inscription on the cake, tried to wolf down the whole thing. Professor Mom, concerned about the possible metamorphosis of her son into a blimp, exclaimed, "Hold on! It is your birthday, so why not take $\dfrac{8^{-\frac{4}{3}} + 2^{-2}}{16^{-\frac{3}{4}} + 2^{-1}}$ of the cake? I'll eat half of what's left over." How much of the cake did the professor eat?

SECTION P.4 *Polynomials*

Objectives

❶ Understand the vocabulary of polynomials.

❷ Add and subtract polynomials.

❸ Multiply polynomials.

❹ Use FOIL in polynomial multiplication.

❺ Use special products in polynomial multiplication.

❻ Perform operations with polynomials in several variables.

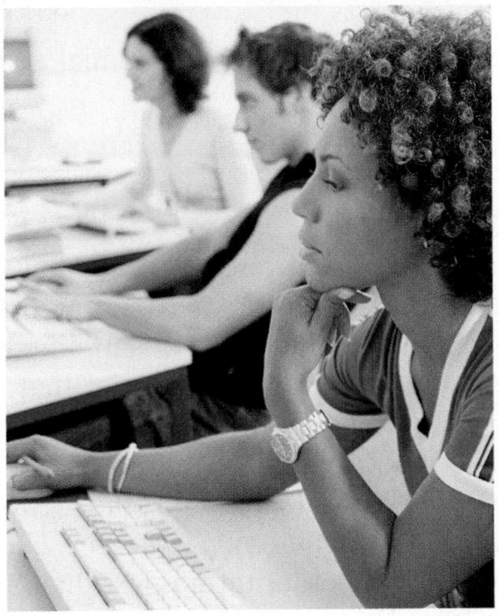

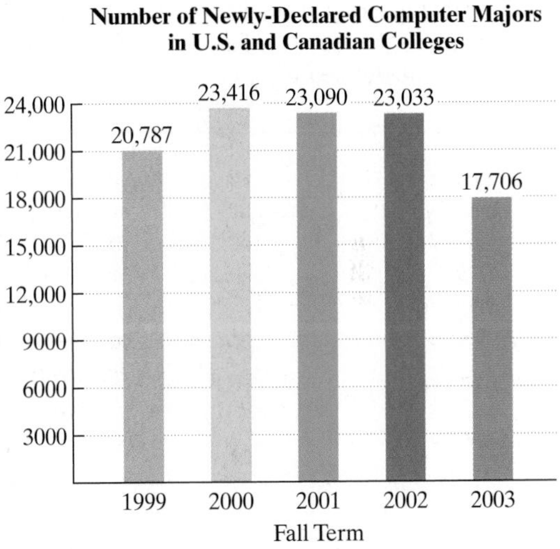

Figure P.11

Source: Computing Research Association

Tech firms might be rebounding from the dot-com bust, but enrollment in college computer programs keeps falling. In the past, a computer degree meant "instant riches, or at least a well-paying, secure job," says San Jose computer science chair David Hayes. "Now, the perception is jobs are going overseas and people are being laid off."

The bar graph in Figure P.11 shows the number of newly-declared computer science and computer engineering majors for the fall term in U.S. and Canadian colleges from 1999 through 2003. The data can be modeled by the formula

$$N = -365x^4 + 2728x^3 - 7106x^2 + 7372x + 20{,}787,$$

where N is the number of newly-declared computer majors for the fall term x years after 1999.

The algebraic expression on the right side of the equation,

$$-365x^4 + 2728x^3 - 7106x^2 + 7372x + 20{,}787,$$

is an example of a *polynomial*. A **polynomial** is a single term or the sum of two or more terms containing variables in the numerator with whole-number exponents. This particular polynomial contains five terms. Equations containing polynomials are used in such diverse areas as science, business, medicine, psychology, and sociology. In this section, we review basic ideas about polynomials and their operations.

1 Understand the vocabulary of polynomials.

How We Describe Polynomials

Consider the polynomial

$$7x^3 - 9x^2 + 13x - 6.$$

We can express this polynomial as

$$7x^3 + (-9x^2) + 13x + (-6).$$

The polynomial contains four terms. It is customary to write the terms in the order of descending powers of the variable. This is the **standard form** of a polynomial.

Some polynomials contain only one variable. Each term of such a polynomial in x is of the form ax^n. If $a \neq 0$, the **degree** of ax^n is n. For example, the degree of the term $7x^3$ is 3.

Study Tip

We can express 0 in many ways, including $0x$, $0x^2$, and $0x^3$. It is impossible to assign a single exponent on the variable. This is why 0 has no defined degree.

The Degree of ax^n

If $a \neq 0$, the degree of ax^n is n. The degree of a nonzero constant is 0. The constant 0 has no defined degree.

Here is an example of a polynomial and the degree of each of its four terms:

$$6x^4 - 3x^3 + 2x - 5.$$

degree 4 | degree 3 | degree 1 | degree of nonzero constant: 0

Notice that the exponent on x for the term $2x$ is understood to be $1{:}2x^1$. For this reason, the degree of $2x$ is 1. You can think of -5 as $-5x^0$; thus, its degree is 0.

A polynomial which when simplified has exactly one term is called a **monomial**. A **binomial** is a polynomial that has two terms, each with a different exponent. A **trinomial** is a polynomial with three terms, each with a different exponent. Simplified polynomials with four or more terms have no special names.

The **degree of a polynomial** is the greatest of the degrees of all its terms. For example, $4x^2 + 3x$ is a binomial of degree 2 because the degree of the first term is 2, and the degree of the other term is less than 2. Also, $7x^5 - 2x^2 + 4$ is a trinomial of degree 5 because the degree of the first term is 5, and the degrees of the other terms are less than 5.

Up to now, we have used x to represent the variable in a polynomial. However, any letter can be used. For example,

- $7x^5 - 3x^3 + 8$ is a polynomial (in x) of degree 5. Because there are three terms, the polynomial is a trinomial.
- $6y^3 + 4y^2 - y + 3$ is a polynomial (in y) of degree 3. Because there are four terms, the polynomial has no special name.
- $z^7 + \sqrt{2}$ is a polynomial (in z) of degree 7. Because there are two terms, the polynomial is a binomial.

Not every algebraic expression is a polynomial. Algebraic expressions whose variables do not contain whole number exponents in numerators such as

$$3x^{-2} + 7 \quad \text{and} \quad 5x^{\frac{3}{2}} + 9x^{\frac{1}{2}} + 2$$

are not polynomials. Furthermore, a quotient of polynomials such as

$$\frac{x^2 + 2x + 5}{x^3 - 7x^2 + 9x - 3}$$

is not a polynomial because the form of a polynomial involves only addition and subtraction of terms, not division.

We can tie together the threads of our discussion with the formal definition of a polynomial in one variable. In this definition, the coefficients of the terms are represented by a_n (read "a sub n"), a_{n-1} (read "a sub n minus 1"), a_{n-2}, and so on. The small letters to the lower right of each a are called **subscripts** and are *not exponents*. Subscripts are used to distinguish one constant from another when a large and undetermined number of such constants are needed.

Definition of a Polynomial in x

A **polynomial in x** is an algebraic expression of the form

$$a_nx^n + a_{n-1}x^{n-1} + a_{n-2}x^{n-2} + \cdots + a_1x + a_0,$$

where $a_n, a_{n-1}, a_{n-2}, \ldots, a_1$, and a_0 are real numbers, $a_n \neq 0$, and n is a non-negative integer. The polynomial is of **degree n**, a_n is the **leading coefficient**, and a_0 is the **constant term**.

2 Add and subtract polynomials.

Adding and Subtracting Polynomials

Polynomials are added and subtracted by combining like terms. For example, we can combine the monomials $-9x^3$ and $13x^3$ using addition as follows:

$$-9x^3 + 13x^3 = (-9 + 13)x^3 = 4x^3.$$

These like terms both contain x to the third power.

Add coefficients and keep the same variable factor, x^3.

EXAMPLE 1 Adding and Subtracting Polynomials

Perform the indicated operations and simplify:

a. $(-9x^3 + 7x^2 - 5x + 3) + (13x^3 + 2x^2 - 8x - 6)$

b. $(7x^3 - 8x^2 + 9x - 6) - (2x^3 - 6x^2 - 3x + 9)$.

Solution

a. $(-9x^3 + 7x^2 - 5x + 3) + (13x^3 + 2x^2 - 8x - 6)$

$= (-9x^3 + 13x^3) + (7x^2 + 2x^2) + (-5x - 8x) + (3 - 6)$ Group like terms.

$= 4x^3 + 9x^2 + (-13x) + (-3)$ Combine like terms.

$= 4x^3 + 9x^2 - 13x - 3$ Simplify.

b. $(7x^3 - 8x^2 + 9x - 6) - (2x^3 - 6x^2 - 3x + 9)$

Change the sign of each coefficient.

$= (7x^3 - 8x^2 + 9x - 6) + (-2x^3 + 6x^2 + 3x - 9)$ Rewrite subtraction as addition of the additive inverse.

$= (7x^3 - 2x^3) + (-8x^2 + 6x^2) + (9x + 3x) + (-6 - 9)$ Group like terms.

$= 5x^3 + (-2x^2) + 12x + (-15)$ Combine like terms.

$= 5x^3 - 2x^2 + 12x - 15$ Simplify.

Study Tip

You can also arrange like terms in columns and combine vertically:

$$\begin{array}{r} 7x^3 - 8x^2 + 9x - 6 \\ \underline{-2x^3 + 6x^2 + 3x - 9} \\ 5x^3 - 2x^2 + 12x - 15 \end{array}$$

The like terms can be combined by adding their coefficients and keeping the same variable factor.

Check Point 1 Perform the indicated operations and simplify:

a. $(-17x^3 + 4x^2 - 11x - 5) + (16x^3 - 3x^2 + 3x - 15)$

b. $(13x^3 - 9x^2 - 7x + 1) - (-7x^3 + 2x^2 - 5x + 9)$.

Multiply polynomials.

Multiplying Polynomials

The product of two monomials is obtained by using properties of exponents. For example,

$$(-8x^6)(5x^3) = -8 \cdot 5x^{6+3} = -40x^9.$$

> Multiply coefficients and add exponents.

Furthermore, we can use the distributive property to multiply a monomial and a polynomial that is not a monomial. For example,

$$3x^4(2x^3 - 7x + 3) = 3x^4 \cdot 2x^3 - 3x^4 \cdot 7x + 3x^4 \cdot 3 = 6x^7 - 21x^5 + 9x^4.$$

> Monomial   Trinomial

How do we multiply two polynomials if neither is a monomial? For example, consider

$$(2x + 3)(x^2 + 4x + 5).$$

> Binomial   Trinomial

One way to perform this multiplication is to distribute $2x$ throughout the trinomial

$$2x(x^2 + 4x + 5)$$

and 3 throughout the trinomial

$$3(x^2 + 4x + 5).$$

Then combine the like terms that result.

> ### Multiplying Polynomials When Neither Is a Monomial
> Multiply each term of one polynomial by each term of the other polynomial. Then combine like terms.

Study Tip

Don't confuse adding and multiplying monomials.

Addition:
$$5x^4 + 6x^4 = 11x^4$$

Multiplication:
$$(5x^4)(6x^4) = (5 \cdot 6)(x^4 \cdot x^4)$$
$$= 30x^{4+4}$$
$$= 30x^8$$

Only like terms can be added or subtracted, but unlike terms may be multiplied.

Addition:
$5x^4 + 3x^2$ cannot be simplified.

Multiplication:
$$(5x^4)(3x^2) = (5 \cdot 3)(x^4 \cdot x^2)$$
$$= 15x^{4+2}$$
$$= 15x^6$$

EXAMPLE 2   Multiplying a Binomial and a Trinomial

Multiply: $(2x + 3)(x^2 + 4x + 5).$

Solution

$(2x + 3)(x^2 + 4x + 5)$

$= 2x(x^2 + 4x + 5) + 3(x^2 + 4x + 5)$ Multiply the trinomial by each term of the binomial.

$= 2x \cdot x^2 + 2x \cdot 4x + 2x \cdot 5 + 3x^2 + 3 \cdot 4x + 3 \cdot 5$ Use the distributive property.

$= 2x^3 + 8x^2 + 10x + 3x^2 + 12x + 15$ Multiply monomials: Multiply coefficients and add exponents.

$= 2x^3 + 11x^2 + 22x + 15$ Combine like terms: $8x^2 + 3x^2 = 11x^2$ and $10x + 12x = 22x.$

Another method for solving Example 2 is to use a vertical format similar to that used for multiplying whole numbers.

$$\begin{array}{rrrrl} & x^2 + & 4x + & 5 & \\ & & 2x + & 3 & \\ \hline & 3x^2 + & 12x + & 15 & 3(x^2 + 4x + 5) \\ 2x^3 + & 8x^2 + & 10x & & 2x(x^2 + 4x + 5) \\ \hline 2x^3 + & 11x^2 + & 22x + & 15 & \text{Combine like terms.} \end{array}$$

Write like terms in the same column.

Check Point 2 Multiply: $(5x - 2)(3x^2 - 5x + 4)$.

4 Use FOIL in polynomial multiplication.

The Product of Two Binomials: FOIL

Frequently, we need to find the product of two binomials. One way to perform this multiplication is to distribute each term in the first binomial through the second binomial. For example, we can find the product of the binomials $3x + 2$ and $4x + 5$ as follows:

$$\begin{aligned}(3x + 2)(4x + 5) &= 3x(4x + 5) + 2(4x + 5) \\ &= 3x(4x) + 3x(5) + 2(4x) + 2(5) \\ &= 12x^2 + 15x + 8x + 10.\end{aligned}$$

Distribute $3x$ over $4x + 5$. Distribute 2 over $4x + 5$.

We can also find the product of $3x + 2$ and $4x + 5$ using a method called FOIL, which is based on our work shown above. Any two binomials can be quickly multiplied by using the FOIL method, in which **F** represents the product of the **first** terms in each binomial, **O** represents the product of the **outside** terms, **I** represents the product of the **inside** terms, and **L** represents the product of the **last**, or second, terms in each binomial. For example, we can use the FOIL method to find the product of the binomials $3x + 2$ and $4x + 5$ as follows:

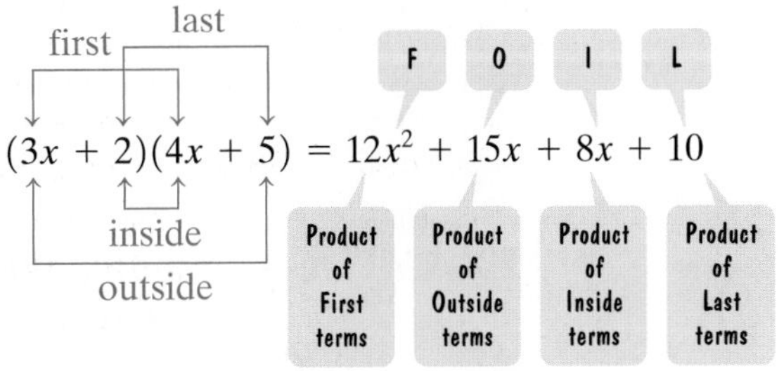

$$(3x + 2)(4x + 5) = 12x^2 + 15x + 8x + 10$$

$$= 12x^2 + 23x + 10 \qquad \text{Combine like terms.}$$

In general, here's how to use the FOIL method to find the product of $ax + b$ and $cx + d$:

Using the FOIL Method to Multiply Binomials

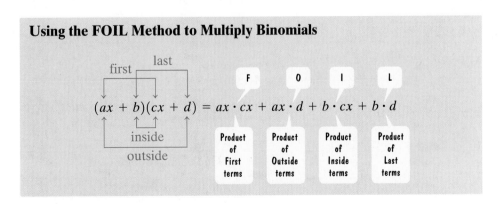

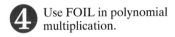

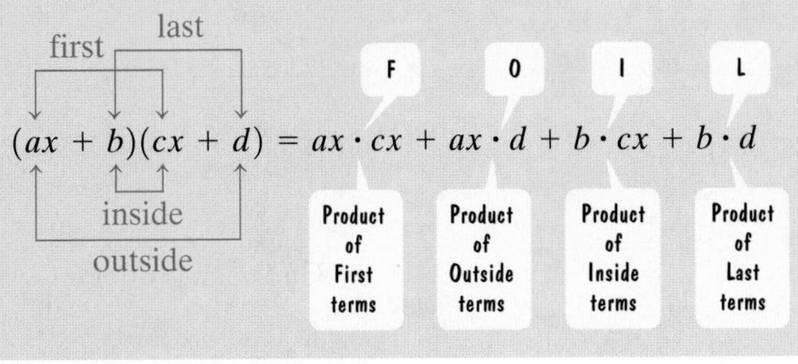

$$(ax + b)(cx + d) = ax \cdot cx + ax \cdot d + b \cdot cx + b \cdot d$$

EXAMPLE 3 Using the FOIL Method

Multiply: $(3x + 4)(5x - 3)$.

Solution

first, last, inside, outside

F O I L

$$(3x + 4)(5x - 3) = 3x \cdot 5x + 3x(-3) + 4 \cdot 5x + 4(-3)$$
$$= 15x^2 - 9x + 20x - 12$$
$$= 15x^2 + 11x - 12 \quad \text{Combine like terms.}$$

Check Point 3 Multiply: $(7x - 5)(4x - 3)$.

5 Use special products in polynomial multiplication.

Multiplying the Sum and Difference of Two Terms

We can use the FOIL method to multiply $A + B$ and $A - B$ as follows:

F O I L

$$(A + B)(A - B) = A^2 - AB + AB - B^2 = A^2 - B^2.$$

Notice that the outside and inside products have a sum of 0 and the terms cancel. The FOIL multiplication provides us with a quick rule for multiplying the sum and difference of two terms, referred to as a special-product formula.

The Product of the Sum and Difference of Two Terms

$$(A + B)(A - B) = A^2 - B^2$$

The product of the sum and the difference of the same two terms is the square of the first term minus the square of the second term.

EXAMPLE 4 Finding the Product of the Sum and Difference of Two Terms

Find each product:

a. $(4y + 3)(4y - 3)$ **b.** $(5a^4 + 6)(5a^4 - 6)$.

Solution Use the special-product formula shown.

$$(A + B)(A - B) = A^2 - B^2$$

First term squared − Second term squared = Product

a. $(4y + 3)(4y - 3) = (4y)^2 - 3^2 = 16y^2 - 9$

b. $(5a^4 + 6)(5a^4 - 6) = (5a^4)^2 - 6^2 = 25a^8 - 36$

Check Point 4 Find each product:

a. $(7x + 8)(7x - 8)$ **b.** $(2y^3 - 5)(2y^3 + 5)$.

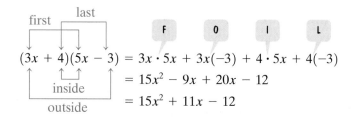

The Square of a Binomial

Let us find $(A + B)^2$, the square of a binomial sum. To do so, we begin with the FOIL method and look for a general rule.

F O I L

$$(A + B)^2 = (A + B)(A + B) = A \cdot A + A \cdot B + A \cdot B + B \cdot B$$
$$= A^2 + 2AB + B^2$$

This result implies the following rule, which is another example of a special-product formula:

Study Tip

Caution! The square of a sum is *not* the sum of the squares.

$(A + B)^2 \neq A^2 + B^2$

The middle term $2AB$ is missing.

$(x + 3)^2 \neq x^2 + 9$

Incorrect!

Show that $(x + 3)^2$ and $x^2 + 9$ are not equal by substituting 5 for x in each expression and simplifying.

The Square of a Binomial Sum

$$(A + B)^2 = A^2 + 2AB + B^2$$

The square of a binomial sum is first term squared plus 2 times the product of the terms plus last term squared.

EXAMPLE 5 Finding the Square of a Binomial Sum

Square each binomial:

a. $(x + 3)^2$ **b.** $(3x + 7)^2$.

Solution Use the special-product formula shown.

$$(A + B)^2 = A^2 + 2AB + B^2$$

	(First Term)^2	+	2 · Product of the Terms	+	(Last Term)^2	= Product
a. $(x + 3)^2 =$	x^2	+	$2 \cdot x \cdot 3$	+	3^2	$= x^2 + 6x + 9$
b. $(3x + 7)^2 =$	$(3x)^2$	+	$2(3x)(7)$	+	7^2	$= 9x^2 + 42x + 49$

Check Point 5 Square each binomial:

a. $(x + 10)^2$ **b.** $(5x + 4)^2$.

Using the FOIL method on $(A - B)^2$, the square of a binomial difference, we obtain the following rule:

The Square of a Binomial Difference

$$(A - B)^2 = A^2 - 2AB + B^2$$

The square of a binomial difference is first term squared minus 2 times the product of the terms plus last term squared.

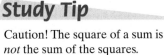

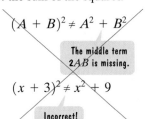

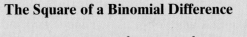

EXAMPLE 6 Finding the Square of a Binomial Difference

Square each binomial using the preceding rule:

a. $(x - 4)^2$ **b.** $(5y - 6)^2$.

Solution Use the special-product formula shown.

$$(A - B)^2 = A^2 - 2AB + B^2$$

	(First Term)^2	−	2 · Product of the Terms	+	(Last Term)^2	= Product
a. $(x - 4)^2 =$	x^2	−	$2 \cdot x \cdot 4$	+	4^2	$= x^2 - 8x + 16$
b. $(5y - 6)^2 =$	$(5y)^2$	−	$2(5y)(6)$	+	6^2	$= 25y^2 - 60y + 36$

Check Point 6 Square each binomial:

a. $(x - 9)^2$ **b.** $(7x - 3)^2$.

Special Products

There are several products that occur so frequently that it's convenient to memorize the form, or pattern, of these formulas.

Special Products

Let A and B represent real numbers, variables, or algebraic expressions.

Special Product	Example
Sum and Difference of Two Terms	
$(A + B)(A - B) = A^2 - B^2$	$(2x + 3)(2x - 3) = (2x)^2 - 3^2 = 4x^2 - 9$
Squaring a Binomial	
$(A + B)^2 = A^2 + 2AB + B^2$	$(y + 5)^2 = y^2 + 2 \cdot y \cdot 5 + 5^2 = y^2 + 10y + 25$
$(A - B)^2 = A^2 - 2AB + B^2$	$(3x - 4)^4 = (3x)^2 - 2 \cdot 3x \cdot 4 + 4^2 = 9x^2 - 24x + 16$
Cubing a Binomial	
$(A + B)^3 = A^3 + 3A^2B + 3AB^2 + B^3$	$(x + 4)^3 = x^3 + 3x^2(4) + 3x(4)^2 + 4^3 = x^3 + 12x^2 + 48x + 64$
$(A - B)^3 = A^3 - 3A^2B + 3AB^2 - B^3$	$(x - 2)^3 = x^3 - 3x^2(2) + 3x(2)^2 - 2^3 = x^3 - 6x^2 + 12x - 8$

Study Tip

Although it's convenient to memorize these forms, the FOIL method can be used on all five examples in the box. To cube $x + 4$, you can first square $x + 4$ using FOIL and then multiply this result by $x + 4$. In short, you do not necessarily have to utilize these special formulas. What is the advantage of knowing and using these forms?

6 Perform operations with polynomials in several variables.

Polynomials in Several Variables

The next time you visit the lumber yard and go rummaging through piles of wood, think *polynomials*, although polynomials a bit different from those we have encountered so far. The forestry industry uses a polynomial in two variables to determine the number of board feet that can be manufactured from a tree with a diameter of x inches and a length of y feet. This polynomial is

$$\frac{1}{4}x^2y - 2xy + 4y.$$

In general, a **polynomial in two variables**, x and y, contains the sum of one or more monomials in the form $ax^n y^m$. The constant, a, is the **coefficient**. The exponents, n and m, represent whole numbers. The **degree** of the monomial $ax^n y^m$ is $n + m$. We'll use the polynomial from the forestry industry to illustrate these ideas.

The coefficients are $\frac{1}{4}$, -2, and 4.

$$\frac{1}{4}x^2y \quad - 2xy \quad + 4y$$

Degree of monomial: $2 + 1 = 3$

Degree of monomial: $1 + 1 = 2$

Degree of monomial $(4x^0y^1)$: $0 + 1 = 1$

The **degree of a polynomial in two variables** is the highest degree of all its terms. For the preceding polynomial, the degree is 3.

Polynomials containing two or more variables can be added, subtracted, and multiplied just like polynomials that contain only one variable. For example, we can add the monomials $-7xy^2$ and $13xy^2$ as follows:

$$-7xy^2 + 13xy^2 = (-7 + 13)xy^2 = 6xy^2.$$

These like terms both contain the variable factors x and y^2.

Add coefficients and keep the same variable factors, xy^2.

EXAMPLE 7 Subtracting Polynomials in Two Variables

Subtract:

$$(5x^3 - 9x^2y + 3xy^2 - 4) - (3x^3 - 6x^2y - 2xy^2 + 3).$$

Solution

$$(5x^3 - 9x^2y + 3xy^2 - 4) - (3x^3 - 6x^2y - 2xy^2 + 3)$$

Change the sign of each coefficient.

$$= (5x^3 - 9x^2y + 3xy^2 - 4) + (-3x^3 + 6x^2y + 2xy^2 - 3)$$ Add the opposite of the polynomial being subtracted.

$$= (5x^3 - 3x^3) + (-9x^2y + 6x^2y) + (3xy^2 + 2xy^2) + (-4 - 3)$$ Group like terms.

$$= 2x^3 - 3x^2y + 5xy^2 - 7$$ Combine like terms by adding coefficients and keeping the same variable factors.

Check Point 7 Subtract: $(x^3 - 4x^2y + 5xy^2 - y^3) - (x^3 - 6x^2y + y^3)$.

EXAMPLE 8 Multiplying Polynomials in Two Variables

Multiply: **a.** $(x + 4y)(3x - 5y)$ **b.** $(5x + 3y)^2$.

Solution We will perform the multiplication in part (a) using the FOIL method. We will multiply in part (b) using the formula for the square of a binomial sum, $(A + B)^2$.

a. $(x + 4y)(3x - 5y)$ *Multiply these binomials using the FOIL method.*

F O I L

$$= (x)(3x) + (x)(-5y) + (4y)(3x) + (4y)(-5y)$$
$$= 3x^2 - 5xy + 12xy - 20y^2$$
$$= 3x^2 + 7xy - 20y^2$$ *Combine like terms.*

$$(A + B)^2 = A^2 + 2 \cdot A \cdot B + B^2$$

b. $(5x + 3y)^2 = (5x)^2 + 2(5x)(3y) + (3y)^2$
$$= 25x^2 + 30xy + 9y^2$$

Check Point 8 Multiply:
a. $(7x - 6y)(3x - y)$ **b.** $(2x + 4y)^2$.

EXERCISE SET P.4

Practice Exercises

In Exercises 1–4, is the algebraic expression a polynomial? If it is, write the polynomial in standard form.

1. $2x + 3x^2 - 5$
2. $2x + 3x^{-1} - 5$
3. $\frac{2x + 3}{x}$
4. $x^2 - x^3 + x^4 - 5$

In Exercises 5–8, find the degree of the polynomial.

5. $3x^2 - 5x + 4$
6. $-4x^3 + 7x^2 - 11$
7. $x^2 - 4x^3 + 9x - 12x^4 + 63$
8. $x^2 - 8x^3 + 15x^4 + 91$

In Exercises 9–14, perform the indicated operations. Write the resulting polynomial in standard form and indicate its degree.

9. $(-6x^3 + 5x^2 - 8x + 9) + (17x^3 + 2x^2 - 4x - 13)$
10. $(-7x^3 + 6x^2 - 11x + 13) + (19x^3 - 11x^2 + 7x - 17)$
11. $(17x^3 - 5x^2 + 4x - 3) - (5x^3 - 9x^2 - 8x + 11)$
12. $(18x^4 - 2x^3 - 7x + 8) - (9x^4 - 6x^3 - 5x + 7)$
13. $(5x^2 - 7x - 8) + (2x^2 - 3x + 7) - (x^2 - 4x - 3)$
14. $(8x^2 + 7x - 5) - (3x^2 - 4x) - (-6x^3 - 5x^2 + 3)$

In Exercises 15–58, find each product.

15. $(x + 1)(x^2 - x + 1)$
16. $(x + 5)(x^2 - 5x + 25)$
17. $(2x - 3)(x^2 - 3x + 5)$
18. $(2x - 1)(x^2 - 4x + 3)$
19. $(x + 7)(x + 3)$
20. $(x + 8)(x + 5)$
21. $(x - 5)(x + 3)$
22. $(x - 1)(x + 2)$
23. $(3x + 5)(2x + 1)$
24. $(7x + 4)(3x + 1)$
25. $(2x - 3)(5x + 3)$
26. $(2x - 5)(7x + 2)$
27. $(5x^2 - 4)(3x^2 - 7)$
28. $(7x^2 - 2)(3x^2 - 5)$
29. $(8x^3 + 3)(x^2 - 5)$
30. $(7x^3 + 5)(x^2 - 2)$
31. $(x + 3)(x - 3)$
32. $(x + 5)(x - 5)$
33. $(3x + 2)(3x - 2)$
34. $(2x + 5)(2x - 5)$
35. $(5 - 7x)(5 + 7x)$
36. $(4 - 3x)(4 + 3x)$
37. $(4x^2 + 5x)(4x^2 - 5x)$
38. $(3x^2 + 4x)(3x^2 - 4x)$
39. $(1 - y^5)(1 + y^5)$
40. $(2 - y^5)(2 + y^5)$
41. $(x + 2)^2$
42. $(x + 5)^2$
43. $(2x + 3)^2$
44. $(3x + 2)^2$
45. $(x - 3)^2$
46. $(x - 4)^2$
47. $(4x^2 - 1)^2$
48. $(5x^2 - 3)^2$
49. $(7 - 2x)^2$
50. $(9 - 5x)^2$
51. $(x + 1)^3$
52. $(x + 2)^3$
53. $(2x + 3)^3$
54. $(3x + 4)^3$
55. $(x - 3)^3$
56. $(x - 1)^3$
57. $(3x - 4)^3$
58. $(2x - 3)^3$

In Exercises 59–66, perform the indicated operations. Indicate the degree of the resulting polynomial.

59. $(5x^2y - 3xy) + (2x^2y - xy)$
60. $(-2x^2y + xy) + (4x^2y + 7xy)$
61. $(4x^2y + 8xy + 11) + (-2x^2y + 5xy + 2)$
62. $(7x^4y^2 - 5x^2y^2 + 3xy) + (-18x^4y^2 - 6x^2y^2 - xy)$
63. $(x^3 + 7xy - 5y^2) - (6x^3 - xy + 4y^2)$
64. $(x^4 - 7xy - 5y^3) - (6x^4 - 3xy + 4y^3)$
65. $(3x^4y^2 + 5x^3y - 3y) - (2x^4y^2 - 3x^3y - 4y + 6x)$
66. $(5x^4y^2 + 6x^3y - 7y) - (3x^4y^2 - 5x^3y - 6y + 8x)$

In Exercises 67–82, find each product.

67. $(x + 5y)(7x + 3y)$
68. $(x + 9y)(6x + 7y)$
69. $(x - 3y)(2x + 7y)$
70. $(3x - y)(2x + 5y)$
71. $(3xy - 1)(5xy + 2)$
72. $(7x^2y + 1)(2x^2y - 3)$
73. $(7x + 5y)^2$
74. $(9x + 7y)^2$
75. $(x^2y^2 - 3)^2$
76. $(x^2y^2 - 5)^2$
77. $(x - y)(x^2 + xy + y^2)$
78. $(x + y)(x^2 - xy + y^2)$
79. $(3x + 5y)(3x - 5y)$
80. $(7x + 3y)(7x - 3y)$
81. $(7xy^2 - 10y)(7xy^2 + 10y)$
82. $(3xy^2 - 4y)(3xy^2 + 4y)$

Practice Plus

In Exercises 83–90, perform the indicated operation or operations.

83. $(3x + 4y)^2 - (3x - 4y)^2$

84. $(5x + 2y)^2 - (5x - 2y)^2$

85. $(5x - 7)(3x - 2) - (4x - 5)(6x - 1)$

86. $(3x + 5)(2x - 9) - (7x - 2)(x - 1)$

87. $(2x + 5)(2x - 5)(4x^2 + 25)$

88. $(3x + 4)(3x - 4)(9x^2 + 16)$

89. $\dfrac{(2x - 7)^5}{(2x - 7)^3}$ **90.** $\dfrac{(5x - 3)^6}{(5x - 3)^4}$

Application Exercises

The bar graph shows the number of people in the United States, in millions, who do yoga.

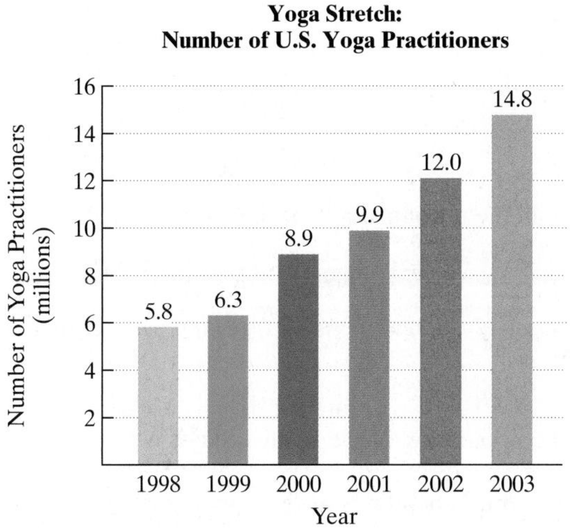

Source: *Yoga Journal*

Here are four mathematical models for the data shown in the graph. In each formula, N represents the number of U.S. yoga practitioners, in millions, x years after 1998.

Model 1 $N = 1.8x + 5.1$

Model 2 $N = 5.6(1.2)^x$

Model 3 $N = 0.17x^2 + 0.95x + 5.68$

Model 4 $N = 0.09x^2 + 0.01x^3 + 1.1x + 5.64$

Use these models to solve Exercises 91–96.

91. Which model uses a polynomial that is not in standard form? Rewrite the model in standard form.

92. If x is any real number from 0 to 5, inclusive, which model does not use a polynomial?

93. Which model best describes the data for 2000?

94. Which model best describes the data for 1998?

95. How well does the model of degree 2 describe the data for 2003?

96. How well does the polynomial model that is not in standard form describe the data for 2002?

In Exercises 97–98, write a polynomial in standard form that models, or represents, the volume of the open box.

97.

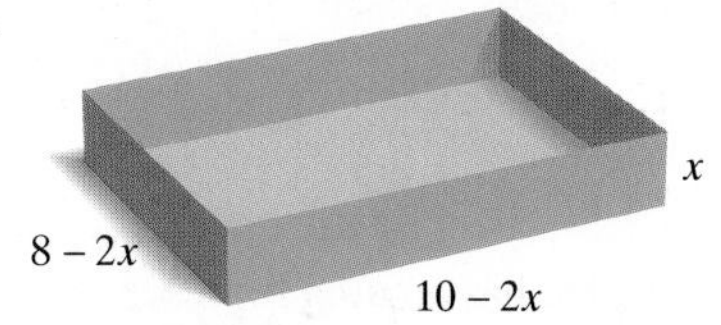

98.

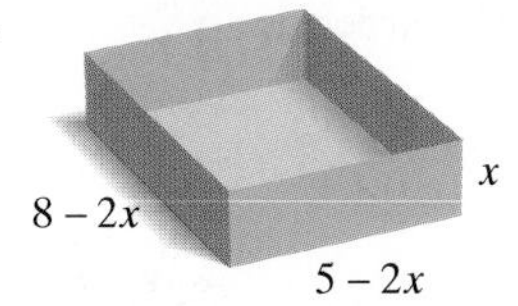

In Exercises 99–100, write a polynomial in standard form that models, or represents, the area of the shaded region.

99.

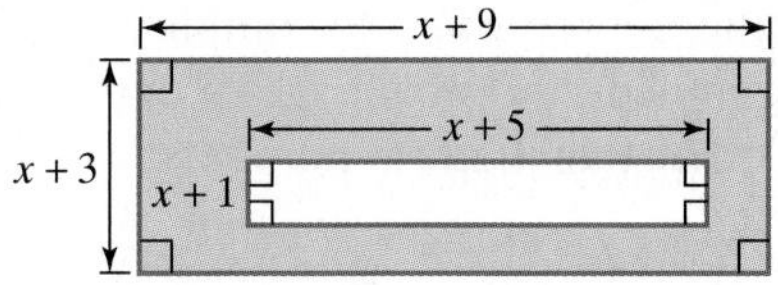

100.

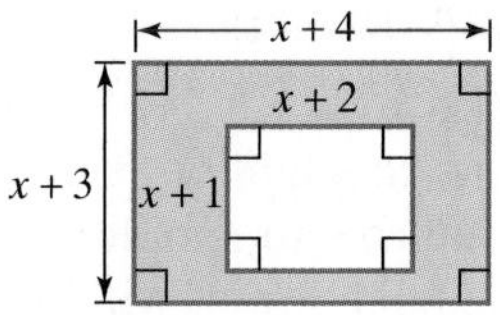

Writing in Mathematics

101. What is a polynomial in x?

102. Explain how to subtract polynomials.

103. Explain how to multiply two binomials using the FOIL method. Give an example with your explanation.

104. Explain how to find the product of the sum and difference of two terms. Give an example with your explanation.

105. Explain how to square a binomial difference. Give an example with your explanation.

106. Explain how to find the degree of a polynomial in two variables.

107. In the section opener, we used the mathematical model

$$N = -365x^4 + 2728x^3 - 7106x^2 + 7372x + 20{,}787$$

to describe the number of newly-declared computer majors, N, for the fall term x years after 1999. Use a calculator to determine these numbers from 1999 through 2003. Compare your results with the data shown in Figure P.11 on page 45. Describe what you observe.

Critical Thinking Exercises

108. Which one of the following is true?

a. $(3x^3 + 2)(3x^3 - 2) = 9x^9 - 4$

b. $(x - 5)^2 = x^2 - 5x + 25$

c. $(x + 1)^2 = x^2 + 1$

d. Suppose a square garden has an area represented by $9x^2$ square feet. If one side is made 7 feet longer and the other side is made 2 feet shorter, then the trinomial that represents the area of the larger garden is $9x^2 + 15x - 14$ square feet.

In Exercises 109–111, perform the indicated operations.

109. $[(7x + 5) + 4y][(7x + 5) - 4y]$

110. $[(3x + y) + 1]^2$

111. $(x^n + 2)(x^n - 2) - (x^n - 3)^2$

112. Express the area of the plane figure shown as a polynomial in standard form.

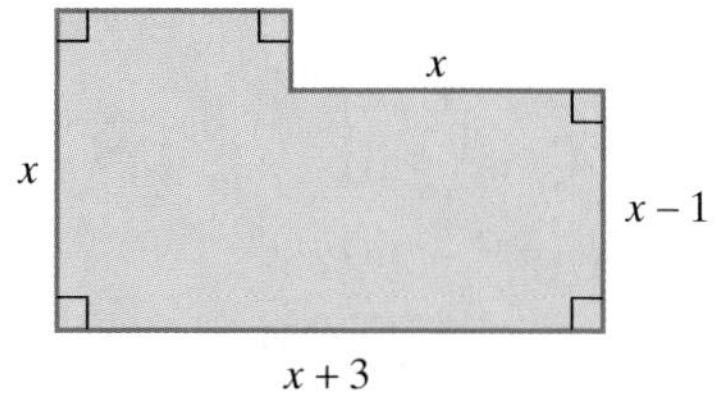

CHAPTER P
MID-CHAPTER CHECK POINT

What You Know: We defined the real numbers $[\{x|x \text{ is rational}\} \cup \{x|x \text{ is irrational}\}]$ and graphed them as points on a number line. We reviewed the basic rules of algebra, using these properties to simplify algebraic expressions. We expanded our knowledge of exponents to include exponents other than natural numbers:

$$b^0 = 1;\quad b^{-n} = \frac{1}{b^n};\quad \frac{1}{b^{-n}} = b^n;\quad b^{\frac{1}{n}} = \sqrt[n]{b};$$

$$b^{\frac{m}{n}} = \left(\sqrt[n]{b}\right)^m = \sqrt[n]{b^m};\quad b^{-\frac{m}{n}} = \frac{1}{b^{\frac{m}{n}}}.$$

We used properties of exponents to simplify exponential expressions and properties of radicals to simplify radical expressions. Finally, we performed operations with polynomials. We used a number of fast methods for finding products of polynomials, including the FOIL method for multiplying binomials, a special-product formula for the product of the sum and difference of two terms $[(A + B)(A - B) = A^2 - B^2]$, and special-product formulas for squaring binomials $[(A + B)^2 = A^2 + 2AB = B^2;$ $(A - B)^2 = A^2 - 2AB + B^2]$.

In Exercises 1–25, simplify the given expression or perform the indicated operation (and simplify, if possible), whichever is appropriate.

1. $(3x + 5)(4x - 7)$

2. $(3x + 5) - (4x - 7)$

3. $\sqrt{6} + 9\sqrt{6}$

4. $3\sqrt{12} - \sqrt{27}$

5. $7x + 3[9 - (2x - 6)]$

6. $(8x - 3)^2$

7. $\left(x^{\frac{1}{3}}y^{-\frac{1}{2}}\right)^6$

8. $\left(\frac{2}{7}\right)^0 - 32^{-\frac{2}{5}}$

9. $(2x - 5) - (x^2 - 3x + 1)$

10. $(2x - 5)(x^2 - 3x + 1)$

11. $x^3 + x^3 - x^3 \cdot x^3$

12. $(9a - 10b)(2a + b)$

13. $\{a, c, d, e\} \cup \{c, d, f, h\}$

14. $\{a, c, d, e\} \cap \{c, d, f, h\}$

15. $(3x^2y^3 - xy + 4y^2) - (-2x^2y^3 - 3xy + 5y^2)$

16. $\frac{24x^2y^{13}}{-2x^5y^{-2}}$

17. $\left(\frac{1}{3}x^{-5}y^4\right)(18x^{-2}y^{-1})$

18. $\sqrt[12]{x^4}$

19. $\frac{24 \times 10^3}{2 \times 10^6}$ (Express the answer in scientific notation.)

20. $\frac{\sqrt[3]{32}}{\sqrt[3]{2}}$

21. $(x^3 + 2)(x^3 - 2)$

22. $(x^2 + 2)^2$

23. $\sqrt{50} \cdot \sqrt{6}$

24. $\frac{11}{7 - \sqrt{3}}$

25. $\frac{11}{\sqrt{3}}$

26. List all the rational numbers in this set:

$$\left\{-11, -\frac{3}{7}, 0, 0.45, \sqrt{23}, \sqrt{25}\right\}.$$

In Exercises 27–28, rewrite each expression without absolute value bars.

27. $|2 - \sqrt{13}|$

28. $x^2|x|$ if $x < 0$

29. If the population of the United States is 2.9×10^8 and each person spends about \$120 per year on ice cream, express the total annual spending on ice cream in scientific notation.

30. A human brain contains 3×10^{10} neurons and a gorilla brain contains 7.5×10^9 neurons. How many times as many neurons are in the brain of a human as in the brain of a gorilla?

31. In 2003, 28.5 million U.S. adults browsed Internet personals and 17.4 million posted online personal ads. The bar graph shows the amount spent in the United States, in millions of dollars, on online dating.

Shopping for a Date: Millions of Dollars Spent in the U.S. on Online Dating

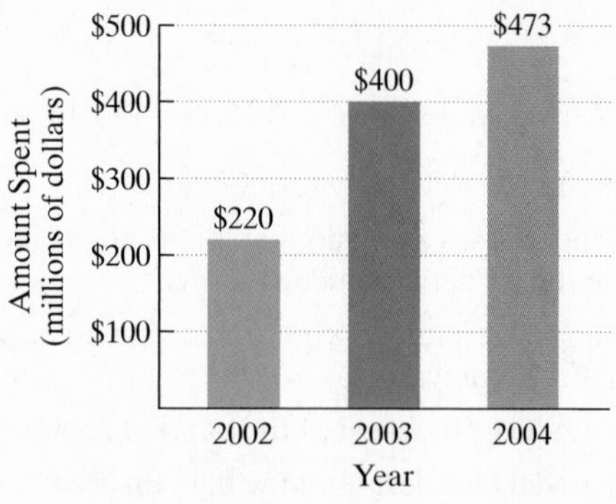

Source: Jupiter Research

Here are three mathematical models for the data shown in the graph. In each formula, D represents the amount spent on online dating, in millions of dollars, x years after 2002.

Model 1 $D = 236(1.5)^x$

Model 2 $D = 127x + 239$

Model 3 $D = -54x^2 + 234x + 220$

a. Which model best describes the data for 2004?

b. According to the polynomial model of degree 1, how much will Americans spend on online dating in 2008?

SECTION P.5 Factoring Polynomials

Objectives

1. Factor out the greatest common factor of a polynomial.
2. Factor by grouping.
3. Factor trinomials.
4. Factor the difference of squares.
5. Factor perfect square trinomials.
6. Factor the sum and difference of two cubes.
7. Use a general strategy for factoring polynomials.
8. Factor algebraic expressions containing fractional and negative exponents.

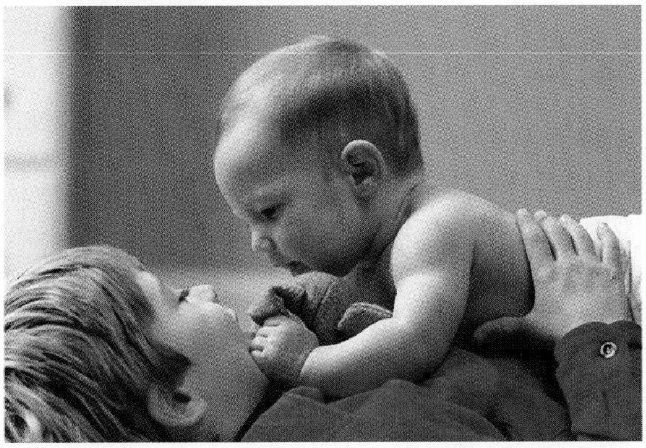

A two-year-old boy is asked, "Do you have a brother?" He answers, "Yes." "What is your brother's name?" "Tom." Asked if Tom has a brother, the two-year-old replies, "No." The child can go in the direction from self to brother, but he cannot reverse this direction and move from brother back to self.

As our intellects develop, we learn to reverse the direction of our thinking. Reversibility of thought is found throughout algebra. For example, we can multiply polynomials and show that

$$5x(2x + 3) = 10x^2 + 15x.$$

We can also reverse this process and express the resulting polynomial as

$$10x^2 + 15x = 5x(2x + 3).$$

Factoring a polynomial containing the sum of monomials means finding an equivalent expression that is a product.

Factoring $10x^2 + 15x$

Sum of monomials

Equivalent expression that is a product

$$10x^2 + 15x = 5x(2x + 3)$$

The factors of $10x^2 + 15x$ are $5x$ and $2x + 3$.

In this section, we will be **factoring over the set of integers**, meaning that the coefficients in the factors are integers. Polynomials that cannot be factored using integer coefficients are called **irreducible over the integers**, or **prime**.

The goal in factoring a polynomial is to use one or more factoring techniques until each of the polynomial's factors, except possibly for a monomial factor, is prime or irreducible. In this situation, the polynomial is said to be **factored completely**.

We will now discuss basic techniques for factoring polynomials.

1 Factor out the greatest common factor of a polynomial.

Common Factors

In any factoring problem, the first step is to look for the *greatest common factor*. The **greatest common factor**, abbreviated GCF, is an expression of the highest degree that divides each term of the polynomial. The distributive property in the reverse direction

$$ab + ac = a(b + c)$$

can be used to factor out the greatest common factor.

EXAMPLE 1   Factoring out the Greatest Common Factor

Factor: **a.** $18x^3 + 27x^2$ **b.** $x^2(x + 3) + 5(x + 3)$.

Solution

a. First, determine the greatest common factor.

> 9 is the greatest integer that divides 18 and 27.

$$18x^3 + 27x^2$$

> x^2 is the greatest expression that divides x^3 and x^2.

The GCF of the two terms of the polynomial is $9x^2$.

$$18x^3 + 27x^2$$
$$= 9x^2(2x) + 9x^2(3) \qquad \text{Express each term as the product of the GCF and its other factor.}$$
$$= 9x^2(2x + 3) \qquad \text{Factor out the GCF.}$$

b. In this situation, the greatest common factor is the common binomial factor $(x + 3)$. We factor out this common factor as follows:

$$x^2(x + 3) + 5(x + 3) = (x + 3)(x^2 + 5). \qquad \text{Factor out the common binomial factor.}$$

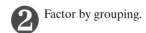 Factor: **a.** $10x^3 - 4x^2$ **b.** $2x(x - 7) + 3(x - 7)$.

2 Factor by grouping.

Factoring by Grouping

Some polynomials have only a greatest common factor of 1. However, by a suitable grouping of the terms, it still may be possible to factor. This process, called **factoring by grouping**, is illustrated in Example 2.

EXAMPLE 2   Factoring by Grouping

Factor: $x^3 + 4x^2 + 3x + 12$.

Solution   There is no factor other than 1 common to all terms. However, we can group terms that have a common factor:

$$\boxed{x^3 + 4x^2} \quad + \quad \boxed{3x + 12}.$$

> Common factor is x^2.   Common factor is 3.

We now factor the given polynomial as follows:

$$x^3 + 4x^2 + 3x + 12$$
$$= (x^3 + 4x^2) + (3x + 12) \qquad \text{Group terms with common factors.}$$
$$= x^2(x + 4) + 3(x + 4) \qquad \text{Factor out the greatest common factor from the grouped terms. The remaining two terms have x + 4 as a common binomial factor.}$$
$$= (x + 4)(x^2 + 3). \qquad \text{Factor out the GCF, x + 4.}$$

Discovery

In Example 2, group the terms as follows:

$$(x^3 + 3x) + (4x^2 + 12).$$

Factor out the greatest common factor from each group and complete the factoring process. Describe what happens. What can you conclude?

Thus, $x^3 + 4x^2 + 3x + 12 = (x + 4)(x^2 + 3)$. Check the factorization by multiplying the right side of the equation using the FOIL method. Because the factorization is correct, you should obtain the original polynomial.

Check Point 2 Factor: $x^3 + 5x^2 - 2x - 10$.

3 Factor trinomials.

Factoring Trinomials

To factor a trinomial of the form $ax^2 + bx + c$, a little trial and error may be necessary.

A Strategy for Factoring $ax^2 + bx + c$

Assume, for the moment, that there is no greatest common factor.

1. Find two First terms whose product is ax^2:

$$(\square x + \quad)(\square x + \quad) = ax^2 + bx + c.$$

2. Find two Last terms whose product is c:

$$(x + \square)(x + \square) = ax^2 + bx + c.$$

3. By trial and error, perform steps 1 and 2 until the sum of the Outside product and the Inside product is bx:

$$(\square x + \square)(\square x + \square) = ax^2 + bx + c.$$

I

O

Sum of O + I

If no such combination exists, the polynomial is prime.

Study Tip

The *error* part of the factoring strategy plays an important role in the process. If you do not get the correct factorization the first time, this is not a bad thing. This error is often helpful in leading you to the correct factorization.

EXAMPLE 3 Factoring Trinomials Whose Leading Coefficients Are 1

Factor: **a.** $x^2 + 6x + 8$ **b.** $x^2 + 3x - 18$.

Solution

a. The factors of the first term are x and x:

$$x^2 + 6x + 8 = (x \qquad)(x \qquad).$$

Factors of 8	8, 1	4, 2	−8, −1	−4, −2
Sum of Factors	9	6	−9	−6

This is the desired sum.

To find the second term of each factor, we must find two integers whose product is 8 and whose sum is 6. From the table in the margin, we see that 4 and 2 are the required integers. Thus,

$$x^2 + 6x + 8 = (x + 4)(x + 2) \text{ or } (x + 2)(x + 4).$$

Factors of -18	18, −1	−18, 1	9, −2	−9, 2	6, −3	−6, 3
Sum of Factors	17	−17	7	−7	3	−3

This is the desired sum.

b. We begin with

$$x^2 + 3x - 18 = (x \quad)(x \quad).$$

To find the second term of each factor, we must find two integers whose product is -18 and whose sum is 3. From the table in the margin, we see that 6 and -3 are the required integers. Thus,

$$x^2 + 3x - 18 = (x + 6)(x - 3) \quad \text{or} \quad (x - 3)(x + 6).$$

Check Point 3 Factor:

a. $x^2 + 13x + 40$ **b.** $x^2 - 5x - 14$.

EXAMPLE 4 Factoring a Trinomial Whose Leading Coefficient Is Not 1

Factor: $8x^2 - 10x - 3$.

Solution

Step 1 Find two First terms whose product is $8x^2$.

$$8x^2 - 10x - 3 \stackrel{?}{=} (8x \quad)(x \quad)$$
$$8x^2 - 10x - 3 \stackrel{?}{=} (4x \quad)(2x \quad)$$

Step 2 Find two Last terms whose product is -3. The possible factorizations are $1(-3)$ and $-1(3)$.

Step 3 Try various combinations of these factors. The correct factorization of $8x^2 - 10x - 3$ is the one in which the sum of the Outside and Inside products is equal to $-10x$. Here is a list of the possible factorizations:

Possible Factorizations of $8x^2 - 10x - 3$	Sum of Outside and Inside Products (Should Equal $-10x$)
$(8x + 1)(x - 3)$	$-24x + x = -23x$
$(8x - 3)(x + 1)$	$8x - 3x = 5x$
$(8x - 1)(x + 3)$	$24x - x = 23x$
$(8x + 3)(x - 1)$	$-8x + 3x = -5x$
$(4x + 1)(2x - 3)$	$-12x + 2x = -10x$
$(4x - 3)(2x + 1)$	$4x - 6x = -2x$
$(4x - 1)(2x + 3)$	$12x - 2x = 10x$
$(4x + 3)(2x - 1)$	$-4x + 6x = 2x$

This is the required middle term.

Thus,

$$8x^2 - 10x - 3 = (4x + 1)(2x - 3) \quad \text{or} \quad (2x - 3)(4x + 1).$$

Show that either of these factorizations is correct by multiplying the factors using the FOIL method. You should obtain the original trinomial.

Check Point 4 Factor: $6x^2 + 19x - 7$.

Study Tip

Here are some suggestions for reducing the list of possible factorizations for $ax^2 + bx + c$:

1. If b is relatively small, avoid the larger factors of a.
2. If c is positive, the signs in both binomial factors must match the sign of b.
3. If the trinomial has no common factor, no binomial factor can have a common factor.
4. Reversing the signs in the binomial factors reverses the sign of bx, the middle term.

EXAMPLE 5 Factoring a Trinomial in Two Variables

Factor: $2x^2 - 7xy + 3y^2$.

Solution

Step 1 Find two First terms whose product is $2x^2$.

$$2x^2 - 7xy + 3y^2 = (2x \quad)(x \quad)$$

Step 2 Find two Last terms whose product is $3y^2$. The possible factorizations are $(y)(3y)$ and $(-y)(-3y)$.

Step 3 Try various combinations of these factors. The correct factorization of $2x^2 - 7xy + 3y^2$ is the one in which the sum of the Outside and Inside products is equal to $-7xy$. Here is a list of possible factorizations:

Possible Factorizations of $2x^2 - 7xy + 3y^2$	Sum of Outside and Inside Products (Should Equal $-7xy$)
$(2x + 3y)(x + y)$	$2xy + 3xy = 5xy$
$(2x + y)(x + 3y)$	$6xy + xy = 7xy$
$(2x - 3y)(x - y)$	$-2xy - 3xy = -5xy$
$(2x - y)(x - 3y)$	$-6xy - xy = -7xy$

This is the required middle term.

Thus,

$$2x^2 - 7xy + 3y^2 = (2x - y)(x - 3y) \quad \text{or} \quad (x - 3y)(2x - y).$$

Use FOIL multiplication to check either of these factorizations.

Check Point 5 Factor: $3x^2 - 13xy + 4y^2$.

4 Factor the difference of squares.

Factoring the Difference of Two Squares

A method for factoring the difference of two squares is obtained by reversing the special product for the sum and difference of two terms.

The Difference of Two Squares

If A and B are real numbers, variables, or algebraic expressions, then

$$A^2 - B^2 = (A + B)(A - B).$$

In words: The difference of the squares of two terms factors as the product of a sum and a difference of those terms.

EXAMPLE 6 Factoring the Difference of Two Squares

Factor: **a.** $x^2 - 4$ **b.** $81x^2 - 49$.

Solution We must express each term as the square of some monomial. Then we use the formula for factoring $A^2 - B^2$.

a. $x^2 - 4 = x^2 - 2^2 = (x + 2)(x - 2)$

$A^2 - B^2 = (A + B)\ (A - B)$

b. $81x^2 - 49 = (9x)^2 - 7^2 = (9x + 7)(9x - 7)$

Check Point 6 Factor: **a.** $x^2 - 81$ **b.** $36x^2 - 25$.

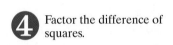

We have seen that a polynomial is factored completely when it is written as the product of prime polynomials. To be sure that you have factored completely, check to see whether any factors with more than one term in the factored polynomial can be factored further. If so, continue factoring.

EXAMPLE 7   A Repeated Factorization

Factor completely: $x^4 - 81$.

Solution

$$x^4 - 81 = (x^2)^2 - 9^2$$
$$= (x^2 + 9)(x^2 - 9)$$
$$= (x^2 + 9)(x^2 - 3^2)$$
$$= (x^2 + 9)(x + 3)(x - 3)$$

Express as the difference of two squares.

The factors are the sum and the difference of the expressions being squared.

The factor $x^2 - 9$ is the difference of two squares and can be factored.

The factors of $x^2 - 9$ are the sum and the difference of the expressions being squared.

Study Tip

Factoring $x^4 - 81$ as
$$(x^2 + 9)(x^2 - 9)$$
is not a complete factorization. The second factor, $x^2 - 9$, is itself a difference of two squares and can be factored.

Are you tempted to further factor $x^2 + 9$, the sum of two squares, in Example 7? Resist the temptation! **The sum of two squares, $A^2 + B^2$, with no common factor other than 1 is a prime polynomial over the integers.**

Check Point 7 Factor completely: $81x^4 - 16$.

⑤ Factor perfect square trinomials.

Factoring Perfect Square Trinomials

Our next factoring technique is obtained by reversing the special products for squaring binomials. The trinomials that are factored using this technique are called **perfect square trinomials**.

Factoring Perfect Square Trinomials

Let A and B be real numbers, variables, or algebraic expressions.

1. $A^2 + 2AB + B^2 = (A + B)^2$

Same sign

2. $A^2 - 2AB + B^2 = (A - B)^2$

Same sign

The two items in the box show that perfect square trinomials come in two forms: one in which the coefficient of the middle term is positive and one in which the coefficient of the middle term is negative. Here's how to recognize a perfect square trinomial:

1. The first and last terms are squares of monomials or integers.
2. The middle term is twice the product of the expressions being squared in the first and last terms.

EXAMPLE 8   Factoring Perfect Square Trinomials

Factor: **a.** $x^2 + 6x + 9$ **b.** $25x^2 - 60x + 36$.

Solution

a. $x^2 + 6x + 9 = x^2 + 2 \cdot x \cdot 3 + 3^2 = (x + 3)^2$

The middle term has a positive sign.

$A^2 \ + \ 2AB \ + \ B^2 \ = \ (A \ + \ B)^2$

b. We suspect that $25x^2 - 60x + 36$ is a perfect square trinomial because $25x^2 = (5x)^2$ and $36 = 6^2$. The middle term can be expressed as twice the product of $5x$ and 6.

$$25x^2 - 60x + 36 = (5x)^2 - 2 \cdot 5x \cdot 6 + 6^2 = (5x - 6)^2$$

$$A^2 - 2AB + B^2 = (A - B)^2$$

Check Point 8 Factor: **a.** $x^2 + 14x + 49$ **b.** $16x^2 - 56x + 49$.

6 Factor the sum and difference of two cubes.

Factoring the Sum and Difference of Two Cubes

We can use the following formulas to factor the sum or the difference of two cubes:

Factoring the Sum and Difference of Two Cubes

1. Factoring the Sum of Two Cubes

$$A^3 + B^3 = (A + B)(A^2 - AB + B^2)$$

Same signs Opposite signs

2. Factoring the Difference of Two Cubes

$$A^3 - B^3 = (A - B)(A^2 + AB + B^2)$$

Same signs Opposite signs

EXAMPLE 9 Factoring Sums and Differences of Two Cubes

Factor: **a.** $x^3 + 8$ **b.** $64x^3 - 125$.

Solution

a. To factor $x^3 + 8$, we must express each term as the cube of some monomial. Then we use the formula for factoring $A^3 + B^3$.

$$x^3 + 8 = x^3 + 2^3 = (x + 2)(x^2 - x \cdot 2 + 2^2) = (x + 2)(x^2 - 2x + 4)$$

$$A^3 + B^3 = (A + B)(A^2 - AB + B^2)$$

b. To factor $64x^3 - 125$, we must express each term as the cube of some monomial. Then use the formula for factoring $A^3 - B^3$.

$$64x^3 - 125 = (4x)^3 - 5^3 = (4x - 5)[(4x)^2 + (4x)(5) + 5^2]$$

$$A^3 - B^3 = (A - B)(A^2 + AB + B^2)$$

$$= (4x - 5)(16x^2 + 20x + 25)$$

Check Point 9 Factor: **a.** $x^3 + 1$ **b.** $125x^3 - 8$.

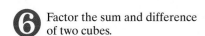

7 Use a general strategy for factoring polynomials.

A Strategy for Factoring Polynomials

It is important to practice factoring a wide variety of polynomials so that you can quickly select the appropriate technique. The polynomial is factored completely when all its polynomial factors, except possibly for monomial factors, are prime. Because of the commutative property, the order of the factors does not matter.

A Strategy for Factoring a Polynomial

1. If there is a common factor, factor out the GCF.
2. Determine the number of terms in the polynomial and try factoring as follows:
   a. If there are two terms, can the binomial be factored by one of the following special forms?

   Difference of two squares: $A^2 - B^2 = (A + B)(A - B)$

   Sum of two cubes: $A^3 + B^3 = (A + B)(A^2 - AB + B^2)$

   Difference of two cubes: $A^3 - B^3 = (A - B)(A^2 + AB + B^2)$

   b. If there are three terms, is the trinomial a perfect square trinomial? If so, factor by one of the following special forms:

   $$A^2 + 2AB + B^2 = (A + B)^2$$
   $$A^2 - 2AB + B^2 = (A - B)^2$$

   If the trinomial is not a perfect square trinomial, try factoring by trial and error.

   c. If there are four or more terms, try factoring by grouping.
3. Check to see if any factors with more than one term in the factored polynomial can be factored further. If so, factor completely.

EXAMPLE 10 Factoring a Polynomial

Factor: $2x^3 + 8x^2 + 8x$.

Solution

Step 1 If there is a common factor, factor out the GCF. Because $2x$ is common to all terms, we factor it out.

$$2x^3 + 8x^2 + 8x = 2x(x^2 + 4x + 4) \quad \text{Factor out the GCF.}$$

Step 2 Determine the number of terms and factor accordingly. The factor $x^2 + 4x + 4$ has three terms and is a perfect square trinomial. We factor using $A^2 + 2AB + B^2 = (A + B)^2$.

$$2x^3 + 8x^2 + 8x = 2x(x^2 + 4x + 4)$$
$$= 2x(x^2 + 2 \cdot x \cdot 2 + 2^2)$$

$A^2 + 2AB + B^2$

$$= 2x(x + 2)^2 \quad A^2 + 2AB + B^2 = (A + B)^2$$

Step 3 Check to see if factors can be factored further. In this problem, they cannot. Thus,

$$2x^3 + 8x^2 + 8x = 2x(x + 2)^2.$$

Check Point 10 Factor: $3x^3 - 30x^2 + 75x$.

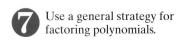

EXAMPLE 11 Factoring a Polynomial

Factor: $x^2 - 25a^2 + 8x + 16$.

Solution

Step 1 If there is a common factor, factor out the GCF. Other than 1 or -1, there is no common factor.

Step 2 Determine the number of terms and factor accordingly. There are four terms. We try factoring by grouping. Grouping into two groups of two terms does not result in a common binomial factor. Let's try grouping as a difference of squares.

$$x^2 - 25a^2 + 8x + 16$$

$$= (x^2 + 8x + 16) - 25a^2$$ Rearrange terms and group as a perfect square trinomial minus $25a^2$ to obtain a difference of squares.

$$= (x + 4)^2 - (5a)^2$$ Factor the perfect square trinomial.

$$= (x + 4 + 5a)(x + 4 - 5a)$$ Factor the difference of squares. The factors are the sum and difference of the expressions being squared.

Step 3 Check to see if factors can be factored further. In this case, they cannot, so we have factored completely.

Check Point 11 Factor: $x^2 - 36a^2 + 20x + 100$.

8 Factor algebraic expressions containing fractional and negative exponents.

Factoring Algebraic Expressions Containing Fractional and Negative Exponents

Although expressions containing fractional and negative exponents are not polynomials, they can be simplified using factoring techniques.

EXAMPLE 12 Factoring Involving Fractional and Negative Exponents

Factor and simplify: $x(x + 1)^{-\frac{3}{4}} + (x + 1)^{\frac{1}{4}}$.

Solution The greatest common factor is $x + 1$ with the *smallest exponent* in the two terms. Thus, the greatest common factor is $(x + 1)^{-\frac{3}{4}}$.

$$x(x + 1)^{-\frac{3}{4}} + (x + 1)^{\frac{1}{4}}$$

$$= (x + 1)^{-\frac{3}{4}}x + (x + 1)^{-\frac{3}{4}}(x + 1)$$ Express each term as the product of the greatest common factor and its other factor.

$$= (x + 1)^{-\frac{3}{4}}[x + (x + 1)]$$ Factor out the greatest common factor.

$$= \frac{2x + 1}{(x + 1)^{\frac{3}{4}}}$$ $b^{-n} = \frac{1}{b^n}$

Check Point 12 Factor and simplify: $x(x - 1)^{-\frac{1}{2}} + (x - 1)^{\frac{1}{2}}$.

EXERCISE SET P.5

Practice Exercises

In Exercises 1–10, factor out the greatest common factor.

1. $18x + 27$
2. $16x - 24$
3. $3x^2 + 6x$
4. $4x^2 - 8x$
5. $9x^4 - 18x^3 + 27x^2$
6. $6x^4 - 18x^3 + 12x^2$
7. $x(x + 5) + 3(x + 5)$
8. $x(2x + 1) + 4(2x + 1)$
9. $x^2(x - 3) + 12(x - 3)$
10. $x^2(2x + 5) + 17(2x + 5)$

In Exercises 11–16, factor by grouping.

11. $x^3 - 2x^2 + 5x - 10$
12. $x^3 - 3x^2 + 4x - 12$
13. $x^3 - x^2 + 2x - 2$
14. $x^3 + 6x^2 - 2x - 12$
15. $3x^3 - 2x^2 - 6x + 4$
16. $x^3 - x^2 - 5x + 5$

In Exercises 17–38, factor each trinomial, or state that the trinomial is prime.

17. $x^2 + 5x + 6$
18. $x^2 + 8x + 15$
19. $x^2 - 2x - 15$
20. $x^2 - 4x - 5$
21. $x^2 - 8x + 15$
22. $x^2 - 14x + 45$
23. $3x^2 - x - 2$
24. $2x^2 + 5x - 3$
25. $3x^2 - 25x - 28$
26. $3x^2 - 2x - 5$
27. $6x^2 - 11x + 4$
28. $6x^2 - 17x + 12$
29. $4x^2 + 16x + 15$
30. $8x^2 + 33x + 4$
31. $9x^2 - 9x + 2$
32. $9x^2 + 5x - 4$
33. $20x^2 + 27x - 8$
34. $15x^2 - 19x + 6$
35. $2x^2 + 3xy + y^2$
36. $3x^2 + 4xy + y^2$
37. $6x^2 - 5xy - 6y^2$
38. $6x^2 - 7xy - 5y^2$

In Exercises 39–48, factor the difference of two squares.

39. $x^2 - 100$
40. $x^2 - 144$
41. $36x^2 - 49$
42. $64x^2 - 81$
43. $9x^2 - 25y^2$
44. $36x^2 - 49y^2$
45. $x^4 - 16$
46. $x^4 - 1$
47. $16x^4 - 81$
48. $81x^4 - 1$

In Exercises 49–56, factor each perfect square trinomial.

49. $x^2 + 2x + 1$
50. $x^2 + 4x + 4$
51. $x^2 - 14x + 49$
52. $x^2 - 10x + 25$
53. $4x^2 + 4x + 1$
54. $25x^2 + 10x + 1$
55. $9x^2 - 6x + 1$
56. $64x^2 - 16x + 1$

In Exercises 57–64, factor using the formula for the sum or difference of two cubes.

57. $x^3 + 27$
58. $x^3 + 64$
59. $x^3 - 64$
60. $x^3 - 27$
61. $8x^3 - 1$
62. $27x^3 - 1$
63. $64x^3 + 27$
64. $8x^3 + 125$

In Exercises 65–92, factor completely, or state that the polynomial is prime.

65. $3x^3 - 3x$
66. $5x^3 - 45x$
67. $4x^2 - 4x - 24$
68. $6x^2 - 18x - 60$
69. $2x^4 - 162$
70. $7x^4 - 7$
71. $x^3 + 2x^2 - 9x - 18$
72. $x^3 + 3x^2 - 25x - 75$
73. $2x^2 - 2x - 112$
74. $6x^2 - 6x - 12$
75. $x^3 - 4x$
76. $9x^3 - 9x$
77. $x^2 + 64$
78. $x^2 + 36$
79. $x^3 + 2x^2 - 4x - 8$
80. $x^3 + 2x^2 - x - 2$
81. $y^5 - 81y$
82. $y^5 - 16y$
83. $20y^4 - 45y^2$
84. $48y^4 - 3y^2$
85. $x^2 - 12x + 36 - 49y^2$
86. $x^2 - 10x + 25 - 36y^2$
87. $9b^2x - 16y - 16x + 9b^2y$
88. $16a^2x - 25y - 25x + 16a^2y$
89. $x^2y - 16y + 32 - 2x^2$
90. $12x^2y - 27y - 4x^2 + 9$
91. $2x^3 - 8a^2x + 24x^2 + 72x$
92. $2x^3 - 98a^2x + 28x^2 + 98x$

In Exercises 93–102, factor and simplify each algebraic expression.

93. $x^{\frac{3}{2}} - x^{\frac{1}{2}}$
94. $x^{\frac{3}{4}} - x^{\frac{1}{4}}$
95. $4x^{-\frac{2}{3}} + 8x^{\frac{1}{3}}$
96. $12x^{-\frac{3}{4}} + 6x^{\frac{1}{4}}$
97. $(x + 3)^{\frac{1}{2}} - (x + 3)^{\frac{3}{2}}$
98. $(x^2 + 4)^{\frac{3}{2}} + (x^2 + 4)^{\frac{7}{2}}$
99. $(x + 5)^{-\frac{1}{2}} - (x + 5)^{-\frac{3}{2}}$
100. $(x^2 + 3)^{-\frac{2}{3}} + (x^2 + 3)^{-\frac{5}{3}}$
101. $(4x - 1)^{\frac{1}{2}} - \frac{1}{3}(4x - 1)^{\frac{3}{2}}$
102. $-8(4x + 3)^{-2} + 10(5x + 1)(4x + 3)^{-1}$

Practice Plus

In Exercises 103–114, factor completely.

103. $10x^2(x + 1) - 7x(x + 1) - 6(x + 1)$
104. $12x^2(x - 1) - 4x(x - 1) - 5(x - 1)$
105. $6x^4 + 35x^2 - 6$
106. $7x^4 + 34x^2 - 5$
107. $y^7 + y$
108. $(y + 1)^3 + 1$
109. $x^4 - 5x^2y^2 + 4y^4$
110. $x^4 - 10x^2y^2 + 9y^4$
111. $(x - y)^4 - 4(x - y)^2$
112. $(x + y)^4 - 100(x + y)^2$
113. $2x^2 - 7xy^2 + 3y^4$
114. $3x^2 + 5xy^2 + 2y^4$

Application Exercises

115. Your computer store is having an incredible sale. The price on one model is reduced by 40%. Then the sale price is reduced by another 40%. If x is the computer's original price, the sale price can be represented by

$$(x - 0.4x) - 0.4(x - 0.4x).$$

a. Factor out $(x - 0.4x)$ from each term. Then simplify the resulting expression.

b. Use the simplified expression from part (a) to answer these questions. With a 40% reduction followed by a 40% reduction, is the computer selling at 20% of its original price? If not, at what percentage of the original price is it selling?

116. Your local electronics store is having an end-of-the-year sale. The price on a large-screen television had been reduced by 30%. Now the sale price is reduced by another 30%. If x is the television's original price, the sale price can be represented by

$$(x - 0.3x) - 0.3(x - 0.3x).$$

a. Factor out $(x - 0.3x)$ from each term. Then simplify the resulting expression.

b. Use the simplified expression from part (a) to answer these questions. With a 30% reduction followed by a 30% reduction, is the television selling at 40% of its original price? If not, at what percentage of the original price is it selling?

In Exercises 117–120,

a. *Write an expression for the area of the shaded region.*

b. *Write the expression in factored form.*

117.

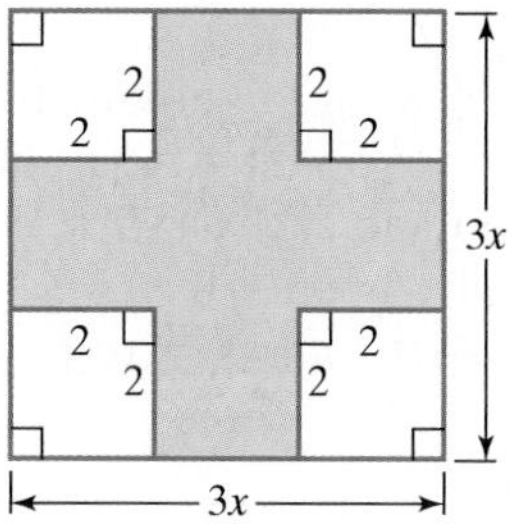

118.

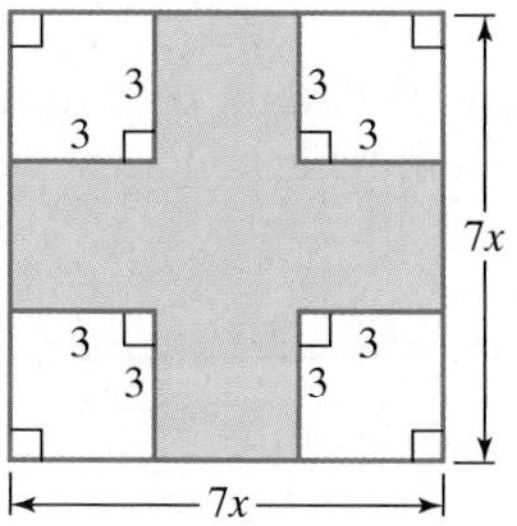

119.

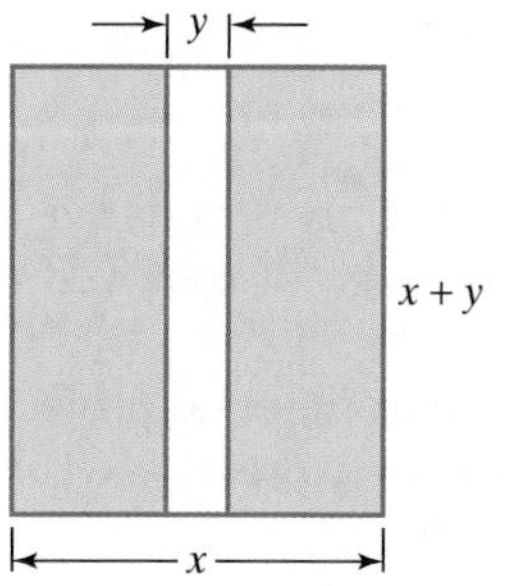

120.

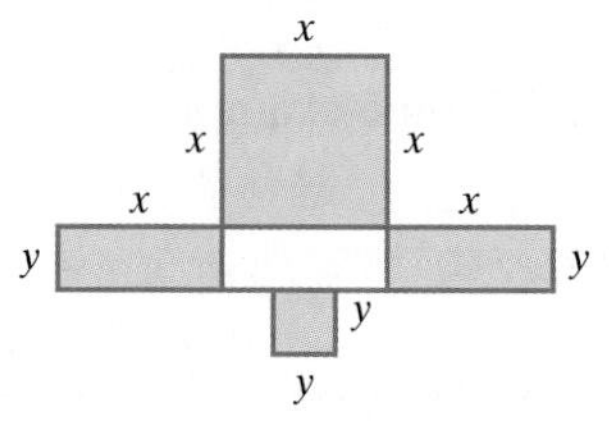

In Exercises 121–122, find the formula for the volume of the region outside the smaller rectangular solid and inside the larger rectangular solid. Then express the volume in factored form.

121.

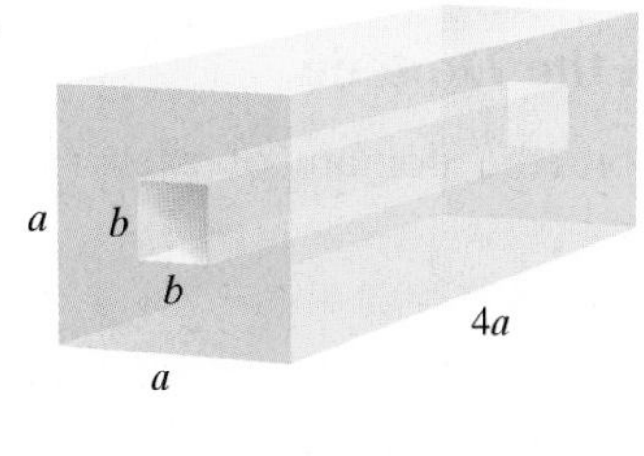

122.

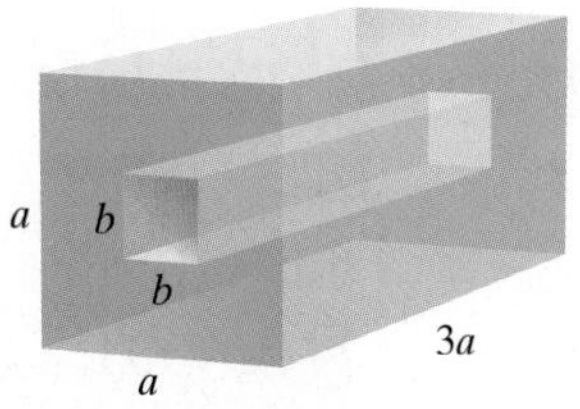

Writing in Mathematics

123. Using an example, explain how to factor out the greatest common factor of a polynomial.

124. Suppose that a polynomial contains four terms. Explain how to use factoring by grouping to factor the polynomial.

125. Explain how to factor $3x^2 + 10x + 8$.

126. Explain how to factor the difference of two squares. Provide an example with your explanation.

127. What is a perfect square trinomial and how is it factored?

128. Explain how to factor $x^3 + 1$.

129. What does it mean to factor completely?

Critical Thinking Exercises

130. Which one of the following is true?

a. Because $x^2 + 1$ is irreducible over the integers, it follows that $x^3 + 1$ is also irreducible.

b. One correct factored form for $x^2 - 4x + 3$ is $x(x - 4) + 3$.

c. $x^3 - 64 = (x - 4)^3$

d. None of the above is true.

In Exercises 131–134, factor completely.

131. $x^{2n} + 6x^n + 8$

132. $-x^2 - 4x + 5$

133. $x^4 - y^4 - 2x^3y + 2xy^3$

134. $(x - 5)^{-\frac{1}{2}}(x + 5)^{-\frac{1}{2}} - (x + 5)^{\frac{1}{2}}(x - 5)^{-\frac{3}{2}}$

In Exercises 135–136, find all integers b so that the trinomial can be factored.

135. $x^2 + bx + 15$

136. $x^2 + 4x + b$

Group Exercise

137. Divide the group in half. Without looking at any factoring problems in the book, each group should create five factoring problems. Make sure that some of your problems require at least two factoring strategies. Next, exchange problems with the other half of the group. Work to factor the five problems. After completing the factorizations, evaluate the factoring problems that you were given. Are they too easy? Too difficult? Can the polynomials really be factored? Share your responses with the half of the group that wrote the problems. Finally, grade each other's work in factoring the polynomials. Each factoring problem is worth 20 points. You may award partial credit. If you take off points, explain why points are deducted and how you decided to take off a particular number of points for the error(s) that you found.

SECTION P.6 Rational Expressions

Objectives

1. Specify numbers that must be excluded from the domain of a rational expression.
2. Simplify rational expressions.
3. Multiply rational expressions.
4. Divide rational expressions.
5. Add and subtract rational expressions.
6. Simplify complex rational expressions.

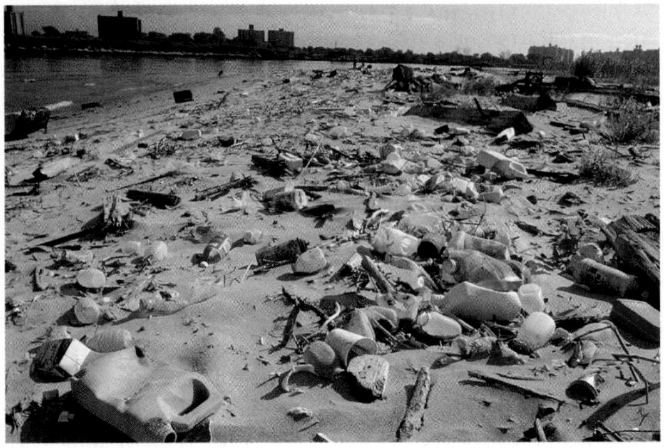

How can we describe the costs of reducing environmental pollution? We often use algebraic expressions involving quotients of polynomials. For example, the algebraic expression

$$\frac{250x}{100 - x}$$

describes the cost, in millions of dollars, to remove x percent of the pollutants that are discharged into a river. Removing a modest percentage of pollutants, say 40%, is far less costly than removing a substantially greater percentage, such as 95%. We see this by evaluating the algebraic expression for $x = 40$ and $x = 95$.

Evaluating $\frac{250x}{100 - x}$ for

$x = 40$: Cost is $\frac{250(40)}{100 - 40} \approx 167.$

$x = 95$: Cost is $\frac{250(95)}{100 - 95} = 4750.$

Discovery

What happens if you try substituting 100 for x in

$$\frac{250x}{100 - x}?$$

What does this tell you about the cost of cleaning up all of the river's pollutants?

The cost increases from approximately \$167 million to a possibly prohibitive \$4750 million, or \$4.75 billion. Costs spiral upward as the percentage of removed pollutants increases.

Many algebraic expressions that describe costs of environmental projects are examples of *rational expressions*. First we will define rational expressions. Then we will review how to perform operations with such expressions.

1 Specify numbers that must be excluded from the domain of a rational expression.

Rational Expressions

A **rational expression** is the quotient of two polynomials. Some examples are

$$\frac{x - 2}{4}, \quad \frac{4}{x - 2}, \quad \frac{x}{x^2 - 1}, \quad \text{and} \quad \frac{x^2 + 1}{x^2 + 2x - 3}.$$

The set of real numbers for which an algebraic expression is defined is the **domain** of the expression. Because rational expressions indicate division and division by zero is undefined, we must exclude numbers from a rational expression's domain that make the denominator zero.

EXAMPLE 1 Excluding Numbers from the Domain

Find all the numbers that must be excluded from the domain of each rational expression:

a. $\frac{4}{x - 2}$ **b.** $\frac{x}{x^2 - 1}.$

Solution To determine the numbers that must be excluded from each domain, examine the denominators.

a. $\frac{4}{x-2}$ **b.** $\frac{x}{x^2-1} = \frac{x}{(x+1)(x-1)}$

This denominator would equal zero if $x = 2$.

This factor would equal zero if $x = -1$.

This factor would equal zero if $x = 1$.

For the rational expression in part (a), we must exclude 2 from the domain. For the rational expression in part (b), we must exclude both -1 and 1 from the domain. These excluded numbers are often written to the right of a rational expression.

$$\frac{4}{x-2}, x \neq 2 \qquad \frac{x}{x^2-1}, x \neq -1, x \neq 1$$

Check Point 1 Find all the numbers that must be excluded from the domain of each rational expression:

a. $\frac{7}{x+5}$ **b.** $\frac{x}{x^2-36}$.

❷ Simplify rational expressions.

Simplifying Rational Expressions

A rational expression is **simplified** if its numerator and denominator have no common factors other than 1 or -1. The following procedure can be used to simplify rational expressions:

Simplifying Rational Expressions

1. Factor the numerator and the denominator completely.
2. Divide both the numerator and the denominator by any common factors.

EXAMPLE 2 Simplifying Rational Expressions

Simplify: **a.** $\frac{x^3+x^2}{x+1}$ **b.** $\frac{x^2+6x+5}{x^2-25}$.

Solution

a. $\frac{x^3+x^2}{x+1} = \frac{x^2(x+1)}{x+1}$ Factor the numerator. Because the denominator is $x + 1$, $x \neq -1$.

$= \frac{x^2\overset{1}{\cancel{(x+1)}}}{\underset{1}{\cancel{x+1}}}$ Divide out the common factor, $x + 1$.

$= x^2, x \neq -1$ Denominators of 1 need not be written because $\frac{a}{1} = a$.

b. $\frac{x^2+6x+5}{x^2-25} = \frac{(x+5)(x+1)}{(x+5)(x-5)}$ Factor the numerator and denominator. Because the denominator is $(x+5)(x-5)$, $x \neq -5$ and $x \neq 5$.

$= \frac{\overset{1}{\cancel{(x+5)}}(x+1)}{\underset{1}{\cancel{(x+5)}}(x-5)}$ Divide out the common factor, $x + 5$.

$= \frac{x+1}{x-5}, \quad x \neq -5, \quad x \neq 5$

Check Point 2 Simplify: **a.** $\frac{x^3+3x^2}{x+3}$ **b.** $\frac{x^2-1}{x^2+2x+1}$.

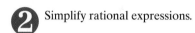

3 Multiply rational expressions.

Multiplying Rational Expressions

The product of two rational expressions is the product of their numerators divided by the product of their denominators. Here is a step-by-step procedure for multiplying rational expressions:

Multiplying Rational Expressions

1. Factor all numerators and denominators completely.
2. Divide numerators and denominators by common factors.
3. Multiply the remaining factors in the numerators and multiply the remaining factors in the denominators.

EXAMPLE 3 Multiplying Rational Expressions

Multiply and simplify:

$$\frac{x-7}{x-1}\cdot\frac{x^2-1}{3x-21}.$$

Solution

$$\frac{x-7}{x-1}\cdot\frac{x^2-1}{3x-21}$$ This is the given multiplication problem.

$$=\frac{x-7}{x-1}\cdot\frac{(x+1)(x-1)}{3(x-7)}$$ Factor as many numerators and denominators as possible. Because the denominator has factors of $x-1$ and $x-7$, $x \neq 1$ and $x \neq 7$.

$$=\frac{\overset{1}{\cancel{x-7}}}{\underset{1}{\cancel{x-1}}}\cdot\frac{(x+1)\overset{1}{\cancel{(x-1)}}}{3\underset{1}{\cancel{(x-7)}}}$$ Divide numerators and denominators by common factors.

$$=\frac{x+1}{3}, x \neq 1, x \neq 7$$ Multiply the remaining factors in the numerators and denominators.

These excluded numbers from the domain must also be excluded from the simplified expression's domain.

Check Point 3 Multiply and simplify:

$$\frac{x+3}{x^2-4}\cdot\frac{x^2-x-6}{x^2+6x+9}.$$

4 Divide rational expressions.

Dividing Rational Expressions

The quotient of two rational expressions is the product of the first expression and the multiplicative inverse, or reciprocal, of the second expression. The reciprocal is found by interchanging the numerator and the denominator. Thus, **we find the quotient of two rational expressions by inverting the divisor and multiplying.**

EXAMPLE 4 Dividing Rational Expressions

Divide and simplify:

$$\frac{x^2-2x-8}{x^2-9}\div\frac{x-4}{x+3}.$$

Solution

$$\frac{x^2-2x-8}{x^2-9} \div \frac{x-4}{x+3}$$ This is the given division problem.

$$= \frac{x^2-2x-8}{x^2-9} \cdot \frac{x+3}{x-4}$$ Invert the divisor and multiply.

$$= \frac{(x-4)(x+2)}{(x+3)(x-3)} \cdot \frac{x+3}{x-4}$$ Factor as many numerators and denominators as possible. For nonzero denominators, $x \neq -3$, $x \neq 3$, and $x \neq 4$.

$$= \frac{\overset{1}{\cancel{(x-4)}}(x+2)}{\underset{1}{\cancel{(x+3)}}(x-3)} \cdot \frac{\overset{1}{\cancel{(x+3)}}}{\underset{1}{\cancel{(x-4)}}}$$ Divide numerators and denominators by common factors.

$$= \frac{x+2}{x-3}, x \neq -3, x \neq 3, x \neq 4$$ Multiply the remaining factors in the numerators and the denominators.

Check Point 4 Divide and simplify:

$$\frac{x^2-2x+1}{x^3+x} \div \frac{x^2+x-2}{3x^2+3}.$$

5 Add and subtract rational expressions.

Adding and Subtracting Rational Expressions with the Same Denominator

We add or subtract rational expressions with the same denominator by (1) adding or subtracting the numerators, (2) placing this result over the common denominator, and (3) simplifying, if possible.

EXAMPLE 5 Subtracting Rational Expressions with the Same Denominator

Subtract: $\frac{5x+1}{x^2-9} - \frac{4x-2}{x^2-9}$.

Solution

$$\frac{5x+1}{x^2-9} - \frac{4x-2}{x^2-9} = \frac{5x+1-(4x-2)}{x^2-9}$$ Subtract numerators and include parentheses to indicate that both terms are subtracted. Place this difference over the common denominator.

$$= \frac{5x+1-4x+2}{x^2-9}$$ Remove parentheses and then change the sign of each term.

$$= \frac{x+3}{x^2-9}$$ Combine like terms.

$$= \frac{\overset{1}{\cancel{x+3}}}{\underset{1}{\cancel{(x+3)}}(x-3)}$$ Factor and simplify ($x \neq -3$ and $x \neq 3$).

$$= \frac{1}{x-3}, x \neq -3, x \neq 3$$

Study Tip

Example 5 shows that when a numerator is being subtracted, we must subtract every term in that expression.

Check Point 5 Subtract: $\frac{x}{x+1} - \frac{3x+2}{x+1}$.

Adding and Subtracting Rational Expressions with Different Denominators

Rational expressions that have no common factors in their denominators can be added or subtracted using one of the following properties:

$$\frac{a}{b}+\frac{c}{d}=\frac{ad+bc}{bd} \qquad \frac{a}{b}-\frac{c}{d}=\frac{ad-bc}{bd},\ b\neq 0,\ d\neq 0.$$

The denominator, bd, is the product of the factors in the two denominators. Because we are looking at rational expressions that have no common factors in their denominators, the product bd gives the least common denominator.

EXAMPLE 6 Subtracting Rational Expressions Having No Common Factors in Their Denominators

Subtract: $\frac{x+2}{2x-3}-\frac{4}{x+3}$.

Solution We need to find the least common denominator. This is the product of the distinct factors in each denominator, namely $(2x-3)(x+3)$. We can therefore use the subtraction property given previously as follows:

$$\frac{a}{b} - \frac{c}{d} = \frac{ad-bc}{bd}$$

$$\frac{x+2}{2x-3}-\frac{4}{x+3}=\frac{(x+2)(x+3)-(2x-3)4}{(2x-3)(x+3)}$$

Observe that $a = x + 2$, $b = 2x - 3$, $c = 4$, and $d = x + 3$.

$$=\frac{x^2+5x+6-(8x-12)}{(2x-3)(x+3)}$$

Multiply.

$$=\frac{x^2+5x+6-8x+12}{(2x-3)(x+3)}$$

Remove parentheses and then change the sign of each term.

$$=\frac{x^2-3x+18}{(2x-3)(x+3)},\ x\neq\frac{3}{2},\ x\neq -3$$

Combine like terms in the numerator.

Check Point 6 Add: $\frac{3}{x+1}+\frac{5}{x-1}$.

The **least common denominator**, or LCD, of several rational expressions is a polynomial consisting of the product of all prime factors in the denominators, with each factor raised to the greatest power of its occurrence in any denominator. When adding and subtracting rational expressions that have different denominators with one or more common factors in the denominators, it is efficient to find the least common denominator first.

Finding the Least Common Denominator

1. Factor each denominator completely.
2. List the factors of the first denominator.
3. Add to the list in step 2 any factors of the second denominator that do not appear in the list.
4. Form the product of each different factor from the list in step 3. This product is the least common denominator.

EXAMPLE 7 Finding the Least Common Denominator

Find the least common denominator of

$$\frac{7}{5x^2 + 15x} \quad \text{and} \quad \frac{9}{x^2 + 6x + 9}.$$

Solution

Step 1 Factor each denominator completely.

$$5x^2 + 15x = 5x(x + 3)$$
$$x^2 + 6x + 9 = (x + 3)^2$$

Step 2 List the factors of the first denominator.

$$5, x, (x + 3)$$

Step 3 Add any unlisted factors from the second denominator. The second denominator is $(x + 3)^2$ or $(x + 3)(x + 3)$. One factor of $x + 3$ is already in our list, but the other factor is not. We add a second factor of $x + 3$ to the list. We have

$$5, x, (x + 3), (x + 3).$$

Step 4 The least common denominator is the product of all factors in the final list. Thus,

$$5x(x + 3)(x + 3), \quad \text{or} \quad 5x(x + 3)^2$$

is the least common denominator.

Check Point 7 Find the least common denominator of

$$\frac{3}{x^2 - 6x + 9} \quad \text{and} \quad \frac{7}{x^2 - 9}.$$

Finding the least common denominator for two (or more) rational expressions is the first step needed to add or subtract the expressions.

Adding and Subtracting Rational Expressions That Have Different Denominators

1. Find the LCD of the rational expressions.
2. Rewrite each rational expression as an equivalent expression whose denominator is the LCD. To do so, multiply the numerator and the denominator of each rational expression by any factor(s) needed to convert the denominator into the LCD.
3. Add or subtract numerators, placing the resulting expression over the LCD.
4. If possible, simplify the resulting rational expression.

EXAMPLE 8 Adding Rational Expressions with Different Denominators

Add: $\frac{x+3}{x^2+x-2}+\frac{2}{x^2-1}$.

Solution

Step 1 Find the least common denominator. Start by factoring the denominators.

$$x^2+x-2=(x+2)(x-1)$$
$$x^2-1=(x+1)(x-1)$$

The factors of the first denominator are $x+2$ and $x-1$. The only factor from the second denominator that is not listed is $x+1$. Thus, the least common denominator is

$$(x+2)(x-1)(x+1).$$

Step 2 Write equivalent expressions with the LCD as denominators. We must rewrite each rational expression with a denominator of $(x+2)(x-1)(x+1)$. We do so by multiplying both the numerator and the denominator of each rational expression by any factor(s) needed to convert the expression's denominator into the LCD.

$$\frac{x+3}{(x+2)(x-1)}\cdot\frac{x+1}{x+1}=\frac{(x+3)(x+1)}{(x+2)(x-1)(x+1)}$$

Multiply the numerator and denominator by $x+1$ to get $(x+2)(x-1)(x+1)$, the LCD.

$$\frac{2}{(x+1)(x-1)}\cdot\frac{x+2}{x+2}=\frac{2(x+2)}{(x+2)(x+1)(x-1)}$$

Multiply the numerator and denominator by $x+2$ to get $(x+2)(x+1)(x-1)$, the LCD.

Because $\frac{x+1}{x+1}=1$ and $\frac{x+2}{x+2}=1$, we are not changing the value of either rational expression, only its appearance.

Now we are ready to perform the indicated addition.

$$\frac{x+3}{x^2+x-2}+\frac{2}{x^2-1}$$

This is the given problem.

$$=\frac{x+3}{(x+2)(x-1)}+\frac{2}{(x+1)(x-1)}$$

Factor the denominators. The LCD is $(x+2)(x+1)(x-1)$.

$$=\frac{(x+3)(x+1)}{(x+2)(x-1)(x+1)}+\frac{2(x+2)}{(x+2)(x-1)(x+1)}$$

Rewrite equivalent expressions with the LCD.

Step 3 Add numerators, putting this sum over the LCD.

$$=\frac{(x+3)(x+1)+2(x+2)}{(x+2)(x-1)(x+1)}$$

$$=\frac{x^2+4x+3+2x+4}{(x+2)(x-1)(x+1)}$$

Perform the multiplications in the numerator.

$$=\frac{x^2+6x+7}{(x+2)(x-1)(x+1)},\ x\neq-2,\ x\neq1,\ x\neq-1$$

Combine like terms in the numerator: $4x+2x=6x$ and $3+4=7$.

Step 4 If necessary, simplify. Because the numerator is prime, no further simplification is possible.

Check Point 8 Subtract: $\frac{x}{x^2-10x+25}-\frac{x-4}{2x-10}$.

6 Simplify complex rational expressions.

Complex Rational Expressions

Complex rational expressions, also called **complex fractions**, have numerators or denominators containing one or more rational expressions. Here are two examples of such expressions:

$$\frac{1+\frac{1}{x}}{1-\frac{1}{x}}$$

Separate rational expressions occur in the numerator and the denominator.

$$\frac{\frac{1}{x+h}-\frac{1}{x}}{h}.$$

Separate rational expressions occur in the numerator.

One method for simplifying a complex rational expression is to combine its numerator into a single expression and combine its denominator into a single expression. Then perform the division by inverting the denominator and multiplying.

EXAMPLE 9 Simplifying a Complex Rational Expression

Simplify: $\dfrac{1+\frac{1}{x}}{1-\frac{1}{x}}$.

Solution

Step 1 Add to get a single rational expression in the numerator.

$$1+\frac{1}{x}=\frac{1}{1}+\frac{1}{x}=\frac{1\cdot x}{1\cdot x}+\frac{1}{x}=\frac{x}{x}+\frac{1}{x}=\frac{x+1}{x}$$

The LCD is $1 \cdot x$, or x.

Step 2 Subtract to get a single rational expression in the denominator.

$$1-\frac{1}{x}=\frac{1}{1}-\frac{1}{x}=\frac{1\cdot x}{1\cdot x}-\frac{1}{x}=\frac{x}{x}-\frac{1}{x}=\frac{x-1}{x}$$

The LCD is $1 \cdot x$, or x.

Step 3 Perform the division indicated by the main fraction bar: Invert and multiply. If possible, simplify.

$$\frac{1+\frac{1}{x}}{1-\frac{1}{x}}=\frac{\frac{x+1}{x}}{\frac{x-1}{x}}=\frac{x+1}{x}\cdot\frac{x}{x-1}=\frac{x+1}{\overset{}{\underset{1}{\cancel{x}}}}\cdot\frac{\overset{1}{\cancel{x}}}{x-1}=\frac{x+1}{x-1}$$

Invert and multiply.

Check Point 9 Simplify: $\dfrac{\frac{1}{x}-\frac{3}{2}}{\frac{1}{x}+\frac{3}{4}}$.

A second method for simplifying a complex rational expression is to find the least common denominator of all the rational expressions in its numerator and denominator. Then multiply each term in its numerator and denominator by this least common denominator. Because we are multiplying by a form of 1, we will obtain an equivalent expression that does not contain fractions in its numerator or denominator. Here we use this method to simplify the complex rational expression in Example 9.

$$\frac{1+\frac{1}{x}}{1-\frac{1}{x}} = \frac{\left(1+\frac{1}{x}\right)}{\left(1-\frac{1}{x}\right)}\cdot\frac{x}{x}$$

The least common denominator of all the rational expressions is x. Multiply the numerator and denominator by x. Because $\frac{x}{x} = 1$, we are not changing the complex fraction ($x \neq 0$).

$$= \frac{1\cdot x + \frac{1}{x}\cdot x}{1\cdot x - \frac{1}{x}\cdot x}$$

Use the distributive property. Be sure to distribute x to every term.

$$= \frac{x+1}{x-1},\ x \neq 0,\ x \neq 1$$

Multiply. The complex rational expression is now simplified.

EXAMPLE 10 Simplifying a Complex Rational Expression

Simplify: $\dfrac{\frac{1}{x+h} - \frac{1}{x}}{h}$.

Solution We will use the method of multiplying each of the three terms, $\frac{1}{x+h}$, $\frac{1}{x}$, and h by the least common denominator. The least common denominator is $x(x+h)$.

$$\frac{\frac{1}{x+h} - \frac{1}{x}}{h}$$

$$= \frac{\left(\frac{1}{x+h} - \frac{1}{x}\right)x(x+h)}{hx(x+h)}$$

Multiply the numerator and denominator by $x(x+h)$, $h \neq 0$, $x \neq 0$, $x \neq -h$.

$$= \frac{\frac{1}{x+h}\cdot x(x+h) - \frac{1}{x}\cdot x(x+h)}{hx(x+h)}$$

Use the distributive property in the numerator.

$$= \frac{x-(x+h)}{hx(x+h)}$$

Simplify: $\frac{1}{\cancel{x+h}}\, x(\cancel{x+h}) = x$ and $\frac{1}{\cancel{x}}\cdot\cancel{x}(x+h) = x+h$.

$$= \frac{x-x-h}{hx(x+h)}$$

Subtract in the numerator.

$$= \frac{-\overset{1}{\cancel{h}}}{\underset{1}{\cancel{h}}x(x+h)}$$

Simplify: $x - x - h = -h$.

$$= -\frac{1}{x(x+h)},\ h \neq 0,\ x \neq 0,\ x \neq -h$$

Divide the numerator and denominator by h.

Check Point 10 Simplify: $\dfrac{\frac{1}{x+7} - \frac{1}{x}}{7}$.

EXERCISE SET P.6

Practice Exercises

In Exercises 1–6, find all numbers that must be excluded from the domain of each rational expression.

1. $\frac{7}{x-3}$

2. $\frac{13}{x+9}$

3. $\frac{x+5}{x^2-25}$

4. $\frac{x+7}{x^2-49}$

5. $\frac{x-1}{x^2+11x+10}$

6. $\frac{x-3}{x^2+4x-45}$

In Exercises 7–14, simplify each rational expression. Find all numbers that must be excluded from the domain of the simplified rational expression.

7. $\frac{3x-9}{x^2-6x+9}$

8. $\frac{4x-8}{x^2-4x+4}$

9. $\frac{x^2-12x+36}{4x-24}$

10. $\frac{x^2-8x+16}{3x-12}$

11. $\frac{y^2+7y-18}{y^2-3y+2}$

12. $\frac{y^2-4y-5}{y^2+5y+4}$

13. $\frac{x^2+12x+36}{x^2-36}$

14. $\frac{x^2-14x+49}{x^2-49}$

In Exercises 15–32, multiply or divide as indicated.

15. $\frac{x-2}{3x+9} \cdot \frac{2x+6}{2x-4}$

16. $\frac{6x+9}{3x-15} \cdot \frac{x-5}{4x+6}$

17. $\frac{x^2-9}{x^2} \cdot \frac{x^2-3x}{x^2+x-12}$

18. $\frac{x^2-4}{x^2-4x+4} \cdot \frac{2x-4}{x+2}$

19. $\frac{x^2-5x+6}{x^2-2x-3} \cdot \frac{x^2-1}{x^2-4}$

20. $\frac{x^2+5x+6}{x^2+x-6} \cdot \frac{x^2-9}{x^2-x-6}$

21. $\frac{x^3-8}{x^2-4} \cdot \frac{x+2}{3x}$

22. $\frac{x^2+6x+9}{x^3+27} \cdot \frac{1}{x+3}$

23. $\frac{x+1}{3} \div \frac{3x+3}{7}$

24. $\frac{x+5}{7} \div \frac{4x+20}{9}$

25. $\frac{x^2-4}{x} \div \frac{x+2}{x-2}$

26. $\frac{x^2-4}{x-2} \div \frac{x+2}{4x-8}$

27. $\frac{4x^2+10}{x-3} \div \frac{6x^2+15}{x^2-9}$

28. $\frac{x^2+x}{x^2-4} \div \frac{x^2-1}{x^2+5x+6}$

29. $\frac{x^2-25}{2x-2} \div \frac{x^2+10x+25}{x^2+4x-5}$

30. $\frac{x^2-4}{x^2+3x-10} \div \frac{x^2+5x+6}{x^2+8x+15}$

31. $\frac{x^2+x-12}{x^2+x-30} \cdot \frac{x^2+5x+6}{x^2-2x-3} \div \frac{x+3}{x^2+7x+6}$

32. $\frac{x^3-25x}{4x^2} \cdot \frac{2x^2-2}{x^2-6x+5} \div \frac{x^2+5x}{7x+7}$

In Exercises 33–54, add or subtract as indicated.

33. $\frac{4x+1}{6x+5} + \frac{8x+9}{6x+5}$

34. $\frac{3x+2}{3x+4} + \frac{3x+6}{3x+4}$

35. $\frac{x^2-2x}{x^2+3x} + \frac{x^2+x}{x^2+3x}$

36. $\frac{x^2-4x}{x^2-x-6} + \frac{4x-4}{x^2-x-6}$

37. $\frac{4x-10}{x-2} - \frac{x-4}{x-2}$

38. $\frac{2x+3}{3x-6} - \frac{3-x}{3x-6}$

39. $\frac{x^2+3x}{x^2+x-12} - \frac{x^2-12}{x^2+x-12}$

40. $\frac{x^2-4x}{x^2-x-6} - \frac{x-6}{x^2-x-6}$

41. $\frac{3}{x+4} + \frac{6}{x+5}$

42. $\frac{8}{x-2} + \frac{2}{x-3}$

43. $\frac{3}{x+1} - \frac{3}{x}$

44. $\frac{4}{x} - \frac{3}{x+3}$

45. $\frac{2x}{x+2} + \frac{x+2}{x-2}$

46. $\frac{3x}{x-3} - \frac{x+4}{x+2}$

47. $\frac{x+5}{x-5} + \frac{x-5}{x+5}$

48. $\frac{x+3}{x-3} + \frac{x-3}{x+3}$

49. $\frac{4}{x^2+6x+9} + \frac{4}{x+3}$

50. $\frac{3}{5x+2} + \frac{5x}{25x^2-4}$

51. $\frac{3x}{x^2+3x-10} - \frac{2x}{x^2+x-6}$

52. $\frac{x}{x^2-2x-24} - \frac{x}{x^2-7x+6}$

53. $\frac{4x^2+x-6}{x^2+3x+2} - \frac{3x}{x+1} + \frac{5}{x+2}$

54. $\frac{6x^2+17x-40}{x^2+x-20} + \frac{3}{x-4} - \frac{5x}{x+5}$

In Exercises 55–68, simplify each complex rational expression.

55. $\frac{\frac{x}{3}-1}{x-3}$

56. $\frac{\frac{x}{4}-1}{x-4}$

57. $\frac{1+\frac{1}{x}}{3-\frac{1}{x}}$

58. $\frac{8+\frac{1}{x}}{4-\frac{1}{x}}$

59. $\frac{\frac{1}{x}+\frac{1}{y}}{x+y}$

60. $\frac{1-\frac{1}{x}}{xy}$

61. $\frac{x-\frac{x}{x+3}}{x+2}$

62. $\frac{x-3}{x-\frac{3}{x-2}}$

63. $\frac{\frac{3}{x-2}-\frac{4}{x+2}}{\frac{7}{x^2-4}}$

64. $\frac{\frac{x}{x-2}+1}{\frac{3}{x^2-4}+1}$

65. $\dfrac{\dfrac{1}{x+1}}{\dfrac{1}{x^2-2x-3}+\dfrac{1}{x-3}}$

66. $\dfrac{\dfrac{6}{x^2+2x-15}-\dfrac{1}{x-3}}{\dfrac{1}{x+5}+1}$

67. $\dfrac{\dfrac{1}{(x+h)^2}-\dfrac{1}{x^2}}{h}$

68. $\dfrac{\dfrac{x+h}{x+h+1}-\dfrac{x}{x+1}}{h}$

Practice Plus

In Exercises 69–76, perform the indicated operations. Simplify the result, if possible.

69. $\left(\dfrac{2x+3}{x+1}\cdot\dfrac{x^2+4x-5}{2x^2+x-3}\right)-\dfrac{2}{x+2}$

70. $\dfrac{1}{x^2-2x-8}\div\left(\dfrac{1}{x-4}-\dfrac{1}{x+2}\right)$

71. $\left(2-\dfrac{6}{x+1}\right)\left(1+\dfrac{3}{x-2}\right)$

72. $\left(4-\dfrac{3}{x+2}\right)\left(1+\dfrac{5}{x-1}\right)$

73. $\dfrac{y^{-1}-(y+5)^{-1}}{5}$

74. $\dfrac{y^{-1}-(y+2)^{-1}}{2}$

75. $\left(\dfrac{1}{a^3-b^3}\cdot\dfrac{ac+ad-bc-bd}{1}\right)-\dfrac{c-d}{a^2+ab+b^2}$

76. $\dfrac{ab}{a^2+ab+b^2}+\left(\dfrac{ac-ad-bc+bd}{ac-ad+bc-bd}\div\dfrac{a^3-b^3}{a^3+b^3}\right)$

Application Exercises

77. The rational expression

$$\frac{130x}{100-x}$$

describes the cost, in millions of dollars, to inoculate x percent of the population against a particular strain of flu.

a. Evaluate the expression for $x = 40$, $x = 80$, and $x = 90$. Describe the meaning of each evaluation in terms of percentage inoculated and cost.

b. For what value of x is the expression undefined?

c. What happens to the cost as x approaches 100%? How can you interpret this observation?

78. Doctors use the rational expression

$$\frac{DA}{A+12}$$

to determine the dosage of a drug prescribed for children. In this expression, A = child's age and D = adult dosage. What is the difference in the child's dosage for a 7-year-old child and a 3-year-old child? Express the answer as a single rational expression in terms of D. Then describe what your answer means in terms of the variables in the rational expression.

79. The bar graph shows the total number of crimes in the United States, in millions, from 1995 through 2002.

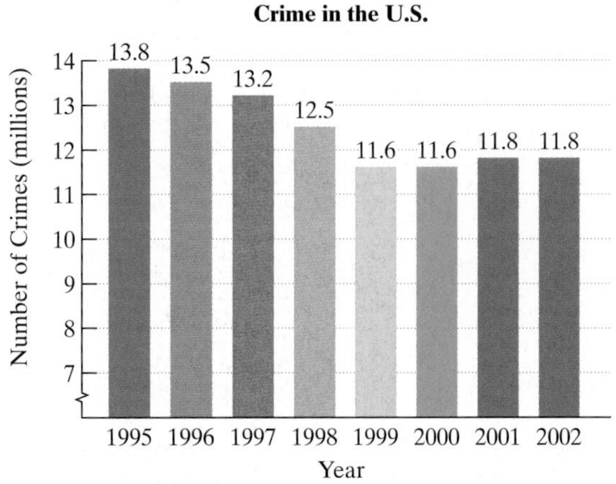

Source: FBI

The polynomial $3.6t + 260$ describes the U.S. population, in millions, t years after 1994. The polynomial $-0.3t + 14$ describes the number of crimes in the United States, in millions, t years after 1994.

a. Write a rational expression that describes the crime rate in the United States t years after 1994.

b. According to the rational expression in part (a), what was the crime rate in 2002? Round to two decimal places. How many crimes does this indicate per 100,000 inhabitants?

c. According to the FBI, there were 4119 crimes per 100,000 U.S. inhabitants in 2002. How well does the rational expression that you evaluated in part (b) model this number?

80. The average rate on a round-trip commute having a one-way distance d is given by the complex rational expression

$$\frac{2d}{\dfrac{d}{r_1}+\dfrac{d}{r_2}},$$

in which r_1 and r_2 are the average rates on the outgoing and return trips, respectively. Simplify the expression. Then find your average rate if you drive to campus averaging 40 miles per hour and return home on the same route averaging 30 miles per hour. Explain why the answer is not 25 miles per hour.

In Exercises 81–82, express the perimeter of each rectangle as a single rational expression.

81.

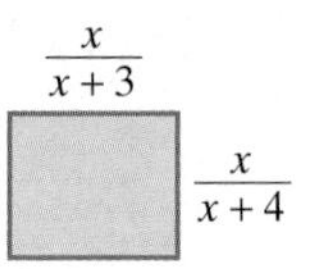

82.

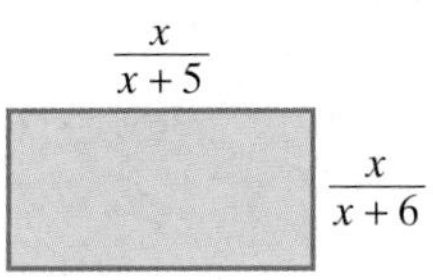

Writing in Mathematics

83. What is a rational expression?

84. Explain how to determine which numbers must be excluded from the domain of a rational expression.

85. Explain how to simplify a rational expression.

86. Explain how to multiply rational expressions.

87. Explain how to divide rational expressions.

88. Explain how to add or subtract rational expressions with the same denominators.

89. Explain how to add rational expressions having no common factors in their denominators. Use $\frac{3}{x+5} + \frac{7}{x+2}$ in your explanation.

90. Explain how to find the least common denominator for denominators of $x^2 - 100$ and $x^2 - 20x + 100$.

91. Describe two ways to simplify $\dfrac{\frac{3}{x} + \frac{2}{x^2}}{\frac{1}{x^2} + \frac{2}{x}}$.

Explain the error in Exercises 92–94. Then rewrite the right side of the equation to correct the error that now exists.

92. $\frac{1}{a} + \frac{1}{b} = \frac{1}{a+b}$

93. $\frac{1}{x} + 7 = \frac{1}{x+7}$

94. $\frac{a}{x} + \frac{a}{b} = \frac{a}{x+b}$

95. A politician claims that each year the crime rate in the United States is decreasing. Explain how to use the polynomials in Exercise 79 to verify this claim.

Critical Thinking Exercises

96. Which one of the following is true?

a. $\frac{a}{b} + \frac{a}{c} = \frac{a}{b+c}$

b. $6 + \frac{1}{x} = \frac{7}{x}$

c. $\frac{1}{x+3} + \frac{x+3}{2} = \frac{1}{\cancel{(x+3)}} + \frac{\cancel{(x+3)}}{2} = 1 + \frac{1}{2} = \frac{3}{2}$

d. $\frac{x^2 - 25}{x - 5} = x - 5$

e. None of the above is true.

In Exercises 97–99, perform the indicated operations.

97. $\frac{1}{x^n - 1} - \frac{1}{x^n + 1} - \frac{1}{x^{2n} - 1}$

98. $\left(1 - \frac{1}{x}\right)\left(1 - \frac{1}{x+1}\right)\left(1 - \frac{1}{x+2}\right)\left(1 - \frac{1}{x+3}\right)$

99. $(x - y)^{-1} + (x - y)^{-2}$

100. In one short sentence, five words or less, explain what

$$\frac{\frac{1}{x} + \frac{1}{x^2} + \frac{1}{x^3}}{\frac{1}{x^4} + \frac{1}{x^5} + \frac{1}{x^6}}$$

does to each number x.

Chapter P
Summary, Review, and Test

Summary: Basic Formulas

Definition of Absolute Value

$$|x| = \begin{cases} x & \text{if } x \geq 0 \\ -x & \text{if } x < 0 \end{cases}$$

Distance between Points *a* and *b* on a Number Line

$$|a - b| \quad \text{or} \quad |b - a|$$

Properties of Algebra

Commutative	$a + b = b + a$ $ab = ba$
Associative	$(a + b) + c = a + (b + c)$ $(ab)c = a(bc)$
Distributive	$a(b + c) = ab + ac$
Identity	$a + 0 = a$ $a \cdot 1 = a$
Inverse	$a + (-a) = 0$ $a \cdot \frac{1}{a} = 1, a \neq 0$

Properties of Exponents

$$b^{-n} = \frac{1}{b^n}, \quad b^0 = 1, \quad b^m \cdot b^n = b^{m+n},$$

$$(b^m)^n = b^{mn}, \quad \frac{b^m}{b^n} = b^{m-n}, \quad (ab)^n = a^n b^n, \quad \left(\frac{a}{b}\right)^n = \frac{a^n}{b^n}$$

Product and Quotient Rules for nth Roots

$$\sqrt[n]{ab} = \sqrt[n]{a} \cdot \sqrt[n]{b}, \quad \sqrt[n]{\frac{a}{b}} = \frac{\sqrt[n]{a}}{\sqrt[n]{b}}$$

Rational Exponents

$$a^{\frac{1}{n}} = \sqrt[n]{a}, \quad a^{-\frac{1}{n}} = \frac{1}{a^{\frac{1}{n}}} = \frac{1}{\sqrt[n]{a}},$$

$$a^{\frac{m}{n}} = (\sqrt[n]{a})^m = \sqrt[n]{a^m}, \quad a^{-\frac{m}{n}} = \frac{1}{a^{\frac{m}{n}}}$$

Special Products

$$(A + B)(A - B) = A^2 - B^2$$
$$(A + B)^2 = A^2 + 2AB + B^2$$
$$(A - B)^2 = A^2 - 2AB + B^2$$
$$(A + B)^3 = A^3 + 3A^2B + 3AB^2 + B^3$$
$$(A - B)^3 = A^3 - 3A^2B + 3AB^2 - B^3$$

Factoring Formulas

$$A^2 - B^2 = (A + B)(A - B)$$
$$A^2 + 2AB + B^2 = (A + B)^2$$
$$A^2 - 2AB + B^2 = (A - B)^2$$
$$A^3 + B^3 = (A + B)(A^2 - AB + B^2)$$
$$A^3 - B^3 = (A - B)(A^2 + AB + B^2)$$

Review Exercises

You can use these review exercises, like the review exercises at the end of each chapter, to test your understanding of the chapter's topics. However, you can also use these exercises as a prerequisite test to check your mastery of the fundamental algebra skills needed in this book.

P.1

In Exercises 1–2, evaluate each algebraic expression for the given value or values of the variable(s).

1. $3 + 6(x - 2)^3$ for $x = 4$

2. $x^2 - 5(x - y)$ for $x = 6$ and $y = 2$

3. You are riding along an expressway traveling x miles per hour. The formula

$$S = 0.015x^2 + x + 10$$

models the recommended safe distance, S, in feet, between your car and other cars on the expressway. What is the recommended safe distance when your speed is 60 miles per hour?

In Exercises 4–7, let $A = \{a, b, c\}$, $B = \{a, c, d, e\}$, *and* $C = \{a, d, f, g\}$. *Find the indicated set.*

4. $A \cap B$ **5.** $A \cup B$

6. $A \cup C$ **7.** $C \cap A$

8. Consider the set:

$$\left\{-17, -\tfrac{9}{13}, 0, 0.75, \sqrt{2}, \pi, \sqrt{81}\right\}.$$

List all numbers from the set that are **a.** natural numbers, **b.** whole numbers, **c.** integers, **d.** rational numbers, **e.** irrational numbers, **f.** real numbers.

In Exercises 9–11, rewrite each expression without absolute value bars.

9. $|-103|$ **10.** $|\sqrt{2} - 1|$

11. $|3 - \sqrt{17}|$

12. Express the distance between the numbers -17 and 4 using absolute value. Then evaluate the absolute value.

In Exercises 13–18, state the name of the property illustrated.

13. $3 + 17 = 17 + 3$

14. $(6 \cdot 3) \cdot 9 = 6 \cdot (3 \cdot 9)$

15. $\sqrt{3}(\sqrt{5} + \sqrt{3}) = \sqrt{15} + 3$

16. $(6 \cdot 9) \cdot 2 = 2 \cdot (6 \cdot 9)$

17. $\sqrt{3}(\sqrt{5} + \sqrt{3}) = (\sqrt{5} + \sqrt{3})\sqrt{3}$

18. $(3 \cdot 7) + (4 \cdot 7) = (4 \cdot 7) + (3 \cdot 7)$

In Exercises 19–22, simplify each algebraic expression.

19. $5(2x - 3) + 7x$

20. $\frac{1}{5}(5x) + [(3y) + (-3y)] - (-x)$

21. $3(4y - 5) - (7y + 2)$ **22.** $8 - 2[3 - (5x - 1)]$

23. The bar graph shows the number of endangered animal species in the United States for six selected years. The data can be modeled by the formulas $E = 10x + 166$ and $E = 0.04x^2 + 9.2x + 169$, in which E represents the number of endangered species x years after 1980. Which formula best describes the actual number of endangered animal species in 2000?

Endangered Animal Species in the U.S.

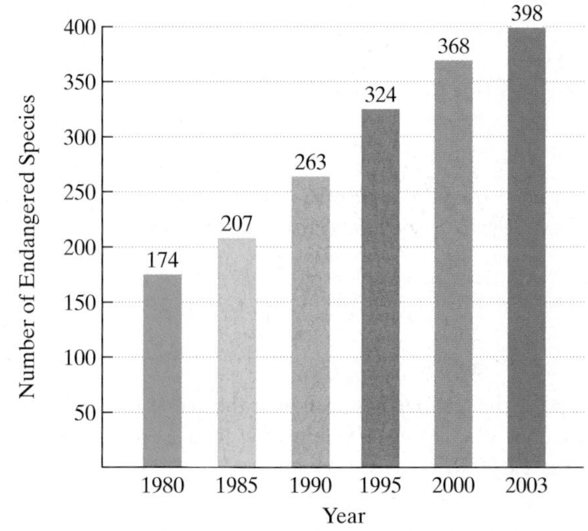

Source: U.S. Fish and Wildlife Service

P.2

Evaluate each exponential expression in Exercises 24–27.

24. $(-3)^3(-2)^2$ **25.** $2^{-4} + 4^{-1}$

26. $5^{-3} \cdot 5$ **27.** $\frac{3^3}{3^6}$

Simplify each exponential expression in Exercises 28–31.

28. $(-2x^4y^3)^3$ **29.** $(-5x^3y^2)(-2x^{-11}y^{-2})$

30. $(2x^3)^{-4}$ **31.** $\frac{7x^5y^6}{28x^{15}y^{-2}}$

In Exercises 32–33, write each number in decimal notation.

32. 3.74×10^4 **33.** 7.45×10^{-5}

In Exercises 34–35, write each number in scientific notation.

34. 3,590,000 **35.** 0.00725

In Exercises 36–37, perform the indicated operation and write the answer in decimal notation.

36. $(3 \times 10^3)(1.3 \times 10^2)$ **37.** $\frac{6.9 \times 10^3}{3 \times 10^5}$

38. If you earned \$1 million per year ($\$10^6$), how long would it take to accumulate \$1 billion ($\$10^9$)?

39. If the population of the United States is 2.9×10^8 and each person spends about \$150 per year going to the movies (or renting movies), express the total annual spending on movies in scientific notation.

P.3

Use the product rule to simplify the expressions in Exercises 40–43. In Exercises 42–43, assume that variables represent nonnegative real numbers.

40. $\sqrt{300}$

41. $\sqrt{12x^2}$

42. $\sqrt{10x}\cdot\sqrt{2x}$

43. $\sqrt{r^3}$

Use the quotient rule to simplify the expressions in Exercises 44–45.

44. $\sqrt{\dfrac{121}{4}}$

45. $\dfrac{\sqrt{96x^3}}{\sqrt{2x}}$ (Assume that $x > 0$.)

In Exercises 46–48, add or subtract terms whenever possible.

46. $7\sqrt{5} + 13\sqrt{5}$

47. $2\sqrt{50} + 3\sqrt{8}$

48. $4\sqrt{72} - 2\sqrt{48}$

In Exercises 49–52, rationalize the denominator.

49. $\dfrac{30}{\sqrt{5}}$

50. $\dfrac{\sqrt{2}}{\sqrt{3}}$

51. $\dfrac{5}{6 + \sqrt{3}}$

52. $\dfrac{14}{\sqrt{7} - \sqrt{5}}$

Evaluate each expression in Exercises 53–56 or indicate that the root is not a real number.

53. $\sqrt[3]{125}$

54. $\sqrt[5]{-32}$

55. $\sqrt[4]{-125}$

56. $\sqrt[4]{(-5)^4}$

Simplify the radical expressions in Exercises 57–61.

57. $\sqrt[3]{81}$

58. $\sqrt[3]{y^5}$

59. $\sqrt[4]{8}\cdot\sqrt[4]{10}$

60. $4\sqrt[3]{16} + 5\sqrt[3]{2}$

61. $\dfrac{\sqrt[4]{32x^5}}{\sqrt[4]{16x}}$ (Assume that $x > 0$.)

In Exercises 62–67, evaluate each expression.

62. $16^{\frac{1}{2}}$

63. $25^{-\frac{1}{2}}$

64. $125^{\frac{1}{3}}$

65. $27^{-\frac{1}{3}}$

66. $64^{\frac{2}{3}}$

67. $27^{-\frac{4}{3}}$

In Exercises 68–70, simplify using properties of exponents.

68. $\left(5x^{\frac{2}{3}}\right)\left(4x^{\frac{1}{4}}\right)$

69. $\dfrac{15x^{\frac{3}{4}}}{5x^{\frac{1}{2}}}$

70. $(125x^6)^{\frac{2}{3}}$

71. Simplify by reducing the index of the radical: $\sqrt[6]{y^3}$.

P.4

In Exercises 72–73, perform the indicated operations. Write the resulting polynomial in standard form and indicate its degree.

72. $(-6x^3 + 7x^2 - 9x + 3) + (14x^3 + 3x^2 - 11x - 7)$

73. $(13x^4 - 8x^3 + 2x^2) - (5x^4 - 3x^3 + 2x^2 - 6)$

In Exercises 74–80, find each product.

74. $(3x - 2)(4x^2 + 3x - 5)$

75. $(3x - 5)(2x + 1)$

76. $(4x + 5)(4x - 5)$

77. $(2x + 5)^2$

78. $(3x - 4)^2$

79. $(2x + 1)^3$

80. $(5x - 2)^3$

In Exercises 81–82, perform the indicated operations. Indicate the degree of the resulting polynomial.

81. $(7x^2 - 8xy + y^2) + (-8x^2 - 9xy - 4y^2)$

82. $(13x^3y^2 - 5x^2y - 9x^2) - (-11x^3y^2 - 6x^2y + 3x^2 - 4)$

In Exercises 83–87, find each product.

83. $(x + 7y)(3x - 5y)$

84. $(3x - 5y)^2$

85. $(3x^2 + 2y)^2$

86. $(7x + 4y)(7x - 4y)$

87. $(a - b)(a^2 + ab + b^2)$

P.5

In Exercises 88–104, factor completely, or state that the polynomial is prime.

88. $15x^3 + 3x^2$

89. $x^2 - 11x + 28$

90. $15x^2 - x - 2$

91. $64 - x^2$

92. $x^2 + 16$

93. $3x^4 - 9x^3 - 30x^2$

94. $20x^7 - 36x^3$

95. $x^3 - 3x^2 - 9x + 27$

96. $16x^2 - 40x + 25$

97. $x^4 - 16$

98. $y^3 - 8$

99. $x^3 + 64$

100. $3x^4 - 12x^2$

101. $27x^3 - 125$

102. $x^5 - x$

103. $x^3 + 5x^2 - 2x - 10$

104. $x^2 + 18x + 81 - y^2$

In Exercises 105–107, factor and simplify each algebraic expression.

105. $16x^{-\frac{3}{4}} + 32x^{\frac{1}{4}}$

106. $(x^2 - 4)(x^2 + 3)^{\frac{1}{2}} - (x^2 - 4)^2(x^2 + 3)^{\frac{3}{2}}$

107. $12x^{-\frac{1}{2}} + 6x^{-\frac{3}{2}}$

P.6

In Exercises 108–110, simplify each rational expression. Also, list all numbers that must be excluded from the domain.

108. $\dfrac{x^3 + 2x^2}{x + 2}$

109. $\dfrac{x^2 + 3x - 18}{x^2 - 36}$

110. $\dfrac{x^2 + 2x}{x^2 + 4x + 4}$

In Exercises 111–113, multiply or divide as indicated.

111. $\dfrac{x^2 + 6x + 9}{x^2 - 4}\cdot\dfrac{x + 3}{x - 2}$

112. $\dfrac{6x + 2}{x^2 - 1} \div \dfrac{3x^2 + x}{x - 1}$

113. $\dfrac{x^2 - 5x - 24}{x^2 - x - 12} \div \dfrac{x^2 - 10x + 16}{x^2 + x - 6}$

In Exercises 114–117, add or subtract as indicated.

114. $\dfrac{2x - 7}{x^2 - 9} - \dfrac{x - 10}{x^2 - 9}$

115. $\dfrac{3x}{x + 2} + \dfrac{x}{x - 2}$

116. $\dfrac{x}{x^2 - 9} + \dfrac{x - 1}{x^2 - 5x + 6}$

117. $\dfrac{4x - 1}{2x^2 + 5x - 3} - \dfrac{x + 3}{6x^2 + x - 2}$

In Exercises 118–120, simplify each complex rational expression.

118. $\dfrac{\dfrac{1}{x} - \dfrac{1}{2}}{\dfrac{1}{3} - \dfrac{x}{6}}$

119. $\dfrac{3 + \dfrac{12}{x}}{1 - \dfrac{16}{x^2}}$

120. $\dfrac{3 - \dfrac{1}{x + 3}}{3 + \dfrac{1}{x + 3}}$

Chapter P Test

In Exercises 1–18, simplify the given expression or perform the indicated operation (and simplify, if possible), whichever is appropriate.

1. $5(2x^2 - 6x) - (4x^2 - 3x)$

2. $7 + 2[3(x + 1) - 2(3x - 1)]$

3. $\{1, 2, 5\} \cap \{5, a\}$

4. $\{1, 2, 5\} \cup \{5, a\}$

5. $(2x^2y^3 - xy + y^2) - (-4x^2y^3 - 5xy - y^2)$

6. $\dfrac{30x^3y^4}{6x^9y^{-4}}$

7. $\sqrt{6r}\sqrt{3r}$ (Assume that $r \geq 0$.)

8. $4\sqrt{50} - 3\sqrt{18}$

9. $\dfrac{3}{5 + \sqrt{2}}$

10. $\sqrt[3]{16x^4}$

11. $\dfrac{x^2 + 2x - 3}{x^2 - 3x + 2}$

12. $\dfrac{5 \times 10^{-6}}{20 \times 10^{-8}}$ (Express the answer in scientific notation.)

13. $(2x - 5)(x^2 - 4x + 3)$

14. $(5x + 3y)^2$

15. $\dfrac{2x + 8}{x - 3} \div \dfrac{x^2 + 5x + 4}{x^2 - 9}$

16. $\dfrac{x}{x + 3} + \dfrac{5}{x - 3}$

17. $\dfrac{2x + 3}{x^2 - 7x + 12} - \dfrac{2}{x - 3}$

18. $\dfrac{\frac{1}{x} - \frac{1}{3}}{\frac{1}{x}}$

In Exercises 19–24, factor completely, or state that the polynomial is prime.

19. $x^2 - 9x + 18$

20. $x^3 + 2x^2 + 3x + 6$

21. $25x^2 - 9$

22. $36x^2 - 84x + 49$

23. $y^3 - 125$

24. $x^2 + 10x + 25 - 9y^2$

25. Factor and simplify:

$$x(x + 3)^{-\frac{3}{5}} + (x + 3)^{\frac{2}{5}}.$$

26. List all the rational numbers in this set:

$$\left\{-7, -\tfrac{4}{5}, 0, 0.25, \sqrt{3}, \sqrt{4}, \tfrac{22}{7}, \pi\right\}.$$

In Exercises 27–28, state the name of the property illustrated.

27. $3(2 + 5) = 3(5 + 2)$

28. $6(7 + 4) = 6 \cdot 7 + 6 \cdot 4$

29. Express in scientific notation: 0.00076.

30. Evaluate: $27^{-\frac{5}{3}}$.

31. In 2003, world population was approximately 6.3×10^9. By some projections, world population will double by 2040. Express the population at that time in scientific notation.

32. Your life expectancy is related to the year when you were born. The bar graph shows life expectancy in the United States by year of birth.

Life Expectancy in the United States, by Year of Birth

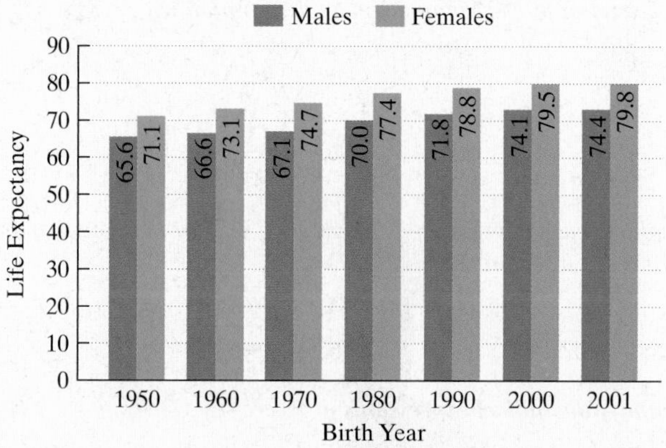

Source: U.S. Bureau of the Census

Here are two mathematical models for the data shown in the graph. In each formula, E represents life expectancy for Americans born t years after 1950.

Model 1 $E = 0.17t + 71$

Model 2 $E = 0.18t + 65$

a. Which model describes the data for the men and which model describes the data for the women?

b. According to the model that describes the data for the men, what is the life expectancy for U.S. men born in 2000? How well does the model describe the life expectancy shown in the graph?

Equations and Inequalities

CHAPTER 1

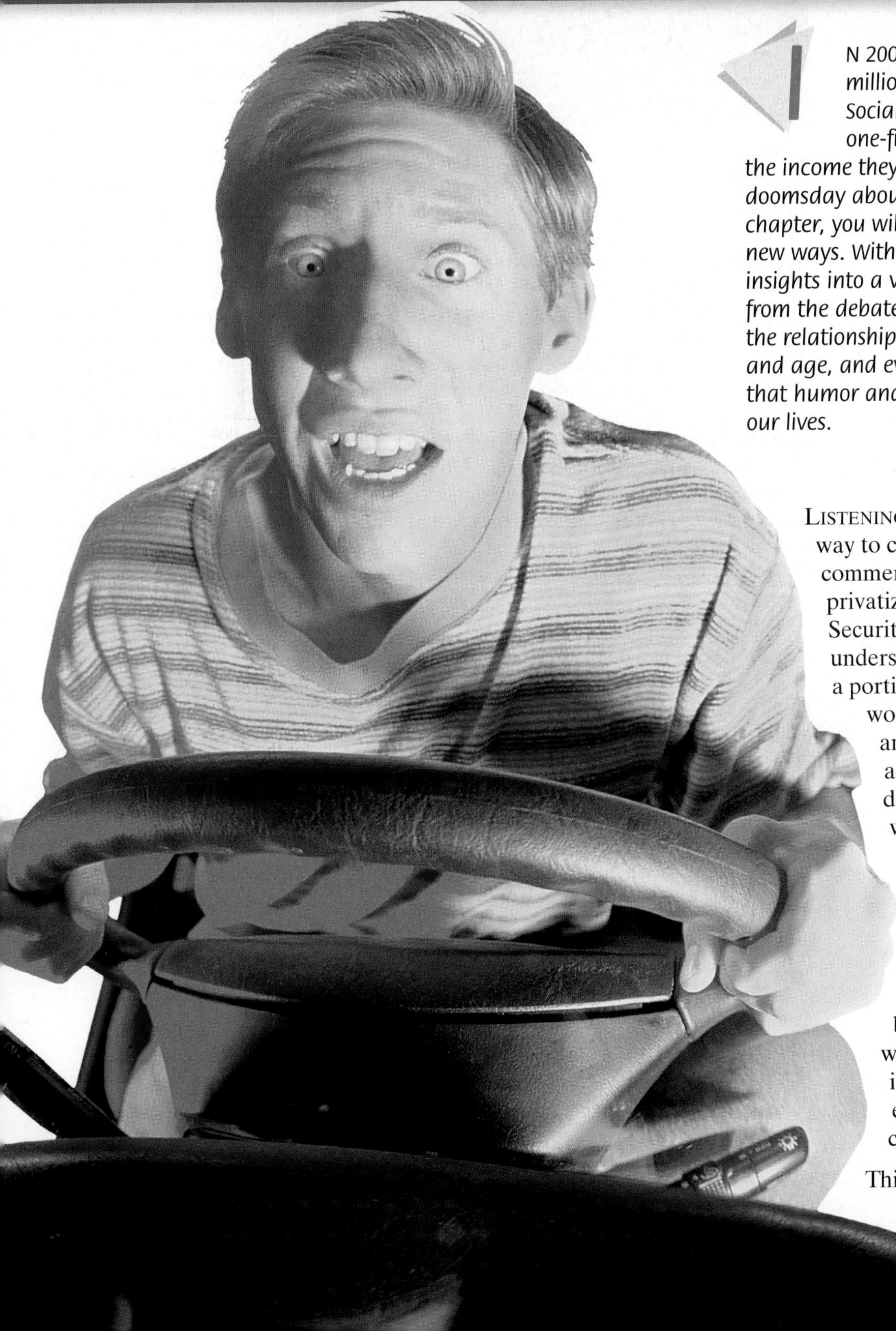

IN 2004, APPROXIMATELY 48 million Americans received Social Security benefits. For one-fifth of retirees, this was all the income they had. Is Social Security doomsday about to arrive? In this chapter, you will learn to use formulas in new ways. With these skills, you will gain insights into a variety of issues, ranging from the debate about Social Security to the relationship between blood pressure and age, and even the positive benefits that humor and laughter can have on our lives.

LISTENING TO THE RADIO ON THE way to campus, you hear political commentators debating the privatization of the Social Security system. As you understand it, the idea is to take a portion of the tax every worker pays into the system and put it into a savings account that workers can decide how to invest. You wonder if a crisis and bankruptcy are fast approaching, and if private accounts will do anything to slow that process. How can mathematical models be used to determine when the money paid out in benefits will begin to exceed the amount collected in taxes?

This problem appears as Exercises 35–37 in the Chapter 1 Test.

SECTION 1.1 Graphs and Graphing Utilities

Objectives

1. Plot points in the rectangular coordinate system.
2. Graph equations in the rectangular coordinate system.
3. Interpret information about a graphing utility's viewing rectangle or table.
4. Use a graph to determine intercepts.
5. Interpret information given by graphs.

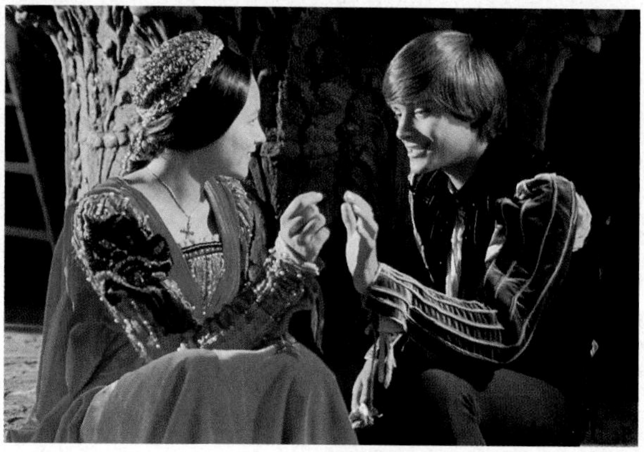

The beginning of the seventeenth century was a time of innovative ideas and enormous intellectual progress in Europe. English theatergoers enjoyed a succession of exciting new plays by Shakespeare. William Harvey proposed the radical notion that the heart was a pump for blood rather than the center of emotion. Galileo, with his new-fangled invention called the telescope, supported the theory of Polish astronomer Copernicus that the sun, not the Earth, was the center of the solar system. Monteverdi was writing the world's first grand operas. French mathematicians Pascal and Fermat invented a new field of mathematics called probability theory.

Into this arena of intellectual electricity stepped French aristocrat René Descartes (1596–1650). Descartes (pronounced "day cart"), propelled by the creativity surrounding him, developed a new branch of mathematics that brought together algebra and geometry in a unified way—a way that visualized numbers as points on a graph, equations as geometric figures, and geometric figures as equations. This new branch of mathematics, called *analytic geometry*, established Descartes as one of the founders of modern thought and among the most original mathematicians and philosophers of any age. We begin this section by looking at Descartes's deceptively simple idea, called the **rectangular coordinate system** or (in his honor) the **Cartesian coordinate system**.

1 Plot points in the rectangular coordinate system.

Points and Ordered Pairs

Descartes used two number lines that intersect at right angles at their zero points, as shown in Figure 1.1. The horizontal number line is the ***x*-axis**. The vertical number line is the ***y*-axis**. The point of intersection of these axes is their zero points, called the **origin**. Positive numbers are shown to the right and above the origin. Negative numbers are shown to the left and below the origin. The axes divide the plane into four quarters, called **quadrants**. The points located on the axes are not in any quadrant.

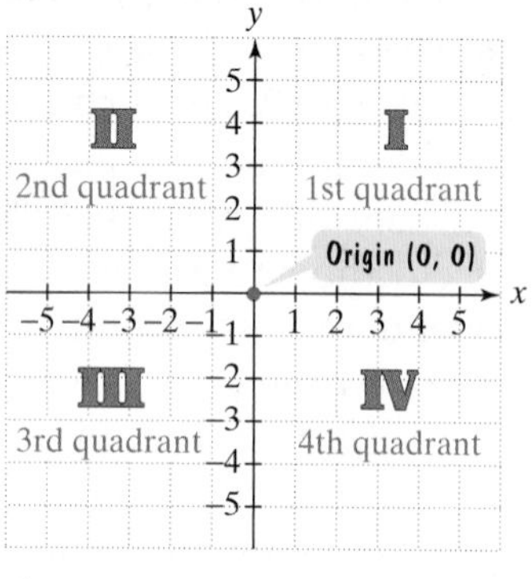

Figure 1.1 The rectangular coordinate system

Each point in the rectangular coordinate system corresponds to an **ordered pair** of real numbers, (x, y). Examples of such pairs are $(-5, 3)$ and $(3, -5)$. The first number in each pair, called the ***x*-coordinate**, denotes the distance and direction from the origin along the x-axis. The second number in each pair, called the ***y*-coordinate**, denotes vertical distance and direction along a line parallel to the y-axis or along the y-axis itself.

Figure 1.2 shows how we **plot**, or locate, the points corresponding to the ordered pairs $(-5, 3)$ and $(3, -5)$. We plot $(-5, 3)$ by going 5 units from 0 to the left along the x-axis. Then we go 3 units up parallel to the y-axis. We plot $(3, -5)$ by going 3 units from 0 to the right along the x-axis and 5 units down parallel to the y-axis. The phrase "the points corresponding to the ordered pairs $(-5, 3)$ and $(3, -5)$" is often abbreviated as "the points $(-5, 3)$ and $(3, -5)$."

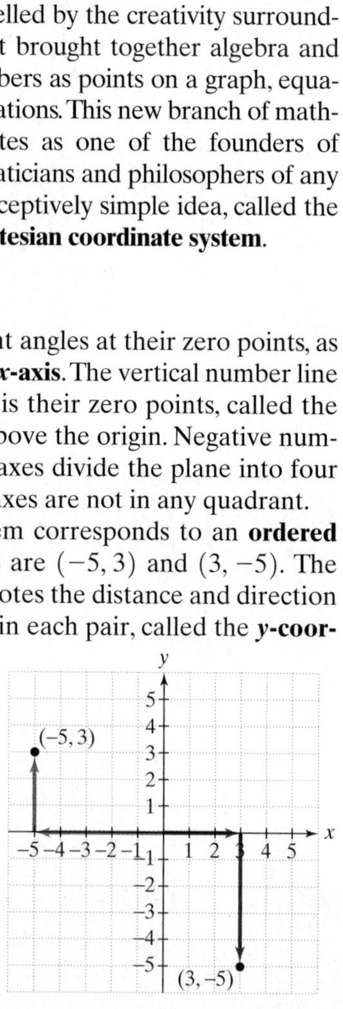

Figure 1.2 Plotting $(-5, 3)$ and $(3, -5)$

Study Tip

The phrase *ordered pair* is used because order is important. The order in which coordinates appear makes a difference in a point's location. This is illustrated in Figure 1.2.

EXAMPLE 1   Plotting Points in the Rectangular Coordinate System

Plot the points: $A(-3, 5)$, $B(2, -4)$, $C(5, 0)$, $D(-5, -3)$, $E(0, 4)$ and $F(0, 0)$.

Solution   See Figure 1.3. We move from the origin and plot the points in the following way:

$A(-3, 5)$:   3 units left, 5 units up

$B(2, -4)$:   2 units right, 4 units down

$C(5, 0)$:   5 units right, 0 units up or down

$D(-5, -3)$:   5 units left, 3 units down

$E(0, 4)$:   0 units right or left, 4 units up

$F(0, 0)$:   0 units right or left, 0 units up or down

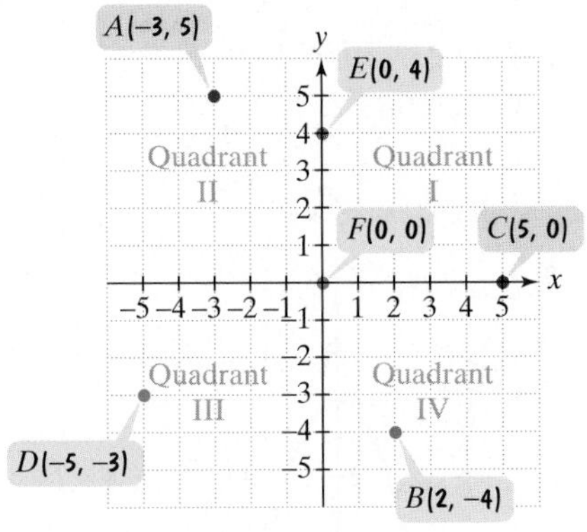

Figure 1.3 Plotting points

Check Point 1   Plot the points: $A(-2, 4)$, $B(4, -2)$, $C(-3, 0)$, and $D(0, -3)$.

② Graph equations in the rectangular coordinate system.

Graphs of Equations

A relationship between two quantities can be expressed as an **equation in two variables**, such as

$$y = 4 - x^2.$$

A **solution of an equation in two variables**, x and y, is an ordered pair of real numbers with the following property: When the x-coordinate is substituted for x and the y-coordinate is substituted for y in the equation, we obtain a true statement. For example, consider the equation $y = 4 - x^2$ and the ordered pair $(3, -5)$. When 3 is substituted for x and -5 is substituted for y, we obtain the statement $-5 = 4 - 3^2$, or $-5 = 4 - 9$, or $-5 = -5$. Because this statement is true, the ordered pair $(3, -5)$ is a solution of the equation $y = 4 - x^2$. We also say that $(3, -5)$ **satisfies** the equation.

We can generate as many ordered-pair solutions as desired to $y = 4 - x^2$ by substituting numbers for x and then finding the corresponding values for y. For example, suppose we let $x = 3$:

Start with x.   Compute y.   Form the ordered pair (x, y).

x	$y = 4 - x^2$	Ordered Pair (x, y)
3	$y = 4 - 3^2 = 4 - 9 = -5$	$(3, -5)$

Let $x = 3$. $(3, -5)$ is a solution of $y = 4 - x^2$.

The **graph of an equation in two variables** is the set of all points whose coordinates satisfy the equation. One method for graphing such equations is the **point-plotting method**. First, we find several ordered pairs that are solutions of the equation. Next, we plot these ordered pairs as points in the rectangular coordinate system. Finally, we connect the points with a smooth curve or line. This often gives us a picture of all ordered pairs that satisfy the equation.

EXAMPLE 2 Graphing an Equation Using the Point-Plotting Method

Graph $y = 4 - x^2$. Select integers for x, starting with -3 and ending with 3.

Solution For each value of x, we find the corresponding value for y.

	Start with x.	Compute y.	Form the ordered pair (x, y).

We selected integers from -3 to 3, inclusive, to include three negative numbers, 0, and three positive numbers. We also wanted to keep the resulting computations for y relatively simple.

x	$y = 4 - x^2$	Ordered Pair (x, y)
-3	$y = 4 - (-3)^2 = 4 - 9 = -5$	$(-3, -5)$
-2	$y = 4 - (-2)^2 = 4 - 4 = 0$	$(-2, 0)$
-1	$y = 4 - (-1)^2 = 4 - 1 = 3$	$(-1, 3)$
0	$y = 4 - 0^2 = 4 - 0 = 4$	$(0, 4)$
1	$y = 4 - 1^2 = 4 - 1 = 3$	$(1, 3)$
2	$y = 4 - 2^2 = 4 - 4 = 0$	$(2, 0)$
3	$y = 4 - 3^2 = 4 - 9 = -5$	$(3, -5)$

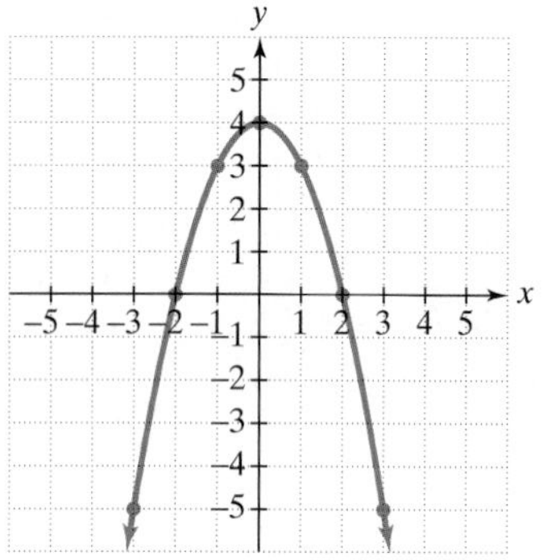

Figure 1.4 The graph of $y = 4 - x^2$

Now we plot the seven points and join them with a smooth curve, as shown in Figure 1.4. The graph of $y = 4 - x^2$ is a curve where the part of the graph to the right of the y-axis is a reflection of the part to the left of it and vice versa. The arrows on the left and the right of the curve indicate that it extends indefinitely in both directions.

Check Point 2 Graph $y = 4 - x$. Select integers for x, starting with -3 and ending with 3.

Study Tip

In Chapters 2 and 3, we will be studying graphs of equations in two variables in which

$$y = \text{a polynomial in } x.$$

Do not be concerned that we have not yet learned techniques, other than plotting points, for graphing such equations. As you solve some of the equations in this chapter, we will display graphs simply to enhance your visual understanding of your work. For now, think of graphs of first-degree polynomials as lines and graphs of second-degree polynomials as symmetric U-shaped cups that open upward or downward.

EXAMPLE 3 Graphing an Equation Using the Point-Plotting Method

Graph $y = |x|$. Select integers for x, starting with -3 and ending with 3.

Solution For each value of x, we find the corresponding value for y.

x	$y = |x|$	Ordered Pair (x, y)
-3	$y = |-3| = 3$	$(-3, 3)$
-2	$y = |-2| = 2$	$(-2, 2)$
-1	$y = |-1| = 1$	$(-1, 1)$
0	$y = |0| = 0$	$(0, 0)$
1	$y = |1| = 1$	$(1, 1)$
2	$y = |2| = 2$	$(2, 2)$
3	$y = |3| = 3$	$(3, 3)$

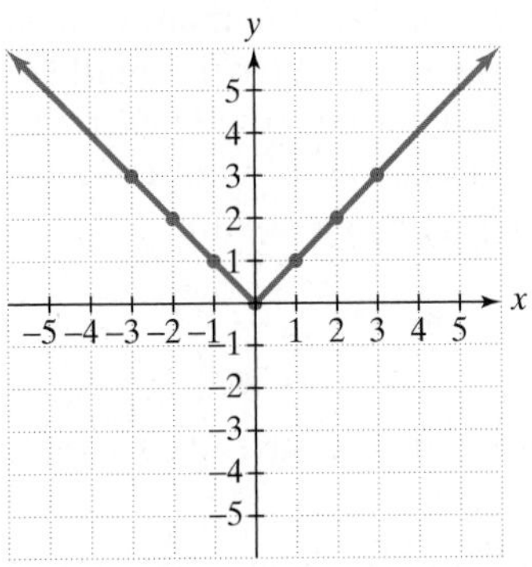

Figure 1.5 The graph of
$y = |x|$

3 Interpret information about a graphing utility's viewing rectangle or table.

We plot the points and connect them, resulting in the graph shown in Figure 1.5. The graph is V-shaped and centered at the origin. For every point (x, y) on the graph, the point $(-x, y)$ is also on the graph. This shows that the absolute value of a positive number is the same as the absolute value of its opposite.

Check Point 3 Graph $y = |x + 1|$. Select integers for x, starting with -4 and ending with 2.

Graphing Equations and Creating Tables Using a Graphing Utility

Graphing calculators or graphing software packages for computers are referred to as **graphing utilities** or graphers. A graphing utility is a powerful tool that quickly generates the graph of an equation in two variables. Figures 1.6(a) and 1.6(b) show two such graphs for the equations in Examples 2 and 3.

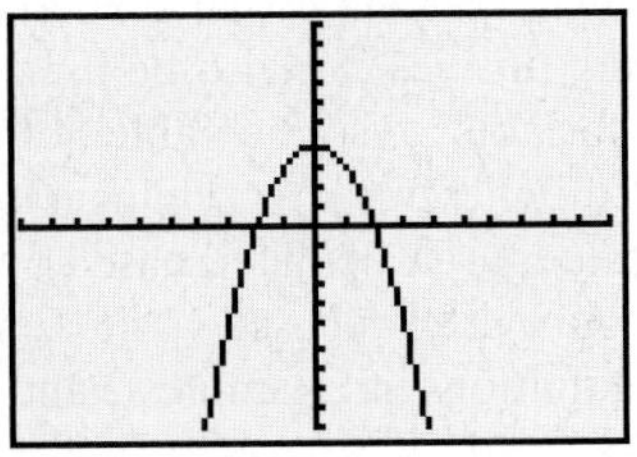

Figure 1.6(a) The graph of
$y = 4 - x^2$

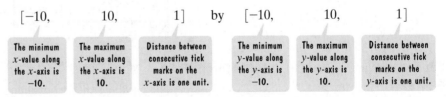

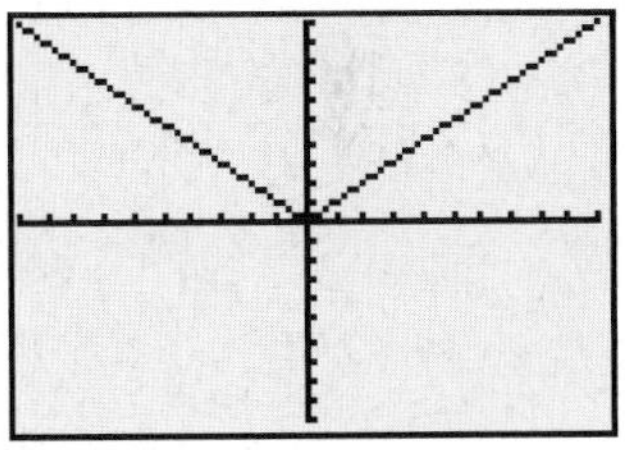

Figure 1.6(b) The graph of
$y = |x|$

What differences do you notice between these graphs and the graphs that we drew by hand? They do seem a bit "jittery." Arrows do not appear on the left and right ends of the graphs. Furthermore, numbers are not given along the axes. For both graphs in Figure 1.6, the x-axis extends from -10 to 10 and the y-axis also extends from -10 to 10. The distance represented by each consecutive tick mark is one unit. We say that the **viewing rectangle**, or the **viewing window**, is $[-10, 10, 1]$ by $[-10, 10, 1]$.

To graph an equation in x and y using a graphing utility, enter the equation and specify the size of the viewing rectangle. The size of the viewing rectangle sets minimum and maximum values for both the x- and y-axes. Enter these values, as well as the values between consecutive tick marks, on the respective axes. The $[-10, 10, 1]$ by $[-10, 10, 1]$ viewing rectangle used in Figure 1.6 is called the **standard viewing rectangle**.

EXAMPLE 4   Understanding the Viewing Rectangle

What is the meaning of a $[-2, 3, 0.5]$ by $[-10, 20, 5]$ viewing rectangle?

Solution We begin with $[-2, 3, 0.5]$, which describes the x-axis. The minimum x-value is -2 and the maximum x-value is 3. The distance between consecutive tick marks is 0.5.

Study Tip

Even if you are not using a graphing utility in the course, read this part of the section. Knowing about viewing rectangles will enable you to understand the graphs that we display in the technology boxes throughout the book.

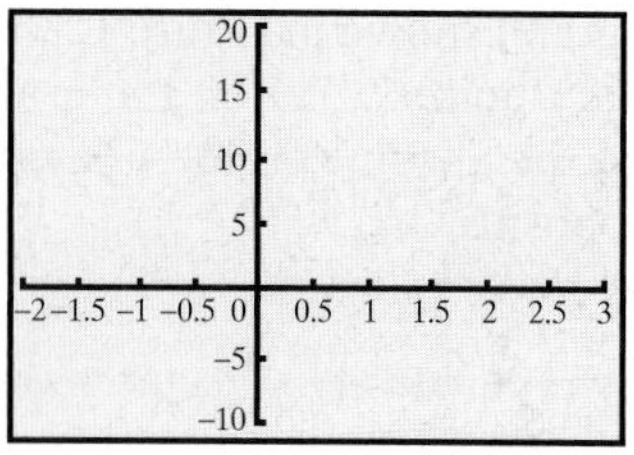

Figure 1.7 A $[-2, 3, 0.5]$ by $[-10, 20, 5]$ viewing rectangle

Next, consider $[-10, 20, 5]$, which describes the y-axis. The minimum y-value is -10 and the maximum y-value is 20. The distance between consecutive tick marks is 5.

Figure 1.7 illustrates a $[-2, 3, 0.5]$ by $[-10, 20, 5]$ viewing rectangle. To make things clearer, we've placed numbers by each tick mark. These numbers do not appear on the axes when you use a graphing utility to graph an equation.

Check Point 4 What is the meaning of a $[-100, 100, 50]$ by $[-100, 100, 10]$ viewing rectangle? Create a figure like the one in Figure 1.7 that illustrates this viewing rectangle.

On most graphing utilities, the display screen is two-thirds as high as it is wide. By using a square setting, you can equally space the x and y tick marks. (This does not occur in the standard viewing rectangle.) Graphing utilities can also *zoom in* and *zoom out.* When you zoom in, you see a smaller portion of the graph, but you do so in greater detail. When you zoom out, you see a larger portion of the graph. Thus, zooming out may help you to develop a better understanding of the overall character of the graph. With practice, you will become more comfortable with graphing equations in two variables using your graphing utility. You will also develop a better sense of the size of the viewing rectangle that will reveal needed information about a particular graph.

Graphing utilities can also be used to create tables showing solutions of equations in two variables. Use the Table Setup function to choose the starting value of x and to input the increment, or change, between the consecutive x-values. The corresponding y-values are calculated based on the equation(s) in two variables in the [Y=] screen. In Figure 1.8, we used a TI-83 Plus to create a table for $y = 4 - x^2$ and $y = |x|$, the equations in Examples 2 and 3.

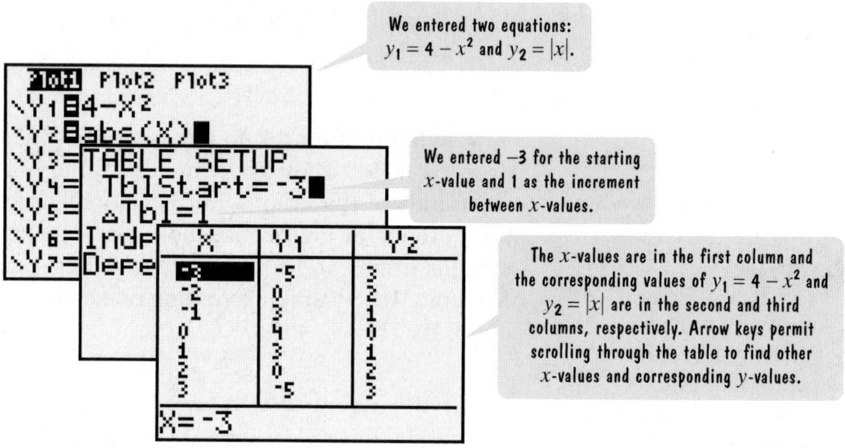

Figure 1.8 Creating a table for $y_1 = 4 - x^2$ and $y_2 = |x|$

4 Use a graph to determine intercepts.

Intercepts

An ***x*-intercept** of a graph is the x-coordinate of a point where the graph intersects the x-axis. For example, look at the graph of $y = 4 - x^2$ in Figure 1.9. The graph crosses the x-axis at $(-2, 0)$ and $(2, 0)$. Thus, the x-intercepts are -2 and 2. **The *y*-coordinate corresponding to an *x*-intercept is always zero.**

A ***y*-intercept** of a graph is the y-coordinate of a point where the graph intersects the y-axis. The graph of $y = 4 - x^2$ in Figure 1.9 shows that the graph crosses the y-axis at $(0, 4)$. Thus, the y-intercept is 4. **The *x*-coordinate corresponding to a *y*-intercept is always zero.**

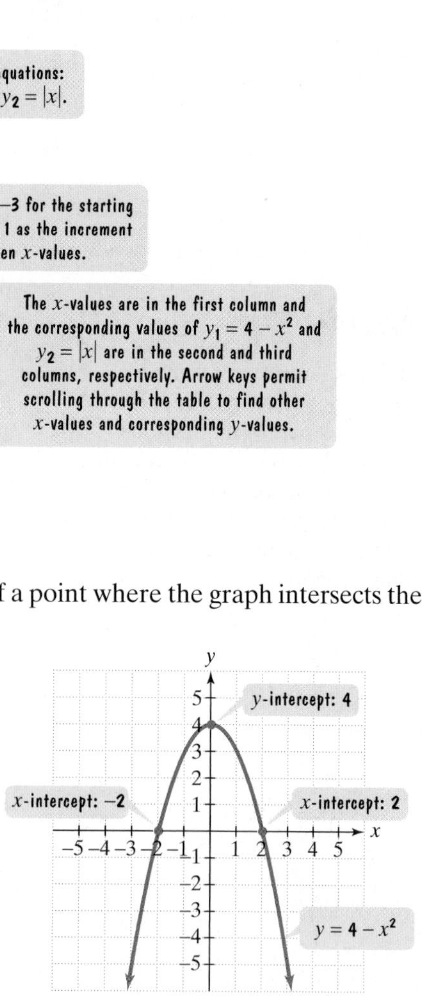

Figure 1.9 Intercepts of $y = 4 - x^2$

Study Tip

Mathematicians tend to use two ways to describe intercepts. Did you notice that we are using single numbers? If a is an x-intercept of a graph, then the graph passes through the point $(a, 0)$. If b is a y-intercept of a graph, then the graph passes through the point $(0, b)$.

Some books state that the x-intercept is the *point* $(a, 0)$ and the x-intercept is *at a* on the x-axis. Similarly, the y-intercept is the *point* $(0, b)$ and the y-intercept is *at b* on the y-axis. In these descriptions, the intercepts are the actual points where the graph intersects the axes.

Although we'll describe intercepts as single numbers, we'll immediately state the point on the x- or y-axis that the graph passes through. Here's the important thing to keep in mind:

x-intercept: The corresponding value of y is 0.

y-intercept: The corresponding value of x is 0.

EXAMPLE 5 Identifying Intercepts

Identify the x- and y-intercepts.

a.

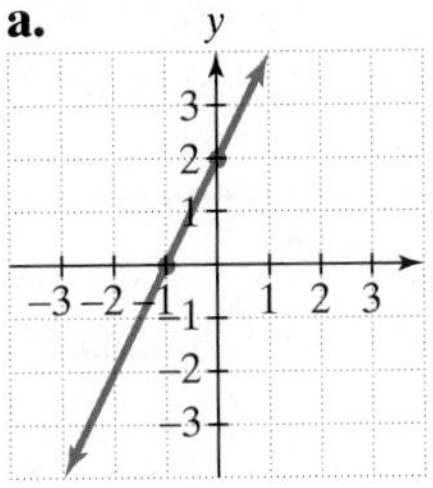

b.

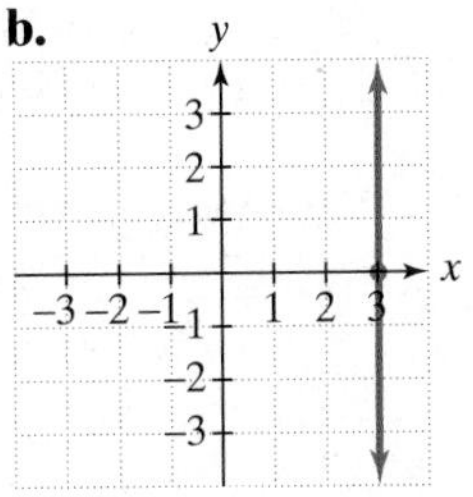

c.

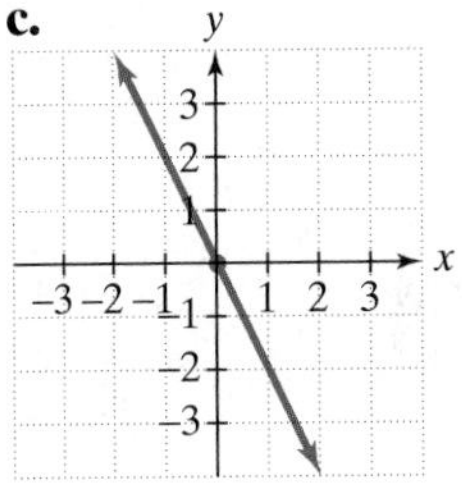

Solution

a. The graph crosses the x-axis at $(-1, 0)$. Thus, the x-intercept is -1. The graph crosses the y-axis at $(0, 2)$. Thus, the y-intercept is 2.

b. The graph crosses the x-axis at $(3, 0)$, so the x-intercept is 3. This vertical line does not cross the y-axis. Thus, there is no y-intercept.

c. This graph crosses the x- and y-axes at the same point, the origin. Because the graph crosses both axes at $(0, 0)$, the x-intercept is 0 and the y-intercept is 0.

Check Point 5 Identify the x- and y-intercepts.

a.

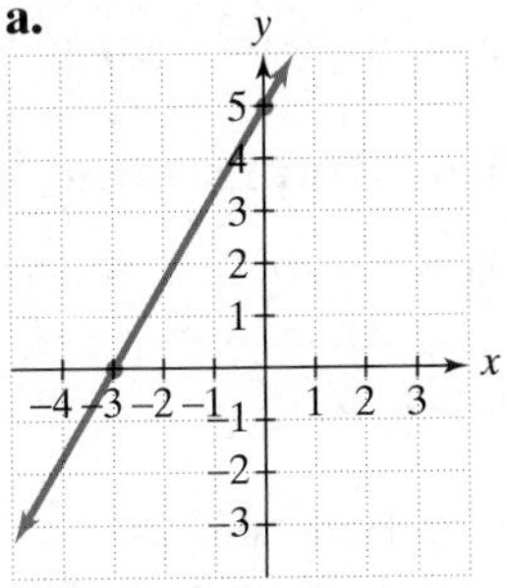

b.

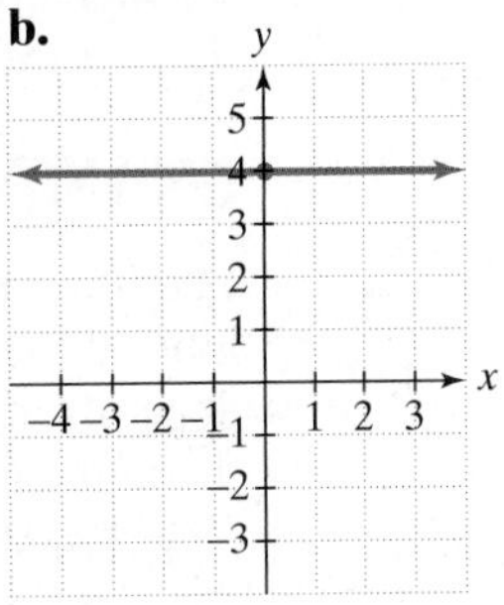

c.

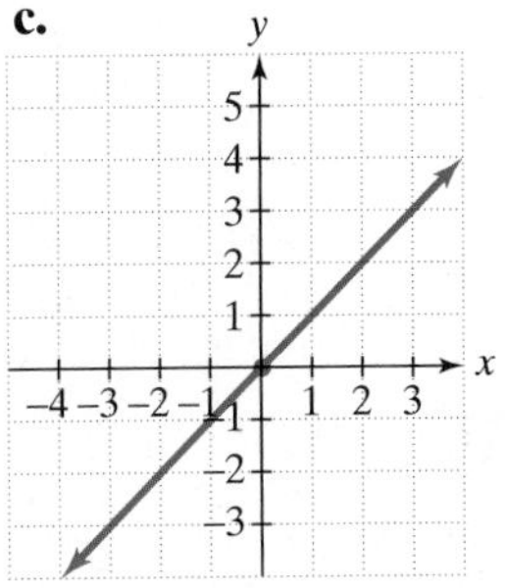

Figure 1.10 illustrates that a graph may have no intercepts or several intercepts.

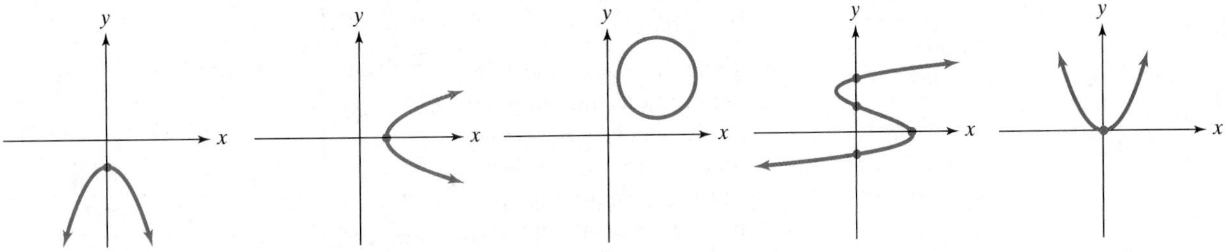

No x-intercept One y-intercept | One x-intercept No y-intercept | No intercepts | One x-intercept Three y-intercepts | The same x-intercept and y-intercept

Figure 1.10

5 Interpret information given by graphs.

Interpreting Information Given by Graphs

Line graphs are often used to illustrate trends over time. Some measure of time, such as months or years, frequently appears on the horizontal axis. Amounts are generally listed on the vertical axis. Points are drawn to represent the given information. The graph is formed by connecting the points with line segments.

Figure 1.11 is an example of a typical line graph. The graph shows the average age at which women in the United States married for the first time from 1890 through 2003. The years are listed on the horizontal axis and the ages are listed on the vertical axis. The symbol ⩘ on the vertical axis shows that there is a break in values between 0 and 20. Thus, the first tick mark on the vertical axis represents an average age of 20.

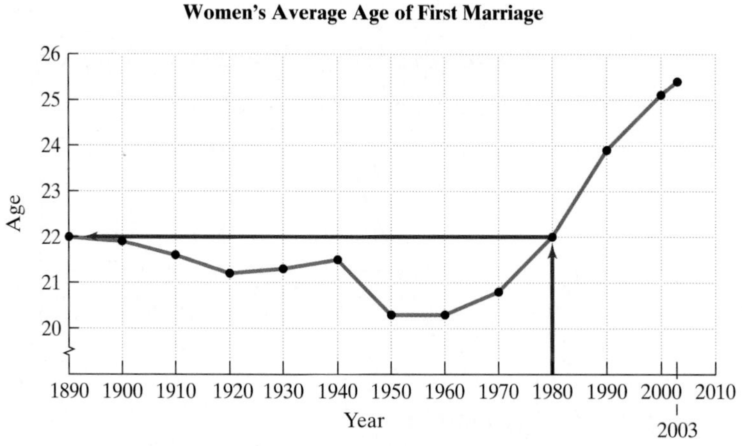

Figure 1.11

Source: U.S. Census Bureau

A line graph displays information in the first quadrant of a rectangular coordinate system. By identifying points on line graphs and their coordinates, you can interpret specific information given by the graph.

For example, the red lines in Figure 1.11 show how to find the average age at which women married for the first time in 1980.

Step 1 Locate 1980 on the horizontal axis.

Step 2 Locate the point on the line graph above 1980.

Step 3 Read across to the corresponding age on the vertical axis.

The age appears to be exactly at 22. Thus, in 1980, women in the United States married for the first time at an average age of 22.

EXAMPLE 6 Interpreting Information Given by a Graph

The line graph in Figure 1.12 shows the percentage of federal prisoners in the United States sentenced for drug offenses from 1970 through 2003.

a. For the period shown, estimate the maximum percentage of federal prisoners sentenced for drug offenses. When did this occur?

b. Table 1.1 shows the number, in thousands, of federal prisoners in the United States for four selected years. Estimate the number of federal prisoners sentenced for drug offenses for the year in part (a).

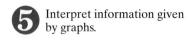

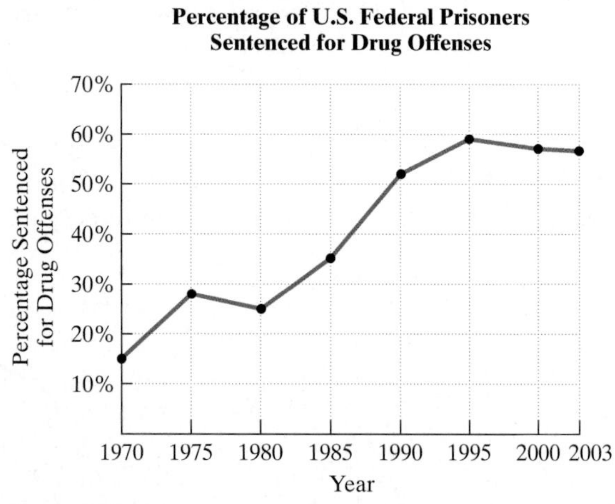

Figure 1.12

Source: Frank Schmalleger, *Criminal Justice Today,* 7th Edition, Prentice Hall, 2003.

Table 1.1 Number of U.S. Federal Prisoners

Year	Federal Prisoners
1990	58,838
1995	89,538
2000	133,921
2003	159,275

Source: Bureau of Justice Statistics

Solution

a. The maximum fraction of federal prisoners sentenced for drug offenses can be found by locating the highest point on the graph. This point lies above 1995 on the horizontal axis. Read across to the corresponding percent on the vertical axis. The number falls below 60% by approximately one unit. It appears that 59% is a reasonable estimate. Thus, the maximum fraction of federal prisoners sentenced for drug offenses is approximately 59%. This occurred in 1995.

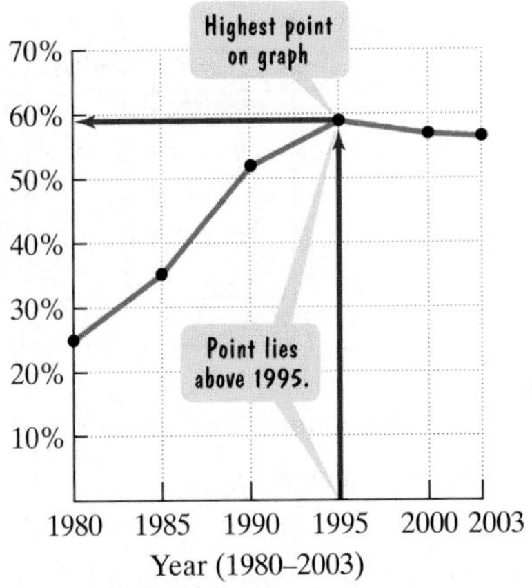

b. Table 1.1 shows that there were 89,538 federal prisoners in 1995. The number sentenced for drug offenses can be determined as follows:

$$\underset{\text{for drug offenses}}{\text{Number sentenced}} = 0.59 \times 89{,}538.$$

To estimate the number sentenced for drug offenses, we round the percent to 60% and the total federal prison population to 90,000.

$$\text{Number sentenced for drug offenses} \approx 60\% \text{ of } 90{,}000$$
$$= 0.6 \times 90{,}000$$
$$= 54{,}000$$

In 1995, approximately 54,000 federal prisoners were sentenced for drug offenses.

Check Point 6 Use Figure 1.12 and Table 1.1 to estimate the number of federal prisoners sentenced for drug offenses in 2003.

EXERCISE SET 1.1

 Practice Exercises

In Exercises 1–12, plot the given point in a rectangular coordinate system.

1. $(1, 4)$ **2.** $(2, 5)$ **3.** $(-2, 3)$

4. $(-1, 4)$ **5.** $(-3, -5)$ **6.** $(-4, -2)$

7. $(4, -1)$ **8.** $(3, -2)$ **9.** $(-4, 0)$

10. $(0, -3)$ **11.** $\left(\frac{7}{2}, -\frac{3}{2}\right)$ **12.** $\left(-\frac{5}{2}, \frac{3}{2}\right)$

Graph each equation in Exercises 13–28. Let $x = -3, -2, -1, 0, 1, 2,$ and 3.

13. $y = x^2 - 2$ **14.** $y = x^2 + 2$ **15.** $y = x - 2$

16. $y = x + 2$ **17.** $y = 2x + 1$ **18.** $y = 2x - 4$

19. $y = -\frac{1}{2}x$ **20.** $y = -\frac{1}{2}x + 2$ **21.** $y = 2|x|$

22. $y = -2|x|$ **23.** $y = |x| + 1$ **24.** $y = |x| - 1$

25. $y = 9 - x^2$ **26.** $y = -x^2$ **27.** $y = x^3$

28. $y = x^3 - 1$

In Exercises 29–32, match the viewing rectangle with the correct figure. Then label the tick marks in the figure to illustrate this viewing rectangle.

29. $[-5, 5, 1]$ by $[-5, 5, 1]$ **30.** $[-10, 10, 2]$ by $[-4, 4, 2]$

31. $[-20, 80, 10]$ by $[-30, 70, 10]$

32. $[-40, 40, 20]$ by $[-1000, 1000, 100]$

a.
b.
c.
d.

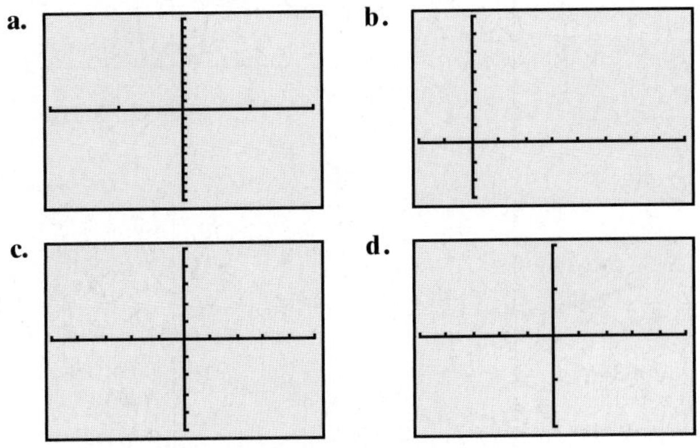

The table of values was generated by a graphing utility with a TABLE feature. Use the table to solve Exercises 33–40.

X	Y1	Y2
-3	9	5
-2	4	4
-1	1	3
0	0	2
1	1	1
2	4	0
3	9	-1

X= -3

33. Which equation corresponds to Y_2 in the table?

a. $y_2 = x + 8$ **b.** $y_2 = x - 2$

c. $y_2 = 2 - x$ **d.** $y_2 = 1 - 2x$

34. Which equation corresponds to Y_1 in the table?

a. $y_1 = -3x$ **b.** $y_1 = x^2$

c. $y_1 = -x^2$ **d.** $y_1 = 2 - x$

35. Does the graph of Y_2 pass through the origin?

36. Does the graph of Y_1 pass through the origin?

37. At which point does the graph of Y_2 cross the x-axis?

38. At which point does the graph of Y_2 cross the y-axis?

39. At which points do the graphs of Y_1 and Y_2 intersect?

40. For which values of x is $Y_1 = Y_2$?

In Exercises 41–46, use the graph to **a.** *determine the x-intercepts, if any;* **b.** *determine the y-intercepts, if any. For each graph, tick marks along the axes represent one unit each.*

41.

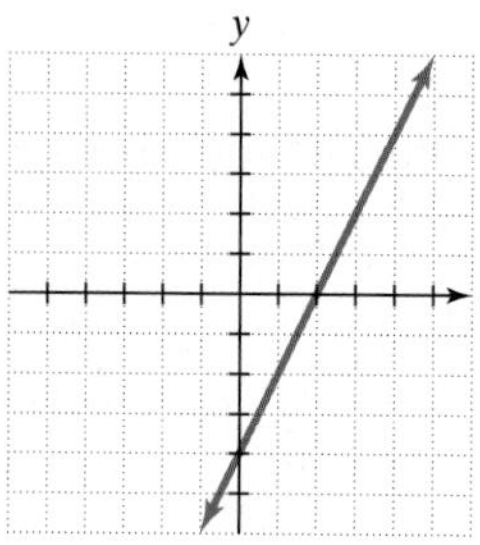

42.

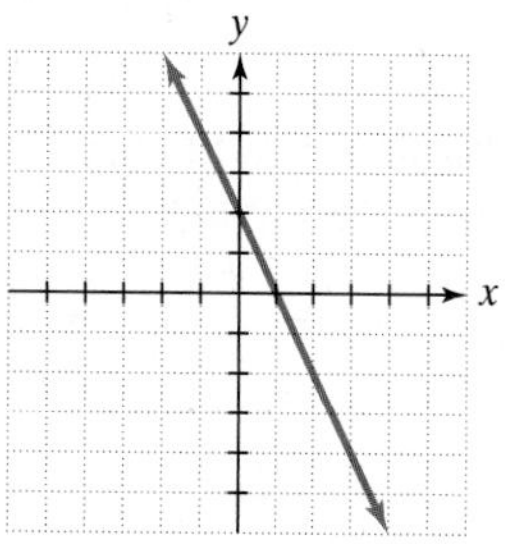

43.

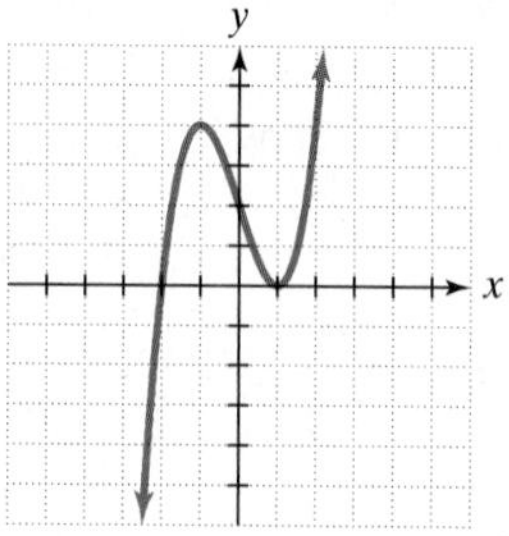

44.

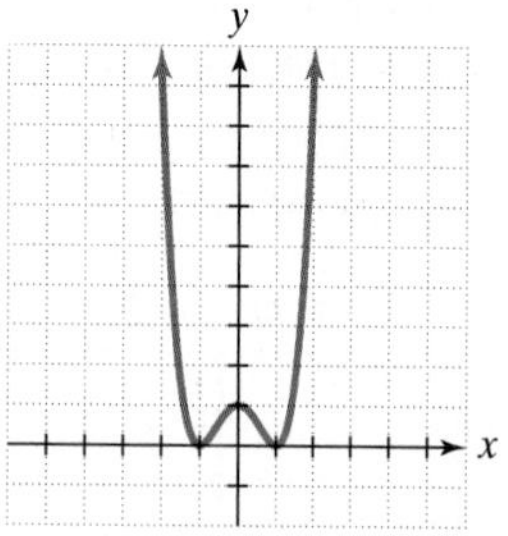

45.

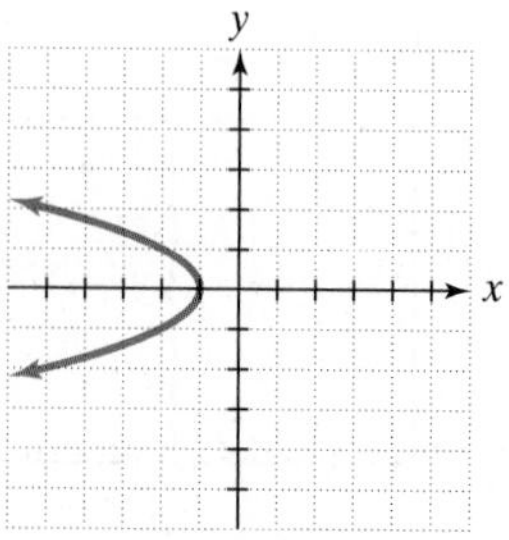

46.

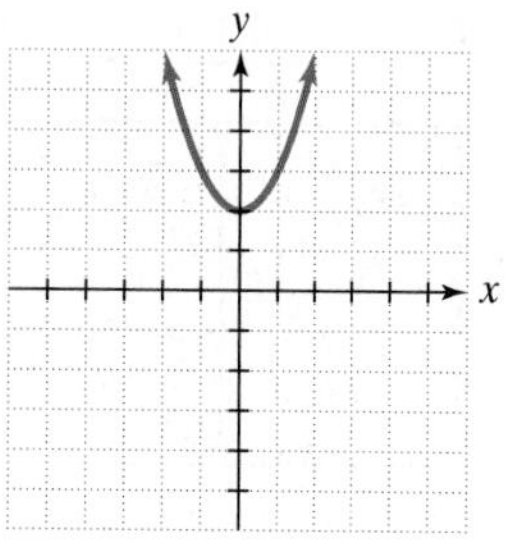

Practice Plus

In Exercises 47–50, write each English sentence as an equation in two variables. Then graph the equation.

47. The y-value is four more than twice the x-value.

48. The y-value is the difference between four and twice the x-value.

49. The y-value is three decreased by the square of the x-value.

50. The y-value is two more than the square of the x-value.

In Exercises 51–54, graph each equation.

51. $y = 5$ (Let $x = -3, -2, -1, 0, 1, 2,$ and 3.)

52. $y = -1$ (Let $x = -3, -2, -1, 0, 1, 2,$ and 3.)

53. $y = \frac{1}{x}$ (Let $x = -2, -1, -\frac{1}{2}, -\frac{1}{3}, \frac{1}{3}, \frac{1}{2}, 1,$ and 2.)

54. $y = -\frac{1}{x}$ (Let $x = -2, -1, -\frac{1}{2}, -\frac{1}{3}, \frac{1}{3}, \frac{1}{2}, 1,$ and 2.)

Application Exercises

We live in an era of democratic aspiration. The number of democracies worldwide is on the rise. The line graph shows the number of democracies worldwide, in four-year periods, from 1973 through 2001, including 2002. Use the graph to solve Exercises 55–60.

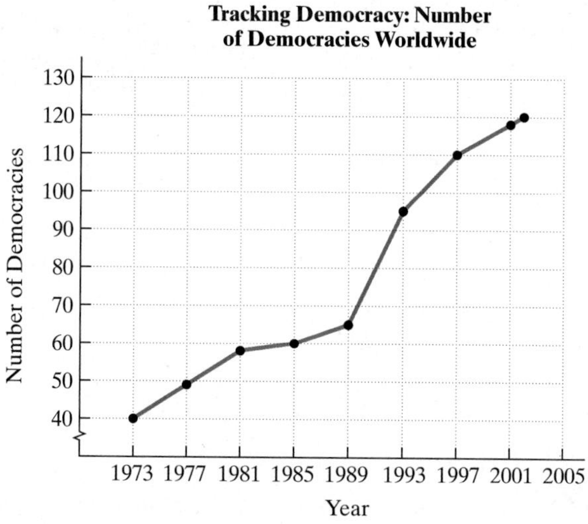

Source: The Freedom House

55. Find an estimate for the number of democracies in 1989.

56. How many more democracies were there in 2002 than in 1973?

57. In which four-year period did the number of democracies increase at the greatest rate?

58. In which four-year period did the number of democracies increase at the slowest rate?

59. In which year were there 49 democracies?

60. In which year were there 110 democracies?

Medical researchers have found that the desirable heart rate, R, in beats per minute, for beneficial exercise is approximated by the mathematical models

$$R = 165 - 0.75A \quad \text{for men}$$

$$R = 143 - 0.65A \quad \text{for women}$$

where A is the person's age. Use these mathematical models to solve Exercises 61–62.

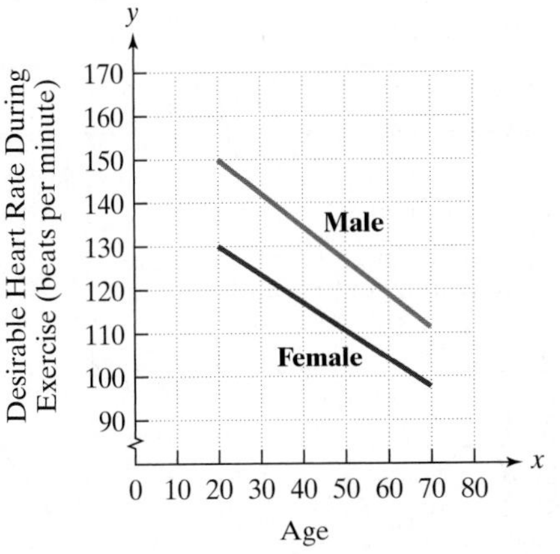

61. What is the desirable heart rate during exercise for a 40-year-old man? Identify your computation as an appropriate point on the blue graph.

62. What is the desirable heart rate during exercise for a 40-year-old woman? Identify your computation as an appropriate point on the red graph.

Autism is a neurological disorder that impedes language and derails social and emotional development. New findings suggest that the condition is not a sudden calamity that strikes children at the age of 2 or 3, but a developmental problem linked to abnormally rapid brain growth during infancy. The graphs show that the heads of severely autistic children start out smaller than average and then go through a period of explosive growth. Exercises 63–64 involve mathematical models for the data shown by the graphs.

Developmental Differences between Healthy Children and Severe Autistics

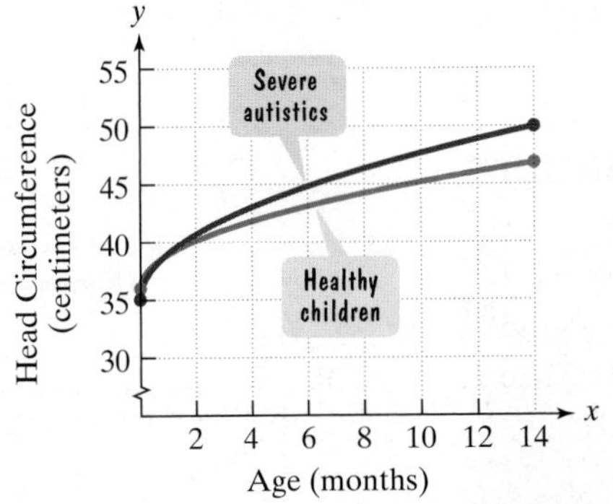

Source: The Journal of the American Medical Association

63. The data for one of the two groups shown by the graphs can be modeled by

$$y = 2.9\sqrt{x} + 36,$$

where y is the head circumference, in centimeters, at age x months, $0 \le x \le 14$.

a. According to the model, what is the head circumference at birth?

b. According to the model, what is the head circumference at 9 months?

c. According to the model, what is the head circumference at 14 months? Use a calculator and round to the nearest tenth of a centimeter.

d. Use the values that you obtained in parts (a) through (c) and the graphs shown above to determine whether the given model describes healthy children or severe autistics.

64. The data for one of the two groups shown by the graphs can be modeled by

$$y = 4\sqrt{x} + 35,$$

where y is the head circumference, in centimeters, at age x months, $0 \le x \le 14$.

a. According to the model, what is the head circumference at birth?

b. According to the model, what is the head circumference at 9 months?

c. According to the model, what is the head circumference at 14 months? Use a calculator and round to the nearest centimeter.

d. Use the values that you obtained in parts (a) through (c) and the graphs shown in the previous column to determine whether the given model describes healthy children or severe autistics.

Writing in Mathematics

65. What is the rectangular coordinate system?

66. Explain how to plot a point in the rectangular coordinate system. Give an example with your explanation.

67. Explain why $(5, -2)$ and $(-2, 5)$ do not represent the same point.

68. Explain how to graph an equation in the rectangular coordinate system.

69. What does a $[-20, 2, 1]$ by $[-4, 5, 0.5]$ viewing rectangle mean?

Technology Exercises

70. Use a graphing utility to verify each of your hand-drawn graphs in Exercises 13–28. Experiment with the size of the viewing rectangle to make the graph displayed by the graphing utility resemble your hand-drawn graph as much as possible.

71. The stated intent of the 1994 "don't ask, don't tell" policy was to reduce the number of discharges of gay men and lesbians from the military. The equation

$$y = 45.48x^2 - 334.35x + 1237.9$$

describes the number of service members, y, discharged from the military for homosexuality x years after 1990. Graph the equation in a $[0, 10, 1]$ by $[0, 2200, 100]$ viewing rectangle. Then describe something about the relationship between x and y that is revealed by looking at the graph that is not obvious from the equation. What does the graph reveal about the success or lack of success of "don't ask, don't tell"?

Critical Thinking Exercises

72. Which one of the following is true?

a. If the coordinates of a point satisfy the inequality $xy > 0$, then (x, y) must be in quadrant I.

b. The ordered pair $(2, 5)$ satisfies $3y - 2x = -4$.

c. If a point is on the x-axis, it is neither up nor down, so $x = 0$.

d. None of the above is true.

In Exercises 73–76, match the story with the correct figure. The figures are labeled (a), (b), (c), and (d).

73. As the blizzard got worse, the snow fell harder and harder.

74. The snow fell more and more softly.

75. It snowed hard, but then it stopped. After a short time, the snow started falling softly.

76. It snowed softly, and then it stopped. After a short time, the snow started falling hard.

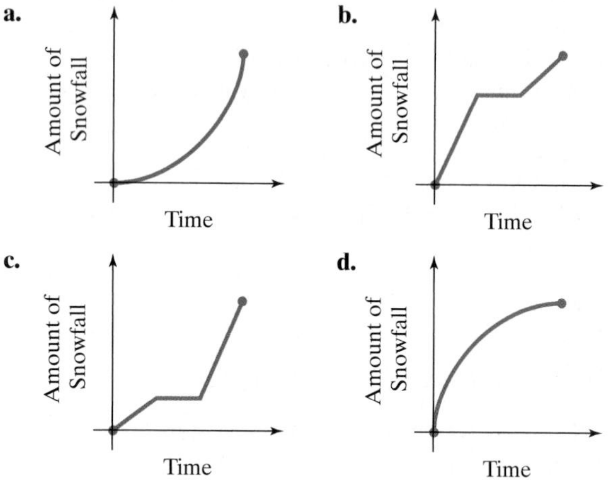

In Exercises 77–78, select the graph that best illustrates each story.

77. An airplane flew from Miami to San Francisco.

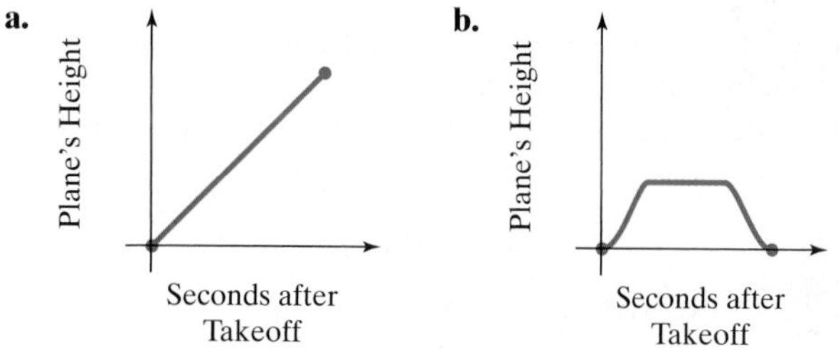

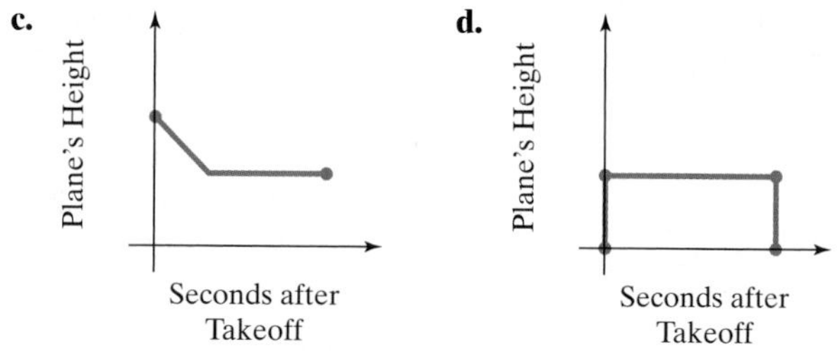

78. At noon, you begin to breathe in.

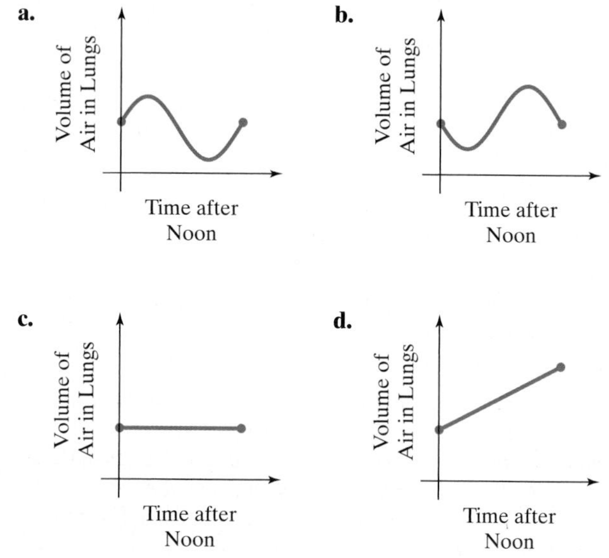

SECTION 1.2 Linear Equations and Rational Equations

Objectives

1. Solve linear equations in one variable.
2. Solve linear equations containing fractions.
3. Solve rational equations with variables in the denominators.
4. Recognize identities, conditional equations, and inconsistent equations.

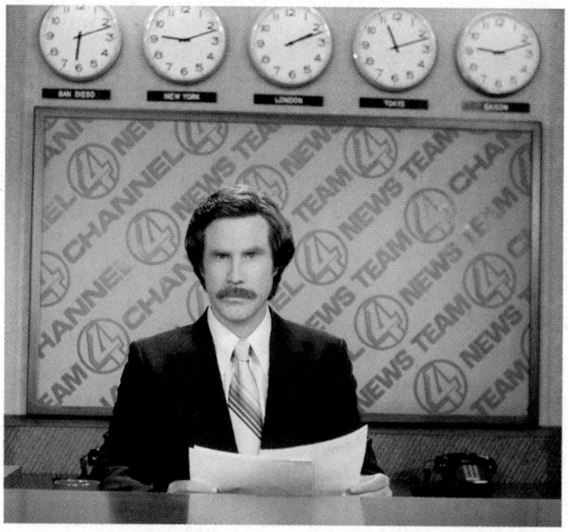

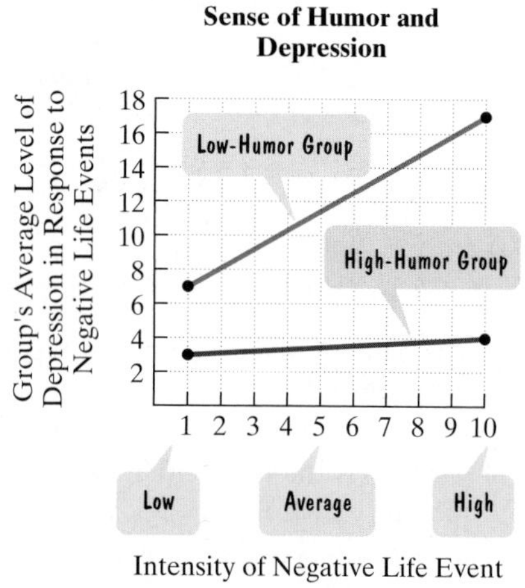

Figure 1.13

Source: Steven Davis and Joseph Palladino, *Psychology*, 3rd Edition, Prentice Hall, 2003.

The belief that humor and laughter can have positive benefits on our lives is not new. The graphs in Figure 1.13 indicate that persons with a low sense of humor have higher levels of depression in response to negative life events than those with a high sense of humor. These graphs can be modeled by the following formulas:

Low-Humor Group High-Humor Group

$$D = \frac{10}{9}x + \frac{53}{9} \qquad D = \frac{1}{9}x + \frac{26}{9}.$$

In each formula, x represents the intensity of a negative life event (from 1, low, to 10, high) and D is the level of depression in response to that event.

Suppose that the low-humor group averages a level of depression of 10 in response to a negative life event. We can determine the intensity of that event by substituting 10 for D in the low-humor model:

$$10 = \frac{10}{9}x + \frac{53}{9}.$$

The two sides of an equation can be reversed. So, we can also express this equation as

$$\frac{10}{9}x + \frac{53}{9} = 10.$$

Notice that the highest exponent on the variable is 1. Such an equation is called a *linear equation in one variable*. In this section, we will study how to solve linear equations.

1 Solve linear equations in one variable.

Solving Linear Equations in One Variable

We begin with a general definition of a linear equation in one variable.

Definition of a Linear Equation

A **linear equation in one variable x** is an equation that can be written in the form

$$ax + b = 0,$$

where a and b are real numbers, and $a \neq 0$.

An example of a linear equation in one variable is

$$4x + 12 = 0.$$

Solving an equation in x involves determining all values of x that result in a true statement when substituted into the equation. Such values are **solutions**, or **roots**, of the equation. For example, substitute -3 for x in $4x + 12 = 0$. We obtain

$$4(-3) + 12 = 0, \quad \text{or} \quad -12 + 12 = 0.$$

This simplifies to the true statement $0 = 0$. Thus, -3 is a solution of the equation $4x + 12 = 0$. We also say that -3 **satisfies** the equation $4x + 12 = 0$, because when we substitute -3 for x, a true statement results. The set of all such solutions is called the equation's **solution set**. For example, the solution set of the equation $4x + 12 = 0$ is $\{-3\}$.

Two or more equations that have the same solution set are called **equivalent equations**. For example, the equations

$$4x + 12 = 0 \quad \text{and} \quad 4x = -12 \quad \text{and} \quad x = -3$$

are equivalent equations because the solution set for each is $\{-3\}$. To solve a linear equation in x, we transform the equation into an equivalent equation one or more times. Our final equivalent equation should be of the form

$$x = \text{a number}.$$

The solution set of this equation is the set consisting of the number.

To generate equivalent equations, we will use the following principles:

Generating Equivalent Equations

An equation can be transformed into an equivalent equation by one or more of the following operations:

	Example
1. Simplify an expression by removing grouping symbols and combining like terms.	• $3(x - 6) = 6x - x$ $3x - 18 = 5x$
2. Add (or subtract) the same real number or variable expression on *both* sides of the equation.	• $3x - 18 = 5x$ $3x - 18 - 3x = 5x - 3x$ (Subtract $3x$ from both sides of the equation.) $-18 = 2x$
3. Multiply (or divide) on *both* sides of the equation by the same *nonzero* quantity.	• $-18 = 2x$ $\frac{-18}{2} = \frac{2x}{2}$ (Divide both sides of the equation by 2.) $-9 = x$
4. Interchange the two sides of the equation.	• $-9 = x$ $x = -9$

If you look closely at the equations in the box, you will notice that we have solved the equation $3(x - 6) = 6x - x$. The final equation, $x = -9$, with x isolated by itself on the left side, shows that $\{-9\}$ is the solution set. The idea in solving a linear equation is to get the variable by itself on one side of the equal sign and a number by itself on the other side.

EXAMPLE 1 Solving a Linear Equation

Solve and check: $2x + 3 = 17$.

Solution Our goal is to obtain an equivalent equation with x isolated on one side and a number on the other side.

$2x + 3 = 17$	This is the given equation.
$2x + 3 - 3 = 17 - 3$	Subtract 3 from both sides.
$2x = 14$	Simplify.
$\frac{2x}{2} = \frac{14}{2}$	Divide both sides by 2.
$x = 7$	Simplify.

Now we check the proposed solution, 7, by replacing x with 7 in the original equation.

$2x + 3 = 17$	This is the original equation.
$2 \cdot 7 + 3 \stackrel{?}{=} 17$	Substitute 7 for x. The question mark indicates that we do not yet know if the two sides are equal.
$14 + 3 \stackrel{?}{=} 17$	Multiply: $2 \cdot 7 = 14$.
$17 = 17$ (This statement is true.)	Add: $14 + 3 = 17$.

Because the check results in a true statement, we conclude that the solution set of the given equation is $\{7\}$.

Check Point 1 Solve and check: $4x + 5 = 29$.

Study Tip

We simplify algebraic expressions. We solve algebraic equations. Notice the differences between the procedures:

Simplifying an Algebraic Expression

Simplify: $3(x - 7) - (5x - 11)$.

This is not an equation. There is no equal sign.

Solution
$$\begin{aligned} &3(x - 7) - (5x - 11) \\ &= 3x - 21 - 5x + 11 \\ &= (3x - 5x) + (-21 + 11) \\ &= -2x + (-10) \\ &= -2x - 10 \end{aligned}$$

Stop! Further simplification is not possible. Avoid the common error of setting $-2x - 10$ equal to 0.

Solving an Algebraic Equation

Solve: $3(x - 7) - (5x - 11) = 14$.

This is an equation. There is an equal sign.

Solution
$$\begin{aligned} 3(x - 7) - (5x - 11) &= 14 \\ 3x - 21 - 5x + 11 &= 14 \\ -2x - 10 &= 14 \\ -2x - 10 + 10 &= 14 + 10 \\ -2x &= 24 \\ \frac{-2x}{-2} &= \frac{24}{-2} \\ x &= -12 \end{aligned}$$

Add 10 to both sides.

Divide both sides by −2.

The solution set is $\{-12\}$.

Here is a step-by-step procedure for solving a linear equation in one variable. Not all of these steps are necessary to solve every equation.

Solving a Linear Equation

1. Simplify the algebraic expression on each side by removing grouping symbols and combining like terms.
2. Collect all the variable terms on one side and all the numbers, or constant terms, on the other side.
3. Isolate the variable and solve.
4. Check the proposed solution in the original equation.

EXAMPLE 2 Solving a Linear Equation

Solve and check: $2(x - 3) - 17 = 13 - 3(x + 2)$.

Solution

Step 1 Simplify the algebraic expression on each side.

Do not begin with 13 − 3. Multiplication (the distributive property) is applied before subtraction.

$$\begin{aligned} 2(x - 3) - 17 &= 13 - 3(x + 2) && \text{This is the given equation.} \\ 2x - 6 - 17 &= 13 - 3x - 6 && \text{Use the distributive property.} \\ 2x - 23 &= -3x + 7 && \text{Combine like terms.} \end{aligned}$$

Step 2 Collect variable terms on one side and constant terms on the other side. We will collect variable terms on the left by adding $3x$ to both sides. We will collect the numbers on the right by adding 23 to both sides.

$$\begin{aligned} 2x - 23 + 3x &= -3x + 7 + 3x && \text{Add } 3x \text{ to both sides.} \\ 5x - 23 &= 7 && \text{Simplify: } 2x + 3x = 5x. \\ 5x - 23 + 23 &= 7 + 23 && \text{Add 23 to both sides.} \\ 5x &= 30 && \text{Simplify.} \end{aligned}$$

Step 3 Isolate the variable and solve. We isolate the variable, x, by dividing both sides of $5x = 30$ by 5.

$$\frac{5x}{5} = \frac{30}{5} \qquad \text{Divide both sides by 5.}$$

$$x = 6 \qquad \text{Simplify.}$$

Step 4 Check the proposed solution in the original equation. Substitute 6 for x in the original equation.

$$2(x - 3) - 17 = 13 - 3(x + 2) \qquad \text{This is the original equation.}$$

$$2(6 - 3) - 17 \stackrel{?}{=} 13 - 3(6 + 2) \qquad \text{Substitute 6 for x.}$$

$$2(3) - 17 \stackrel{?}{=} 13 - 3(8) \qquad \text{Simplify inside parentheses.}$$

$$6 - 17 \stackrel{?}{=} 13 - 24 \qquad \text{Multiply.}$$

$$-11 = -11 \qquad \text{Subtract.}$$

The true statement $-11 = -11$ verifies that the solution set is $\{6\}$.

Technology

You can use a graphing utility to check the solution of a linear equation. Enter each side of the equation separately under y_1 and y_2. Then use the table or the graphs to locate the x-value for which the y-values are the same. This x-value is the solution.

Let's verify our work in Example 2 and show that 6 is the solution of

$$2(x - 3) - 17 = 13 - 3(x + 2).$$

Enter $y_1 = 2(x - 3) - 17$ in the $\boxed{y=}$ screen.   Enter $y_2 = 13 - 3(x + 2)$ in the $\boxed{y=}$ screen.

Numeric Check

Display a table for y_1 and y_2.

Graphic Check

Display graphs for y_1 and y_2 and use the intersection feature. The solution is the x-coordinate of the intersection point.

X	Y1	Y2
1	-21	4
2	-19	1
3	-17	-2
4	-15	-5
5	-13	-8
6	-11	-11
7	-9	-14

X=6

When $x = 6$, y_1 and y_2 have the same value, namely -11. This verifies that 6 is the solution of $2(x - 3) = 13 - 3(x + 2)$.

$y_1 = 2(x - 3) - 17$

$y_2 = 13 - 3(x + 2)$

x-coordinate of intersection point is 6.

Intersection
X=6   Y=-11

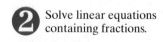

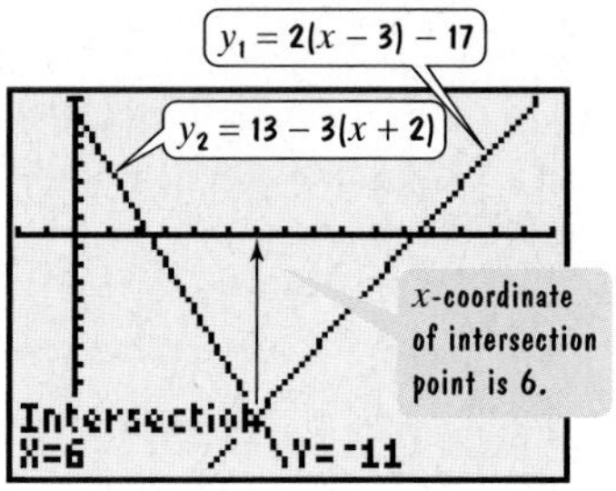

$[-2, 16, 1]$ by $[-14, 8, 1]$

Choose a large enough viewing rectangle so that you can see the intersection point.

Check Point 2 Solve and check: $4(2x + 1) = 29 + 3(2x - 5)$.

Solve linear equations containing fractions.

Linear Equations with Fractions

Equations are easier to solve when they do not contain fractions. How do we remove fractions from an equation? We begin by multiplying both sides of the equation by the least common denominator of any fractions in the equation. The least common denominator is the smallest number that all denominators will divide into. Multiplying every term on both sides of the equation by the least common denominator will eliminate the fractions in the equation. Example 3 shows how we "clear an equation of fractions."

EXAMPLE 3 Solving a Linear Equation Involving Fractions

Solve and check: $\frac{x+2}{4} - \frac{x-1}{3} = 2.$

Solution The fractional terms have denominators of 4 and 3. The smallest number that is divisible by 4 and 3 is 12. We begin by multiplying both sides of the equation by 12, the least common denominator.

$$\frac{x+2}{4} - \frac{x-1}{3} = 2$$ This is the given equation.

$$12\left(\frac{x+2}{4} - \frac{x-1}{3}\right) = 12 \cdot 2$$ Multiply both sides by 12.

$$12\left(\frac{x+2}{4}\right) - 12\left(\frac{x-1}{3}\right) = 24$$ Use the distributive property and multiply each term on the left by 12.

$$\overset{3}{\cancel{12}}\left(\frac{x+2}{\underset{1}{\cancel{4}}}\right) - \overset{4}{\cancel{12}}\left(\frac{x-1}{\underset{1}{\cancel{3}}}\right) = 24$$ Divide out common factors in each multiplication on the left.

$$3(x+2) - 4(x-1) = 24$$ The fractions are now cleared.

$$3x + 6 - 4x + 4 = 24$$ Use the distributive property.

$$-x + 10 = 24$$ Combine like terms: $3x - 4x = -x$ and $6 + 4 = 10$.

$$-x + 10 - 10 = 24 - 10$$ Subtract 10 from both sides.

$$-x = 14$$ Simplify.

We're not finished. A negative sign should not precede the variable.

Isolate x by multiplying or dividing both sides of this equation by -1.

$$\frac{-x}{-1} = \frac{14}{-1}$$ Divide both sides by -1.

$$x = -14$$ Simplify.

Check the proposed solution. Substitute –14 for x in the original equation. You should obtain $2 = 2$. This true statement verifies that the solution set is $\{-14\}$.

Check Point 3 Solve and check: $\frac{x-3}{4} = \frac{5}{14} - \frac{x+5}{7}.$

3 Solve rational equations with variables in the denominators.

Rational Equations

A **rational equation** is an equation containing one or more rational expressions. In Example 3, we solved a rational equation with constants in the denominators. This rational equation was a linear equation. Now, let's consider a rational equation such as

$$\frac{1}{x} = \frac{1}{5} + \frac{3}{2x}.$$

Can you see how this rational equation differs from the rational equation that we solved earlier? The variable, x, appears in two of the denominators. Although this rational equation is not a linear equation, the solution procedure still involves multiplying each side by the least common denominator. However, we must avoid any values of the variable that make a denominator zero. For example, examine the denominators in the equation

$$\frac{1}{x} = \frac{1}{5} + \frac{3}{2x}.$$

This denominator would equal zero if $x = 0$. This denominator would equal zero if $x = 0$.

We see that x cannot equal zero. With this in mind, let's solve the equation.

EXAMPLE 4 Solving a Rational Equation

Solve: $\frac{1}{x} = \frac{1}{5} + \frac{3}{2x}$.

Solution The denominators are x, 5, and $2x$. The least common denominator is $10x$. We begin by multiplying both sides of the equation by $10x$. We will also write the restriction that x cannot equal zero to the right of the equation.

$$\frac{1}{x} = \frac{1}{5} + \frac{3}{2x}, \quad x \neq 0$$ This is the given equation.

$$10x \cdot \frac{1}{x} = 10x\left(\frac{1}{5} + \frac{3}{2x}\right)$$ Multiply both sides by 10x.

$$10x \cdot \frac{1}{x} = 10x \cdot \frac{1}{5} + 10x \cdot \frac{3}{2x}$$ Use the distributive property. Be sure to multiply each term by 10x.

$$\cancel{10x} \cdot \frac{1}{\cancel{x}} = \overset{2}{\cancel{10}}x \cdot \frac{1}{\underset{1}{\cancel{5}}} + \overset{5}{\cancel{10}}\cancel{x} \cdot \frac{3}{\underset{1}{\cancel{2}\cancel{x}}}$$ Divide out common factors in the multiplications.

$$10 = 2x + 15$$ Complete the multiplications.

Observe that the resulting equation,

$$10 = 2x + 15,$$

is now cleared of fractions. With the variable term, $2x$, already on the right, we will collect constant terms on the left by subtracting 15 from both sides.

$$10 - 15 = 2x + 15 - 15$$ Subtract 15 from both sides.

$$-5 = 2x$$ Simplify.

Finally, we isolate the variable, x, in $-5 = 2x$ by dividing both sides by 2.

$$\frac{-5}{2} = \frac{2x}{2}$$ Divide both sides by 2.

$$-\frac{5}{2} = x$$ Simplify.

We check our solution by substituting $-\frac{5}{2}$ into the original equation or by using a calculator. With a calculator, evaluate each side of the equation for $x = -\frac{5}{2}$, or for $x = -2.5$. Note that the original restriction that $x \neq 0$ is met. The solution set is $\left\{-\frac{5}{2}\right\}$.

Check Point 4 Solve: $\frac{5}{2x} = \frac{17}{18} - \frac{1}{3x}$.

EXAMPLE 5 Solving a Rational Equation

Solve: $\frac{x}{x-3} = \frac{3}{x-3} + 9.$

Solution We must avoid any values of the variable x that make a denominator zero.

$$\frac{x}{x-3} = \frac{3}{x-3} + 9$$

These denominators are zero if $x = 3$.

We see that x cannot equal 3. With denominators of $x - 3$, $x - 3$, and 1, the least common denominator is $x - 3$. We multiply both sides of the equation by $x - 3$. We also write the restriction that x cannot equal 3 to the right of the equation.

$$\frac{x}{x-3} = \frac{3}{x-3} + 9, \quad x \neq 3$$ This is the given equation.

$$(x-3)\cdot\frac{x}{x-3} = (x-3)\left(\frac{3}{x-3} + 9\right)$$ Multiply both sides by $x - 3$.

$$(x-3)\cdot\frac{x}{x-3} = (x-3)\cdot\frac{3}{x-3} + (x-3)\cdot 9$$ Use the distributive property.

$$\cancel{(x-3)}\cdot\frac{x}{\cancel{x-3}} = \cancel{(x-3)}\cdot\frac{3}{\cancel{x-3}} + 9(x-3)$$ Divide out common factors in two of the multiplications.

$$x = 3 + 9(x-3)$$ Simplify.

The resulting equation is cleared of fractions. We now solve for x.

$$x = 3 + 9x - 27$$ Use the distributive property.

$$x = 9x - 24$$ Combine numerical terms.

$$x - 9x = 9x - 24 - 9x$$ Subtract $9x$ from both sides.

$$-8x = -24$$ Simplify.

$$\frac{-8x}{-8} = \frac{-24}{-8}$$ Divide both sides by -8.

$$x = 3$$ Simplify.

Study Tip

Reject any proposed solution that causes any denominator in an equation to equal 0.

The proposed solution, 3, is *not* a solution because of the restriction that $x \neq 3$. There is *no solution to this equation.* The solution set for this equation contains no elements. The solution set is $\varnothing$, the empty set.

Check Point 5 Solve: $\frac{x}{x-2} = \frac{2}{x-2} - \frac{2}{3}.$

EXAMPLE 6 Solving a Rational Equation to Determine When Two Equations Are Equal

Consider the equations

$$y_1 = \frac{3}{x+6} + \frac{1}{x-2} \quad \text{and} \quad y_2 = \frac{4}{x^2 + 4x - 12}.$$

Find all values of x for which $y_1 = y_2$.

Solution Because we are interested in one or more values of x that cause y_1 and y_2 to be equal, we set the expressions that define y_1 and y_2 equal to each other:

$$\frac{3}{x+6} + \frac{1}{x-2} = \frac{4}{x^2 + 4x - 12}.$$

To identify values of x that make denominators zero, let's factor $x^2 + 4x - 12$, the denominator on the right. This factorization is also necessary in identifying the least common denominator.

$$\frac{3}{x + 6} + \frac{1}{x - 2} = \frac{4}{(x + 6)(x - 2)}$$

| This denominator is zero if $x = -6$. | This denominator is zero if $x = 2$. | This denominator is zero if $x = -6$ or $x = 2$. |

We see that x cannot equal -6 or 2. The least common denominator is $(x + 6)(x - 2)$.

$$\frac{3}{x + 6} + \frac{1}{x - 2} = \frac{4}{(x + 6)(x - 2)}, \quad x \neq -6, \quad x \neq 2$$

This is the given equation with a denominator factored.

$$(x + 6)(x - 2)\left(\frac{3}{x + 6} + \frac{1}{x - 2}\right) = (x + 6)(x - 2) \cdot \frac{4}{(x + 6)(x - 2)}$$

Multiply both sides by $(x + 6)(x - 2)$, the LCD.

$$\cancel{(x + 6)}(x - 2) \cdot \frac{3}{\cancel{x + 6}} + (x + 6)\cancel{(x - 2)} \cdot \frac{1}{\cancel{x - 2}} = \cancel{(x + 6)}\cancel{(x - 2)} \cdot \frac{4}{\cancel{(x + 6)}\cancel{(x - 2)}}$$

Use the distributive property and divide out common factors.

$$3(x - 2) + 1(x + 6) = 4$$

Simplify. This equation is cleared of fractions.

$$3x - 6 + x + 6 = 4$$

Use the distributive property.

$$4x = 4$$

Combine like terms.

$$\frac{4x}{4} = \frac{4}{4}$$

Divide both sides by 4.

$$x = 1$$

Simplify. This is not part of the restriction that $x \neq -6$ and $x \neq 2$.

The value of x for which $y_1 = y_2$ is 1.

Check

Is $y_1 = y_2$ when $x = 1$? We use the given equations

$$y_1 = \frac{3}{x + 6} + \frac{1}{x - 2} \quad \text{and} \quad y_2 = \frac{4}{x^2 + 4x - 12}$$

to answer the question.

Checking by Hand

Substitute 1 for x in y_1 and y_2.

$$y_1 = \frac{3}{1 + 6} + \frac{1}{1 - 2} = \frac{3}{7} + \frac{1}{-1}$$

$$= \frac{3}{7} - 1 = \frac{3}{7} - \frac{7}{7} = -\frac{4}{7}$$

$$y_2 = \frac{4}{1^2 + 4 \cdot 1 - 12} = \frac{4}{1 + 4 - 12}$$

$$= \frac{4}{-7} = -\frac{4}{7}$$

When $x = 1$, y_1 and y_2 have the same value, namely, $-\frac{4}{7}$.

Checking with a Graphing Utility

Display a table showing values for y_1 and y_2. Enter the equations as y_1 and y_2, and be careful with parentheses.

| $y_1 = 3 \div (x + 6) + 1 \div (x - 2)$ | $y_2 = 4 \div (x^2 + 4x - 12)$ |

X	Y₁	Y₂
-3	.8	-.2667
-2	.5	-.25
-1	.26667	-.2667
0	0	-.3333
1	-.5714	-.5714
2	ERROR	ERROR
3	1.3333	.44444

X=1

No matter how far up or down you scroll, $y_1 = y_2$ only when $x = 1$.

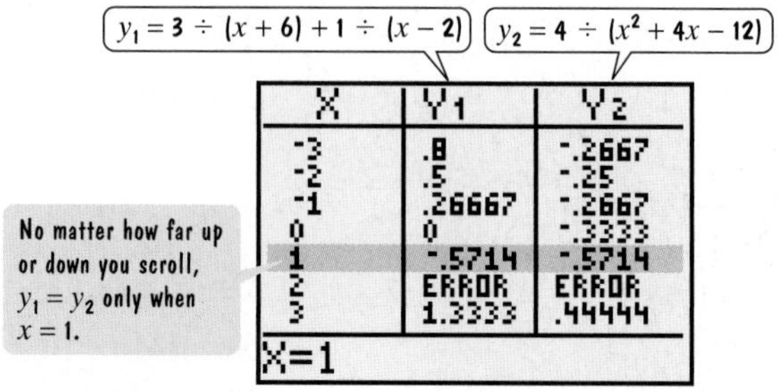

Check Point 6 Consider the equations

$$y_1 = \frac{1}{x+4} + \frac{1}{x-4} \quad \text{and} \quad y_2 = \frac{22}{x^2 - 16}.$$

Find all values of x for which $y_1 = y_2$ and check.

4 Recognize identities, conditional equations, and inconsistent equations.

Types of Equations

We tend to place things in categories, allowing us to order and structure the world. For example, you can categorize yourself by your age group, your ethnicity, your academic major, or your gender. Equations can be placed into categories that depend on their solution sets.

An equation that is true for all real numbers for which both sides are defined is called an **identity**. An example of an identity is

$$x + 3 = x + 2 + 1.$$

Every number plus 3 is equal to that number plus 2 plus 1. Therefore, the solution set to this equation is the set of all real numbers, expressed as $\{x|x \text{ is a real number}\}$.

Another example of an identity is

$$\frac{2x}{x} = 2.$$

Because division by 0 is undefined, this equation is true for all real number values of x except 0. The solution set is the set of nonzero real numbers, expressed as $\{x|x \text{ is a real number and } x \neq 0\}$.

An equation that is not an identity, but that is true for at least one real number, is called a **conditional equation**. The equation $2x + 3 = 17$ is an example of a conditional equation. The equation is not an identity and is true only if x is 7.

An **inconsistent equation** is an equation that is not true for even one real number. An example of an inconsistent equation is

$$x = x + 7.$$

There is no number that is equal to itself plus 7. The equation's solution set is $\varnothing$, the empty set. Some inconsistent equations are less obvious than this. Consider the equation in Example 5,

$$\frac{x}{x-3} = \frac{3}{x-3} + 9.$$

This equation is not true for any real number and has no solution. Thus, it is inconsistent.

Study Tip

If you are concerned by the vocabulary of equation types, keep in mind that there are three possible situations. We can state these situations informally as follows:

1. $x =$ a real number — conditional equation
2. $x =$ all real numbers — identity
3. $x =$ no real numbers. — inconsistent equation

EXAMPLE 7 Categorizing an Equation

Solve and determine whether the equation

$$2(x + 1) = 2x + 3$$

is an identity, a conditional equation, or an inconsistent equation.

Solution Begin by applying the distributive property on the left side. We obtain

$$2x + 2 = 2x + 3$$

Does something look strange? Can doubling a number and increasing the product by 2 give the same result as doubling the same number and increasing the product by 3? No. Let's continue solving the equation by subtracting $2x$ from both sides.

$$2x - 2x + 2 = 2x - 2x + 3$$

$$2 = 3$$

Keep reading. $2 = 3$ is not the solution.

The original equation, $2(x + 1) = 2x + 3$, is equivalent to the statement $2 = 3$, which is false for every value of x. The equation is inconsistent and has no solution. The solution set is $\varnothing$, the empty set.

Technology

Consider the graphs of $y_1 = 2(x + 1) = 2x + 2$ and $y_2 = 2x + 3$. The graphs appear to be parallel lines with no intersection point. This verifies that the equation

$$2(x + 1) = 2x + 3$$

has no solution and is inconsistent.

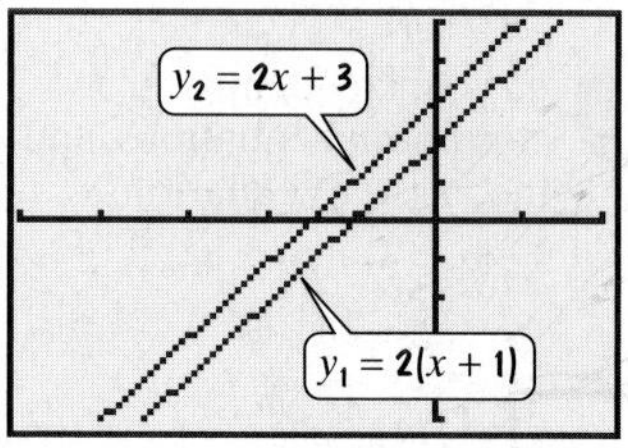

[−5, 2, 1] by [−5, 5, 1]

Check Point 7 Solve and determine whether the equation

$$4x - 7 = 4(x - 1) + 3$$

is an identity, a conditional equation, or an inconsistent equation.

EXERCISE SET 1.2

Practice Exercises

In Exercises 1–16, solve and check each linear equation.

1. $7x - 5 = 72$
2. $6x - 3 = 63$
3. $11x - (6x - 5) = 40$
4. $5x - (2x - 10) = 35$
5. $2x - 7 = 6 + x$
6. $3x + 5 = 2x + 13$
7. $7x + 4 = x + 16$
8. $13x + 14 = 12x - 5$
9. $3(x - 2) + 7 = 2(x + 5)$
10. $2(x - 1) + 3 = x - 3(x + 1)$
11. $3(x - 4) - 4(x - 3) = x + 3 - (x - 2)$
12. $2 - (7x + 5) = 13 - 3x$
13. $16 = 3(x - 1) - (x - 7)$
14. $5x - (2x + 2) = x + (3x - 5)$
15. $25 - [2 + 5y - 3(y + 2)] = -3(2y - 5) - [5(y - 1) - 3y + 3]$
16. $45 - [4 - 2y - 4(y + 7)] = -4(1 + 3y) - [4 - 3(y + 2) - 2(2y - 5)]$

Exercises 17–30 contain linear equations with constants in denominators. Solve each equation.

17. $\frac{x}{3} = \frac{x}{2} - 2$
18. $\frac{x}{5} = \frac{x}{6} + 1$
19. $20 - \frac{x}{3} = \frac{x}{2}$
20. $\frac{x}{5} - \frac{1}{2} = \frac{x}{6}$
21. $\frac{3x}{5} = \frac{2x}{3} + 1$
22. $\frac{x}{2} = \frac{3x}{4} + 5$
23. $\frac{3x}{5} - x = \frac{x}{10} - \frac{5}{2}$
24. $2x - \frac{2x}{7} = \frac{x}{2} + \frac{17}{2}$
25. $\frac{x + 3}{6} = \frac{3}{8} + \frac{x - 5}{4}$
26. $\frac{x + 1}{4} = \frac{1}{6} + \frac{2 - x}{3}$
27. $\frac{x}{4} = 2 + \frac{x - 3}{3}$
28. $5 + \frac{x - 2}{3} = \frac{x + 3}{8}$
29. $\frac{x + 1}{3} = 5 - \frac{x + 2}{7}$
30. $\frac{3x}{5} - \frac{x - 3}{2} = \frac{x + 2}{3}$

Exercises 31–50 contain rational equations with variables in denominators. For each equation, **a.** *Write the value or values of the variable that make a denominator zero. These are the restrictions on the variable.* **b.** *Keeping the restrictions in mind, solve the equation.*

31. $\frac{4}{x} = \frac{5}{2x} + 3$
32. $\frac{5}{x} = \frac{10}{3x} + 4$
33. $\frac{2}{x} + 3 = \frac{5}{2x} + \frac{13}{4}$
34. $\frac{7}{2x} - \frac{5}{3x} = \frac{22}{3}$
35. $\frac{2}{3x} + \frac{1}{4} = \frac{11}{6x} - \frac{1}{3}$
36. $\frac{5}{2x} - \frac{8}{9} = \frac{1}{18} - \frac{1}{3x}$
37. $\frac{x - 2}{2x} + 1 = \frac{x + 1}{x}$
38. $\frac{4}{x} = \frac{9}{5} - \frac{7x - 4}{5x}$
39. $\frac{1}{x - 1} + 5 = \frac{11}{x - 1}$
40. $\frac{3}{x + 4} - 7 = \frac{-4}{x + 4}$
41. $\frac{8x}{x + 1} = 4 - \frac{8}{x + 1}$
42. $\frac{2}{x - 2} = \frac{x}{x - 2} - 2$
43. $\frac{3}{2x - 2} + \frac{1}{2} = \frac{2}{x - 1}$
44. $\frac{3}{x + 3} = \frac{5}{2x + 6} + \frac{1}{x - 2}$
45. $\frac{3}{x + 2} + \frac{2}{x - 2} = \frac{8}{(x + 2)(x - 2)}$

46. $\dfrac{5}{x+2} + \dfrac{3}{x-2} = \dfrac{12}{(x+2)(x-2)}$

47. $\dfrac{2}{x+1} - \dfrac{1}{x-1} = \dfrac{2x}{x^2-1}$

48. $\dfrac{4}{x+5} + \dfrac{2}{x-5} = \dfrac{32}{x^2-25}$

49. $\dfrac{1}{x-4} - \dfrac{5}{x+2} = \dfrac{6}{x^2-2x-8}$

50. $\dfrac{6}{x+3} - \dfrac{5}{x-2} = \dfrac{-20}{x^2+x-6}$

In Exercises 51–56, find all values of x satisfying the given conditions.

51. $y_1 = 5(2x-8) - 2$, $y_2 = 5(x-3) + 3$, and $y_1 = y_2$.

52. $y_1 = 7(3x-2) + 5$, $y_2 = 6(2x-1) + 24$, and $y_1 = y_2$.

53. $y_1 = \dfrac{x-3}{5}$, $y_2 = \dfrac{x-5}{4}$, and $y_1 - y_2 = 1$.

54. $y_1 = \dfrac{x+1}{4}$, $y_2 = \dfrac{x-2}{3}$, and $y_1 - y_2 = -4$.

55. $y_1 = \dfrac{5}{x+4}$, $y_2 = \dfrac{3}{x+3}$, $y_3 = \dfrac{12x+19}{x^2+7x+12}$, and $y_1 + y_2 = y_3$.

56. $y_1 = \dfrac{2x-1}{x^2+2x-8}$, $y_2 = \dfrac{2}{x+4}$, $y_3 = \dfrac{1}{x-2}$, and $y_1 + y_2 = y_3$.

In Exercises 57–60, find all values of x such that $y = 0$.

57. $y = 4[x - (3-x)] - 7(x+1)$

58. $y = 2[3x - (4x-6)] - 5(x-6)$

59. $y = \dfrac{x+6}{3x-12} - \dfrac{5}{x-4} - \dfrac{2}{3}$

60. $y = \dfrac{1}{5x+5} - \dfrac{3}{x+1} + \dfrac{7}{5}$

In Exercises 61–68, determine whether each equation is an identity, a conditional equation, or an inconsistent equation.

61. $4(x-7) = 4x - 28$

62. $4(x-7) = 4x + 28$

63. $2x + 3 = 2x - 3$

64. $\dfrac{7x}{x} = 7$

65. $4x + 5x = 8x$

66. $8x + 2x = 9x$

67. $\dfrac{2x}{x-3} = \dfrac{6}{x-3} + 4$

68. $\dfrac{3}{x-3} = \dfrac{x}{x-3} + 3$

The equations in Exercises 69–80 combine the types of equations we have discussed in this section. Solve each equation. Then state whether the equation is an identity, a conditional equation, or an inconsistent equation.

69. $\dfrac{x+5}{2} - 4 = \dfrac{2x-1}{3}$

70. $\dfrac{x+2}{7} = 5 - \dfrac{x+1}{3}$

71. $\dfrac{2}{x-2} = 3 + \dfrac{x}{x-2}$

72. $\dfrac{6}{x+3} + 2 = \dfrac{-2x}{x+3}$

73. $8x - (3x+2) + 10 = 3x$

74. $2(x+2) + 2x = 4(x+1)$

75. $\dfrac{2}{x} + \dfrac{1}{2} = \dfrac{3}{4}$

76. $\dfrac{3}{x} - \dfrac{1}{6} = \dfrac{1}{3}$

77. $\dfrac{4}{x-2} + \dfrac{3}{x+5} = \dfrac{7}{(x+5)(x-2)}$

78. $\dfrac{1}{x-1} = \dfrac{1}{(2x+3)(x-1)} + \dfrac{4}{2x+3}$

79. $\dfrac{4x}{x+3} - \dfrac{12}{x-3} = \dfrac{4x^2+36}{x^2-9}$

80. $\dfrac{4}{x^2+3x-10} - \dfrac{1}{x^2+x-6} = \dfrac{3}{x^2-x-12}$

In Exercises 81–84, use the Y= *screen to write the equation being solved. Then use the table to solve the equation.*

81.

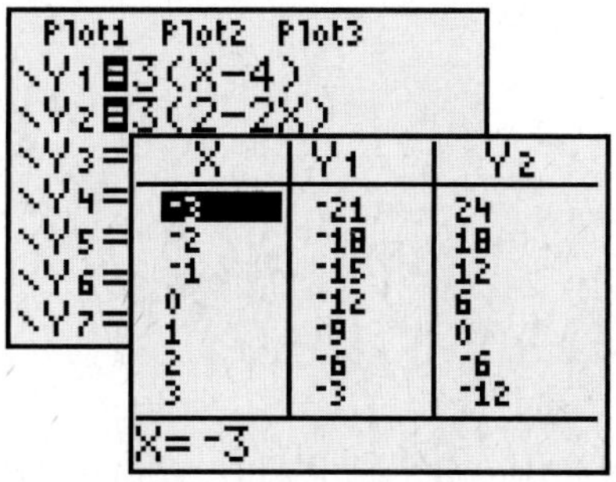

82.

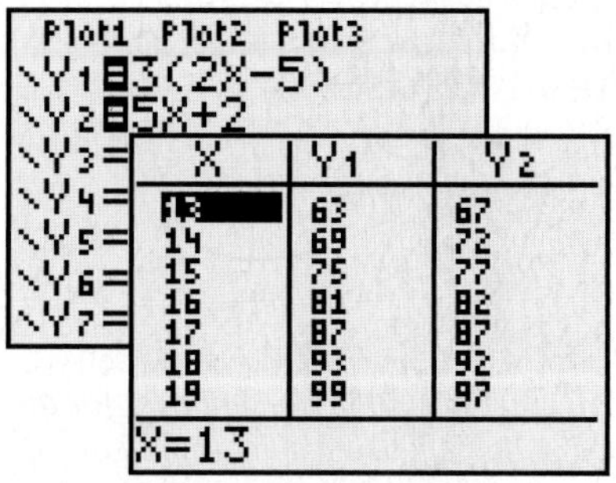

83.

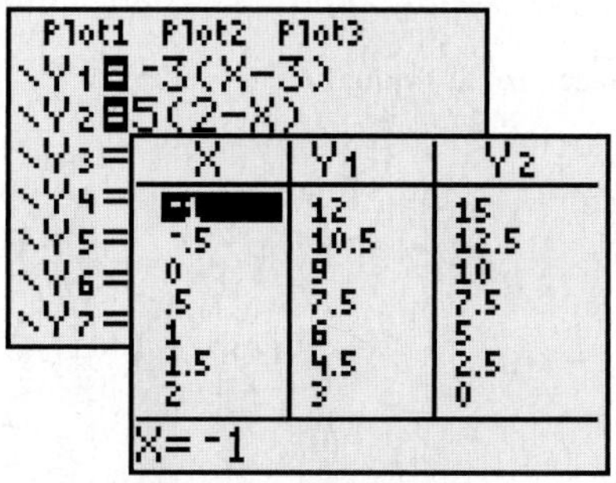

84.

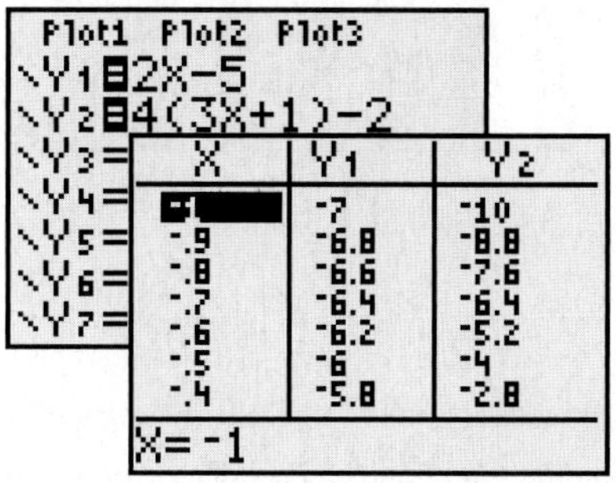

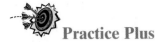

Practice Plus

85. Evaluate $x^2 - x$ for the value of x satisfying $4(x - 2) + 2 = 4x - 2(2 - x)$.

86. Evaluate $x^2 - x$ for the value of x satisfying $2(x - 6) = 3x + 2(2x - 1)$.

87. Evaluate $x^2 - (xy - y)$ for x satisfying $\frac{3(x + 3)}{5} = 2x + 6$ and y satisfying $-2y - 10 = 5y + 18$.

88. Evaluate $x^2 - (xy - y)$ for x satisfying $\frac{13x - 6}{4} = 5x + 2$ and y satisfying $5 - y = 7(y + 4) + 1$.

In Exercises 89–96, solve each equation.

89. $[(3 + 6)^2 \div 3] \cdot 4 = -54x$

90. $2^3 - [4(5 - 3)^3] = -8x$

91. $5 - 12x = 8 - 7x - [6 \div 3(2 + 5^3) + 5x]$

92. $2(5x + 58) = 10x + 4(21 \div 3.5 - 11)$

93. $0.7x + 0.4(20) = 0.5(x + 20)$

94. $0.5(x + 2) = 0.1 + 3(0.1x + 0.3)$

95. $4x + 13 - \{2x - [4(x - 3) - 5]\} = 2(x - 6)$

96. $-2\{7 - [4 - 2(1 - x) + 3]\} = 10 - [4x - 2(x - 3)]$

Application Exercises

The bar graph shows the average cost of tuition and fees at public four-year colleges in the United States. The data can be modeled by the formula

$$T = 165x + 2771,$$

where T represents the average cost of tuition and fees for the school year ending x years after 1996. Use the formula to solve Exercises 97–98.

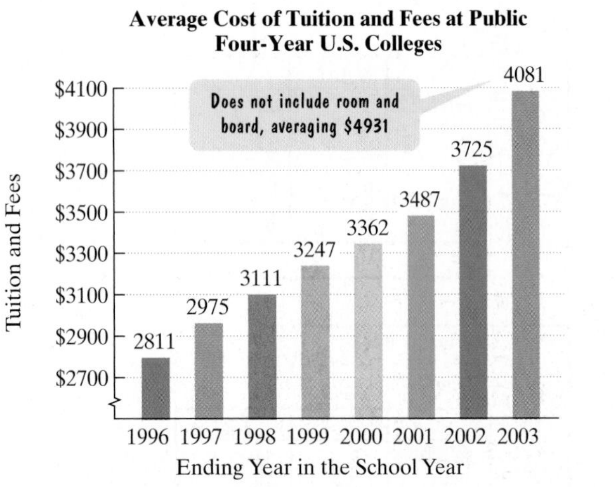

Source: The College Board

97. When will tuition and fees at public U.S. colleges average $4421?

98. When will tuition and fees at public U.S. colleges average $4751?

In the section opener, we used two formulas to model the level of depression, D, in response to the intensity of a negative life event, x, from 1, low, to 10, high:

Low-Humor Group: $D = \frac{10}{9}x + \frac{53}{9}$ High-Humor Group: $D = \frac{1}{9}x + \frac{26}{9}$.

Use these formulas to solve Exercises 99–100.

99. If the high-humor group averages a level of depression of 3.5, or $\frac{7}{2}$, in response to a negative life event, what is the intensity of that event? How is the solution shown on the line graph in Figure 1.13 on page 94?

100. If the low-humor group averages a level of depression of 10 in response to a negative life event, what is the intensity of that event? How is the solution shown on the line graph in Figure 1.13 on page 94?

Two formulas that approximate the dosage of a drug prescribed for children are

$$\text{Young's rule: } C = \frac{DA}{A + 12}$$

$$\text{and Cowling's rule: } C = \frac{D(A + 1)}{24}.$$

In each formula, A = the child's age, in years, D = an adult dosage, and C = the proper child's dosage. The formulas apply for ages 2 through 13, inclusive. Use the formulas to solve Exercises 101–102.

101. When the adult dosage is 1000 milligrams, a child is given 500 milligrams. Using Young's rule, what is that child's age?

102. When the adult dosage is 1000 milligrams, a child is given 300 milligrams. Using Young's rule, what is that child's age? Round to the nearest year.

The graphs illustrate Young's rule and Cowling's rule when the dosage of a drug prescribed for an adult is 1000 milligrams. Use the graphs to solve Exercises 103–106.

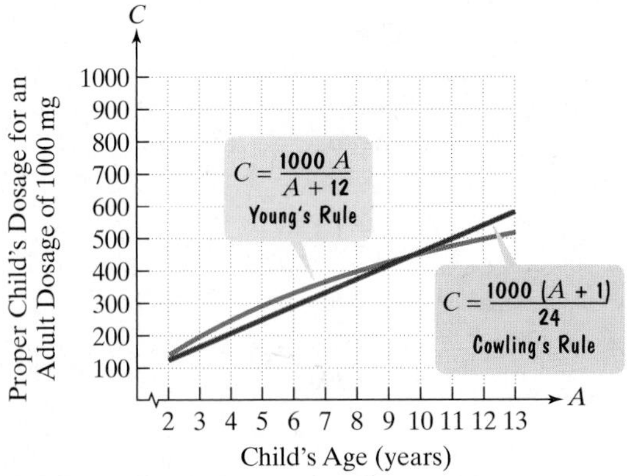

103. Identify your solution to Exercise 101 as a point on the appropriate graph.

104. Identify your solution to Exercise 102 as a point on the appropriate graph.

105. Does either formula consistently give a smaller dosage than the other? If so, which one?

106. Is there an age at which the dosage given by one formula becomes greater than the dosage given by the other? If so, what is a reasonable estimate of that age?

Formulas with rational expressions are often used to model learning. Many of these formulas model the proportion of correct responses in terms of the number of trials of a particular task. One such model, called a learning curve, is

$$P = \frac{0.9x - 0.4}{0.9x + 0.1},$$

where P is the proportion of correct responses after x trials. If P = 0, there are no correct responses. If P = 1, all responses are correct. The graph of the rational formula is shown. Use the formula to solve Exercises 107–108.

A Learning Curve

$P = \dfrac{0.9x - 0.4}{0.9x + 0.1}$

Number of Learning Trials

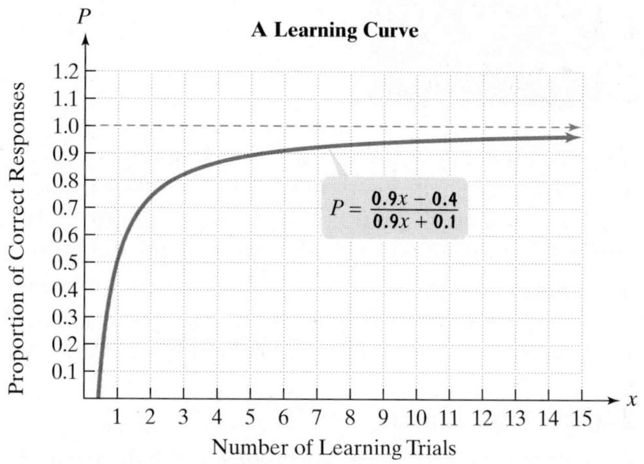

107. How many learning trials are necessary for 0.95 of the responses to be correct? Identify your solution as a point on the graph.

108. How many learning trials are necessary for 0.5 of the responses to be correct? Identify your solution as a point on the graph.

109. A company wants to increase the 10% peroxide content of its product by adding pure peroxide (100% peroxide). If *x* liters of pure peroxide are added to 500 liters of its 10% solution, the concentration, *C*, of the new mixture is given by

$$C = \frac{x + 0.1(500)}{x + 500}.$$

How many liters of pure peroxide should be added to produce a new product that is 28% peroxide?

110. Suppose that *x* liters of pure acid are added to 200 liters of a 35% acid solution.

  a. Write a formula that gives the concentration, *C*, of the new mixture. (*Hint*: See Exercise 109.)

  b. How many liters of pure acid should be added to produce a new mixture that is 74% acid?

Writing in Mathematics

111. What is a linear equation in one variable? Give an example of this type of equation.

112. Suppose that you solve $\dfrac{x}{5} - \dfrac{x}{2} = 1$ by multiplying both sides by 20, rather than the least common denominator of 5 and 2 (namely, 10). Describe what happens. If you get the correct solution, why do you think we clear the equation of fractions by multiplying by the *least* common denominator?

113. Suppose you are an algebra teacher grading the following solution on an examination:

$$-3(x - 6) = 2 - x$$
$$-3x - 18 = 2 - x$$
$$-2x - 18 = 2$$
$$-2x = -16$$
$$x = 8.$$

You should note that 8 checks, and the solution set is {8}. The student who worked the problem therefore wants full credit. Can you find any errors in the solution? If full credit is 10 points, how many points should you give the student? Justify your position.

114. Explain how to find restrictions on the variable in a rational equation.

115. Why should restrictions on the variable in a rational equation be listed before you begin solving the equation?

116. What is an identity? Give an example.

117. What is a conditional equation? Give an example.

118. What is an inconsistent equation? Give an example.

119. Describe the trend shown by the graph in Exercises 107–108 in terms of learning new tasks. What happens initially and what happens as time increases?

Technology Exercises

In Exercises 120–123, use your graphing utility to enter each side of the equation separately under y_1 and y_2. Then use the utility's TABLE *or* GRAPH *feature to solve the equation.*

120. $5x + 2(x - 1) = 3x + 10$

121. $2x + 3(x - 4) = 4x - 7$

122. $\dfrac{x - 3}{5} - 1 = \dfrac{x - 5}{4}$

123. $\dfrac{2x - 1}{3} - \dfrac{x - 5}{6} = \dfrac{x - 3}{4}$

Critical Thinking Exercises

124. Which one of the following is true?

  a. The equation $-7x = x$ has no solution.

  b. The equations $\dfrac{x}{x - 4} = \dfrac{4}{x - 4}$ and $x = 4$ are equivalent.

  c. The equations $3y - 1 = 11$ and $3y - 7 = 5$ are equivalent.

  d. If *a* and *b* are any real numbers, then $ax + b = 0$ always has one number in its solution set.

125. If *x* represents a number, write an English sentence about the number that results in an inconsistent equation.

126. Find *b* such that $\dfrac{7x + 4}{b} + 13 = x$ has a solution set given by {−6}.

127. Find *b* such that $\dfrac{4x - b}{x - 5} = 3$ has a solution set given by Ø.

SECTION 1.3 Models and Applications

Objectives

1. Use linear equations to solve problems.
2. Solve a formula for a variable.

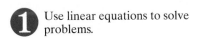

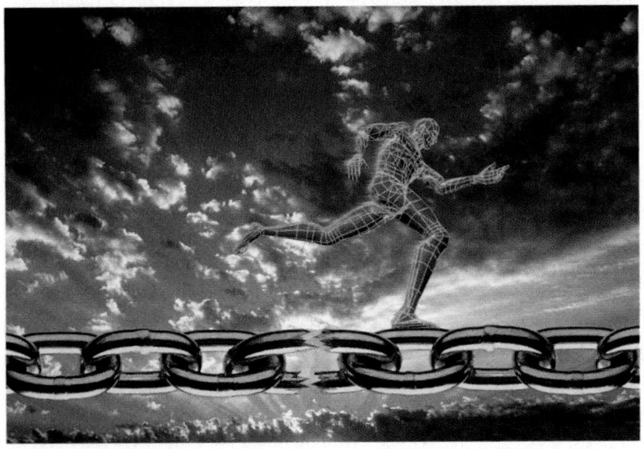

The human race is undeniably becoming a faster race. Since the beginning of the past century, track-and-field records have fallen in everything from sprints to miles to marathons. The performance arc is clearly rising, but no one knows how much higher it can climb. At some point, even the best-trained body simply has to up and quit. The question is, just where is that point, and is it possible for athletes, trainers, and genetic engineers to push it higher? In this section, you will learn a problem-solving strategy that uses linear equations to determine if anyone will ever run a 3-minute mile.

1 Use linear equations to solve problems.

Problem Solving with Linear Equations

We have seen that a model is a mathematical representation of a real-world situation. In this section, we will be solving problems that are presented in English. This means that we must obtain models by translating from the ordinary language of English into the language of algebraic equations. To translate, however, we must understand the English prose and be familiar with the forms of algebraic language. Here are some general steps we will follow in solving word problems:

Study Tip

When solving word problems, particularly problems involving geometric figures, drawing a picture of the situation is often helpful. Label x on your drawing and, where appropriate, label other parts of the drawing in terms of x.

Strategy for Solving Word Problems

Step 1 Read the problem carefully. Attempt to state the problem in your own words and state what the problem is looking for. Let x (or any variable) represent one of the quantities in the problem.

Step 2 If necessary, write expressions for any other unknown quantities in the problem in terms of x.

Step 3 Write an equation in x that models the verbal conditions of the problem.

Step 4 Solve the equation and answer the problem's question.

Step 5 Check the solution *in the original wording* of the problem, not in the equation obtained from the words.

EXAMPLE 1 Walk It Off

Experts concerned with fitness and health suggest that we should walk 10,000 steps per day, about 5 miles. Depending on stride length, each mile ranges between 2000 and 2500 steps. The graph in Figure 1.14 at the top of the next page shows the number of steps it takes to burn off various foods. (The data are based on a body weight of 150 to 165 pounds.)

The number of steps needed to burn off a cheeseburger exceeds the number needed to burn off a 12-ounce soda by 4140. The number needed to burn off a doughnut exceeds the number needed to burn off a 12-ounce soda by 2300. If you chow down a cheeseburger, doughnut, and 12-ounce soda, a 16,790-step walk is needed to burn off the calories (and perhaps alleviate the guilt). Determine the number of steps it takes to burn off a cheeseburger, a doughnut, and a 12-ounce soda.

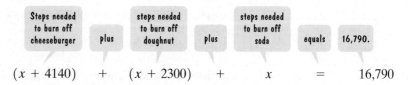

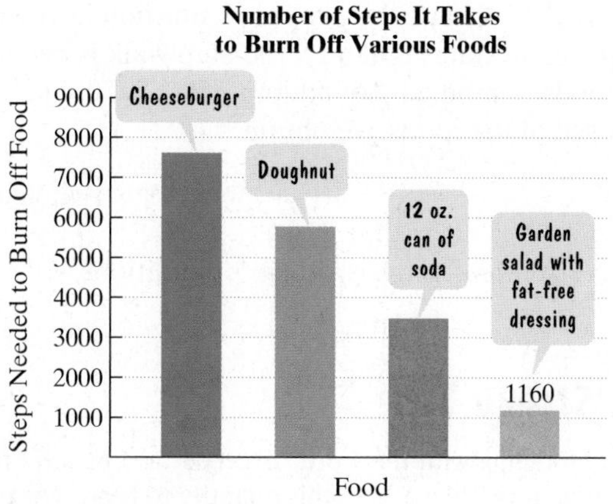

Figure 1.14
Source: The Step Diet Book

Solution

Step 1 Let x represent one of the quantities. We know something about the number of steps needed to burn off a cheeseburger and a doughnut: The numbers exceed that of a 12-ounce soda by 4140 and 2300, respectively. We will let

$x =$ the number of steps needed to burn off a 12-ounce soda.

Step 2 Represent other quantities in terms of x. Because the number of steps needed to burn off a cheeseburger exceeds the number needed to burn off a 12-ounce soda by 4140, let

$x + 4140 =$ the number of steps needed to burn off a cheeseburger.

Because the number of steps needed to burn off a doughnut exceeds the number needed to burn off a 12-ounce soda by 2300, let

$x + 2300 =$ the number of steps needed to burn off a doughnut.

Step 3 Write an equation in x that models the conditions. A 16,790-step walk is needed to burn off a "meal" consisting of a cheeseburger, a doughnut, and a 12-ounce soda.

Steps needed to burn off cheeseburger | plus | steps needed to burn off doughnut | plus | steps needed to burn off soda | equals | 16,790.

$$(x + 4140) \quad + \quad (x + 2300) \quad + \quad x \quad = \quad 16{,}790$$

Step 4 Solve the equation and answer the question.

$$(x + 4140) + (x + 2300) + x = 16{,}790$$ This is the equation that models the problem's conditions.

$$3x + 6440 = 16{,}790$$ Remove parentheses, regroup, and combine like terms.

$$3x = 10{,}350$$ Subtract 6440 from both sides.

$$x = 3450$$ Divide both sides by 3.

Thus,

the number of steps needed to burn off a 12-ounce soda $= x = 3450$.

the number of steps needed to burn off a cheeseburger
$= x + 4140 = 3450 + 4140 = 7590$.

the number of steps needed to burn off a doughnut
$= x + 2300 = 3450 + 2300 = 5750$.

It takes 7590 steps to burn off a cheeseburger, 5750 steps to burn off a doughnut, and 3450 steps to burn off a 12-ounce soda.

Step 5 Check the proposed solution in the original wording of the problem. The problem states that a 16,790-step walk is needed to burn off the calories in the three foods combined. By adding 7590, 5750, and 3450, the numbers that we found for each of the foods, we obtain

$$7590 + 5750 + 3450 = 16{,}790,$$

as specified by the problem's conditions.

Study Tip

Modeling with the word "exceeds" can be a bit tricky. It's helpful to identify the smaller quantity. Then add to this quantity to represent the larger quantity. For example, suppose that Tim's height exceeds Tom's height by a inches. Tom is the shorter person. If Tom's height is represented by x, then Tim's height is represented by $x + a$.

Check Point 1 Basketball, bicycle riding, and football are the three sports and recreational activities in the United States with the greatest number of medically treated injuries. In 2004, the number of injuries from basketball exceeded those from football by 0.6 million. The number of injuries from bicycling exceeded those from football by 0.3 million. Combined, basketball, bicycling, and football accounted for 3.9 million injuries. Determine the number of medically treated injuries from each of these recreational activities in 2004.

(*Source:* U.S. Consumer Product Safety Commission)

Mile Records

1886	4:12.3	1958	3:54.5
1923	4:10.4	1966	3:51.3
1933	4:07.6	1979	3:48.9
1945	4:01.3	1985	3:46.3
1954	3:59.4	1999	3:43.1

Source: U.S.A. Track and Field

EXAMPLE 2 Will Anyone Ever Run a Three-Minute Mile?

One yardstick for measuring how steadily—if slowly—athletic performance has improved is the mile run. In 1923, the record for the mile was a comparatively sleepy 4 minutes, 10.4 seconds. In 1954, Roger Bannister of Britain cracked the 4-minute mark, coming in at 3 minutes, 59.4 seconds. In the half-century since, about 0.3 second per year has been shaved off Bannister's record. If this trend continues, by which year will someone run a 3-minute mile?

Solution In solving this problem, we will express time for the mile run in seconds. Our interest is in a time of 3 minutes, or 180 seconds.

Step 1 Let x represent one of the quantities. Here is the critical information in the problem:

- In 1954, the record was 3 minutes, 59.4 seconds, or 239.4 seconds.
- The record has decreased by 0.3 second per year since then.

We are interested in when the record will be 180 seconds. Let

x = the number of years after 1954 when someone will run a 3-minute mile.

Step 2 Represent other quantities in terms of x. There are no other unknown quantities to find, so we can skip this step.

Step 3 Write an equation in x that models the conditions.

The 1954 record time | decreased by | 0.3 second per year for x years | equals | the 3-minute, or 180-second, mile.

$$239.4 \quad - \quad 0.3x \quad = \quad 180$$

A Poky Species

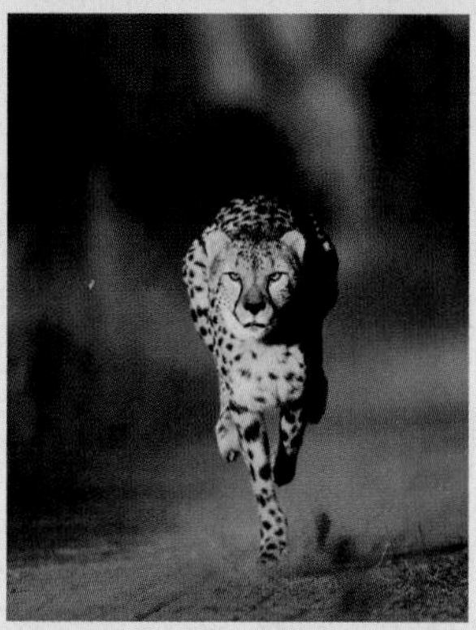

For a species that prides itself on its athletic prowess, human beings are a pretty poky group. Lions can sprint at up to 50 miles per hour; cheetahs move even faster, flooring it to a sizzling 70 miles per hour. But most humans—with our willowy spines and awkward, upright gait—would have trouble cracking 20 miles per hour with a tail wind, a flat track, and a good pair of running shoes.

Step 4 Solve the equation and answer the question.

$$239.4 - 0.3x = 180$$ This is the equation that models the problem's conditions.

$$239.4 - 239.4 - 0.3x = 180 - 239.4$$ Subtract 239.4 from both sides.

$$-0.3x = -59.4$$ Simplify.

$$\frac{-0.3x}{-0.3} = \frac{-59.4}{-0.3}$$ Divide both sides by -0.3.

$$x = 198$$ Simplify.

Using current trends, by 198 years (gasp!) after 1954, or in 2152, someone will run a 3-minute mile.

Step 5 Check the proposed solution in the original wording of the problem. The problem states that the record time should be 180 seconds. Do we obtain 180 seconds if we decrease the 1954 record time, 239.4 seconds, by 0.3 second per year for 198 years, our proposed solution?

$$239.4 - 0.3(198) = 239.4 - 59.4 = 180$$

This verifies that, using current trends, the 3-minute mile will be run 198 years after 1954.

Check Point 2 Got organic milk? Although organic milk accounts for only 1.2% of the market, consumption is increasing. In 2004, Americans purchased 40.7 million gallons of organic milk, increasing at a rate of 5.6 million gallons per year. If this trend continues, when will Americans purchase 79.9 million gallons of organic milk?

(*Source*: National Dairy Council)

EXAMPLE 3 Selecting a Long-Distance Carrier

You are choosing between two long-distance telephone plans. Plan A has a monthly fee of \$20 with a charge of \$0.05 per minute for all long-distance calls. Plan B has a monthly fee of \$5 with a charge of \$0.10 per minute for all long-distance calls. For how many minutes of long-distance calls will the costs for the two plans be the same?

Solution

Step 1 Let x represent one of the quantities. Let

x = the number of minutes of long-distance calls for which the two plans cost the same.

Step 2 Represent other quantities in terms of x. There are no other unknown quantities, so we can skip this step.

Step 3 Write an equation in x that models the conditions. The monthly cost for plan A is the monthly fee, \$20, plus the per minute charge, \$0.05, times the number of minutes of long-distance calls, x. The monthly cost for plan B is the monthly fee, \$5, plus the per-minute charge, \$0.10, times the number of minutes of long-distance calls, x.

The monthly cost for plan A	must equal	the monthly cost for plan B.
$20 + 0.05x$	$=$	$5 + 0.10x$

Step 4   Solve the equation and answer the question.

$$20 + 0.05x = 5 + 0.10x$$ This is the equation that models the problem's conditions.

$$20 = 5 + 0.05x$$ Subtract 0.05x from both sides.

$$15 = 0.05x$$ Subtract 5 from both sides.

$$\frac{15}{0.05} = \frac{0.05x}{0.05}$$ Divide both sides by 0.05.

$$300 = x$$ Simplify.

Because x represents the number of minutes of long-distance calls for which the two plans cost the same, the costs will be the same for 300 minutes of long-distance calls.

Step 5   Check the proposed solution in the original wording of the problem. The problem states that the costs for the two plans should be the same. Let's see if they are the same with 300 minutes of long-distance calls:

$$\text{Cost for plan A} = \$20 + \$0.05(300) = \$20 + \$15 = \$35$$

Monthly fee   Per-minute charge

$$\text{Cost for plan B} = \$5 + \$0.10(300) = \$5 + \$30 = \$35.$$

With 300 minutes, or 5 hours, of long-distance chatting, both plans cost $35 for the month. Thus, the proposed solution, 300 minutes, satisfies the problem's conditions.

Technology

We can use a graphing utility to numerically or graphically verify our work in Example 3.

The monthly cost for plan A   must equal   the monthly cost for plan B.

$$20 + 0.05x = 5 + 0.10x$$

Enter $y_1 = 20 + .05x$.   Enter $y_2 = 5 + .10x$.

Numeric Check

Display a table for y_1 and y_2.

When $x = 300$, y_1 and y_2 have the same value, 35. With 300 minutes of calls, costs are the same, $35, for both plans.

Graphic Check

Display graphs for y_1 and y_2. Use the intersection feature.

$y_1 = 20 + 0.05x$

$y_2 = 5 + 0.10x$

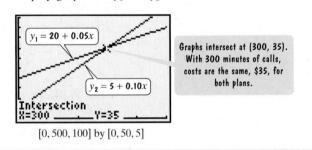

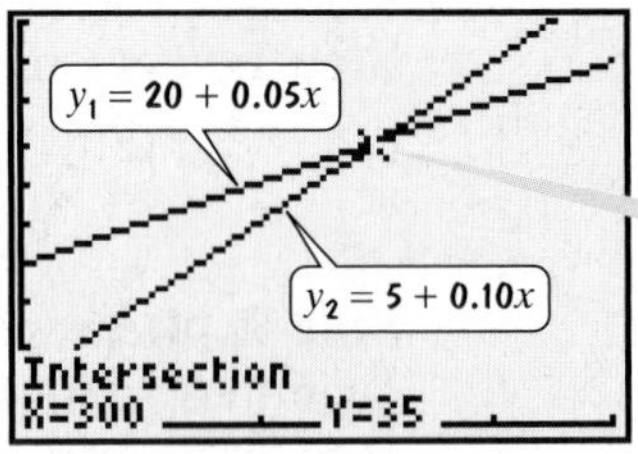

Graphs intersect at (300, 35). With 300 minutes of calls, costs are the same, $35, for both plans.

[0, 500, 100] by [0, 50, 5]

 Check Point 3   You are choosing between two long-distance telephone plans. Plan A has a monthly fee of $15 with a charge of $0.08 per minute for all long-distance calls. Plan B has a monthly fee of $3 with a charge of $0.12 per minute for all long-distance calls. For how many minutes of long-distance calls will the costs for the two plans be the same?

EXAMPLE 4 Education Pays Off

The graph in Figure 1.15 shows that for the period from 1982 through 2002, those with the most education had the fastest growth in wages. In 2002, the median annual income for people with a college degree was $52,000. This is a 160% increase over the median income in 1982. What were people with a college degree earning in 1982?

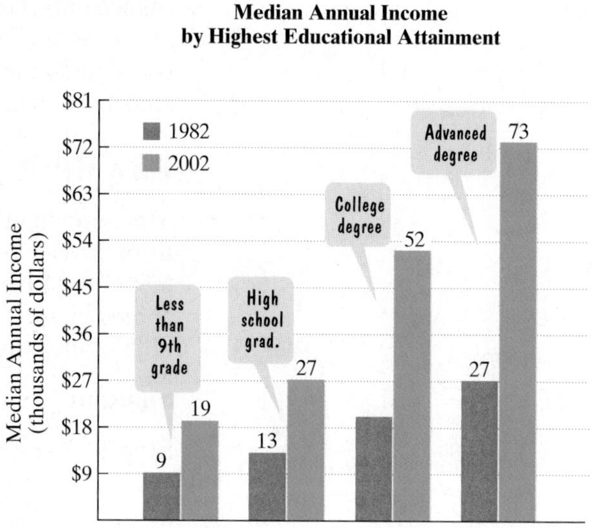

Figure 1.15

Source: U.S. Census Bureau

Solution

Step 1 Let x represent one of the quantities. We will let

x = the median income of people with a college degree in 1982.

Step 2 Represent other quantities in terms of x. There are no other unknown quantities to find, so we can skip this step.

Step 3 Write an equation in x that models the conditions. The median income in 1982 plus the 160% increase is the median income in 2002, $52,000.

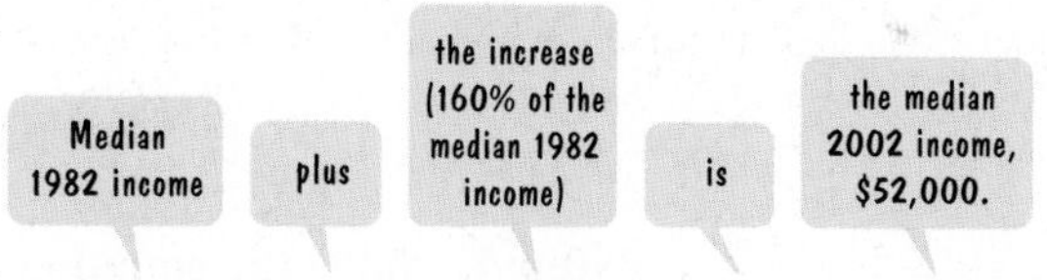

$$x + 1.6x = 52{,}000$$

Study Tip

Observe that the 1982 income, x, increased by 160% is $x + 1.6x$ and *not* $x + 1.6$ or $x + 160$.

Step 4 Solve the equation and answer the question.

$$x + 1.6x = 52{,}000$$ This is the equation that models the problem's conditions.

$$2.6x = 52{,}000$$ Combine like terms: $x + 1.6x = 1x + 1.6x = 2.6x$.

$$\frac{2.6x}{2.6} = \frac{52{,}000}{2.6}$$ Divide both sides by 2.6.

$$x = 20{,}000$$

In 1982, people with a college degree were earning $20,000.

Step 5 Check the proposed solution in the original wording of the problem. The 1982 income, $20,000, plus the 160% increase should equal the 2002 income given in the original wording, $52,000:

$$20{,}000 + 160\% \text{ of } 20{,}000 = 20{,}000 + 1.6(20{,}000) = 20{,}000 + 32{,}000 = 52{,}000.$$

This verifies that in 1982, college graduates were earning $20,000.

Check Point 4 After a 30% price reduction, you purchase a new computer for $840. What was the computer's price before the reduction?

Our next example is about simple interest. Simple interest involves interest calculated only on the amount of money that we invest, called the **principal**. The formula $I = Pr$ is used to find the simple interest, I, earned for one year when the principal, P, is invested at an annual interest rate, r. Dual investment problems involve different amounts of money in two or more investments, each paying a different rate.

EXAMPLE 5 Solving a Dual Investment Problem

Your grandmother needs your help. She has \$50,000 to invest. Part of this money is to be invested in noninsured bonds paying 15% annual interest. The rest of this money is to be invested in a government-insured certificate of deposit paying 7% annual interest. She told you that she requires \$6000 per year in extra income from both of these investments. How much money should be placed in each investment?

Solution

Step 1 Let x represent one of the quantities. We will let

$x =$ the amount invested in the noninsured bonds at 15%.

Step 2 Represent other quantities in terms of x. The other quantity that we seek is the amount invested at 7% in the certificate of deposit. Because the total amount Grandma has to invest is \$50,000 and we already used up x,

$50{,}000 - x =$ the amount invested in the certificate of deposit at 7%.

Step 3 Write an equation in x that models the conditions. Because Grandma requires \$6000 in total interest, the interest for the two investments combined must be \$6000. Interest is Pr or rP for each investment.

Interest from the 15% investment	plus	interest from the 7% investment	is	\$6000.
$0.15x$	$+$	$0.07(50{,}000 - x)$	$=$	6000
rate times principal		rate times principal		

Step 4 Solve the equation and answer the question.

$0.15x + 0.07(50{,}000 - x) = 6000$	This is the equation that models the problem's conditions.
$0.15x + 3500 - 0.07x = 6000$	Use the distributive property.
$0.08x + 3500 = 6000$	Combine like terms.
$0.08x = 2500$	Subtract 3500 from both sides.
$\dfrac{0.08x}{0.08} = \dfrac{2500}{0.08}$	Divide both sides by 0.08.
$x = 31{,}250$	Simplify.

Thus,

the amount invested at 15% $= x = 31{,}250$.

the amount invested at 7% $= 50{,}000 - 31{,}250 = 18{,}750$.

Grandma should invest \$31,250 at 15% and \$18,750 at 7%.

Step 5 Check the proposed solution in the original wording of the problem. The problem states that the total interest from the dual investments should be \$6000. Can Grandma count on \$6000 interest? The interest earned on \$31,250 at 15% is (\$31,250) (0.15), or \$4687.50. The interest earned on \$18,750 at 7% is (\$18,750) (0.07),

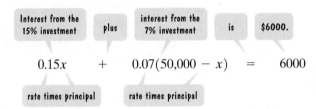

or $1312.50. The total interest is $4687.50 + $1312.50, or $6000, exactly as it should be. You've made your grandmother happy. (Now if you would just visit her more often …)

Check Point **5** You inherited $5000 with the stipulation that for the first year the money had to be invested in two funds paying 9% and 11% annual interest. How much did you invest at each rate if the total interest earned for the year was $487?

Solving geometry problems usually requires a knowledge of basic geometric ideas and formulas. Formulas for area, perimeter, and volume are given in Table 1.2.

Table 1.2   Common Formulas for Area, Perimeter, and Volume

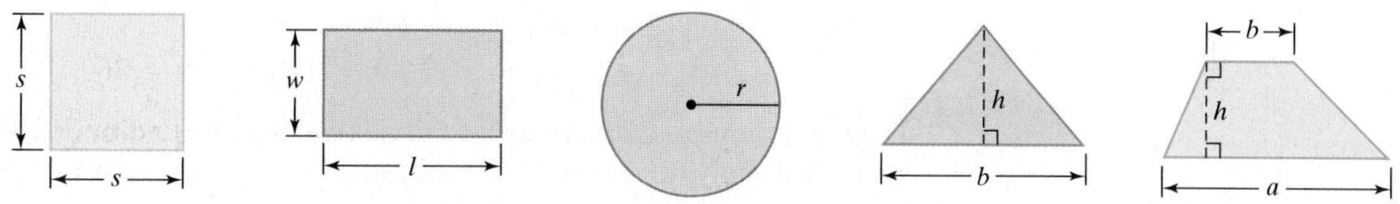

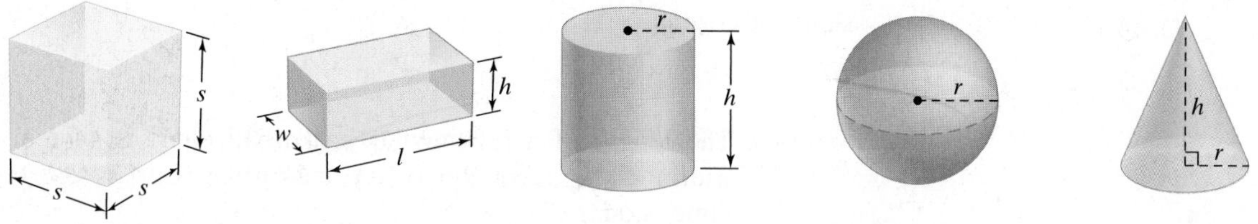

We will be using the formula for the perimeter of a rectangle, $P = 2l + 2w$, in our next example. The formula states that a rectangle's perimeter is the sum of twice its length and twice its width.

EXAMPLE 6   Finding the Dimensions of an American Football Field

The length of an American football field is 200 feet more than the width. If the perimeter of the field is 1040 feet, what are its dimensions?

Solution

Step 1   Let x represent one of the quantities. We know something about the length; the length is 200 feet more than the width. We will let

$$x = \text{the width.}$$

Step 2   Represent other quantities in terms of x. Because the length is 200 feet more than the width, we add 200 to the width to represent the length. Thus,

$$x + 200 = \text{the length.}$$

Figure 1.16 An American football field

Figure 1.16 illustrates an American football field and its dimensions.

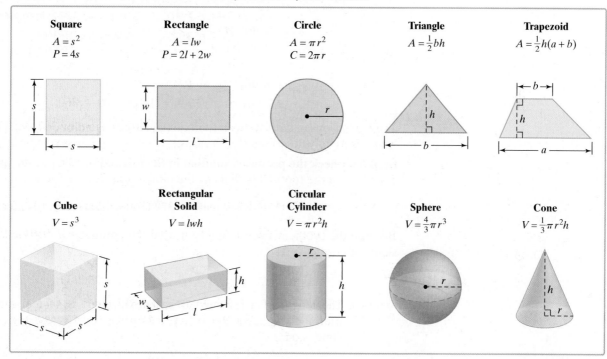

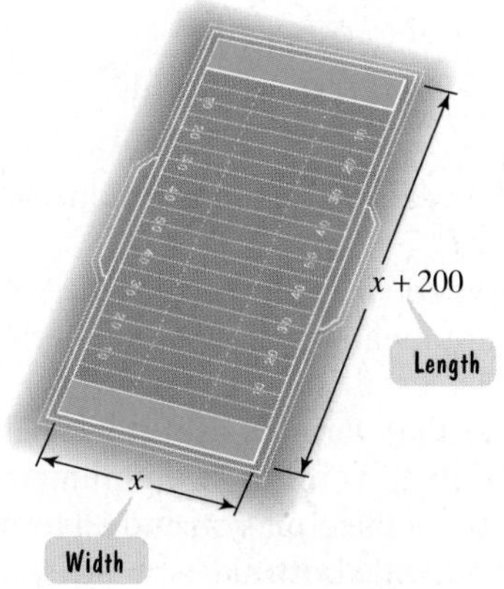

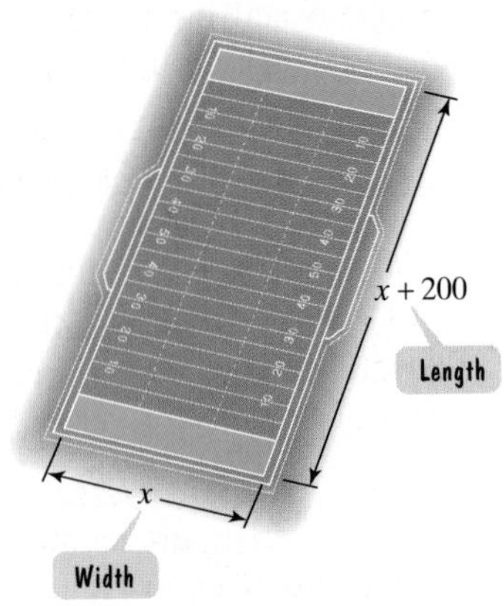

Figure 1.16 (repeated)

Step 3 Write an equation in x that models the conditions. Because the perimeter of the field is 1040 feet,

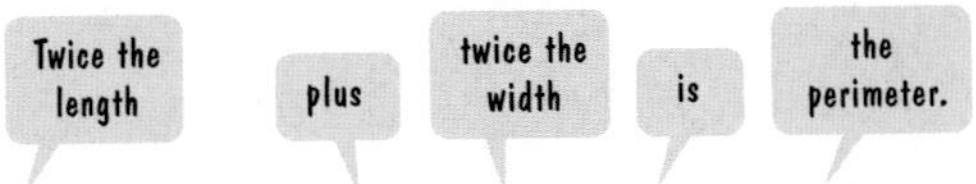

$$2(x + 200) \quad + \quad 2x \quad = \quad 1040.$$

Step 4 Solve the equation and answer the question.

$$2(x + 200) + 2x = 1040 \quad \text{This is the equation that models the problem's conditions.}$$
$$2x + 400 + 2x = 1040 \quad \text{Apply the distributive property.}$$
$$4x + 400 = 1040 \quad \text{Combine like terms: } 2x + 2x = 4x.$$
$$4x = 640 \quad \text{Subtract 400 from both sides.}$$
$$x = 160 \quad \text{Divide both sides by 4.}$$

Thus,

$$\text{width} = x = 160.$$
$$\text{length} = x + 200 = 160 + 200 = 360.$$

The dimensions of an American football field are 160 feet by 360 feet. (The 360-foot length is usually described as 120 yards.)

Step 5 Check the proposed solution in the original wording of the problem. The perimeter of the football field using the dimensions that we found is

$$2(160 \text{ feet}) + 2(360 \text{ feet}) = 320 \text{ feet} + 720 \text{ feet} = 1040 \text{ feet}.$$

Because the problem's wording tells us that the perimeter is 1040 feet, our dimensions are correct.

Check Point 6 The length of a rectangular basketball court is 44 feet more than the width. If the perimeter of the basketball court is 288 feet, what are its dimensions?

2 Solve a formula for a variable.

Solving a Formula for One of Its Variables

We know that solving an equation is the process of finding the number (or numbers) that make the equation a true statement. All of the equations we have solved contained only one letter, x.

By contrast, formulas contain two or more letters, representing two or more variables. An example is the formula for the perimeter of a rectangle:

$$2l + 2w = P.$$

We say that this formula is solved for the variable P because P is alone on one side of the equation and the other side does not contain a P.

Solving a formula for a variable means rewriting the formula so that the variable is isolated on one side of the equation. It does not mean obtaining a numerical value for that variable.

To solve a formula for one of its variables, treat that variable as if it were the only variable in the equation. Think of the other variables as if they were numbers. Isolate all terms with the specified variable on one side of the equation and all terms without the specified variable on the other side. Then divide both sides by the same nonzero quantity to get the specified variable alone. The next example shows how to do this.

EXAMPLE 7 Solving a Formula for a Variable

Solve the formula $2l + 2w = P$ for l.

Solution First, isolate $2l$ on the left by subtracting $2w$ from both sides. Then solve for l by dividing both sides by 2.

We need to isolate l.

$$2l + 2w = P \quad \text{This is the given formula.}$$

$$2l + 2w - 2w = P - 2w \quad \text{Isolate } 2l \text{ by subtracting } 2w \text{ from both sides.}$$

$$2l = P - 2w \quad \text{Simplify.}$$

$$\frac{2l}{2} = \frac{P - 2w}{2} \quad \text{Solve for } l \text{ by dividing both sides by 2.}$$

$$l = \frac{P - 2w}{2} \quad \text{Simplify.}$$

Check Point 7 Solve the formula $2l + 2w = P$ for w.

EXAMPLE 8 Solving a Formula for a Variable That Occurs Twice

The formula

$$A = P + Prt$$

describes the amount, A, that a principal of P dollars is worth after t years when invested at a simple annual interest rate, r. Solve this formula for P.

Solution Notice that all the terms with P already occur on the right side of the formula.

We need to isolate P.

$$A = P + Prt$$

We can factor P from the two terms on the right to convert the two occurrences of P into one.

$$A = P + Prt \quad \text{This is the given formula.}$$

$$A = P(1 + rt) \quad \text{Factor out } P \text{ on the right side of the equation.}$$

$$\frac{A}{1 + rt} = \frac{P(1 + rt)}{1 + rt} \quad \text{Divide both sides by } 1 + rt.$$

$$\frac{A}{1 + rt} = P \quad \text{Simplify: } \frac{P\cancel{(1 + rt)}}{1\cancel{(1 + rt)}} = \frac{P}{1} = P.$$

Equivalently,

$$P = \frac{A}{1 + rt}.$$

Check Point 8 Solve the formula $P = C + MC$ for C.

Study Tip

You cannot solve $A = P + Prt$ for P by subtracting Prt from both sides and writing

$$A - Prt = P.$$

When a formula is solved for a specified variable, that variable must be isolated on one side. The variable P occurs on both sides of

$$A - Prt = P.$$

EXERCISE SET 1.3

Practice Exercises

Use the five-step strategy for solving word problems to find the number or numbers described in Exercises 1–10.

1. When five times a number is decreased by 4, the result is 26. What is the number?
2. When two times a number is decreased by 3, the result is 11. What is the number?
3. When a number is decreased by 20% of itself, the result is 20. What is the number?
4. When a number is decreased by 30% of itself, the result is 28. What is the number?
5. When 60% of a number is added to the number, the result is 192. What is the number?
6. When 80% of a number is added to the number, the result is 252. What is the number?
7. 70% of what number is 224?
8. 70% of what number is 252?
9. One number exceeds another by 26. The sum of the numbers is 64. What are the numbers?
10. One number exceeds another by 24. The sum of the numbers is 58. What are the numbers?

Practice Plus

In Exercises 11–18, find all values of x satisfying the given conditions.

11. $y_1 = 13x - 4$, $y_2 = 5x + 10$, and y_1 exceeds y_2 by 2.
12. $y_1 = 10x + 6$, $y_2 = 12x - 7$, and y_1 exceeds y_2 by 3.
13. $y_1 = 10(2x - 1)$, $y_2 = 2x + 1$, and y_1 is 14 more than 8 times y_2.
14. $y_1 = 9(3x - 5)$, $y_2 = 3x - 1$, and y_1 is 51 less than 12 times y_2.
15. $y_1 = 2x + 6$, $y_2 = x + 8$, $y_3 = x$, and the difference between 3 times y_1 and 5 times y_2 is 22 less than y_3.
16. $y_1 = 2.5$, $y_2 = 2x + 1$, $y_3 = x$, and the difference between 2 times y_1 and 3 times y_2 is 8 less than 4 times y_3.
17. $y_1 = \frac{1}{x}$, $y_2 = \frac{1}{2x}$, $y_3 = \frac{1}{x-1}$, and the sum of 3 times y_1 and 4 times y_2 is the product of 4 and y_3.
18. $y_1 = \frac{1}{x}$, $y_2 = \frac{1}{x^2 - x}$, $y_3 = \frac{1}{x-1}$, and the difference between 6 times y_1 and 3 times y_2 is the product of 7 and y_3.

Application Exercises

19. Each day, the number of births in the world exceeds the number of deaths by 229 thousand. The combined number of births and deaths is 521 thousand. Determine the number of births and the number of deaths per day.

Daily Growth of World Population

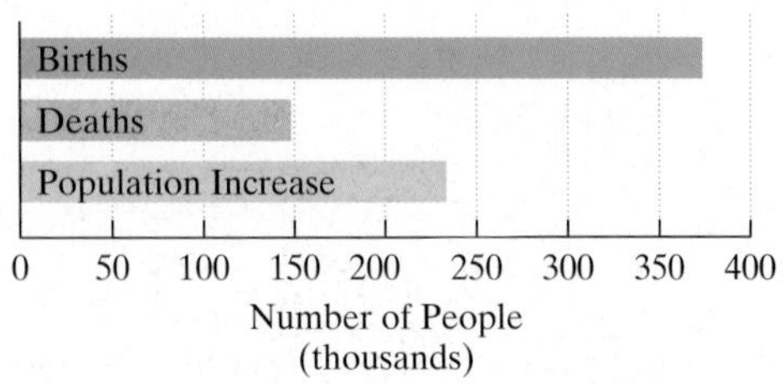

Source: "Population Update" 2000

20. Americans say keep the penny. A survey asked a random sample of U.S. adults if they favored abolishing the penny. The circle graph shows the results.

Do You Favor Abolishing the Penny?

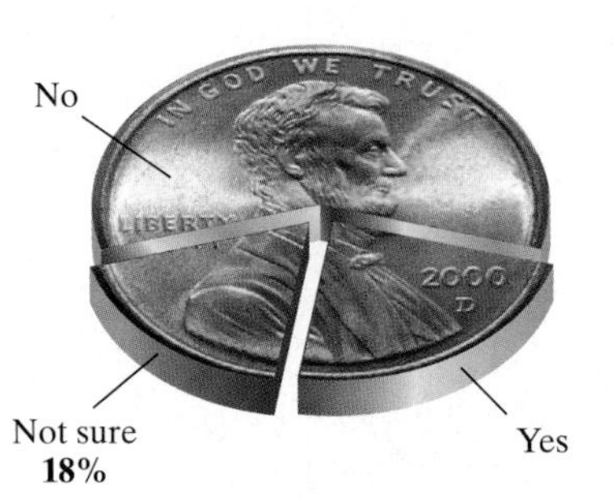

Source: Harris poll of 2136 adults

A total of 82% of those polled responded yes or no. The no responses (keep the penny) exceeded the yes responses (abolish the penny) by 36%. Determine the percentage who responded yes and the percentage who responded no.

21. The bar graph shows the number of Internet users, in millions, for the countries with the most users. The number of Internet users in Japan exceeds China by 10 million and the number of Internet users in the United States exceeds China by 123 million. There are a total of 271 million Internet users in the United States, Japan, and China. Determine the number of users, in millions, in each country.

Countries with the Most Internet Users

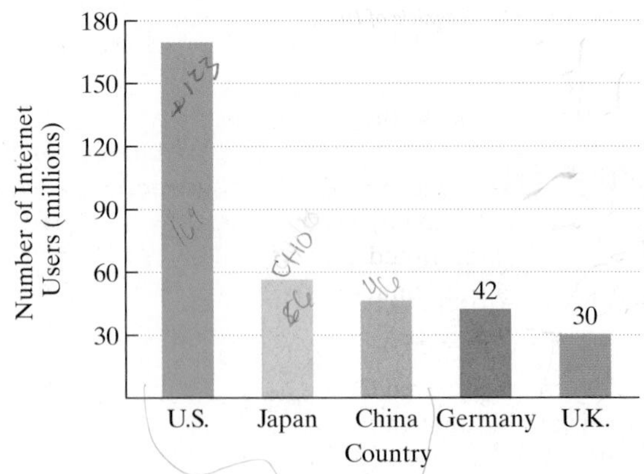

Source: *CyberAtlas*

22. The bar graph at the top of the next page shows the percentage of global energy used by the countries consuming the most energy. The percentage of global energy used by China exceeds Russia by 6% and the percentage of global energy used by the United States exceeds Russia by 16.4%. Combined, the United States, China, and Russia consume 40.4% of the world's energy. Determine the percentage of global energy used by each country.

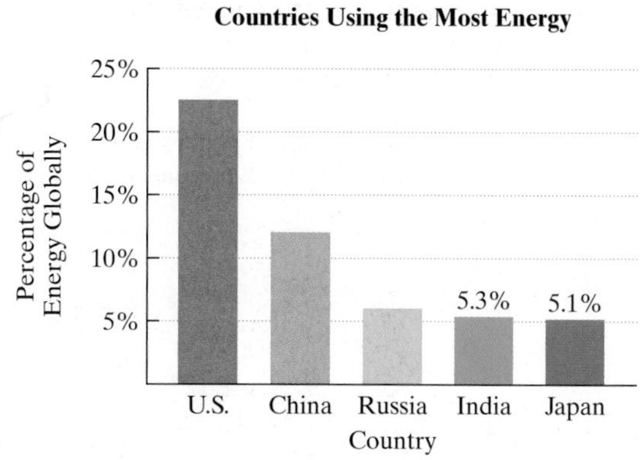

Source: World Bank Group

23. The circle graph shows the political ideology of U.S. college freshmen. The percentage of liberals exceeds twice that of conservatives by 4.4%. Find the percentage of liberals and the percentage of conservatives. (*Hint*: You'll need to use the percents displayed on the graph to determine the combined percentage of liberals and conservatives.)

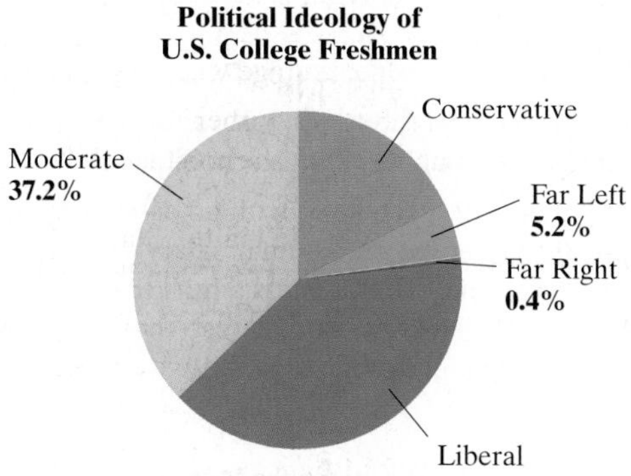

Source: *The Chronicle of Higher Education*

24. The bar graph shows the breakdown of the 7485 hate crimes reported in the United States in 2004. The number of hate crimes based on race exceeded three times the number based on sexual orientation by 127. Find the number of hate crimes reported in the United States in 2004 based on race and based on sexual orientation.

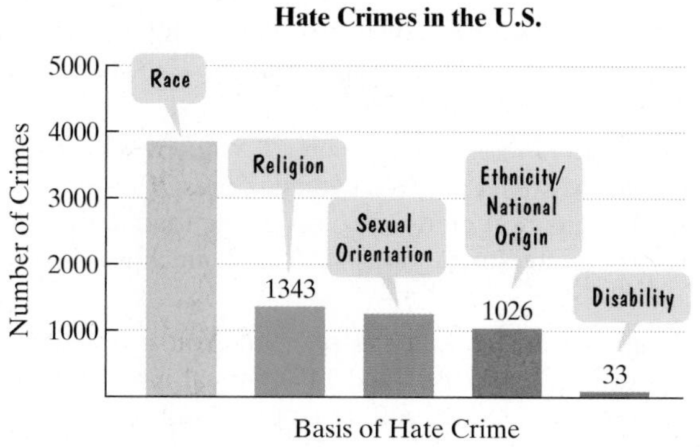

Source: F.B.I.

According to one mathematical model, the average life expectancy for American men born in 1900 was 55 years. Life expectancy has increased by about 0.2 year for each birth year after 1900. Use this information to solve Exercises 25–26.

25. If this trend continues, for which birth year will the average life expectancy be 85 years?

26. If this trend continues, for which birth year will the average life expectancy be 91 years?

The graph shows the number of Americans without health insurance from 2000 through 2003.

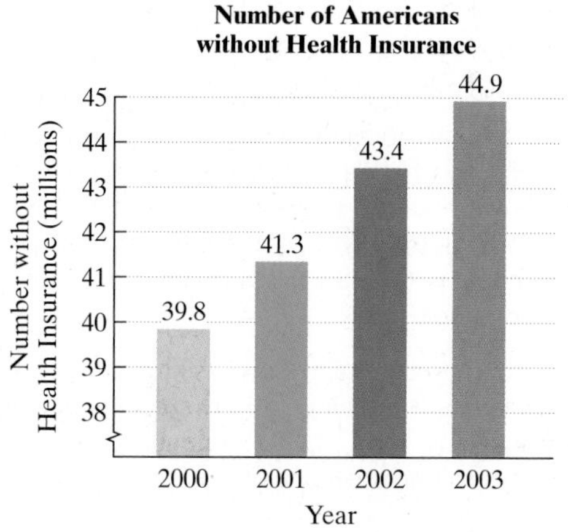

Source: U.S. Census Bureau

In 2000, there were 39.8 million Americans without health insurance. This number has increased at an average rate of 1.7 million people per year. Use this description to solve Exercises 27–28.

27. a. Write a formula that models the number of Americans without health insurance, y, x years after 2000.

b. Use the formula in part (a) to determine when the number of Americans without health insurance will exceed the number in 2003 by 8.5 million.

c. Graph the formula in part (a) and show the solution to part (b) on the graph.

28. a. Write a formula that models the number of Americans without health insurance, y, x years after 2000.

b. Use the formula in part (a) to determine when the number of Americans without health insurance will exceed the number in 2003 by 10.2 million.

c. Graph the formula in part (a) and show the solution to part (b) on the graph.

In 2003, the price of a BMW 7 Series was approximately $80,500 with a depreciation of $8705 per year. Use this information to solve Exercises 29–30.

29. After how many years will the car's value be $19,565?

30. After how many years will the car's value be $36,975?

31. You are choosing between two health clubs. Club A offers membership for a fee of $40 plus a monthly fee of $25. Club B offers membership for a fee of $15 plus a monthly fee of $30. After how many months will the total cost at each health club be the same? What will be the total cost for each club?

32. Video Store A charges $9 to rent a video game for one week. Although only members can rent from the store, membership is free. Video Store B charges only $4 to rent a video game for one week. Only members can rent from the store and membership is $50 per year. After how many video-game rentals will the total amount spent at each store be the same? What will be the total amount spent at each store?

33. The bus fare in a city is $1.25. People who use the bus have the option of purchasing a monthly coupon book for $15.00. With the coupon book, the fare is reduced to $0.75. Determine the number of times in a month the bus must be used so that the total monthly cost without the coupon book is the same as the total monthly cost with the coupon book.

34. A coupon book for a bridge costs $30 per month. The toll for the bridge is normally $5.00, but it is reduced to $3.50 for people who have purchased the coupon book. Determine the number of times in a month the bridge must be crossed so that the total monthly cost without the coupon book is the same as the total monthly cost with the coupon book.

35. In 2005, there were 13,300 students at college A, with a projected enrollment increase of 1000 students per year. In the same year, there were 26,800 students at college B, with a projected enrollment decline of 500 students per year.

a. According to these projections, when will the colleges have the same enrollment? What will be the enrollment in each college at that time?

b. Use the following table to numerically check your work in part (a). What equations were entered for Y_1 and Y_2 to obtain this table?

X	Y_1	Y_2
7	20300	23300
8	21300	22800
9	22300	22300
10	23300	21800
11	24300	21300
12	25300	20800
13	26300	20300

X=7

36. In 2000, the population of Greece was 10,600,000, with projections of a population decrease of 28,000 people per year. In the same year, the population of Belgium was 10,200,000, with projections of a population decrease of 12,000 people per year. (*Source*: United Nations) According to these projections, when will the two countries have the same population? What will be the population at that time?

37. After a 20% reduction, you purchase a television for $336. What was the television's price before the reduction?

38. After a 30% reduction, you purchase a dictionary for $30.80. What was the dictionary's price before the reduction?

39. Including 8% sales tax, an inn charges $162 per night. Find the inn's nightly cost before the tax is added.

40. Including 5% sales tax, an inn charges $252 per night. Find the inn's nightly cost before the tax is added.

The graph shows average yearly earnings in the United States by highest educational attainment. Use the relevant information shown in the graph to solve Exercises 41–42.

Average Earnings of Full-Time Workers in the U.S., by Highest Educational Attainment

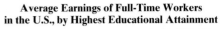

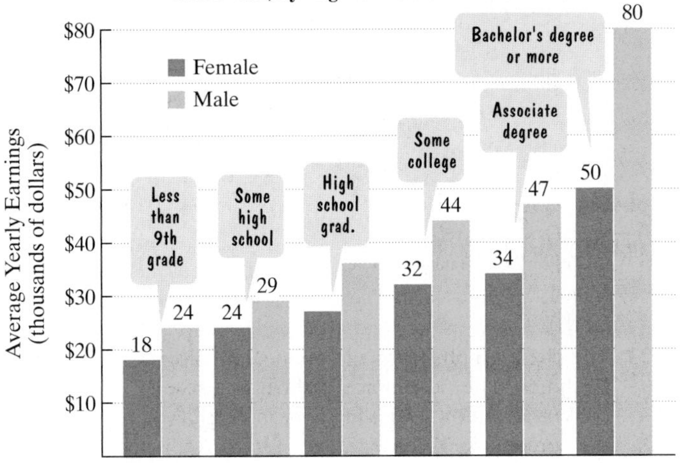

Source: U.S. Census Bureau

41. The annual salary for men with some college is an increase of 22% over the annual salary for men whose highest educational attainment is a high school degree. What is the annual salary, to the nearest thousand dollars, for men whose highest educational attainment is a high school degree?

42. The annual salary for women with an associate degree is an increase of 26% over the annual salary for women whose highest educational attainment is a high school degree. What is the annual salary, to the nearest thousand dollars, for women whose highest educational attainment is a high school degree?

Exercises 43–44 involve markup, the amount added to the dealer's cost of an item to arrive at the selling price of that item.

43. The selling price of a refrigerator is $584. If the markup is 25% of the dealer's cost, what is the dealer's cost of the refrigerator?

44. The selling price of a scientific calculator is $15. If the markup is 25% of the dealer's cost, what is the dealer's cost of the calculator?

45. You invested $7000 in two accounts paying 6% and 8% annual interest, respectively. If the total interest earned for the year was $520, how much was invested at each rate?

46. You invested $11,000 in stocks and bonds, paying 5% and 8% annual interest, respectively. If the total interest earned for the year was $730, how much was invested in stocks and how much was invested in bonds?

47. Things did not go quite as planned. You invested $8000, part of it in stock that paid 12% annual interest. However, the rest of the money suffered a 5% loss. If the total annual income from both investments was $620, how much was invested at each rate?

48. Things did not go quite as planned. You invested $12,000, part of it in stock that paid 14% annual interest. However, the rest of the money suffered a 6% loss. If the total annual income from both investments was $680, how much was invested at each rate?

49. A rectangular soccer field is twice as long as it is wide. If the perimeter of the soccer field is 300 yards, what are its dimensions?

50. A rectangular swimming pool is three times as long as it is wide. If the perimeter of the pool is 320 feet, what are its dimensions?

51. The length of the rectangular tennis court at Wimbledon is 6 feet longer than twice the width. If the court's perimeter is 228 feet, what are the court's dimensions?

52. The length of a rectangular pool is 6 meters less than twice the width. If the pool's perimeter is 126 meters, what are its dimensions?

53. The rectangular painting in the figure shown measures 12 inches by 16 inches and contains a frame of uniform width around the four edges. The perimeter of the rectangle formed by the painting and its frame is 72 inches. Determine the width of the frame.

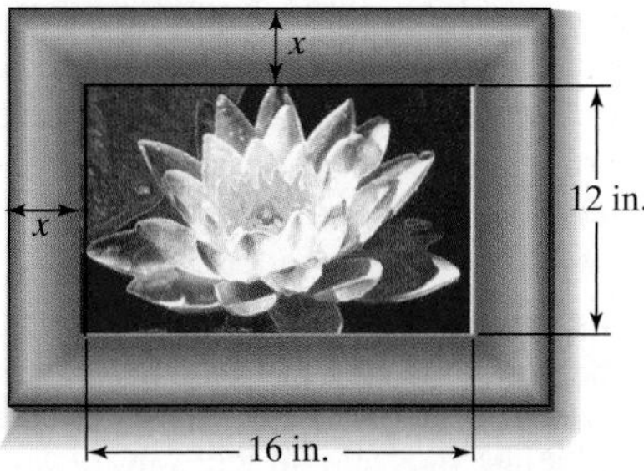

54. The rectangular swimming pool in the figure shown measures 40 feet by 60 feet and contains a path of uniform width around the four edges. The perimeter of the rectangle formed by the pool and the surrounding path is 248 feet. Determine the width of the path.

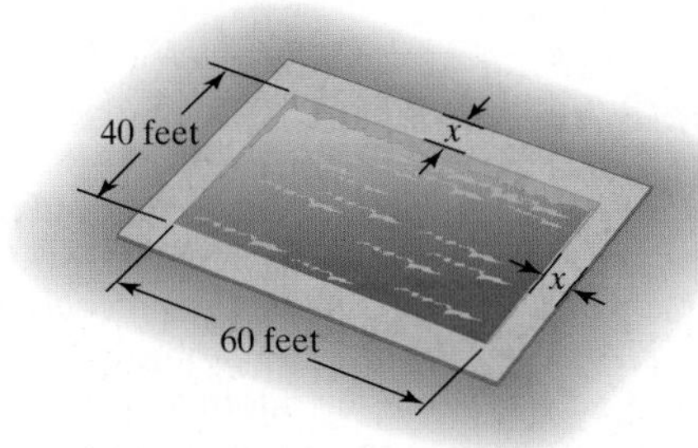

55. An automobile repair shop charged a customer \$448, listing \$63 for parts and the remainder for labor. If the cost of labor is \$35 per hour, how many hours of labor did it take to repair the car?

56. A repair bill on a sailboat came to \$1603, including \$532 for parts and the remainder for labor. If the cost of labor is \$63 per hour, how many hours of labor did it take to repair the sailboat?

57. An HMO pamphlet contains the following recommended weight for women: "Give yourself 100 pounds for the first 5 feet plus 5 pounds for every inch over 5 feet tall." Using this description, what height corresponds to a recommended weight of 135 pounds?

58. A job pays an annual salary of \$33,150, which includes a holiday bonus of \$750. If paychecks are issued twice a month, what is the gross amount for each paycheck?

59. Answer the question in the following *Peanuts* cartoon strip. (*Note:* You may not use the answer given in the cartoon!)

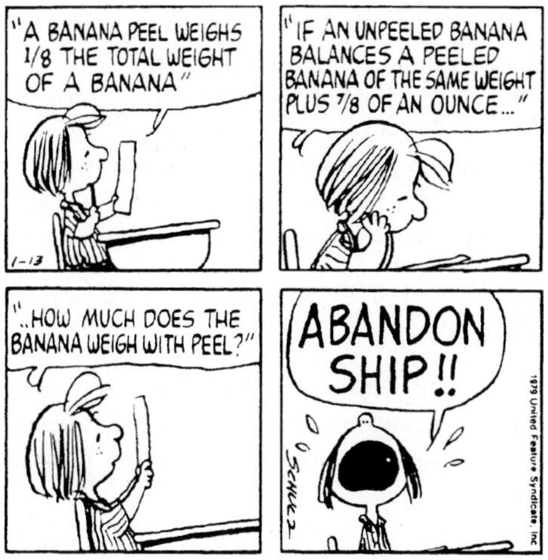

PEANUTS reprinted by permission of United Features Syndicate, Inc.

60. For a long-distance person-to-person telephone call, a telephone company charges \$0.43 for the first minute, \$0.32 for each additional minute, and a \$2.10 service charge. If the cost of a call is \$5.73, how long did the person talk?

In Exercises 61–80, solve each formula for the specified variable. Do you recognize the formula? If so, what does it describe?

61. $A = lw$ for w

62. $D = RT$ for R

63. $A = \frac{1}{2}bh$ for b

64. $V = \frac{1}{3}Bh$ for B

65. $I = Prt$ for P

66. $C = 2\pi r$ for r

67. $E = mc^2$ for m

68. $V = \pi r^2 h$ for h

69. $T = D + pm$ for p

70. $P = C + MC$ for M

71. $A = \frac{1}{2}h(a + b)$ for a

72. $A = \frac{1}{2}h(a + b)$ for b

73. $S = P + Prt$ for r

74. $S = P + Prt$ for t

75. $B = \dfrac{F}{S - V}$ for S

76. $S = \dfrac{C}{1 - r}$ for r

77. $IR + Ir = E$ for I

78. $A = 2lw + 2lh + 2wh$ for h

79. $\dfrac{1}{p} + \dfrac{1}{q} = \dfrac{1}{f}$ for f

80. $\dfrac{1}{R} = \dfrac{1}{R_1} + \dfrac{1}{R_2}$ for R_1

Writing in Mathematics

81. In your own words, describe a step-by-step approach for solving algebraic word problems.

82. Write an original word problem that can be solved using a linear equation. Then solve the problem.

83. Explain what it means to solve a formula for a variable.

84. Did you have difficulties solving some of the problems that were assigned in this exercise set? Discuss what you did if this happened to you. Did your course of action enhance your ability to solve algebraic word problems?

85. The mile records in Example 2 on page 110 are a yardstick for measuring how athletes are getting better and better. Do you think that there is a limit to human performance? Explain your answer. If so, when might we reach it?

86. The bar graph in Exercises 41–42 shows average earnings of U.S. men and women, by highest educational attainment. Describe the trend shown by the graph. Discuss any aspects of the data that surprised you.

Technology Exercises

87. Use a graphing utility to numerically or graphically verify your work in any one exercise from Exercises 31–34. For assistance on how to do this, refer to the Technology box on page 112.

88. The formula $y = 0.2x + 55$ models the average life expectancy, y, of American men born x years after 1900. Graph the formula in a $[0, 200, 20]$ by $[0, 100, 10]$ viewing rectangle. Then use the TRACE or ZOOM feature to verify your answer in Exercise 25 or 26.

89. A tennis club offers two payment options. Members can pay a monthly fee of \$30 plus \$5 per hour for court rental time. The second option has no monthly fee, but court time costs \$7.50 per hour.

a. Write a mathematical model representing total monthly costs for each option for x hours of court rental time.

b. Use a graphing utility to graph the two models in a $[0, 15, 1]$ by $[0, 120, 20]$ viewing rectangle.

c. Use your utility's trace or intersection feature to determine where the two graphs intersect. Describe what the coordinates of this intersection point represent in practical terms.

d. Verify part (c) using an algebraic approach by setting the two models equal to one another and determining how many hours one has to rent the court so that the two plans result in identical monthly costs.

Critical Thinking Exercises

90. At the north campus of a performing arts school, 10% of the students are music majors. At the south campus, 90% of the students are music majors. The campuses are merged into one east campus. If 42% of the 1000 students at the east campus are music majors, how many students did the north and south campuses have before the merger?

91. The price of a dress is reduced by 40%. When the dress still does not sell, it is reduced by 40% of the reduced price. If the price of the dress after both reductions is \$72, what was the original price?

92. In a film, the actor Charles Coburn plays an elderly "uncle" character criticized for marrying a woman when he is 3 times her age. He wittily replies, "Ah, but in 20 years time I shall only be twice her age." How old is the "uncle" and the woman?

93. Suppose that we agree to pay you 8¢ for every problem in this chapter that you solve correctly and fine you 5¢ for every problem done incorrectly. If at the end of 26 problems we do not owe each other any money, how many problems did you solve correctly?

94. It was wartime when the Ricardos found out Mrs. Ricardo was pregnant. Ricky Ricardo was drafted and made out a will, deciding that \$14,000 in a savings account was to be divided between his wife and his child-to-be. Rather strangely, and certainly with gender bias, Ricky stipulated that if the child were a boy, he would get twice the amount of the mother's portion. If it were a girl, the mother would get twice the amount the girl was to receive. We'll never know what Ricky was thinking of, for (as fate would have it) he did not return from war. Mrs. Ricardo gave birth to twins—a boy and a girl. How was the money divided?

95. A thief steals a number of rare plants from a nursery. On the way out, the thief meets three security guards, one after another. To each security guard, the thief is forced to give one-half the plants that he still has, plus 2 more. Finally, the thief leaves the nursery with 1 lone palm. How many plants were originally stolen?

96. Solve for C: $V = C - \frac{C - S}{L}N.$

Group Exercise

97. One of the best ways to learn how to *solve* a word problem in algebra is to *design* word problems of your own. Creating a word problem makes you very aware of precisely how much information is needed to solve the problem. You must also focus on the best way to present information to a reader and on how much information to give. As you write your problem, you gain skills that will help you solve problems created by others.

The group should design five different word problems that can be solved using linear equations. All of the problems should be on different topics. For example, the group should not have more than one problem on simple interest. The group should turn in both the problems and their algebraic solutions.

SECTION 1.4 Complex Numbers

Objectives

1. Add and subtract complex numbers.
2. Multiply complex numbers.
3. Divide complex numbers.
4. Perform operations with square roots of negative numbers.

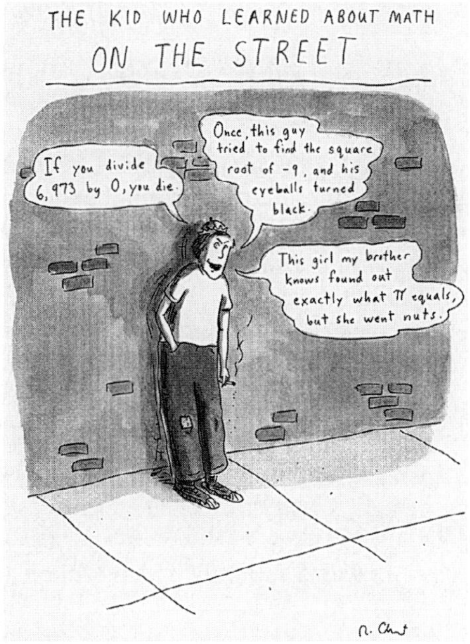

Who is this kid warning us about our eyeballs turning black if we attempt to find the square root of -9? Don't believe what you hear on the street. Although square roots of negative numbers are not real numbers, they do play a significant role in algebra. In this section, we move beyond the real numbers and discuss square roots with negative radicands.

The Imaginary Unit *i*

In the next section, we will study equations whose solutions may involve the square roots of negative numbers. Because the square of a real number is never negative, there is no real number x such that $x^2 = -1$. To provide a setting in which such equations have solutions, mathematicians invented an expanded system of numbers, the complex numbers. The *imaginary number i*, defined to be a solution of the equation $x^2 = -1$, is the basis of this new set.

The Imaginary Unit *i*

The **imaginary unit *i*** is defined as

$$i = \sqrt{-1}, \text{ where } i^2 = -1.$$

Using the imaginary unit i, we can express the square root of any negative number as a real multiple of i. For example,

$$\sqrt{-25} = \sqrt{-1}\sqrt{25} = i\sqrt{25} = 5i.$$

We can check this result by squaring $5i$ and obtaining -25.

$$(5i)^2 = 5^2 i^2 = 25(-1) = -25$$

A new system of numbers, called *complex numbers*, is based on adding multiples of i, such as $5i$, to the real numbers.

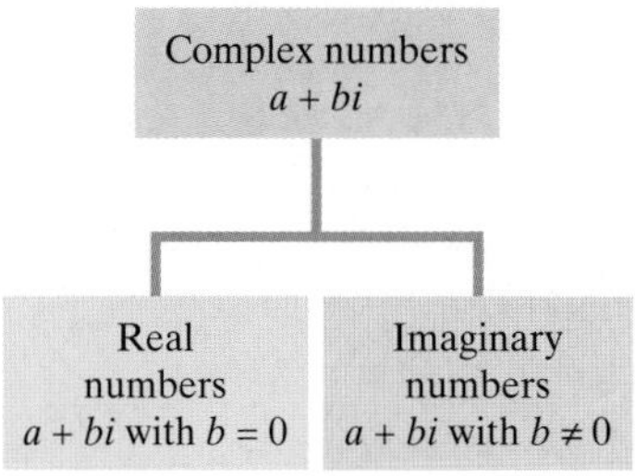

Figure 1.17 The complex number system

Complex Numbers and Imaginary Numbers

The set of all numbers in the form

$$a + bi$$

with real numbers a and b, and i, the imaginary unit, is called the set of **complex numbers**. The real number a is called the **real part** and the real number b is called the **imaginary part** of the complex number $a + bi$. If $b \neq 0$, then the complex number is called an **imaginary number** (Figure 1.17). An imaginary number in the form bi is called a **pure imaginary number**.

Here are some examples of complex numbers. Each number can be written in the form $a + bi$.

$-4 + 6i$		$2i = 0 + 2i$		$3 = 3 + 0i$	
a, the real part, is -4.	b, the imaginary part, is 6.	a, the real part, is 0.	b, the imaginary part, is 2.	a, the real part, is 3.	b, the imaginary part, is 0.

Can you see that b, the imaginary part, is not zero in the first two complex numbers? Because $b \neq 0$, these complex numbers are imaginary numbers. Furthermore, the imaginary number $2i$ is a pure imaginary number. By contrast, the imaginary part of the complex number on the right is zero. This complex number is not an imaginary number. The number 3, or $3 + 0i$, is a real number.

A complex number is said to be **simplified** if it is expressed in the **standard form** $a + bi$. If b is a radical, we usually write i before b. For example, we write $7 + i\sqrt{5}$ rather than $7 + \sqrt{5}i$, which could easily be confused with $7 + \sqrt{5i}$.

Expressed in standard form, two complex numbers are equal if and only if their real parts are equal and their imaginary parts are equal.

Equality of Complex Numbers

$a + bi = c + di$ if and only if $a = c$ and $b = d$.

1 Add and subtract complex numbers.

Operations with Complex Numbers

The form of a complex number $a + bi$ is like the binomial $a + bx$. Consequently, we can add, subtract, and multiply complex numbers using the same methods we used for binomials, remembering that $i^2 = -1$.

Adding and Subtracting Complex Numbers

1. $(a + bi) + (c + di) = (a + c) + (b + d)i$
   In words, this says that you add complex numbers by adding their real parts, adding their imaginary parts, and expressing the sum as a complex number.
2. $(a + bi) - (c + di) = (a - c) + (b - d)i$
   In words, this says that you subtract complex numbers by subtracting their real parts, subtracting their imaginary parts, and expressing the difference as a complex number.

EXAMPLE 1 Adding and Subtracting Complex Numbers

Perform the indicated operations, writing the result in standard form:

a. $(5 - 11i) + (7 + 4i)$ **b.** $(-5 + i) - (-11 - 6i)$.

Solution

a. $(5 - 11i) + (7 + 4i)$

$= 5 - 11i + 7 + 4i$ Remove the parentheses.

$= 5 + 7 - 11i + 4i$ Group real and imaginary terms.

$= (5 + 7) + (-11 + 4)i$ Add real parts and add imaginary parts.

$= 12 - 7i$ Simplify.

b. $(-5 + i) - (-11 - 6i)$

$= -5 + i + 11 + 6i$ Remove the parentheses. Change signs of real and imaginary parts in the complex number being subtracted.

$= -5 + 11 + i + 6i$ Group real and imaginary terms.

$= (-5 + 11) + (1 + 6)i$ Add real parts and add imaginary parts.

$= 6 + 7i$ Simplify.

Study Tip

The following examples, using the same integers as in Example 1, show how operations with complex numbers are just like operations with polynomials.

a. $(5 - 11x) + (7 + 4x)$
$= 12 - 7x$

b. $(-5 + x) - (-11 - 6x)$
$= -5 + x + 11 + 6x$
$= 6 + 7x$

Check Point 1 Add or subtract as indicated:

a. $(5 - 2i) + (3 + 3i)$ **b.** $(2 + 6i) - (12 - i)$.

Multiplication of complex numbers is performed the same way as multiplication of polynomials, using the distributive property and the FOIL method. After completing the multiplication, we replace any occurrences of i^2 with -1. This idea is illustrated in the next example.

2 Multiply complex numbers.

EXAMPLE 2 Multiplying Complex Numbers

Find the products:

a. $4i(3 - 5i)$ **b.** $(7 - 3i)(-2 - 5i)$.

Solution

a. $4i(3 - 5i)$

$= 4i \cdot 3 - 4i \cdot 5i$ Distribute $4i$ throughout the parentheses.

$= 12i - 20i^2$ Multiply.

$= 12i - 20(-1)$ Replace i^2 with -1.

$= 20 + 12i$ Simplify to $12i + 20$ and write in standard form.

b. $(7 - 3i)(-2 - 5i)$

F O I L

$= -14 - 35i + 6i + 15i^2$ Use the FOIL method.

$= -14 - 35i + 6i + 15(-1)$ $i^2 = -1$

$= -14 - 15 - 35i + 6i$ Group real and imaginary terms.

$= -29 - 29i$ Combine real and imaginary terms.

Check Point 2 Find the products:

a. $7i(2 - 9i)$ **b.** $(5 + 4i)(6 - 7i)$.

3 Divide complex numbers.

Complex Conjugates and Division

It is possible to multiply complex numbers and obtain a real number. This occurs when we multiply $a + bi$ and $a - bi$.

F O I L

$(a + bi)(a - bi) = a^2 - abi + abi - b^2i^2$ Use the FOIL method.

$= a^2 - b^2(-1)$ $i^2 = -1$

$= a^2 + b^2$ Notice that this product eliminates i.

For the complex number $a + bi$, we define its *complex conjugate* to be $a - bi$. The multiplication of complex conjugates results in a real number.

Conjugate of a Complex Number

The **complex conjugate** of the number $a + bi$ is $a - bi$, and the complex conjugate of $a - bi$ is $a + bi$. The multiplication of complex conjugates gives a real number.

$$(a + bi)(a - bi) = a^2 + b^2$$
$$(a - bi)(a + bi) = a^2 + b^2$$

Complex conjugates are used to divide complex numbers. By multiplying the numerator and the denominator of the division by the complex conjugate of the denominator, you will obtain a real number in the denominator.

EXAMPLE 3 Using Complex Conjugates to Divide Complex Numbers

Divide and express the result in standard form: $\frac{7 + 4i}{2 - 5i}$.

Solution The complex conjugate of the denominator, $2 - 5i$, is $2 + 5i$. Multiplication of both the numerator and the denominator by $2 + 5i$ will eliminate i from the denominator.

$$\frac{7+4i}{2-5i} = \frac{(7+4i)}{(2-5i)} \cdot \frac{(2+5i)}{(2+5i)}$$
Multiply the numerator and the denominator by the complex conjugate of the denominator.

F O I L

$$= \frac{14 + 35i + 8i + 20i^2}{2^2 + 5^2}$$
Use the FOIL method in the numerator and $(a - bi)(a + bi) = a^2 + b^2$ in the denominator.

$$= \frac{14 + 43i + 20(-1)}{29}$$
Combine imaginary terms and replace i^2 with -1.

$$= \frac{-6 + 43i}{29}$$
Combine real terms in the numerator: $14 + 20(-1) = 14 - 20 = -6$.

$$= -\frac{6}{29} + \frac{43}{29}i$$
Express the answer in standard form.

Observe that the quotient is expressed in the standard form $a + bi$, with $a = -\frac{6}{29}$ and $b = \frac{43}{29}$.

Check Point 3 Divide and express the result in standard form: $\frac{5 + 4i}{4 - i}$.

4 Perform operations with square roots of negative numbers.

Roots of Negative Numbers

The square of $4i$ and the square of $-4i$ both result in -16:

$$(4i)^2 = 16i^2 = 16(-1) = -16 \qquad (-4i)^2 = 16i^2 = 16(-1) = -16.$$

Consequently, in the complex number system -16 has two square roots, namely, $4i$ and $-4i$. We call $4i$ the **principal square root** of -16.

Principal Square Root of a Negative Number

For any positive real number b, the **principal square root** of the negative number $-b$ is defined by

$$\sqrt{-b} = i\sqrt{b}.$$

Consider the multiplication problem

$$5i \cdot 2i = 10i^2 = 10(-1) = -10.$$

This problem can also be given in terms of principal square roots of negative numbers:

$$\sqrt{-25} \cdot \sqrt{-4}.$$

Because the product rule for radicals only applies to real numbers, multiplying radicands is incorrect. **When performing operations with square roots of negative numbers, begin by expressing all square roots in terms of *i*.** Then perform the indicated operation.

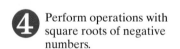

Correct:

$$\sqrt{-25}\cdot\sqrt{-4} = i\sqrt{25}\cdot i\sqrt{4} = 5i\cdot 2i = 10i^2 = 10(-1) = -10$$

Incorrect:

$$\sqrt{-25}\cdot\sqrt{-4} = \sqrt{-25\cdot -4} = \sqrt{100} = 10$$

EXAMPLE 4 Operations Involving Square Roots of Negative Numbers

Perform the indicated operations and write the result in standard form:

a. $\sqrt{-18} - \sqrt{-8}$ **b.** $\left(-1 + \sqrt{-5}\right)^2$ **c.** $\dfrac{-25 + \sqrt{-50}}{15}$.

Solution Begin by expressing all square roots of negative numbers in terms of i.

a. $\sqrt{-18} - \sqrt{-8} = i\sqrt{18} - i\sqrt{8} = i\sqrt{9\cdot 2} - i\sqrt{4\cdot 2}$
$= 3i\sqrt{2} - 2i\sqrt{2} = i\sqrt{2}$

$(A + B)^2 = A^2 + 2AB + B^2$

b. $(-1 + \sqrt{-5})^2 = (-1 + i\sqrt{5})^2 = (-1)^2 + 2(-1)(i\sqrt{5}) + (i\sqrt{5})^2$
$= 1 - 2i\sqrt{5} + 5i^2$
$= 1 - 2i\sqrt{5} + 5(-1)$
$= -4 - 2i\sqrt{5}$

c. $\dfrac{-25 + \sqrt{-50}}{15}$

$= \dfrac{-25 + i\sqrt{50}}{15}$ $\sqrt{-b} = i\sqrt{b}$

$= \dfrac{-25 + 5i\sqrt{2}}{15}$ $\sqrt{50} = \sqrt{25\cdot 2} = 5\sqrt{2}$

$= \dfrac{-25}{15} + \dfrac{5i\sqrt{2}}{15}$ Write the complex number in standard form.

$= -\dfrac{5}{3} + i\dfrac{\sqrt{2}}{3}$ Simplify.

Complex Numbers on a Postage Stamp

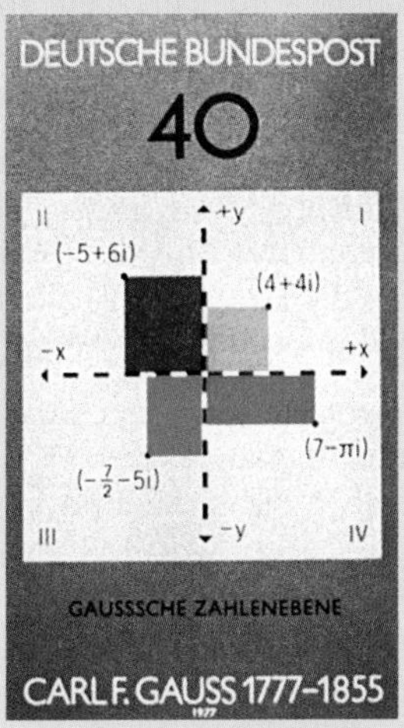

This stamp honors the work done by the German mathematician Carl Friedrich Gauss (1777–1855) with complex numbers. Gauss represented complex numbers as points in the plane.

Check Point 4 Perform the indicated operations and write the result in standard form:

a. $\sqrt{-27} + \sqrt{-48}$ **b.** $\left(-2 + \sqrt{-3}\right)^2$ **c.** $\dfrac{-14 + \sqrt{-12}}{2}$.

EXERCISE SET 1.4

Practice Exercises

In Exercises 1–8, add or subtract as indicated and write the result in standard form.

1. $(7 + 2i) + (1 - 4i)$
2. $(-2 + 6i) + (4 - i)$
3. $(3 + 2i) - (5 - 7i)$
4. $(-7 + 5i) - (-9 - 11i)$
5. $6 - (-5 + 4i) - (-13 - i)$
6. $7 - (-9 + 2i) - (-17 - i)$
7. $8i - (14 - 9i)$
8. $15i - (12 - 11i)$

In Exercises 9–20, find each product and write the result in standard form.

9. $-3i(7i - 5)$
10. $-8i(2i - 7)$
11. $(-5 + 4i)(3 + i)$
12. $(-4 - 8i)(3 + i)$
13. $(7 - 5i)(-2 - 3i)$
14. $(8 - 4i)(-3 + 9i)$
15. $(3 + 5i)(3 - 5i)$
16. $(2 + 7i)(2 - 7i)$
17. $(-5 + i)(-5 - i)$
18. $(-7 - i)(-7 + i)$
19. $(2 + 3i)^2$
20. $(5 - 2i)^2$

In Exercises 21–28, divide and express the result in standard form.

21. $\dfrac{2}{3-i}$ **22.** $\dfrac{3}{4+i}$

23. $\dfrac{2i}{1+i}$ **24.** $\dfrac{5i}{2-i}$

25. $\dfrac{8i}{4-3i}$ **26.** $\dfrac{-6i}{3+2i}$

27. $\dfrac{2+3i}{2+i}$ **28.** $\dfrac{3-4i}{4+3i}$

In Exercises 29–44, perform the indicated operations and write the result in standard form.

29. $\sqrt{-64}-\sqrt{-25}$ **30.** $\sqrt{-81}-\sqrt{-144}$

31. $5\sqrt{-16}+3\sqrt{-81}$ **32.** $5\sqrt{-8}+3\sqrt{-18}$

33. $\left(-2+\sqrt{-4}\right)^2$ **34.** $\left(-5-\sqrt{-9}\right)^2$

35. $\left(-3-\sqrt{-7}\right)^2$ **36.** $\left(-2+\sqrt{-11}\right)^2$

37. $\dfrac{-8+\sqrt{-32}}{24}$ **38.** $\dfrac{-12+\sqrt{-28}}{32}$

39. $\dfrac{-6-\sqrt{-12}}{48}$ **40.** $\dfrac{-15-\sqrt{-18}}{33}$

41. $\sqrt{-8}\left(\sqrt{-3}-\sqrt{5}\right)$ **42.** $\sqrt{-12}\left(\sqrt{-4}-\sqrt{2}\right)$

43. $\left(3\sqrt{-5}\right)\left(-4\sqrt{-12}\right)$ **44.** $\left(3\sqrt{-7}\right)\left(2\sqrt{-8}\right)$

Practice Plus

In Exercises 45–50, perform the indicated operation(s) and write the result in standard form.

45. $(2-3i)(1-i)-(3-i)(3+i)$

46. $(8+9i)(2-i)-(1-i)(1+i)$

47. $(2+i)^2-(3-i)^2$

48. $(4-i)^2-(1+2i)^2$

49. $5\sqrt{-16}+3\sqrt{-81}$

50. $5\sqrt{-8}+3\sqrt{-18}$

51. Evaluate x^2-2x+2 for $x=1+i$.

52. Evaluate x^2-2x+5 for $x=1-2i$.

53. Evaluate $\dfrac{x^2+19}{2-x}$ for $x=3i$.

54. Evaluate $\dfrac{x^2+11}{3-x}$ for $x=4i$.

Application Exercises

Complex numbers are used in electronics to describe the current in an electric circuit. Ohm's law relates the current in a circuit, I, in amperes, the voltage of the circuit, E, in volts, and the resistance of the circuit, R, in ohms, by the formula $E = IR$. Use this formula to solve Exercises 55–56.

55. Find E, the voltage of a circuit, if $I=(4-5i)$ amperes and $R=(3+7i)$ ohms.

56. Find E, the voltage of a circuit, if $I=(2-3i)$ amperes and $R=(3+5i)$ ohms.

57. The mathematician Girolamo Cardano is credited with the first use (in 1545) of negative square roots in solving the now-famous problem, "Find two numbers whose sum is 10 and whose product is 40." Show that the complex numbers $5+i\sqrt{15}$ and $5-i\sqrt{15}$ satisfy the conditions of the problem. (Cardano did not use the symbolism $i\sqrt{15}$ or even $\sqrt{-15}$. He wrote R.m 15 for $\sqrt{-15}$, meaning "radix minus 15." He regarded the numbers 5 + R.m 15 and 5 − R.m 15 as "fictitious" or "ghost numbers," and considered the problem "manifestly impossible." But in a mathematically adventurous spirit, he exclaimed, "Nevertheless, we will operate.")

Writing in Mathematics

58. What is i?

59. Explain how to add complex numbers. Provide an example with your explanation.

60. Explain how to multiply complex numbers and give an example.

61. What is the complex conjugate of $2+3i$? What happens when you multiply this complex number by its complex conjugate?

62. Explain how to divide complex numbers. Provide an example with your explanation.

63. Explain each of the three jokes in the cartoon on page 123.

64. A stand-up comedian uses algebra in some jokes, including one about a telephone recording that announces "You have just reached an imaginary number. Please multiply by i and dial again." Explain the joke.

Explain the error in Exercises 65–66.

65. $\sqrt{-9}+\sqrt{-16}=\sqrt{-25}=i\sqrt{25}=5i$

66. $\left(\sqrt{-9}\right)^2=\sqrt{-9}\cdot\sqrt{-9}=\sqrt{81}=9$

Critical Thinking Exercises

67. Which one of the following is true?

a. Some irrational numbers are not complex numbers.

b. $(3+7i)(3-7i)$ is an imaginary number.

c. $\dfrac{7+3i}{5+3i}=\dfrac{7}{5}$

d. In the complex number system, x^2+y^2 (the sum of two squares) can be factored as $(x+yi)(x-yi)$.

In Exercises 68–70, perform the indicated operations and write the result in standard form.

68. $\dfrac{4}{(2+i)(3-i)}$

69. $\dfrac{1+i}{1+2i}+\dfrac{1-i}{1-2i}$

70. $\dfrac{8}{1+\dfrac{2}{i}}$

SECTION 1.5 Quadratic Equations

Objectives

1. Solve quadratic equations by factoring.
2. Solve quadratic equations by the square root property.
3. Solve quadratic equations by completing the square.
4. Solve quadratic equations using the quadratic formula.
5. Use the discriminant to determine the number and type of solutions.
6. Determine the most efficient method to use when solving a quadratic equation.
7. Solve problems modeled by quadratic equations.

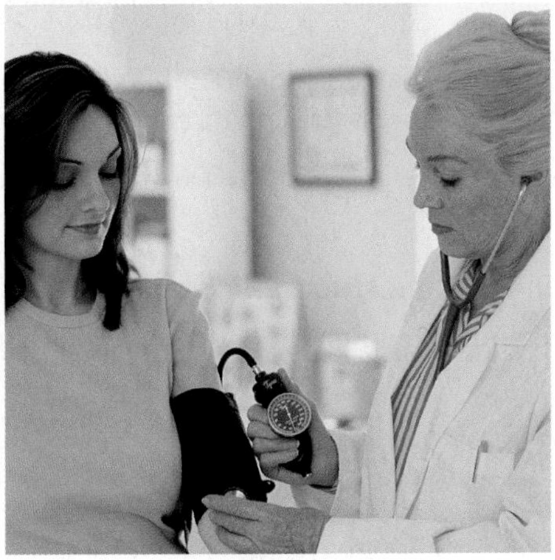

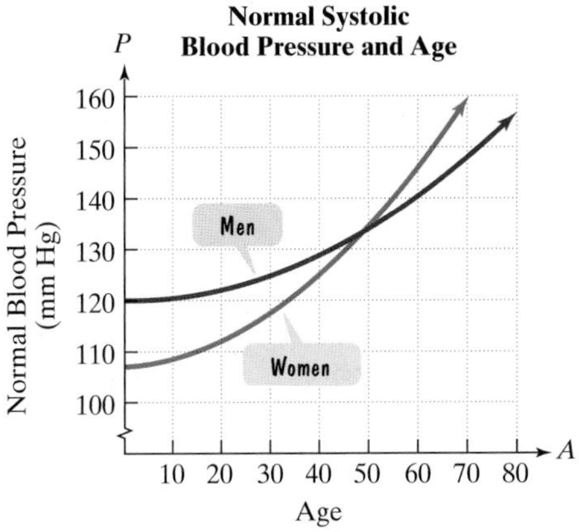

Figure 1.18

Until fairly recently, many doctors believed that your blood pressure was theirs to know and yours to worry about. Today, however, people are encouraged to find out their blood pressure. That pumped-up cuff that squeezes against your upper arm measures blood pressure in millimeters (mm) of mercury (Hg). Blood pressure is given in two numbers: systolic pressure over diastolic pressure, such as 120 over 80. Systolic pressure is the pressure of blood against the artery walls when the heart contracts. Diastolic pressure is the pressure of blood against the artery walls when the heart is at rest.

The graphs in Figure 1.18 illustrate how systolic pressure increases with age as the arteries become less elastic. The blue graph representing women's normal systolic blood pressure is narrower than the red graph representing men's normal systolic blood pressure. Up to approximately age 50, women's normal systolic blood pressure is lower than men's, although it is increasing at a faster rate. After age 50, women's normal systolic blood pressure is higher than men's.

Normal systolic blood pressure is modeled by the following formulas:

Men

$$P = 0.006A^2 - 0.02A + 120$$

Women

$$P = 0.01A^2 + 0.05A + 107.$$

In each formula, P is the normal systolic blood pressure, in millimeters of mercury, at age A.

Suppose we are interested in the age of a man with a normal systolic blood pressure of 125 millimeters of mercury. We can use the red graph in Figure 1.19 to approximate the value of A for which $P = 125$. Locate 125 on the vertical axis and then move to the right to the red graph and locate the point for which 125 is the second coordinate. From this point, we look to the horizontal axis to find the corresponding first coordinate. A reasonable estimate is 31. Thus, $P = 125$ for $A \approx 31$. We see that 31 is the approximate age of a man whose normal systolic blood pressure is 125 mm Hg.

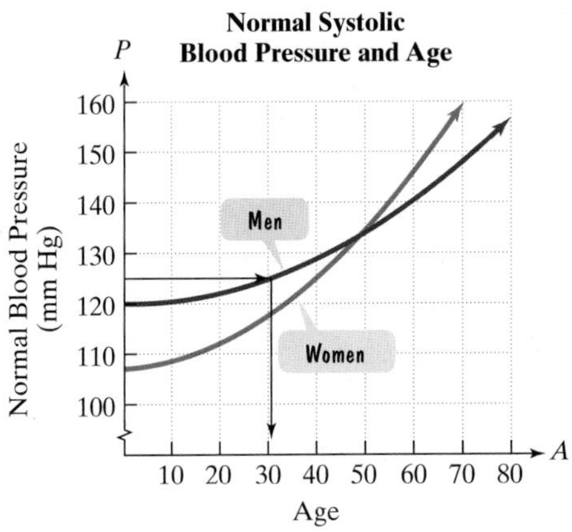

Figure 1.19

We can obtain this information algebraically by using the formula that models male blood pressure, $P = 0.006A^2 - 0.02A + 120$, and setting P equal to 125:

$$0.006A^2 - 0.02A + 120 = 125.$$

This equation is an example of a *quadratic equation*. By subtracting 125 from both sides, we can express the equation in *general form*:

$$0.006A^2 - 0.02A - 5 = 0.$$

Definition of a Quadratic Equation

A **quadratic equation** in x is an equation that can be written in the **general form**

$$ax^2 + bx + c = 0,$$

where a, b, and c are real numbers, with $a \neq 0$. A quadratic equation in x is also called a **second-degree polynomial equation** in x.

In this section, we study a number of methods for solving quadratic equations. We also look at applications of these equations, returning to the blood pressure application later in the section.

1 Solve quadratic equations by factoring.

Solving Quadratic Equations by Factoring

Here is an example of a quadratic equation in general form:

$$x^2 - 7x + 10 = 0.$$

$a = 1$ $b = -7$ $c = 10$

We can factor the left side of this equation. We obtain $(x - 5)(x - 2) = 0$. If a quadratic equation has zero on one side and a factored expression on the other side, it can be solved using the **zero-product principle**.

The Zero-Product Principle

If the product of two algebraic expressions is zero, then at least one of the factors is equal to zero.

If $AB = 0$, then $A = 0$ or $B = 0$.

For example, consider the equation $(x - 5)(x - 2) = 0$. According to the zero-product principle, this product can be zero only if at least one of the factors is zero. We set each individual factor equal to zero and solve each resulting equation for x.

$$(x - 5)(x - 2) = 0$$
$$x - 5 = 0 \quad \text{or} \quad x - 2 = 0$$
$$x = 5 \qquad\qquad x = 2$$

We can check each of these proposed solutions, 5 and 2, in the original quadratic equation, $x^2 - 7x + 10 = 0$. Substitute each one separately for x into the equation.

Check 5:

$$x^2 - 7x + 10 = 0$$
$$5^2 - 7 \cdot 5 + 10 \stackrel{?}{=} 0$$
$$25 - 35 + 10 \stackrel{?}{=} 0$$
$$0 = 0, \quad \text{true}$$

Check 2:

$$x^2 - 7x + 10 = 0$$
$$2^2 - 7 \cdot 2 + 10 \stackrel{?}{=} 0$$
$$4 - 14 + 10 \stackrel{?}{=} 0$$
$$0 = 0, \quad \text{true}$$

The resulting true statements indicate that the solutions are 2 and 5. The solution set is $\{2, 5\}$. Note that with a quadratic equation, we can have two solutions, compared to the conditional linear equation that had one.

Solving a Quadratic Equation by Factoring

1. If necessary, rewrite the equation in the general form $ax^2 + bx + c = 0$, moving all terms to one side, thereby obtaining zero on the other side.
2. Factor completely.
3. Apply the zero-product principle, setting each factor containing a variable equal to zero.
4. Solve the equations in step 3.
5. Check the solutions in the original equation.

EXAMPLE 1   Solving Quadratic Equations by Factoring

Solve by factoring:

   a. $4x^2 - 2x = 0$ **b.** $2x^2 + 7x = 4.$

Solution

   a. We begin with $4x^2 - 2x = 0$.

Step 1   Move all terms to one side and obtain zero on the other side. All terms are already on the left and zero is on the other side, so we can skip this step.

Step 2   Factor. We factor out $2x$ from the two terms on the left side.

$$4x^2 - 2x = 0 \qquad \text{This is the given equation.}$$
$$2x(2x - 1) = 0 \qquad \text{Factor.}$$

Steps 3 and 4   Set each factor equal to zero and solve the resulting equations.

$$2x = 0 \quad \text{or} \quad 2x - 1 = 0$$
$$x = 0 \qquad\qquad 2x = 1$$
$$x = \tfrac{1}{2}$$

Step 5   Check the solutions in the original equation.

$$\begin{array}{cc}
\textbf{Check 0:} & \textbf{Check } \tfrac{1}{2}\textbf{:} \\
4x^2 - 2x = 0 & 4x^2 - 2x = 0 \\
4 \cdot 0^2 - 2 \cdot 0 \overset{?}{=} 0 & 4\left(\tfrac{1}{2}\right)^2 - 2\left(\tfrac{1}{2}\right) \overset{?}{=} 0 \\
0 - 0 \overset{?}{=} 0 & 4\left(\tfrac{1}{4}\right) - 2\left(\tfrac{1}{2}\right) \overset{?}{=} 0 \\
0 = 0, \quad \text{true} & 1 - 1 \overset{?}{=} 0 \\
& 0 = 0, \quad \text{true}
\end{array}$$

The solution set is $\left\{0, \tfrac{1}{2}\right\}$.

   b. Next, we solve $2x^2 + 7x = 4$.

Step 1   Move all terms to one side and obtain zero on the other side. Subtract 4 from both sides and write the equation in general form.

$$2x^2 + 7x = 4 \qquad \text{This is the given equation.}$$
$$2x^2 + 7x - 4 = 4 - 4 \qquad \text{Subtract 4 from both sides.}$$
$$2x^2 + 7x - 4 = 0 \qquad \text{Simplify.}$$

Step 2   Factor.

$$2x^2 + 7x - 4 = 0$$
$$(2x - 1)(x + 4) = 0$$

Steps 3 and 4   Set each factor equal to zero and solve the resulting equations.

$$2x - 1 = 0 \quad \text{or} \quad x + 4 = 0$$
$$2x = 1 \qquad\qquad x = -4$$
$$x = \tfrac{1}{2}$$

Step 5 Check the solutions in the original equation.

Check $\frac{1}{2}$:

$$2x^2 + 7x = 4$$
$$2\left(\tfrac{1}{2}\right)^2 + 7\left(\tfrac{1}{2}\right) \stackrel{?}{=} 4$$
$$\tfrac{1}{2} + \tfrac{7}{2} \stackrel{?}{=} 4$$
$$4 = 4, \quad \text{true}$$

Check −4:

$$2x^2 + 7x = 4$$
$$2(-4)^2 + 7(-4) \stackrel{?}{=} 4$$
$$32 + (-28) \stackrel{?}{=} 4$$
$$4 = 4, \quad \text{true}$$

The solution set is $\left\{-4, \frac{1}{2}\right\}$.

Check Point 1 Solve by factoring:

a. $3x^2 - 9x = 0$ **b.** $2x^2 + x = 1$.

Technology

You can use a graphing utility to check the real solutions of a quadratic equation. **The real solutions of $ax^2 + bx + c = 0$ correspond to the x-intercepts of the graph of $y = ax^2 + bx + c$.** For example, to check the solutions of $2x^2 + 7x = 4$, or $2x^2 + 7x - 4 = 0$, graph $y = 2x^2 + 7x - 4$. The cuplike U-shaped graph is shown on the right. Note that it is important to have all nonzero terms on one side of the quadratic equation before entering it into the graphing utility. The x-intercepts are -4 and $\frac{1}{2}$, and the graph of $y = 2x^2 + 7x - 4$ passes through $(-4, 0)$ and $\left(\frac{1}{2}, 0\right)$. This verifies that $\left\{-4, \frac{1}{2}\right\}$ is the solution set of $2x^2 + 7x - 4 = 0$.

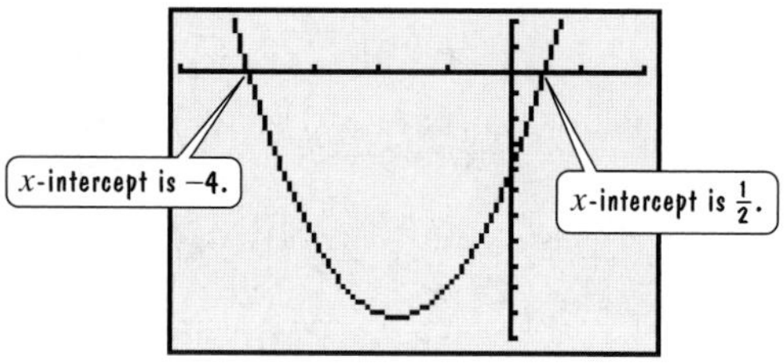

[−5, 2, 1] by [−11, 2, 1]

2 Solve quadratic equations by the square root property.

Solving Quadratic Equations by the Square Root Property

Quadratic equations of the form $u^2 = d$, where u is an algebraic expression, and d is a nonzero real number, can be solved by the *square root property*. First, isolate the squared expression u^2 on one side of the equation and the number d on the other side. Then take the square root of both sides. Remember, there are two numbers whose square is d. One number is $\sqrt{d}$ and one is $-\sqrt{d}$.

We can use factoring to verify that $u^2 = d$ has these two solutions.

$$u^2 = d \quad \text{This is the given equation.}$$
$$u^2 - d = 0 \quad \text{Move all terms to one side and obtain zero on the other side.}$$
$$\left(u + \sqrt{d}\right)\left(u - \sqrt{d}\right) = 0 \quad \text{Factor.}$$
$$u + \sqrt{d} = 0 \quad \text{or} \quad u - \sqrt{d} = 0 \quad \text{Set each factor equal to zero.}$$
$$u = -\sqrt{d} \qquad u = \sqrt{d} \quad \text{Solve the resulting equations.}$$

Because the solutions differ only in sign, we can write them in abbreviated notation as $u = \pm\sqrt{d}$. We read this as "u equals positive or negative the square root of d" or "u equals plus or minus the square root of d."

Now that we have verified these solutions, we can solve $u^2 = d$ directly by taking square roots. This process is called the **square root property**.

The Square Root Property

If u is an algebraic expression and d is a nonzero real number, then $u^2 = d$ has exactly two solutions:

$$\text{If } u^2 = d, \text{ then } u = \sqrt{d} \text{ or } u = -\sqrt{d}.$$

Equivalently,

$$\text{If } u^2 = d, \text{ then } u = \pm\sqrt{d}.$$

EXAMPLE 2 Solving Quadratic Equations by the Square Root Property

Solve by the square root property:

a. $3x^2 - 15 = 0$ **b.** $9x^2 + 25 = 0$ **c.** $(x - 2)^2 = 6$.

Solution To apply the square root property, we need a squared expression by itself on one side of the equation.

$$3x^2 - 15 = 0 \qquad 9x^2 + 25 = 0 \qquad (x - 2)^2 = 6$$

We want x^2 by itself. We want x^2 by itself. The squared expression is by itself.

a.

$3x^2 - 15 = 0$	This is the original equation.
$3x^2 = 15$	Add 15 to both sides.
$x^2 = 5$	Divide both sides by 3.
$x = \sqrt{5}$ or $x = -\sqrt{5}$	Apply the square root property. Equivalently, $x = \pm\sqrt{5}$.

By checking both proposed solutions in the original equation, we can confirm that the solution set is $\{-\sqrt{5}, \sqrt{5}\}$ or $\{\pm\sqrt{5}\}$.

b.

$9x^2 + 25 = 0$	This is the original equation.
$9x^2 = -25$	Subtract 25 from both sides.
$x^2 = -\frac{25}{9}$	Divide both sides by 9.
$x = \pm\sqrt{-\frac{25}{9}}$	Apply the square root property.
$x = \pm i\sqrt{\frac{25}{9}} = \pm\frac{5}{3}i$	Express solutions in terms of i.

Because the equation has an x^2-term and no x-term, we can check both proposed solutions, $\pm\frac{5}{3}i$, at once.

Check $\frac{5}{3}i$ and $-\frac{5}{3}i$:

$$9x^2 + 25 = 0$$

$$9\left(\pm\frac{5}{3}i\right)^2 + 25 \stackrel{?}{=} 0$$

$$9\left(\frac{25}{9}i^2\right) + 25 \stackrel{?}{=} 0$$

$$25i^2 + 25 \stackrel{?}{=} 0$$

$i^2 = -1$

$$25(-1) + 25 \stackrel{?}{=} 0$$

$$0 = 0, \quad \text{true}$$

The solutions are $-\frac{5}{3}i$ and $\frac{5}{3}i$. The solution set is $\left\{-\frac{5}{3}i, \frac{5}{3}i\right\}$ or $\left\{\pm\frac{5}{3}i\right\}$.

c.

$(x - 2)^2 = 6$	This is the original equation.
$x - 2 = \pm\sqrt{6}$	Apply the square root property.
$x = 2 \pm \sqrt{6}$	Add 2 to both sides.

By checking both values in the original equation, we can confirm that the solution set is $\{2 + \sqrt{6}, 2 - \sqrt{6}\}$ or $\{2 \pm \sqrt{6}\}$.

Technology

The graph of

$$y = 9x^2 + 25$$

has no x-intercepts. This shows that

$$9x^2 + 25 = 0$$

has no real solutions. Example 2(b) on the previous page algebraically established that the solutions are imaginary numbers.

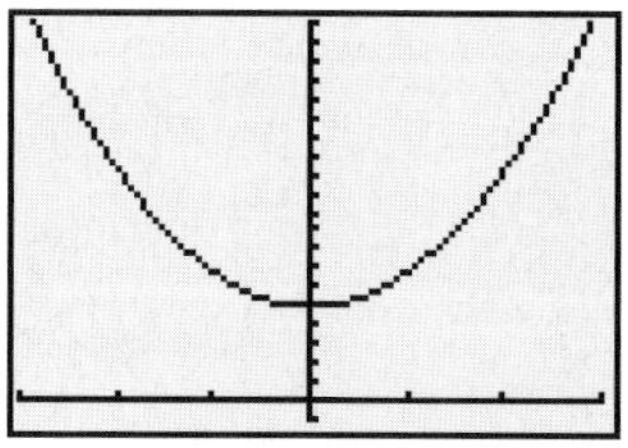

$[-3, 3, 1]$ by $[-5, 100, 5]$

Check Point 2 Solve by the square root property:

a. $3x^2 - 21 = 0$ **b.** $5x^2 + 45 = 0$ **c.** $(x + 5)^2 = 11$.

③ Solve quadratic equations by completing the square.

Completing the Square

How do we solve an equation in the form $ax^2 + bx + c = 0$ if the trinomial $ax^2 + bx + c$ cannot be factored? We cannot use the zero-product principle in such a case. However, we can convert the equation into an equivalent equation that can be solved using the square root property. This is accomplished by **completing the square**.

Visualizing Completing the Square

This figure, with area $x^2 + 8x$, is not a complete square. The bottom-right corner is missing.

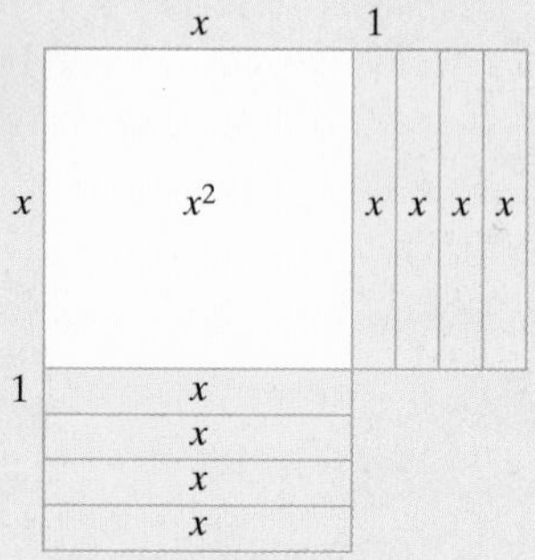

Area: $x^2 + 8x$

Add 16 square units to the missing portion and you, literally, complete the square.

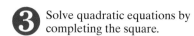

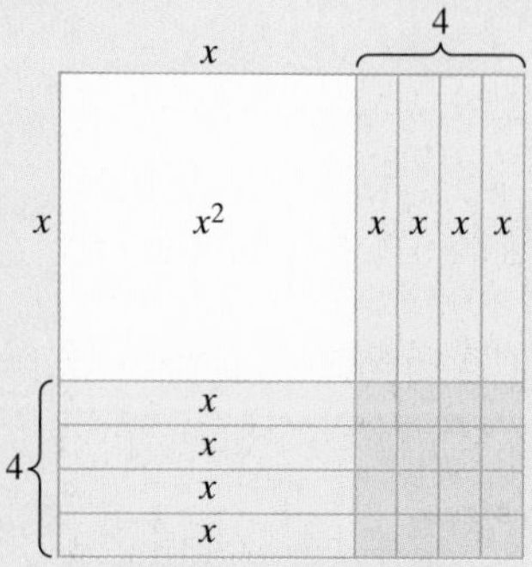

Area: $x^2 + 8x + 16 = (x + 4)^2$

Completing the Square

If $x^2 + bx$ is a binomial, then by adding $\left(\dfrac{b}{2}\right)^2$, which is the square of half the coefficient of x, a perfect square trinomial will result. That is,

$$x^2 + bx + \left(\frac{b}{2}\right)^2 = \left(x + \frac{b}{2}\right)^2$$

EXAMPLE 3   Completing the Square

What term should be added to each binomial so that it becomes a perfect square trinomial? Write and factor the trinomial.

a. $x^2 + 8x$ **b.** $x^2 - 7x$ **c.** $x^2 + \dfrac{3}{5}x$

Solution To complete the square, we must add a term to each binomial. The term that should be added is the square of half the coefficient of x.

$$x^2 + 8x \qquad x^2 - 7x \qquad x^2 + \frac{3}{5}x$$

Add $\left(\frac{8}{2}\right)^2 = 4^2$.   Add $\left(\frac{-7}{2}\right)^2$, or $\frac{49}{4}$,   Add $\left(\frac{1}{2} \cdot \frac{3}{5}\right)^2 = \left(\frac{3}{10}\right)^2$.
Add 16 to complete   to complete   Add $\frac{9}{100}$ to complete
the square.   the square.   the square.

a. The coefficient of the x-term in $x^2 + 8x$ is 8. Half of 8 is 4, and $4^2 = 16$. Add 16. The result is a perfect square trinomial.

$$x^2 + 8x + 16 = (x + 4)^2$$

b. The coefficient of the x-term in $x^2 - 7x$ is -7. Half of -7 is $-\dfrac{7}{2}$, and $\left(-\dfrac{7}{2}\right)^2 = \dfrac{49}{4}$. Add $\dfrac{49}{4}$. The result is a perfect square trinomial.

$$x^2 - 7x + \frac{49}{4} = \left(x - \frac{7}{2}\right)^2$$

c. The coefficient of the x-term in $x^2 + \frac{3}{5}x$ is $\frac{3}{5}$. Half of $\frac{3}{5}$ is $\frac{1}{2} \cdot \frac{3}{5}$, or $\frac{3}{10}$, and $\left(\frac{3}{10}\right)^2 = \frac{9}{100}$. Add $\frac{9}{100}$. The result is a perfect square trinomial.

$$x^2 + \frac{3}{5}x + \frac{9}{100} = \left(x + \frac{3}{10}\right)^2$$

Study Tip

You may not be accustomed to factoring perfect square trinomials in which fractions are involved. The constant in the factorization is always half the coefficient of x.

$$x^2 - 7x + \frac{49}{4} = \left(x - \frac{7}{2}\right)^2$$

Half the coefficient of x, -7, is $-\frac{7}{2}$.

$$x^2 + \frac{3}{5}x + \frac{9}{100} = \left(x + \frac{3}{10}\right)^2$$

Half the coefficient of x, $\frac{3}{5}$, is $\frac{3}{10}$.

Check Point 3 What term should be added to each binomial so that it becomes a perfect square trinomial? Write and factor the trinomial.

a. $x^2 + 6x$ **b.** $x^2 - 5x$ **c.** $x^2 + \frac{2}{3}x$

We can solve any quadratic equation by completing the square. If the coefficient of the x^2-term is one, we add the square of half the coefficient of x to both sides of the equation. **When you add a constant term to one side of the equation to complete the square, be certain to add the same constant to the other side of the equation.** These ideas are illustrated in Example 4.

EXAMPLE 4 Solving a Quadratic Equation by Completing the Square

Solve by completing the square: $x^2 - 6x + 4 = 0$.

Solution We begin by subtracting 4 from both sides. This is done to isolate the binomial $x^2 - 6x$, so that we can complete the square.

$$x^2 - 6x + 4 = 0 \quad \text{This is the original equation.}$$

$$x^2 - 6x = -4 \quad \text{Subtract 4 from both sides.}$$

Next, we work with $x^2 - 6x = -4$ and complete the square. Find half the coefficient of the x-term and square it. The coefficient of the x-term is -6. Half of -6 is -3 and $(-3)^2 = 9$. Thus, we add 9 to both sides of the equation.

$$x^2 - 6x + 9 = -4 + 9 \quad \text{Add 9 to both sides to complete the square.}$$

$$(x - 3)^2 = 5 \quad \text{Factor and simplify.}$$

$$x - 3 = \sqrt{5} \quad \text{or} \quad x - 3 = -\sqrt{5} \quad \text{Apply the square root property.}$$

$$x = 3 + \sqrt{5} \qquad x = 3 - \sqrt{5} \quad \text{Add 3 to both sides in each equation.}$$

The solutions are $3 \pm \sqrt{5}$, and the solution set is $\{3 + \sqrt{5}, 3 - \sqrt{5}\}$, or $\{3 \pm \sqrt{5}\}$.

Check Point 4 Solve by completing the square: $x^2 + 4x - 1 = 0$.

If the coefficient of the x^2-term in a quadratic equation is not 1, you must divide each side of the equation by this coefficient before completing the square. For example, to solve $9x^2 - 6x - 4 = 0$ by completing the square, first divide every term by 9:

$$\frac{9x^2}{9} - \frac{6x}{9} - \frac{4}{9} = \frac{0}{9}$$
$$x^2 - \frac{6}{9}x - \frac{4}{9} = 0$$
$$x^2 - \frac{2}{3}x - \frac{4}{9} = 0.$$

Now that the coefficient of the x^2-term is 1, we can solve by completing the square.

EXAMPLE 5 Solving a Quadratic Equation by Completing the Square

Solve by completing the square: $9x^2 - 6x - 4 = 0$.

Solution

$$9x^2 - 6x - 4 = 0$$ This is the original equation.

$$x^2 - \frac{2}{3}x - \frac{4}{9} = 0$$ Divide both sides by 9.

$$x^2 - \frac{2}{3}x = \frac{4}{9}$$ Add $\frac{4}{9}$ to both sides to isolate the binomial.

$$x^2 - \frac{2}{3}x + \frac{1}{9} = \frac{4}{9} + \frac{1}{9}$$ Complete the square: Half of $-\frac{2}{3}$ is $-\frac{2}{6}$, or $-\frac{1}{3}$, and $\left(-\frac{1}{3}\right)^2 = \frac{1}{9}$.

$$\left(x - \frac{1}{3}\right)^2 = \frac{5}{9}$$ Factor and simplify.

$$x - \frac{1}{3} = \sqrt{\frac{5}{9}} \quad \text{or} \quad x - \frac{1}{3} = -\sqrt{\frac{5}{9}}$$ Apply the square root property.

$$x - \frac{1}{3} = \frac{\sqrt{5}}{3} \qquad x - \frac{1}{3} = -\frac{\sqrt{5}}{3}$$ $\sqrt{\frac{5}{9}} = \frac{\sqrt{5}}{\sqrt{9}} = \frac{\sqrt{5}}{3}$

$$x = \frac{1}{3} + \frac{\sqrt{5}}{3} \qquad x = \frac{1}{3} - \frac{\sqrt{5}}{3}$$ Add $\frac{1}{3}$ to both sides and solve for x.

$$x = \frac{1 + \sqrt{5}}{3} \qquad x = \frac{1 - \sqrt{5}}{3}$$ Express solutions with a common denominator.

The solutions are $\frac{1 \pm \sqrt{5}}{3}$, and the solution set is $\left\{\frac{1 \pm \sqrt{5}}{3}\right\}$.

Check Point 5 Solve by completing the square: $2x^2 + 3x - 4 = 0$.

Technology

Obtain a decimal approximation for each solution of $9x^2 - 6x - 4 = 0$, the equation in Example 5.

$$\frac{1 + \sqrt{5}}{3} \approx 1.1$$
$$\frac{1 - \sqrt{5}}{3} \approx -0.4$$

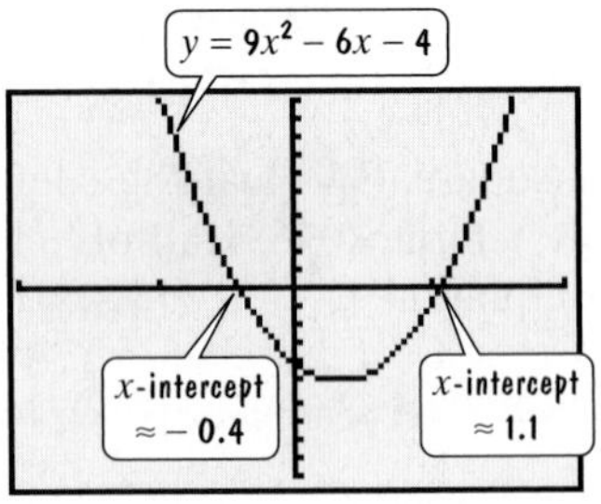

[–2, 2, 1] by [–10, 10, 1]

The x-intercepts of $y = 9x^2 - 6x - 4$ verify the solutions.

4 Solve quadratic equations using the quadratic formula.

Solving Quadratic Equations Using the Quadratic Formula

We can use the method of completing the square to derive a formula that can be used to solve all quadratic equations. The derivation given on the next page also shows a particular quadratic equation, $3x^2 - 2x - 4 = 0$, to specifically illustrate each of the steps.

Deriving the Quadratic Formula

General Form of a Quadratic Equation	Comment	A Specific Example
$ax^2 + bx + c = 0, a > 0$	This is the given equation.	$3x^2 - 2x - 4 = 0$
$x^2 + \frac{b}{a}x + \frac{c}{a} = 0$	Divide both sides by a so that the coefficient of x^2 is 1.	$x^2 - \frac{2}{3}x - \frac{4}{3} = 0$
$x^2 + \frac{b}{a}x = -\frac{c}{a}$	Isolate the binomial by adding $-\frac{c}{a}$ on both sides of the equation.	$x^2 - \frac{2}{3}x = \frac{4}{3}$
$x^2 + \frac{b}{a}x + \left(\frac{b}{2a}\right)^2 = -\frac{c}{a} + \left(\frac{b}{2a}\right)^2$ (half)^2 $x^2 + \frac{b}{a}x + \frac{b^2}{4a^2} = -\frac{c}{a} + \frac{b^2}{4a^2}$	Complete the square. Add the square of half the coefficient of x to both sides.	$x^2 - \frac{2}{3}x + \left(-\frac{1}{3}\right)^2 = \frac{4}{3} + \left(-\frac{1}{3}\right)^2$ (half)^2 $x^2 - \frac{2}{3}x + \frac{1}{9} = \frac{4}{3} + \frac{1}{9}$
$\left(x + \frac{b}{2a}\right)^2 = -\frac{c}{a} \cdot \frac{4a}{4a} + \frac{b^2}{4a^2}$	Factor on the left side and obtain a common denominator on the right side.	$\left(x - \frac{1}{3}\right)^2 = \frac{4}{3} \cdot \frac{3}{3} + \frac{1}{9}$
$\left(x + \frac{b}{2a}\right)^2 = \frac{-4ac + b^2}{4a^2}$ $\left(x + \frac{b}{2a}\right)^2 = \frac{b^2 - 4ac}{4a^2}$	Add fractions on the right side.	$\left(x - \frac{1}{3}\right)^2 = \frac{12 + 1}{9}$ $\left(x - \frac{1}{3}\right)^2 = \frac{13}{9}$
$x + \frac{b}{2a} = \pm\sqrt{\frac{b^2 - 4ac}{4a^2}}$	Apply the square root property.	$x - \frac{1}{3} = \pm\sqrt{\frac{13}{9}}$
$x + \frac{b}{2a} = \pm\frac{\sqrt{b^2 - 4ac}}{2a}$	Take the square root of the quotient, simplifying the denominator.	$x - \frac{1}{3} = \pm\frac{\sqrt{13}}{3}$
$x = \frac{-b}{2a} \pm \frac{\sqrt{b^2 - 4ac}}{2a}$	Solve for x by subtracting $\frac{b}{2a}$ from both sides.	$x = \frac{1}{3} \pm \frac{\sqrt{13}}{3}$
$x = \frac{-b \pm \sqrt{b^2 - 4ac}}{2a}$	Combine fractions on the right side.	$x = \frac{1 \pm \sqrt{13}}{3}$

The formula shown at the bottom of the left column is called the *quadratic formula*. A similar proof shows that the same formula can be used to solve quadratic equations if a, the coefficient of the x^2-term, is negative.

The Quadratic Formula

The solutions of a quadratic equation in general form $ax^2 + bx + c = 0$, with $a \neq 0$, are given by the **quadratic formula**

$$x = \frac{-b \pm \sqrt{b^2 - 4ac}}{2a}.$$

x equals negative b plus or minus the square root of $b^2 - 4ac$, all divided by $2a$.

To use the quadratic formula, write the quadratic equation in general form if necessary. Then determine the numerical values for a (the coefficient of the x^2-term), b (the coefficient of the x-term), and c (the constant term). Substitute the values of a, b, and c into the quadratic formula and evaluate the expression. The $\pm$ sign indicates that there are two solutions of the equation.

To Die at Twenty

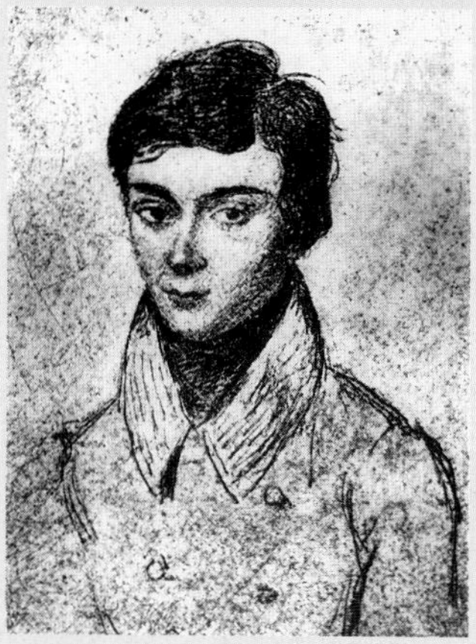

Can the equations

$$7x^5 + 12x^3 - 9x + 4 = 0$$

and

$$8x^6 - 7x^5 + 4x^3 - 19 = 0$$

be solved using a formula similar to the quadratic formula? The first equation has five solutions and the second has six solutions, but they cannot be found using a formula. How do we know? In 1832, a 20-year-old Frenchman, Evariste Galois, wrote down a proof showing that there is no general formula to solve equations when the exponent on the variable is 5 or greater. Galois was jailed as a political activist several times while still a teenager. The day after his brilliant proof he fought a duel over a woman. The duel was a political setup. As he lay dying, Galois told his brother, Alfred, of the manuscript that contained his proof: "Mathematical manuscripts are in my room. On the table. Take care of my work. Make it known. Important. Don't cry, Alfred. I need all my courage—to die at twenty." (Our source is Leopold Infeld's biography of Galois, *Whom the Gods Love*. Some historians, however, dispute the story of Galois's ironic death the very day after his algebraic proof. Mathematical truths seem more reliable than historical ones!)

EXAMPLE 6 Solving a Quadratic Equation Using the Quadratic Formula

Solve using the quadratic formula: $2x^2 - 6x + 1 = 0$.

Solution The given equation is in general form. Begin by identifying the values for a, b, and c.

$$2x^2 - 6x + 1 = 0$$

$a = 2$ $b = -6$ $c = 1$

Substituting these values into the quadratic formula and simplifying gives the equation's solutions.

$$x = \frac{-b \pm \sqrt{b^2 - 4ac}}{2a}$$ Use the quadratic formula.

$$= \frac{-(-6) \pm \sqrt{(-6)^2 - 4(2)(1)}}{2 \cdot 2}$$ Substitute the values for a, b, and c: $a = 2$, $b = -6$, and $c = 1$.

$$= \frac{6 \pm \sqrt{36 - 8}}{4}$$ $-(-6) = 6$, $(-6)^2 = (-6)(-6) = 36$, and $4(2)(1) = 8$.

$$= \frac{6 \pm \sqrt{28}}{4}$$ Complete the subtraction under the radical.

$$= \frac{6 \pm 2\sqrt{7}}{4}$$ $\sqrt{28} = \sqrt{4 \cdot 7} = \sqrt{4}\sqrt{7} = 2\sqrt{7}$

$$= \frac{2(3 \pm \sqrt{7})}{4}$$ Factor out 2 from the numerator.

$$= \frac{3 \pm \sqrt{7}}{2}$$ Divide the numerator and denominator by 2.

The solution set is $\left\{\frac{3 + \sqrt{7}}{2}, \frac{3 - \sqrt{7}}{2}\right\}$ or $\left\{\frac{3 \pm \sqrt{7}}{2}\right\}$.

Technology

You can use a graphing utility to verify that the solutions of $2x^2 - 6x + 1 = 0$ are $\frac{3 \pm \sqrt{7}}{2}$. Begin by entering $y_1 = 2x^2 - 6x + 1$ in the [Y=] screen. Then evaluate this equation at each of the proposed solutions.

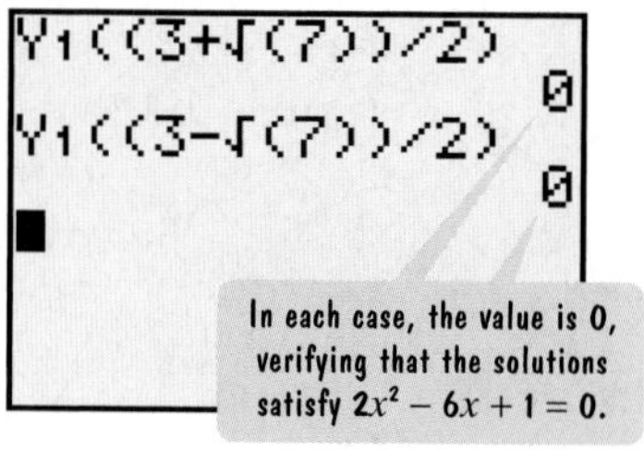
Y1((3+√(7))/2)
0
Y1((3−√(7))/2)
0
In each case, the value is 0, verifying that the solutions satisfy 2x² − 6x + 1 = 0.

Check Point 6 Solve using the quadratic formula:

$$2x^2 + 2x - 1 = 0.$$

EXAMPLE 7 Solving a Quadratic Equation Using the Quadratic Formula

Solve using the quadratic formula: $3x^2 - 2x + 4 = 0$.

Solution The given equation is in general form. Begin by identifying the values for *a*, *b*, and *c*.

$$3x^2 - 2x + 4 = 0$$

$a = 3 \quad b = -2 \quad c = 4$

$x = \dfrac{-b \pm \sqrt{b^2 - 4ac}}{2a}$ Use the quadratic formula.

$= \dfrac{-(-2) \pm \sqrt{(-2)^2 - 4(3)(4)}}{2(3)}$ Substitute the values for a, b, and c: $a = 3$, $b = -2$, and $c = 4$,

$= \dfrac{2 \pm \sqrt{4 - 48}}{6}$ $-(-2) = 2$ and $(-2)^2 = (-2)(-2) = 4$.

$= \dfrac{2 \pm \sqrt{-44}}{6}$ Subtract under the radical. Because the number under the radical sign is negative, the solutions will not be real numbers.

$= \dfrac{2 \pm 2i\sqrt{11}}{6}$ $\sqrt{-44} = \sqrt{4(11)(-1)} = 2i\sqrt{11}$

$= \dfrac{2(1 \pm i\sqrt{11})}{6}$ Factor 2 from the numerator.

$= \dfrac{1 \pm i\sqrt{11}}{3}$ Divide numerator and denominator by 2.

$= \dfrac{1}{3} \pm i\dfrac{\sqrt{11}}{3}$ Write the complex numbers in standard form.

The solutions are complex conjugates, and the solution set is $\left\{\frac{1}{3} + i\frac{\sqrt{11}}{3}, \frac{1}{3} - i\frac{\sqrt{11}}{3}\right\}$ or $\left\{\frac{1}{3} \pm i\frac{\sqrt{11}}{3}\right\}$.

Study Tip

Checking irrational and imaginary solutions can be time-consuming. The solutions given by the quadratic formula are always correct, unless you have made a careless error. Checking for computational errors or errors in simplification is sufficient.

If $ax^2 + bx + c = 0$ has imaginary solutions, the graph of $y = ax^2 + bx + c$ will not have *x*-intercepts. This is illustrated by the imaginary solutions of $3x^2 - 2x + 4 = 0$ in Example 7 and the graph of $y = 3x^2 - 2x + 4$ in Figure 1.20.

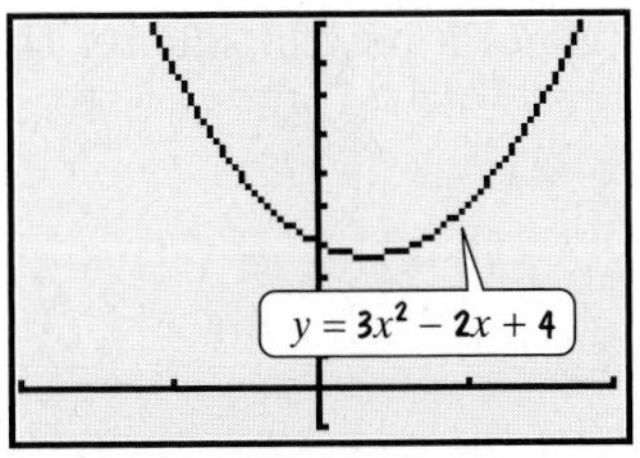

[−2, 2, 1] by [−1, 10, 1]

Figure 1.20 This graph has no *x*-intercepts

Check Point 7 Solve using the quadratic formula:

$$x^2 - 2x + 2 = 0.$$

5 Use the discriminant to determine the number and type of solutions.

The Discriminant

The quantity $b^2 - 4ac$, which appears under the radical sign in the quadratic formula, is called the **discriminant**. Table 1.3 on the next page shows how the discriminant of the quadratic equation $ax^2 + bx + c = 0$ determines the number and type of solutions.

Table 1.3 The Discriminant and the Kinds of Solutions to $ax^2 + bx + c = 0$

Discriminant $b^2 - 4ac$	Kinds of Solutions to $ax^2 + bx + c = 0$	Graph of $y = ax^2 + bx + c$
$b^2 - 4ac > 0$	**Two unequal real solutions;** If a, b, and c are rational numbers and the discriminant is a perfect square, the solutions are rational. If the discriminant is not a perfect square, the solutions are irrational.	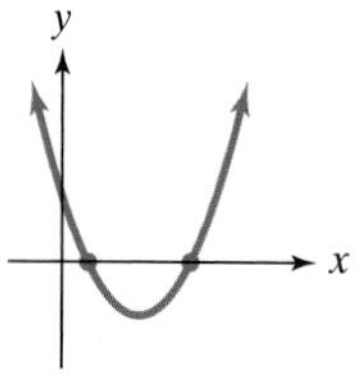 Two x-intercepts
$b^2 - 4ac = 0$	**One solution (a repeated solution) that is a real number;** If a, b, and c are rational numbers, the repeated solution is also a rational number.	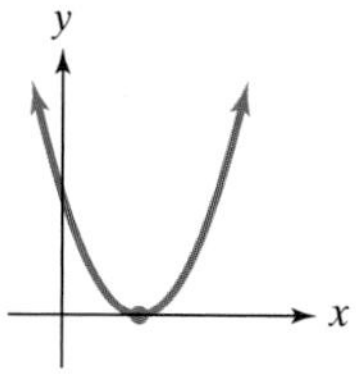 One x-intercept
$b^2 - 4ac < 0$	**No real solution; two imaginary solutions;** The solutions are complex conjugates.	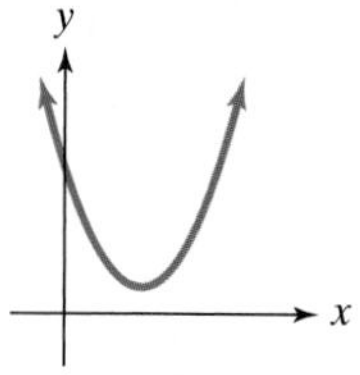 No x-intercepts

EXAMPLE 8 Using the Discriminant

For each equation, compute the discriminant. Then determine the number and type of solutions:

a. $3x^2 + 4x - 5 = 0$ **b.** $9x^2 - 6x + 1 = 0$ **c.** $3x^2 - 8x + 7 = 0.$

Solution Begin by identifying the values for a, b, and c in each equation. Then compute $b^2 - 4ac$, the discriminant.

a. $3x^2 + 4x - 5 = 0$

$a = 3$ $b = 4$ $c = -5$

Substitute and compute the discriminant:

$$b^2 - 4ac = 4^2 - 4 \cdot 3(-5) = 16 - (-60) = 16 + 60 = 76.$$

The discriminant, 76, is a positive number that is not a perfect square. Thus, there are two irrational solutions.

b. $9x^2 - 6x + 1 = 0$

$a = 9$ $b = -6$ $c = 1$

Substitute and compute the discriminant:

$$b^2 - 4ac = (-6)^2 - 4 \cdot 9 \cdot 1 = 36 - 36 = 0.$$

The discriminant, 0, shows that there is only one real solution. This real solution is a rational number.

c. $3x^2 - 8x + 7 = 0$

$a = 3$ $b = -8$ $c = 7$

$$b^2 - 4ac = (-8)^2 - 4 \cdot 3 \cdot 7 = 64 - 84 = -20$$

The negative discriminant, -20, shows that there are two imaginary solutions. (These solutions are complex conjugates of each other.)

Check Point 8 For each equation, compute the discriminant. Then determine the number and type of solutions:

a. $x^2 + 6x + 9 = 0$ **b.** $2x^2 - 7x - 4 = 0$ **c.** $3x^2 - 2x + 4 = 0$.

6 Determine the most efficient method to use when solving a quadratic equation.

Determining Which Method to Use

All quadratic equations can be solved by the quadratic formula. However, if an equation is in the form $u^2 = d$, such as $x^2 = 5$ or $(2x + 3)^2 = 8$, it is faster to use the square root property, taking the square root of both sides. If the equation is not in the form $u^2 = d$, write the quadratic equation in general form $(ax^2 + bx + c = 0)$. Try to solve the equation by factoring. If $ax^2 + bx + c$ cannot be factored, then solve the quadratic equation by the quadratic formula.

Because we used the method of completing the square to derive the quadratic formula, we no longer need it for solving quadratic equations. However, we will use completing the square later in the book to help graph circles and other kinds of equations.

Table 1.4 summarizes our observations about which technique to use when solving a quadratic equation.

Table 1.4 Determining the Most Efficient Technique to Use When Solving a Quadratic Equation

Description and Form of the Quadratic Equation	Most Efficient Solution Method	Example
$ax^2 + bx + c = 0$ and $ax^2 + bx + c$ can be factored easily.	Factor and use the zero-product principle.	$3x^2 + 5x - 2 = 0$ $(3x - 1)(x + 2) = 0$ $3x - 1 = 0$ or $x + 2 = 0$ $x = \frac{1}{3}$ $\quad x = -2$
$ax^2 + bx = 0$ The quadratic equation has no constant term. $(c = 0)$	Factor and use the zero-product principle.	$6x^2 + 9x = 0$ $3x(2x + 3) = 0$ $3x = 0$ or $2x + 3 = 0$ $x = 0$ $\quad 2x = -3$ $x = -\frac{3}{2}$
$ax^2 + c = 0$ The quadratic equation has no x-term. $(b = 0)$	Solve for x^2 and apply the square root property.	$7x^2 - 4 = 0$ $7x^2 = 4$ $x^2 = \frac{4}{7}$ $x = \pm\frac{2}{\sqrt{7}} = \pm\frac{2}{\sqrt{7}} \cdot \frac{\sqrt{7}}{\sqrt{7}} = \pm\frac{2\sqrt{7}}{7}$
$u^2 = d$; u is a first-degree polynomial.	Use the square root property.	$(x + 4)^2 = 5$ $x + 4 = \pm\sqrt{5}$ $x = -4 \pm \sqrt{5}$
$ax^2 + bx + c = 0$ and $ax^2 + bx + c$ cannot be factored or the factoring is too difficult.	Use the quadratic formula: $x = \frac{-b \pm \sqrt{b^2 - 4ac}}{2a}$.	$x^2 - 2x - 6 = 0$ $a = 1$ $\quad b = -2$ $\quad c = -6$ $x = \frac{-(-2) \pm \sqrt{(-2)^2 - 4(1)(-6)}}{2}$ $= \frac{2 \pm \sqrt{4 - (-24)}}{2}$ $= \frac{2 \pm \sqrt{28}}{2} = \frac{2 \pm \sqrt{4}\sqrt{7}}{2}$ $= \frac{2 \pm 2\sqrt{7}}{2} = \frac{2(1 \pm \sqrt{7})}{2}$ $= 1 \pm \sqrt{7}$

7 Solve problems modeled by quadratic equations.

Applications

EXAMPLE 9 Blood Pressure and Age

We opened this section with graphs (Figure 1.18, repeated on the right) showing that a person's normal systolic pressure, measured in millimeters of mercury (mm Hg), depends on his or her age. The formula

$$P = 0.006A^2 - 0.02A + 120$$

models a man's normal systolic pressure, P, at age A. Find the age, to the nearest year, of a man whose normal systolic blood pressure is 125 mm Hg.

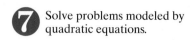

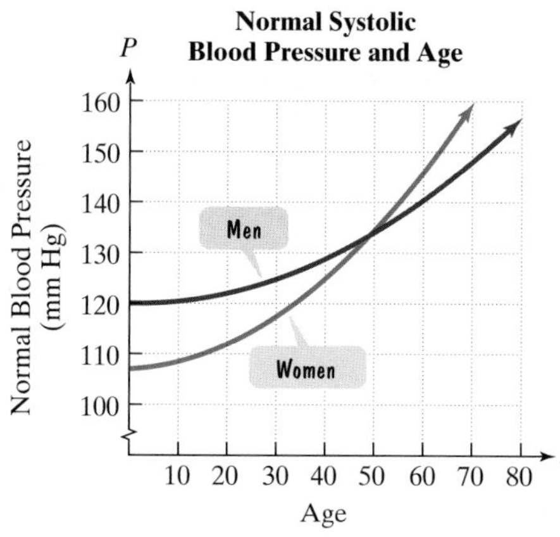

Figure 1.18 (repeated)

Solution We are interested in the age of a man with a normal systolic blood pressure of 125 millimeters of mercury. Thus, we substitute 125 for P in the given formula for men. Then we solve for A, the man's age.

$$P = 0.006A^2 - 0.02A + 120$$ This is the given formula for men.

$$125 = 0.006A^2 - 0.02A + 120$$ Substitute 125 for P.

$$0 = 0.006A^2 - 0.02A - 5$$ Subtract 125 from both sides and write the quadratic equation in general form.

$a = 0.006$ $b = -0.02$ $c = -5$

Because the trinomial on the right side of the equation is prime, we solve using the quadratic formula.

Notice that the variable is A, rather than the usual x.

$$A = \frac{-b \pm \sqrt{b^2 - 4ac}}{2a}$$ Use the quadratic formula.

$$= \frac{-(-0.02) \pm \sqrt{(-0.02)^2 - 4(0.006)(-5)}}{2(0.006)}$$ Substitute the values for a, b, and c: $a = 0.006$, $b = -0.02$, and $c = -5$.

$$= \frac{0.02 \pm \sqrt{0.1204}}{0.012}$$ Use a calculator to simplify the radicand.

$$\approx \frac{0.02 \pm 0.347}{0.012}$$ Use a calculator: $\sqrt{0.1204} \approx 0.347$.

$$A \approx \frac{0.02 + 0.347}{0.012} \quad \text{or} \quad A \approx \frac{0.02 - 0.347}{0.012}$$

$$A \approx 31 \qquad\qquad A \approx -27$$ Use a calculator and round to the nearest integer.

Reject this solution. Age cannot be negative.

The positive solution indicates that 31 is the approximate age of a man whose normal systolic blood pressure is 125 mm Hg. The solution can be visualized as the point (31, 125) on the red graph representing men in Figure 1.18. Take a moment to locate this point on the graph.

Technology

On most calculators, here is how to approximate

$$\frac{0.02 + \sqrt{0.1204}}{0.012}.$$

Many Scientific Calculators

[(] .02 [+] .1204 [√] [)] [÷] .012 [=]

Many Graphing Calculators

[(] .02 [+] [√] .1204 [)] [÷] .012 [ENTER]

If your calculator displays an open parenthesis after √, you'll need to enter another closed parenthesis here.

Check Point 9 The formula $P = 0.01A^2 + 0.05A + 107$ models a woman's normal systolic blood pressure, P, at age A. Use this formula to find the age, to the nearest year, of a woman whose normal systolic blood pressure is 115 mm Hg. Use the blue graph in Figure 1.18 to verify your solution.

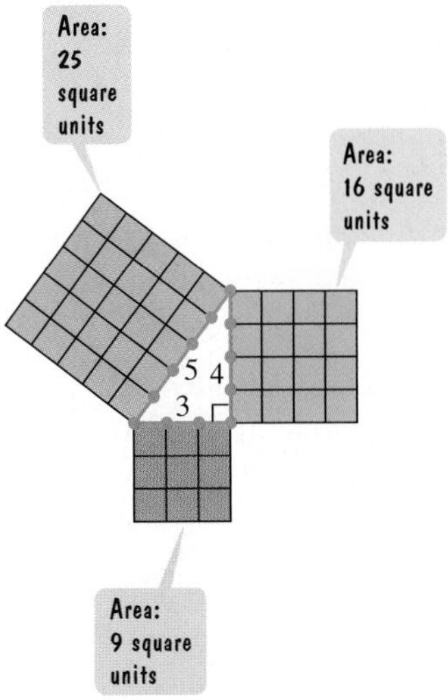

Figure 1.21 The area of the large square equals the sum of the areas of the smaller squares.

In our next example, we will be using the *Pythagorean Theorem* to obtain a verbal model. The ancient Greek philosopher and mathematician Pythagoras (approximately 582–500 B.C.) founded a school whose motto was "All is number." Pythagoras is best remembered for his work with the **right triangle**, a triangle with one angle measuring 90°. The side opposite the 90° angle is called the **hypotenuse**. The other sides are called **legs**. Pythagoras found that if he constructed squares on each of the legs, as well as a larger square on the hypotenuse, the sum of the areas of the smaller squares is equal to the area of the larger square. This is illustrated in Figure 1.21.

This relationship is usually stated in terms of the lengths of the three sides of a right triangle and is called the **Pythagorean Theorem**.

The Pythagorean Theorem

The sum of the squares of the lengths of the legs of a right triangle equals the square of the length of the hypotenuse.

If the legs have lengths a and b, and the hypotenuse has length c, then

$$a^2 + b^2 = c^2.$$

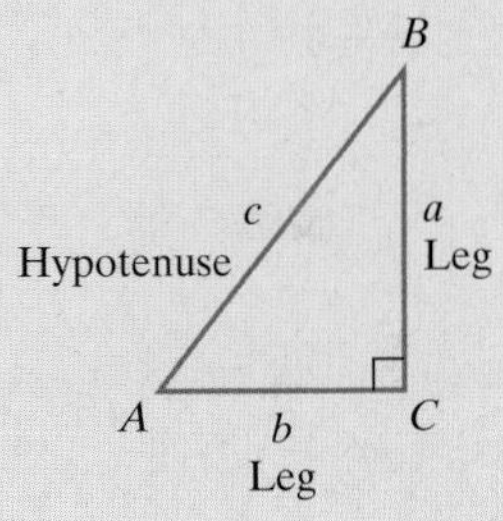

EXAMPLE 10 Using the Pythagorean Theorem

In a 25-inch television set, the length of the screen's diagonal is 25 inches. If the screen's height is 15 inches, what is its width?

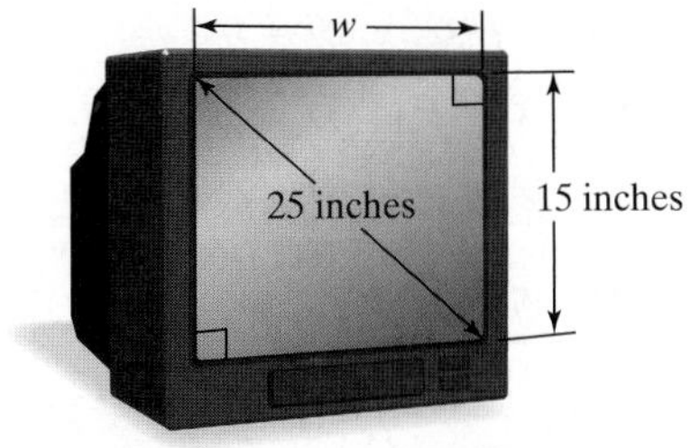

Figure 1.22 A right triangle is formed by the television's height, width, and diagonal.

Solution Figure 1.22 shows a right triangle that is formed by the height, width, and diagonal. We can find w, the screen's width, using the Pythagorean Theorem.

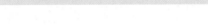

(Leg)²	plus	(Leg)²	equals	(Hypotenuse)².
w^2	$+$	15^2	$=$	25^2

The equation $w^2 + 15^2 = 25^2$ can be solved most efficiently by the square root property.

$$w^2 + 15^2 = 25^2$$ This is the equation resulting from the Pythagorean Theorem.

$$w^2 + 225 = 625$$ Square 15 and 25.

$$w^2 = 400$$ Isolate w^2 by subtracting 225 from both sides.

$$w = \sqrt{400} \quad \text{or} \quad w = -\sqrt{400}$$ Apply the square root property.

$$w = 20 \qquad w = -20$$ Simplify.

Because w represents the width of the television's screen, this dimension must be positive. We reject -20. Thus, the width of the television is 20 inches.

Check Point 10 What is the width of a 15-inch television set whose height is 9 inches?

EXERCISE SET 1.5

Practice Exercises

Solve each equation in Exercises 1–14 by factoring.

1. $x^2 - 3x - 10 = 0$ **2.** $x^2 - 13x + 36 = 0$

3. $x^2 = 8x - 15$ **4.** $x^2 = -11x - 10$

5. $6x^2 + 11x - 10 = 0$ **6.** $9x^2 + 9x + 2 = 0$

7. $3x^2 - 2x = 8$ **8.** $4x^2 - 13x = -3$

9. $3x^2 + 12x = 0$ **10.** $5x^2 - 20x = 0$

11. $2x(x - 3) = 5x^2 - 7x$ **12.** $16x(x - 2) = 8x - 25$

13. $7 - 7x = (3x + 2)(x - 1)$

14. $10x - 1 = (2x + 1)^2$

Solve each equation in Exercises 15–34 by the square root property.

15. $3x^2 = 27$ **16.** $5x^2 = 45$

17. $5x^2 + 1 = 51$ **18.** $3x^2 - 1 = 47$

19. $2x^2 - 5 = -55$ **20.** $2x^2 - 7 = -15$

21. $(x + 2)^2 = 25$ **22.** $(x - 3)^2 = 36$

23. $3(x - 4)^2 = 15$ **24.** $3(x + 4)^2 = 21$

25. $(x + 3)^2 = -16$ **26.** $(x - 1)^2 = -9$

27. $(x - 3)^2 = -5$ **28.** $(x + 2)^2 = -7$

29. $(3x + 2)^2 = 9$ **30.** $(4x - 1)^2 = 16$

31. $(5x - 1)^2 = 7$ **32.** $(8x - 3)^2 = 5$

33. $(3x - 4)^2 = 8$ **34.** $(2x + 8)^2 = 27$

In Exercises 35–46, determine the constant that should be added to the binomial so that it becomes a perfect square trinomial. Then write and factor the trinomial.

35. $x^2 + 12x$ **36.** $x^2 + 16x$

37. $x^2 - 10x$ **38.** $x^2 - 14x$

39. $x^2 + 3x$ **40.** $x^2 + 5x$

41. $x^2 - 7x$ **42.** $x^2 - 9x$

43. $x^2 - \frac{2}{3}x$ **44.** $x^2 + \frac{4}{5}x$

45. $x^2 - \frac{1}{3}x$ **46.** $x^2 - \frac{1}{4}x$

Solve each equation in Exercises 47–64 by completing the square.

47. $x^2 + 6x = 7$ **48.** $x^2 + 6x = -8$

49. $x^2 - 2x = 2$ **50.** $x^2 + 4x = 12$

51. $x^2 - 6x - 11 = 0$ **52.** $x^2 - 2x - 5 = 0$

53. $x^2 + 4x + 1 = 0$ **54.** $x^2 + 6x - 5 = 0$

55. $x^2 - 5x + 6 = 0$ **56.** $x^2 + 7x - 8 = 0$

57. $x^2 + 3x - 1 = 0$ **58.** $x^2 - 3x - 5 = 0$

59. $2x^2 - 7x + 3 = 0$ **60.** $2x^2 + 5x - 3 = 0$

61. $4x^2 - 4x - 1 = 0$ **62.** $2x^2 - 4x - 1 = 0$

63. $3x^2 - 2x - 2 = 0$ **64.** $3x^2 - 5x - 10 = 0$

Solve each equation in Exercises 65–74 using the quadratic formula.

65. $x^2 + 8x + 15 = 0$ **66.** $x^2 + 8x + 12 = 0$

67. $x^2 + 5x + 3 = 0$ **68.** $x^2 + 5x + 2 = 0$

69. $3x^2 - 3x - 4 = 0$ **70.** $5x^2 + x - 2 = 0$

71. $4x^2 = 2x + 7$ **72.** $3x^2 = 6x - 1$

73. $x^2 - 6x + 10 = 0$ **74.** $x^2 - 2x + 17 = 0$

In Exercises 75–82, compute the discriminant. Then determine the number and type of solutions for the given equation.

75. $x^2 - 4x - 5 = 0$ **76.** $4x^2 - 2x + 3 = 0$

77. $2x^2 - 11x + 3 = 0$ **78.** $2x^2 + 11x - 6 = 0$

79. $x^2 - 2x + 1 = 0$ **80.** $3x^2 = 2x - 1$

81. $x^2 - 3x - 7 = 0$ **82.** $3x^2 + 4x - 2 = 0$

Solve each equation in Exercises 83–108 by the method of your choice.

83. $2x^2 - x = 1$ **84.** $3x^2 - 4x = 4$

85. $5x^2 + 2 = 11x$ **86.** $5x^2 = 6 - 13x$

87. $3x^2 = 60$ **88.** $2x^2 = 250$

89. $x^2 - 2x = 1$ **90.** $2x^2 + 3x = 1$

91. $(2x + 3)(x + 4) = 1$ **92.** $(2x - 5)(x + 1) = 2$

93. $(3x - 4)^2 = 16$ **94.** $(2x + 7)^2 = 25$

95. $3x^2 - 12x + 12 = 0$ **96.** $9 - 6x + x^2 = 0$

97. $4x^2 - 16 = 0$ **98.** $3x^2 - 27 = 0$

99. $x^2 - 6x + 13 = 0$ **100.** $x^2 - 4x + 29 = 0$

101. $x^2 = 4x - 7$ **102.** $5x^2 = 2x - 3$

103. $2x^2 - 7x = 0$ **104.** $2x^2 + 5x = 3$

105. $\frac{1}{x} + \frac{1}{x + 2} = \frac{1}{3}$ **106.** $\frac{1}{x} + \frac{1}{x + 3} = \frac{1}{4}$

107. $\frac{2x}{x - 3} + \frac{6}{x + 3} = -\frac{28}{x^2 - 9}$

108. $\frac{3}{x - 3} + \frac{5}{x - 4} = \frac{x^2 - 20}{x^2 - 7x + 12}$

In Exercises 109–114, find the x-intercept(s) of the graph of each equation. Use the x-intercepts to match the equation with its graph. The graphs are shown in [−10, 10, 1] by [−10, 10, 1] viewing rectangles and labeled (a) through (f).

109. $y = x^2 - 4x - 5$ **110.** $y = x^2 - 6x + 7$

111. $y = -(x + 1)^2 + 4$ **112.** $y = -(x + 3)^2 + 1$

113. $y = x^2 - 2x + 2$ **114.** $y = x^2 + 6x + 9$

a.

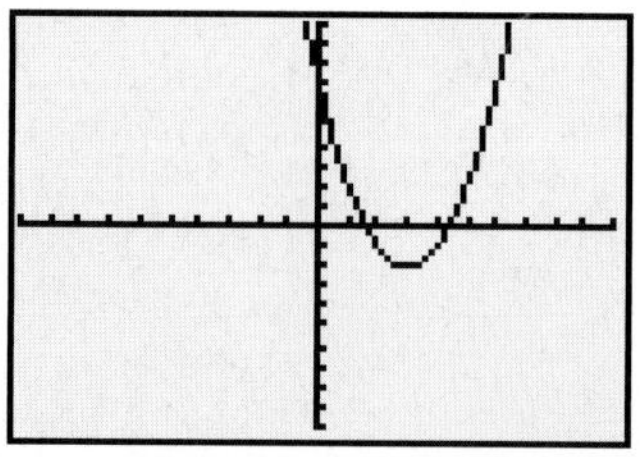

b.

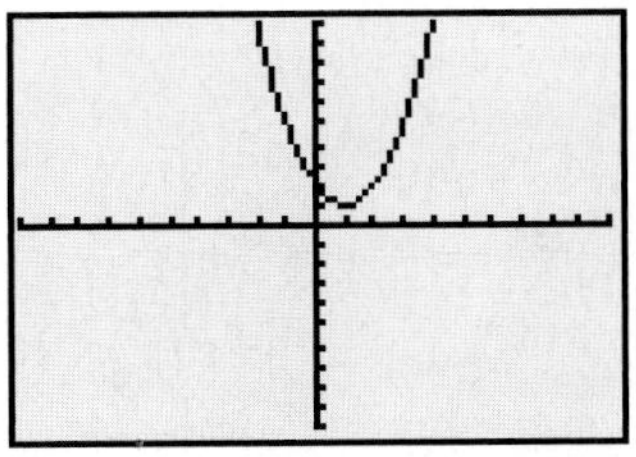

c.

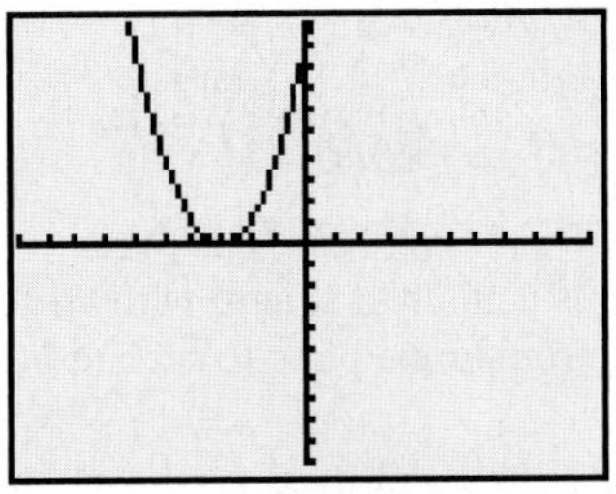

d.

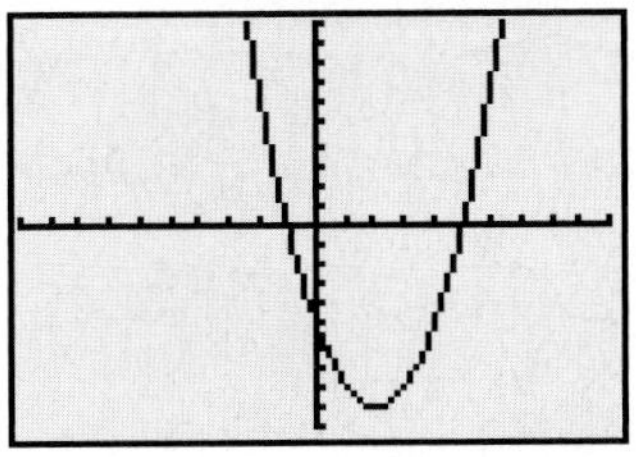

e.

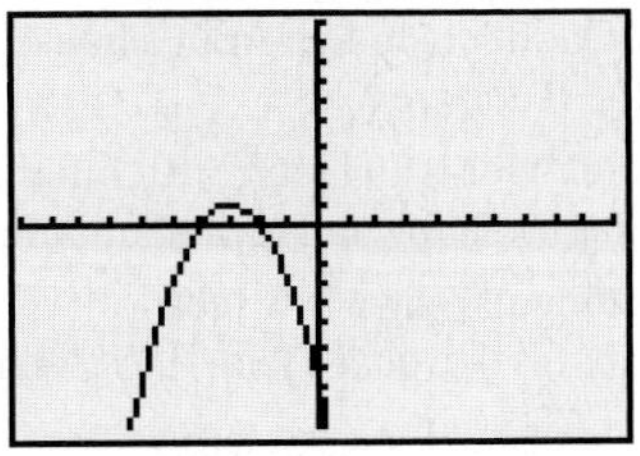

f.

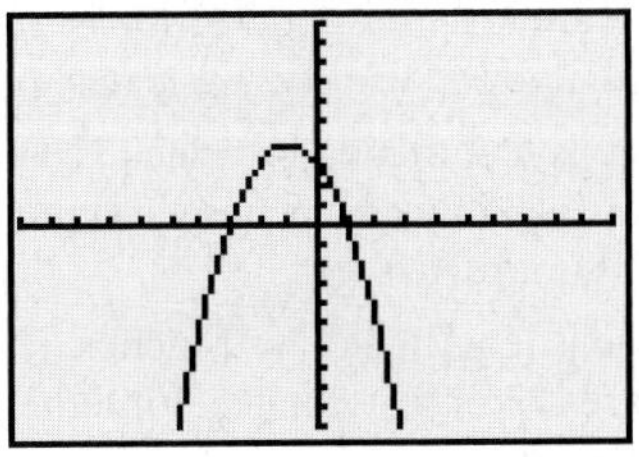

In Exercises 115–122, find all values of x satisfying the given conditions.

115. $y = 2x^2 - 3x$ and $y = 2$.

116. $y = 5x^2 + 3x$ and $y = 2$.

117. $y_1 = x - 1$, $y_2 = x + 4$, and $y_1 y_2 = 14$.

118. $y_1 = x - 3$, $y_2 = x + 8$, and $y_1 y_2 = -30$.

119. $y_1 = \frac{2x}{x + 2}$, $y_2 = \frac{3}{x + 4}$, and $y_1 + y_2 = 1$.

120. $y_1 = \frac{3}{x - 1}$, $y_2 = \frac{8}{x}$, and $y_1 + y_2 = 3$.

121. $y_1 = 2x^2 + 5x - 4$, $y_2 = -x^2 + 15x - 10$, and $y_1 - y_2 = 0$.

122. $y_1 = -x^2 + 4x - 2$, $y_2 = -3x^2 + x - 1$, and $y_1 - y_2 = 0$.

Practice Plus

In Exercises 123–124, list all numbers that must be excluded from the domain of each rational expression.

123. $\frac{3}{2x^2 + 4x - 9}$ **124.** $\frac{7}{2x^2 - 8x + 5}$

125. When the sum of 6 and twice a positive number is subtracted from the square of the number, 0 results. Find the number.

126. When the sum of 1 and twice a negative number is subtracted from twice the square of the number, 0 results. Find the number.

In Exercises 127–130, solve each equation by the method of your choice.

127. $\frac{1}{x^2 - 3x + 2} = \frac{1}{x + 2} + \frac{5}{x^2 - 4}$

128. $\frac{x - 1}{x - 2} + \frac{x}{x - 3} = \frac{1}{x^2 - 5x + 6}$

129. $\sqrt{2}x^2 + 3x - 2\sqrt{2} = 0$

130. $\sqrt{3}x^2 + 6x + 7\sqrt{3} = 0$

Application Exercises

A driver's age has something to do with his or her chance of getting into a fatal car crash. The bar graph shows the number of fatal vehicle crashes per 100 million miles driven for drivers of various age groups. For example, 25-year-old drivers are involved in 4.1 fatal crashes per 100 million miles driven. Thus, when a group of 25-year-old Americans have driven a total of 100 million miles, approximately 4 have been in accidents in which someone died.

Age of U.S. Drivers and Fatal Crashes

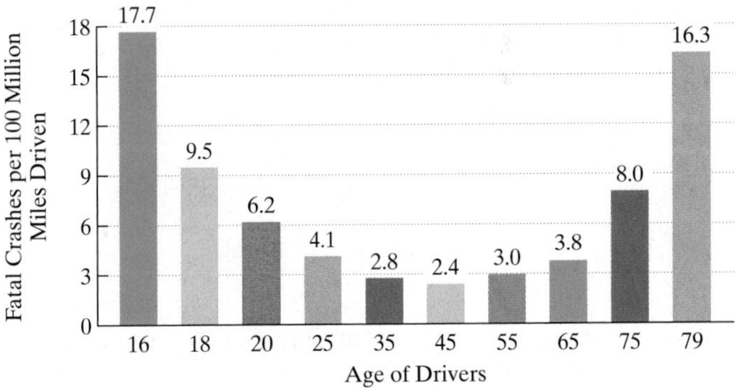

Source: Insurance Institute for Highway Safety

The number of fatal vehicle crashes per 100 million miles, y, for drivers of age x can be modeled by the formula

$$y = 0.013x^2 - 1.19x + 28.24.$$

Use the formula to solve Exercises 131–132.

131. What age groups are expected to be involved in 3 fatal crashes per 100 million miles driven? How well does the formula model the trend in the actual data shown in the bar graph?

132. What age groups are expected to be involved in 10 fatal crashes per 100 million miles driven? How well does the formula model the trend in the actual data shown in the bar graph ?

Throwing events in track and field include the shot put, the discus throw, the hammer throw, and the javelin throw. The distance that an athlete can achieve depends on the initial velocity of the object thrown and the angle above the horizontal at which the object leaves the hand.

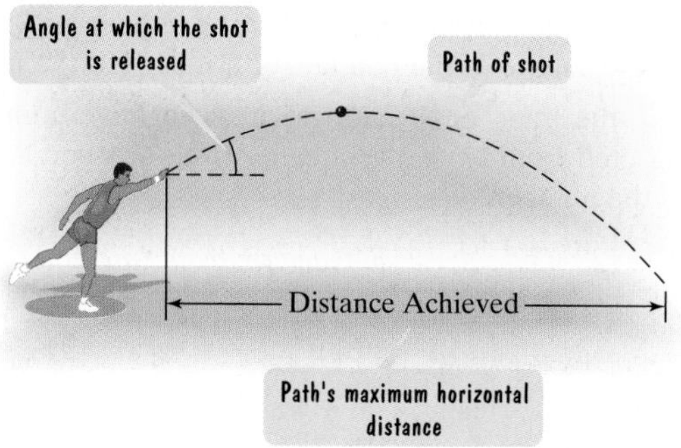

In Exercises 133–134, an athlete whose event is the shot put releases the shot with the same initial velocity, but at different angles.

133. When the shot is released at an angle of 35°, its path can be modeled by the formula

$$y = -0.01x^2 + 0.7x + 6.1,$$

in which x is the shot's horizontal distance, in feet, and y is its height, in feet. This formula is shown by one of the graphs, (a) or (b), in the figure. Use the formula to determine the shot's maximum distance. Use a calculator and round to the nearest tenth of a foot. Which graph, (a) or (b), shows the shot's path?

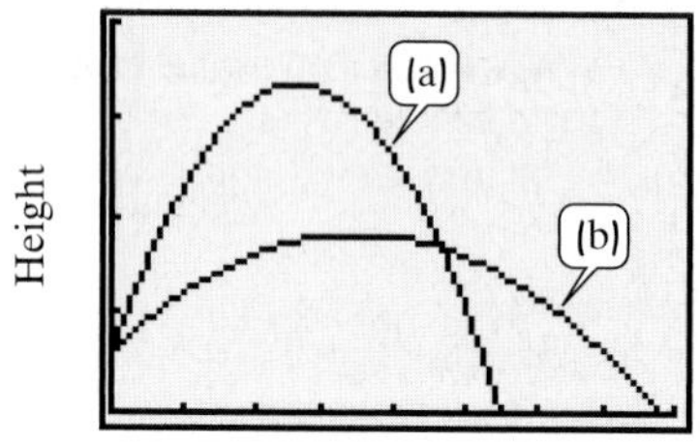

Horizontal Distance
[0, 80, 10] by [0, 40, 10]

134. When the shot is released at an angle of 65°, its path can be modeled by the formula

$$y = -0.04x^2 + 2.1x + 6.1,$$

in which x is the shot's horizontal distance, in feet, and y is its height, in feet. This formula is shown by one of the graphs, (a) or (b), in the figure in Exercise 133. Use the formula to determine the shot's maximum distance. Use a calculator and round to the nearest tenth of a foot. Which graph, (a) or (b), shows the shot's path?

Use the Pythagorean Theorem and the square root property to solve Exercises 135–138. Express answers in simplified radical form. Then find a decimal approximation to the nearest tenth.

135. The figure at the top of the next column shows that the doorway into a room is 4 feet wide and 8 feet high. What is the length of the longest rectangular panel that can be taken through this doorway diagonally?

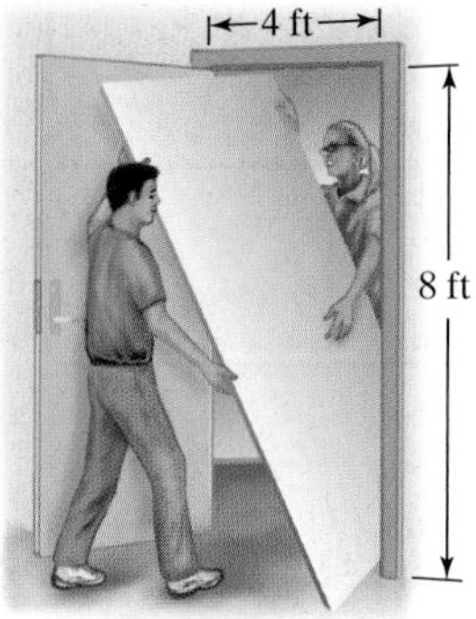

136. A baseball diamond is actually a square with 90-foot sides. What is the distance from home plate to second base?

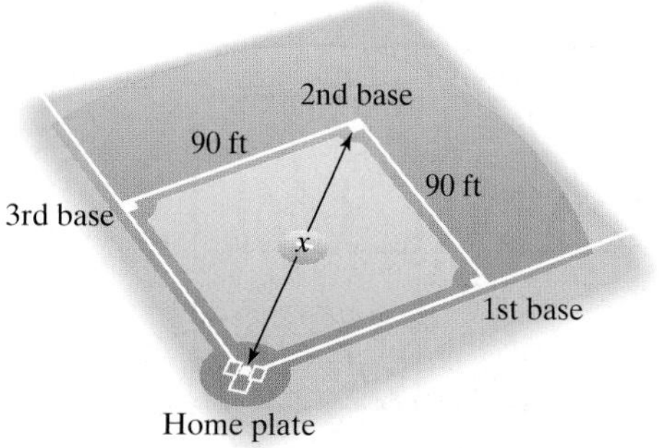

137. The base of a 20-foot ladder is 15 feet from a house. How far up the house does the ladder reach?

138. The base of a 30-foot ladder is 10 feet from a building. If the ladder reaches the flat roof, how tall is the building?

139. The length of a rectangular sign is 3 feet longer than the width. If the sign's area is 54 square feet, find its length and width.

140. A rectangular parking lot has a length that is 3 yards greater than the width. The area of the parking lot is 180 square yards. Find the length and the width.

141. Each side of a square is lengthened by 3 inches. The area of this new, larger square is 64 square inches. Find the length of a side of the original square.

142. Each side of a square is lengthened by 2 inches. The area of this new, larger square is 36 square inches. Find the length of a side of the original square.

143. A pool measuring 10 meters by 20 meters is surrounded by a path of uniform width, as shown in the figure. If the area of the pool and the path combined is 600 square meters, what is the width of the path?

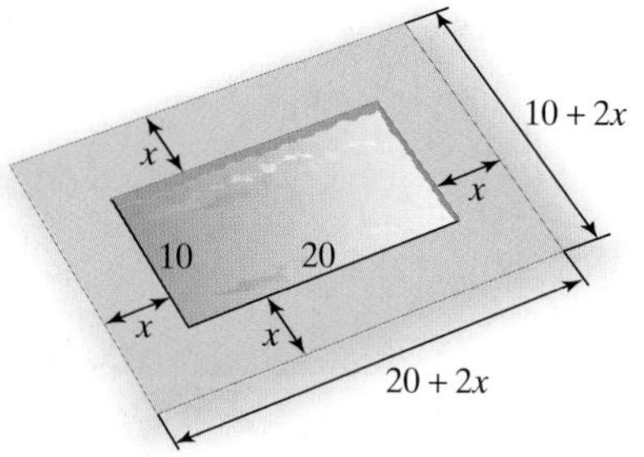

144. A vacant rectangular lot is being turned into a community vegetable garden measuring 15 meters by 12 meters. A path of uniform width is to surround the garden, as shown in the figure. If the area of the garden and path combined is 378 square meters, find the width of the path.

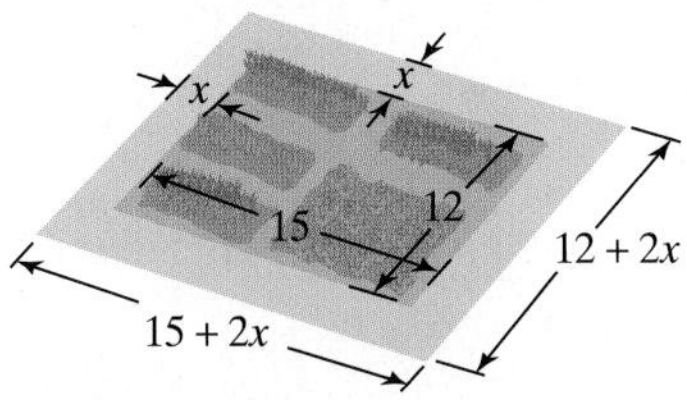

145. A machine produces open boxes using square sheets of metal. The figure illustrates that the machine cuts equal-sized squares measuring 2 inches on a side from the corners and then shapes the metal into an open box by turning up the sides. If each box must have a volume of 200 cubic inches, find the length and width of the open box.

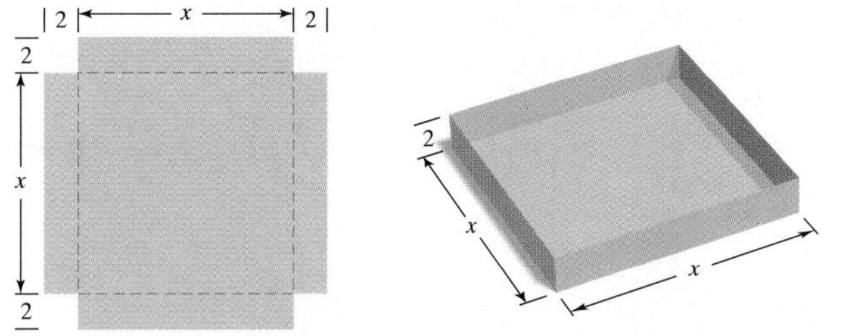

146. A machine produces open boxes using square sheets of metal. The machine cuts equal-sized squares measuring 3 inches on a side from the corners and then shapes the metal into an open box by turning up the sides. If each box must have a volume of 75 cubic inches, find the length and width of the open box.

147. A rain gutter is made from sheets of aluminum that are 20 inches wide. As shown in the figure, the edges are turned up to form right angles. Determine the depth of the gutter that will allow a cross-sectional area of 13 square inches. Show that there are two different solutions to the problem. Round to the nearest tenth of an inch.

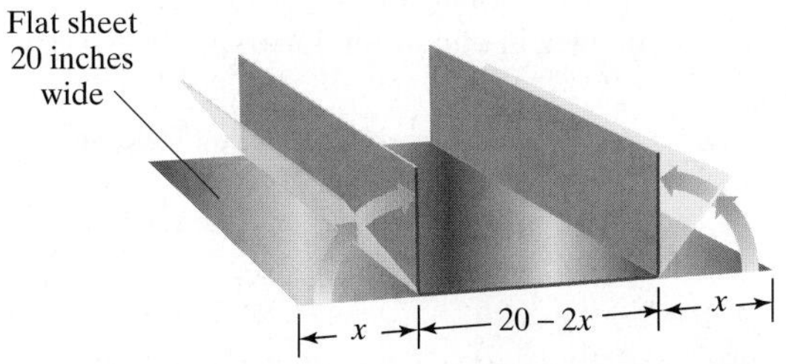

148. A piece of wire is 8 inches long. The wire is cut into two pieces and then each piece is bent into a square. Find the length of each piece if the sum of the areas of these squares is to be 2 square inches.

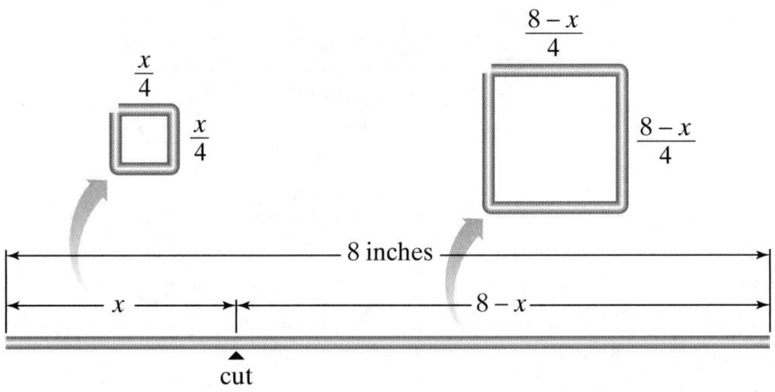

Writing in Mathematics

149. What is a quadratic equation?

150. Explain how to solve $x^2 + 6x + 8 = 0$ using factoring and the zero-product principle.

151. Explain how to solve $x^2 + 6x + 8 = 0$ by completing the square.

152. Explain how to solve $x^2 + 6x + 8 = 0$ using the quadratic formula.

153. How is the quadratic formula derived?

154. What is the discriminant and what information does it provide about a quadratic equation?

155. If you are given a quadratic equation, how do you determine which method to use to solve it?

156. Describe the relationship between the solutions of $ax^2 + bx + c = 0$ and the graph of $y = ax^2 + bx + c$.

157. If a quadratic equation $ax^2 + bx + c = 0$ has imaginary solutions, how is this shown on the graph of $y = ax^2 + bx + c$?

Technology Exercises

158. Use a graphing utility and x-intercepts to verify any of the real solutions that you obtained for three of the quadratic equations in Exercises 65–74.

159. Use a graphing utility to graph $y = ax^2 + bx + c$ related to any five of the quadratic equations, $ax^2 + bx + c = 0$, in Exercises 75–82. How does each graph illustrate what you determined algebraically using the discriminant?

Critical Thinking Exercises

160. Which one of the following is true?

a. The equation $(2x - 3)^2 = 25$ is equivalent to $2x - 3 = 5$.

b. Every quadratic equation has two distinct numbers in its solution set.

c. A quadratic equation whose coefficients are real numbers can never have a solution set containing one real number and one imaginary number.

d. The equation $ax^2 + c = 0$ cannot be solved by the quadratic formula.

161. Write a quadratic equation in general form whose solution set is $\{-3, 5\}$.

162. Solve for t: $s = -16t^2 + v_0 t$.

163. A rectangular swimming pool is 12 meters long and 8 meters wide. A tile border of uniform width is to be built around the pool using 120 square meters of tile. The tile is from a discontinued stock (so no additional materials are available) and all 120 square meters are to be used. How wide should the border be? Round to the nearest tenth of a meter. If zoning laws require at least a 2-meter-wide border around the pool, can this be done with the available tile?

CHAPTER 1
MID-CHAPTER CHECK POINT

What You Know: We used the rectangular coordinate system to represent ordered pairs of real numbers and to graph equations in two variables. We saw that linear equations can be written in the form $ax + b = 0, a \neq 0$, and quadratic equations can be written in the general form $ax^2 + bx + c = 0, a \neq 0$. We solved linear equations and saw that some linear equations have no solution, whereas others have all real numbers as solutions. We solved quadratic equations using factoring, the square root property, completing the square, and the quadratic formula. We saw that the discriminant of $ax^2 + bx + c = 0$, $b^2 - 4ac$, determines the number and type of solutions. We performed operations with complex numbers and used the imaginary unit i ($i = \sqrt{-1}$, where $i^2 = -1$) to represent solutions of quadratic equations with negative discriminants. Only real solutions correspond to x-intercepts. We also solved rational equations by multiplying both sides by the least common denominator and clearing fractions. We developed a strategy for solving a variety of applied problems, using equations to model verbal conditions.

In Exercises 1–12, solve each equation.

1. $-5 + 3(x + 5) = 2(3x - 4)$
2. $5x^2 - 2x = 7$
3. $\frac{x-3}{5} - 1 = \frac{x-5}{4}$
4. $3x^2 - 6x - 2 = 0$
5. $4x - 2(1 - x) = 3(2x + 1) - 5$
6. $5x^2 + 1 = 37$
7. $x(2x - 3) = -4$
8. $\frac{3x}{4} - \frac{x}{3} + 1 = \frac{4x}{5} - \frac{3}{20}$
9. $(x + 3)^2 = 24$
10. $\frac{1}{x^2} - \frac{4}{x} + 1 = 0$
11. $3x + 1 - (x - 5) = 2x - 4$
12. $\frac{2x}{x^2 + 6x + 8} = \frac{x}{x+4} - \frac{2}{x+2}$

In Exercises 13–17, find the x-intercepts of the graph of each equation.

13. $y = x^2 + 6x + 2$
14. $y = 4(x + 1) - 3x - (6 - x)$
15. $y = 2x^2 + 26$
16. $y = \frac{x^2}{3} + \frac{x}{2} - \frac{2}{3}$
17. $y = x^2 - 5x + 8$

In Exercises 18–19, find all values of x satisfying the given conditions.

18. $y_1 = 3(2x - 5) - 2(4x + 1)$, $y_2 = -5(x + 3) - 2$, and $y_1 = y_2$.
19. $y_1 = 2x + 3$, $y_2 = x + 2$, and $y_1 y_2 = 10$.
20. Solve by completing the square: $x^2 + 10x - 3 = 0$.

In Exercises 21–22, without solving the equation, determine the number and type of solutions.

21. $2x^2 + 5x + 4 = 0$
22. $10x(x + 4) = 15x - 15$

In Exercises 23–25, graph each equation in a rectangular coordinate system.

23. $y = 2x - 1$
24. $y = 1 - |x|$
25. $y = x^2 + 2$
26. Solve for n: $L = a + (n - 1)d$.
27. Solve for l: $A = 2lw + 2lh + 2wh$.
28. Solve for f_1: $f = \frac{f_1 f_2}{f_1 + f_2}$.
29. The bar graph shows the defense budget, in billions of dollars, for the countries with the largest defense budgets. The defense budget of Russia exceeds Japan's by $4 billion, and the defense budget of the United States exceeds Japan's by $251 billion. The combined defense budgets of the United States, Russia, and Japan are $375 billion. Determine the defense budget, in billions of dollars, for each country.

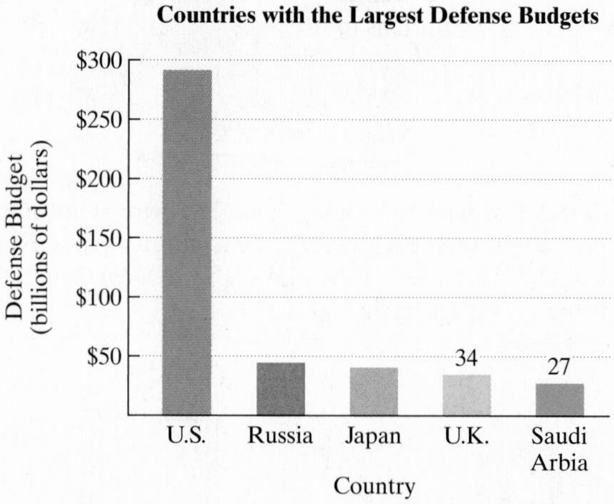

Source: Russell Ash, *The Top 10 of Everything 2004*

30. The average weight for female infants at birth is 7 pounds, with a monthly weight gain of 1.5 pounds. After how many months does a baby girl weigh 16 pounds?

31. You invested \$25,000 in two accounts paying 8% and 9% annual interest, respectively. At the end of the year, the total interest from these investments was \$2135. How much was invested at each rate?

32. Photo Shop A charges \$1.60 to develop a roll of film plus \$0.11 for each print. Photo Shop B charges \$1.20 to develop a roll of film plus \$0.13 per print. For how many prints will the amount spent at each photo shop be the same? What will be that amount?

33. In 2002, the average weight for an American woman aged 20 through 29 was 157 pounds. This is a 22% increase over the average weight in 1960. What was the average weight, to the nearest pound, for an American woman aged 20 through 29 in 1960? (*Source:* National Center for Health Statistics)

34. You invested \$4000. On part of this investment, you earned 4% interest. On the remainder of the investment, you lost 3%. Combining earnings and losses, the annual income from these investments was \$55. How much was invested at each rate?

In Exercises 35–36, find the dimensions of each rectangle.

35. The rectangle's length exceeds twice its width by 5 feet. The perimeter is 46 feet.

36. The rectangle's length is 1 foot shorter than twice its width. The area is 28 square feet.

37. A vertical pole is supported by three wires. Each wire is 13 yards long and is anchored in the ground 5 yards from the base of the pole. How far up the pole will the wires be attached?

38. The graph shows the rapid growth of multinational corporations from 1970 through 2001, including the starting dates of some notable corporations. The data can be modeled by the equation

$$N = 62.2x^2 + 7000,$$

where N represents the number of multinational corporations in the world x years after 1970. Use this model to determine in which year there were 46,000 multinational corporations. How well does the model describe the actual number of corporations for that year shown in the graph?

Number of Multinational Corporations in the World

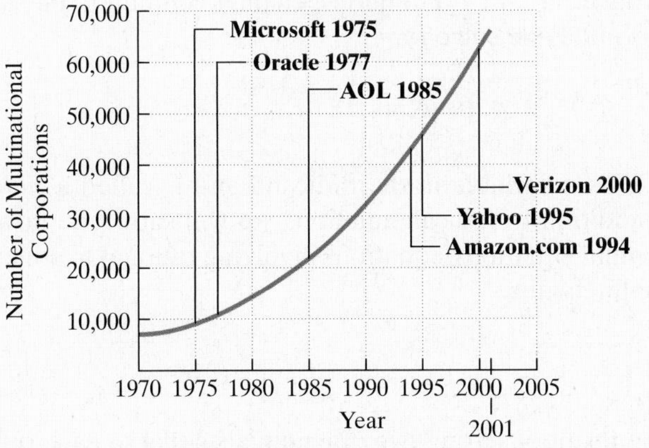

Source: Medard Gabel, *Global Inc.*, The New Press, 2003

39. The bar graph shows the percentage of foreign-born Americans from 1930 through 2003. The data can be modeled by the equation

$$P = 0.0049x^2 - 0.359x + 11.78,$$

where P is the percentage of the U.S. population that was foreign-born x years after 1930. If the trend shown by the data continues, in which year will 15% of the U.S. population be foreign-born?

Percentage of U.S. Population That Was Foreign-Born, 1930-2003

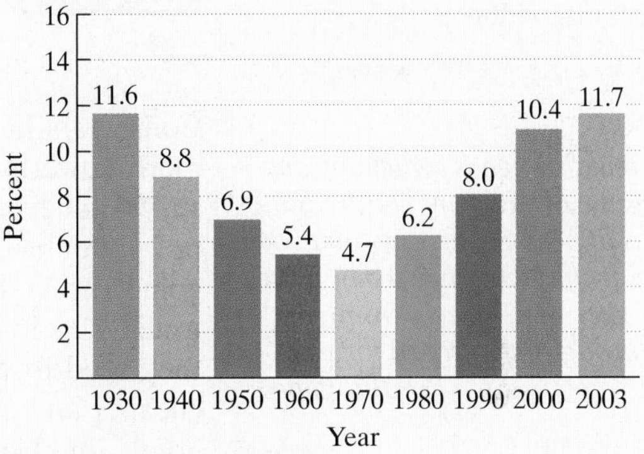

Source: U.S. Census Bureau

In Exercises 40–45, perform the indicated operations and write the result in standard form.

40. $(6 - 2i) - (7 - i)$

41. $3i(2 + i)$

42. $(1 + i)(4 - 3i)$

43. $\dfrac{1 + i}{1 - i}$

44. $\sqrt{-75} - \sqrt{-12}$

45. $(2 - \sqrt{-3})^2$

SECTION 1.6 Other Types of Equations

Objectives

1. Solve polynomial equations by factoring.
2. Solve radical equations.
3. Solve equations with rational exponents.
4. Solve equations that are quadratic in form.
5. Solve equations involving absolute value.

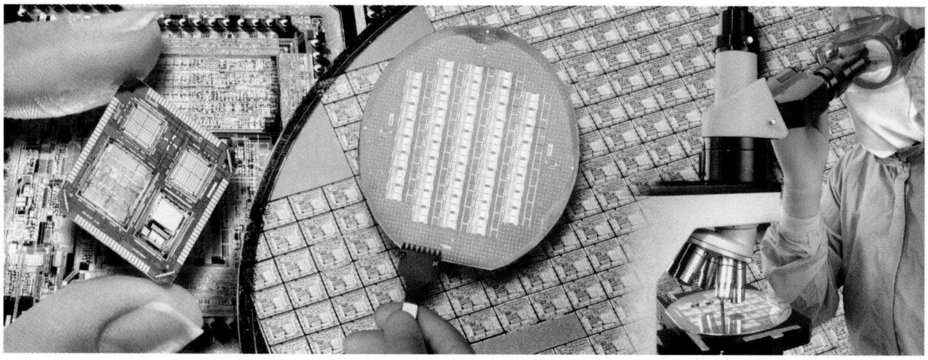

"A breakthrough is something that changes the behavior of hundreds of millions of people where, if you took it away from them, they'd say, 'You can't take that away from me.' Breakthroughs are critical for us."
Bill Gates

Computers have become faster and more powerful. Moore's Law, which states that every 18 months computer power doubles at no extra cost, is still going strong. But customers are not trading up for new models the way they used to, despite being lured by cheaper and cheaper prices. The bar graph in Figure 1.23 shows the number of personal computers sold in the United States each year from 1996 through 2002. Does the trend shown by the graph suggest a new idea afoot in the land, a philosophy of "good enough" when it comes to high tech? Or will a new wave of innovation reverse this trend?

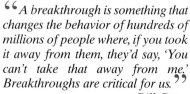

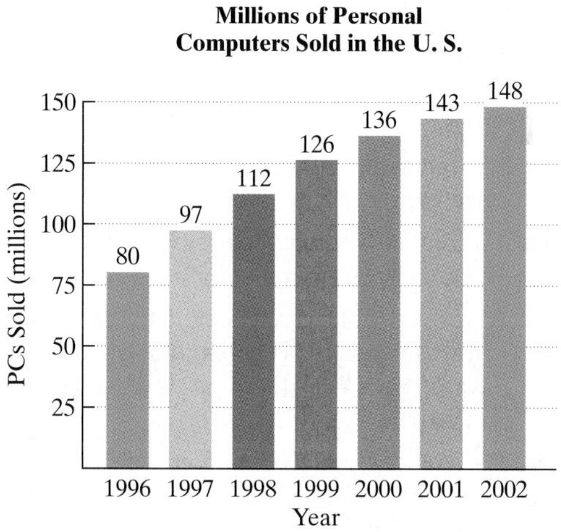

Figure 1.23
Source: Newsweek

The bar graph indicates that PC sales are increasing for the period shown. However, the rate of increase is slowing down. For this reason, a square-root formula is appropriate for modeling the data. The formula

$$P = 28\sqrt{t} + 80$$

models the number of personal computers sold in the United States, P, in millions, t years after 1996.

If trends from 1996 through 2002 continue and there are no new breakthroughs to spur sales, in what year will 192 million personal computers be sold? Substitute 192 for P in the formula and solve for t:

$$192 = 28\sqrt{t} + 80.$$

The resulting equation contains a variable in the radicand and is called a *radical equation*. In this section, in addition to radical equations, we will show you how to solve certain kinds of polynomial equations, equations involving rational exponents, and equations involving absolute value.

1 Solve polynomial equations by factoring.

Polynomial Equations

A **polynomial equation** is the result of setting two polynomials equal to each other. The equation is in **general form** if one side is 0 and the polynomial on the other side is in descending powers of the variable. The **degree of a polynomial equation** is the same as the highest degree of any term in the equation. Here are examples of three polynomial equations:

$$3x + 5 = 14 \qquad 2x^2 + 7x = 4 \qquad x^3 + x^2 = 4x + 4.$$

This equation is of degree 1 because 1 is the highest degree.

This equation is of degree 2 because 2 is the highest degree.

This equation is of degree 3 because 3 is the highest degree.

Notice that a polynomial equation of degree 1 is a linear equation. A polynomial equation of degree 2 is a quadratic equation.

Some polynomial equations of degree 3 or higher can be solved by moving all terms to one side, thereby obtaining 0 on the other side. Once the equation is in general form, factor and then set each factor equal to 0.

EXAMPLE 1 Solving a Polynomial Equation by Factoring

Solve by factoring: $3x^4 = 27x^2$.

Solution

Step 1 Move all terms to one side and obtain zero on the other side. Subtract $27x^2$ from both sides.

$$3x^4 = 27x^2 \quad \text{This is the given equation.}$$

$$3x^4 - 27x^2 = 27x^2 - 27x^2 \quad \text{Subtract } 27x^2 \text{ from both sides.}$$

$$3x^4 - 27x^2 = 0 \quad \text{Simplify.}$$

Step 2 Factor. We can factor $3x^2$ from each term.

$$3x^4 - 27x^2 = 0$$

$$3x^2(x^2 - 9) = 0$$

Steps 3 and 4 Set each factor equal to zero and solve the resulting equations.

$$3x^2 = 0 \quad \text{or} \quad x^2 - 9 = 0$$

$$x^2 = 0 \qquad x^2 = 9$$

$$x = \pm\sqrt{0} \qquad x = \pm\sqrt{9}$$

$$x = 0 \qquad x = \pm 3$$

Step 5 Check the solutions in the original equation. Check the three solutions, 0, –3, and 3, by substituting them into the original equation. Can you verify that the solution set is $\{-3, 0, 3\}$?

Study Tip

In solving $3x^4 = 27x^2$, be careful not to divide both sides by x^2. If you do, you'll lose 0 as a solution. In general, do not divide both sides of an equation by a variable because that variable might take on the value 0 and you cannot divide by 0.

Check Point 1 Solve by factoring: $4x^4 = 12x^2$.

EXAMPLE 2 Solving a Polynomial Equation by Factoring

Solve by factoring: $x^3 + x^2 = 4x + 4$.

Solution

Step 1 Move all terms to one side and obtain zero on the other side. Subtract $4x$ and subtract 4 from both sides.

$$x^3 + x^2 = 4x + 4 \quad \text{This is the given equation.}$$

$$x^3 + x^2 - 4x - 4 = 4x + 4 - 4x - 4 \quad \text{Subtract 4x and 4 from both sides.}$$

$$x^3 + x^2 - 4x - 4 = 0 \quad \text{Simplify.}$$

Technology

A graphing utility's TABLE feature can be used to numerically verify that $\{-2, -1, 2\}$ is the solution set of

$$x^3 + x^2 = 4x + 4$$

Enter $y_1 = x^3 + x^2$.   Enter $y_2 = 4x + 4$.

y_1 and y_2 are equal when $x = -2$, $x = -1$, and $x = 2$.	X	Y1	Y2
	-4	-48	-12
	-3	-18	-8
	-2	-4	-4
	-1	0	0
	0	0	4
	1	2	8
	2	12	12

X= -4

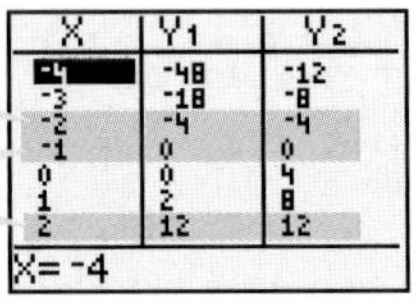

Step 2   Factor. Use factoring by grouping. Group terms that have a common factor.

$$\boxed{x^3 + x^2} + \boxed{-4x - 4} = 0$$

Common factor is x^2.   Common factor is -4.

$$x^2(x + 1) - 4(x + 1) = 0 \qquad \text{Factor } x^2 \text{ from the first two terms and } -4 \text{ from the last two terms.}$$

$$(x + 1)(x^2 - 4) = 0 \qquad \text{Factor out the common binomial, } x + 1, \text{ from each term.}$$

Steps 3 and 4   Set each factor equal to zero and solve the resulting equations.

$$x + 1 = 0 \qquad \text{or} \qquad x^2 - 4 = 0$$
$$x = -1 \qquad\qquad\qquad x^2 = 4$$
$$x = \pm\sqrt{4} = \pm 2$$

Step 5   Check the solutions in the original equation. Check the three solutions, $-1, -2$ and 2, by substituting them into $x^3 + x^2 = 4x + 4$, the original equation. Can you verify that the solution set is $\{-2, -1, 2\}$?

Discovery

Suggest a method involving intersecting graphs that can be used with a graphing utility to verify that $\{-2, -1, 2\}$ is the solution set of

$$x^3 + x^2 = 4x + 4.$$

Apply this method to verify the solution set.

Technology

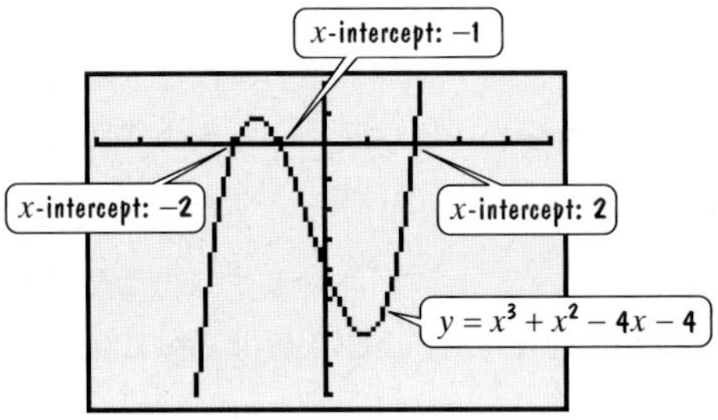

x-intercept: −1
x-intercept: −2
x-intercept: 2
$y = x^3 + x^2 - 4x - 4$
$[-5, 5, 1]$ by $[-8, 2, 1]$

You can use a graphing utility to check the solutions of $x^3 + x^2 - 4x - 4 = 0$. Graph $y = x^3 + x^2 - 4x - 4$, as shown on the left. The x-intercepts are $-2, -1$, and 2, corresponding to the equation's solutions.

Check Point 2   Solve by factoring: $2x^3 + 3x^2 = 8x + 12$.

② Solve radical equations.

Radical Equations

A **radical equation** is an equation in which the variable occurs in a square root, cube root, or any higher root. An example of a radical equation is

$$\sqrt{x} = 9.$$

We solve the equation by squaring both sides:

Squaring both sides eliminates the square root.

$$\left(\sqrt{x}\right)^2 = 9^2$$
$$x = 81.$$

The proposed solution, 81, can be checked in the original equation, $\sqrt{x} = 9$. Because $\sqrt{81} = 9$, the solution is 81 and the solution set is $\{81\}$.

In general, we solve radical equations with square roots by squaring both sides of the equation. We solve radical equations with nth roots by raising both sides of the equation to the nth power. Unfortunately, if n is even, all the solutions of the equation raised to the even power may not be solutions of the original equation. Consider, for example, the equation

$$x = 4.$$

If we square both sides, we obtain

$$x^2 = 16.$$
$$x = \pm\sqrt{16} = \pm 4$$

This new equation has two solutions, -4 and 4. By contrast, only 4 is a solution of the original equation, $x = 4$. For this reason, **when raising both sides of an equation to an even power, always check proposed solutions in the original equation.**

Here is a general method for solving radical equations with nth roots:

Solving Radical Equations Containing *n*th Roots

1. If necessary, arrange terms so that one radical is isolated on one side of the equation.
2. Raise both sides of the equation to the nth power to eliminate the nth root.
3. Solve the resulting equation. If this equation still contains radicals, repeat steps 1 and 2.
4. Check all proposed solutions in the original equation.

Extra solutions may be introduced when you raise both sides of a radical equation to an even power. Such solutions, which are not solutions of the given equation, are called **extraneous solutions** or **extraneous roots**.

EXAMPLE 3 Solving a Radical Equation

Solve: $\sqrt{2x - 1} + 2 = x$.

Solution

Step 1 Isolate a radical on one side. We isolate the radical, $\sqrt{2x-1}$, by subtracting 2 from both sides.

$$\sqrt{2x - 1} + 2 = x \quad \text{This is the given equation.}$$
$$\sqrt{2x - 1} = x - 2 \quad \text{Subtract 2 from both sides.}$$

Study Tip

Be sure to square *both sides* of an equation. Do *not* square each term.

Correct:

$$\left(\sqrt{2x - 1}\right)^2 = (x - 2)^2$$

Incorrect!

$$\left(\sqrt{2x - 1}\right)^2 = x^2 - 2^2$$

Step 2 Raise both sides to the *n*th power. Because n, the index, is 2, we square both sides.

$$\left(\sqrt{2x - 1}\right)^2 = (x - 2)^2$$
$$2x - 1 = x^2 - 4x + 4 \quad \text{Simplify. Use the formula } (A - B)^2 = A^2 - 2AB + B^2 \text{ on the right side.}$$

Step 3 Solve the resulting equation. Because of the x^2-term, the resulting equation is a quadratic equation. We can obtain 0 on the left side by subtracting $2x$ and adding 1 on both sides.

$$2x - 1 = x^2 - 4x + 4 \quad \text{The resulting equation is quadratic.}$$
$$0 = x^2 - 6x + 5 \quad \text{Write in general form, subtracting 2x and adding 1 on both sides.}$$
$$0 = (x - 1)(x - 5) \quad \text{Factor.}$$
$$x - 1 = 0 \quad \text{or} \quad x - 5 = 0 \quad \text{Set each factor equal to 0.}$$
$$x = 1 \qquad x = 5 \quad \text{Solve the resulting equations.}$$

Step 4   Check the proposed solutions in the original equation.

Check 1:

$$\sqrt{2x - 1} + 2 = x$$
$$\sqrt{2 \cdot 1 - 1} + 2 \overset{?}{=} 1$$
$$\sqrt{1} + 2 \overset{?}{=} 1$$
$$1 + 2 \overset{?}{=} 1$$
$$3 = 1, \quad \text{false}$$

Check 5:

$$\sqrt{2x - 1} + 2 = x$$
$$\sqrt{2 \cdot 5 - 1} + 2 \overset{?}{=} 5$$
$$\sqrt{9} + 2 \overset{?}{=} 5$$
$$3 + 2 \overset{?}{=} 5$$
$$5 = 5, \quad \text{true}$$

Thus, 1 is an extraneous solution. The only solution is 5, and the solution set is $\{5\}$.

Technology

A graphing utility's $\boxed{\text{TABLE}}$ feature provides a numeric check that 1 is not a solution and 5 is a solution of $\sqrt{2x - 1} + 2 = x$.

When $x = 1$, y_1 and y_2 are not equal. 1 is an extraneous solution.

When $x = 5$, y_1 and y_2 are equal.

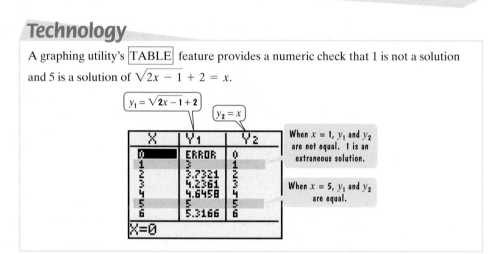

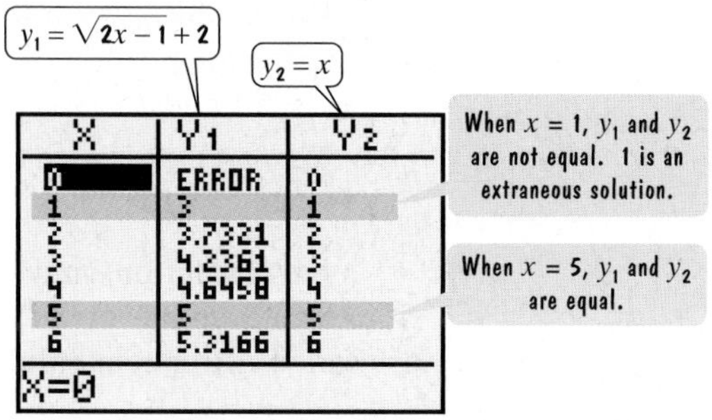

Check Point 3   Solve: $\sqrt{x + 3} + 3 = x$.

The solution of radical equations with two or more square root expressions involves isolating a radical, squaring both sides, and then repeating this process. Let's consider an equation containing two square root expressions.

EXAMPLE 4   Solving an Equation That Has Two Radicals

Solve: $\sqrt{3x + 1} - \sqrt{x + 4} = 1$.

Solution

Step 1   Isolate a radical on one side. We can isolate the radical $\sqrt{3x + 1}$ by adding $\sqrt{x + 4}$ to both sides. We obtain

$$\sqrt{3x + 1} = \sqrt{x + 4} + 1.$$

Step 2   Square both sides.

$$\left(\sqrt{3x + 1}\right)^2 = \left(\sqrt{x + 4} + 1\right)^2$$

Squaring the expression on the right side of the equation can be a bit tricky. We have to use the formula

$$(A + B)^2 = A^2 + 2AB + B^2.$$

Focusing on just the right side, here is how the squaring is done:

$$(A + B)^2 = A^2 + 2 \cdot A \cdot B + B^2$$

$$\left(\sqrt{x + 4} + 1\right)^2 = \left(\sqrt{x + 4}\right)^2 + 2 \cdot \sqrt{x + 4} \cdot 1 + 1^2 = x + 4 + 2\sqrt{x + 4} + 1.$$

Now let's return to squaring both sides.

$$\left(\sqrt{3x+1}\right)^2 = \left(\sqrt{x+4}+1\right)^2$$ Square both sides of the equation with an isolated radical.

$$3x + 1 = x + 4 + 2\sqrt{x+4} + 1$$ $\left(\sqrt{3x+1}\right)^2 = 3x + 1$; square the right side using the formula for $(A + B)^2$.

$$3x + 1 = x + 5 + 2\sqrt{x+4}$$ Combine numerical terms on the right side: $4 + 1 = 5$.

Can you see that the resulting equation still contains a radical, namely $\sqrt{x+4}$? Thus, we need to repeat the first two steps.

Repeat Step 1 Isolate a radical on one side. We isolate $2\sqrt{x+4}$, the radical term, by subtracting $x + 5$ from both sides. We obtain

$$3x + 1 = x + 5 + 2\sqrt{x+4}$$ This is the equation from our last step.

$$2x - 4 = 2\sqrt{x+4}.$$ Subtract x and subtract 5 from both sides.

Although we can simplify the equation by dividing both sides by 2, this sort of simplification is not always helpful. Thus, we will work with the equation in this form.

Repeat Step 2 Square both sides.

Be careful in squaring both sides. Use $(A - B)^2 = A^2 - 2AB + B^2$ to square the left side. Use $(AB)^2 = A^2B^2$ to square the right side.

$$(2x - 4)^2 = \left(2\sqrt{x+4}\right)^2$$ Square both sides.

$$4x^2 - 16x + 16 = 4(x + 4)$$ Square both 2 and $\sqrt{x+4}$ on the right side.

Step 3 Solve the resulting equation. We solve this quadratic equation by writing it in general form.

$$4x^2 - 16x + 16 = 4x + 16$$ Use the distributive property.

$$4x^2 - 20x = 0$$ Subtract $4x + 16$ from both sides.

$$4x(x - 5) = 0$$ Factor.

$$4x = 0 \quad \text{or} \quad x - 5 = 0$$ Set each factor equal to zero.

$$x = 0 \qquad\qquad x = 5$$ Solve for x.

Step 4 Check the proposed solutions in the original equation.

Check 0:

$$\sqrt{3x+1} - \sqrt{x+4} = 1$$
$$\sqrt{3\cdot 0+1} - \sqrt{0+4} \stackrel{?}{=} 1$$
$$\sqrt{1} - \sqrt{4} \stackrel{?}{=} 1$$
$$1 - 2 \stackrel{?}{=} 1$$
$$-1 = 1, \quad \text{false}$$

Check 5:

$$\sqrt{3x+1} - \sqrt{x+4} = 1$$
$$\sqrt{3\cdot 5+1} - \sqrt{5+4} \stackrel{?}{=} 1$$
$$\sqrt{16} - \sqrt{9} \stackrel{?}{=} 1$$
$$4 - 3 \stackrel{?}{=} 1$$
$$1 = 1, \quad \text{true}$$

The check indicates that 0 is not a solution. It is an extraneous solution brought about by squaring each side of the equation. The only solution is 5, and the solution set is $\{5\}$.

Technology

The graph of

$$y = \sqrt{3x+1} - \sqrt{x+4} - 1$$

has only one x-intercept at 5. This verifies that the solution set of $\sqrt{3x+1} - \sqrt{x+4} = 1$ is $\{5\}$.

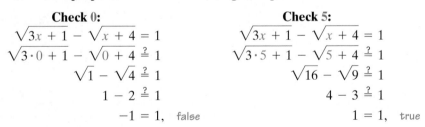

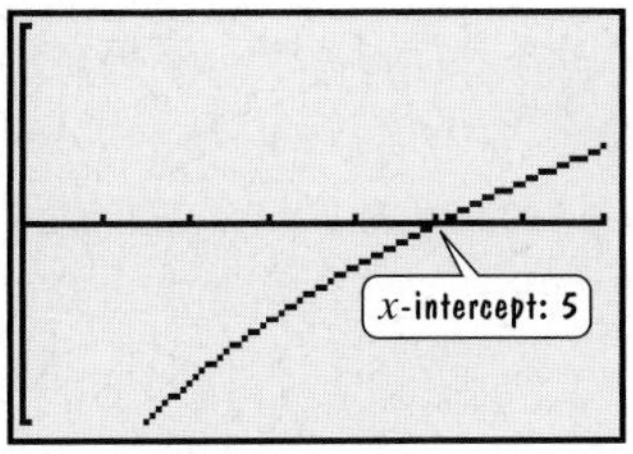

[0, 7, 1] by [−1, 1, 1]

Check Point 4 Solve: $\sqrt{x+5} - \sqrt{x-3} = 2$.

3 Solve equations with rational exponents.

Equations with Rational Exponents

We know that expressions with rational exponents represent radicals:

$$a^{\frac{m}{n}} = (\sqrt[n]{a})^m = \sqrt[n]{a^m}.$$

For example, the radical equation $3\sqrt[4]{x^3} - 6 = 0$ can be expressed as $3x^{\frac{3}{4}} - 6 = 0$.

A radical equation with rational exponents can be solved by

1. isolating the expression with the rational exponent, and
2. raising both sides of the equation to a power that is the reciprocal of the rational exponent.

Solving Radical Equations of the Form $x^{\frac{m}{n}} = k$

Assume that m and n are positive integers, $\frac{m}{n}$ is in lowest terms, and k is a real number.

1. Isolate the expression with the rational exponent.
2. Raise both sides of the equation to the $\frac{n}{m}$ power.

If m is even:	**If m is odd:**
$x^{\frac{m}{n}} = k$	$x^{\frac{m}{n}} = k$
$\left(x^{\frac{m}{n}}\right)^{\frac{n}{m}} = \pm k^{\frac{n}{m}}$	$\left(x^{\frac{m}{n}}\right)^{\frac{n}{m}} = k^{\frac{n}{m}}$
$x = \pm k^{\frac{n}{m}}$	$x = k^{\frac{n}{m}}$

It is incorrect to insert the $\pm$ symbol when the numerator of the exponent is odd. An odd index has only one root.

3. Check all proposed solutions in the original equation to find out if they are actual solutions or extraneous solutions.

EXAMPLE 5 Solving Equations Involving Rational Exponents

Solve:

a. $3x^{\frac{3}{4}} - 6 = 0$ **b.** $x^{\frac{2}{3}} - \frac{3}{4} = -\frac{1}{2}$.

Solution

a. Our goal is to isolate $x^{\frac{3}{4}}$. Then we can raise both sides of the equation to the $\frac{4}{3}$ power because $\frac{4}{3}$ is the reciprocal of $\frac{3}{4}$.

$3x^{\frac{3}{4}} - 6 = 0$ This is the given equation; we will isolate $x^{\frac{3}{4}}$.

$3x^{\frac{3}{4}} = 6$ Add 6 to both sides.

$\frac{3x^{\frac{3}{4}}}{3} = \frac{6}{3}$ Divide both sides by 3.

$x^{\frac{3}{4}} = 2$ Simplify.

$\left(x^{\frac{3}{4}}\right)^{\frac{4}{3}} = 2^{\frac{4}{3}}$ Raise both sides to the $\frac{4}{3}$ power. Because $\frac{m}{n} = \frac{3}{4}$ and m is odd, we do not use the $\pm$ symbol.

$x = 2^{\frac{4}{3}}$ Simplify the left side: $\left(x^{\frac{3}{4}}\right)^{\frac{4}{3}} = x^{\frac{3\cdot4}{4\cdot3}} = x^{\frac{12}{12}} = x^1 = x$.

The proposed solution is $2^{\frac{4}{3}}$. Complete the solution process by checking this value in the given equation.

$3x^{\frac{3}{4}} - 6 = 0$ This is the original equation.

$3\left(2^{\frac{4}{3}}\right)^{\frac{3}{4}} - 6 \stackrel{?}{=} 0$ Substitute the proposed solution.

$3\cdot2 - 6 \stackrel{?}{=} 0$ $\left(2^{\frac{4}{3}}\right)^{\frac{3}{4}} = 2^{\frac{4\cdot3}{3\cdot4}} = 2^{\frac{12}{12}} = 2^1 = 2$

$0 = 0$, true $3\cdot2 - 6 = 6 - 6 = 0$

The solution is $2^{\frac{4}{3}} = \sqrt[3]{2^4} \approx 2.52$. The solution set is $\left\{2^{\frac{4}{3}}\right\}$.

b. To solve $x^{\frac{2}{3}} - \frac{3}{4} = -\frac{1}{2}$, our goal is to isolate $x^{\frac{2}{3}}$. Then we can raise both sides of the equation to the $\frac{3}{2}$ power because $\frac{3}{2}$ is the reciprocal of $\frac{2}{3}$.

$x^{\frac{2}{3}} - \frac{3}{4} = -\frac{1}{2}$ This is the given equation.

$x^{\frac{2}{3}} = \frac{1}{4}$ Add $\frac{3}{4}$ to both sides. $\frac{3}{4} - \frac{1}{2} = \frac{3}{4} - \frac{2}{4} = \frac{1}{4}$.

$\left(x^{\frac{2}{3}}\right)^{\frac{3}{2}} = \pm\left(\frac{1}{4}\right)^{\frac{3}{2}}$ Raise both sides to the $\frac{3}{2}$ power. Because $\frac{m}{n} = \frac{2}{3}$ and m is even, the $\pm$ symbol is necessary.

$x = \pm\frac{1}{8}$ $\left(\frac{1}{4}\right)^{\frac{3}{2}} = \left(\sqrt{\frac{1}{4}}\right)^3 = \left(\frac{1}{2}\right)^3 = \frac{1}{8}$

Take a moment to verify that the solution set is $\left\{-\frac{1}{8}, \frac{1}{8}\right\}$.

Check Point 5 Solve: **a.** $5x^{\frac{3}{2}} - 25 = 0$ **b.** $x^{\frac{2}{3}} - 8 = -4$.

4 Solve equations that are quadratic in form.

Equations That Are Quadratic in Form

Some equations that are not quadratic can be written as quadratic equations using an appropriate substitution. Here are some examples:

Given Equation	Substitution	New Equation
$x^4 - 8x^2 - 9 = 0$ or $(x^2)^2 - 8x^2 - 9 = 0$	$u = x^2$	$u^2 - 8u - 9 = 0$
$5x^{\frac{2}{3}} + 11x^{\frac{1}{3}} + 2 = 0$ or $5\left(x^{\frac{1}{3}}\right)^2 + 11x^{\frac{1}{3}} + 2 = 0$	$u = x^{\frac{1}{3}}$	$5u^2 + 11u + 2 = 0$

An equation that is **quadratic in form** is one that can be expressed as a quadratic equation using an appropriate substitution. Both of the preceding given equations are quadratic in form.

Equations that are quadratic in form contain an expression to a power, the same expression to that power squared, and a constant term. By letting u equal the expression to the power, a quadratic equation in u will result. Now it's easy. Solve this quadratic equation for u. Finally, use your substitution to find the values for the variable in the given equation. Example 6 shows how this is done.

EXAMPLE 6 Solving an Equation Quadratic in Form

Solve: $x^4 - 8x^2 - 9 = 0$.

Solution Notice that the equation contains an expression to a power, x^2, the same expression to that power squared, x^4 or $(x^2)^2$, and a constant term, -9. We let u equal the expression to the power. Thus,

$$\text{let } u = x^2.$$

Now we write the given equation as a quadratic equation in u and solve for u.

$x^4 - 8x^2 - 9 = 0$ This is the given equation.

$(x^2)^2 - 8x^2 - 9 = 0$ The given equation contains x^2 and x^2 squared.

$u^2 - 8u - 9 = 0$ Replace x^2 with u.

$(u - 9)(u + 1) = 0$ Factor.

$u - 9 = 0$ or $u + 1 = 0$ Apply the zero-product principle.

$u = 9$ $\quad u = -1$ Solve for u.

We're not done! Why not? We were asked to solve for x and we have values for u. We use the original substitution, $u = x^2$, to solve for x. Replace u with x^2 in each equation shown, namely $u = 9$ and $u = -1$.

$$\begin{array}{ll} x^2 = 9 & x^2 = -1 \\ x = \pm\sqrt{9} & x = \pm\sqrt{-1} \\ x = \pm 3 & x = \pm i \end{array}$$

The solution set is $\{-3, 3, -i, i\}$. The graph in the technology box shows that only the real solutions, -3 and 3, appear as x-intercepts.

Technology

The graph of

$$y = x^4 - 8x^2 - 9$$

has x-intercepts at -3 and 3. This verifies that the real solutions of

$$x^4 - 8x^2 - 9 = 0$$

are -3 and 3. The imaginary solutions, $-i$ and i, are not shown as intercepts.

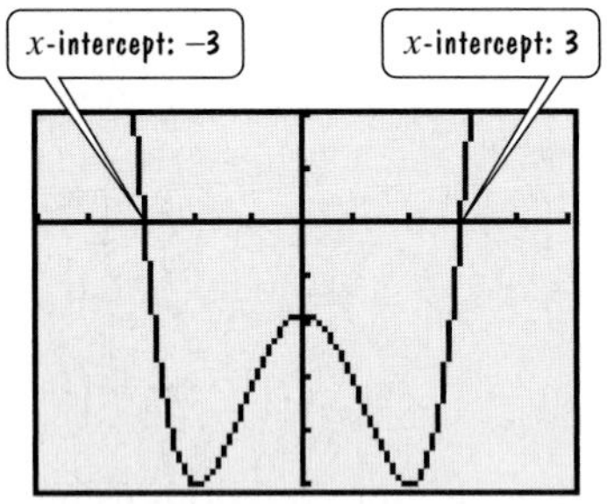

[−5, 5, 1] by [−25, 10, 5]

Check Point 6 Solve: $x^4 - 5x^2 + 6 = 0$.

EXAMPLE 7 Solving an Equation Quadratic in Form

Solve: $5x^{\frac{2}{3}} + 11x^{\frac{1}{3}} + 2 = 0$.

Solution Notice that the equation contains an expression to a power, $x^{\frac{1}{3}}$, the same expression to that power squared, $x^{\frac{2}{3}}$ or $\left(x^{\frac{1}{3}}\right)^2$, and a constant term, 2. We let u equal the expression to the power. Thus,

$$\text{let } u = x^{\frac{1}{3}}.$$

Now we write the given equation as a quadratic equation in u and solve for u.

$$5x^{\frac{2}{3}} + 11x^{\frac{1}{3}} + 2 = 0 \quad \text{This is the given equation.}$$

$$5\left(x^{\frac{1}{3}}\right)^2 + 11x^{\frac{1}{3}} + 2 = 0 \quad \text{The given equation contains } x^{\frac{1}{3}} \text{ and } x^{\frac{1}{3}} \text{ squared.}$$

$$5u^2 + 11u + 2 = 0 \quad \text{Replace } x^{\frac{1}{3}} \text{ with } u.$$

$$(5u + 1)(u + 2) = 0 \quad \text{Factor.}$$

$$5u + 1 = 0 \quad \text{or} \quad u + 2 = 0 \quad \text{Set each factor equal to 0.}$$

$$5u = -1 \qquad u = -2 \quad \text{Solve for } u.$$

$$u = -\frac{1}{5}$$

Use the original substitution, $u = x^{\frac{1}{3}}$, to solve for x. Replace u with $x^{\frac{1}{3}}$ in each of the preceding equations, namely $u = -\frac{1}{5}$ and $u = -2$.

$$x^{\frac{1}{3}} = -\frac{1}{5} \qquad x^{\frac{1}{3}} = -2 \quad \text{Replace } u \text{ with } x^{\frac{1}{3}}.$$

$$\left(x^{\frac{1}{3}}\right)^3 = \left(-\frac{1}{5}\right)^3 \qquad \left(x^{\frac{1}{3}}\right)^3 = (-2)^3 \quad \text{Solve for } x \text{ by cubing both sides of each equation.}$$

$$x = -\frac{1}{125} \qquad x = -8$$

Check these values to verify that the solution set is $\{-8, -\frac{1}{125}\}$.

Check Point 7 Solve: $3x^{\frac{2}{3}} - 11x^{\frac{1}{3}} - 4 = 0$.

5 Solve equations involving absolute value.

Equations Involving Absolute Value

We have seen that the absolute value of x, denoted $|x|$, describes the distance of x from zero on a number line. Now consider an **absolute value equation**, such as

$$|x| = 2.$$

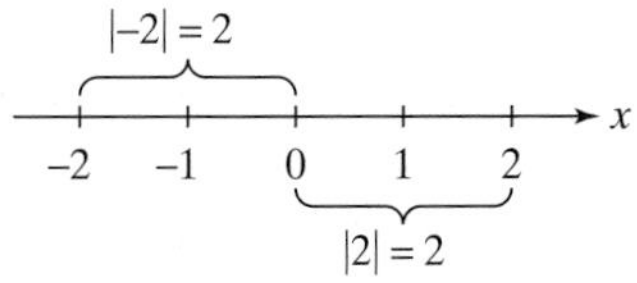

Figure 1.24

This means that we must determine real numbers whose distance from the origin on a number line is 2. Figure 1.24 shows that there are two numbers such that $|x| = 2$, namely, 2 and -2. We write $x = 2$ or $x = -2$. This observation can be generalized as follows:

> **Rewriting an Absolute Value Equation without Absolute Value Bars**
> If c is a positive real number and X represents any algebraic expression, then $|X| = c$ is equivalent to $X = c$ or $X = -c$.

Technology

You can use a graphing utility to verify the solution of an absolute value equation. Consider, for example,

$$|2x - 3| = 11$$

Graph $y_1 = |2x - 3|$ and $y_2 = 11$. The graphs are shown in a $[-10, 10, 1]$ by $[-1, 15, 1]$ viewing rectangle. The x-coordinates of the intersection points are -4 and 7, verifying that $\{-4, 7\}$ is the solution set.

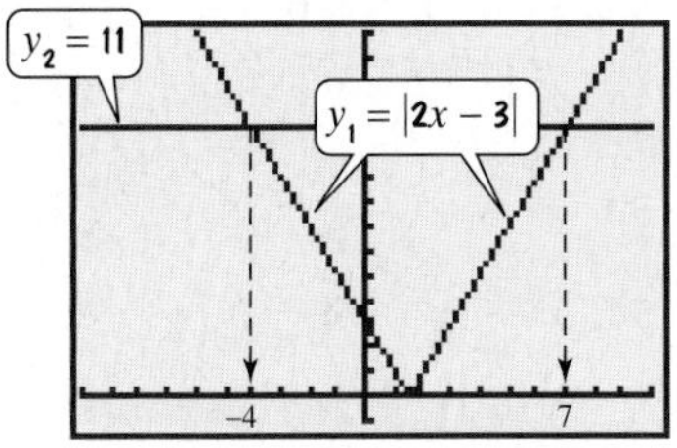

$[-10, 10, 1]$ by $[-1, 15, 1]$

EXAMPLE 8 Solving an Equation Involving Absolute Value

Solve: $|2x - 3| = 11$.

Solution

$$|2x - 3| = 11$$ This is the given equation.

$$2x - 3 = 11 \quad \text{or} \quad 2x - 3 = -11$$ Rewrite the equation without absolute value bars.

$$2x = 14 \qquad\qquad 2x = -8$$ Add 3 to both sides of each equation.

$$x = 7 \qquad\qquad x = -4$$ Divide both sides of each equation by 2.

Check 7:	**Check −4:**	
$\lvert 2x - 3\rvert = 11$	$\lvert 2x - 3\rvert = 11$	This is the original equation.
$\lvert 2(7) - 3\rvert \stackrel{?}{=} 11$	$\lvert 2(-4) - 3\rvert \stackrel{?}{=} 11$	Substitute the proposed solutions.
$\lvert 14 - 3\rvert \stackrel{?}{=} 11$	$\lvert -8 - 3\rvert \stackrel{?}{=} 11$	Perform operations inside the absolute value bars.
$\lvert 11\rvert \stackrel{?}{=} 11$	$\lvert -11\rvert \stackrel{?}{=} 11$	
$11 = 11$, true	$11 = 11$, true	These true statements indicate that 7 and −4 are solutions.

The solution set is $\{-4, 7\}$.

Check Point 8 Solve: $|2x - 1| = 5$.

EXAMPLE 9 Solving an Equation Involving Absolute Value

Solve: $5|1 - 4x| - 15 = 0$.

Solution

$$5|1 - 4x| - 15 = 0$$ This is the given equation.

We need to isolate $|1 - 4x|$, the absolute value expression.

$$5|1 - 4x| = 15$$ Add 15 to both sides.

$$|1 - 4x| = 3$$ Divide both sides by 5.

$$1 - 4x = 3 \quad \text{or} \quad 1 - 4x = -3$$ Rewrite $|X| = c$ as $X = c$ or $X = -c$.

$$-4x = 2 \qquad\qquad -4x = -4$$ Subtract 1 from both sides of each equation.

$$x = -\frac{1}{2} \qquad\qquad x = 1$$ Divide both sides of each equation by −4.

Take a moment to check $-\frac{1}{2}$ and 1, the proposed solutions, in the original equation, $5|1 - 4x| - 15 = 0$. In each case, you should obtain the true statement $0 = 0$. The solution set is $\{-\frac{1}{2}, 1\}$. These solutions appear as x-intercepts for the graph of $y = 5|1 - 4x| - 15$, as shown in Figure 1.25.

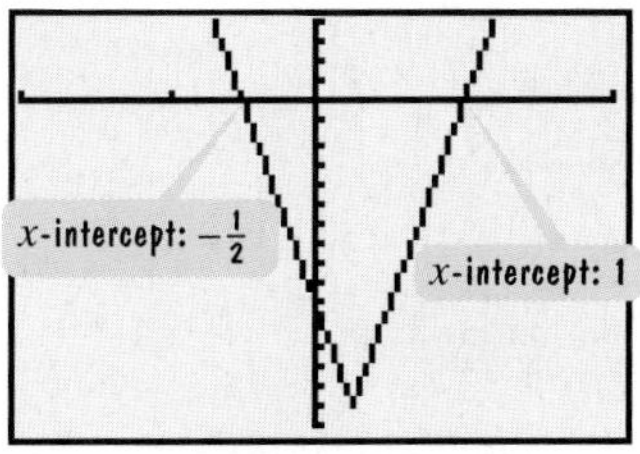

$[-2, 2, 1]$ by $[-16, 4, 1]$

Figure 1.25 The graph of $y = 5|1 - 4x| - 15$

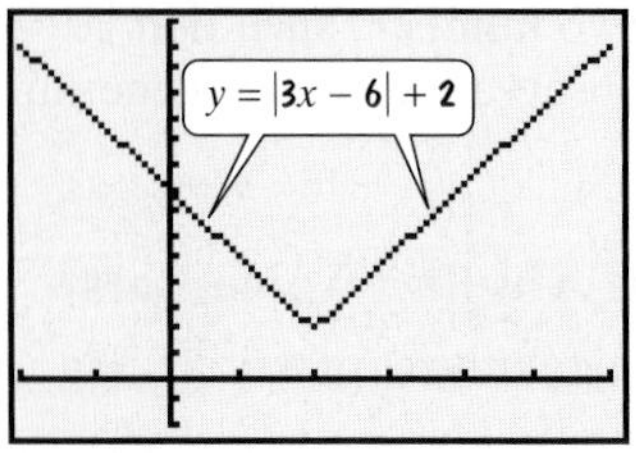

[−2, 6, 1] by [−2, 15, 1]

Figure 1.26 An absolute value equation whose graph has no x-intercepts

Check Point 9 Solve: $4|1 - 2x| - 20 = 0$.

The absolute value of a number is never negative. Thus, if X is an algebraic expression and c is a negative number, then $|X| = c$ has no solution. For example, the equation $|3x - 6| = -2$ has no solution because $|3x - 6|$ cannot be negative. The solution set is $\varnothing$, the empty set. The graph of $y = |3x - 6| + 2$, shown in Figure 1.26, lies above the x-axis and has no x-intercepts.

The absolute value of 0 is 0. Thus, if X is an algebraic expression and $|X| = 0$, the solution is found by solving $X = 0$. For example, the solution of $|x - 2| = 0$ is obtained by solving $x - 2 = 0$. The solution is 2 and the solution set is $\{2\}$.

EXERCISE SET 1.6

Practice Exercises

Solve each polynomial equation in Exercises 1–10 by factoring and then using the zero-product principle.

1. $3x^4 - 48x^2 = 0$ **2.** $5x^4 - 20x^2 = 0$

3. $3x^3 + 2x^2 = 12x + 8$ **4.** $4x^3 - 12x^2 = 9x - 27$

5. $2x - 3 = 8x^3 - 12x^2$ **6.** $x + 1 = 9x^3 + 9x^2$

7. $4y^3 - 2 = y - 8y^2$ **8.** $9y^3 + 8 = 4y + 18y^2$

9. $2x^4 = 16x$ **10.** $3x^4 = 81x$

Solve each radical equation in Exercises 11–30. Check all proposed solutions.

11. $\sqrt{3x + 18} = x$ **12.** $\sqrt{20 - 8x} = x$

13. $\sqrt{x + 3} = x - 3$ **14.** $\sqrt{x + 10} = x - 2$

15. $\sqrt{2x + 13} = x + 7$ **16.** $\sqrt{6x + 1} = x - 1$

17. $x - \sqrt{2x + 5} = 5$ **18.** $x - \sqrt{x + 11} = 1$

19. $\sqrt{2x + 19} - 8 = x$ **20.** $\sqrt{2x + 15} - 6 = x$

21. $\sqrt{3x + 10} = x + 4$ **22.** $\sqrt{x} - 3 = x - 9$

23. $\sqrt{x + 8} - \sqrt{x - 4} = 2$

24. $\sqrt{x + 5} - \sqrt{x - 3} = 2$

25. $\sqrt{x - 5} - \sqrt{x - 8} = 3$

26. $\sqrt{2x - 3} - \sqrt{x - 2} = 1$

27. $\sqrt{2x + 3} + \sqrt{x - 2} = 2$

28. $\sqrt{x + 2} + \sqrt{3x + 7} = 1$

29. $\sqrt{3\sqrt{x + 1}} = \sqrt{3x - 5}$

30. $\sqrt{1 + 4\sqrt{x}} = 1 + \sqrt{x}$

Solve each equation with rational exponents in Exercises 31–40. Check all proposed solutions.

31. $x^{\frac{3}{2}} = 8$ **32.** $x^{\frac{3}{2}} = 27$

33. $(x - 4)^{\frac{3}{2}} = 27$ **34.** $(x + 5)^{\frac{3}{2}} = 8$

35. $6x^{\frac{5}{2}} - 12 = 0$ **36.** $8x^{\frac{5}{3}} - 24 = 0$

37. $(x - 4)^{\frac{2}{3}} = 16$ **38.** $(x + 5)^{\frac{2}{3}} = 4$

39. $(x^2 - x - 4)^{\frac{3}{4}} - 2 = 6$ **40.** $(x^2 - 3x + 3)^{\frac{3}{2}} - 1 = 0$

Solve each equation in Exercises 41–60 by making an appropriate substitution.

41. $x^4 - 5x^2 + 4 = 0$ **42.** $x^4 - 13x^2 + 36 = 0$

43. $9x^4 = 25x^2 - 16$ **44.** $4x^4 = 13x^2 - 9$

45. $x - 13\sqrt{x} + 40 = 0$ **46.** $2x - 7\sqrt{x} - 30 = 0$

47. $x^{-2} - x^{-1} - 20 = 0$ **48.** $x^{-2} - x^{-1} - 6 = 0$

49. $x^{\frac{2}{3}} - x^{\frac{1}{3}} - 6 = 0$ **50.** $2x^{\frac{2}{3}} + 7x^{\frac{1}{3}} - 15 = 0$

51. $x^{\frac{3}{2}} - 2x^{\frac{3}{4}} + 1 = 0$ **52.** $x^{\frac{2}{5}} + x^{\frac{1}{5}} - 6 = 0$

53. $2x - 3x^{\frac{1}{2}} + 1 = 0$ **54.** $x + 3x^{\frac{1}{2}} - 4 = 0$

55. $(x - 5)^2 - 4(x - 5) - 21 = 0$

56. $(x + 3)^2 + 7(x + 3) - 18 = 0$

57. $(x^2 - x)^2 - 14(x^2 - x) + 24 = 0$

58. $(x^2 - 2x)^2 - 11(x^2 - 2x) + 24 = 0$

59. $\left(y - \frac{8}{y}\right)^2 + 5\left(y - \frac{8}{y}\right) - 14 = 0$

60. $\left(y - \frac{10}{y}\right)^2 + 6\left(y - \frac{10}{y}\right) - 27 = 0$

In Exercises 61–78, solve each absolute value equation or indicate that the equation has no solution.

61. $|x| = 8$ **62.** $|x| = 6$

63. $|x - 2| = 7$ **64.** $|x + 1| = 5$

65. $|2x - 1| = 5$ **66.** $|2x - 3| = 11$

67. $2|3x - 2| = 14$ **68.** $3|2x - 1| = 21$

69. $7|5x| + 2 = 16$ **70.** $7|3x| + 2 = 16$

71. $2\left|4 - \frac{5}{2}x\right| + 6 = 18$ **72.** $4\left|1 - \frac{3}{4}x\right| + 7 = 10$

73. $|x + 1| + 5 = 3$ **74.** $|x + 1| + 6 = 2$

75. $|2x - 1| + 3 = 3$ **76.** $|3x - 2| + 4 = 4$

Hint for Exercises 77–78: Absolute value expressions are equal when the expressions inside the absolute value bars are equal to or opposites of each other.

77. $|3x - 1| = |x + 5|$ **78.** $|2x - 7| = |x + 3|$

In Exercises 79–84, find the x-intercepts of the graph of each equation. Then use the x-intercepts to match the equation with its graph. [The graphs are labeled (a) through (f).]

79. $y = \sqrt{x+2} + \sqrt{x-1} - 3$

80. $y = \sqrt{x-4} + \sqrt{x+4} - 4$

81. $y = x^{\frac{1}{3}} + 2x^{\frac{1}{6}} - 3$

82. $y = x^{-2} - x^{-1} - 6$

83. $y = (x+2)^2 - 9(x+2) + 20$

84. $y = 2(x+2)^2 + 5(x+2) - 3$

a.

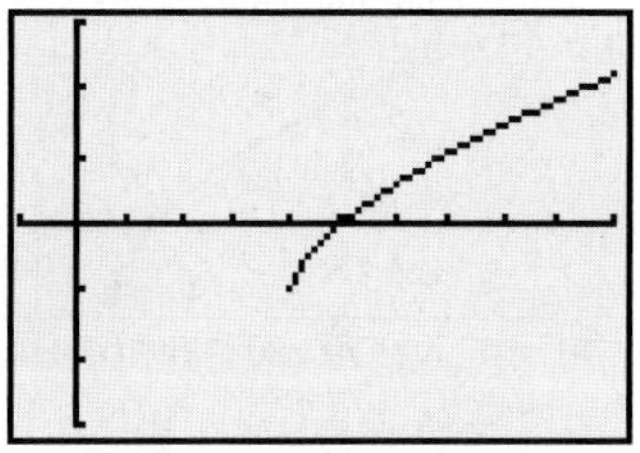

[−1, 10, 1] by [−3, 3, 1]

b.

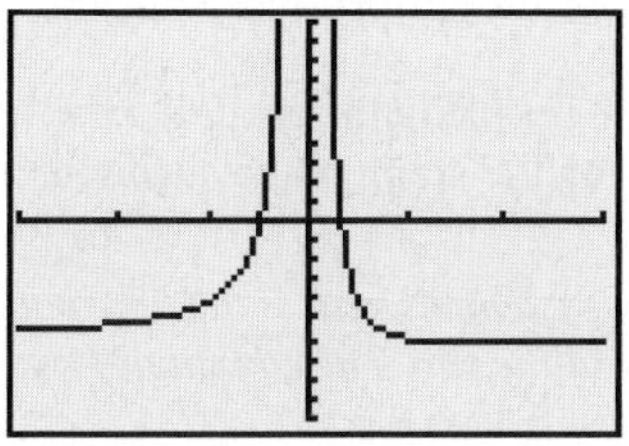

[−3, 3, 1] by [−10, 10, 1]

c.

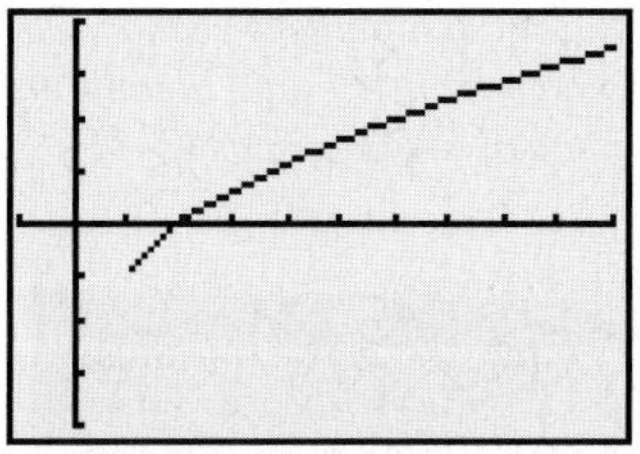

[−1, 10, 1] by [−4, 4, 1]

d.

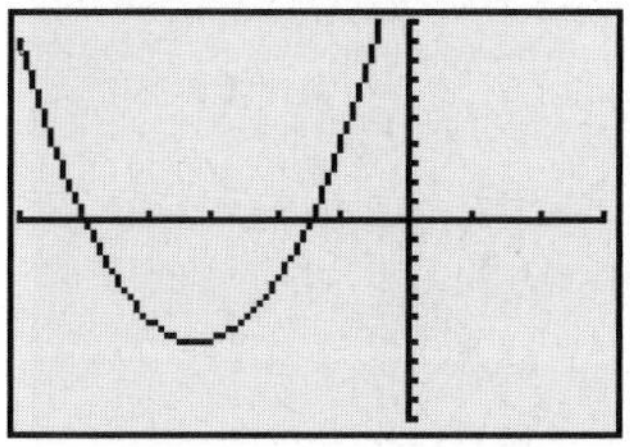

[−6, 3, 1] by [−10, 10, 1]

e.

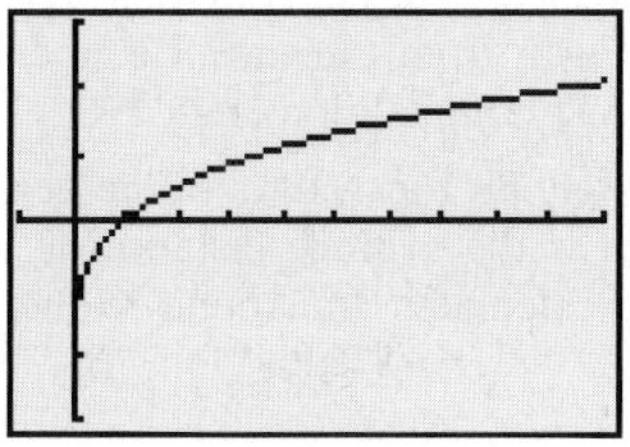

[−1, 10, 1] by [−3, 3, 1]

f.

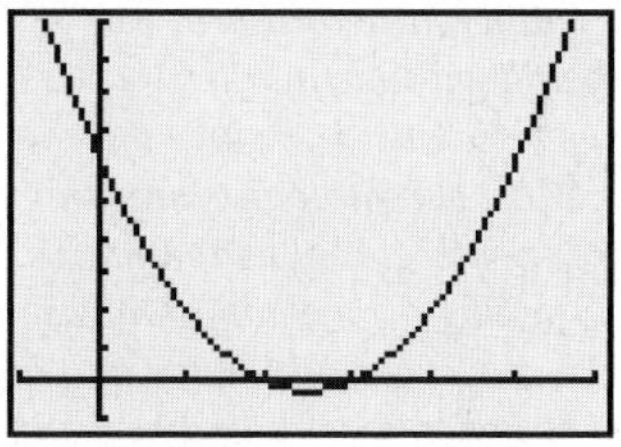

[−1, 6, 1] by [−1, 10, 1]

In Exercises 85–94, find all values of x satisfying the given conditions.

85. $y = |5 - 4x|$ and $y = 11$.

86. $y = |2 - 3x|$ and $y = 13$.

87. $y = x + \sqrt{x+5}$ and $y = 7$.

88. $y = x - \sqrt{x-2}$ and $y = 4$.

89. $y = 2x^3 + x^2 - 8x + 2$ and $y = 6$.

90. $y = x^3 + 4x^2 - x + 6$ and $y = 10$.

91. $y = (x+4)^{\frac{3}{2}}$ and $y = 8$.

92. $y = (x-5)^{\frac{3}{2}}$ and $y = 125$.

93. $y_1 = (x^2 - 1)^2$, $y_2 = 2(x^2 - 1)$, and y_1 exceeds y_2 by 3.

94. $y_1 = 6\left(\frac{2x}{x-3}\right)^2$, $y_2 = 5\left(\frac{2x}{x-3}\right)$, and y_1 exceeds y_2 by 6.

Practice Plus

In Exercises 95–98, solve each equation.

95. $|x^2 + 2x - 36| = 12$

96. $|x^2 + 6x + 1| = 8$

97. $x(x+1)^3 - 42(x+1)^2 = 0$

98. $x(x-2)^3 - 35(x-2)^2 = 0$

99. If 5 times a number is decreased by 4, the principal square root of this difference is 2 less than the number. Find the number(s).

100. If a number is decreased by 3, the principal square root of this difference is 5 less than the number. Find the number(s).

101. Solve for V: $r = \sqrt{\frac{3V}{\pi h}}$.

102. Solve for A: $r = \sqrt{\frac{A}{4\pi}}$.

In Exercises 103–104, list all numbers that must be excluded from the domain of each expression.

103. $\frac{|x-1| - 3}{|x+2| - 14}$

104. $\frac{x^3 - 2x^2 - 9x + 18}{x^3 + 3x^2 - x - 3}$

Application Exercises

The formula

$$P = 28\sqrt{t} + 80$$

models the number of personal computers sold in the United States, P, in millions, t years after 1996. Use the model to solve Exercises 105–106.

105. If trends indicated by this model continue, in what year will 192 million personal computers be sold?

106. According to the model, in what year were 143 million personal computers sold? Round to the nearest year. How well does this describe the data shown for this year in Figure 1.23 on page 150?

In 2002, the average surface temperature on Earth was 57.9°F, approximately 1.4° higher than it was one hundred years ago. Worldwide temperatures have risen only 9°F since the end of the last ice age 12,000 years ago. Most climatologists are convinced that over the next one hundred years, global temperatures will continue to increase, possibly setting off a chain of devastating events beginning with a rise in sea levels worldwide and ending with the destruction of water supplies, forests, and agriculture in many parts of the world. The graph shows global annual average temperatures from 1880 through 2002, with projections from 2002 through 2100.

Global Annual Average Temperatures and Projections through 2100

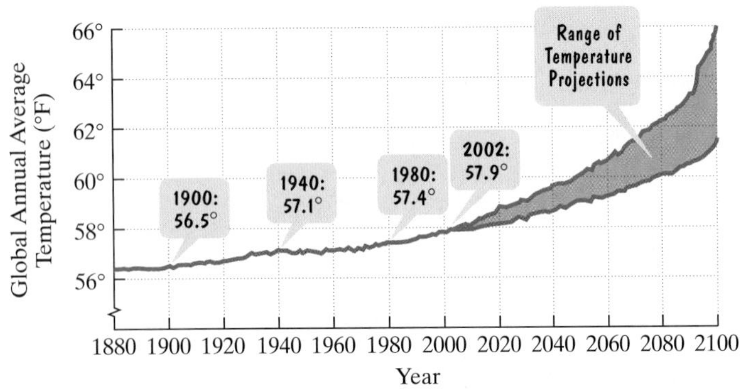

Source: National Oceanic and Atmospheric Administration

The temperature projections shown in the graph can be modeled by two equations:

$$H = 0.083x + 57.9$$

Models temperatures at the high end of the range

$$L = 0.36\sqrt{x} + 57.9.$$

Models temperatures at the low end of the range

In these equations, H and L describe projected global annual average temperatures, in degrees Fahrenheit, x years after 2002, where $0 \le x \le 98$. Use the models to solve Exercises 107–110.

107. Use H and L to determine the temperatures at the high and low end of the range of projected global average temperatures for 2100. Round to the nearest tenth of a degree.

108. Use H and L to determine the temperatures at the high and low end of the range of projected global average temperatures for 2080. Round to the nearest tenth of a degree.

109. Use H and L to determine by which year the projected global average temperature will exceed the 2002 average of 57.9° by one degree. Round to the nearest year.

110. Use H and L to determine by which year the projected global average temperature will exceed the 2002 average of 57.9° by two degrees.

Out of a group of 50,000 births, the number of people, y, surviving to age x is modeled by the equation

$$y = 5000\sqrt{100 - x}.$$

The graph of the equation is shown. Use the equation to solve Exercises 111–112.

Number of Survivors, by Age, from a Group of 50,000 Births

$y = 5000\sqrt{100 - x}$

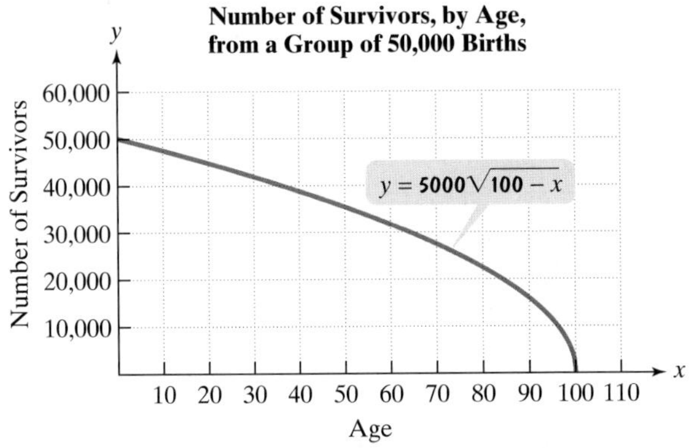

111. To what age will 40,000 people in the group survive? Identify the solution as a point on the graph of the equation.

112. To what age will 35,000 people in the group survive? Identify the solution as a point on the graph of the equation.

For each planet in our solar system, its year is the time it takes the planet to revolve once around the sun. The equation

$$E = 0.2x^{\frac{3}{2}}$$

models the number of Earth days in a planet's year, E, where x is the average distance of the planet from the sun, in millions of kilometers. Use the equation to solve Exercises 113–114.

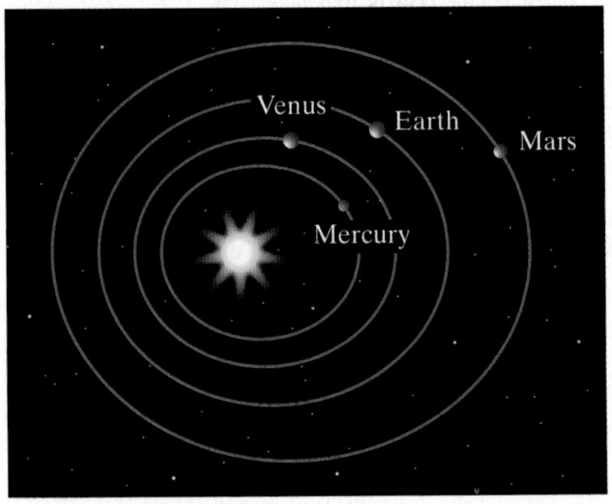

113. We, of course, have 365 Earth days in our year. What is the average distance of Earth from the sun? Use a calculator and round to the nearest million kilometers.

114. There are approximately 88 Earth days in the year of the planet Mercury. What is the average distance of Mercury from the sun? Use a calculator and round to the nearest million kilometers.

Use the Pythagorean Theorem to solve Exercises 115–116.

115. Two vertical poles of lengths 6 feet and 8 feet, respectively, stand 10 feet apart. A cable reaches from the top of one pole to some point on the ground between the poles and then to the top of the other pole. Where should this point be located to use 18 feet of cable?

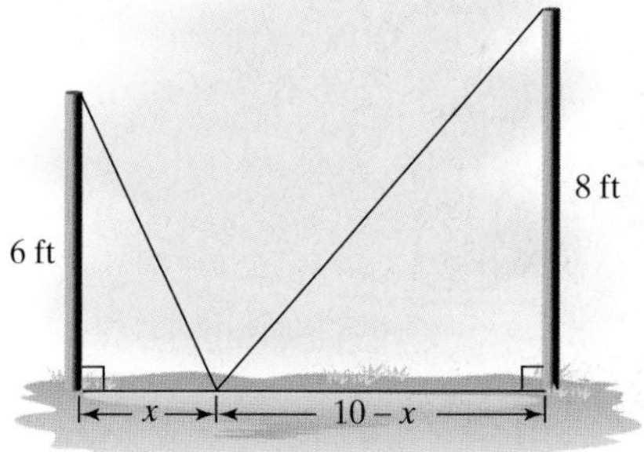

116. Towns A and B are located 6 miles and 3 miles, respectively, from a major expressway. The point on the expressway closest to town A is 12 miles from the point on the expressway closest to town B. Two new roads are to be built from A to the expressway and then to B.

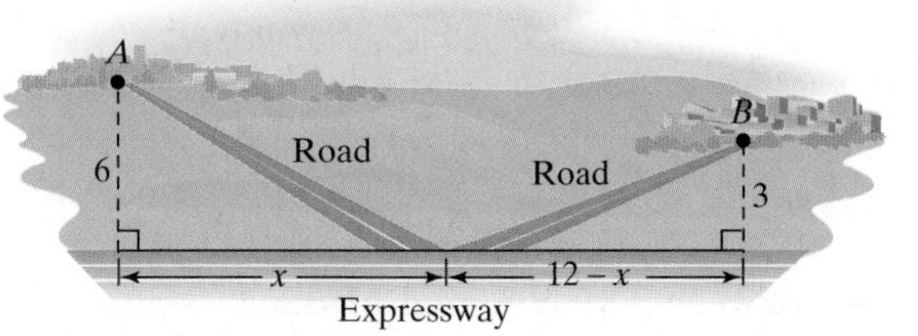

a. Express the combined lengths of the new roads in terms of where the roads are positioned from A, shown as x in the figure.

b. Where should the roads be positioned from A if their combined lengths is 15 miles?

Writing in Mathematics

117. Without actually solving the equation, give a general description of how to solve $x^3 - 5x^2 - x + 5 = 0$.

118. In solving $\sqrt{3x+4} - \sqrt{2x+4} = 2$, why is it a good idea to isolate a radical term? What if we don't do this and simply square each side? Describe what happens.

119. What is an extraneous solution to a radical equation?

120. Explain how to recognize an equation that is quadratic in form. Provide two original examples with your explanation.

121. Describe two methods for solving this equation: $x - 5\sqrt{x} + 4 = 0$.

122. Explain how to solve an equation involving absolute value.

123. Explain why the procedure that you explained in Exercise 122 does not apply to the equation $|x - 2| = -3$. What is the solution set for this equation?

124. Describe the trend shown by the graph in Exercises 111–112. When is the rate of decrease most rapid? What does this mean about survival rate by age?

Technology Exercises

In Exercises 125–127, use a graphing utility and the graph's x-intercepts to solve each equation. Check by direct substitution. A viewing rectangle is given.

125. $x^3 + 3x^2 - x - 3 = 0$

$[-6, 6, 1]$ by $[-6, 6, 1]$

126. $-x^4 + 4x^3 - 4x^2 = 0$

$[-6, 6, 1]$ by $[-9, 2, 1]$

127. $\sqrt{2x+13} - x - 5 = 0$

$[-5, 5, 1]$ by $[-5, 5, 1]$

128. Use a graphing utility to obtain the graph of the equation in Exercises 111–112. Then use the TRACE feature to trace along the curve until you reach the point that visually shows the solution to Exercise 111 or 112.

Critical Thinking Exercises

129. Which one of the following is true?

a. Squaring both sides of $\sqrt{y+4} + \sqrt{y-1} = 5$ leads to $y + 4 + y - 1 = 25$, an equation with no radicals.

b. The equation $(x^2 - 2x)^9 - 5(x^2 - 2x)^3 + 6 = 0$ is quadratic in form and can be solved by letting

$$u = (x^2 - 2x)^3.$$

c. If a radical equation has two proposed solutions and one of these values is not a solution, the other value is also not a solution.

d. None of these statements is true.

130. Solve: $\sqrt{6x-2} = \sqrt{2x+3} - \sqrt{4x-1}$.

131. Solve *without* squaring both sides:

$$5 - \frac{2}{x} = \sqrt{5 - \frac{2}{x}}.$$

132. Solve for x: $\sqrt[3]{x\sqrt{x}} = 9$.

133. Solve for x: $x^{\frac{5}{6}} + x^{\frac{2}{3}} - 2x^{\frac{1}{2}} = 0$.

SECTION 1.7 Linear Inequalities and Absolute Value Inequalities

Objectives

1. Use interval notation.
2. Find intersections and unions of intervals.
3. Solve linear inequalities.
4. Recognize inequalities with no solution or all real numbers as solutions.
5. Solve compound inequalities.
6. Solve absolute value inequalities.

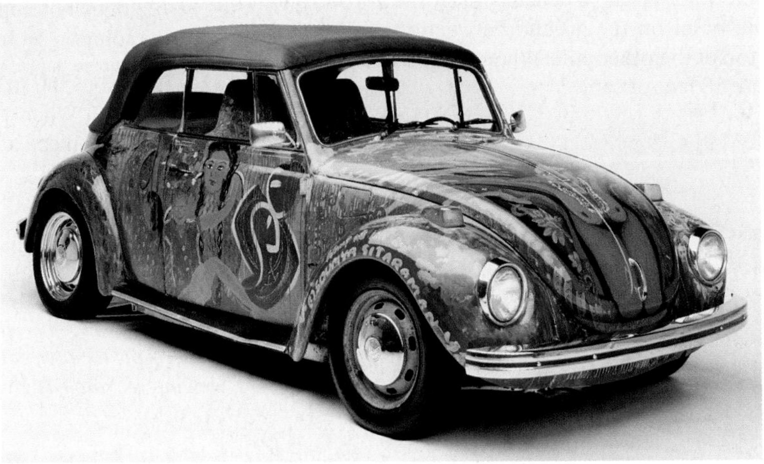

Rent-a-Heap, a car rental company, charges \$125 per week plus \$0.20 per mile to rent one of their cars. Suppose you are limited by how much money you can spend for the week: You can spend at most \$335. If we let x represent the number of miles you drive the heap in a week, we can write an inequality that models the given conditions:

The weekly charge of \$125 | plus | the charge of \$0.20 per mile for x miles | must be less than or equal to | \$335.

$$125 + 0.20x \leq 335.$$

Placing an inequality symbol between a polynomial of degee 1 and a constant results in a *linear inequality in one variable*. In this section, we will study how to solve linear inequalities such as the one shown above. **Solving an inequality** is the process of finding the set of numbers that make the inequality a true statement. These numbers are called the **solutions** of the inequality and we say that they **satisfy** the inequality. The set of all solutions is called the **solution set** of the inequality. Set-builder notation and a new notation, called *interval notation*, are used to represent solution sets. We begin this section by looking at interval notation.

1 Use interval notation.

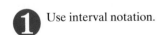

Interval Notation

Subsets of real numbers can be represented using **interval notation**. Suppose that a and b are two real numbers such that $a < b$.

Interval Notation	Graph
The **open interval** (a, b) represents the set of real numbers between, but not including, a and b. $$(a, b) = \{x \mid a < x < b\}$$ x is greater than a $(a < x)$ and x is less than b $(x < b)$.	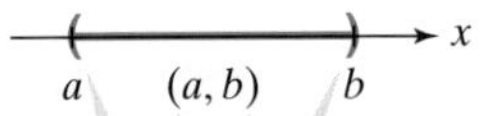 The parentheses in the graph and in interval notation indicate that a and b, the endpoints, are excluded from the interval.

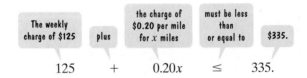

(continued)

Interval Notation	Graph
The **closed interval** $[a, b]$ represents the set of real numbers between, and including, a and b. $$[a, b] = \{x \mid a \leq x \leq b\}$$ x is greater than or equal to a $(a \leq x)$ and x is less than or equal to b $(x \leq b)$.	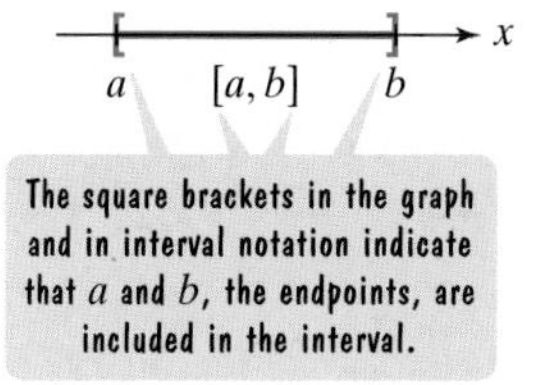
The **infinite interval** (a, ∞) represents the set of real numbers that are greater than a. $$(a, \infty) = \{x \mid x > a\}$$ The infinity symbol does not represent a real number. It indicates that the interval extends indefinitely to the right.	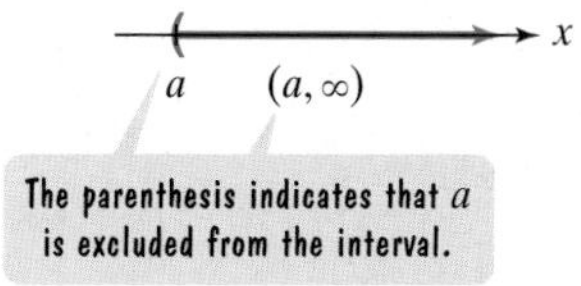
The **infinite interval** $(-\infty, b]$ represents the set of real numbers that are less than or equal to b. $$(-\infty, b] = \{x \mid x \leq b\}$$ The negative infinity symbol indicates that the interval extends indefinitely to the left.	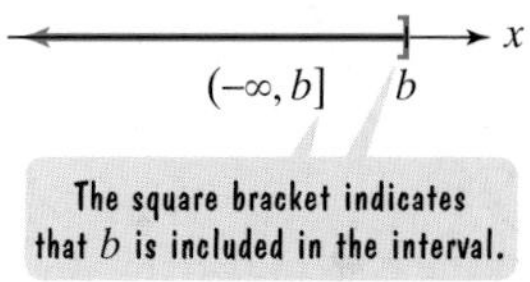

Parentheses and Brackets in Interval Notation

Parentheses indicate endpoints that are not included in an interval. Square brackets indicate endpoints that are included in an interval.

Table 1.5 lists nine possible types of intervals used to describe subsets of real numbers.

Table 1.5 Intervals on the Real Number Line

Let a and b be real numbers such that $a < b$.		
Interval Notation	**Set-Builder Notation**	**Graph**
(a, b)	$\{x \mid a < x < b\}$	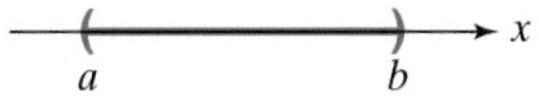
$[a, b]$	$\{x \mid a \leq x \leq b\}$	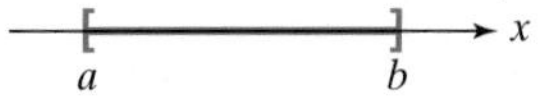
$[a, b)$	$\{x \mid a \leq x < b\}$	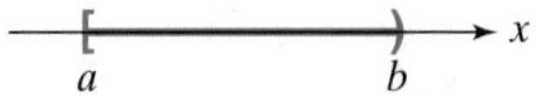
$(a, b]$	$\{x \mid a < x \leq b\}$	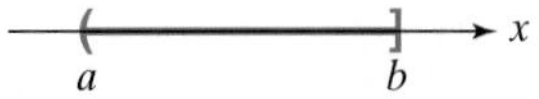
(a, ∞)	$\{x \mid x > a\}$	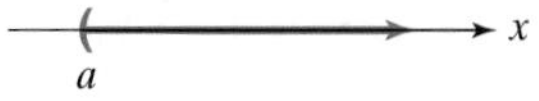
$[a, \infty)$	$\{x \mid x \geq a\}$	
$(-\infty, b)$	$\{x \mid x < b\}$	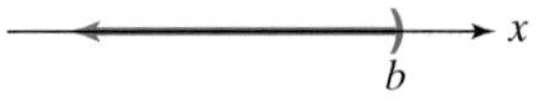
$(-\infty, b]$	$\{x \mid x \leq b\}$	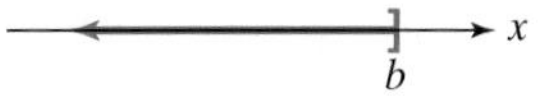
$(-\infty, \infty)$	$\{x \mid x$ is a real number$\}$ or $\mathbb{R}$ (set of all real numbers)	

EXAMPLE 1 Using Interval Notation

Express each interval in set-builder notation and graph:

a. $(-1, 4]$ **b.** $[2.5, 4]$ **c.** $(-4, \infty)$.

Solution

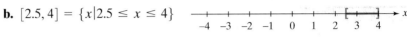

a. $(-1, 4] = \{x \mid -1 < x \leq 4\}$

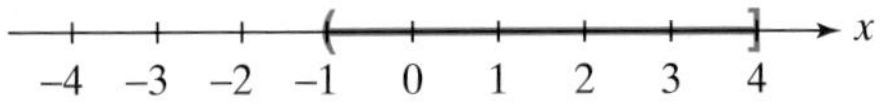

b. $[2.5, 4] = \{x \mid 2.5 \leq x \leq 4\}$

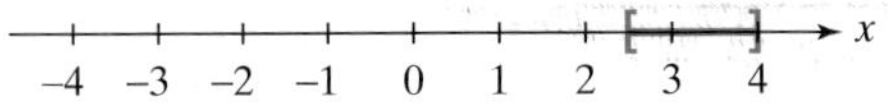

c. $(-4, \infty) = \{x \mid x > -4\}$

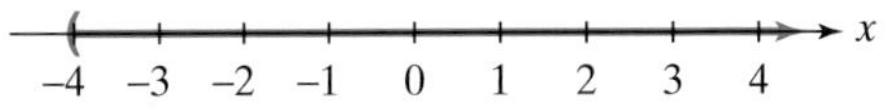

Check Point 1 Express each interval in set-builder notation and graph:

a. $[-2, 5)$ **b.** $[1, 3.5]$ **c.** $(-\infty, -1)$.

2 Find intersections and unions of intervals.

Intersections and Unions of Intervals

In Chapter P, we learned how to find intersections and unions of sets. Recall that $A \cap B$ (A intersection B) is the set of elements common to both set A and set B. By contrast, $A \cup B$ (A union B) is the set of elements in set A or in set B or in both sets.

Because intervals represent sets, it is possible to find their intersections and unions. Graphs are helpful in this process.

Finding Intersections and Unions of Intervals

1. Graph each interval on a number line.
2. **a.** To find the intersection, take the portion of the number line that the two graphs have in common.

   b. To find the union, take the portion of the number line representing the total collection of numbers in the two graphs.

EXAMPLE 2 Finding Intersections and Unions of Intervals

Use graphs to find each set:

a. $(1, 4) \cap [2, 8]$ **b.** $(1, 4) \cup [2, 8]$.

Solution

a. $(1, 4) \cap [2, 8]$, the intersection of the intervals $(1, 4)$ and $[2, 8]$, consists of the numbers that are in both intervals.

Graph $(1, 4)$:

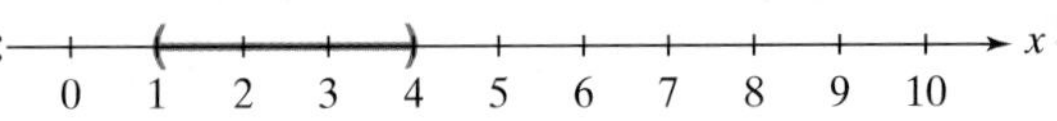

$\{x \mid 1 < x < 4\}$

Graph $[2, 8]$:

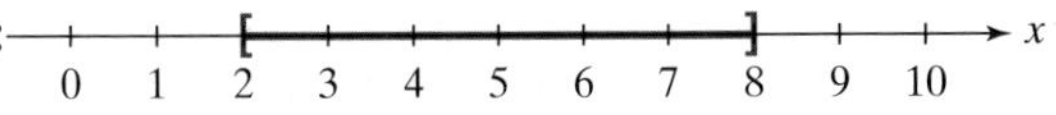

$\{x \mid 2 \leq x \leq 8\}$

To find $(1, 4) \cap [2, 8]$, take the portion of the number line that the two graphs have in common.

Numbers in both $(1, 4)$ and $[2, 8]$:

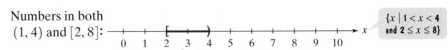

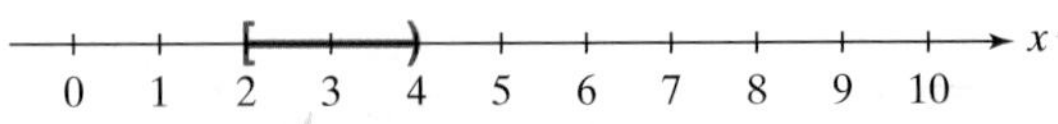

$\{x \mid 1 < x < 4$ and $2 \leq x \leq 8\}$

The numbers common to both intervals are those that are greater than or equal to 2 and less than 4: $[2, 4)$.

Thus, $(1, 4) \cap [2, 8] = [2, 4)$.

b. $(1, 4) \cup [2, 8]$, the union of the intervals $(1, 4)$ and $[2, 8]$, consists of the numbers that are in either one interval or the other (or both).

Graph $(1, 4)$: $\{x \mid 1 < x < 4\}$

Graph $[2, 8]$: $\{x \mid 2 \leq x \leq 8\}$

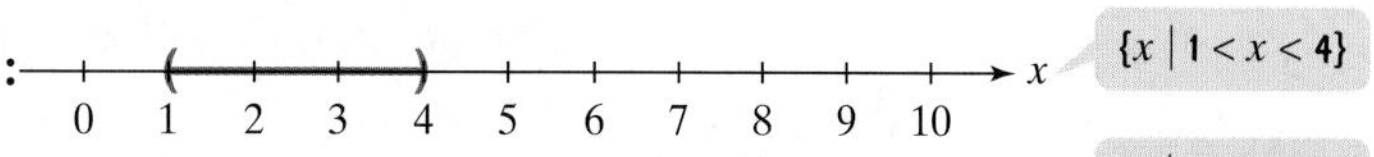

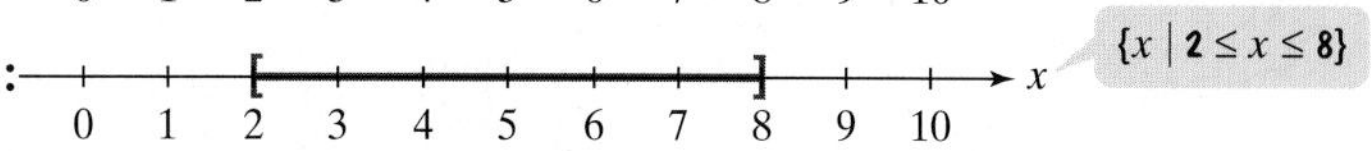

To find $(1, 4) \cup [2, 8]$, take the portion of the number line representing the total collection of numbers in the two graphs.

Numbers in either
$(1, 4)$ or $[2, 8]$ or both: $\{x \mid 1 < x < 4$ or $2 \leq x \leq 8\}$

The numbers in either one interval or the other (or both) are those that are greater than 1 and less than or equal to 8: $(1, 8]$.

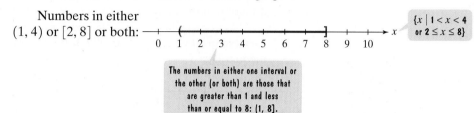

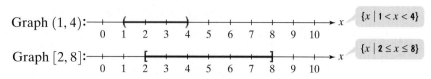

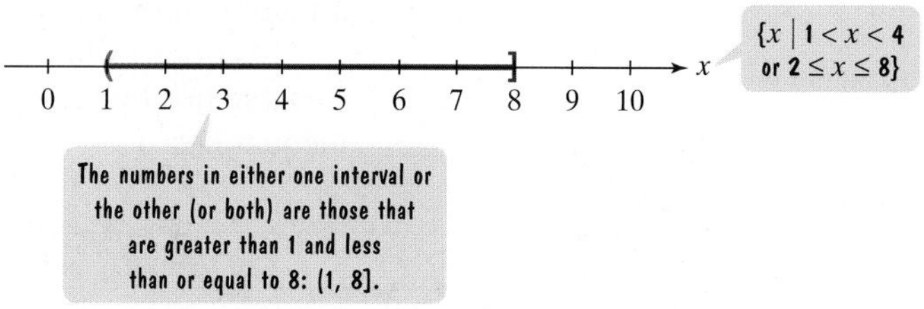

Thus, $(1, 4) \cup [2, 8] = (1, 8]$.

Check Point 2 Use graphs to find each set:
   a. $[1, 3] \cap (2, 6)$ **b.** $[1, 3] \cup (2, 6)$.

❸ Solve linear inequalities.

Solving Linear Inequalities in One Variable

We know that a linear equation in x can be expressed as $ax + b = 0$. A **linear inequality in x** can be written in one of the following forms: $ax + b < 0, ax + b \leq 0$, $ax + b > 0, ax + b \geq 0$. In each form, $a \neq 0$. Back to our question that opened this section: How many miles can you drive your Rent-a-Heap car if you can spend at most \$335 per week? We answer the question by solving

$$0.20x + 125 \leq 335$$

for x. The solution procedure is nearly identical to that for solving

$$0.20x + 125 = 335.$$

Our goal is to get x by itself on the left side. We do this by subtracting 125 from both sides to isolate $0.20x$:

$0.20x + 125 \leq 335$	This is the given inequality.
$0.20x + 125 - 125 \leq 335 - 125$	Subtract 125 from both sides.
$0.20x \leq 210.$	Simplify.

Finally, we isolate x from $0.20x$ by dividing both sides of the inequality by 0.20:

$\dfrac{0.20x}{0.20} \leq \dfrac{210}{0.20}$	Divide both sides by 0.20.
$x \leq 1050.$	Simplify.

With at most \$335 per week to spend, you can travel at most 1050 miles.

We started with the inequality $0.20x + 125 \leq 335$ and obtained the inequality $x \leq 1050$ in the final step. Both of these inequalities have the same solution set, namely $\{x \mid x \leq 1050\}$. Inequalities such as these, with the same solution set, are said to be **equivalent**.

We isolated x from $0.20x$ by dividing both sides of $0.20x \leq 210$ by 0.20, a positive number. Let's see what happens if we divide both sides of an inequality by a negative number. Consider the inequality $10 < 14$. Divide 10 and 14 by -2:

$$\frac{10}{-2} = -5 \quad \text{and} \quad \frac{14}{-2} = -7.$$

Because -5 lies to the right of -7 on the number line, -5 is greater than -7:

$$-5 > -7.$$

Notice that the direction of the inequality symbol is reversed:

$$10 < 14$$
$$\updownarrow$$
$$-5 > -7.$$

Dividing by -2 changes the direction of the inequality symbol.

In general, **when we multiply or divide both sides of an inequality by a negative number, the direction of the inequality symbol is reversed**. When we reverse the direction of the inequality symbol, we say that we change the *sense* of the inequality.

We can isolate a variable in a linear inequality the same way we can isolate a variable in a linear equation. The following properties are used to create equivalent inequalities:

Properties of Inequalities

Property	The Property in Words	Example
The Addition Property of Inequality If $a < b$, then $a + c < b + c$. If $a < b$, then $a - c < b - c$.	If the same quantity is added to or subtracted from both sides of an inequality, the resulting inequality is equivalent to the original one.	$2x + 3 < 7$ Subtract 3: $2x + 3 - 3 < 7 - 3.$ Simplify: $2x < 4.$
The Positive Multiplication Property of Inequality If $a < b$ and c is positive, then $ac < bc$. If $a < b$ and c is positive, then $\frac{a}{c} < \frac{b}{c}$.	If we multiply or divide both sides of an inequality by the same positive quantity, the resulting inequality is equivalent to the original one.	$2x < 4$ Divide by 2: $\frac{2x}{2} < \frac{4}{2}.$ Simplify: $x < 2.$
The Negative Multiplication Property of Inequality If $a < b$ and c is negative, then $ac > bc$. If $a < b$ and c is negative, $\frac{a}{c} > \frac{b}{c}$.	If we multiply or divide both sides of an inequality by the same negative quantity and reverse the direction of the inequality symbol, the resulting inequality is equivalent to the original one.	$-4x < 20$ Divide by -4 and reverse the sense of the inequality: $\frac{-4x}{-4} > \frac{20}{-4}.$ Simplify: $x > -5.$

EXAMPLE 3 Solving a Linear Inequality

Solve and graph the solution set on a number line:

$$3 - 2x \le 11.$$

Solution

$3 - 2x \le 11$	This is the given inequality.
$3 - 2x - 3 \le 11 - 3$	Subtract 3 from both sides.
$-2x \le 8$	Simplify.
$\frac{-2x}{-2} \ge \frac{8}{-2}$	Divide both sides by -2 and change the sense of the inequality.
$x \ge -4$	Simplify.

The solution set consists of all real numbers that are greater than or equal to -4, expressed as $\{x \mid x \ge -4\}$ in set-builder notation. The interval notation for this solution set is $[-4, \infty)$. The graph of the solution set is shown as follows:

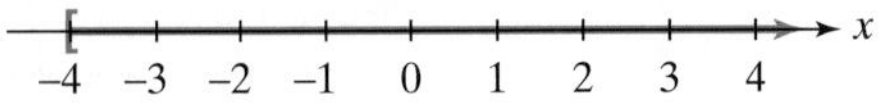

Discovery

As a partial check, select one number from the solution set for the inequality in Example 3. Substitute that number into the original inequality. Perform the resulting computations. You should obtain a true statement.

Is it possible to perform a partial check using a number that is not in the solution set? What should happen in this case? Try doing this.

Check Point 3 Solve and graph the solution set on a number line:

$$2 - 3x \le 5.$$

EXAMPLE 4 Solving a Linear Inequality

Solve and graph the solution set on a number line:

$$-2x - 4 > x + 5.$$

Solution

Step 1 Simplify each side. Because each side is already simplified, we can skip this step.

Step 2 Collect variable terms on one side and constant terms on the other side. We will collect variable terms on the left and constant terms on the right.

$-2x - 4 > x + 5$	This is the given inequality.
$-2x - 4 - x > x + 5 - x$	Subtract x from both sides.
$-3x - 4 > 5$	Simplify.
$-3x - 4 + 4 > 5 + 4$	Add 4 to both sides.
$-3x > 9$	Simplify.

Step 3 Isolate the variable and solve. We isolate the variable, x, by dividing both sides by -3. Because we are dividing by a negative number, we must reverse the inequality symbol.

$\frac{-3x}{-3} < \frac{9}{-3}$	Divide both sides by -3 and change the sense of the inequality.
$x < -3$	Simplify.

Step 4 Express the solution set in set-builder or interval notation and graph the set on a number line. The solution set consists of all real numbers that are less than -3, expressed in set-builder notation as $\{x \mid x < -3\}$. The interval notation for this solution set is $(-\infty, -3)$. The graph of the solution set is shown as follows:

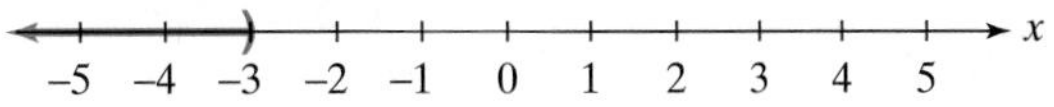

–5 –4 –3 –2 –1 0 1 2 3 4 5
x

Study Tip

You can solve

$$-2x - 4 > x + 5$$

by isolating x on the right side. Add $2x$ to both sides.

$$-2x - 4 + 2x > x + 5 + 2x$$
$$-4 > 3x + 5$$

Now subtract 5 from both sides.

$$-4 - 5 > 3x + 5 - 5$$
$$-9 > 3x$$

Finally, divide both sides by 3.

$$\frac{-9}{3} > \frac{3x}{3}$$
$$-3 > x$$

This last inequality means the same thing as $x < -3$.

Technology

You can use a graphing utility to check the solution set of a linear inequality. Enter each side of the inequality separately under y_1 and y_2. Then use the table or the graphs. To use the table, first locate the x-value for which the y-values are the same. Then scroll up or down to locate x values for which y_1 is greater than y_2 or for which y_1 is less than y_2. To use the graphs, locate the intersection point and then find the x-values for which the graph of y_1 lies above the graph of y_2 ($y_1 > y_2$) or for which the graph of y_1 lies below the graph of y_2 ($y_1 < y_2$).

Let's verify our work in Example 4 and show that $(-\infty, -3)$ is the solution set of

$$-2x - 4 > x + 5.$$

Enter $y_1 = -2x - 4$ in the $\boxed{y=}$ screen.

Enter $y_2 = x + 5$ in the $\boxed{y=}$ screen.

We are looking for values of x for which y_1 is greater than y_2.

Numeric Check

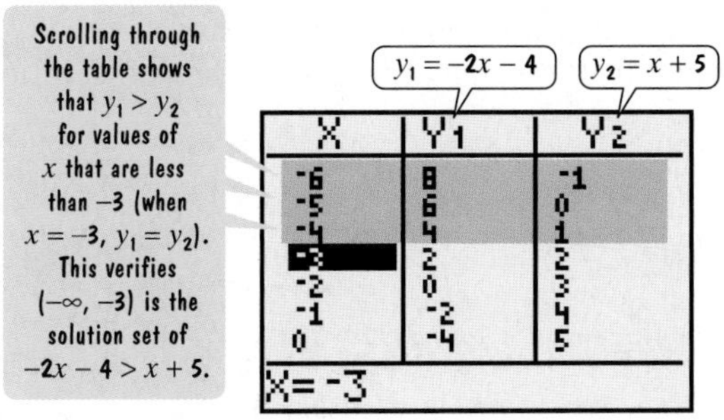

Scrolling through the table shows that $y_1 > y_2$ for values of x that are less than -3 (when $x = -3$, $y_1 = y_2$). This verifies $(-\infty, -3)$ is the solution set of $-2x - 4 > x + 5$.
$y_1 = -2x - 4$
$y_2 = x + 5$
X Y1 Y2
-6 8 -1
-5 6 0
-4 4 1
-3 2 2
-2 0 3
-1 -2 4
0 -4 5
X= -3

Graphic Check

Display the graphs for y_1 and y_2. Use the intersection feature. The solution set is the set of x-values for which the graph of y_1 lies above the graph of y_2.

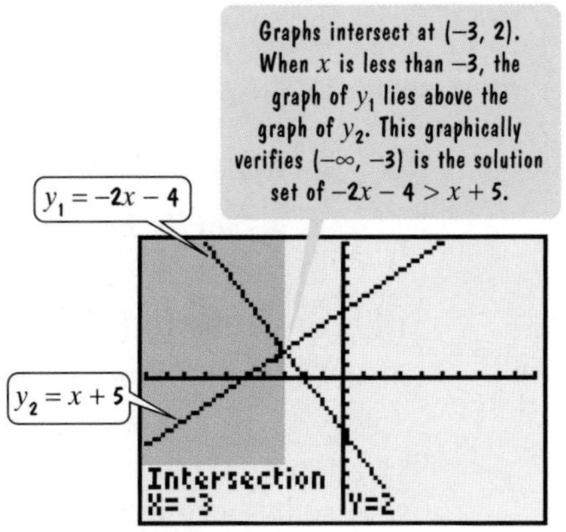

Graphs intersect at $(-3, 2)$. When x is less than -3, the graph of y_1 lies above the graph of y_2. This graphically verifies $(-\infty, -3)$ is the solution set of $-2x - 4 > x + 5$.
$y_1 = -2x - 4$
$y_2 = x + 5$
Intersection
X= -3
Y=2

[–10, 10, 1] by [–10, 10, 1]

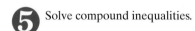

Check Point 4 Solve and graph the solution set on a number line: $3x + 1 > 7x - 15$.

4 Recognize inequalities with no solution or all real numbers as solutions.

Inequalities with Unusual Solution Sets

We have seen that some equations have no solution. This is also true for some inequalities. An example of such an inequality is

$$x > x + 1.$$

There is no number that is greater than itself plus 1. This inequality has no solution and its solution set is $\emptyset$, the empty set.

By contrast, some inequalities are true for all real numbers. An example of such an inequality is

$$x < x + 1.$$

Every real number is less than itself plus 1. The solution set is $\{x|x \text{ is a real number}\}$ or $\mathbb{R}$. In interval notation, the solution set is $(-\infty, \infty)$.

If you attempt to solve an inequality that has no solution, you will eliminate the variable and obtain a false statement, such as $0 > 1$. If you attempt to solve an inequality that is true for all real numbers, you will eliminate the variable and obtain a true statement, such as $0 < 1$.

EXAMPLE 5 Solving Linear Inequalities

Solve each inequality:

a. $2(x + 4) > 2x + 3$ **b.** $x + 7 \le x - 2$.

Solution

a.

$2(x + 4) > 2x + 3$ This is the given inequality.

$2x + 8 > 2x + 3$ Apply the distributive property.

$2x + 8 - 2x > 2x + 3 - 2x$ Subtract 2x from both sides.

$8 > 3$ Simplify. The statement $8 > 3$ is true.

The inequality $8 > 3$ is true for all values of x. Because this inequality is equivalent to the original inequality, the original inequality is true for all real numbers. The solution set is

$$\{x|x \text{ is a real number}\} \text{ or } \mathbb{R} \text{ or } (-\infty, \infty).$$

b.

$x + 7 \le x - 2$ This is the given inequality.

$x + 7 - x \le x - 2 - x$ Subtract x from both sides.

$7 \le -2$ Simplify. The statement $7 \le -2$ is false.

The inequality $7 \le -2$ is false for all values of x. Because this inequality is equivalent to the original inequality, the original inequality has no solution. The solution set is $\emptyset$.

Technology

The graphs of

$y_1 = 2(x + 4)$ and $y_2 = 2x + 3$

are parallel lines. The graph of y_1 is always above the graph of y_2. Every value of x satisfies the inequality $y_1 > y_2$. Thus, the solution set of the inequality

$$2(x + 4) > 2x + 3$$

is $(-\infty, \infty)$.

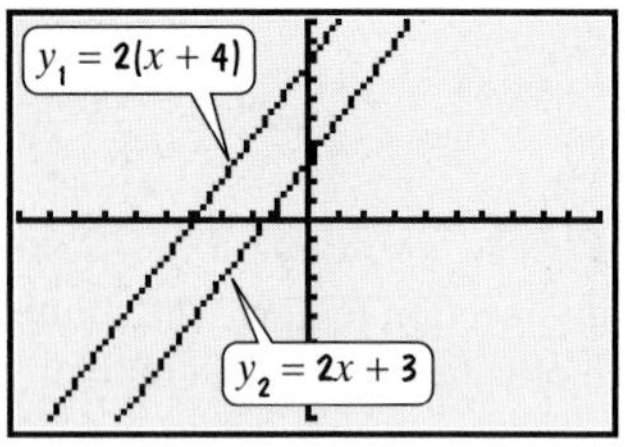

[−10, 10, 1] by [−10, 10, 1]

Check Point 5 Solve each inequality:

a. $3(x + 1) > 3x + 2$ **b.** $x + 1 \le x - 1$.

5 Solve compound inequalities.

Solving Compound Inequalities

We now consider two inequalities such as

$$-3 < 2x + 1 \quad \text{and} \quad 2x + 1 \le 3$$

expressed as a **compound inequality**

$$-3 < 2x + 1 \le 3.$$

The word "and" does not appear when the inequality is written in the shorter form, although intersection is implied. The shorter form enables us to solve both inequalities at once. By performing the same operation on all three parts of the inequality, our goal is to **isolate x in the middle.**

Technology

To check Example 6, graph each part of

$$-3 < 2x + 1 \leq 3.$$

Enter $y_1 = -3$.	Enter $y_2 = 2x + 1$.	Enter $y_3 = 3$.

The figure shows that the graph of $y_2 = 2x + 1$ lies above the graph of $y_1 = -3$ and on or below the graph of $y_3 = 3$ when x is in the interval $(-2, 1]$.

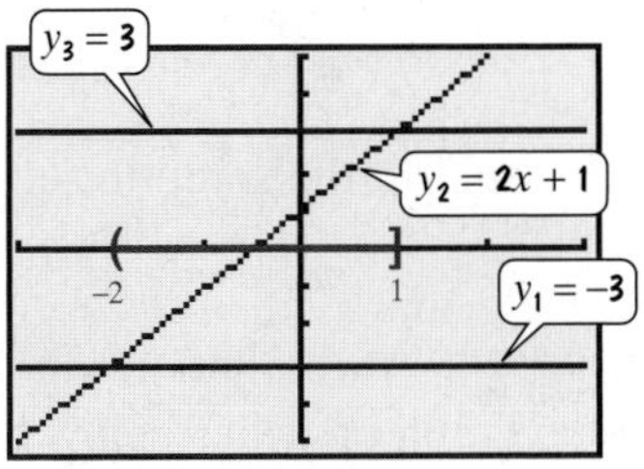

[−3, 3, 1] by [−5, 5, 1]

EXAMPLE 6 Solving a Compound Inequality

Solve and graph the solution set on a number line:

$$-3 < 2x + 1 \leq 3.$$

Solution We would like to isolate x in the middle. We can do this by first subtracting 1 from all three parts of the compound inequality. Then we isolate x from $2x$ by dividing all three parts of the inequality by 2.

$$-3 < 2x + 1 \leq 3 \quad \text{This is the given inequality.}$$

$$-3 - 1 < 2x + 1 - 1 \leq 3 - 1 \quad \text{Subtract 1 from all three parts.}$$

$$-4 < 2x \leq 2 \quad \text{Simplify.}$$

$$\frac{-4}{2} < \frac{2x}{2} \leq \frac{2}{2} \quad \text{Divide each part by 2.}$$

$$-2 < x \leq 1 \quad \text{Simplify.}$$

The solution set consists of all real numbers greater than -2 and less than or equal to 1, represented by $\{x|-2 < x \leq 1\}$ in set-builder notation and $(-2, 1]$ in interval notation. The graph is shown as follows:

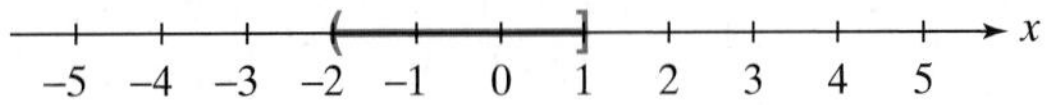

Check Point 6 Solve and graph the solution set on a number line: $1 \leq 2x + 3 < 11$.

6 Solve absolute value inequalities.

Solving Inequalities with Absolute Value

We know that $|x|$ describes the distance of x from zero on a real number line. We can use this geometric interpretation to solve an inequality such as

$$|x| < 2.$$

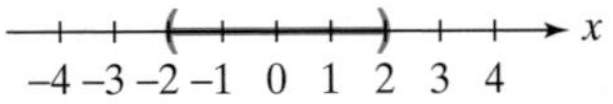

Figure 1.27 $|x| < 2$, so $-2 < x < 2$.

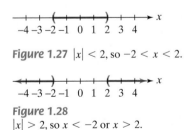

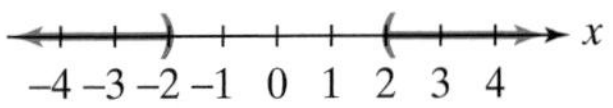

Figure 1.28 $|x| > 2$, so $x < -2$ or $x > 2$.

This means that the distance of x from 0 is *less than* 2, as shown in Figure 1.27. The interval shows values of x that lie less than 2 units from 0. Thus, x can lie between -2 and 2. That is, x is greater than -2 and less than 2. We write $(-2, 2)$ or $\{x|-2 < x < 2.\}$

Some absolute value inequalities use the "greater than" symbol. For example, $|x| > 2$ means that the distance of x from 0 is *greater than* 2, as shown in Figure 1.28. Thus, x can be less than -2 *or* greater than 2. We write $x < -2$ or $x > 2$.

These observations suggest the following principles for solving inequalities with absolute value.

Study Tip

In the $|X| < c$ case, we have one compound inequality to solve. In the $|X| > c$ case, we have two separate inequalities to solve.

Solving an Absolute Value Inequality

If X is an algebraic expression and c is a positive number,

1. The solutions of $|X| < c$ are the numbers that satisfy $-c < X < c$.
2. The solutions of $|X| > c$ are the numbers that satisfy $X < -c$ or $X > c$.

These rules are valid if $<$ is replaced by $\leq$ and $>$ is replaced by $\geq$.

EXAMPLE 7 Solving an Absolute Value Inequality

Solve and graph the solution set on a number line: $|x - 4| < 3$.

Solution We rewrite the inequality without absolute value bars.

$|X| < c$ means $-c < X < c$.

$$|x - 4| < 3 \quad \text{means} \quad -3 < x - 4 < 3.$$

We solve the compound inequality by adding 4 to all three parts.

$$-3 < x - 4 < 3$$
$$-3 + 4 < x - 4 + 4 < 3 + 4$$
$$1 < x < 7$$

The solution set is all real numbers greater than 1 and less than 7, denoted by $\{x|1 < x < 7\}$ or $(1, 7)$. The graph of the solution set is shown as follows:

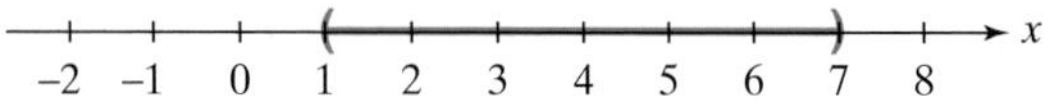

We can use the rectangular coordinate system to visualize the solution set of

$$|x - 4| < 3.$$

Figure 1.29 shows the graphs of $y_1 = |x - 4|$ and $y_2 = 3$. The solution set of $|x - 4| < 3$ consists of all values of x for which the blue graph of y_1 lies below the red graph of y_2. These x-values make up the interval $(1, 7)$, which is the solution set.

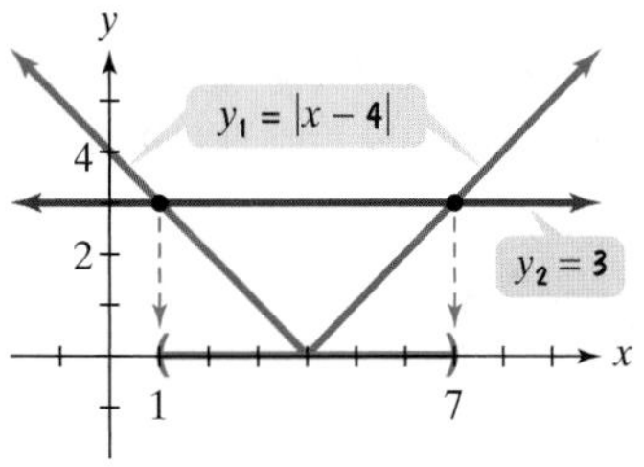

Figure 1.29 The solution set of $|x - 4| < 3$ is $(1, 7)$.

Check Point 7 Solve and graph the solution set on a number line: $|x - 2| < 5$.

EXAMPLE 8 Solving an Absolute Value Inequality

Solve and graph the solution set on a number line: $-2|3x + 5| + 7 \geq -13$.

Solution

$-2|3x + 5| + 7 \geq -13$ — This is the given inequality.

We need to isolate $|3x + 5|$, the absolute value expression.

$-2|3x + 5| + 7 - 7 \geq -13 - 7$ — Subtract 7 from both sides.

$-2|3x + 5| \geq -20$ — Simplify.

$\dfrac{-2|3x + 5|}{-2} \leq \dfrac{-20}{-2}$ — Divide both sides by -2 and change the sense of the inequality.

$|3x + 5| \leq 10$ — Simplify.

$-10 \leq 3x + 5 \leq 10$ — Rewrite without absolute value bars: $|X| \leq c$ means $-c \leq X \leq c$.

Now we need to isolate x in the middle.

$-10 - 5 \leq 3x + 5 - 5 \leq 10 - 5$ — Subtract 5 from all three parts.

$-15 \leq 3x \leq 5$ — Simplify.

$\dfrac{-15}{3} \leq \dfrac{3x}{3} \leq \dfrac{5}{3}$ — Divide each part by 3.

$-5 \leq x \leq \dfrac{5}{3}$ — Simplify.

The solution set is $\left\{x \middle| -5 \le x \le \frac{5}{3}\right\}$ in set-builder notation and $\left[-5, \frac{5}{3}\right]$ in interval notation. The graph is shown as follows:

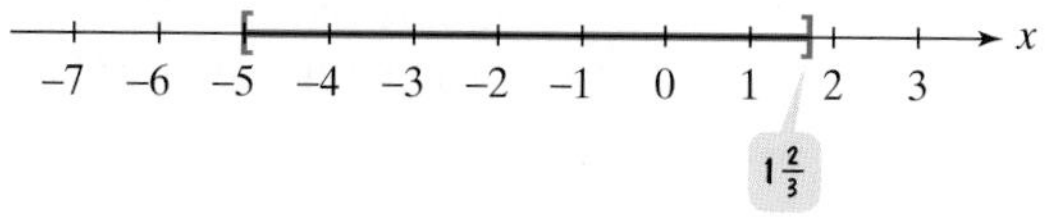

Check Point 8 Solve and graph the solution set on a number line: $-3|5x - 2| + 20 \ge -19$.

EXAMPLE 9 Solving an Absolute Value Inequality

Solve and graph the solution set on a number line: $7 < |5 - 2x|$.

Solution We begin by expressing the inequality with the absolute value expression on the left side:

$$|5 - 2x| > 7.$$

$c < |X|$ means the same thing as $|X| > c$. In both cases, the inequality symbol points to c.

We rewrite this inequality without absolute value bars.

$|X| > c$ means $X < -c$ or $X > c$.

$$|5 - 2x| > 7 \text{ means } 5 - 2x < -7 \text{ or } 5 - 2x > 7.$$

We solve each of these inequalities separately. Then we take the union of their solution sets.

$5 - 2x < -7$	or	$5 - 2x > 7$	These are the inequalities without absolute value bars.
$5 - 5 - 2x < -7 - 5$		$5 - 5 - 2x > 7 - 5$	Subtract 5 from both sides.
$-2x < -12$		$-2x > 2$	Simplify.
$\frac{-2x}{-2} > \frac{-12}{-2}$		$\frac{-2x}{-2} < \frac{2}{-2}$	Divide both sides by −2 and change the sense of the inequality.
$x > 6$		$x < -1$	Simplify.

The solution set consists of all numbers that are less than −1 or greater than 6. The solution set is $\{x | x < -1 \text{ or } x > 6\}$, or, in interval notation $(-\infty, -1) \cup (6, \infty)$. The graph of the solution set is shown as follows:

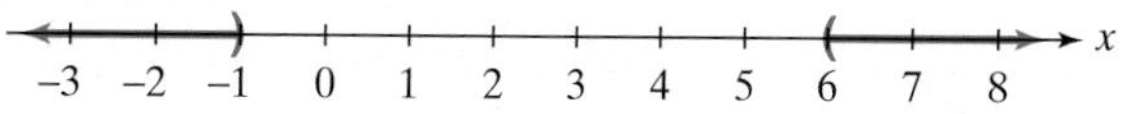

Study Tip

The graph of the solution set for $|X| > c$ will be divided into two intervals whose union cannot be represented as a single interval. The graph of the solution set for $|X| < c$ will be a single interval. Avoid the common error of rewriting $|X| > c$ as $-c < X > c$.

Check Point 9 Solve and graph the solution set on a number line: $18 < |6 - 3x|$.

Applications

Our next example shows how to use an inequality to select the better deal between two pricing options. We use our strategy for solving word problems, translating from the verbal conditions of the problem to a linear inequality.

EXAMPLE 10   Selecting the Better Deal

Acme Car rental agency charges $4 a day plus $0.15 per mile, whereas Interstate rental agency charges $20 a day and $0.05 per mile. How many miles must be driven to make the daily cost of an Acme rental a better deal than an Interstate rental?

Solution

Step 1   Let x represent one of the quantities. We are looking for the number of miles that must be driven in a day to make Acme the better deal. Thus,

let x = the number of miles driven in a day.

Step 2   Represent other quantities in terms of x. We are not asked to find another quantity, so we can skip this step.

Step 3   Write an inequality in x that models the conditions. Acme is a better deal than Interstate if the daily cost of Acme is less than the daily cost of Interstate.

$$4 \quad + \quad 0.15 \cdot x \qquad\qquad < \qquad\qquad 20 \quad + \quad 0.05 \cdot x$$

Technology

The graphs of the daily cost models for the car rental agencies

$$y_1 = 4 + 0.15x$$
$$\text{and } y_2 = 20 + 0.05x$$

are shown in a $[0, 300, 10]$ by $[0, 40, 4]$ viewing rectangle. The graphs intersect at $(160, 28)$. To the left of $x = 160$, the graph of Acme's daily cost lies below that of Interstate's daily cost. This shows that for fewer than 160 miles per day, Acme offers the better deal.

Step 4   Solve the inequality and answer the question.

$$4 + 0.15x < 20 + 0.05x$$
This is the inequality that models the verbal conditions.

$$4 + 0.15x - 0.05x < 20 + 0.05x - 0.05x$$
Subtract 0.05x from both sides.

$$4 + 0.1x < 20$$
Simplify.

$$4 + 0.1x - 4 < 20 - 4$$
Subtract 4 from both sides.

$$0.1x < 16$$
Simplify.

$$\frac{0.1x}{0.1} < \frac{16}{0.1}$$
Divide both sides by 0.1.

$$x < 160$$
Simplify.

Thus, driving fewer than 160 miles per day makes Acme the better deal.

Step 5   Check the proposed solution in the original wording of the problem. One way to do this is to take a mileage less than 160 miles per day to see if Acme is the better deal. Suppose that 150 miles are driven in a day.

$$\text{Cost for Acme} = 4 + 0.15(150) = 26.50$$
$$\text{Cost for Interstate} = 20 + 0.05(150) = 27.50$$

Acme has a lower daily cost, making Acme the better deal.

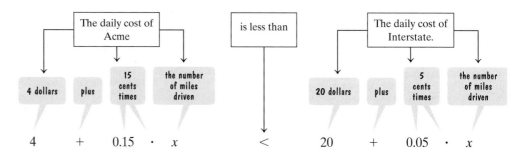

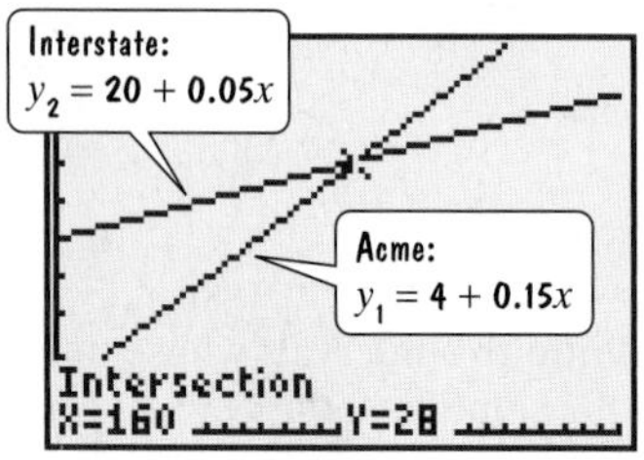

Check Point 10 A car can be rented from Basic Rental for $260 per week with no extra charge for mileage. Continental charges $80 per week plus 25 cents for each mile driven to rent the same car. How many miles must be driven in a week to make the rental cost for Basic Rental a better deal than Continental's?

EXERCISE SET 1.7

Practice Exercises

In Exercises 1–14, express each interval in set-builder notation and graph the interval on a number line.

1. $(1, 6]$
2. $(-2, 4]$
3. $[-5, 2)$
4. $[-4, 3)$
5. $[-3, 1]$
6. $[-2, 5]$
7. $(2, \infty)$
8. $(3, \infty)$
9. $[-3, \infty)$
10. $[-5, \infty)$
11. $(-\infty, 3)$
12. $(-\infty, 2)$
13. $(-\infty, 5.5)$
14. $(-\infty, 3.5]$

In Exercises 15–26, use graphs to find each set.

15. $(-3, 0) \cap [-1, 2]$
16. $(-4, 0) \cap [-2, 1]$
17. $(-3, 0) \cup [-1, 2]$
18. $(-4, 0) \cup [-2, 1]$
19. $(-\infty, 5) \cap [1, 8)$
20. $(-\infty, 6) \cap [2, 9)$
21. $(-\infty, 5) \cup [1, 8)$
22. $(-\infty, 6) \cup [2, 9)$
23. $[3, \infty) \cap (6, \infty)$
24. $[2, \infty) \cap (4, \infty)$
25. $[3, \infty) \cup (6, \infty)$
26. $[2, \infty) \cup (4, \infty)$

In all exercises, other than $\varnothing$, use interval notation to express solution sets and graph each solution set on a number line.

In Exercises 27–50, solve each linear inequality.

27. $5x + 11 < 26$
28. $2x + 5 < 17$
29. $3x - 7 \geq 13$
30. $8x - 2 \geq 14$
31. $-9x \geq 36$
32. $-5x \leq 30$
33. $8x - 11 \leq 3x - 13$
34. $18x + 45 \leq 12x - 8$
35. $4(x + 1) + 2 \geq 3x + 6$
36. $8x + 3 > 3(2x + 1) + x + 5$
37. $2x - 11 < -3(x + 2)$
38. $-4(x + 2) > 3x + 20$
39. $1 - (x + 3) \geq 4 - 2x$
40. $5(3 - x) \leq 3x - 1$
41. $\frac{x}{4} - \frac{3}{2} \leq \frac{x}{2} + 1$
42. $\frac{3x}{10} + 1 \geq \frac{1}{5} - \frac{x}{10}$
43. $1 - \frac{x}{2} > 4$
44. $7 - \frac{4}{5}x < \frac{3}{5}$
45. $\frac{x - 4}{6} \geq \frac{x - 2}{9} + \frac{5}{18}$
46. $\frac{4x - 3}{6} + 2 \geq \frac{2x - 1}{12}$
47. $4(3x - 2) - 3x < 3(1 + 3x) - 7$
48. $3(x - 8) - 2(10 - x) > 5(x - 1)$
49. $5(x - 2) - 3(x + 4) \geq 2x - 20$
50. $6(x - 1) - (4 - x) \geq 7x - 8$

In Exercises 51–58, solve each compound inequality.

51. $6 < x + 3 < 8$
52. $7 < x + 5 < 11$
53. $-3 \leq x - 2 < 1$
54. $-6 < x - 4 \leq 1$
55. $-11 < 2x - 1 \leq -5$
56. $3 \leq 4x - 3 < 19$
57. $-3 \leq \frac{2}{3}x - 5 < -1$
58. $-6 \leq \frac{1}{2}x - 4 < -3$

In Exercises 59–94, solve each absolute value inequality.

59. $|x| < 3$
60. $|x| < 5$
61. $|x - 1| \leq 2$
62. $|x + 3| \leq 4$
63. $|2x - 6| < 8$
64. $|3x + 5| < 17$
65. $|2(x - 1) + 4| \leq 8$
66. $|3(x - 1) + 2| \leq 20$
67. $\left|\frac{2x + 6}{3}\right| < 2$
68. $\left|\frac{3(x - 1)}{4}\right| < 6$
69. $|x| > 3$
70. $|x| > 5$
71. $|x - 1| \geq 2$
72. $|x + 3| \geq 4$
73. $|3x - 8| > 7$
74. $|5x - 2| > 13$
75. $\left|\frac{2x + 2}{4}\right| \geq 2$
76. $\left|\frac{3x - 3}{9}\right| \geq 1$
77. $\left|3 - \frac{2}{3}x\right| > 5$
78. $\left|3 - \frac{3}{4}x\right| > 9$
79. $3|x - 1| + 2 \geq 8$
80. $5|2x + 1| - 3 \geq 9$
81. $-2|x - 4| \geq -4$
82. $-3|x + 7| \geq -27$
83. $-4|1 - x| < -16$
84. $-2|5 - x| < -6$
85. $3 \leq |2x - 1|$
86. $9 \leq |4x + 7|$
87. $5 > |4 - x|$
88. $2 > |11 - x|$
89. $1 < |2 - 3x|$
90. $4 < |2 - x|$
91. $12 < \left|-2x + \frac{6}{7}\right| + \frac{3}{7}$
92. $1 < \left|x - \frac{11}{3}\right| + \frac{7}{3}$
93. $4 + \left|3 - \frac{x}{3}\right| \geq 9$
94. $\left|2 - \frac{x}{2}\right| - 1 \leq 1$

In Exercises 95–102, use interval notation to represent all values of x satisfying the given conditions.

95. $y_1 = \frac{x}{2} + 3,\ y_2 = \frac{x}{3} + \frac{5}{2},$ and $y_1 \leq y_2$.
96. $y_1 = \frac{2}{3}(6x - 9) + 4,\ y_2 = 5x + 1,$ and $y_1 > y_2$.
97. $y = 1 - (x + 3) + 2x$ and y is at least 4.
98. $y = 2x - 11 + 3(x + 2)$ and y is at most 0.
99. $y = |3x - 4| + 2$ and $y < 8$.
100. $y = |2x - 5| + 1$ and $y > 9$.
101. $y = 7 - \left|\frac{x}{2} + 2\right|$ and y is at most 4.
102. $y = 8 - |5x + 3|$ and y is at least 6.

Practice Plus

In Exercises 103–104, use the graph of $y = |4 - x|$ to solve each inequality.

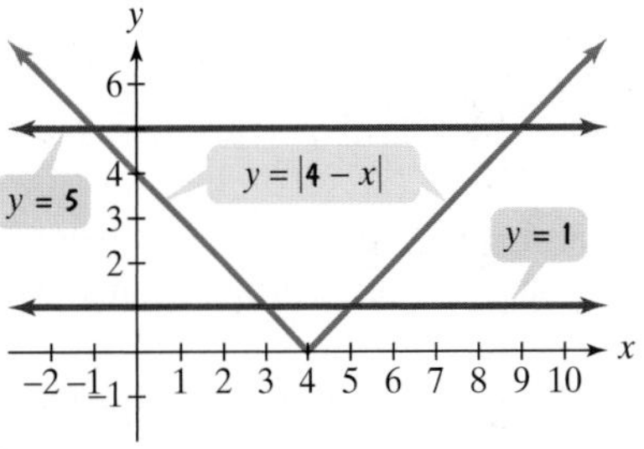

103. $|4 - x| < 5$
104. $|4 - x| \geq 5$

In Exercises 105–106, use the table to solve each inequality.

105. $-2 \le 5x + 3 < 13$

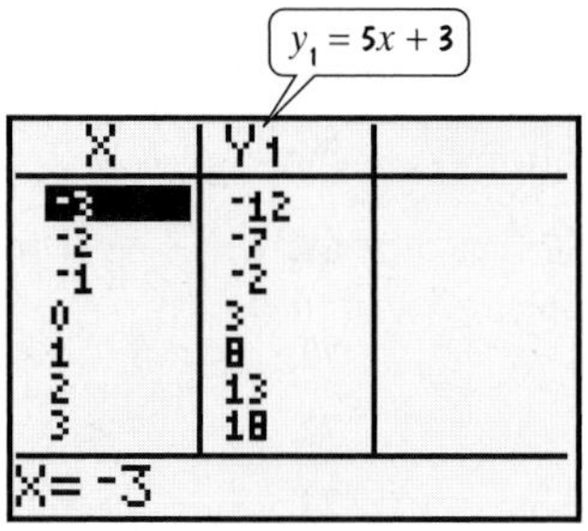

106. $-3 < 2x - 5 \le 3$

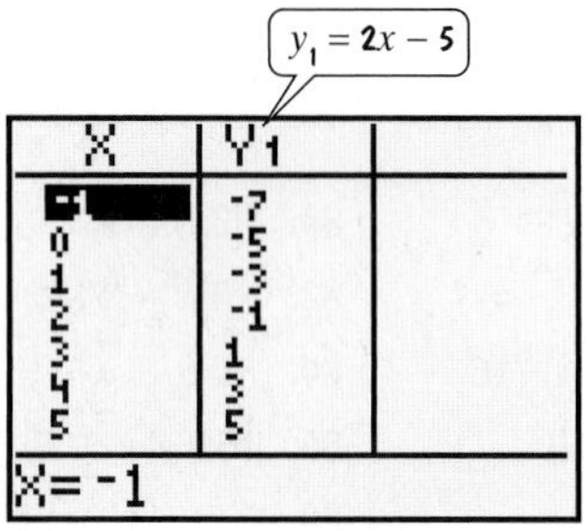

107. When 3 times a number is subtracted from 4, the absolute value of the difference is at least 5. Use interval notation to express the set of all numbers that satisfy this condition.

108. When 4 times a number is subtracted from 5, the absolute value of the difference is at most 13. Use interval notation to express the set of all numbers that satisfy this condition.

Application Exercises

The graphs show that the three components of love, namely passion, intimacy, and commitment, progress differently over time. Passion peaks early in a relationship and then declines. By contrast, intimacy and commitment build gradually. Use the graphs to solve Exercises 109–116.

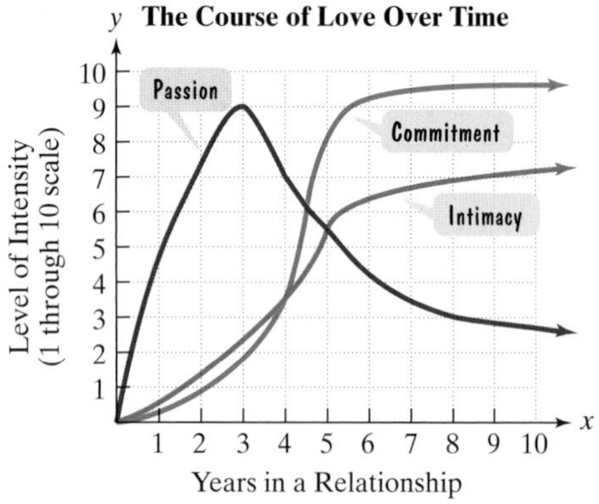

Source: R. J. Sternberg. A Triangular Theory of Love, *Psychological Review*, 93, 119–135.

109. Use interval notation to write an inequality that expresses for which years in a relationship intimacy is greater than commitment.

110. Use interval notation to write an inequality that expresses for which years in a relationship passion is greater than or equal to intimacy.

111. What is the relationship between passion and intimacy on the interval $[5, 7)$?

112. What is the relationship between intimacy and commitment on the interval $[4, 7)$?

113. What is the relationship between passion and commitment for $\{x|6 < x < 8\}$?

114. What is the relationship between passion and commitment for $\{x|7 < x < 9\}$?

115. What is the maximum level of intensity for passion? After how many years in a relationship does this occur?

116. After approximately how many years do levels of intensity for commitment exceed the maximum level of intensity for passion?

117. The percentage, P, of U.S. voters who used electronic voting systems, such as optical scans, in national elections can be modeled by the formula

$$P = 3.1x + 25.8,$$

where x is the number of years after 1994. In which years will more than 63% of U.S. voters use electronic systems?

118. The percentage, P, of U.S. voters who used punch cards or lever machines in national elections can be modeled by the formula

$$P = -2.5x + 63.1,$$

where x is the number of years after 1994. In which years will fewer than 38.1% of U.S. voters use punch cards or lever machines?

119. A basic cellular phone plan costs \$20 per month for 60 calling minutes. Additional time costs \$0.40 per minute. The formula

$$C = 20 + 0.40(x - 60)$$

gives the monthly cost for this plan, C, for x calling minutes, where $x > 60$. How many calling minutes are possible for a monthly cost of at least \$28 and at most \$40?

120. The formula for converting Fahrenheit temperature, F, to Celsius temperature, C, is

$$C = \frac{5}{9}(F - 32).$$

If Celsius temperature ranges from 15° to 35°, inclusive, what is the range for the Fahrenheit temperature? Use interval notation to express this range.

121. If a coin is tossed 100 times, we would expect approximately 50 of the outcomes to be heads. It can be demonstrated that a coin is unfair if h, the number of outcomes that result in heads, satisfies $\left|\frac{h - 50}{5}\right| \ge 1.645$. Describe the number of outcomes that determine an unfair coin that is tossed 100 times.

In Exercises 122–133, use the strategy for solving word problems, translating from the verbal conditions of the problem to a linear inequality.

122. A truck can be rented from Basic Rental for \$50 per day plus \$0.20 per mile. Continental charges \$20 per day plus \$0.50 per mile to rent the same truck. How many miles must be driven in a day to make the rental cost for Basic Rental a better deal than Continental's?

123. You are choosing between two long-distance telephone plans. Plan A has a monthly fee of \$15 with a charge of \$0.08 per minute for all long-distance calls. Plan B has a monthly fee of \$3 with a charge of \$0.12 per minute for all long-distance calls. How many minutes of long-distance calls in a month make plan A the better deal?

124. A city commission has proposed two tax bills. The first bill requires that a homeowner pay \$1800 plus 3% of the assessed home value in taxes. The second bill requires taxes of \$200 plus 8% of the assessed home value. What price range of home assessment would make the first bill a better deal?

125. A local bank charges \$8 per month plus 5¢ per check. The credit union charges \$2 per month plus 8¢ per check. How many checks should be written each month to make the credit union a better deal?

126. A company manufactures and sells blank audiocassette tapes. The weekly fixed cost is \$10,000 and it costs \$0.40 to produce each tape. The selling price is \$2.00 per tape. How many tapes must be produced and sold each week for the company to generate a profit?

127. A company manufactures and sells personalized stationery. The weekly fixed cost is \$3000 and it costs \$3.00 to produce each package of stationery. The selling price is \$5.50 per package. How many packages of stationery must be produced and sold each week for the company to generate a profit?

128. An elevator at a construction site has a maximum capacity of 2800 pounds. If the elevator operator weighs 265 pounds and each cement bag weighs 65 pounds, how many bags of cement can be safely lifted on the elevator in one trip?

129. An elevator at a construction site has a maximum capacity of 3000 pounds. If the elevator operator weighs 245 pounds and each cement bag weighs 95 pounds, how many bags of cement can be safely lifted on the elevator in one trip?

130. To earn an A in a course, you must have a final average of at least 90%. On the first four examinations, you have grades of 86%, 88%, 92%, and 84%. If the final examination counts as two grades, what must you get on the final to earn an A in the course?

131. On two examinations, you have grades of 86 and 88. There is an optional final examination, which counts as one grade. You decide to take the final in order to get a course grade of A, meaning a final average of at least 90.

a. What must you get on the final to earn an A in the course?

b. By taking the final, if you do poorly, you might risk the B that you have in the course based on the first two exam grades. If your final average is less than 80, you will lose your B in the course. Describe the grades on the final that will cause this to happen.

132. Parts for an automobile repair cost \$175. The mechanic charges \$34 per hour. If you receive an estimate for at least \$226 and at most \$294 for fixing the car, what is the time interval that the mechanic will be working on the job?

133. The toll to a bridge is \$3.00. A three-month pass costs \$7.50 and reduces the toll to \$0.50. A six-month pass costs \$30 and permits crossing the bridge for no additional fee. How many crossings per three-month period does it take for the three-month pass to be the best deal?

Writing in Mathematics

134. When graphing the solutions of an inequality, what does a parenthesis signify? What does a bracket signify?

135. Describe ways in which solving a linear inequality is similar to solving a linear equation.

136. Describe ways in which solving a linear inequality is different than solving a linear equation.

137. What is a compound inequality and how is it solved?

138. Describe how to solve an absolute value inequality involving the symbol $<$. Give an example.

139. Describe how to solve an absolute value inequality involving the symbol $>$. Give an example.

140. Explain why $|x| < -4$ has no solution.

141. Describe the solution set of $|x| > -4$.

Technology Exercises

In Exercises 142–143, solve each inequality using a graphing utility. Graph each side separately. Then determine the values of x for which the graph for the left side lies above the graph for the right side.

142. $-3(x - 6) > 2x - 2$

143. $-2(x + 4) > 6x + 16$

144. Use a graphing utility's TABLE feature to verify your work in Exercises 142–143.

145. A bank offers two checking account plans. Plan A has a base service charge of \$4.00 per month plus 10¢ per check. Plan B charges a base service charge of \$2.00 per month plus 15¢ per check.

a. Write models for the total monthly costs for each plan if x checks are written.

b. Use a graphing utility to graph the models in the same $[0, 50, 10]$ by $[0, 10, 1]$ viewing rectangle.

c. Use the graphs (and the intersection feature) to determine for what number of checks per month plan A will be better than plan B.

d. Verify the result of part (c) algebraically by solving an inequality.

Critical Thinking Exercises

146. Which one of the following is true?

a. The first step in solving $|2x - 3| > -7$ is to rewrite the inequality as $2x - 3 > -7$ or $2x - 3 < 7$.

b. The smallest real number in the solution set of $2x > 6$ is 4.

c. All irrational numbers satisfy $|x - 4| > 0$.

d. None of these statements is true.

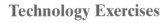

147. What's wrong with this argument? Suppose x and y represent two real numbers, where $x > y$.

$2 > 1$	This is a true statement.
$2(y - x) > 1(y - x)$	Multiply both sides by $y - x$.
$2y - 2x > y - x$	Use the distributive property.
$y - 2x > -x$	Subtract y from both sides.
$y > x$	Add $2x$ to both sides.

The final inequality, $y > x$, is impossible because we were initially given $x > y$.

148. Write an absolute value inequality for which the interval shown is the solution.

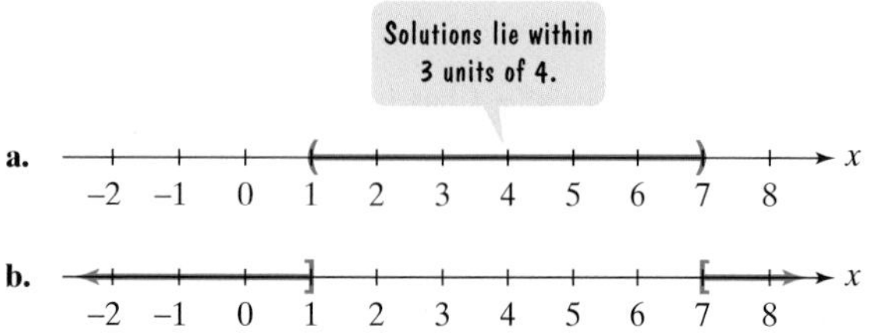

149. Here are two inequalities that describe the range of monthly average temperatures, T, in degrees Fahrenheit for two American cities:

Model 1: $|T - 57| < 7$

Model 2: $|T - 50| < 22$.

Which model describes Albany, New York, and which model describes San Francisco, California?

Group Exercise

150. Each group member should research one situation that provides two different pricing options. These can involve areas such as public transportation options (with or without coupon books), cell phone plans, long-distance telephone plans, or anything of interest. Be sure to bring in all the details for each option. At a second group meeting, select the two pricing situations that are most interesting and relevant. Using each situation, write a word problem about selecting the better of the two options. The word problem should be one that can be solved using a linear inequality. The group should turn in the two problems and their solutions.

Chapter 1 Summary, Review, and Test

Summary

DEFINITIONS AND CONCEPTS	EXAMPLES
1.1 Graphs and Graphing Utilities	
a. The rectangular coordinate system consists of a horizontal number line, the x-axis, and a vertical number line, the y-axis, intersecting at their zero points, the origin. Each point in the system corresponds to an ordered pair of real numbers (x, y). The first number in the pair is the x-coordinate; the second number is the y-coordinate. See Figure 1.1 on page 84.	Ex. 1, p. 85
b. An ordered pair is a solution of an equation in two variables if replacing the variables by the corresponding coordinates results in a true statement. The ordered pair is said to satisfy the equation. The graph of the equation is the set of all points whose coordinates satisfy the equation. One method for graphing an equation is to plot ordered-pair solutions and connect them with a smooth curve or line.	Ex. 2, p. 86; Ex. 3, p. 86
c. An x-intercept of a graph is the x-coordinate of a point where the graph intersects the x-axis. The y-coordinate corresponding to an x-intercept is always zero. A y-intercept of a graph is the y-coordinate of a point where the graph intersects the y-axis. The x-coordinate corresponding to a y-intercept is always zero.	Ex. 5, p. 89
1.2 Linear Equations and Rational Equations	
a. A linear equation in one variable x can be written in the form $ax + b = 0, a \neq 0$.	
b. The procedure for solving a linear equation is given in the box on page 97.	Ex. 1, p. 96; Ex. 2, p. 97
c. If an equation contains fractions, begin by multiplying both sides by the least common denominator, thereby clearing fractions.	Ex. 3, p. 99
d. A rational equation is an equation containing one or more rational expressions. If an equation contains rational expressions with variable denominators, avoid in the solution set any values of the variable that make a denominator zero.	Ex. 4, p. 100; Ex. 5, p. 101; Ex. 6, p. 101
e. An identity is an equation that is true for all real numbers for which both sides are defined. A conditional equation is not an identity and is true for at least one real number. An inconsistent equation is an equation that is not true for even one real number.	Ex. 7, p. 103

1.3 Models and Applications	
a. A five-step procedure for solving word problems using equations that model verbal conditions is given in the box on page 108.	Ex. 1, p. 108; Ex. 2, p. 110; Ex. 3, p. 111; Ex. 4, p. 113; Ex. 5, p. 114; Ex. 6, p. 115
b. Solving a formula for a variable means rewriting the formula so that the variable is isolated on one side of the equation.	Ex. 7, p. 117; Ex. 8, p. 117
1.4 Complex Numbers	
a. The imaginary unit i is defined as $i = \sqrt{-1}$, where $i^2 = -1$. The set of numbers in the form $a + bi$ is called the set of complex numbers; a is the real part and b is the imaginary part. If $b = 0$, the complex number is a real number. If $b \neq 0$, the complex number is an imaginary number. Complex numbers in the form bi are called pure imaginary numbers.	
b. Rules for adding and subtracting complex numbers are given in the box on page 124.	Ex. 1, p. 124
c. To multiply complex numbers, multiply as if they are polynomials. After completing the multiplication, replace i^2 with -1 and simplify.	Ex. 2, p. 125
d. The complex conjugate of $a + bi$ is $a - bi$ and vice versa. The multiplication of complex conjugates gives a real number: $(a + bi)(a - bi) = a^2 + b^2$.	
e. To divide complex numbers, multiply the numerator and the denominator by the complex conjugate of the denominator.	Ex. 3, p. 126
f. When performing operations with square roots of negative numbers, begin by expressing all square roots in terms of i. The principal square root of $-b$ is defined by $\sqrt{-b} = i\sqrt{b}$.	Ex. 4, p. 127
1.5 Quadratic Equations	
a. A quadratic equation in x can be written in the general form $ax^2 + bx + c = 0, a \neq 0$.	
b. The procedure for solving a quadratic equation by factoring and the zero-product principle is given in the box on page 131.	Ex. 1, p. 131
c. The procedure for solving a quadratic equation by the square root property is given in the box on page 132.	Ex. 2, p. 133
d. All quadratic equations can be solved by completing the square. Isolate the binomial with the two variable terms on one side of the equation. If the coefficient of the x^2-term is not one, divide each side of the equation by this coefficient. Then add the square of half the coefficient of x to both sides.	Ex. 4, p. 135; Ex. 5, p. 136
e. All quadratic equations can be solved by the quadratic formula $x = \frac{-b \pm \sqrt{b^2 - 4ac}}{2a}$. The formula is derived by completing the square of the equation $ax^2 + bx + c = 0$.	Ex. 6, p. 138; Ex. 7, p. 139
f. The discriminant, $b^2 - 4ac$, indicates the number and type of solutions to the quadratic equation $ax^2 + bx + c = 0$, shown in Table 1.3 on page 140.	Ex. 8, p. 140
g. Table 1.4 on page 141 shows the most efficient technique to use when solving a quadratic equation.	Ex. 9, p. 142; Ex. 10, p. 143
1.6 Other Types of Equations	
a. Some polynomial equations of degree 3 or greater can be solved by moving all terms to one side, obtaining zero on the other side, factoring, and using the zero-product principle. Factoring by grouping is often used.	Ex. 1, p. 151; Ex. 2, p. 151
b. A radical equation is an equation in which the variable occurs in a square root, cube root, and so on. A radical equation can be solved by isolating the radical and raising both sides of the equation to a power equal to the radical's index. When raising both sides to an even power, check all proposed solutions in the original equation. Eliminate extraneous solutions from the solution set.	Ex. 3, p. 153; Ex. 4, p. 154
c. A radical equation with rational exponents can be solved by isolating the expression with the rational exponent and raising both sides of the equation to a power that is the reciprocal of the rational exponent. See the details in the box on page 156.	Ex. 5, p. 156

d. An equation is quadratic in form if it can be written in the form $au^2 + bu + c = 0$, where u is an algebraic expression and $a \neq 0$. Solve for u and use the substitution that resulted in this equation to find the values for the variable in the given equation.	Ex. 6, p. 157; Ex. 7, p. 158
e. Absolute value equations in the form $|X| = c, c > 0$, can be solved by rewriting the equation without absolute value bars: $X = c$ or $X = -c$.	Ex. 8, p. 159; Ex. 9, p. 159

1.7 Linear Inequalities and Absolute Value Inequalities

a. Solution sets of inequalities are expressed using set-builder notation and interval notation. In interval notation, parentheses indicate endpoints that are not included in an interval. Square brackets indicate endpoints that are included in an interval. See Table 1.5 on page 165.	Ex. 1, p. 166
b. A procedure for finding intersections and unions of intervals is given in the box on page 166.	Ex. 2, p. 166
c. A linear inequality in one variable x can be expressed as $ax + b \leq c, ax + b < c, ax + b \geq c$, or $ax + b > c, a \neq 0$.	
d. A linear inequality is solved using a procedure similar to solving a linear equation. However, when multiplying or dividing by a negative number, change the sense of the inequality.	Ex. 3, p. 168; Ex. 4, p. 169; Ex. 5, p. 170
e. A compound inequality with three parts can be solved by isolating the variable in the middle.	Ex. 6, p. 171
f. Inequalities involving absolute value can be solved by rewriting the inequalities without absolute value bars. The ways to do this are shown in the box on page 171.	Ex. 7, p. 172; Ex. 8, p. 172; Ex. 9, p. 173

Review Exercises

1.1

Graph each equation in Exercises 1–4. Let $x = -3, -2, -1, 0, 1, 2,$ and 3.

1. $y = 2x - 2$ **2.** $y = x^2 - 3$

3. $y = x$ **4.** $y = |x| - 2$

5. What does a $[-20, 40, 10]$ by $[-5, 5, 1]$ viewing rectangle mean? Draw axes with tick marks and label the tick marks to illustrate this viewing rectangle.

In Exercises 6–8, use the graph and determine the x-intercepts, if any, and the y-intercepts, if any. For each graph, tick marks along the axes represent one unit each.

6.
7.
8.
Afghanistan accounts for 76% of the world's illegal opium production. Opium-poppy cultivation nets big money in a country where most people earn less than $1 per day. (Source: Newsweek) The line graph shows opium-poppy cultivation, in thousands of acres, in Afghanistan from 1990 through 2004. Use the graph to solve Exercises 9–14.

Afghanistan's Opium Crop

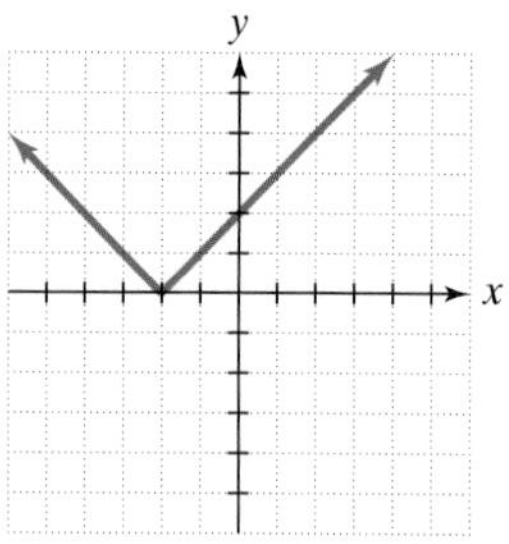

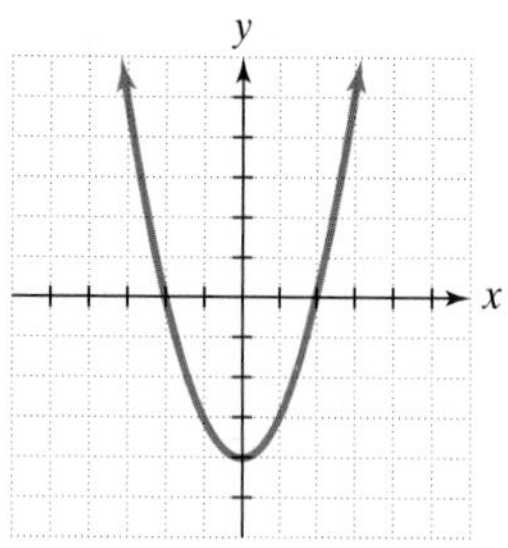

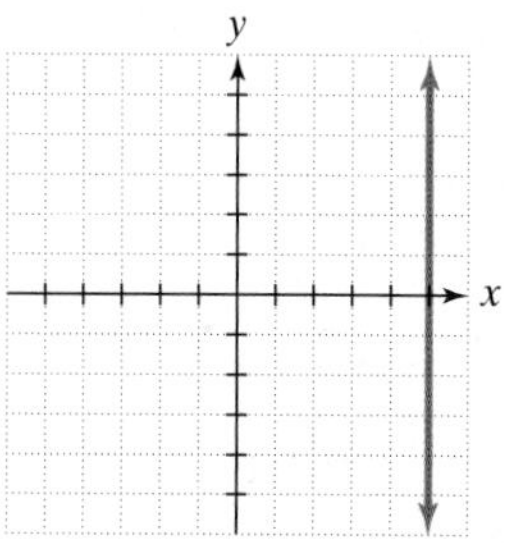

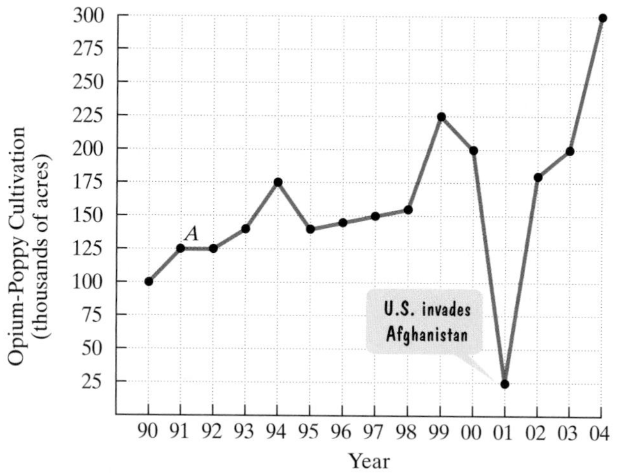

Source: U.N. Office on Drugs and Crime

9. What are the coordinates of point A? What does this mean in terms of the information given by the graph?

10. In which year were 150 thousand acres used for opium-poppy cultivation?

11. For the period shown, when did opium cultivation reach a minimum? How many thousands of acres were used to cultivate the illegal crop?

12. For the period shown, when did opium cultivation reach a maximum? How many thousands of acres were used to cultivate the illegal crop?

13. Between which two years did opium cultivation not change?

14. Between which two years did opium cultivation increase at the greatest rate? What is a reasonable estimate of the increase, in thousands of acres, used to cultivate the illegal crop during this period?

1.2

In Exercises 15–35, solve each equation. Then state whether the equation is an identity, a conditional equation, or an inconsistent equation.

15. $2x - 5 = 7$

16. $5x + 20 = 3x$

17. $7(x - 4) = x + 2$

18. $1 - 2(6 - x) = 3x + 2$

19. $2(x - 4) + 3(x + 5) = 2x - 2$

20. $2x - 4(5x + 1) = 3x + 17$

21. $7x + 5 = 5(x + 3) + 2x$

22. $7x + 13 = 2(2x - 5) + 3x + 23$

23. $\frac{2x}{3} = \frac{x}{6} + 1$

24. $\frac{x}{2} - \frac{1}{10} = \frac{x}{5} + \frac{1}{2}$

25. $\frac{2x}{3} = 6 - \frac{x}{4}$

26. $\frac{x}{4} = 2 - \frac{x - 3}{3}$

27. $\frac{3x + 1}{3} - \frac{13}{2} = \frac{1 - x}{4}$

28. $\frac{9}{4} - \frac{1}{2x} = \frac{4}{x}$

29. $\frac{7}{x - 5} + 2 = \frac{x + 2}{x - 5}$

30. $\frac{1}{x - 1} - \frac{1}{x + 1} = \frac{2}{x^2 - 1}$

31. $\frac{5}{x + 3} + \frac{1}{x - 2} = \frac{8}{x^2 + x - 6}$

32. $\frac{1}{x + 5} = 0$

33. $\frac{4}{x + 2} + \frac{3}{x} = \frac{10}{x^2 + 2x}$

34. $3 - 5(2x + 1) - 2(x - 4) = 0$

35. $\frac{x + 2}{x + 3} + \frac{1}{x^2 + 2x - 3} - 1 = 0$

1.3

In Exercises 36–43, use the five-step strategy for solving word problems.

36. The fast-food chains may be touting their "new and improved" salads, but how do they measure up in terms of calories?

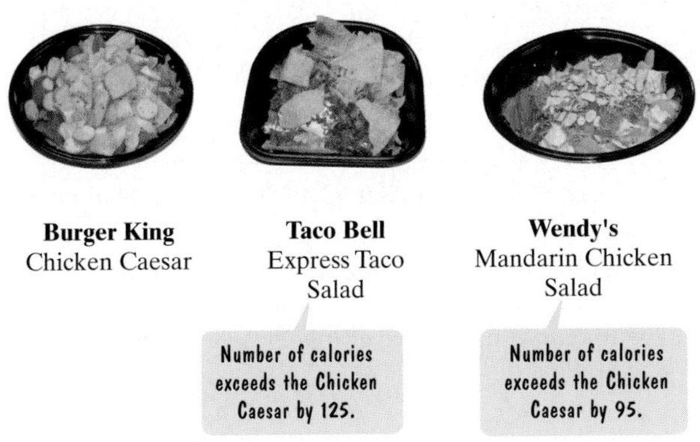

Source: Newsweek

Combined, the three salads contain 1705 calories. Determine the number of calories in each salad.

37. The bar graph shows that in 1970, 37.4% of U.S. adults smoked cigarettes. For the period from 1970 through 2002, the percentage of smokers among U.S. adults decreased at an average rate of 0.5% per year. If this trend continues, when will only 18.4% of U.S. adults smoke cigarettes?

Butt Out: Percentage of Cigarette Smokers Among U.S. Adults

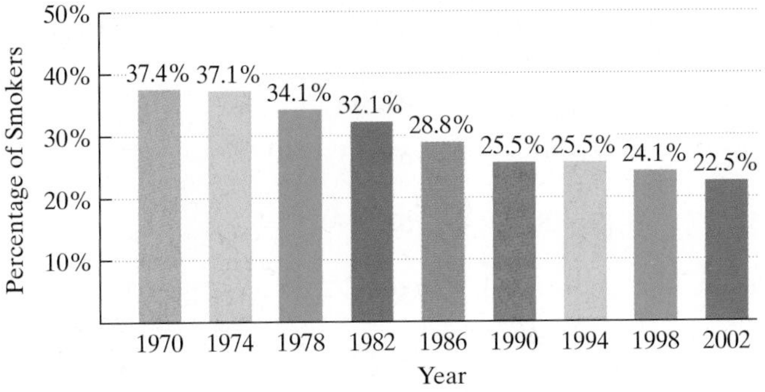

Source: Centers for Disease Control and Prevention

38. You are choosing between two long-distance telephone plans. One plan has a monthly fee of $15 with a charge of $0.05 per minute. The other plan has a monthly fee of $5 with a charge of $0.07 per minute. For how many minutes of long-distance calls will the costs for the two plans be the same?

39. After a 20% price reduction, a cordless phone sold for $48. What was the phone's price before the reduction?

40. A salesperson earns $300 per week plus 5% commission of sales. How much must be sold to earn $800 in a week?

41. You invested $9000 in two funds paying 4% and 7% annual interest, respectively. At the end of the year, the total interest from these investments was $555. How much was invested at each rate?

42. You invested $8000 in two funds paying 2% and 5% annual interest, respectively. At the end of the year, the interest from the 5% investment exceeded the interest from the 2% investment by $85. How much money was invested at each rate?

43. The length of a rectangular field is 6 yards less than triple the width. If the perimeter of the field is 340 yards, what are its dimensions?

44. In 2007, there were 14,100 students at college A, with a projected enrollment increase of 1500 students per year. In the same year, there were 41,700 students at college B, with a projected enrollment decline of 800 students per year.

a. Let x represent the number of years after 2007. Write, but do not solve, an equation that can be used to find how many years after 2007 the colleges will have the same enrollment.

b. The following table is based on your equation in part (a). Y_1 represents one side of the equation and Y_2 represents the other side of the equation. Use the table to answer the following questions: In which year will the colleges have the same enrollment? What will be the enrollment in each college at that time?

X	Y1	Y2
7	24600	36100
8	26100	35300
9	27600	34500
10	29100	33700
11	30600	32900
12	32100	32100
13	33600	31300

X=7

In Exercises 45–47, solve each formula for the specified variable.

45. $vt + gt^2 = s$ for g **46.** $T = gr + gvt$ for g

47. $T = \frac{A - P}{Pr}$ for P

1.4

In Exercises 48–57, perform the indicated operations and write the result in standard form.

48. $(8 - 3i) - (17 - 7i)$ **49.** $4i(3i - 2)$

50. $(7 - i)(2 + 3i)$ **51.** $(3 - 4i)^2$

52. $(7 + 8i)(7 - 8i)$ **53.** $\frac{6}{5 + i}$

54. $\frac{3 + 4i}{4 - 2i}$ **55.** $\sqrt{-32} - \sqrt{-18}$

56. $(-2 + \sqrt{-100})^2$ **57.** $\frac{4 + \sqrt{-8}}{2}$

1.5

Solve each equation in Exercises 58–59 by factoring.

58. $2x^2 + 15x = 8$ **59.** $5x^2 + 20x = 0$

Solve each equation in Exercises 60–63 by the square root property.

60. $2x^2 - 3 = 125$ **61.** $\frac{x^2}{2} + 5 = -3$

62. $(x + 3)^2 = -10$ **63.** $(3x - 4)^2 = 18$

In Exercises 64–65, determine the constant that should be added to the binomial so that it becomes a perfect square trinomial. Then write and factor the trinomial.

64. $x^2 + 20x$ **65.** $x^2 - 3x$

Solve each equation in Exercises 66–67 by completing the square.

66. $x^2 - 12x + 27 = 0$ **67.** $3x^2 - 12x + 11 = 0$

Solve each equation in Exercises 68–70 using the quadratic formula.

68. $x^2 = 2x + 4$ **69.** $x^2 - 2x + 19 = 0$

70. $2x^2 = 3 - 4x$

In Exercises 71–72, without solving the given quadratic equation, determine the number and type of solutions.

71. $x^2 - 4x + 13 = 0$ **72.** $9x^2 = 2 - 3x$

Solve each equation in Exercises 73–81 by the method of your choice.

73. $2x^2 - 11x + 5 = 0$ **74.** $(3x + 5)(x - 3) = 5$

75. $3x^2 - 7x + 1 = 0$ **76.** $x^2 - 9 = 0$

77. $(x - 3)^2 - 25 = 0$ **78.** $3x^2 - x + 2 = 0$

79. $3x^2 - 10x = 8$ **80.** $(x + 2)^2 + 4 = 0$

81. $\frac{5}{x + 1} + \frac{x - 1}{4} = 2$

82. The formula $W = 3t^2$ models the weight of a human fetus, W, in grams, after t weeks, where $0 \le t \le 39$. After how many weeks does the fetus weigh 588 grams?

83. In 1945, 35.4% of taxes collected by the U.S. Treasury came from corporate income taxes. Since then, corporations have worked hard to convince lawmakers that they shouldn't pay taxes. The bar graph shows the percentage of federal taxes from corporate income taxes for selected years from 1985 through 2003. The data can be modeled by the formula

$$P = -0.035x^2 + 0.65x + 7.6,$$

where P represents the percentage of federal taxes from corporations x years after 1985. If these trends continue, by which year (to the nearest year) will corporations pay no taxes?

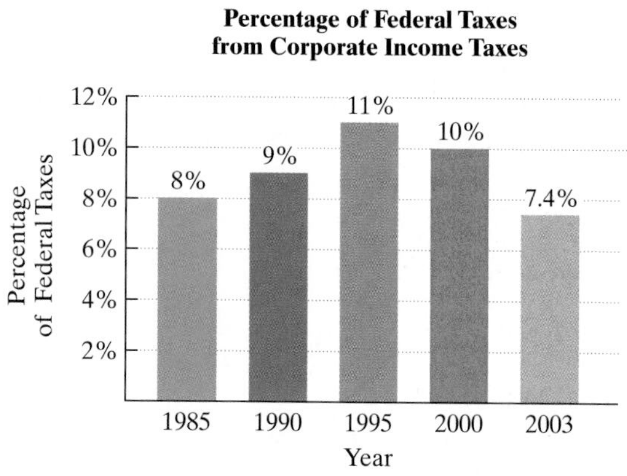

Source: White House Office of Management and Budget

84. An architect is allowed 15 square yards of floor space to add a small bedroom to a house. Because of the room's design in relationship to the existing structure, the width of the rectangular floor must be 7 yards less than two times the length. Find the length and width of the rectangular floor that the architect is permitted.

85. A building casts a shadow that is double the length of its height. If the distance from the end of the shadow to the top of the building is 300 meters, how high is the building? Round to the nearest meter.

1.6

Solve each polynomial equation in Exercises 86–87.

86. $2x^4 = 50x^2$ **87.** $2x^3 - x^2 - 18x + 9 = 0$

Solve each radical equation in Exercises 88–89.

88. $\sqrt{2x - 3} + x = 3$ **89.** $\sqrt{x - 4} + \sqrt{x + 1} = 5$

Solve the equations with rational exponents in Exercises 90–91.

90. $3x^{\frac{3}{4}} - 24 = 0$ **91.** $(x - 7)^{\frac{2}{3}} = 25$

Solve each equation in Exercises 92–93 by making an appropriate substitution.

92. $x^4 - 5x^2 + 4 = 0$ **93.** $x^{\frac{1}{2}} + 3x^{\frac{1}{4}} - 10 = 0$

Solve the equations containing absolute value in Exercises 94–95.

94. $|2x + 1| = 7$ **95.** $2|x - 3| - 6 = 10$

Solve each equation in Exercises 96–102 by the method of your choice.

96. $3x^{\frac{4}{3}} - 5x^{\frac{2}{3}} + 2 = 0$ **97.** $2\sqrt{x - 1} = x$

98. $|2x - 5| - 3 = 0$ **99.** $x^3 + 2x^2 = 9x + 18$

100. $\sqrt{8 - 2x} - x = 0$ **101.** $x^3 + 3x^2 - 2x - 6 = 0$

102. $-4|x + 1| + 12 = 0$

103. By 2010, India could become the world's most HIV-afflicted country. The bar graph shows the increase in the country's HIV infections from 1998 through 2001. The formula $N = 0.3\sqrt{x} + 3.4$ models the number of HIV infections in India, N, in millions, x years after 1998.

AIDS in India

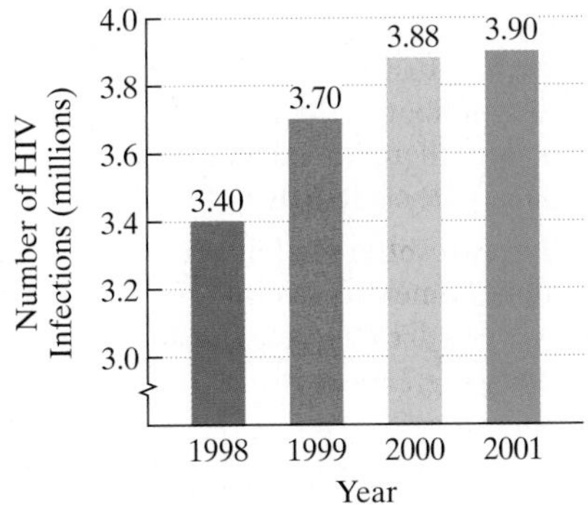

Source: UNAIDS

If trends indicated by the data continue, use the model to determine when the number of HIV infections in India will reach 4.3 million.

1.7

In Exercises 104–106, express each interval in set-builder notation and graph the interval on a number line.

104. $[-3, 5)$ **105.** $(-2, \infty)$ **106.** $(-\infty, 0]$

In Exercises 107–110, use graphs to find each set.

107. $(-2, 1] \cap [-1, 3)$ **108.** $(-2, 1] \cup [-1, 3)$

109. $[1, 3) \cap (0, 4)$ **110.** $[1, 3) \cup (0, 4)$

In Exercises 111–121, solve each inequality. Other than ∅, use interval notation to express solution sets and graph each solution set on a number line.

111. $-6x + 3 \le 15$ **112.** $6x - 9 \ge -4x - 3$

113. $\frac{x}{3} - \frac{3}{4} - 1 > \frac{x}{2}$ **114.** $6x + 5 > -2(x - 3) - 25$

115. $3(2x - 1) - 2(x - 4) \ge 7 + 2(3 + 4x)$

116. $5(x - 2) - 3(x + 4) \ge 2x - 20$

117. $7 < 2x + 3 \le 9$ **118.** $|2x + 3| \le 15$

119. $\left|\frac{2x + 6}{3}\right| > 2$ **120.** $|2x + 5| - 7 \ge -6$

121. $-4|x + 2| + 5 \le -7$

In Exercises 122–123, use interval notation to represent all values of x satisfying the given conditions.

122. $y_1 = -10 - 3(2x + 1)$, $y_2 = 8x + 1$, and $y_1 > y_2$.

123. $y = 3 - |2x - 5|$ and y is at least -6.

124. A car rental agency rents a certain car for \$40 per day with unlimited mileage or \$24 per day plus \$0.20 per mile. How far can a customer drive this car per day for the \$24 option to cost no more than the unlimited mileage option?

125. To receive a B in a course, you must have an average of at least 80% but less than 90% on five exams. Your grades on the first four exams were 95%, 79%, 91%, and 86%. What range of grades on the fifth exam will result in a B for the course?

126. A retiree requires an annual income of at least \$9000 from an investment paying 7.5% annual interest. How much should the retiree invest to achieve the desired return?

Chapter 1 Test

In Exercises 1–23, solve each equation or inequality. Other than ∅, use interval notation to express solution sets of inequalities and graph these solution sets on a number line.

1. $7(x - 2) = 4(x + 1) - 21$

2. $-10 - 3(2x + 1) - 8x - 1 = 0$

3. $\frac{2x - 3}{4} = \frac{x - 4}{2} - \frac{x + 1}{4}$

4. $\frac{2}{x - 3} - \frac{4}{x + 3} = \frac{8}{x^2 - 9}$

5. $2x^2 - 3x - 2 = 0$ **6.** $(3x - 1)^2 = 75$

7. $(x + 3)^2 + 25 = 0$ **8.** $x(x - 2) = 4$

9. $4x^2 = 8x - 5$ **10.** $x^3 - 4x^2 - x + 4 = 0$

11. $\sqrt{x - 3} + 5 = x$ **12.** $\sqrt{8 - 2x} - x = 0$

13. $\sqrt{x + 4} + \sqrt{x - 1} = 5$ **14.** $5x^{\frac{3}{2}} - 10 = 0$

15. $x^{\frac{2}{3}} - 9x^{\frac{1}{3}} + 8 = 0$ **16.** $\left|\frac{2}{3}x - 6\right| = 2$

17. $-3|4x - 7| + 15 = 0$ **18.** $\frac{1}{x^2} - \frac{4}{x} + 1 = 0$

19. $\frac{2x}{x^2 + 6x + 8} + \frac{2}{x + 2} = \frac{x}{x + 4}$

20. $3(x + 4) \ge 5x - 12$ **21.** $\frac{x}{6} + \frac{1}{8} \le \frac{x}{2} - \frac{3}{4}$

22. $-3 \le \frac{2x + 5}{3} < 6$ **23.** $|3x + 2| \ge 3$

In Exercises 24–25, use interval notation to represent all values of x satisfying the given conditions.

24. $y = 2x - 5$, and y is at least -3 and no more than 7.

25. $y = \left|\frac{2 - x}{4}\right|$ and y is at least 1.

In Exercises 26–27, use graphs to find each set.

26. $[-1, 2) \cup (0, 5]$ **27.** $[-1, 2) \cap (0, 5]$

In Exercises 28–29, solve each formula for the specified variable.

28. $V = \frac{1}{3}lwh$ for h **29.** $y - y_1 = m(x - x_1)$ for x

In Exercises 30–31, graph each equation in a rectangular coordinate system.

30. $y = 2 - |x|$ **31.** $y = x^2 - 4$

In Exercises 32–34, perform the indicated operations and write the result in standard form.

32. $(6 - 7i)(2 + 5i)$ **33.** $\frac{5}{2 - i}$

34. $2\sqrt{-49} + 3\sqrt{-64}$

Without changes, the graphs show projections for the amount being paid in Social Security benefits and the amount going into the system. All data are expressed in billions of dollars.

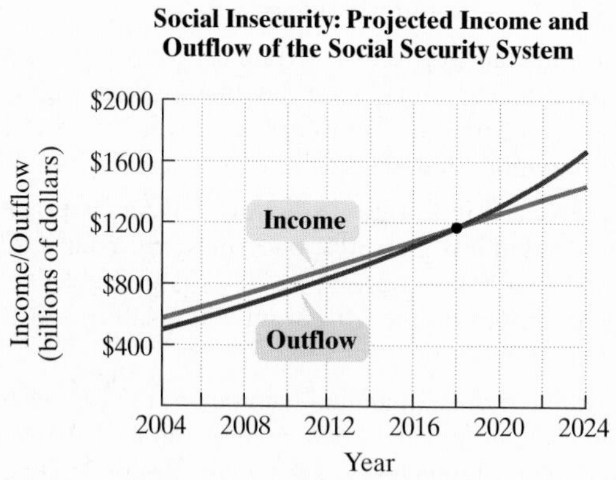

Source: 2004 Social Security Trustees Report

Exercises 35–37 are based on the data shown by the graphs.

35. In 2004, the system's income was \$575 billion, projected to increase at an average rate of \$43 billion per year. In which year will the system's income be \$1177 billion?

36. The data for the system's outflow can be modeled by the formula

$$B = 0.07x^2 + 47.4x + 500,$$

where B represents the amount paid in benefits, in billions of dollars, x years after 2004. According to this model, when will the amount paid in benefits be \$1177 billion? Round to the nearest year.

37. How well do your answers to Exercises 35 and 36 model the data shown by the graphs?

38. From 2002 through 2004, there were 2598 books categorized as U.S. politics and government. The number of books in 2003 exceeded the number in 2002 by 62 and the number in 2004 exceeded the number in 2002 by 190. How many books on U.S. politics were there for each of the three years? (*Source:* Andrew Grabois, R. R. Bowker)

39. The costs for two different kinds of heating systems for a three-bedroom home are given in the following table. After how many years will total costs for solar heating and electric heating be the same? What will be the cost at that time?

System	Cost to Install	Operating Cost/Year
Solar	\$29,700	\$150
Electric	\$5000	\$1100

40. You invested \$10,000 in two accounts paying 8% and 10% annual interest, respectively. At the end of the year, the total interest from these investments was \$940. How much was invested at each rate?

41. The length of a rectangular carpet is 4 feet greater than twice its width. If the area is 48 square feet, find the carpet's length and width.

42. A vertical pole is to be supported by a wire that is 26 feet long and anchored 24 feet from the base of the pole. How far up the pole should the wire be attached?

43. After a 60% reduction, a jacket sold for \$20. What was the jacket's price before the reduction?

44. You are choosing between two telephone plans for local calls. Plan A charges \$25 per month for unlimited calls. Plan B has a monthly fee of \$13 with a charge of \$0.06 per local call. How many local telephone calls in a month make plan A the better deal?

CHAPTER 2

Functions and Graphs

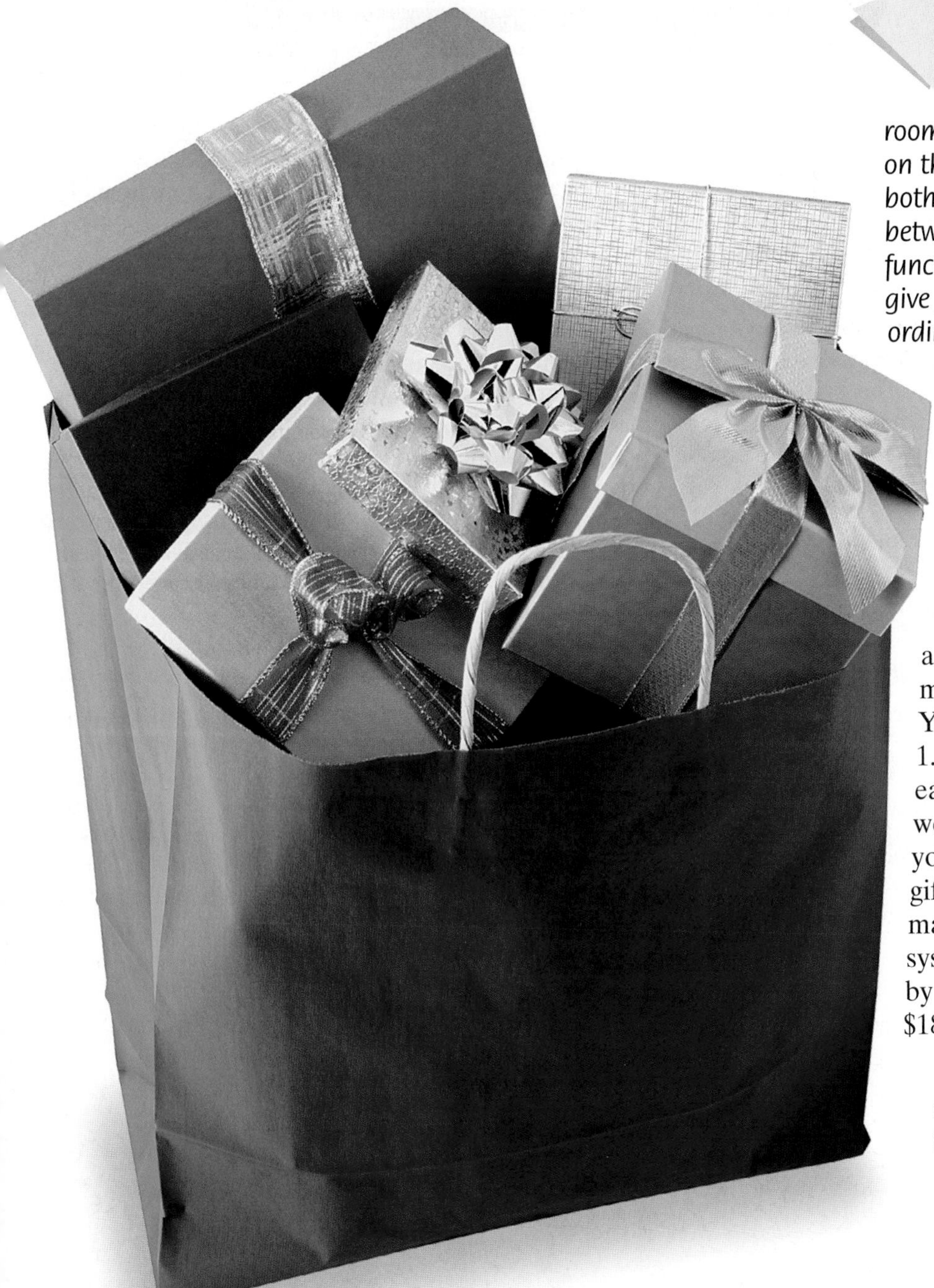

THE COST OF MAILING A *package depends on its weight. The probability that you and another person in a room share the same birthday depends on the number of people in the room. In both these situations, the relationship between variables can be described by a function. Understanding this concept will give you a new perspective on many ordinary situations.*

'TIS THE SEASON AND YOU'VE waited until the last minute to mail your holiday gifts. Your only option is overnight express mail. You realize that the cost of mailing a gift depends on its weight, but the mailing costs seem somewhat odd. Your packages that weigh 1.1 pounds, 1.5 pounds, and 2 pounds cost \$15.75 each to send overnight. Packages that weigh 2.01 pounds and 3 pounds cost you \$18.50 each. Finally, your heaviest gift is barely over 3 pounds and its mailing cost is \$21.25. What sort of system is this in which costs increase by \$2.75, stepping from \$15.75 to \$18.50 and from \$18.50 to \$21.25?

Graphs that ascend in steps are discussed on page 210 in Section 2.2.

SECTION 2.1 Basics of Functions and Their Graphs

Objectives

1. Find the domain and range of a relation.
2. Determine whether a relation is a function.
3. Determine whether an equation represents a function.
4. Evaluate a function.
5. Graph functions by plotting points.
6. Use the vertical line test to identify functions.
7. Obtain information about a function from its graph.
8. Identify the domain and range of a function from its graph.
9. Identify intercepts from a function's graph.

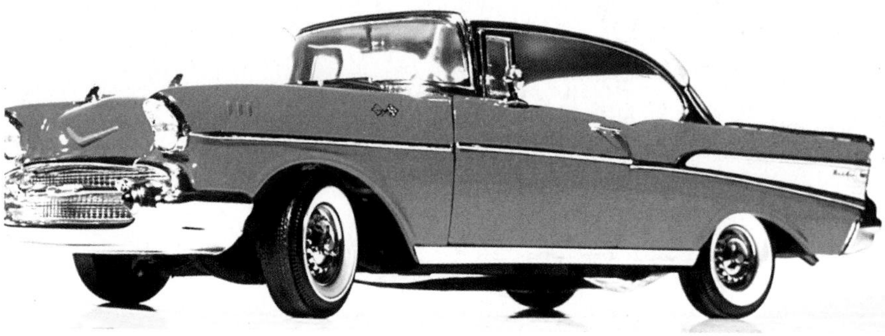

Have you ever seen a gas-guzzling car from the 1950s, with its huge fins and overstated design? The worst year for automobile fuel efficiency was 1958, when cars averaged a dismal 12.4 miles per gallon. There is a formula that approximately describes fuel efficiency of U.S. cars over time. The formula is

$$y = 0.0075x^2 - 0.2672x + 14.8.$$

The variable x represents the number of years after 1940. The variable y represents the average number of miles per gallon for U.S. automobiles.

The mathematical model for fuel efficiency indicates that miles per gallon depend on the number of years after 1940. For each value of x, or the number of years after 1940, the model gives precisely one value of y, or the average number of miles per gallon for U.S. automobiles. Under these conditions, we say that fuel efficiency is a *function* of time.

In this section, you will be introduced to the basics of functions and their graphs. Much of our work in this course will be devoted to the important topic of functions and how they model your world.

1 Find the domain and range of a relation.

Relations

Studies show that exercise can promote good long-term health no matter how much you weigh. A brisk half-hour walk each day is enough to get the benefits. Combined with a healthy diet, it also helps to stave off obesity. How many calories does your workout burn? The graph in Figure 2.1 shows the calories burned per hour in six activities.

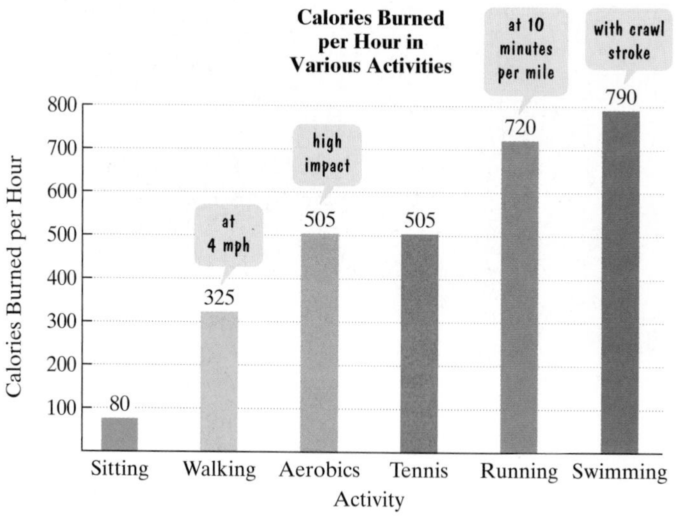

Figure 2.1 Counting calories
Source: fitresource.com

The information shown in the bar graph indicates a correspondence between the activities and calories burned per hour. We can write this correspondence using a set of ordered pairs:

$$\{(\text{sitting}, 80), (\text{walking}, 325), (\text{aerobics}, 505), (\text{tennis}, 505), (\text{running}, 720), (\text{swimming}, 790)\}.$$

These braces indicate that we are representing a set.

The mathematical term for a set of ordered pairs is a *relation*.

Definition of a Relation

A **relation** is any set of ordered pairs. The set of all first components of the ordered pairs is called the **domain** of the relation and the set of all second components is called the **range** of the relation.

EXAMPLE 1 Finding the Domain and Range of a Relation

Find the domain and the range of the relation:

$$\{(\text{sitting}, 80), (\text{walking}, 325), (\text{aerobics}, 505), (\text{tennis}, 505), (\text{running}, 720), (\text{swimming}, 790)\}.$$

Solution The domain is the set of all first components. Thus, the domain is

$$\{\text{sitting, walking, aerobics, tennis, running, swimming}\}.$$

The range is the set of all second components. Thus, the range is

$$\{80, 325, 505, 720, 790\}.$$

Although both aerobics and tennis burn 505 calories per hour, it is not necessary to list 505 twice.

Check Point 1 Find the domain and the range of the relation:

$$\{(5, 12.8), (10, 16.2), (15, 18.9), (20, 20.7), (25, 21.8)\}.$$

As you worked Check Point 1, did you wonder if there was a rule that assigned the "inputs" in the domain to the "outputs" in the range? For example, for the ordered pair (15, 18.9), how does the output 18.9 depend on the input 15? Think paid vacation days! The first number in each ordered pair is the number of years that a full-time employee has been employed by a medium to large U.S. company. The second number is the average number of paid vacation days each year. Consider, for example, the ordered pair (15, 18.9).

(15, 18.9)

After 15 years, | workers average 18.9 paid vacation days per year.

The relation in the vacation-days example can be pictured as follows:

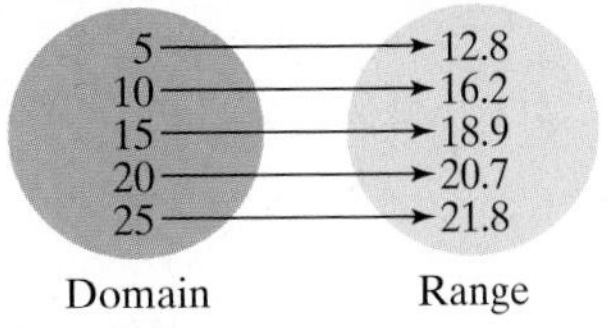

The five points in Figure 2.2 are another way to visually represent the relation.

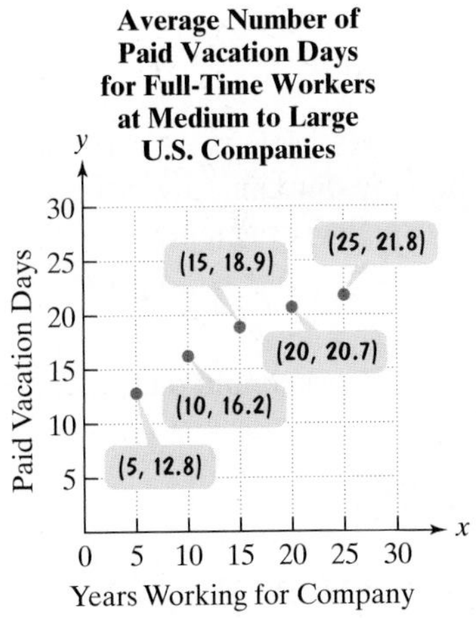

Figure 2.2 The graph of a relation showing a correspondence between years with a company and paid vacation days

Source: Bureau of Labor Statistics

2 Determine whether a relation is a function.

Activity	Calories Burned per hour
Sitting	80
Walking	325
Aerobics	505
Tennis	505
Running	720
Swimming	790

Functions

Shown in the margin are the calories burned per hour for the activities in the bar graph in Figure 2.1 on page 186. We've used this information to define two relations. Figure 2.3(a) shows a correspondence between activities and calories burned. Figure 2.3(b) shows a correspondence between calories burned and activities.

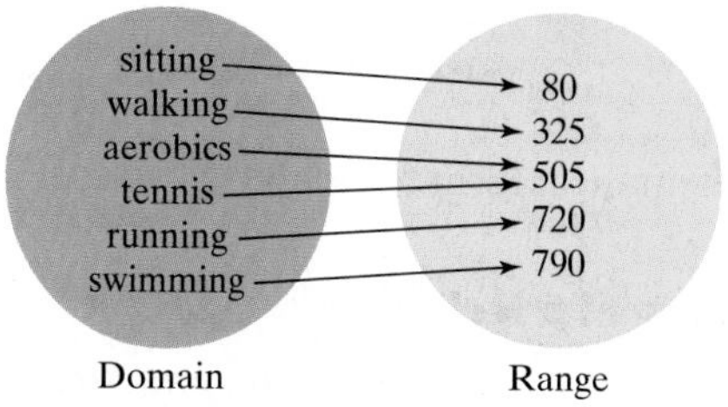

Figure 2.3(a) Activities correspond to calories burned.

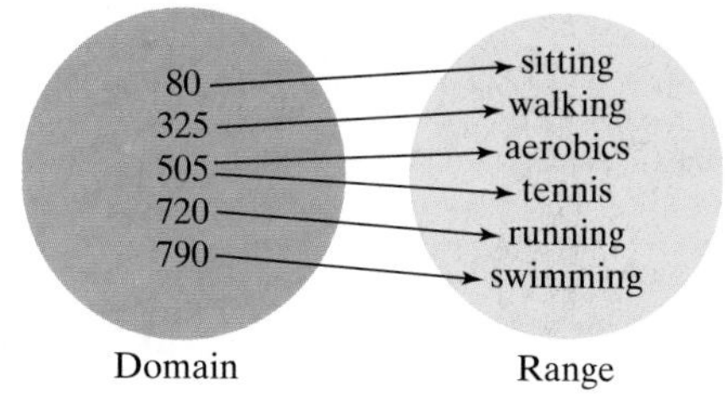

Figure 2.3(b) Calories burned correspond to activities.

A relation in which each member of the domain corresponds to exactly one member of the range is a **function**. Can you see that the relation in Figure 2.3(a) is a function? Each activity in the domain corresponds to exactly one number representing calories burned per hour in the range. If we know the activity, we know the calories burned per hour. Notice that more than one element in the domain can correspond to the same element in the range: Aerobics and tennis both burn 505 calories per hour.

Is the relation in Figure 2.3(b) a function? Does each member of the domain correspond to precisely one member of the range? This relation is not a function because there is a member of the domain that corresponds to two members of the range:

(505, aerobics) (505, tennis).

The member of the domain, 505, corresponds to both aerobics and tennis. If we know the calories burned per hour, 505, we cannot be sure of the activity. Because **a function is a relation in which no two ordered pairs have the same first component and different second components**, the ordered pairs (505, aerobics) and (505, tennis) are not ordered pairs of a function.

Same first component

(505, aerobics) (505, tennis)

Different second components

Definition of a Function

A **function** is a correspondence from a first set, called the **domain**, to a second set, called the **range**, such that each element in the domain corresponds to *exactly one* element in the range.

Can you see that the correspondence between years worked and paid vacation days from Check Point 1 is a function?

Each element in the domain

$\{(5, 12.8), (10, 16.2), (15, 18.9), (20, 20.7), (25, 21.8)\}$

corresponds to exactly one element in the range.

However, Example 2 illustrates that not every correspondence between sets is a function.

EXAMPLE 2 Determining Whether a Relation Is a Function

Determine whether each relation is a function:

a. $\{(1, 6), (2, 6), (3, 8), (4, 9)\}$ **b.** $\{(6, 1), (6, 2), (8, 3), (9, 4)\}$.

Solution We begin by making a figure for each relation that shows the domain and the range (Figure 2.4).

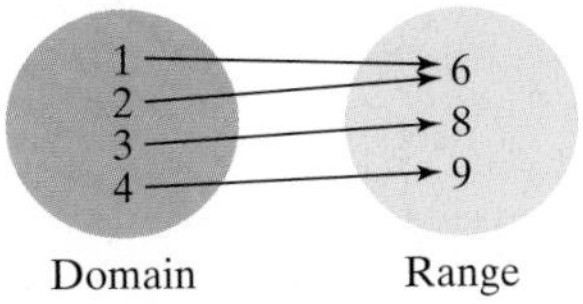

Figure 2.4(a)

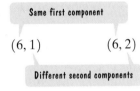

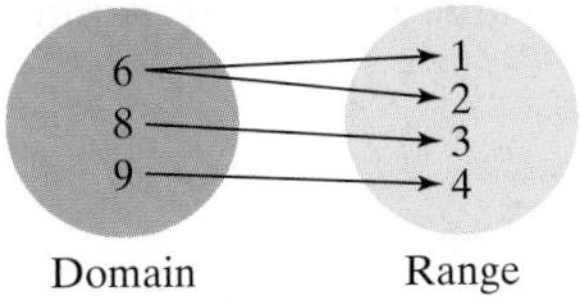

Figure 2.4(b)

a. Figure 2.4(a) shows that every element in the domain corresponds to exactly one element in the range. The element 1 in the domain corresponds to the element 6 in the range. Furthermore, 2 corresponds to 6, 3 corresponds to 8, and 4 corresponds to 9. No two ordered pairs in the given relation have the same first component and different second components. Thus, the relation is a function.

b. Figure 2.4(b) shows that 6 corresponds to both 1 and 2. If any element in the domain corresponds to more than one element in the range, the relation is not a function. This relation is not a function; two ordered pairs have the same first component and different second components.

Same first component

$(6, 1)$ $(6, 2)$

Different second components

Look at Figure 2.4(a) again. The fact that 1 and 2 in the domain correspond to the same number, 6, in the range does not violate the definition of a function. **A function can have two different first components with the same second component.** By contrast, a relation is not a function when two different ordered pairs have the same first component and different second components. Thus, the relation in Figure 2.4(b) is not a function.

Study Tip

If a relation is a function, reversing the components in each of its ordered pairs may result in a relation that is not a function.

Check Point 2 Determine whether each relation is a function:

a. $\{(1, 2), (3, 4), (5, 6), (5, 8)\}$ **b.** $\{(1, 2), (3, 4), (6, 5), (8, 5)\}$.

3 Determine whether an equation represents a function.

Functions as Equations

Functions are usually given in terms of equations rather than as sets of ordered pairs. For example, here is an equation that models paid vacation days each year as a function of years working for a company:

$$y = -0.016x^2 + 0.93x + 8.5.$$

The variable x represents years working for a company. The variable y represents the average number of vacation days each year. The variable y is a function of the variable x. For each value of x, there is one and only one value of y. The variable x is called the **independent variable** because it can be assigned any value from the domain. Thus, x can be assigned any positive integer representing the number of years working for a company. The variable y is called the **dependent variable** because its value depends on x. Paid vacation days depend on years working for a company. The value of the dependent variable, y, is calculated after selecting a value for the independent variable, x.

We have seen that not every set of ordered pairs defines a function. Similarly, not all equations with the variables x and y define a function. If an equation is solved for y and more than one value of y can be obtained for a given x, then the equation does not define y as a function of x.

EXAMPLE 3 Determining Whether an Equation Represents a Function

Determine whether each equation defines y as a function of x:

a. $x^2 + y = 4$ **b.** $x^2 + y^2 = 4$.

Solution Solve each equation for y in terms of x. If two or more values of y can be obtained for a given x, the equation is not a function.

a.

$$x^2 + y = 4 \quad \text{This is the given equation.}$$
$$x^2 + y - x^2 = 4 - x^2 \quad \text{Solve for y by subtracting } x^2 \text{ from both sides.}$$
$$y = 4 - x^2 \quad \text{Simplify.}$$

From this last equation we can see that for each value of x, there is one and only one value of y. For example, if $x = 1$, then $y = 4 - 1^2 = 3$. The equation defines y as a function of x.

b.

$$x^2 + y^2 = 4 \quad \text{This is the given equation.}$$
$$x^2 + y^2 - x^2 = 4 - x^2 \quad \text{Isolate } y^2 \text{ by subtracting } x^2 \text{ from both sides.}$$
$$y^2 = 4 - x^2 \quad \text{Simplify.}$$
$$y = \pm\sqrt{4 - x^2} \quad \text{Apply the square root property: If } u^2 = d, \text{ then } u = \pm\sqrt{d}.$$

The $\pm$ in this last equation shows that for certain values of x (all values between -2 and 2), there are two values of y. For example, if $x = 1$, then $y = \pm\sqrt{4 - 1^2} = \pm\sqrt{3}$. For this reason, the equation does not define y as a function of x.

Check Point 3 Solve each equation for y and then determine whether the equation defines y as a function of x:

a. $2x + y = 6$ **b.** $x^2 + y^2 = 1$.

4 Evaluate a function.

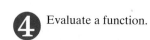

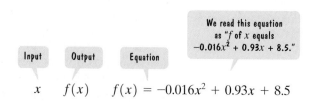

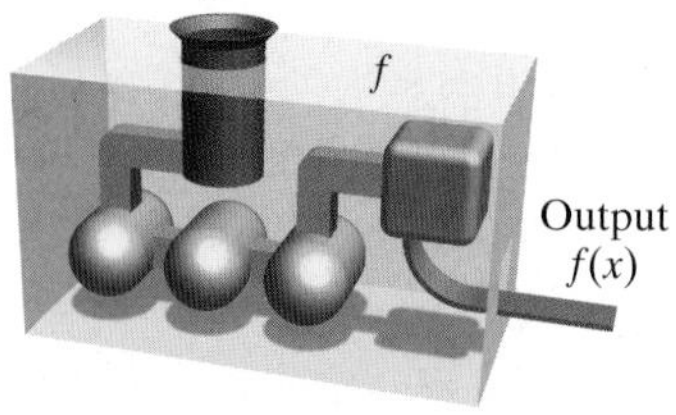

Figure 2.5 A "function machine" with inputs and outputs

Function Notation

If an equation in x and y gives one and only one value of y for each value of x, then the variable y is a function of the variable x. When an equation represents a function, the function is often named by a letter such as f, g, h, F, G, or H. Any letter can be used to name a function. Suppose that f names a function. Think of the domain as the set of the function's inputs and the range as the set of the function's outputs. As shown in Figure 2.5, input is represented by x and the output by $f(x)$. The special notation $f(x)$, read "f of x" or "f at x," represents the **value of the function at the number x.**

Let's make this clearer by considering a specific example. We know that the equation

$$y = -0.016x^2 + 0.93x + 8.5$$

defines y as a function of x. We'll name the function f. Now, we can apply our new function notation.

Study Tip

The notation $f(x)$ does *not* mean "f times x." The notation describes the value of the function at x.

Input: x Output: $f(x)$ Equation: $f(x) = -0.016x^2 + 0.93x + 8.5$

We read this equation as "f of x equals $-0.016x^2 + 0.93x + 8.5$."

Technology

Graphing utilities can be used to evaluate functions. The screens below show the evaluation of

$f(x) = -0.016x^2 + 0.93x + 8.5$

at 10 on a TI-83 Plus graphing calculator. The function f is named Y_1.

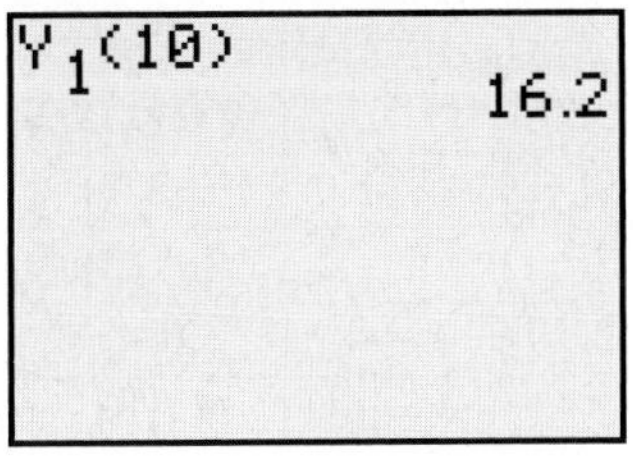

Suppose we are interested in finding $f(10)$, the function's output when the input is 10. To find the value of the function at 10, we substitute 10 for x. We are **evaluating the function** at 10.

$$f(x) = -0.016x^2 + 0.93x + 8.5 \qquad \text{This is the given function.}$$
$$f(10) = -0.016(10)^2 + 0.93(10) + 8.5 \qquad \text{Replace each occurrence of } x \text{ with 10.}$$
$$= -0.016(100) + 0.93(10) + 8.5 \qquad \text{Evaluate the exponential expression: } 10^2 = 100.$$
$$= -1.6 + 9.3 + 8.5 \qquad \text{Perform the multiplications.}$$
$$= 16.2 \qquad \text{Add from left to right.}$$

The statement $f(10) = 16.2$, read "f of 10 equals 16.2," tells us that the value of the function at 10 is 16.2. When the function's input is 10, its output is 16.2. (After 10 years, workers average 16.2 vacation days each year.) To find other function values, such as $f(15)$, $f(20)$, or $f(23)$, substitute the specified input values for x into the function's equation.

If a function is named f and x represents the independent variable, the notation $f(x)$ corresponds to the y-value for a given x. Thus,

$$f(x) = -0.016x^2 + 0.93x + 8.5 \quad \text{and} \quad y = -0.016x^2 + 0.93x + 8.5$$

define the same function. This function may be written as

$$y = f(x) = -0.016x^2 + 0.93x + 8.5.$$

EXAMPLE 4 Evaluating a Function

If $f(x) = x^2 + 3x + 5$, evaluate each of the following:

a. $f(2)$ **b.** $f(x + 3)$ **c.** $f(-x)$.

Solution We substitute 2, $x + 3$, and $-x$ for x in the equation for f. When replacing x with a variable or an algebraic expression, you might find it helpful to think of the function's equation as

$$f(\boxed{x}) = \boxed{x}^2 + 3\boxed{x} + 5.$$

a. We find $f(2)$ by substituting 2 for x in the equation.

$$f(\boxed{2}) = \boxed{2}^2 + 3 \cdot \boxed{2} + 5 = 4 + 6 + 5 = 15$$

Thus, $f(2) = 15$.

b. We find $f(x + 3)$ by substituting $x + 3$ for x in the equation.

$$f(\boxed{x+3}) = \boxed{(x+3)}^2 + 3\boxed{(x+3)} + 5$$

Equivalently,

$$f(x + 3) = (x + 3)^2 + 3(x + 3) + 5$$
$$= x^2 + 6x + 9 + 3x + 9 + 5 \qquad \text{Square } x + 3 \text{ using } (A + B)^2 = A^2 + 2AB + B^2. \text{ Distribute 3 throughout the parentheses.}$$
$$= x^2 + 9x + 23. \qquad \text{Combine like terms.}$$

c. We find $f(-x)$ by substituting $-x$ for x in the equation.

$$f(\boxed{-x}) = \boxed{(-x)}^2 + 3\boxed{(-x)} + 5$$

Equivalently,

$$f(-x) = (-x)^2 + 3(-x) + 5$$
$$= x^2 - 3x + 5.$$

Discovery

Using $f(x) = x^2 + 3x + 5$ and the answers in parts (b) and (c):

1. Is $f(x + 3)$ equal to $f(x) + f(3)$?
2. Is $f(-x)$ equal to $-f(x)$?

Check Point 4 If $f(x) = x^2 - 2x + 7$, evaluate each of the following:

a. $f(-5)$ **b.** $f(x + 4)$ **c.** $f(-x)$.

Graph functions by plotting points.

Graphs of Functions

The **graph of a function** is the graph of its ordered pairs. For example, the graph of $f(x) = 2x$ is the set of points (x, y) in the rectangular coordinate system satisfying $y = 2x$. Similarly, the graph of $g(x) = 2x + 4$ is the set of points (x, y) in the rectangular coordinate system satisfying the equation $y = 2x + 4$. In the next example, we graph both of these functions in the same rectangular coordinate system.

EXAMPLE 5   Graphing Functions

Graph the functions $f(x) = 2x$ and $g(x) = 2x + 4$ in the same rectangular coordinate system. Select integers for x, starting with -2 and ending with 2.

Solution   We begin by setting up a partial table of coordinates for each function. Then, we plot the five points in each table and connect them, as shown in Figure 2.6. The graph of each function is a straight line. Do you see a relationship between the two graphs? The graph of g is the graph of f shifted vertically up by 4 units.

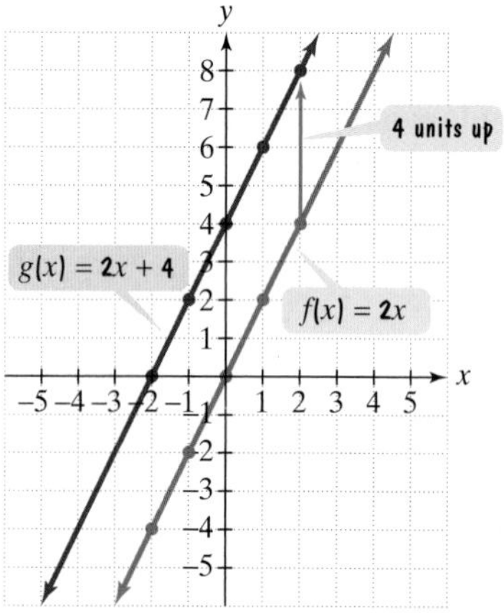

Figure 2.6

x	$f(x) = 2x$	(x, y) or $(x, f(x))$
-2	$f(-2) = 2(-2) = -4$	$(-2, -4)$
-1	$f(-1) = 2(-1) = -2$	$(-1, -2)$
0	$f(0) = 2 \cdot 0 = 0$	$(0, 0)$
1	$f(1) = 2 \cdot 1 = 2$	$(1, 2)$
2	$f(2) = 2 \cdot 2 = 4$	$(2, 4)$

Choose x.   Compute $f(x)$ by evaluating f at x.   Form the ordered pair.

x	$g(x) = 2x + 4$	(x, y) or $(x, g(x))$
-2	$g(-2) = 2(-2) + 4 = 0$	$(-2, 0)$
-1	$g(-1) = 2(-1) + 4 = 2$	$(-1, 2)$
0	$g(0) = 2 \cdot 0 + 4 = 4$	$(0, 4)$
1	$g(1) = 2 \cdot 1 + 4 = 6$	$(1, 6)$
2	$g(2) = 2 \cdot 2 + 4 = 8$	$(2, 8)$

Choose x.   Compute $g(x)$ by evaluating g at x.   Form the ordered pair.

The graphs in Example 5 are straight lines. All functions with equations of the form $f(x) = mx + b$ graph as straight lines. Such functions, called **linear functions**, will be discussed in detail in Section 2.3.

Technology

We can use a graphing utility to check the tables and the graphs in Example 5 for the functions

$$f(x) = 2x \qquad \text{and} \qquad g(x) = 2x + 4.$$

Enter $y_1 = 2x$ in the $\boxed{y=}$ screen.

Enter $y_2 = 2x + 4$ in the $\boxed{y=}$ screen.

We entered -2 for the starting x-value and 1 as an increment between x-values to check our tables in Example 5.

Checking Tables

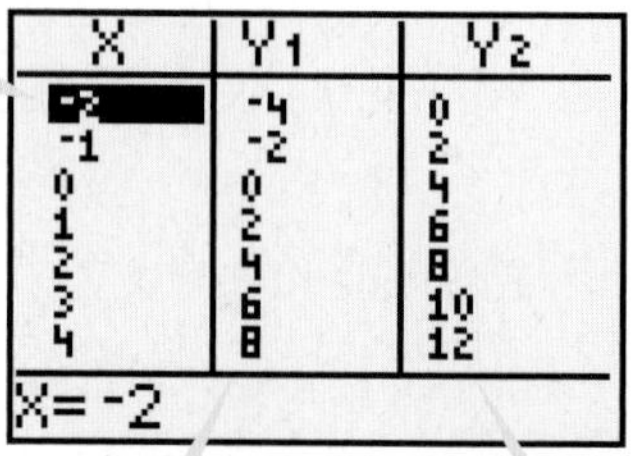

X	Y₁	Y₂
-2	-4	0
-1	-2	2
0	0	4
1	2	6
2	4	8
3	6	10
4	8	12

X=-2

Use the first five ordered pairs (x, y_1) to check the first table.

Use the first five ordered pairs (x, y_2) to check the second table.

Checking Graphs

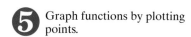

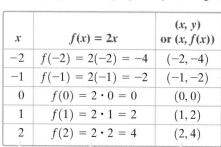

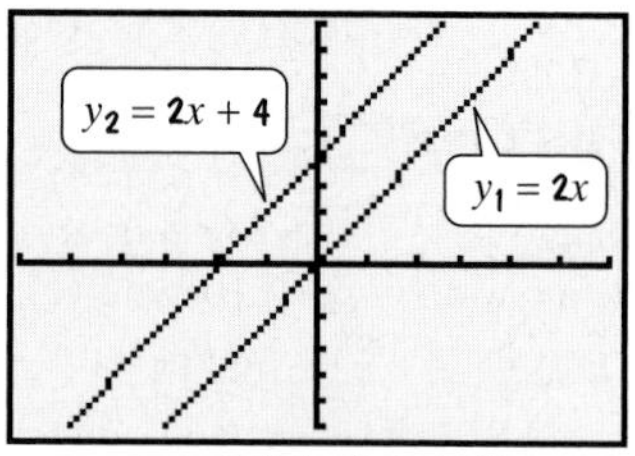

$[-6, 6, 1]$ by $[-6, 9, 1]$

We selected this viewing rectangle, or window, to match Figure 2.6.

Check Point 5 Graph the functions $f(x) = 2x$ and $g(x) = 2x - 3$ in the same rectangular coordinate system. Select integers for x, starting with -2 and ending with 2. How is the graph of g related to the graph of f?

6 Use the vertical line test to identify functions.

The Vertical Line Test

Not every graph in the rectangular coordinate system is the graph of a function. The definition of a function specifies that no value of x can be paired with two or more different values of y. Consequently, if a graph contains two or more different points with the same first coordinate, the graph cannot represent a function. This is illustrated in Figure 2.7. Observe that points sharing a common first coordinate are vertically above or below each other.

This observation is the basis of a useful test for determining whether a graph defines y as a function of x. The test is called the **vertical line test**.

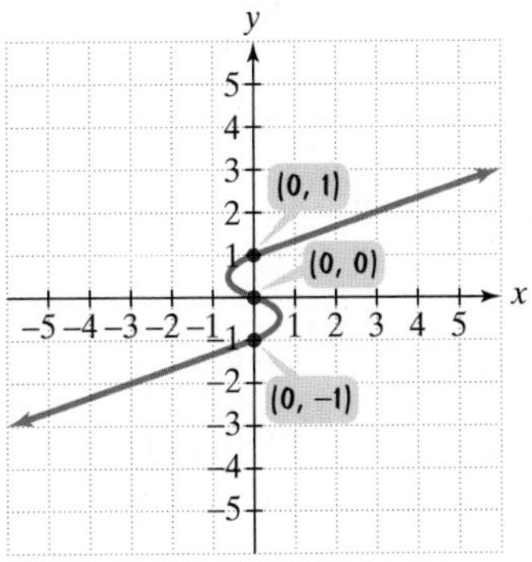

Figure 2.7 y is not a function of x because 0 is paired with three values of y, namely, 1, 0, and -1.

The Vertical Line Test for Functions

If any vertical line intersects a graph in more than one point, the graph does not define y as a function of x.

EXAMPLE 6 Using the Vertical Line Test

Use the vertical line test to identify graphs in which y is a function of x.

a.
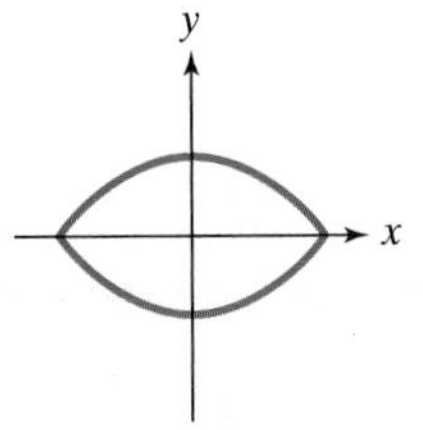

b.
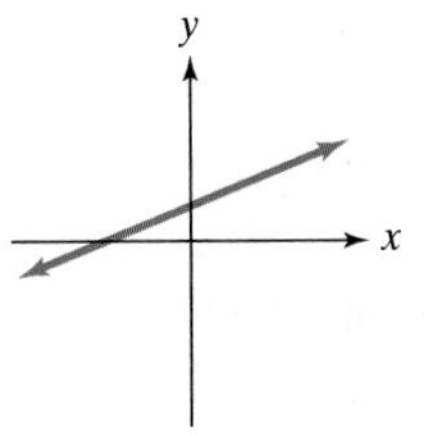

c.
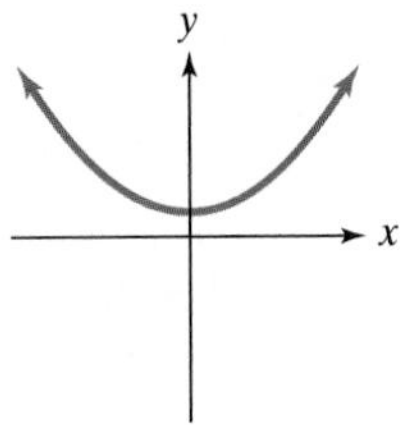

d.
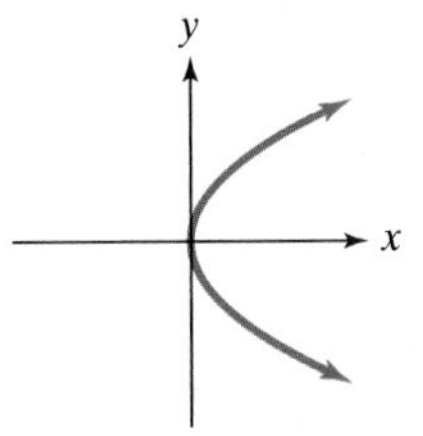

Solution y is a function of x for the graphs in (b) and (c).

a.
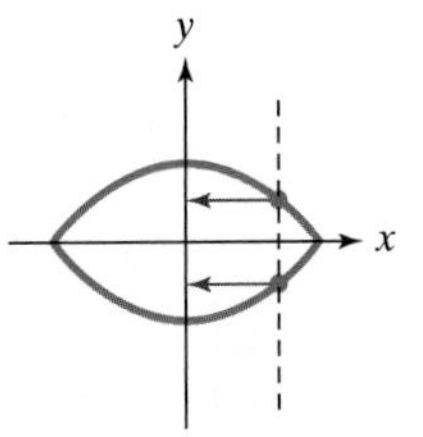

y **is not a function of** x. Two values of y correspond to an x-value.

b.
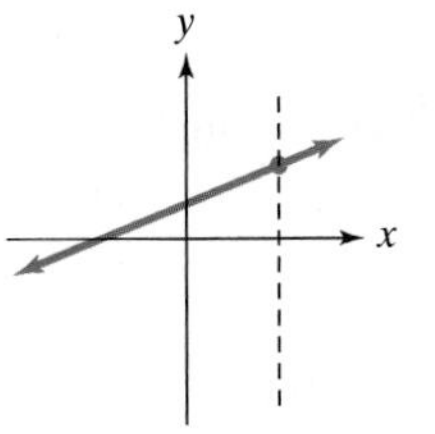

y **is a function of** x.

c.
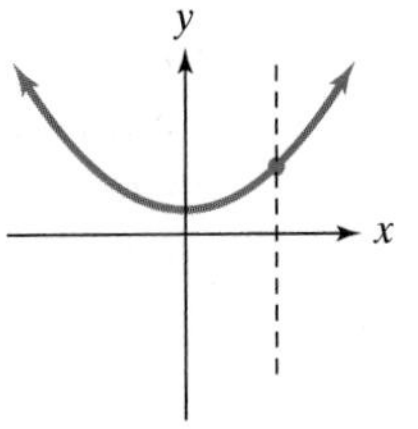

y **is a function of** x.

d.
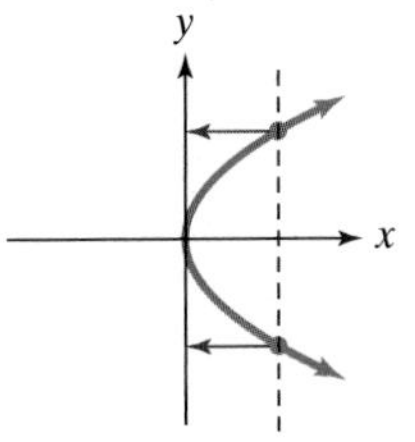

y **is not a function of** x. Two values of y correspond to an x-value.

Check Point 6 Use the vertical line test to identify graphs in which y is a function of x.

a.
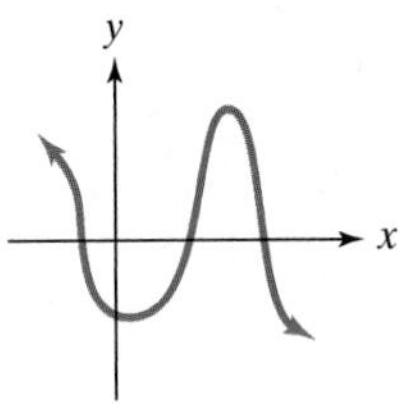

b.
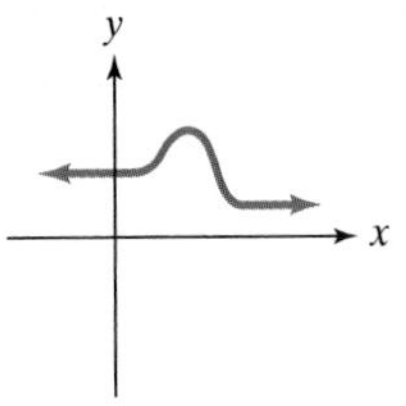

c.
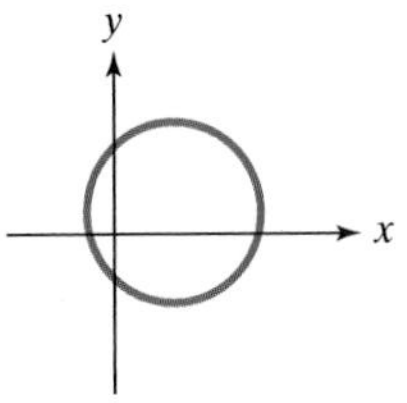

7 Obtain information about a function from its graph.

Obtaining Information from Graphs

You can obtain information about a function from its graph. At the right or left of a graph, you will find closed dots, open dots, or arrows.

- A closed dot indicates that the graph does not extend beyond this point and the point belongs to the graph.
- An open dot indicates that the graph does not extend beyond this point and the point does not belong to the graph.
- An arrow indicates that the graph extends indefinitely in the direction in which the arrow points.

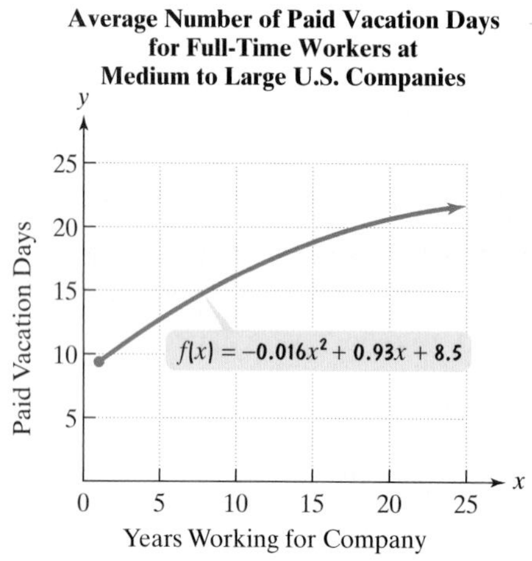

Figure 2.8

Source: Bureau of Labor Statistics

EXAMPLE 7 Analyzing the Graph of a Function

The function

$$f(x) = -0.016x^2 + 0.93x + 8.5$$

models the average number of paid vacation days each year, $f(x)$, for full-time workers at medium to large U.S. companies after x years. The graph of f is shown in Figure 2.8.

a. Explain why f represents the graph of a function.

b. Use the graph to find a reasonable estimate of $f(5)$.

c. For what value of x is $f(x) = 20$?

d. Describe the general trend shown by the graph.

Solution

a. No vertical line intersects the graph of f more than once. By the vertical line test, f represents the graph of a function.

b. To find $f(5)$, or f of 5, we locate 5 on the x-axis. The figure shows the point on the graph of f for which 5 is the first coordinate. From this point, we look to the y-axis to find the corresponding y-coordinate. A reasonable estimate of the y-coordinate is 13. Thus, $f(5) \approx 13$. After 5 years, a worker can expect approximately 13 paid vacation days.

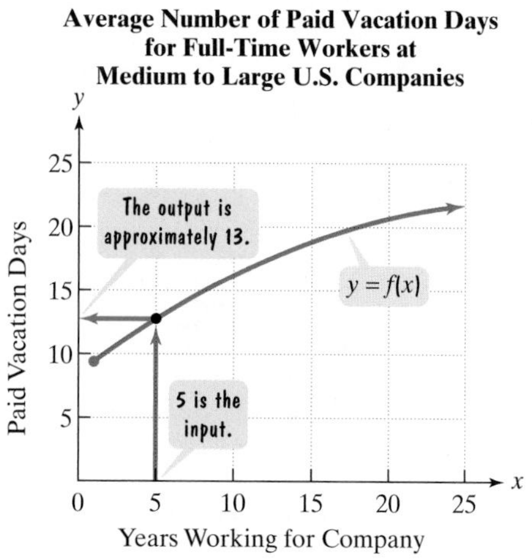

c. To find the value of x for which $f(x) = 20$, we locate 20 on the y-axis. The figure shows that there is one point on the graph of f for which 20 is the second coordinate. From this point, we look to the x-axis to find the corresponding x-coordinate. A reasonable estimate of the x-coordinate is 18. Thus, $f(x) = 20$ for $x \approx 18$. A worker with 20 paid vacation days has been with the company approximately 18 years.

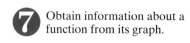

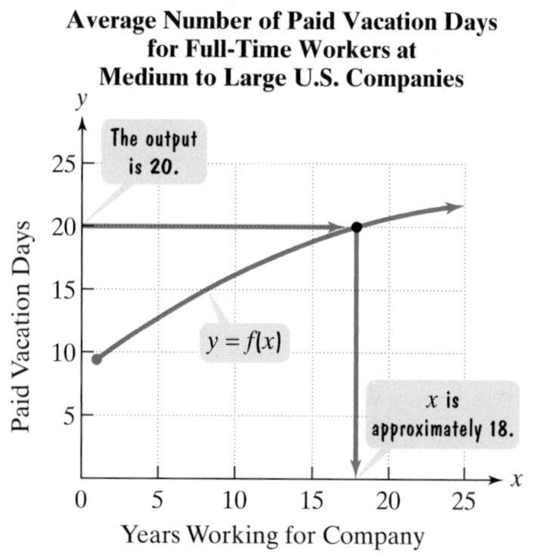

d. The graph of f in Figure 2.8 is rising from left to right. This shows that paid vacation days increase as time with the company increases. However, the rate of increase is slowing down as the graph moves to the right. This means that the increase in paid vacation days takes place more slowly the longer an employee is with the company.

Check Point 7
a. Use the graph of f in Figure 2.8 to find a reasonable estimate of $f(10)$.
b. For what value of x is $f(x) = 15$? Round to the nearest whole number.

8 Identify the domain and range of a function from its graph.

Identifying Domain and Range from a Function's Graph

Study Tip

Throughout this discussion, we will be using interval notation. Recall that square brackets indicate endpoints that are included in an interval. Parentheses indicate endpoints that are not included in an interval. For more detail on interval notation, see Section 1.7, pages 164–166.

Figure 2.9 illustrates how the graph of a function is used to determine the function's domain and its range.

Domain: set of inputs

Found on the x-axis

Range: set of outputs

Found on the y-axis

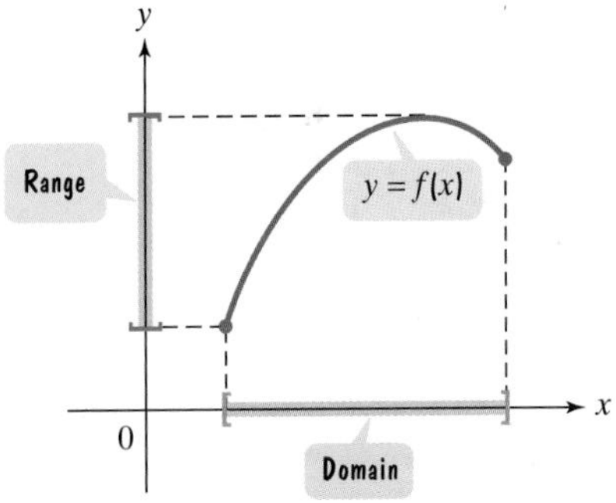

Figure 2.9 Domain and range of f

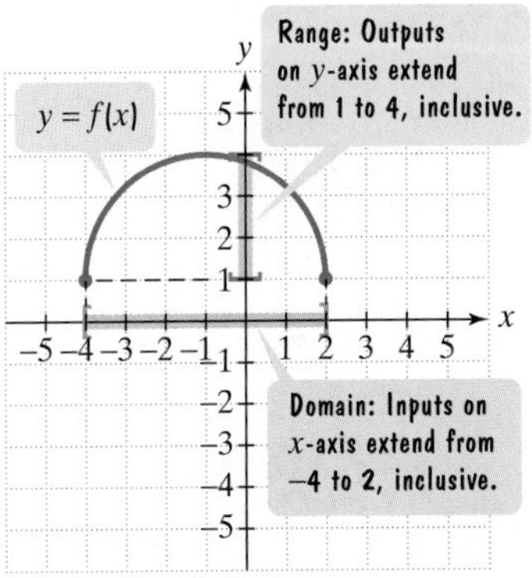

Figure 2.10 Domain and range of f

Let's apply these ideas to the graph of the function shown in Figure 2.10. To find the domain, look for all the inputs on the x-axis that correspond to points on the graph. Can you see that they extend from -4 to 2, inclusive? The function's domain can be represented as follows:

Using Set-Builder Notation

$\{x \mid -4 \leq x \leq 2\}$

The set of all x | such that | x is greater than or equal to -4 and less than or equal to 2.

Using Interval Notation

$[-4, 2]$.

The square brackets indicate -4 and 2 are included. Note the square brackets on the x-axis in Figure 2.10.

To find the range, look for all the outputs on the y-axis that correspond to points on the graph. They extend from 1 to 4, inclusive. The function's range can be represented as follows:

Using Set-Builder Notation

$\{y \mid 1 \leq y \leq 4\}$

The set of all y | such that | y is greater than or equal to 1 and less than or equal to 4.

Using Interval Notation

$[1, 4]$.

The square brackets indicate 1 and 4 are included. Note the square brackets on the y-axis in Figure 2.10.

EXAMPLE 8 Identifying the Domain and Range of a Function from Its Graph

Use the graph of each function to identify its domain and its range.

a.

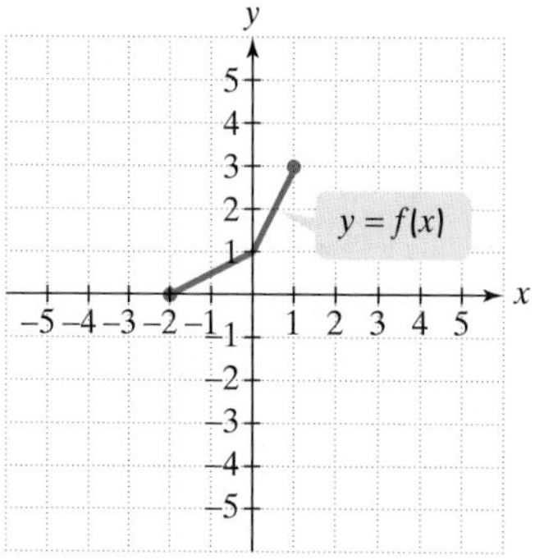

b.

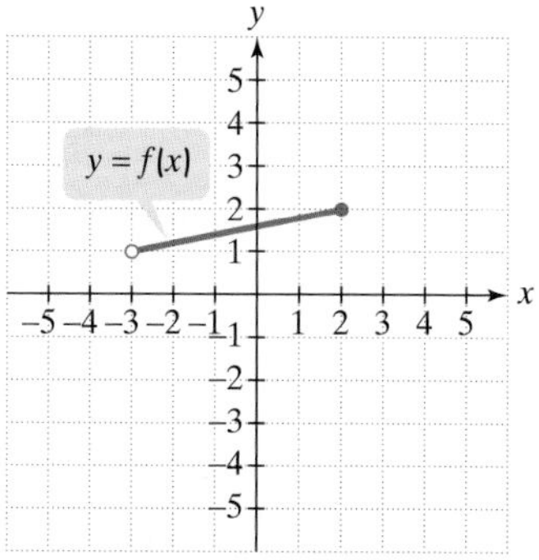

c.

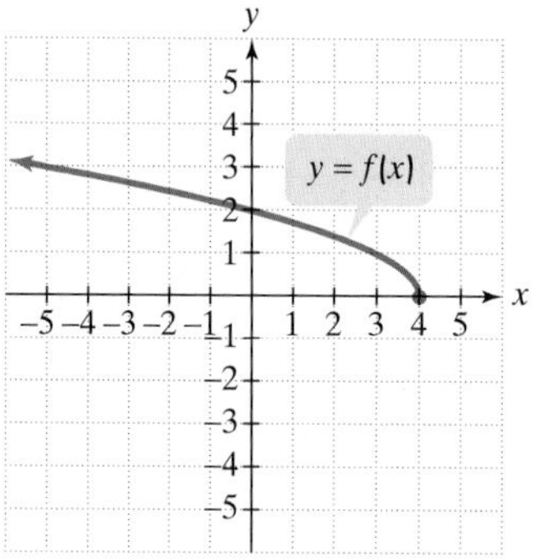

d.

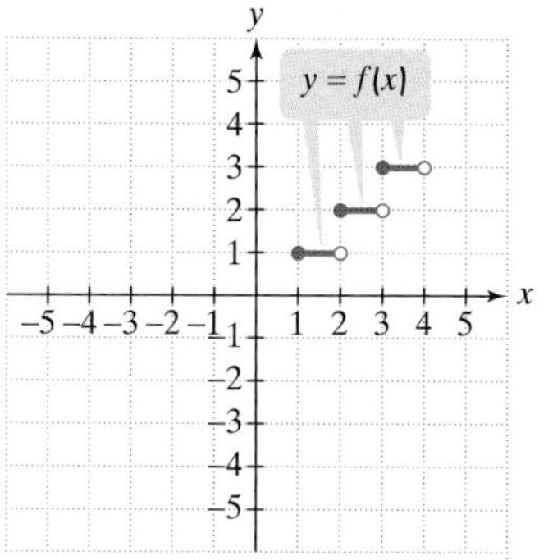

Solution For the graph of each function, the domain is highlighted in blue on the x-axis and the range is highlighted in green on the y-axis.

a.

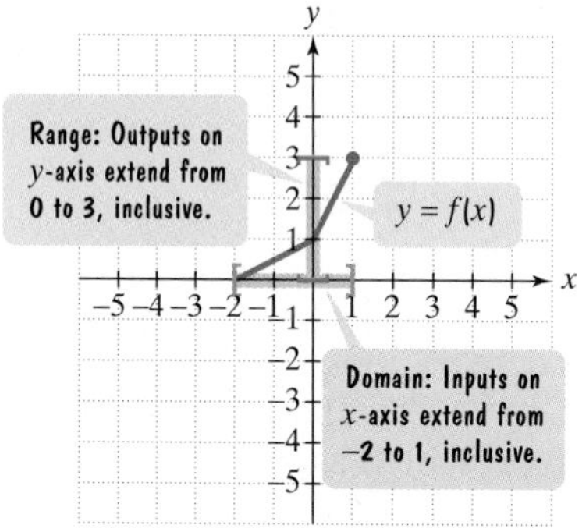

Domain $= \{x|-2 \le x \le 1\}$ or $[-2, 1]$

Range $= \{y|0 \le y \le 3\}$ or $[0, 3]$

b.

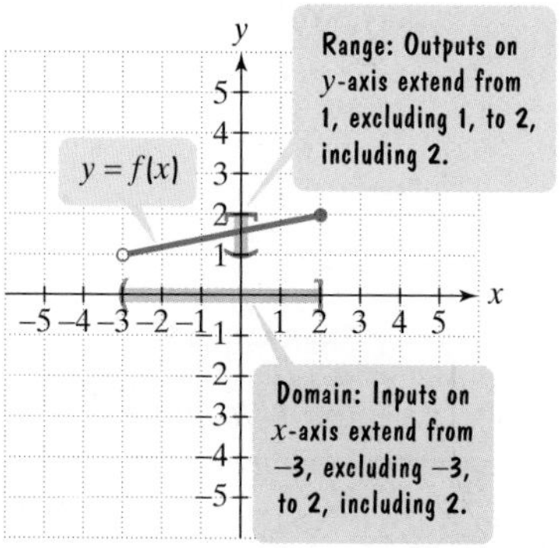

Domain $= \{x|-3 < x \le 2\}$ or $(-3, 2]$

Range $= \{y|1 < y \le 2\}$ or $(1, 2]$

c.

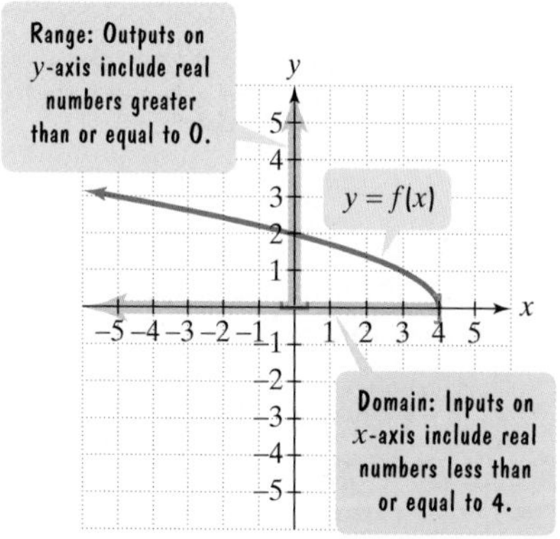

Domain $= \{x|x \le 4\}$ or $(-\infty, 4]$

Range $= \{y|y \ge 0\}$ or $[0, \infty)$

d.

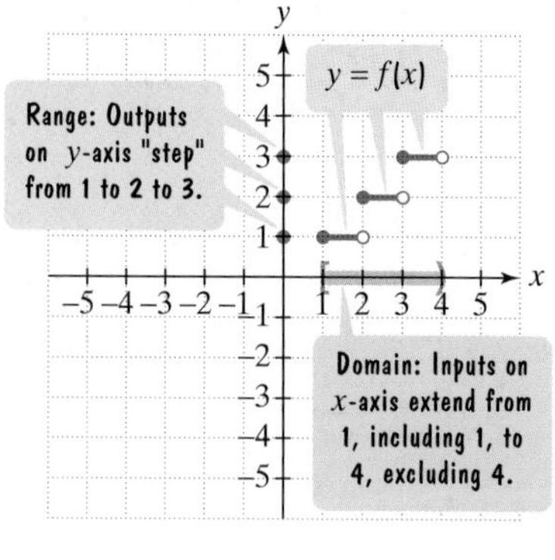

Domain $= \{x|1 \le x < 4\}$ or $[1, 4)$

Range $= \{y|y = 1, 2, 3\}$

Check Point 8 Use the graph of each function to identify its domain and its range.

a.

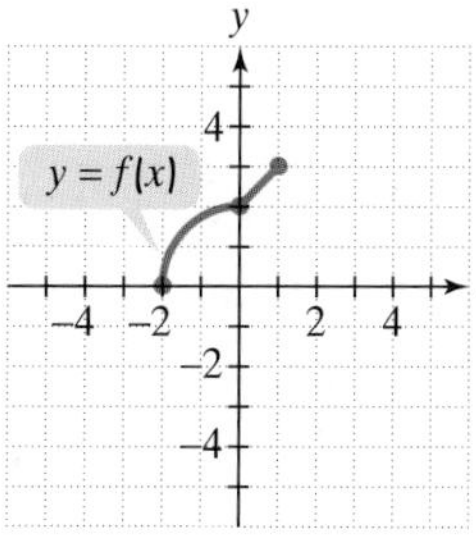

b.

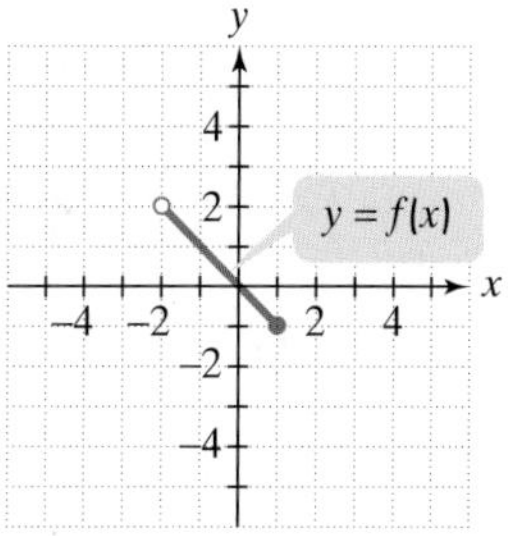

c.

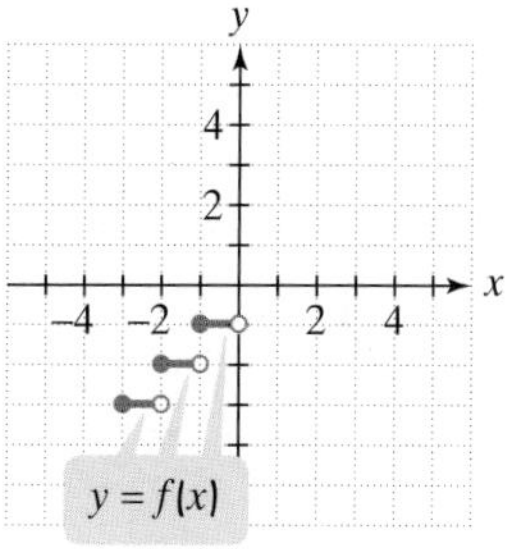

9 Identify intercepts from a function's graph.

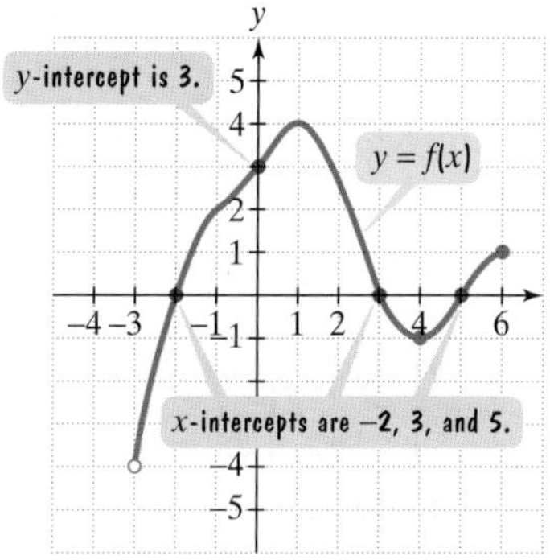

Figure 2.11 Identifying intercepts

Identifying Intercepts from a Function's Graph

Figure 2.11 illustrates how we can identify intercepts from a function's graph. To find the x-intercepts, look for the points at which the graph crosses the x-axis. There are three such points: $(-2, 0)$, $(3, 0)$, and $(5, 0)$. Thus, the x-intercepts are -2, 3, and 5. We express this in function notation by writing $f(-2) = 0$, $f(3) = 0$, and $f(5) = 0$. We say that -2, 3, and 5 are the *zeros of the function*. The **zeros of a function**, f, are the x-values for which $f(x) = 0$.

To find the y-intercept, look for the point at which the graph crosses the y-axis. This occurs at $(0, 3)$. Thus, the y-intercept is 3. We express this in function notation by writing $f(0) = 3$.

By the definition of a function, for each value of x we can have at most one value for y. What does this mean in terms of intercepts? **A function can have more than one x-intercept but at most one y-intercept.**

EXERCISE SET 2.1

Practice Exercises

In Exercises 1–10, determine whether each relation is a function. Give the domain and range for each relation.

1. $\{(1, 2), (3, 4), (5, 5)\}$ **2.** $\{(4, 5), (6, 7), (8, 8)\}$

3. $\{(3, 4), (3, 5), (4, 4), (4, 5)\}$

4. $\{(5, 6), (5, 7), (6, 6), (6, 7)\}$

5. $\{(3, -2), (5, -2), (7, 1), (4, 9)\}$

6. $\{(10, 4), (-2, 4), (-1, 1), (5, 6)\}$

7. $\{(-3, -3), (-2, -2), (-1, -1), (0, 0)\}$

8. $\{(-7, -7), (-5, -5), (-3, -3), (0, 0)\}$

9. $\{(1, 4), (1, 5), (1, 6)\}$

10. $\{(4, 1), (5, 1), (6, 1)\}$

In Exercises 11–26, determine whether each equation defines y as a function of x.

11. $x + y = 16$ **12.** $x + y = 25$

13. $x^2 + y = 16$ **14.** $x^2 + y = 25$

15. $x^2 + y^2 = 16$ **16.** $x^2 + y^2 = 25$

17. $x = y^2$ **18.** $4x = y^2$

19. $y = \sqrt{x + 4}$ **20.** $y = -\sqrt{x + 4}$

21. $x + y^3 = 8$ **22.** $x + y^3 = 27$

23. $xy + 2y = 1$ **24.** $xy - 5y = 1$

25. $|x| - y = 2$ **26.** $|x| - y = 5$

In Exercises 27–38, evaluate each function at the given values of the independent variable and simplify.

27. $f(x) = 4x + 5$

a. $f(6)$ **b.** $f(x + 1)$ **c.** $f(-x)$

28. $f(x) = 3x + 7$

a. $f(4)$ **b.** $f(x + 1)$ **c.** $f(-x)$

29. $g(x) = x^2 + 2x + 3$

a. $g(-1)$ **b.** $g(x + 5)$ **c.** $g(-x)$

30. $g(x) = x^2 - 10x - 3$

a. $g(-1)$ **b.** $g(x + 2)$ **c.** $g(-x)$

31. $h(x) = x^4 - x^2 + 1$

a. $h(2)$ **b.** $h(-1)$

c. $h(-x)$ **d.** $h(3a)$

32. $h(x) = x^3 - x + 1$

a. $h(3)$ **b.** $h(-2)$

c. $h(-x)$ **d.** $h(3a)$

33. $f(r) = \sqrt{r + 6} + 3$

a. $f(-6)$ **b.** $f(10)$ **c.** $f(x - 6)$

34. $f(r) = \sqrt{25 - r} - 6$

a. $f(16)$ **b.** $f(-24)$ **c.** $f(25 - 2x)$

35. $f(x) = \dfrac{4x^2 - 1}{x^2}$

a. $f(2)$ **b.** $f(-2)$ **c.** $f(-x)$

36. $f(x) = \dfrac{4x^3 + 1}{x^3}$

a. $f(2)$ **b.** $f(-2)$ **c.** $f(-x)$

37. $f(x) = \dfrac{x}{|x|}$

a. $f(6)$ **b.** $f(-6)$ **c.** $f(r^2)$

38. $f(x) = \dfrac{|x + 3|}{x + 3}$

a. $f(5)$ **b.** $f(-5)$ **c.** $f(-9 - x)$

In Exercises 39–50, graph the given functions, f and g, in the same rectangular coordinate system. Select integers for x, starting with −2 and ending with 2. Once you have obtained your graphs, describe how the graph of g is related to the graph of f.

39. $f(x) = x, g(x) = x + 3$

40. $f(x) = x, g(x) = x - 4$

41. $f(x) = -2x, g(x) = -2x - 1$

42. $f(x) = -2x, g(x) = -2x + 3$

43. $f(x) = x^2, g(x) = x^2 + 1$

44. $f(x) = x^2, g(x) = x^2 - 2$

45. $f(x) = |x|, g(x) = |x| - 2$

46. $f(x) = |x|, g(x) = |x| + 1$

47. $f(x) = x^3, g(x) = x^3 + 2$

48. $f(x) = x^3, g(x) = x^3 - 1$

49. $f(x) = 3, g(x) = 5$

50. $f(x) = -1, g(x) = 4$

In Exercises 51–54, graph the given square root functions, f and g, in the same rectangular coordinate system. Use the integer values of x given to the right of each function to obtain ordered pairs. Because only nonnegative numbers have square roots that are real numbers, be sure that each graph appears only for values of x that cause the expression under the radical sign to be greater than or equal to zero. Once you have obtained your graphs, describe how the graph of g is related to the graph of f.

51. $f(x) = \sqrt{x}$ $(x = 0, 1, 4, 9)$ and
$g(x) = \sqrt{x} - 1$ $(x = 0, 1, 4, 9)$

52. $f(x) = \sqrt{x}$ $(x = 0, 1, 4, 9)$ and
$g(x) = \sqrt{x} + 2$ $(x = 0, 1, 4, 9)$

53. $f(x) = \sqrt{x}$ $(x = 0, 1, 4, 9)$ and
$g(x) = \sqrt{x - 1}$ $(x = 1, 2, 5, 10)$

54. $f(x) = \sqrt{x}$ $(x = 0, 1, 4, 9)$ and
$g(x) = \sqrt{x + 2}$ $(x = -2, -1, 2, 7)$

In Exercises 55–64, use the vertical line test to identify graphs in which y is a function of x.

55.

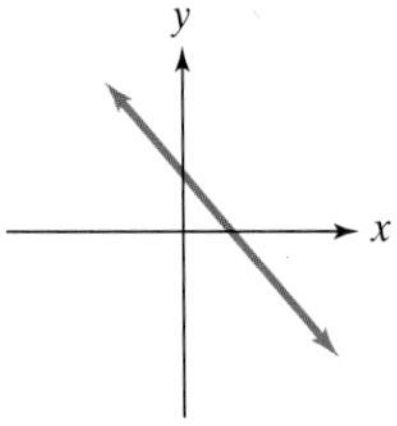

56.

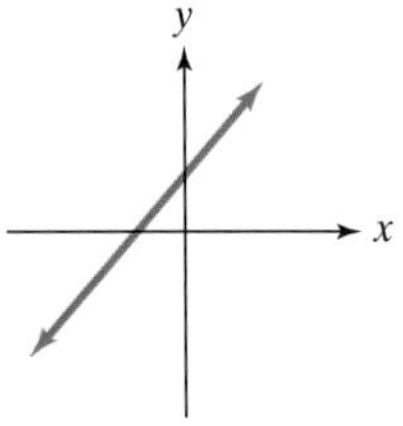

57.

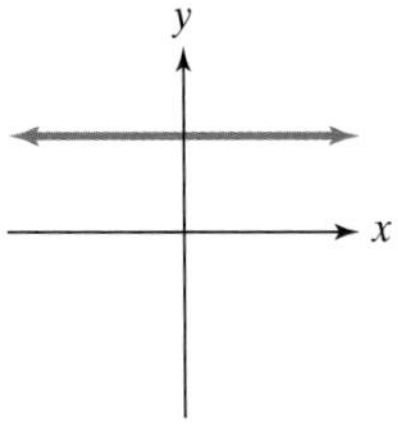

58.

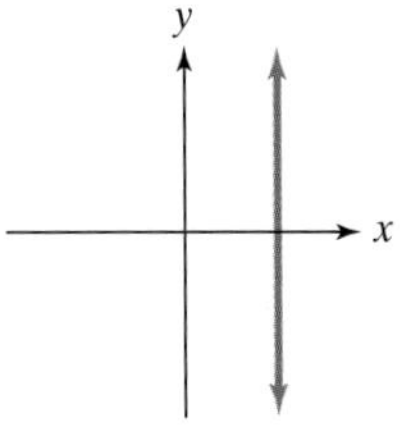

59.

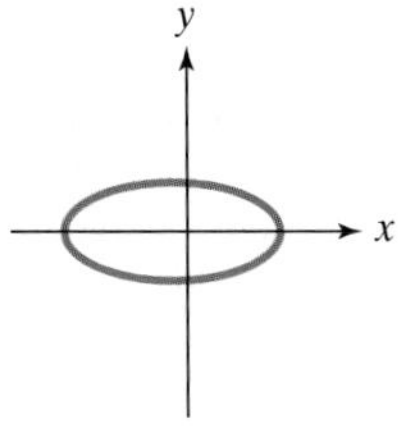

60.

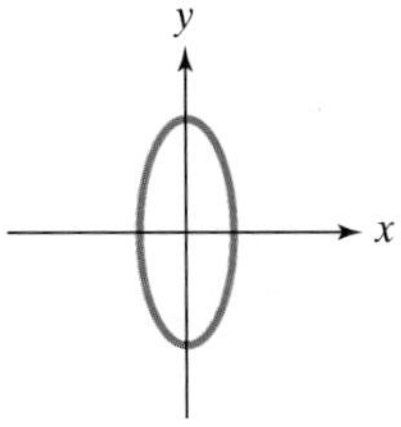

61.

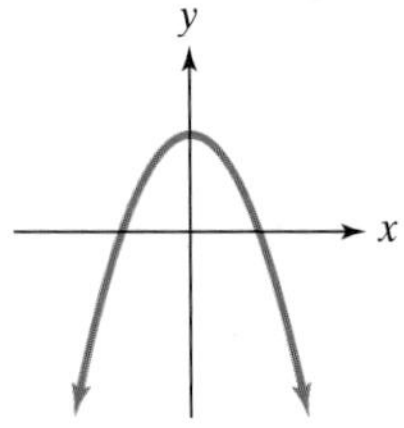

62.

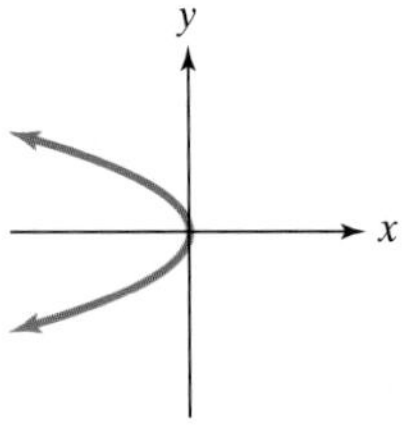

63.

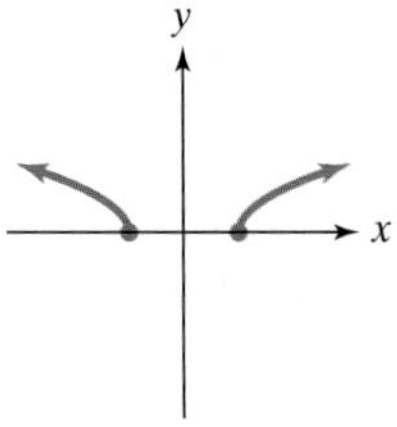

64.

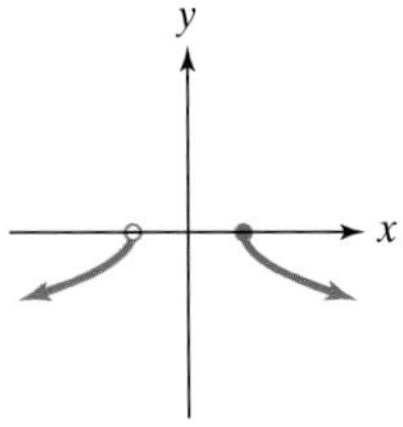

In Exercises 65–70, use the graph of f to find each indicated function value.

65. $f(-2)$

66. $f(2)$

67. $f(4)$

68. $f(-4)$

69. $f(-3)$

70. $f(-1)$

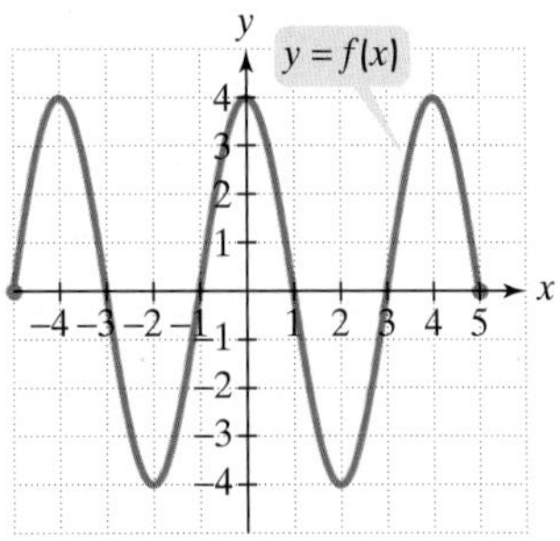

Use the graph of g to solve Exercises 71–76.

71. Find $g(-4)$.

72. Find $g(2)$.

73. Find $g(-10)$.

74. Find $g(10)$.

75. For what value of x is $g(x) = 1$?

76. For what value of x is $g(x) = -1$?

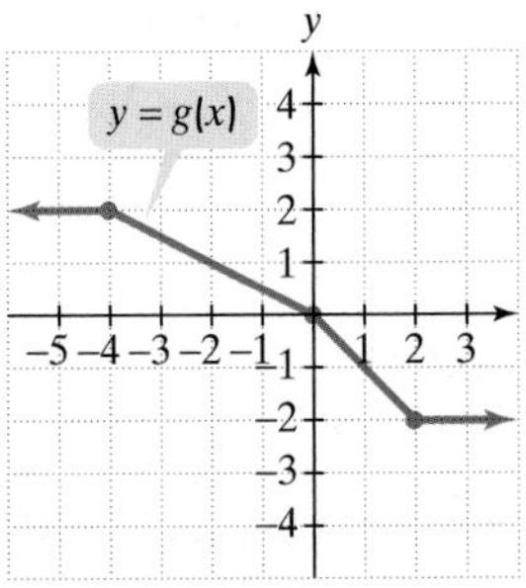

In Exercises 77–92, use the graph to determine **a.** *the function's domain;* **b.** *the function's range;* **c.** *the x-intercepts, if any;* **d.** *the y-intercept, if any; and* **e.** *the function values indicated below the graphs.*

77.

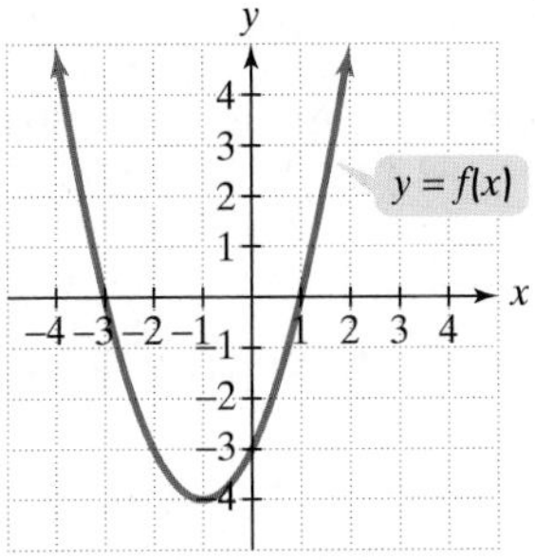

$f(-2) = ?$ $f(2) = ?$

78.

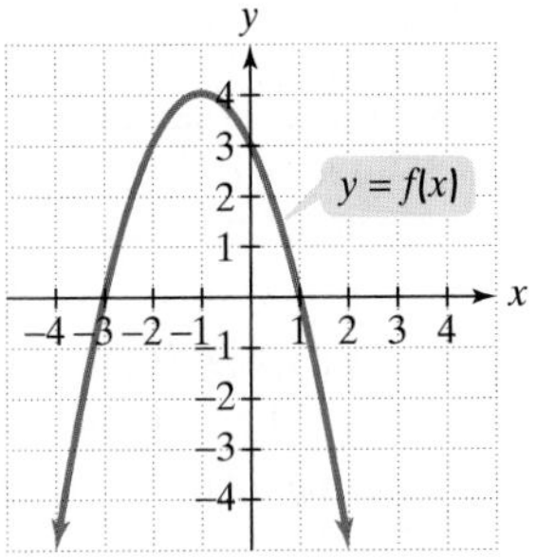

$f(-2) = ?$ $f(2) = ?$

79.

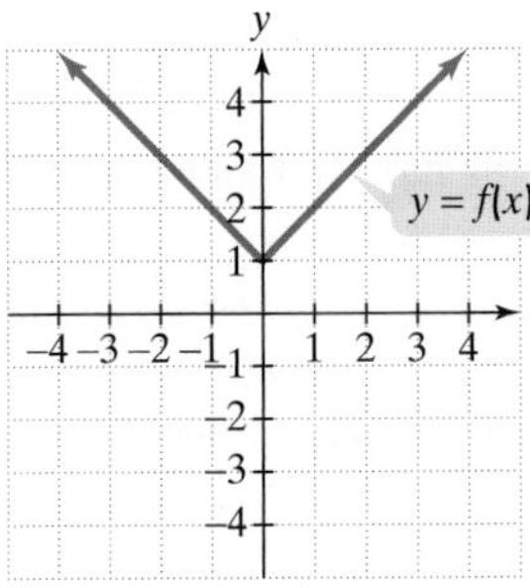

$f(-1) = ?$ $f(3) = ?$

80.

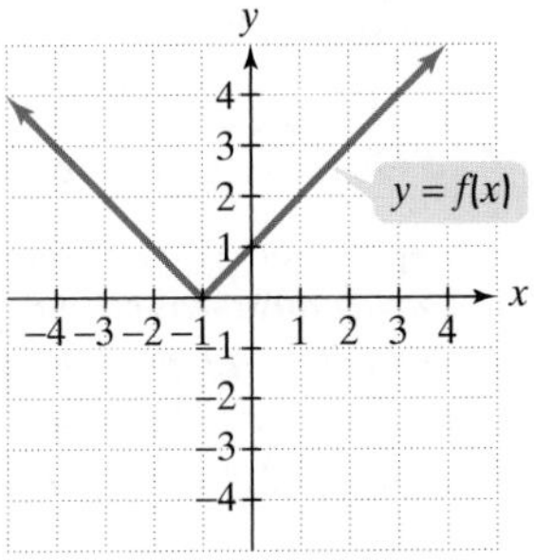

$f(-4) = ?$ $f(3) = ?$

81.

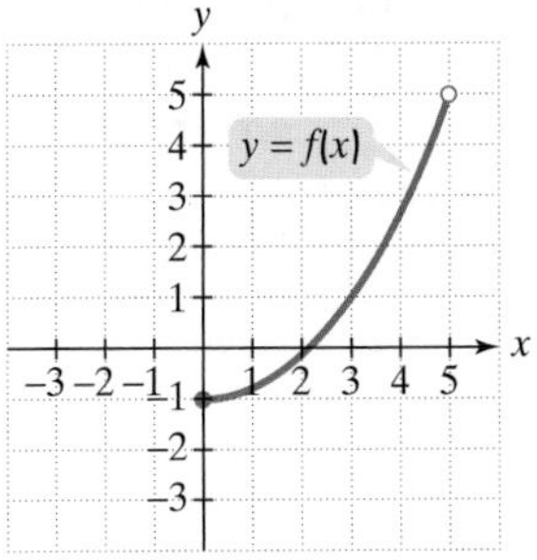

$f(3) = ?$

82.

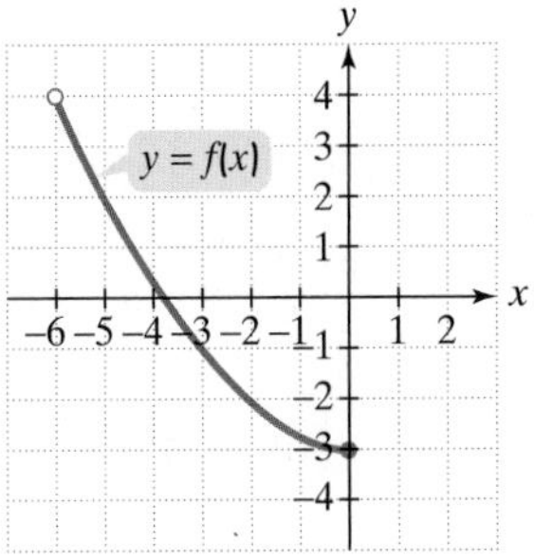

$f(-5) = ?$

83.

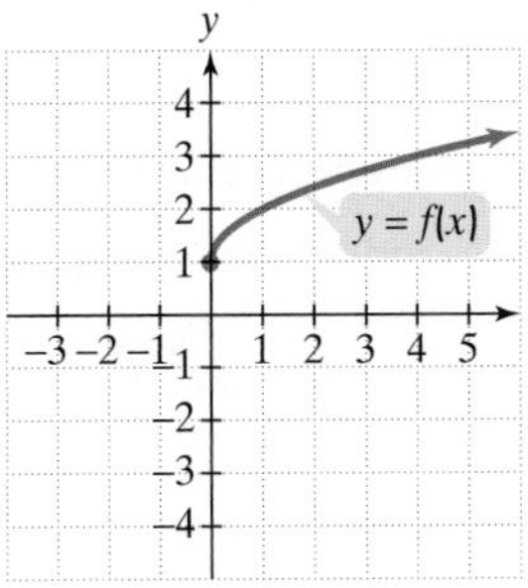

$f(4) = ?$

84.

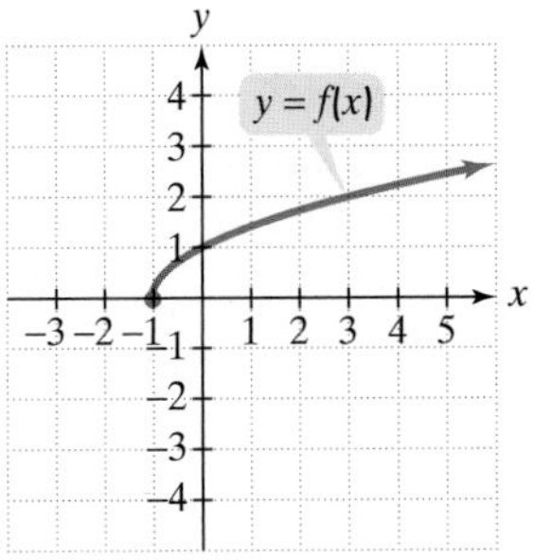

$f(3) = ?$

85.

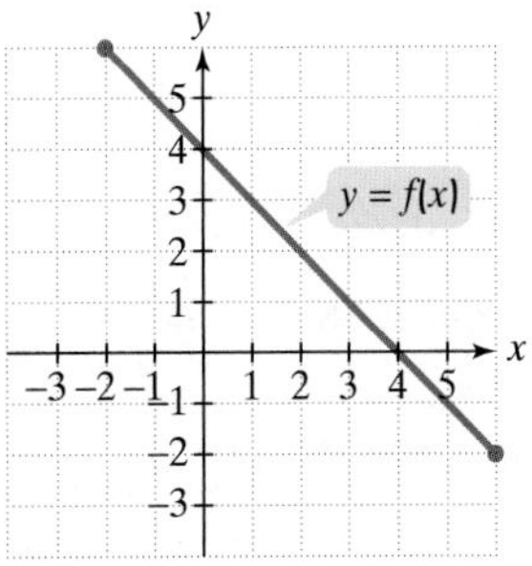

$f(-1) = ?$

86.

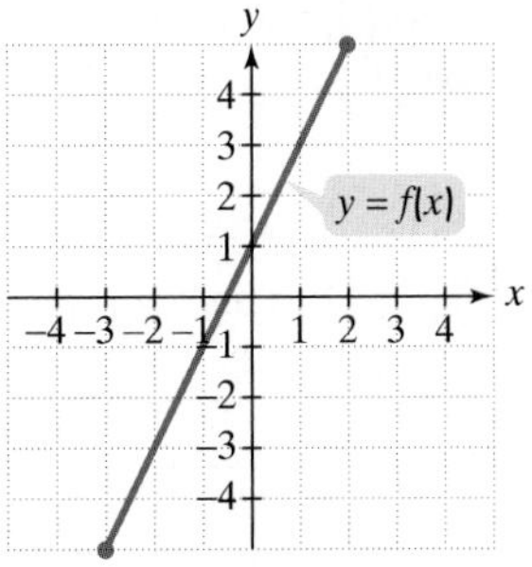

$f(-2) = ?$

87.

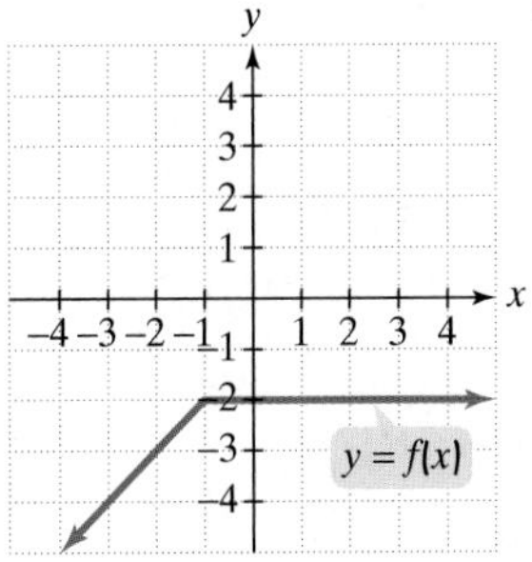

$f(-4) = ?$ $f(4) = ?$

88.

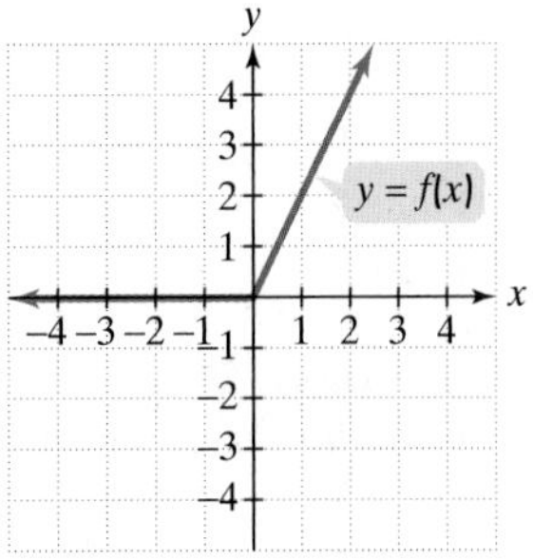

$f(-2) = ?$ $f(2) = ?$

89.

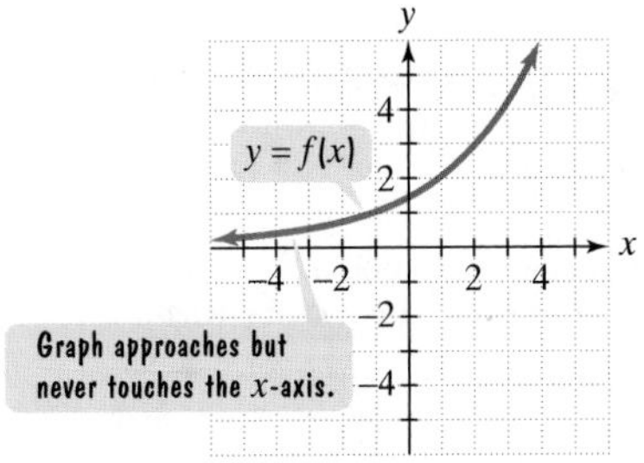

$f(4) = ?$

90.

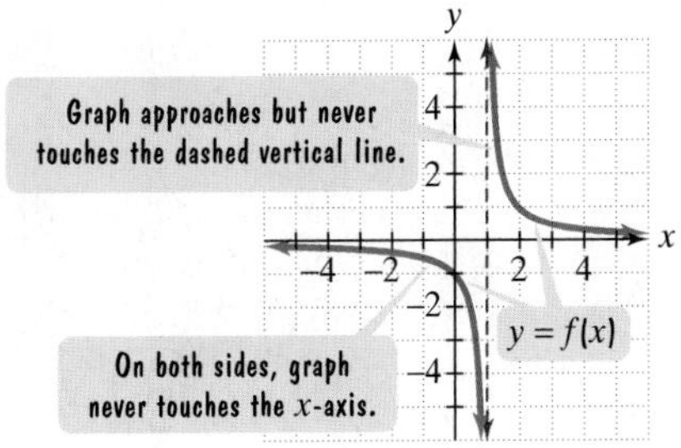

$f(2) = ?$

91.

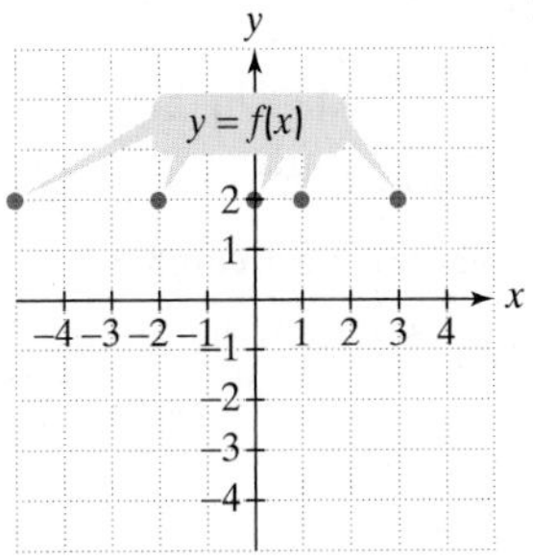

$f(-5) + f(3) = ?$

92.

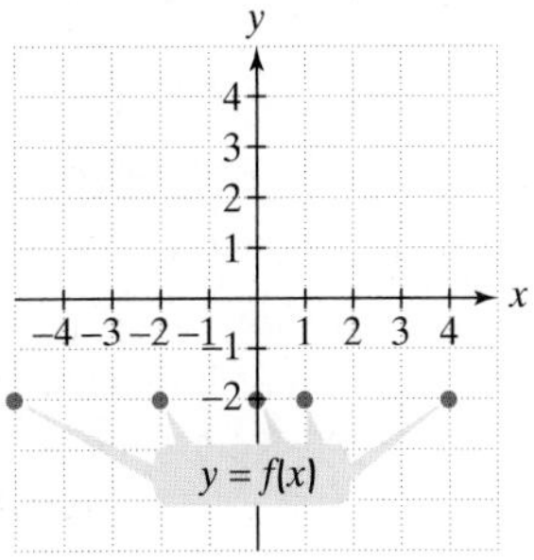

$f(-5) + f(4) = ?$

Practice Plus

In Exercises 93–94, let $f(x) = x^2 - x + 4$ *and* $g(x) = 3x - 5$.

93. Find $g(1)$ and $f(g(1))$. **94.** Find $g(-1)$ and $f(g(-1))$.

In Exercises 95–96, let f and g be defined by the following table:

x	$f(x)$	$g(x)$
−2	6	0
−1	3	4
0	−1	1
1	−4	−3
2	0	−6

95. Find $\sqrt{f(-1) - f(0)} - [g(2)]^2 + f(-2) \div g(2) \cdot g(-1)$.

96. Find $|f(1) - f(0)| - [g(1)]^2 + g(1) \div f(-1) \cdot g(2)$.

In Exercises 97–98, find $f(-x) - f(x)$ *for the given function f. Then simplify the expression.*

97. $f(x) = x^3 + x - 5$ **98.** $f(x) = x^2 - 3x + 7$

Application Exercises

99. The bar graph shows the percentage of children in the world's leading industrial countries who daydream about being rich.

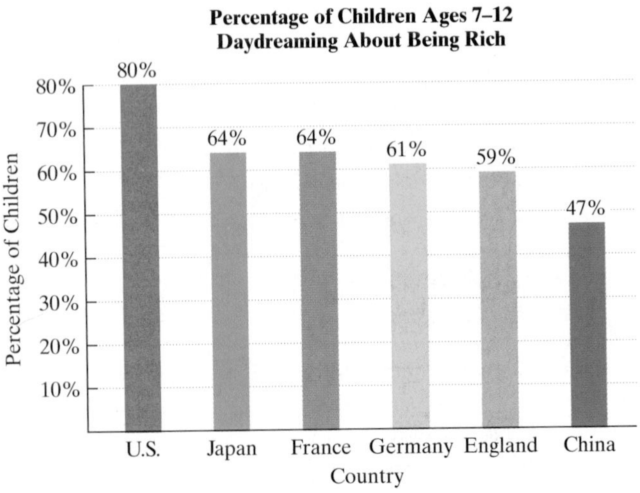

Source: Roper Starch Worldwide for A.B.C. Research

a. Write a set of six ordered pairs in which countries correspond to the percentage of children daydreaming about being rich. Each ordered pair should be in the form

(country, percent).

b. Is the relation in part (a) a function? Explain your answer.

c. Write a set of six ordered pairs in which the percentage of children daydreaming about being rich corresponds to countries. Each ordered pair should be in the form

(percent, country).

d. Is the relation in part (c) a function? Explain your answer.

100. The bar graph shows the breakdown of political ideologies in the United States.

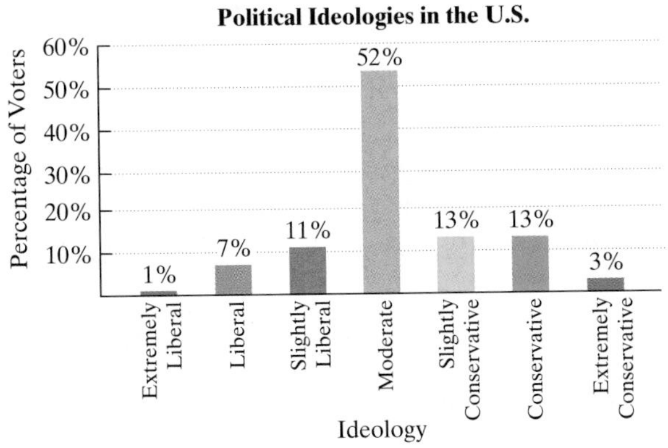

Source: Center for Political Studies, University of Michigan

a. Write a set of seven ordered pairs in which political ideologies correspond to percentages. Each ordered pair should be in the form

(ideology, percent).

Use EL, L, SL, M, SC, C, and EC to represent the respective ideologies from left to right.

b. Is the relation in part (a) a function? Explain your answer.

c. Write a set of seven ordered pairs in which percentages correspond to political ideologies. Each ordered pair should be in the form

(percent, ideology).

d. Is the relation in part (c) a function? Explain your answer.

The male minority? The graphs show enrollment in U.S. colleges, with projections through 2009. The trend indicated by the graphs is among the hottest topics of debate among college-admissions officers. Some private liberal arts colleges have quietly begun special efforts to recruit men—including admissions preferences for them.

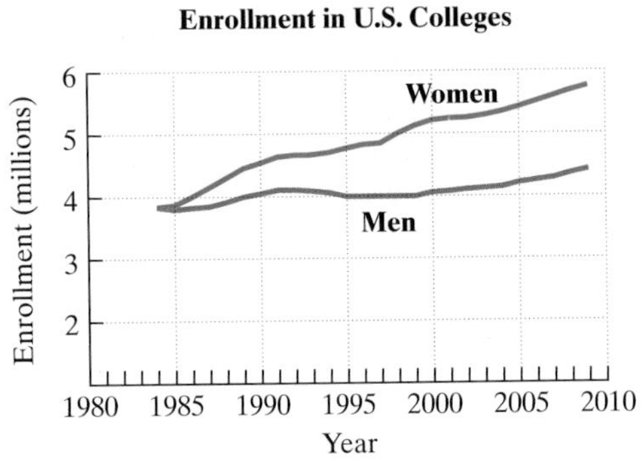

Source: Department of Education

The function

$$W(x) = 0.07x + 4.1$$

models the number of women, $W(x)$, in millions, enrolled in U.S. colleges x years after 1984. The function

$$M(x) = 0.01x + 3.9$$

models the number of men, $M(x)$, in millions, enrolled in U.S. colleges x years after 1984. Use these functions to solve Exercises 101–104.

101. Find and interpret $W(16)$. Identify this information as a point on the graph for women.

102. Find and interpret $M(16)$. Identify this information as a point on the graph for men.

103. Find and interpret $W(20) - M(20)$.

104. Find and interpret $W(25) - M(25)$.

The wage gap is used to compare the status of women's earnings relative to men's. The wage gap is expressed as a percent and is calculated by dividing the median, or middlemost, annual earnings for women by the median annual earnings for men. The line graph shows the wage gap for selected years from 1960 through 2003.

Median Women's Earnings as a Percentage of Median Men's Earnings in the U.S.

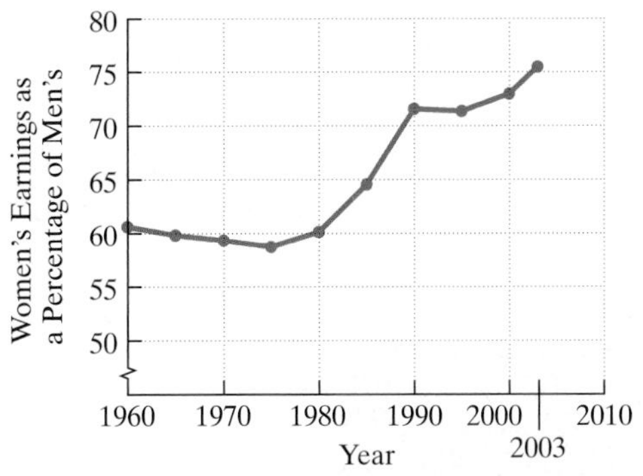

Source: U.S. Women's Bureau

The function

$$P(x) = 0.012x^2 - 0.16x + 60$$

models median women's earnings as a percentage of median men's earnings, $P(x)$, x years after 1960. Use the graph and this function to solve Exercises 105–106.

105. a. Use the graph to estimate, to the nearest percent, women's earnings as a percentage of men's in 2000.

b. Use the function to find women's earnings as a percentage of men's in 2000.

c. In 2000, median annual earnings for U.S. women and men were \$27,355 and \$37,339, respectively. What were women's earnings as a percentage of men's? Use a calculator and round to the nearest tenth of a percent. How well do your answers in parts (a) and (b) model the actual data?

106. a. Use the graph to estimate, to the nearest percent, women's earnings as a percentage of men's in 2003.

b. Use the function to find women's earnings as a percentage of men's in 2003. Round to the nearest tenth of a percent.

c. In 2003, median annual earnings for U.S. women and men were \$30,724 and \$40,668, respectively. What were women's earnings as a percentage of men's? Use a calculator and round to the nearest tenth of a percent. How well do your answers in parts (a) and (b) model the actual data?

In Exercises 107–110, you will be developing functions that model given conditions.

107. A company that manufactures bicycles has a fixed cost of \$100,000. It costs \$100 to produce each bicycle. The total cost for the company is the sum of its fixed cost and variable costs. Write the total cost, C, as a function of the number of bicycles produced, x. Then find and interpret $C(90)$.

108. A car was purchased for \$22,500. The value of the car decreases by \$3200 per year for the first six years. Write a function that describes the value of the car, V, after x years, where $0 \leq x \leq 7$. Then find and interpret $V(3)$.

109. You commute to work a distance of 40 miles and return on the same route at the end of the day. Your average rate on the return trip is 30 miles per hour faster than your average rate on the outgoing trip. Write the total time, T, in hours, devoted to your outgoing and return trips as a function of your rate on the outgoing trip, x. Then find and interpret $T(30)$. Hint:

$$\text{Time traveled} = \frac{\text{Distance traveled}}{\text{Rate of travel}}.$$

110. A chemist working on a flu vaccine needs to mix a 10% sodium-iodine solution with a 60% sodium-iodine solution to obtain a 50-milliliter mixture. Write the amount of sodium iodine in the mixture, S, in milliliters, as a function of the number of milliliters of the 10% solution used, x. Then find and interpret $S(30)$.

Writing in Mathematics

111. What is a relation? Describe what is meant by its domain and its range.

112. Explain how to determine whether a relation is a function. What is a function?

113. How do you determine if an equation in x and y defines y as a function of x?

114. Does $f(x)$ mean f times x when referring to a function f? If not, what does $f(x)$ mean? Provide an example with your explanation.

115. What is the graph of a function?

116. Explain how the vertical line test is used to determine whether a graph represents a function.

117. Explain how to identify the domain and range of a function from its graph.

118. For people filing a single return, federal income tax is a function of adjusted gross income because for each value of adjusted gross income there is a specific tax to be paid. By contrast, the price of a house is not a function of the lot size on which the house sits because houses on same-sized lots can sell for many different prices.

a. Describe an everyday situation between variables that is a function.

b. Describe an everyday situation between variables that is not a function.

119. Do you believe that the trend shown by the graphs for Exercises 101–104 should be reversed by providing admissions preferences for men? Explain your position on this issue.

Technology Exercise

120. Use a graphing utility to verify any five pairs of graphs that you drew by hand in Exercises 39–54.

Critical Thinking Exercises

121. Which one of the following is true based on the graph of f in the figure?

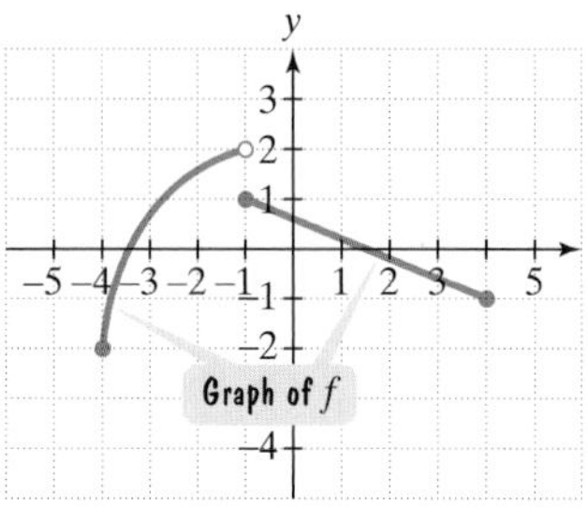

a. The domain of f is $[-4, 1) \cup (1, 4]$.

b. The range of f is $[-2, 2]$.

c. $f(-1) - f(4) = 2$

d. $f(0) = 2.1$

122. If $f(x) = 3x + 7$, find $\dfrac{f(a + h) - f(a)}{h}$.

123. Give an example of a relation with the following characteristics: The relation is a function containing two ordered pairs. Reversing the components in each ordered pair results in a relation that is not a function.

124. If $f(x + y) = f(x) + f(y)$ and $f(1) = 3$, find $f(2)$, $f(3)$, and $f(4)$. Is $f(x + y) = f(x) + f(y)$ for all functions?

SECTION 2.2 More on Functions and Their Graphs

Objectives

1. Find and simplify a function's difference quotient.
2. Understand and use piecewise functions.
3. Identify intervals on which a function increases, decreases, or is constant.
4. Use graphs to locate relative maxima or minima.
5. Identify even or odd functions and recognize their symmetries.
6. Graph step functions.

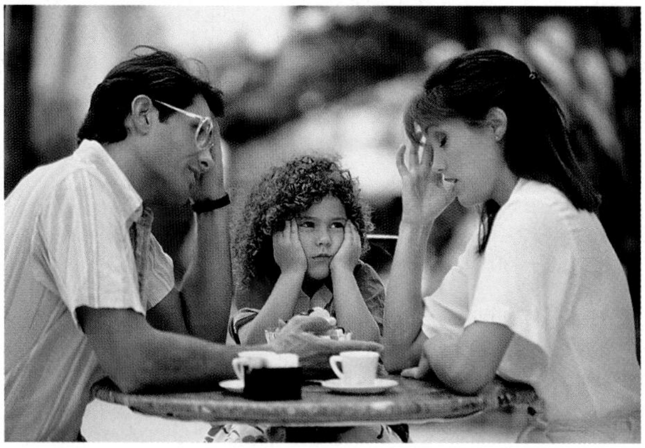

"Our relationship is just going through a phase." When that phase reaches a point where married couples cannot agree on anything else, they agree on a divorce, frequently arranged so that lawyers can live happily ever after. The graph in Figure 2.12 shows the percent distribution of divorces in the United States by number of years of marriage.

Percent Distribution of Divorces by Number of Years of Marriage

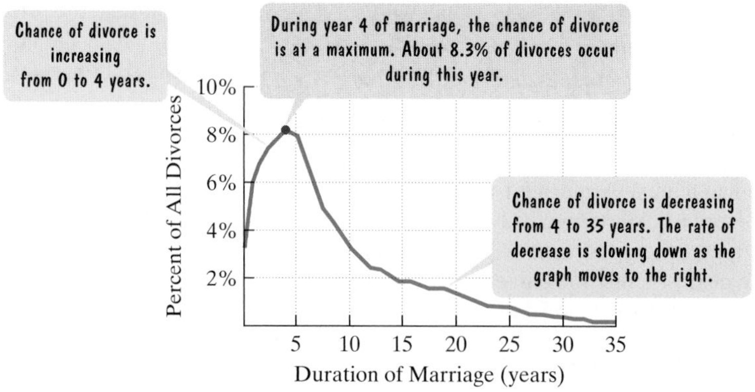

Figure 2.12
Source: Divorce Center

You are probably familiar with the words and phrases used to describe the graph in Figure 2.12:

increasing   decreasing   maximum   slowing rate of decrease

In this section, you will enhance your intuitive understanding of ways of describing graphs by viewing these descriptions from the perspective of functions.

1 Find and simplify a function's difference quotient.

Functions and Difference Quotients

In the next section, we will be studying the average rate of change of a function. A ratio, called the *difference quotient*, plays an important role in understanding the rate at which functions change.

Definition of a Difference Quotient

The expression

$$\frac{f(x+h)-f(x)}{h}$$

for $h \neq 0$ is called the **difference quotient**.

EXAMPLE 1 Evaluating and Simplifying a Difference Quotient

If $f(x) = 2x^2 - x + 3$, find and simplify each expression:

a. $f(x+h)$ **b.** $\frac{f(x+h)-f(x)}{h}, h \neq 0.$

Solution

a. We find $f(x+h)$ by replacing x with $x+h$ each time that x appears in the equation.

$$f(x) = 2x^2 \quad - \quad x \quad + \quad 3$$

Replace x with $x+h$. Replace x with $x+h$. Replace x with $x+h$. Copy the 3. There is no x in this term.

$$\begin{aligned} f(x+h) &= 2(x+h)^2 - (x+h) + 3 \\ &= 2(x^2 + 2xh + h^2) - x - h + 3 \\ &= 2x^2 + 4xh + 2h^2 - x - h + 3 \end{aligned}$$

b. Using our result from part (a), we obtain the following:

This is $f(x+h)$ from part (a). This is $f(x)$ from the given equation.

$$\frac{f(x+h)-f(x)}{h} = \frac{\boxed{2x^2 + 4xh + 2h^2 - x - h + 3} - (2x^2 - x + 3)}{h}$$

$$= \frac{2x^2 + 4xh + 2h^2 - x - h + 3 - 2x^2 + x - 3}{h}$$ Remove parentheses and change the sign of each term in the parentheses.

$$= \frac{(2x^2 - 2x^2) + (-x + x) + (3 - 3) + 4xh + 2h^2 - h}{h}$$ Group like terms.

$$= \frac{4xh + 2h^2 - 1h}{h}$$ Simplify.

We wrote $-h$ as $-1h$ to avoid possible errors in the next factoring step.

$$= \frac{h(4x + 2h - 1)}{h}$$ Factor h from the numerator.

$$= 4x + 2h - 1$$ Divide out identical factors of h in the numerator and denominator.

Check Point 1 If $f(x) = -2x^2 + x + 5$, find and simplify each expression:

a. $f(x+h)$ **b.** $\frac{f(x+h)-f(x)}{h}, h \neq 0.$

2 Understand and use piecewise functions.

Piecewise Functions

A cellular phone company offers the following plan:

- \$20 per month buys 60 minutes.
- Additional time costs \$0.40 per minute.

We can represent this plan mathematically by writing the total monthly cost, C, as a function of the number of calling minutes, t.

$$C(t) = \begin{cases} 20 & \text{if } 0 \le t \le 60 \\ 20 + 0.40(t - 60) & \text{if } t > 60 \end{cases}$$

The cost is \$20 for up to and including 60 calling minutes.

The cost is \$20 plus \$0.40 per minute for additional time for more than 60 calling minutes.

\$20 for first 60 minutes

\$0.40 per minute

times the number of calling minutes exceeding 60

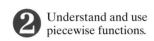

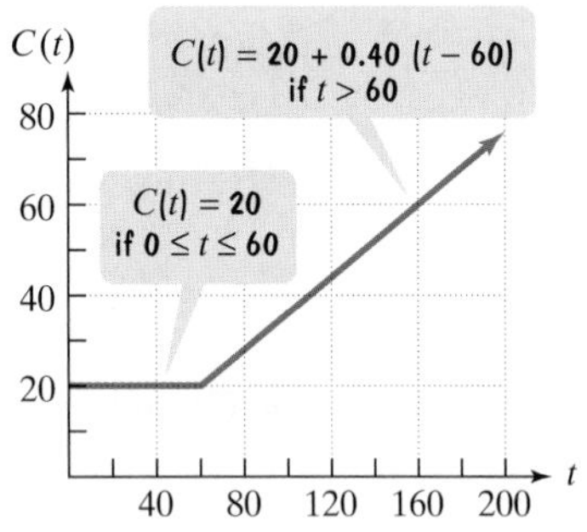

Figure 2.13

A function that is defined by two (or more) equations over a specified domain is called a **piecewise function**. Many cellular phone plans can be represented with piecewise functions. The graph of the piecewise function described above is shown in Figure 2.13.

EXAMPLE 2 Evaluating a Piecewise Function

Use the function that describes the cellular phone plan

$$C(t) = \begin{cases} 20 & \text{if } 0 \le t \le 60 \\ 20 + 0.40(t - 60) & \text{if } t > 60 \end{cases}$$

to find and interpret each of the following:

a. $C(30)$ **b.** $C(100)$.

Solution

a. To find $C(30)$, we let $t = 30$. Because 30 lies between 0 and 60, we use the first line of the piecewise function.

$C(t) = 20$ This is the function's equation for $0 \le t \le 60$.

$C(30) = 20$ Replace t with 30. Regardless of this function's input, the constant output is 20.

This means that with 30 calling minutes, the monthly cost is \$20. This can be visually represented by the point $(30, 20)$ on the first piece of the graph in Figure 2.13.

b. To find $C(100)$, we let $t = 100$. Because 100 is greater than 60, we use the second line of the piecewise function.

$C(t) = 20 + 0.40(t - 60)$ This is the function's equation for $t > 60$.

$C(100) = 20 + 0.40(100 - 60)$ Replace t with 100.

$= 20 + 0.40(40)$ Subtract within parentheses: $100 - 60 = 40$.

$= 20 + 16$ Multiply: $0.40(40) = 16$.

$= 36$ Add: $20 + 16 = 36$.

Thus, $C(100) = 36$. This means that with 100 calling minutes, the monthly cost is \$36. This can be visually represented by the point $(100, 36)$ on the second piece of the graph in Figure 2.13.

Check Point 2 Use the function in Example 2 to find and interpret each of the following:

a. $C(40)$ **b.** $C(80)$.

Identify solutions on the graph in Figure 2.13.

3 Identify intervals on which a function increases, decreases, or is constant.

Increasing and Decreasing Functions

Too late for that flu shot now! It's only 8 A.M. and you're feeling lousy. Your temperature is 101°F. Fascinated by the way that algebra models the world (your author is projecting a bit here), you decide to construct graphs showing your body temperature as a function of the time of day. You decide to let x represent the number of hours after 8 A.M. and $f(x)$ your temperature at time x.

At 8 A.M. your temperature is 101°F and you are not feeling well. However, your temperature starts to decrease. It reaches normal (98.6°F) by 11 A.M. Feeling energized, you construct the graph shown on the right, indicating decreasing temperature for $\{x|0 < x < 3\}$, or on the interval $(0, 3)$.

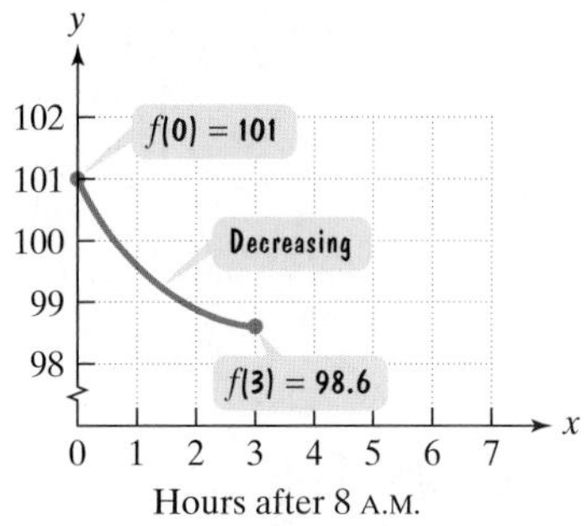

Temperature decreases on (0, 3), reaching 98.6° by 11 A.M.

Did creating that first graph drain you of your energy? Your temperature starts to rise after 11 A.M. By 1 P.M., 5 hours after 8 A.M., your temperature reaches 100°F. However, you keep plotting points on your graph. At the right, we can see that your temperature increases for $\{x|3 < x < 5\}$, or on the interval $(3, 5)$.

The graph of f is decreasing to the left of $x = 3$ and increasing to the right of $x = 3$. Thus, your temperature 3 hours after 8 A.M. was at its lowest point. Your relative minimum temperature was 98.6°.

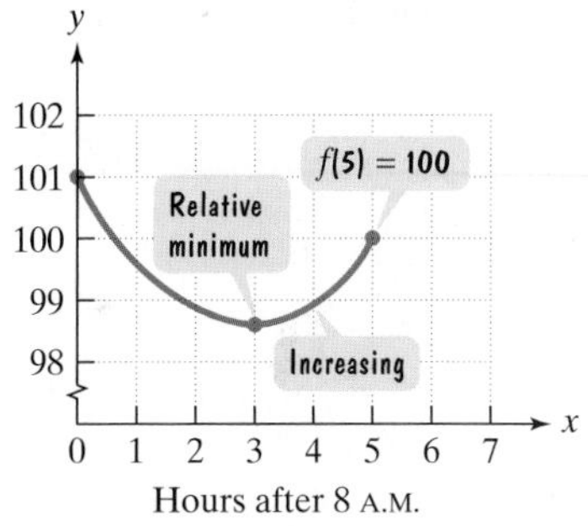

Temperature increases on (3, 5).

By 3 P.M., your temperature is no worse than it was at 1 P.M.: It is still 100°F. (Of course, it's no better, either.) Your temperature remained the same, or constant, for $\{x|5 < x < 7\}$, or on the interval $(5, 7)$.

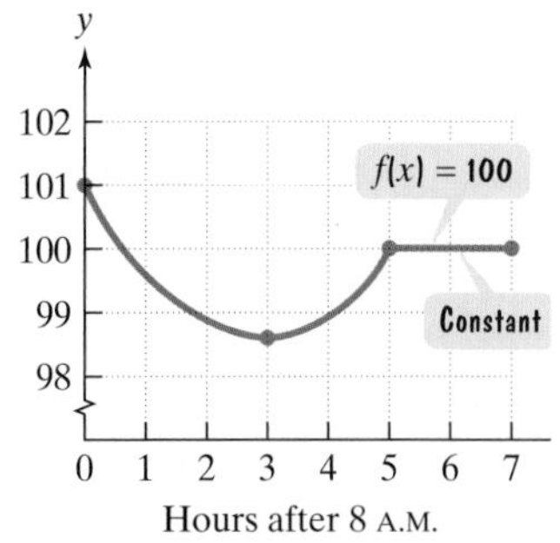

The time-temperature flu scenario illustrates that a function f is increasing when its graph rises from left to right, decreasing when its graph falls from left to right, and remains constant when it neither rises nor falls. Let's now provide a more precise algebraic description for these intuitive concepts.

Increasing, Decreasing, and Constant Functions

1. A function is **increasing** on an open interval, I, if for any x_1 and x_2 in the interval, where $x_1 < x_2$, then $f(x_1) < f(x_2)$.
2. A function is **decreasing** on an open interval, I, if for any x_1 and x_2 in the interval, where $x_1 < x_2$, then $f(x_1) > f(x_2)$.
3. A function is **constant** on an open interval, I, if for any x_1 and x_2 in the interval, where $x_1 < x_2$, then $f(x_1) = f(x_2)$.

Increasing

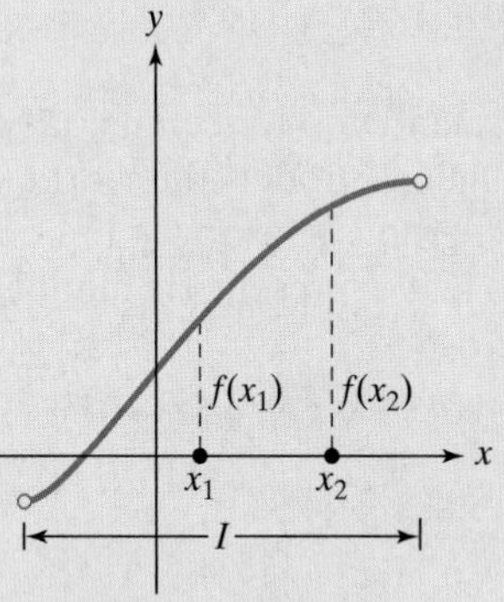

(1) For $x_1 < x_2$ in I,
$f(x_1) < f(x_2)$;
f is increasing on I.

Decreasing

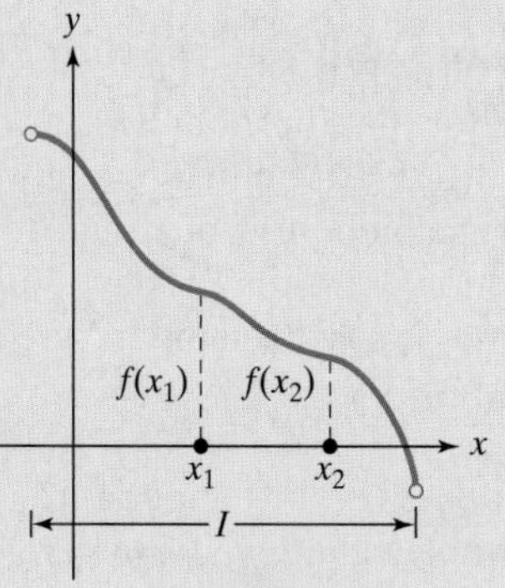

(2) For $x_1 < x_2$ in I,
$f(x_1) > f(x_2)$;
f is decreasing on I.

Constant

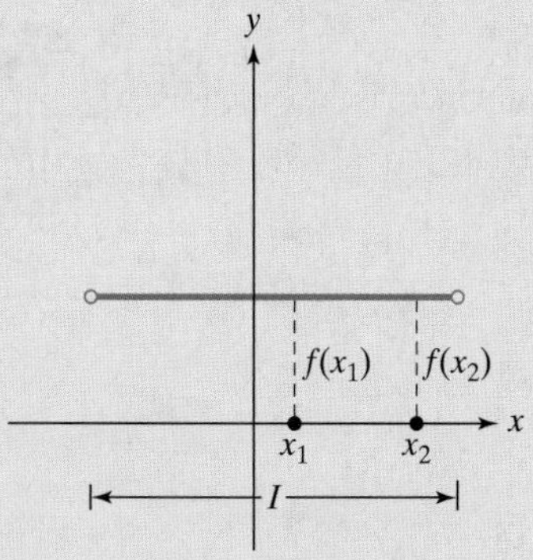

(3) For $x_1 < x_2$ in I,
$f(x_1) = f(x_2)$;
f is constant on I.

Study Tip

The open intervals describing where functions increase, decrease, or are constant, use x-coordinates and not the y-coordinates.

EXAMPLE 3 Intervals on Which a Function Increases, Decreases, or Is Constant

State the intervals on which each given function is increasing, decreasing, or constant.

a.

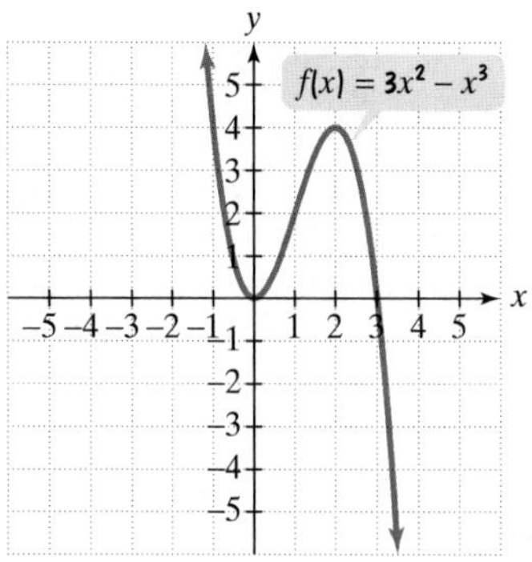

b.

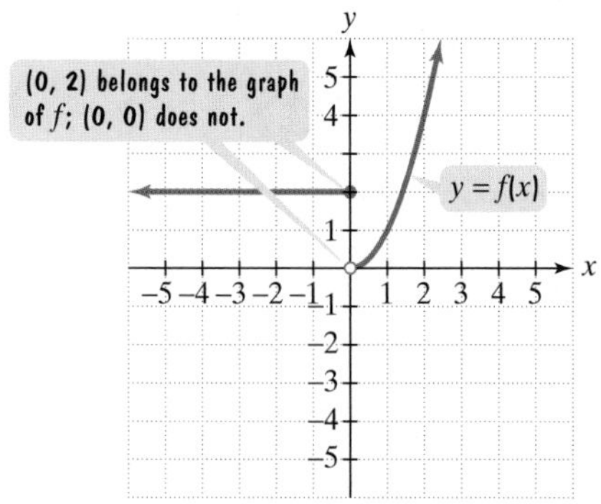

Solution

a. The function is decreasing on the interval $(-\infty, 0)$, increasing on the interval $(0, 2)$, and decreasing on the interval $(2, \infty)$.

b. Although the function's equations are not given, the graph indicates that the function is defined in two pieces. The part of the graph to the left of the y-axis shows that the function is constant on the interval $(-\infty, 0)$. The part to the right of the y-axis shows that the function is increasing on the interval $(0, \infty)$.

Check Point 3 State the intervals on which the given function is increasing, decreasing, or constant.

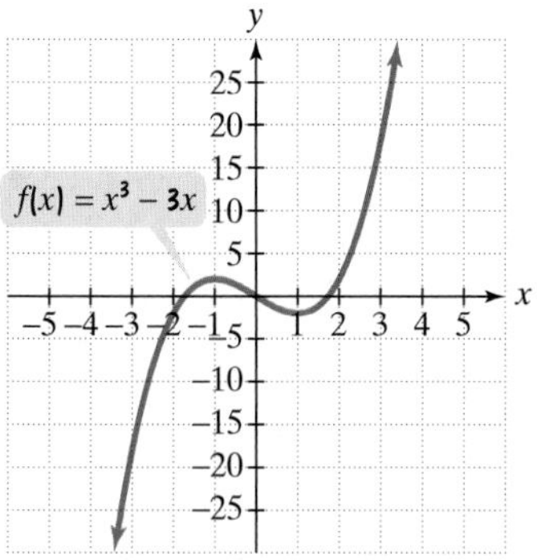

4 Use graphs to locate relative maxima or minima.

Relative Maxima and Relative Minima

The points at which a function changes its increasing or decreasing behavior can be used to find the *relative maximum* or *relative minimum* values of the function. For example, consider the function with which we opened the previous section:

$$f(x) = 0.0075x^2 - 0.2672x + 14.8.$$

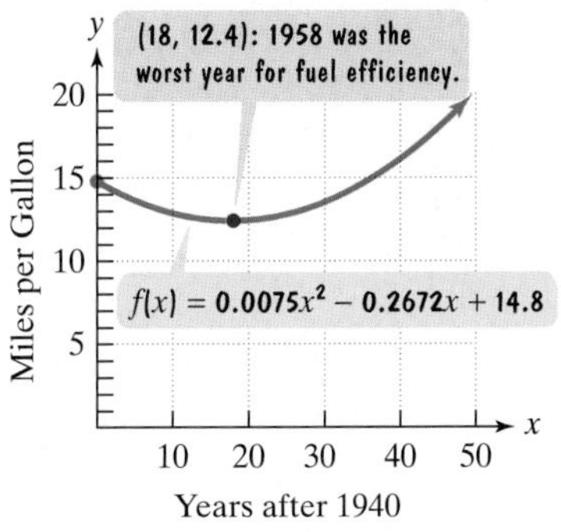

Figure 2.14 Fuel efficiency of U.S. automobiles over time

Recall that the function models the average number of miles per gallon of U.S. automobiles, $f(x)$, x years after 1940. The graph of this function is shown as a continuous curve in Figure 2.14. (It can also be shown as a series of points, each point representing a year and miles per gallon for that year.)

The graph of f is decreasing to the left of $x = 18$ and increasing to the right of $x = 18$. Thus, 18 years after 1940, in 1958, fuel efficiency was at a minimum. We say that the relative minimum fuel efficiency is $f(18)$, or approximately 12.4 miles per gallon. Mathematicians use the word "relative" to suggest that relative to an open interval about 18, the value $f(18)$ is smallest.

Definitions of Relative Maximum and Relative Minimum

1. A function value $f(a)$ is a **relative maximum** of f if there exists an open interval about a such that $f(a) > f(x)$ for all x in the open interval.
2. A function value $f(b)$ is a **relative minimum** of f if there exists an open interval about b such that $f(b) < f(x)$ for all x in the open interval.

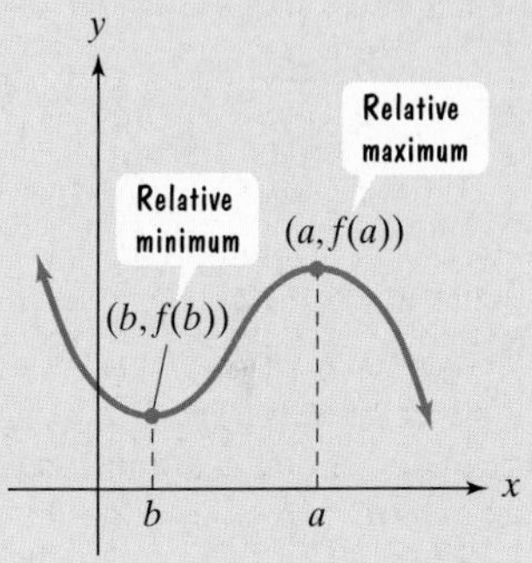

Study Tip

The word *local* is sometimes used instead of *relative* when describing maxima or minima. If f has a relative, or local, maximum at a, $f(a)$ is greater than the values of f near a. If f has a relative, or local, minimum at b, $f(b)$ is less than the values of f near b.

If the graph of a function is given, we can often visually locate the number(s) at which the function has a relative maximum or a relative minimum. For example, the graph of f in Figure 2.15 shows that

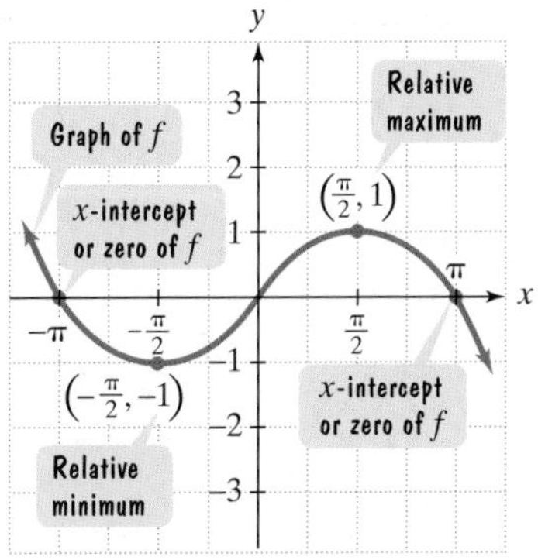

Figure 2.15 Using a graph to locate where f has a relative maximum or minimum

- f has a relative maximum at $\frac{\pi}{2}$.

  The relative maximum is $f\left(\frac{\pi}{2}\right) = 1$.

- f has a relative minimum at $-\frac{\pi}{2}$.

  The relative minimum is $f\left(-\frac{\pi}{2}\right) = -1$.

Notice that f does not have a relative maximum or minimum at $-\pi$ and π, the x-intercepts, or zeros, of the function.

5 Identify even or odd functions and recognize their symmetries.

Even and Odd Functions and Symmetry

Is beauty in the eye of the beholder? Or are there certain objects (or people) that are so well balanced and proportioned that they are universally pleasing to the eye? What constitutes an attractive human face? In Figure 2.16, we've drawn lines between paired features and marked the midpoints. Notice how the features line up almost perfectly. Each half of the face is a mirror image of the other half through the white vertical line.

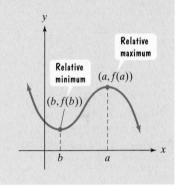

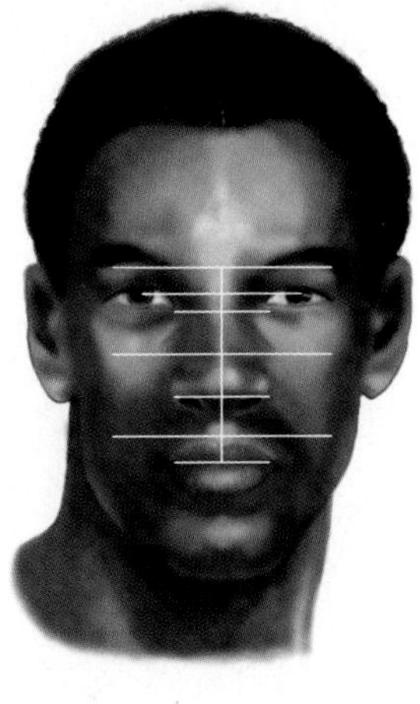

Figure 2.16 To most people, an attractive face is one in which each half is an almost perfect mirror image of the other half.

Did you know that graphs of some equations exhibit exactly the kind of symmetry shown by the attractive face in Figure 2.16? The word *symmetry* comes from the Greek *symmetria*, meaning "the same measure." We can identify graphs with symmetry by looking at a function's equation and determining if the function is *even* or *odd*.

Definition of Even and Odd Functions

The function f is an **even function** if

$$f(-x) = f(x) \quad \text{for all } x \text{ in the domain of } f.$$

The right side of the equation of an even function does not change if x is replaced with $-x$.

The function f is an **odd function** if

$$f(-x) = -f(x) \quad \text{for all } x \text{ in the domain of } f.$$

Every term in the right side of the equation of an odd function changes its sign if x is replaced with $-x$.

EXAMPLE 4 Identifying Even or Odd Functions

Determine whether each of the following functions is even, odd, or neither:

a. $f(x) = x^3 - 6x$ **b.** $g(x) = x^4 - 2x^2$ **c.** $h(x) = x^2 + 2x + 1.$

Solution In each case, replace x with $-x$ and simplify. If the right side of the equation stays the same, the function is even. If every term on the right changes sign, the function is odd.

a. We use the given function's equation, $f(x) = x^3 - 6x$, to find $f(-x)$.

Use $f(x) = x^3 - 6x$.

Replace x with $-x$.

$$f(-x) = (-x)^3 - 6(-x) = (-x)(-x)(-x) - 6(-x) = -x^3 + 6x$$

There are two terms on the right side of the given equation, $f(x) = x^3 - 6x$, and each term changed its sign when we replaced x with $-x$. Because $f(-x) = -f(x)$, f is an odd function.

b. We use the given function's equation, $g(x) = x^4 - 2x^2$, to find $g(-x)$.

Use $g(x) = x^4 - 2x^2$.

Replace x with $-x$.

$$g(-x) = (-x)^4 - 2(-x)^2 = (-x)(-x)(-x)(-x) - 2(-x)(-x)$$
$$= x^4 - 2x^2$$

The right side of the equation of the given function, $g(x) = x^4 - 2x^2$, did not change when we replaced x with $-x$. Because $g(-x) = g(x)$, g is an even function.

c. We use the given function's equation, $h(x) = x^2 + 2x + 1$, to find $h(-x)$.

Use $h(x) = x^2 + 2x + 1$.

Replace x with $-x$.

$$h(-x) = (-x)^2 + 2(-x) + 1 = x^2 - 2x + 1$$

The right side of the equation of the given function, $h(x) = x^2 + 2x + 1$, changed when we replaced x with $-x$. Thus, $h(-x) \neq h(x)$, so h is not an even function. The sign of *each* of the three terms in the equation for $h(x)$ did not change when we replaced x with $-x$. Only the second term changed signs. Thus, $h(-x) \neq -h(x)$, so h is not an odd function. We conclude that h is neither an even nor an odd function.

Check Point 4 Determine whether each of the following functions is even, odd, or neither:

a. $f(x) = x^2 + 6$ **b.** $g(x) = 7x^3 - x$ **c.** $h(x) = x^5 + 1$.

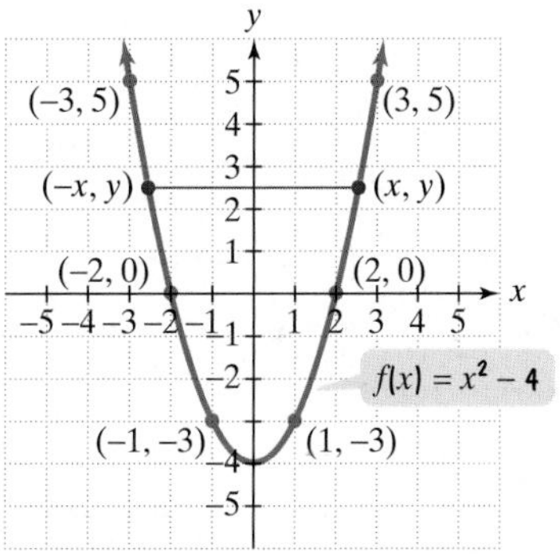

Figure 2.17 y-axis symmetry with $f(-x) = f(x)$

Now, let's see what even and odd functions tell us about a function's graph. Begin with the even function $f(x) = x^2 - 4$, shown in Figure 2.17. The function is even because

$$f(-x) = (-x)^2 - 4 = x^2 - 4 = f(x).$$

Examine the pairs of points shown, such as $(3, 5)$ and $(-3, 5)$. Notice that we obtain the same y-coordinate whenever we evaluate the function at a value of x and the value of its opposite, $-x$. Like the attractive face, each half of the graph is a mirror image of the other half through the y-axis. If we were to fold the paper along the y-axis, the two halves of the graph would coincide. This causes the graph to be *symmetric with respect to the y-axis*. A graph is **symmetric with respect to the y-axis** if, for every point (x, y) on the graph, the point $(-x, y)$ is also on the graph. All even functions have graphs with this kind of symmetry.

Even Functions and y-Axis Symmetry

The graph of an even function in which $f(-x) = f(x)$ is symmetric with respect to the y-axis.

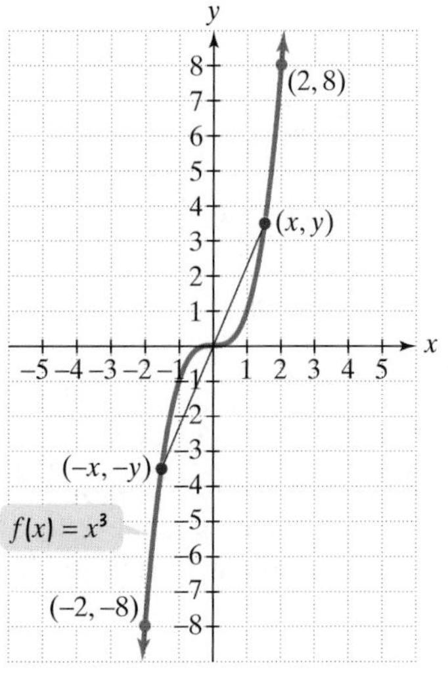

Figure 2.18 Origin symmetry with $f(-x) = -f(x)$

Now, consider the graph of the function $f(x) = x^3$, shown in Figure 2.18. The function is odd because

$$f(-x) = (-x)^3 = (-x)(-x)(-x) = -x^3 = -f(x).$$

Although the graph in Figure 2.18 is not symmetric with respect to the y-axis, it is symmetric in another way. Look at the pairs of points, such as $(2, 8)$ and $(-2, -8)$. For each point (x, y) on the graph, the point $(-x, -y)$ is also on the graph. The points $(2, 8)$ and $(-2, -8)$ are reflections of one another about the origin. This means that

- the points are the same distance from the origin, and
- the points lie on a line through the origin.

A graph is **symmetric with respect to the origin** if, for every point (x, y) on the graph, the point $(-x, -y)$ is also on the graph. Observe that the first- and third-quadrant portions of $f(x) = x^3$ are reflections of one another with respect to the origin. Notice that $f(x)$ and $f(-x)$ have opposite signs, so that $f(-x) = -f(x)$. All odd functions have graphs with origin symmetry.

Odd Functions and Origin Symmetry

The graph of an odd function in which $f(-x) = -f(x)$ is symmetric with respect to the origin.

6 Graph step functions.

Table 2.1 Cost of First-Class Mail (Effective June 30, 2002)

Weight Not Over	Cost
1 ounce	$0.37
2 ounces	0.60
3 ounces	0.83
4 ounces	1.06
5 ounces	1.29

Source: U.S. Postal Service

Step Functions

Have you ever mailed a letter that seemed heavier than usual? Perhaps you worried that the letter would not have enough postage. Costs for mailing a letter weighing up to 5 ounces are given in Table 2.1. If your letter weighs an ounce or less, the cost is $0.37. If your letter weighs 1.05 ounces, 1.50 ounces, 1.90 ounces, or 2.00 ounces, the cost "steps" to $0.60. The cost does not take on any value between $0.37 and $0.60. If your letter weighs 2.05 ounces, 2.50 ounces, 2.90 ounces, or 3 ounces, the cost "steps" to $0.83. Cost increases are $0.23 per step.

Now, let's see what the graph of the function that models this situation looks like. Let

x = the weight of the letter, in ounces, and

$y = f(x)$ = the cost of mailing a letter weighing x ounces.

The graph is shown in Figure 2.19. Notice how it consists of a series of steps that jump vertically 0.23 unit at each integer. The graph is constant between each pair of consecutive integers.

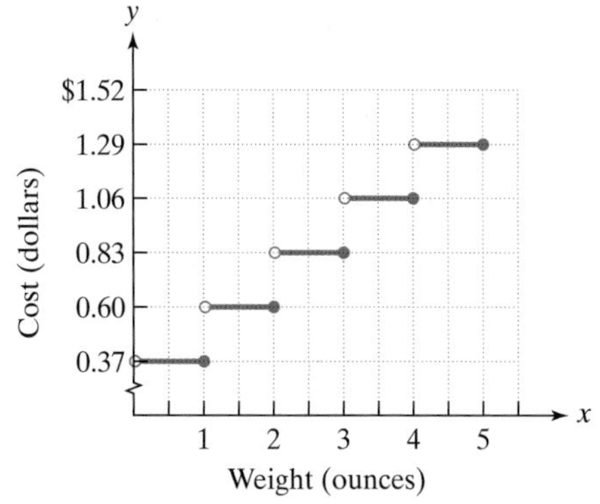

Figure 2.19

Mathematicians have defined functions that describe situations where function values graphically form discontinuous steps. One such function is called the **greatest integer function**, symbolized by $\text{int}(x)$ or $[\![x]\!]$. And what is $\text{int}(x)$?

$$\text{int}(x) = \text{the greatest integer that is less than or equal to } x$$

For example,

$$\text{int}(1) = 1, \quad \text{int}(1.3) = 1, \quad \text{int}(1.5) = 1, \quad \text{int}(1.9) = 1.$$

1 is the greatest integer that is less than or equal to 1, 1.3, 1.5, and 1.9.

Here are some additional examples:

$$\text{int}(2) = 2, \quad \text{int}(2.3) = 2, \quad \text{int}(2.5) = 2, \quad \text{int}(2.9) = 2.$$

2 is the greatest integer that is less than or equal to 2, 2.3, 2.5, and 2.9.

Notice how we jumped from 1 to 2 in the function values for $\text{int}(x)$. In particular,

$$\text{If } 1 \le x < 2, \quad \text{then} \quad \text{int}(x) = 1.$$
$$\text{If } 2 \le x < 3, \quad \text{then} \quad \text{int}(x) = 2.$$

The graph of $f(x) = \text{int}(x)$ is shown in Figure 2.20. The graph of the greatest integer function jumps vertically one unit at each integer. However, the graph is constant between each pair of consecutive integers. The rightmost horizontal step shown in the graph illustrates that

$$\text{If } 5 \le x < 6, \quad \text{then} \quad \text{int}(x) = 5.$$

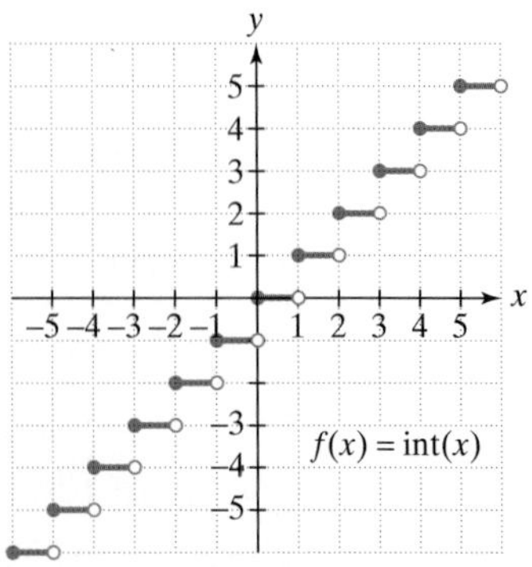

Figure 2.20 The graph of the greatest integer function

In general,

$$\text{If } n \le x < n + 1, \text{ where } n \text{ is an integer,} \quad \text{then} \quad \text{int}(x) = n.$$

By contrast to the graph for the cost of first-class mail, the graph of the greatest integer function includes the point on the left of each horizontal step, but does not include the point on the right. The domain of $f(x) = \text{int}(x)$ is the set of all real numbers, $(-\infty, \infty)$. The range is the set of all integers.

Technology

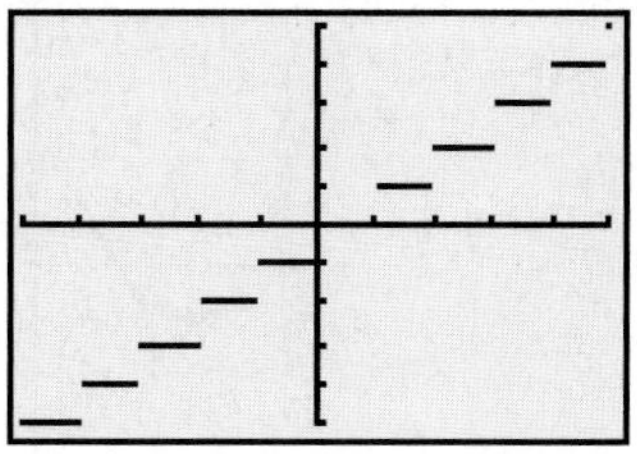

The graph of $f(x) = \text{int}(x)$, shown on the left, was obtained with a graphing utility. By graphing in "dot" mode, we can see the discontinuities at the integers. By looking at the graph, it is impossible to tell that, for each step, the point on the left is included and the point on the right is not. We must trace along the graph to obtain such information.

EXERCISE SET 2.2

Practice Exercises

In Exercises 1–22, find and simplify the difference quotient

$$\frac{f(x + h) - f(x)}{h}, h \neq 0$$

for the given function.

1. $f(x) = 4x$
2. $f(x) = 7x$
3. $f(x) = 3x + 7$
4. $f(x) = 6x + 1$
5. $f(x) = x^2$
6. $f(x) = 2x^2$
7. $f(x) = x^2 - 4x + 3$
8. $f(x) = x^2 - 5x + 8$
9. $f(x) = 2x^2 + x - 1$
10. $f(x) = 3x^2 + x + 5$
11. $f(x) = -x^2 + 2x + 4$
12. $f(x) = -x^2 - 3x + 1$
13. $f(x) = -2x^2 + 5x + 7$
14. $f(x) = -3x^2 + 2x - 1$
15. $f(x) = -2x^2 - x + 3$
16. $f(x) = -3x^2 + x - 1$
17. $f(x) = 6$
18. $f(x) = 7$
19. $f(x) = \frac{1}{x}$
20. $f(x) = \frac{1}{2x}$
21. $f(x) = \sqrt{x}$
22. $f(x) = \sqrt{x - 1}$

In Exercises 23–28, evaluate each piecewise function at the given values of the independent variable.

23. $f(x) = \begin{cases} 3x + 5 & \text{if } x < 0 \\ 4x + 7 & \text{if } x \geq 0 \end{cases}$

    a. $f(-2)$ **b.** $f(0)$ **c.** $f(3)$

24. $f(x) = \begin{cases} 6x - 1 & \text{if } x < 0 \\ 7x + 3 & \text{if } x \geq 0 \end{cases}$

    a. $f(-3)$ **b.** $f(0)$ **c.** $f(4)$

25. $g(x) = \begin{cases} x + 3 & \text{if } x \geq -3 \\ -(x + 3) & \text{if } x < -3 \end{cases}$

    a. $g(0)$ **b.** $g(-6)$ **c.** $g(-3)$

26. $g(x) = \begin{cases} x + 5 & \text{if } x \geq -5 \\ -(x + 5) & \text{if } x < -5 \end{cases}$

    a. $g(0)$ **b.** $g(-6)$ **c.** $g(-5)$

27. $h(x) = \begin{cases} \dfrac{x^2 - 9}{x - 3} & \text{if } x \neq 3 \\ 6 & \text{if } x = 3 \end{cases}$

    a. $h(5)$ **b.** $h(0)$ **c.** $h(3)$

28. $h(x) = \begin{cases} \dfrac{x^2 - 25}{x - 5} & \text{if } x \neq 5 \\ 10 & \text{if } x = 5 \end{cases}$

    a. $h(7)$ **b.** $h(0)$ **c.** $h(5)$

In Exercises 29–40, use the graph to determine

a. *intervals on which the function is increasing, if any.*
b. *intervals on which the function is decreasing, if any.*
c. *intervals on which the function is constant, if any.*

29.

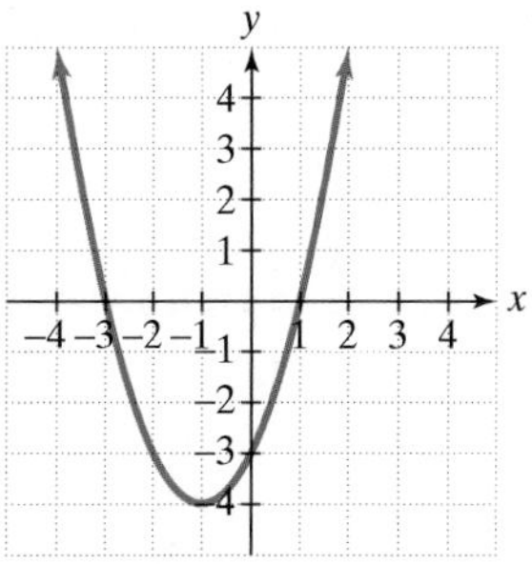

30.

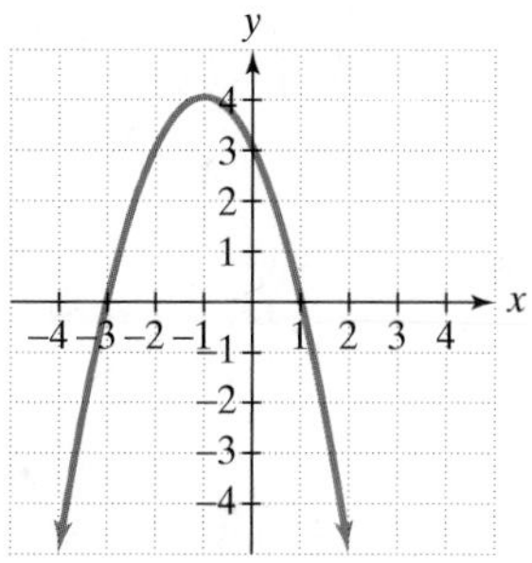

31.

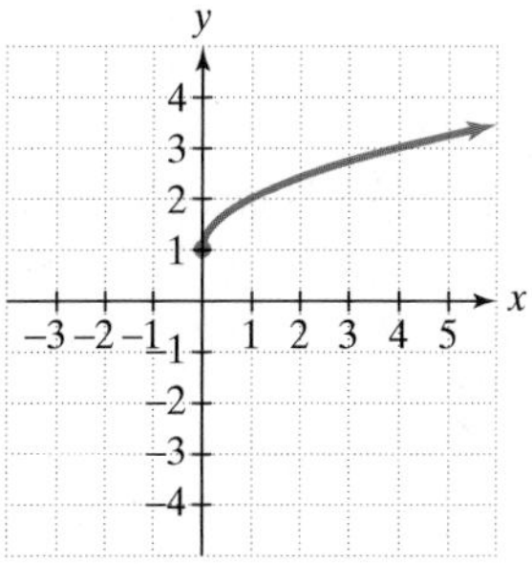

32.

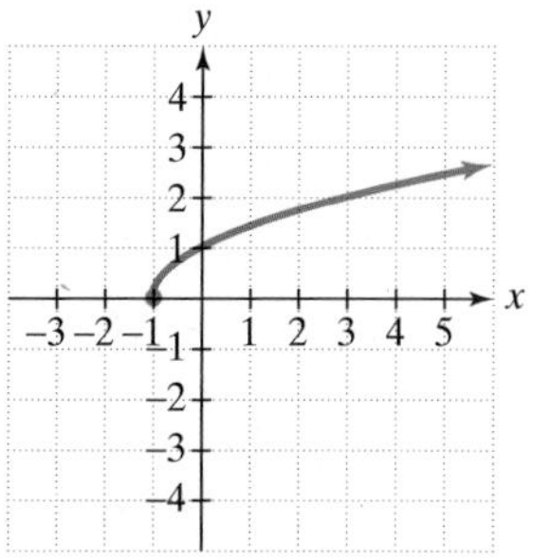

33.

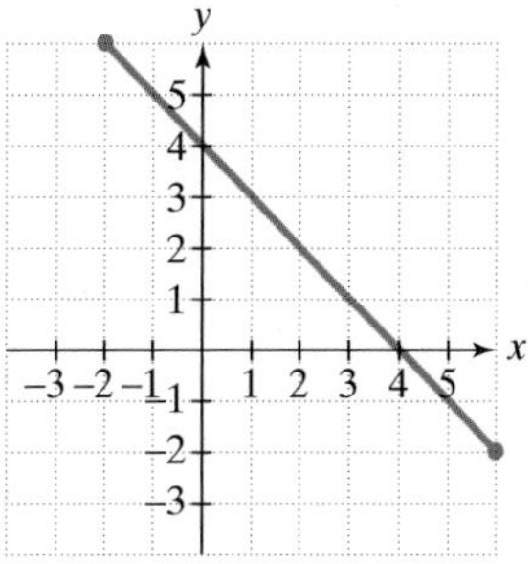

34.

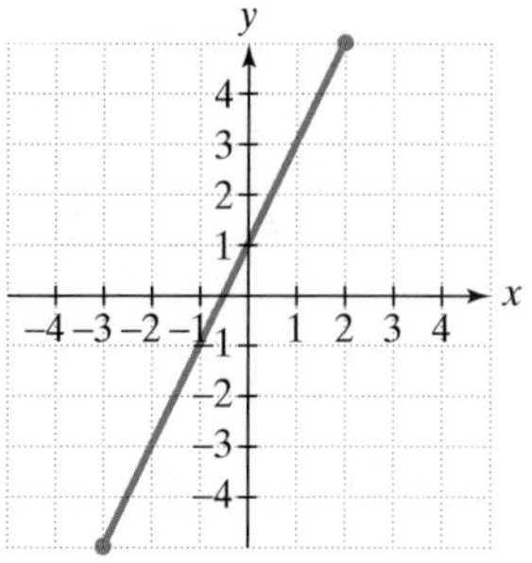

35.

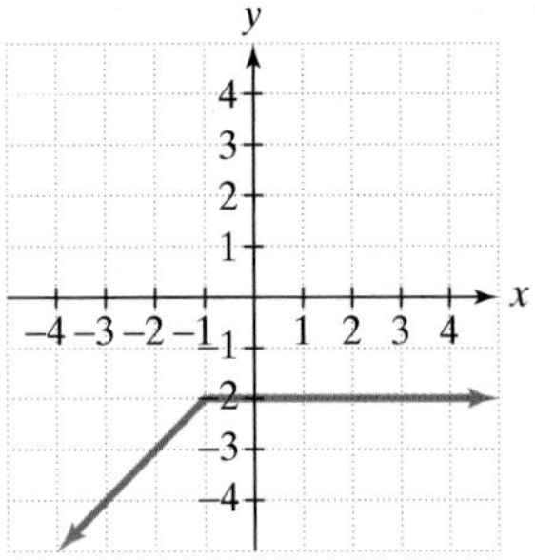

36.

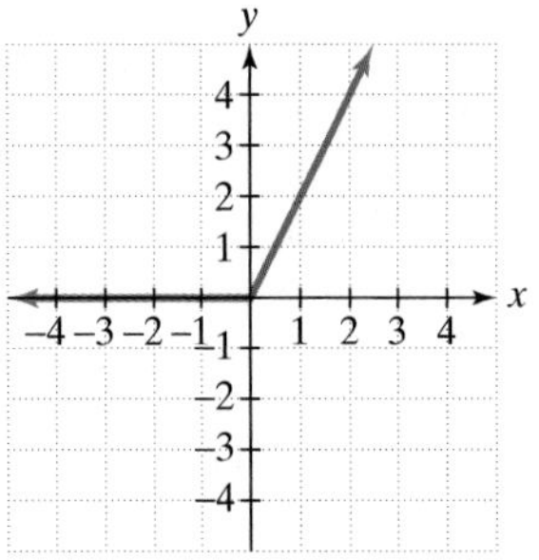

37.

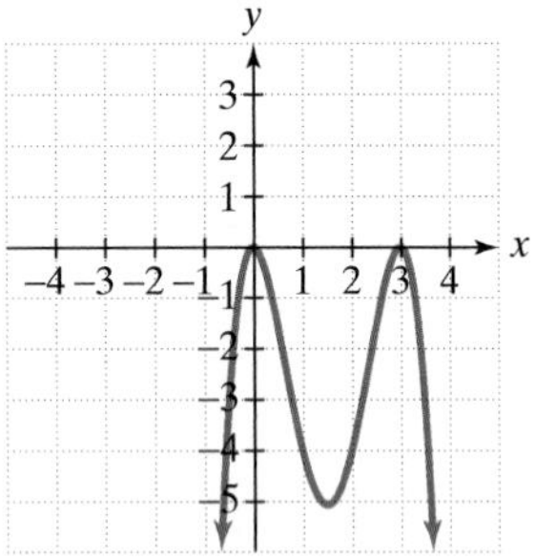

38.

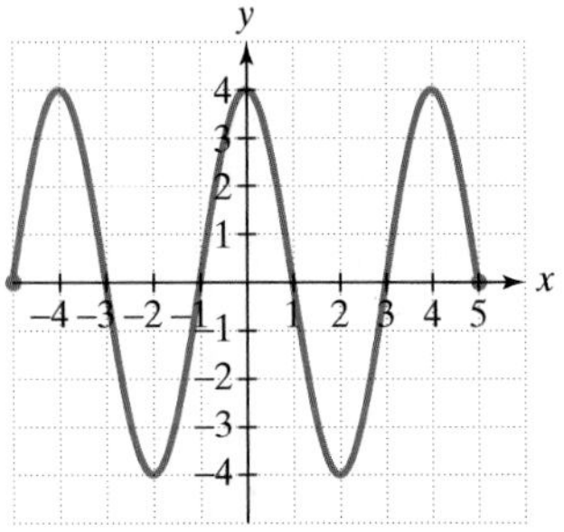

39. **40.**

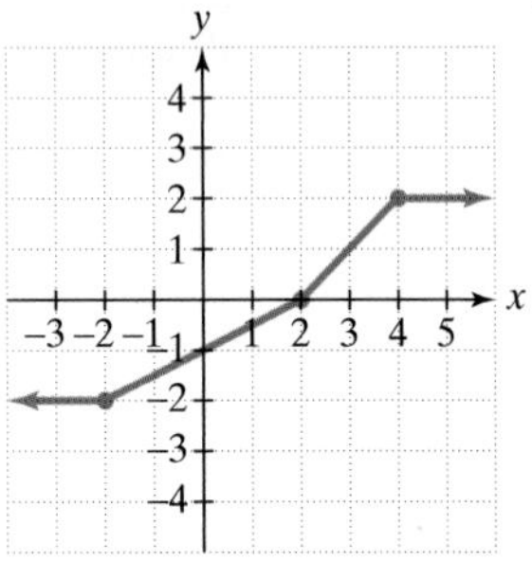

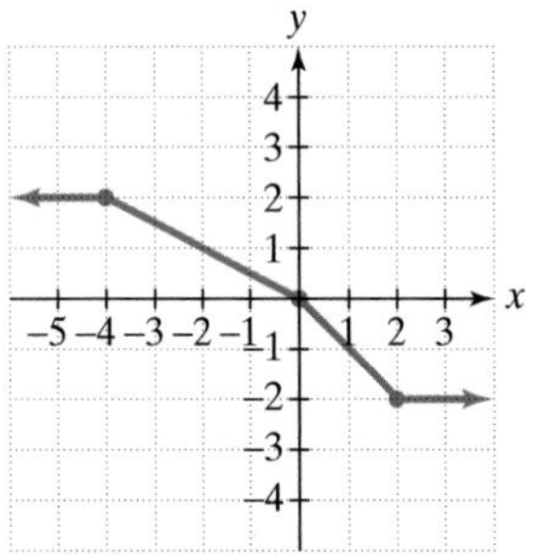

In Exercises 41–44, the graph of a function f is given. Use the graph to find each of the following:

    a. The numbers, if any, at which f has a relative maximum. What are these relative maxima?

    b. The numbers, if any, at which f has a relative minimum. What are these relative minima?

41.

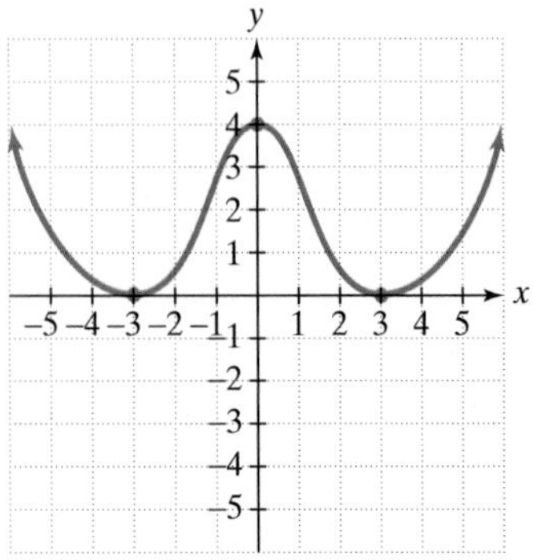

42.

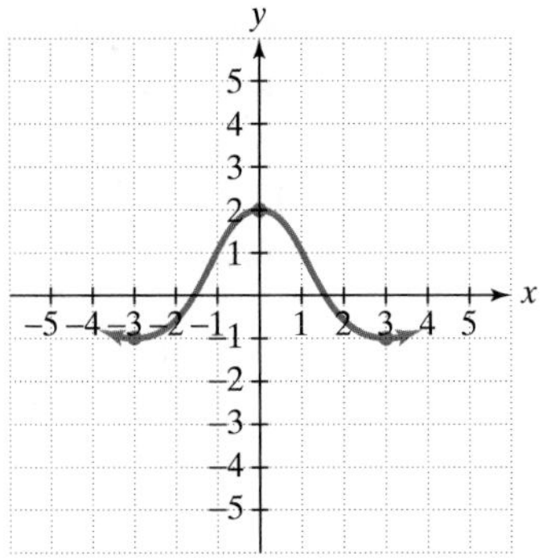

43.

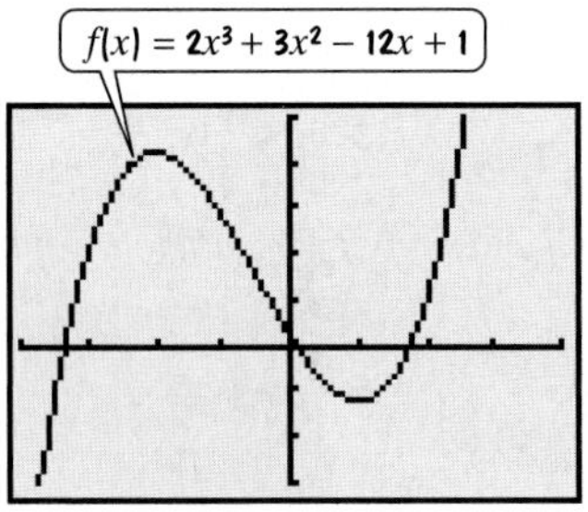

$$f(x) = 2x^3 + 3x^2 - 12x + 1$$

[−4, 4, 1] by [−15, 25, 5]

44.

$$f(x) = 2x^3 - 15x^2 + 24x + 19$$

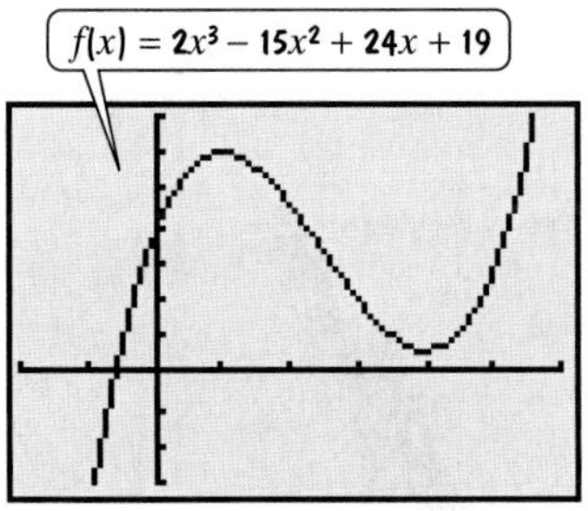

[−2, 6, 1] by [−15, 35, 5]

In Exercises 45–56, determine whether each function is even, odd, or neither.

45. $f(x) = x^3 + x$ **46.** $f(x) = x^3 - x$

47. $g(x) = x^2 + x$ **48.** $g(x) = x^2 - x$

49. $h(x) = x^2 - x^4$ **50.** $h(x) = 2x^2 + x^4$

51. $f(x) = x^2 - x^4 + 1$ **52.** $f(x) = 2x^2 + x^4 + 1$

53. $f(x) = \frac{1}{5}x^6 - 3x^2$ **54.** $f(x) = 2x^3 - 6x^5$

55. $f(x) = x\sqrt{1 - x^2}$ **56.** $f(x) = x^2\sqrt{1 - x^2}$

In Exercises 57–60, use possible symmetry to determine whether each graph is the graph of an even function, an odd function, or a function that is neither even nor odd.

57.

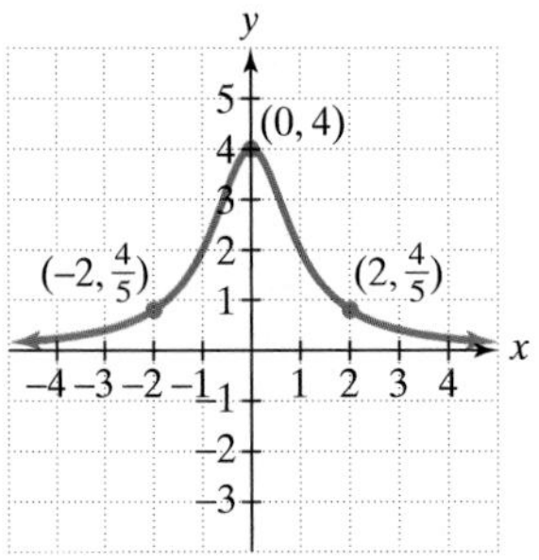

58.

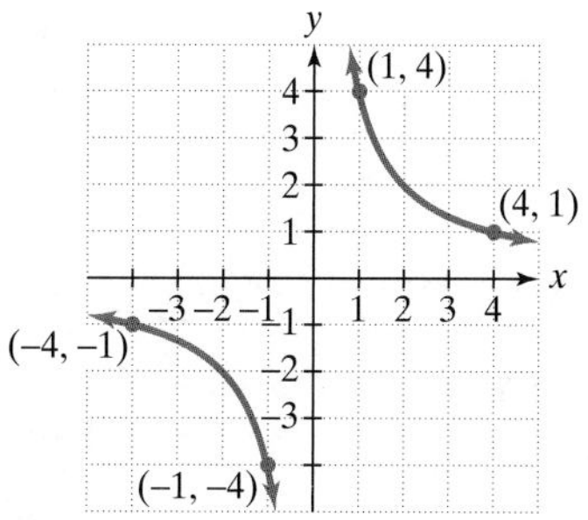

59.

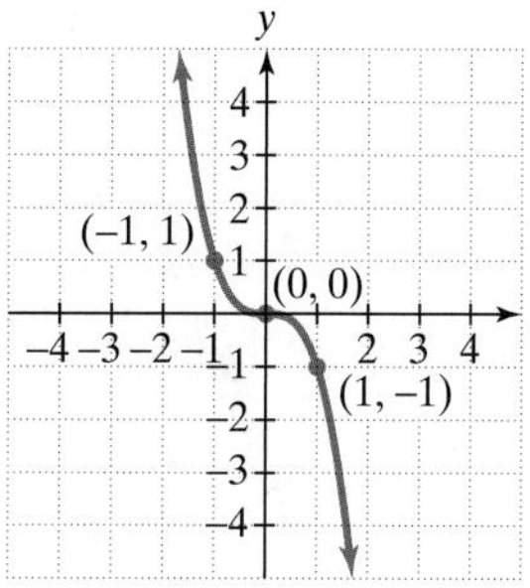

60.

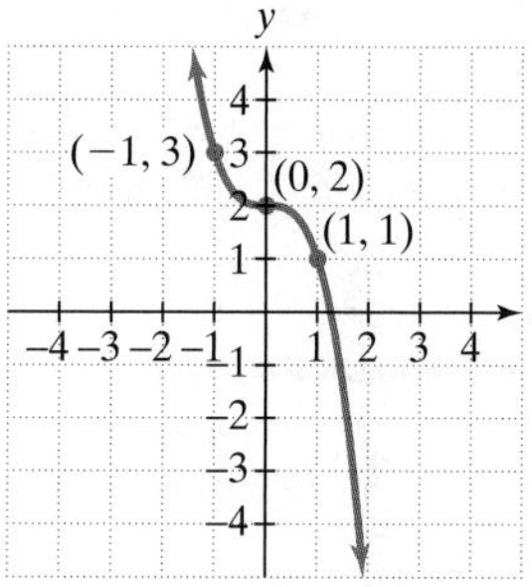

61. Use the graph of f to determine each of the following. Where applicable, use interval notation.

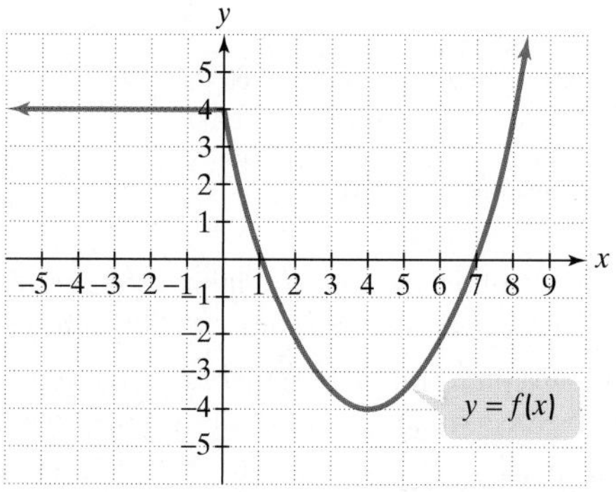

a. the domain of f
b. the range of f
c. the x-intercepts
d. the y-intercept
e. intervals on which f is increasing
f. intervals on which f is decreasing
g. intervals on which f is constant
h. the number at which f has a relative minimum
i. the relative minimum of f
j. $f(-3)$
k. the values of x for which $f(x) = -2$
l. Is f even, odd, or neither?

62. Use the graph of f to determine each of the following. Where applicable, use interval notation.

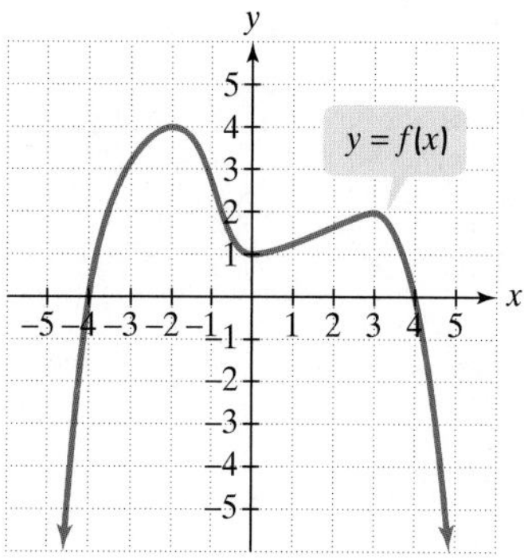

a. the domain of f
b. the range of f
c. the x-intercepts
d. the y-intercept
e. intervals on which f is increasing
f. intervals on which f is decreasing
g. values of x for which $f(x) \leq 0$
h. the numbers at which f has a relative maximum
i. the relative maxima of f
j. $f(-2)$
k. the values of x for which $f(x) = 0$
l. Is f even, odd, or neither?

63. Use the graph of f to determine each of the following. Where applicable, use interval notation.

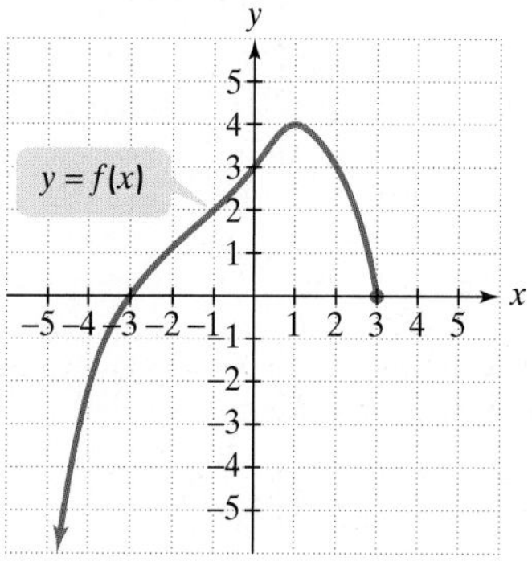

a. the domain of f
b. the range of f
c. the zeros of f
d. $f(0)$
e. intervals on which f is increasing
f. intervals on which f is decreasing
g. values of x for which $f(x) \leq 0$
h. any relative maxima and the numbers at which they occur
i. the value of x for which $f(x) = 4$
j. Is $f(-1)$ positive or negative?

64. Use the graph of f to determine each of the following. Where applicable, use interval notation.

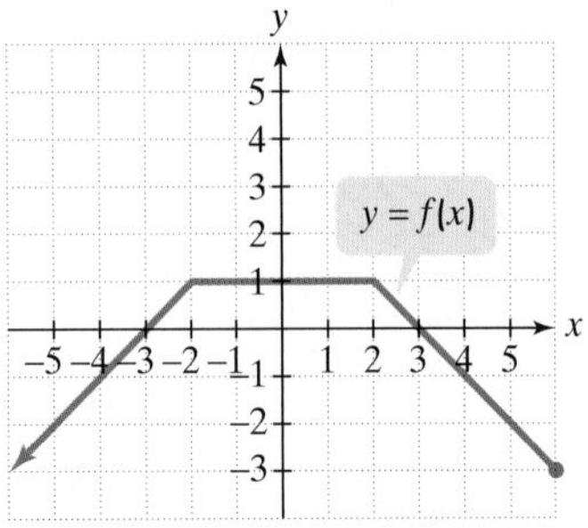

a. the domain of f
b. the range of f
c. the zeros of f
d. $f(0)$
e. intervals on which f is increasing
f. intervals on which f is decreasing
g. intervals on which f is constant
h. values of x for which $f(x) > 0$
i. values of x for which $f(x) = -2$
j. Is $f(4)$ positive or negative?
k. Is f even, odd, or neither?
l. Is $f(2)$ a relative maximum?

In Exercises 65–70, if $f(x) = int(x)$, find each function value.

65. $f(1.06)$ **66.** $f(2.99)$ **67.** $f\left(\frac{1}{3}\right)$
68. $f(-1.5)$ **69.** $f(-2.3)$ **70.** $f(-99.001)$

Practice Plus

In Exercises 71–72, let f be defined by the following graph:

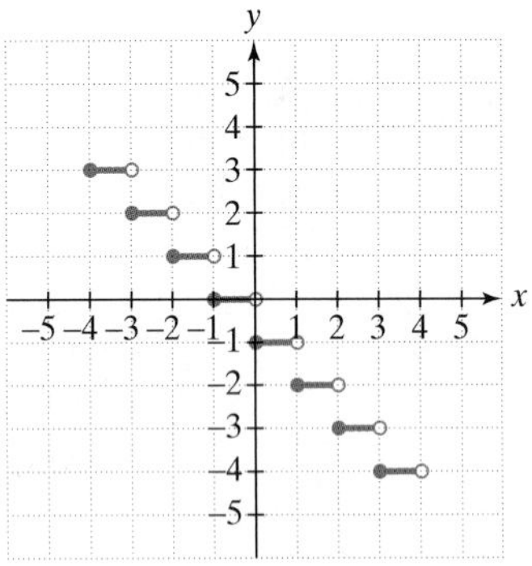

71. Find
$$\sqrt{f(-1.5) + f(-0.9)} - [f(\pi)]^2 + f(-3) \div f(1) \cdot f(-\pi).$$

72. Find
$$\sqrt{f(-2.5) - f(1.9)} - [f(-\pi)]^2 + f(-3) \div f(1) \cdot f(\pi).$$

A cellular phone company offers the following plans. Also given are the piecewise functions that describe these plans. Use this information to solve Exercises 73–74.

Plan A

- \$30 per month buys 120 minutes.
- Additional time costs \$0.30 per minute.

$$C(t) = \begin{cases} 30 & \text{if } 0 \le t \le 120 \\ 30 + 0.30(t - 120) & \text{if } t > 120 \end{cases}$$

Plan B

- \$40 per month buys 200 minutes.
- Additional time costs \$0.30 per minute.

$$C(t) = \begin{cases} 40 & \text{if } 0 \le t \le 200 \\ 40 + 0.30(t - 200) & \text{if } t > 200 \end{cases}$$

73. Simplify the algebraic expression in the second line of the piecewise function for plan A. Then use point-plotting to graph the function.

74. Simplify the algebraic expression in the second line of the piecewise function for plan B. Then use point-plotting to graph the function.

In Exercises 75–76, write a piecewise function that describes each cellular phone billing plan. Then graph the function.

75. \$50 per month buys 400 minutes. Additional time costs \$0.30 per minute.

76. \$60 per month buys 450 minutes. Additional time costs \$0.35 per minute.

Application Exercises

The figure shows the percentage of Jewish Americans in the U.S. population, $f(x)$, x years after 1900. Use the graph to solve Exercises 77–84.

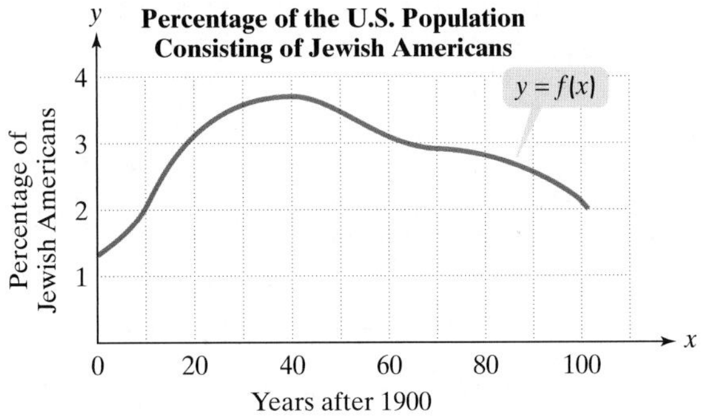

Source: American Jewish Yearbook

77. Use the graph to find a reasonable estimate of $f(60)$. What does this mean in terms of the variables in this situation?

78. Use the graph to find a reasonable estimate of $f(100)$. What does this mean in terms of the variables in this situation?

79. For what value or values of x is $f(x) = 3$? Round to the nearest year. What does this mean in terms of the variables in this situation?

80. For what value or values of x is $f(x) = 2.5$? Round to the nearest year. What does this mean in terms of the variables in this situation?

81. In which year did the percentage of Jewish Americans in the U.S. population reach a maximum? What is a reasonable estimate of the percentage for that year?

82. In which year was the percentage of Jewish Americans in the U.S. population at a minimum? What is a reasonable estimate of the percentage for that year?

83. Explain why f represents the graph of a function.

84. Describe the general trend shown by the graph.

The function

$$f(x) = 0.4x^2 - 36x + 1000$$

models the number of accidents, $f(x)$, per 50 million miles driven as a function of a driver's age, x, in years, where x includes drivers from ages 16 through 74, inclusive. The graph of f is shown. Use the graph of f, and possibly the equation, to solve Exercises 85–88.

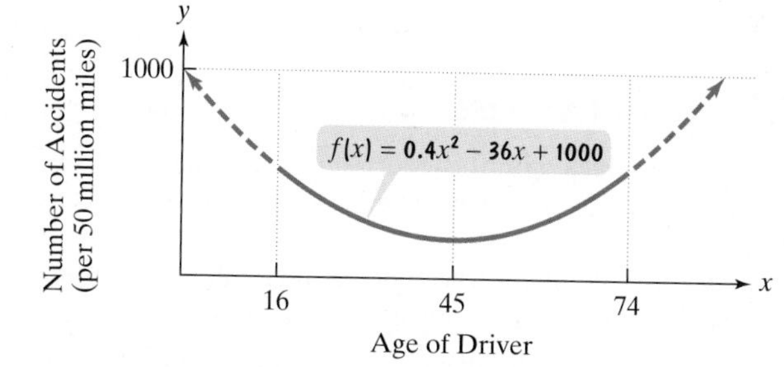

85. State the intervals on which the function is increasing and decreasing. Describe what this means in terms of the variables modeled by the function.

86. For what value of x does the graph reach its lowest point? Use the equation for f to find the minimum value of y. Describe the practical significance of this minimum value.

87. Use the graph to identify two different ages for which drivers have the same number of accidents. Use the equation for f to find the number of accidents for drivers at each of these ages.

88. Use the equation for f to find and interpret $f(50)$. Identify this information as a point on the graph of f.

The graph shows cigarette consumption per U.S. adult from 1910 through 2003. The data can be modeled by the piecewise function

$$f(x) = \begin{cases} 61.9x + 132 & \text{if } 0 \le x \le 30 \\ -2.2x^2 + 256x - 3503 & \text{if } 30 < x \le 93, \end{cases}$$

where x represents years after 1910 and $f(x)$ represents cigarette consumption per U.S. adult. Use this information to solve Exercises 89–92.

Cigarette Consumption per U.S. Adult

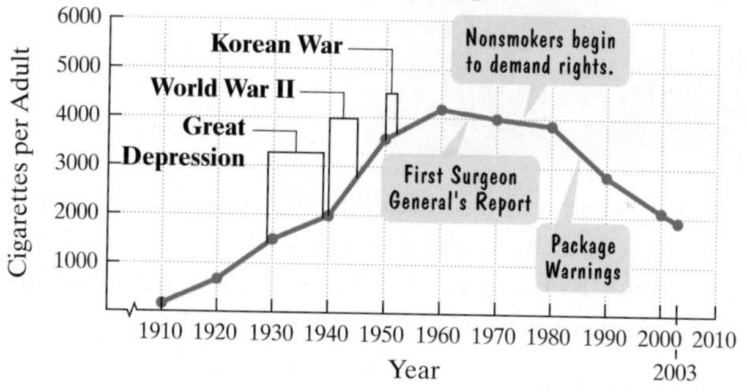

Source: U.S. Department of Health and Human Services

89. Use the piecewise function that models the data to find cigarette consumption in 1940. How well does the function describe the actual consumption for that year shown by the line graph?

90. Use the piecewise function that models the data to find cigarette consumption in 1990. How well does the function describe the actual consumption for that year shown by the line graph?

91. For the period shown, in which year was cigarette consumption at a maximum? Use the graph to find a reasonable estimate of consumption for that year. How well does the piecewise function model this estimate?

92. For the period shown, in which year was cigarette consumption at a minimum? Use the graph to find a reasonable estimate of consumption for that year. How well does the piecewise function model this estimate?

93. The cost of a telephone call between two cities is \$0.10 for the first minute and \$0.05 for each additional minute or portion of a minute. Draw a graph of the cost, C, in dollars, of the phone call as a function of time, t, in minutes, on the interval $(0, 5]$.

94. A cargo service charges a flat fee of \$4 plus \$1 for each pound or fraction of a pound to mail a package. Let $C(x)$ represent the cost to mail a package that weighs x pounds. Graph the cost function on the interval $(0, 5]$.

Writing in Mathematics

95. Explain how to find the difference quotient, $\frac{f(x+h) - f(x)}{h}$, if a function's equation is given.

96. What is a piecewise function?

97. What does it mean if a function f is increasing on an interval?

98. Suppose that a function f is increasing on (a, b), decreasing on (b, c), and defined at b. Describe what occurs at $x = b$. What does the function value $f(b)$ represent?

99. If you are given a function's equation, how do you determine if the function is even, odd, or neither?

100. If you are given a function's graph, how do you determine if the function is even, odd, or neither?

101. What is a step function? Give an example of an everyday situation that can be modeled using such a function. Do not use the cost-of-mail example.

102. Explain how to find $\text{int}(-3.000004)$.

Technology Exercises

103. The function

$$f(x) = -0.00002x^3 + 0.008x^2 - 0.3x + 6.95$$

models the number of annual physician visits, $f(x)$, by a person of age x. Graph the function in a $[0, 100, 5]$ by $[0, 40, 2]$ viewing rectangle. What does the shape of the graph indicate about the relationship between one's age and the number of annual physician visits? Use the TRACE or minimum function capability to find the coordinates of the minimum point on the graph of the function. What does this mean?

In Exercises 104–109, use a graphing utility to graph each function. Use a $[-5, 5, 1]$ by $[-5, 5, 1]$ viewing rectangle. Then find the intervals on which the function is increasing, decreasing, or constant.

104. $f(x) = x^3 - 6x^2 + 9x + 1$ **105.** $g(x) = |4 - x^2|$

106. $h(x) = |x - 2| + |x + 2|$ **107.** $f(x) = x^{\frac{1}{3}}(x - 4)$

108. $g(x) = x^{\frac{2}{3}}$ **109.** $h(x) = 2 - x^{\frac{2}{5}}$

110. a. Graph the functions $f(x) = x^n$ for $n = 2, 4$, and 6 in a $[-2, 2, 1]$ by $[-1, 3, 1]$ viewing rectangle.

b. Graph the functions $f(x) = x^n$ for $n = 1, 3$, and 5 in a $[-2, 2, 1]$ by $[-2, 2, 1]$ viewing rectangle.

c. If n is even, where is the graph of $f(x) = x^n$ increasing and where is it decreasing?

d. If n is odd, what can you conclude about the graph of $f(x) = x^n$ in terms of increasing or decreasing behavior?

e. Graph all six functions in a $[-1, 3, 1]$ by $[-1, 3, 1]$ viewing rectangle. What do you observe about the graphs in terms of how flat or how steep they are?

Critical Thinking Exercises

111. Sketch the graph of f using the following properties. (More than one correct graph is possible.) f is a piecewise function that is decreasing on $(-\infty, 2)$, $f(2) = 0$, f is increasing on $(2, \infty)$, and the range of f is $[0, \infty)$.

112. Define a piecewise function on the intervals $(-\infty, 2]$, $(2, 5)$, and $[5, \infty)$ that does not "jump" at 2 or 5 such that one piece is a constant function, another piece is an increasing function, and the third piece is a decreasing function.

113. Suppose that $h(x) = \frac{f(x)}{g(x)}$. The function f can be even, odd, or neither. The same is true for the function g.

a. Under what conditions is h definitely an even function?

b. Under what conditions is h definitely an odd function?

114. Take another look at the cost of first-class mail and its graph (Table 2.1 and Figure 2.19 on page 210. Change the description of the heading in the left column of Table 2.1 so that the graph includes the point on the left of each horizontal step, but does not include the point on the right.

Group Exercise

115. (For assistance with this exercise, refer to the discussion of piecewise functions on page 204, as well as to Exercises 73–74.)

Group members who have cellular phone plans should describe the total monthly cost of the plan as follows:

$_____ per month buys _____ minutes. Additional time costs $ _____ per minute.

(For simplicity, ignore off-peak rates, roaming charges, etc.) The group should select any three plans, from "basic" to "premier." For each plan selected, write a piecewise function that describes the plan and graph the function. Graph the three functions in the same rectangular coordinate system. Now examine the graphs. For any given number of calling minutes, the best plan is the one whose graph is lowest at that point. Compare the three calling plans. Over how many minutes does one plan become better than another? (You can check out cellular phone plans by visiting www.point.com.)

SECTION 2.3 Linear Functions and Slope

Objectives

1. Calculate a line's slope.
2. Write the point-slope form of the equation of a line.
3. Write and graph the slope-intercept form of the equation of a line.
4. Graph horizontal or vertical lines.
5. Recognize and use the general form of a line's equation.
6. Use intercepts to graph the general form of a line's equation.
7. Model data with linear functions and make predictions.

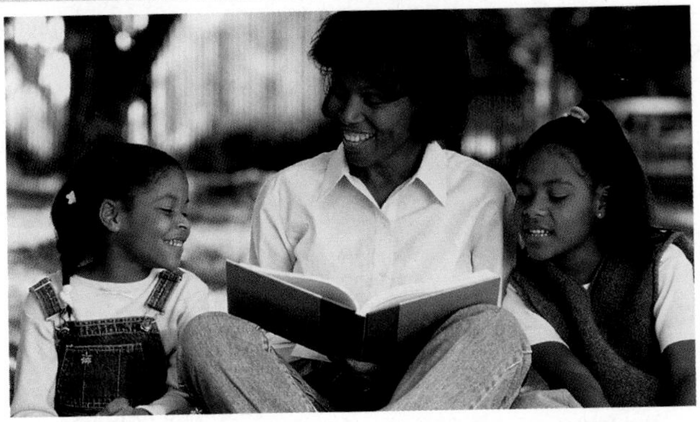

Is there a relationship between literacy and child mortality? As the percentage of adult females who are literate increases, does the mortality of children under five decrease? Figure 2.21, based on data from the United Nations, indicates that this is, indeed, the case. Each point in the figure represents one country.

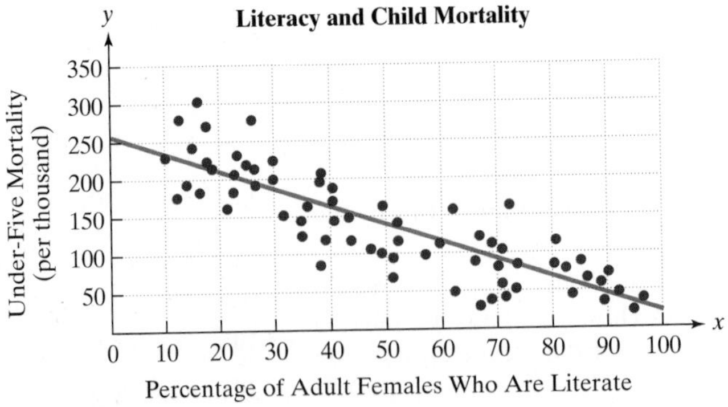

Figure 2.21
Source: United Nations

Data presented in a visual form as a set of points is called a **scatter plot**. Also shown in Figure 2.21 is a line that passes through or near the points. A line that best fits the data points in a scatter plot is called a **regression line**. By writing the equation of this line, we can obtain a model for the data and make predictions about child mortality based on the percentage of literate adult females in a country.

Data often fall on or near a line. In this section, we will use functions to model such data and make predictions. We begin with a discussion of a line's steepness.

1 Calculate a line's slope.

The Slope of a Line

Mathematicians have developed a useful measure of the steepness of a line, called the *slope* of the line. Slope compares the vertical change (the **rise**) to the horizontal change (the **run**) when moving from one fixed point to another along the line. To calculate the slope of a line, we use a ratio that compares the change in y (the rise) to the corresponding change in x (the run).

Slope and the Streets of San Francisco

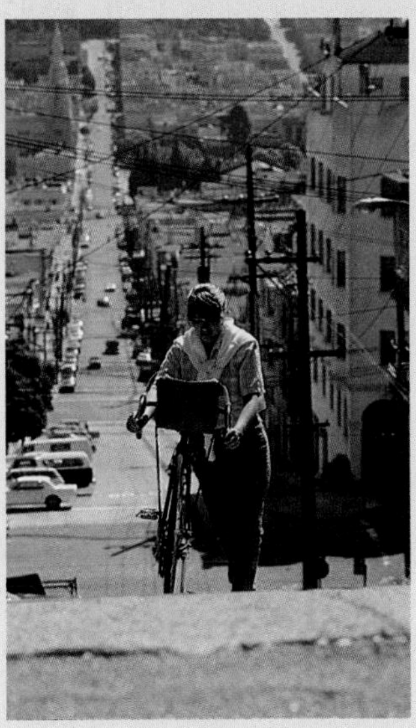

San Francisco's Filbert Street has a slope of 0.613, meaning that for every horizontal distance of 100 feet, the street ascends 61.3 feet vertically. With its 31.5° angle of inclination, the street is too steep to pave and is only accessible by wooden stairs.

Definition of Slope

The **slope** of the line through the distinct points (x_1, y_1) and (x_2, y_2) is

$$\frac{\text{Change in } y}{\text{Change in } x} = \frac{\text{Rise}}{\text{Run}} = \frac{y_2 - y_1}{x_2 - x_1}$$

where $x_2 - x_1 \neq 0$.

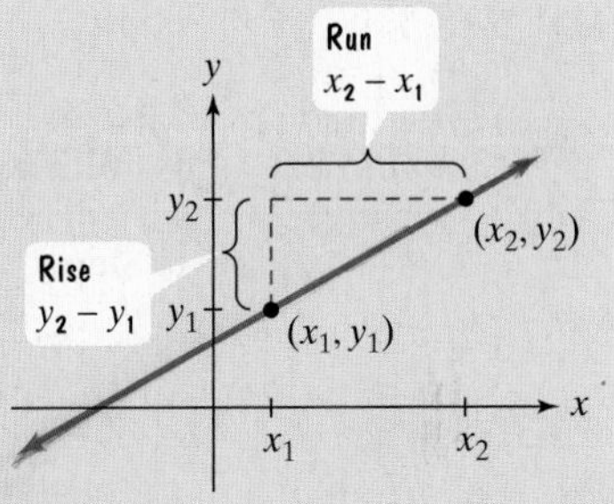

It is common notation to let the letter m represent the slope of a line. The letter m is used because it is the first letter of the French verb *monter*, meaning to rise, or to ascend.

EXAMPLE 1 Using the Definition of Slope

Find the slope of the line passing through each pair of points:

a. $(-3, -1)$ and $(-2, 4)$ **b.** $(-3, 4)$ and $(2, -2)$.

Solution

a. Let $(x_1, y_1) = (-3, -1)$ and $(x_2, y_2) = (-2, 4)$. We obtain the slope as follows:

$$m = \frac{\text{Change in } y}{\text{Change in } x} = \frac{y_2 - y_1}{x_2 - x_1} = \frac{4 - (-1)}{-2 - (-3)} = \frac{5}{1} = 5.$$

The situation is illustrated in Figure 2.22(a). The slope of the line is 5, indicating that there is a vertical change, a rise, of 5 units for each horizontal change, a run, of 1 unit. The slope is positive, and the line rises from left to right.

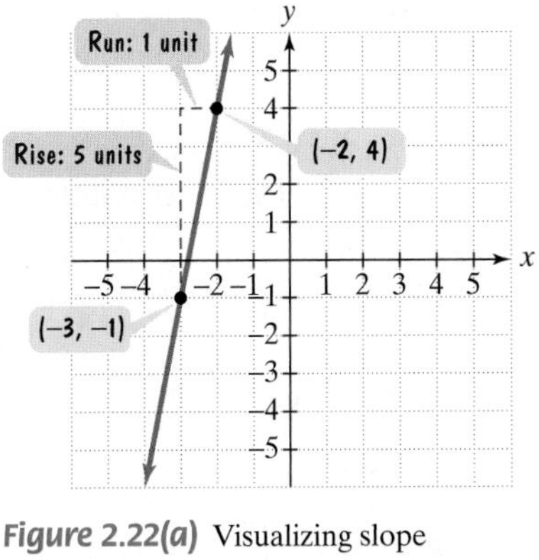

Figure 2.22(a) Visualizing slope

Study Tip

When computing slope, it makes no difference which point you call (x_1, y_1) and which point you call (x_2, y_2). If we let $(x_1, y_1) = (-2, 4)$ and $(x_2, y_2) = (-3, -1)$, the slope is still 5:

$$m = \frac{\text{Change in } y}{\text{Change in } x} = \frac{y_2 - y_1}{x_2 - x_1} = \frac{-1 - 4}{-3 - (-2)} = \frac{-5}{-1} = 5.$$

However, you should not subtract in one order in the numerator $(y_2 - y_1)$ and then in a different order in the denominator $(x_1 - x_2)$. The slope is *not* -5:

$$\frac{-1 - 4}{-2 - (-3)} = \frac{-5}{1} = -5. \quad \textit{Incorrect}$$

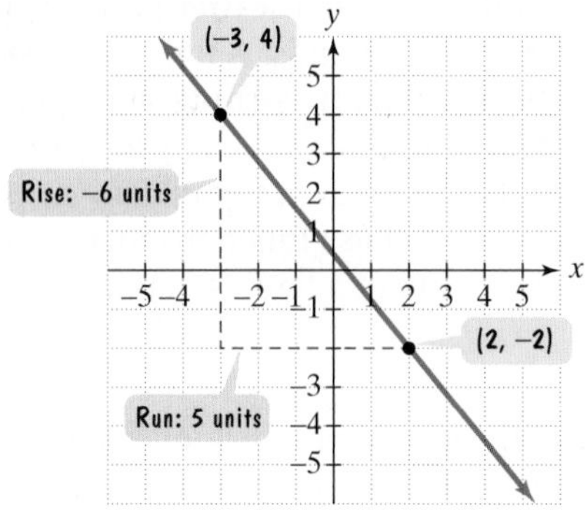

Figure 2.22(b)

b. We can let $(x_1, y_1) = (-3, 4)$ and $(x_2, y_2) = (2, -2)$. The slope of the line shown in Figure 2.22(b) is computed as follows:

$$m = \frac{\text{Change in } y}{\text{Change in } x} = \frac{y_2 - y_1}{x_2 - x_1} = \frac{-2 - 4}{2 - (-3)} = \frac{-6}{5} = -\frac{6}{5}.$$

The slope of the line is $-\frac{6}{5}$. For every vertical change of -6 units (6 units down), there is a corresponding horizontal change of 5 units. The slope is negative and the line falls from left to right.

Check Point 1 Find the slope of the line passing through each pair of points:

a. $(-3, 4)$ and $(-4, -2)$ **b.** $(4, -2)$ and $(-1, 5)$.

Example 1 illustrates that a line with a positive slope is rising from left to right and a line with a negative slope is falling from left to right. By contrast, a horizontal line neither rises nor falls and has a slope of zero. A vertical line has no horizontal change, so $x_2 - x_1 = 0$ in the formula for slope. Because we cannot divide by zero, the slope of a vertical line is undefined. This discussion is summarized in Table 2.2.

Table 2.2 Possibilities for a Line's Slope

Positive Slope	Negative Slope	Zero Slope	Undefined Slope
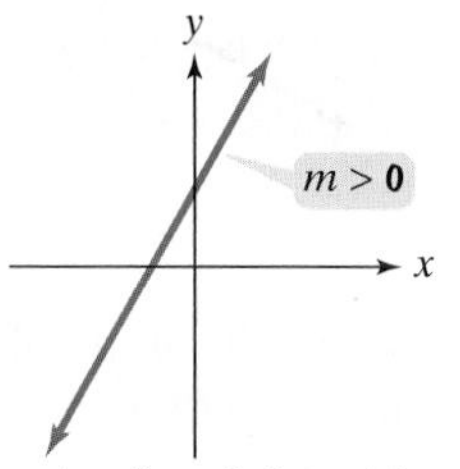	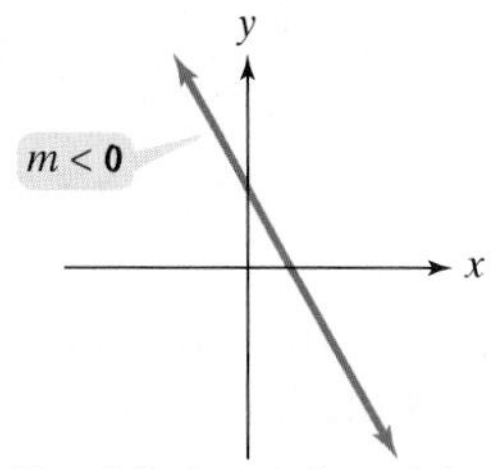	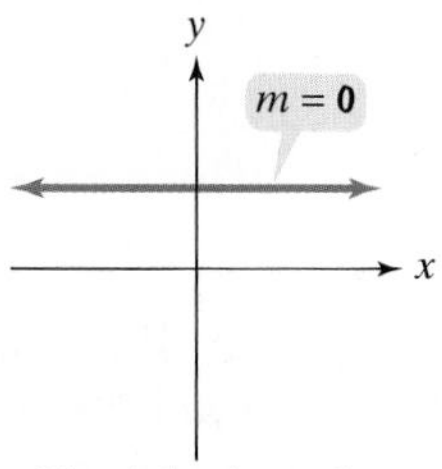	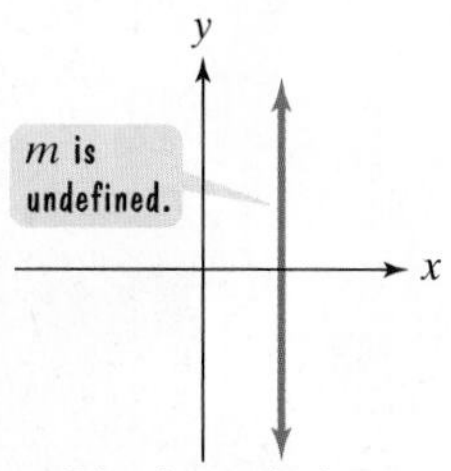
Line rises from left to right.	Line falls from left to right.	Line is horizontal.	Line is vertical.

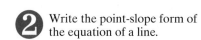

Study Tip

Always be clear in the way you use language, especially in mathematics. For example, it's not a good idea to say that a line has "no slope." This could mean that the slope is zero or that the slope is undefined.

2 Write the point-slope form of the equation of a line.

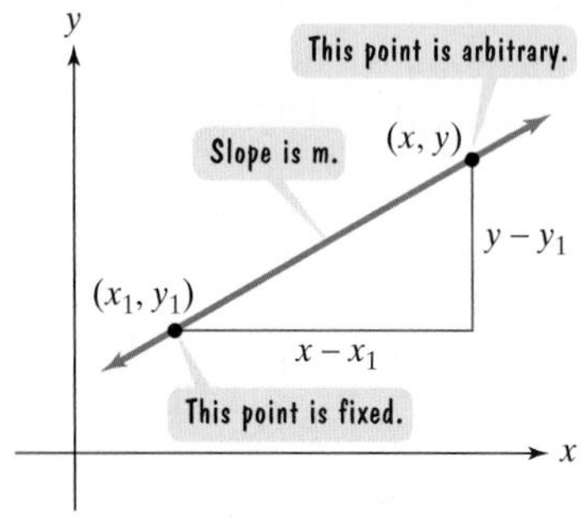

Figure 2.23 A line passing through (x_1, y_1) with slope m

The Point-Slope Form of the Equation of a Line

We can use the slope of a line to obtain various forms of the line's equation. For example, consider a nonvertical line that has slope m and that contains the point (x_1, y_1). Now, let (x, y) represent any other point on the line, shown in Figure 2.23. Keep in mind that the point (x, y) is arbitrary and is not in one fixed position. By contrast, the point (x_1, y_1) is fixed.

Regardless of where the point (x, y) is located, the steepness of the line in Figure 2.23 remains the same. Thus, the ratio for the slope stays a constant m. This means that for all points along the line

$$m = \frac{\text{Change in } y}{\text{Change in } x} = \frac{y - y_1}{x - x_1}.$$

We can clear the fraction by multiplying both sides by $x - x_1$, the least common denominator.

$$m = \frac{y - y_1}{x - x_1} \quad \text{This is the slope of the line in Figure 2.23.}$$

$$m(x - x_1) = \frac{y - y_1}{x - x_1} \cdot x - x_1 \quad \text{Multiply both sides by } x - x_1.$$

$$m(x - x_1) = y - y_1 \quad \text{Simplify: } \frac{y - y_1}{\cancel{x - x_1}} \cdot \cancel{x - x_1} = y - y_1.$$

Now, if we reverse the two sides, we obtain the *point-slope form* of the equation of a line.

Point-Slope Form of the Equation of a Line

The **point-slope form of the equation** of a nonvertical line with slope m that passes through the point (x_1, y_1) is

$$y - y_1 = m(x - x_1).$$

For example, the point-slope form of the equation of the line passing through $(1, 5)$ with slope 2; $(m = 2)$ is

$$y - 5 = 2(x - 1).$$

We will soon be expressing the equation of a nonvertical line in function notation. To do so, we need to solve the point-slope form of a line's equation for y. Example 2 illustrates how to isolate y on one side of the equal sign.

EXAMPLE 2 Writing the Point-Slope Form of the Equation of a Line

Write the point-slope form of the equation of the line with slope 4 that passes through the point $(-1, 3)$. Then solve the equation for y.

Solution We use the point-slope form of the equation of a line with $m = 4$, $x_1 = -1$, and $y_1 = 3$.

$$y - y_1 = m(x - x_1)$$ This is the point-slope form of the equation.

$$y - 3 = 4[x - (-1)]$$ Substitute the given values.

$$y - 3 = 4(x + 1)$$ We now have the point-slope form of the equation of the given line.

We can solve this equation for y by first applying the distributive property on the right side.

$$y - 3 = 4x + 4$$

Finally, we add 3 to both sides.

$$y = 4x + 7$$

Check Point 2 Write the point-slope form of the equation of the line with slope 6 that passes through the point $(2, -5)$. Then solve the equation for y.

EXAMPLE 3 Writing the Point-Slope Form of the Equation of a Line

Write the point-slope form of the equation of the line passing through the points $(4, -3)$ and $(-2, 6)$. (See Figure 2.24.) Then solve the equation for y.

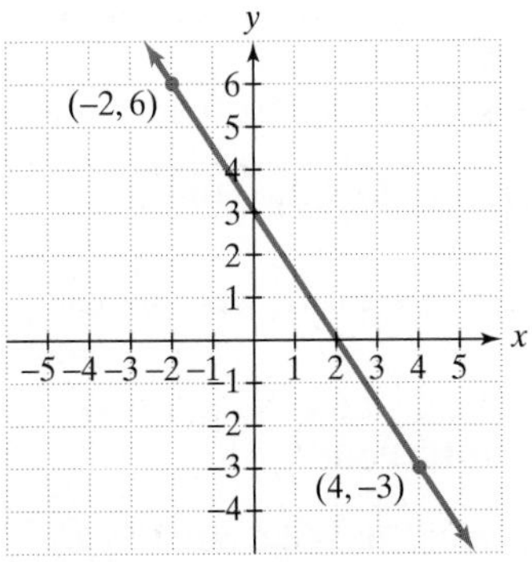

Figure 2.24 Write the point-slope form of the equation of this line.

Solution To use the point-slope form, we need to find the slope. The slope is the change in the y-coordinates divided by the corresponding change in the x-coordinates.

$$m = \frac{6 - (-3)}{-2 - 4} = \frac{9}{-6} = -\frac{3}{2}$$ This is the definition of slope using $(4, -3)$ and $(-2, 6)$.

We can take either point on the line to be (x_1, y_1). Let's use $(x_1, y_1) = (4, -3)$. Now, we are ready to write the point-slope form of the equation.

$$y - y_1 = m(x - x_1)$$ This is the point-slope form of the equation.

$$y - (-3) = -\tfrac{3}{2}(x - 4)$$ Substitute: $(x_1, y_1) = (4, -3)$ and $m = -\frac{3}{2}$.

$$y + 3 = -\tfrac{3}{2}(x - 4)$$ Simplify.

We now have the point-slope form of the equation of the line shown in Figure 2.24. Now, we solve this equation for y.

Discovery

You can use either point for (x_1, y_1) when you write a line's point-slope equation. Rework Example 3 using $(-2, 6)$ for (x_1, y_1). Once you solve for y, you should still obtain

$$y = -\frac{3}{2}x + 3.$$

We need to isolate y.

$$y + 3 = -\frac{3}{2}(x - 4) \quad \text{This is the point-slope form of the equation.}$$

$$y + 3 = -\frac{3}{2}x + 6 \quad \text{Use the distributive property.}$$

$$y = -\frac{3}{2}x + 3 \quad \text{Subtract 3 from both sides.}$$

Check Point 3 Write the point-slope form of the equation of the line passing through the points $(-2, -1)$ and $(-1, -6)$. Then solve the equation for y.

3 Write and graph the slope-intercept form of the equation of a line.

The Slope-Intercept Form of the Equation of a Line

Let's write the point-slope form of the equation of a nonvertical line with slope m and y-intercept b. The line is shown in Figure 2.25. Because the y-intercept is b, the line passes through $(0, b)$. We use the point-slope form with $x_1 = 0$ and $y_1 = b$.

$$y - y_1 = m(x - x_1)$$

Let $y_1 = b$. Let $x_1 = 0$.

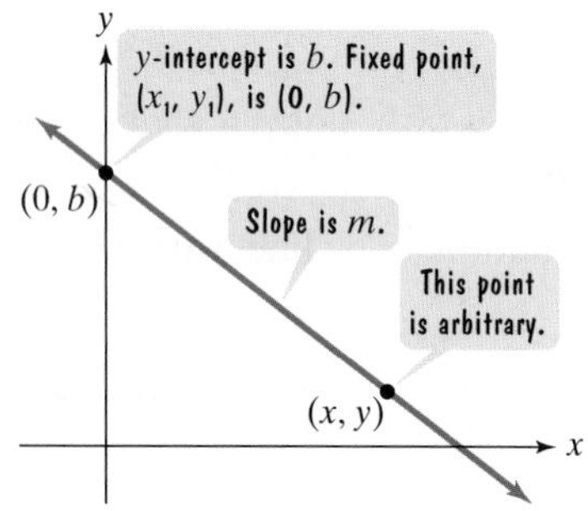

Figure 2.25 A line with slope m and y-intercept b

We obtain

$$y - b = m(x - 0).$$

Simplifying on the right side gives us

$$y - b = mx.$$

Finally, we solve for y by adding b to both sides.

$$y = mx + b$$

Thus, if a line's equation is written with y isolated on one side, the x-coefficient is the line's slope and the constant term is the y-intercept. This form of a line's equation is called the *slope-intercept form* of the line.

Slope-Intercept Form of the Equation of a Line

The **slope-intercept form of the equation** of a nonvertical line with slope m and y-intercept b is

$$y = mx + b.$$

The slope-intercept form of a line's equation, $y = mx + b$, can be expressed in function notation by replacing y with $f(x)$:

$$f(x) = mx + b.$$

We have seen that functions in this form are called **linear functions**. Thus, in the equation of a linear function, the x-coefficient is the line's slope and the constant term is the y-intercept. Here are two examples:

$$y = 2x - 4$$

The slope is 2. The y-intercept is -4.

$$f(x) = \frac{1}{2}x + 2.$$

The slope is $\frac{1}{2}$. The y-intercept is 2.

If a linear function's equation is in slope-intercept form, we can use the y-intercept and the slope to obtain its graph.

Graphing $y = mx + b$ Using the Slope and y-Intercept

1. Plot the point containing the y-intercept on the y-axis. This is the point $(0, b)$.
2. Obtain a second point using the slope, m. Write m as a fraction, and use rise over run, starting at the point containing the y-intercept, to plot this point.
3. Use a straightedge to draw a line through the two points. Draw arrowheads at the ends of the line to show that the line continues indefinitely in both directions.

EXAMPLE 4 Graphing Using the Slope and y-Intercept

Graph the linear function: $f(x) = -\frac{3}{2}x + 2.$

Solution The equation of the line is in the form $f(x) = mx + b$. We can find the slope, m, by identifying the coefficient of x. We can find the y-intercept, b, by identifying the constant term.

$$f(x) = -\frac{3}{2}x + 2$$

The slope is $-\frac{3}{2}$.

The y-intercept is 2.

Now that we have identified the slope and the y-intercept, we use the three-step procedure to graph the equation.

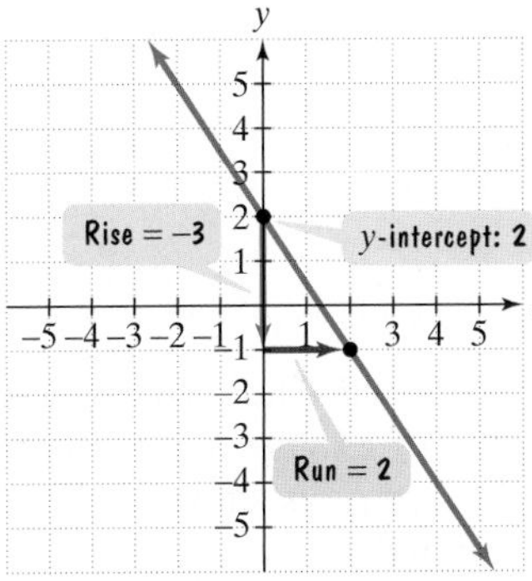

Figure 2.26 The graph of $f(x) = -\frac{3}{2}x + 2$

Step 1 **Plot the point containing the y-intercept on the y-axis.** The y-intercept is 2. We plot $(0, 2)$, shown in Figure 2.26.

Step 2 **Obtain a second point using the slope, m. Write m as a fraction, and use rise over run, starting at the point containing the y-intercept, to plot this point.** The slope, $-\frac{3}{2}$, is already written as a fraction.

$$m = -\frac{3}{2} = \frac{-3}{2} = \frac{\text{Rise}}{\text{Run}}$$

We plot the second point on the line by starting at $(0, 2)$, the first point. Based on the slope, we move 3 units *down* (the rise) and 2 units to the *right* (the run). This puts us at a second point on the line, $(2, -1)$, shown in Figure 2.26.

Step 3 **Use a straightedge to draw a line through the two points.** The graph of the linear function $f(x) = -\frac{3}{2}x + 2$ is shown as a blue line in Figure 2.26.

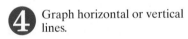

Check Point 4 Graph the linear function: $f(x) = \frac{3}{5}x + 1.$

4 Graph horizontal or vertical lines.

Equations of Horizontal and Vertical Lines

Some things change very little. For example, from 1997 through 2003, the federal minimum wage remained constant at \$5.15 per hour, indicated by the green bars in Figure 2.27. These bars show the minimum wage before it was adjusted for inflation. Also shown in the figure is a blue horizontal line segment that passes through the tops of the seven green bars.

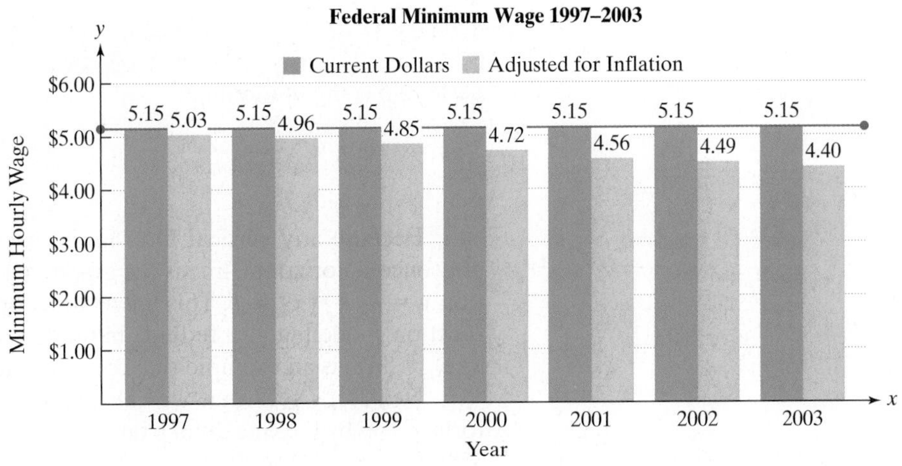

Figure 2.27

Source: www.dol.gov/esa/public/minwage

We can use $y = mx + b$, the slope-intercept form of a line's equation, to obtain an equation that models the federal minimum wage, y, in current dollars, in year x, where x is between 1997 and 2003, inclusive. The horizontal blue line segment in Figure 2.27 on the previous page provides the values for m and b:

$$y = mx + b.$$

Because the line segment is horizontal, $m = 0$.

The y-intercept is shown as \$5.15, so $b = 5.15$.

Thus, an equation that models the federal minimum wage between 1997 and 2003, inclusive, is

$$y = 0x + 5.15, \quad \text{or } y = 5.15.$$

The federal minimum wage remained constant at \$5.15 per hour. Using function notation, we can write

$$f(x) = 5.15.$$

For all year inputs, x, in the domain from 1997 through 2003, inclusive,

the output is a constant 5.15.

In general, if a line is horizontal, its slope is zero: $m = 0$. Thus, the equation $y = mx + b$ becomes $y = b$, where b is the y-intercept. For example, the graph of $y = -4$ is a horizontal line with a y-intercept of -4. The graph is shown in Figure 2.28. Three of the points along the line are shown and labeled.

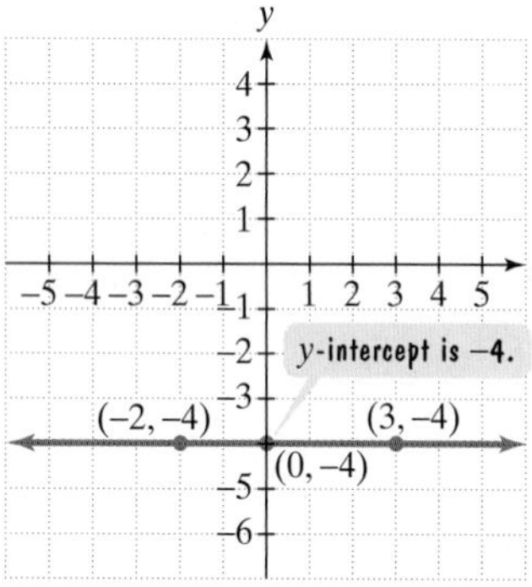

Figure 2.28 The graph of $y = -4$ or $f(x) = -4$

x	$y = -4$	(x, y)
−2	−4	(−2, −4)
0	−4	(0, −4)
3	−4	(3, −4)

For all choices of x,

y is a constant −4.

No matter what the x-coordinate is, the corresponding y-coordinate for every point on the line in Figure 2.28 is -4.

Equation of a Horizontal Line

A horizontal line is given by an equation of the form

$$y = b,$$

where b is the y-intercept.

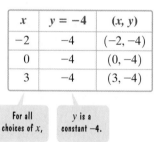

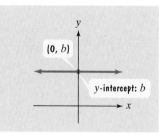

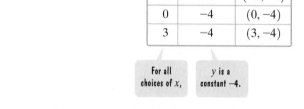

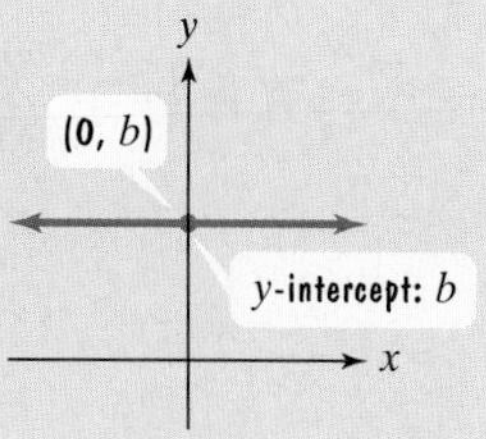

Because any vertical line can intersect the graph of a horizontal line $y = b$ only once, a horizontal line is the graph of a function. Thus, we can express the equation $y = b$ as $f(x) = b$. This linear function is often called a **constant function**. The function modeling the federal minimum wage from 1997 through 2003, namely $f(x) = 5.15$, is an example of a constant function.

Next, let's see what we can discover about the graph of an equation of the form $x = a$ by looking at an example.

EXAMPLE 5   Graphing a Vertical Line

Graph the linear equation: $x = 2$.

Solution   All ordered pairs that are solutions of $x = 2$ have a value of x that is always 2. Any value can be used for y. Let's select three of the possible values for y: -2, 0, and 3.

$x = 2$	y	(x, y)
2	-2	$(2, -2)$
2	0	$(2, 0)$
2	3	$(2, 3)$

The table shows that three ordered pairs that are solutions of $x = 2$ are $(2, -2)$, $(2, 0)$, and $(2, 3)$. Drawing a line that passes through the three points gives the vertical line shown in Figure 2.29.

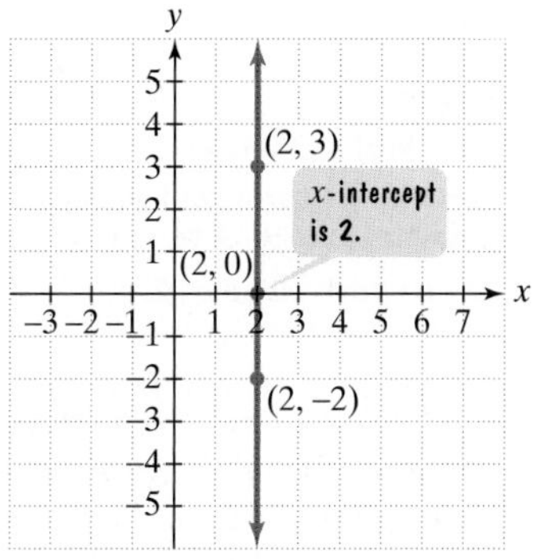

Figure 2.29 The graph of $x = 2$

Equation of a Vertical Line

A vertical line is given by an equation of the form

$$x = a,$$

where a is the x-intercept.

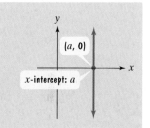

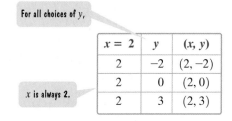

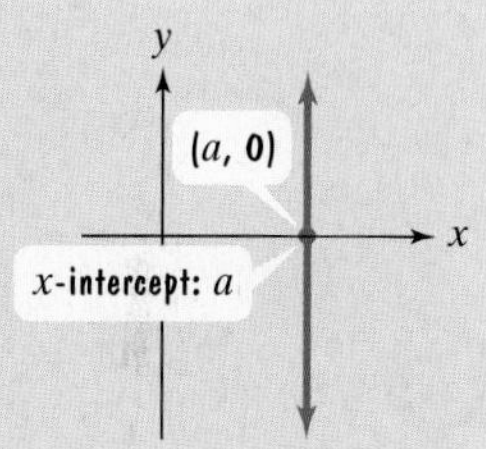

Does a vertical line represent the graph of a linear function? No. Look at the graph of $x = 2$ in Figure 2.29. A vertical line drawn through $(2, 0)$ intersects the graph infinitely many times. This shows that infinitely many outputs are associated with the input 2. **No vertical line is a linear function.**

Check Point 5   Graph the linear equation: $x = -3$.

⑤ Recognize and use the general form of a line's equation.

The General Form of the Equation of a Line

The vertical line whose equation is $x = 5$ cannot be written in slope-intercept form, $y = mx + b$, because its slope is undefined. However, every line has an equation that can be expressed in the form $Ax + By + C = 0$. For example, $x = 5$ can be expressed as $1x + 0y - 5 = 0$, or $x - 5 = 0$. The equation $Ax + By + C = 0$ is called the *general form* of the equation of a line.

General Form of the Equation of a Line

Every line has an equation that can be written in the **general form**

$$Ax + By + C = 0,$$

where A, B, and C are real numbers, and A and B are not both zero.

If the equation of a line is given in general form, it is possible to find the slope, m, and the y-intercept, b, for the line. We solve the equation for y, transforming it into the slope–intercept form $y = mx + b$. In this form, the coefficient of x is the slope of the line and the constant term is its y-intercept.

EXAMPLE 6 Finding the Slope and the y-Intercept

Find the slope and the y-intercept of the line whose equation is $3x + 2y - 4 = 0$.

Solution The equation is given in general form. We begin by rewriting it in the form $y = mx + b$. We need to solve for y.

Our goal is to isolate y.

$$3x + 2y - 4 = 0$$ This is the given equation.

$$2y = -3x + 4$$ Isolate the term containing y by adding $-3x + 4$ to both sides.

$$\frac{2y}{2} = \frac{-3x + 4}{2}$$ Divide both sides by 2.

$$y = -\frac{3}{2}x + 2$$ On the right, divide each term in the numerator by 2 to obtain slope-intercept form.

slope ($-\frac{3}{2}$) y-intercept (2)

The coefficient of x, $-\frac{3}{2}$, is the slope and the constant term, 2, is the y-intercept. This is the form of the equation that we graphed in Figure 2.26 on page 221.

Check Point 6 Find the slope and the y-intercept of the line whose equation is $3x + 6y - 12 = 0$. Then use the y-intercept and the slope to graph the equation.

6 Use intercepts to graph the general form of a line's equation.

Using Intercepts to Graph $Ax + By + C = 0$

Example 6 and Check Point 6 illustrate that one way to graph the general form of a line's equation is to convert to slope-intercept form, $y = mx + b$. Then use the slope and the y-intercept to obtain the graph.

A second method for graphing $Ax + By + C = 0$ uses intercepts. This method does not require rewriting the general form in a different form.

Using Intercepts to Graph $Ax + By + C = 0$

1. Find the x-intercept. Let $y = 0$ and solve for x. Plot the point containing the x-intercept on the x-axis.
2. Find the y-intercept. Let $x = 0$ and solve for y. Plot the point containing the y-intercept on the y-axis.
3. Use a straightedge to draw a line through the two points containing the intercepts. Draw arrowheads at the ends of the line to show that the line continues indefinitely in both directions.

EXAMPLE 7 Using Intercepts to Graph a Linear Equation

Graph using intercepts: $4x - 3y - 6 = 0$.

Solution

Step 1 Find the x-intercept. Let $y = 0$ and solve for x.

$$4x - 3 \cdot 0 - 6 = 0$$ Replace y with 0 in $4x - 3y - 6 = 0$.

$$4x - 6 = 0$$ Simplify.

$$4x = 6$$ Add 6 to both sides.

$$x = \frac{6}{4} = \frac{3}{2}$$ Divide both sides by 4.

The x-intercept is $\frac{3}{2}$, so the line passes through $\left(\frac{3}{2}, 0\right)$ or $(1.5, 0)$, as shown in Figure 2.30.

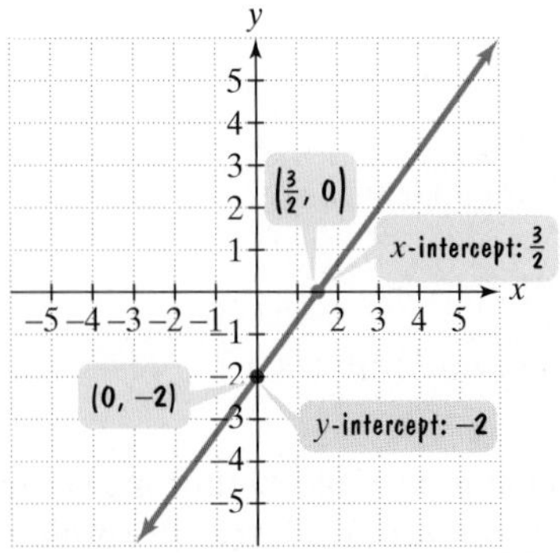

Figure 2.30 The graph of $4x - 3y - 6 = 0$

Step 2   Find the y-intercept. Let x = 0 and solve for y.

$$4 \cdot 0 - 3y - 6 = 0 \qquad \text{Replace x with 0 in } 4x - 3y - 6 = 0.$$
$$-3y - 6 = 0 \qquad \text{Simplify.}$$
$$-3y = 6 \qquad \text{Add 6 to both sides.}$$
$$y = -2 \qquad \text{Divide both sides by } -3.$$

The y-intercept is −2, so the line passes through $(0, -2)$, as shown in Figure 2.30.

Step 3   Graph the equation by drawing a line through the two points containing the intercepts. The graph of $4x - 3y - 6 = 0$ is shown in Figure 2.30.

Check Point 7 Graph using intercepts: $3x - 2y - 6 = 0$.

We've covered a lot of territory. Let's take a moment to summarize the various forms for equations of lines.

Equations of Lines

1. Point-slope form: $y - y_1 = m(x - x_1)$
2. Slope-intercept form: $y = mx + b$ or $f(x) = mx + b$
3. Horizontal line: $y = b$
4. Vertical line: $x = a$
5. General form: $Ax + By + C = 0$

Model data with linear functions and make predictions.

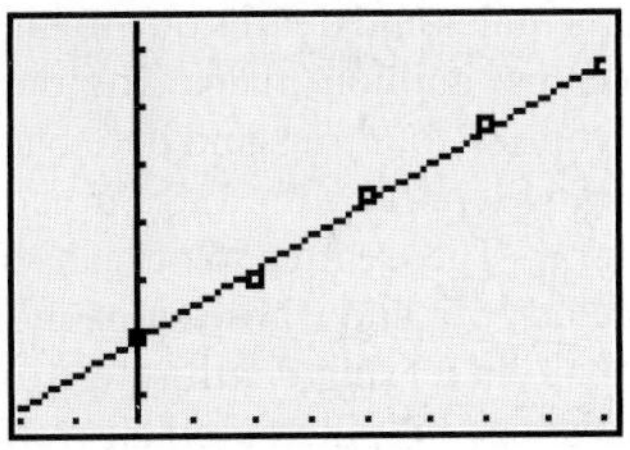

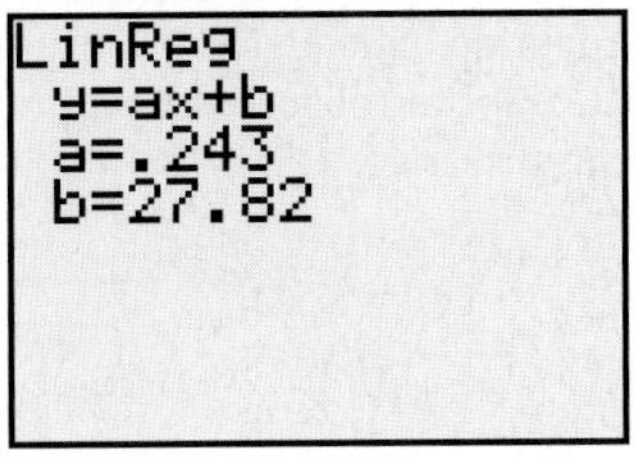

Applications

Linear functions are useful for modeling data that fall on or near a line. For example, the bar graph in Figure 2.31(a) gives the median age of the U.S. population in the indicated year. (The median age is the age in the middle when all the ages of the U.S. population are arranged from youngest to oldest.) The data are displayed as a set of five points in a rectangular coordinate system in Figure 2.31(b).

Technology

You can use a graphing utility to obtain a model for a scatter plot in which the data points fall on or near a straight line. After entering the data in Figure 2.31(b), a graphing utility displays a scatter plot of the data and the regression line, that is, the line that best fits the data.

[−10, 40, 5] by [25, 39, 2]

Also displayed is the regression line's equation.

The Graying of America: Median Age of the U.S. Population

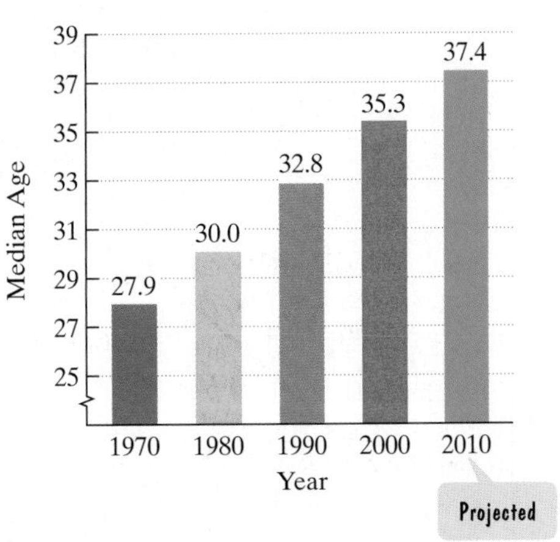

Figure 2.31(a)

Figure 2.31(b)

Source: U.S. Census Bureau

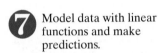

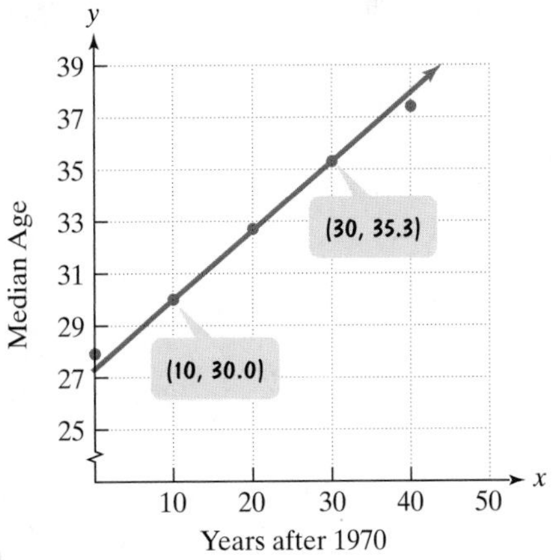

Also shown on the scatter plot in Figure 2.31(b) is a line that passes through or near the five points. By writing the equation of this line, we can obtain a model of the data and make predictions about the median age of the U.S. population in the future.

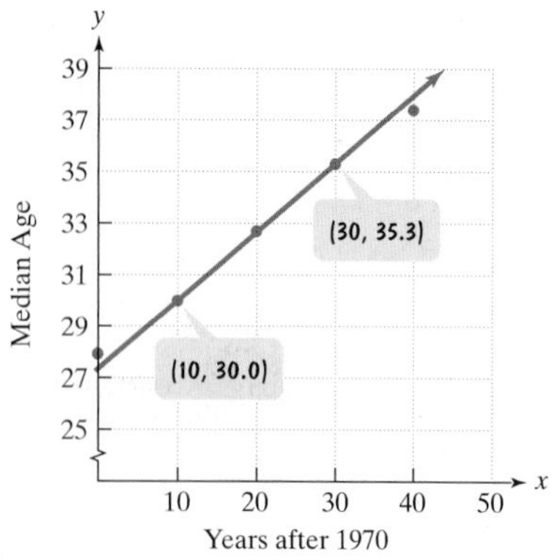

Figure 2.31(b) (repeated)

EXAMPLE 8 Modeling the Graying of America

Write the slope-intercept equation of the line shown in Figure 2.31(b). Use the equation to predict the median age of the U.S. population in 2020.

Solution The line in Figure 2.31(b) passes through (10, 30.0) and (30, 35.3). We start by finding its slope.

$$m = \frac{\text{Change in } y}{\text{Change in } x} = \frac{35.3 - 30.0}{30 - 10} = \frac{5.3}{20} = 0.265$$

The slope indicates that each year the median age of the U.S. population is increasing by 0.265 years.

Now, we write the line's slope-intercept equation.

$$y - y_1 = m(x - x_1)$$ Begin with the point-slope form.

$$y - 30.0 = 0.265(x - 10)$$ Either ordered pair can be (x_1, y_1). Let $(x_1, y_1) = (10, 30.0)$. From above, $m = 0.265$.

$$y - 30.0 = 0.265x - 2.65$$ Apply the distributive property.

$$y = 0.265x + 27.35$$ Add 30 to both sides and solve for y.

A linear function that models the median age of the U.S. population, $f(x)$, x years after 1970 is

$$f(x) = 0.265x + 27.35.$$

Now, let's use this function to predict the median age in 2020. Because 2020 is 50 years after 1970, we substitute 50 for x and evaluate the function at 50.

$$f(50) = 0.265(50) + 27.35 = 40.6$$

Our model predicts that the median age of the U.S. population in 2020 will be 40.6.

Check Point 8 Use the data points (10, 30.0) and (20, 32.8) from Figure 2.31(b) to write a slope-intercept equation that models the median age of the U.S. population x years after 1970. Use this model to predict the median age in 2020.

Cigarettes and Lung Cancer

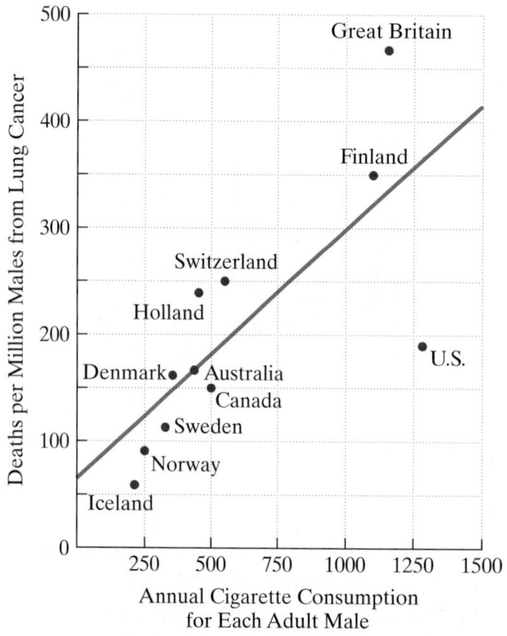

This scatter plot shows a relationship between cigarette consumption among males and deaths due to lung cancer per million males. The data are from 11 countries and date back to a 1964 report by the U.S. Surgeon General. The scatter plot can be modeled by a line whose slope indicates an increasing death rate from lung cancer with increased cigarette consumption. At that time, the tobacco industry argued that in spite of this regression line, tobacco use is not the cause of cancer. Recent data do, indeed, show a causal effect between tobacco use and numerous diseases.

Source: Smoking and Health, Washington, D.C., 1964

EXERCISE SET 2.3

Practice Exercises

In Exercises 1–10, find the slope of the line passing through each pair of points or state that the slope is undefined. Then indicate whether the line through the points rises, falls, is horizontal, or is vertical.

1. $(4, 7)$ and $(8, 10)$
2. $(2, 1)$ and $(3, 4)$
3. $(-2, 1)$ and $(2, 2)$
4. $(-1, 3)$ and $(2, 4)$
5. $(4, -2)$ and $(3, -2)$
6. $(4, -1)$ and $(3, -1)$
7. $(-2, 4)$ and $(-1, -1)$
8. $(6, -4)$ and $(4, -2)$
9. $(5, 3)$ and $(5, -2)$
10. $(3, -4)$ and $(3, 5)$

In Exercises 11–38, use the given conditions to write an equation for each line in point-slope form and slope-intercept form.

11. Slope $= 2$, passing through $(3, 5)$
12. Slope $= 4$, passing through $(1, 3)$
13. Slope $= 6$, passing through $(-2, 5)$
14. Slope $= 8$, passing through $(4, -1)$
15. Slope $= -3$, passing through $(-2, -3)$
16. Slope $= -5$, passing through $(-4, -2)$
17. Slope $= -4$, passing through $(-4, 0)$
18. Slope $= -2$, passing through $(0, -3)$
19. Slope $= -1$, passing through $\left(-\frac{1}{2}, -2\right)$
20. Slope $= -1$, passing through $\left(-4, -\frac{1}{4}\right)$
21. Slope $= \frac{1}{2}$, passing through the origin
22. Slope $= \frac{1}{3}$, passing through the origin
23. Slope $= -\frac{2}{3}$, passing through $(6, -2)$
24. Slope $= -\frac{3}{5}$, passing through $(10, -4)$
25. Passing through $(1, 2)$ and $(5, 10)$
26. Passing through $(3, 5)$ and $(8, 15)$
27. Passing through $(-3, 0)$ and $(0, 3)$
28. Passing through $(-2, 0)$ and $(0, 2)$
29. Passing through $(-3, -1)$ and $(2, 4)$
30. Passing through $(-2, -4)$ and $(1, -1)$
31. Passing through $(-3, -2)$ and $(3, 6)$
32. Passing through $(-3, 6)$ and $(3, -2)$
33. Passing through $(-3, -1)$ and $(4, -1)$
34. Passing through $(-2, -5)$ and $(6, -5)$
35. Passing through $(2, 4)$ with x-intercept $= -2$
36. Passing through $(1, -3)$ with x-intercept $= -1$
37. x-intercept $= -\frac{1}{2}$ and y-intercept $= 4$
38. x-intercept $= 4$ and y-intercept $= -2$

In Exercises 39–48, give the slope and y-intercept of each line whose equation is given. Then graph the linear function.

39. $y = 2x + 1$
40. $y = 3x + 2$
41. $f(x) = -2x + 1$
42. $f(x) = -3x + 2$
43. $f(x) = \frac{3}{4}x - 2$
44. $f(x) = \frac{3}{4}x - 3$
45. $y = -\frac{3}{5}x + 7$
46. $y = -\frac{2}{5}x + 6$
47. $g(x) = -\frac{1}{2}x$
48. $g(x) = -\frac{1}{3}x$

In Exercises 49–58, graph each equation in a rectangular coordinate system.

49. $y = -2$
50. $y = 4$
51. $x = -3$
52. $x = 5$
53. $y = 0$
54. $x = 0$
55. $f(x) = 1$
56. $f(x) = 3$
57. $3x - 18 = 0$
58. $3x + 12 = 0$

In Exercises 59–66,
- **a.** *Rewrite the given equation in slope-intercept form.*
- **b.** *Give the slope and y-intercept.*
- **c.** *Use the slope and y-intercept to graph the linear function.*

59. $3x + y - 5 = 0$
60. $4x + y - 6 = 0$
61. $2x + 3y - 18 = 0$
62. $4x + 6y + 12 = 0$
63. $8x - 4y - 12 = 0$
64. $6x - 5y - 20 = 0$
65. $3y - 9 = 0$
66. $4y + 28 = 0$

In Exercises 67–72, use intercepts to graph each equation.

67. $6x - 2y - 12 = 0$
68. $6x - 9y - 18 = 0$
69. $2x + 3y + 6 = 0$
70. $3x + 5y + 15 = 0$
71. $8x - 2y + 12 = 0$
72. $6x - 3y + 15 = 0$

Practice Plus

In Exercises 73–76, find the slope of the line passing through each pair of points or state that the slope is undefined. Assume that all variables represent positive real numbers. Then indicate whether the line through the points rises, falls, is horizontal, or is vertical.

73. $(0, a)$ and $(b, 0)$
74. $(-a, 0)$ and $(0, -b)$
75. (a, b) and $(a, b + c)$
76. $(a - b, c)$ and $(a, a + c)$

In Exercises 77–78, give the slope and y-intercept of each line whose equation is given. Assume that $B \neq 0$.

77. $Ax + By = C$ **78.** $Ax = By - C$

In Exercises 79–80, find the value of y if the line through the two given points is to have the indicated slope.

79. $(3, y)$ and $(1, 4)$, $m = -3$

80. $(-2, y)$ and $(4, -4)$, $m = \frac{1}{3}$

In Exercises 81–82, graph each linear function.

81. $3x - 4f(x) - 6 = 0$ **82.** $6x - 5f(x) - 20 = 0$

83. If one point on a line is $(3, -1)$ and the line's slope is -2, find the y-intercept.

84. If one point on a line is $(2, -6)$ and the line's slope is $-\frac{3}{2}$, find the y-intercept.

Use the figure to make the lists in Exercises 85–86.

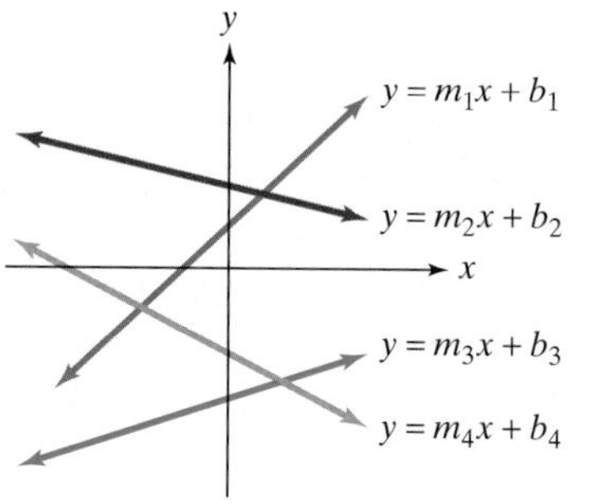

85. List the slopes $m_1, m_2, m_3,$ and m_4 in order of decreasing size.

86. List the y-intercepts $b_1, b_2, b_3,$ and b_4 in order of decreasing size.

Application Exercises

Though increasing numbers of Americans are obese, fewer are trimming down by watching what they eat. The bar graph shows the percentage of American adults on weight-loss diets for four selected years. The data are displayed as two sets of four points each, one scatter plot for the percentage of dieting women and one for the percentage of dieting men. Also shown in each scatter plot is a line that passes through or near the four points. Use these lines to solve Exercises 87–88.

Percentage of American Adults on Weight-Loss Diets

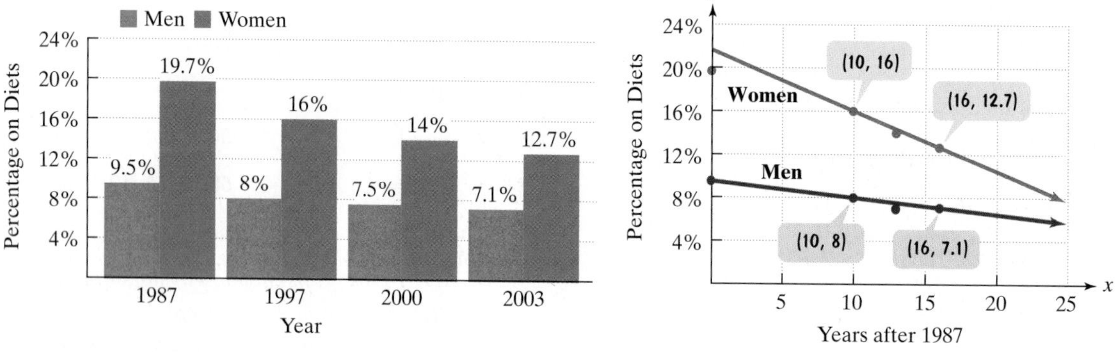

Source: Mediamark Research, American Demographics

87. In this exercise, you will use the blue line for the women shown on the scatter plot to develop a model for the percentage of dieting American women.

a. Use the two points whose coordinates are shown by the voice balloons to find the point-slope form of the equation of the line that models the percentage of adult women on diets, y, x years after 1987.

b. Write the equation in part (a) in slope-intercept form. Use function notation.

c. Use the linear function to predict the percentage of adult women on weight-loss diets in 2007.

88. In this exercise, you will use the red line for men shown on the scatter plot to develop a model for the percentage of dieting American men.

a. Use the two points whose coordinates are shown by the voice balloons to find the point-slope form of the equation of the line that models the percentage of adult men on diets, y, x years after 1987.

b. Write the equation in part (a) in slope-intercept form. Use function notation.

c. Use the linear function to predict the percentage of adult men on weight-loss diets in 2007.

89. The bar graph shows life expectancies for Americans born in seven selected years.

Life Expectancy in the U.S. by Birth Year

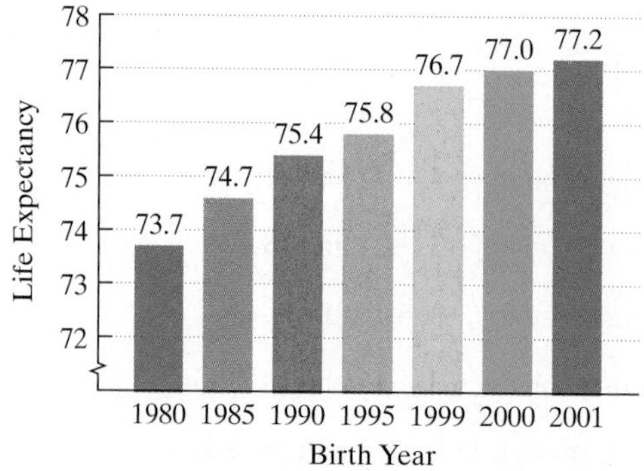

Source: National Center for Health Statistics

a. Let x represent the number of birth years after 1980 and let y represent life expectancy. Create a scatter plot that displays the data as a set of seven points in a rectangular coordinate system.

b. Draw a line through the two points that show life expectancies for 1985 and 2000. Use the coordinates of these points to write the line's equation in point-slope form and slope-intercept form. Round the slope to two decimal places.

c. Write a linear function that models life expectancy, $E(x)$, for Americans born x years after 1980. Then use this function to predict the life expectancy of an American born in 2020.

90. The bar graph shows the number of global HIV/AIDS cases, in millions, from 1999 through 2003.

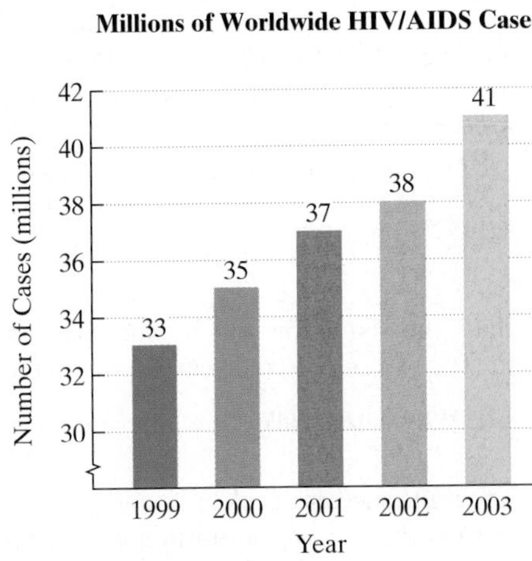

Source: UNAIDS

a. Let x represent the number of years after 1999 and let y represent the number of HIV/AIDS cases worldwide, in millions. Create a scatter plot that displays the data as a set of five points in a rectangular coordinate system.

b. Draw a line through the two points that show the number of cases in 2000 and 2003. Use the coordinates of these points to write the line's equation in point-slope form and slope-intercept form.

c. Write a linear function that models the number of HIV/AIDS cases worldwide, $A(x)$, in millions, x years after 1999. Then use this function to predict the number of cases in 2010.

91. Shown, again, is the scatter plot that indicates a relationship between the percentage of adult females in a country who are literate and the mortality of children under five. Also

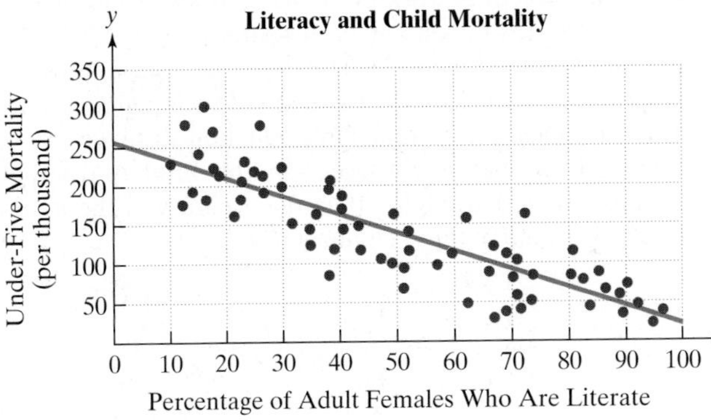

Source: United Nations

shown is a line that passes through or near the points. Find a linear function that models the data by finding the slope-intercept form of the line's equation. Use the function to make a prediction about child mortality based on the percentage of adult females in a country who are literate.

92. Just as money doesn't buy happiness for individuals, the two don't necessarily go together for countries either. However, the scatter plot does show a relationship between a country's annual per capita income and the percentage of people in that country who call themselves "happy."

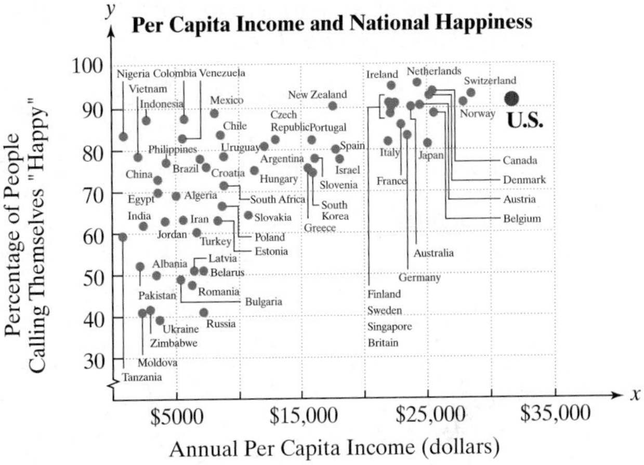

Source: Richard Layard, *Happiness: Lessons from a New Science*, Penguin, 2005

Draw a line that fits the data so that the spread of the data points around the line is as small as possible. Use the coordinates of two points along your line to write the slope-intercept form of its equation. Express the equation in function notation and use the linear function to make a prediction about national happiness based on per capita income.

Writing in Mathematics

93. What is the slope of a line and how is it found?

94. Describe how to write the equation of a line if two points along the line are known.

95. Explain how to derive the slope-intercept form of a line's equation, $y = mx + b$, from the point-slope form

$$y - y_1 = m(x - x_1).$$

96. Explain how to graph the equation $x = 2$. Can this equation be expressed in slope-intercept form? Explain.

97. Explain how to use the general form of a line's equation to find the line's slope and y-intercept.

98. Explain how to use intercepts to graph the general form of a line's equation.

99. Take another look at the scatter plot in Exercise 91. Although there is a relationship between literacy and child mortality, we cannot conclude that increased literacy causes child mortality to decrease. Offer two or more possible explanations for the data in the scatter plot.

Technology Exercises

Use a graphing utility to graph each equation in Exercises 100–103. Then use the $\boxed{\text{TRACE}}$ *feature to trace along the line and find the coordinates of two points. Use these points to compute the line's slope. Check your result by using the coefficient of x in the line's equation.*

100. $y = 2x + 4$ **101.** $y = -3x + 6$

102. $y = -\frac{1}{2}x - 5$ **103.** $y = \frac{3}{4}x - 2$

104. Is there a relationship between alcohol from moderate wine consumption and heart disease death rate? The table gives data from 19 developed countries.

France

Country	A	B	C	D	E	F	G
Liters of alcohol from drinking wine, per person per year (x)	2.5	3.9	2.9	2.4	2.9	0.8	9.1
Deaths from heart disease, per 100,000 people per year (y)	211	167	131	191	220	297	71

U.S.

Country	H	I	J	K	L	M	N	O	P	Q	R	S
(x)	0.8	0.7	7.9	1.8	1.9	0.8	6.5	1.6	5.8	1.3	1.2	2.7
(y)	211	300	107	167	266	227	86	207	115	285	199	172

Source: New York Times, December 28, 1994

a. Use the statistical menu of your graphing utility to enter the 19 ordered pairs of data items shown in the table.

b. Use the $\boxed{\text{DRAW}}$ menu and the scatter plot capability to draw a scatter plot of the data.

c. Select the linear regression option. Use your utility to obtain values for a and b for the equation of the regression line, $y = ax + b$. You may also be given a **correlation coefficient**, r. Values of r close to 1 indicate that the points can be described by a linear relationship and the regression line has a positive slope. Values of r close to -1 indicate that the points can be described by a linear relationship and the regression line has a negative slope. Values of r close to 0 indicate no linear relationship between the variables. In this case, a linear model does not accurately describe the data.

d. Use the appropriate sequence (consult your manual) to graph the regression equation on top of the points in the scatter plot.

Critical Thinking Exercises

105. Which one of the following is true?

a. A linear function with nonnegative slope has a graph that rises from left to right.

b. Every line in the rectangular coordinate system has an equation that can be expressed in slope-intercept form.

c. The graph of the linear function $5x + 6y = 30$ is a line passing through the point $(6, 0)$ with slope $-\frac{5}{6}$.

d. The graph of $x = 7$ in the rectangular coordinate system is the single point $(7, 0)$.

In Exercises 106–107, find the coefficients that must be placed in each shaded area so that the function's graph will be a line satisfying the specified conditions.

106. ▢ $x +$ ▢ $y = 12$; x-intercept $= -2$; y-intercept $= 4$

107. ▢ $x +$ ▢ $y = 12$; y-intercept $= -6$; slope $= \dfrac{1}{2}$

108. Prove that the equation of a line passing through $(a, 0)$ and $(0, b)$ $(a \neq 0, b \neq 0)$ can be written in the form $\dfrac{x}{a} + \dfrac{y}{b} = 1$. Why is this called the *intercept form* of a line?

109. Excited about the success of celebrity stamps, post office officials were rumored to have put forth a plan to institute two new types of thermometers. On these new scales, $°E$ represents degrees Elvis and $°M$ represents degrees Madonna. If it is known that $40°E = 25°M$, $280°E = 125°M$, and degrees Elvis is linearly related to degrees Madonna, write an equation expressing E in terms of M.

Group Exercise

110. In Example 8 on page 226, we used the data in Figure 2.31 on page 225 to develop a linear function that modeled the graying of America. For this group exercise, you might find it helpful to pattern your work after Figure 2.31 and the solution to Example 8. Group members should begin by consulting an almanac, newspaper, magazine, or the Internet to find data that appear to lie approximately on or near a line. Working by hand or using a graphing utility, group members should construct scatter plots for the data that were assembled. If working by hand, draw a line that approximately fits the data in each scatter plot and then write its equation as a function in slope-intercept form. If using a graphing utility, obtain the equation of each regression line. Then use each linear function's equation to make predictions about what might occur in the future. Are there circumstances that might affect the accuracy of the prediction? List some of these circumstances.

SECTION 2.4 More on Slope

Objectives

1. Find slopes and equations of parallel and perpendicular lines.
2. Interpret slope as rate of change.
3. Find a function's average rate of change.

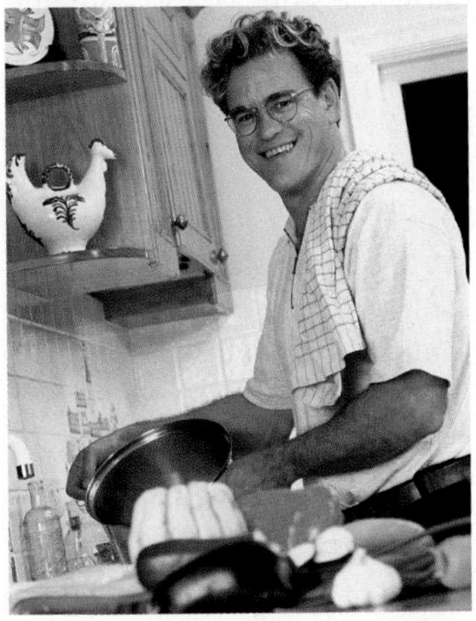

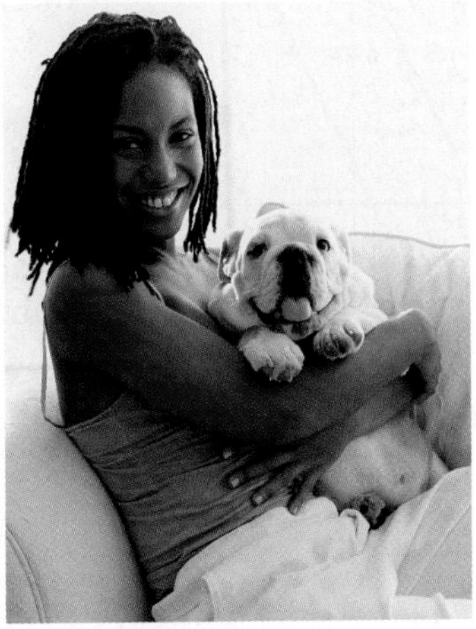

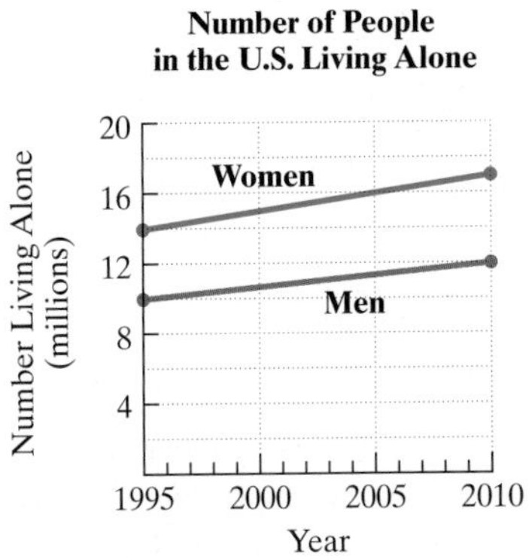

Figure 2.32

Source: Forrester Research

A best guess at the look of our nation in the next decades indicates that the number of men and women living alone will increase each year. Figure 2.32 shows that by 2010, approximately 12 million men and 17 million women will be living alone.

By looking at Figure 2.32, can you tell that the green graph representing women has a greater slope than the blue graph representing men? This indicates a greater yearly rate of change in the millions of women living alone than in the millions of men living alone. In this section, you will learn to interpret slope as a rate of change. You will also explore the relationships between slopes of parallel and perpendicular lines.

1 Find slopes and equations of parallel and perpendicular lines.

Parallel and Perpendicular Lines

Two nonintersecting lines that lie in the same plane are **parallel**. If two lines do not intersect, the ratio of the vertical change to the horizontal change is the same for each line. Because two parallel lines have the same "steepness," they must have the same slope.

Slope and Parallel Lines

1. If two nonvertical lines are parallel, then they have the same slope.
2. If two distinct nonvertical lines have the same slope, then they are parallel.
3. Two distinct vertical lines, both with undefined slopes, are parallel.

EXAMPLE 1 Writing Equations of a Line Parallel to a Given Line

Write an equation of the line passing through $(-3, 1)$ and parallel to the line whose equation is $y = 2x + 1$. Express the equation in point-slope form and slope-intercept form.

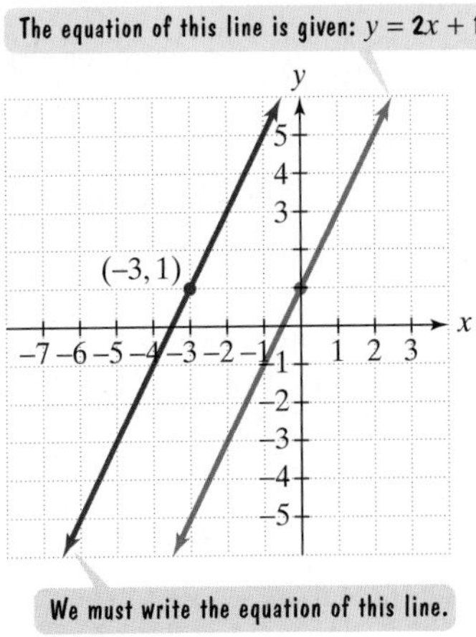

Figure 2.33

Solution The situation is illustrated in Figure 2.33. We are looking for the equation of the red line shown on the left. How do we obtain this equation? Notice that the line passes through the point $(-3, 1)$. Using the point-slope form of the line's equation, we have $x_1 = -3$ and $y_1 = 1$.

$$y - y_1 = m(x - x_1)$$

$y_1 = 1$ $\quad$ $x_1 = -3$

The equation of this line is given: $y = 2x + 1$.

$(-3, 1)$

We must write the equation of this line.

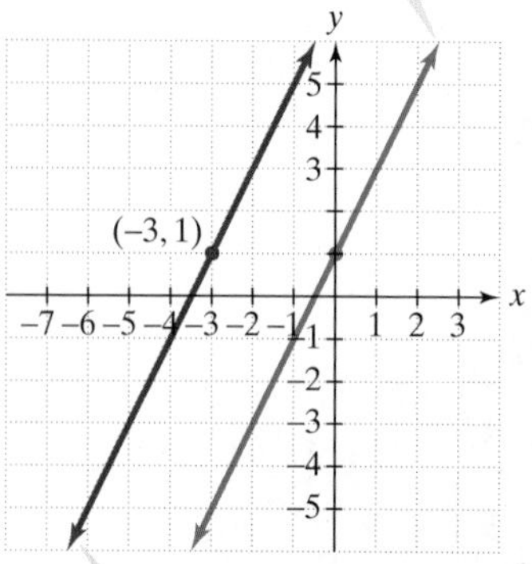

Figure 2.33 (repeated)

With $(x_1, y_1) = (-3, 1)$, the only thing missing from the equation of the red line is m, the slope. Do we know anything about the slope of either line in Figure 2.33? The answer is yes; we know the slope of the blue line on the right, whose equation is given.

$$y = 2x + 1$$

The slope of the blue line on the right in Figure 2.33 is 2.

Parallel lines have the same slope. Because the slope of the blue line is 2, the slope of the red line, the line whose equation we must write, is also 2: $m = 2$. We now have values for x_1, y_1, and m for the red line.

$$y - y_1 = m(x - x_1)$$

$y_1 = 1$ $m = 2$ $x_1 = -3$

The point-slope form of the red line's equation is

$$y - 1 = 2[x - (-3)] \text{ or}$$
$$y - 1 = 2(x + 3).$$

Solving for y, we obtain the slope-intercept form of the equation.

$$y - 1 = 2x + 6 \qquad \text{Apply the distributive property.}$$
$$y = 2x + 7 \qquad \text{Add 1 to both sides. This is the slope-intercept}$$

form, $y = mx + b$, of the equation. Using function notation, the equation is $f(x) = 2x + 7$.

Check Point 1 Write an equation of the line passing through $(-2, 5)$ and parallel to the line whose equation is $y = 3x + 1$. Express the equation in point-slope form and slope-intercept form.

Two lines that intersect at a right angle (90°) are said to be **perpendicular**, shown in Figure 2.34. The relationship between the slopes of perpendicular lines is not as obvious as the relationship between parallel lines. Figure 2.34 shows line AB, with slope $\frac{c}{d}$. Rotate line AB through 90° to the left to obtain line $A'B'$, perpendicular to line AB. The figure indicates that the rise and the run of the new line are reversed from the original line, but the rise is now negative. This means that the slope of the new line is $-\frac{d}{c}$. Notice that the product of the slopes of the two perpendicular lines is -1:

$$\left(\frac{c}{d}\right)\left(-\frac{d}{c}\right) = -1.$$

This relationship holds for all perpendicular lines and is summarized in the following box:

B'

Slope $= -\frac{d}{c}$

$-d$

A' d

c B

A Slope $= \frac{c}{d}$

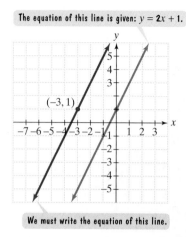

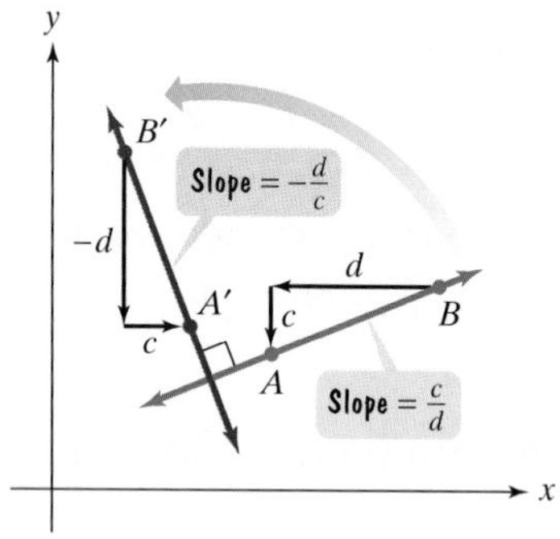

Figure 2.34 Slopes of perpendicular lines

Slope and Perpendicular Lines

1. If two nonvertical lines are perpendicular, then the product of their slopes is -1.
2. If the product of the slopes of two lines is -1, then the lines are perpendicular.
3. A horizontal line having zero slope is perpendicular to a vertical line having undefined slope.

An equivalent way of stating this relationship is to say that **one line is perpendicular to another line if its slope is the *negative reciprocal* of the slope of the other line.** For example, if a line has slope 5, any line having slope $-\frac{1}{5}$ is perpendicular to it. Similarly, if a line has slope $-\frac{3}{4}$, any line having slope $\frac{4}{3}$ is perpendicular to it.

EXAMPLE 2 Writing Equations of a Line Perpendicular to a Given Line

a. Find the slope of any line that is perpendicular to the line whose equation is $x + 4y - 8 = 0$.

b. Write the equation of the line passing through $(3, -5)$ and perpendicular to the line whose equation is $x + 4y - 8 = 0$. Express the equation in general form.

Solution

a. We begin by writing the equation of the given line, $x + 4y - 8 = 0$, in slope-intercept form. Solve for y.

$$x + 4y - 8 = 0 \quad \text{This is the given equation.}$$

$$4y = -x + 8 \quad \text{To isolate the } y\text{-term, subtract } x \text{ and add 8 on both sides.}$$

$$y = -\frac{1}{4}x + 2 \quad \text{Divide both sides by 4.}$$

Slope is $-\frac{1}{4}$.

The given line has slope $-\frac{1}{4}$. Any line perpendicular to this line has a slope that is the negative reciprocal of $-\frac{1}{4}$. Thus, the slope of any perpendicular line is 4.

b. Let's begin by writing the point-slope form of the perpendicular line's equation. Because the line passes through the point $(3, -5)$, we have $x_1 = 3$ and $y_1 = -5$. In part (a), we determined that the slope of any line perpendicular to $x + 4y - 8 = 0$ is 4, so the slope of this particular perpendicular line must also be 4: $m = 4$.

$$y - y_1 = m(x - x_1)$$

$y_1 = -5$ $m = 4$ $x_1 = 3$

The point-slope form of the perpendicular line's equation is

$$y - (-5) = 4(x - 3) \text{ or}$$
$$y + 5 = 4(x - 3).$$

How can we express this equation in general form $(Ax + By + C = 0)$? We need to obtain zero on one side of the equation. Let's do this and keep A, the coefficient of x, positive.

$$y + 5 = 4(x - 3) \quad \text{This is the point-slope form of the line's equation.}$$

$$y + 5 = 4x - 12 \quad \text{Apply the distributive property.}$$

$$y - y + 5 - 5 = 4x - y - 12 - 5 \quad \text{To obtain 0 on the left, subtract } y \text{ and subtract 5 on both sides.}$$

$$0 = 4x - y - 17 \quad \text{Simplify.}$$

The general form of the perpendicular line's equation is $4x - y - 17 = 0$.

Check Point 2

a. Find the slope of any line that is perpendicular to the line whose equation is $x + 3y - 12 = 0$.

b. Write the equation of the line passing through $(-2, -6)$ and perpendicular to the line whose equation is $x + 3y - 12 = 0$. Express the equation in general form.

2 Interpret slope as rate of change.

Slope as Rate of Change

Slope is defined as the ratio of a change in y to a corresponding change in x. It describes how fast y is changing with respect to x. For a linear function, slope may be interpreted as the rate of change of the dependent variable per unit change in the independent variable.

Our next example shows how slope can be interpreted as a rate of change in an applied situation. When calculating slope in applied problems, keep track of the units in the numerator and the denominator.

EXAMPLE 3 Slope as a Rate of Change

The line graphs for the number of women and men living alone are shown again in Figure 2.35. Find the slope of the line segment for the women. Describe what this slope represents.

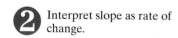

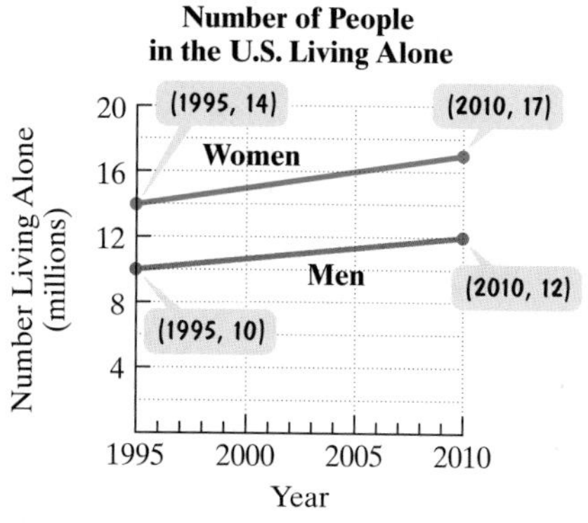

Figure 2.35
Source: Forrester Research

Solution We let x represent a year and y the number of women living alone in that year. The two points shown on the line segment for women have the following coordinates:

$$(1995, 14) \quad \text{and} \quad (2010, 17).$$

In 1995, 14 million U.S. women lived alone.

In 2010, 17 million U.S. women are projected to live alone.

Now we compute the slope:

$$m = \frac{\text{Change in } y}{\text{Change in } x} = \frac{17 - 14}{2010 - 1995} = \frac{3}{15} = \frac{1}{5} = \frac{0.2 \text{ million women}}{\text{year}}.$$

The unit in the numerator is *million women.*

The unit in the denominator is *year.*

The slope indicates that the number of U.S. women living alone is projected to increase by 0.2 million each year. The rate of change is 0.2 million women per year.

Check Point 3 Use the graph in Figure 2.35 to find the slope of the line segment for the men. Express the slope correct to two decimal places and describe what it represents.

In Check Point 3 did you find that the slope of the line segment for the men is different from that of the women? The rate of change for men living alone is not equal to the rate of change for women living alone. Because of these different slopes, if you extend the line segments in Figure 2.35, the resulting lines will intersect. They are not parallel.

3 Find a function's average rate of change.

The Average Rate of Change of a Function

If the graph of a function is not a straight line, the **average rate of change** between any two points is the slope of the line containing the two points. This line is called a **secant line**. For example, Figure 2.36 shows the graph of a particular man's height, in inches, as a function of his age, in years. Two points on the graph are labeled: (13, 57) and (18, 76). At age 13, this man was 57 inches tall and at age 18, he was 76 inches tall.

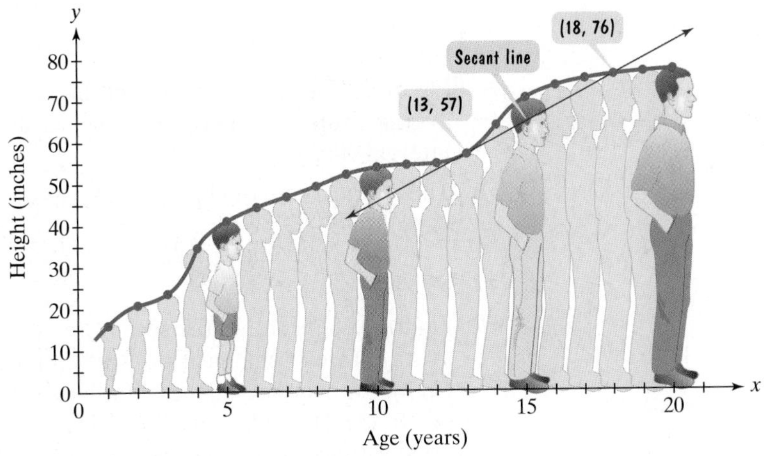

Figure 2.36 Height as a function of age

The man's average growth rate between ages 13 and 18 is the slope of the secant line containing (13, 57) and (18, 76):

$$m = \frac{\text{Change in } y}{\text{Change in } x} = \frac{76 - 57}{18 - 13} = \frac{19}{5} = 3\frac{4}{5}.$$

This man's average rate of change, or average growth rate, from age 13 to age 18 was $3\frac{4}{5}$, or 3.8, inches per year.

The Average Rate of Change of a Function

Let $(x_1, f(x_1))$ and $(x_2, f(x_2))$ be distinct points on the graph of a function f. (See Figure 2.37.) The **average rate of change of f** from x_1 to x_2 is

$$\frac{f(x_2) - f(x_1)}{x_2 - x_1}.$$

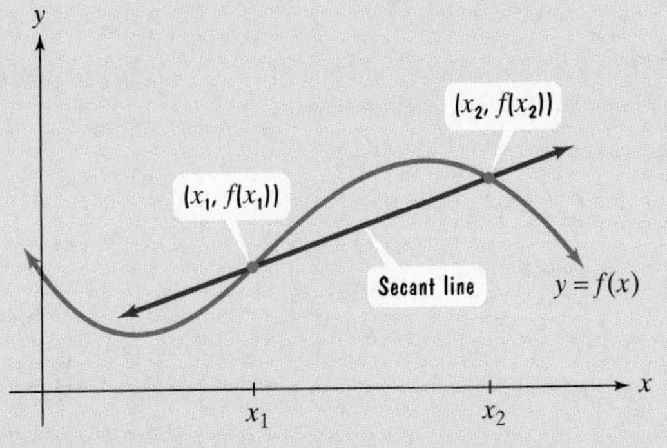

Figure 2.37

EXAMPLE 4 Finding the Average Rate of Change

Find the average rate of change of $f(x) = x^2$ from

a. $x_1 = 0$ to $x_2 = 1$ **b.** $x_1 = 1$ to $x_2 = 2$ **c.** $x_1 = -2$ to $x_2 = 0$.

Solution

a. The average rate of change of $f(x) = x^2$ from $x_1 = 0$ to $x_2 = 1$ is

$$\frac{f(x_2) - f(x_1)}{x_2 - x_1} = \frac{f(1) - f(0)}{1 - 0} = \frac{1^2 - 0^2}{1} = 1.$$

Figure 2.38(a) shows the secant line of $f(x) = x^2$ from $x_1 = 0$ to $x_2 = 1$. The average rate of change is positive and the function is increasing on the interval $(0, 1)$.

b. The average rate of change of $f(x) = x^2$ from $x_1 = 1$ to $x_2 = 2$ is

$$\frac{f(x_2) - f(x_1)}{x_2 - x_1} = \frac{f(2) - f(1)}{2 - 1} = \frac{2^2 - 1^2}{1} = 3.$$

Figure 2.38(b) shows the secant line of $f(x) = x^2$ from $x_1 = 1$ to $x_2 = 2$. The average rate of change is positive and the function is increasing on the interval $(1, 2)$. Can you see that the graph rises more steeply on the interval $(1, 2)$ than on $(0, 1)$? This is because the average rate of change from $x_1 = 1$ to $x_2 = 2$ is greater than the average rate of change from $x_1 = 0$ to $x_2 = 1$.

c. The average rate of change of $f(x) = x^2$ from $x_1 = -2$ to $x_2 = 0$ is

$$\frac{f(x_2) - f(x_1)}{x_2 - x_1} = \frac{f(0) - f(-2)}{0 - (-2)} = \frac{0^2 - (-2)^2}{2} = \frac{-4}{2} = -2.$$

Figure 2.38(c) shows the secant line of $f(x) = x^2$ from $x_1 = -2$ to $x_2 = 0$. The average rate of change is negative and the function is decreasing on the interval $(-2, 0)$.

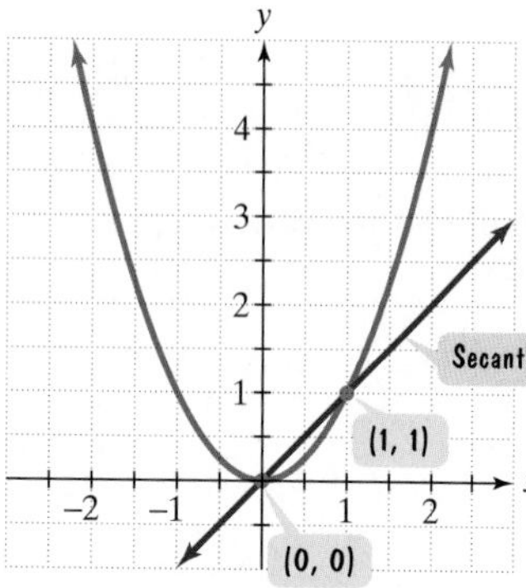

Figure 2.38(a) The secant line of $f(x) = x^2$ from $x_1 = 0$ to $x_2 = 1$

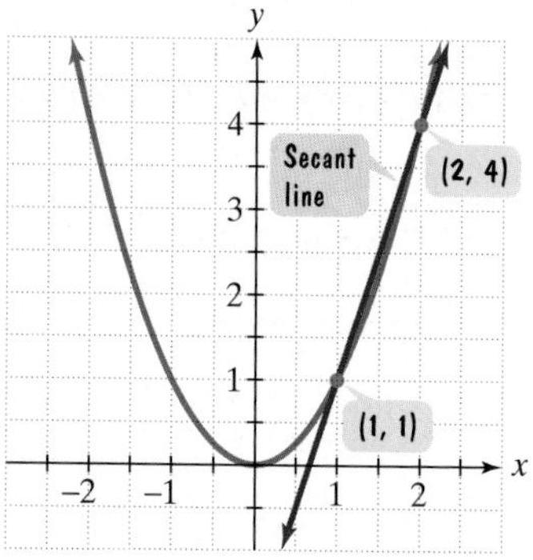

Figure 2.38(b) The secant line of $f(x) = x^2$ from $x_1 = 1$ to $x_2 = 2$

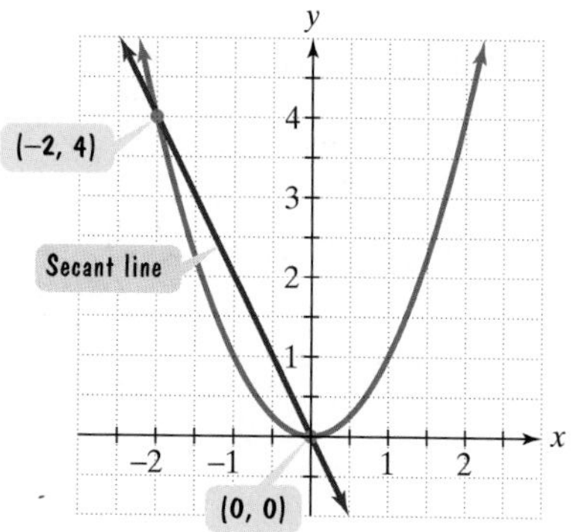

Figure 2.38(c) The secant line of $f(x) = x^2$ from $x_1 = -2$ to $x_2 = 0$

Check Point 4 Find the average rate of change of $f(x) = x^3$ from

a. $x_1 = 0$ to $x_2 = 1$ **b.** $x_1 = 1$ to $x_2 = 2$ **c.** $x_1 = -2$ to $x_2 = 0$.

Suppose we are interested in the average rate of change of f from $x_1 = x$ to $x_2 = x + h$. In this case, the average rate of change is

$$\frac{f(x_2) - f(x_1)}{x_2 - x_1} = \frac{f(x + h) - f(x)}{x + h - x} = \frac{f(x + h) - f(x)}{h}.$$

Do you recognize the last expression? It is the difference quotient that you used in Section 2.2. Thus, the difference quotient gives the average rate of change of a function from x to $x + h$. In the difference quotient, h is thought of as a number very close to 0. In this way, the average rate of change can be found for a very short interval.

EXAMPLE 5 Finding the Average Rate of Change

When a person receives a drug injected into a muscle, the concentration of the drug in the body, measured in milligrams per 100 milliliters, is a function of the time elapsed after the injection, measured in hours. Figure 2.39 shows the graph of such a function, where x represents hours after the injection and $f(x)$ is the drug's concentration at time x. Find the average rate of change in the drug's concentration between 3 and 7 hours.

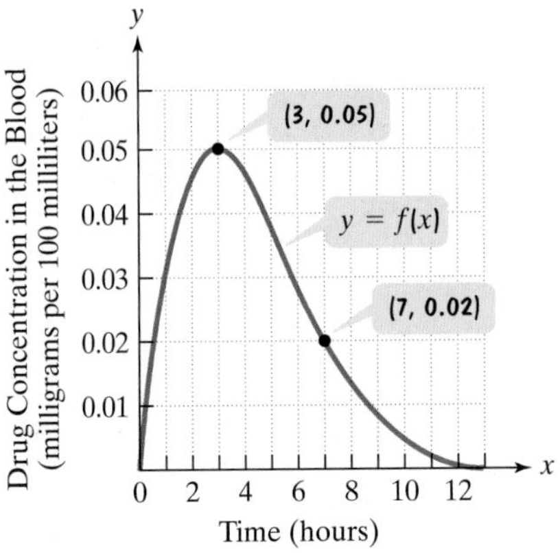

Figure 2.39 Concentration of a drug as a function of time

Solution At 3 hours, the drug's concentration is 0.05 and at 7 hours, the concentration is 0.02. The average rate of change in its concentration between 3 and 7 hours is

$$\frac{f(x_2) - f(x_1)}{x_2 - x_1} = \frac{f(7) - f(3)}{7 - 3} = \frac{0.02 - 0.05}{7 - 3} = \frac{-0.03}{4} = -0.0075.$$

The average rate of change is -0.0075. This means that the drug's concentration is decreasing at an average rate of 0.0075 milligrams per 100 milliliters per hour.

Study Tip

Units used to describe x and y tend to "pile up" when expressing the rate of change of y with respect to x. The unit used to express the rate of change of y with respect to x is

the unit used to describe y **per** the unit used to describe x.

In Figure 2.39, y, or drug concentration, is described in milligrams per 100 milliliters.

In Figure 2.39, x, or time, is described in hours.

In Figure 2.39, the rate of change is described in terms of milligrams per 100 milliliters per hour.

Check Point 5 Use Figure 2.39 to find the average rate of change in the drug's concentration between 1 hour and 3 hours.

How Calculus Studies Change

Calculus allows motion and change to be analyzed by "freezing the frame" of a continuous changing process, instant by instant. For example, Figure 2.40 shows a male's changing height over intervals of time. Over the period of time from P to D, his average rate of growth is his change in height—that is, his height at time D minus his height at time P—divided by the change in time from P to D. This is the slope of secant line PD.

The secant lines PD, PC, PB, and PA shown in Figure 2.40 have slopes that show average growth rates for successively shorter periods of time. Calculus makes these time frames so small that they approach a single point—that is, a single instant in time. This point is shown as point P in Figure 2.40. The slope of the line that touches the graph at P gives the male's growth rate at one instant in time, P.

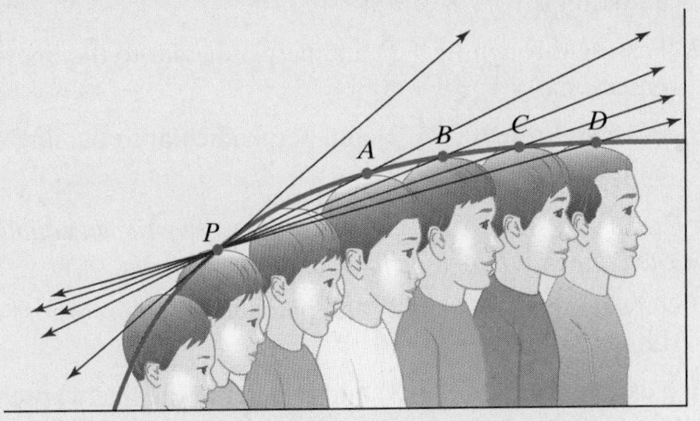

Figure 2.40 Analyzing continuous growth over intervals of time and at an instant in time

EXERCISE SET 2.4

Practice Exercises

In Exercises 1–4, write an equation for line L in point-slope form and slope-intercept form.

1.

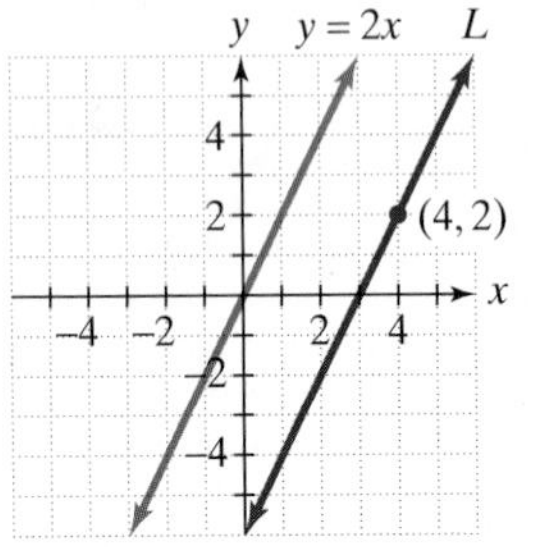

L is parallel to $y = 2x$.

2.

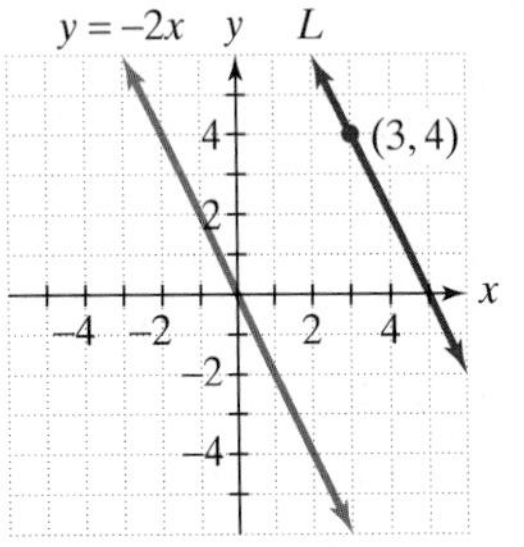

L is parallel to $y = -2x$.

3.

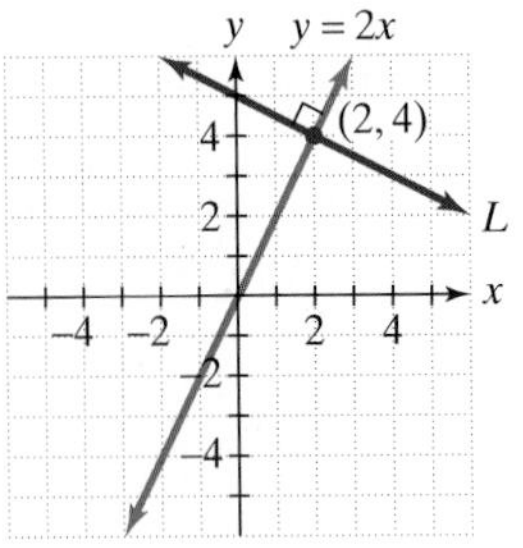

L is perpendicular to $y = 2x$.

4.

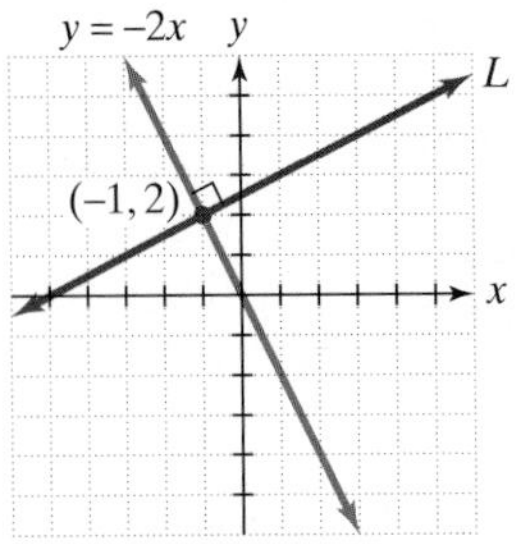

L is perpendicular to $y = -2x$.

In Exercises 5–8, use the given conditions to write an equation for each line in point-slope form and slope-intercept form.

5. Passing through $(-8, -10)$ and parallel to the line whose equation is $y = -4x + 3$
6. Passing through $(-2, -7)$ and parallel to the line whose equation is $y = -5x + 4$
7. Passing through $(2, -3)$ and perpendicular to the line whose equation is $y = \frac{1}{5}x + 6$
8. Passing through $(-4, 2)$ and perpendicular to the line whose equation is $y = \frac{1}{3}x + 7$

In Exercises 9–12, use the given conditions to write an equation for each line in point-slope form and general form.

9. Passing through $(-2, 2)$ and parallel to the line whose equation is $2x - 3y - 7 = 0$
10. Passing through $(-1, 3)$ and parallel to the line whose equation is $3x - 2y - 5 = 0$
11. Passing through $(4, -7)$ and perpendicular to the line whose equation is $x - 2y - 3 = 0$
12. Passing through $(5, -9)$ and perpendicular to the line whose equation is $x + 7y - 12 = 0$

In Exercises 13–18, find the average rate of change of the function from x_1 to x_2.

13. $f(x) = 3x$ from $x_1 = 0$ to $x_2 = 5$
14. $f(x) = 6x$ from $x_1 = 0$ to $x_2 = 4$
15. $f(x) = x^2 + 2x$ from $x_1 = 3$ to $x_2 = 5$
16. $f(x) = x^2 - 2x$ from $x_1 = 3$ to $x_2 = 6$
17. $f(x) = \sqrt{x}$ from $x_1 = 4$ to $x_2 = 9$
18. $f(x) = \sqrt{x}$ from $x_1 = 9$ to $x_2 = 16$

Practice Plus

In Exercises 19–24, write the slope-intercept equation of a function f whose graph satisfies the given conditions.

19. The graph of f passes through $(-1, 5)$ and is perpendicular to the line whose equation is $x = 6$.
20. The graph of f passes through $(-2, 6)$ and is perpendicular to the line whose equation is $x = -4$.
21. The graph of f passes through $(-6, 4)$ and is perpendicular to the line that has an x-intercept of 2 and a y-intercept of -4.
22. The graph of f passes through $(-5, 6)$ and is perpendicular to the line that has an x-intercept of 3 and a y-intercept of -9.
23. The graph of f is perpendicular to the line whose equation is $3x - 2y - 4 = 0$ and has the same y-intercept as this line.
24. The graph of f is perpendicular to the line whose equation is $4x - y - 6 = 0$ and has the same y-intercept as this line.

Application Exercises

In Exercises 25–28, a linear function that models data is described. Find the slope of each model. Then describe what this means in terms of the rate of change of the dependent variable per unit change in the independent variable.

25. The linear function $f(x) = 0.01x + 57.7$ models the global average temperature of Earth, $f(x)$, in degrees Fahrenheit, x years after 1995.
26. The linear function $f(x) = 2x + 10$ models the amount, $f(x)$, in billions of dollars, that the drug industry spent on marketing information about drugs to doctors x years after 2000. (*Source:* IMS Health)
27. The linear function $f(x) = -0.52x + 24.7$ models the percentage of U.S. adults who smoked cigarettes, $f(x)$, x years after 1997. (*Source:* National Center for Health Statistics)
28. The linear function $f(x) = -0.28x + 1.7$ models the percentage of U.S. taxpayers who were audited by the IRS, $f(x)$, x years after 1996. (*Source:* IRS)

The bar graph shows the average amount that U.S. consumers spent on four pieces of the entertainment pie from 2002 through 2004.

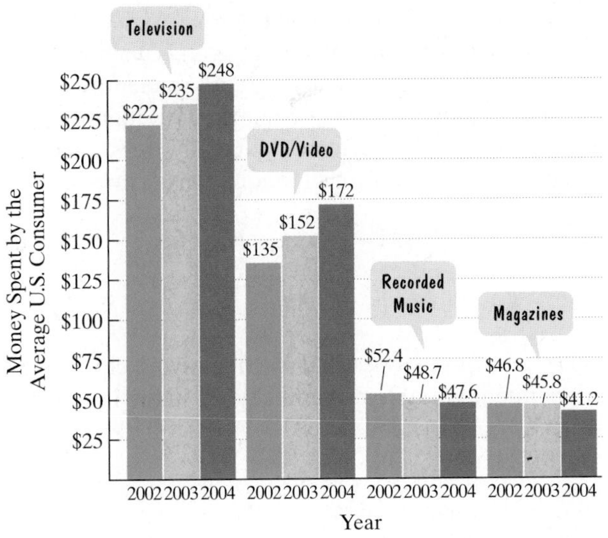

Source: Veronis, Suhler, Stevenson, Communications Industry Forecast and Report

In Exercises 29–32, find a linear function in slope-intercept form that models the given description. Each function should model the average amount, $f(x)$, that U.S. consumers spent on the mode of entertainment x years after 2002.

29. In 2002, the average U.S. consumer spent \$222 on television (broadcast, cable, and satellite) and this amount has increased at an average rate of \$13 per year since then.

30. In 2002, the average U.S. consumer spent \$135 on DVDs and videos, and this amount has increased at an average rate of \$18.50 per year since then.

31. In 2002, the average U.S. consumer spent \$52.40 on recorded music and this amount has decreased at an average rate of \$2.40 per year since then.

32. In 2002, the average U.S. consumer spent \$46.80 on magazines and this amount has decreased at an average rate of \$2.80 per year since then.

The graph shows the percentage of sales of recorded music in the United States for rock and rap/hip-hop from 1997 through 2003. Use the information shown to solve Exercises 33–34.

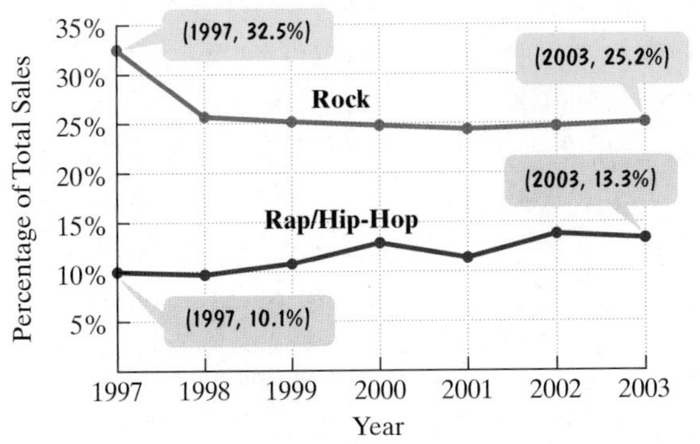

Source: RIAA

33. Find the average rate of change in the percentage of total sales of rock from 1997 through 2003. Round to the nearest hundredth of a percent.

34. Find the average rate of change in the percentage of total sales of rap/hip-hop from 1997 through 2003. Round to the nearest hundredth of a percent.

Writing in Mathematics

35. If two lines are parallel, describe the relationship between their slopes.

36. If two lines are perpendicular, describe the relationship between their slopes.

37. If you know a point on a line and you know the equation of a line perpendicular to this line, explain how to write the line's equation.

38. A formula in the form $y = mx + b$ models the cost, y, of a four-year college x years after 2005. Would you expect m to be positive, negative, or zero? Explain your answer.

39. What is a secant line?

40. What is the average rate of change of a function?

Technology Exercise

41. a. Why are the lines whose equations are $y = \frac{1}{3}x + 1$ and $y = -3x - 2$ perpendicular?

b. Use a graphing utility to graph the equations in a $[-10, 10, 1]$ by $[-10, 10, 1]$ viewing rectangle. Do the lines appear to be perpendicular?

c. Now use the zoom square feature of your utility. Describe what happens to the graphs. Explain why this is so.

Critical Thinking Exercises

In Exercises 42–43, draw a graph that illustrates each data description.

42. From 1971 through 1980, the percentage of Americans who were obese held constant at 15%. After 1980, this percentage increased at a rate of 0.8% per year.
(*Source:* National Center for Health Statistics)

43. In 1970, the daily calories in the U.S. food supply was 3300 per person. From 1970 through 1975, daily calories in the food supply decreased at a rate of 20 calories per person per year. From 1975 through the present, this number has increased at a rate of 28 calories per person per year.
(*Source:* USDA, Economic Research Service)

44. What is the slope of a line that is perpendicular to the line whose equation is $Ax + By + C = 0$, $A \neq 0$ and $B \neq 0$?

45. Determine the value of A so that the line whose equation is $Ax + y - 2 = 0$ is perpendicular to the line containing the points $(1, -3)$ and $(-2, 4)$.

CHAPTER 2
MID-CHAPTER CHECK POINT

What You Know: We learned that a function is a relation in which no two ordered pairs have the same first component and different second components. We represented functions as equations and used function notation. We graphed functions and applied the vertical line test to identify graphs of functions. We determined the domain and range of a function from its graph, using inputs on the x-axis for the domain and outputs on the y-axis for the range. We used graphs to identify intervals on which functions increase, decrease, or are constant, as well as to locate relative maxima or minima. We identified even functions [$f(-x) = f(x)$: y-axis symmetry] and odd functions [$f(-x) = -f(x)$: origin symmetry]. Finally, we studied linear functions and slope, using slope (change in y divided by change in x) to develop various forms for equations of lines:

Point-slope form	Slope-intercept form	Horizontal line	Vertical line	General form
$y - y_1 = m(x - x_1)$	$y = f(x) = mx + b$	$y = f(x) = b$	$x = a$	$Ax + By + C = 0$

We saw that parallel lines have the same slope and that perpendicular lines have slopes that are negative reciprocals. For linear functions, slope was interpreted as the rate of change of the dependent variable per unit change in the independent variable. For nonlinear functions, the slope of the secant line between $(x_1, f(x_1))$ and $(x_2, f(x_2))$ described the average rate of change of f from x_1 to x_2: $\dfrac{f(x_2) - f(x_1)}{x_2 - x_1}$.

In Exercises 1–6, determine whether each relation is a function. Give the domain and range for each relation.

1. $\{(2, 6), (1, 4), (2, -6)\}$

2. $\{(0, 1), (2, 1), (3, 4)\}$

3.

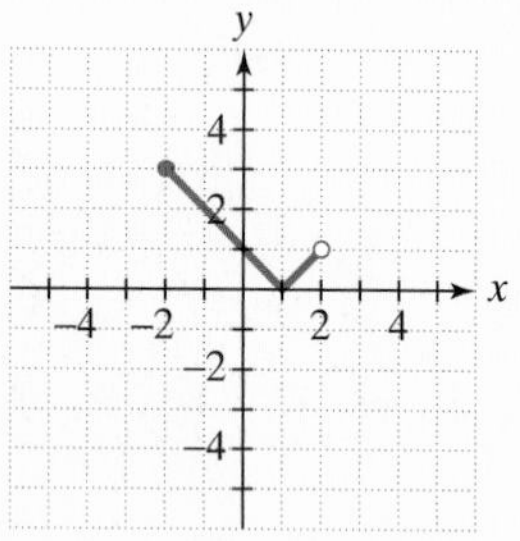

4.

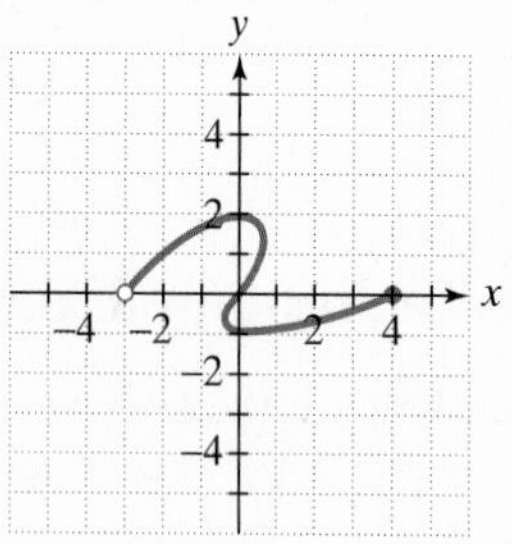

5.

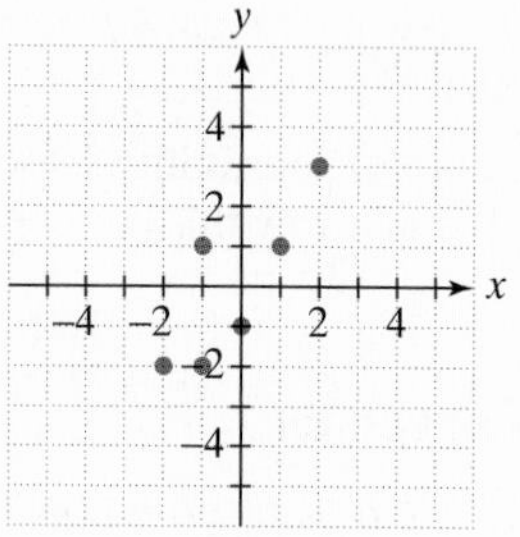

6.

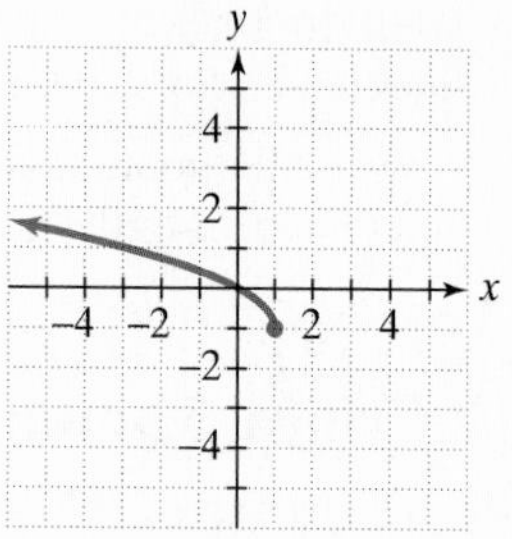

In Exercises 7–8, determine whether each equation defines y as a function of x.

7. $x^2 + y = 5$

8. $x + y^2 = 5$

Use the graph of f to solve Exercises 9–24. Where applicable, use interval notation.

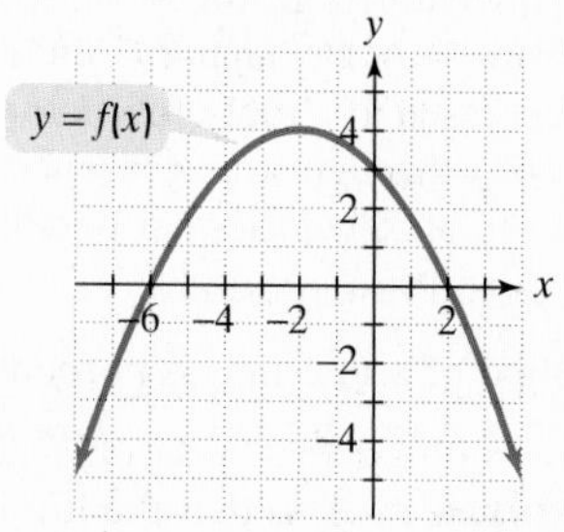

9. Explain why f represents the graph of a function.

10. Find the domain of f.

11. Find the range of f.

12. Find the x-intercept(s).

13. Find the y-intercept.

14. Find the interval(s) on which f is increasing.

15. Find the interval(s) on which f is decreasing.

16. At what number does f have a relative maximum?

17. What is the relative maximum of f?

18. Find $f(-4)$.

19. For what value or values of x is $f(x) = -2$?

20. For what value or values of x is $f(x) = 0$?

21. For what values of x is $f(x) > 0$?

22. Is $f(100)$ positive or negative?

23. Is f even, odd, or neither?

24. Find the average rate of change of f from $x_1 = -4$ to $x_2 = 4$.

In Exercises 25–36, graph each equation in a rectangular coordinate system.

25. $y = -2x$

26. $y = -2$

27. $x + y = -2$

28. $y = \frac{1}{3}x - 2$

29. $x = 3.5$

30. $4x - 2y = 8$

31. $f(x) = x^2 - 4$

32. $f(x) = x - 4$

33. $f(x) = |x| - 4$

34. $5y = -3x$

35. $5y = 20$

36. $f(x) = \begin{cases} -1 & \text{if } x \le 0 \\ 1 & \text{if } x > 0 \end{cases}$

37. Let $f(x) = -2x^2 + x - 5$.

a. Find $f(-x)$. Is f even, odd, or neither?

b. Find $\dfrac{f(x + h) - f(x)}{h}$, $h \neq 0$.

38. Let $C(x) = \begin{cases} 30 & \text{if} \quad 0 \le t \le 200 \\ 30 + 0.40(t - 200) & \text{if} \quad t > 200 \end{cases}$.

a. Find $C(150)$. **b.** Find $C(250)$.

In Exercises 39–42, write a function in slope-intercept form whose graph satisfies the given conditions.

39. Slope $= -2$, passing through $(-4, 3)$

40. Passing through $(-1, -5)$ and $(2, 1)$

41. Passing through $(3, -4)$ and parallel to the line whose equation is $3x - y - 5 = 0$

42. Passing through $(-4, -3)$ and perpendicular to the line whose equation is $2x - 5y - 10 = 0$

43. Determine whether the line through $(2, -4)$ and $(7, 0)$ is parallel to a second line through $(-4, 2)$ and $(1, 6)$.

44. The graph shows the percentage of U.S. colleges that offered distance learning by computer for selected years from 1995 through 2002.

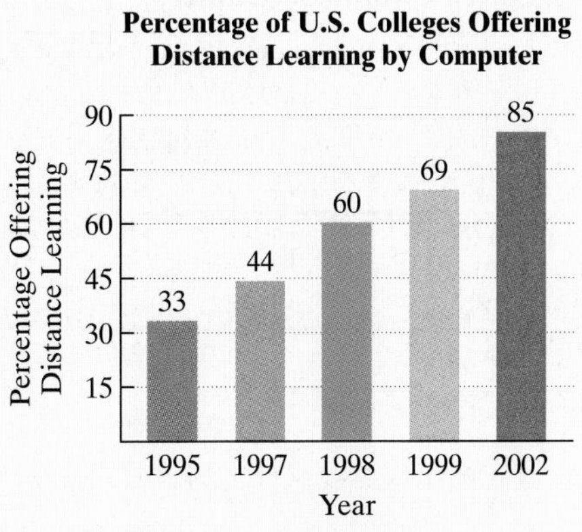

Source: International Data Corporation

The data can be modeled by the linear function $f(x) = 7.8x + 33$, where x is the number of years after 1995 and $f(x)$ is the percentage of U.S. colleges offering distance learning. Find the slope of this function and describe its meaning as a rate of change.

45. Find the average rate of change of $f(x) = 3x^2 - x$ from $x_1 = -1$ to $x_2 = 2$.

SECTION 2.5 Transformations of Functions

Objectives

1. Recognize graphs of common functions.
2. Use vertical shifts to graph functions.
3. Use horizontal shifts to graph functions.
4. Use reflections to graph functions.
5. Use vertical stretching and shrinking to graph functions.
6. Use horizontal stretching and shrinking to graph functions.
7. Graph functions involving a sequence of transformations.

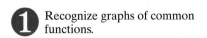

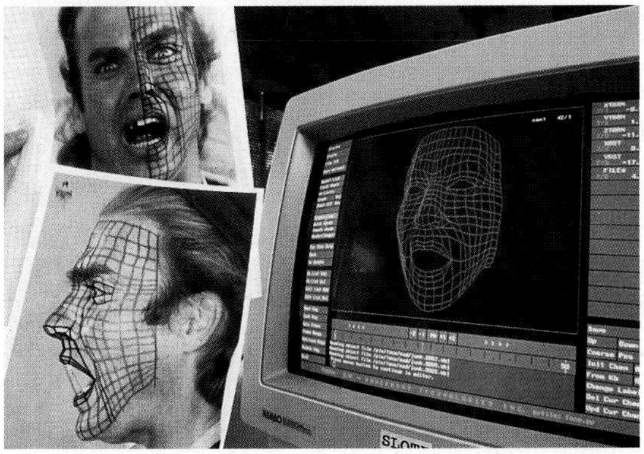

Have you seen *Terminator 2, The Mask*, or *The Matrix*? These were among the first films to use spectacular effects in which a character or object having one shape was transformed in a fluid fashion into a quite different shape. The name for such a transformation is **morphing**. The effect allows a real actor to be seamlessly transformed into a computer-generated animation. The animation can be made to perform impossible feats before it is morphed back to the conventionally filmed image.

Like transformed movie images, the graph of one function can be turned into the graph of a different function. To do this, we need to rely on a function's equation. Knowing that a graph is a transformation of a familiar graph makes graphing easier.

1 Recognize graphs of common functions.

Graphs of Common Functions

Table 2.3 on the next page gives names to seven frequently encountered functions in algebra. The table shows each function's graph and lists characteristics of the function. Study the shape of each graph and take a few minutes to verify the function's characteristics from its graph. Knowing these graphs is essential for analyzing their transformations into more complicated graphs.

Table 2.3 Algebra's Common Graphs

Constant Function

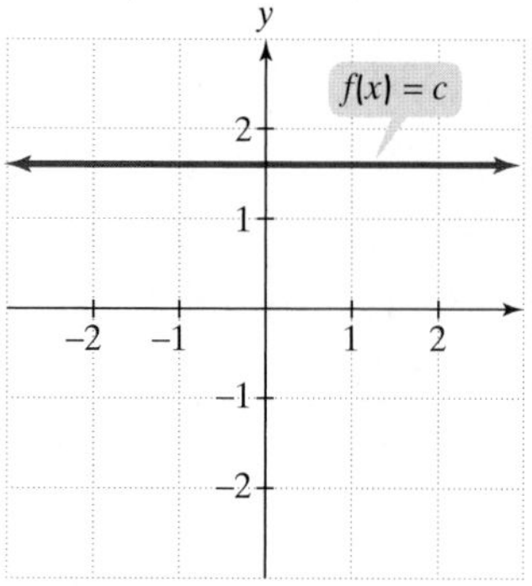

- Domain: $(-\infty, \infty)$
- Range: the single number c
- Constant on $(-\infty, \infty)$
- Even function

Identity Function

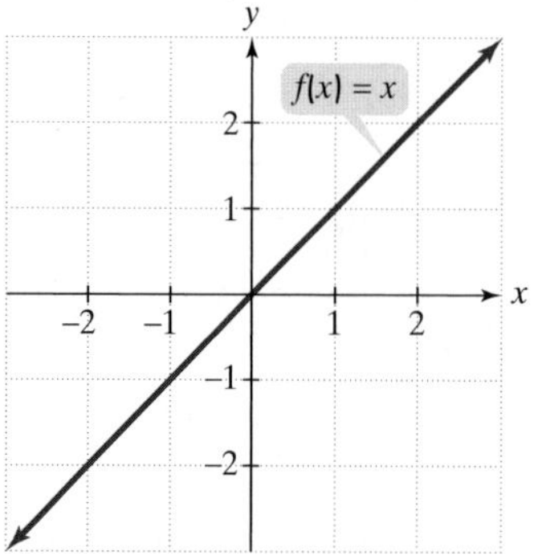

- Domain: $(-\infty, \infty)$
- Range: $(-\infty, \infty)$
- Increasing on $(-\infty, \infty)$
- Odd function

Absolute Value Function

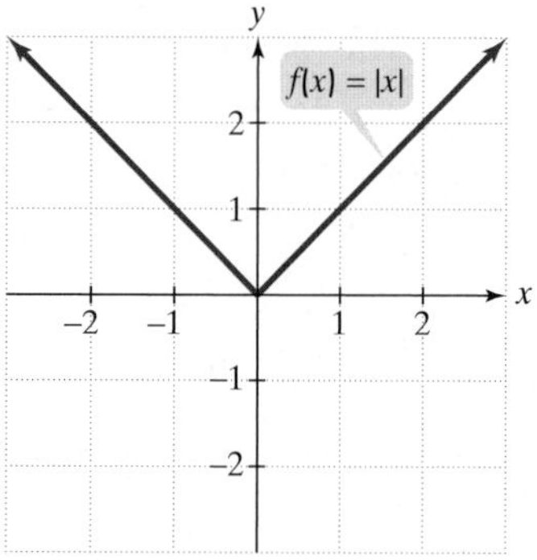

- Domain: $(-\infty, \infty)$
- Range: $[0, \infty)$
- Decreasing on $(-\infty, 0)$ and increasing on $(0, \infty)$
- Even function

Standard Quadratic Function

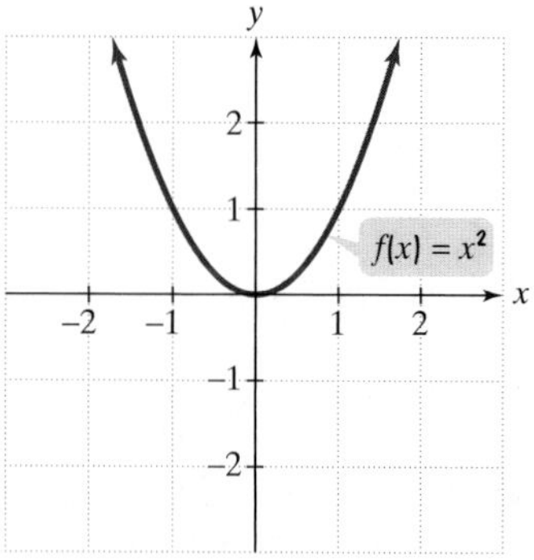

- Domain: $(-\infty, \infty)$
- Range: $[0, \infty)$
- Decreasing on $(-\infty, 0)$ and increasing on $(0, \infty)$
- Even function

Square Root Function

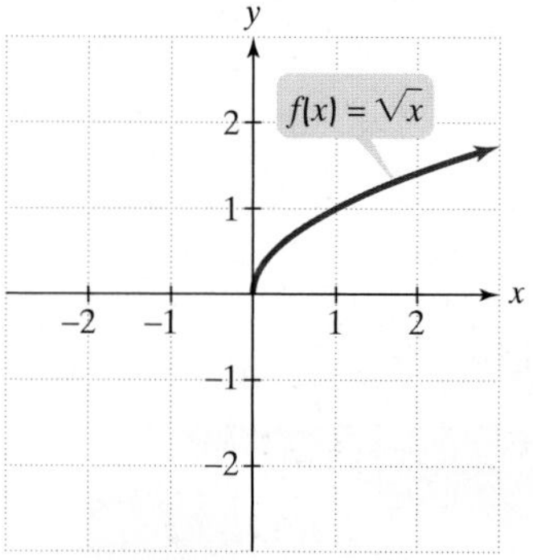

- Domain: $[0, \infty)$
- Range: $[0, \infty)$
- Increasing on $(0, \infty)$
- Neither even nor odd

Standard Cubic Function

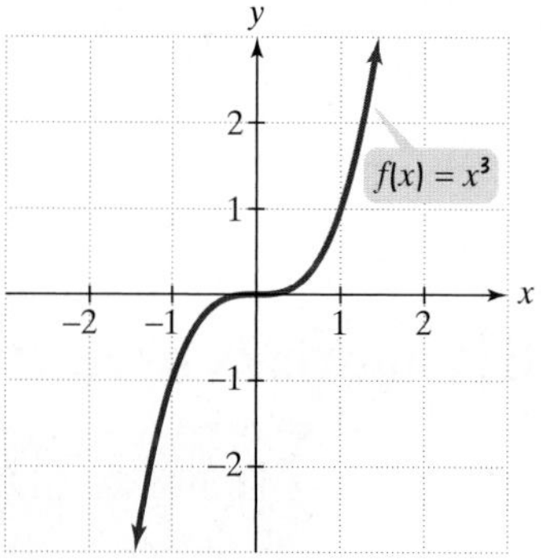

- Domain: $(-\infty, \infty)$
- Range: $(-\infty, \infty)$
- Increasing on $(-\infty, \infty)$
- Odd function

Cube Root Function

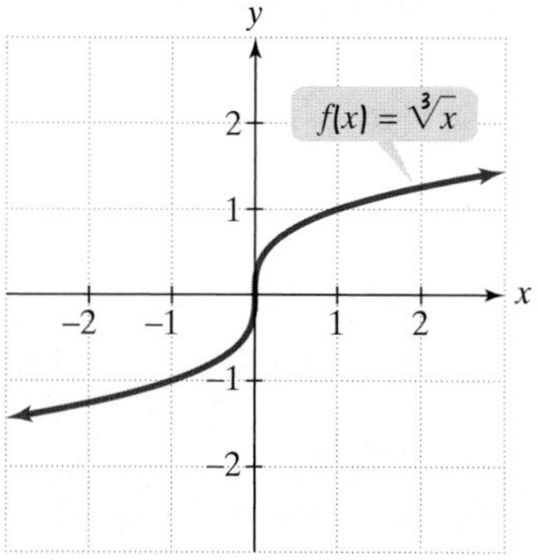

- Domain: $(-\infty, \infty)$
- Range: $(-\infty, \infty)$
- Increasing on $(-\infty, \infty)$
- Odd function

Discovery

The study of how changing a function's equation can affect its graph can be explored with a graphing utility. Use your graphing utility to verify the hand-drawn graphs as you read this section.

2 Use vertical shifts to graph functions.

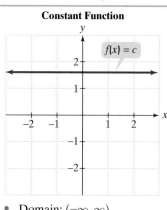

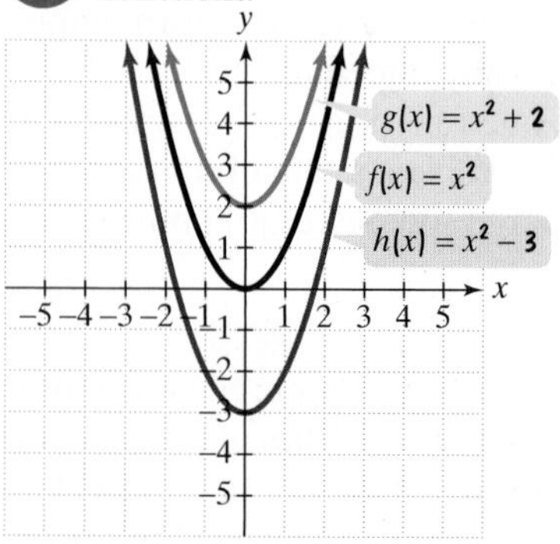

Figure 2.41 Vertical shifts

Vertical Shifts

Let's begin by looking at three graphs whose shapes are the same. Figure 2.41 shows the graphs. The black graph in the middle is the standard quadratic function, $f(x) = x^2$. Now, look at the blue graph on the top. The equation of this graph, $g(x) = x^2 + 2$, adds 2 to the right side of $f(x) = x^2$. The y-coordinate of each point of g is 2 more than the corresponding y-coordinate of each point of f. What effect does this have on the graph of f? It shifts the graph vertically up by 2 units.

$$g(x) = x^2 + 2 = f(x) + 2$$

The graph of g shifts the graph of f up 2 units.

Finally, look at the red graph on the bottom in Figure 2.41. The equation of this graph, $h(x) = x^2 - 3$, subtracts 3 from the right side of $f(x) = x^2$. The y-coordinate of each point of h is 3 less than the corresponding y-coordinate of each point of f. What effect does this have on the graph of f? It shifts the graph vertically down by 3 units.

$$h(x) = x^2 - 3 = f(x) - 3$$

The graph of h | shifts the graph of f down 3 units.

In general, if c is positive, $y = f(x) + c$ shifts the graph of f upward c units and $y = f(x) - c$ shifts the graph of f downward c units. These are called **vertical shifts** of the graph of f.

Vertical Shifts

Let f be a function and c a positive real number.

- The graph of $y = f(x) + c$ is the graph of $y = f(x)$ shifted c units vertically upward.
- The graph of $y = f(x) - c$ is the graph of $y = f(x)$ shifted c units vertically downward.

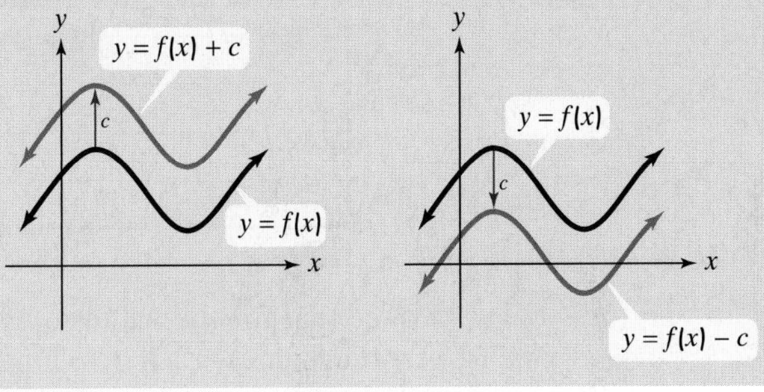

Study Tip

To keep track of transformations, identify a number of points on the given function's graph. Then analyze what happens to the coordinates of these points with each transformation.

EXAMPLE 1 Vertical Shift Down

Use the graph of $f(x) = |x|$ to obtain the graph of $g(x) = |x| - 4$.

Solution The graph of $g(x) = |x| - 4$ has the same shape as the graph of $f(x) = |x|$. However, it is shifted down vertically 4 units.

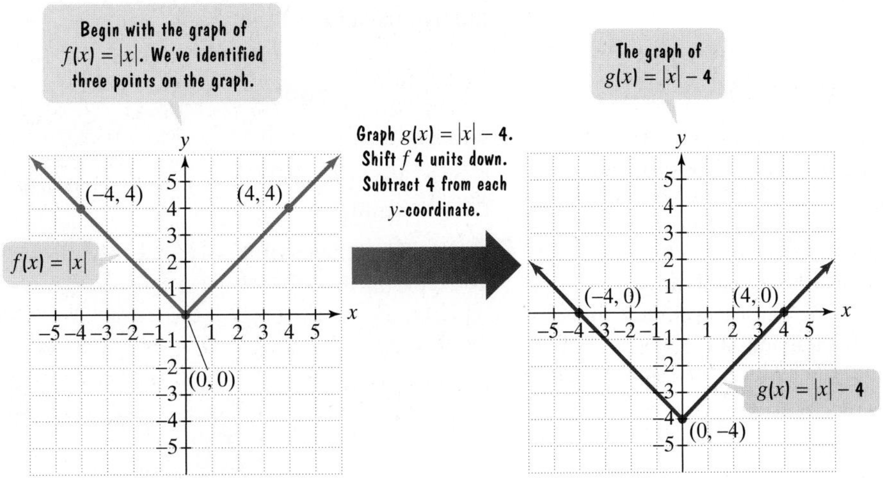

Check Point 1 Use the graph of $f(x) = |x|$ to obtain the graph of $g(x) = |x| + 3$.

3 Use horizontal shifts to graph functions.

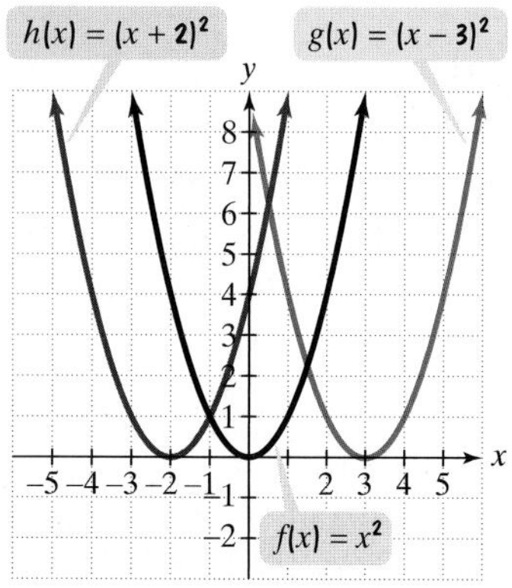

Figure 2.42 Horizontal shifts

Horizontal Shifts

We return to the graph of $f(x) = x^2$, the standard quadratic function. In Figure 2.42, the graph of function f is in the middle of the three graphs. By contrast to the vertical shift situation, this time there are graphs to the left and to the right of the graph of f. Look at the blue graph on the right. The equation of this graph, $g(x) = (x - 3)^2$, subtracts 3 from each value of x before squaring it. What effect does this have on the graph of $f(x) = x^2$? It shifts the graph horizontally to the right by 3 units.

$$g(x) = (x - 3)^2 = f(x - 3)$$

The graph of g shifts the graph of f 3 units to the right.

Does it seem strange that *subtracting* 3 in the domain causes a shift of 3 units to the *right*? Perhaps a partial table of coordinates for each function will numerically convince you of this shift.

x	$f(x) = x^2$
-2	$(-2)^2 = 4$
-1	$(-1)^2 = 1$
0	$0^2 = 0$
1	$1^2 = 1$
2	$2^2 = 4$

x	$g(x) = (x - 3)^2$
1	$(1 - 3)^2 = (-2)^2 = 4$
2	$(2 - 3)^2 = (-1)^2 = 1$
3	$(3 - 3)^2 = 0^2 = 0$
4	$(4 - 3)^2 = 1^2 = 1$
5	$(5 - 3)^2 = 2^2 = 4$

Notice that for the values of $f(x)$ and $g(x)$ to be the same, the values of x used in graphing g must each be 3 units greater than those used to graph f. For this reason, the graph of g is the graph of f shifted 3 units to the right.

Now, look at the red graph on the left in Figure 2.42. The equation of this graph, $h(x) = (x + 2)^2$, adds 2 to each value of x before squaring it. What effect does this have on the graph of $f(x) = x^2$? It shifts the graph horizontally to the left by 2 units.

$$h(x) = (x + 2)^2 = f(x + 2)$$

The graph of h shifts the graph of f 2 units to the left.

In general, if c is positive, $y = f(x + c)$ shifts the graph of f to the left c units and $y = f(x - c)$ shifts the graph of f to the right c units. These are called **horizontal shifts** of the graph of f.

Study Tip

On a number line, if x represents a number and c is positive, then $x + c$ lies c units to the right of x and $x - c$ lies c units to the left of x. This orientation does not apply to horizontal shifts: $f(x + c)$ causes a shift of c units to the left and $f(x - c)$ causes a shift of c units to the right.

Horizontal Shifts

Let f be a function and c a positive real number.

- The graph of $y = f(x + c)$ is the graph of $y = f(x)$ shifted to the left c units.
- The graph of $y = f(x - c)$ is the graph of $y = f(x)$ shifted to the right c units.

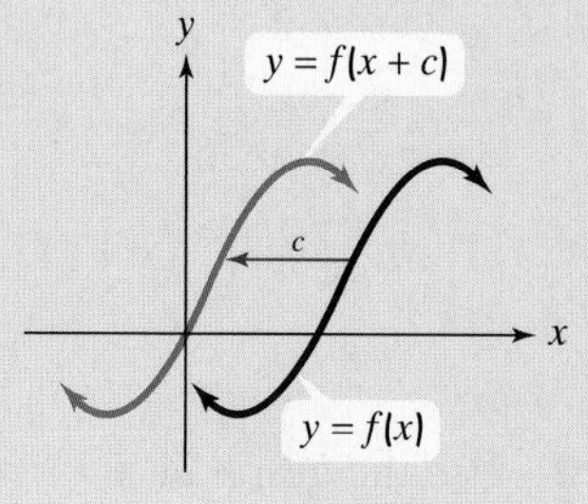

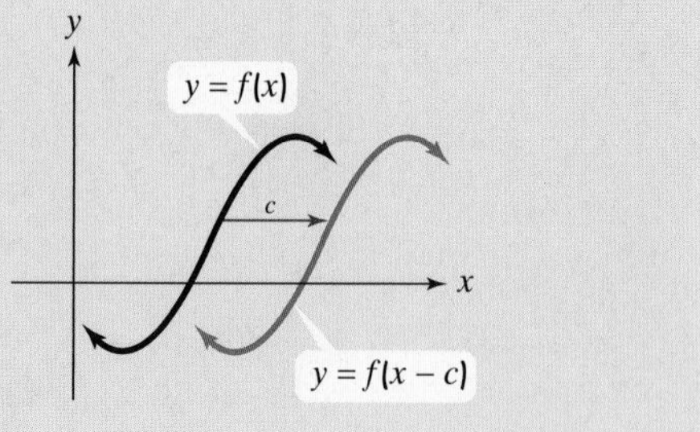

EXAMPLE 2 Horizontal Shift to the Left

Use the graph of $f(x) = \sqrt{x}$ to obtain the graph of $g(x) = \sqrt{x+5}$.

Solution Compare the equations for $f(x) = \sqrt{x}$ and $g(x) = \sqrt{x+5}$. The equation for g adds 5 to each value of x before taking the square root.

$$y = g(x) = \sqrt{x+5} = f(x+5)$$

The graph of g | shifts the graph of f 5 units to the left.

The graph of $g(x) = \sqrt{x+5}$ has the same shape as the graph of $f(x) = \sqrt{x}$. However, it is shifted horizontally to the left 5 units.

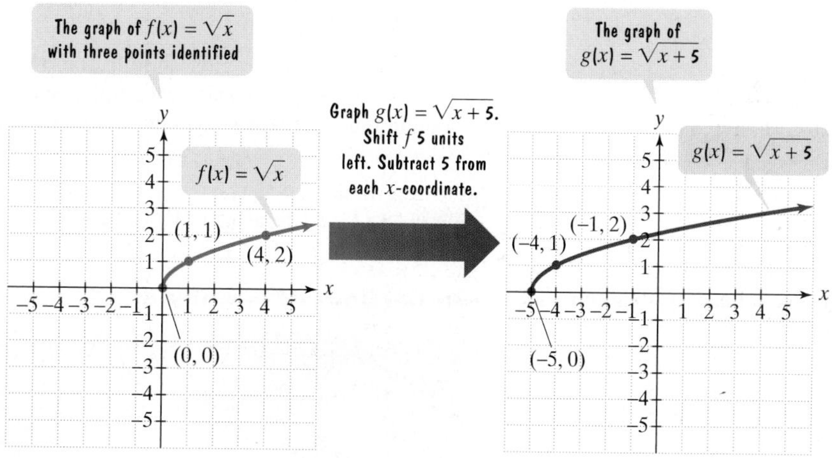

Check Point 2 Use the graph of $f(x) = \sqrt{x}$ to obtain the graph of $g(x) = \sqrt{x-4}$.

Some functions can be graphed by combining horizontal and vertical shifts. These functions will be variations of a function whose equation you know how to graph, such as the standard quadratic function, the standard cubic function, the square root function, the cube root function, or the absolute value function.

In our next example, we will use the graph of the standard quadratic function, $f(x) = x^2$, to obtain the graph of $h(x) = (x+1)^2 - 3$. We will graph three functions:

$$f(x) = x^2 \qquad g(x) = (x+1)^2 \qquad h(x) = (x+1)^2 - 3.$$

Start by graphing the standard quadratic function. | Shift the graph of f horizontally one unit to the left. | Shift the graph of g vertically down 3 units.

EXAMPLE 3 Combining Horizontal and Vertical Shifts

Use the graph of $f(x) = x^2$ to obtain the graph of $h(x) = (x + 1)^2 - 3$.

Solution

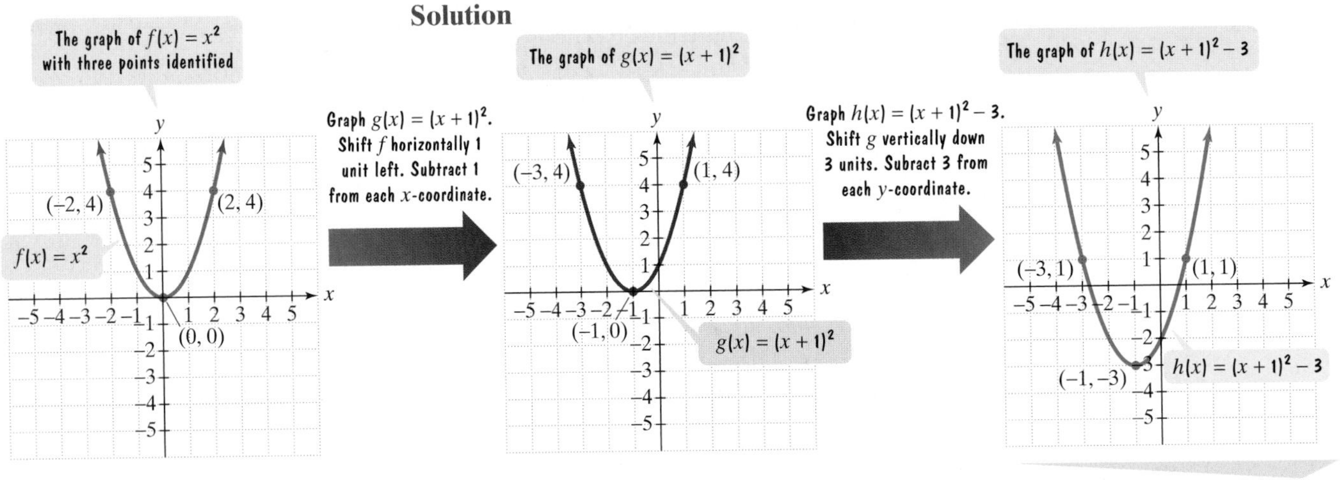

The graph of $f(x) = x^2$ with three points identified
(−2, 4)
(2, 4)
(0, 0)
$f(x) = x^2$
Graph $g(x) = (x + 1)^2$. Shift f horizontally 1 unit left. Subtract 1 from each x-coordinate.
The graph of $g(x) = (x + 1)^2$
(−3, 4)
(1, 4)
(−1, 0)
$g(x) = (x + 1)^2$
Graph $h(x) = (x + 1)^2 - 3$. Shift g vertically down 3 units. Subtract 3 from each y-coordinate.
The graph of $h(x) = (x + 1)^2 - 3$
(−3, 1)
(1, 1)
(−1, −3)
$h(x) = (x + 1)^2 - 3$

Discovery

Work Example 3 by first shifting the graph of $f(x) = x^2$ three units down, graphing $g(x) = x^2 - 3$. Now, shift this graph one unit left to graph $h(x) = (x + 1)^2 - 3$. Did you obtain the last graph shown in the solution of Example 3? What can you conclude?

Check Point 3 Use the graph of $f(x) = \sqrt{x}$ to obtain the graph of $h(x) = \sqrt{x - 1} - 2$.

4 Use reflections to graph functions.

Reflections of Graphs

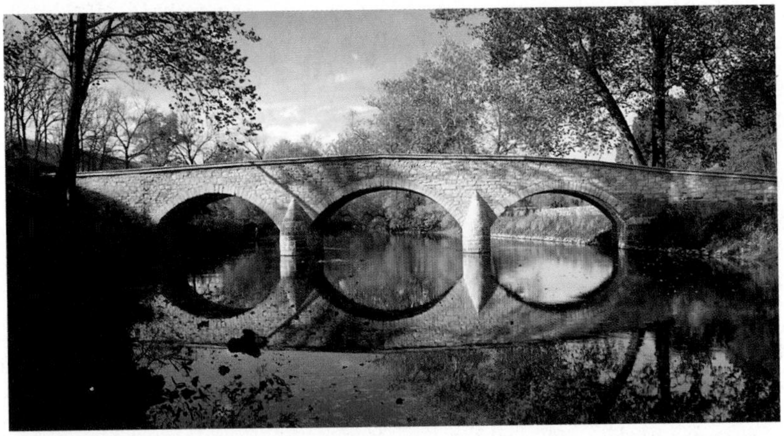

This photograph shows a reflection of an old bridge in a Maryland river. This perfect reflection occurs because the surface of the water is absolutely still. A mild breeze rippling the water's surface would distort the reflection.

Is it possible for graphs to have mirror-like qualities? Yes. Figure 2.43 shows the graphs of $f(x) = x^2$ and $g(x) = -x^2$. The graph of g is a **reflection about the x-axis** of the graph of f. For corresponding values of x, the y-coordinates of g are the opposites of the y-coordinates of f. In general, the graph of $y = -f(x)$ reflects the graph of f about the x-axis. Thus, the graph of g is a reflection of the graph of f about the x-axis because

$$g(x) = -x^2 = -f(x).$$

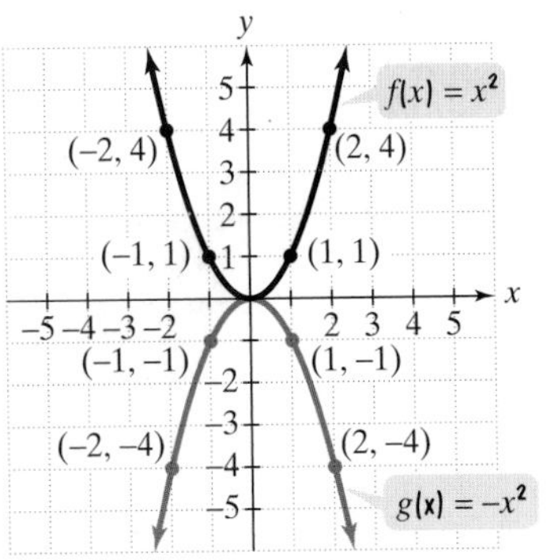

$f(x) = x^2$
(−2, 4)
(2, 4)
(−1, 1)
(1, 1)
(−1, −1)
(1, −1)
(−2, −4)
(2, −4)
$g(x) = -x^2$

Figure 2.43 Reflection about the x-axis

Reflection about the x-Axis

The graph of $y = -f(x)$ is the graph of $y = f(x)$ reflected about the x-axis.

EXAMPLE 4 Reflection about the x-Axis

Use the graph of $f(x) = \sqrt[3]{x}$ to obtain the graph of $g(x) = -\sqrt[3]{x}$.

Solution Compare the equations for $f(x) = \sqrt[3]{x}$ and $g(x) = -\sqrt[3]{x}$. The graph of g is a reflection about the x-axis of the graph of f because

$$g(x) = -\sqrt[3]{x} = -f(x).$$

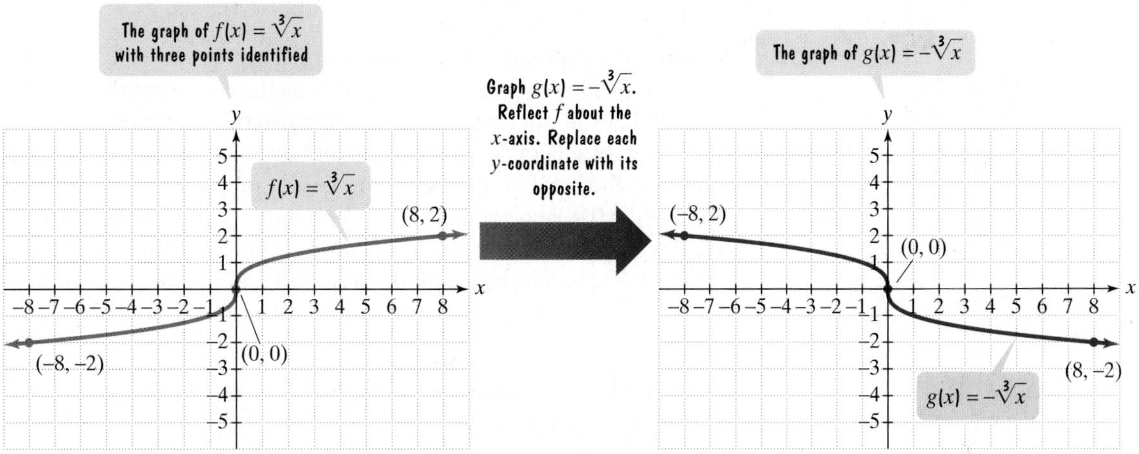

Check Point 4 Use the graph of $f(x) = |x|$ to obtain the graph of $g(x) = -|x|$.

It is also possible to reflect graphs about the y-axis.

Reflection about the y-Axis

The graph of $y = f(-x)$ is the graph of $y = f(x)$ reflected about the y-axis.

For corresponding values of y, the x-coordinates of $y = f(-x)$ are the opposite of those of $y = f(x)$.

EXAMPLE 5 Reflection about the y-Axis

Use the graph of $f(x) = \sqrt{x}$ to obtain the graph of $h(x) = \sqrt{-x}$.

Solution Compare the equations for $f(x) = \sqrt{x}$ and $h(x) = \sqrt{-x}$. The graph of h is a reflection about the y-axis of the graph of f because

$$h(x) = \sqrt{-x} = f(-x).$$

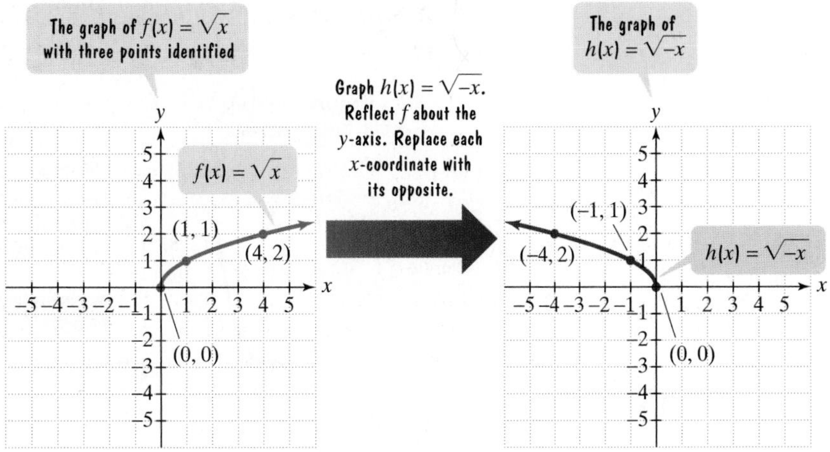

Check Point 5 Use the graph of $f(x) = \sqrt[3]{x}$ to obtain the graph of $h(x) = \sqrt[3]{-x}$.

5 Use vertical stretching and shrinking to graph functions.

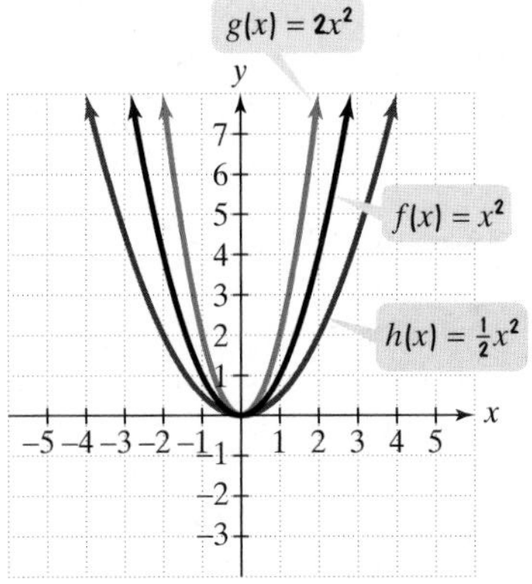

Figure 2.44 Vertically stretching and shrinking $f(x) = x^2$

Vertical Stretching and Shrinking

Morphing does much more than move an image horizontally, vertically, or about an axis. An object having one shape is transformed into a different shape. Horizontal shifts, vertical shifts, and reflections do not change the basic shape of a graph. Graphs remain rigid and proportionally the same when they undergo these transformations. How can we shrink and stretch graphs, thereby altering their basic shapes?

Look at the three graphs in Figure 2.44. The black graph in the middle is the graph of the standard quadratic function, $f(x) = x^2$. Now, look at the blue graph on the top. The equation of this graph is $g(x) = 2x^2$, or $g(x) = 2f(x)$. Thus, for each x, the y-coordinate of g is 2 times as large as the corresponding y-coordinate on the graph of f. The result is a narrower graph because the values of y are rising faster. We say that the graph of g is obtained by vertically *stretching* the graph of f. Now, look at the red graph on the bottom. The equation of this graph is $h(x) = \frac{1}{2}x^2$, or $h(x) = \frac{1}{2}f(x)$. Thus, for each x, the y-coordinate of h is one-half as large as the corresponding y-coordinate on the graph of f. The result is a wider graph because the values of y are rising more slowly. We say that the graph of h is obtained by vertically *shrinking* the graph of f.

These observations can be summarized as follows:

Vertically Stretching and Shrinking Graphs

Let f be a function and c a positive real number.

- If $c > 1$, the graph of $y = cf(x)$ is the graph of $y = f(x)$ vertically stretched by multiplying each of its y-coordinates by c.
- If $0 < c < 1$, the graph of $y = cf(x)$ is the graph of $y = f(x)$ vertically shrunk by multiplying each of its y-coordinates by c.

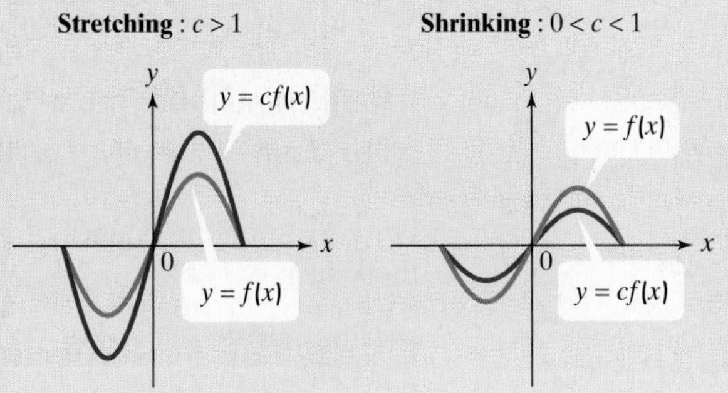

EXAMPLE 6 Vertically Shrinking a Graph

Use the graph of $f(x) = x^3$ to obtain the graph of $h(x) = \frac{1}{2}x^3$.

Solution The graph of $h(x) = \frac{1}{2}x^3$ is obtained by vertically shrinking the graph of $f(x) = x^3$.

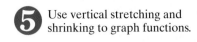

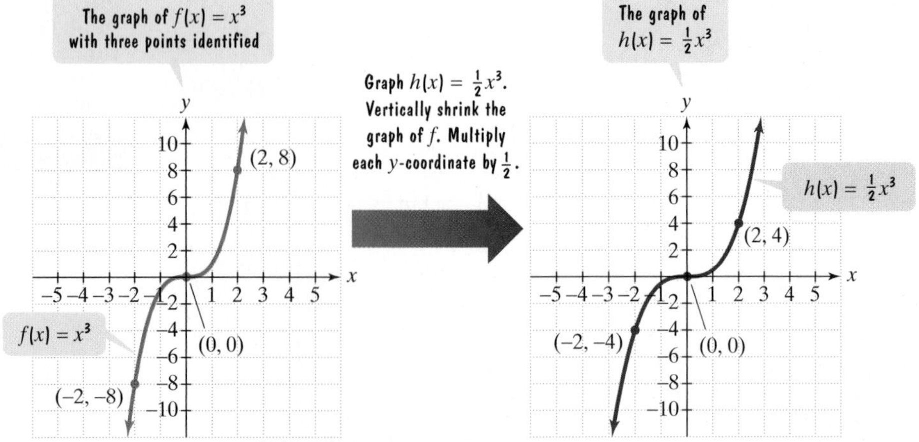

Check Point 6 Use the graph of $f(x) = |x|$ to obtain the graph of $g(x) = 2|x|$.

6 Use horizontal stretching and shrinking to graph functions.

Horizontal Stretching and Shrinking

It is also possible to horizontally stretch and shrink graphs.

Horizontally Stretching and Shrinking Graphs

Let f be a function and c a positive real number.

- If $c > 1$, the graph of $y = f(cx)$ is the graph of $y = f(x)$ horizontally shrunk by dividing each of its x-coordinates by c.
- If $0 < c < 1$, the graph of $y = f(cx)$ is the graph of $y = f(x)$ horizontally stretched by dividing each of its x-coordinates by c.

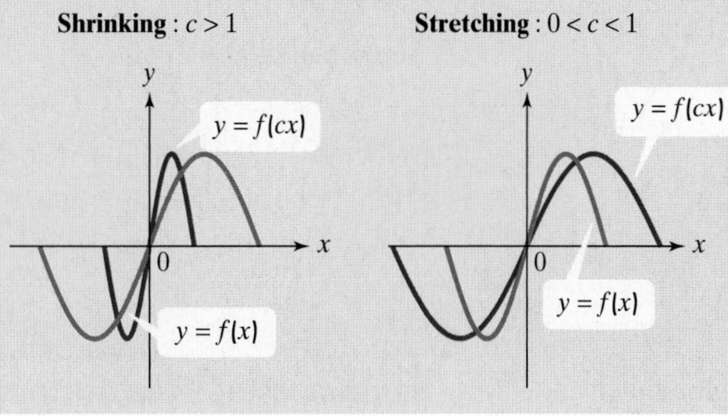

EXAMPLE 7 Horizontally Stretching and Shrinking a Graph

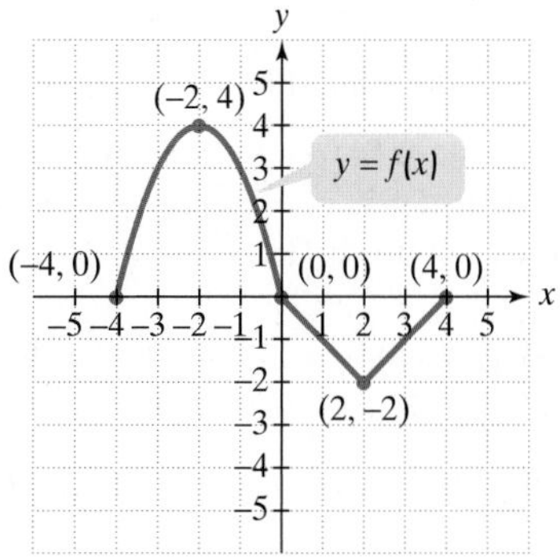

Figure 2.45

Use the graph of $y = f(x)$ in Figure 2.45 to obtain each of the following graphs:

a. $g(x) = f(2x)$ **b.** $h(x) = f\left(\frac{1}{2}x\right)$.

Solution

a. The graph of $g(x) = f(2x)$ is obtained by horizontally shrinking the graph of $y = f(x)$.

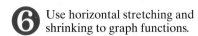

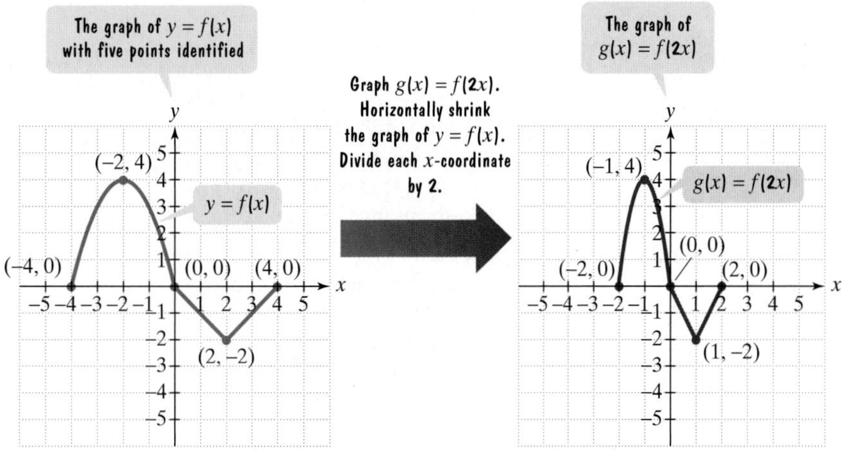

b. The graph of $h(x) = f\left(\frac{1}{2}x\right)$ is obtained by horizontally stretching the graph of $y = f(x)$.

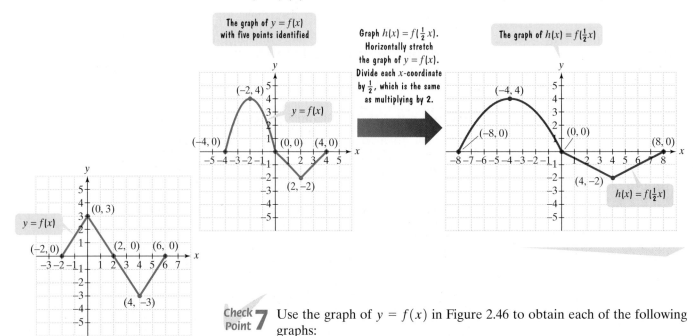

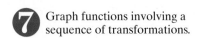

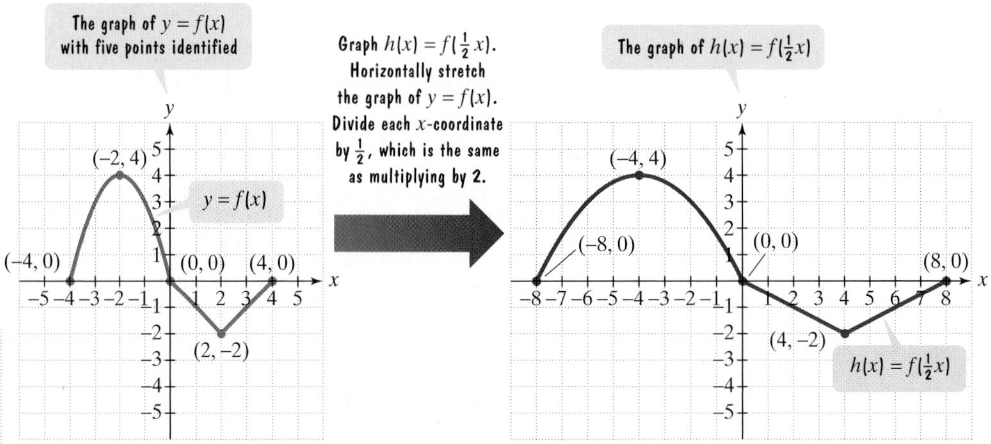

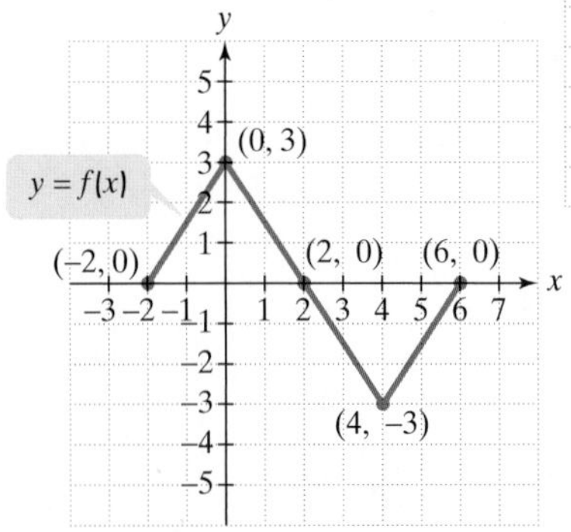

Figure 2.46

Check Point 7 Use the graph of $y = f(x)$ in Figure 2.46 to obtain each of the following graphs:

a. $g(x) = f(2x)$ **b.** $h(x) = f\left(\frac{1}{2}x\right)$.

7 Graph functions involving a sequence of transformations.

Sequences of Transformations

Table 2.4 summarizes the procedures for transforming the graph of $y = f(x)$.

Table 2.4 Summary of Transformations
In each case, c represents a positive real number.

To Graph:	Draw the Graph of f and:	Changes in the Equation of $y = f(x)$
Vertical shifts $y = f(x) + c$ $y = f(x) - c$	 Raise the graph of f by c units. Lower the graph of f by c units.	 c is added to $f(x)$. c is subtracted from $f(x)$.
Horizontal shifts $y = f(x + c)$ $y = f(x - c)$	 Shift the graph of f to the left c units. Shift the graph of f to the right c units.	 x is replaced with $x + c$. x is replaced with $x - c$.
Reflection about the x-axis $y = -f(x)$	Reflect the graph of f about the x-axis.	$f(x)$ is multiplied by -1.
Reflection about the y-axis $y = f(-x)$	Reflect the graph of f about the y-axis.	x is replaced with $-x$.
Vertical stretching or shrinking $y = cf(x), c > 1$ $y = cf(x), 0 < c < 1$	 Multiply each y-coordinate of $y = f(x)$ by c, vertically stretching the graph of f. Multiply each y-coordinate of $y = f(x)$ by c, vertically shrinking the graph of f.	 $f(x)$ is multiplied by $c, c > 1$. $f(x)$ is multiplied by $c, 0 < c < 1$.
Horizontal stretching or shrinking $y = f(cx), c > 1$ $y = f(cx), 0 < c < 1$	 Divide each x-coordinate of $y = f(x)$ by c, horizontally shrinking the graph of f. Divide each x-coordinate of $y = f(x)$ by c, horizontally stretching the graph of f.	 x is replaced with $cx, c > 1$. x is replaced with $cx, 0 < c < 1$.

A function involving more than one transformation can be graphed by performing transformations in the following order:

1. Horizontal shifting **2.** Stretching or shrinking
3. Reflecting **4.** Vertical shifting

EXAMPLE 8 Graphing Using a Sequence of Transformations

Use the graph of $y = f(x)$ given in Figure 2.45 of Example 7 on page 249, and repeated below, to graph $y = -\frac{1}{2}f(x - 1) + 3$.

Solution Our graphs will evolve in the following order:

1. Horizontal shifting: Graph $y = f(x - 1)$ by shifting the graph of $y = f(x)$ 1 unit to the right.
2. Shrinking: Graph $y = \frac{1}{2}f(x - 1)$ by shrinking the previous graph by a factor of $\frac{1}{2}$.
3. Reflecting: Graph $y = -\frac{1}{2}f(x - 1)$ by reflecting the previous graph about the x-axis.
4. Vertical shifting: Graph $y = -\frac{1}{2}f(x - 1) + 3$ by shifting the previous graph up 3 units.

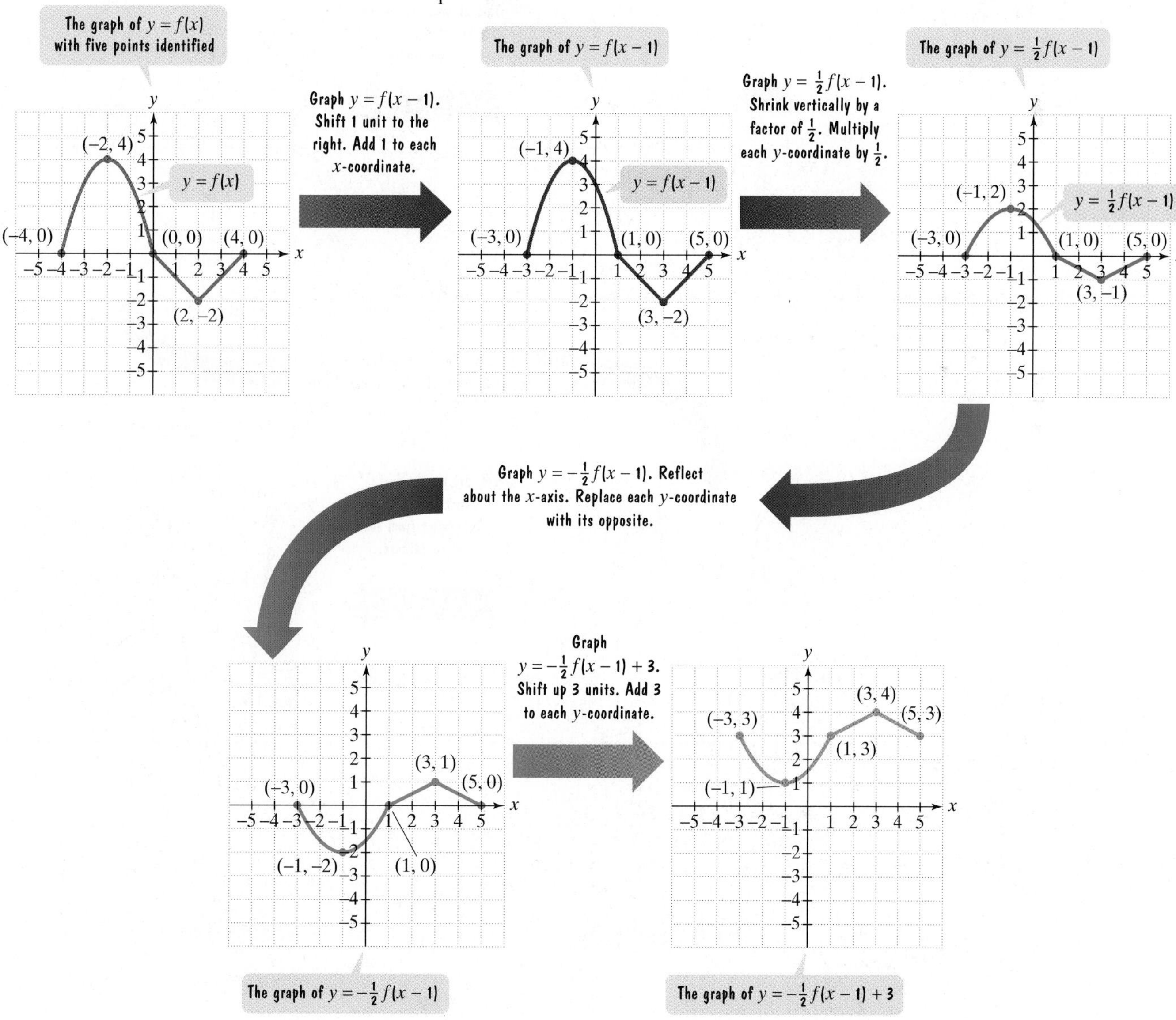

Check Point 8 Use the graph of $y = f(x)$ given in Figure 2.46 of Check Point 7 to graph $y = -\frac{1}{3}f(x + 1) - 2$.

EXAMPLE 9 Graphing Using a Sequence of Transformations

Use the graph of $f(x) = x^2$ to graph $g(x) = 2(x + 3)^2 - 1$.

Solution Our graphs will evolve in the following order:

1. Horizontal shifting: Graph $y = (x + 3)^2$ by shifting the graph of $f(x) = x^2$ three units to the left.
2. Stretching: Graph $y = 2(x + 3)^2$ by stretching the previous graph by a factor of 2.
3. Vertical shifting: Graph $g(x) = 2(x + 3)^2 - 1$ by shifting the previous graph down 1 unit.

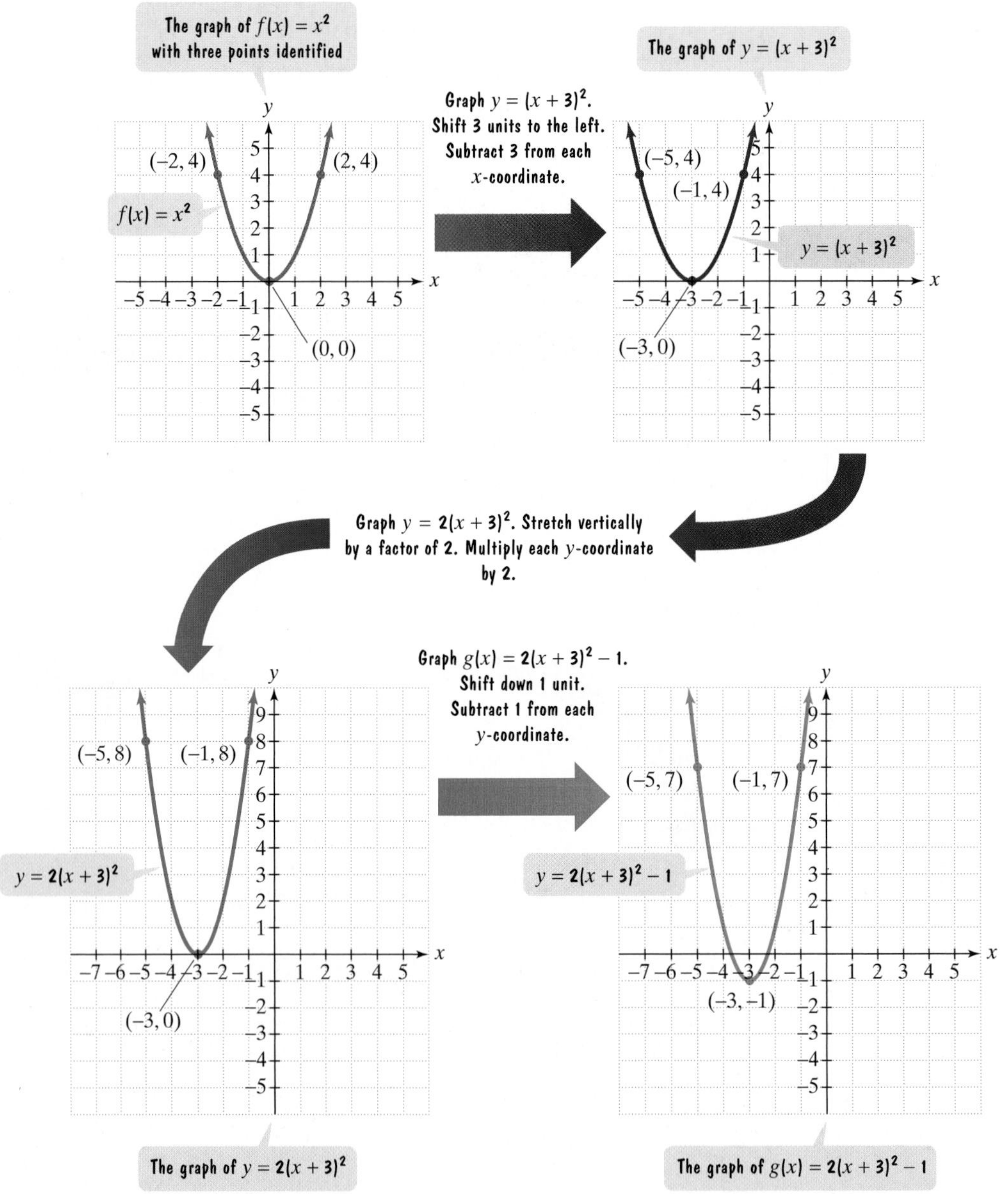
The graph of $f(x) = x^2$ with three points identified
(−2, 4)
(2, 4)
(0, 0)
$f(x) = x^2$
Graph $y = (x + 3)^2$. Shift 3 units to the left. Subtract 3 from each x-coordinate.
The graph of $y = (x + 3)^2$
(−5, 4)
(−1, 4)
(−3, 0)
$y = (x + 3)^2$
Graph $y = 2(x + 3)^2$. Stretch vertically by a factor of 2. Multiply each y-coordinate by 2.
(−5, 8)
(−1, 8)
(−3, 0)
$y = 2(x + 3)^2$
Graph $g(x) = 2(x + 3)^2 - 1$. Shift down 1 unit. Subtract 1 from each y-coordinate.
(−5, 7)
(−1, 7)
(−3, −1)
$y = 2(x + 3)^2 - 1$
The graph of $y = 2(x + 3)^2$
The graph of $g(x) = 2(x + 3)^2 - 1$

Check Point 9 Use the graph of $f(x) = x^2$ to graph $g(x) = 2(x - 1)^2 + 3$.

EXERCISE SET 2.5

Practice Exercises

In Exercises 1–16, use the graph of $y = f(x)$ to graph each function g.

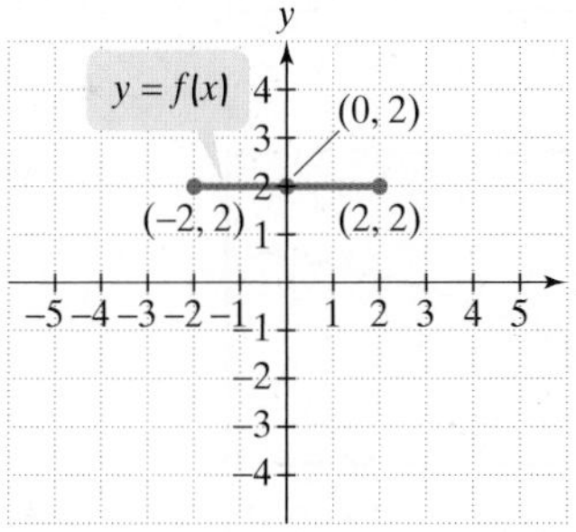

1. $g(x) = f(x) + 1$
2. $g(x) = f(x) - 1$
3. $g(x) = f(x + 1)$
4. $g(x) = f(x - 1)$
5. $g(x) = f(x - 1) - 2$
6. $g(x) = f(x + 1) + 2$
7. $g(x) = f(-x)$
8. $g(x) = -f(x)$
9. $g(x) = -f(x) + 3$
10. $g(x) = f(-x) + 3$
11. $g(x) = \frac{1}{2}f(x)$
12. $g(x) = 2f(x)$
13. $g(x) = f\left(\frac{1}{2}x\right)$
14. $g(x) = f(2x)$
15. $g(x) = -f\left(\frac{1}{2}x\right) + 1$
16. $g(x) = -f(2x) - 1$

In Exercises 17–32, use the graph of $y = f(x)$ to graph each function g.

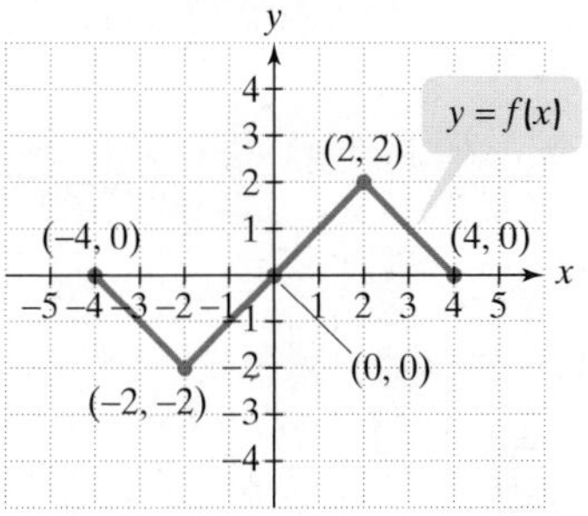

17. $g(x) = f(x) - 1$
18. $g(x) = f(x) + 1$
19. $g(x) = f(x - 1)$
20. $g(x) = f(x + 1)$
21. $g(x) = f(x - 1) + 2$
22. $g(x) = f(x + 1) - 2$
23. $g(x) = -f(x)$
24. $g(x) = f(-x)$
25. $g(x) = f(-x) + 1$
26. $g(x) = -f(x) + 1$
27. $g(x) = 2f(x)$
28. $g(x) = \frac{1}{2}f(x)$
29. $g(x) = f(2x)$
30. $g(x) = f\left(\frac{1}{2}x\right)$
31. $g(x) = 2f(x + 2) + 1$
32. $g(x) = 2f(x + 2) - 1$

In Exercises 33–44, use the graph of $y = f(x)$ to graph each function g.

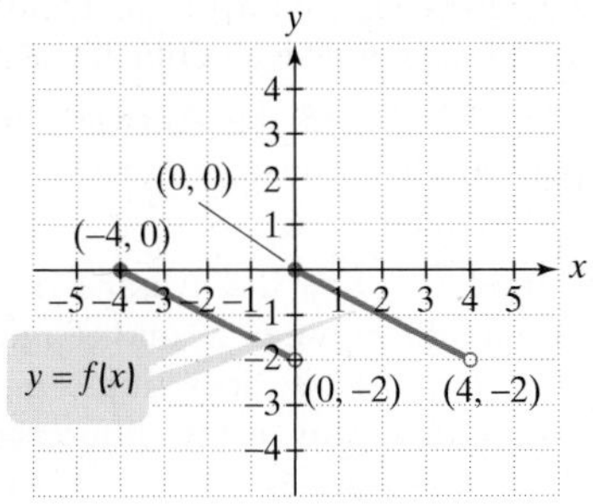

33. $g(x) = f(x) + 2$
34. $g(x) = f(x) - 2$
35. $g(x) = f(x + 2)$
36. $g(x) = f(x - 2)$
37. $g(x) = -f(x + 2)$
38. $g(x) = -f(x - 2)$
39. $g(x) = -\frac{1}{2}f(x + 2)$
40. $g(x) = -\frac{1}{2}f(x - 2)$
41. $g(x) = -\frac{1}{2}f(x + 2) - 2$
42. $g(x) = -\frac{1}{2}f(x - 2) + 2$
43. $g(x) = \frac{1}{2}f(2x)$
44. $g(x) = 2f\left(\frac{1}{2}x\right)$

In Exercises 45–52, use the graph of $y = f(x)$ to graph each function g.

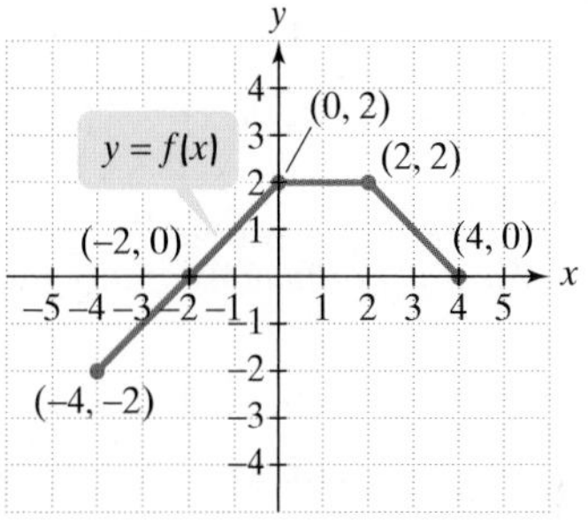

45. $g(x) = f(x - 1) - 1$
46. $g(x) = f(x + 1) + 1$
47. $g(x) = -f(x - 1) + 1$
48. $g(x) = -f(x + 1) - 1$
49. $g(x) = 2f\left(\frac{1}{2}x\right)$
50. $g(x) = \frac{1}{2}f(2x)$
51. $g(x) = \frac{1}{2}f(x + 1)$
52. $g(x) = 2f(x - 1)$

In Exercises 53–66, begin by graphing the standard quadratic function, $f(x) = x^2$. Then use transformations of this graph to graph the given function.

53. $g(x) = x^2 - 2$
54. $g(x) = x^2 - 1$
55. $g(x) = (x - 2)^2$
56. $g(x) = (x - 1)^2$
57. $h(x) = -(x - 2)^2$
58. $h(x) = -(x - 1)^2$
59. $h(x) = (x - 2)^2 + 1$
60. $h(x) = (x - 1)^2 + 2$
61. $g(x) = 2(x - 2)^2$
62. $g(x) = \frac{1}{2}(x - 1)^2$
63. $h(x) = 2(x - 2)^2 - 1$
64. $h(x) = \frac{1}{2}(x - 1)^2 - 1$
65. $h(x) = -2(x + 1)^2 + 1$
66. $h(x) = -2(x + 2)^2 + 1$

In Exercises 67–80, begin by graphing the square root function, $f(x) = \sqrt{x}$. Then use transformations of this graph to graph the given function.

67. $g(x) = \sqrt{x} + 2$
68. $g(x) = \sqrt{x} + 1$
69. $g(x) = \sqrt{x + 2}$
70. $g(x) = \sqrt{x + 1}$
71. $h(x) = -\sqrt{x + 2}$
72. $h(x) = -\sqrt{x + 1}$
73. $h(x) = \sqrt{-x + 2}$
74. $h(x) = \sqrt{-x + 1}$
75. $g(x) = \frac{1}{2}\sqrt{x + 2}$
76. $g(x) = 2\sqrt{x + 1}$
77. $h(x) = \sqrt{x + 2} - 2$
78. $h(x) = \sqrt{x + 1} - 1$
79. $g(x) = 2\sqrt{x + 2} - 2$
80. $g(x) = 2\sqrt{x + 1} - 1$

In Exercises 81–94, begin by graphing the absolute value function, $f(x) = |x|$. Then use transformations of this graph to graph the given function.

81. $g(x) = |x| + 4$
82. $g(x) = |x| + 3$
83. $g(x) = |x + 4|$
84. $g(x) = |x + 3|$
85. $h(x) = |x + 4| - 2$
86. $h(x) = |x + 3| - 2$
87. $h(x) = -|x + 4|$
88. $h(x) = -|x + 3|$

89. $g(x) = -|x + 4| + 1$ **90.** $g(x) = -|x + 4| + 2$

91. $h(x) = 2|x + 4|$ **92.** $h(x) = 2|x + 3|$

93. $g(x) = -2|x + 4| + 1$ **94.** $g(x) = -2|x + 3| + 2$

In Exercises 95–106, begin by graphing the standard cubic function, $f(x) = x^3$. Then use transformations of this graph to graph the given function.

95. $g(x) = x^3 - 3$ **96.** $g(x) = x^3 - 2$

97. $g(x) = (x - 3)^3$ **98.** $g(x) = (x - 2)^3$

99. $h(x) = -x^3$ **100.** $h(x) = -(x - 2)^3$

101. $h(x) = \frac{1}{2}x^3$ **102.** $h(x) = \frac{1}{4}x^3$

103. $r(x) = (x - 3)^3 + 2$ **104.** $r(x) = (x - 2)^3 + 1$

105. $h(x) = \frac{1}{2}(x - 3)^3 - 2$ **106.** $h(x) = \frac{1}{2}(x - 2)^3 - 1$

In Exercises 107–118, begin by graphing the cube root function, $f(x) = \sqrt[3]{x}$. Then use transformations of this graph to graph the given function.

107. $g(x) = \sqrt[3]{x} + 2$ **108.** $g(x) = \sqrt[3]{x} - 2$

109. $g(x) = \sqrt[3]{x + 2}$ **110.** $g(x) = \sqrt[3]{x - 2}$

111. $h(x) = \frac{1}{2}\sqrt[3]{x + 2}$ **112.** $h(x) = \frac{1}{2}\sqrt[3]{x - 2}$

113. $r(x) = \frac{1}{2}\sqrt[3]{x + 2} - 2$ **114.** $r(x) = \frac{1}{2}\sqrt[3]{x - 2} + 2$

115. $h(x) = -\sqrt[3]{x + 2}$ **116.** $h(x) = -\sqrt[3]{x - 2}$

117. $g(x) = \sqrt[3]{-x - 2}$ **118.** $g(x) = \sqrt[3]{-x + 2}$

 ### Practice Plus

In Exercises 119–122, use transformations of the graph of the greatest integer function, $f(x) = \text{int}(x)$, to graph each function. (The graph of $f(x) = \text{int}(x)$ is shown in Figure 2.20 on page 210.)

119. $g(x) = 2\,\text{int}\,(x + 1)$ **120.** $g(x) = 3\,\text{int}\,(x - 1)$

121. $h(x) = \text{int}(-x) + 1$ **122.** $h(x) = \text{int}(-x) - 1$

In Exercises 123–126, write a possible equation for the function whose graph is shown. Each graph shows a transformation of a common function.

123.

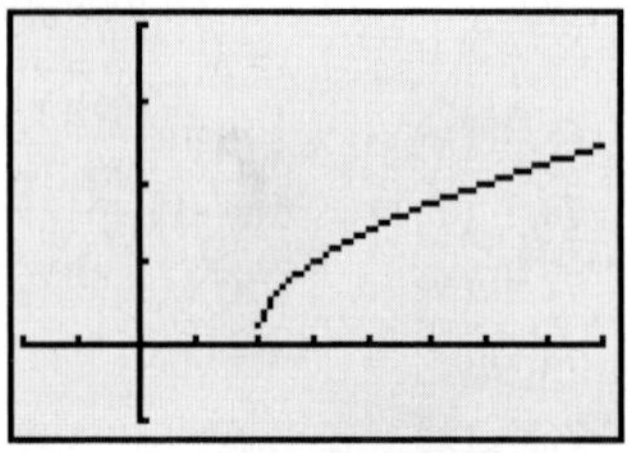

[−2, 8, 1] by [−1, 4, 1]

124.

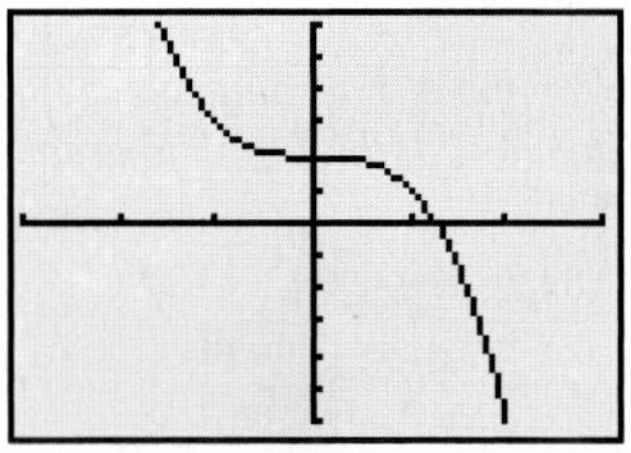

[−3, 3, 1] by [−6, 6, 1]

125.

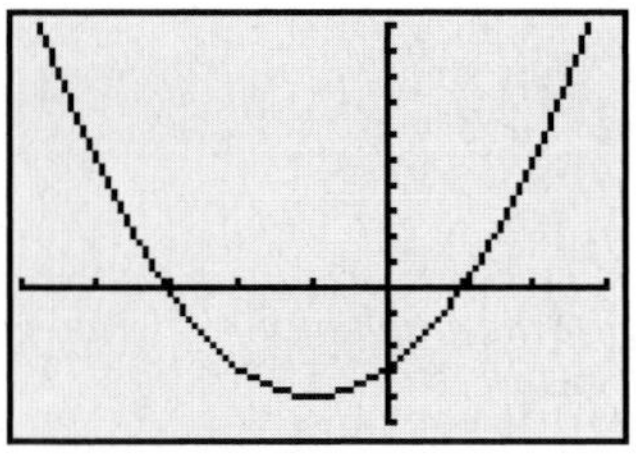

[−5, 3, 1] by [−5, 10, 1]

126.

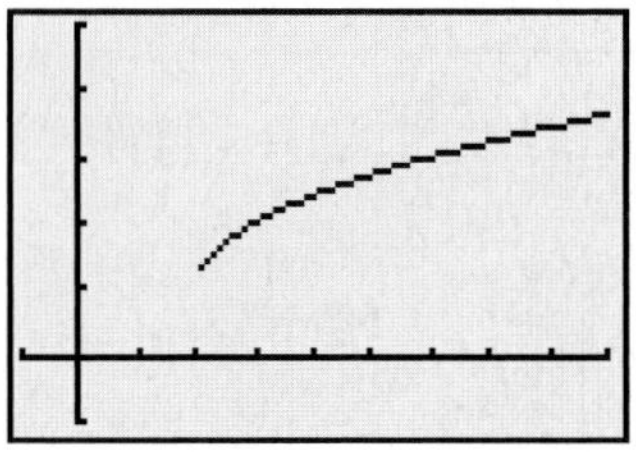

[−1, 9, 1] by [−1, 5, 1]

Application Exercises

127. The function $f(x) = 2.9\sqrt{x} + 20.1$ models the median height, $f(x)$, in inches, of boys who are x months of age. The graph of f is shown.

Boys' Heights

$f(x) = 2.9\sqrt{x} + 20.1$

Median Height (inches)

Age (months)

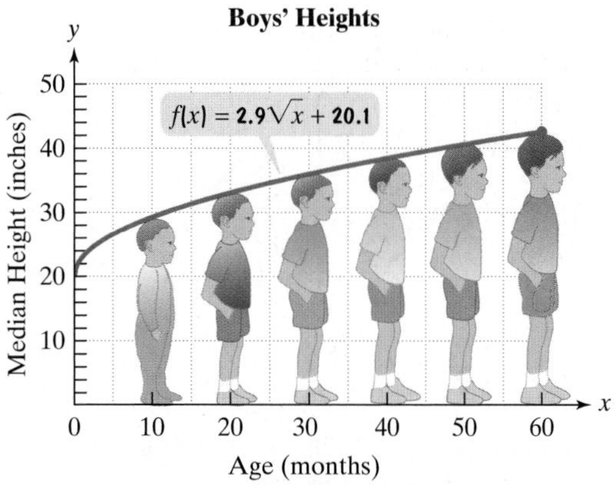

Source: Laura Walther Nathanson, *The Portable Pediatrician for Parents*

a. Describe how the graph can be obtained using transformations of the square root function $f(x) = \sqrt{x}$.

b. According to the model, what is the median height of boys who are 48 months, or four years, old? Use a calculator and round to the nearest tenth of an inch. The actual median height for boys at 48 months is 40.8 inches. How well does the model describe the actual height?

c. Use the model to find the average rate of change, in inches per month, between birth and 10 months. Round to the nearest tenth.

d. Use the model to find the average rate of change, in inches per month, between 50 and 60 months. Round to the nearest tenth. How does this compare with your answer in part (c)? How is this difference shown by the graph?

128. The function $f(x) = 3.1\sqrt{x} + 19$ models the median height, $f(x)$, in inches, of girls who are x months of age. The graph of f is shown.

Girls' Heights

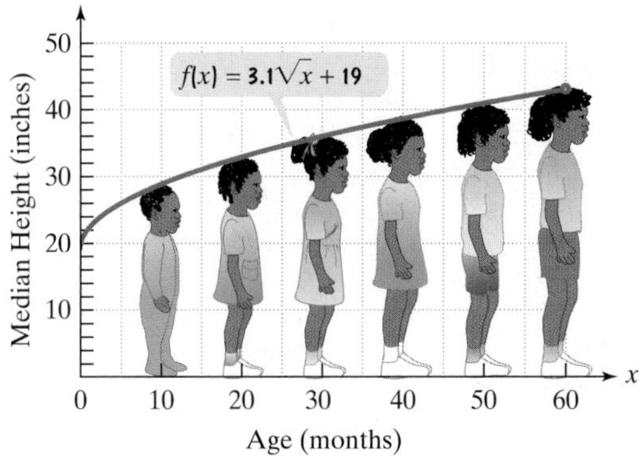

Source: Laura Walther Nathanson, *The Portable Pediatrician for Parents*

a. Describe how the graph can be obtained using transformations of the square root function $f(x) = \sqrt{x}$.

b. According to the model, what is the median height of girls who are 48 months, or four years, old? Use a calculator and round to the nearest tenth of an inch. The actual median height for girls at 48 months is 40.2 inches. How well does the model describe the actual height?

c. Use the model to find the average rate of change, in inches per month, between birth and 10 months. Round to the nearest tenth.

d. Use the model to find the average rate of change, in inches per month, between 50 and 60 months. Round to the nearest tenth. How does this compare with your answer in part (c)? How is this difference shown by the graph?

Writing in Mathematics

129. What must be done to a function's equation so that its graph is shifted vertically upward?

130. What must be done to a function's equation so that its graph is shifted horizontally to the right?

131. What must be done to a function's equation so that its graph is reflected about the x-axis?

132. What must be done to a function's equation so that its graph is reflected about the y-axis?

133. What must be done to a function's equation so that its graph is stretched vertically?

134. What must be done to a function's equation so that its graph is shrunk horizontally?

Technology Exercises

135. a. Use a graphing utility to graph $f(x) = x^2 + 1$.

b. Graph $f(x) = x^2 + 1$, $g(x) = f(2x)$, $h(x) = f(3x)$, and $k(x) = f(4x)$ in the same viewing rectangle.

c. Describe the relationship among the graphs of f, g, h, and k, with emphasis on different values of x for points on all four graphs that give the same y-coordinate.

d. Generalize by describing the relationship between the graph of f and the graph of g, where $g(x) = f(cx)$ for $c > 1$.

e. Try out your generalization by sketching the graphs of $f(cx)$ for $c = 1$, $c = 2$, $c = 3$, and $c = 4$ for a function of your choice.

136. a. Use a graphing utility to graph $f(x) = x^2 + 1$.

b. Graph $f(x) = x^2 + 1$, $g(x) = f\left(\frac{1}{2}x\right)$, and $h(x) = f\left(\frac{1}{4}x\right)$ in the same viewing rectangle.

c. Describe the relationship among the graphs of f, g, and h, with emphasis on different values of x for points on all three graphs that give the same y-coordinate.

d. Generalize by describing the relationship between the graph of f and the graph of g, where $g(x) = f(cx)$ for $0 < c < 1$.

e. Try out your generalization by sketching the graphs of $f(cx)$ for $c = 1$, and $c = \frac{1}{2}$, and $c = \frac{1}{4}$ for a function of your choice.

Critical Thinking Exercises

137. Which one of the following is true?

a. If $f(x) = |x|$ and $g(x) = |x + 3| + 3$, then the graph of g is a translation of the graph of f three units to the right and three units upward.

b. If $f(x) = -\sqrt{x}$ and $g(x) = \sqrt{-x}$, then f and g have identical graphs.

c. If $f(x) = x^2$ and $g(x) = 5(x^2 - 2)$, then the graph of g can be obtained from the graph of f by stretching f five units followed by a downward shift of two units.

d. If $f(x) = x^3$ and $g(x) = -(x - 3)^3 - 4$, then the graph of g can be obtained from the graph of f by moving f three units to the right, reflecting about the x-axis, and then moving the resulting graph down four units.

In Exercises 138–141, functions f and g are graphed in the same rectangular coordinate system. If g is obtained from f through a sequence of transformations, find an equation for g.

138.

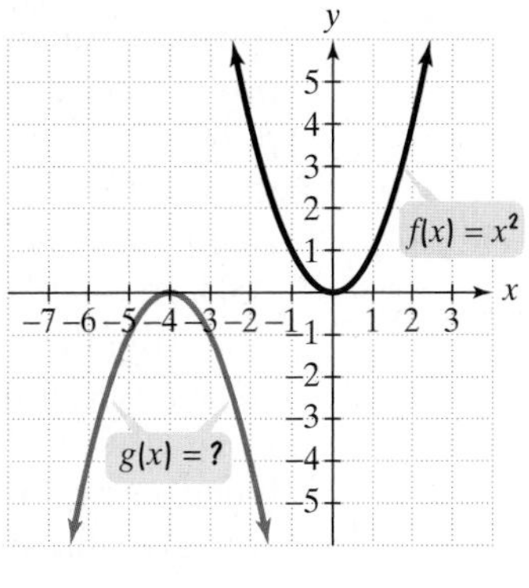

139.

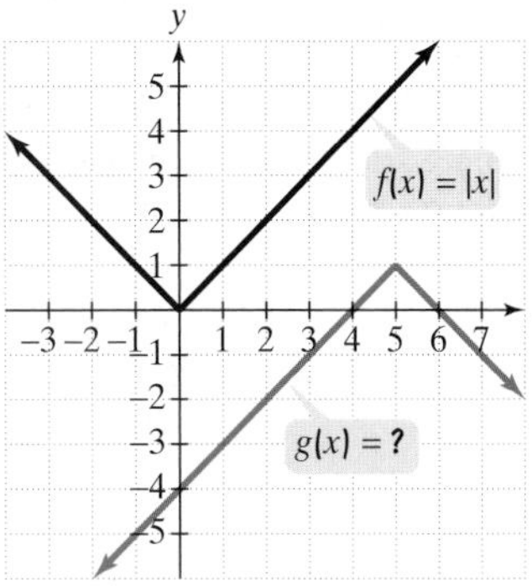

140.

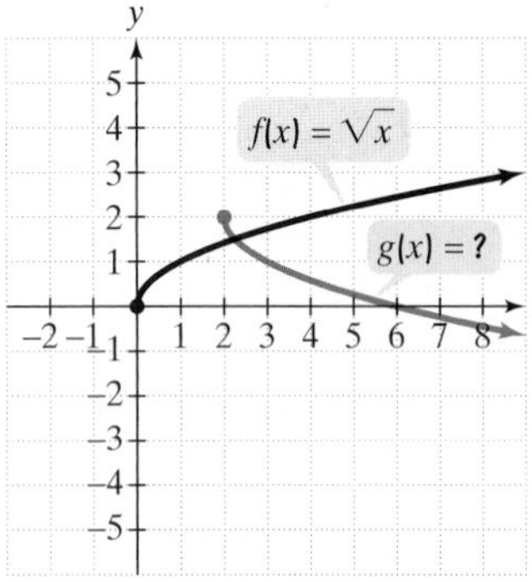

141.

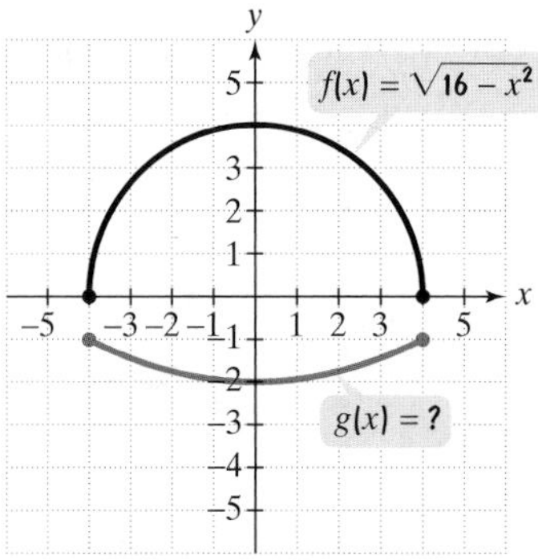

For Exercises 142–145, assume that (a, b) is a point on the graph of f. What is the corresponding point on the graph of each of the following functions?

142. $y = f(-x)$ **143.** $y = 2f(x)$

144. $y = f(x - 3)$ **145.** $y = f(x) - 3$

SECTION 2.6 Combinations of Functions; Composite Functions

Objectives

1. Find the domain of a function.
2. Combine functions using the algebra of functions, specifying domains.
3. Form composite functions.
4. Determine domains for composite functions.
5. Write functions as compositions.

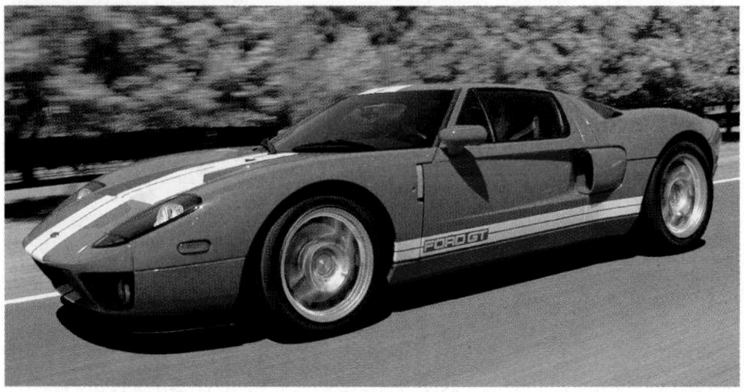

America's big three automakers, GM, Ford, and Chrysler, know that consumers have always been willing to pay more for cool design and a hot car. In 2004, Detroit gave consumers that opportunity, unleashing 40 new cars or updated models. The Big Three's discovery of the automobile in 2004 might seem odd, but from 1990 through 2003, Detroit focused much of its energy and money on SUVs and trucks, which commanded high prices and high profits, and saw less competition from imports. The line graphs in Figure 2.47 show the number, in millions, of cars and SUVs sold by the Big Three from 1990 through 2003. In this section, we will look at these data from the perspective of functions. By considering total sales of cars and SUVs, you will see that functions can be combined using procedures that will remind you of combining algebraic expressions.

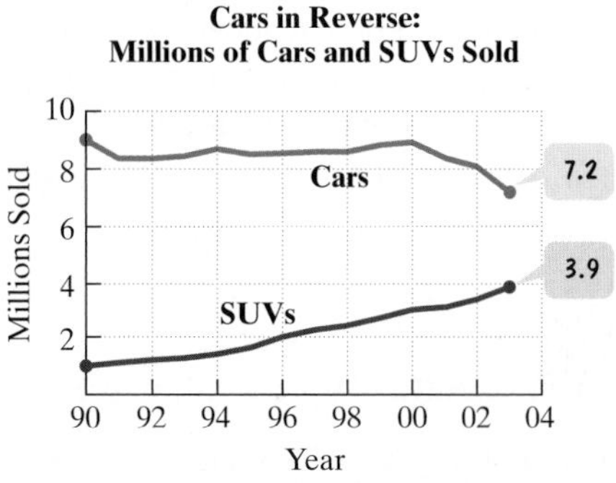

Figure 2.47 Sales by America's Big Three automakers

Study Tip

Throughout this section, we will be using the intersection of sets expressed in interval notation. Recall that the intersection of sets A and B, written $A \cap B$, is the set of elements common to both set A and set B. When sets A and B are in interval notation, to find the intersection, graph each interval and take the portion of the number line that the two graphs have in common. We will also be using notation involving the union of sets A and B, $A \cup B$, meaning the set of elements in A or in B or in both. For more detail, see Section P.1, pages 5–6 and Section 1.7, pages 166–167.

1 Find the domain of a function.

The Domain of a Function

We begin with two functions that model the data in Figure 2.47.

$$C(x) = -0.14x + 9 \qquad S(x) = 0.22x + 1$$

Car sales, $C(x)$, in millions, x years after 1990

SUV sales, $S(x)$, in millions, x years after 1990

How far beyond 1990 should we extend these models? The trend in Figure 2.47 shows decreasing car sales and increasing SUV sales. Because the three automakers focused on SUVs from 1990 through 2003 and launched a fleet of new cars in 2004, the trend shown by the data changed in 2004. Thus, we should not extend the models beyond 2003. Because x represents the number of years after 1990,

$$\text{Domain of } C = \{x \mid x = 0, 1, 2, 3, \ldots, 13\}$$

and

$$\text{Domain of } S = \{x \mid x = 0, 1, 2, 3, \ldots, 13\}.$$

Functions that model data often have their domains explicitly given with the function's equation. However, for most functions, only an equation is given and the domain is not specified. In cases like this, the domain of a function f is the largest set of real numbers for which the value of $f(x)$ is a real number. For example, consider the function

$$f(x) = \frac{1}{x - 3}.$$

Because division by 0 is undefined (and not a real number), the denominator, $x - 3$, cannot be 0. Thus, x cannot equal 3. The domain of the function consists of all real numbers other than 3, represented by

$$\text{Domain of } f = \{x \mid x \text{ is a real number and } x \neq 3\}.$$

Using interval notation,

$$\text{Domain of } f = (-\infty, 3) \cup (3, \infty).$$

All real numbers less than 3 | or | All real numbers greater than 3

Now consider a function involving a square root:

$$g(x) = \sqrt{x - 3}.$$

Because only nonnegative numbers have square roots that are real numbers, the expression under the square root sign, $x - 3$, must be nonnegative. We can use inspection to see that $x - 3 \geq 0$ if $x \geq 3$. The domain of g consists of all real numbers that are greater than or equal to 3:

$$\text{Domain of } g = \{x \mid x \geq 3\} \text{ or } [3, \infty).$$

Finding a Function's Domain

If a function f does not model data or verbal conditions, its domain is the largest set of real numbers for which the value of $f(x)$ is a real number. Exclude from a function's domain real numbers that cause division by zero and real numbers that result in a square root of a negative number.

EXAMPLE 1   Finding the Domain of a Function

Find the domain of each function:

a. $f(x) = x^2 - 7x$ **b.** $g(x) = \dfrac{3x + 2}{x^2 - 2x - 3}$ **c.** $h(x) = \sqrt{3x + 12}$.

Solution

a. The function $f(x) = x^2 - 7x$ contains neither division nor a square root. For every real number, x, the algebraic expression $x^2 - 7x$ represents a real number. Thus, the domain of f is the set of all real numbers.

$$\text{Domain of } f = (-\infty, \infty)$$

b. The function $g(x) = \dfrac{3x + 2}{x^2 - 2x - 3}$ contains division. Because division by 0 is undefined, we must exclude from the domain the values of x that cause the denominator, $x^2 - 2x - 3$, to be 0. We can identify these values by setting $x^2 - 2x - 3$ equal to 0.

$$x^2 - 2x - 3 = 0 \quad \text{\small Set the function's denominator equal to 0.}$$
$$(x + 1)(x - 3) = 0 \quad \text{\small Factor.}$$
$$x + 1 = 0 \quad \text{or} \quad x - 3 = 0 \quad \text{\small Set each factor equal to 0.}$$
$$x = -1 \qquad\qquad x = 3 \quad \text{\small Solve the resulting equations.}$$

We must exclude -1 and 3 from the domain of g.

$$\text{Domain of } g = (-\infty, -1) \cup (-1, 3) \cup (3, \infty)$$

c. The function $h(x) = \sqrt{3x + 12}$ contains an even root. Because only nonnegative numbers have real square roots, the quantity under the radical sign, $3x + 12$, must be greater than or equal to 0.

$$3x + 12 \geq 0 \quad \text{\small Set the function's radicand greater than}$$
$$\text{\small or equal to 0.}$$
$$3x \geq -12 \quad \text{\small Subtract 12 from both sides.}$$
$$x \geq -4 \quad \text{\small Divide both sides by 3. Division by a positive}$$
$$\text{\small number preserves the sense of the inequality.}$$

The domain of h consists of all real numbers greater than or equal to -4.

$$\text{Domain of } h = [-4, \infty)$$

The domain is highlighted on the x-axis in Figure 2.48.

$h(x) = \sqrt{3x + 12}$

Domain $[-4, \infty)$

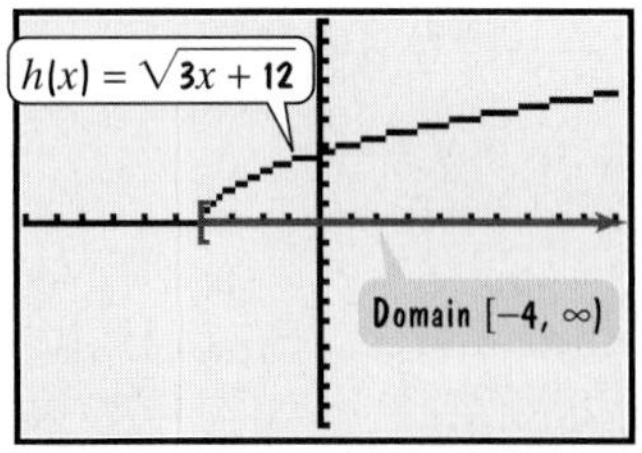

$[-10, 10, 1]$ by $[-10, 10, 1]$

Figure 2.48

Check Point 1   Find the domain of each function:

a. $f(x) = x^2 + 3x - 17$ **b.** $g(x) = \dfrac{5x}{x^2 - 49}$

c. $h(x) = \sqrt{9x - 27}$.

The Algebra of Functions

We return to the functions that model millions of car and SUV sales from 1990 through 2003:

$$C(x) = -0.14x + 9 \qquad S(x) = 0.22x + 1.$$

Car sales, $C(x)$, in millions, x years after 1990

SUV sales, $S(x)$, in millions, x years after 1990

How can we use these functions to find total sales of cars and SUVs in 2003? Because 2003 is 13 years after 1990 and 13 is in the domain of each function, we need to find the sum of two function values:

$$C(13) + S(13).$$

Here is how it's done:

$$C(13) = -0.14(13) + 9 = 7.18 \qquad S(13) = 0.22(13) + 1 = 3.86$$

Substitute 13 for x in $C(x) = -0.14x + 9$.

7.18 million cars were sold in 2003.

Substitute 13 for x in $S(x) = 0.22x + 1$.

3.86 million SUVs were sold in 2003.

$$C(13) + S(13) = 7.18 + 3.86 = 11.04.$$

Thus, a total of 11.04 million cars and SUVs were sold in 2003.

There is a second way that we can obtain this number. We can first add the functions C and S to obtain a new function, $C + S$. To do so, we add the terms to the right of the equal sign for $C(x)$ to the terms to the right of the equal sign for $S(x)$:

$$\begin{aligned}(C + S)(x) &= C(x) + S(x)\\ &= (-0.14x + 9) + (0.22x + 1) &&\text{Add terms for } C(x) \text{ and } S(x).\\ &= 0.08x + 10. &&\text{Combine like terms.}\end{aligned}$$

Thus,

$$(C + S)(x) = 0.08x + 10.$$

Total car and SUV sales, in millions, x years after 1990

Do you see how we can use this new function to find total car and SUV sales in 2003? Substitute 13 for x in the equation for $C + S$:

$$(C + S)(13) = 0.08(13) + 10 = 11.04.$$

Substitute 13 for x in $(C + S)(x) = 0.08x + 10$.

As we found above, a total of 11.04 million cars and SUVs were sold in 2003.

The domain of the new function, $C + S$, consists of the numbers x that are in the domain of C **and** in the domain of S. If D_C represents the domain of C and D_S represents the domain of S, the domain of $C + S$ is $D_C \cap D_S$. Because both C and S model data from 1990 through 2003,

$$\text{Domain of } C + S = \{0, 1, 2, 3, \ldots, 13\}.$$

The function that models total car and SUV sales illustrates that functions can be added algebraically. We can also combine functions using subtraction, multiplication, and division by performing operations with the algebraic expressions that appear on the right side of the equations. The domain for each of these functions consists of all real numbers that are common to the domains of the functions being combined. Furthermore, when combining functions using division, values that make the divisor zero must be excluded from the domain.

The following definitions summarize our discussion:

2 Combine functions using the algebra of functions, specifying domains.

The Algebra of Functions: Sum, Difference, Product, and Quotient of Functions

Let f and g be two functions. The **sum** $f + g$, the **difference** $f - g$, the **product** fg, and the **quotient** $\frac{f}{g}$ are functions whose domains are the set of all real numbers common to the domains of f and g $(D_f \cap D_g)$, defined as follows:

1. Sum: $(f + g)(x) = f(x) + g(x)$
2. Difference: $(f - g)(x) = f(x) - g(x)$
3. Product: $(fg)(x) = f(x) \cdot g(x)$
4. Quotient: $\left(\frac{f}{g}\right)(x) = \frac{f(x)}{g(x)}$, provided $g(x) \neq 0$.

EXAMPLE 2 Combining Functions

Let $f(x) = 2x - 1$ and $g(x) = x^2 + x - 2$. Find each of the following functions:

a. $(f + g)(x)$ **b.** $(f - g)(x)$ **c.** $(fg)(x)$ **d.** $\left(\frac{f}{g}\right)(x)$.

Determine the domain for each function.

Solution

a. $(f + g)(x) = f(x) + g(x)$ — This is the definition of the sum $f + g$.

$= (2x - 1) + (x^2 + x - 2)$ — Substitute the given functions.

$= x^2 + 3x - 3$ — Remove parentheses and combine like terms.

b. $(f - g)(x) = f(x) - g(x)$ — This is the definition of the difference $f - g$.

$= (2x - 1) - (x^2 + x - 2)$ — Substitute the given functions.

$= 2x - 1 - x^2 - x + 2$ — Remove parentheses and change the sign of each term in the second set of parentheses.

$= -x^2 + x + 1$ — Combine like terms and arrange terms in descending powers of x.

c. $(fg)(x) = f(x) \cdot g(x)$ — This is the definition of the product fg.

$= (2x - 1)(x^2 + x - 2)$ — Substitute the given functions.

$= 2x(x^2 + x - 2) - 1(x^2 + x - 2)$ — Multiply each term in the second factor by $2x$ and -1, respectively.

$= 2x^3 + 2x^2 - 4x - x^2 - x + 2$ — Use the distributive property.

$= 2x^3 + (2x^2 - x^2) + (-4x - x) + 2$ — Rearrange terms so that like terms are adjacent.

$= 2x^3 + x^2 - 5x + 2$ — Combine like terms.

d. $\left(\frac{f}{g}\right)(x) = \frac{f(x)}{g(x)}$ — This is the definition of the quotient $\frac{f}{g}$.

$= \frac{2x - 1}{x^2 + x - 2}$ — Substitute the given functions. This rational expression cannot be simplified.

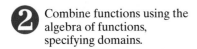

Because the equations for f and g do not involve division or contain even roots, the domain of both f and g is the set of all real numbers. Thus, the domain of $f + g, f - g$, and fg is the set of all real numbers, $(-\infty, \infty)$.

The function $\frac{f}{g}$ contains division. We must exclude from its domain values of x that cause the denominator, $x^2 + x - 2$, to be 0. Let's identify these values.

$$x^2 + x - 2 = 0 \quad \text{Set the denominator of } \frac{f}{g} \text{ equal to 0.}$$
$$(x + 2)(x - 1) = 0 \quad \text{Factor.}$$
$$x + 2 = 0 \quad \text{or} \quad x - 1 = 0 \quad \text{Set each factor equal to 0.}$$
$$x = -2 \qquad\qquad x = 1 \quad \text{Solve the resulting equations.}$$

We must exclude -2 and 1 from the domain of $\frac{f}{g}$.

$$\text{Domain of } \frac{f}{g} = (-\infty, -2) \cup (-2, 1) \cup (1, \infty)$$

Check Point 2 Let $f(x) = x - 5$ and $g(x) = x^2 - 1$. Find each of the following functions:

a. $(f + g)(x)$ **b.** $(f - g)(x)$ **c.** $(fg)(x)$ **d.** $\left(\dfrac{f}{g}\right)(x)$.

Determine the domain for each function.

EXAMPLE 3   Adding Functions and Determining the Domain

Let $f(x) = \sqrt{x + 3}$ and $g(x) = \sqrt{x - 2}$. Find each of the following:

a. $(f + g)(x)$ **b.** the domain of $f + g$.

Solution

a. $(f + g)(x) = f(x) + g(x) = \sqrt{x + 3} + \sqrt{x - 2}$

b. The domain of $f + g$ is the set of all real numbers that are common to the domain of f and the domain of g. Thus, we must find the domains of f and g before finding their intersection. We will do so for f first.

Note that $f(x) = \sqrt{x + 3}$ is a function involving the square root of $x + 3$. Because the square root of a negative quantity is not a real number, the value of $x + 3$ must be nonnegative. Thus, the domain of f is all x such that $x + 3 \geq 0$. Equivalently, the domain is $\{x | x \geq -3\}$, or $[-3, \infty)$.

Likewise, $g(x) = \sqrt{x - 2}$ is also a square root function. Because the square root of a negative quantity is not a real number, the value of $x - 2$ must be nonnegative. Thus, the domain of g is all x such that $x - 2 \geq 0$. Equivalently, the domain is $\{x | x \geq 2\}$, or $[2, \infty)$.

Now, we can use a number line to determine $D_f \cap D_g$, the domain of $f + g$. Figure 2.49 shows the domain of f in blue and the domain of g in red. Can you see that all real numbers greater than or equal to 2 are common to both domains? This is shown in purple on the number line. Thus, the domain of $f + g$ is $[2, \infty)$.

Domain of f

Domain of g

Domain of $f + g$

-3   2

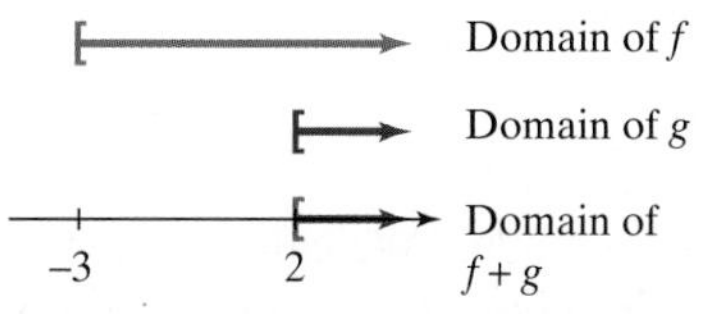

Figure 2.49 Finding the domain of the sum $f + g$

Technology

Domain $[2, \infty)$

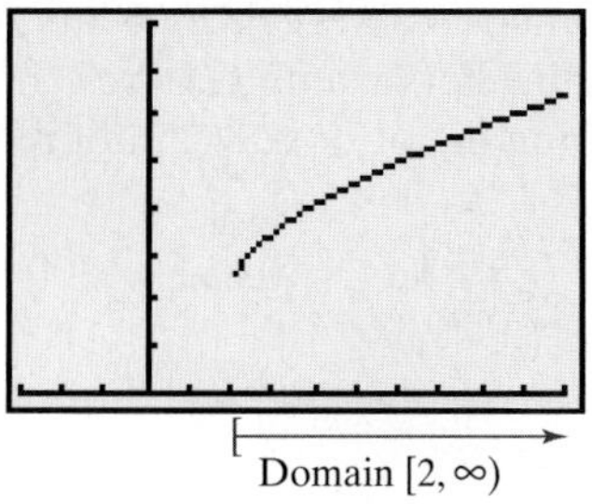

The graph on the left is the graph of

$$y = \sqrt{x + 3} + \sqrt{x - 2}$$

in a $[-3, 10, 1]$ by $[0, 8, 1]$ viewing rectangle. The graph reveals what we discovered algebraically in Example 3(b). The domain of this function is $[2, \infty)$.

Check Point 3 Let $f(x) = \sqrt{x-3}$ and $g(x) = \sqrt{x+1}$. Find each of the following:

a. $(f+g)(x)$ **b.** the domain of $f+g$.

3 Form composite functions.

Composite Functions

There is another way of combining two functions. To help understand this new combination, suppose that your local computer store is having a sale. The models that are on sale cost either \$300 less than the regular price or 85% of the regular price. If x represents the computer's regular price, the discounts can be described with the following functions:

$$f(x) = x - 300 \qquad g(x) = 0.85x.$$

The computer is on sale for \$300 less than its regular price.

The computer is on sale for 85% of its regular price.

At the store, you bargain with the salesperson. Eventually, she makes an offer you can't refuse. The sale price will be 85% of the regular price followed by a \$300 reduction:

$$0.85x - 300.$$

85% of the regular price

followed by a \$300 reduction

In terms of the functions f and g, this offer can be obtained by taking the output of $g(x) = 0.85x$, namely $0.85x$, and using it as the input of f:

$$f(x) = x - 300$$

Replace x with $0.85x$, the output of $g(x) = 0.85x$.

$$f(0.85x) = 0.85x - 300.$$

Because $0.85x$ is $g(x)$, we can write this last equation as

$$f(g(x)) = 0.85x - 300.$$

We read this equation as "f of g of x is equal to $0.85x - 300$." We call $f(g(x))$ the **composition of the function *f* with *g***, or a **composite function**. This composite function is written $f \circ g$. Thus,

$$(f \circ g)(x) = f(g(x)) = 0.85x - 300.$$

This can be read "f of g of x" or "f composed with g of x."

Like all functions, we can evaluate $f \circ g$ for a specified value of x in the function's domain. For example, here's how to find the value of the composite function describing the offer you cannot refuse at 1400:

$$(f \circ g)(x) = 0.85x - 300$$

Replace x with 1400.

$$(f \circ g)(1400) = 0.85(1400) - 300 = 1190 - 300 = 890.$$

This means that a computer that regularly sells for \$1400 is on sale for \$890 subject to both discounts. We can use a partial table of coordinates for each of the discount functions, g and f, to numerically verify this result.

Computer's regular price | 85% of the regular price

x	$g(x) = 0.85x$
1200	1020
1300	1105
1400	1190

85% of the regular price | \$300 reduction

x	$f(x) = x - 300$
1020	720
1105	805
1190	890

Using these tables, we can find $(f \circ g)(1400)$:

$$(f \circ g)(1400) = f(g(1400)) = f(1190) = 890.$$

The table for g shows that $g(1400) = 1190$.

The table for f shows that $f(1190) = 890$.

This verifies that a computer that regularly sells for \$1400 is on sale for \$890 subject to both discounts.

Before you run out to buy a computer, let's generalize our discussion of the computer's double discount and define the composition of any two functions.

The Composition of Functions

The **composition of the function f with g** is denoted by $f \circ g$ and is defined by the equation

$$(f \circ g)(x) = f(g(x)).$$

The **domain of the composite function $f \circ g$** is the set of all x such that

1. x is in the domain of g and
2. $g(x)$ is in the domain of f.

The composition of f with g, $f \circ g$, is pictured as a machine with inputs and outputs in Figure 2.50. The diagram indicates that the output of g, or $g(x)$, becomes the input for "machine" f. If $g(x)$ is not in the domain of f, it cannot be input into machine f, and so $g(x)$ must be discarded.

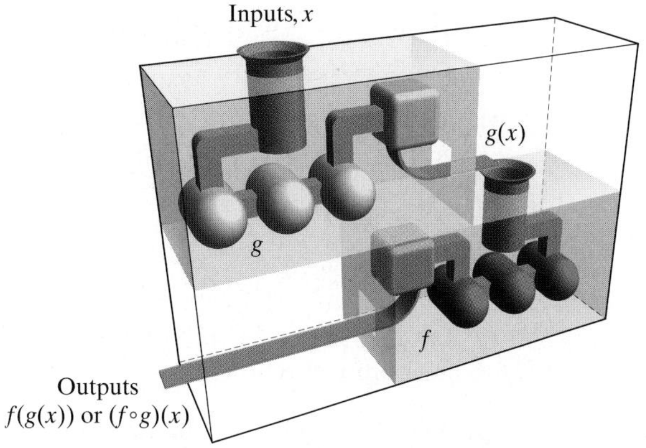

Figure 2.50 Inputting one function into a second function

EXAMPLE 4   Forming Composite Functions

Given $f(x) = 3x - 4$ and $g(x) = x^2 - 2x + 6$, find each of the following composite functions:

    a. $(f \circ g)(x)$ **b.** $(g \circ f)(x)$.

Solution

    a. We begin with $(f \circ g)(x)$, the composition of f with g. Because $(f \circ g)(x)$ means $f(g(x))$, we must replace each occurrence of x in the equation for f with $g(x)$.

$$f(x) = 3x - 4 \qquad \text{This is the given equation for } f.$$

Replace x with $g(x)$.

$$(f \circ g)(x) = f(g(x)) = 3g(x) - 4$$

$$= 3(x^2 - 2x + 6) - 4 \qquad \begin{array}{l} \text{Because} \\ g(x) = x^2 - 2x + 6, \\ \text{replace } g(x) \text{ with } x^2 - 2x + 6. \end{array}$$

$$= 3x^2 - 6x + 18 - 4 \qquad \text{Use the distributive property.}$$

$$= 3x^2 - 6x + 14 \qquad \text{Simplify.}$$

Thus, $(f \circ g)(x) = 3x^2 - 6x + 14$.

    b. Next, we find $(g \circ f)(x)$, the composition of g with f. Because $(g \circ f)(x)$ means $g(f(x))$, we must replace each occurrence of x in the equation for g with $f(x)$.

$$g(x) = x^2 - 2x + 6 \qquad \text{This is the equation for } g.$$

Replace x with $f(x)$.

$$(g \circ f)(x) = g(f(x)) = (f(x))^2 - 2f(x) + 6$$

$$= (3x - 4)^2 - 2(3x - 4) + 6 \qquad \begin{array}{l} \text{Because } f(x) = 3x - 4, \\ \text{replace } f(x) \text{ with } 3x - 4. \end{array}$$

$$= 9x^2 - 24x + 16 - 6x + 8 + 6 \qquad \begin{array}{l} \text{Use } (A - B)^2 = \\ A^2 - 2AB + B^2 \text{ to} \\ \text{square } 3x - 4. \end{array}$$

$$= 9x^2 - 30x + 30 \qquad \begin{array}{l} \text{Simplify:} \\ -24x - 6x = -30x \\ \text{and } 16 + 8 + 6 = 30. \end{array}$$

Thus, $(g \circ f)(x) = 9x^2 - 30x + 30$. **Notice that $(f \circ g)(x)$ is not the same function as $(g \circ f)(x)$.**

Check Point 4 Given $f(x) = 5x + 6$ and $g(x) = 2x^2 - x - 1$, find each of the following composite functions:

    a. $(f \circ g)(x)$ **b.** $(g \circ f)(x)$.

4 Determine domains for composite functions.

We need to be careful in determining the domain for a composite function.

Excluding Values from the Domain of $(f \circ g)(x) = f(g(x))$

The following values must be excluded from the input x:

- If x is not in the domain of g, it must not be in the domain of $f \circ g$.
- Any x for which $g(x)$ is not in the domain of f must not be in the domain of $f \circ g$.

EXAMPLE 5 Forming a Composite Function and Finding Its Domain

Given $f(x) = \frac{2}{x-1}$ and $g(x) = \frac{3}{x}$, find each of the following:

a. $(f \circ g)(x)$ **b.** the domain of $f \circ g$.

Solution

a. Because $(f \circ g)(x)$ means $f(g(x))$, we must replace x in $f(x) = \frac{2}{x-1}$ with $g(x)$.

Study Tip

The procedure for simplifying complex fractions can be found in Section P.6, pages 75–76.

$$(f \circ g)(x) = f(g(x)) = \frac{2}{g(x)-1} = \frac{2}{\frac{3}{x}-1} = \frac{2}{\frac{3}{x}-1} \cdot \frac{x}{x} = \frac{2x}{3-x}$$

$g(x) = \frac{3}{x}$

Simplify the complex fraction by multiplying by $\frac{x}{x}$, or 1.

Thus, $(f \circ g)(x) = \frac{2x}{3-x}$.

b. We determine values to exclude from the domain of $(f \circ g)(x)$ in two steps.

Rules for Excluding Numbers from the Domain of $(f \circ g)(x) = f(g(x))$	Applying the Rules to $f(x) = \frac{2}{x-1}$ and $g(x) = \frac{3}{x}$
If x is not in the domain of g, it must not be in the domain of $f \circ g$.	Because $g(x) = \frac{3}{x}$, 0 is not in the domain of g. Thus, 0 must be excluded from the domain of $f \circ g$.
Any x for which $g(x)$ is not in the domain of f must not be in the domain of $f \circ g$.	Because $f(g(x)) = \frac{2}{g(x)-1}$, we must exclude from the domain of $f \circ g$ any x for which $g(x) = 1$. $\frac{3}{x} = 1$ Set $g(x)$ equal to 1. $3 = x$ Multiply both sides by x. 3 must be excluded from the domain of $f \circ g$.

We see that 0 and 3 must be excluded from the domain of $f \circ g$. The domain of $f \circ g$ is

$$(-\infty, 0) \cup (0, 3) \cup (3, \infty).$$

Check Point 5 Given $f(x) = \frac{4}{x+2}$ and $g(x) = \frac{1}{x}$, find each of the following:

a. $(f \circ g)(x)$ **b.** the domain of $f \circ g$.

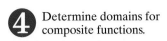

5 Write functions as compositions.

Decomposing Functions

When you form a composite function, you "compose" two functions to form a new function. It is also possible to reverse this process. That is, you can "decompose" a given function and express it as a composition of two functions. Although there is more than one way to do this, there is often a "natural" selection that comes to mind first. For example, consider the function h defined by

$$h(x) = (3x^2 - 4x + 1)^5.$$

The function h takes $3x^2 - 4x + 1$ and raises it to the power 5. A natural way to write h as a composition of two functions is to raise the function $g(x) = 3x^2 - 4x + 1$ to the power 5. Thus, if we let

$$f(x) = x^5 \text{ and } g(x) = 3x^2 - 4x + 1, \text{ then}$$
$$(f \circ g)(x) = f(g(x)) = f(3x^2 - 4x + 1) = (3x^2 - 4x + 1)^5.$$

EXAMPLE 6 Writing a Function as a Composition

Express $h(x)$ as a composition of two functions:

$$h(x) = \sqrt[3]{x^2 + 1}.$$

Solution The function h takes $x^2 + 1$ and takes its cube root. A natural way to write h as a composition of two functions is to take the cube root of the function $g(x) = x^2 + 1$. Thus, we let

$$f(x) = \sqrt[3]{x} \text{ and } g(x) = x^2 + 1.$$

We can check this composition by finding $(f \circ g)(x)$. This should give the original function, namely $h(x) = \sqrt[3]{x^2 + 1}$.

$$(f \circ g)(x) = f(g(x)) = f(x^2 + 1) = \sqrt[3]{x^2 + 1} = h(x)$$

Study Tip

Suppose the form of function h is $h(x) = (\text{algebraic expression})^{\text{power}}$. Function h can be expressed as a composition, $f \circ g$, using

$f(x) = x^{\text{power}}$
$g(x) = \text{algebraic expression}.$

Check Point 6 Express $h(x)$ as a composition of two functions:

$$h(x) = \sqrt{x^2 + 5}.$$

EXERCISE SET 2.6

Practice Exercises

In Exercises 1–30, find the domain of each function.

1. $f(x) = 3(x - 4)$
2. $f(x) = 2(x + 5)$
3. $g(x) = \dfrac{3}{x - 4}$
4. $g(x) = \dfrac{2}{x + 5}$
5. $f(x) = x^2 - 2x - 15$
6. $f(x) = x^2 + x - 12$
7. $g(x) = \dfrac{3}{x^2 - 2x - 15}$
8. $g(x) = \dfrac{2}{x^2 + x - 12}$
9. $f(x) = \dfrac{1}{x + 7} + \dfrac{3}{x - 9}$
10. $f(x) = \dfrac{1}{x + 8} + \dfrac{3}{x - 10}$
11. $g(x) = \dfrac{1}{x^2 + 1} - \dfrac{1}{x^2 - 1}$
12. $g(x) = \dfrac{1}{x^2 + 4} - \dfrac{1}{x^2 - 4}$
13. $h(x) = \dfrac{4}{\frac{3}{x} - 1}$
14. $h(x) = \dfrac{5}{\frac{4}{x} - 1}$
15. $f(x) = \dfrac{1}{\frac{4}{x - 1} - 2}$
16. $f(x) = \dfrac{1}{\frac{4}{x - 2} - 3}$
17. $f(x) = \sqrt{x - 3}$
18. $f(x) = \sqrt{x + 2}$
19. $g(x) = \dfrac{1}{\sqrt{x - 3}}$
20. $g(x) = \dfrac{1}{\sqrt{x + 2}}$
21. $g(x) = \sqrt{5x + 35}$
22. $g(x) = \sqrt{7x - 70}$
23. $f(x) = \sqrt{24 - 2x}$
24. $f(x) = \sqrt{84 - 6x}$
25. $h(x) = \sqrt{x - 2} + \sqrt{x + 3}$
26. $h(x) = \sqrt{x - 3} + \sqrt{x + 4}$
27. $g(x) = \dfrac{\sqrt{x - 2}}{x - 5}$
28. $g(x) = \dfrac{\sqrt{x - 3}}{x - 6}$

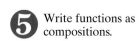

29. $f(x) = \dfrac{2x + 7}{x^3 - 5x^2 - 4x + 20}$

30. $f(x) = \dfrac{7x + 2}{x^3 - 2x^2 - 9x + 18}$

In Exercises 31–48, find $f + g$, $f - g$, fg, and $\frac{f}{g}$. Determine the domain for each function.

31. $f(x) = 2x + 3,\ g(x) = x - 1$

32. $f(x) = 3x - 4,\ g(x) = x + 2$

33. $f(x) = x - 5,\ g(x) = 3x^2$

34. $f(x) = x - 6,\ g(x) = 5x^2$

35. $f(x) = 2x^2 - x - 3,\ g(x) = x + 1$

36. $f(x) = 6x^2 - x - 1,\ g(x) = x - 1$

37. $f(x) = 3 - x^2,\ g(x) = x^2 + 2x - 15$

38. $f(x) = 5 - x^2,\ g(x) = x^2 + 4x - 12$

39. $f(x) = \sqrt{x},\ g(x) = x - 4$

40. $f(x) = \sqrt{x},\ g(x) = x - 5$

41. $f(x) = 2 + \dfrac{1}{x},\ g(x) = \dfrac{1}{x}$

42. $f(x) = 6 - \dfrac{1}{x},\ g(x) = \dfrac{1}{x}$

43. $f(x) = \dfrac{5x + 1}{x^2 - 9},\ g(x) = \dfrac{4x - 2}{x^2 - 9}$

44. $f(x) = \dfrac{3x + 1}{x^2 - 25},\ g(x) = \dfrac{2x - 4}{x^2 - 25}$

45. $f(x) = \sqrt{x + 4},\ g(x) = \sqrt{x - 1}$

46. $f(x) = \sqrt{x + 6},\ g(x) = \sqrt{x - 3}$

47. $f(x) = \sqrt{x - 2},\ g(x) = \sqrt{2 - x}$

48. $f(x) = \sqrt{x - 5},\ g(x) = \sqrt{5 - x}$

In Exercises 49–64, find

a. $(f \circ g)(x)$; **b.** $(g \circ f)(x)$; **c.** $(f \circ g)(2)$.

49. $f(x) = 2x,\ g(x) = x + 7$

50. $f(x) = 3x,\ g(x) = x - 5$

51. $f(x) = x + 4,\ g(x) = 2x + 1$

52. $f(x) = 5x + 2,\ g(x) = 3x - 4$

53. $f(x) = 4x - 3,\ g(x) = 5x^2 - 2$

54. $f(x) = 7x + 1,\ g(x) = 2x^2 - 9$

55. $f(x) = x^2 + 2,\ g(x) = x^2 - 2$

56. $f(x) = x^2 + 1,\ g(x) = x^2 - 3$

57. $f(x) = 4 - x,\ g(x) = 2x^2 + x + 5$

58. $f(x) = 5x - 2,\ g(x) = -x^2 + 4x - 1$

59. $f(x) = \sqrt{x},\ g(x) = x - 1$

60. $f(x) = \sqrt{x},\ g(x) = x + 2$

61. $f(x) = 2x - 3,\ g(x) = \dfrac{x + 3}{2}$

62. $f(x) = 6x - 3,\ g(x) = \dfrac{x + 3}{6}$

63. $f(x) = \dfrac{1}{x},\ g(x) = \dfrac{1}{x}$

64. $f(x) = \dfrac{2}{x},\ g(x) = \dfrac{2}{x}$

In Exercises 65–72, find

a. $(f \circ g)(x)$; **b.** *the domain of* $f \circ g$.

65. $f(x) = \dfrac{2}{x + 3},\ g(x) = \dfrac{1}{x}$

66. $f(x) = \dfrac{5}{x + 4},\ g(x) = \dfrac{1}{x}$

67. $f(x) = \dfrac{x}{x + 1},\ g(x) = \dfrac{4}{x}$

68. $f(x) = \dfrac{x}{x + 5},\ g(x) = \dfrac{6}{x}$

69. $f(x) = \sqrt{x},\ g(x) = x - 2$

70. $f(x) = \sqrt{x},\ g(x) = x - 3$

71. $f(x) = x^2 + 4,\ g(x) = \sqrt{1 - x}$

72. $f(x) = x^2 + 1,\ g(x) = \sqrt{2 - x}$

In Exercises 73–80, express the given function h as a composition of two functions f and g so that $h(x) = (f \circ g)(x)$.

73. $h(x) = (3x - 1)^4$

74. $h(x) = (2x - 5)^3$

75. $h(x) = \sqrt[3]{x^2 - 9}$

76. $h(x) = \sqrt{5x^2 + 3}$

77. $h(x) = |2x - 5|$

78. $h(x) = |3x - 4|$

79. $h(x) = \dfrac{1}{2x - 3}$

80. $h(x) = \dfrac{1}{4x + 5}$

Practice Plus

Use the graphs of f and g to solve Exercises 81–88.

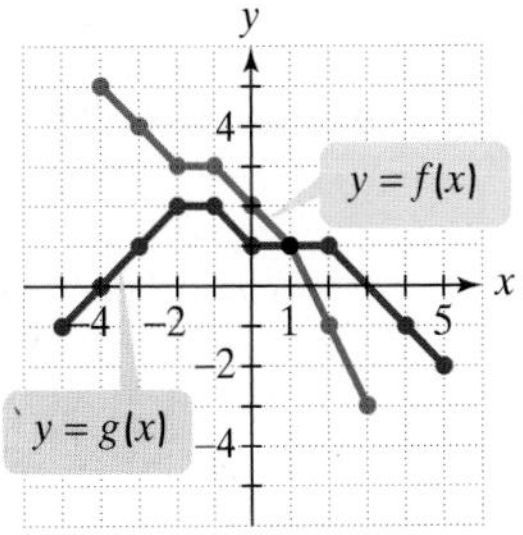

81. Find $(f + g)(-3)$.

82. Find $(g - f)(-2)$.

83. Find $(fg)(2)$.

84. Find $\left(\dfrac{g}{f}\right)(3)$.

85. Find the domain of $f + g$.

86. Find the domain of $\dfrac{f}{g}$.

87. Graph $f + g$.

88. Graph $f - g$.

In Exercises 89–92, use the graphs of f and g to evaluate each composite function.

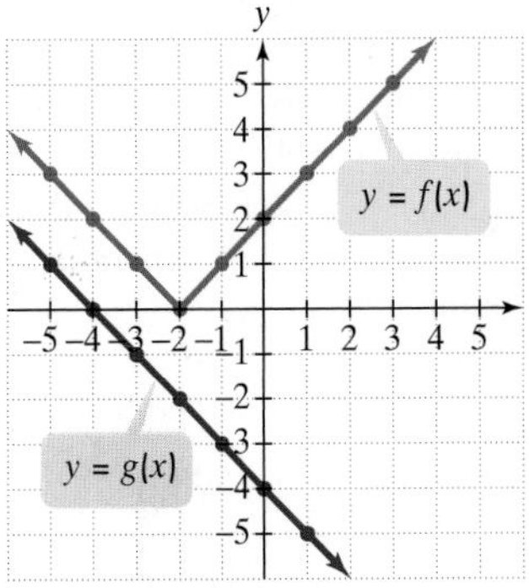

89. $(f \circ g)(-1)$

90. $(f \circ g)(1)$

91. $(g \circ f)(0)$

92. $(g \circ f)(-1)$

In Exercises 93–94, find all values of x satisfying the given conditions.

93. $f(x) = 2x - 5$, $g(x) = x^2 - 3x + 8$, and $(f \circ g)(x) = 7$.

94. $f(x) = 1 - 2x$, $g(x) = 3x^2 + x - 1$, and $(f \circ g)(x) = -5$.

Application Exercises

The table shows the total number of births and the total number of deaths in the United States from 1995 through 2003.

Births and Deaths in the U.S.

Year	Births	Deaths
1995	3,899,589	2,312,132
1996	3,891,494	2,314,690
1997	3,880,894	2,314,245
1998	3,941,553	2,337,256
1999	3,959,417	2,391,399
2000	4,058,814	2,403,351
2001	4,025,933	2,416,425
2002	4,022,000	2,436,000
2003	4,093,000	2,423,000

Source: Department of Health and Human Services

The data can be modeled by the following functions:

Number of births $B(x) = 26{,}208x + 3{,}869{,}910$

Number of deaths $D(x) = 17{,}964x + 2{,}300{,}198.$

In each function, x represents the number of years after 1995. Assume that the functions apply only to the years shown in the table. Use these functions to solve Exercises 95–98.

95. Find the domain of B.

96. Find the domain of D.

97. **a.** Find $(B - D)(x)$. What does this function represent?

b. Use the function in part (a) to find $(B - D)(8)$. What does this mean in terms of the U.S. population and to which year does this apply?

c. Use the data shown in the table to find $(B - D)(8)$. How well does the difference of functions used in part (b) model this number?

98. **a.** Find $(B - D)(x)$. What does this function represent?

b. Use the function in part (a) to find $(B - D)(6)$. What does this mean in terms of the U.S. population and to which year does this apply?

c. Use the data shown in the table to find $(B - D)(6)$. How well does the difference of functions used in part (b) model this number?

Consider the following functions:

$f(x)$ = *population of the world's more developed regions in year x*

$g(x)$ = *population of the world's less developed regions in year x*

$h(x)$ = *total world population in year x.*

Use these functions and the graphs shown to answer Exercises 99–102.

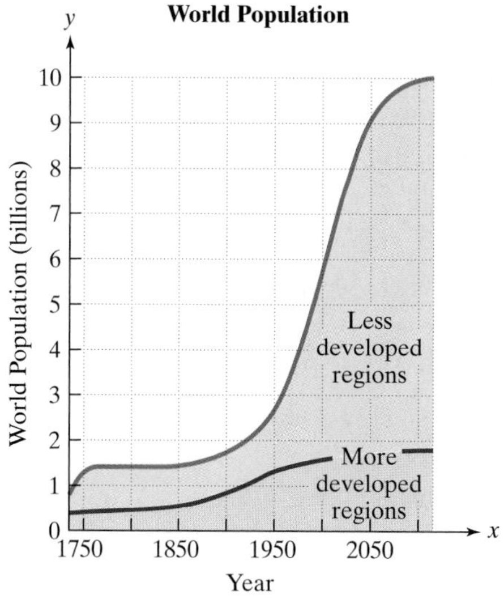

Source: Population Reference Bureau

99. What does the function $f + g$ represent?

100. What does the function $h - g$ represent?

101. Use the graph to estimate $(f + g)(2000)$.

102. Use the graph to estimate $(h - g)(2000)$.

103. A company that sells radios has yearly fixed costs of \$600,000. It costs the company \$45 to produce each radio. Each radio will sell for \$65. The company's costs and revenue are modeled by the following functions:

$C(x) = 600{,}000 + 45x$ This function models the company's costs.

$R(x) = 65x.$ This function models the company's revenue.

Find and interpret $(R - C)(20{,}000)$, $(R - C)(30{,}000)$, and $(R - C)(40{,}000)$.

104. A department store has two locations in a city. From 2000 through 2004, the profits for each of the store's two branches are modeled by the functions $f(x) = -0.44x + 13.62$ and $g(x) = 0.51x + 11.14$. In each model, x represents the number of years after 2000, and f and g represent the profit, in millions of dollars.

a. What is the slope of f? Describe what this means.

b. What is the slope of g? Describe what this means.

c. Find $f + g$. What is the slope of this function? What does this mean?

105. The regular price of a computer is x dollars. Let $f(x) = x - 400$ and $g(x) = 0.75x$.

a. Describe what the functions f and g model in terms of the price of the computer.

b. Find $(f \circ g)(x)$ and describe what this models in terms of the price of the computer.

c. Repeat part (b) for $(g \circ f)(x)$.

d. Which composite function models the greater discount on the computer, $f \circ g$ or $g \circ f$? Explain.

106. The regular price of a pair of jeans is x dollars. Let $f(x) = x - 5$ and $g(x) = 0.6x$.
- **a.** Describe what functions f and g model in terms of the price of the jeans.
- **b.** Find $(f \circ g)(x)$ and describe what this models in terms of the price of the jeans.
- **c.** Repeat part (b) for $(g \circ f)(x)$.
- **d.** Which composite function models the greater discount on the jeans, $f \circ g$ or $g \circ f$? Explain.

Writing in Mathematics

107. If a function is defined by an equation, explain how to find its domain.

108. If equations for f and g are given, explain how to find $f - g$.

109. If equations for two functions are given, explain how to obtain the quotient function and its domain.

110. Describe a procedure for finding $(f \circ g)(x)$. What is the name of this function?

111. Describe the values of x that must be excluded from the domain of $(f \circ g)(x)$.

112. We opened the section with two functions that modeled the data in Figure 2.47 on page 256. Car sales, $C(x)$, in millions, x years after 1990 were modeled by $C(x) = -0.14x + 9$. SUV sales, $S(x)$, in millions, x years after 1990 were modeled by $S(x) = 0.22x + 1$. Explain how these models were obtained from Figure 2.47.

Technology Exercises

113. The function $f(t) = -0.14t^2 + 0.51t + 31.6$ models the U.S. population ages 65 and older, $f(t)$, in millions, t years after 1990. The function $g(t) = 0.54t^2 + 12.64t + 107.1$ models the total yearly cost of Medicare, $g(t)$, in billions of dollars, t years after 1990. Graph the function $\frac{g}{f}$ in a $[0, 15, 1]$ by $[0, 60, 10]$ viewing rectangle. What does the shape of the graph indicate about the per capita costs of Medicare for the U.S. population ages 65 and over with increasing time?

114. Graph $y_1 = x^2 - 2x$, $y_2 = x$, and $y_3 = y_1 \div y_2$ in the same $[-10, 10, 1]$ by $[-10, 10, 1]$ viewing rectangle. Then use the TRACE feature to trace along y_3. What happens at $x = 0$? Explain why this occurs.

115. Graph $y_1 = \sqrt{2 - x}$, $y_2 = \sqrt{x}$, and $y_3 = \sqrt{2 - y_2}$ in the same $[-4, 4, 1]$ by $[0, 2, 1]$ viewing rectangle. If y_1 represents f and y_2 represents g, use the graph of y_3 to find the domain of $f \circ g$. Then verify your observation algebraically.

Critical Thinking Exercises

116. Which one of the following is true?
- **a.** If $f(x) = x^2 - 4$ and $g(x) = \sqrt{x^2 - 4}$, then $(f \circ g)(x) = -x^2$ and $(f \circ g)(5) = -25$.
- **b.** There can never be two functions f and g, where $f \neq g$, for which $(f \circ g)(x) = (g \circ f)(x)$.
- **c.** If $f(7) = 5$ and $g(4) = 7$, then $(f \circ g)(4) = 35$.
- **d.** If $f(x) = \sqrt{x}$ and $g(x) = 2x - 1$, then $(f \circ g)(5) = g(2)$.

117. Prove that if f and g are even functions, then fg is also an even function.

118. Define two functions f and g so that $f \circ g = g \circ f$.

119. Use the graphs given in Exercises 99–102 to create a graph that shows the population, in billions, of less developed regions from 1950 through 2050.

SECTION 2.7 *Inverse Functions*

Objectives

1. Verify inverse functions.
2. Find the inverse of a function.
3. Use the horizontal line test to determine if a function has an inverse function.
4. Use the graph of a one-to-one function to graph its inverse function.
5. Find the inverse of a function and graph both functions on the same axes.

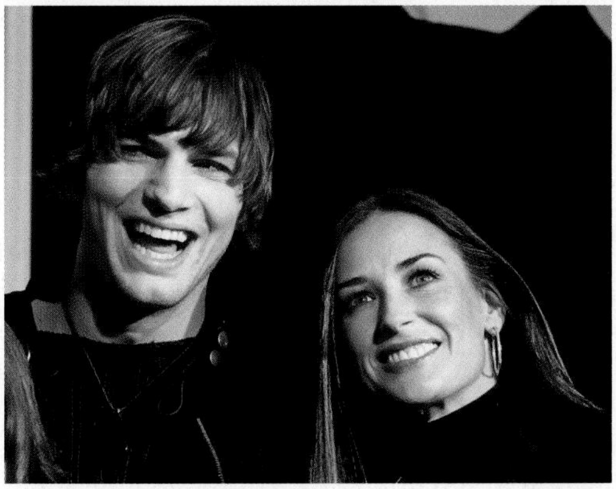

In most societies, women say they prefer to marry men who are older than themselves, whereas men say they prefer women who are younger. Evolutionary psychologists attribute these preferences to female concern with a partner's material resources and male concern with a partner's fertility (*Source:* David M. Buss, *Psychological Inquiry*, 6, 1–30). When the man is considerably older than the woman, people rarely comment. However, when the woman is older, as in the relationship between actors Ashton Kutcher and Demi Moore, people take notice.

Figure 2.51 shows the preferred age in a mate in five selected countries. We can focus on the data for the women and define a function.

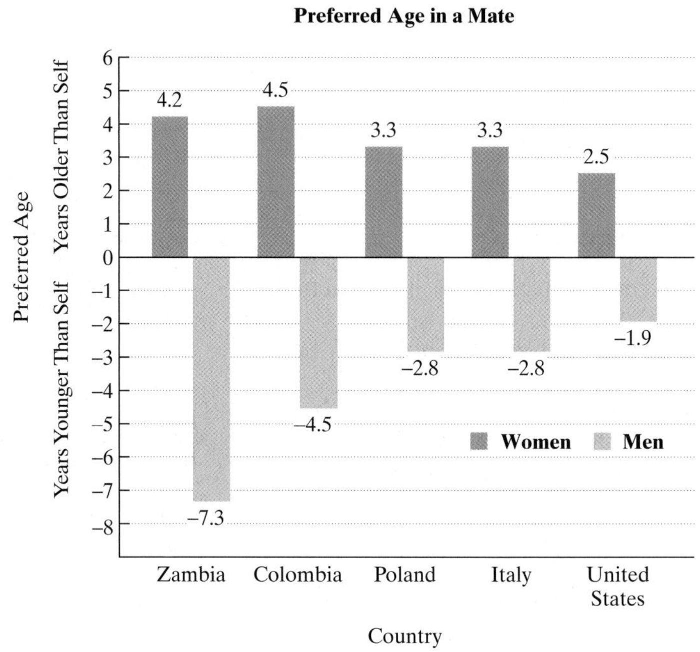

Figure 2.51
Source: Carole Wade and Carol Tavris, *Psychology*, 6th Edition, Prentice Hall, 2000

Let the domain of the function be the set of the five countries shown in the graph. Let the range be the set of the average number of years women in each of the respective countries prefer men who are older than themselves. The function can be written as follows:

$$f: \{(\text{Zambia}, 4.2), (\text{Colombia}, 4.5), (\text{Poland}, 3.3), (\text{Italy}, 3.3), (\text{U.S.}, 2.5)\}.$$

Now let's "undo" f by interchanging the first and second components in each of its ordered pairs. Switching the inputs and outputs of f, we obtain the following relation:

Same first component

$$\text{Undoing } f: \{(4.2, \text{Zambia}), (4.5, \text{Colombia}), (3.3, \text{Poland}), (3.3, \text{Italy}), (2.5, \text{U.S.})\}.$$

Different second components

Can you see that this relation is not a function? Two of its ordered pairs have the same first component and different second components. This violates the definition of a function.

If a function f is a set of ordered pairs, (x, y), then the changes produced by f can be "undone" by reversing the components of all the ordered pairs. The resulting relation, (y, x), may or may not be a function. In this section, we will develop these ideas by studying functions whose compositions have a special "undoing" relationship.

Inverse Functions

Here are two functions that describe situations related to the price of a computer, x:

$$f(x) = x - 300 \qquad g(x) = x + 300.$$

The function f subtracts \$300 from the computer's price and the function g adds \$300 to the computer's price. Let's see what $f(g(x))$ does. Put $g(x)$ into f:

$$f(x) = x - 300$$ This is the given equation for f.

Replace x with $g(x)$.

$$f(g(x)) = g(x) - 300$$
$$= x + 300 - 300$$ Because $g(x) = x + 300$, replace $g(x)$ with $x + 300$.
$$= x.$$ This is the computer's original price.

By putting $g(x)$ into f and finding $f(g(x))$, we see that the computer's price, x, went through two changes: the first, an increase; the second, a decrease:

$$x + 300 - 300.$$

The final price of the computer, x, is identical to its starting price, x.

In general, if the changes made to x by a function g are undone by the changes made by a function f, then

$$f(g(x)) = x.$$

Assume, also, that this "undoing" takes place in the other direction:

$$g(f(x)) = x.$$

Under these conditions, we say that each function is the *inverse function* of the other. The fact that g is the inverse of f is expressed by renaming g as f^{-1}, read "f-inverse." For example, the inverse functions

$$f(x) = x - 300 \qquad g(x) = x + 300$$

are usually named as follows:

$$f(x) = x - 300 \qquad f^{-1}(x) = x + 300.$$

We can use partial tables of coordinates for f and f^{-1} to gain numerical insight into the relationship between a function and its inverse function.

Computer's regular price — \$300 reduction

x	$f(x) = x - 300$
1200	900
1300	1000
1400	1100

Ordered pairs for f: (1200, 900), (1300, 1000), (1400, 1100)

Price with \$300 reduction — \$300 price increase

x	$f^{-1}(x) = x + 300$
900	1200
1000	1300
1100	1400

Ordered pairs for f^{-1}: (900, 1200), (1000, 1300), (1100, 1400)

The tables illustrate that if a function f is the set of ordered pairs (x, y), then its inverse, f^{-1}, is the set of ordered pairs (y, x). Using these tables, we can see how one function's changes to x are undone by the other function:

$$(f^{-1} \circ f)(1300) = f^{-1}(f(1300)) = f^{-1}(1000) = 1300.$$

The table for f shows that $f(1300) = 1000$.

The table for f^{-1} shows that $f^{-1}(1000) = 1300$.

The final price of the computer, \$1300, is identical to its starting price, \$1300.

With these ideas in mind, we present the formal definition of the inverse of a function:

Study Tip

The notation f^{-1} represents the inverse function of f. The -1 is *not* an exponent. The notation f^{-1} does *not* mean $\frac{1}{f}$:

$$f^{-1} \neq \frac{1}{f}.$$

Definition of the Inverse of a Function

Let f and g be two functions such that

$$f(g(x)) = x \qquad \text{for every } x \text{ in the domain of } g$$

and

$$g(f(x)) = x \qquad \text{for every } x \text{ in the domain of } f.$$

The function g is the **inverse of the function** f and is denoted by f^{-1} (read "f-inverse"). Thus, $f(f^{-1}(x)) = x$ and $f^{-1}(f(x)) = x$. The domain of f is equal to the range of f^{-1}, and vice versa.

1 Verify inverse functions.

EXAMPLE 1 Verifying Inverse Functions

Show that each function is the inverse of the other:

$$f(x) = 3x + 2 \quad \text{and} \quad g(x) = \frac{x-2}{3}.$$

Solution To show that f and g are inverses of each other, we must show that $f(g(x)) = x$ and $g(f(x)) = x$. We begin with $f(g(x))$.

$$f(x) = 3x + 2$$ This is the equation for f.

Replace x with $g(x)$.

$$f(g(x)) = 3g(x) + 2 = 3\left(\frac{x-2}{3}\right) + 2 = x - 2 + 2 = x$$

$g(x) = \frac{x-2}{3}$

Next, we find $g(f(x))$.

$$g(x) = \frac{x-2}{3}$$ This is the equation for g.

Replace x with $f(x)$.

$$g(f(x)) = \frac{f(x)-2}{3} = \frac{(3x+2)-2}{3} = \frac{3x}{3} = x$$

$f(x) = 3x + 2$

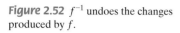

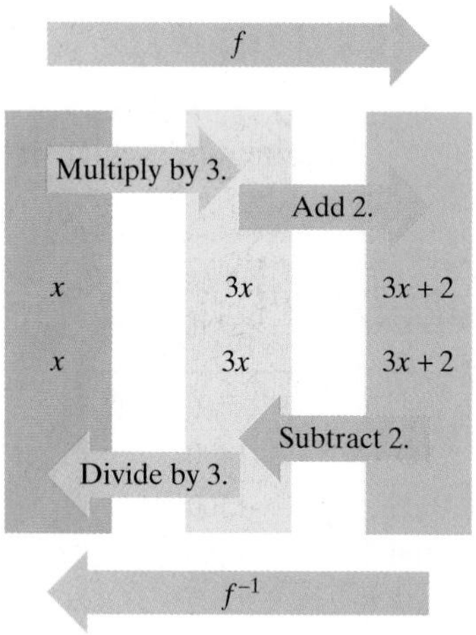

Figure 2.52 f^{-1} undoes the changes produced by f.

Because g is the inverse of f (and vice versa), we can use inverse notation and write

$$f(x) = 3x + 2 \quad \text{and} \quad f^{-1}(x) = \frac{x-2}{3}.$$

Notice how f^{-1} undoes the changes produced by f: f changes x by *multiplying* by 3 and *adding* 2, and f^{-1} undoes this by *subtracting* 2 and *dividing* by 3. This "undoing" process is illustrated in Figure 2.52.

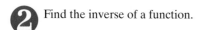

Check Point 1 Show that each function is the inverse of the other:

$$f(x) = 4x - 7 \quad \text{and} \quad g(x) = \frac{x+7}{4}.$$

2 Find the inverse of a function.

Finding the Inverse of a Function

The definition of the inverse of a function tells us that the domain of f is equal to the range of f^{-1}, and vice versa. This means that if the function f is the set of ordered pairs (x, y), then the inverse of f is the set of ordered pairs (y, x). If a function is defined by an equation, we can obtain the equation for f^{-1}, the inverse of f, by interchanging the role of x and y in the equation for the function f.

Finding the Inverse of a Function

The equation for the inverse of a function f can be found as follows:

1. Replace $f(x)$ with y in the equation for $f(x)$.

2. Interchange x and y.

3. Solve for y. If this equation does not define y as a function of x, the function f does not have an inverse function and this procedure ends. If this equation does define y as a function of x, the function f has an inverse function.

4. If f has an inverse function, replace y in step 3 by $f^{-1}(x)$. We can verify our result by showing that $f(f^{-1}(x)) = x$ and $f^{-1}(f(x)) = x$.

Study Tip

The procedure for finding a function's inverse uses a *switch-and-solve* strategy. Switch x and y, and then solve for y.

EXAMPLE 2   Finding the Inverse of a Function

Find the inverse of $f(x) = 7x - 5$.

Solution

Step 1   Replace $f(x)$ with y:
$$y = 7x - 5.$$

Step 2   Interchange x and y:
$$x = 7y - 5. \qquad \text{This is the inverse function.}$$

Step 3   Solve for y:
$$x + 5 = 7y \qquad \text{Add 5 to both sides.}$$
$$\frac{x + 5}{7} = y. \qquad \text{Divide both sides by 7.}$$

Discovery

In Example 2, we found that if $f(x) = 7x - 5$, then
$$f^{-1}(x) = \frac{x + 5}{7}.$$
Verify this result by showing that
$$f(f^{-1}(x)) = x$$
and
$$f^{-1}(f(x)) = x.$$

Step 4   Replace y with $f^{-1}(x)$:
$$f^{-1}(x) = \frac{x + 5}{7}. \qquad \text{The equation is written with } f^{-1} \text{ on the left.}$$

Thus, the inverse of $f(x) = 7x - 5$ is $f^{-1}(x) = \dfrac{x + 5}{7}$.

The inverse function, f^{-1}, undoes the changes produced by f. f changes x by multiplying by 7 and subtracting 5. f^{-1} undoes this by adding 5 and dividing by 7.

Check Point 2 Find the inverse of $f(x) = 2x + 7$.

EXAMPLE 3   Finding the Inverse of a Function

Find the inverse of $f(x) = x^3 + 1$.

Solution

Step 1   Replace $f(x)$ with y: $y = x^3 + 1$.

Step 2   Interchange x and y: $x = y^3 + 1$.

Step 3   Solve for y:

Our goal is to isolate y. Because $\sqrt[3]{y^3} = y$, we will take the cube root of both sides of the equation.

$$x - 1 = y^3 \qquad \text{Subtract 1 from both sides.}$$
$$\sqrt[3]{x - 1} = \sqrt[3]{y^3} \qquad \text{Take the cube root on both sides.}$$
$$\sqrt[3]{x - 1} = y. \qquad \text{Simplify.}$$

Step 4   Replace y with $f^{-1}(x)$: $f^{-1}(x) = \sqrt[3]{x - 1}$.

Thus, the inverse of $f(x) = x^3 + 1$ is $f^{-1}(x) = \sqrt[3]{x - 1}$.

Check Point 3 Find the inverse of $f(x) = 4x^3 - 1$.

EXAMPLE 4 Finding the Inverse of a Function

Find the inverse of $f(x) = \frac{5}{x} + 4$.

Solution

Step 1 Replace $f(x)$ with y:

$$y = \frac{5}{x} + 4.$$

Step 2 Interchange x and y:

$$x = \frac{5}{y} + 4.$$

Our goal is to isolate y. To get y out of the denominator, we will multiply both sides of the equation by y, $y \neq 0$.

Step 3 Solve for y:

$$x = \frac{5}{y} + 4$$ This is the equation from step 2.

$$xy = \left(\frac{5}{y} + 4\right)y$$ Multiply both sides by y, $y \neq 0$.

$$xy = \frac{5}{y} \cdot y + 4y$$ Use the distributive property.

$$xy = 5 + 4y$$ Simplify: $\frac{5}{y} \cdot y = 5$.

$$xy - 4y = 5$$ Subtract $4y$ from both sides.

$$y(x - 4) = 5$$ Factor out y from $xy - 4y$ to obtain a single occurrence of y.

$$\frac{y(x-4)}{x-4} = \frac{5}{x-4}$$ Divide both sides by $x - 4$.

$$y = \frac{5}{x-4}.$$ Simplify.

Step 4 Replace y with $f^{-1}(x)$:

$$f^{-1}(x) = \frac{5}{x-4}.$$

Thus, the inverse of $f(x) = \frac{5}{x} + 4$ is $f^{-1}(x) = \frac{5}{x-4}$.

Check Point 4 Find the inverse of $f(x) = \frac{3}{x} - 1$.

3 Use the horizontal line test to determine if a function has an inverse function.

The Horizontal Line Test and One-to-One Functions

Let's see what happens if we try to find the inverse of the standard quadratic function, $f(x) = x^2$.

Step 1 Replace *f*(*x*) with *y*: $y = x^2$.

Step 2 Interchange *x* and *y*: $x = y^2$.

Step 3 Solve for *y*: We apply the square root property to solve $y^2 = x$ for y. We obtain

$$y = \pm\sqrt{x}.$$

The $\pm$ in this last equation shows that for certain values of x (all positive real numbers), there are two values of y. Because this equation does not represent y as a function of x, the standard quadratic function $f(x) = x^2$ does not have an inverse function.

We can use a few of the solutions of $y = x^2$ to illustrate numerically that this function does not have an inverse:

Four solutions of $y = x^2$. $(-2, 4),\ (-1, 1),\ (1, 1),\ (2, 4),$

Interchange x and y in each ordered pair. $(4, -2),\ (1, -1),\ (1, 1),\ (4, 2).$

The input 1 is associated with two outputs, −1 and 1.

The input 4 is associated with two outputs, −2 and 2.

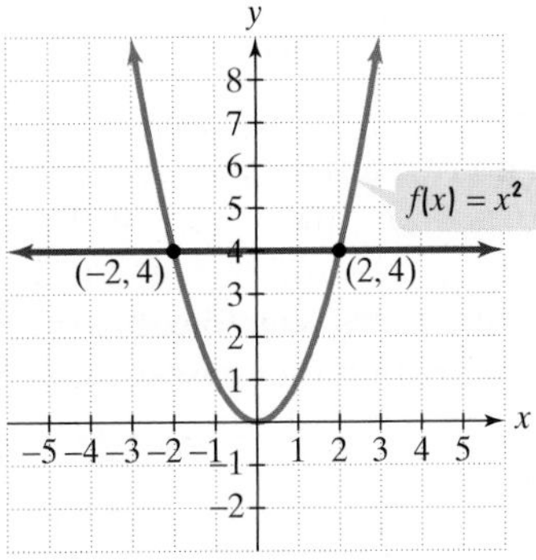

Figure 2.53 The horizontal line intersects the graph twice.

A function provides exactly one output for each input. Thus, the ordered pairs in the bottom row do not define a function.

Can we look at the graph of a function and tell if it represents a function with an inverse? Yes. The graph of the standard quadratic function $f(x) = x^2$ is shown in Figure 2.53. Four units above the x-axis, a horizontal line is drawn. This line intersects the graph at two of its points, $(-2, 4)$ and $(2, 4)$. Inverse functions have ordered pairs with the coordinates reversed. We just saw what happened when we interchanged x and y. We obtained $(4, -2)$ and $(4, 2)$, and these ordered pairs do not define a function.

If any horizontal line, such as the one in Figure 2.53, intersects a graph at two or more points, the set of these points will not define a function when their coordinates are reversed. This suggests the **horizontal line test** for inverse functions.

Discovery

How might you restrict the domain of $f(x) = x^2$, graphed in Figure 2.53, so that the remaining portion of the graph passes the horizontal line test?

The Horizontal Line Test For Inverse Functions

A function f has an inverse that is a function, f^{-1}, if there is no horizontal line that intersects the graph of the function f at more than one point.

EXAMPLE 5 Applying the Horizontal Line Test

Which of the following graphs represent functions that have inverse functions?

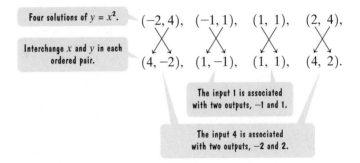

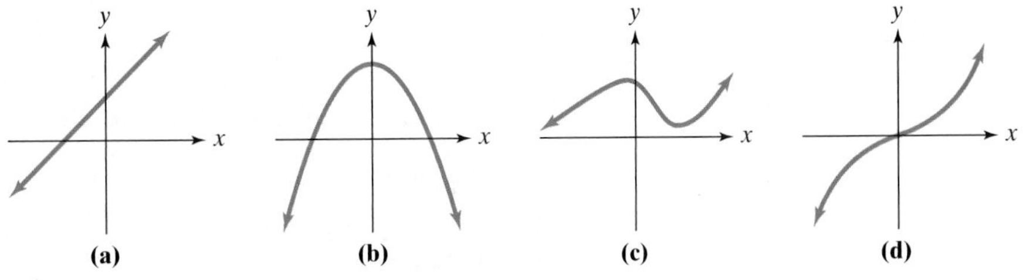

Solution Notice that horizontal lines can be drawn in graphs (b) and (c) that intersect the graphs more than once. These graphs do not pass the horizontal line test. These are not the graphs of functions with inverse functions. By contrast, no horizontal line can be drawn in graphs (a) and (d) that intersects the graphs more than once. These graphs pass the horizontal line test. Thus, the graphs in parts (a) and (d) represent functions that have inverse functions.

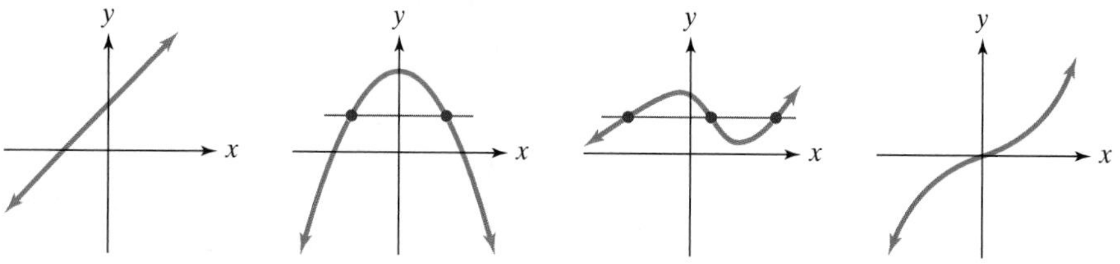

(a) (b) (c) (d)

Check Point 5 Which of the following graphs represent functions that have inverse functions?

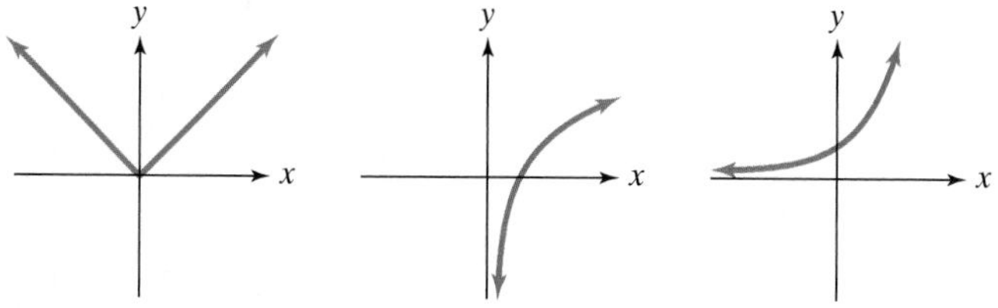

(a) (b) (c)

A function passes the horizontal line test when no two different ordered pairs have the same second component. This means that if $x_1 \neq x_2$, then $f(x_1) \neq f(x_2)$. Such a function is called a **one-to-one function**. Thus, **a one-to-one function is a function in which no two different ordered pairs have the same second component.** Only one-to-one functions have inverse functions. Any function that passes the horizontal line test is a one-to-one function. Any one-to-one function has a graph that passes the horizontal line test.

4 Use the graph of a one-to-one function to graph its inverse function.

Graphs of f and f^{-1}

There is a relationship between the graph of a one-to-one function, f, and its inverse, f^{-1}. Because inverse functions have ordered pairs with the coordinates interchanged, if the point (a, b) is on the graph of f then the point (b, a) is on the graph of f^{-1}. The points (a, b) and (b, a) are symmetric with respect to the line $y = x$. Thus, **the graph of f^{-1} is a reflection of the graph of f about the line $y = x$.** This is illustrated in Figure 2.54.

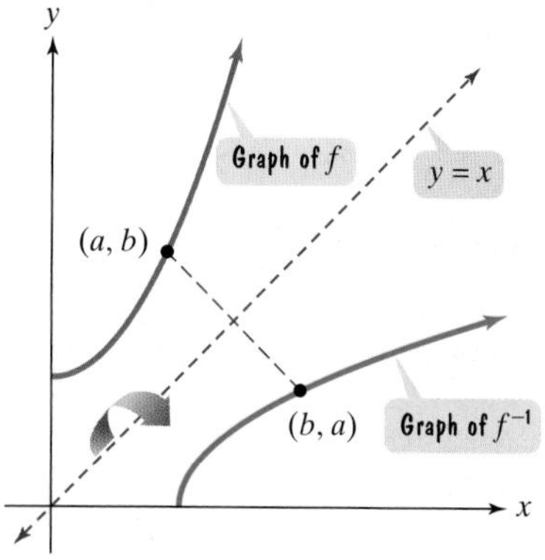

Figure 2.54 The graph of f^{-1} is a reflection of the graph of f about $y = x$.

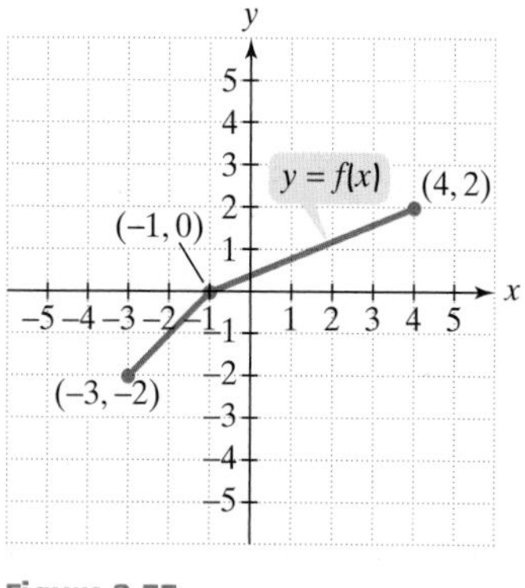

Figure 2.55

EXAMPLE 6 Graphing the Inverse Function

Use the graph of f in Figure 2.55 to draw the graph of its inverse function.

Solution We begin by noting that no horizontal line intersects the graph of f at more than one point, so f does have an inverse function. Because the points $(-3, -2)$, $(-1, 0)$, and $(4, 2)$ are on the graph of f, the graph of the inverse function, f^{-1}, has points with these ordered pairs reversed. Thus, $(-2, -3)$, $(0, -1)$, and $(2, 4)$ are on the graph of f^{-1}. We can use these points to graph f^{-1}. The graph of f^{-1} is shown in green in Figure 2.56. Note that the green graph of f^{-1} is the reflection of the blue graph of f about the line $y = x$.

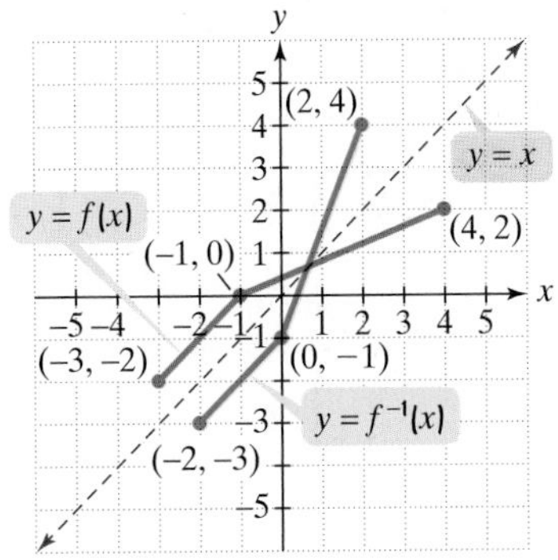

Figure 2.56 The graphs of f and f^{-1}

Check Point 6 The graph of function f consists of two line segments, one segment from $(-2, -2)$ to $(-1, 0)$ and a second segment from $(-1, 0)$ to $(1, 2)$. Graph f and use the graph to draw the graph of its inverse function.

5 Find the inverse of a function and graph both functions on the same axes.

In our final example, we will first find f^{-1}. Then we will graph f and f^{-1} in the same rectangular coordinate system.

EXAMPLE 7 Finding the Inverse of a Domain-Restricted Function

Find the inverse of $f(x) = x^2 - 1$ if $x \geq 0$. Graph f and f^{-1} in the same rectangular coordinate system.

Solution The graph of $f(x) = x^2 - 1$ is the graph of the standard quadratic function shifted vertically down 1 unit. Figure 2.57 shows the function's graph. This graph fails the horizontal line test, so the function $f(x) = x^2 - 1$ does not have an inverse function. By restricting the domain to $x \geq 0$, as given, we obtain a new function whose graph is shown in red in Figure 2.57. This red portion of the graph is increasing on the interval $(0, \infty)$ and passes the horizontal line test. This tells us that $f(x) = x^2 - 1$ has an inverse function if we restrict its domain to $x \geq 0$. We use our four-step procedure to find this inverse function. Begin with $f(x) = x^2 - 1, x \geq 0$.

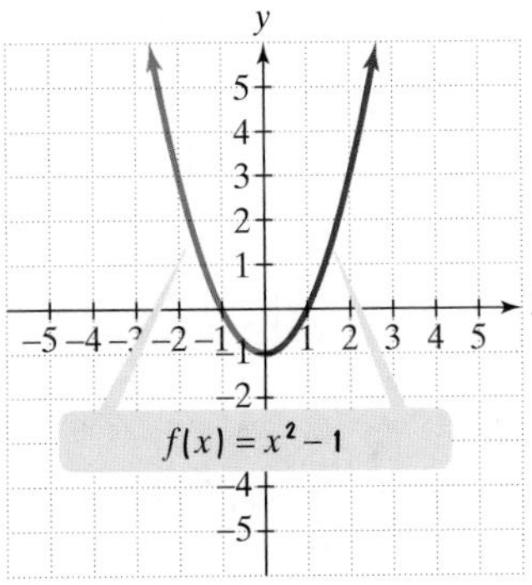

Figure 2.57

Step 1 Replace $f(x)$ with y: $y = x^2 - 1, x \geq 0$.

Step 2 Interchange x and y: $x = y^2 - 1, y \geq 0$.

Step 3 Solve for y:

$$x = y^2 - 1, y \geq 0 \quad \text{This is the equation from step 2.}$$

$$x + 1 = y^2 \quad \text{Add 1 to both sides.}$$

$$\sqrt{x + 1} = y \quad \text{Apply the square root property.}$$

Because $y \geq 0$, take only the principal square root and not the negative square root.

Step 4 Replace y with $f^{-1}(x)$: $f^{-1}(x) = \sqrt{x + 1}$.

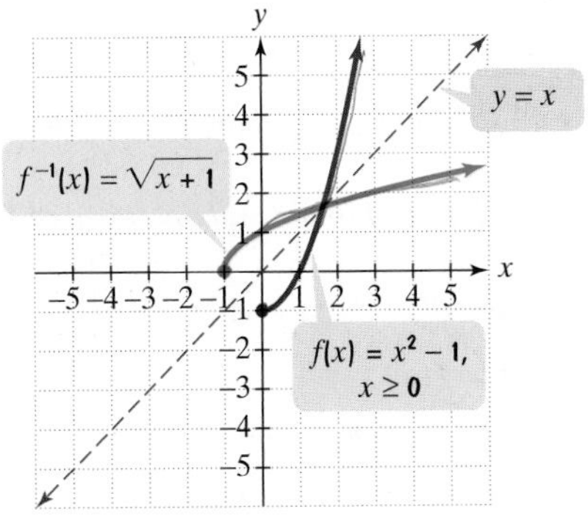

Figure 2.58

Thus, the inverse of $f(x) = x^2 - 1, x \geq 0$, is $f^{-1}(x) = \sqrt{x + 1}$. The graphs of f and f^{-1} are shown in Figure 2.58. We obtained the graph of $f^{-1}(x) = \sqrt{x + 1}$ by shifting the graph of the square root function, $y = \sqrt{x}$, horizontally to the left 1 unit. Note that the green graph of f^{-1} is the reflection of the red graph of f about the line $y = x$.

Check Point 7 Find the inverse of $f(x) = x^2 + 1$ if $x \geq 0$. Graph f and f^{-1} in the same rectangular coordinate system.

EXERCISE SET 2.7

Practice Exercises

In Exercises 1–10, find $f(g(x))$ and $g(f(x))$ and determine whether each pair of functions f and g are inverses of each other.

1. $f(x) = 4x$ and $g(x) = \frac{x}{4}$

2. $f(x) = 6x$ and $g(x) = \frac{x}{6}$

3. $f(x) = 3x + 8$ and $g(x) = \frac{x - 8}{3}$

4. $f(x) = 4x + 9$ and $g(x) = \frac{x - 9}{4}$

5. $f(x) = 5x - 9$ and $g(x) = \frac{x + 5}{9}$

6. $f(x) = 3x - 7$ and $g(x) = \frac{x + 3}{7}$

7. $f(x) = \frac{3}{x - 4}$ and $g(x) = \frac{3}{x} + 4$

8. $f(x) = \frac{2}{x - 5}$ and $g(x) = \frac{2}{x} + 5$

9. $f(x) = -x$ and $g(x) = -x$

10. $f(x) = \sqrt[3]{x - 4}$ and $g(x) = x^3 + 4$

The functions in Exercises 11–28 are all one-to-one. For each function,

a. *Find an equation for $f^{-1}(x)$, the inverse function.*

b. *Verify that your equation is correct by showing that $f(f^{-1}(x)) = x$ and $f^{-1}(f(x)) = x$.*

11. $f(x) = x + 3$

12. $f(x) = x + 5$

13. $f(x) = 2x$

14. $f(x) = 4x$

15. $f(x) = 2x + 3$

16. $f(x) = 3x - 1$

17. $f(x) = x^3 + 2$

18. $f(x) = x^3 - 1$

19. $f(x) = (x + 2)^3$

20. $f(x) = (x - 1)^3$

21. $f(x) = \frac{1}{x}$

22. $f(x) = \frac{2}{x}$

23. $f(x) = \sqrt{x}$

24. $f(x) = \sqrt[3]{x}$

25. $f(x) = \frac{7}{x} - 3$

26. $f(x) = \frac{4}{x} + 9$

27. $f(x) = \frac{2x + 1}{x - 3}$

28. $f(x) = \frac{2x - 3}{x + 1}$

Which graphs in Exercises 29–34 represent functions that have inverse functions?

29.

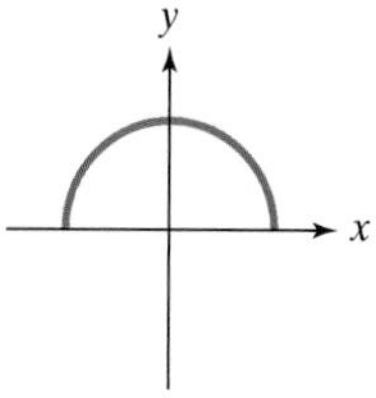

30.

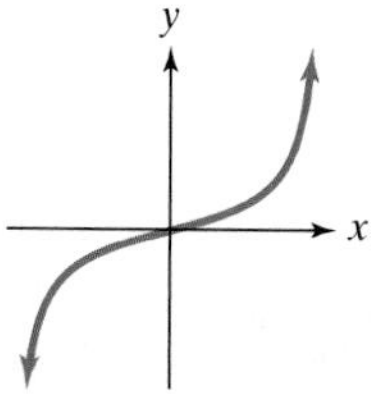

31.

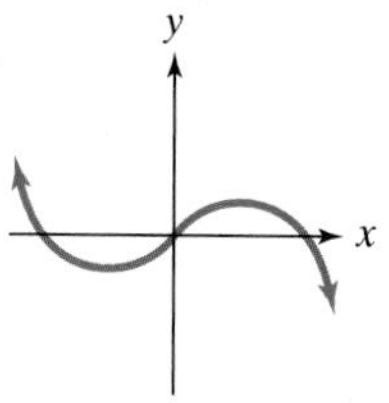

32.

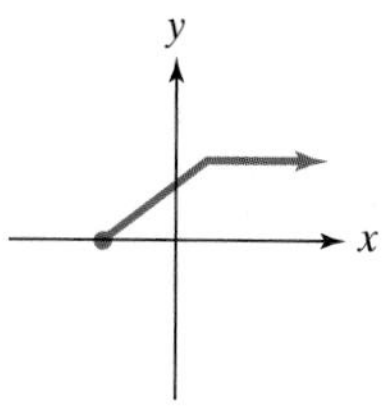

33.

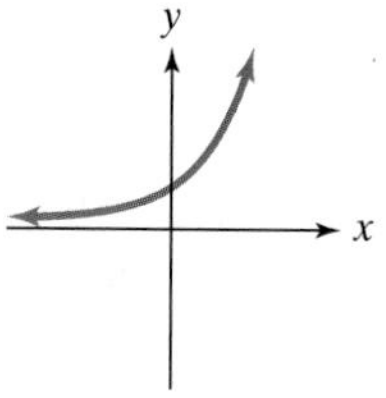

34.

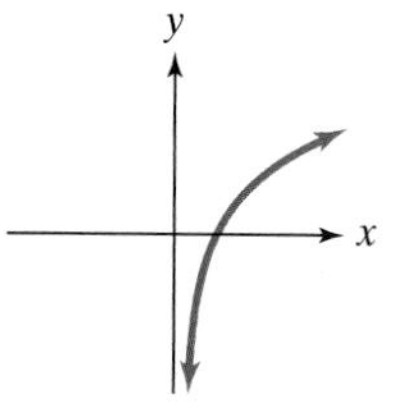

In Exercises 35–38, use the graph of f to draw the graph of its inverse function.

35.

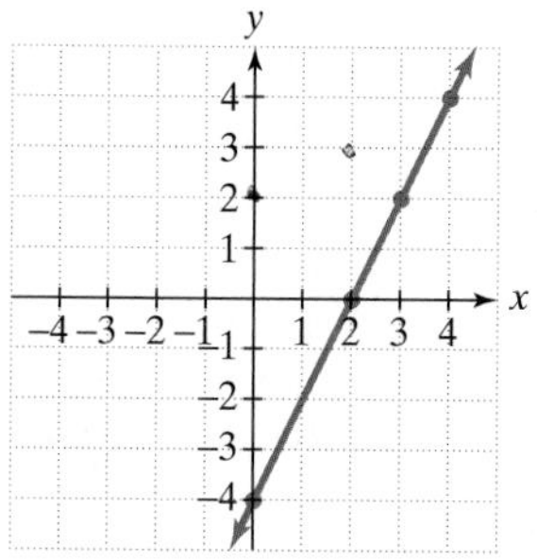

36.

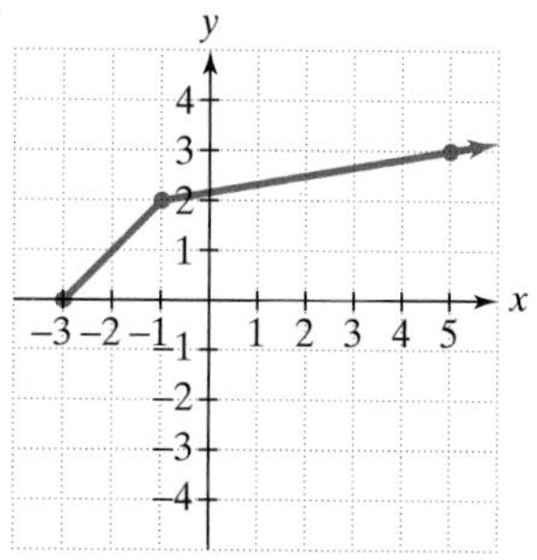

37.

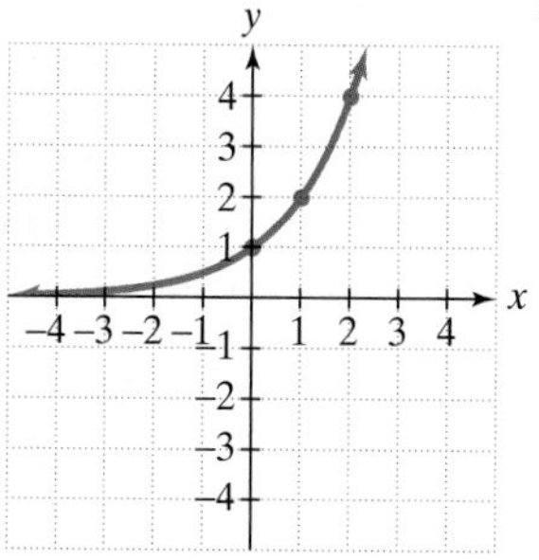

38.

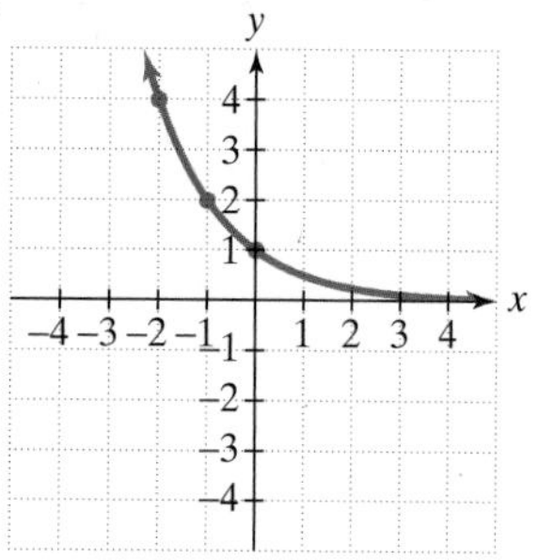

In Exercises 39–52,

a. *Find an equation for $f^{-1}(x)$.*

b. *Graph f and f^{-1} in the same rectangular coordinate system.*

c. *Use interval notation to give the domain and the range of f and f^{-1}.*

39. $f(x) = 2x - 1$

40. $f(x) = 2x - 3$

41. $f(x) = x^2 - 4, x \geq 0$

42. $f(x) = x^2 - 1, x \leq 0$

43. $f(x) = (x - 1)^2, x \leq 1$

44. $f(x) = (x - 1)^2, x \geq 1$

45. $f(x) = x^3 - 1$

46. $f(x) = x^3 + 1$

47. $f(x) = (x + 2)^3$

48. $f(x) = (x - 2)^3$

(Hint for Exercises 49–52: To solve for a variable involving an nth root, raise both sides of the equation to the nth power: $(\sqrt[n]{y})^n = y$.)

49. $f(x) = \sqrt{x - 1}$

50. $f(x) = \sqrt{x} + 2$

51. $f(x) = \sqrt[3]{x} + 1$

52. $f(x) = \sqrt[3]{x - 1}$

Practice Plus

In Exercises 53–58, f and g are defined by the following tables. Use the tables to evaluate each composite function.

x	$f(x)$
−1	1
0	4
1	5
2	−1

x	$g(x)$
−1	0
1	1
4	2
10	−1

53. $f(g(1))$ **54.** $f(g(4))$ **55.** $(g \circ f)(-1)$

56. $(g \circ f)(0)$ **57.** $f^{-1}(g(10))$ **58.** $f^{-1}(g(1))$

In Exercises 59–64, let

$$f(x) = 2x - 5$$
$$g(x) = 4x - 1$$
$$h(x) = x^2 + x + 2.$$

Evaluate the indicated function without finding an equation for the function.

59. $(f \circ g)(0)$ **60.** $(g \circ f)(0)$ **61.** $f^{-1}(1)$

62. $g^{-1}(7)$ **63.** $g(f[h(1)])$ **64.** $f(g[h(1)])$

Application Exercises

65. Refer to Figure 2.51 on page 270. Recall that the bar graphs in the figure show the preferred age in a mate in five selected countries.

a. Consider a function, f, whose domain is the set of the five countries shown in the graph. Let the range be the set of the average number of years men in each of the respective countries prefer women who are younger than themselves. Write the function f as a set of ordered pairs.

b. Write the relation that is the inverse of f as a set of ordered pairs. Is this relation a function? Explain your answer.

66. The bar graph shows the number of days in a school year for the six countries with the longest school years. (To put these numbers in perspective, a school year in the United States is 180 days.)

Countries with the Longest School Years

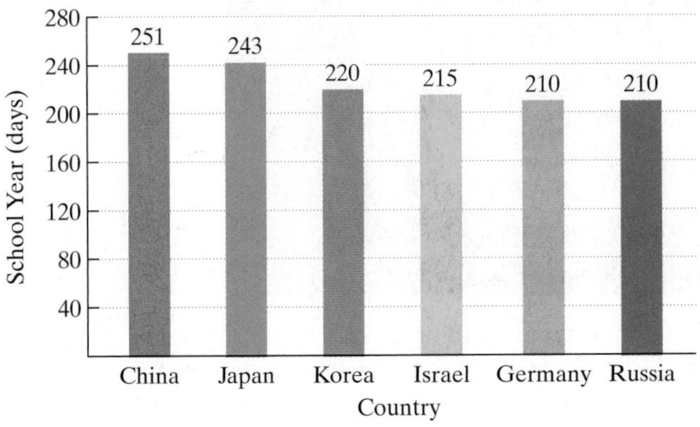

Source: UNESCO

a. Consider a function, f, whose domain is the set of six countries shown. Let the range be the number of days in the school year for each respective country. Write the function f as a set of ordered pairs.

b. Write the relation that is the inverse of f as a set of ordered pairs. Is this relation a function? Explain your answer.

67. The graph represents the probability of two people in the same room sharing a birthday as a function of the number of people in the room. Call the function f.

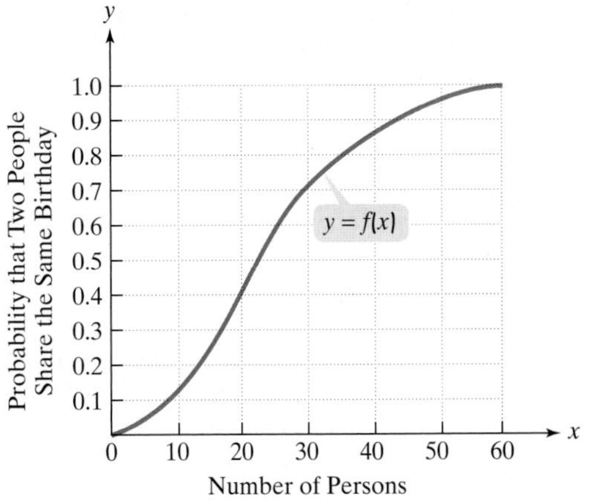

a. Explain why f has an inverse that is a function.

b. Describe in practical terms the meaning of $f^{-1}(0.25)$, $f^{-1}(0.5)$, and $f^{-1}(0.7)$.

68. A study of 900 working women in Texas showed that their feelings changed throughout the day. As the graph indicates, the women felt better as time passed, except for a blip (that's slang for relative maximum) at lunchtime.

Average Level of Happiness at Different Times of Day

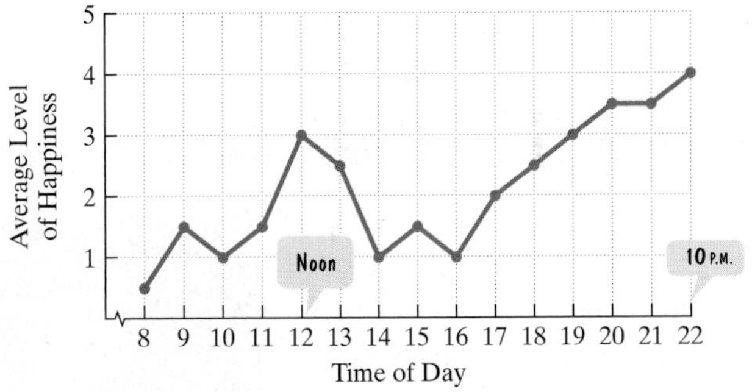

Source: D. Kahneman et. al., "A Survey Method for Characterizing Daily Life Experience," *Science*.

a. Does the graph have an inverse that is a function? Explain your answer.

b. Identify two or more times of day when the average happiness level is 3. Express your answers as ordered pairs.

c. Do the ordered pairs in part (b) indicate that the graph represents a one-to-one function? Explain your answer.

69. The formula

$$y = f(x) = \frac{9}{5}x + 32$$

is used to convert from x degrees Celsius to y degrees Fahrenheit. The formula

$$y = g(x) = \frac{5}{9}(x - 32)$$

is used to convert from x degrees Fahrenheit to y degrees Celsius. Show that f and g are inverse functions.

Writing in Mathematics

70. Explain how to determine if two functions are inverses of each other.
71. Describe how to find the inverse of a one-to-one function.
72. What is the horizontal line test and what does it indicate?
73. Describe how to use the graph of a one-to-one function to draw the graph of its inverse function.
74. How can a graphing utility be used to visually determine if two functions are inverses of each other?
75. What explanations can you offer for the trends shown by the graph in Exercise 68?

Technology Exercises

In Exercises 76–83, use a graphing utility to graph the function. Use the graph to determine whether the function has an inverse that is a function (that is, whether the function is one-to-one).

76. $f(x) = x^2 - 1$
77. $f(x) = \sqrt[3]{2 - x}$
78. $f(x) = \frac{x^3}{2}$
79. $f(x) = \frac{x^4}{4}$
80. $f(x) = \text{int}(x - 2)$
81. $f(x) = |x - 2|$
82. $f(x) = (x - 1)^3$
83. $f(x) = -\sqrt{16 - x^2}$

In Exercises 84–86, use a graphing utility to graph f and g in the same viewing rectangle. In addition, graph the line $y = x$ and visually determine if f and g are inverses.

84. $f(x) = 4x + 4, g(x) = 0.25x - 1$
85. $f(x) = \frac{1}{x} + 2, g(x) = \frac{1}{x - 2}$
86. $f(x) = \sqrt[3]{x} - 2, g(x) = (x + 2)^3$

Critical Thinking Exercises

87. Which one of the following is true?
    - **a.** The inverse of $\{(1, 4), (2, 7)\}$ is $\{(2, 7), (1, 4)\}$.
    - **b.** The function $f(x) = 5$ is one-to-one.
    - **c.** If $f(x) = 3x$, then $f^{-1}(x) = \frac{1}{3x}$.
    - **d.** The domain of f is the same as the range of f^{-1}.
88. If $f(x) = 3x$ and $g(x) = x + 5$, find $(f \circ g)^{-1}(x)$ and $(g^{-1} \circ f^{-1})(x)$.
89. Show that
$$f(x) = \frac{3x - 2}{5x - 3}$$
is its own inverse.
90. *Freedom 7* was the spacecraft that carried the first American into space in 1961. Total flight time was 15 minutes and the spacecraft reached a maximum height of 116 miles. Consider a function, s, that expresses *Freedom 7*'s height, $s(t)$, in miles, after t minutes. Is s a one-to-one function? Explain your answer.
91. If $f(2) = 6$, and f is one-to-one, find x satisfying $8 + f^{-1}(x - 1) = 10$.

Group Exercise

92. In Tom Stoppard's play *Arcadia*, the characters dream and talk about mathematics, including ideas involving graphing, composite functions, symmetry, and lack of symmetry in things that are tangled, mysterious, and unpredictable. Group members should read the play. Present a report on the ideas discussed by the characters that are related to concepts that we studied in this chapter. Bring in a copy of the play and read appropriate excerpts.

SECTION 2.8 Distance and Midpoint Formulas; Circles

Objectives

1. Find the distance between two points.
2. Find the midpoint of a line segment.
3. Write the standard form of a circle's equation.
4. Give the center and radius of a circle whose equation is in standard form.
5. Convert the general form of a circle's equation to standard form.

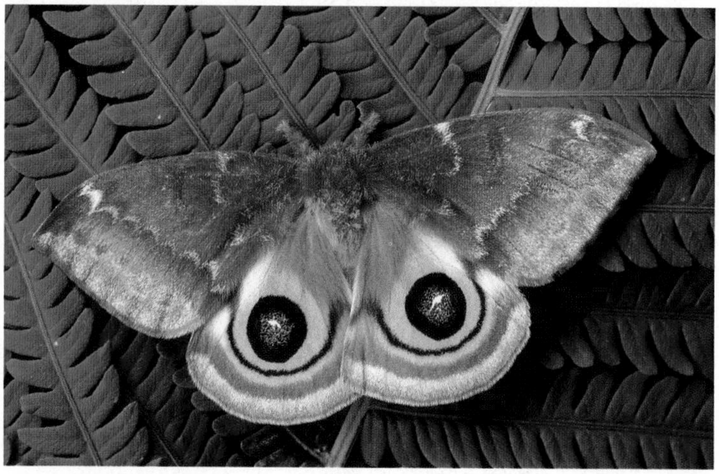

It's a good idea to know your way around a circle. Clocks, angles, maps, and compasses are based on circles. Circles occur everywhere in nature: in ripples on water, patterns on a moth's wings, and cross sections of trees. Some consider the circle to be the most pleasing of all shapes.

The rectangular coordinate system gives us a unique way of knowing a circle. It enables us to translate a circle's geometric definition into an algebraic equation. To do this, we must first develop a formula for the distance between any two points in rectangular coordinates.

1 Find the distance between two points.

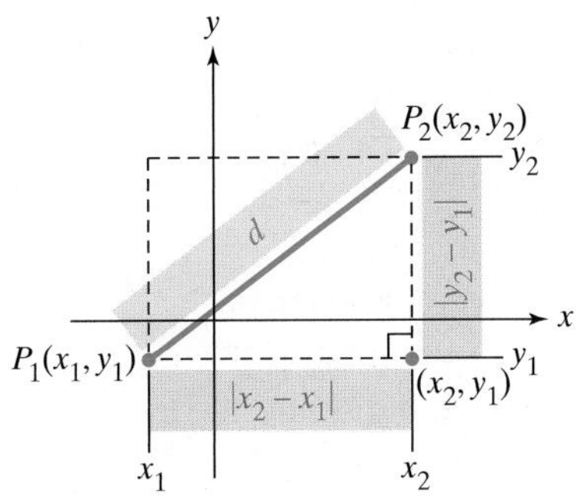

Figure 2.59

The Distance Formula

Using the Pythagorean Theorem, we can find the distance between the two points $P_1(x_1, y_1)$ and $P_2(x_2, y_2)$ in the rectangular coordinate system. The two points are illustrated in Figure 2.59.

The distance that we need to find is represented by d and shown in blue. Notice that the distance between two points on the dashed horizontal line is the absolute value of the difference between the x-coordinates of the two points. This distance, $|x_2 - x_1|$, is shown in pink. Similarly, the distance between two points on the dashed vertical line is the absolute value of the difference between the y-coordinates of the two points. This distance, $|y_2 - y_1|$, is also shown in pink.

Because the dashed lines are horizontal and vertical, a right triangle is formed. Thus, we can use the Pythagorean Theorem to find the distance d. Squaring the lengths of the triangle's sides results in positive numbers, so absolute value notation is not necessary.

$$d^2 = (x_2 - x_1)^2 + (y_2 - y_1)^2$$ Apply the Pythagorean Theorem to the right triangle in Figure 2.59.

$$d = \pm\sqrt{(x_2 - x_1)^2 + (y_2 - y_1)^2}$$ Apply the square root property.

$$d = \sqrt{(x_2 - x_1)^2 + (y_2 - y_1)^2}$$ Because distance is nonnegative, write only the principal square root.

This result is called the **distance formula**.

The Distance Formula

The distance, d, between the points (x_1, y_1) and (x_2, y_2) in the rectangular coordinate system is

$$d = \sqrt{(x_2 - x_1)^2 + (y_2 - y_1)^2}.$$

To compute the distance between two points, find the square of the difference between the x-coordinates plus the square of the difference between the y-coordinates. The principal square root of this sum is the distance.

When using the distance formula, it does not matter which point you call (x_1, y_1) and which you call (x_2, y_2).

EXAMPLE 1 Using the Distance Formula

Find the distance between $(-1, 4)$ and $(3, -2)$.

Solution We will let $(x_1, y_1) = (-1, 4)$ and $(x_2, y_2) = (3, -2)$.

$$d = \sqrt{(x_2 - x_1)^2 + (y_2 - y_1)^2}$$ Use the distance formula.

$$= \sqrt{(-2 - 4)^2 + [3 - (-1)]^2}$$ Substitute the given values.

$$= \sqrt{(-6)^2 + 4^2}$$ Perform operations inside grouping symbols: $-2 - 4 = -6$ and $3 - (-1) = 3 + 1 = 4$.

$$= \sqrt{36 + 16}$$ Square -6 and 4.

Caution: This does not equal $\sqrt{36} + \sqrt{16}$.

$$= \sqrt{52}$$ Add.

$$= \sqrt{4 \cdot 13} = 2\sqrt{13} \approx 7.21$$ $\sqrt{52} = \sqrt{4 \cdot 13} = \sqrt{4}\sqrt{13} = 2\sqrt{13}$

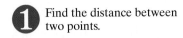

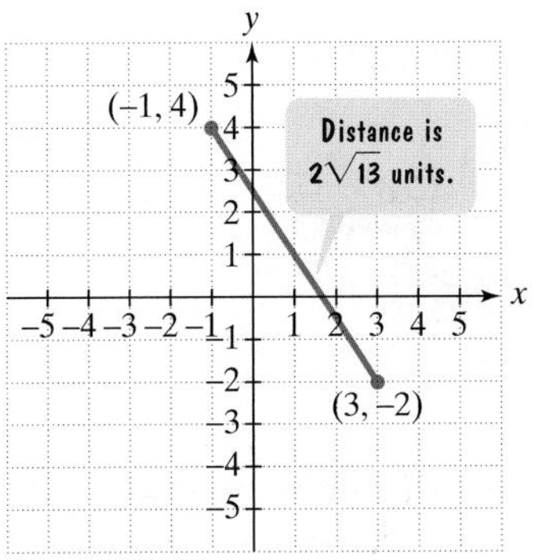

Figure 2.60 Finding the distance between two points

The distance between the given points is $2\sqrt{13}$ units, or approximately 7.21 units. The situation is illustrated in Figure 2.60.

Check Point 1 Find the distance between $(-4, 9)$ and $(1, -3)$.

2 Find the midpoint of a line segment.

The Midpoint Formula

The distance formula can be used to derive a formula for finding the midpoint of a line segment between two given points. The formula is given as follows:

The Midpoint Formula

Consider a line segment whose endpoints are (x_1, y_1) and (x_2, y_2). The coordinates of the segment's midpoint are

$$\left(\frac{x_1 + x_2}{2}, \frac{y_1 + y_2}{2}\right).$$

To find the midpoint, take the average of the two x-coordinates and the average of the two y-coordinates.

Study Tip

The midpoint formula requires finding the *sum* of coordinates. By contrast, the distance formula requires finding the *difference* of coordinates:

Midpoint: Sum of coordinates

$$\left(\frac{x_1 + x_2}{2}, \frac{y_1 + y_2}{2}\right)$$

Distance: Difference of coordinates

$$\sqrt{(x_2 - x_1)^2 + (y_2 - y_1)^2}$$

It's easy to confuse the two formulas. Be sure to use addition, not subtraction, when applying the midpoint formula.

EXAMPLE 2 Using the Midpoint Formula

Find the midpoint of the line segment with endpoints $(1, -6)$ and $(-8, -4)$.

Solution To find the coordinates of the midpoint, we average the coordinates of the endpoints.

$$\text{Midpoint} = \left(\frac{1 + (-8)}{2}, \frac{-6 + (-4)}{2}\right) = \left(\frac{-7}{2}, \frac{-10}{2}\right) = \left(-\frac{7}{2}, -5\right)$$

Average the x-coordinates. Average the y-coordinates.

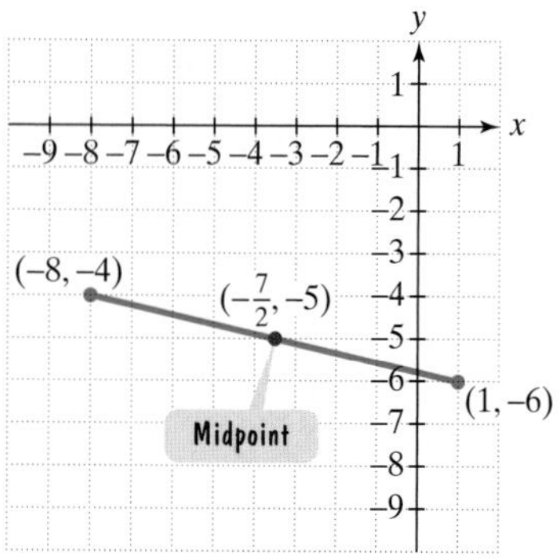

Figure 2.61 Finding a line segment's midpoint

Figure 2.61 illustrates that the point $\left(-\frac{7}{2}, -5\right)$ is midway between the points $(1, -6)$ and $(-8, -4)$.

Check Point 2 Find the midpoint of the line segment with endpoints $(1, 2)$ and $(7, -3)$.

Circles

Our goal is to translate a circle's geometric definition into an equation. We begin with this geometric definition.

Definition of a Circle

A **circle** is the set of all points in a plane that are equidistant from a fixed point, called the **center**. The fixed distance from the circle's center to any point on the circle is called the **radius**.

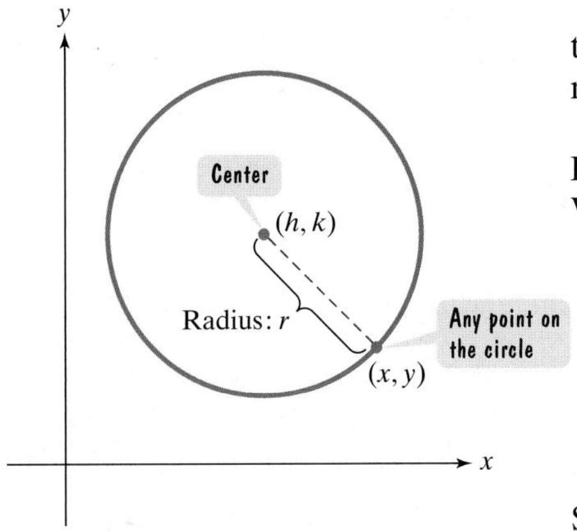

Figure 2.62 A circle centered at (h, k) with radius r

Figure 2.62 is our starting point for obtaining a circle's equation. We've placed the circle into a rectangular coordinate system. The circle's center is (h, k) and its radius is r. We let (x, y) represent the coordinates of any point on the circle.

What does the geometric definition of a circle tell us about point (x, y) in Figure 2.62? The point is on the circle if and only if its distance from the center is r. We can use the distance formula to express this idea algebraically:

The distance between (x, y) and (h, k) | is always | r.

$$\sqrt{(x-h)^2+(y-k)^2} = r.$$

Squaring both sides of $\sqrt{(x-h)^2+(y-k)^2} = r$ yields the *standard form of the equation of a circle.*

3 Write the standard form of a circle's equation.

The Standard Form of the Equation of a Circle

The **standard form of the equation of a circle** with center (h, k) and radius r is

$$(x-h)^2+(y-k)^2=r^2.$$

EXAMPLE 3 Finding the Standard Form of a Circle's Equation

Write the standard form of the equation of the circle with center $(0, 0)$ and radius 2. Graph the circle.

Solution The center is $(0, 0)$. Because the center is represented as (h, k) in the standard form of the equation, $h = 0$ and $k = 0$. The radius is 2, so we will let $r = 2$ in the equation.

$$(x-h)^2+(y-k)^2=r^2 \quad \text{This is the standard form of a circle's equation.}$$

$$(x-0)^2+(y-0)^2=2^2 \quad \text{Substitute 0 for } h\text{, 0 for } k\text{, and 2 for } r.$$

$$x^2+y^2=4 \quad \text{Simplify.}$$

The standard form of the equation of the circle is $x^2 + y^2 = 4$. Figure 2.63 shows the graph.

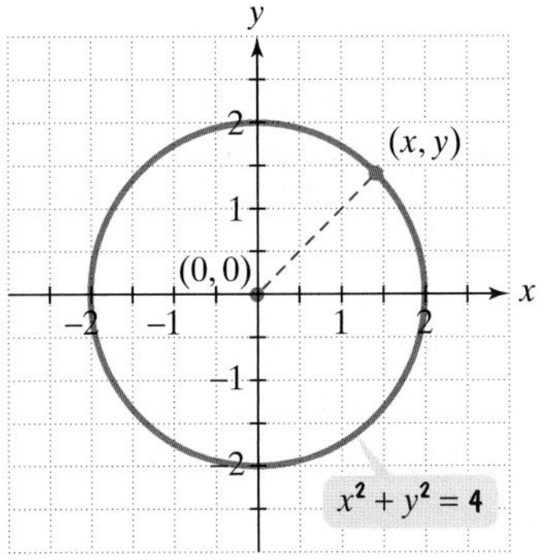

Figure 2.63 The graph of $x^2 + y^2 = 4$

Check Point 3 Write the standard form of the equation of the circle with center $(0, 0)$ and radius 4.

Technology

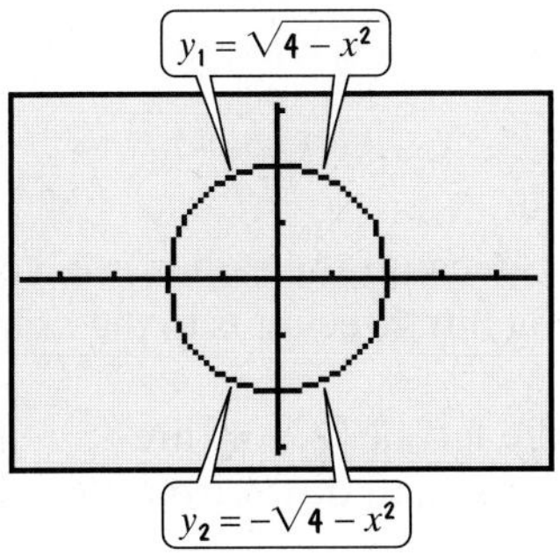

To graph a circle with a graphing utility, first solve the equation for y.

$$x^2+y^2=4$$

$$y^2=4-x^2$$

$$y=\pm\sqrt{4-x^2} \quad y \text{ is not a function of } x.$$

Graph the two equations

$$y_1=\sqrt{4-x^2} \quad \text{and} \quad y_2=-\sqrt{4-x^2}$$

in the same viewing rectangle. The graph of $y_1 = \sqrt{4-x^2}$ is the top semicircle because y is always positive. The graph of $y_2 = -\sqrt{4-x^2}$ is the bottom semicircle because y is always negative. Use a ZOOM SQUARE setting so that the circle looks like a circle. (Many graphing utilities have problems connecting the two semicircles because the segments directly across horizontally from the center become nearly vertical.)

Example 3 and Check Point 3 involved circles centered at the origin. The standard form of the equation of all such circles is $x^2 + y^2 = r^2$, where r is the circle's radius. Now, let's consider a circle whose center is not at the origin.

EXAMPLE 4 Finding the Standard Form of a Circle's Equation

Write the standard form of the equation of the circle with center $(-2, 3)$ and radius 4.

Solution The center is $(-2, 3)$. Because the center is represented as (h, k) in the standard form of the equation, $h = -2$ and $k = 3$. The radius is 4, so we will let $r = 4$ in the equation.

$$(x - h)^2 + (y - k)^2 = r^2 \quad \text{This is the standard form of a circle's equation.}$$

$$[x - (-2)]^2 + (y - 3)^2 = 4^2 \quad \text{Substitute } -2 \text{ for } h, 3 \text{ for } k, \text{ and } 4 \text{ for } r.$$

$$(x + 2)^2 + (y - 3)^2 = 16 \quad \text{Simplify.}$$

The standard form of the equation of the circle is $(x + 2)^2 + (y - 3)^2 = 16$.

Check Point 4 Write the standard form of the equation of the circle with center $(5, -6)$ and radius 10.

4 Give the center and radius of a circle whose equation is in standard form.

EXAMPLE 5 Using the Standard Form of a Circle's Equation to Graph the Circle

a. Find the center and radius of the circle whose equation is

$$(x - 2)^2 + (y + 4)^2 = 9.$$

b. Graph the equation.

c. Use the graph to identify the relation's domain and range.

Solution

a. We begin by finding the circle's center, (h, k), and its radius, r. We can find the values for h, k, and r by comparing the given equation to the standard form of the equation of a circle, $(x - h)^2 + (y - k)^2 = r^2$.

$$(x - 2)^2 + (y + 4)^2 = 9$$

$$(x - 2)^2 + (y - (-4))^2 = 3^2$$

This is $(x - h)^2$, with $h = 2$. This is $(y - k)^2$, with $k = -4$. This is r^2, with $r = 3$.

We see that $h = 2$, $k = -4$, and $r = 3$. Thus, the circle has center $(h, k) = (2, -4)$ and a radius of 3 units.

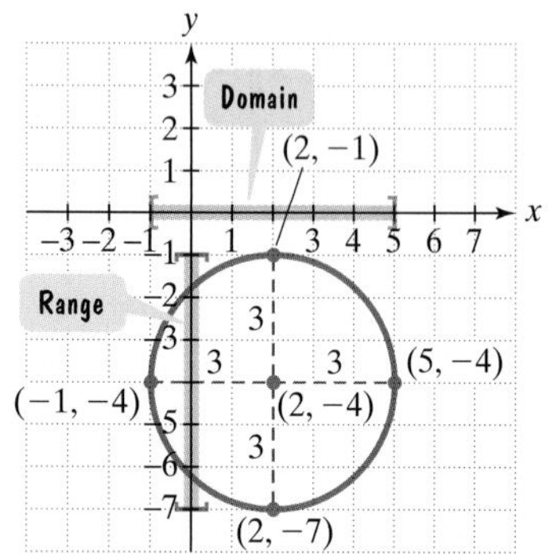

Figure 2.64 The graph of $(x - 2)^2 + (y + 4)^2 = 9$

b. To graph this circle, first plot the center $(2, -4)$. Because the radius is 3, you can locate at least four points on the circle by going out three units to the right, to the left, up, and down from the center.

The points three units to the right and to the left of $(2, -4)$ are $(5, -4)$ and $(-1, -4)$, respectively. The points three units up and down from $(2, -4)$ are $(2, -1)$ and $(2, -7)$, respectively.

Using these points, we obtain the graph in Figure 2.64.

c. The four points that we located on the circle can be used to determine the relation's domain and range. The points $(-1, -4)$ and $(5, -4)$ show that values of x extend from -1 to 5, inclusive:

$$\text{Domain} = [-1, 5].$$

The points $(2, -7)$ and $(2, -1)$ show that values of y extend from -7 to -1, inclusive:

$$\text{Range} = [-7, -1].$$

Check Point 5 **a.** Find the center and radius of the circle whose equation is

$$(x + 3)^2 + (y - 1)^2 = 4.$$

b. Graph the equation.

c. Use the graph to identify the relation's domain and range.

If we square $x - 2$ and $y + 4$ in the standard form of the equation from Example 5, we obtain another form for the circle's equation.

$(x - 2)^2 + (y + 4)^2 = 9$	This is the standard form of the equation from Example 5.
$x^2 - 4x + 4 + y^2 + 8y + 16 = 9$	Square $x - 2$ and $y + 4$.
$x^2 + y^2 - 4x + 8y + 20 = 9$	Combine constants and rearrange terms.
$x^2 + y^2 - 4x + 8y + 11 = 0$	Subtract 9 from both sides.

This result suggests that an equation in the form $x^2 + y^2 + Dx + Ey + F = 0$ can represent a circle. This is called the *general form of the equation of a circle.*

The General Form of the Equation of a Circle

The **general form of the equation of a circle** is

$$x^2 + y^2 + Dx + Ey + F = 0$$

where D, E, and F are real numbers.

5 Convert the general form of a circle's equation to standard form.

We can convert the general form of the equation of a circle to the standard form $(x - h)^2 + (y - k)^2 = r^2$. We do so by completing the square on x and y. Let's see how this is done.

EXAMPLE 6 Converting the General Form of a Circle's Equation to Standard Form and Graphing the Circle

Write in standard form and graph: $x^2 + y^2 + 4x - 6y - 23 = 0$.

Study Tip

To review completing the square, see Section 1.5, pages 134–136.

Solution Because we plan to complete the square on both x and y, let's rearrange the terms so that x-terms are arranged in descending order, y-terms are arranged in descending order, and the constant term appears on the right.

$x^2 + y^2 + 4x - 6y - 23 = 0$	This is the given equation.
$(x^2 + 4x \quad\quad) + (y^2 - 6y \quad\quad) = 23$	Rewrite in anticipation of completing the square.
$(x^2 + 4x + 4) + (y^2 - 6y + 9) = 23 + 4 + 9$	Complete the square on x: $\frac{1}{2} \cdot 4 = 2$ and $2^2 = 4$, so add 4 to both sides. Complete the square on y: $\frac{1}{2}(-6) = -3$ and $(-3)^2 = 9$, so add 9 to both sides.
$(x + 2)^2 + (y - 3)^2 = 36$	Factor on the left and add on the right.

Remember that numbers added on the left side must also be added on the right side.

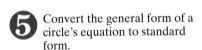

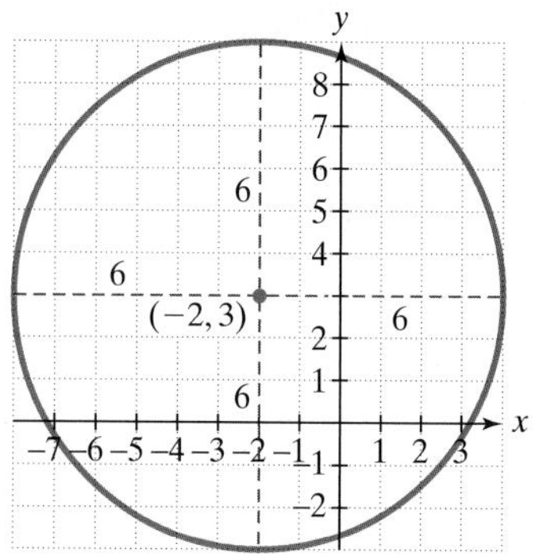

Figure 2.65 The graph of $(x + 2)^2 + (y - 3)^2 = 36$

This last equation, $(x + 2)^2 + (y - 3)^2 = 36$, is in standard form. We can identify the circle's center and radius by comparing this equation to the standard form of the equation of a circle, $(x - h)^2 + (y - k)^2 = r^2$.

$$(x + 2)^2 + (y - 3)^2 = 36$$

$$(x - (-2))^2 + (y - 3)^2 = 6^2$$

This is $(x - h)^2$, with $h = -2$. This is $(y - k)^2$, with $k = 3$. This is r^2, with $r = 6$.

We use the center, $(h, k) = (-2, 3)$, and the radius, $r = 6$, to graph the circle. The graph is shown in Figure 2.65.

Technology

To graph $x^2 + y^2 + 4x - 6y - 23 = 0$, rewrite the equation as a quadratic equation in y.

$$y^2 - 6y + (x^2 + 4x - 23) = 0$$

Now solve for y using the quadratic formula, with $a = 1$, $b = -6$, and $c = x^2 + 4x - 23$.

$$y = \frac{-b \pm \sqrt{b^2 - 4ac}}{2a} = \frac{-(-6) \pm \sqrt{(-6)^2 - 4 \cdot 1(x^2 + 4x - 23)}}{2 \cdot 1} = \frac{6 \pm \sqrt{36 - 4(x^2 + 4x - 23)}}{2}$$

Because we will enter these equations, there is no need to simplify. Enter

$$y_1 = \frac{6 + \sqrt{36 - 4(x^2 + 4x - 23)}}{2}$$

and

$$y_2 = \frac{6 - \sqrt{36 - 4(x^2 + 4x - 23)}}{2}.$$

Use a ZOOM SQUARE setting. The graph is shown on the right.

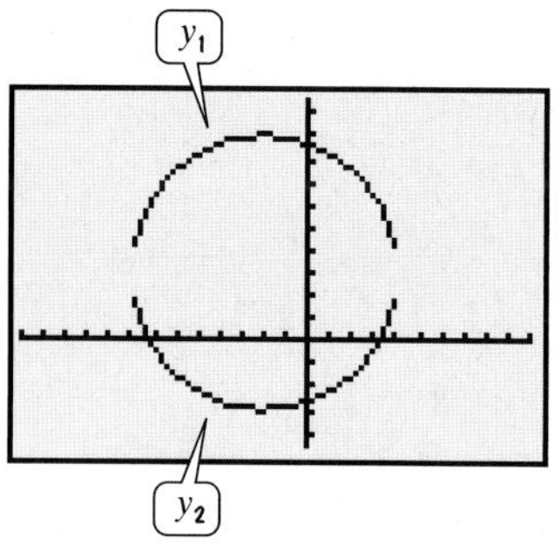

Check Point 6 Write in standard form and graph:

$$x^2 + y^2 + 4x - 4y - 1 = 0.$$

EXERCISE SET 2.8

Practice Exercises

In Exercises 1–18, find the distance between each pair of points. If necessary, round answers to two decimals places.

1. $(2, 3)$ and $(14, 8)$
2. $(5, 1)$ and $(8, 5)$
3. $(4, -1)$ and $(-6, 3)$
4. $(2, -3)$ and $(-1, 5)$
5. $(0, 0)$ and $(-3, 4)$
6. $(0, 0)$ and $(3, -4)$
7. $(-2, -6)$ and $(3, -4)$
8. $(-4, -1)$ and $(2, -3)$
9. $(0, -3)$ and $(4, 1)$
10. $(0, -2)$ and $(4, 3)$
11. $(3.5, 8.2)$ and $(-0.5, 6.2)$
12. $(2.6, 1.3)$ and $(1.6, -5.7)$
13. $(0, -\sqrt{3})$ and $(\sqrt{5}, 0)$
14. $(0, -\sqrt{2})$ and $(\sqrt{7}, 0)$
15. $(3\sqrt{3}, \sqrt{5})$ and $(-\sqrt{3}, 4\sqrt{5})$
16. $(2\sqrt{3}, \sqrt{6})$ and $(-\sqrt{3}, 5\sqrt{6})$
17. $\left(\frac{7}{3}, \frac{1}{5}\right)$ and $\left(\frac{1}{3}, \frac{6}{5}\right)$
18. $\left(-\frac{1}{4}, -\frac{1}{7}\right)$ and $\left(\frac{3}{4}, \frac{6}{7}\right)$

In Exercises 19–30, find the midpoint of each line segment with the given endpoints.

19. $(6, 8)$ and $(2, 4)$
20. $(10, 4)$ and $(2, 6)$
21. $(-2, -8)$ and $(-6, -2)$
22. $(-4, -7)$ and $(-1, -3)$
23. $(-3, -4)$ and $(6, -8)$
24. $(-2, -1)$ and $(-8, 6)$
25. $\left(-\frac{7}{2}, \frac{3}{2}\right)$ and $\left(-\frac{5}{2}, -\frac{11}{2}\right)$
26. $\left(-\frac{2}{5}, \frac{7}{15}\right)$ and $\left(-\frac{2}{5}, -\frac{4}{15}\right)$
27. $(8, 3\sqrt{5})$ and $(-6, 7\sqrt{5})$
28. $(7\sqrt{3}, -6)$ and $(3\sqrt{3}, -2)$
29. $(\sqrt{18}, -4)$ and $(\sqrt{2}, 4)$
30. $(\sqrt{50}, -6)$ and $(\sqrt{2}, 6)$

In Exercises 31–40, write the standard form of the equation of the circle with the given center and radius.

31. Center $(0, 0)$, $r = 7$ **32.** Center $(0, 0)$, $r = 8$

33. Center $(3, 2)$, $r = 5$ **34.** Center $(2, -1)$, $r = 4$

35. Center $(-1, 4)$, $r = 2$ **36.** Center $(-3, 5)$, $r = 3$

37. Center $(-3, -1)$, $r = \sqrt{3}$

38. Center $(-5, -3)$, $r = \sqrt{5}$

39. Center $(-4, 0)$, $r = 10$ **40.** Center $(-2, 0)$, $r = 6$

In Exercises 41–48, give the center and radius of the circle described by the equation and graph each equation. Use the graph to identify the relation's domain and range.

41. $x^2 + y^2 = 16$ **42.** $x^2 + y^2 = 49$

43. $(x - 3)^2 + (y - 1)^2 = 36$

44. $(x - 2)^2 + (y - 3)^2 = 16$

45. $(x + 3)^2 + (y - 2)^2 = 4$

46. $(x + 1)^2 + (y - 4)^2 = 25$

47. $(x + 2)^2 + (y + 2)^2 = 4$

48. $(x + 4)^2 + (y + 5)^2 = 36$

In Exercises 49–60, complete the square and write the equation in standard form. Then give the center and radius of each circle and graph the equation.

49. $x^2 + y^2 + 6x + 2y + 6 = 0$

50. $x^2 + y^2 + 8x + 4y + 16 = 0$

51. $x^2 + y^2 - 10x - 6y - 30 = 0$

52. $x^2 + y^2 - 4x - 12y - 9 = 0$

53. $x^2 + y^2 + 8x - 2y - 8 = 0$

54. $x^2 + y^2 + 12x - 6y - 4 = 0$

55. $x^2 - 2x + y^2 - 15 = 0$

56. $x^2 + y^2 - 6y - 7 = 0$

57. $x^2 + y^2 - x + 2y + 1 = 0$

58. $x^2 + y^2 + x + y - \frac{1}{2} = 0$

59. $x^2 + y^2 + 3x - 2y - 1 = 0$

60. $x^2 + y^2 + 3x + 5y + \frac{9}{4} = 0$

Practice Plus

In Exercises 61–62, a line segment through the center of each circle intersects the circle at the points shown.

a. *Find the coordinates of the circle's center.*

b. *Find the radius of the circle.*

c. *Use your answers from parts (a) and (b) to write the standard form of the circle's equation.*

61.

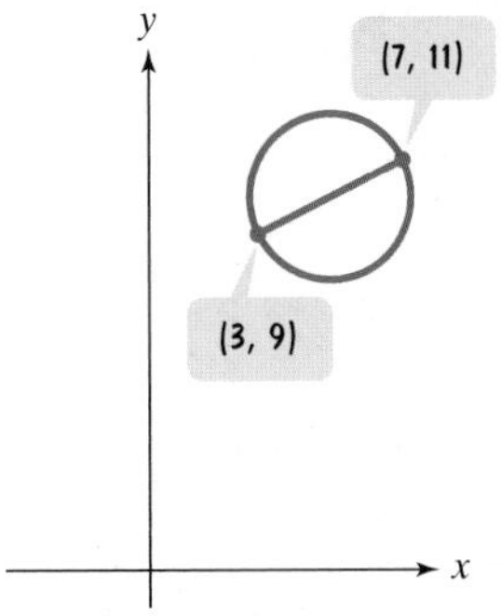

62.

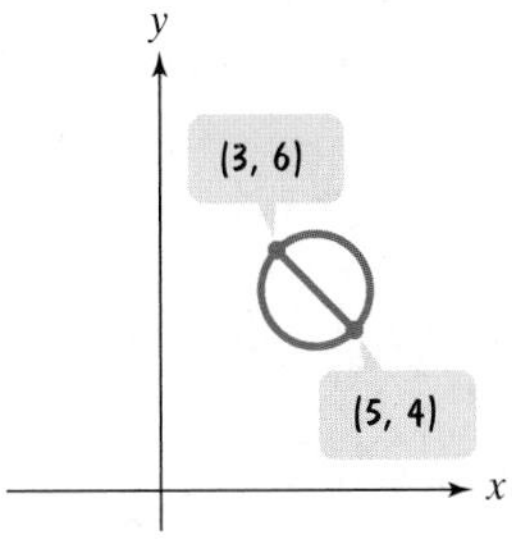

In Exercises 63–66, graph both equations in the same rectangular coordinate system and find all points of intersection. Then show that these ordered pairs satisfy the equations.

63. $x^2 + y^2 = 16$
$x - y = 4$

64. $x^2 + y^2 = 9$
$x - y = 3$

65. $(x - 2)^2 + (y + 3)^2 = 4$
$y = x - 3$

66. $(x - 3)^2 + (y + 1)^2 = 9$
$y = x - 1$

Application Exercises

67. A rectangular coordinate system with coordinates in miles is placed on the map in the figure shown. Bangkok has coordinates $(-115, 170)$ and Phnom Penh has coordinates $(65, 70)$. How long will it take a plane averaging 400 miles per hour to fly directly from one city to the other? Round to the nearest tenth of an hour. Approximately how many minutes is the flight?

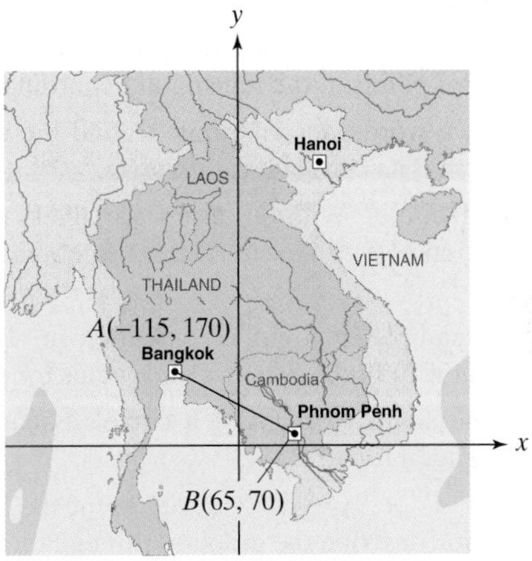

68. The Ferris wheel in the figure has a radius of 68 feet. The clearance between the wheel and the ground is 14 feet. The rectangular coordinate system shown has its origin on the ground directly below the center of the wheel. Use the coordinate system to write the equation of the circular wheel.

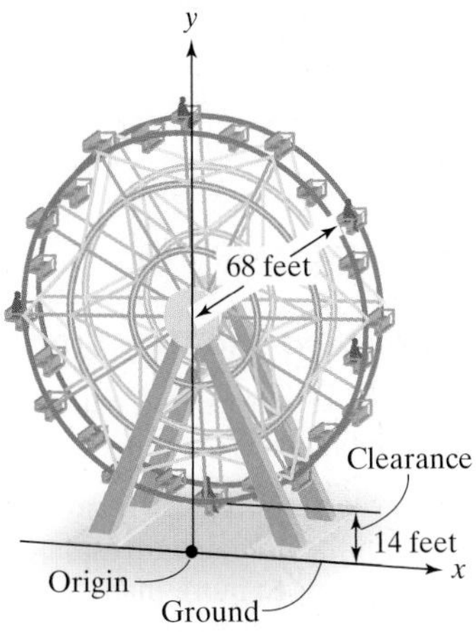

69. A rectangular coordinate system with coordinates in miles is placed with the origin at the center of Los Angeles. The figure indicates that the University of Southern California is located 2.4 miles west and 2.7 miles south of central Los Angeles. A seismograph on the campus shows that a small earthquake occurred. The quake's epicenter is estimated to be approximately 30 miles from the university. Write the standard form of the equation for the set of points that could be the epicenter of the quake.

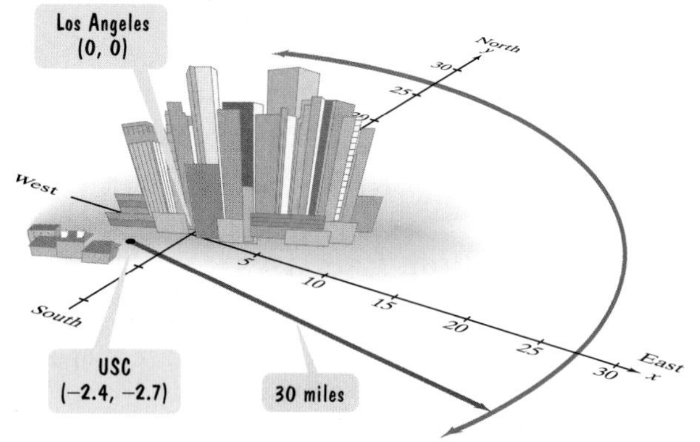

Writing in Mathematics

70. In your own words, describe how to find the distance between two points in the rectangular coordinate system.

71. In your own words, describe how to find the midpoint of a line segment if its endpoints are known.

72. What is a circle? Without using variables, describe how the definition of a circle can be used to obtain a form of its equation.

73. Give an example of a circle's equation in standard form. Describe how to find the center and radius for this circle.

74. How is the standard form of a circle's equation obtained from its general form?

75. Does $(x-3)^2 + (y-5)^2 = 0$ represent the equation of a circle? If not, describe the graph of this equation.

76. Does $(x-3)^2 + (y-5)^2 = -25$ represent the equation of a circle? What sort of set is the graph of this equation?

Technology Exercises

In Exercises 77–79, use a graphing utility to graph each circle whose equation is given.

77. $x^2 + y^2 = 25$

78. $(y+1)^2 = 36 - (x-3)^2$

79. $x^2 + 10x + y^2 - 4y - 20 = 0$

Critical Thinking Exercises

80. Which one of the following is true?

a. The equation of the circle whose center is at the origin with radius 16 is $x^2 + y^2 = 16$.

b. The graph of $(x-3)^2 + (y+5)^2 = 36$ is a circle with radius 6 centered at $(-3, 5)$.

c. The graph of $(x-4) + (y+6) = 25$ is a circle with radius 5 centered at $(4, -6)$.

d. None of the above is true.

81. Show that the points $A(1, 1+d)$, $B(3, 3+d)$, and $C(6, 6+d)$ are collinear (lie along a straight line) by showing that the distance from A to B plus the distance from B to C equals the distance from A to C.

82. Prove the midpoint formula by using the following procedure.

a. Show that the distance between (x_1, y_1) and $\left(\frac{x_1+x_2}{2}, \frac{y_1+y_2}{2}\right)$ is equal to the distance between (x_2, y_2) and $\left(\frac{x_1+x_2}{2}, \frac{y_1+y_2}{2}\right)$.

b. Use the procedure from Exercise 81 and the distances from part (a) to show that the points (x_1, y_1), $\left(\frac{x_1+x_2}{2}, \frac{y_1+y_2}{2}\right)$, and (x_2, y_2) are collinear.

83. Find the area of the donut-shaped region bounded by the graphs of $(x-2)^2 + (y+3)^2 = 25$ and $(x-2)^2 + (y+3)^2 = 36$.

84. A **tangent line** to a circle is a line that intersects the circle at exactly one point. The tangent line is perpendicular to the radius of the circle at this point of contact. Write the point-slope form of the equation of a line tangent to the circle whose equation is $x^2 + y^2 = 25$ at the point $(3, -4)$.

Chapter 2 Summary, Review, and Test

Summary

DEFINITIONS AND CONCEPTS	EXAMPLES
2.1 Basics of Functions and Their Graphs	
a. A relation is any set of ordered pairs. The set of first components is the domain of the relation and the set of second components is the range.	Ex. 1, p. 187
b. A function is a correspondence from a first set, called the domain, to a second set, called the range, such that each element in the domain corresponds to exactly one element in the range. If any element in a relation's domain corresponds to more than one element in the range, the relation is not a function.	Ex. 2, p. 189

DEFINITIONS AND CONCEPTS	EXAMPLES
c. Functions are usually given in terms of equations involving x and y, in which x is the independent variable and y is the dependent variable. If an equation is solved for y and more than one value of y can be obtained for a given x, then the equation does not define y as a function of x. If an equation defines a function, the value of the function at x, $f(x)$, often replaces y.	Ex. 3, p. 190; Ex. 4, p. 191
d. The graph of a function is the graph of its ordered pairs.	Ex. 5, p. 192
e. The vertical line test for functions: If any vertical line intersects a graph in more than one point, the graph does not define y as a function of x.	Ex. 6, p. 193
f. The graph of a function can be used to determine the function's domain and its range. To find the domain, look for all the inputs on the x-axis that correspond to points on the graph. To find the range, look for all the outputs on the y-axis that correspond to points on the graph.	Ex. 8, p. 196
g. The zeros of a function, f, are the values of x for which $f(x) = 0$. At these values, the graph of f has x-intercepts. A function can have more than one x-intercept but at most one y-intercept.	Figure 2.11, p. 197

2.2 More on Functions and Their Graphs

a. The difference quotient is $\dfrac{f(x+h) - f(x)}{h}, h \neq 0.$	Ex. 1, p. 203
b. Piecewise functions are defined by two (or more) equations over a specified domain.	Ex. 2, p. 204
c. A function is increasing on intervals where its graph rises, decreasing on intervals where it falls, and constant on intervals where it neither rises nor falls. Precise definitions are given in the box on page 206.	Ex. 3, p. 206
d. If the graph of a function is given, we can often visually locate the number(s) at which the function has a relative maximum or relative minimum. Precise definitions are given in the box on page 207.	Figure 2.15, p. 207
e. The graph of an even function in which $f(-x) = f(x)$ is symmetric with respect to the y-axis. The graph of an odd function in which $f(-x) = -f(x)$ is symmetric with respect to the origin.	Ex. 4, p. 208
f. The graph of $f(x) = \text{int}(x)$, where $\text{int}(x)$ is the greatest integer that is less than or equal to x, has function values that form discontinuous steps, shown in Figure 2.20 on page 210. If $n \le x < n + 1$, where n is an integer, then $\text{int}(x) = n$.	

2.3 Linear Functions and Slope

a. The slope, m, of the line through (x_1, y_1) and (x_2, y_2) is $m = \dfrac{y_2 - y_1}{x_2 - x_1}$.	Ex. 1, p. 217
b. Equations of lines include point-slope form, $y - y_1 = m(x - x_1)$, slope-intercept form, $y = mx + b$, and general form, $Ax + By + C = 0$. The equation of a horizontal line is $y = b$; a vertical line is $x = a$. A vertical line is not a linear function.	Ex. 2, p. 219; Ex. 3, p. 219; Ex. 5, p. 223
c. Linear functions in the form $f(x) = mx + b$ can be graphed using the slope, m, and the y-intercept, b. (See the box on page 220.) Linear equations in the general form $Ax + By + C = 0$ can be solved for y and graphed using the slope and the y-intercept. Intercepts can also be used to graph $Ax + By + C = 0$. (See the box on page 224.)	Ex. 4, p. 221; Ex. 6, p. 224; Ex. 7, p. 224

2.4 More on Slope

a. Parallel lines have equal slopes. Perpendicular lines have slopes that are negative reciprocals.	Ex. 1, p. 231; Ex. 2, p. 233
b. The slope of a linear function is the rate of change of the dependent variable per unit change of the independent variable.	Ex. 3, p. 234
c. The average rate of change of f from x_1 to x_2 is $\dfrac{f(x_2) - f(x_1)}{x_2 - x_1}.$	Ex. 4, p. 235; Ex. 5, p. 237

DEFINITIONS AND CONCEPTS	EXAMPLES
2.5 *Transformations of Functions*	
a. Table 2.3 on page 242 shows the graphs of the constant function, $f(x) = c$, the identity function, $f(x) = x$, the absolute value function, $f(x) = \lvert x \rvert$, the standard quadratic function, $f(x) = x^2$, the square root function, $f(x) = \sqrt{x}$, the standard cubic function, $f(x) = x^3$, and the cube root function, $f(x) = \sqrt[3]{x}$. The table also lists characteristics of each function.	
b. Table 2.4 on page 250 summarizes how to graph a function using vertical shifts, $y = f(x) \pm c$, horizontal shifts, $y = f(x \pm c)$, reflections about the x-axis, $y = -f(x)$, reflections about the y-axis, $y = f(-x)$, vertical stretching, $y = cf(x), c > 1$, vertical shrinking, $y = cf(x), 0 < c < 1$, horizontal shrinking, $y = f(cx), c > 1$, and horizontal stretching, $y = f(cx), 0 < c < 1$.	Ex. 1, p. 243; Ex. 2, p. 245; Ex. 3, p. 246; Ex. 4, p. 247; Ex. 5, p. 247; Ex. 6, p. 248; Ex. 7, p. 249
c. A function involving more than one transformation can be graphed in the following order: (1) horizontal shifting; (2) stretching or shrinking; (3) reflecting; (4) vertical shifting.	Ex. 8, p. 251; Ex. 9, p. 252
2.6 *Combinations of Functions; Composite Functions*	
a. If a function f does not model data or verbal conditions, its domain is the largest set of real numbers for which the value of $f(x)$ is a real number. Exclude from a function's domain real numbers that cause division by zero and real numbers that result in a square root of a negative number.	Ex. 1, p. 258
b. When functions are given as equations, they can be added, subtracted, multiplied, or divided by performing operations with the algebraic expressions that appear on the right side of the equations. Definitions for the sum $f + g$, the difference $f - g$, the product fg, and the quotient $\frac{f}{g}$ functions, with domains $D_f \cap D_g$, and $g(x) \neq 0$ for the quotient function, are given in the box on page 260.	Ex. 2, p. 260; Ex. 3, p. 261
c. The composition of functions f and g, $f \circ g$, is defined by $(f \circ g)(x) = f(g(x))$. The domain of the composite function $f \circ g$ is given in the box on page 263. This composite function is obtained by replacing each occurrence of x in the equation for f with $g(x)$.	Ex. 4, p. 264; Ex. 5, p. 265
2.7 *Inverse Functions*	
a. If $f(g(x)) = x$ and $g(f(x)) = x$, function g is the inverse of function f, denoted f^{-1} and read "f–inverse." Thus, to show that f and g are inverses of each other, one must show that $f(g(x)) = x$ and $g(f(x)) = x$.	Ex. 1, p. 272
b. The procedure for finding a function's inverse uses a switch-and-solve strategy. Switch x and y, and then solve for y. The procedure is given in the box on page 273.	Ex. 2, p. 273; Ex. 3, p. 273; Ex. 4, p. 274
c. The horizontal line test for inverse functions: A function f has an inverse that is a function, f^{-1}, if there is no horizontal line that intersects the graph of the function f at more than one point.	Ex. 5, p. 275
d. A one-to-one function is one in which no two different ordered pairs have the same second component. Only one-to-one functions have inverse functions.	
e. If the point (a, b) is on the graph of f, then the point (b, a) is on the graph of f^{-1}. The graph of f^{-1} is a reflection of the graph of f about the line $y = x$.	Ex. 6, p. 276; Ex. 7, p. 277
2.8 *Distance and Midpoint Formulas; Circles*	
a. The distance, d, between the points (x_1, y_1) and (x_2, y_2) is given by $d = \sqrt{(x_2 - x_1)^2 + (y_2 - y_1)^2}$.	Ex. 1, p. 281
b. The midpoint of the line segment whose endpoints are (x_1, y_1) and (x_2, y_2) is the point with coordinates $\left(\frac{x_1 + x_2}{2}, \frac{y_1 + y_2}{2}\right)$.	Ex. 2, p. 282
c. The standard form of the equation of a circle with center (h, k) and radius r is $(x - h)^2 + (y - k)^2 = r^2$.	Ex. 3, p. 283; Ex. 4, p. 284; Ex. 5, p. 284
d. The general form of the equation of a circle is $x^2 + y^2 + Dx + Ey + F = 0$.	
e. To convert from the general form to the standard form of a circle's equation, complete the square on x and y.	Ex. 6, p. 285

Review Exercises

2.1 and 2.2

In Exercises 1–3, determine whether each relation is a function. Give the domain and range for each relation.

1. $\{(2, 7), (3, 7), (5, 7)\}$

2. $\{(1, 10), (2, 500), (13, \pi)\}$

3. $\{(12, 13), (14, 15), (12, 19)\}$

In Exercises 4–6, determine whether each equation defines y as a function of x.

4. $2x + y = 8$

5. $3x^2 + y = 14$

6. $2x + y^2 = 6$

In Exercises 7–10, evaluate each function at the given values of the independent variable and simplify.

7. $f(x) = 5 - 7x$

a. $f(4)$ **b.** $f(x + 3)$ **c.** $f(-x)$

8. $g(x) = 3x^2 - 5x + 2$

a. $g(0)$ **b.** $g(-2)$

c. $g(x - 1)$ **d.** $g(-x)$

9. $g(x) = \begin{cases} \sqrt{x - 4} & \text{if } x \geq 4 \\ 4 - x & \text{if } x < 4 \end{cases}$

a. $g(13)$ **b.** $g(0)$ **c.** $g(-3)$

10. $f(x) = \begin{cases} \dfrac{x^2 - 1}{x - 1} & \text{if } x \neq 1 \\ 12 & \text{if } x = 1 \end{cases}$

a. $f(-2)$ **b.** $f(1)$ **c.** $f(2)$

In Exercises 11–16, use the vertical line test to identify graphs in which y is a function of x.

11.

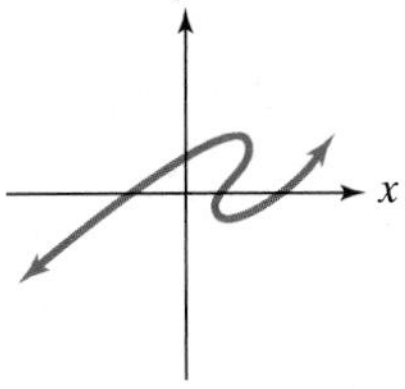

12.

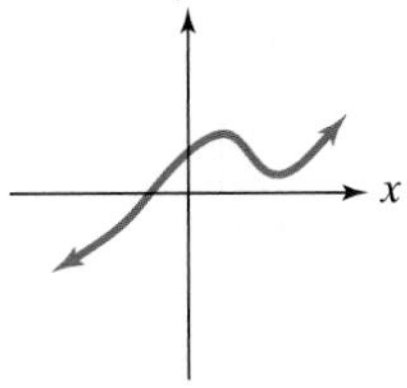

13.

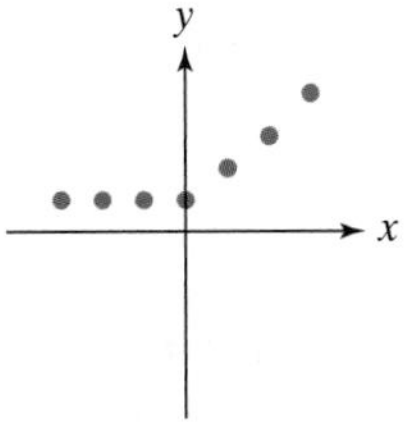

14.

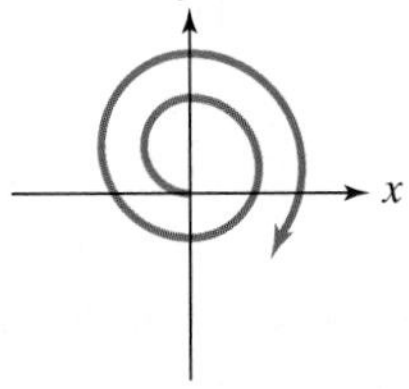

15.

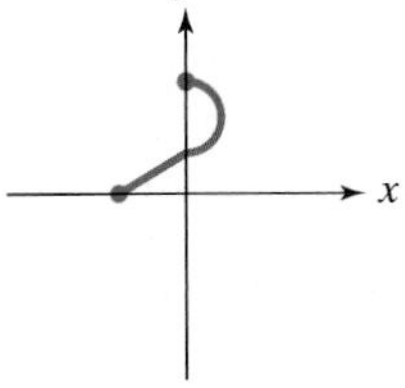

16.

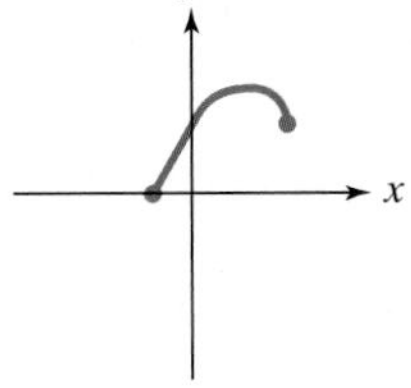

In Exercises 17–18, find and simplify the difference quotient

$$\frac{f(x + h) - f(x)}{h}, \quad h \neq 0$$

for the given function.

17. $f(x) = 8x - 11$

18. $f(x) = -2x^2 + x + 10$

In Exercises 19–21, use the graph to determine **a.** *the function's domain;* **b.** *the function's range;* **c.** *the x-intercepts, if any;* **d.** *the y-intercept, if any;* **e.** *intervals on which the function is increasing, decreasing, or constant; and* **f.** *the function values indicated below the graphs.*

19.

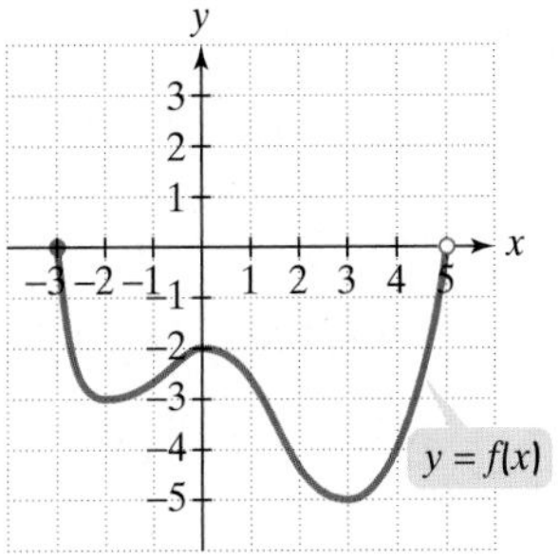

$f(-2) = ?$ $f(3) = ?$

20.

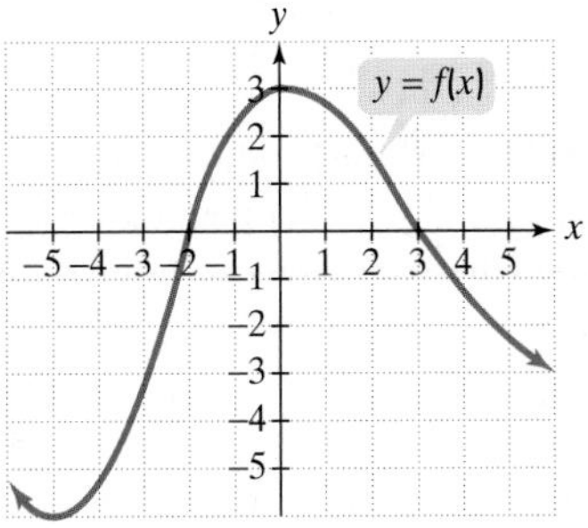

$f(-2) = ?$ $f(6) = ?$

21.

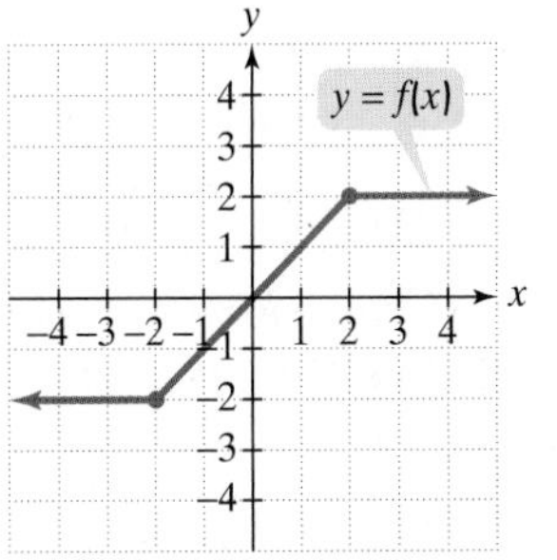

$f(-9) = ?$ $f(14) = ?$

In Exercises 22–23, find each of the following:

a. *The numbers, if any, at which f has a relative maximum. What are these relative maxima?*

b. *The numbers, if any, at which f has a relative minimum. What are these relative minima?*

22. Use the graph in Exercise 19.

23. Use the graph in Exercise 20.

In Exercises 24–26, determine whether each function is even, odd, or neither. State each function's symmetry. If you are using a graphing utility, graph the function and verify its possible symmetry.

24. $f(x) = x^3 - 5x$

25. $f(x) = x^4 - 2x^2 + 1$

26. $f(x) = 2x\sqrt{1 - x^2}$

27. The graph shows the height, in meters, of an eagle in terms of its time, in seconds, in flight.

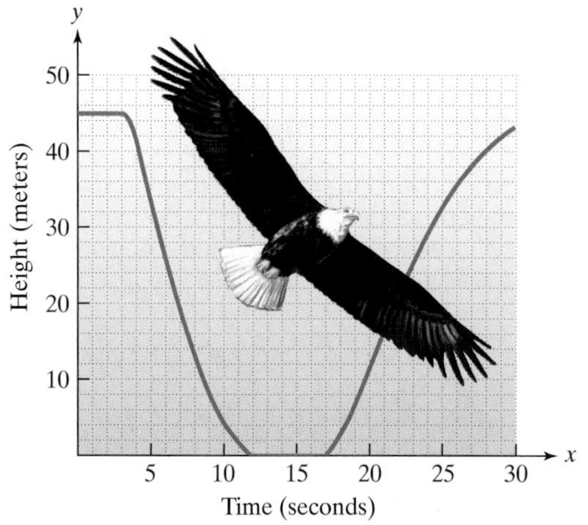

a. Is the eagle's height a function of time? Use the graph to explain why or why not.

b. On which interval is the function decreasing? Describe what this means in practical terms.

c. On which intervals is the function constant? What does this mean for each of these intervals?

d. On which interval is the function increasing? What does this mean?

28. A cargo service charges a flat fee of \$5 plus \$1.50 for each pound or fraction of a pound. Graph shipping cost, $C(x)$, in dollars, as a function of weight, x, in pounds, for $0 < x \leq 5$.

2.3 and 2.4

In Exercises 29–32, find the slope of the line passing through each pair of points or state that the slope is undefined. Then indicate whether the line through the points rises, falls, is horizontal, or is vertical.

29. $(3, 2)$ and $(5, 1)$

30. $(-1, -2)$ and $(-3, -4)$

31. $\left(-3, \frac{1}{4}\right)$ and $\left(6, \frac{1}{4}\right)$

32. $(-2, 5)$ and $(-2, 10)$

In Exercises 33–36, use the given conditions to write an equation for each line in point-slope form and slope-intercept form.

33. Passing through $(-3, 2)$ with slope -6

34. Passing through $(1, 6)$ and $(-1, 2)$

35. Passing through $(4, -7)$ and parallel to the line whose equation is $3x + y - 9 = 0$

36. Passing through $(-3, 6)$ and perpendicular to the line whose equation is $y = \frac{1}{3}x + 4$

37. Write the general form of the equation of the line passing through $(-12, -1)$ and perpendicular to the line whose equation is $6x - y - 4 = 0$.

In Exercises 38–41, give the slope and y-intercept of each line whose equation is given. Then graph the line.

38. $y = \frac{2}{5}x - 1$

39. $f(x) = -4x + 5$

40. $2x + 3y + 6 = 0$

41. $2y - 8 = 0$

42. Graph using intercepts: $2x - 5y - 10 = 0$.

43. Graph: $2x - 10 = 0$.

44. You can click a mouse and bet the house. The points in the graph show the dizzying growth of online gambling. With more than 1800 sites, the industry has become the Web's biggest moneymaker.

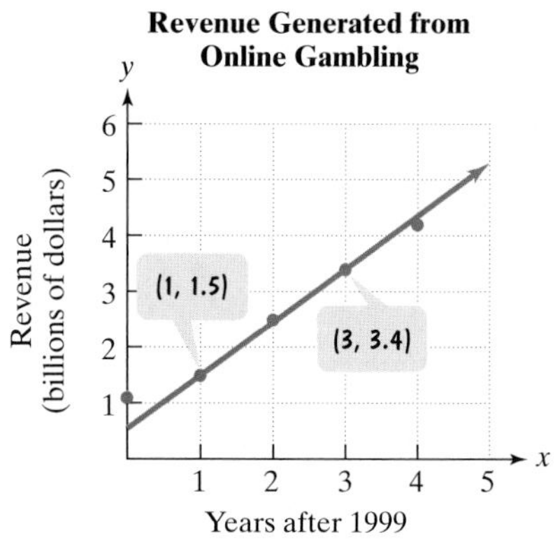

Source: Newsweek

a. Use the two points whose coordinates are shown by the voice balloons to find the point-slope form of the equation of the line that models revenue from online gambling, y, in billions of dollars, x years after 1999.

b. Write the equation in part (a) in slope-intercept form.

c. In 2003, nearly \$3.5 billion was lost on Internet bets, triggering a sharp backlash that threatened to shut down Internet wagering. If this crackdown on the industry is not successful, use your slope-intercept model to predict the billions of dollars in revenue from online gambling in 2009.

45. The graph shows new AIDS diagnoses among the general U.S. population, y, for year x, where $1999 \leq x \leq 2003$.

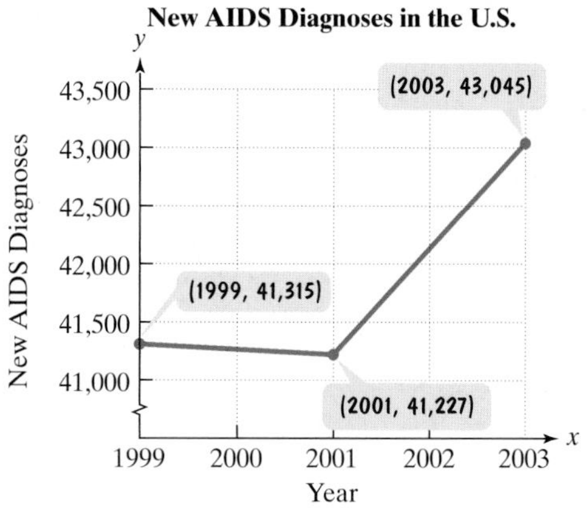

Source: Centers for Disease Control

a. Find the slope of the line passing through (1999, 41,315) and (2001, 41,227). Then express the slope as a rate of change with the proper units attached.

b. Find the slope of the line passing through (2001, 41,227) and (2003, 43,045). Then express the slope as a rate of change.

c. Draw a line passing through (1999, 41,315) and (2003, 43,045) and find its slope. Is the slope the average of the slopes of the lines that you found in parts (a) and (b)? Explain your answer.

46. Find the average rate of change of $f(x) = x^2 - 4x$ from $x_1 = 5$ to $x_2 = 9$.

2.5

In Exercises 47–51, use the graph of $y = f(x)$ to graph each function g.

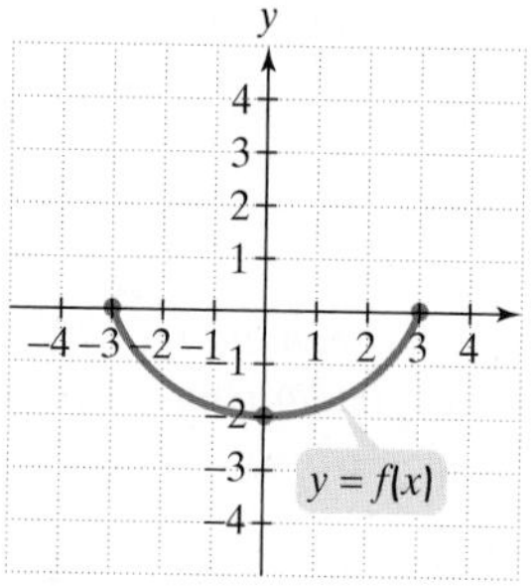

47. $g(x) = f(x + 2) + 3$
48. $g(x) = \frac{1}{2}f(x - 1)$
49. $g(x) = -f(2x)$
50. $g(x) = 2f\left(\frac{1}{2}x\right)$
51. $g(x) = -f(-x) - 1$

In Exercises 52–55, begin by graphing the standard quadratic function, $f(x) = x^2$. Then use transformations of this graph to graph the given function.

52. $g(x) = x^2 + 2$
53. $h(x) = (x + 2)^2$
54. $r(x) = -(x + 1)^2$
55. $y(x) = \frac{1}{2}(x - 1)^2 + 1$

In Exercises 56–58, begin by graphing the square root function, $f(x) = \sqrt{x}$. Then use transformations of this graph to graph the given function.

56. $g(x) = \sqrt{x + 3}$
57. $h(x) = \sqrt{3 - x}$
58. $r(x) = 2\sqrt{x + 2}$

In Exercises 59–61, begin by graphing the absolute value function, $f(x) = |x|$. Then use transformations of this graph to graph the given function.

59. $g(x) = |x + 2| - 3$
60. $h(x) = -|x - 1| + 1$
61. $r(x) = \frac{1}{2}|x + 2|$

In Exercises 62–64, begin by graphing the standard cubic function, $f(x) = x^3$. Then use transformations of this graph to graph the given function.

62. $g(x) = \frac{1}{2}(x - 1)^3$
63. $h(x) = -(x + 1)^3$
64. $r(x) = \frac{1}{4}x^3 - 1$

In Exercises 65–67, begin by graphing the cube root function, $f(x) = \sqrt[3]{x}$. Then use transformations of this graph to graph the given function.

65. $g(x) = \sqrt[3]{x + 2} - 1$
66. $h(x) = -\sqrt[3]{2x}$
67. $r(x) = -2\sqrt[3]{-x}$

2.6

In Exercises 68–73, find the domain of each function.

68. $f(x) = x^2 + 6x - 3$
69. $g(x) = \dfrac{4}{x - 7}$
70. $h(x) = \sqrt{8 - 2x}$
71. $f(x) = \dfrac{x}{x^2 + 4x - 21}$
72. $g(x) = \dfrac{\sqrt{x - 2}}{x - 5}$
73. $f(x) = \sqrt{x - 1} + \sqrt{x + 5}$

In Exercises 74–76, find $f + g$, $f - g$, fg, and $\frac{f}{g}$. Determine the domain for each function.

74. $f(x) = 3x - 1, \quad g(x) = x - 5$
75. $f(x) = x^2 + x + 1, \quad g(x) = x^2 - 1$
76. $f(x) = \sqrt{x + 7}, \quad g(x) = \sqrt{x - 2}$

*In Exercises 77–78, find **a.** $(f \circ g)(x)$; **b.** $(g \circ f)(x)$; **c.** $(f \circ g)(3)$.*

77. $f(x) = x^2 + 3, \quad g(x) = 4x - 1$
78. $f(x) = \sqrt{x}, \quad g(x) = x + 1$

*In Exercises 79–80, find **a.** $(f \circ g)(x)$; **b.** the domain of $(f \circ g)$.*

79. $f(x) = \dfrac{x + 1}{x - 2}, \quad g(x) = \dfrac{1}{x}$
80. $f(x) = \sqrt{x - 1}, \quad g(x) = x + 3$

In Exercises 81–82, express the given function h as a composition of two functions f and g so that $h(x) = (f \circ g)(x)$.

81. $h(x) = (x^2 + 2x - 1)^4$
82. $h(x) = \sqrt[3]{7x + 4}$

2.7

In Exercises 83–84, find $f(g(x))$ and $g(f(x))$ and determine whether each pair of functions f and g are inverses of each other.

83. $f(x) = \dfrac{3}{5}x + \dfrac{1}{2}$ and $g(x) = \dfrac{5}{3}x - 2$
84. $f(x) = 2 - 5x$ and $g(x) = \dfrac{2 - x}{5}$

The functions in Exercises 85–87 are all one-to-one. For each function,

a. *Find an equation for $f^{-1}(x)$, the inverse function.*

b. *Verify that your equation is correct by showing that $f(f^{-1}(x)) = x$ and $f^{-1}(f(x)) = x$.*

85. $f(x) = 4x - 3$
86. $f(x) = 8x^3 + 1$
87. $f(x) = \dfrac{2}{x} + 5$

Which graphs in Exercises 88–91 represent functions that have inverse functions?

88.

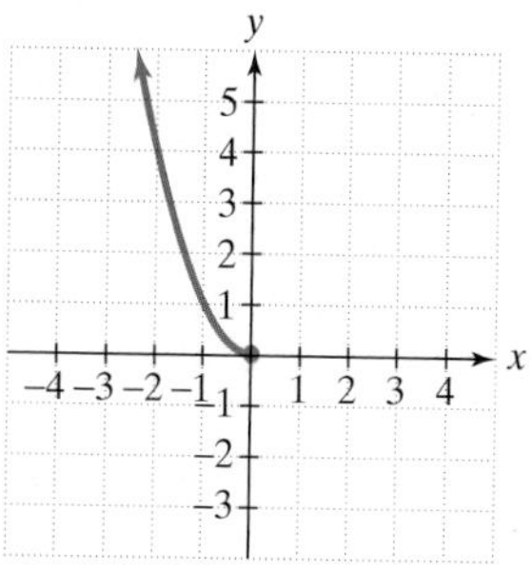

89.

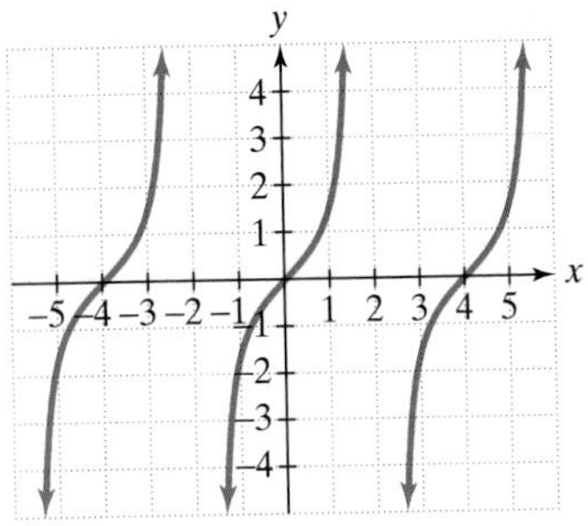

90.

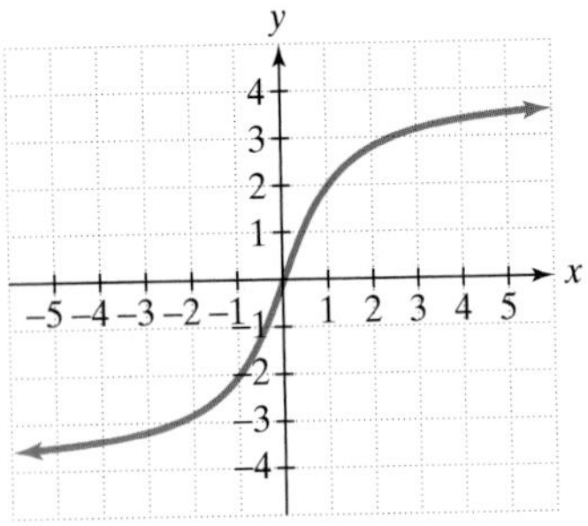

91.

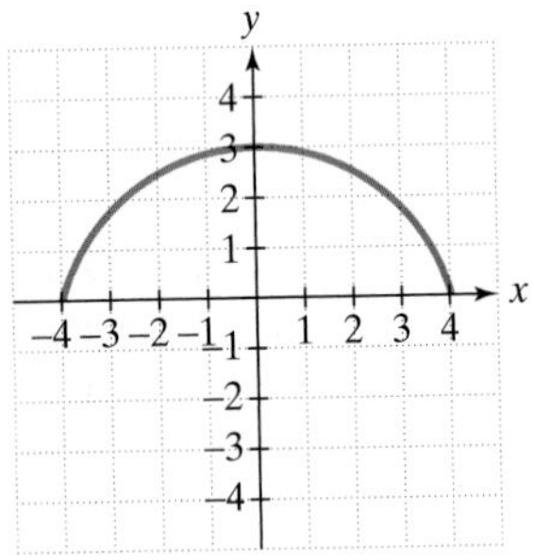

92. Use the graph of f in the figure shown to draw the graph of its inverse function.

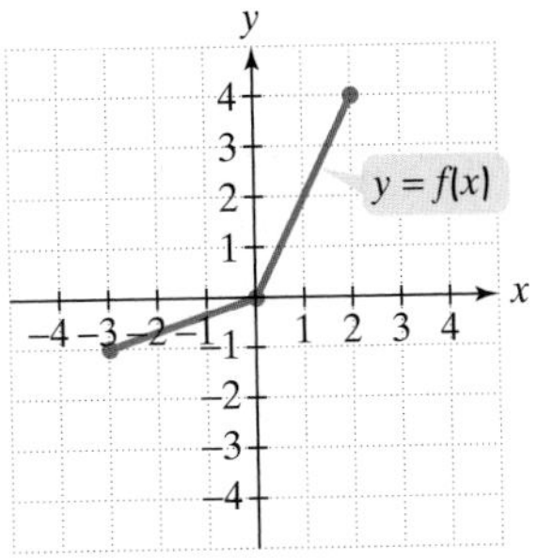

In Exercises 93–94, find an equation for $f^{-1}(x)$. Then graph f and f^{-1} in the same rectangular coordinate system.

93. $f(x) = 1 - x^2, x \geq 0$ **94.** $f(x) = \sqrt{x} + 1$

2.8

In Exercises 95–96, find the distance between each pair of points. If necessary, round answers to two decimal places.

95. $(-2, 3)$ and $(3, -9)$ **96.** $(-4, 3)$ and $(-2, 5)$

In Exercises 97–98, find the midpoint of each line segment with the given endpoints.

97. $(2, 6)$ and $(-12, 4)$ **98.** $(4, -6)$ and $(-15, 2)$

In Exercises 99–100, write the standard form of the equation of the circle with the given center and radius.

99. Center $(0, 0), r = 3$ **100.** Center $(-2, 4), r = 6$

In Exercises 101–103, give the center and radius of each circle and graph its equation. Use the graph to identify the relation's domain and range.

101. $x^2 + y^2 = 1$ **102.** $(x + 2)^2 + (y - 3)^2 = 9$

103. $x^2 + y^2 - 4x + 2y - 4 = 0$

Chapter 2 Test

1. List by letter all relations that are not functions.

a. $\{(7, 5), (8, 5), (9, 5)\}$

b. $\{(5, 7), (5, 8), (5, 9)\}$

c.

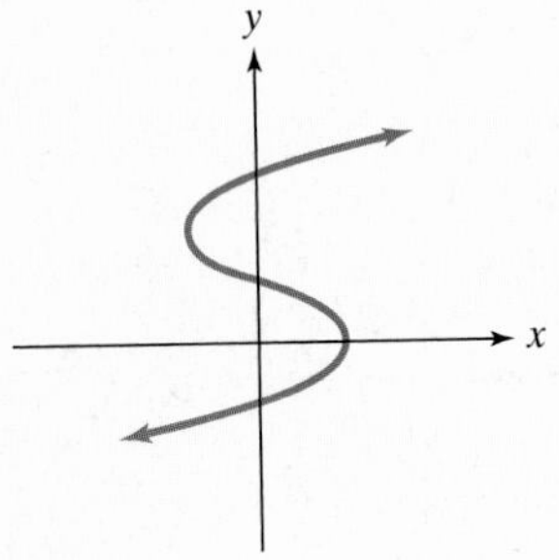

d. $x^2 + y^2 = 100$

e.

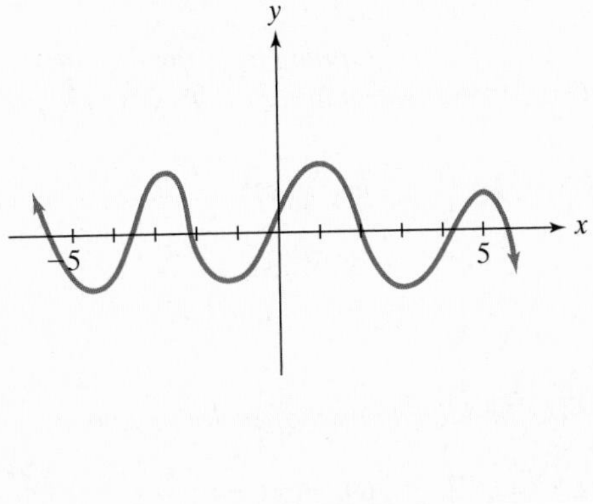

2. Use the graph of $y = f(x)$ to solve this exercise.

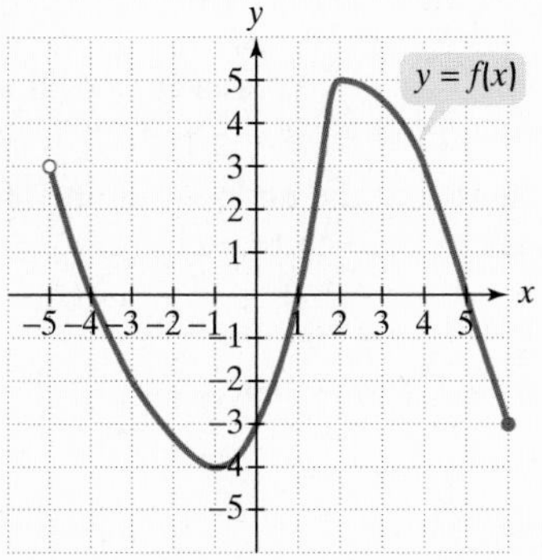

a. What is $f(4) - f(-3)$?
b. What is the domain of f?
c. What is the range of f?
d. On which interval or intervals is f increasing?
e. On which interval or intervals is f decreasing?
f. For what number does f have a relative maximum? What is the relative maximum?
g. For what number does f have a relative minimum? What is the relative minimum?
h. What are the x-intercepts?
i. What is the y-intercept?

3. Use the graph of $y = f(x)$ to solve this exercise.

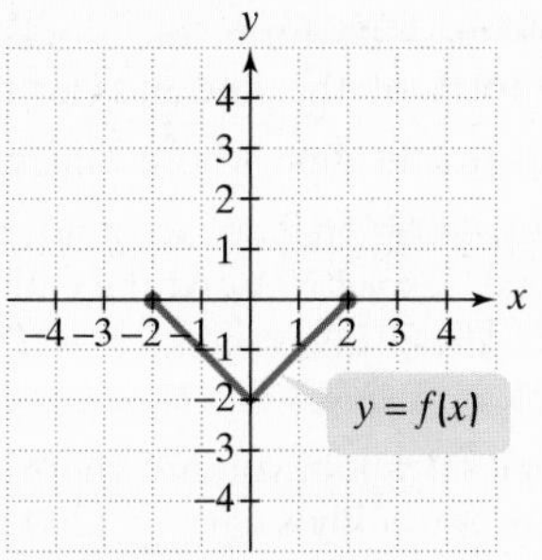

a. What are the zeros of f?
b. Find the value(s) of x for which $f(x) = -1$.
c. Find the value(s) of x for which $f(x) = -2$.
d. Is f even, odd, or neither?
e. Does f have an inverse function?
f. Is $f(0)$ a relative maximum, a relative minimum, or neither?
g. Graph $g(x) = f(x + 1) - 1$.
h. Graph $h(x) = \frac{1}{2}f\left(\frac{1}{2}x\right)$.
i. Graph $r(x) = -f(-x) + 1$.
j. Find the average rate of change of f from $x_1 = -2$ to $x_2 = 1$.

In Exercises 4–15, graph each equation in a rectangular coordinate system. If two functions are indicated, graph both in the same system. Then use your graphs to identify each relation's domain and range.

4. $x + y = 4$
5. $x^2 + y^2 = 4$
6. $f(x) = 4$
7. $f(x) = -\frac{1}{3}x + 2$
8. $(x + 2)^2 + (y - 1)^2 = 9$
9. $f(x) = \begin{cases} 2 & \text{if } x \le 0 \\ -1 & \text{if } x > 0 \end{cases}$
10. $x^2 + y^2 + 4x - 6y - 3 = 0$
11. $f(x) = |x|$ and $g(x) = \frac{1}{2}|x + 1| - 2$
12. $f(x) = x^2$ and $g(x) = -(x - 1)^2 + 4$
13. $f(x) = 2x - 4$ and f^{-1}
14. $f(x) = x^3 - 1$ and f^{-1}
15. $f(x) = x^2 - 1,\ x \ge 0$, and f^{-1}

In Exercises 16–23, let $f(x) = x^2 - x - 4$ and $g(x) = 2x - 6$.

16. Find $f(x - 1)$.
17. Find $\dfrac{f(x + h) - f(x)}{h}$.
18. Find $(g - f)(x)$.
19. Find $\left(\dfrac{f}{g}\right)(x)$ and its domain.
20. Find $(f \circ g)(x)$.
21. Find $(g \circ f)(x)$.
22. Find $g(f(-1))$.
23. Find $f(-x)$. Is f even, odd, or neither?

In Exercises 24–25, use the given conditions to write an equation for each line in point-slope form and slope-intercept form.

24. Passing through $(2, 1)$ and $(-1, -8)$
25. Passing through $(-4, 6)$ and perpendicular to the line whose equation is $y = -\frac{1}{4}x + 5$
26. Write the general form of the equation of the line passing through $(-7, -10)$ and parallel to the line whose equation is $4x + 2y - 5 = 0$.
27. When adjusted for inflation, the federal minimum wage from 1997 through 2003 continued to decrease. The points in the scatter plot show the minimum hourly inflation-adjusted wage for this period. Also shown is a line that passes through or near the points.

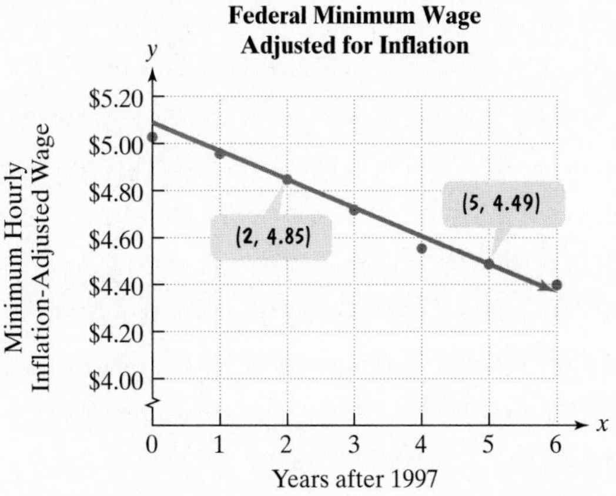

Source: www.dul.gov/esa/pudha/minwage

a. Use the two points whose coordinates are shown by the voice balloons to find the point-slope form of the equation of the line that models the minimum hourly inflation-adjusted wage, y, x years after 1997.
b. Write the equation in part (a) in slope-intercept form. Use function notation.
c. Use the linear function to predict the minimum hourly inflation-adjusted wage in 2007.

28. Find the average rate of change of $f(x) = 3x^2 - 5$ from $x_1 = 6$ to $x_2 = 10$.

29. If $g(x) = \begin{cases} \sqrt{x-3} & \text{if } x \geq 3 \\ 3 - x & \text{if } x < 3 \end{cases}$, find $g(-1)$ and $g(7)$.

In Exercises 30–31, find the domain of each function.

30. $f(x) = \dfrac{3}{x+5} + \dfrac{7}{x-1}$

31. $f(x) = 3\sqrt{x+5} + 7\sqrt{x-1}$

32. If $f(x) = \dfrac{7}{x-4}$ and $g(x) = \dfrac{2}{x}$, find $(f \circ g)(x)$ and the domain of $f \circ g$.

33. Express $h(x) = (2x+3)^7$ as a composition of two functions f and g so that $h(x) = (f \circ g)(x)$.

34. Find the length and the midpoint of the line segment whose endpoints are $(2, -2)$ and $(5, 2)$.

Cumulative Review Exercises (Chapters 1–2)

Use the graph of $y = f(x)$ to solve Exercises 1–5.

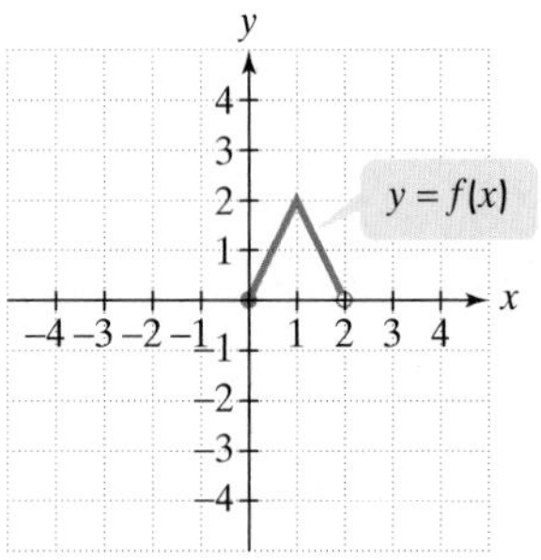

1. Find the domain and the range of f.

2. For what value(s) of x is $f(x) = 1$?

3. Find the relative maximum.

4. Graph $g(x) = f(x-1) + 1$.

5. Graph $h(x) = -2f\left(\frac{1}{2}x\right)$.

In Exercises 6–10, solve each equation or inequality.

6. $(x+3)(x-4) = 8$

7. $3(4x - 1) = 4 - 6(x - 3)$

8. $\sqrt{x} + 2 = x$

9. $x^{\frac{2}{3}} - x^{\frac{1}{3}} - 6 = 0$

10. $\dfrac{x}{2} - 3 \leq \dfrac{x}{4} + 2$

In Exercises 11–14, graph each equation in a rectangular coordinate system. If two functions are indicated, graph both in the same system. Then use your graphs to identify each relation's domain and range.

11. $3x - 6y - 12 = 0$

12. $(x-2)^2 + (y+1)^2 = 4$

13. $f(x) = \sqrt[3]{x}$ and $g(x) = \sqrt[3]{x-3} + 4$

14. $f(x) = \sqrt{x-3} + 2$ and f^{-1}

In Exercises 15–16, let $f(x) = 4 - x^2$ and $g(x) = x + 5$.

15. Find $\dfrac{f(x+h) - f(x)}{h}$ and simplify.

16. Find all values of x satisfying $(f \circ g)(x) = 0$.

17. Write the point-slope form, the slope-intercept form, and the general form of the line passing through $(-2, 5)$ and perpendicular to the line whose equation is $y = -\frac{1}{4}x + \frac{1}{3}$.

18. You invested \$6000 in two accounts paying 7% and 9% annual interest, respectively. At the end of the year, the total interest from these investments was \$510. How much was invested at each rate?

19. For a summer sales job, you are choosing between two pay arrangements: a weekly salary of \$200 plus 5% commission on sales, or a straight 15% commission. For how many dollars of sales will the earnings be the same regardless of the pay arrangement?

20. The length of a rectangular garden is 2 feet more than twice its width. If 22 feet of fencing is needed to enclose the garden, what are its dimensions?

Polynomial and Rational Functions

CHAPTER 3

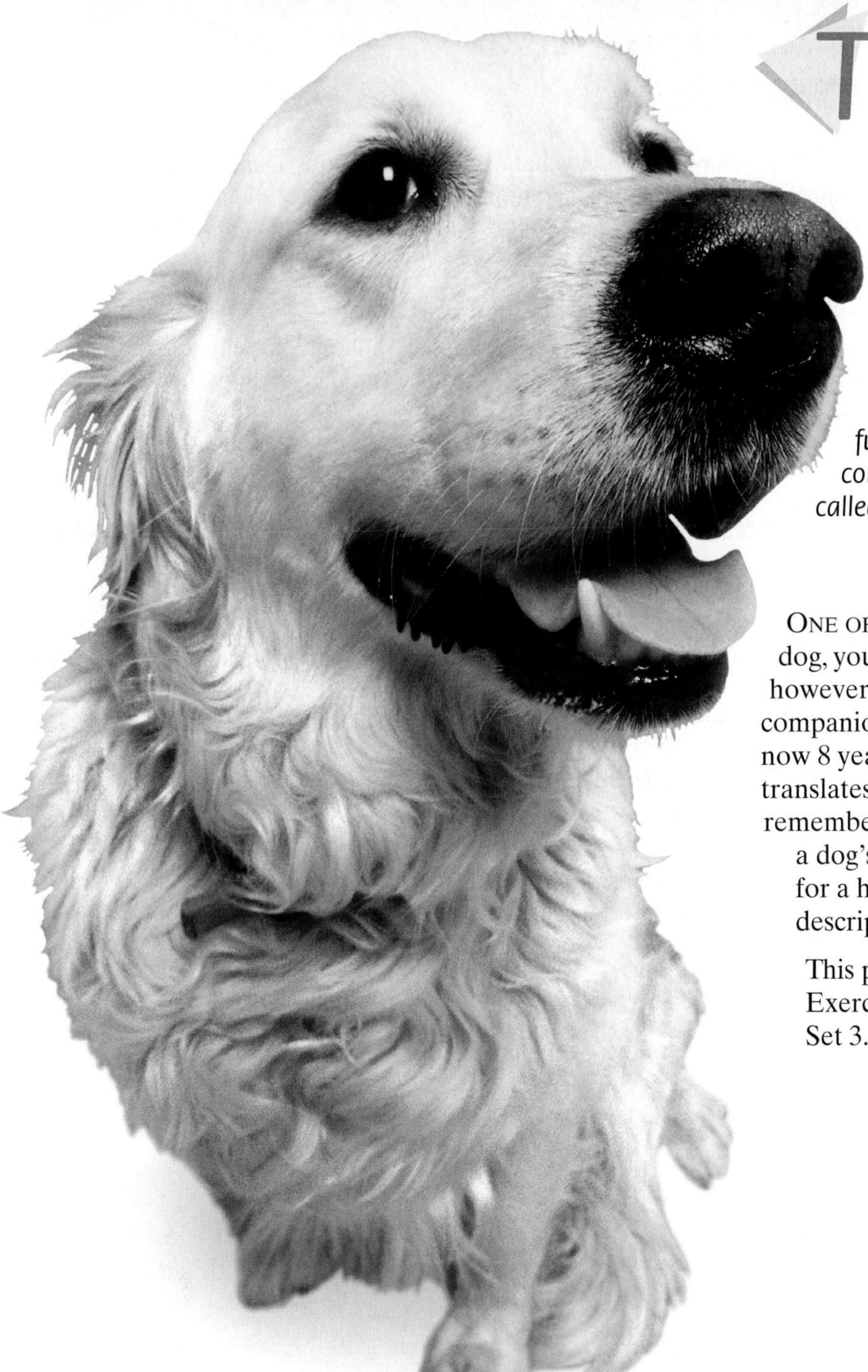

THERE IS A FUNCTION THAT models the age in human years, $H(x)$, of a dog that is x years old:

$$H(x) = -0.001618x^4 + 0.077326x^3 - 1.2367x^2 + 11.460x + 2.914.$$

The function contains variables to powers that are whole numbers and is an example of a ***polynomial function****. In this chapter, we study polynomial functions and functions that consist of quotients of polynomials, called* ***rational functions****.*

ONE OF THE JOYS OF YOUR LIFE IS YOUR dog, your very special buddy. Lately, however, you've noticed that your companion is slowing down a bit. He's now 8 years old and you wonder how this translates into human years. You remember something about every year of a dog's life being equal to seven years for a human. Is there a more accurate description?

This problem appears as Exercises 63–64 in Exercise Set 3.4.

SECTION 3.1 Quadratic Functions

Objectives

1. Recognize characteristics of parabolas.
2. Graph parabolas.
3. Determine a quadratic function's minimum or maximum value.
4. Solve problems involving a quadratic function's minimum or maximum value.

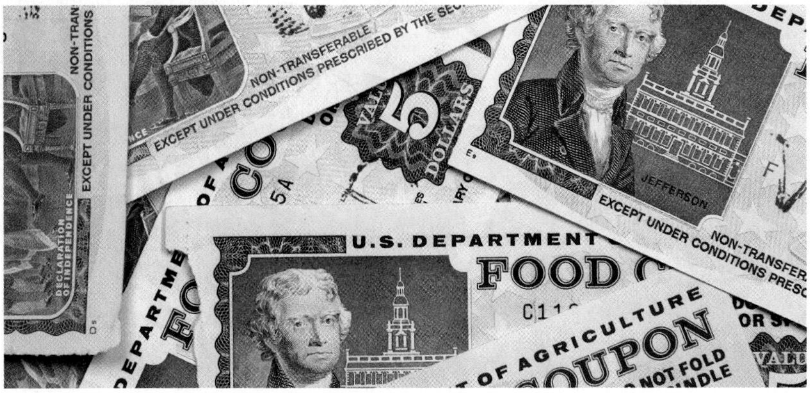

The Food Stamp Program is the first line of defense against hunger for millions of American families. The program provides benefits for eligible participants to purchase approved food items at approved food stores. Over half of all participants are children; one out of six is a low-income older adult. The function

$$f(x) = 0.22x^2 - 0.50x + 7.68$$

models the number of households, $f(x)$, in millions, participating in the program x years after 1999. For example, to find the number of households receiving food stamps in 2005, substitute 6 for x because 2005 is 6 years after 1999:

$$f(6) = 0.22(6)^2 - 0.50(6) + 7.68 = 12.6.$$

Thus, in 2005, 12.6 million households received food stamps.

The function $f(x) = 0.22x^2 - 0.50x + 7.68$ is an example of a quadratic function. We have seen that a **quadratic function** is any function of the form

$$f(x) = ax^2 + bx + c,$$

where a, b, and c are real numbers, with $a \neq 0$. A quadratic function is a polynomial function whose greatest exponent is 2. In this section, we study quadratic functions and their graphs.

1 Recognize characteristics of parabolas.

Graphs of Quadratic Functions

The graph of any quadratic function is called a **parabola**. Parabolas are shaped like cups, as shown in Figure 3.1. If the coefficient of x^2 (the value of a in $ax^2 + bx + c$) is positive, the parabola opens upward. If the coefficient of x^2 is negative, the graph opens downward. The **vertex** (or turning point) of the parabola is the lowest point on the graph when it opens upward and the highest point on the graph when it opens downward.

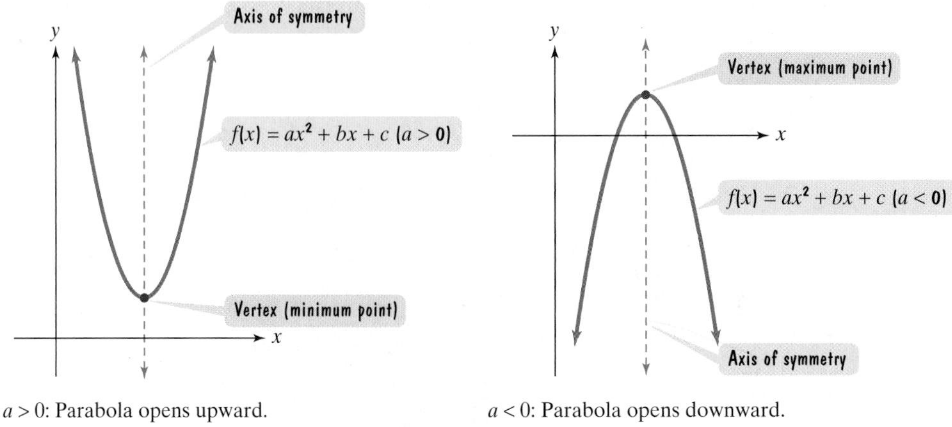

Figure 3.1 Characteristics of graphs of quadratic functions

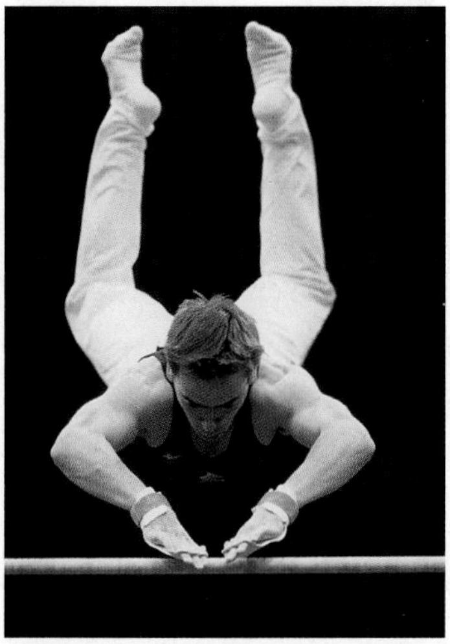

Look at the unusual image of the word "mirror" shown below. The artist, Scott Kim, has created the image so that the two halves of the whole are mirror images of each other. A parabola shares this kind of symmetry, in which a line through the vertex divides the figure in half. Parabolas are symmetric with respect to this line, called the **axis of symmetry**. If a parabola is folded along its axis of symmetry, the two halves match exactly.

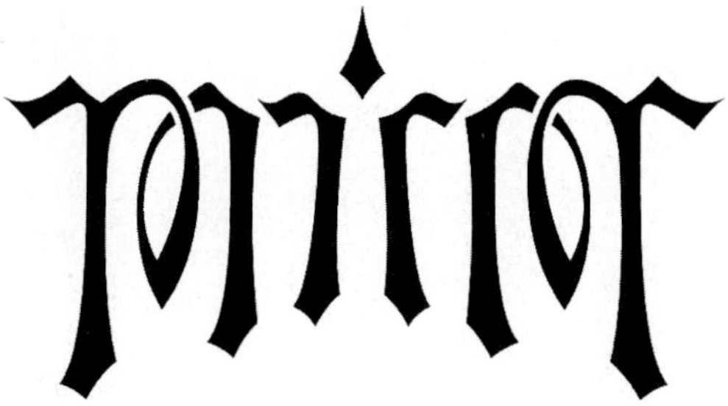

Graphing Quadratic Functions in Standard Form

2 Graph parabolas.

In our earlier work with transformations, we applied a series of transformations to the graph of $f(x) = x^2$. The graph of this function is a parabola. The vertex for this parabola is $(0, 0)$. In Figure 3.2(a), the graph of $f(x) = ax^2$ for $a > 0$ is shown in black; it opens *upward*. In Figure 3.2(b), the graph of $f(x) = ax^2$ for $a < 0$ is shown in black; it opens *downward*.

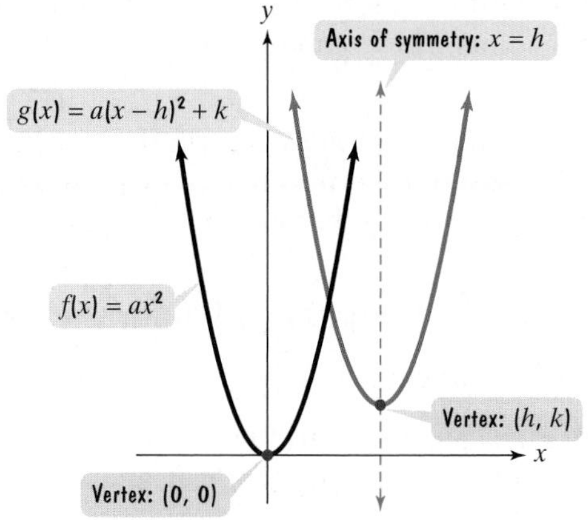

Figure 3.2(a)
$a > 0$: Parabola opens upward.

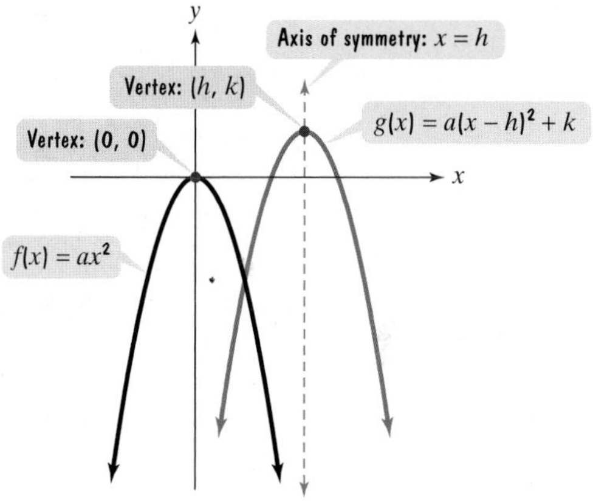

Figure 3.2(b)
$a < 0$: Parabola opens downward.

Transformations of $f(x) = ax^2$

Figure 3.2(a) and 3.2(b) also show the graph of $g(x) = a(x - h)^2 + k$ in blue. Compare these graphs to those of $f(x) = ax^2$. Observe that h determines the horizontal shift and k determines the vertical shift of the graph of $f(x) = ax^2$:

$$g(x) = a(x - h)^2 + k.$$

If $h > 0$, the graph of $f(x) = ax^2$ is shifted h units to the right.

If $k > 0$, the graph of $y = a(x - h)^2$ is shifted k units up.

Consequently, the vertex $(0, 0)$ on the black graph of $f(x) = ax^2$ moves to the point (h, k) on the blue graph of $g(x) = a(x - h)^2 + k$. The axis of symmetry is the vertical line whose equation is $x = h$.

The form of the expression for g is convenient because it immediately identifies the vertex of the parabola as (h, k). This is the **standard form** of a quadratic function.

The Standard Form of a Quadratic Function

The quadratic function

$$f(x) = a(x - h)^2 + k, \qquad a \neq 0$$

is in **standard form**. The graph of f is a parabola whose vertex is the point (h, k). The parabola is symmetric with respect to the line $x = h$. If $a > 0$, the parabola opens upward; if $a < 0$, the parabola opens downward.

The sign of a in $f(x) = a(x - h)^2 + k$ determines whether the parabola opens upward or downward. Furthermore, if $|a|$ is small, the parabola opens more flatly than if $|a|$ is large. Here is a general procedure for graphing parabolas whose equations are in standard form:

Graphing Quadratic Functions with Equations in Standard Form

To graph $f(x) = a(x - h)^2 + k$,

1. Determine whether the parabola opens upward or downward. If $a > 0$, it opens upward. If $a < 0$, it opens downward.
2. Determine the vertex of the parabola. The vertex is (h, k).
3. Find any x-intercepts by solving $f(x) = 0$. The function's real zeros are the x-intercepts.
4. Find the y-intercept by computing $f(0)$.
5. Plot the intercepts, the vertex, and additional points as necessary. Connect these points with a smooth curve that is shaped like a cup.

In the graphs that follow, we will show each axis of symmetry as a dashed vertical line. Because this vertical line passes through the vertex, (h, k), its equation is $x = h$. The line is dashed because it is not part of the parabola.

EXAMPLE 1 Graphing a Quadratic Function in Standard Form

Graph the quadratic function $f(x) = -2(x - 3)^2 + 8$.

Solution We can graph this function by following the steps in the preceding box. We begin by identifying values for a, h, and k.

Standard form: $f(x) = a(x - h)^2 + k$

$a = -2$ $\quad h = 3$ $\quad k = 8$

Given function: $f(x) = -2(x - 3)^2 + 8$

Step 1 Determine how the parabola opens. Note that a, the coefficient of x^2, is -2. Thus, $a < 0$; this negative value tells us that the parabola opens downward.

Step 2 Find the vertex. The vertex of the parabola is at (h, k). Because $h = 3$ and $k = 8$, the parabola has its vertex at $(3, 8)$.

Step 3 Find the x-intercepts by solving $f(x) = 0$. Replace $f(x)$ with 0 in $f(x) = -2(x - 3)^2 + 8$.

$0 = -2(x - 3)^2 + 8$ — Find x-intercepts, setting f(x) equal to 0.

$2(x - 3)^2 = 8$ — Solve for x. Add $2(x - 3)^2$ to both sides of the equation.

$(x - 3)^2 = 4$ — Divide both sides by 2.

$x - 3 = \sqrt{4}$ or $x - 3 = -\sqrt{4}$ — Apply the square root property.

$x - 3 = 2$ $\qquad x - 3 = -2$ — $\sqrt{4} = 2$

$x = 5$ $\qquad x = 1$ — Add 3 to both sides in each equation.

The x-intercepts are 1 and 5. The parabola passes through $(1, 0)$. and $(5, 0)$.

Vertex: (3, 8)

x-intercept: 1   x-intercept: 5

y-intercept: −10   Axis of symmetry: x = 3

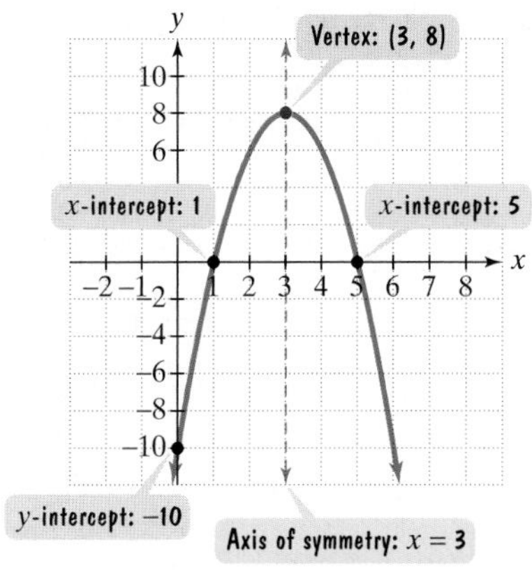

Figure 3.3 The graph of
$f(x) = -2(x - 3)^2 + 8$

Step 4   Find the y-intercept by computing f(0). Replace x with 0 in $f(x) = -2(x - 3)^2 + 8$.

$$f(0) = -2(0 - 3)^2 + 8 = -2(-3)^2 + 8 = -2(9) + 8 = -10$$

The y-intercept is -10. The parabola passes through $(0, -10)$.

Step 5   Graph the parabola. With a vertex at $(3, 8)$, x-intercepts at 1 and 5, and a y-intercept at -10, the graph of f is shown in Figure 3.3. The axis of symmetry is the vertical line whose equation is $x = 3$.

Check Point 1   Graph the quadratic function $f(x) = -(x - 1)^2 + 4$.

EXAMPLE 2   Graphing a Quadratic Function in Standard Form

Graph the quadratic function $f(x) = (x + 3)^2 + 1$.

Solution   We begin by finding values for a, h, and k.

$$f(x) = a(x - h)^2 + k \quad \text{Standard form of quadratic function}$$
$$f(x) = (x + 3)^2 + 1 \quad \text{Given function}$$
$$f(x) = 1(x - (-3))^2 + 1$$

$a = 1 \qquad h = -3 \qquad k = 1$

Step 1   Determine how the parabola opens. Note that a, the coefficient of x^2, is 1. Thus, $a > 0$; this positive value tells us that the parabola opens upward.

Step 2   Find the vertex. The vertex of the parabola is at (h, k). Because $h = -3$ and $k = 1$, the parabola has its vertex at $(-3, 1)$.

Step 3   Find the x-intercepts by solving f(x) = 0. Replace $f(x)$ with 0 in $f(x) = (x + 3)^2 + 1$. Because the vertex is at $(-3, 1)$, which lies above the x-axis, and the parabola opens upward, it appears that this parabola has no x-intercepts. We can verify this observation algebraically.

$$0 = (x + 3)^2 + 1 \qquad \text{Find possible x-intercepts, setting } f(x) \text{ equal to 0.}$$

$$-1 = (x + 3)^2 \qquad \text{Solve for x. Subtract 1 from both sides.}$$

$$x + 3 = \sqrt{-1} \quad \text{or} \quad x + 3 = -\sqrt{-1} \qquad \text{Apply the square root property.}$$
$$x + 3 = i \qquad\qquad x + 3 = -i \qquad \sqrt{-1} = i$$
$$x = -3 + i \qquad\qquad x = -3 - i \qquad \text{The solutions are } -3 \pm i.$$

Because this equation has no real solutions, the parabola has no x-intercepts.

Step 4   Find the y-intercept by computing f(0). Replace x with 0 in $f(x) = (x + 3)^2 + 1$.

$$f(0) = (0 + 3)^2 + 1 = 3^2 + 1 = 9 + 1 = 10$$

The y-intercept is 10. The parabola passes through $(0, 10)$.

Step 5   Graph the parabola. With a vertex at $(-3, 1)$, no x-intercepts, and a y-intercept at 10, the graph of f is shown in Figure 3.4. The axis of symmetry is the vertical line whose equation is $x = -3$.

Axis of symmetry: x = −3

y-intercept: 10

Vertex: (−3, 1)

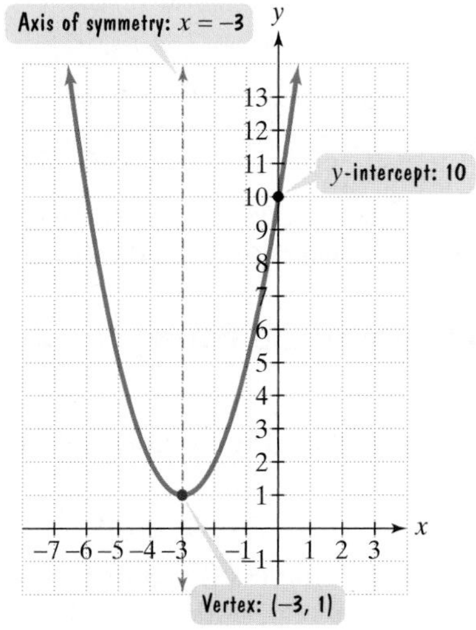

Figure 3.4 The graph of
$f(x) = (x + 3)^2 + 1$

Check Point 2   Graph the quadratic function $f(x) = (x - 2)^2 + 1$.

Graphing Quadratic Functions in the Form $f(x) = ax^2 + bx + c$

Quadratic functions are frequently expressed in the form $f(x) = ax^2 + bx + c$. How can we identify the vertex of a parabola whose equation is in this form? Completing the square provides the answer to this question.

$$f(x) = ax^2 + bx + c$$

$$= a\left(x^2 + \frac{b}{a}x\right) + c \qquad \text{Factor out } a \text{ from } ax^2 + bx.$$

$$= a\left(x^2 + \frac{b}{a}x + \frac{b^2}{4a^2}\right) + c - a\left(\frac{b^2}{4a^2}\right)$$

Complete the square by adding the square of half the coefficient of x.

By completing the square, we added $a \cdot \frac{b^2}{4a^2}$. To avoid changing the function's equation, we must subtract this term.

$$= a\left(x + \frac{b}{2a}\right)^2 + c - \frac{b^2}{4a} \qquad \text{Write the trinomial as the square of a binomial and simplify the constant term.}$$

Compare this form of the equation with a quadratic function's **standard form**.

Standard form $\quad f(x) = a(x - h)^2 + k$

$h = -\frac{b}{2a} \qquad k = c - \frac{b^2}{4a}$

Equation under discussion $\quad f(x) = a\left(x - \left(-\frac{b}{2a}\right)\right)^2 + c - \frac{b^2}{4a}$

The important part of this observation is that h, the x-coordinate of the vertex, is $-\frac{b}{2a}$. The y-coordinate can be found by evaluating the function at $-\frac{b}{2a}$.

The Vertex of a Parabola Whose Equation Is $f(x) = ax^2 + bx + c$

Consider the parabola defined by the quadratic function $f(x) = ax^2 + bx + c$. The parabola's vertex is $\left(-\frac{b}{2a}, f\left(-\frac{b}{2a}\right)\right)$.

We can apply our five-step procedure and graph parabolas in the form $f(x) = ax^2 + bx + c$. The only step that is different is how we determine the vertex.

EXAMPLE 3 Graphing a Quadratic Function in the Form $f(x) = ax^2 + bx + c$

Graph the quadratic function $f(x) = -x^2 - 2x + 1$. Use the graph to identify the function's domain and its range.

Solution

Step 1 Determine how the parabola opens. Note that a, the coefficient of x^2, is -1. Thus, $a < 0$; this negative value tells us that the parabola opens downward.

Step 2 Find the vertex. We know that the x-coordinate of the vertex is $x = -\frac{b}{2a}$. We identify a, b, and c in $f(x) = ax^2 + bx + c$.

$$f(x) = -x^2 - 2x + 1$$

$a = -1 \qquad b = -2 \qquad c = 1$

Substitute the values of a and b into the equation for the x-coordinate:

$$x = -\frac{b}{2a} = -\frac{-2}{2(-1)} = -\left(\frac{-2}{-2}\right) = -1.$$

The x-coordinate of the vertex is -1 and the vertex is at $(-1, f(-1))$. We substitute -1 for x in the equation of the function, $f(x) = -x^2 - 2x + 1$, to find the y-coordinate:

$$f(-1) = -(-1)^2 - 2(-1) + 1 = -1 + 2 + 1 = 2.$$

The vertex is at $(-1, 2)$.

Step 3   Find the x-intercepts by solving $f(x) = 0$. Replace $f(x)$ with 0 in $f(x) = -x^2 - 2x + 1$. We obtain $0 = -x^2 - 2x + 1$. This equation cannot be solved by factoring. We will use the quadratic formula to solve it.

$$-x^2 - 2x + 1 = 0$$

$$a = -1 \qquad b = -2 \qquad c = 1$$

$$x = \frac{-b \pm \sqrt{b^2 - 4ac}}{2a} = \frac{-(-2) \pm \sqrt{(-2)^2 - 4(-1)(1)}}{2(-1)} = \frac{2 \pm \sqrt{4 - (-4)}}{-2}$$

To locate the x-intercepts, we need decimal approximations. Thus, there is no need to simplify the radical form of the solutions.

$$x = \frac{2 + \sqrt{8}}{-2} \approx -2.4 \qquad \text{or} \qquad x = \frac{2 - \sqrt{8}}{-2} \approx 0.4$$

The x-intercepts are approximately -2.4 and 0.4. The parabola passes through $(-2.4, 0)$ and $(0.4, 0)$.

Step 4   Find the y-intercept by computing $f(0)$. Replace x with 0 in $f(x) = -x^2 - 2x + 1$.

$$f(0) = -0^2 - 2 \cdot 0 + 1 = 1$$

The y-intercept is 1. The parabola passes through $(0, 1)$.

Step 5   Graph the parabola. With a vertex at $(-1, 2)$, x-intercepts at approximately -2.4 and 0.4, and a y-intercept at 1, the graph of f is shown in Figure 3.5(a). The axis of symmetry is the vertical line whose equation is $x = -1$.

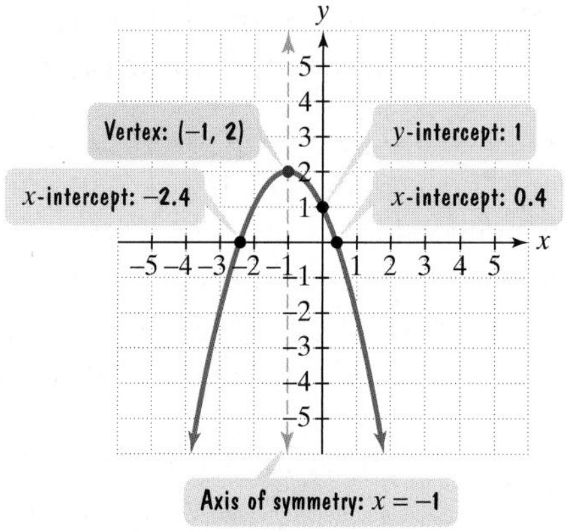

Figure 3.5(a) The graph of $f(x) = -x^2 - 2x + 1$

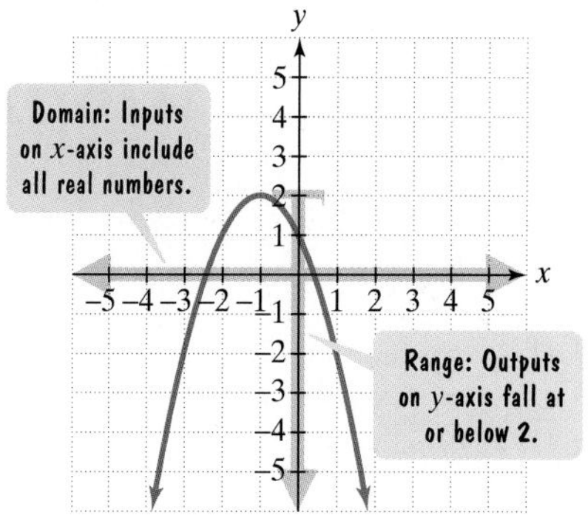

Figure 3.5(b) Determining the domain and range of $f(x) = -x^2 - 2x + 1$

Now we are ready to determine the domain and range of $f(x) = -x^2 - 2x + 1$. We can use the parabola, shown again in Figure 3.5(b), to do so. To find the domain, look for all the inputs on the x-axis that correspond to points on the graph. As the graph widens and continues to fall at both ends, can you see that these inputs include all real numbers?

Domain of f is $\{x | x \text{ is a real number}\}$ or $(-\infty, \infty)$.

To find the range, look for all the outputs on the y-axis that correspond to points on the graph. Figure 3.5(b) shows that the parabola's vertex, $(-1, 2)$, is the highest point on the graph. Because the y-coordinate of the vertex is 2, outputs on the y-axis fall at or below 2.

$$\text{Range of } f \text{ is } \{y \mid y \leq 2\} \text{ or } (-\infty, 2].$$

Check Point 3 Graph the quadratic function $f(x) = -x^2 + 4x + 1$. Use the graph to identify the function's domain and its range.

3 Determine a quadratic function's minimum or maximum value.

Minimum and Maximum Values of Quadratic Functions

Consider the quadratic function $f(x) = ax^2 + bx + c$. If $a > 0$, the parabola opens upward and the vertex is its lowest point. If $a < 0$, the parabola opens downward and the vertex is its highest point. The x-coordinate of the vertex is $-\frac{b}{2a}$. Thus, we can find the minimum or maximum value of f by evaluating the quadratic function at $x = -\frac{b}{2a}$.

Minimum and Maximum: Quadratic Functions

Consider the quadratic function $f(x) = ax^2 + bx + c$.

1. If $a > 0$, then f has a minimum that occurs at $x = -\frac{b}{2a}$. This minimum value is $f\left(-\frac{b}{2a}\right)$.
2. If $a < 0$, then f has a maximum that occurs at $x = -\frac{b}{2a}$. This maximum value is $f\left(-\frac{b}{2a}\right)$.

In each case, the value of x gives the location of the minimum or maximum value. The value of y, or $f\left(-\frac{b}{2a}\right)$, gives that minimum or maximum value.

EXAMPLE 4 Obtaining Information about a Quadratic Function from Its Equation

Consider the quadratic function $f(x) = -3x^2 + 6x - 13$.

a. Determine, without graphing, whether the function has a minimum value or a maximum value.

b. Find the minimum or maximum value and determine where it occurs.

c. Identify the function's domain and its range.

Solution We begin by identifying a, b, and c in the function's equation:

$$f(x) = -3x^2 + 6x - 13.$$

$a = -3$ $b = 6$ $c = -13$

a. Because $a < 0$, the function has a maximum value.

b. The maximum value occurs at

$$x = -\frac{b}{2a} = -\frac{6}{2(-3)} = -\frac{6}{-6} = -(-1) = 1.$$

The maximum value occurs at $x = 1$ and the maximum value of $f(x) = -3x^2 + 6x - 13$ is

$$f(1) = -3 \cdot 1^2 + 6 \cdot 1 - 13 = -3 + 6 - 13 = -10.$$

We see that the maximum is -10 at $x = 1$.

c. Like all quadratic functions, the domain is $(-\infty, \infty)$. Because the function's maximum value is -10, the range includes all real numbers at or below -10. The range is $(-\infty, -10]$.

We can use the graph of $f(x) = -3x^2 + 6x - 13$ to visualize the results of Example 4. Figure 3.6 shows the graph in a $[-6, 6, 1]$ by $[-50, 20, 10]$ viewing rectangle. The maximum function feature verifies that the function's maximum is -10 at $x = 1$. Notice that x gives the location of the maximum and y gives the maximum value. Notice, too, that the maximum value is -10 and not the ordered pair $(1, -10)$.

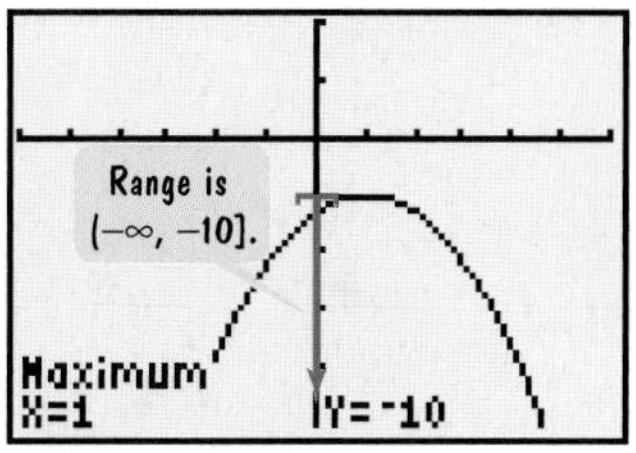

[−6, 6, 1] by [−50, 20, 10]

Figure 3.6

Check Point 4 Repeat parts (a) through (c) of Example 4 using the quadratic function $f(x) = 4x^2 - 16x + 1000$.

4 Solve problems involving a quadratic function's minimum or maximum value.

Applications of Quadratic Functions

When did the minimum number of households participate in the food stamp program? What is the age of a driver having the least number of car accidents? If you throw a baseball vertically upward, after how many seconds will it reach its maximum height and what is that height? The answers to these questions involve finding the maximum or minimum value of a quadratic function, as well as where this value occurs.

EXAMPLE 5 The Food Stamp Program

Figure 3.7 shows the number of U.S. households, in millions, participating in the Food Stamp Program from 1999 through 2004. The function

$$f(x) = 0.22x^2 - 0.50x + 7.68$$

models the number of households, $f(x)$, in millions, participating in the program x years after 1999. According to this function, in which year was the number of participants at a minimum? How many households received food stamps for that year? How well does this model the data shown in Figure 3.7?

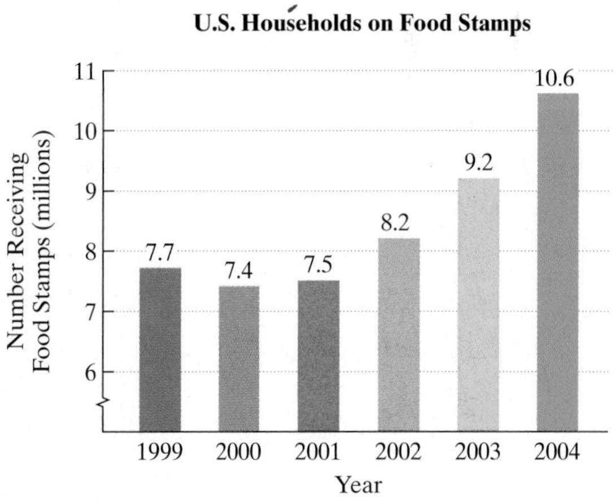

Figure 3.7

Source: Food Stamp Program

Solution We begin by identifying a, b, and c in the function's equation:

$$f(x) = 0.22x^2 - 0.50x + 7.68.$$

$a = 0.22$ $b = -0.50$ $c = 7.68$

Because $a > 0$, the function has a minimum value. The minimum value occurs at

$$x = -\frac{b}{2a} = -\frac{(-0.50)}{2(0.22)} = \frac{0.50}{0.44} \approx 1.$$

This means that the number of households receiving food stamps was at a minimum approximately 1 year after 1999, in 2000. Using the model $f(x) = 0.22x^2 - 0.50x + 7.68$, the number of households, in millions, for that year was

$$f(1) = 0.22(1)^2 - 0.50(1) + 7.68 = 7.4.$$

In 2000, the number of households receiving food stamps was at a minimum of 7.4 million. Because this is precisely what is shown in Figure 3.7 on the previous page, the function models the data extremely well.

Technology

Because of the decreasing-increasing cuplike shape of the data in Figure 3.7, a quadratic function is an appropriate model. We entered the data using

(number of years after 1999, millions of participants).

Upon entering the QUADratic REGression program, we obtain the results shown in the screen on the right. Thus, the quadratic function of best fit is

$$f(x) = 0.22x^2 - 0.50x + 7.68,$$

where x represents the number of years after 1999 and $f(x)$ represents the number of U.S. households, in millions, on food stamps.

Data:
(0, 7.7), (1,7.4), (2, 7.5), (3, 8.2), (4, 9.2), (5, 10.6)

```
QuadReg
 y=ax²+bx+c
 a=.2160714286
 b=-.4917857143
 c=7.682142857
```

Check Point 5 The function $f(x) = 0.4x^2 - 36x + 1000$ models the number of accidents, $f(x)$, per 50 million miles driven, for drivers x years old, where $16 \le x \le 74$. What is the age of a driver having the least number of car accidents? What is the minimum number of car accidents per 50 million miles driven?

Quadratic functions can also be modeled from verbal conditions. Once we have obtained a quadratic function, we can then use the x-coordinate of the vertex to determine its maximum or minimum value. Here is a step-by-step strategy for solving these kinds of problems:

Strategy for Solving Problems Involving Maximizing or Minimizing Quadratic Functions

1. Read the problem carefully and decide which quantity is to be maximized or minimized.
2. Use the conditions of the problem to express the quantity as a function in one variable.
3. Rewrite the function in the form $f(x) = ax^2 + bx + c$.
4. Calculate $-\frac{b}{2a}$. If $a > 0$, f has a minimum at $x = -\frac{b}{2a}$. This minimum value is $f\left(-\frac{b}{2a}\right)$. If $a < 0$, f has a maximum at $x = -\frac{b}{2a}$. This maximum value is $f\left(-\frac{b}{2a}\right)$.
5. Answer the question posed in the problem.

EXAMPLE 6 Minimizing a Product

Among all pairs of numbers whose difference is 10, find a pair whose product is as small as possible. What is the minimum product?

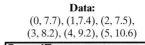

Solution

Step 1   Decide what must be maximized or minimized. We must minimize the product of two numbers. Calling the numbers x and y, and calling the product P, we must minimize

$$P = xy.$$

Step 2   Express this quantity as a function in one variable. In the formula $P = xy$, P is expressed in terms of two variables, x and y. However, because the difference of the numbers is 10, we can write

$$x - y = 10.$$

We can solve this equation for y in terms of x (or vice versa), substitute the result into $P = xy$, and obtain P as a function of one variable.

$$-y = -x + 10 \qquad \text{Subtract x from both sides of } x - y = 10.$$

$$y = x - 10 \qquad \text{Multiply both sides of the equation by } -1 \text{ and solve for y.}$$

Now we substitute $x - 10$ for y in $P = xy$.

$$P = xy = x(x - 10).$$

Because P is now a function of x, we can write

$$P(x) = x(x - 10).$$

Step 3   Write the function in the form $f(x) = ax^2 + bx + c$. We apply the distributive property to obtain

$$P(x) = x(x - 10) = x^2 - 10x.$$

$$\boxed{a = 1} \quad \boxed{b = -10}$$

Step 4   Calculate $-\dfrac{b}{2a}$. If $a > 0$, the function has a minimum at this value. The voice balloons show that $a = 1$ and $b = -10$.

$$x = -\frac{b}{2a} = -\frac{-10}{2(1)} = -(-5) = 5$$

This means that the product, P, of two numbers whose difference is 10 is a minimum when one of the numbers, x, is 5.

Step 5   Answer the question posed by the problem. The problem asks for the two numbers and the minimum product. We found that one of the numbers, x, is 5. Now we must find the second number, y.

$$y = x - 10 = 5 - 10 = -5$$

The number pair whose difference is 10 and whose product is as small as possible is $5, -5$. The minimum product is $5(-5)$, or -25.

Check Point 6 Among all pairs of numbers whose difference is 8, find a pair whose product is as small as possible. What is the minimum product?

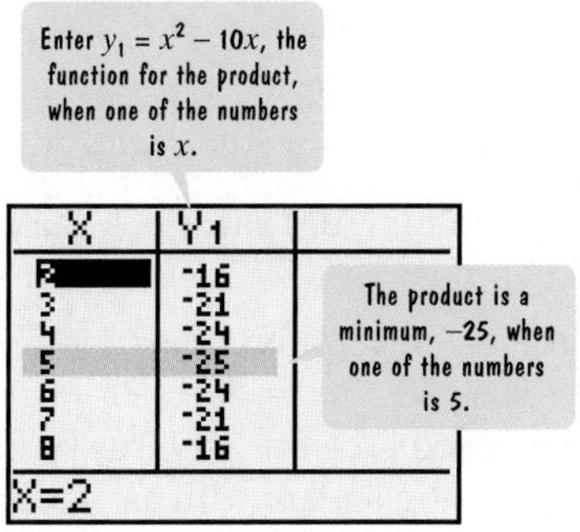

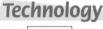

EXAMPLE 7 Maximizing Area

You have 100 yards of fencing to enclose a rectangular region. Find the dimensions of the rectangle that maximize the enclosed area. What is the maximum area?

Solution

Step 1 Decide what must be maximized or minimized. We must maximize area. What we do not know are the rectangle's dimensions, x and y.

Step 2 Express this quantity as a function in one variable. Because we must maximize area, we have $A = xy$. We need to transform this into a function in which A is represented by one variable. Because you have 100 yards of fencing, the perimeter of the rectangle is 100 yards. This means that

$$2x + 2y = 100.$$

We can solve this equation for y in terms of x, substitute the result into $A = xy$, and obtain A as a function in one variable. We begin by solving for y.

$$2y = 100 - 2x \quad \text{Subtract } 2x \text{ from both sides.}$$

$$y = \frac{100 - 2x}{2} \quad \text{Divide both sides by 2.}$$

$$y = 50 - x \quad \text{Divide each term in the numerator by 2.}$$

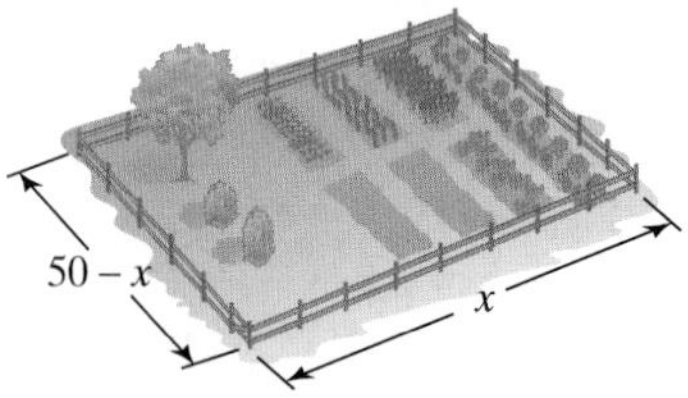

Figure 3.8 What value of x will maximize the rectangle's area?

Now we substitute $50 - x$ for y in $A = xy$.

$$A = xy = x(50 - x)$$

The rectangle and its dimensions are illustrated in Figure 3.8. Because A is now a function of x, we can write

$$A(x) = x(50 - x).$$

This function models the area, $A(x)$, of any rectangle whose perimeter is 100 yards in terms of one of its dimensions, x.

Step 3 Write the function in the form $f(x) = ax^2 + bx + c$. We apply the distributive property to obtain

$$A(x) = x(50 - x) = 50x - x^2 = -x^2 + 50x.$$

$a = -1$ $b = 50$

Step 4 Calculate $-\frac{b}{2a}$. If $a < 0$, the function has a maximum at this value. The voice balloons show that $a = -1$ and $b = 50$.

$$x = -\frac{b}{2a} = -\frac{50}{2(-1)} = 25$$

This means that the area, $A(x)$, of a rectangle with perimeter 100 yards is a maximum when one of the rectangle's dimensions, x, is 25 yards.

Step 5 Answer the question posed by the problem. We found that $x = 25$. Figure 3.8 shows that the rectangle's other dimension is $50 - x = 50 - 25 = 25$. The dimensions of the rectangle that maximize the enclosed area are 25 yards by 25 yards. The rectangle that gives the maximum area is actually a square with an area of 25 yards $\cdot$ 25 yards, or 625 square yards.

Technology

The graph of the area function

$$A(x) = x(50 - x)$$

was obtained with a graphing utility using a [0, 50, 2] by [0, 700, 25] viewing rectangle. The maximum function feature verifies that a maximum area of 625 square yards occurs when one of the dimensions is 25 yards.

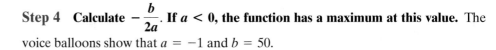

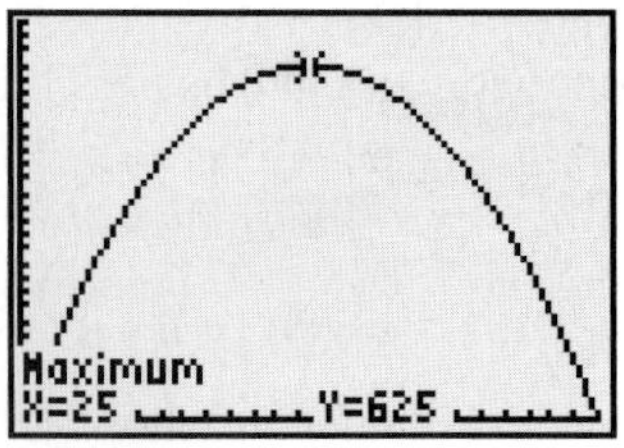

Check Point 7 You have 120 feet of fencing to enclose a rectangular region. Find the dimensions of the rectangle that maximize the enclosed area. What is the maximum area?

EXERCISE SET 3.1

Practice Exercises

In Exercises 1–4, the graph of a quadratic function is given. Write the function's equation, selecting from the following options.

$f(x) = (x + 1)^2 - 1$ ✓

$g(x) = (x + 1)^2 + 1$ ✓

$h(x) = (x - 1)^2 + 1$ ✓

$j(x) = (x - 1)^2 - 1$

1.
2.
3.
4.
In Exercises 5–8, the graph of a quadratic function is given. Write the function's equation, selecting from the following options.

$f(x) = x^2 + 2x + 1$ ✓

$g(x) = x^2 - 2x + 1$

$h(x) = x^2 - 1$ ✓

$j(x) = -x^2 - 1$ ✓

5.
6.
7.
8.
In Exercises 9–16, find the coordinates of the vertex for the parabola defined by the given quadratic function.

9. $f(x) = 2(x - 3)^2 + 1$

10. $f(x) = -3(x - 2)^2 + 12$

11. $f(x) = -2(x + 1)^2 + 5$

12. $f(x) = -2(x + 4)^2 - 8$

13. $f(x) = 2x^2 - 8x + 3$

14. $f(x) = 3x^2 - 12x + 1$

15. $f(x) = -x^2 - 2x + 8$

16. $f(x) = -2x^2 + 8x - 1$

In Exercises 17–38, use the vertex and intercepts to sketch the graph of each quadratic function. Give the equation of the parabola's axis of symmetry. Use the graph to determine the function's domain and range.

17. $f(x) = (x - 4)^2 - 1$

18. $f(x) = (x - 1)^2 - 2$

19. $f(x) = (x - 1)^2 + 2$

20. $f(x) = (x - 3)^2 + 2$

21. $y - 1 = (x - 3)^2$

22. $y - 3 = (x - 1)^2$

23. $f(x) = 2(x + 2)^2 - 1$

24. $f(x) = \frac{5}{4} - \left(x - \frac{1}{2}\right)^2$

25. $f(x) = 4 - (x - 1)^2$

26. $f(x) = 1 - (x - 3)^2$

27. $f(x) = x^2 - 2x - 3$

28. $f(x) = x^2 - 2x - 15$

29. $f(x) = x^2 + 3x - 10$

30. $f(x) = 2x^2 - 7x - 4$

31. $f(x) = 2x - x^2 + 3$

32. $f(x) = 5 - 4x - x^2$

33. $f(x) = x^2 + 6x + 3$

34. $f(x) = x^2 + 4x - 1$

35. $f(x) = 2x^2 + 4x - 3$

36. $f(x) = 3x^2 - 2x - 4$

37. $f(x) = 2x - x^2 - 2$

38. $f(x) = 6 - 4x + x^2$

In Exercises 39–44, an equation of a quadratic function is given.

  a. Determine, without graphing, whether the function has a minimum value or a maximum value.

  b. Find the minimum or maximum value and determine where it occurs.

  c. Identify the function's domain and its range.

39. $f(x) = 3x^2 - 12x - 1$

40. $f(x) = 2x^2 - 8x - 3$

41. $f(x) = -4x^2 + 8x - 3$

42. $f(x) = -2x^2 - 12x + 3$

43. $f(x) = 5x^2 - 5x$

44. $f(x) = 6x^2 - 6x$

Practice Plus

In Exercises 45–48, give the domain and the range of each quadratic function whose graph is described.

45. The vertex is $(-1, -2)$ and the parabola opens up.

46. The vertex is $(-3, -4)$ and the parabola opens down.

47. Maximum $= -6$ at $x = 10$

48. Minimum $= 18$ at $x = -6$

In Exercises 49–52, write an equation in standard form of the parabola that has the same shape as the graph of $f(x) = 2x^2$, but with the given point as the vertex.

49. $(5, 3)$

50. $(7, 4)$

51. $(-10, -5)$

52. $(-8, -6)$

In Exercises 53–56, write an equation in standard form of the parabola that has the same shape as the graph of $f(x) = 3x^2$ or $g(x) = -3x^2$, but with the given maximum or minimum.

53. Maximum $= 4$ at $x = -2$

54. Maximum $= -7$ at $x = 5$

55. Minimum $= 0$ at $x = 11$

56. Minimum $= 0$ at $x = 9$

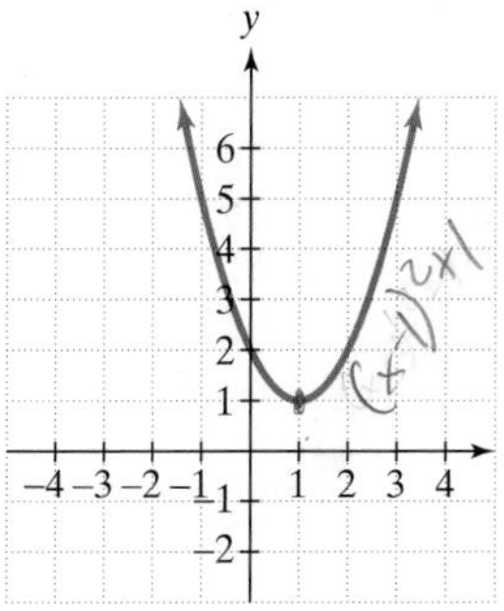

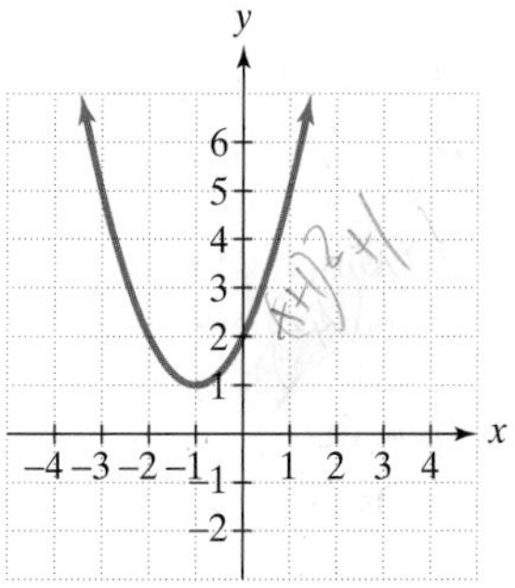

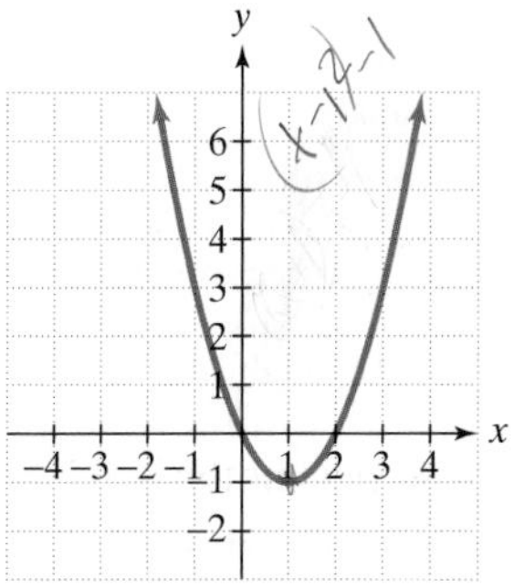

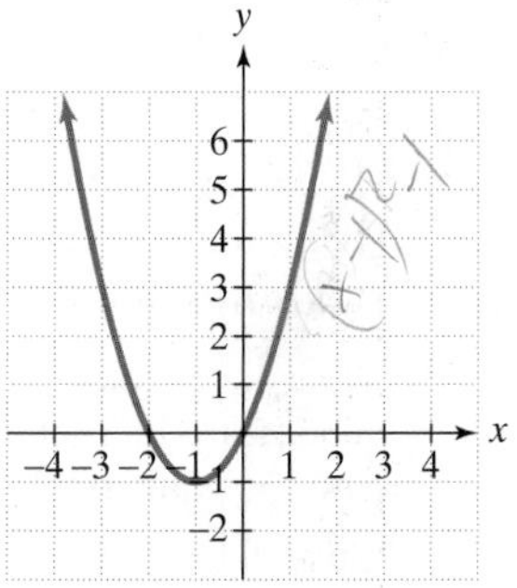

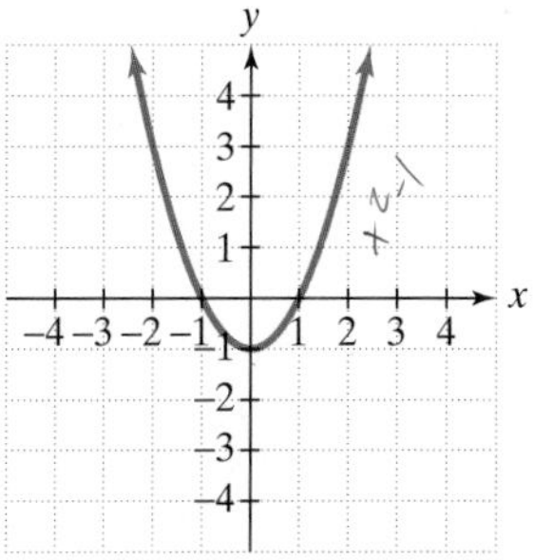

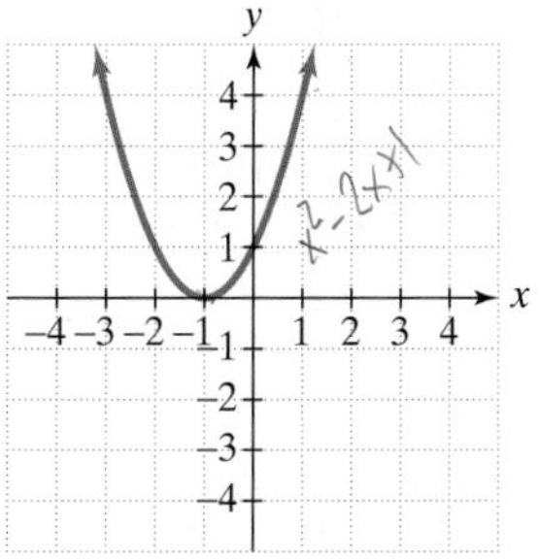

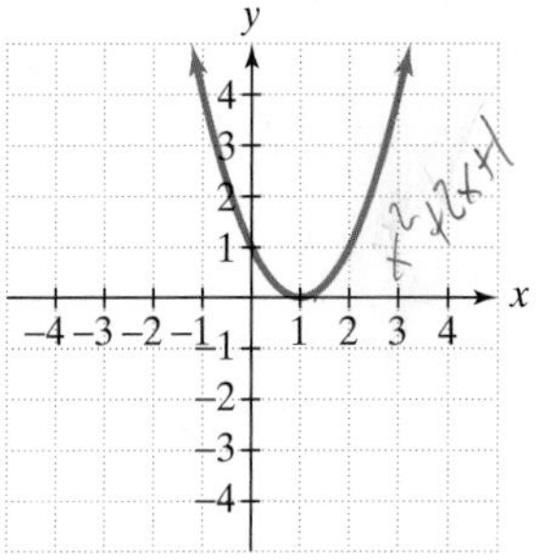

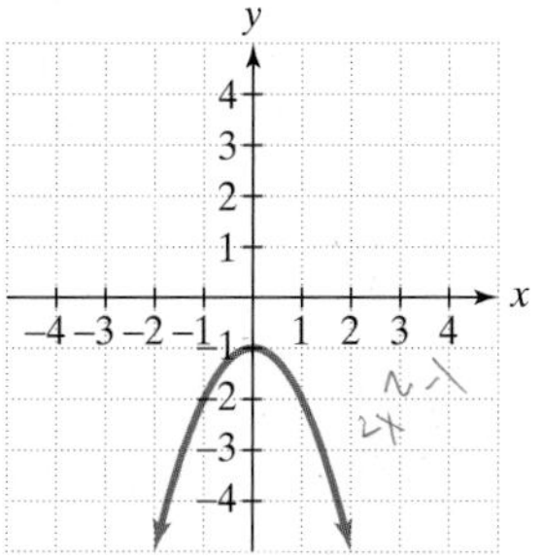

Application Exercises

57. The graph shows per capita U.S. adult wine consumption (in gallons per person) for selected years from 1980 through 2003. The function

$$f(x) = 0.005x^2 - 0.104x + 2.626$$

models U.S. wine consumption, $f(x)$, in gallons per person, x years after 1980. According to this function, in which year was wine consumption at a minimum? Round to the nearest year. What does the function give for per capita consumption, to the nearest tenth of a gallon, for that year? How well does this model the data shown in the graph?

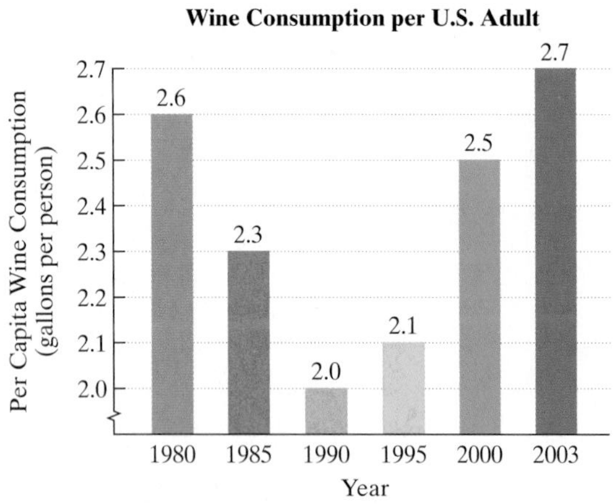

Source: Adams Business Media

58. After declining in the late 1990s, the number of gang-related murders across the United States has increased in recent years. The graph shows the number of gang-related homicides in the United States. The function

$$f(x) = 33x^2 - 255x + 1230$$

models the number of gang-related homicides across the nation, $f(x)$, x years after 1995. According to this function, in which year was the number of homicides at a minimum? Round to the nearest year. What does the function give for the number of gang-related murders for that year? How well does this model the data shown in the graph?

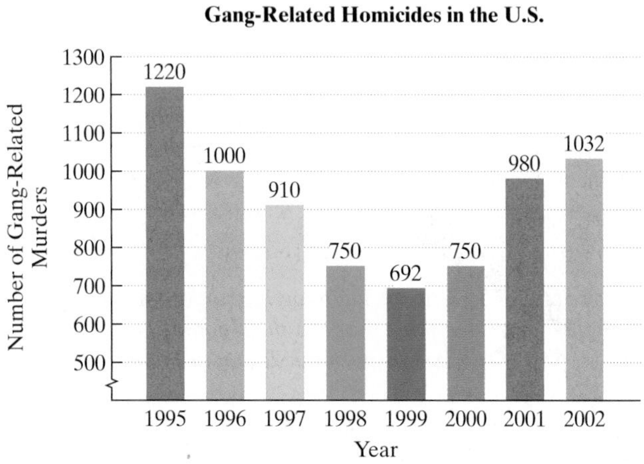

Source: Professor James Alan Fox, Northeastern University

59. A person standing close to the edge on the top of a 200-foot building throws a baseball vertically upward. The quadratic function

$$s(t) = -16t^2 + 64t + 200$$

models the ball's height above the ground, $s(t)$, in feet, t seconds after it was thrown.

a. After how many seconds does the ball reach its maximum height? What is the maximum height?

b. How many seconds does it take until the ball finally hits the ground? Round to the nearest tenth of a second.

c. Find $s(0)$ and describe what this means.

d. Use your results from parts (a) through (c) to graph the quadratic function. Begin the graph with $t = 0$ and end with the value of t for which the ball hits the ground.

60. A person standing close to the edge on the top of a 160-foot building throws a baseball vertically upward. The quadratic function

$$s(t) = -16t^2 + 64t + 160$$

models the ball's height above the ground, $s(t)$, in feet, t seconds after it was thrown.

a. After how many seconds does the ball reach its maximum height? What is the maximum height?

b. How many seconds does it take until the ball finally hits the ground? Round to the nearest tenth of a second.

c. Find $s(0)$ and describe what this means.

d. Use your results from parts (a) through (c) to graph the quadratic function. Begin the graph with $t = 0$ and end with the value of t for which the ball hits the ground.

61. Among all pairs of numbers whose sum is 16, find a pair whose product is as large as possible. What is the maximum product?

62. Among all pairs of numbers whose sum is 20, find a pair whose product is as large as possible. What is the maximum product?

63. Among all pairs of numbers whose difference is 16, find a pair whose product is as small as possible. What is the minimum product?

64. Among all pairs of numbers whose difference is 24, find a pair whose product is as small as possible. What is the minimum product?

65. You have 600 feet of fencing to enclose a rectangular plot that borders on a river. If you do not fence the side along the river, find the length and width of the plot that will maximize the area. What is the largest area that can be enclosed?

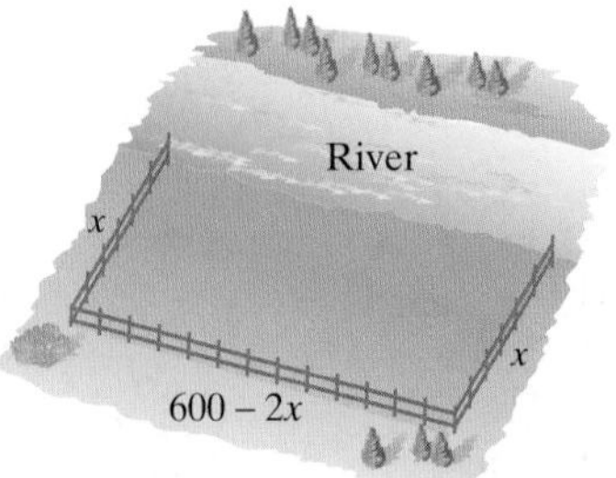

66. You have 200 feet of fencing to enclose a rectangular plot that borders on a river. If you do not fence the side along the river, find the length and width of the plot that will maximize the area. What is the largest area that can be enclosed?

$200 - 2x$

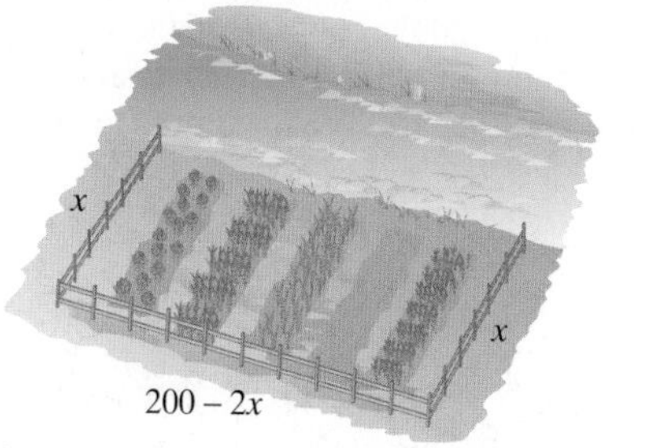

67. You have 50 yards of fencing to enclose a rectangular region. Find the dimensions of the rectangle that maximize the enclosed area. What is the maximum area?

68. You have 80 yards of fencing to enclose a rectangular region. Find the dimensions of the rectangle that maximize the enclosed area. What is the maximum area?

69. A rectangular playground is to be fenced off and divided in two by another fence parallel to one side of the playground. Six hundred feet of fencing is used. Find the dimensions of the playground that maximize the total enclosed area. What is the maximum area?

70. A rectangular playground is to be fenced off and divided in two by another fence parallel to one side of the playground. Four hundred feet of fencing is used. Find the dimensions of the playground that maximize the total enclosed area. What is the maximum area?

71. A rain gutter is made from sheets of aluminum that are 20 inches wide by turning up the edges to form right angles. Determine the depth of the gutter that will maximize its cross-sectional area and allow the greatest amount of water to flow. What is the maximum cross-sectional area?

Flat sheet
20 inches
wide

| x | $20 - 2x$ | x |

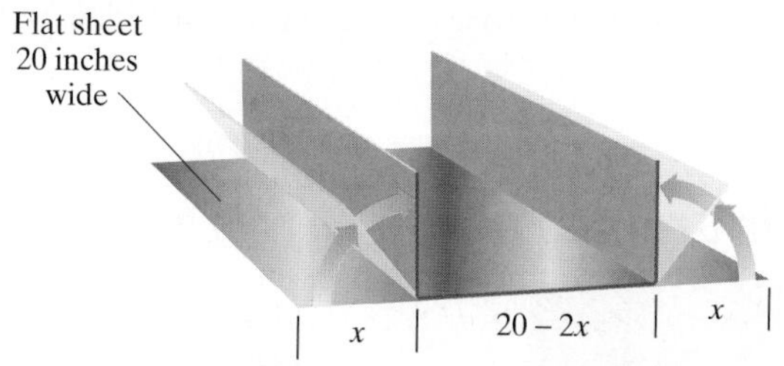

72. A rain gutter is made from sheets of aluminum that are 12 inches wide by turning up the edges to form right angles. Determine the depth of the gutter that will maximize its cross-sectional area and allow the greatest amount of water to flow. What is the maximum cross-sectional area?

73. Hunky Beef, a local sandwich store, has a fixed weekly cost of $525.00, and variable costs for making a roast beef sandwich are $0.55.

a. Let x represent the number of roast beef sandwiches made and sold each week. Write the weekly cost function, C, for Hunky Beef. (*Hint:* The cost function is the sum of fixed and variable costs.)

b. The function $R(x) = -0.001x^2 + 3x$ describes the money, in dollars, that Hunky Beef takes in each week from the sale of x roast beef sandwiches. Use this revenue function and the cost function from part (a) to write the store's weekly profit function, P. (*Hint:* The profit function is the difference between revenue and cost functions.)

c. Use the store's profit function to determine the number of roast beef sandwiches it should make and sell each week to maximize profit. What is the maximum weekly profit?

Writing in Mathematics

74. What is a quadratic function?

75. What is a parabola? Describe its shape.

76. Explain how to decide whether a parabola opens upward or downward.

77. Describe how to find a parabola's vertex if its equation is expressed in standard form. Give an example.

78. Describe how to find a parabola's vertex if its equation is in the form $f(x) = ax^2 + bx + c$. Use $f(x) = x^2 - 6x + 8$ as an example.

79. A parabola that opens upward has its vertex at (1, 2). Describe as much as you can about the parabola based on this information. Include in your discussion the number of x-intercepts (if any) for the parabola.

Technology Exercises

80. Use a graphing utility to verify any five of your hand-drawn graphs in Exercises 17–38.

81. **a.** Use a graphing utility to graph $y = 2x^2 - 82x + 720$ in a standard viewing rectangle. What do you observe?

b. Find the coordinates of the vertex for the given quadratic function.

c. The answer to part (b) is $(20.5, -120.5)$. Because the leading coefficient, 2, of the given function is positive, the vertex is a minimum point on the graph. Use this fact to help find a viewing rectangle that will give a relatively complete picture of the parabola. With an axis of symmetry at $x = 20.5$, the setting for x should extend past this, so try Xmin = 0 and Xmax = 30. The setting for y should include (and probably go below) the y-coordinate of the graph's minimum y-value, so try Ymin = -130. Experiment with Ymax until your utility shows the parabola's major features.

d. In general, explain how knowing the coordinates of a parabola's vertex can help determine a reasonable viewing rectangle on a graphing utility for obtaining a complete picture of the parabola.

In Exercises 82–85, find the vertex for each parabola. Then determine a reasonable viewing rectangle on your graphing utility and use it to graph the quadratic function.

82. $y = -0.25x^2 + 40x$ **83.** $y = -4x^2 + 20x + 160$

84. $y = 5x^2 + 40x + 600$ **85.** $y = 0.01x^2 + 0.6x + 100$

86. The following data show fuel efficiency, in miles per gallon, for all U.S. automobiles in the indicated year.

x (Years after 1940)	y (Average Number of Miles per Gallon for U.S. Automobiles)
1940: 0	14.8
1950: 10	13.9
1960: 20	13.4
1970: 30	13.5
1980: 40	15.9
1990: 50	20.2
2000: 60	22.0

Source: U.S. Department of Transportation

a. Use a graphing utility to draw a scatter plot of the data. Explain why a quadratic function is appropriate for modeling these data.
b. Use the quadratic regression feature to find the quadratic function that best fits the data.
c. Use the model in part (b) to determine the worst year for automobile fuel efficiency. What was the average number of miles per gallon for that year?
d. Use a graphing utility to draw a scatter plot of the data and graph the quadratic function of best fit on the scatter plot.

Critical Thinking Exercises

87. Which one of the following is true?

a. No quadratic functions have a range of $(-\infty, \infty)$.
b. The vertex of the parabola described by $f(x) = 2(x - 5)^2 - 1$ is at $(5, 1)$.
c. The graph of $f(x) = -2(x + 4)^2 - 8$ has one y-intercept and two x-intercepts.
d. The maximum value of y for the quadratic function $f(x) = -x^2 + x + 1$ is 1.

In Exercises 88–89, find the axis of symmetry for each parabola whose equation is given. Use the axis of symmetry to find a second point on the parabola whose y-coordinate is the same as the given point.

88. $f(x) = 3(x + 2)^2 - 5$; $(-1, -2)$

89. $f(x) = (x - 3)^2 + 2$; $(6, 11)$

In Exercises 90–91, write the equation of each parabola in standard form.

90. Vertex: $(-3, -4)$; The graph passes through the point $(1, 4)$.

91. Vertex: $(-3, -1)$; The graph passes through the point $(-2, -3)$.

92. A rancher has 1000 feet of fencing to construct six corrals, as shown in the figure. Find the dimensions that maximize the enclosed area. What is the maximum area?

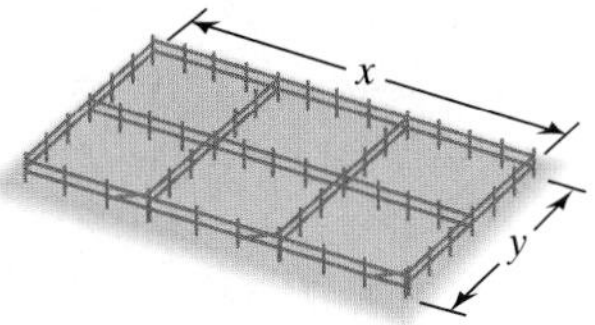

93. The annual yield per lemon tree is fairly constant at 320 pounds when the number of trees per acre is 50 or fewer. For each additional tree over 50, the annual yield per tree for all trees on the acre decreases by 4 pounds due to overcrowding. Find the number of trees that should be planted on an acre to produce the maximum yield. How many pounds is the maximum yield?

Group Exercise

94. Each group member should consult an almanac, newspaper, magazine, or the Internet to find data that initially increase and then decrease, or vice versa, and therefore can be modeled by a quadratic function. Group members should select the two sets of data that are most interesting and relevant. For each data set selected,

a. Use the quadratic regression feature of a graphing utility to find the quadratic function that best fits the data.
b. Use the equation of the quadratic function to make a prediction from the data. What circumstances might affect the accuracy of your prediction?
c. Use the equation of the quadratic function to write and solve a problem involving maximizing or minimizing the function.

SECTION 3.2 Polynomial Functions and Their Graphs

Objectives

1. Identify polynomial functions.
2. Recognize characteristics of graphs of polynomial functions.
3. Determine end behavior.
4. Use factoring to find zeros of polynomial functions.
5. Identify zeros and their multiplicities.
6. Use the Intermediate Value Theorem.
7. Understand the relationship between degree and turning points.
8. Graph polynomial functions.

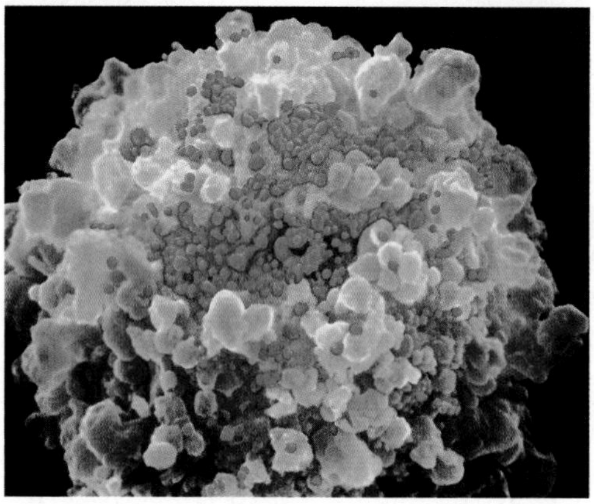

Magnified 6000 times, this color-scanned image shows a T-lymphocyte blood cell (green) infected with the HIV virus (red). Depletion of the number of T-cells causes destruction of the immune system.

In 1980, U.S. doctors diagnosed 41 cases of a rare form of cancer, Kaposi's sarcoma, that involved skin lesions, pneumonia, and severe immunological deficiencies. All cases involved gay men ranging in age from 26 to 51. By the end of 2002, approximately 890,000 Americans, straight and gay, male and female, old and young, were infected with the HIV virus.

Modeling AIDS-related data and making predictions about the epidemic's havoc is serious business. Figure 3.9 shows the number of AIDS cases diagnosed in the United States from 1983 through 2002.

AIDS Cases Diagnosed in the U.S., 1983–2002

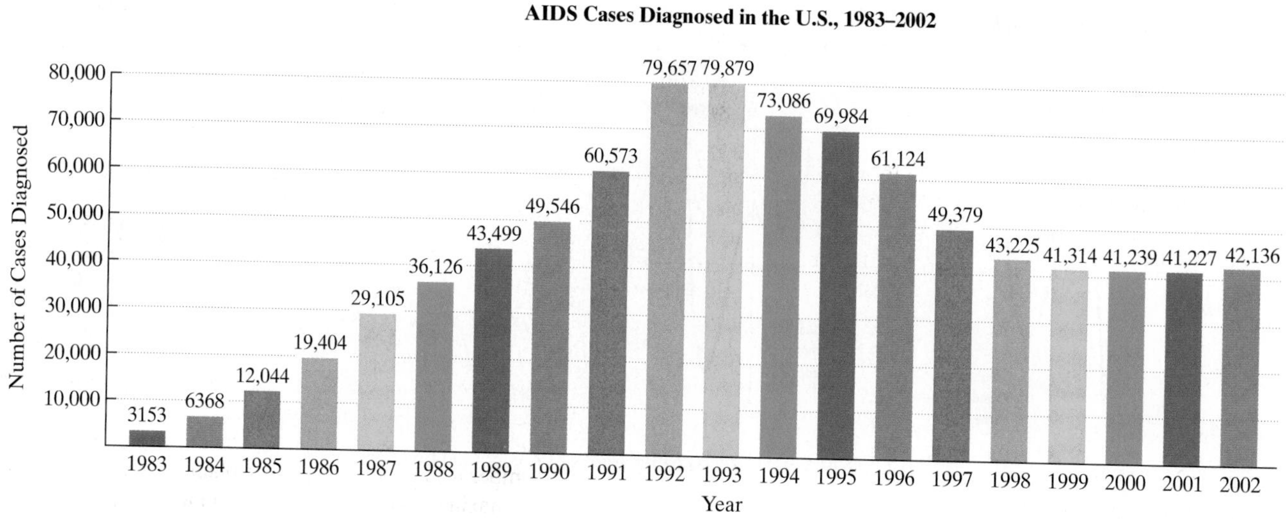

Figure 3.9

Source: Department of Health and Human Services

1 Identify polynomial functions.

Changing circumstances and unforeseen events can result in models for AIDS-related data that are not particularly useful over long periods of time. For example, the function

$$f(x) = -49x^3 + 806x^2 + 3776x + 2503$$

models the number of AIDS cases diagnosed in the United States x years after 1983. The model was obtained using a portion of the data shown in Figure 3.9, namely cases diagnosed from 1983 through 1991, inclusive. Figure 3.10 shows the graph of f from 1983 through 1991. This function is an example of a *polynomial function of degree 3*.

$f(x) = -49x^3 + 806x^2 + 3776x + 2503$

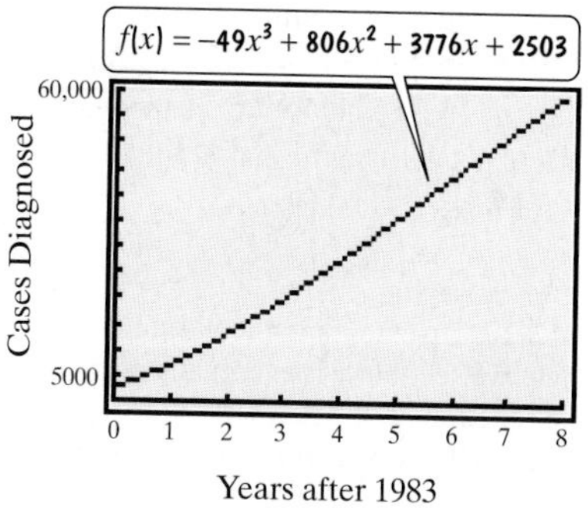

[0, 8, 1] by [0, 60,000, 5000]

Figure 3.10 The graph of a function modeling the number of AIDS cases from 1983 through 1991

Definition of a Polynomial Function

Let n be a nonnegative integer and let $a_n, a_{n-1}, \ldots, a_2, a_1, a_0$ be real numbers, with $a_n \neq 0$. The function defined by

$$f(x) = a_n x^n + a_{n-1} x^{n-1} + \cdots + a_2 x^2 + a_1 x + a_0$$

is called a **polynomial function of degree n**. The number a_n, the coefficient of the variable to the highest power, is called the **leading coefficient**.

A constant function $f(x) = c$, where $c \neq 0$, is a polynomial function of degree 0. A linear function $f(x) = mx + b$, where $m \neq 0$, is a polynomial function of degree 1. A quadratic function $f(x) = ax^2 + bx + c$, where $a \neq 0$, is a polynomial function of degree 2. In this section, we focus on polynomial functions of degree 3 or higher.

2 Recognize characteristics of graphs of polynomial functions.

Smooth, Continuous Graphs

Polynomial functions of degree 2 or higher have graphs that are *smooth* and *continuous*. By **smooth**, we mean that the graphs contain only rounded curves with no sharp corners. By **continuous**, we mean that the graphs have no breaks and can be drawn without lifting your pencil from the rectangular coordinate system. These ideas are illustrated in Figure 3.11 on the next page.

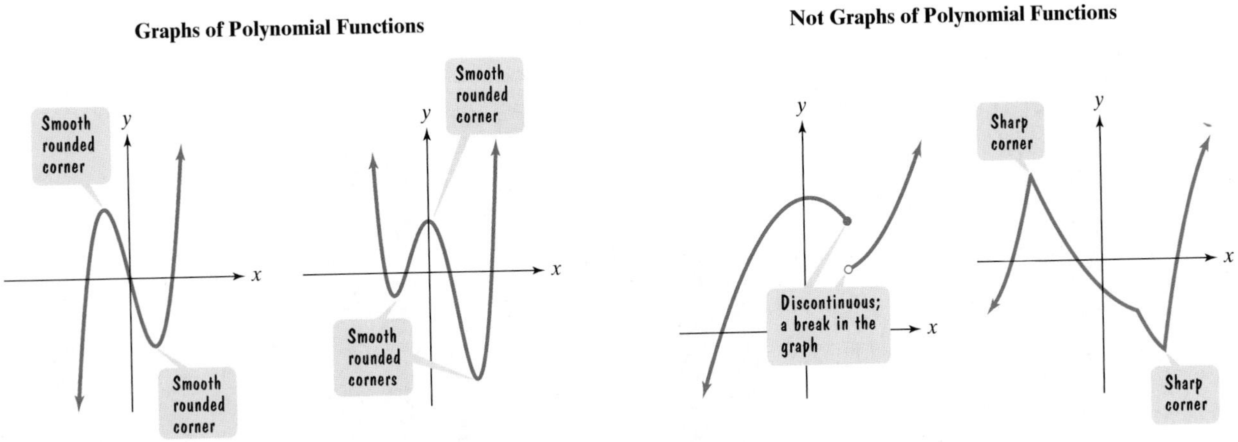

Figure 3.11 Recognizing graphs of polynomial functions

3 Determine end behavior.

End Behavior of Polynomial Functions

Figure 3.12 shows the graph of the function

$$f(x) = -49x^3 + 806x^2 + 3776x + 2503,$$

which models the number of U.S. AIDS cases from 1983 through 1991. Look what happens to the graph when we extend the year up through 2005. By year 21 (2004), the values of y are negative and the function no longer models AIDS cases. We've added an arrow to the graph at the far right to emphasize that it continues to decrease without bound. It is this far-right *end behavior* of the graph that makes it inappropriate for modeling AIDS cases into the future.

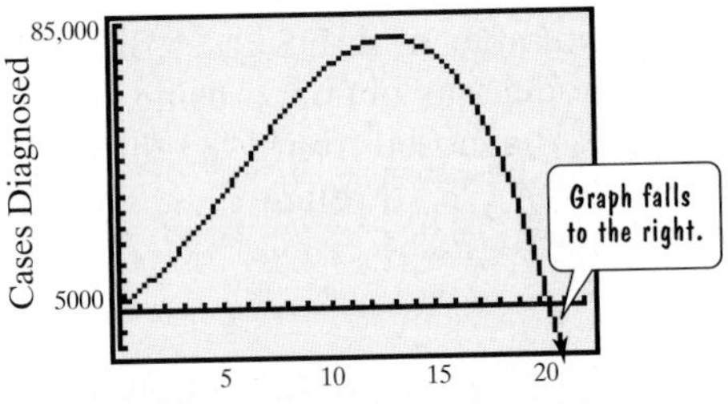

[0, 22, 1] by [−10,000, 85,000, 5000]

Figure 3.12 By extending the viewing rectangle, we see that y is eventually negative and the function no longer models the number of AIDS cases. Model breakdown occurs by 2004.

The behavior of a graph of a function to the far left or the far right is called its **end behavior**. Although the graph of a polynomial function may have intervals where it increases or decreases, the graph will eventually rise or fall without bound as it moves far to the left or far to the right.

How can you determine whether the graph of a polynomial function goes up or down at each end? The end behavior of a polynomial function

$$f(x) = a_n x^n + a_{n-1}x^{n-1} + \cdots + a_1 x + a_0$$

depends upon the leading term $a_n x^n$, because when $|x|$ is large, the other terms are relatively insignificant in size. In particular, the sign of the leading coefficient, a_n, and the degree, n, of the polynomial function reveal its end behavior. In terms of end behavior, only the term of highest degree counts, as summarized by the **Leading Coefficient Test**.

Study Tip

Odd-degree polynomial functions have graphs with opposite behavior at each end. Even-degree polynomial functions have graphs with the same behavior at each end.

The Leading Coefficient Test

As x increases or decreases without bound, the graph of the polynomial function

$$f(x) = a_n x^n + a_{n-1}x^{n-1} + a_{n-2}x^{n-2} + \cdots + a_1 x + a_0 \quad (a_n \neq 0)$$

eventually rises or falls. In particular,

1. For n odd:

If the leading coefficient is positive, the graph falls to the left and rises to the right.

If the leading coefficient is negative, the graph rises to the left and falls to the right.

2. For n even:

If the leading coefficient is positive, the graph rises to the left and to the right.

If the leading coefficient is negative, the graph falls to the left and to the right.

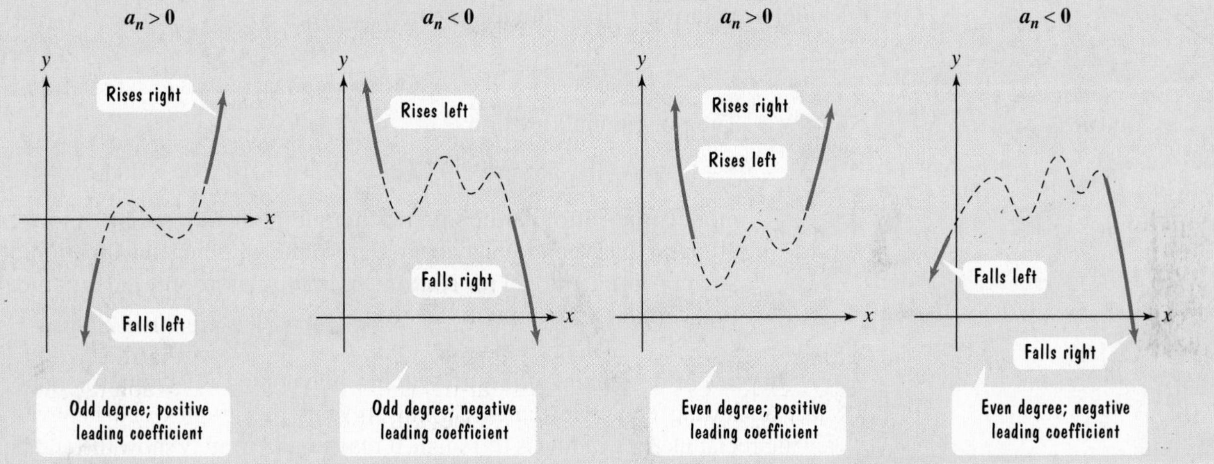

EXAMPLE 1 Using the Leading Coefficient Test

Use the Leading Coefficient Test to determine the end behavior of the graph of

$$f(x) = x^3 + 3x^2 - x - 3.$$

Solution We begin by identifying the sign of the leading coefficient and the degree of the polynomial.

$$f(x) = x^3 + 3x^2 - x - 3$$

The leading coefficient, 1, is positive.

The degree of the polynomial, 3, is odd.

The degree of the function f is 3, which is odd. Odd-degree polynomial functions have graphs with opposite behavior at each end. The leading coefficient, 1, is positive. Thus, the graph falls to the left and rises to the right. The graph of f is shown in Figure 3.13.

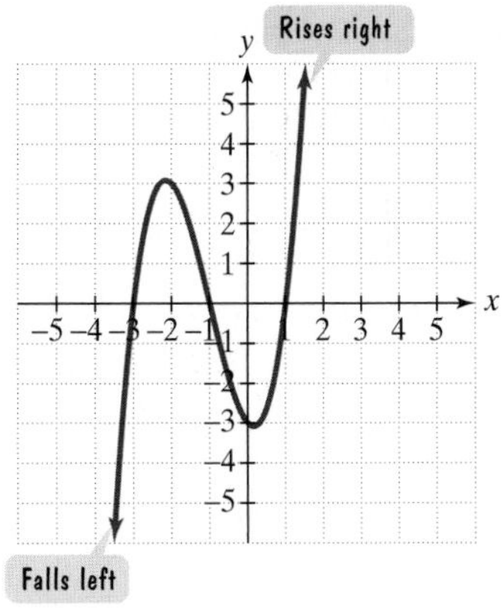

Figure 3.13 The graph of $f(x) = x^3 + 3x^2 - x - 3$

Check Point 1 Use the Leading Coefficient Test to determine the end behavior of the graph of $f(x) = x^4 - 4x^2$.

EXAMPLE 2 Using the Leading Coefficient Test

Use end behavior to explain why

$$f(x) = -49x^3 + 806x^2 + 3776x + 2503$$

is only an appropriate model for AIDS cases for a limited time period.

Solution We begin by identifying the sign of the leading coefficient and the degree of the polynomial.

$$f(x) = -49x^3 + 806x^2 + 3776x + 2503$$

The leading coefficient, −49, is negative.

The degree of the polynomial, 3, is odd.

The degree of f is 3, which is odd. Odd-degree polynomial functions have graphs with opposite behavior at each end. The leading coefficient, −49, is negative. Thus, the graph rises to the left and falls to the right. The fact that the graph falls to the right indicates that at some point the number of AIDS cases will be negative, an impossibility. If a function has a graph that decreases without bound over time, it will not be capable of modeling nonnegative phenomena over long time periods. Model breakdown will eventually occur.

Check Point 2 The polynomial function

$$f(x) = -0.27x^3 + 9.2x^2 - 102.9x + 400$$

models the ratio of students to computers in U.S. public schools x years after 1980. Use end behavior to determine whether this function could be an appropriate model for computers in the classroom well into the twenty-first century. Explain your answer.

If you use a graphing utility to graph a polynomial function, it is important to select a viewing rectangle that accurately reveals the graph's end behavior. If the viewing rectangle, or window, is too small, it may not accurately show the end behavior.

EXAMPLE 3 Using the Leading Coefficient Test

The graph of $f(x) = -x^4 + 8x^3 + 4x^2 + 2$ was obtained with a graphing utility using a $[-8, 8, 1]$ by $[-10, 10, 1]$ viewing rectangle. The graph is shown in Figure 3.14(a). Does the graph show the end behavior of the function?

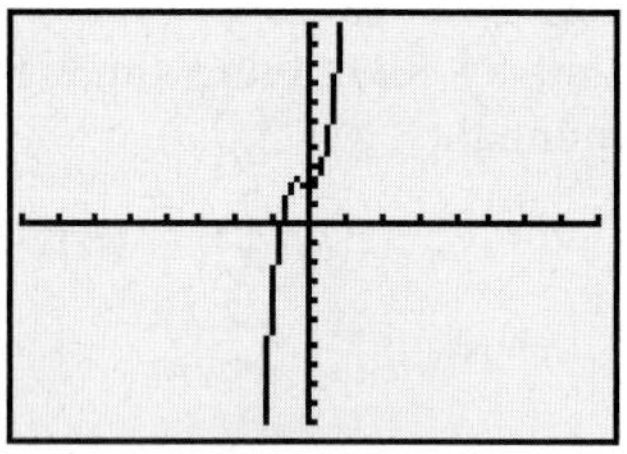

[−8, 8, 1] by [−10, 10, 1]

Figure 3.14(a)

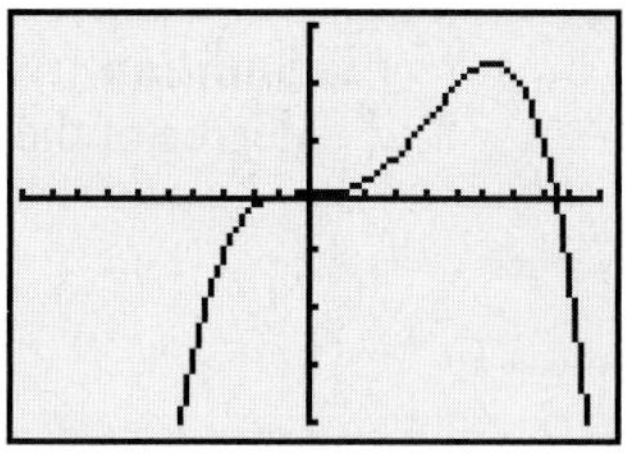

[−10, 10, 1] by [−1000, 750, 250]

Figure 3.14(b)

Solution We begin by identifying the sign of the leading coefficient and the degree of the polynomial.

$$f(x) = -x^4 + 8x^3 + 4x^2 + 2$$

The leading coefficient, −1, is negative.

The degree of the polynomial, 4, is even.

The degree of f is 4, which is even. Even-degree polynomial functions have graphs with the same behavior at each end. The leading coefficient, −1, is negative. Thus, the graph should fall to the left and fall to the right. The graph in Figure 3.14(a) is falling to the left, but it is not falling to the right. Therefore, the graph is not complete enough to show end behavior. A more complete graph of the function is shown in a larger viewing rectangle in Figure 3.14(b).

Check Point 3 The graph of $f(x) = x^3 + 13x^2 + 10x - 4$ is shown in a standard viewing rectangle in Figure 3.15. Use the Leading Coefficient Test to determine whether the graph shows the end behavior of the function. Explain your answer.

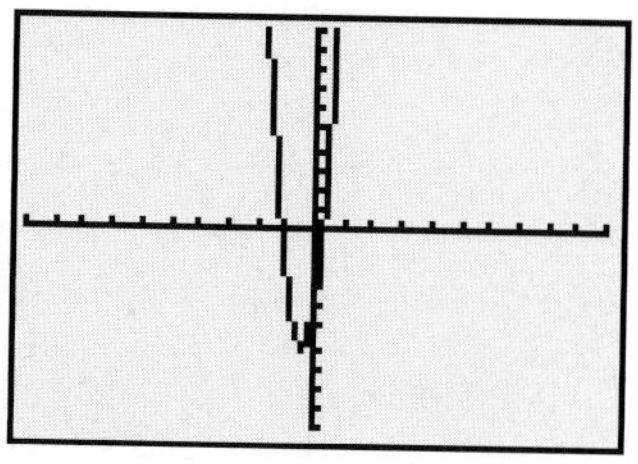

Figure 3.15

④ Use factoring to find zeros of polynomial functions.

Zeros of Polynomial Functions

If f is a polynomial function, then the values of x for which $f(x)$ is equal to 0 are called the **zeros** of f. These values of x are the **roots**, or **solutions**, of the polynomial equation $f(x) = 0$. Each real root of the polynomial equation appears as an x-intercept of the graph of the polynomial function.

EXAMPLE 4   Finding Zeros of a Polynomial Function

Find all zeros of $f(x) = x^3 + 3x^2 - x - 3$.

Solution   By definition, the zeros are the values of x for which $f(x)$ is equal to 0. Thus, we set $f(x)$ equal to 0:

$$f(x) = x^3 + 3x^2 - x - 3 = 0.$$

We solve the polynomial equation $x^3 + 3x^2 - x - 3 = 0$ for x as follows:

$x^3 + 3x^2 - x - 3 = 0$ This is the equation needed to find the function's zeros.

$x^2(x + 3) - 1(x + 3) = 0$ Factor x^2 from the first two terms and -1 from the last two terms.

$(x + 3)(x^2 - 1) = 0$ A common factor of $x + 3$ is factored from the expression.

$x + 3 = 0$ or $x^2 - 1 = 0$ Set each factor equal to 0.

$x = -3$ $x^2 = 1$ Solve for x.

$x = \pm 1$ Remember that if $x^2 = d$, then $x^2 = \pm\sqrt{d}$.

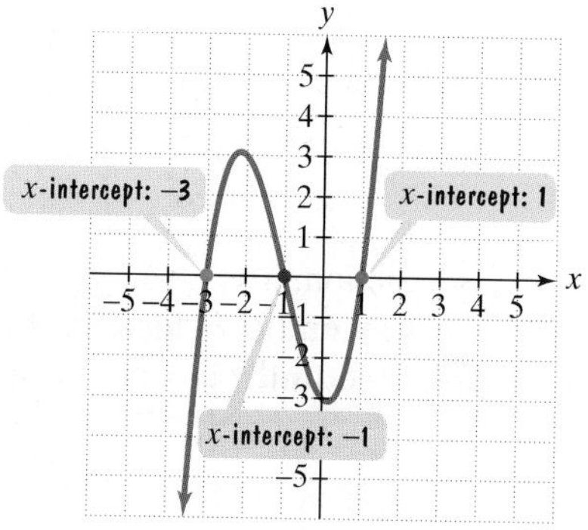

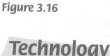

Figure 3.16

The zeros of f are -3, -1, and 1. The graph of f in Figure 3.16 shows that each zero is an x-intercept. The graph passes through the points $(-3, 0)$, $(-1, 0)$, and $(1, 0)$.

Technology

A graphing utility can be used to verify that -3, -1, and 1 are the three real zeros of $f(x) = x^3 + 3x^2 - x - 3$.

Numeric Check
Display a table for the function.

Enter $y_1 = x^3 + 3x^2 - x - 3$.

y_1 is equal to 0 when $x = -3$, $x = -1$, and $x = 1$.

-3, -1, and 1 are the real zeros.

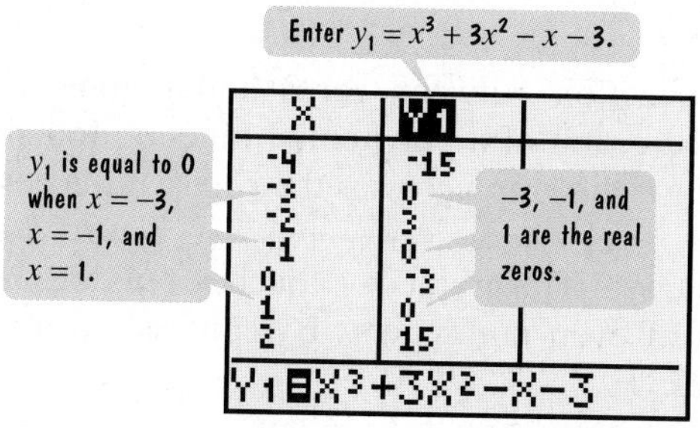

Graphic Check
Display a graph for the function. The x-intercepts indicate that -3, -1, and 1 are the real zeros.

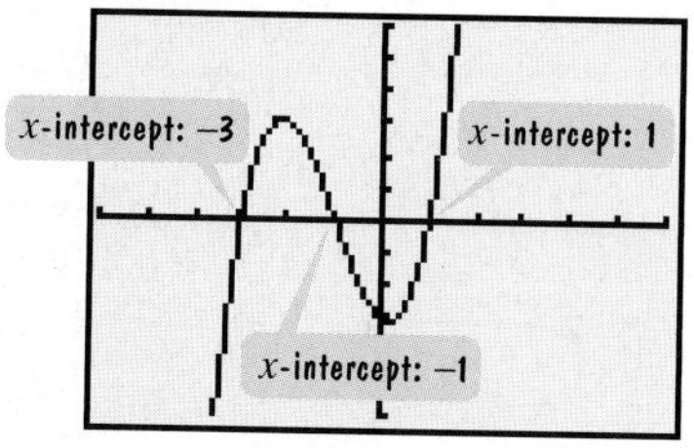

[−6, 6, 1] by [−6, 6, 1]

The utility's ZERO feature on the graph of f also verifies that -3, -1, and 1 are the function's real zeros.

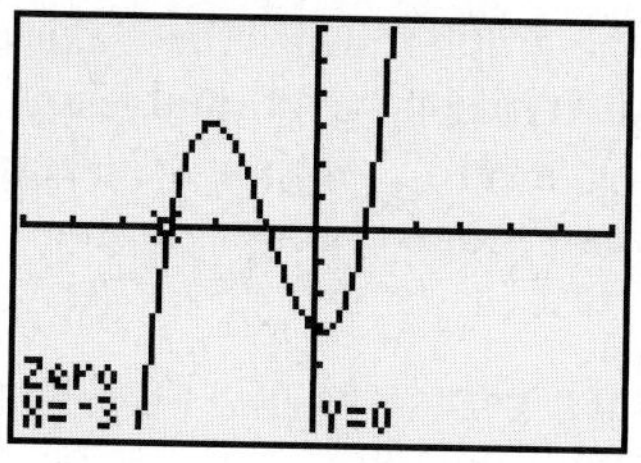

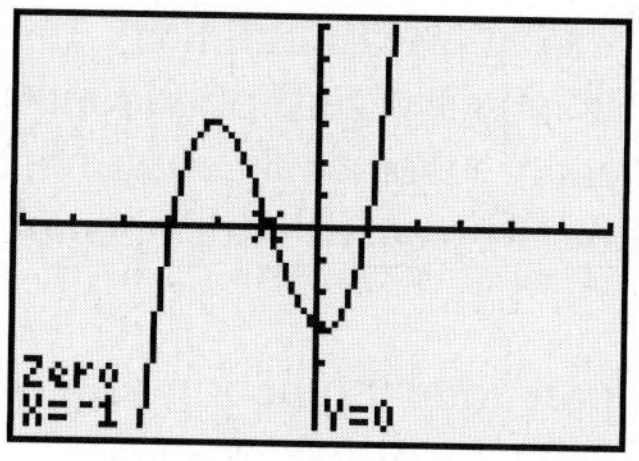

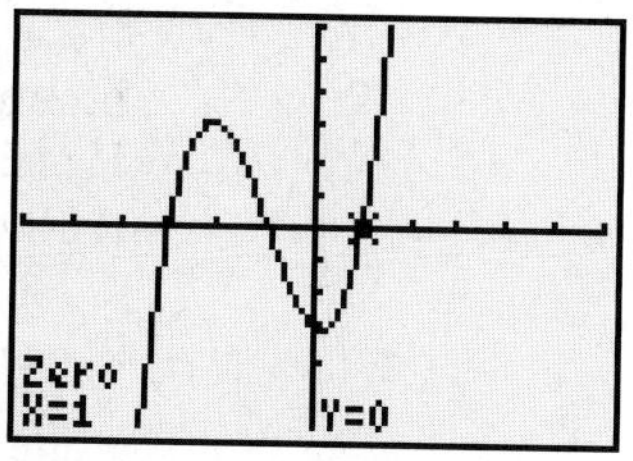

Check Point 4 Find all zeros of $f(x) = x^3 + 2x^2 - 4x - 8$.

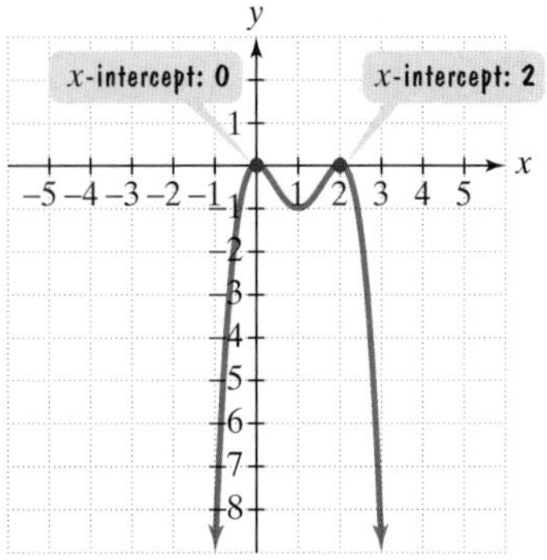

Figure 3.17 The zeros of $f(x) = -x^4 + 4x^3 - 4x^2$, namely 0 and 2, are the x-intercepts for the graph of f.

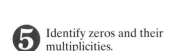

EXAMPLE 5 Finding Zeros of a Polynomial Function

Find all zeros of $f(x) = -x^4 + 4x^3 - 4x^2$.

Solution We find the zeros of f by setting $f(x)$ equal to 0 and solving the resulting equation.

$$-x^4 + 4x^3 - 4x^2 = 0 \quad \text{We now have a polynomial equation.}$$

$$x^4 - 4x^3 + 4x^2 = 0 \quad \text{Multiply both sides by } -1. \text{ This step is optional.}$$

$$x^2(x^2 - 4x + 4) = 0 \quad \text{Factor out } x^2.$$

$$x^2(x - 2)^2 = 0 \quad \text{Factor completely.}$$

$$x^2 = 0 \quad \text{or} \quad (x - 2)^2 = 0 \quad \text{Set each factor equal to 0.}$$

$$x = 0 \qquad x = 2 \quad \text{Solve for } x.$$

The zeros of $f(x) = -x^4 + 4x^3 - 4x^2$ are 0 and 2. The graph of f, shown in Figure 3.17, has x-intercepts at 0 and 2. The graph passes through the points $(0, 0)$ and $(2, 0)$.

Check Point 5 Find all zeros of $f(x) = x^4 - 4x^2$.

5 Identify zeros and their multiplicities.

Multiplicities of Zeros

We can use the results of factoring to express a polynomial as a product of factors. For instance, in Example 5, we can use our factoring to express the function's equation as follows:

$$f(x) = -x^4 + 4x^3 - 4x^2 = -(x^4 - 4x^3 + 4x^2) = -x^2(x - 2)^2.$$

The factor x occurs twice: $x^2 = x \cdot x$.

The factor $(x - 2)$ occurs twice: $(x - 2)^2 = (x - 2)(x - 2)$.

Notice that each factor occurs twice. In factoring the equation for the polynomial function f, if the same factor $x - r$ occurs k times, but not $k + 1$ times, we call r a **zero with multiplicity k**. For the polynomial function

$$f(x) = -x^2(x - 2)^2,$$

0 and 2 are both zeros with multiplicity 2.

Multiplicity provides another connection between zeros and graphs. The multiplicity of a zero tells us whether the graph of a polynomial function touches the x-axis at the zero and turns around, or if the graph crosses the x-axis at the zero. For example, look again at the graph of $f(x) = -x^4 + 4x^3 - 4x^2$ in Figure 3.17. Each zero, 0 and 2, is a zero with multiplicity 2. The graph of f touches, but does not cross, the x-axis at each of these zeros of even multiplicity. By contrast, a graph crosses the x-axis at zeros of odd multiplicity.

Multiplicity and x-Intercepts

If r is a zero of **even multiplicity**, then the graph **touches** the x-axis **and turns around** at r. If r is a zero of **odd multiplicity**, then the graph **crosses** the x-axis at r. Regardless of whether the multiplicity of a zero is even or odd, graphs tend to flatten out at zeros with multiplicity greater than one.

If a polynomial function's equation is expressed as a product of linear factors, we can quickly identify zeros and their multiplicities.

EXAMPLE 6   Finding Zeros and Their Multiplicities

Find the zeros of $f(x) = (x + 1)(2x - 3)^2$ and give the multiplicity of each zero. State whether the graph crosses the x-axis or touches the x-axis and turns around at each zero.

Solution   We find the zeros of f by setting $f(x)$ equal to 0:

$$(x + 1)(2x - 3)^2 = 0.$$

Set each factor equal to 0.

$$\boxed{\begin{array}{c} x + 1 = 0 \\ x = -1 \end{array}} \qquad \boxed{\begin{array}{c} 2x - 3 = 0 \\ x = \frac{3}{2} \end{array}}$$

$$(x + 1)^1(2x - 3)^2 = 0$$

This exponent is **1**. Thus, the multiplicity of −1 is **1**.

This exponent is **2**. Thus, the multiplicity of $\frac{3}{2}$ is **2**.

−1 is a zero of odd multiplicity. Graph crosses x-axis.

$\frac{3}{2}$ is a zero of even multiplicity. Graph touches x-axis, flattens, and turns around.

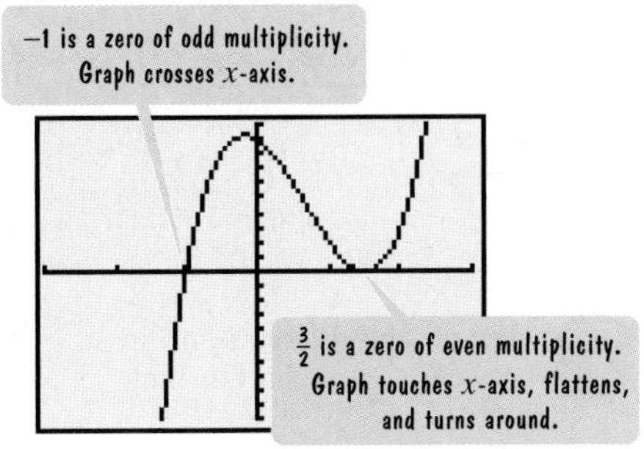

$[-3, 3, 1]$ by $[-10, 10, 1]$

Figure 3.18 The graph of $f(x) = (x + 1)(2x - 3)^2$

The zeros of $f(x) = (x + 1)(2x - 3)^2$ are −1, with multiplicity 1, and $\frac{3}{2}$, with multiplicity 2. Because the multiplicity of −1 is odd, the graph crosses the x-axis at this zero. Because the multiplicity of $\frac{3}{2}$ is even, the graph touches the x-axis and turns around at this zero. These relationships are illustrated by the graph of f in Figure 3.18.

Check Point 6   Find the zeros of $f(x) = -4\left(x + \frac{1}{2}\right)^2(x - 5)^3$ and give the multiplicity of each zero. State whether the graph crosses the x-axis or touches the x-axis and turns around at each zero.

⑥ Use the Intermediate Value Theorem.

The Intermediate Value Theorem

The *Intermediate Value Theorem* tells us of the existence of real zeros. The idea behind the theorem is illustrated in Figure 3.19. The figure shows that if $(a, f(a))$ lies below the x-axis and $(b, f(b))$ lies above the x-axis, the smooth, continuous graph of a polynomial function f must cross the x-axis at some value c between a and b. This value is a real zero for the function.

These observations are summarized in the **Intermediate Value Theorem**.

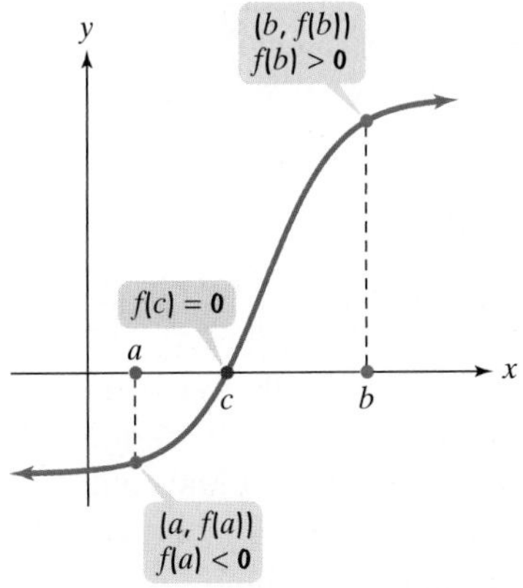

Figure 3.19 The graph must cross the x-axis at some value between a and b.

> **The Intermediate Value Theorem for Polynomials**
>
> Let f be a polynomial function with real coefficients. If $f(a)$ and $f(b)$ have opposite signs, then there is at least one value of c between a and b for which $f(c) = 0$. Equivalently, the equation $f(x) = 0$ has at least one real root between a and b.

EXAMPLE 7   Using the Intermediate Value Theorem

Show that the polynomial function $f(x) = x^3 - 2x - 5$ has a real zero between 2 and 3.

Solution   Let us evaluate f at 2 and at 3. If $f(2)$ and $f(3)$ have opposite signs, then there is at least one real zero between 2 and 3. Using $f(x) = x^3 - 2x - 5$, we obtain

$$f(2) = 2^3 - 2 \cdot 2 - 5 = 8 - 4 - 5 = -1$$

$f(2)$ is negative.

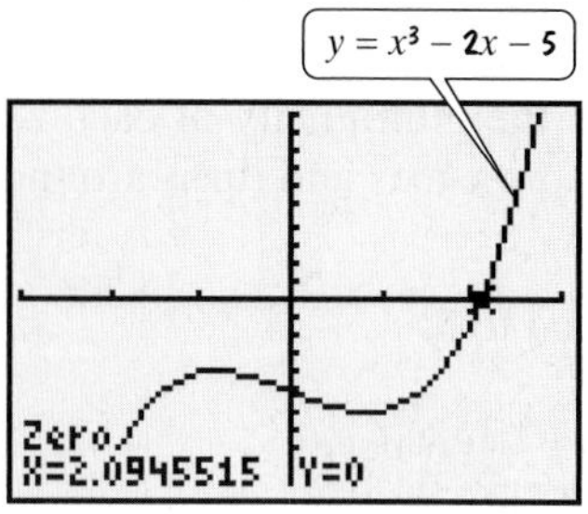

[−3, 3, 1] by [−10, 10, 1]

Figure 3.20

and

$$f(3) = 3^3 - 2 \cdot 3 - 5 = 27 - 6 - 5 = 16.$$

$f(3)$ is positive.

Because $f(2) = -1$ and $f(3) = 16$, the sign change shows that the polynomial function has a real zero between 2 and 3. This zero is actually irrational and is approximated using a graphing utility's ZERO feature as 2.0945515 in Figure 3.20.

Check Point 7 Show that the polynomial function $f(x) = 3x^3 - 10x + 9$ has a real zero between −3 and −2.

7 Understand the relationship between degree and turning points.

Turning Points of Polynomial Functions

The graph of $f(x) = x^5 - 6x^3 + 8x + 1$ is shown in Figure 3.21. The graph has four smooth **turning points**. At each turning point, the graph changes direction from increasing to decreasing or vice versa. The given equation has 5 as its greatest exponent and is therefore a polynomial function of degree 5. Notice that the graph has four turning points. In general, **if f is a polynomial function of degree n, then the graph of f has at most $n - 1$ turning points.**

Figure 3.21 illustrates that the y-coordinate of each turning point is either a relative maximum or a relative minimum of f. Without the aid of a graphing utility or a knowledge of calculus, it is difficult and often impossible to locate turning points of polynomial functions with degrees greater than 2. If necessary, test values can be taken between the x-intercepts to get a general idea of how high the graph rises or how low the graph falls. For the purpose of graphing in this section, a general estimate is sometimes appropriate and necessary.

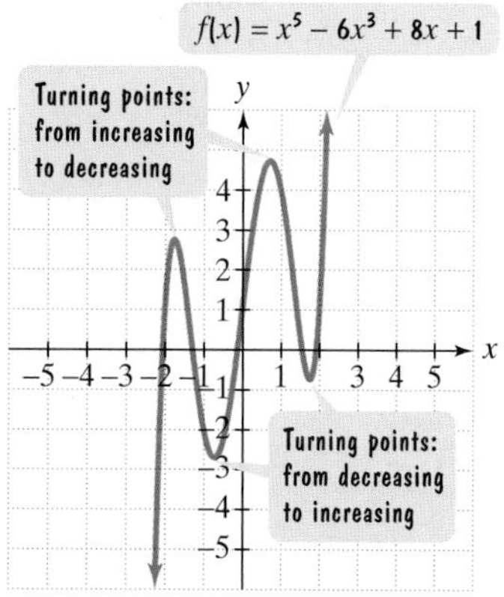

Figure 3.21 Graph with four turning points

8 Graph polynomial functions.

A Strategy for Graphing Polynomial Functions

Here's a general strategy for graphing a polynomial function. A graphing utility is a valuable complement, but not a necessary component, to this strategy. If you are using a graphing utility, some of the steps listed in the following box will help you to select a viewing rectangle that shows the important parts of the graph.

Graphing a Polynomial Function

$$f(x) = a_n x^n + a_{n-1}x^{n-1} + a_{n-2}x^{n-2} + \cdots + a_1 x + a_0, \; a_n \neq 0$$

1. Use the Leading Coefficient Test to determine the graph's end behavior.
2. Find x-intercepts by setting $f(x) = 0$ and solving the resulting polynomial equation. If there is an x-intercept at r as a result of $(x - r)^k$ in the complete factorization of $f(x)$, then
   - **a.** If k is even, the graph touches the x-axis at r and turns around.
   - **b.** If k is odd, the graph crosses the x-axis at r.
   - **c.** If $k > 1$, the graph flattens out at $(r, 0)$.
3. Find the y-intercept by computing $f(0)$.
4. Use symmetry, if applicable, to help draw the graph:
   - **a.** y-axis symmetry: $f(-x) = f(x)$
   - **b.** Origin symmetry: $f(-x) = -f(x)$.
5. Use the fact that the maximum number of turning points of the graph is $n - 1$ to check whether it is drawn correctly.

Study Tip

Remember that, without calculus, it is often impossible to give the exact location of turning points. However, you can obtain additional points satisfying the function to estimate how high the graph rises or how low it falls. To obtain these points, use values of x between (and to the left and right of) the x-intercepts.

EXAMPLE 8 Graphing a Polynomial Function

Graph: $f(x) = x^4 - 2x^2 + 1$.

Solution

Step 1 Determine end behavior. Identify the sign of a_n, the leading coefficient, and the degree, n, of the polynomial function.

$$f(x) = x^4 - 2x^2 + 1$$

The leading coefficient, 1, is positive.

The degree of the polynomial function, 4, is even.

Because the degree, 4, is even, the graph has the same behavior at each end. The leading coefficient, 1, is positive. Thus, the graph rises to the left and rises to the right.

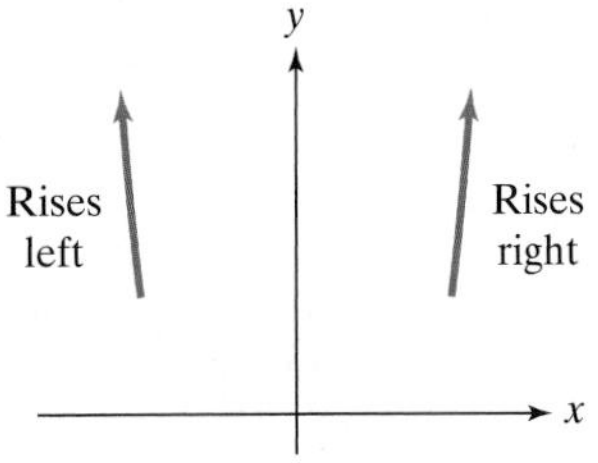

Step 2 Find x-intercepts (zeros of the function) by setting $f(x) = 0$.

$$x^4 - 2x^2 + 1 = 0$$

$$(x^2 - 1)(x^2 - 1) = 0 \quad \text{Factor.}$$

$$(x + 1)(x - 1)(x + 1)(x - 1) = 0 \quad \text{Factor completely.}$$

$$(x + 1)^2(x - 1)^2 = 0 \quad \text{Express the factorization in a more compact form.}$$

$$(x + 1)^2 = 0 \quad \text{or} \quad (x - 1)^2 = 0 \quad \text{Set each factorization equal to 0.}$$

$$x = -1 \qquad x = 1 \quad \text{Solve for x.}$$

We see that -1 and 1 are both repeated zeros with multiplicity 2. Because of the even multiplicity, the graph touches the x-axis at -1 and 1 and turns around. Furthermore, the graph tends to flatten out at these zeros with multiplicity greater than one.

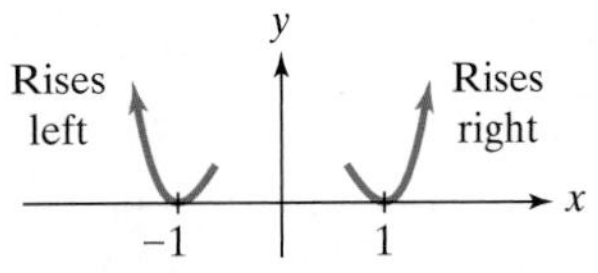

Step 3 Find the y-intercept by computing $f(0)$. We use $f(x) = x^4 - 2x^2 + 1$ and compute $f(0)$.

$$f(0) = 0^4 - 2 \cdot 0^2 + 1 = 1$$

There is a y-intercept at 1, so the graph passes through $(0, 1)$.

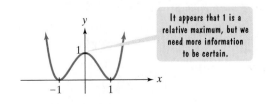

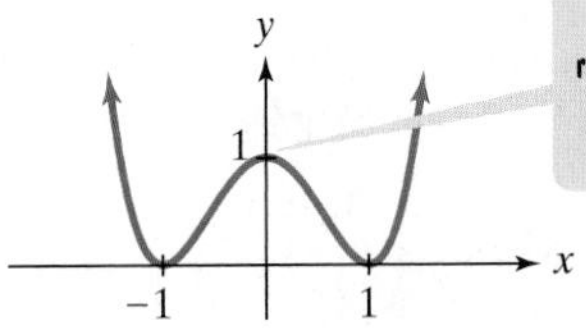

It appears that 1 is a relative maximum, but we need more information to be certain.

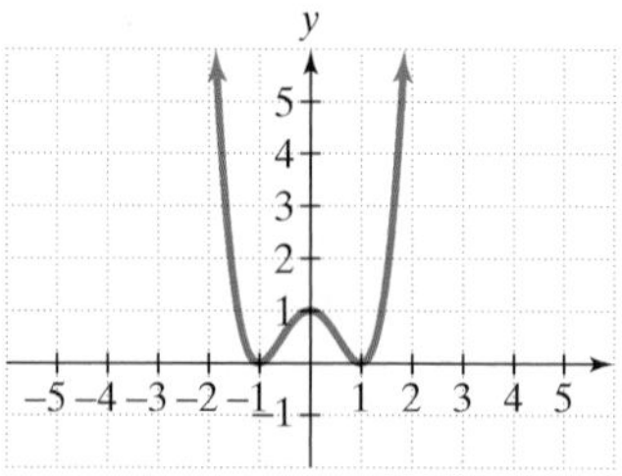

Figure 3.22 The graph of $f(x) = x^4 - 2x^2 + 1$

Step 4 Use possible symmetry to help draw the graph. Our partial graph suggests y-axis symmetry. Let's verify this by finding $f(-x)$.

$$f(x) = x^4 - 2x^2 + 1$$

Replace x with $-x$.

$$f(-x) = (-x)^4 - 2(-x)^2 + 1 = x^4 - 2x^2 + 1$$

Because $f(-x) = f(x)$, the graph of f is symmetric with respect to the y-axis. Figure 3.22 shows the graph of $f(x) = x^4 - 2x^2 + 1$.

Step 5 Use the fact that the maximum number of turning points of the graph is $n - 1$ to check whether it is drawn correctly. Because $n = 4$, the maximum number of turning points is $4 - 1$, or 3. Because the graph in Figure 3.22 has three turning points, we have not violated the maximum number possible. Can you see how this verifies that 1 is indeed a relative maximum and $(0, 1)$ is a turning point? If the graph rose above 1 on either side of $x = 0$, it would have to rise above 1 on the other side as well because of symmetry. This would require additional turning points to smoothly curve back to the x-intercepts. The graph already has three turning points, which is the maximum number for a fourth-degree polynomial function.

Check Point 8 Use the five-step strategy to graph $f(x) = x^3 - 3x^2$.

EXERCISE SET 3.2

Practice Exercises

In Exercises 1–10, determine which functions are polynomial functions. For those that are, identify the degree.

1. $f(x) = 5x^2 + 6x^3$

2. $f(x) = 7x^2 + 9x^4$

3. $g(x) = 7x^5 - \pi x^3 + \frac{1}{5}x$

4. $g(x) = 6x^7 + \pi x^5 + \frac{2}{3}x$

5. $h(x) = 7x^3 + 2x^2 + \frac{1}{x}$

6. $h(x) = 8x^3 - x^2 + \frac{2}{x}$

7. $f(x) = x^{\frac{1}{2}} - 3x^2 + 5$

8. $f(x) = x^{\frac{1}{3}} - 4x^2 + 7$

9. $f(x) = \frac{x^2 + 7}{x^3}$

10. $f(x) = \frac{x^2 + 7}{3}$

In Exercises 11–14, identify which graphs are not those of polynomial functions.

11.

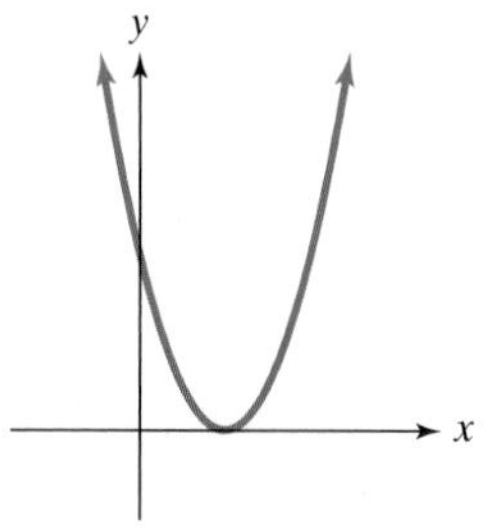

12.

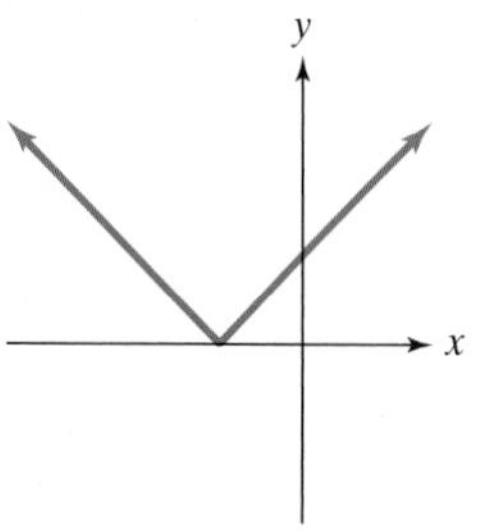

13.

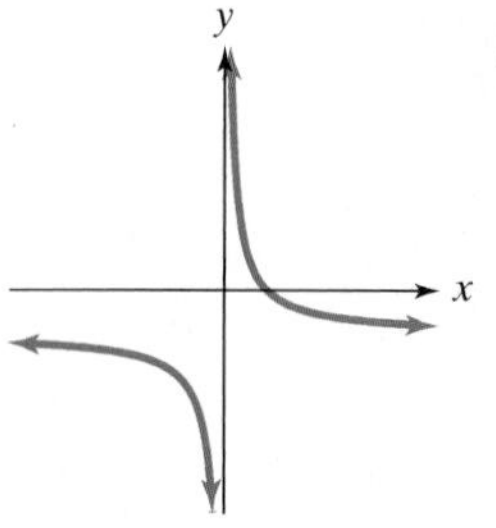

14.

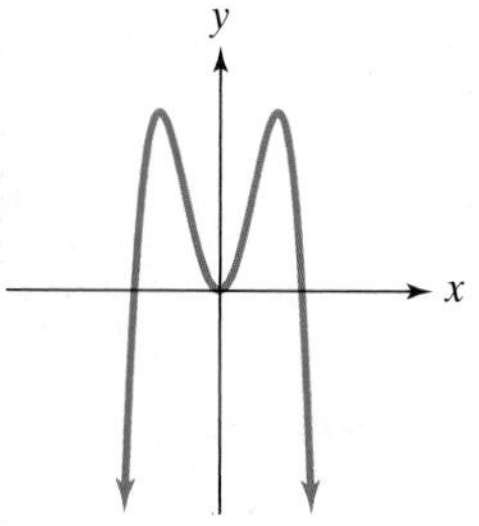

In Exercises 15–18, use the Leading Coefficient Test to determine the end behavior of the graph of the given polynomial function. Then use this end behavior to match the polynomial function with its graph. [The graphs are labeled (a) through (d).]

15. $f(x) = -x^4 + x^2$

16. $f(x) = x^3 - 4x^2$

17. $f(x) = (x - 3)^2$

18. $f(x) = -x^3 - x^2 + 5x - 3$

(a)

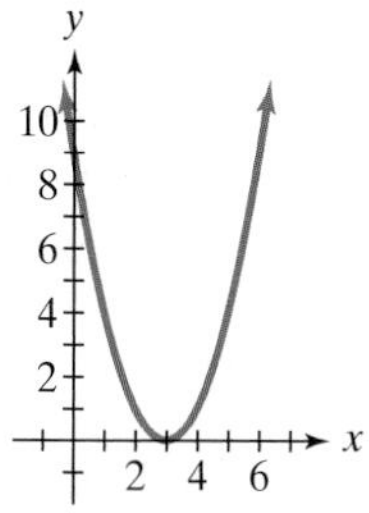

(b)

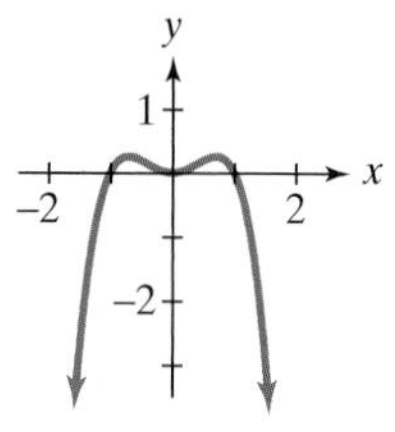

(c)

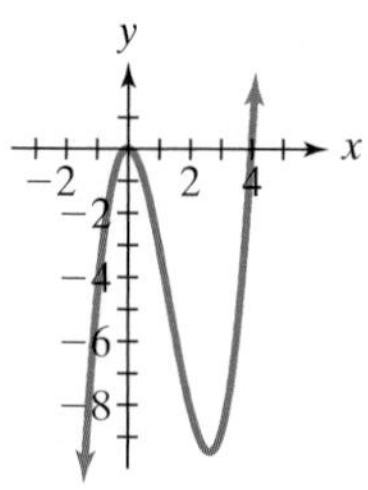

(d)

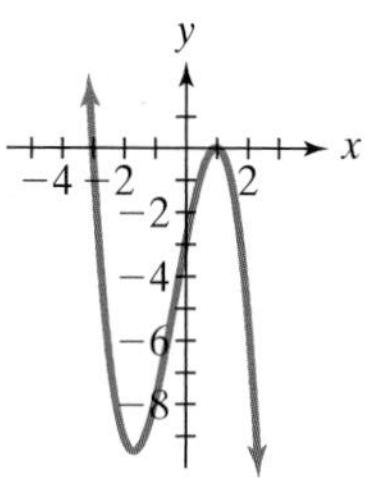

In Exercises 19–24, use the Leading Coefficient Test to determine the end behavior of the graph of the polynomial function.

19. $f(x) = 5x^3 + 7x^2 - x + 9$

20. $f(x) = 11x^3 - 6x^2 + x + 3$

21. $f(x) = 5x^4 + 7x^2 - x + 9$

22. $f(x) = 11x^4 - 6x^2 + x + 3$

23. $f(x) = -5x^4 + 7x^2 - x + 9$

24. $f(x) = -11x^4 - 6x^2 + x + 3$

In Exercises 25–32, find the zeros for each polynomial function and give the multiplicity for each zero. State whether the graph crosses the x-axis, or touches the x-axis and turns around, at each zero.

25. $f(x) = 2(x - 5)(x + 4)^2$

26. $f(x) = 3(x + 5)(x + 2)^2$

27. $f(x) = 4(x - 3)(x + 6)^3$

28. $f(x) = -3\left(x + \frac{1}{2}\right)(x - 4)^3$

29. $f(x) = x^3 - 2x^2 + x$

30. $f(x) = x^3 + 4x^2 + 4x$

31. $f(x) = x^3 + 7x^2 - 4x - 28$

32. $f(x) = x^3 + 5x^2 - 9x - 45$

In Exercises 33–40, use the Intermediate Value Theorem to show that each polynomial has a real zero between the given integers.

33. $f(x) = x^3 - x - 1$; between 1 and 2

34. $f(x) = x^3 - 4x^2 + 2$; between 0 and 1

35. $f(x) = 2x^4 - 4x^2 + 1$; between -1 and 0

36. $f(x) = x^4 + 6x^3 - 18x^2$; between 2 and 3

37. $f(x) = x^3 + x^2 - 2x + 1$; between -3 and -2

38. $f(x) = x^5 - x^3 - 1$; between 1 and 2

39. $f(x) = 3x^3 - 10x + 9$; between -3 and -2

40. $f(x) = 3x^3 - 8x^2 + x + 2$; between 2 and 3

In Exercises 41–64,

a. *Use the Leading Coefficient Test to determine the graph's end behavior.*
b. *Find the x-intercepts. State whether the graph crosses the x-axis, or touches the x-axis and turns around, at each intercept.*
c. *Find the y-intercept.*
d. *Determine whether the graph has y-axis symmetry, origin symmetry, or neither.*
e. *If necessary, find a few additional points and graph the function. Use the maximum number of turning points to check whether it is drawn correctly.*

41. $f(x) = x^3 + 2x^2 - x - 2$
42. $f(x) = x^3 + x^2 - 4x - 4$

43. $f(x) = x^4 - 9x^2$
44. $f(x) = x^4 - x^2$

45. $f(x) = -x^4 + 16x^2$
46. $f(x) = -x^4 + 4x^2$

47. $f(x) = x^4 - 2x^3 + x^2$
48. $f(x) = x^4 - 6x^3 + 9x^2$

49. $f(x) = -2x^4 + 4x^3$
50. $f(x) = -2x^4 + 2x^3$

51. $f(x) = 6x^3 - 9x - x^5$
52. $f(x) = 6x - x^3 - x^5$

53. $f(x) = 3x^2 - x^3$
54. $f(x) = \frac{1}{2} - \frac{1}{2}x^4$

55. $f(x) = -3(x - 1)^2(x^2 - 4)$

56. $f(x) = -2(x - 4)^2(x^2 - 25)$

57. $f(x) = x^2(x - 1)^3(x + 2)$

58. $f(x) = x^3(x + 2)^2(x + 1)$

59. $f(x) = -x^2(x - 1)(x + 3)$

60. $f(x) = -x^2(x + 2)(x - 2)$

61. $f(x) = -2x^3(x - 1)^2(x + 5)$

62. $f(x) = -3x^3(x - 1)^2(x + 3)$

63. $f(x) = (x - 2)^2(x + 4)(x - 1)$

64. $f(x) = (x + 3)(x + 1)^3(x + 4)$

Practice Plus

In Exercises 65–72, complete graphs of polynomial functions whose zeros are integers are shown.

a. *Find the zeros and state whether the multiplicity of each zero is even or odd.*
b. *Write an equation, expressed as the product of factors, of a polynomial function that might have each graph. Use a leading coefficient of 1 or −1, and make the degree of f as small as possible.*
c. *Use both the equation in part (b) and the graph to find the y-intercept.*

65.

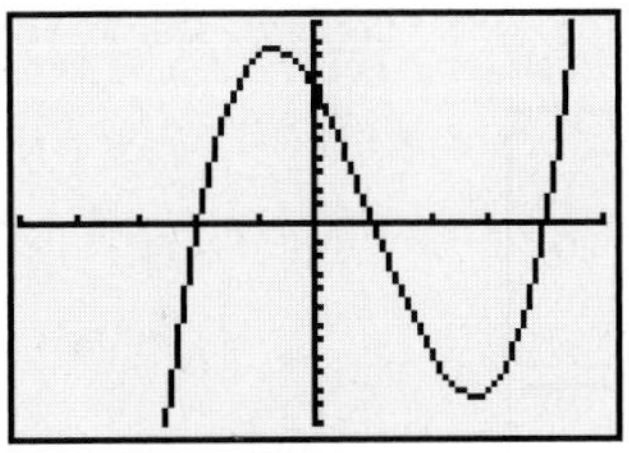

[−5, 5, 1] by [−12, 12, 1]

66.

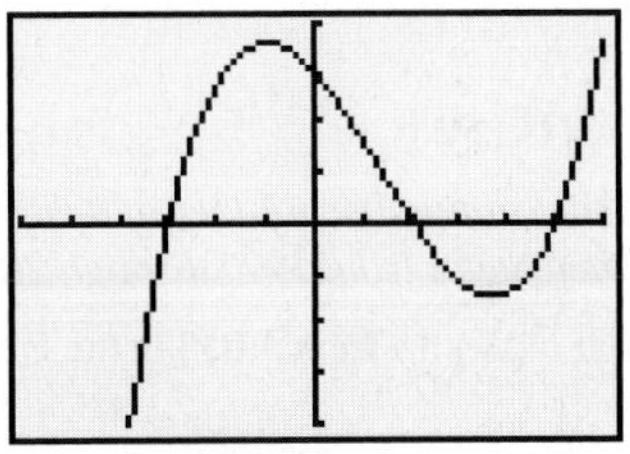

[−6, 6, 1] by [−40, 40, 10]

67.

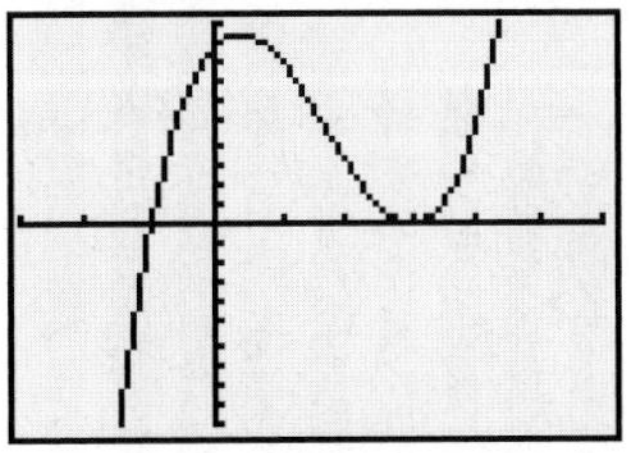

[−3, 6, 1] by [−10, 10, 1]

68.

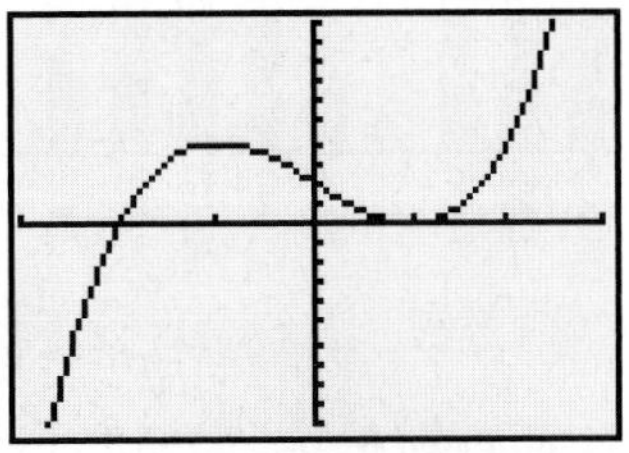

[−3, 3, 1] by [−10, 10, 1]

69.

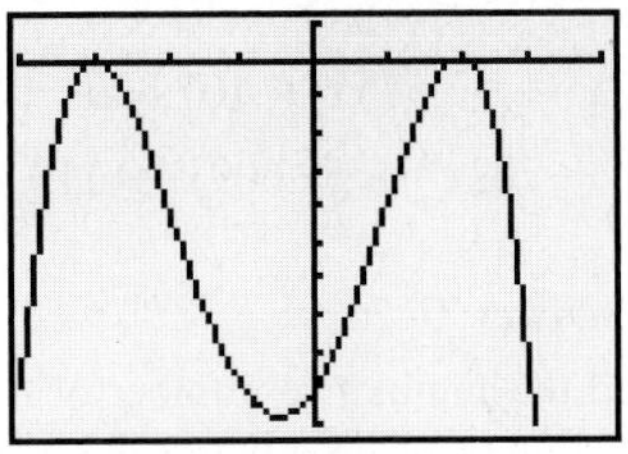

[−4, 4, 1] by [−40, 4, 4]

70.

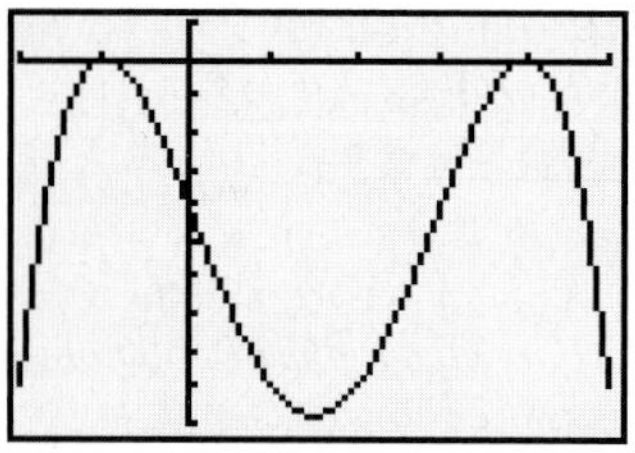

[−2, 5, 1] by [−40, 4, 4]

71.

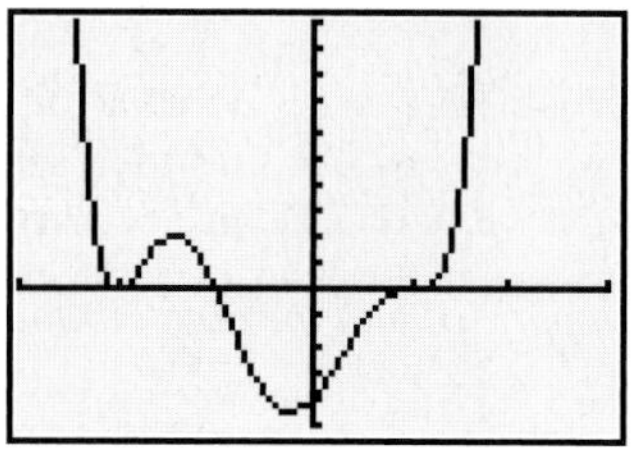

[−3, 3, 1] by [−5, 10, 1]

72.

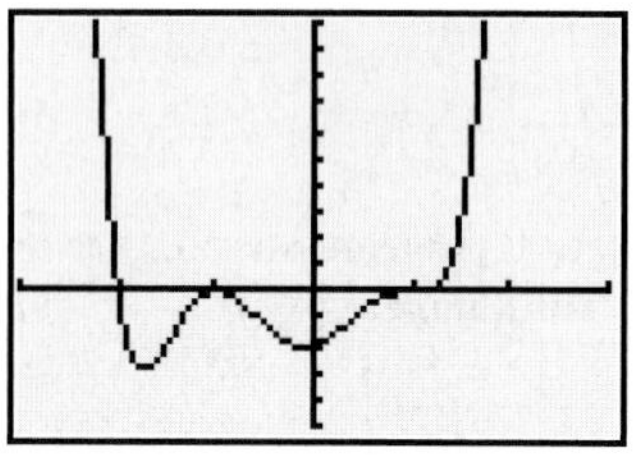

[−3, 3, 1] by [−5, 10, 1]

 Application Exercises

The bar graph shows the cumulative number of deaths from AIDS in the United States from 1990 through 2002.

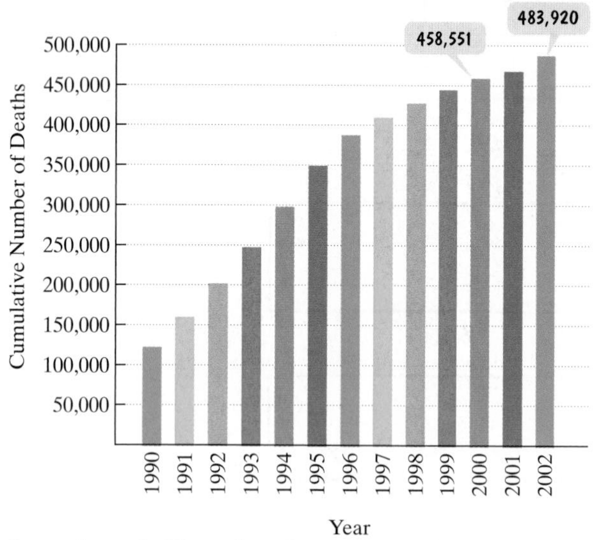

Source: Centers for Disease Control

The data in the bar graph can be modeled by the following second- and third-degree polynomial functions:

Cumulative number of AIDS deaths *x* years after 1990

$f(x) = -2212x^2 + 57{,}575x + 107{,}896$

$g(x) = -84x^3 - 702x^2 + 50{,}609x + 113{,}435.$

Use these functions to solve Exercises 73–76.

73. Use both functions to find the cumulative number of AIDS deaths in 2000. Which function provides a better description for the actual number shown in the bar graph?

74. Use both functions to find the cumulative number of AIDS deaths in 2002. Which function provides a better description for the actual number shown in the bar graph?

75. Use the Leading Coefficient Test to determine the end behavior to the right for the graph of *f*. Will this function be useful in modeling the cumulative number of AIDS deaths over an extended period of time? Explain your answer.

76. Use the Leading Coefficient Test to determine the end behavior to the right for the graph of *g*. Will this function be useful in modeling the cumulative number of AIDS deaths over an extended period of time? Explain your answer.

77. Although it has been more than 50 years since the Supreme Court ruled against school segregation, data from the Civil Rights Project at Harvard University indicate that integration and academic equality remain elusive. The graph shows the percentage of the average African-American student's classmates who were white for the period from 1970 through 2002.

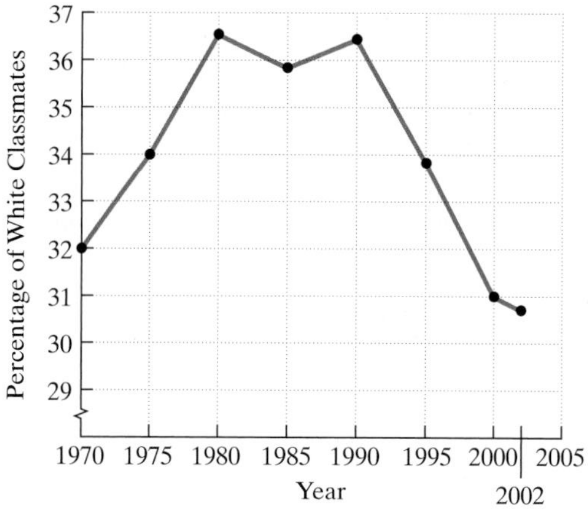

Source: Civil Rights Project, Harvard University

a. For which years was the percentage of white classmates increasing?

b. For which years was the percentage of white classmates decreasing?

c. How many turning points (from increasing to decreasing or from decreasing to increasing) does the graph have for the period shown?

d. Suppose that a polynomial function is used to model the data shown in the graph using

(number of years after 1970, percentage of the average African-American student's classmates who were white).

Use the number of turning points to determine the degree of the polynomial function of best fit.

e. For the model in part (d), should the leading coefficient of the polynomial function be positive or negative? Explain your answer.

78. The graphs show the percentage of husbands and wives with one or more children who said their marriage was going well "all the time" at various stages in their relationships.

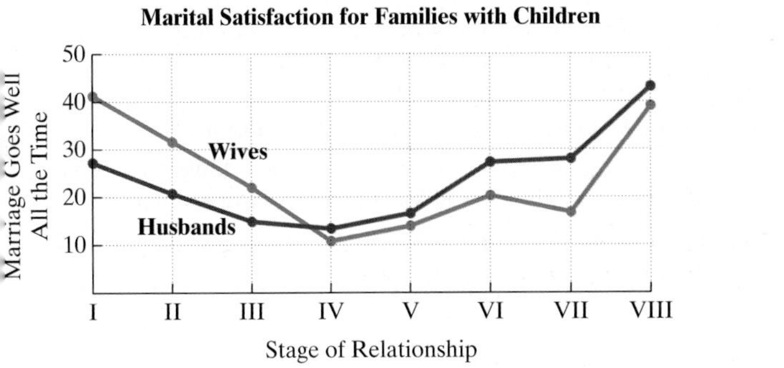

ge I: Beginning families — Stage V: Families with teenagers
ge II: Child-bearing families — Stage VI: Families with adult children leaving home
ge III: Families with preschool children — Stage VII: Families in the middle years
ge IV: Families with school-age children — Stage VIII: Aging families

Source: Rollins, B., & Feldman, H. (1970), Marital satisfaction over the family life cycle. *Journal of Marriage and the Family*, 32, 20–28.

a. Between which stages was marital satisfaction for wives decreasing?

b. Between which stages was marital satisfaction for wives increasing?

c. How many turning points (from decreasing to increasing or from increasing to decreasing) are shown in the graph for wives?

d. Suppose that a polynomial function is used to model the data shown in the graph for wives using

(stage in the relationship, percentage indicating that the marriage was going well all the time).

Use the number of turning points to determine the degree of the polynomial function of best fit.

e. For the model in part (d), should the leading coefficient of the polynomial function be positive or negative? Explain your answer.

Writing in Mathematics

79. What is a polynomial function?

80. What do we mean when we describe the graph of a polynomial function as smooth and continuous?

81. What is meant by the end behavior of a polynomial function?

82. Explain how to use the Leading Coefficient Test to determine the end behavior of a polynomial function.

83. Why is a third-degree polynomial function with a negative leading coefficient not appropriate for modeling nonnegative real-world phenomena over a long period of time?

84. What are the zeros of a polynomial function and how are they found?

85. Explain the relationship between the multiplicity of a zero and whether or not the graph crosses or touches the x-axis at that zero.

86. If f is a polynomial function, and $f(a)$ and $f(b)$ have opposite signs, what must occur between a and b? If $f(a)$ and $f(b)$ have the same sign, does it necessarily mean that this will not occur? Explain your answer.

87. Explain the relationship between the degree of a polynomial function and the number of turning points on its graph.

88. Can the graph of a polynomial function have no x-intercepts? Explain.

89. Can the graph of a polynomial function have no y-intercept? Explain.

90. Describe a strategy for graphing a polynomial function. In your description, mention intercepts, the polynomial's degree, and turning points.

91. The graphs shown in Exercise 78 indicate that marital satisfaction tends to be greatest at the beginning and at the end of the stages in the relationship, with a decline occurring in the middle. What explanations can you offer for this trend?

Technology Exercises

92. Use a graphing utility to verify any five of the graphs that you drew by hand in Exercises 41–64.

Write a polynomial function that imitates the end behavior of each graph in Exercises 93–96. The dashed portions of the graphs indicate that you should focus only on imitating the left and right behavior of the graph and can be flexible about what occurs between the left and right ends. Then use your graphing utility to graph the polynomial function and verify that you imitated the end behavior shown in the given graph.

93.

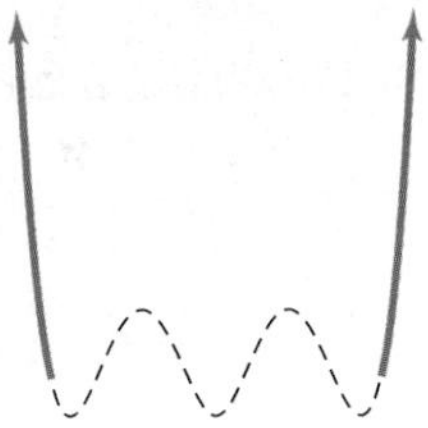

94.

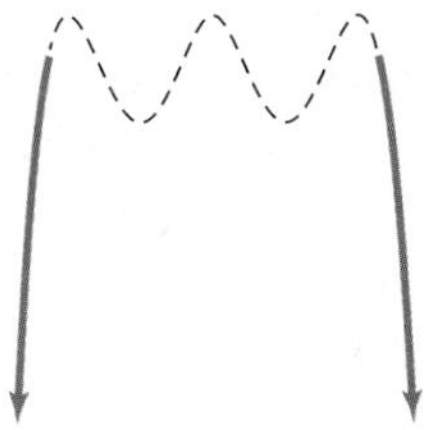

95.

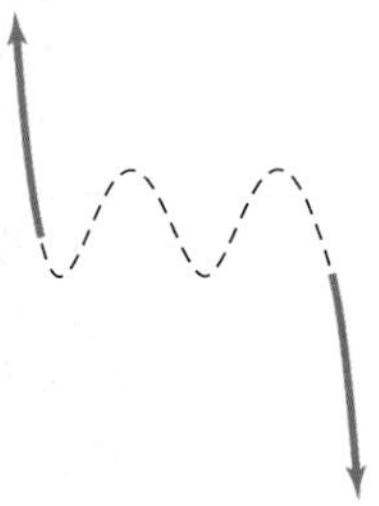

96.

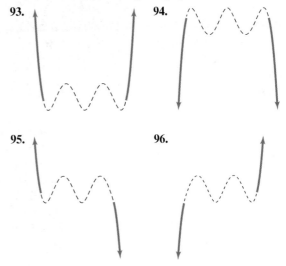

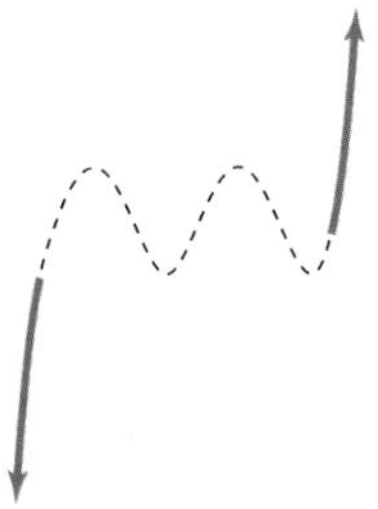

In Exercises 97–100, use a graphing utility with a viewing rectangle large enough to show end behavior to graph each polynomial function.

97. $f(x) = x^3 + 13x^2 + 10x - 4$

98. $f(x) = -2x^3 + 6x^2 + 3x - 1$

99. $f(x) = -x^4 + 8x^3 + 4x^2 + 2$

100. $f(x) = -x^5 + 5x^4 - 6x^3 + 2x + 20$

In Exercises 101–102, use a graphing utility to graph f and g in the same viewing rectangle. Then use the ZOOM OUT *feature to show that f and g have identical end behavior.*

101. $f(x) = x^3 - 6x + 1, \quad g(x) = x^3$

102. $f(x) = -x^4 + 2x^3 - 6x, \quad g(x) = -x^4$

Critical Thinking Exercises

103. Which one of the following is true?

a. If $f(x) = -x^3 + 4x$, then the graph of f falls to the left and falls to the right.

b. A mathematical model that is a polynomial of degree n whose leading term is $a_n x^n$, n odd and $a_n < 0$, is ideally suited to describe nonnegative phenomena over unlimited periods of time.

c. There is more than one third-degree polynomial function with the same three x-intercepts.

d. The graph of a function with origin symmetry can rise to the left and to the right.

Use the descriptions in Exercises 104–105 to write an equation of a polynomial function with the given characteristics. Use a graphing utility to graph your function to see if you are correct. If not, modify the function's equation and repeat this process.

104. Crosses the x-axis at -4, 0, and 3; lies above the x-axis between -4 and 0; lies below the x-axis between 0 and 3

105. Touches the x-axis at 0 and crosses the x-axis at 2; lies below the x-axis between 0 and 2

SECTION 3.3 Dividing Polynomials; Remainder and Factor Theorems

Objectives

1. Use long division to divide polynomials.
2. Use synthetic division to divide polynomials.
3. Evaluate a polynomial using the Remainder Theorem.
4. Use the Factor Theorem to solve a polynomial equation.

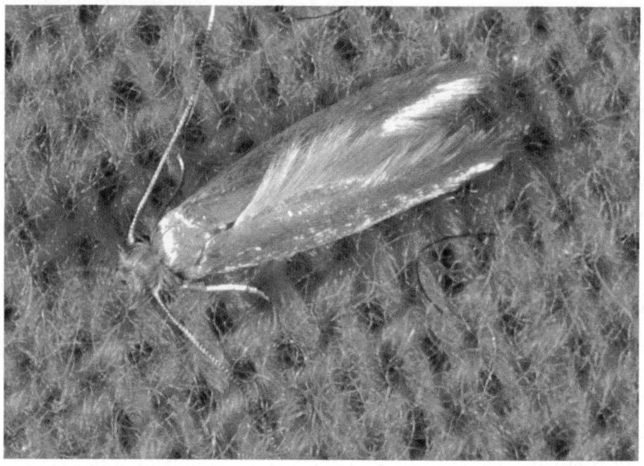

A moth has moved into your closet. She appeared in your bedroom at night, but somehow her relatively stout body escaped your clutches. Within a few weeks, swarms of moths in your tattered wardrobe suggest that Mama Moth was in the family way. There must be at least 200 critters nesting in every crevice of your clothing.

Two hundred plus moth-tykes from one female moth—is this possible? Indeed it is. The number of eggs, $f(x)$, in a female moth is a function of her abdominal width, x, in millimeters, modeled by

$$f(x) = 14x^3 - 17x^2 - 16x + 34, \quad 1.5 \le x \le 3.5.$$

Because there are 200 moths feasting on your favorite sweaters, Mama's abdominal width can be estimated by finding the solutions of the polynomial equation

$$14x^3 - 17x^2 - 16x + 34 = 200.$$

How can we solve such an equation? You might begin by subtracting 200 from both sides to obtain zero on one side. But then what? The factoring that we used in the previous section will not work in this situation.

In Section 3.4, we will present techniques for solving certain kinds of polynomial equations. These techniques will further enhance your ability to manipulate algebraically the polynomial functions that model your world. Because these techniques are based on understanding polynomial division, in this section we look at two methods for dividing polynomials. (We'll return to Mama Moth's abdominal width in the exercise set.)

1 Use long division to divide polynomials.

Long Division of Polynomials and the Division Algorithm

We begin by looking at division by a polynomial containing more than one term, such as

$$x + 3\overline{)x^2 + 10x + 21}.$$

Divisor has two terms and is a binomial.

The polynomial dividend has three terms and is a trinomial.

When a divisor has more than one term, the four steps used to divide whole numbers–**divide, multiply, subtract, bring down the next term**—form the repetitive procedure for polynomial long division.

EXAMPLE 1 Long Division of Polynomials

Divide $x^2 + 10x + 21$ by $x + 3$.

Solution The following steps illustrate how polynomial division is very similar to numerical division.

$$x + 3\overline{)x^2 + 10x + 21}$$

Arrange the terms of the dividend $(x^2 + 10x + 21)$ and the divisor $(x + 3)$ in descending powers of x.

$$\begin{array}{r} x \\ x + 3\overline{)x^2 + 10x + 21} \end{array}$$

Divide x^2 (the first term in the dividend) by x (the first term in the divisor): $\frac{x^2}{x} = x$. Align like terms.

$x(x + 3) = x^2 + 3x$

$$\begin{array}{r} x \\ x + 3\overline{)x^2 + 10x + 21} \\ \underline{x^2 + 3x} \end{array}$$

Multiply each term in the divisor $(x + 3)$ by x, aligning terms of the product under like terms in the dividend.

$$\begin{array}{r} x \\ x + 3\overline{)x^2 + 10x + 21} \\ \underline{\ominus x^2 \oplus\!\!\!\!\ominus 3x} \\ 7x \end{array}$$

Change signs of the polynomial being subtracted.

Subtract $x^2 + 3x$ from $x^2 + 10x$ by changing the sign of each term in the lower expression and adding.

$$\begin{array}{r} x \\ x + 3\overline{)x^2 + 10x + 21} \\ \underline{x^2 + 3x} \downarrow \\ 7x + 21 \end{array}$$

Bring down 21 from the original dividend and add algebraically to form a new dividend.

$$\begin{array}{r} x + 7 \\ x + 3\overline{)x^2 + 10x + 21} \\ \underline{x^2 + 3x} \\ 7x + 21 \end{array}$$

Find the second term of the quotient. Divide the first term of $7x + 21$ by x, the first term of the divisor: $\frac{7x}{x} = 7$.

$7(x + 3) = 7x + 21$

$$\begin{array}{r} x + 7 \\ x + 3\overline{)x^2 + 10x + 21} \\ \underline{x^2 + 3x} \\ 7x + 21 \\ \underline{\ominus 7x \ominus 21} \\ 0 \end{array}$$

Remainder

Multiply the divisor $(x + 3)$ by 7, aligning under like terms in the new dividend. Then subtract to obtain the remainder of 0.

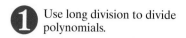

The quotient is $x + 7$. Because the remainder is 0, we can conclude that $x + 3$ is a factor of $x^2 + 10x + 21$ and

$$\frac{x^2 + 10x + 21}{x + 3} = x + 7.$$

Check Point 1 Divide $x^2 + 14x + 45$ by $x + 9$.

Before considering additional examples, let's summarize the general procedure for dividing one polynomial by another.

Long Division of Polynomials

1. **Arrange** the terms of both the dividend and the divisor in descending powers of the variable.
2. **Divide** the first term in the dividend by the first term in the divisor. The result is the first term of the quotient.
3. **Multiply** every term in the divisor by the first term in the quotient. Write the resulting product beneath the dividend with like terms lined up.
4. **Subtract** the product from the dividend.
5. **Bring down** the next term in the original dividend and write it next to the remainder to form a new dividend.
6. Use this new expression as the dividend and repeat this process until the remainder can no longer be divided. This will occur when the degree of the remainder (the highest exponent on a variable in the remainder) is less than the degree of the divisor.

In our next long division, we will obtain a nonzero remainder.

EXAMPLE 2   Long Division of Polynomials

Divide $4 - 5x - x^2 + 6x^3$ by $3x - 2$.

Solution   We begin by writing the dividend in descending powers of x.

$$4 - 5x - x^2 + 6x^3 = 6x^3 - x^2 - 5x + 4$$

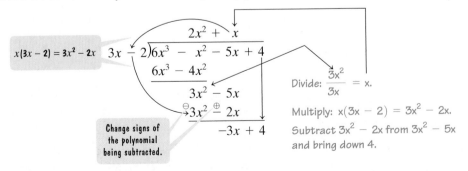

Now we divide $3x^2$ by $3x$ to obtain x, multiply x and the divisor, and subtract.

$x(3x - 2) = 3x^2 - 2x$

$$
\begin{array}{r}
2x^2 + x \\
3x - 2 \overline{\smash{)}6x^3 - x^2 - 5x + 4} \\
6x^3 - 4x^2 \\
\hline
3x^2 - 5x \\
3x^2 - 2x \\
\hline
-3x + 4
\end{array}
$$

Divide: $\dfrac{3x^2}{3x} = x.$

Multiply: $x(3x - 2) = 3x^2 - 2x.$

Subtract $3x^2 - 2x$ from $3x^2 - 5x$ and bring down 4.

Change signs of the polynomial being subtracted.

Now we divide $-3x$ by $3x$ to obtain -1, multiply -1 and the divisor, and subtract.

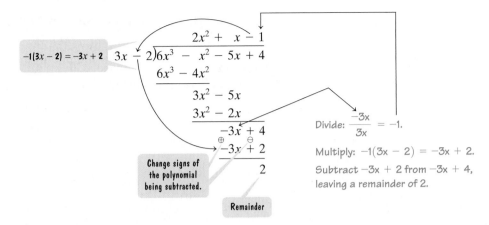

The quotient is $2x^2 + x - 1$ and the remainder is 2. When there is a nonzero remainder, as in this example, list the quotient, plus the remainder above the divisor. Thus,

$$\frac{6x^3 - x^2 - 5x + 4}{3x - 2} = \underbrace{2x^2 + x - 1}_{\text{Quotient}} + \underbrace{\frac{2}{3x - 2}}_{\substack{\text{Remainder} \\ \text{above divisor}}}.$$

An important property of division can be illustrated by clearing fractions in the equation that concluded Example 2. Multiplying both sides of this equation by $3x - 2$ results in the following equation:

$$6x^3 - x^2 - 5x + 4 = \underbrace{(3x - 2)}_{\text{Divisor}}\underbrace{(2x^2 + x - 1)}_{\text{Quotient}} + \underbrace{2}_{\text{Remainder}}.$$

Dividend

Polynomial long division is checked by multiplying the divisor with the quotient and then adding the remainder. This should give the dividend. The process illustrates the **Division Algorithm**.

The Division Algorithm

If $f(x)$ and $d(x)$ are polynomials, with $d(x) \neq 0$, and the degree of $d(x)$ is less than or equal to the degree of $f(x)$, then there exist unique polynomials $q(x)$ and $r(x)$ such that

$$\underbrace{f(x)}_{\text{Dividend}} = \underbrace{d(x)}_{\text{Divisor}} \cdot \underbrace{q(x)}_{\text{Quotient}} + \underbrace{r(x)}_{\text{Remainder}}.$$

The remainder, $r(x)$, equals 0 or it is of degree less than the degree of $d(x)$. If $r(x) = 0$, we say that $d(x)$ **divides evenly** into $f(x)$ and that $d(x)$ and $q(x)$ are **factors** of $f(x)$.

Check Point 2 Divide $7 - 11x - 3x^2 + 2x^3$ by $x - 3$. Express the result in the form quotient, plus remainder divided by divisor.

If a power of x is missing in either a dividend or a divisor, add that power of x with a coefficient of 0 and then divide. In this way, like terms will be aligned as you carry out the long division.

EXAMPLE 3 Long Division of Polynomials

Divide $6x^4 + 5x^3 + 3x - 5$ by $3x^2 - 2x$.

Solution We write the dividend, $6x^4 + 5x^3 + 3x - 5$, as $6x^4 + 5x^3 + 0x^2 + 3x - 5$ to keep all like terms aligned.

Multiply.

$$
\begin{array}{r|l}
 & 2x^2 + 3x + 2 \\
3x^2 - 2x & 6x^4 + 5x^3 + 0x^2 + 3x - 5 \\
 & \ominus 6x^4 \oplus 4x^3 \\
 & \quad 9x^3 + 0x^2 \\
 & \quad \ominus 9x^3 \oplus 6x^2 \\
 & \qquad 6x^2 + 3x \\
 & \qquad \ominus 6x^2 \oplus 4x \\
 & \qquad\quad 7x - 5
\end{array}
$$

$2x^2(3x^2 - 2x) = 6x^4 - 4x^3$

$3x(3x^2 - 2x) = 9x^3 - 6x^2$

$2(3x^2 - 2x) = 6x^2 - 4x$

Remainder

The division process is finished because the degree of $7x - 5$, which is 1, is less than the degree of the divisor $3x^2 - 2x$, which is 2. The answer is

$$\frac{6x^4 + 5x^3 + 3x - 5}{3x^2 - 2x} = 2x^2 + 3x + 2 + \frac{7x - 5}{3x^2 - 2x}.$$

Check Point 3 Divide $2x^4 + 3x^3 - 7x - 10$ by $x^2 - 2x$.

2 Use synthetic division to divide polynomials.

Dividing Polynomials Using Synthetic Division

We can use **synthetic division** to divide polynomials if the divisor is of the form $x - c$. This method provides a quotient more quickly than long division. Let's compare the two methods showing $x^3 + 4x^2 - 5x + 5$ divided by $x - 3$.

Long Division

$$
\begin{array}{r|l}
 & x^2 + 7x + 16 \\
x - 3 & x^3 + 4x^2 - 5x + 5 \\
 & \ominus x^3 \oplus 3x^2 \\
 & \quad 7x^2 - 5x \\
 & \quad \ominus 7x^2 \oplus 21x \\
 & \qquad 16x + 5 \\
 & \qquad \ominus 16x \oplus 48 \\
 & \qquad\qquad 53
\end{array}
$$

Quotient

Dividend

Divisor $x - c$; $c = 3$

Remainder

Synthetic Division

$$
\begin{array}{r|rrrr}
3 & 1 & 4 & -5 & 5 \\
 & & 3 & 21 & 48 \\
\hline
 & 1 & 7 & 16 & 53
\end{array}
$$

Notice the relationship between the polynomials in the long division process and the numbers that appear in synthetic division.

These are the coefficients of the dividend $x^3 + 4x^2 - 5x + 5$.

The divisor is $x - 3$. This is 3, or c, in $x - c$.

$$
\begin{array}{r|rrrr}
3 & 1 & 4 & -5 & 5 \\
 & & 3 & 21 & 48 \\
\hline
 & 1 & 7 & 16 & 53
\end{array}
$$

These are the coefficients of the quotient $x^2 + 7x + 16$.

This is the remainder.

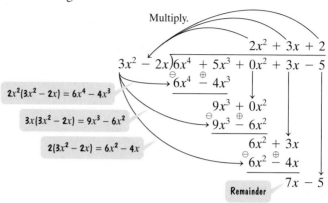

Now let's look at the steps involved in synthetic division.

Synthetic Division

To divide a polynomial by $x - c$:

Example

$x - 3 \overline{)x^3 + 4x^2 - 5x + 5}$

1. Arrange the polynomial in descending powers, with a 0 coefficient for any missing term.

2. Write c for the divisor, $x - c$. To the right, write the coefficients of the dividend.

$3 |\ \ 1\ \ 4\ \ -5\ \ 5$

3. Write the leading coefficient of the dividend on the bottom row.

$3 |\ \ 1\ \ 4\ \ -5\ \ 5$
$\ \ \ \ \ \ \downarrow$ Bring down 1.
$\ \ \ \ \ \ 1$

4. Multiply c (in this case, 3) times the value just written on the bottom row. Write the product in the next column in the second row.

$3 |\ \ 1\ \ 4\ \ -5\ \ 5$
$\ \ \ \ \ \ \ \ \ 3$
$\ \ \ \ \ 1$

Multiply by 3: $3 \cdot 1 = 3$.

5. Add the values in this new column, writing the sum in the bottom row.

$3 |\ \ 1\ \ 4 |\ \ -5\ \ 5$
$\ \ \ \ \ \ \ \ \ 3$ | Add.
$\ \ \ \ \ 1\ \ 7$

6. Repeat this series of multiplications and additions until all columns are filled in.

$3 |\ \ 1\ \ 4\ \ -5 |\ \ 5$
$\ \ \ \ \ \ \ \ \ 3\ \ 21$ | Add.
$\ \ \ \ \ 1\ \ 7\ \ 16$

Multiply by 3: $3 \cdot 7 = 21$.

$3 |\ \ 1\ \ 4\ \ -5\ \ 5 |$
$\ \ \ \ \ \ \ \ \ 3\ \ 21\ \ 48$ | Add.
$\ \ \ \ \ 1\ \ 7\ \ 16\ \ 53$

Multiply by 3: $3 \cdot 16 = 48$.

7. Use the numbers in the last row to write the quotient, plus the remainder above the divisor. **The degree of the first term of the quotient is one less than the degree of the first term of the dividend.** The final value in this row is the remainder.

Written from
1 7 16 53
the last row of the synthetic division

$1x^2 + 7x + 16 + \dfrac{53}{x - 3}$

$x - 3 \overline{)x^3 + 4x^2 - 5x +\ \ 5}$

EXAMPLE 4   Using Synthetic Division

Use synthetic division to divide $5x^3 + 6x + 8$ by $x + 2$.

Solution The divisor must be in the form $x - c$. Thus, we write $x + 2$ as $x - (-2)$. This means that $c = -2$. Writing a 0 coefficient for the missing x^2-term in the dividend, we can express the division as follows:

$$x - (-2) \overline{)5x^3 + 0x^2 + 6x + 8}.$$

Now we are ready to set up the problem so that we can use synthetic division.

Use the coefficients of the dividend
$5x^3 + 0x^2 + 6x + 8$ in descending powers of x.

This is c in
$x - (-2)$. $-2 |\ \ 5\ \ 0\ \ 6\ \ 8$

We begin the synthetic division process by bringing down 5. This is followed by a series of multiplications and additions.

1. Bring down 5.

$$\begin{array}{r|rrrr} -2 & 5 & 0 & 6 & 8 \\ & & & & \\ \hline & 5 & & & \end{array}$$

2. Multiply: $-2(5) = -10$.

$$\begin{array}{r|rrrr} -2 & 5 & 0 & 6 & 8 \\ & & -10 & & \\ \hline & 5 & & & \end{array}$$

Multiply by −2.

3. Add: $0 + (-10) = -10$.

$$\begin{array}{r|rrrr} -2 & 5 & 0 & 6 & 8 \\ & & -10 & & \\ \hline & 5 & -10 & & \end{array}$$

Add.

4. Multiply: $-2(-10) = 20$.

$$\begin{array}{r|rrrr} -2 & 5 & 0 & 6 & 8 \\ & & -10 & 20 & \\ \hline & 5 & -10 & & \end{array}$$

Multiply by −2.

5. Add: $6 + 20 = 26$.

$$\begin{array}{r|rrrr} -2 & 5 & 0 & 6 & 8 \\ & & -10 & 20 & \\ \hline & 5 & -10 & 26 & \end{array}$$

Add.

6. Multiply: $-2(26) = -52$.

$$\begin{array}{r|rrrr} -2 & 5 & 0 & 6 & 8 \\ & & -10 & 20 & -52 \\ \hline & 5 & -10 & 26 & \end{array}$$

Multiply by −2.

7. Add: $8 + (-52) = -44$.

$$\begin{array}{r|rrrr} -2 & 5 & 0 & 6 & 8 \\ & & -10 & 20 & -52 \\ \hline & 5 & -10 & 26 & -44 \end{array}$$

Add.

The numbers in the last row represent the coefficients of the quotient and the remainder. The degree of the first term of the quotient is one less than that of the dividend. Because the degree of the dividend, $5x^3 + 6x + 8$, is 3, the degree of the quotient is 2. This means that the 5 in the last row represents $5x^2$.

$$\begin{array}{r|rrrr} -2 & 5 & 0 & 6 & 8 \\ & & -10 & 20 & -52 \\ \hline & 5 & -10 & 26 & -44 \end{array}$$

The quotient is $5x^2 - 10x + 26$.

The remainder is −44.

Thus,

$$x + 2\overline{)5x^3 + 6x + 8} \quad = \quad 5x^2 - 10x + 26 - \frac{44}{x+2}$$

Check Point 4 Use synthetic division to divide $x^3 - 7x - 6$ by $x + 2$.

3 Evaluate a polynomial using the Remainder Theorem.

The Remainder Theorem

Let's consider the Division Algorithm when the dividend, $f(x)$, is divided by $x - c$. In this case, the remainder must be a constant because its degree is less than one, the degree of $x - c$.

$$f(x) = (x - c)q(x) + r$$

Dividend: $f(x)$; Divisor: $(x - c)$; Quotient: $q(x)$; The remainder, r, is a constant when dividing by $x - c$.

Now let's evaluate f at c.

$f(c) = (c - c)q(c) + r$ Find $f(c)$ by letting $x = c$ in $f(x) = (x - c)q(x) + r$. This will give an expression for r.

$f(c) = 0 \cdot q(c) + r$ $c - c = 0$

$f(c) = r$ $0 \cdot q(c) = 0$ and $0 + r = r$.

What does this last equation mean? If a polynomial is divided by $x - c$, the remainder is the value of the polynomial at c. This result is called the **Remainder Theorem**.

The Remainder Theorem

If the polynomial $f(x)$ is divided by $x - c$, then the remainder is $f(c)$.

Example 5 shows how we can use the Remainder Theorem to evaluate a polynomial function at 2. Rather than substituting 2 for x, we divide the function by $x - 2$. The remainder is $f(2)$.

EXAMPLE 5 Using the Remainder Theorem to Evaluate a Polynomial Function

Given $f(x) = x^3 - 4x^2 + 5x + 3$, use the Remainder Theorem to find $f(2)$.

Solution By the Remainder Theorem, if $f(x)$ is divided by $x - 2$, then the remainder is $f(2)$. We'll use synthetic division to divide.

$$\begin{array}{r|rrrr} 2 & 1 & -4 & 5 & 3 \\ & & 2 & -4 & 2 \\ \hline & 1 & -2 & 1 & 5 \end{array}$$

Remainder

The remainder, 5, is the value of $f(2)$. Thus, $f(2) = 5$. We can verify that this is correct by evaluating $f(2)$ directly. Using $f(x) = x^3 - 4x^2 + 5x + 3$, we obtain

$$f(2) = 2^3 - 4 \cdot 2^2 + 5 \cdot 2 + 3 = 8 - 16 + 10 + 3 = 5.$$

Check Point 5 Given $f(x) = 3x^3 + 4x^2 - 5x + 3$, use the Remainder Theorem to find $f(-4)$.

4 Use the Factor Theorem to solve a polynomial equation.

The Factor Theorem

Let's look again at the Division Algorithm when the divisor is of the form $x - c$.

$$f(x) = (x - c)q(x) + r$$

Dividend, Divisor, Quotient, Constant remainder

By the Remainder Theorem, the remainder r is $f(c)$, so we can substitute $f(c)$ for r:

$$f(x) = (x - c)q(x) + f(c).$$

Notice that if $f(c) = 0$, then

$$f(x) = (x - c)q(x)$$

so that $x - c$ is a factor of $f(x)$. This means that for the polynomial function $f(x)$, if $f(c) = 0$, then $x - c$ is a factor of $f(x)$.

Let's reverse directions and see what happens if $x - c$ is a factor of $f(x)$. This means that

$$f(x) = (x - c)q(x).$$

If we replace x in $f(x) = (x - c)q(x)$ with c, we obtain

$$f(c) = (c - c)q(c) = 0 \cdot q(c) = 0.$$

Thus, if $x - c$ is a factor of $f(x)$, then $f(c) = 0$.

We have proved a result known as the **Factor Theorem**.

> **The Factor Theorem**
>
> Let $f(x)$ be a polynomial.
>
> **a.** If $f(c) = 0$, then $x - c$ is a factor of $f(x)$.
>
> **b.** If $x - c$ is a factor of $f(x)$, then $f(c) = 0$.

The example that follows shows how the Factor Theorem can be used to solve a polynomial equation.

EXAMPLE 6   Using the Factor Theorem

Solve the equation $2x^3 - 3x^2 - 11x + 6 = 0$ given that 3 is a zero of $f(x) = 2x^3 - 3x^2 - 11x + 6$.

Solution   We are given that 3 is a zero of $f(x) = 2x^3 - 3x^2 - 11x + 6$. This means that $f(3) = 0$. Because $f(3) = 0$, the Factor Theorem tells us that $x - 3$ is a factor of $f(x)$. We'll use synthetic division to divide $f(x)$ by $x - 3$.

$$\begin{array}{r|rrrr} 3 & 2 & -3 & -11 & 6 \\ & & 6 & 9 & -6 \\ \hline & 2 & 3 & -2 & 0 \end{array} \qquad \begin{array}{r} 2x^2 + 3x - 2 \\ x - 3\overline{)2x^3 - 3x^2 - 11x + 6} \end{array}$$

The remainder, 0, verifies that $x - 3$ is a factor of $2x^3 - 3x^2 - 11x + 6$.

Equivalently,

$$2x^3 - 3x^2 - 11x + 6 = (x - 3)(2x^2 + 3x - 2).$$

Now we can solve the polynomial equation.

$$2x^3 - 3x^2 - 11x + 6 = 0 \qquad \text{This is the given equation.}$$

$$(x - 3)(2x^2 + 3x - 2) = 0 \qquad \text{Factor using the result from the synthetic division.}$$

$$(x - 3)(2x - 1)(x + 2) = 0 \qquad \text{Factor the trinomial.}$$

$$x - 3 = 0 \quad \text{or} \quad 2x - 1 = 0 \quad \text{or} \quad x + 2 = 0 \qquad \text{Set each factor equal to 0.}$$

$$x = 3 \qquad\qquad x = \tfrac{1}{2} \qquad\qquad x = -2 \qquad \text{Solve for x.}$$

The solution set is $\left\{-2, \frac{1}{2}, 3\right\}$.

Based on the Factor Theorem, the following statements are useful in solving polynomial equations:

1. If $f(x)$ is divided by $x - c$ and the remainder is zero, then c is a zero of f and c is a root of the polynomial equation $f(x) = 0$.

2. If $f(x)$ is divided by $x - c$ and the remainder is zero, then $x - c$ is a factor of $f(x)$.

Check Point 6   Solve the equation $15x^3 + 14x^2 - 3x - 2 = 0$ given that -1 is a zero of $f(x) = 15x^3 + 14x^2 - 3x - 2$.

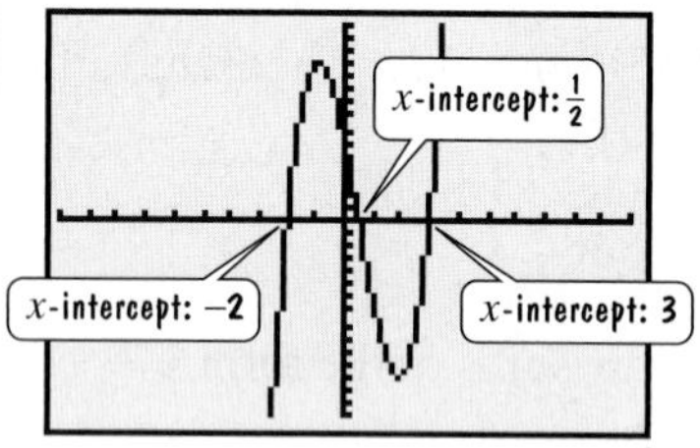

EXERCISE SET 3.3

Practice Exercises

In Exercises 1–16, divide using long division. State the quotient, $q(x)$, and the remainder, $r(x)$.

1. $(x^2 + 8x + 15) \div (x + 5)$

2. $(x^2 + 3x - 10) \div (x - 2)$

3. $(x^3 + 5x^2 + 7x + 2) \div (x + 2)$

4. $(x^3 - 2x^2 - 5x + 6) \div (x - 3)$

5. $(6x^3 + 7x^2 + 12x - 5) \div (3x - 1)$

6. $(6x^3 + 17x^2 + 27x + 20) \div (3x + 4)$

7. $(12x^2 + x - 4) \div (3x - 2)$

8. $(4x^2 - 8x + 6) \div (2x - 1)$

9. $\dfrac{2x^3 + 7x^2 + 9x - 20}{x + 3}$

10. $\dfrac{3x^2 - 2x + 5}{x - 3}$

11. $\dfrac{4x^4 - 4x^2 + 6x}{x - 4}$

12. $\dfrac{x^4 - 81}{x - 3}$

13. $\dfrac{6x^3 + 13x^2 - 11x - 15}{3x^2 - x - 3}$

14. $\dfrac{x^4 + 2x^3 - 4x^2 - 5x - 6}{x^2 + x - 2}$

15. $\dfrac{18x^4 + 9x^3 + 3x^2}{3x^2 + 1}$

16. $\dfrac{2x^5 - 8x^4 + 2x^3 + x^2}{2x^3 + 1}$

In Exercises 17–32, divide using synthetic division.

17. $(2x^2 + x - 10) \div (x - 2)$

18. $(x^2 + x - 2) \div (x - 1)$

19. $(3x^2 + 7x - 20) \div (x + 5)$

20. $(5x^2 - 12x - 8) \div (x + 3)$

21. $(4x^3 - 3x^2 + 3x - 1) \div (x - 1)$

22. $(5x^3 - 6x^2 + 3x + 11) \div (x - 2)$

23. $(6x^5 - 2x^3 + 4x^2 - 3x + 1) \div (x - 2)$

24. $(x^5 + 4x^4 - 3x^2 + 2x + 3) \div (x - 3)$

25. $(x^2 - 5x - 5x^3 + x^4) \div (5 + x)$

26. $(x^2 - 6x - 6x^3 + x^4) \div (6 + x)$

27. $\dfrac{x^5 + x^3 - 2}{x - 1}$

28. $\dfrac{x^7 + x^5 - 10x^3 + 12}{x + 2}$

29. $\dfrac{x^4 - 256}{x - 4}$

30. $\dfrac{x^7 - 128}{x - 2}$

31. $\dfrac{2x^5 - 3x^4 + x^3 - x^2 + 2x - 1}{x + 2}$

32. $\dfrac{x^5 - 2x^4 - x^3 + 3x^2 - x + 1}{x - 2}$

In Exercises 33–40, use synthetic division and the Remainder Theorem to find the indicated function value.

33. $f(x) = 2x^3 - 11x^2 + 7x - 5;\quad f(4)$

34. $f(x) = x^3 - 7x^2 + 5x - 6;\quad f(3)$

35. $f(x) = 3x^3 - 7x^2 - 2x + 5;\quad f(-3)$

36. $f(x) = 4x^3 + 5x^2 - 6x - 4;\quad f(-2)$

37. $f(x) = x^4 + 5x^3 + 5x^2 - 5x - 6;\quad f(3)$

38. $f(x) = x^4 - 5x^3 + 5x^2 + 5x - 6;\quad f(2)$

39. $f(x) = 2x^4 - 5x^3 - x^2 + 3x + 2;\quad f\left(-\dfrac{1}{2}\right)$

40. $f(x) = 6x^4 + 10x^3 + 5x^2 + x + 1;\quad f\left(-\dfrac{2}{3}\right)$

41. Use synthetic division to divide
$$f(x) = x^3 - 4x^2 + x + 6 \text{ by } x + 1.$$
Use the result to find all zeros of f.

42. Use synthetic division to divide
$$f(x) = x^3 - 2x^2 - x + 2 \text{ by } x + 1.$$
Use the result to find all zeros of f.

43. Solve the equation $2x^3 - 5x^2 + x + 2 = 0$ given that 2 is a zero of $f(x) = 2x^3 - 5x^2 + x + 2$.

44. Solve the equation $2x^3 - 3x^2 - 11x + 6 = 0$ given that -2 is a zero of $f(x) = 2x^3 - 3x^2 - 11x + 6$.

45. Solve the equation $12x^3 + 16x^2 - 5x - 3 = 0$ given that $-\dfrac{3}{2}$ is a root.

46. Solve the equation $3x^3 + 7x^2 - 22x - 8 = 0$ given that $-\dfrac{1}{3}$ is a root.

Practice Plus

In Exercises 47–50, use the graph or the table to determine a solution of each equation. Use synthetic division to verify that this number is a solution of the equation. Then solve the polynomial equation.

47. $x^3 + 2x^2 - 5x - 6 = 0$

$y = x^3 + 2x^2 - 5x - 6$

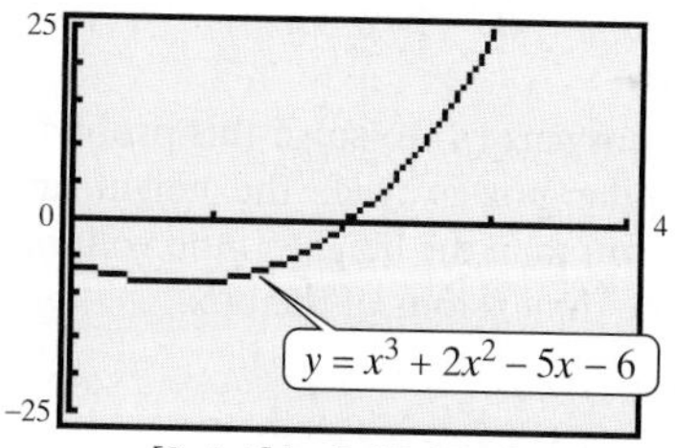

$[0, 4, 1]$ by $[-25, 25, 5]$

48. $2x^3 + x^2 - 13x + 6 = 0$

$y = 2x^3 + x^2 - 13x + 6$

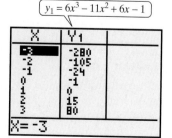

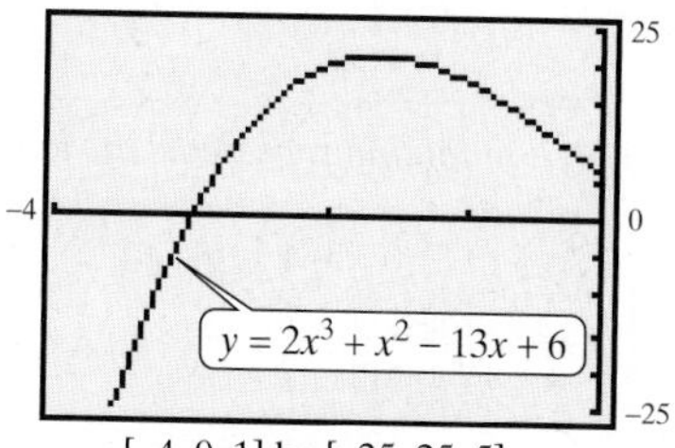

$[-4, 0, 1]$ by $[-25, 25, 5]$

49. $6x^3 - 11x^2 + 6x - 1 = 0$

$y_1 = 6x^3 - 11x^2 + 6x - 1$

50. $2x^3 + 11x^2 - 7x - 6 = 0$

$$y_1 = 2x^3 + 11x^2 - 7x - 6$$

X	Y1
-3	60
-2	36
-1	10
0	-6
1	0
2	40
3	126

X = -3

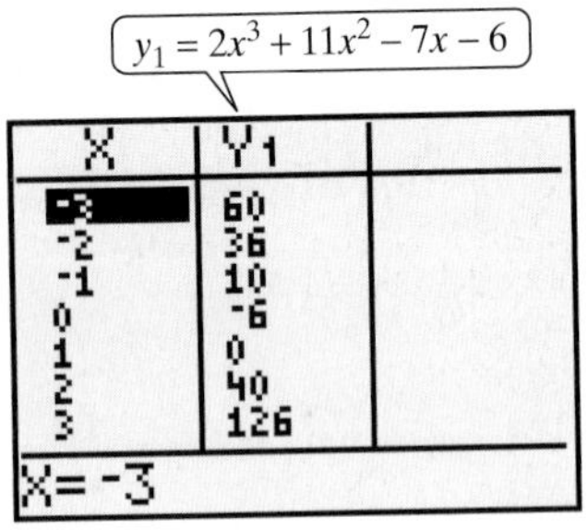

⭐ Application Exercises

51. a. Use synthetic division to show that 3 is a solution of the polynomial equation

$$14x^3 - 17x^2 - 16x - 177 = 0.$$

b. Use the solution from part (a) to solve this problem. The number of eggs, $f(x)$, in a female moth is a function of her abdominal width, x, in millimeters, modeled by

$$f(x) = 14x^3 - 17x^2 - 16x + 34.$$

What is the abdominal width when there are 211 eggs?

52. a. Use synthetic division to show that 2 is a solution of the polynomial equation

$$2h^3 + 14h^2 - 72 = 0.$$

b. Use the solution from part (a) to solve this problem. The width of a rectangular box is twice the height and the length is 7 inches more than the height. If the volume is 72 cubic inches, find the dimensions of the box.

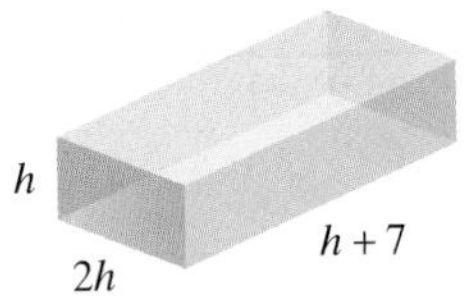

In Exercises 53–54, write a polynomial that represents the length of each rectangle.

53.

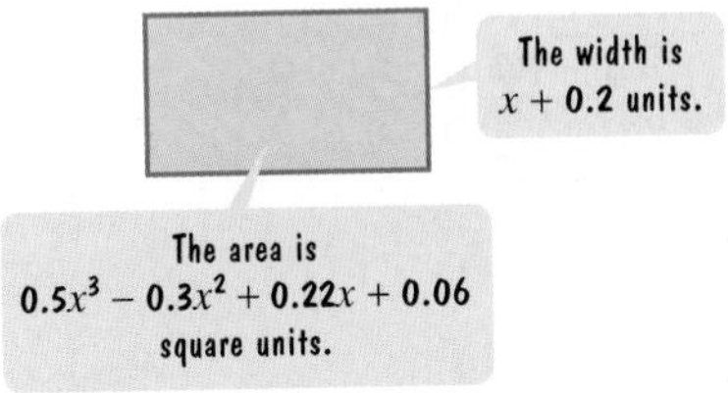

The width is $x + 0.2$ units.

The area is $0.5x^3 - 0.3x^2 + 0.22x + 0.06$ square units.

54.

The width is $x + \frac{3}{4}$ units.

The area is $8x^3 + 6x^2 - 5x + 3$ square units.

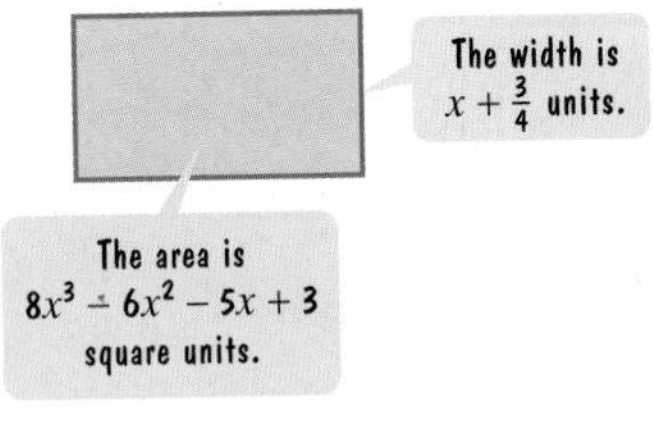

During the 1980s, the controversial economist Arthur Laffer promoted the idea that tax increases lead to a reduction in government revenue. Called supply-side economics, the theory uses functions such as

$$f(x) = \frac{80x - 8000}{x - 110}, 30 \le x \le 100.$$

This function models the government tax revenue, $f(x)$, in tens of billions of dollars, in terms of the tax rate, x. The graph of the function is shown. It illustrates tax revenue decreasing quite dramatically as the tax rate increases. At a tax rate of (gasp) 100%, the government takes all our money and no one has an incentive to work. With no income earned, zero dollars in tax revenue is generated.

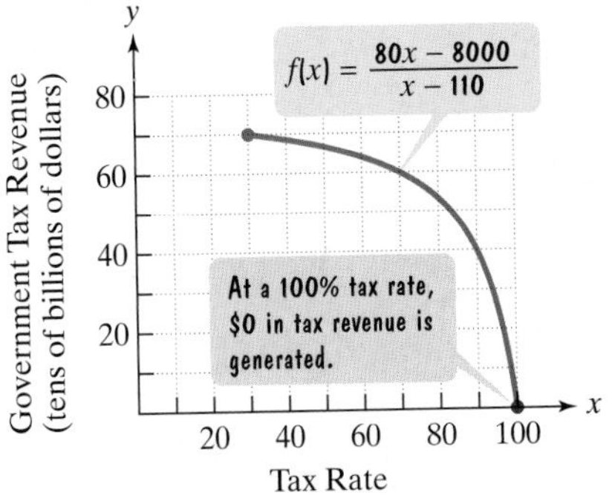

Use function f and its graph to solve Exercises 55–56.

55. a. Find and interpret $f(30)$. Identify the solution as a point on the graph of the function.

b. Rewrite the function by using long division to perform

$$(80x - 8000) \div (x - 110).$$

Then use this new form of the function to find $f(30)$. Do you obtain the same answer as you did in part (a)?

c. Is f a polynomial function? Explain your answer.

56. a. Find and interpret $f(40)$. Identify the solution as a point on the graph of the function.

b. Rewrite the function by using long division to perform

$$(80x - 8000) \div (x - 110).$$

Then use this new form of the function to find $f(40)$. Do you obtain the same answer as you did in part (a)?

c. Is f a polynomial function? Explain your answer.

✏️ Writing in Mathematics

57. Explain how to perform long division of polynomials. Use $2x^3 - 3x^2 - 11x + 7$ divided by $x - 3$ in your explanation.

58. In your own words, state the Division Algorithm.

59. How can the Division Algorithm be used to check the quotient and remainder in a long division problem?

60. Explain how to perform synthetic division. Use the division problem in Exercise 57 to support your explanation.

61. State the Remainder Theorem.

62. Explain how the Remainder Theorem can be used to find $f(-6)$ if $f(x) = x^4 + 7x^3 + 8x^2 + 11x + 5$. What advantage is there to using the Remainder Theorem in this situation rather than evaluating $f(-6)$ directly?

63. How can the Factor Theorem be used to determine if $x - 1$ is a factor of $x^3 - 2x^2 - 11x + 12$?

64. If you know that -2 is a zero of

$$f(x) = x^3 + 7x^2 + 4x - 12,$$

explain how to solve the equation

$$x^3 + 7x^2 + 4x - 12 = 0.$$

Technology Exercise

65. For each equation that you solved in Exercises 43–46, use a graphing utility to graph the polynomial function defined by the left side of the equation. Use end behavior to obtain a complete graph. Then use the graph's x-intercepts to verify your solutions.

Critical Thinking Exercises

66. Which one of the following is true?

a. If a trinomial in x of degree 6 is divided by a trinomial in x of degree 3, the degree of the quotient is 2.

b. Synthetic division could not be used to find the quotient of $10x^3 - 6x^2 + 4x - 1$ and $x - \frac{1}{2}$.

c. Any problem that can be done by synthetic division can also be done by the method for long division of polynomials.

d. If a polynomial long-division problem results in a remainder that is a whole number, then the divisor is a factor of the dividend.

67. Find k so that $4x + 3$ is a factor of

$$20x^3 + 23x^2 - 10x + k.$$

68. When $2x^2 - 7x + 9$ is divided by a polynomial, the quotient is $2x - 3$ and the remainder is 3. Find the polynomial.

69. Find the quotient of $x^{3n} + 1$ and $x^n + 1$.

70. Synthetic division is a process for dividing a polynomial by $x - c$. The coefficient of x is 1. How might synthetic division be used if you are dividing by $2x - 4$?

71. Use synthetic division to show that 5 is a solution of

$$x^4 - 4x^3 - 9x^2 + 16x + 20 = 0.$$

Then solve the polynomial equation.

SECTION 3.4 *Zeros of Polynomial Functions*

Objectives

1. Use the Rational Zero Theorem to find possible rational zeros.
2. Find zeros of a polynomial function.
3. Solve polynomial equations.
4. Use the Linear Factorization Theorem to find polynomials with given zeros.
5. Use Descartes's Rule of Signs.

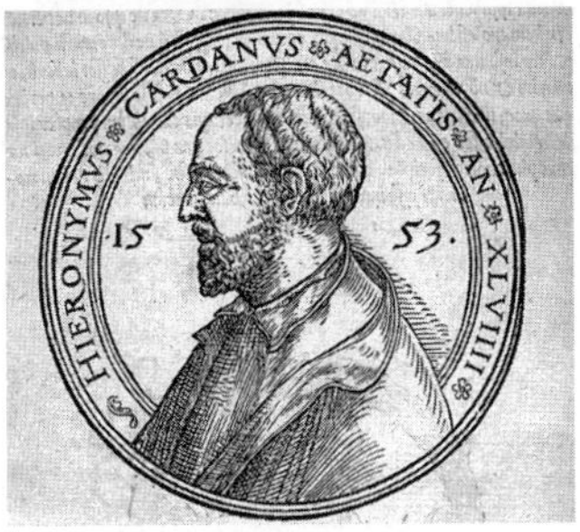

You stole my formula!

Tartaglia's Secret Formula for One Solution of $x^3 + mx = n$

$$x = \sqrt[3]{\sqrt{\left(\frac{n}{2}\right)^2 + \left(\frac{m}{3}\right)^3} + \frac{n}{2}} - \sqrt[3]{\sqrt{\left(\frac{n}{2}\right)^2 + \left(\frac{m}{3}\right)^3} - \frac{n}{2}}$$

Popularizers of mathematics are sharing bizarre stories that are giving math a secure place in popular culture. One episode, able to compete with the wildest fare served up by television talk shows and the tabloids, involves three Italian mathematicians and, of all things, zeros of polynomial functions.

Tartaglia (1499–1557), poor and starving, has found a formula that gives a root for a third-degree polynomial equation. Cardano (1501–1576) begs Tartaglia to reveal the secret formula, wheedling it from him with the promise he will find the impoverished Tartaglia a patron. Then Cardano publishes his famous work *Ars Magna*, in which he presents Tartaglia's formula as his own. Cardano uses his most talented student, Ferrari (1522–1565), who derived a formula for a root of a fourth-degree polynomial equation, to falsely accuse Tartaglia of plagiarism. The dispute becomes violent and Tartaglia is fortunate to escape alive.

The noise from this "You Stole My Formula" episode is quieted by the work of French mathematician Evariste Galois (1811–1832). Galois proved that there is no general formula for finding roots of polynomial equations of degree 5 or higher. There are, however, methods for finding roots. In this section, we study methods for finding zeros of polynomial functions. We begin with a theorem that plays an important role in this process.

Study Tip

Be sure you are familiar with the various kinds of zeros of polynomial functions. Here's a quick example:

$$f(x) = (x + 3)(2x - 1)(x + \sqrt{2})(x - \sqrt{2})(x - 4 + 5i)(x - 4 - 5i).$$

Zeros: -3, $\frac{1}{2}$, $-\sqrt{2}$, $\sqrt{2}$, $4 - 5i$, $4 + 5i$

Rational zeros: -3, $\frac{1}{2}$ — Irrational zeros: $-\sqrt{2}$, $\sqrt{2}$ — Complex imaginary zeros: $4 - 5i$, $4 + 5i$

Real zeros: rational zeros and irrational zeros — Nonreal zeros: complex imaginary zeros

1 Use the Rational Zero Theorem to find possible rational zeros.

The Rational Zero Theorem

The Rational Zero Theorem provides us with a tool that we can use to make a list of all possible rational zeros of a polynomial function. Equivalently, the theorem gives all possible rational roots of a polynomial equation. Not every number in the list will be a zero of the function, but every rational zero of the polynomial function will appear somewhere in the list.

The Rational Zero Theorem

If $f(x) = a_nx^n + a_{n-1}x^{n-1} + \cdots + a_1x + a_0$ has *integer* coefficients and $\frac{p}{q}$ (where $\frac{p}{q}$ is reduced to lowest terms) is a rational zero of f, then p is a factor of the constant term, a_0, and q is a factor of the leading coefficient, a_n.

You can explore the "why" behind the Rational Zero Theorem in Exercise 90 of Exercise Set 3.4. For now, let's see if we can figure out what the theorem tells us about possible rational zeros. To use the theorem, list all the integers that are factors of the constant term, a_0. Then list all the integers that are factors of the leading coefficient, a_n. Finally list all possible rational zeros:

$$\text{Possible rational zeros} = \frac{\text{Factors of the constant term}}{\text{Factors of the leading coefficient}}.$$

EXAMPLE 1 Using the Rational Zero Theorem

List all possible rational zeros of $f(x) = -x^4 + 3x^2 + 4$.

Solution The constant term is 4. We list all of its factors: ±1, ±2, ±4. The leading coefficient is -1. Its factors are ±1.

Factors of the constant term, 4: ±1, ±2, ±4

Factors of the leading coefficient, -1: ±1

Because

$$\text{Possible rational zeros} = \frac{\text{Factors of the constant term}}{\text{Factors of the leading coefficient}},$$

we must take each number in the first row, ±1, ±2, ±4, and divide by each number in the second row, ±1.

$$\text{Possible rational zeros} = \frac{\text{Factors of } 4}{\text{Factors of } -1} = \frac{\pm1, \pm2, \pm4}{\pm1} = \pm1, \quad \pm2, \quad \pm4$$

±1: Divide ±1 by ±1. ±2: Divide ±2 by ±1. ±4: Divide ±4 by ±1.

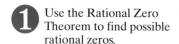

Study Tip

Always keep in mind the relationship among zeros, roots, and x-intercepts. The zeros of a function f are the roots, or solutions, of the equation $f(x) = 0$. Furthermore, the real zeros, or real roots, are the x-intercepts of the graph of f.

There are six possible rational zeros. The graph of $f(x) = -x^4 + 3x^2 + 4$ is shown in Figure 3.23. The x-intercepts are -2 and 2. Thus, -2 and 2 are the actual rational zeros.

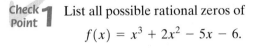

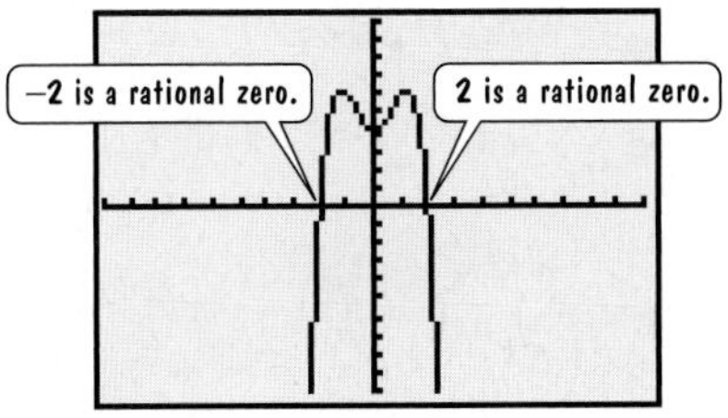

Figure 3.23 The graph of $f(x) = -x^4 + 3x^2 + 4$ shows that -2 and 2 are rational zeros.

Check Point 1 List all possible rational zeros of

$$f(x) = x^3 + 2x^2 - 5x - 6.$$

EXAMPLE 2 Using the Rational Zero Theorem

List all possible rational zeros of $f(x) = 15x^3 + 14x^2 - 3x - 2$.

Solution The constant term is -2 and the leading coefficient is 15.

$$\text{Possible rational zeros} = \frac{\text{Factors of the constant term, } -2}{\text{Factors of the leading coefficient, } 15} = \frac{\pm 1, \pm 2}{\pm 1, \pm 3, \pm 5, \pm 15}$$

$$= \pm 1, \ \pm 2, \ \pm\frac{1}{3}, \ \pm\frac{2}{3}, \ \pm\frac{1}{5}, \ \pm\frac{2}{5}, \ \pm\frac{1}{15}, \ \pm\frac{2}{15}$$

Divide ±1 and ±2 by ±1. | Divide ±1 and ±2 by ±3. | Divide ±1 and ±2 by ±5. | Divide ±1 and ±2 by ±15.

There are 16 possible rational zeros. The actual solution set of

$$15x^3 + 14x^2 - 3x - 2 = 0$$

is $\left\{-1, -\frac{1}{3}, \frac{2}{5}\right\}$, which contains three of the 16 possible zeros.

Check Point 2 List all possible rational zeros of

$$f(x) = 4x^5 + 12x^4 - x - 3.$$

2 Find zeros of a polynomial function.

How do we determine which (if any) of the possible rational zeros are rational zeros of the polynomial function? To find the first rational zero, we can use a trial-and-error process involving synthetic division: If $f(x)$ is divided by $x - c$ and the remainder is zero, then c is a zero of f. After we identify the first rational zero, we use the result of the synthetic division to factor the original polynomial. Then we set each factor equal to zero to identify any additional rational zeros.

EXAMPLE 3 Finding Zeros of a Polynomial Function

Find all zeros of $f(x) = x^3 + 2x^2 - 5x - 6$.

Solution We begin by listing all possible rational zeros.

Possible rational zeros

$$= \frac{\text{Factors of the constant term, } -6}{\text{Factors of the leading coefficient, } 1} = \frac{\pm 1, \pm 2, \pm 3, \pm 6}{\pm 1} = \pm 1, \pm 2, \pm 3, \pm 6$$

Divide the eight numbers in the numerator by ±1.

Now we will use synthetic division to see if we can find a rational zero among the possible rational zeros $\pm 1, \pm 2, \pm 3, \pm 6$. Keep in mind that if $f(x)$ is divided by

$x - c$ and the remainder is zero, then c is a zero of f. Let's start by testing 1. If 1 is not a rational zero, then we will test other possible rational zeros.

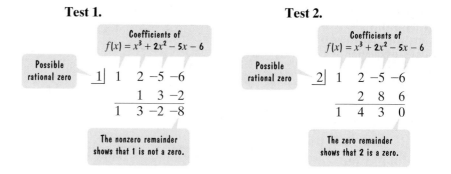

The zero remainder tells us that 2 is a zero of the polynomial function $f(x) = x^3 + 2x^2 - 5x - 6$. Equivalently, 2 is a solution, or root, of the polynomial equation $x^3 + 2x^2 - 5x - 6 = 0$. Thus, $x - 2$ is a factor of the polynomial. The first three numbers in the bottom row of the synthetic division on the right, 1, 4, and 3, give the coefficients of the other factor. This factor is $x^2 + 4x + 3$.

$x^3 + 2x^2 - 5x - 6 = 0$ Finding the zeros of $f(x) = x^3 + 2x^2 - 5x - 6$ is the same as finding the roots of this equation.

$(x - 2)(x^2 + 4x + 3) = 0$ Factor using the result from the synthetic division.

$(x - 2)(x + 3)(x + 1) = 0$ Factor completely.

$x - 2 = 0$ or $x + 3 = 0$ or $x + 1 = 0$ Set each factor equal to zero.

$x = 2$ $x = -3$ $x = -1$ Solve for x.

The solution set is $\{-3, -1, 2\}$. The zeros of f are $-3, -1,$ and 2.

Check Point 3 Find all zeros of
$$f(x) = x^3 + 8x^2 + 11x - 20.$$

Our work in Example 3 involved finding zeros of a third-degree polynomial function. The Rational Zero Theorem is a tool that allows us to rewrite such functions as products of two factors, one linear and one quadratic. Zeros of the quadratic factor are found by factoring, the quadratic formula, or the square root property.

EXAMPLE 4   Finding Zeros of a Polynomial Function

Find all zeros of $f(x) = x^3 + 7x^2 + 11x - 3$.

Solution   We begin by listing all possible rational zeros.

$$\text{Possible rational zeros} = \frac{\text{Factors of the constant term, } -3}{\text{Factors of the leading coefficient, } 1} = \frac{\pm 1, \pm 3}{\pm 1} = \pm 1, \pm 3$$

Now we will use synthetic division to see if we can find a rational zero among the four possible rational zeros.

Test 1.

$1 \mid$ 1 7 11 −3
      1 8 19
   1 8 19 16

Test −1.

$-1 \mid$ 1 7 11 −3
        −1 −6 −5
     1 6 5 −8

Test 3.

$3 \mid$ 1 7 11 −3
      3 30 123
   1 10 41 120

Test −3.

$-3 \mid$ 1 7 11 −3
        −3 −12 3
     1 4 −1 0

The zero remainder on the right tells us that -3 is a zero of the polynomial function $f(x) = x^3 + 7x^2 + 11x - 3$. To find all zeros of f, we proceed as follows:

$$x^3 + 7x^2 + 11x - 3 = 0$$ Finding the zeros of f is the same thing as finding the roots of $f(x) = 0$.

$$(x + 3)(x^2 + 4x - 1) = 0$$ This result is from the last synthetic division. The first three numbers in the bottom row, 1, 4, and -1, give the coefficients of the second factor.

$$x + 3 = 0 \quad \text{or} \quad x^2 + 4x - 1 = 0$$ Set each factor equal to 0.

$$x = -3$$ Solve the linear equation.

We can use the quadratic formula to solve $x^2 + 4x - 1 = 0$.

$$x = \frac{-b \pm \sqrt{b^2 - 4ac}}{2a}$$ We use the quadratic formula because $x^2 + 4x - 1$ cannot be factored.

$$= \frac{-4 \pm \sqrt{4^2 - 4(1)(-1)}}{2(1)}$$ Let $a = 1$, $b = 4$, and $c = -1$.

$$= \frac{-4 \pm \sqrt{20}}{2}$$ Multiply and subtract under the radical: $4^2 - 4(1)(-1) = 16 - (-4) = 16 + 4 = 20$.

$$= \frac{-4 \pm 2\sqrt{5}}{2}$$ $\sqrt{20} = \sqrt{4 \cdot 5} = 2\sqrt{5}$

$$= -2 \pm \sqrt{5}$$ Divide the numerator and the denominator by 2.

The solution set is $\{-3, -2 - \sqrt{5}, -2 + \sqrt{5}\}$. The zeros of f are -3, $-2 - \sqrt{5}$, and $-2 + \sqrt{5}$. Among these three real zeros, one zero is rational and two are irrational.

Check Point 4 Find all zeros of $f(x) = x^3 + x^2 - 5x - 2$.

If the degree of a polynomial function or equation is 4 or higher, it is often necessary to find more than one linear factor by synthetic division.

One way to speed up the process of finding the first zero is to graph the function. Any x-intercept is a zero.

3 Solve polynomial equations.

EXAMPLE 5 Solving a Polynomial Equation

Solve: $x^4 - 6x^2 - 8x + 24 = 0$.

Solution Recall that we refer to the *zeros* of a polynomial function and the *roots* of a polynomial equation. Because we are given an equation, we will use the word "roots," rather than "zeros," in the solution process. We begin by listing all possible rational roots.

$$\text{Possible rational roots} = \frac{\text{Factors of the constant term, 24}}{\text{Factors of the leading coefficient, 1}}$$

$$= \frac{\pm 1, \pm 2, \pm 3, \pm 4, \pm 6, \pm 8, \pm 12, \pm 24}{\pm 1}$$

$$= \pm 1, \pm 2, \pm 3, \pm 4, \pm 6, \pm 8, \pm 12, \pm 24$$

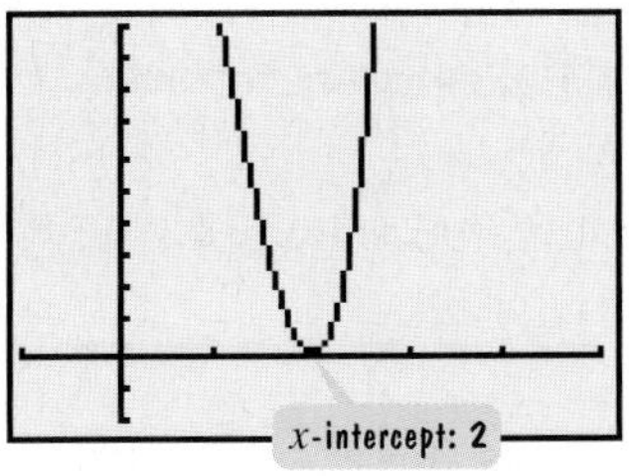

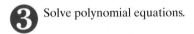

Figure 3.24 The graph of $f(x) = x^4 - 6x^2 - 8x + 24$ in a $[-1, 5, 1]$ by $[-2, 10, 1]$ viewing rectangle

The graph of $f(x) = x^4 - 6x^2 - 8x + 24$ is shown in Figure 3.24. Because the x-intercept is 2, we will test 2 by synthetic division and show that it is a root of the given equation. Without the graph, the procedure would be to start the trial-and-error synthetic division with 1 and proceed until a zero remainder is found, as we did in Example 4.

Now we can rewrite the given equation in factored form.

$$x^4 - 6x^2 - 8x + 24 = 0 \qquad \text{This is the given equation.}$$

$$(x - 2)(x^3 + 2x^2 - 2x - 12) = 0 \qquad \text{This is the result obtained from the synthetic division. The first four numbers in the bottom row, 1, 2, } -2, \text{ and } -12, \text{ give the coefficients of the second factor.}$$

$$x - 2 = 0 \quad \text{or} \quad x^3 + 2x^2 - 2x - 12 = 0 \qquad \text{Set each factor equal to 0.}$$

We can use the same approach to look for rational roots of the polynomial equation $x^3 + 2x^2 - 2x - 12 = 0$, listing all possible rational roots. Without the graph in Figure 3.24, the procedure would be to start testing possible rational roots by trial-and-error synthetic division. However, take a second look at the graph in Figure 3.24. Because the graph turns around at 2, this means that 2 is a root of even multiplicity. Thus, 2 must also be a root of $x^3 + 2x^2 - 2x - 12 = 0$, confirmed by the following synthetic division.

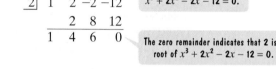

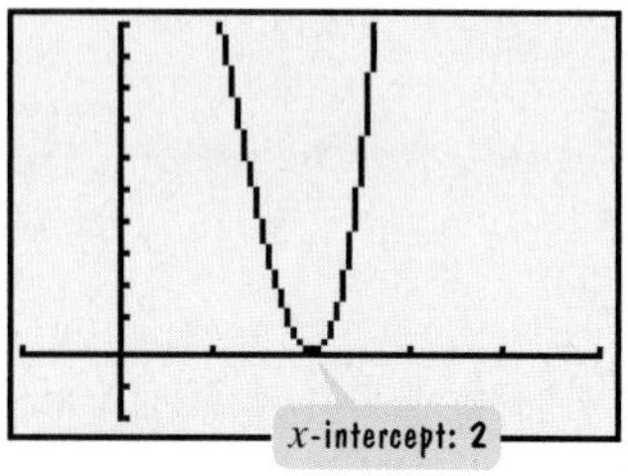

Figure 3.24 (repeated)

Now we can solve the original equation as follows:

$$x^4 - 6x^2 - 8x + 24 = 0 \qquad \text{This is the given equation.}$$

$$(x - 2)(x^3 + 2x^2 - 2x - 12) = 0 \qquad \text{This factorization was obtained from the first synthetic division.}$$

$$(x - 2)(x - 2)(x^2 + 4x + 6) = 0 \qquad \text{This factorization was obtained from the second synthetic division. The first three numbers in the bottom row, 1, 4, and 6, give the coefficients of the third factor.}$$

$$x - 2 = 0 \quad \text{or} \quad x - 2 = 0 \quad \text{or} \quad x^2 + 4x + 6 = 0 \qquad \text{Set each factor equal to 0.}$$
$$x = 2 \qquad\qquad x = 2 \qquad\qquad\qquad \text{Solve the linear equations.}$$

We can use the quadratic formula to solve $x^2 + 4x + 6 = 0$.

$$x = \frac{-b \pm \sqrt{b^2 - 4ac}}{2a} \qquad \text{We use the quadratic formula because } x^2 + 4x + 6 \text{ cannot be factored.}$$

$$= \frac{-4 \pm \sqrt{4^2 - 4(1)(6)}}{2(1)} \qquad \text{Let } a = 1, b = 4, \text{ and } c = 6.$$

$$= \frac{-4 \pm \sqrt{-8}}{2} \qquad \text{Multiply and subtract under the radical:} \\ 4^2 - 4(1)(6) = 16 - 24 = -8.$$

$$= \frac{-4 \pm 2i\sqrt{2}}{2} \qquad \sqrt{-8} = \sqrt{4(2)(-1)} = 2i\sqrt{2}$$

$$= -2 \pm i\sqrt{2} \qquad \text{Simplify.}$$

The solution set of the original equation, $x^4 - 6x^2 - 8x + 24 = 0$, is $\{2, -2 - i\sqrt{2}, -2 + i\sqrt{2}\}$. The graph in Figure 3.24 illustrates that a graphing utility does not reveal the two imaginary roots.

In Example 5, 2 is a repeated root of the equation with multiplicity 2. The example illustrates two general properties:

Properties of Polynomial Equations

1. If a polynomial equation is of degree n, then counting multiple roots separately, the equation has n roots.
2. If $a + bi$ is a root of a polynomial equation with real coefficients $(b \neq 0)$, then the complex imaginary number $a - bi$ is also a root. Complex imaginary roots, if they exist, occur in conjugate pairs.

Check Point 5 Solve: $x^4 - 6x^3 + 22x^2 - 30x + 13 = 0$.

The Fundamental Theorem of Algebra

The fact that a polynomial equation of degree n has n roots is a consequence of a theorem proved in 1799 by a 22-year-old student named Carl Friedrich Gauss in his doctoral dissertation. His result is called the **Fundamental Theorem of Algebra**.

The Fundamental Theorem of Algebra

If $f(x)$ is a polynomial of degree n, where $n \geq 1$, then the equation $f(x) = 0$ has at least one complex root.

Suppose, for example, that $f(x) = 0$ represents a polynomial equation of degree n. By the Fundamental Theorem of Algebra, we know that this equation has at least one complex root; we'll call it c_1. By the Factor Theorem, we know that $x - c_1$ is a factor of $f(x)$. Therefore, we obtain

$$(x - c_1)q_1(x) = 0 \quad \text{The degree of the polynomial } q_1(x) \text{ is } n - 1.$$

$$x - c_1 = 0 \quad \text{or} \quad q_1(x) = 0. \quad \text{Set each factor equal to 0.}$$

If the degree of $q_1(x)$ is at least 1, by the Fundamental Theorem of Algebra, the equation $q_1(x) = 0$ has at least one complex root. We'll call it c_2. The Factor Theorem gives us

$$q_1(x) = 0 \quad \text{The degree of } q_1(x) \text{ is } n - 1.$$

$$(x - c_2)q_2(x) = 0 \quad \text{The degree of } q_2(x) \text{ is } n - 2.$$

$$x - c_2 = 0 \quad \text{or} \quad q_2(x) = 0. \quad \text{Set each factor equal to 0.}$$

Let's see what we have up to this point and then continue the process.

$$f(x) = 0 \quad \text{This is the original polynomial equation of degree } n.$$

$$(x - c_1)q_1(x) = 0 \quad \text{This is the result from our first application of the Fundamental Theorem.}$$

$$(x - c_1)(x - c_2)q_2(x) = 0 \quad \text{This is the result from our second application of the Fundamental Theorem.}$$

By continuing this process, we will obtain the product of n linear factors. Setting each of these linear factors equal to zero results in n complex roots. Thus, if $f(x)$ is a polynomial of degree n, where $n \geq 1$, then $f(x) = 0$ has exactly n roots, where roots are counted according to their multiplicity.

4 Use the Linear Factorization Theorem to find polynomials with given zeros.

The Linear Factorization Theorem

In Example 5, we found that $x^4 - 6x^2 - 8x + 24 = 0$ has $\{2, -2 \pm i\sqrt{2}\}$ as a solution set, where 2 is a repeated root with multiplicity 2. The polynomial can be factored over the complex nonreal numbers as follows:

These are the four zeros.

$$f(x) = x^4 - 6x^2 - 8x + 24$$
$$= [x - (-2 + i\sqrt{2})][x - (-2 - i\sqrt{2})](x - 2)(x - 2).$$

These are the linear factors.

This fourth-degree polynomial has four linear factors. Just as an nth-degree polynomial equation has n roots, an nth-degree polynomial has n linear factors. This is formally stated as the **Linear Factorization Theorem**.

The Linear Factorization Theorem

If $f(x) = a_nx^n + a_{n-1}x^{n-1} + \cdots + a_1x + a_0$, where $n \geq 1$ and $a_n \neq 0$, then

$$f(x) = a_n(x - c_1)(x - c_2)\cdots(x - c_n),$$

where $c_1, c_2, \ldots, c_n$ are complex numbers (possibly real and not necessarily distinct). In words: An nth-degree polynomial can be expressed as the product of a nonzero constant and n linear factors.

Many of our problems involving polynomial functions and polynomial equations dealt with the process of finding zeros and roots. The Linear Factorization Theorem enables us to reverse this process, finding a polynomial function when the zeros are given.

EXAMPLE 6 Finding a Polynomial Function with Given Zeros

Find a fourth-degree polynomial function $f(x)$ with real coefficients that has -2, 2, and i as zeros and such that $f(3) = -150$.

Solution Because i is a zero and the polynomial has real coefficients, the conjugate, $-i$, must also be a zero. We can now use the Linear Factorization Theorem.

$$f(x) = a_n(x - c_1)(x - c_2)(x - c_3)(x - c_4)$$ This is the linear factorization for a fourth-degree polynomial.

$$= a_n(x + 2)(x - 2)(x - i)(x + i)$$ Use the given zeros: $c_1 = -2$, $c_2 = 2$, $c_3 = i$, and, from above, $c_4 = -i$.

$$= a_n(x^2 - 4)(x^2 + 1)$$ Multiply: $(x - i)(x + i) = x^2 - i^2 = x^2 - (-1) = x^2 + 1$.

$$f(x) = a_n(x^4 - 3x^2 - 4)$$ Complete the multiplication.

$$f(3) = a_n(3^4 - 3\cdot 3^2 - 4) = -150$$ To find a_n, use the fact that $f(3) = -150$.

$$a_n(81 - 27 - 4) = -150$$ Solve for a_n.

$$50a_n = -150$$ Simplify: $81 - 27 - 4 = 50$.

$$a_n = -3$$ Divide both sides by 50.

Substituting -3 for a_n in the formula for $f(x)$, we obtain

$$f(x) = -3(x^4 - 3x^2 - 4).$$

Equivalently,

$$f(x) = -3x^4 + 9x^2 + 12.$$

Technology

The graph of $f(x) = -3x^4 + 9x^2 + 12$, shown in a $[-3, 3, 1]$ by $[-200, 20, 20]$ viewing rectangle, verifies that -2 and 2 are real zeros. By tracing along the curve, we can check that $f(3) = -150$.

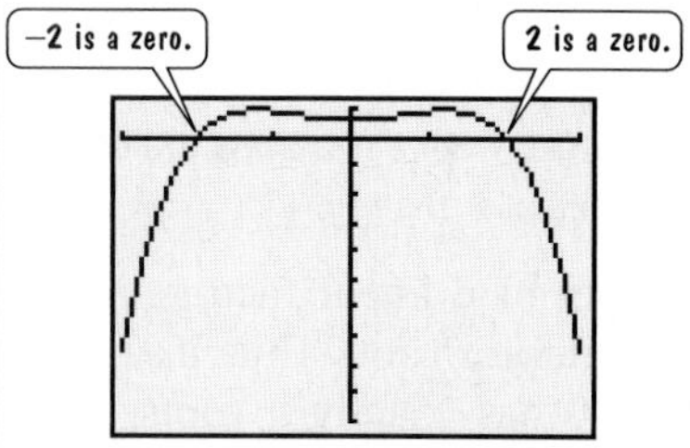

Check Point 6 Find a third-degree polynomial function $f(x)$ with real coefficients that has -3 and i as zeros and such that $f(1) = 8$.

5 Use Descartes's Rule of Sign.

Descartes's Rule of Signs

Because an nth-degree polynomial equation might have roots that are imaginary numbers, we should note that such an equation can have *at most n* real roots. **Descartes's Rule of Signs** provides even more specific information about the number of real zeros that a polynomial can have. The rule is based on considering *variations in sign* between consecutive coefficients. For example, the function $f(x) = 3x^7 - 2x^5 - x^4 + 7x^2 + x - 3$ has three sign changes:

$$f(x) = 3x^7 - 2x^5 - x^4 + 7x^2 + x - 3.$$

sign change sign change sign change

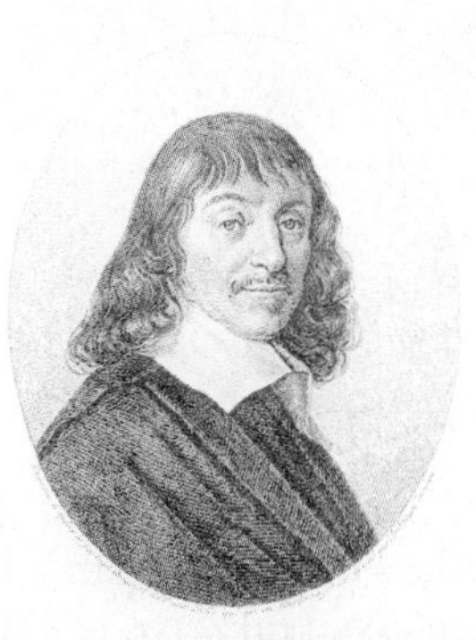

"An equation can have as many true [positive] roots as it contains changes of sign, from plus to minus or from minus to plus." René Descartes (1596–1650) in *La Géométrie* (1637)

Descartes's Rule of Signs

Let $f(x) = a_n x^n + a_{n-1}x^{n-1} + \cdots + a_2 x^2 + a_1 x + a_0$ be a polynomial with real coefficients.

1. The number of *positive real zeros* of f is either
   a. the same as the number of sign changes of $f(x)$
   or
   b. less than the number of sign changes of $f(x)$ by a positive even integer.
   If $f(x)$ has only one variation in sign, then f has exactly one positive real zero.
2. The number of *negative real zeros* of f is either
   a. the same as the number of sign changes of $f(-x)$
   or
   b. less than the number of sign changes of $f(-x)$ by a positive even integer.
   If $f(-x)$ has only one variation in sign, then f has exactly one negative real zero.

Study Tip

The number of real zeros given by Descartes's Rule of Signs includes rational zeros from a list of possible rational zeros, as well as irrational zeros not on the list. It does not include any imaginary zeros.

Table 3.1 illustrates what Descartes's Rule of Signs tells us about the positive real zeros of various polynomial functions.

Table 3.1 Descartes's Rule of Signs and Positive Real Zeros

Polynomial Function	Sign Changes	Conclusion
$f(x) = 3x^7 - 2x^5 - x^4 + 7x^2 + x - 3.$ (sign change, sign change, sign change)	3	There are 3 positive real zeros. or There is $3 - 2 = 1$ positive real zero.
$f(x) = 4x^5 + 2x^4 - 3x^2 + x + 5$ (sign change, sign change)	2	There are 2 positive real zeros. or There are $2 - 2 = 0$ positive real zeros.
$f(x) = -7x^6 - 5x^4 + x + 9$ (sign change)	1	There is 1 positive real zero.

EXAMPLE 7 Using Descartes's Rule of Signs

Determine the possible numbers of positive and negative real zeros of $f(x) = x^3 + 2x^2 + 5x + 4$.

Solution

1. To find possibilities for positive real zeros, count the number of sign changes in the equation for $f(x)$. Because all the coefficients are positive, there are no variations in sign. Thus, there are no positive real zeros.
2. To find possibilities for negative real zeros, count the number of sign changes in the equation for $f(-x)$. We obtain this equation by replacing x with $-x$ in the given function.

$$f(x) = x^3 + 2x^2 + 5x + 4$$

Replace x with $-x$.

$$f(-x) = (-x)^3 + 2(-x)^2 + 5(-x) + 4$$
$$= -x^3 + 2x^2 - 5x + 4$$

Now count the sign changes.

$$f(x) = -x^3 + 2x^2 - 5x + 4x$$

sign change sign change sign change

There are three variations in sign. The number of negative real zeros of f is either equal to the number of sign changes, 3, or is less than this number by an even integer. This means that either there are 3 negative real zeros or there is $3 - 2 = 1$ negative real zero.

What do the results of Example 7 mean in terms of solving

$$x^3 + 2x^2 + 5x + 4 = 0?$$

Without using Descartes's Rule of Signs, we list the possible rational roots as follows:

Possible rational roots

$$= \frac{\text{Factors of the constant term, 4}}{\text{Factors of the leading coefficient, 1}} = \frac{\pm 1, \pm 2, \pm 4}{\pm 1} = \pm 1, \pm 2, \pm 4$$

However, Descartes's Rule of Signs informed us that $f(x) = x^3 + 2x^2 + 5x + 4$ has no positive real zeros. Thus, the polynomial equation $x^3 + 2x^2 + 5x + 4 = 0$ has no positive real roots. This means that we can eliminate the positive numbers from our list of possible rational roots. Possible rational roots include only -1, -2, and -4. We can use synthetic division and test two of the three possible rational roots of $x^3 + 2x^2 + 5x + 4 = 0$ as follows:

Test -1.

$$\begin{array}{r|rrrr} -1 & 1 & 2 & 5 & 4 \\ & & -1 & -1 & -4 \\ \hline & 1 & 1 & 4 & 0 \end{array}$$

The zero remainder shows that -1 is a root.

Test -2.

$$\begin{array}{r|rrrr} -2 & 1 & 2 & 5 & 4 \\ & & -2 & 0 & -10 \\ \hline & 1 & 0 & 5 & -6 \end{array}$$

The nonzero remainder shows that -2 is not a root.

By solving the equation $x^3 + 2x^2 + 5x + 4 = 0$, you will find that this equation of degree 3 has three roots. One root is -1 and the other two roots are imaginary numbers in a conjugate pair. Verify this by completing the solution process.

Check Point 7 Determine the possible numbers of positive and negative real zeros of $f(x) = x^4 - 14x^3 + 71x^2 - 154x + 120$.

EXERCISE SET 3.4

Practice Exercises

In Exercises 1–8, use the Rational Zero Theorem to list all possible rational zeros for each given function.

1. $f(x) = x^3 + x^2 - 4x - 4$
2. $f(x) = x^3 + 3x^2 - 6x - 8$
3. $f(x) = 3x^4 - 11x^3 - x^2 + 19x + 6$
4. $f(x) = 2x^4 + 3x^3 - 11x^2 - 9x + 15$
5. $f(x) = 4x^4 - x^3 + 5x^2 - 2x - 6$
6. $f(x) = 3x^4 - 11x^3 - 3x^2 - 6x + 8$
7. $f(x) = x^5 - x^4 - 7x^3 + 7x^2 - 12x - 12$
8. $f(x) = 4x^5 - 8x^4 - x + 2$

In Exercises 9–16,

a. *List all possible rational zeros.*
b. *Use synthetic division to test the possible rational zeros and find an actual zero.*
c. *Use the quotient from part (b) to find the remaining zeros of the polynomial function.*

9. $f(x) = x^3 + x^2 - 4x - 4$
10. $f(x) = x^3 - 2x^2 - 11x + 12$
11. $f(x) = 2x^3 - 3x^2 - 11x + 6$
12. $f(x) = 2x^3 - 5x^2 + x + 2$
13. $f(x) = x^3 + 4x^2 - 3x - 6$
14. $f(x) = 2x^3 + x^2 - 3x + 1$
15. $f(x) = 2x^3 + 6x^2 + 5x + 2$
16. $f(x) = x^3 - 4x^2 + 8x - 5$

In Exercises 17–24,

a. *List all possible rational roots.*
b. *Use synthetic division to test the possible rational roots and find an actual root.*
c. *Use the quotient from part (b) to find the remaining roots and solve the equation.*

17. $x^3 - 2x^2 - 11x + 12 = 0$
18. $x^3 - 2x^2 - 7x - 4 = 0$
19. $x^3 - 10x - 12 = 0$
20. $x^3 - 5x^2 + 17x - 13 = 0$
21. $6x^3 + 25x^2 - 24x + 5 = 0$
22. $2x^3 - 5x^2 - 6x + 4 = 0$
23. $x^4 - 2x^3 - 5x^2 + 8x + 4 = 0$
24. $x^4 - 2x^2 - 16x - 15 = 0$

In Exercises 25–32, find an nth-degree polynomial function with real coefficients satisfying the given conditions. If you are using a graphing utility, use it to graph the function and verify the real zeros and the given function value.

25. $n = 3$; 1 and $5i$ are zeros; $f(-1) = -104$
26. $n = 3$; 4 and $2i$ are zeros; $f(-1) = -50$
27. $n = 3$; -5 and $4 + 3i$ are zeros; $f(2) = 91$
28. $n = 3$; 6 and $-5 + 2i$ are zeros; $f(2) = -636$
29. $n = 4$; i and $3i$ are zeros; $f(-1) = 20$
30. $n = 4$; -2, $-\frac{1}{2}$, and i are zeros; $f(1) = 18$
31. $n = 4$; -2, 5, and $3 + 2i$ are zeros; $f(1) = -96$
32. $n = 4$; -4, $\frac{1}{3}$, and $2 + 3i$ are zeros; $f(1) = 100$

In Exercises 33–38, use Descartes's Rule of Signs to determine the possible number of positive and negative real zeros for each given function.

33. $f(x) = x^3 + 2x^2 + 5x + 4$
34. $f(x) = x^3 + 7x^2 + x + 7$
35. $f(x) = 5x^3 - 3x^2 + 3x - 1$
36. $f(x) = -2x^3 + x^2 - x + 7$
37. $f(x) = 2x^4 - 5x^3 - x^2 - 6x + 4$
38. $f(x) = 4x^4 - x^3 + 5x^2 - 2x - 6$

In Exercises 39–52, find all zeros of the polynomial function or solve the given polynomial equation. Use the Rational Zero Theorem, Descartes's Rule of Signs, and possibly the graph of the polynomial function shown by a graphing utility as an aid in obtaining the first zero or the first root.

39. $f(x) = x^3 - 4x^2 - 7x + 10$
40. $f(x) = x^3 + 12x^2 + 21x + 10$
41. $2x^3 - x^2 - 9x - 4 = 0$
42. $3x^3 - 8x^2 - 8x + 8 = 0$
43. $f(x) = x^4 - 2x^3 + x^2 + 12x + 8$
44. $f(x) = x^4 - 4x^3 - x^2 + 14x + 10$
45. $x^4 - 3x^3 - 20x^2 - 24x - 8 = 0$
46. $x^4 - x^3 + 2x^2 - 4x - 8 = 0$
47. $f(x) = 3x^4 - 11x^3 - x^2 + 19x + 6$
48. $f(x) = 2x^4 + 3x^3 - 11x^2 - 9x + 15$
49. $4x^4 - x^3 + 5x^2 - 2x - 6 = 0$
50. $3x^4 - 11x^3 - 3x^2 - 6x + 8 = 0$
51. $2x^5 + 7x^4 - 18x^2 - 8x + 8 = 0$
52. $4x^5 + 12x^4 - 41x^3 - 99x^2 + 10x + 24 = 0$

Practice Plus

Exercises 53–60, show incomplete graphs of given polynomial functions.

a. *Find all the zeros of each function.*
b. *Without using a graphing utility, draw a complete graph of the function.*

53. $f(x) = -x^3 + x^2 + 16x - 16$

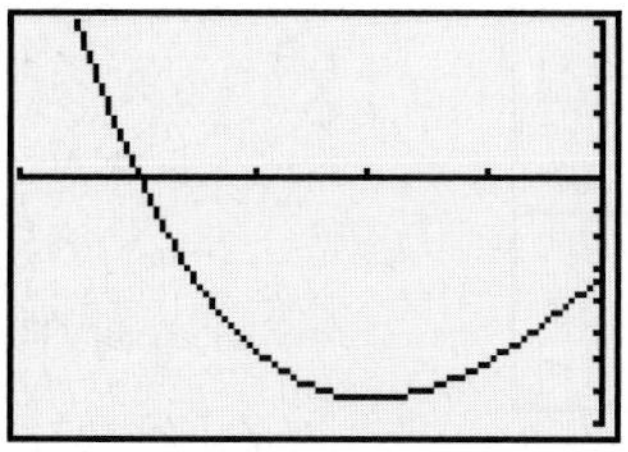

[−5, 0, 1] by [−40, 25, 5]

54. $f(x) = -x^3 + 3x^2 - 4$

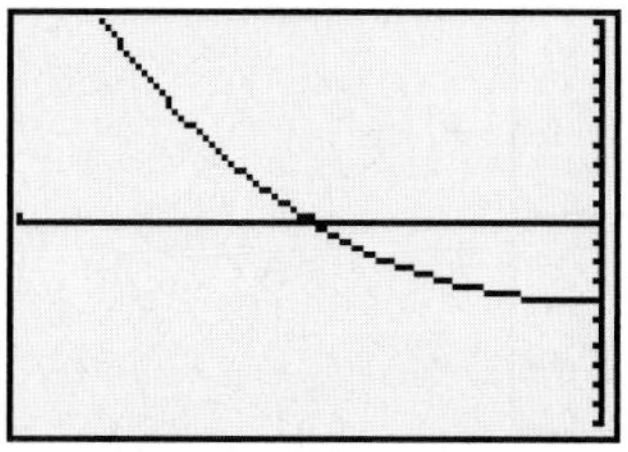

[−2, 0, 1] by [−10, 10, 1]

55. $f(x) = 4x^3 - 8x^2 - 3x + 9$

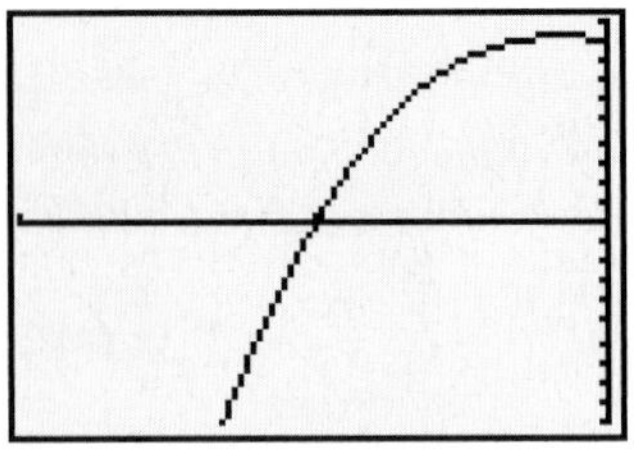

$[-2, 0, 1]$ by $[-10, 10, 1]$

56. $f(x) = 3x^3 + 2x^2 + 2x - 1$

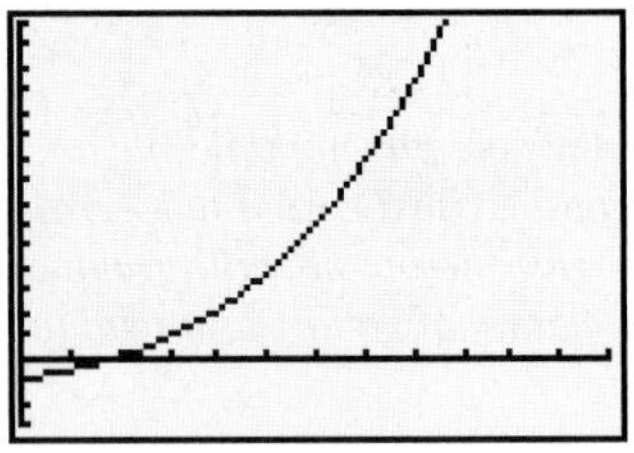

$[0, 2, \frac{1}{6}]$ by $[-3, 15, 1]$

57. $f(x) = 2x^4 - 3x^3 - 7x^2 - 8x + 6$

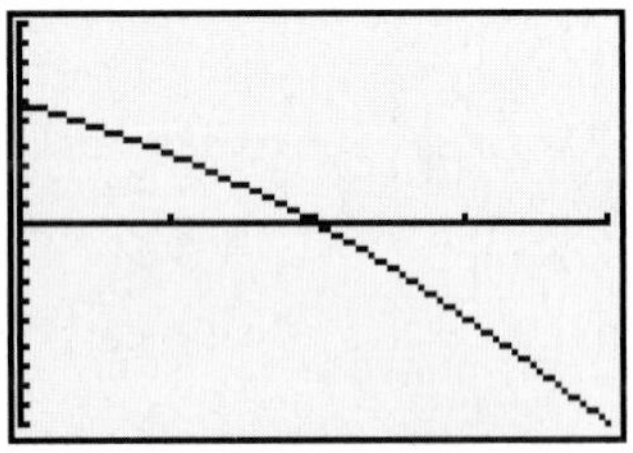

$[0, 1, \frac{1}{4}]$ by $[-10, 10, 1]$

58. $f(x) = 2x^4 + 2x^3 - 22x^2 - 18x + 36$

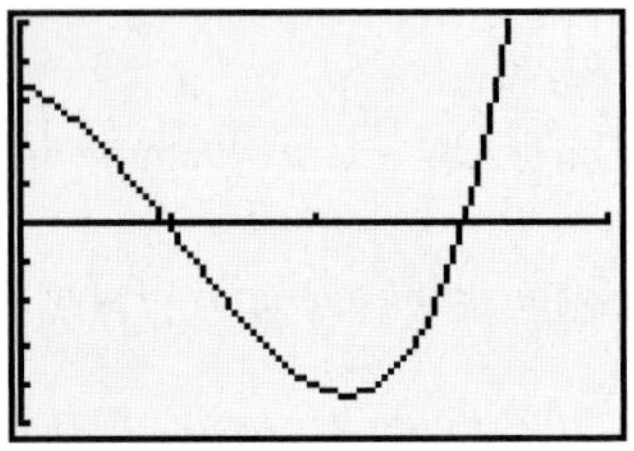

$[0, 4, 1]$ by $[-50, 50, 10]$

59. $f(x) = 3x^5 + 2x^4 - 15x^3 - 10x^2 + 12x + 8$

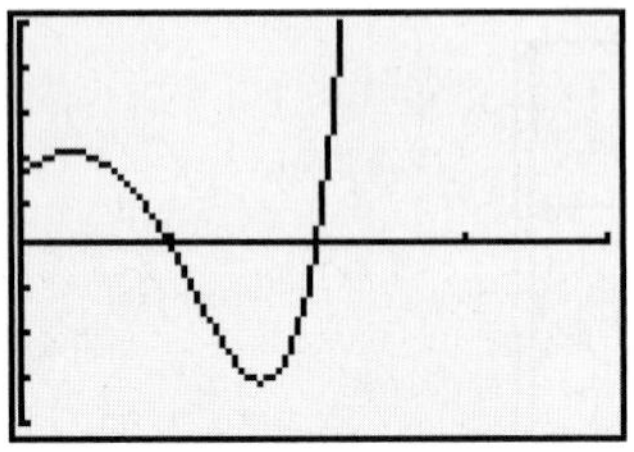

$[0, 4, 1]$ by $[-20, 25, 5]$

60. $f(x) = -5x^4 + 4x^3 - 19x^2 + 16x + 4$

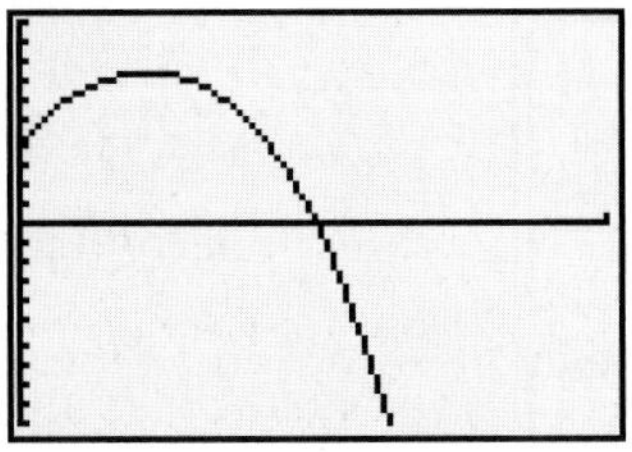

$[0, 2, 1]$ by $[-10, 10, 1]$

Application Exercises

The graphs are based on a study of the percentage of professional works completed in each age decade of life by 738 people who lived to be at least 79. Use the graphs to solve Exercises 61–62.

Age Trends in Professional Productivity

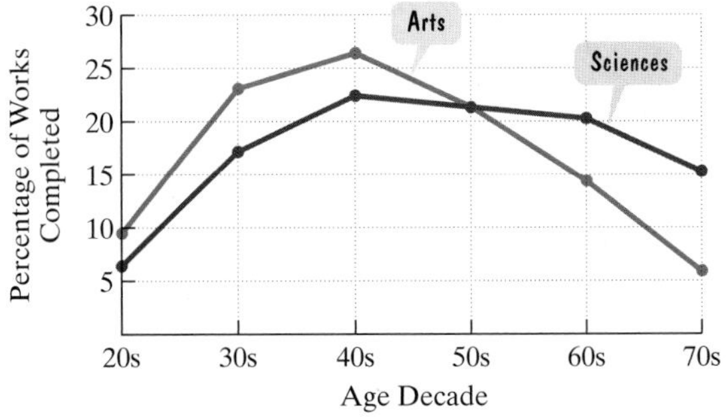

Source: Dennis, W. (1966), Creative productivity between the ages of 20 and 80 years. *Journal of Gerontology*, 21, 1–8.

61. Suppose that a polynomial function f is used to model the data shown in the graph for the arts using

(age decade, percentage of works completed).

a. Use the graph to solve the polynomial equation $f(x) = 27$. Describe what this means in terms of an age decade and productivity.

b. Describe the degree and the leading coefficient of a function f that can be used to model the data in the graph.

62. Suppose that a polynomial function g is used to model the data shown in the graph for the sciences using

(age decade, percentage of works completed).

a. Use the graph to solve the polynomial equation $g(x) = 20$. Find only the meaningful value of x and then describe what this means in terms of an age decade and productivity.

b. Describe the degree and the leading coefficient of a function g that can be used to model the data in the graph.

The polynomial function

$$H(x) = -0.001618x^4 + 0.077326x^3 - 1.2367x^2 + 11.460x + 2.914$$

models the age in human years, $H(x)$, of a dog that is x years old, where $x \geq 1$. Although the coefficients make it difficult to solve equations algebraically using this function, a graph of the function makes approximate solutions possible.

Use the graph shown to solve Exercises 63–64. Round all answers to the nearest year.

Dog's Age in Human Years

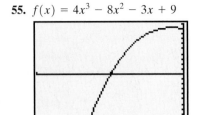

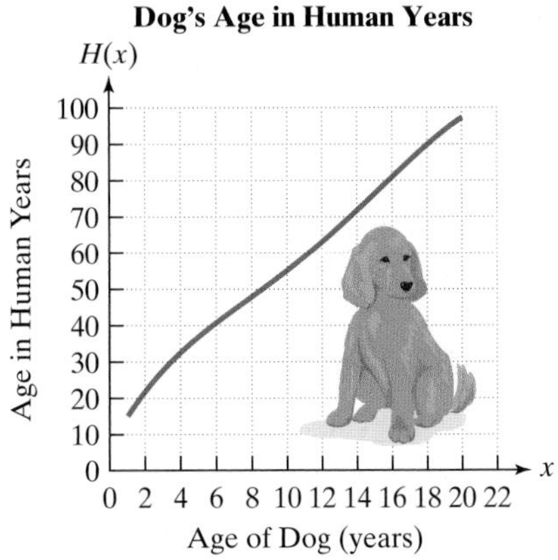

Source: U.C. Davis

63. If you are 25, what is the equivalent age for dogs?

64. If you are 90, what is the equivalent age for dogs?

65. Set up an equation to answer the question in either Exercise 63 or 64. Bring all terms to one side and obtain zero on the other side. What are some of the difficulties involved in solving this equation? Explain how the Intermediate Value Theorem can be used to verify the approximate solution that you obtained from the graph.

66. The concentration of a drug, in parts per million, in a patient's blood x hours after the drug is administered is given by the function

$$f(x) = -x^4 + 12x^3 - 58x^2 + 132x.$$

How many hours after the drug is administered will it be eliminated from the bloodstream?

67. A box with an open top is formed by cutting squares out of the corners of a rectangular piece of cardboard 10 inches by 8 inches and then folding up the sides. If x represents the length of the side of the square cut from each corner of the rectangle, what size square must be cut if the volume of the box is to be 48 cubic inches?

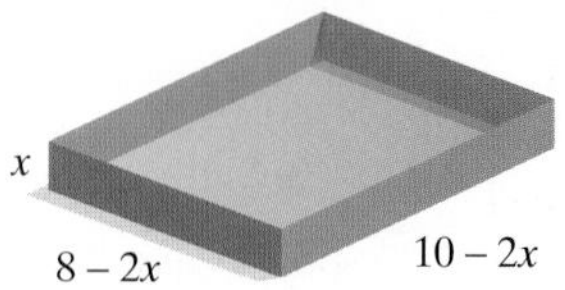

Writing in Mathematics

68. Describe how to find the possible rational zeros of a polynomial function.

69. How does the linear factorization of $f(x)$, that is,

$$f(x) = a_n(x - c_1)(x - c_2)\cdots(x - c_n),$$

show that a polynomial equation of degree n has n roots?

70. Describe how to use Descartes's Rule of Signs to determine the possible number of positive real zeros of a polynomial function.

71. Describe how to use Descartes's Rule of Signs to determine the possible number of negative roots of a polynomial equation.

72. Why must every polynomial equation of degree 3 have at least one real root?

73. Explain why the equation $x^4 + 6x^2 + 2 = 0$ has no rational roots.

74. Suppose $\frac{3}{4}$ is a root of a polynomial equation. What does this tell us about the leading coefficient and the constant term in the equation?

75. Use the graphs for Exercises 61–62 to describe one similarity and one difference between age trends in professional productivity in the arts and the sciences.

Technology Exercises

The equations in Exercises 76–79 have real roots that are rational. Use the Rational Zero Theorem to list all possible rational roots. Then graph the polynomial function in the given viewing rectangle to determine which possible rational roots are actual roots of the equation.

76. $2x^3 - 15x^2 + 22x + 15 = 0$; $[-1, 6, 1]$ by $[-50, 50, 10]$

77. $6x^3 - 19x^2 + 16x - 4 = 0$; $[0, 2, 1]$ by $[-3, 2, 1]$

78. $2x^4 + 7x^3 - 4x^2 - 27x - 18 = 0$; $[-4, 3, 1]$ by $[-45, 45, 15]$

79. $4x^4 + 4x^3 + 7x^2 - x - 2 = 0$; $[-2, 2, 1]$ by $[-5, 5, 1]$

80. Use Descartes's Rule of Signs to determine the possible number of positive and negative real zeros of $f(x) = 3x^4 + 5x^2 + 2$. What does this mean in terms of the graph of f? Verify your result by using a graphing utility to graph f.

81. Use Descartes's Rule of Signs to determine the possible number of positive and negative real zeros of $f(x) = x^5 - x^4 + x^3 - x^2 + x - 8$. Verify your result by using a graphing utility to graph f.

82. Write equations for several polynomial functions of odd degree and graph each function. Is it possible for the graph to have no real zeros? Explain. Try doing the same thing for polynomial functions of even degree. Now is it possible to have no real zeros?

Use a graphing utility to obtain a complete graph for each polynomial function in Exercises 83–86. Then determine the number of real zeros and the number of imaginary zeros for each function.

83. $f(x) = x^3 - 6x - 9$

84. $f(x) = 3x^5 - 2x^4 + 6x^3 - 4x^2 - 24x + 16$

85. $f(x) = 3x^4 + 4x^3 - 7x^2 - 2x - 3$

86. $f(x) = x^6 - 64$

Critical Thinking Exercises

87. Which one of the following is true?

a. The equation $x^3 + 5x^2 + 6x + 1 = 0$ has one positive real root.

b. Descartes's Rule of Signs gives the exact number of positive and negative real roots for a polynomial equation.

c. Every polynomial equation of degree 3 has at least one rational root.

d. None of the above is true.

88. Give an example of a polynomial equation that has no real roots. Describe how you obtained the equation.

89. If the volume of the solid shown in the figure is 208 cubic inches, find the value of x.

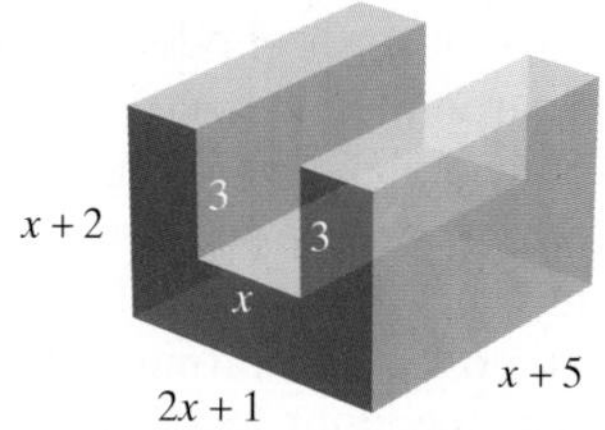

90. In this exercise, we lead you through the steps involved in the proof of the Rational Zero Theorem. Consider the polynomial equation

$$a_nx^n + a_{n-1}x^{n-1} + a_{n-2}x^{n-2} + \cdots + a_1x + a_0 = 0,$$

where $\frac{p}{q}$ is a rational root reduced to lowest terms.

a. Substitute $\frac{p}{q}$ for x in the equation and show that the equation can be written as

$$a_np^n + a_{n-1}p^{n-1}q + a_{n-2}p^{n-2}q^2 + \cdots + a_1pq^{n-1} = -a_0q^n.$$

b. Why is p a factor of the left side of the equation?

c. Because p divides the left side, it must also divide the right side. However, because $\frac{p}{q}$ is reduced to lowest terms, p

cannot divide q. Thus, p and q have no common factors other than -1 and 1. Because p does divide the right side and it is not a factor of q^n, what can you conclude?

d. Rewrite the equation from part (a) with all terms containing q on the left and the term that does not have a factor of q on the right. Use an argument that parallels parts (b) and (c) to conclude that q is a factor of a_n.

In Exercises 91–94, the graph of a polynomial function is given. What is the smallest degree that each polynomial could have?

91.

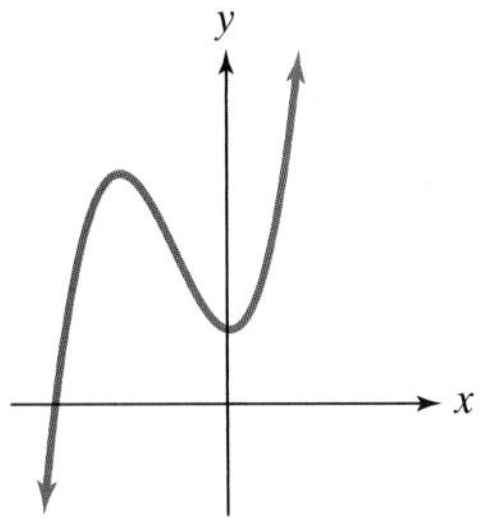

92.

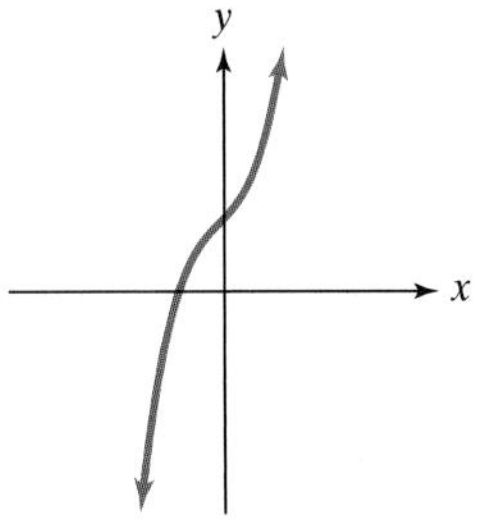

93.

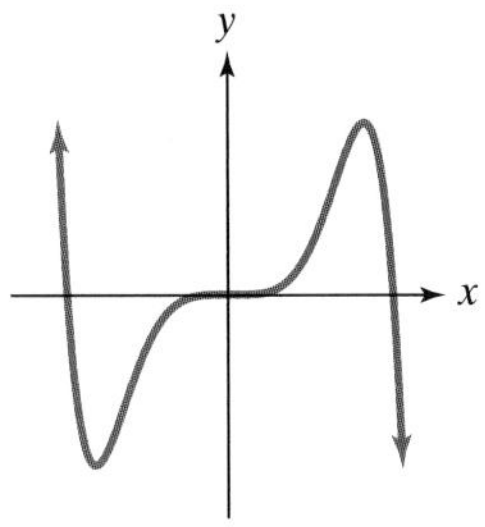

94.

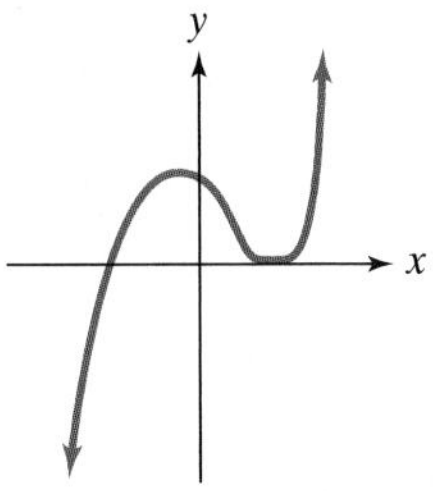

95. Explain why a polynomial function of degree 20 cannot cross the x-axis exactly once.

CHAPTER 3 MID-CHAPTER CHECK POINT

What You Know: We graphed quadratic functions using vertices, intercepts, and additional points, as necessary. We learned that the vertex of $f(x) = a(x - h)^2 + k$ is (h, k) and the vertex of $f(x) = ax^2 + bx + c$ is $\left(-\frac{b}{2a}, f\left(-\frac{b}{2a}\right)\right)$. We used the vertex to solve problems that involved minimizing or maximizing quadratic functions. We learned a number of techniques for finding the zeros of a polynomial function f of degree 3 or higher or, equivalently, finding the roots, or solutions, of the equation $f(x) = 0$. For some functions, the zeros were found by factoring $f(x)$. For other functions, we listed possible rational zeros and used synthetic division and the Factor Theorem to determine the zeros. We saw that graphs cross the x-axis at zeros of odd multiplicity and touch the x-axis and turn around at zeros of even multiplicity. We learned to graph polynomial functions using zeros, the Leading Coefficient Test, intercepts, and symmetry. We checked graphs using the fact that a polynomial function of degree n has a graph with at most $n - 1$ turning points. After finding zeros of polynomial functions, we reversed directions by using the Linear Factorization Theorem to find functions with given zeros.

In Exercises 1–4, graph the given quadratic function. Give each function's domain and range.

1. $f(x) = (x - 3)^2 - 4$ **2.** $f(x) = 5 - (x + 2)^2$

3. $f(x) = -x^2 - 4x + 5$ **4.** $f(x) = 3x^2 - 6x + 1$

In Exercises 5–13, find all zeros of each polynomial function. Then graph the function.

5. $f(x) = (x - 2)^2(x + 1)^3$ **6.** $f(x) = -(x - 2)^2(x + 1)^2$

7. $f(x) = x^3 - x^2 - 4x + 4$ **8.** $f(x) = x^4 - 5x^2 + 4$

9. $f(x) = -(x + 1)^6$ **10.** $f(x) = -6x^3 + 7x^2 - 1$

11. $f(x) = 2x^3 - 2x$ **12.** $f(x) = x^3 - 2x^2 + 26x$

13. $f(x) = -x^3 + 5x^2 - 5x - 3$

In Exercises 14–19, solve each polynomial equation.

14. $x^3 - 3x + 2 = 0$

15. $6x^3 - 11x^2 + 6x - 1 = 0$

16. $(2x + 1)(3x - 2)^3(2x - 7) = 0$

17. $2x^3 + 5x^2 - 200x - 500 = 0$

18. $x^4 - x^3 - 11x^2 = x + 12$

19. $2x^4 + x^3 - 17x^2 - 4x + 6 = 0$

20. A company manufactures and sells bath cabinets. The function

$$P(x) = -x^2 + 150x - 4425$$

models the company's daily profit, $P(x)$, when x cabinets are manufactured and sold per day. How many cabinets should be manufactured and sold per day to maximize the company's profit? What is the maximum daily profit?

21. Among all pairs of numbers whose sum is -18, find a pair whose product is as large as possible. What is the maximum product?

22. The base of a triangle measures 40 inches minus twice the measure of its height. For what measure of the height does the triangle have a maximum area? What is the maximum area?

In Exercises 23–24, divide, using synthetic division if possible.

23. $(6x^4 - 3x^3 - 11x^2 + 2x + 4) \div (3x^2 - 1)$

24. $(2x^4 - 13x^3 + 17x^2 + 18x - 24) \div (x - 4)$

In Exercises 25–26, find an nth-degree polynomial function with real coefficients satisfying the given conditions.

25. $n = 3$; 1 and i are zeros; $f(-1) = 8$

26. $n = 4$; 2 (with multiplicity 2) and $3i$ are zeros; $f(0) = 36$

27. Does $f(x) = x^3 - x - 5$ have a real zero between 1 and 2?

SECTION 3.5 Rational Functions and Their Graphs

Objectives

1. Find the domain of rational functions.
2. Use arrow notation.
3. Identify vertical asymptotes.
4. Identify horizontal asymptotes.
5. Use transformations to graph rational functions.
6. Graph rational functions.
7. Identify slant asymptotes.
8. Solve applied problems involving rational functions.

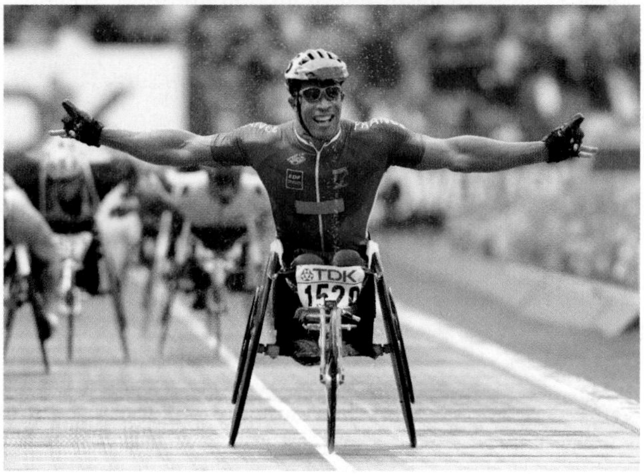

Technology is now promising to bring light, fast, and beautiful wheelchairs to millions of disabled people. The cost of manufacturing these radically different wheelchairs can be modeled by rational functions. In this section, we will see how graphs of these functions illustrate that low prices are possible with high production levels, which are urgently needed in this situation. There are more than half a billion people with disabilities in developing countries; an estimated 20 million need wheelchairs right now.

1 Find the domain of rational functions.

Rational Functions

Rational functions are quotients of polynomial functions. This means that rational functions can be expressed as

$$f(x) = \frac{p(x)}{q(x)},$$

where p and q are polynomial functions and $q(x) \neq 0$. The **domain** of a rational function is the set of all real numbers except the x-values that make the denominator zero. For example, the domain of the rational function

$$f(x) = \frac{x^2 + 7x + 9}{x(x-2)(x+5)}$$

This is $p(x)$.

This is $q(x)$.

is the set of all real numbers except 0, 2, and -5.

EXAMPLE 1 Finding the Domain of a Rational Function

Find the domain of each rational function:

a. $f(x) = \dfrac{x^2 - 9}{x - 3}$ **b.** $g(x) = \dfrac{x}{x^2 - 9}$ **c.** $h(x) = \dfrac{x + 3}{x^2 + 9}$.

Solution Rational functions contain division. Because division by 0 is undefined, we must exclude from the domain of each function values of x that cause the polynomial function in the denominator to be 0.

a. The denominator of $f(x) = \dfrac{x^2 - 9}{x - 3}$ is 0 if $x = 3$. Thus, x cannot equal 3. The domain of f consists of all real numbers except 3. We can express the domain in set-builder or interval notation:

$$\text{Domain of } f = \{x \mid x \neq 3\}$$
$$\text{Domain of } f = (-\infty, 3) \cup (3, \infty).$$

Study Tip

Because the domain of a rational function is the set of all real numbers except those for which the denominator is 0, you can identify such numbers by setting the denominator equal to 0 and solving for x. Exclude the resulting real values of x from the domain.

b. The denominator of $g(x) = \frac{x}{x^2 - 9}$ is 0 if $x = -3$ or $x = 3$. Thus, the domain of g consists of all real numbers except -3 and 3. We can express the domain in set-builder or interval notation:

$$\text{Domain of } g = \{x | x \neq -3, x \neq 3\}$$
$$\text{Domain of } g = (-\infty, -3) \cup (-3, 3) \cup (3, \infty).$$

c. No real numbers cause the denominator of $h(x) = \frac{x + 3}{x^2 + 9}$ to equal 0. The domain of h consists of all real numbers.

$$\text{Domain of } h = (-\infty, \infty)$$

Check Point 1 Find the domain of each rational function:

a. $f(x) = \frac{x^2 - 25}{x - 5}$ **b.** $g(x) = \frac{x}{x^2 - 25}$ **c.** $h(x) = \frac{x + 5}{x^2 + 25}$.

2 Use arrow notation.

The most basic rational function is the **reciprocal function**, defined by $f(x) = \frac{1}{x}$. The denominator of the reciprocal function is zero when $x = 0$, so the domain of f is the set of all real numbers except 0.

Let's look at the behavior of f near the excluded value 0. We start by evaluating $f(x)$ to the left of 0.

x approaches 0 from the left.

x	-1	-0.5	-0.1	-0.01	-0.001
$f(x) = \frac{1}{x}$	-1	-2	-10	-100	-1000

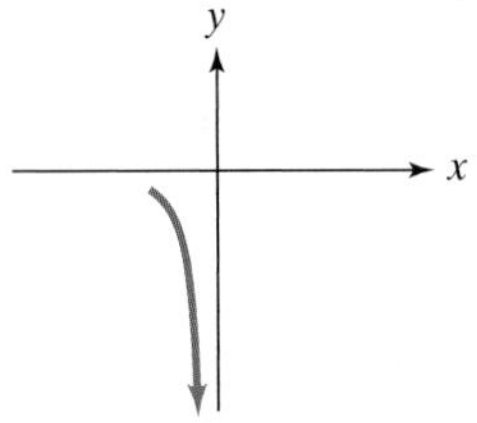

Mathematically, we say that "x approaches 0 from the left." From the table and the accompanying graph, it appears that as x approaches 0 from the left, the function values, $f(x)$, decrease without bound. We say that "$f(x)$ approaches negative infinity." We use a special arrow notation to describe this situation symbolically:

$$\text{As } x \to 0^-, \ f(x) \to -\infty.$$

As x approaches 0 from the left, $f(x)$ approaches negative infinity (that is, the graph falls).

Observe that the minus ($-$) superscript on the 0 ($x \to 0^-$) is read "from the left."

Next, we evaluate $f(x)$ to the right of 0.

x approaches 0 from the right.

x	0.001	0.01	0.1	0.5	1
$f(x) = \frac{1}{x}$	1000	100	10	2	1

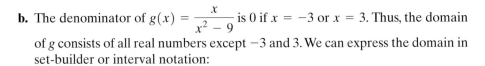

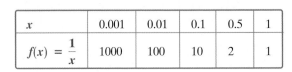

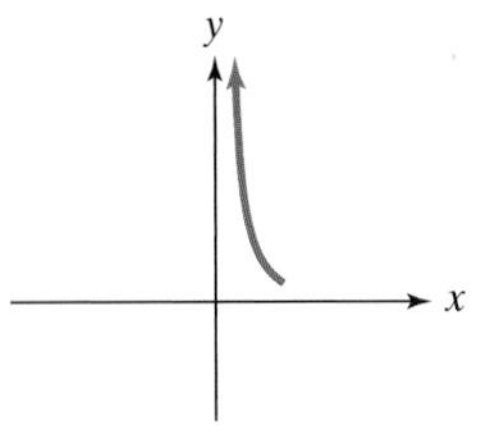

Mathematically, we say that "x approaches 0 from the right." From the table and the accompanying graph, it appears that as x approaches 0 from the right, the function values, $f(x)$, increase without bound. We say that "$f(x)$ approaches infinity." We again use a special arrow notation to describe this situation symbolically:

As $x \to 0^+$, $f(x) \to \infty$.

As x approaches 0 from the right, $f(x)$ approaches infinity (that is, the graph rises).

Observe that the plus (+) superscript on the 0 ($x \to 0^+$) is read "from the right."

Now let's see what happens to the function values, $f(x)$, as x gets farther away from the origin. The following tables suggest what happens to $f(x)$ as x increases or decreases without bound.

x increases without bound:

x	1	10	100	1000
$f(x) = \frac{1}{x}$	1	0.1	0.01	0.001

x decreases without bound:

x	−1	−10	−100	−1000
$f(x) = \frac{1}{x}$	−1	−0.1	−0.01	−0.001

It appears that as x increases or decreases without bound, the function values, $f(x)$, are getting progressively closer to 0.

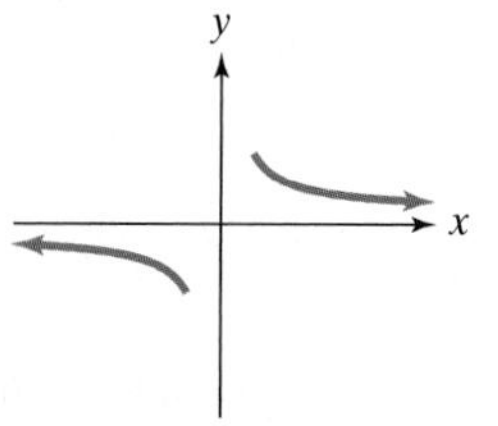

Figure 3.25 $f(x)$ approaches 0 as x increases or decreases without bound.

Figure 3.25 illustrates the end behavior of $f(x) = \frac{1}{x}$ as x increases or decreases without bound. The graph shows that the function values, $f(x)$, are approaching 0. This means that as x increases or decreases without bound, the graph of f is approaching the horizontal line $y = 0$ (that is, the x-axis). We use arrow notation to describe this situation:

As $x \to \infty$, $f(x) \to 0$ and as $x \to -\infty$, $f(x) \to 0$.

As x approaches infinity (that is, increases without bound), $f(x)$ approaches 0.

As x approaches negative infinity (that is, decreases without bound), $f(x)$ approaches 0.

Thus, as x approaches infinity ($x \to \infty$) or as x approaches negative infinity ($x \to -\infty$), the function values are approaching zero: $f(x) \to 0$.

The graph of the reciprocal function $f(x) = \frac{1}{x}$ is shown in Figure 3.26. Unlike the graph of a polynomial function, the graph of the reciprocal function has a break and is composed of two distinct branches.

Study Tip

If x is far from 0, then $\frac{1}{x}$ is close to 0. By contrast, if x is close to 0, then $\frac{1}{x}$ is far from 0.

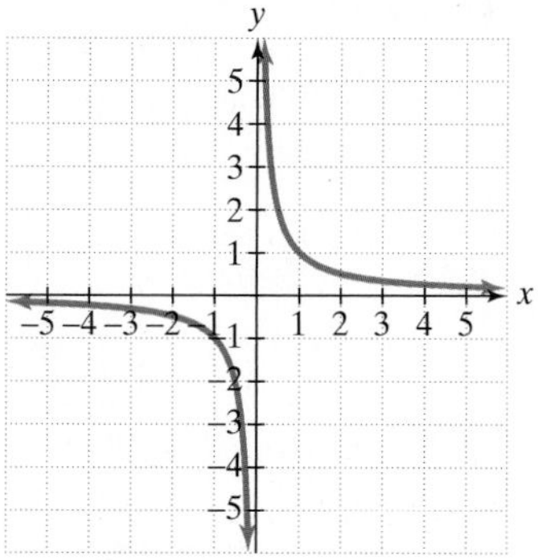

Figure 3.26 The graph of the reciprocal function $f(x) = \frac{1}{x}$

The arrow notation used throughout our discussion of the reciprocal function is summarized in the following box:

Arrow Notation

Symbol	Meaning
$x \to a^+$	x approaches a from the right.
$x \to a^-$	x approaches a from the left.
$x \to \infty$	x approaches infinity; that is, x increases without bound.
$x \to -\infty$	x approaches negative infinity; that is, x decreases without bound.

Another basic rational function is $f(x) = \dfrac{1}{x^2}$. The graph of this even function, with y-axis symmetry and positive function values, is shown in Figure 3.27. Like the reciprocal function, the graph has a break and is composed of two distinct branches.

As $x \to 0^-$, $f(x) \to \infty$.
Function values increase without bound.

As $x \to 0^+$, $f(x) \to \infty$.
Function values increase without bound.

As $x \to -\infty$ (decreases without bound), $f(x) \to 0$.

As $x \to \infty$ (increases without bound), $f(x) \to 0$.

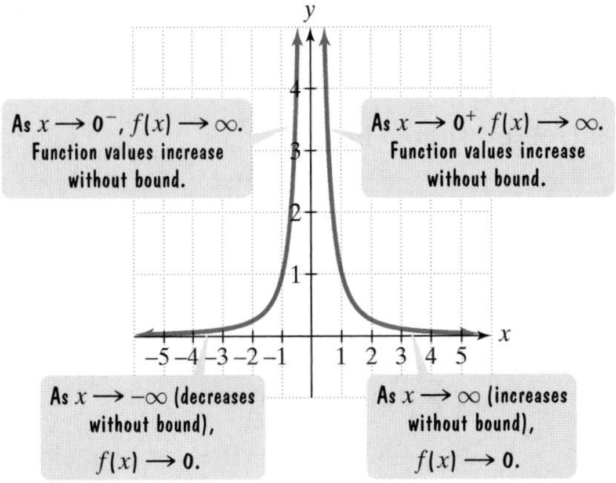

Figure 3.27 The graph of $f(x) = \dfrac{1}{x^2}$

③ Identify vertical asymptotes.

Vertical Asymptotes of Rational Functions

Look again at the graph of $f(x) = \dfrac{1}{x^2}$ in Figure 3.27. The curve approaches, but does not touch, the y-axis. The y-axis, or $x = 0$, is said to be a *vertical asymptote* of the graph. A rational function may have no vertical asymptotes, one vertical asymptote, or several vertical asymptotes. The graph of a rational function never intersects a vertical asymptote. We will use dashed lines to show asymptotes.

Definition of a Vertical Asymptote

The line $x = a$ is a **vertical asymptote** of the graph of a function f if $f(x)$ increases or decreases without bound as x approaches a.

As $x \to a^+$, $f(x) \to \infty$.

As $x \to a^-$, $f(x) \to \infty$.

As $x \to a^+$, $f(x) \to -\infty$.

As $x \to a^-$, $f(x) \to -\infty$.

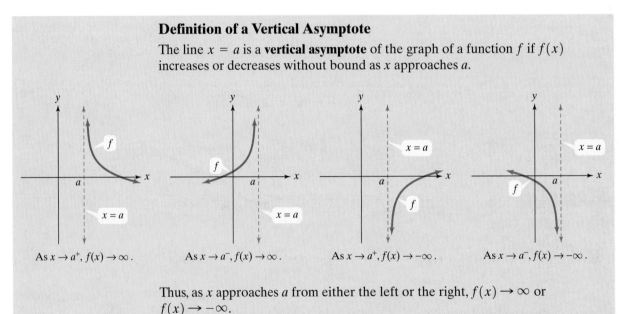

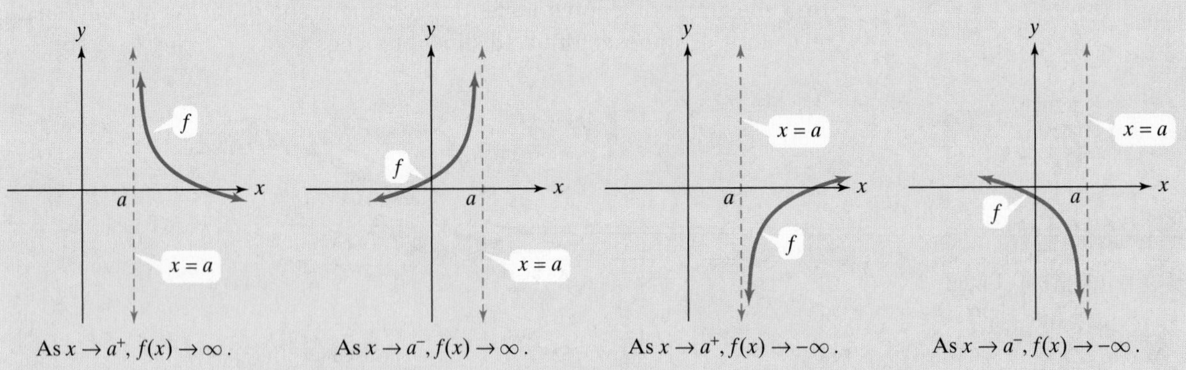

Thus, as x approaches a from either the left or the right, $f(x) \to \infty$ or $f(x) \to -\infty$.

If the graph of a rational function has vertical asymptotes, they can be located using the following theorem:

Locating Vertical Asymptotes

If $f(x) = \dfrac{p(x)}{q(x)}$ is a rational function in which $p(x)$ and $q(x)$ have no common factors and a is a zero of $q(x)$, the denominator, then $x = a$ is a vertical asymptote of the graph of f.

EXAMPLE 2 Finding the Vertical Asymptotes of a Rational Function

Find the vertical asymptotes, if any, of the graph of each rational function:

a. $f(x) = \dfrac{x}{x^2 - 9}$ **b.** $g(x) = \dfrac{x + 3}{x^2 - 9}$ **c.** $h(x) = \dfrac{x + 3}{x^2 + 9}$.

Solution Factoring is usually helpful in identifying zeros of denominators.

a. $f(x) = \dfrac{x}{x^2 - 9} = \dfrac{x}{(x + 3)(x - 3)}$

This factor is 0 if $x = -3$.

This factor is 0 if $x = 3$.

There are no common factors in the numerator and the denominator. The zeros of the denominator are -3 and 3. Thus, the lines $x = -3$ and $x = 3$ are the vertical asymptotes for the graph of f. [See Figure 3.28(a).]

b. We will use factoring to see if there are common factors.

$$g(x) = \frac{x + 3}{x^2 - 9} = \frac{(x + 3)}{(x + 3)(x - 3)} = \frac{1}{x - 3}$$

There is a common factor, $x + 3$, so simplify.

This denominator is 0 if $x = 3$.

The only zero of the denominator of $g(x)$ in simplified form is 3. Thus, the line $x = 3$ is the only vertical asymptote of the graph of g. [See Figure 3.28(b).]

c. We cannot factor the denominator of $h(x)$ over the real numbers.

$$h(x) = \frac{x + 3}{x^2 + 9}$$

No real numbers make this denominator 0.

The denominator has no real zeros. Thus, the graph of h has no vertical asymptotes. [See Figure 3.28(c).]

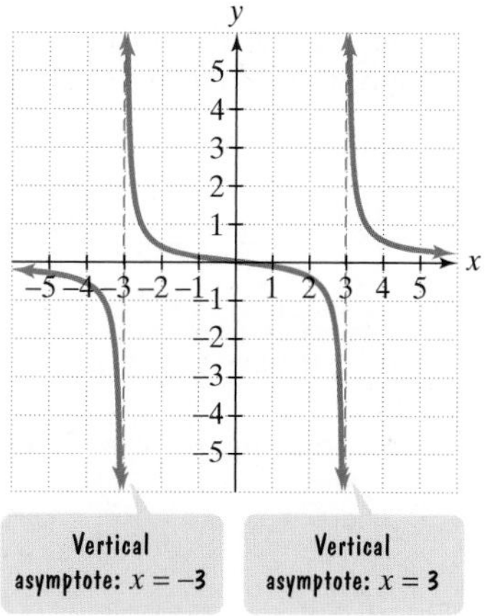

Figure 3.28(a)

The graph of $f(x) = \dfrac{x}{x^2 - 9}$ has two vertical asymptotes.

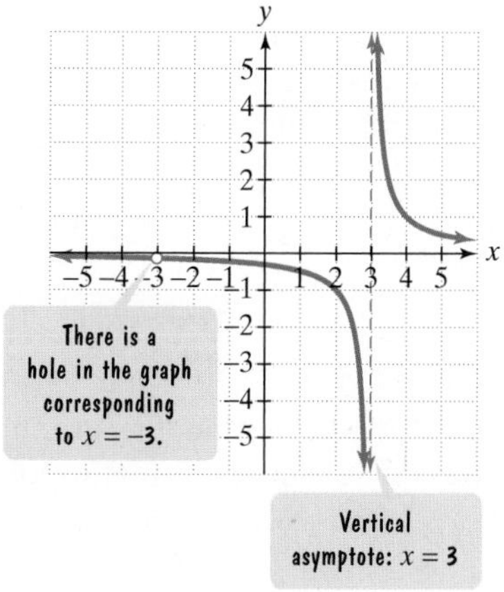

Figure 3.28(b)

The graph of $g(x) = \dfrac{x + 3}{x^2 - 9}$ has one vertical asymptote.

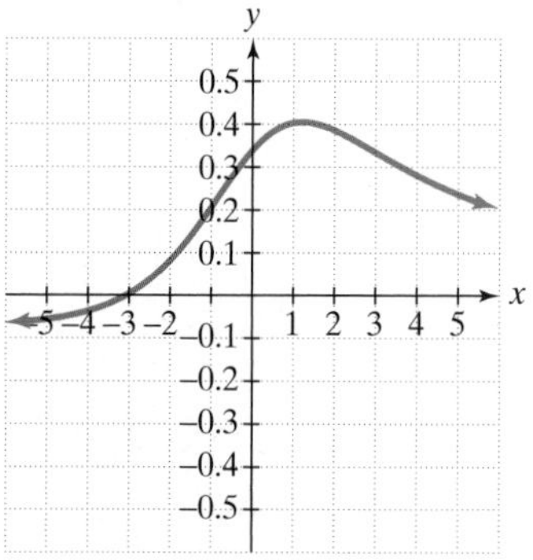

Figure 3.28(c)

The graph of $h(x) = \dfrac{x + 3}{x^2 + 9}$ has no vertical asymptotes.

Check Point 2 Find the vertical asymptotes, if any, of the graph of each rational function:

a. $f(x) = \dfrac{x}{x^2 - 1}$ **b.** $g(x) = \dfrac{x - 1}{x^2 - 1}$ **c.** $h(x) = \dfrac{x - 1}{x^2 + 1}$.

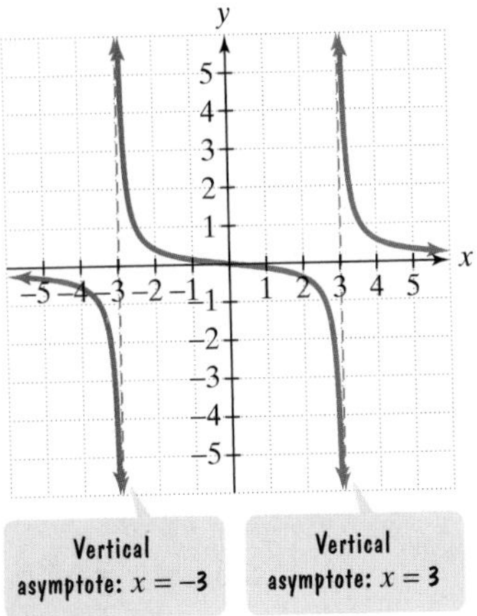
Vertical asymptote: $x = -3$
Vertical asymptote: $x = 3$

Figure 3.28(a) (repeated)

Technology

The graph of the rational function $f(x) = \frac{x}{x^2 - 9}$, drawn by hand in Figure 3.28(a), is graphed below in a $[-5, 5, 1]$ by $[-4, 4, 1]$ viewing rectangle. The graph is shown in connected mode and in dot mode. In connected mode, the graphing utility plots many points and connects the points with curves. In dot mode, the utility plots the same points, but does not connect them.

Connected Mode

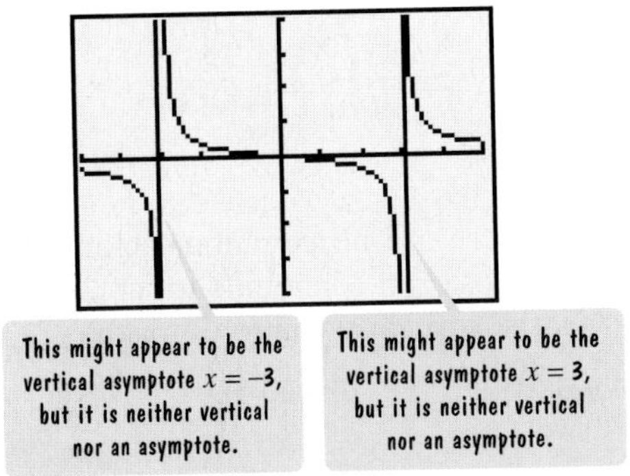
This might appear to be the vertical asymptote $x = -3$, but it is neither vertical nor an asymptote.
This might appear to be the vertical asymptote $x = 3$, but it is neither vertical nor an asymptote.

Dot Mode

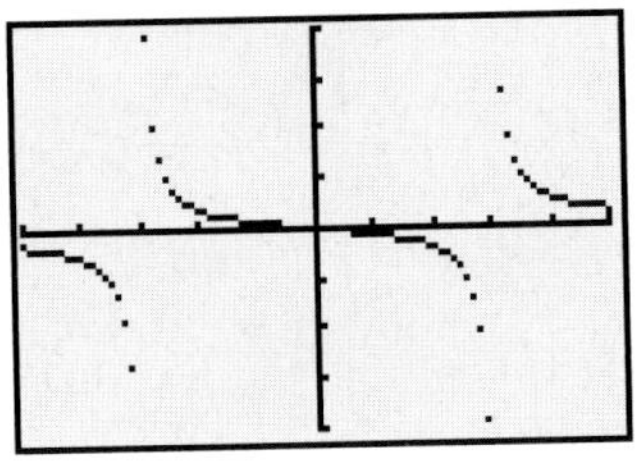

The steep lines in connected mode that are "almost" the vertical asymptotes $x = -3$ and $x = 3$ are not part of the graph and do not represent the vertical asymptotes. The graphing utility has incorrectly connected the last point to the left of $x = -3$ with the first point to the right of $x = -3$. It has also incorrectly connected the last point to the left of $x = 3$ with the first point to the right of $x = 3$. The effect is to create two near-vertical segments that look like asymptotes. This erroneous effect does not appear using dot mode.

A value where the denominator of a rational function is zero does not necessarily result in a vertical asymptote. There is a hole corresponding to $x = a$, and not a vertical asymptote, in the graph of a rational function under the following conditions: The value a causes the denominator to be zero, but there is a reduced form of the function's equation in which a does not cause the denominator to be zero.

Consider, for example, the function

$$f(x) = \frac{x^2 - 4}{x - 2}.$$

Because the denominator is zero when $x = 2$, the function's domain is all real numbers except 2. However, there is a reduced form of the equation in which 2 does not cause the denominator to be zero:

$$f(x) = \frac{x^2 - 4}{x - 2} = \frac{(x + 2)\cancel{(x - 2)}}{\cancel{x - 2}} = x + 2, \ x \neq 2.$$

Denominator is zero at $x = 2$.

In this reduced form, 2 does not result in a zero denominator.

Figure 3.29 shows that the graph has a hole corresponding to $x = 2$. Graphing utilities do not show this feature of the graph.

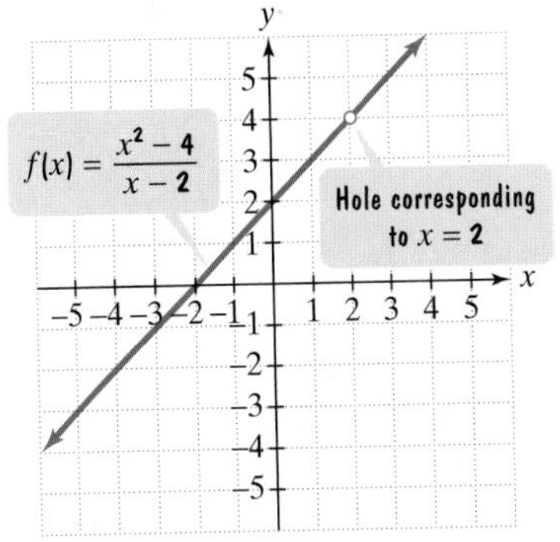
$f(x) = \frac{x^2 - 4}{x - 2}$
Hole corresponding to $x = 2$

Figure 3.29 A graph with a hole corresponding to the denominator's zero

4 Identify horizontal asymptotes.

Horizontal Asymptotes of Rational Functions

Figure 3.26, repeated at the left, shows the graph of the reciprocal function $f(x) = \frac{1}{x}$. As $x \to \infty$ and as $x \to -\infty$, the function values are approaching 0: $f(x) \to 0$. The line $y = 0$ (that is, the x-axis) is a *horizontal asymptote* of the graph. Many, but not all, rational functions have horizontal asymptotes.

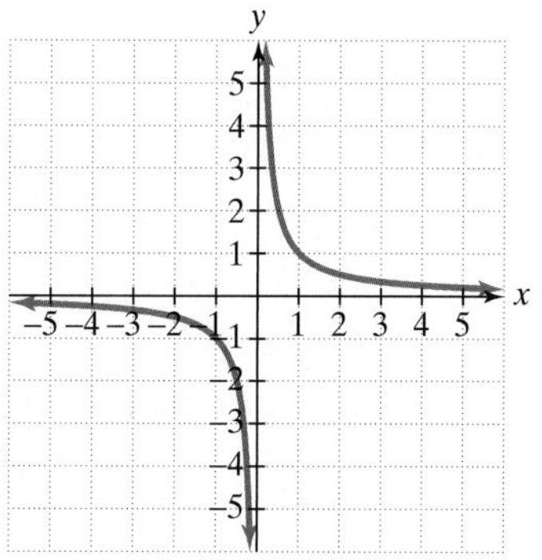

Figure 3.26 The graph of $f(x) = \frac{1}{x}$ (repeated)

Definition of a Horizontal Asymptote

The line $y = b$ is a **horizontal asymptote** of the graph of a function f if $f(x)$ approaches b as x increases or decreases without bound.

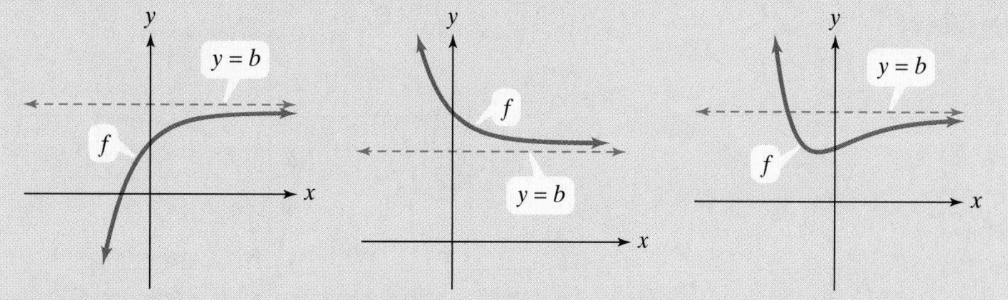

As $x \to \infty, f(x) \to b$. As $x \to \infty, f(x) \to b$. As $x \to \infty, f(x) \to b$.

Recall that a rational function may have several vertical asymptotes. By contrast, it can have at most one horizontal asymptote. Although a graph can never intersect a vertical asymptote, it may cross its horizontal asymptote.

If the graph of a rational function has a horizontal asymptote, it can be located using the following theorem:

Locating Horizontal Asymptotes

Let f be the rational function given by

$$f(x) = \frac{a_n x^n + a_{n-1}x^{n-1} + \cdots + a_1 x + a_0}{b_m x^m + b_{m-1}x^{m-1} + \cdots + b_1 x + b_0}, \quad a_n \neq 0, b_m \neq 0.$$

The degree of the numerator is n. The degree of the denominator is m.

1. If $n < m$, the x-axis, or $y = 0$, is the horizontal asymptote of the graph of f.
2. If $n = m$, the line $y = \frac{a_n}{b_m}$ is the horizontal asymptote of the graph of f.
3. If $n > m$, the graph of f has no horizontal asymptote.

EXAMPLE 3 Finding the Horizontal Asymptote of a Rational Function

Find the horizontal asymptote, if any, of the graph of each rational function:

a. $f(x) = \frac{4x}{2x^2 + 1}$ **b.** $g(x) = \frac{4x^2}{2x^2 + 1}$ **c.** $h(x) = \frac{4x^3}{2x^2 + 1}$.

Solution

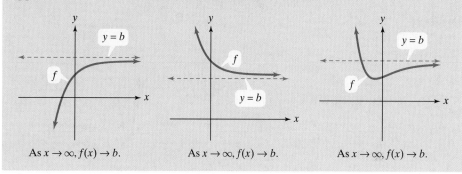

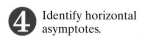

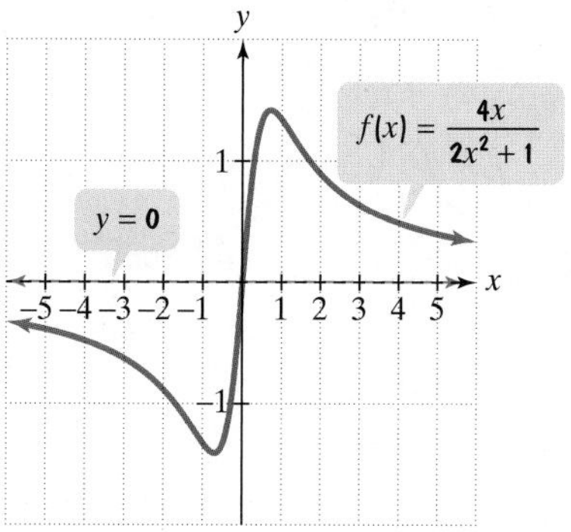

Figure 3.30(a) The horizontal asymptote of the graph is $y = 0$.

a. $f(x) = \frac{4x}{2x^2 + 1}$

The degree of the numerator, 1, is less than the degree of the denominator, 2. Thus, the graph of f has the x-axis as a horizontal asymptote. [See Figure 3.30(a).] The equation of the horizontal asymptote is $y = 0$.

b. $g(x) = \dfrac{4x^2}{2x^2 + 1}$

The degree of the numerator, 2, is equal to the degree of the denominator, 2. The leading coefficients of the numerator and denominator, 4 and 2, are used to obtain the equation of the horizontal asymptote. The equation of the horizontal asymptote is $y = \frac{4}{2}$ or $y = 2$. [See Figure 3.30(b).]

c. $h(x) = \dfrac{4x^3}{2x^2 + 1}$

The degree of the numerator, 3, is greater than the degree of the denominator, 2. Thus, the graph of h has no horizontal asymptote. [See Figure 3.30(c).]

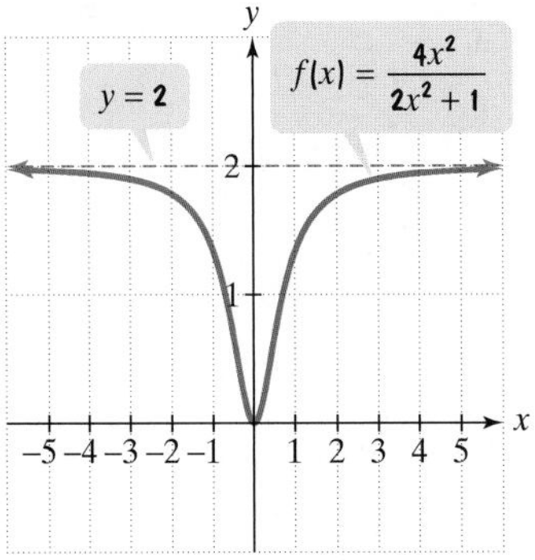

Figure 3.30(b) The horizontal asymptote of the graph is $y = 2$.

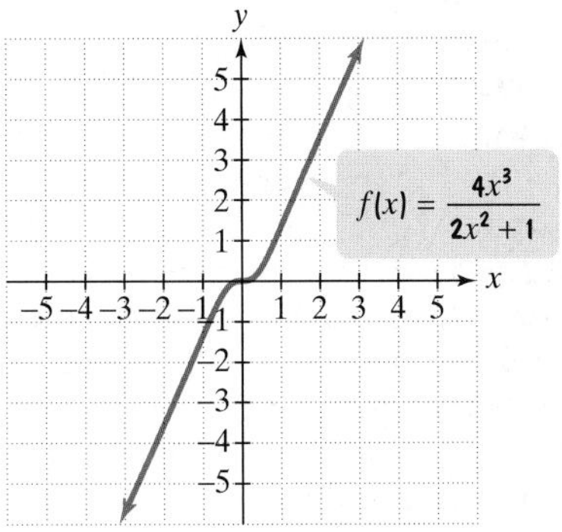

Figure 3.30(c) The graph has no horizontal asymptote.

Check Point 3 Find the horizontal asymptote, if any, of the graph of each rational function:

a. $f(x) = \dfrac{9x^2}{3x^2 + 1}$ **b.** $g(x) = \dfrac{9x}{3x^2 + 1}$ **c.** $h(x) = \dfrac{9x^3}{3x^2 + 1}$.

⑤ Use transformations to graph rational functions.

Using Transformations to Graph Rational Functions

Table 3.2 shows the graphs of two rational functions, $f(x) = \dfrac{1}{x}$ and $f(x) = \dfrac{1}{x^2}$. The dashed green lines indicate the asymptotes.

Table 3.2   Graphs of Common Rational Functions

Some rational functions can be graphed using transformations (horizontal shifting, stretching or shrinking, reflecting, vertical shifting) of these two common graphs.

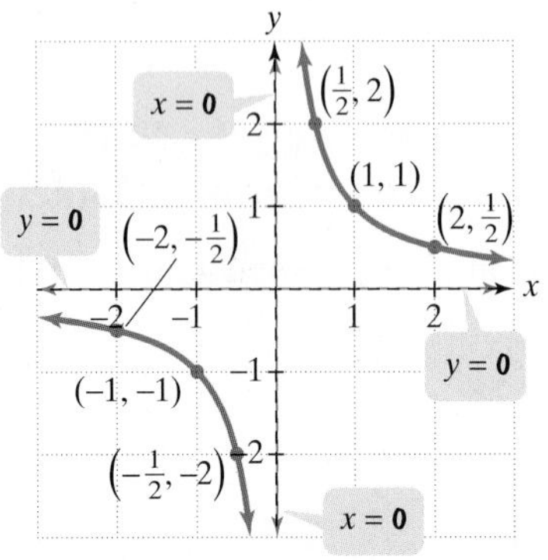

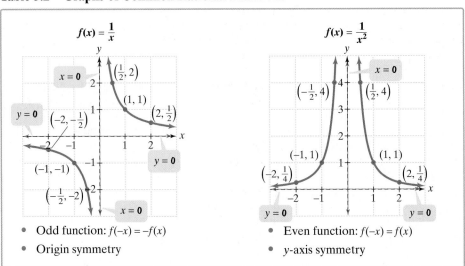

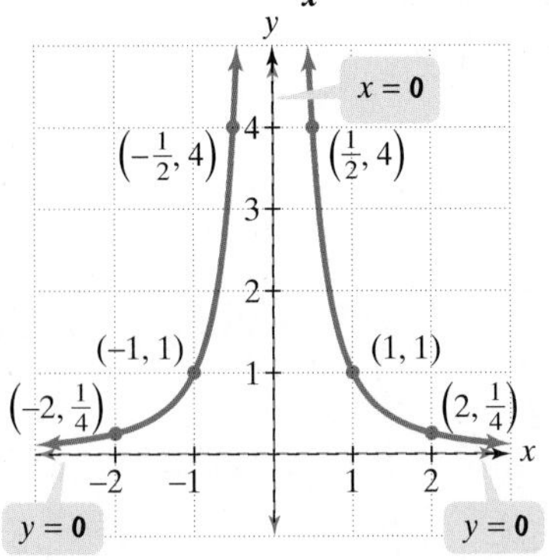

EXAMPLE 4 Using Transformations to Graph a Rational Function

Use the graph of $f(x) = \frac{1}{x^2}$ to graph $g(x) = \frac{1}{(x-2)^2} + 1$.

Solution

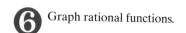

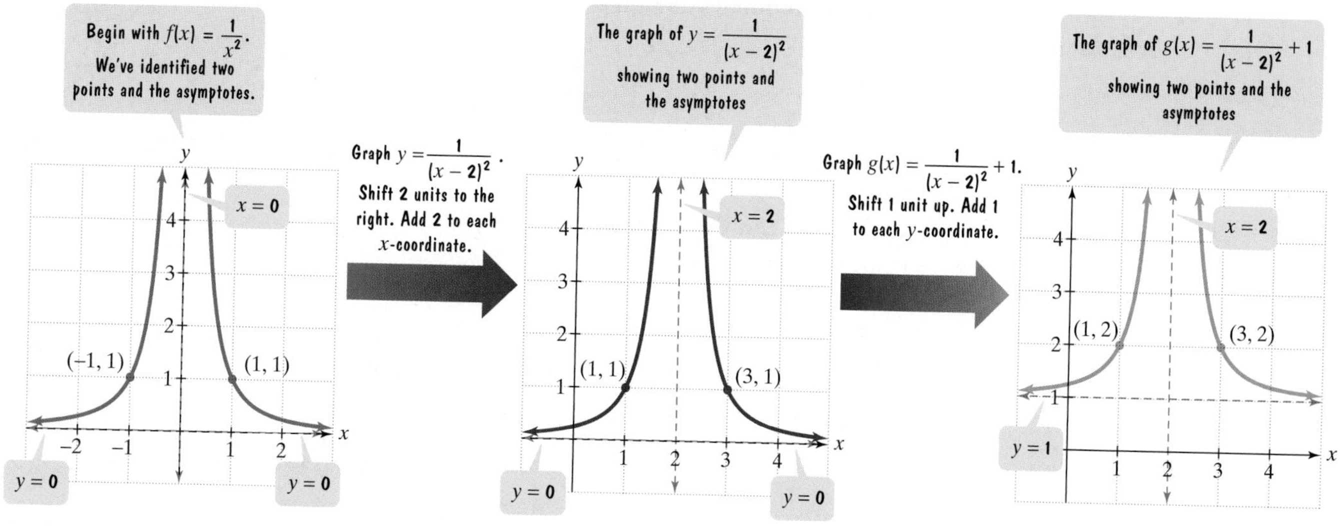

Check Point 4 Use the graph of $f(x) = \frac{1}{x}$ to graph $g(x) = \frac{1}{x+2} - 1$.

6 Graph rational functions.

Graphing Rational Functions

Rational functions that are not transformations of $f(x) = \frac{1}{x}$ or $f(x) = \frac{1}{x^2}$ can be graphed using the following suggestions:

Strategy for Graphing a Rational Function

The following strategy can be used to graph

$$f(x) = \frac{p(x)}{q(x)},$$

where p and q are polynomial functions with no common factors.

1. Determine whether the graph of f has symmetry.

$$f(-x) = f(x)\text{: } y\text{-axis symmetry}$$
$$f(-x) = -f(x)\text{: origin symmetry}$$

2. Find the y-intercept (if there is one) by evaluating $f(0)$.
3. Find the x-intercepts (if there are any) by solving the equation $p(x) = 0$.
4. Find any vertical asymptote(s) by solving the equation $q(x) = 0$.
5. Find the horizontal asymptote (if there is one) using the rule for determining the horizontal asymptote of a rational function.
6. Plot at least one point between and beyond each x-intercept and vertical asymptote.
7. Use the information obtained previously to graph the function between and beyond the vertical asymptotes.

EXAMPLE 5 Graphing a Rational Function

Graph: $f(x) = \frac{2x}{x-1}$.

Solution

Step 1 Determine symmetry.

$$f(-x) = \frac{2(-x)}{-x-1} = \frac{-2x}{-x-1} = \frac{2x}{x+1}$$

Because $f(-x)$ does not equal $f(x)$ or $-f(x)$, the graph has neither y-axis nor origin symmetry.

Step 2 Find the y-intercept. Evaluate $f(0)$.

$$f(0) = \frac{2 \cdot 0}{0-1} = \frac{0}{-1} = 0$$

The y-intercept is 0, so the graph passes through the origin.

Step 3 Find x-intercept(s). This is done by solving $p(x) = 0$.

$$2x = 0 \quad \text{Set the numerator equal to 0.}$$

$$x = 0 \quad \text{Divide both sides by 2.}$$

There is only one x-intercept. This verifies that the graph passes through the origin.

Step 4 Find the vertical asymptote(s). Solve $q(x) = 0$, thereby finding zeros of the denominator.

$$x - 1 = 0 \quad \text{Set the denominator equal to 0.}$$

$$x = 1 \quad \text{Add 1 to both sides.}$$

The equation of the vertical asymptote is $x = 1$.

Step 5 Find the horizontal asymptote. Because the numerator and denominator of $f(x) = \frac{2x}{x-1}$ have the same degree, 1, the leading coefficients of the numerator and denominator, 2 and 1, respectively, are used to obtain the equation of the horizontal asymptote. The equation is

$$y = \frac{2}{1} = 2.$$

The equation of the horizontal asymptote is $y = 2$.

Step 6 Plot points between and beyond each x-intercept and vertical asymptote. With an x-intercept at 0 and a vertical asymptote at $x = 1$, we evaluate the function at -2, -1, $\frac{1}{2}$, 2, and 4.

x	-2	-1	$\frac{1}{2}$	2	4
$f(x) = \frac{2x}{x-1}$	$\frac{4}{3}$	1	-2	4	$\frac{8}{3}$

Figure 3.31 shows these points, the y-intercept, the x-intercept, and the asymptotes.

Step 7 Graph the function. The graph of $f(x) = \frac{2x}{x-1}$ is shown in Figure 3.32.

Technology

The graph of $y = \frac{2x}{x-1}$, obtained using the dot mode in a $[-6, 6, 1]$ by $[-6, 6\ 1]$ viewing rectangle, verifies that our hand-drawn graph is correct.

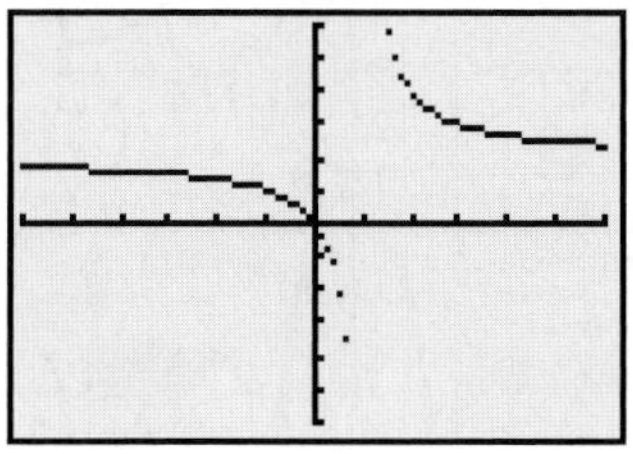

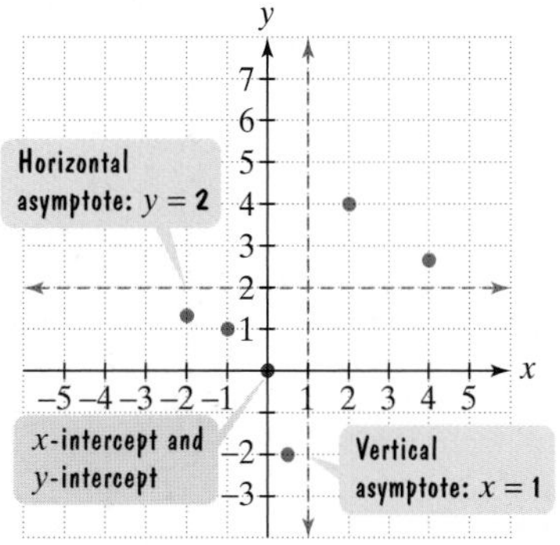

Figure 3.31 Preparing to graph the rational function $f(x) = \frac{2x}{x-1}$

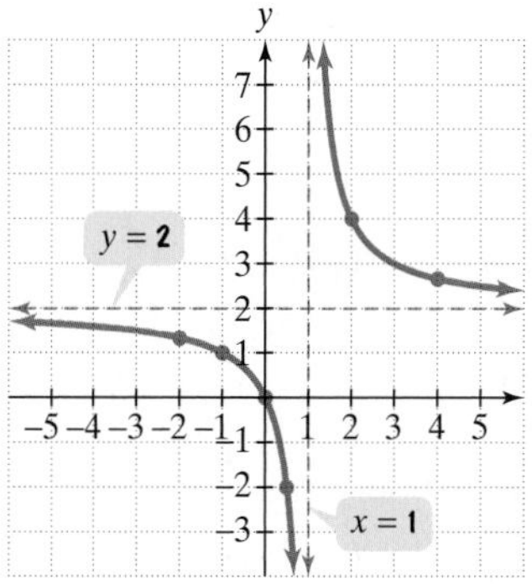

Figure 3.32 The graph of $f(x) = \frac{2x}{x-1}$

Check Point 5 Graph: $f(x) = \frac{3x}{x-2}$.

EXAMPLE 6 Graphing a Rational Function

Graph: $f(x) = \frac{3x^2}{x^2-4}$.

Solution

Step 1 Determine symmetry. $f(-x) = \frac{3(-x)^2}{(-x)^2-4} = \frac{3x^2}{x^2-4} = f(x)$: The graph of f is symmetric with respect to the y-axis.

Step 2 Find the y-intercept. $f(0) = \frac{3 \cdot 0^2}{0^2-4} = \frac{0}{-4} = 0$: The y-intercept is 0, so the graph passes through the origin.

Step 3 Find the x-intercept(s). $3x^2 = 0$, so $x = 0$: The x-intercept is 0, verifying that the graph passes through the origin.

Step 4 Find the vertical asymptote(s). Set $q(x) = 0$.

$$x^2 - 4 = 0 \quad \text{Set the denominator equal to 0.}$$
$$x^2 = 4 \quad \text{Add 4 to both sides.}$$
$$x = \pm 2 \quad \text{Use the square root property.}$$

The vertical asymptotes are $x = -2$ and $x = 2$.

Step 5 Find the horizontal asymptote. Because the numerator and denominator of $f(x) = \frac{3x^2}{x^2-4}$ have the same degree, 2, their leading coefficients, 3 and 1, are used to determine the equation of the horizontal asymptote. The equation is $y = \frac{3}{1} = 3$.

Study Tip

Because the graph has y-axis symmetry, it is not necessary to evaluate the even function at -3 and again at 3.

$$f(-3) = f(3) = \tfrac{27}{5}$$

This also applies to evaluation at -1 and 1.

Step 6 Plot points between and beyond each x-intercept and vertical asymptote. With an x-intercept at 0 and vertical asymptotes at $x = -2$ and $x = 2$, we evaluate the function at -3, -1, 1, 3, and 4.

x	-3	-1	1	3	4
$f(x) = \frac{3x^2}{x^2-4}$	$\frac{27}{5}$	-1	-1	$\frac{27}{5}$	4

Figure 3.33 at the top of the next page shows these points, the y-intercept, the x-intercept, and the asymptotes.

Step 7   Graph the function. The graph of $f(x) = \dfrac{3x^2}{x^2 - 4}$ is shown in Figure 3.34. The y-axis symmetry is now obvious.

Horizontal asymptote: $y = 3$

x-intercept and y-intercept

Vertical asymptote: $x = -2$

Vertical asymptote: $x = 2$

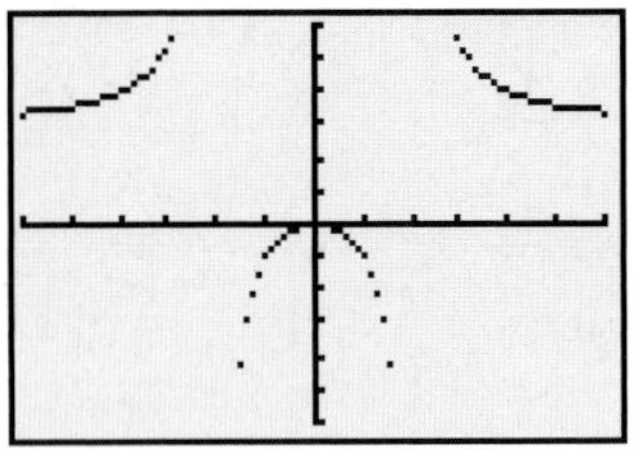

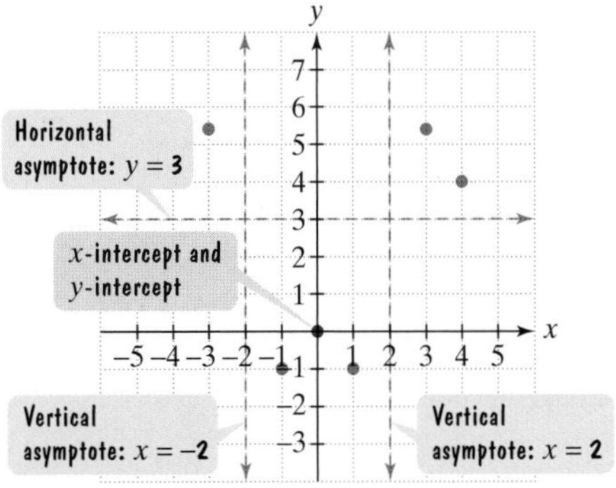

Figure 3.33 Preparing to graph $f(x) = \dfrac{3x^2}{x^2 - 4}$

$y = 3$

$x = -2$

$x = 2$

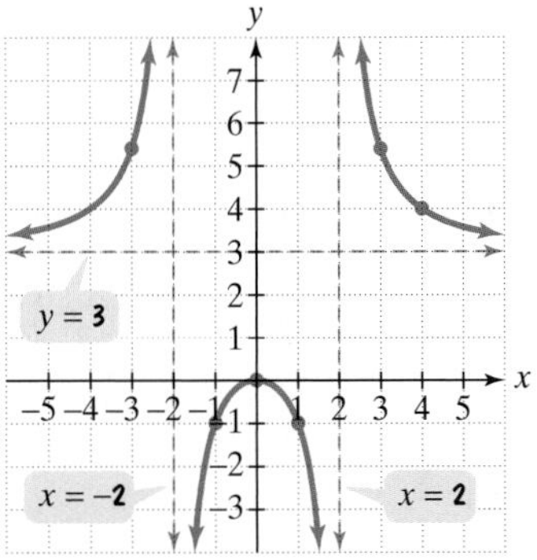

Figure 3.34 The graph of $f(x) = \dfrac{3x^2}{x^2 - 4}$

Check Point 6   Graph: $f(x) = \dfrac{2x^2}{x^2 - 9}$.

Example 7 illustrates that not every rational function has vertical and horizontal asymptotes.

EXAMPLE 7   Graphing a Rational Function

Graph: $f(x) = \dfrac{x^4}{x^2 + 1}$.

Solution

Step 1   Determine symmetry. $f(-x) = \dfrac{(-x)^4}{(-x)^2 + 1} = \dfrac{x^4}{x^2 + 1} = f(x)$: The graph of f is symmetric with respect to the y-axis.

Step 2   Find the y-intercept. $f(0) = \dfrac{0^4}{0^2 + 1} = \dfrac{0}{1} = 0$: The y-intercept is 0.

Step 3   Find the x-intercept(s). $x^4 = 0$, so $x = 0$: The x-intercept is 0.

Step 4   Find the vertical asymptote. Set $q(x) = 0$.

$$x^2 + 1 = 0 \qquad \text{Set the denominator equal to 0.}$$

$$x^2 = -1 \qquad \text{Subtract 1 from both sides.}$$

Although this equation has imaginary roots ($x = \pm i$), there are no real roots. Thus, the graph of f has no vertical asymptotes.

Step 5   Find the horizontal asymptote. Because the degree of the numerator, 4, is greater than the degree of the denominator, 2, there is no horizontal asymptote.

Step 6   Plot points between and beyond each x-intercept and vertical asymptote. With an x-intercept at 0 and no vertical asymptotes, let's look at function values at

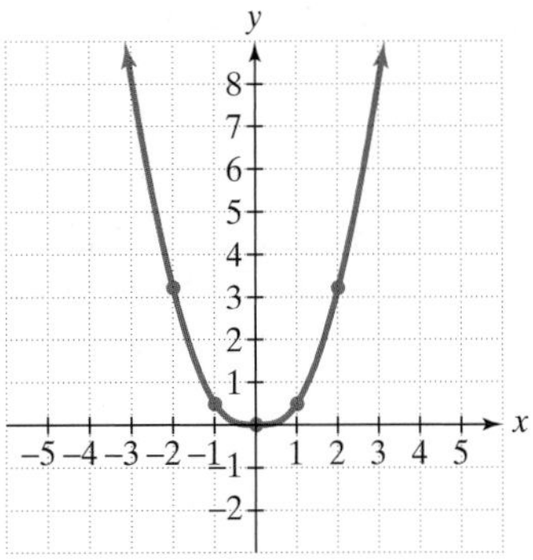

Figure 3.35 The graph of $f(x) = \frac{x^4}{x^2+1}$

$-2, -1, 1,$ and 2. You can evaluate the function at 1 and 2. Use y-axis symmetry to obtain function values at -1 and -2:

$$f(-1) = f(1) \text{ and } f(-2) = f(2).$$

x	-2	-1	1	2
$f(x) = \frac{x^4}{x^2+1}$	$\frac{16}{5}$	$\frac{1}{2}$	$\frac{1}{2}$	$\frac{16}{5}$

Step 7 Graph the function. Figure 3.35 shows the graph of f using the points obtained from the table and y-axis symmetry. Notice that as x approaches infinity or negative infinity ($x \to \infty$ or $x \to -\infty$), the function values, $f(x)$, are getting larger without bound $[f(x) \to \infty]$.

Check Point 7 Graph: $f(x) = \frac{x^4}{x^2+2}$.

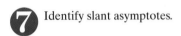

7 Identify slant asymptotes.

Slant Asymptotes

Examine the graph of

$$f(x) = \frac{x^2+1}{x-1},$$

shown in Figure 3.36. Note that the degree of the numerator, 2, is greater than the degree of the denominator, 1. Thus, the graph of this function has no horizontal asymptote. However, the graph has a **slant asymptote**, $y = x + 1$.

The graph of a rational function has a slant asymptote if the degree of the numerator is one more than the degree of the denominator. The equation of the slant asymptote can be found by division. For example, to find the slant asymptote for the graph of $f(x) = \frac{x^2+1}{x-1}$, divide $x - 1$ into $x^2 + 1$:

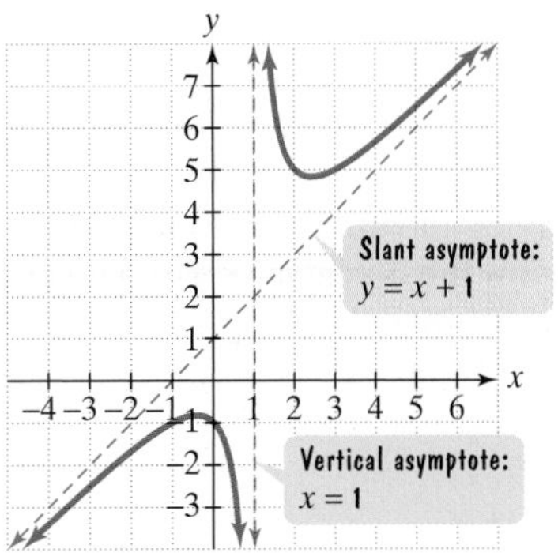

Figure 3.36 The graph of $f(x) = \frac{x^2+1}{x-1}$ with a slant asymptote

$$\begin{array}{r|rrr} 1 & 1 & 0 & 1 \\ & & 1 & 1 \\ \hline & 1 & 1 & 2 \end{array}$$

Remainder

$$x - 1 \overline{)\, x^2 + 0x + 1} \qquad 1x + 1 + \frac{2}{x-1}$$

Observe that

$$f(x) = \frac{x^2+1}{x-1} = \underbrace{x + 1}_{} + \frac{2}{x-1}.$$

The equation of the slant asymptote is $y = x + 1$.

As $|x| \to \infty$, the value of $\frac{2}{x-1}$ is approximately 0. Thus, when $|x|$ is large, the function is very close to $y = x + 1 + 0$. This means that as $x \to \infty$ or as $x \to -\infty$, the graph of f gets closer and closer to the line whose equation is $y = x + 1$. The line $y = x + 1$ is a slant asymptote of the graph.

In general, if $f(x) = \frac{p(x)}{q(x)}$, p and q have no common factors, and the degree of p is one greater than the degree of q, find the slant asymptote by dividing $q(x)$ into $p(x)$. The division will take the form

$$\frac{p(x)}{q(x)} = mx + b + \frac{\text{remainder}}{q(x)}.$$

Slant asymptote: $y = mx + b$

The equation of the slant asymptote is obtained by dropping the term with the remainder. Thus, the equation of the slant asymptote is $y = mx + b$.

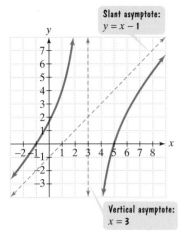

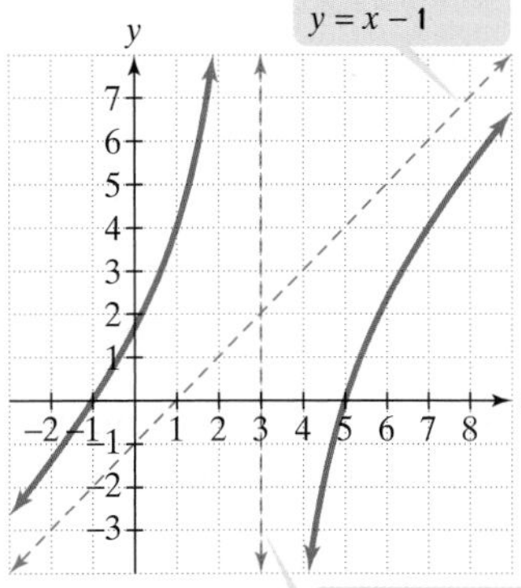

Figure 3.37 The graph of $f(x) = \dfrac{x^2 - 4x - 5}{x - 3}$

EXAMPLE 8 Finding the Slant Asymptote of a Rational Function

Find the slant asymptote of $f(x) = \dfrac{x^2 - 4x - 5}{x - 3}$.

Solution Because the degree of the numerator, 2, is exactly one more than the degree of the denominator, 1, and $x - 3$ is not a factor of $x^2 - 4x - 5$, the graph of f has a slant asymptote. To find the equation of the slant asymptote, divide $x - 3$ into $x^2 - 4x - 5$:

$$\begin{array}{r|rrr} 3 & 1 & -4 & -5 \\ & & 3 & -3 \\ \hline & 1 & -1 & -8 \end{array} \quad \text{Remainder}$$

$$1x - 1 - \frac{8}{x - 3}$$
$$x - 3\overline{)x^2 - 4x - 5}$$

Drop the remainder term and you'll have the equation of the slant asymptote.

The equation of the slant asymptote is $y = x - 1$. Using our strategy for graphing rational functions, the graph of $f(x) = \dfrac{x^2 - 4x - 5}{x - 3}$ is shown in Figure 3.37.

Check Point 8 Find the slant asymptote of $f(x) = \dfrac{2x^2 - 5x + 7}{x - 2}$.

8 Solve applied problems involving rational functions.

Applications

There are numerous examples of asymptotic behavior in functions that describe real-world phenomena. Let's consider an example from the business world. The **cost function**, C, for a business is the sum of its fixed and variable costs:

$$C(x) = (\text{fixed cost}) + cx.$$

Cost per unit times the number of units produced, x.

The **average cost** per unit for a company to produce x units is the sum of its fixed and variable costs divided by the number of units produced. The **average cost function** is a rational function that is denoted by $\overline{C}$. Thus,

Cost of producing x units: fixed plus variable costs

$$\overline{C}(x) = \frac{(\text{fixed cost}) + cx}{x}.$$

Number of units produced

EXAMPLE 9 Average Cost of Producing a Wheelchair

A company is planning to manufacture wheelchairs that are light, fast, and beautiful. The fixed monthly cost will be \$500,000 and it will cost \$400 to produce each radically innovative chair.

a. Write the cost function, C, of producing x wheelchairs.

b. Write the average cost function, $\overline{C}$, of producing x wheelchairs.

c. Find and interpret $\overline{C}(1000)$, $\overline{C}(10{,}000)$, and $\overline{C}(100{,}000)$.

d. What is the horizontal asymptote for the graph of the average cost function, $\overline{C}$? Describe what this represents for the company.

Solution

a. The cost function of producing x wheelchairs, C, is the sum of the fixed cost and the variable costs.

Fixed cost is \$500,000.

Variable cost: \$400 for each wheelchair produced

$$C(x) = 500{,}000 + 400x$$

b. The average cost function of producing x wheelchairs, $\overline{C}$, is the sum of the fixed and variable costs divided by the number of wheelchairs produced.

$$\overline{C}(x) = \frac{500{,}000 + 400x}{x} \quad \text{or} \quad \overline{C}(x) = \frac{400x + 500{,}000}{x}$$

c. We evaluate $\overline{C}$ at 1000, 10,000, and 100,000, interpreting the results.

$$\overline{C}(1000) = \frac{400(1000) + 500{,}000}{1000} = 900$$

The average cost per wheelchair of producing 1000 wheelchairs per month is \$900.

$$\overline{C}(10{,}000) = \frac{400(10{,}000) + 500{,}000}{10{,}000} = 450$$

The average cost per wheelchair of producing 10,000 wheelchairs per month is \$450.

$$\overline{C}(100{,}000) = \frac{400(100{,}000) + 500{,}000}{100{,}000} = 405$$

The average cost per wheelchair of producing 100,000 wheelchairs per month is \$405. Notice that with higher production levels, the cost of producing each wheelchair decreases.

d. We developed the average cost function

$$\overline{C}(x) = \frac{400x + 500{,}000}{x}$$

in which the degree of the numerator, 1, is equal to the degree of the denominator, 1. The leading coefficients of the numerator and denominator, 400 and 1, are used to obtain the equation of the horizontal asymptote. The equation of the horizontal asymptote is

$$y = \frac{400}{1} \quad \text{or} \quad y = 400.$$

The horizontal asymptote is shown in Figure 3.38. This means that the more wheelchairs produced per month, the closer the average cost per wheelchair for the company comes to \$400. The least possible cost per wheelchair is approaching \$400. Competitively low prices take place with high production levels, posing a major problem for small businesses.

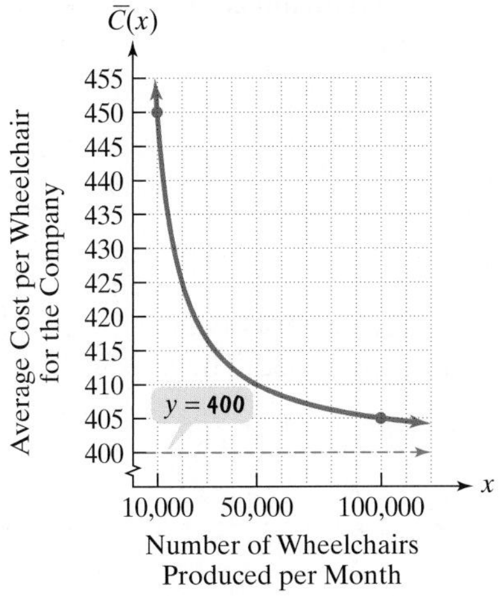

Figure 3.38 As production level increases, the average cost per wheelchair approaches \$400.

Check Point 9 The time: the not-too-distant future. A new company is hoping to replace traditional computers and two-dimensional monitors with its virtual reality system. The fixed monthly cost will be \$600,000 and it will cost \$500 to produce each system.

a. Write the cost function, C, of producing x virtual reality systems.

b. Write the average cost function, $\overline{C}$, of producing x virtual reality systems.

c. Find and interpret $\overline{C}(1000)$, $\overline{C}(10{,}000)$, and $\overline{C}(100{,}000)$.

d. What is the horizontal asymptote for the graph of the average cost function, $\overline{C}$? Describe what this represents for the company.

EXERCISE SET 3.5

Practice Exercises

In Exercises 1–8, find the domain of each rational function.

1. $f(x) = \dfrac{5x}{x - 4}$
2. $f(x) = \dfrac{7x}{x - 8}$
3. $g(x) = \dfrac{3x^2}{(x - 5)(x + 4)}$
4. $g(x) = \dfrac{2x^2}{(x - 2)(x + 6)}$
5. $h(x) = \dfrac{x + 7}{x^2 - 49}$
6. $h(x) = \dfrac{x + 8}{x^2 - 64}$
7. $f(x) = \dfrac{x + 7}{x^2 + 49}$
8. $f(x) = \dfrac{x + 8}{x^2 + 64}$

Use the graph of the rational function in the figure shown to complete each statement in Exercises 9–14.

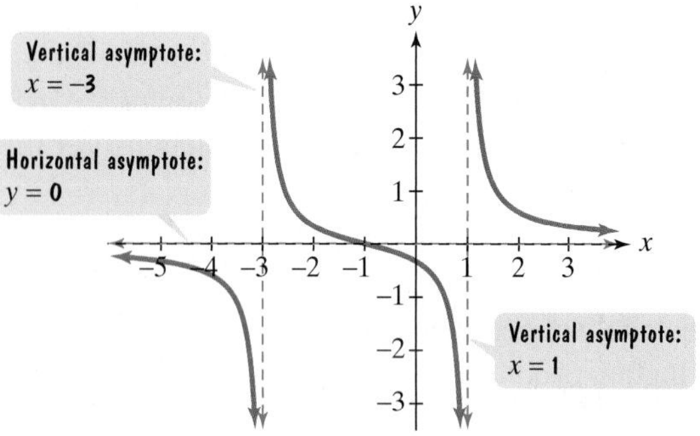

9. As $x \to -3^-$, $f(x) \to$ ____.
10. As $x \to -3^+$, $f(x) \to$ ____.
11. As $x \to 1^-$, $f(x) \to$ ____.
12. As $x \to 1^+$, $f(x) \to$ ____.
13. As $x \to -\infty$, $f(x) \to$ ____.
14. As $x \to \infty$, $f(x) \to$ ____.

Use the graph of the rational function in the figure shown to complete each statement in Exercises 15–20.

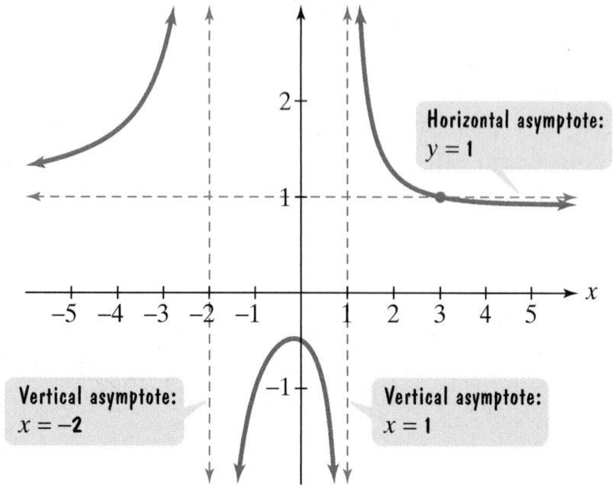

15. As $x \to 1^+$, $f(x) \to$ ____.
16. As $x \to 1^-$, $f(x) \to$ ____.
17. As $x \to -2^+$, $f(x) \to$ ____.
18. As $x \to -2^-$, $f(x) \to$ ____.
19. As $x \to \infty$, $f(x) \to$ ____.
20. As $x \to -\infty$, $f(x) \to$ ____.

In Exercises 21–28, find the vertical asymptotes, if any, of the graph of each rational function.

21. $f(x) = \dfrac{x}{x + 4}$
22. $f(x) = \dfrac{x}{x - 3}$
23. $g(x) = \dfrac{x + 3}{x(x + 4)}$
24. $g(x) = \dfrac{x + 3}{x(x - 3)}$
25. $h(x) = \dfrac{x}{x(x + 4)}$
26. $h(x) = \dfrac{x}{x(x - 3)}$
27. $r(x) = \dfrac{x}{x^2 + 4}$
28. $r(x) = \dfrac{x}{x^2 + 3}$

In Exercises 29–36, find the horizontal asymptote, if any, of the graph of each rational function.

29. $f(x) = \dfrac{12x}{3x^2 + 1}$
30. $f(x) = \dfrac{15x}{3x^2 + 1}$
31. $g(x) = \dfrac{12x^2}{3x^2 + 1}$
32. $g(x) = \dfrac{15x^2}{3x^2 + 1}$
33. $h(x) = \dfrac{12x^3}{3x^2 + 1}$
34. $h(x) = \dfrac{15x^3}{3x^2 + 1}$
35. $f(x) = \dfrac{-2x + 1}{3x + 5}$
36. $f(x) = \dfrac{-3x + 7}{5x - 2}$

In Exercises 37–48, use transformations of $f(x) = \dfrac{1}{x}$ or $f(x) = \dfrac{1}{x^2}$ to graph each rational function.

37. $g(x) = \dfrac{1}{x - 1}$
38. $g(x) = \dfrac{1}{x - 2}$
39. $h(x) = \dfrac{1}{x} + 2$
40. $h(x) = \dfrac{1}{x} + 1$
41. $g(x) = \dfrac{1}{x + 1} - 2$
42. $g(x) = \dfrac{1}{x + 2} - 2$
43. $g(x) = \dfrac{1}{(x + 2)^2}$
44. $g(x) = \dfrac{1}{(x + 1)^2}$
45. $h(x) = \dfrac{1}{x^2} - 4$
46. $h(x) = \dfrac{1}{x^2} - 3$
47. $h(x) = \dfrac{1}{(x - 3)^2} + 1$
48. $h(x) = \dfrac{1}{(x - 3)^2} + 2$

In Exercises 49–70, follow the seven steps on page 359 to graph each rational function.

49. $f(x) = \dfrac{4x}{x - 2}$
50. $f(x) = \dfrac{3x}{x - 1}$
51. $f(x) = \dfrac{2x}{x^2 - 4}$
52. $f(x) = \dfrac{4x}{x^2 - 1}$
53. $f(x) = \dfrac{2x^2}{x^2 - 1}$
54. $f(x) = \dfrac{4x^2}{x^2 - 9}$

55. $f(x) = \dfrac{-x}{x+1}$

56. $f(x) = \dfrac{-3x}{x+2}$

57. $f(x) = -\dfrac{1}{x^2-4}$

58. $f(x) = -\dfrac{2}{x^2-1}$

59. $f(x) = \dfrac{2}{x^2+x-2}$

60. $f(x) = \dfrac{-2}{x^2-x-2}$

61. $f(x) = \dfrac{2x^2}{x^2+4}$

62. $f(x) = \dfrac{4x^2}{x^2+1}$

63. $f(x) = \dfrac{x+2}{x^2+x-6}$

64. $f(x) = \dfrac{x-4}{x^2-x-6}$

65. $f(x) = \dfrac{x^4}{x^2+2}$

66. $f(x) = \dfrac{2x^4}{x^2+1}$

67. $f(x) = \dfrac{x^2+x-12}{x^2-4}$

68. $f(x) = \dfrac{x^2}{x^2+x-6}$

69. $f(x) = \dfrac{3x^2+x-4}{2x^2-5x}$

70. $f(x) = \dfrac{x^2-4x+3}{(x+1)^2}$

In Exercises 71–78, **a.** *Find the slant asymptote of the graph of each rational function and* **b.** *Follow the seven-step strategy and use the slant asymptote to graph each rational function.*

71. $f(x) = \dfrac{x^2-1}{x}$

72. $f(x) = \dfrac{x^2-4}{x}$

73. $f(x) = \dfrac{x^2+1}{x}$

74. $f(x) = \dfrac{x^2+4}{x}$

75. $f(x) = \dfrac{x^2+x-6}{x-3}$

76. $f(x) = \dfrac{x^2-x+1}{x-1}$

77. $f(x) = \dfrac{x^3+1}{x^2+2x}$

78. $f(x) = \dfrac{x^3-1}{x^2-9}$

Practice Plus

In Exercises 79–84, the equation for f is given by the simplified expression that results after performing the indicated operation. Write the equation for f and then graph the function.

79. $\dfrac{5x^2}{x^2-4} \cdot \dfrac{x^2+4x+4}{10x^3}$

80. $\dfrac{x-5}{10x-2} \div \dfrac{x^2-10x+25}{25x^2-1}$

81. $\dfrac{x}{2x+6} - \dfrac{9}{x^2-9}$

82. $\dfrac{2}{x^2+3x+2} - \dfrac{4}{x^2+4x+3}$

83. $\dfrac{1-\dfrac{3}{x+2}}{1+\dfrac{1}{x-2}}$

84. $\dfrac{x-\dfrac{1}{x}}{x+\dfrac{1}{x}}$

In Exercises 85–88, use long division to rewrite the equation for g in the form quotient, plus remainder divided by divisor. Then use this form of the function's equation and transformations of $f(x) = \dfrac{1}{x}$ *to graph g.*

85. $g(x) = \dfrac{2x+7}{x+3}$

86. $g(x) = \dfrac{3x+7}{x+2}$

87. $g(x) = \dfrac{3x-7}{x-2}$

88. $g(x) = \dfrac{2x-9}{x-4}$

Application Exercises

89. A company is planning to manufacture mountain bikes. The fixed monthly cost will be \$100,000 and it will cost \$100 to produce each bicycle.
   a. Write the cost function, C, of producing x mountain bikes.
   b. Write the average cost function, $\overline{C}$, of producing x mountain bikes.
   c. Find and interpret $\overline{C}(500)$, $\overline{C}(1000)$, $\overline{C}(2000)$, and $\overline{C}(4000)$.
   d. What is the horizontal asymptote for the graph of the average cost function, $\overline{C}$? Describe what this means in practical terms.

90. A company that manufactures running shoes has a fixed monthly cost of \$300,000. It costs \$30 to produce each pair of shoes.
   a. Write the cost function, C, of producing x pairs of shoes.
   b. Write the average cost function, $\overline{C}$, of producing x pairs of shoes.
   c. Find and interpret $\overline{C}(1000)$, $\overline{C}(10{,}000)$, and $\overline{C}(100{,}000)$.
   d. What is the horizontal asymptote for the graph of the average cost function, $\overline{C}$? Describe what this represents for the company.

91. The function

$$f(x) = \frac{6.5x^2 - 20.4x + 234}{x^2 + 36}$$

models the pH level, $f(x)$, of the human mouth x minutes after a person eats food containing sugar. The graph of this function is shown in the figure.

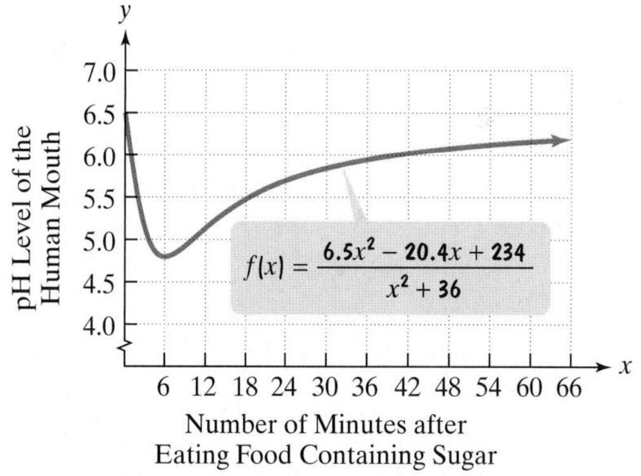

   a. Use the graph to obtain a reasonable estimate, to the nearest tenth, of the pH level of the human mouth 42 minutes after a person eats food containing sugar.
   b. After eating sugar, when is the pH level the lowest? Use the function's equation to determine the pH level, to the nearest tenth, at this time.
   c. According to the graph, what is the normal pH level of the human mouth?
   d. What is the equation of the horizontal asymptote associated with this function? Describe what this means in terms of the mouth's pH level over time.
   e. Use the graph to describe what happens to the pH level during the first hour.

92. A drug is injected into a patient and the concentration of the drug in the bloodstream is monitored. The drug's concentration, $C(t)$, in milligrams per liter, after t hours is modeled by

$$C(t) = \frac{5t}{t^2+1}.$$

The graph of this rational function, obtained with a graphing utility, is shown in the figure.

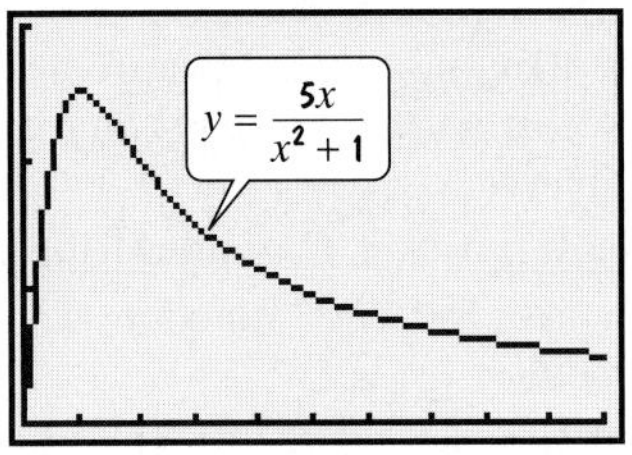

[0, 10, 1] by [0, 3, 1]

a. Use the graph to obtain a reasonable estimate of the drug's concentration after 3 hours. Then verify this estimate algebraically.

b. Use the function's equation, $C(t) = \frac{5t}{t^2 + 1}$, to find the horizontal asymptote for the graph. Describe what this means about the drug's concentration in the patient's bloodstream as time increases.

*Among all deaths from a particular disease, the percentage that are smoking related (21–39 cigarettes per day) is a function of the disease's **incidence ratio**. The incidence ratio describes the number of times more likely smokers are than nonsmokers to die from the disease. The following table shows the incidence ratios for heart disease and lung cancer for two age groups.*

Incidence Ratios

	Heart Disease	Lung Cancer
Ages 55–64	1.9	10
Ages 65–74	1.7	9

Source: Alexander M. Walker, *Observations and Inference*, Epidemiology Resources Inc., 1991.

For example, the incidence ratio of 9 in the table means that smokers between the ages of 65 and 74 are 9 times more likely than nonsmokers in the same age group to die from lung cancer. The rational function

$$P(x) = \frac{100(x - 1)}{x}$$

models the percentage of smoking-related deaths among all deaths from a disease, $P(x)$, in terms of the disease's incidence ratio, x. The graph of the rational function is shown. Use this function to solve Exercises 93–96.

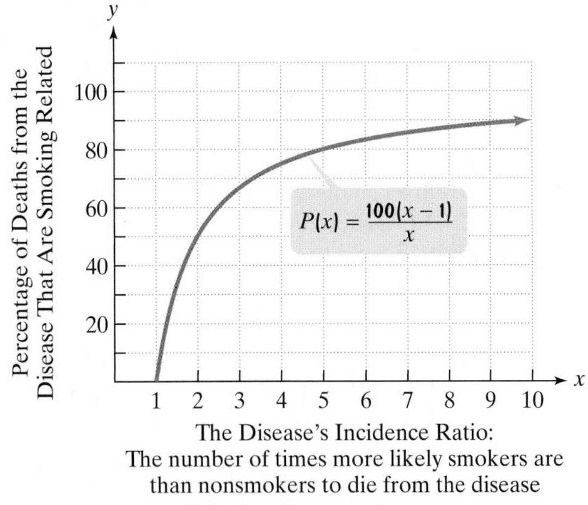

93. Find $P(10)$. Describe what this means in terms of the incidence ratio, 10, given in the table. Identify your solution as a point on the graph.

94. Find $P(9)$. Round to the nearest percent. Describe what this means in terms of the incidence ratio, 9, given in the table. Identify your solution as a point on the graph.

95. What is the horizontal asymptote of the graph? Describe what this means about the percentage of deaths caused by smoking with increasing incidence ratios.

96. According to the model and its graph, is there a disease for which all deaths are caused by smoking? Explain your answer.

97. The graph shows the U.S. population, by gender, for selected years from 1950 through 2002.

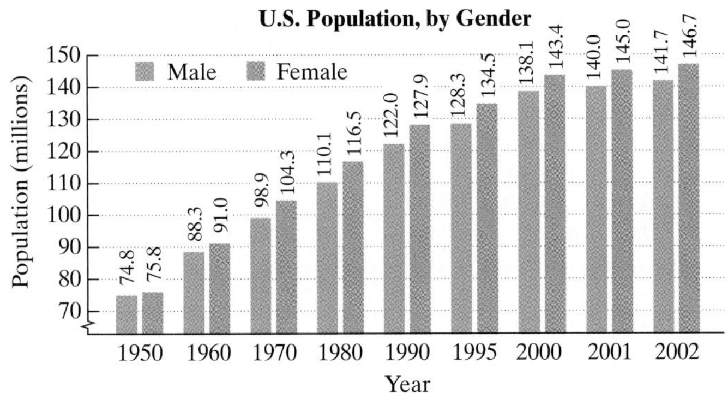

Source: U.S. Census Bureau

a. Write a fraction that shows the ratio of males to females in 1995. Then express the fraction as a decimal, rounded to the nearest thousandth. How many males per 1000 females were there in 1995?

b. How many males per 1000 females were there in 2002?

c. The function $p(x) = 1.256x + 74.2$ models the male U.S. population, $p(x)$, in millions, x years after 1950. The function $q(x) = 1.324x + 76.71$ models the female U.S. population, $q(x)$, in millions, x years after 1950. Write a function that models the ratio of males to females x years after 1950.

d. Use the function that you wrote in part (c) to find the number of males per 1000 females in 1995. How well does the function model the actual number that you determined in part (a)?

e. Use the function that you wrote in part (c) to find the number of males per 1000 females in 2002. How well does the function model the actual number that you determined in part (b)?

f. What is the equation of the horizontal asymptote associated with the function in part (c)? Round to the nearest thousandth. What does this mean about the number of males per 1000 females over time?

Writing in Mathematics

98. What is a rational function?

99. Use everyday language to describe the graph of a rational function f such that as $x \to -\infty$, $f(x) \to 3$.

100. Use everyday language to describe the behavior of a graph near its vertical asymptote if $f(x) \to \infty$ as $x \to -2^-$ and $f(x) \to -\infty$ as $x \to -2^+$.

101. If you are given the equation of a rational function, explain how to find the vertical asymptotes, if any, of the function's graph.

102. If you are given the equation of a rational function, explain how to find the horizontal asymptote, if any, of the function's graph.

103. Describe how to graph a rational function.

104. If you are given the equation of a rational function, how can you tell if the graph has a slant asymptote? If it does, how do you find its equation?

105. Is every rational function a polynomial function? Why or why not? Does a true statement result if the two adjectives *rational* and *polynomial* are reversed? Explain.

106. Although your friend has a family history of heart disease, he smokes, on average, 25 cigarettes per day. He sees the table showing incidence ratios for heart disease (see Exercises 93–96) and feels comfortable that they are less than 2, compared to 9 and 10 for lung cancer. He claims that all family deaths have been from heart disease and decides not to give up smoking. Use the given function and its graph to describe some additional information not given in the table that might influence his decision.

Technology Exercises

107. Use a graphing utility to verify any five of your hand-drawn graphs in Exercises 37–78.

108. Use a graphing utility to graph $y = \frac{1}{x}$, $y = \frac{1}{x^3}$, and $\frac{1}{x^5}$ in the same viewing rectangle. For odd values of n, how does changing n affect the graph of $y = \frac{1}{x^n}$?

109. Use a graphing utility to graph $y = \frac{1}{x^2}$, $y = \frac{1}{x^4}$, and $y = \frac{1}{x^6}$ in the same viewing rectangle. For even values of n, how does changing n affect the graph of $y = \frac{1}{x^n}$?

110. Use a graphing utility to graph

$$f(x) = \frac{x^2 - 4x + 3}{x - 2} \quad \text{and} \quad g(x) = \frac{x^2 - 5x + 6}{x - 2}.$$

What differences do you observe between the graph of f and the graph of g? How do you account for these differences?

111. The rational function

$$f(x) = \frac{27{,}725(x - 14)}{x^2 + 9} - 5x$$

models the number of arrests, $f(x)$, per 100,000 drivers, for driving under the influence of alcohol, as a function of a driver's age, x.

a. Graph the function in a [0, 70, 5] by [0, 400, 20] viewing rectangle.

b. Describe the trend shown by the graph.

c. Use the ZOOM and TRACE features or the maximum function feature of your graphing utility to find the age that corresponds to the greatest number of arrests. How many arrests, per 100,000 drivers, are there for this age group?

Critical Thinking Exercises

112. Which one of the following is true?

a. The graph of a rational function cannot have both a vertical asymptote and a horizontal asymptote.

b. It is not possible to have a rational function whose graph has no y-intercept.

c. The graph of a rational function can have three horizontal asymptotes.

d. The graph of a rational function can never cross a vertical asymptote.

113. Which one of the following is true?

a. The function $f(x) = \frac{1}{\sqrt{x - 3}}$ is a rational function.

b. The x-axis is a horizontal asymptote for the graph of $f(x) = \frac{4x - 1}{x + 3}$.

c. The number of televisions that a company can produce per week after t weeks of production is given by

$$N(t) = \frac{3000t^2 + 30{,}000t}{t^2 + 10t + 25}.$$

Using this model, the company will eventually be able to produce 30,000 televisions in a single week.

d. None of the given statements is true.

In Exercises 114–117, write the equation of a rational function $f(x) = \frac{p(x)}{q(x)}$ having the indicated properties, in which the degrees of p and q are as small as possible. More than one correct function may be possible. Graph your function using a graphing utility to verify that it has the required properties.

114. f has a vertical asymptote given by $x = 3$, a horizontal asymptote $y = 0$, y-intercept at -1, and no x-intercept.

115. f has vertical asymptotes given by $x = -2$ and $x = 2$, a horizontal asymptote $y = 2$, y-intercept at $\frac{9}{2}$, x-intercepts at -3 and 3, and y-axis symmetry.

116. f has a vertical asymptote given by $x = 1$, a slant asymptote whose equation is $y = x$, y-intercept at 2, and x-intercepts at -1 and 2.

117. f has no vertical, horizontal, or slant asymptotes, and no x-intercepts.

SECTION 3.6 Polynomial and Rational Inequalities

Objectives

1. Solve polynomial inequalities.
2. Solve rational inequalities.
3. Solve problems modeled by polynomial or rational inequalities.

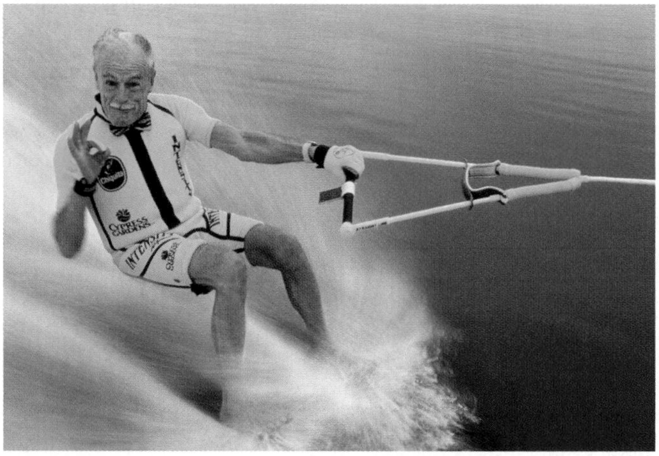

People are going to live longer in the twenty-first century. This will put added pressure on the Social Security and Medicare systems. The bar graph in Figure 3.39 shows the cost of Medicare, in billions of dollars, through 2005.

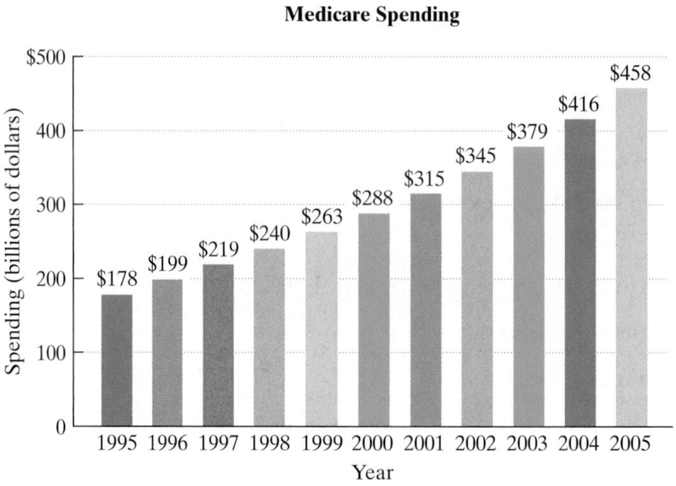

Figure 3.39
Source: Congressional Budget Office

Medicare spending, $f(x)$, in billions of dollars, x years after 1995 can be modeled by the quadratic function

$$f(x) = 1.2x^2 + 15.2x + 181.4.$$

To determine in which years Medicare spending will exceed \$500 billion, we must solve the inequality

$$1.2x^2 + 15.2x + 181.4 > 500.$$

Medicare spending (for $1.2x^2 + 15.2x + 181.4$); exceeds (for $>$); \$500 billion. (for 500)

We begin by subtracting 500 from both sides. This will give us zero on the right:

$$1.2x^2 + 15.2x + 181.4 - 500 > 500 - 500$$
$$1.2x^2 + 15.2x - 318.6 > 0.$$

The form of this inequality is $ax^2 + bx + c > 0$. Such an inequality is called a *polynomial inequality*.

Definition of a Polynomial Inequality

A polynomial inequality is any inequality that can be put into one of the forms

$$f(x) < 0, \quad f(x) > 0, \quad f(x) \le 0, \quad \text{or} \quad f(x) \ge 0,$$

where f is a polynomial function.

In this section, we establish the basic techniques for solving polynomial inequalities. We will use these techniques to solve inequalities involving rational functions.

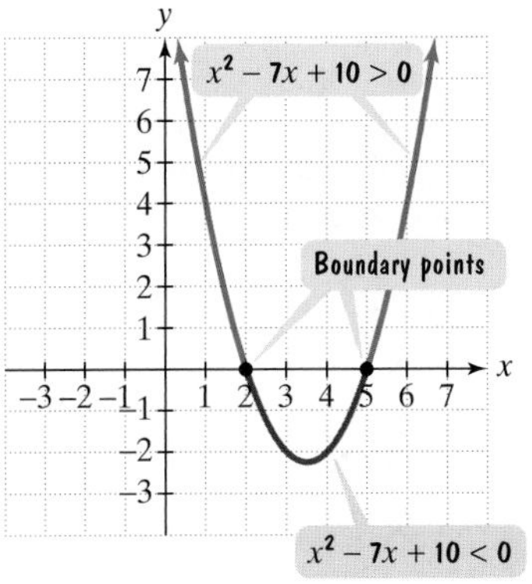

Figure 3.40

Solving Polynomial Inequalities

Graphs can help us visualize the solutions of polynomial inequalities. For example, the graph of $f(x) = x^2 - 7x + 10$ is shown in Figure 3.40. The x-intercepts, 2 and 5, are **boundary points** between where the graph lies above the x-axis, shown in blue, and where the graph lies below the x-axis, shown in red.

Locating the x-intercepts of a polynomial function, f, is an important step in finding the solution set for polynomial inequalities in the form $f(x) < 0$ or $f(x) > 0$. We use the x-intercepts of f as boundary points that divide the real number line into intervals. On each interval, the graph of f is either above the x-axis $[f(x) > 0]$ or below the x-axis $[f(x) < 0]$. For this reason, x-intercepts play a fundamental role in solving polynomial inequalities. The x-intercepts are found by solving the equation $f(x) = 0$.

1 Solve polynomial inequalities.

Procedure for Solving Polynomial Inequalities

1. Express the inequality in the form
$$f(x) < 0 \quad \text{or} \quad f(x) > 0,$$
where f is a polynomial function.
2. Solve the equation $f(x) = 0$. The real solutions are the **boundary points.**
3. Locate these boundary points on a number line, thereby dividing the number line into intervals.
4. Choose one representative number, called a **test value**, within each interval and evaluate f at that number.
   - **a.** If the value of f is positive, then $f(x) > 0$ for all numbers, x, in the interval.
   - **b.** If the value of f is negative, then $f(x) < 0$ for all numbers, x, in the interval.
5. Write the solution set, selecting the interval or intervals that satisfy the given inequality.

This procedure is valid if $<$ is replaced by $\le$ or $>$ is replaced by $\ge$. However, if the inequality involves $\le$ or $\ge$, include the boundary points [the solutions of $f(x) = 0$] in the solution set.

EXAMPLE 1 Solving a Polynomial Inequality

Solve and graph the solution set on a real number line: $2x^2 + x > 15$.

Solution

Step 1 Express the inequality in the form $f(x) < 0$ or $f(x) > 0$. We begin by rewriting the inequality so that 0 is on the right side.

$2x^2 + x > 15$ This is the given inequality.

$2x^2 + x - 15 > 15 - 15$ Subtract 15 from both sides.

$2x^2 + x - 15 > 0$ Simplify.

This inequality is equivalent to the one we wish to solve. It is in the form $f(x) > 0$, where $f(x) = 2x^2 + x - 15$.

Step 2   Solve the equation $f(x) = 0$. We find the x-intercepts of $f(x) = 2x^2 + x - 15$ by solving the equation $2x^2 + x - 15 = 0$.

$$2x^2 + x - 15 = 0 \qquad \text{This polynomial equation is a quadratic equation.}$$

$$(2x - 5)(x + 3) = 0 \qquad \text{Factor.}$$

$$2x - 5 = 0 \quad \text{or} \quad x + 3 = 0 \qquad \text{Set each factor equal to 0.}$$

$$x = \tfrac{5}{2} \qquad\qquad x = -3 \qquad \text{Solve for x.}$$

The x-intercepts of f are -3 and $\frac{5}{2}$. We will use these x-intercepts as boundary points on a number line.

Step 3   Locate the boundary points on a number line and separate the line into intervals. The number line with the boundary points is shown as follows:

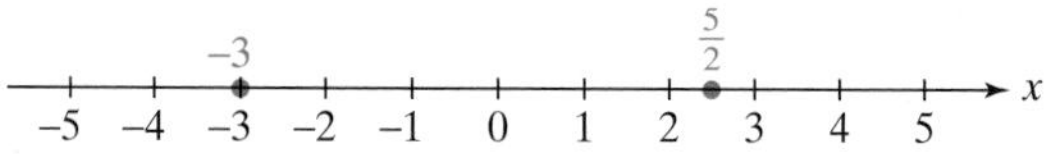

The boundary points divide the number line into three intervals:

$$(-\infty, \quad -3) \quad \left(-3, \tfrac{5}{2}\right) \quad \left(\tfrac{5}{2}, \quad \infty\right).$$

Step 4   Choose one test value within each interval and evaluate f at that number.

Interval	Test Value	Substitute into $f(x) = 2x^2 + x - 15$	Conclusion
$(-\infty, -3)$	-4	$f(-4) = 2(-4)^2 + (-4) - 15$ $= 13$, positive	$f(x) > 0$ for all x in $(-\infty, -3)$.
$\left(-3, \dfrac{5}{2}\right)$	0	$f(0) = 2 \cdot 0^2 + 0 - 15$ $= -15$, negative	$f(x) < 0$ for all x in $\left(-3, \dfrac{5}{2}\right)$.
$\left(\dfrac{5}{2}, \infty\right)$	3	$f(3) = 2 \cdot 3^2 + 3 - 15$ $= 6$, positive	$f(x) > 0$ for all x in $\left(\dfrac{5}{2}, \infty\right)$.

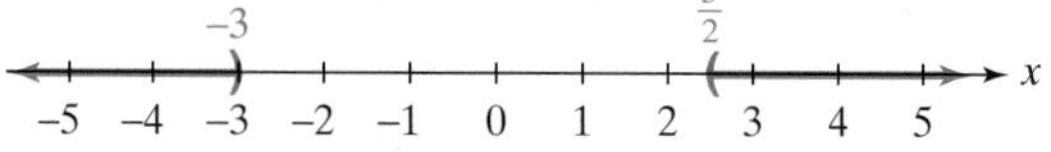

Technology

The solution set for
$$2x^2 + x > 15$$
or, equivalently,
$$2x^2 + x - 15 > 0$$
can be verified with a graphing utility. The graph of $f(x) = 2x^2 + x - 15$ was obtained using a $[-10, 10, 1]$ by $[-16, 6, 1]$ viewing rectangle. The graph lies above the x-axis, representing $>$, for all x in $(-\infty, -3)$ or $\left(\frac{5}{2}, \infty\right)$.

Step 5   Write the solution set, selecting the interval or intervals that satisfy the given inequality. We are interested in solving $2x^2 + x - 15 > 0$, where $f(x) = 2x^2 + x - 15$. Based on our work in step 4, we see that $f(x) > 0$ for all x in $(-\infty, -3)$ or $\left(\frac{5}{2}, \infty\right)$. Thus, the solution set of the given inequality, $2x^2 + x > 15$, or, equivalently, $2x^2 + x - 15 > 0$, is

$$(-\infty, -3) \cup \left(\tfrac{5}{2}, \infty\right) \text{ or } \left\{x \,\middle|\, x < -3 \text{ or } x > \tfrac{5}{2}\right\}.$$

The graph of the solution set on a number line is shown as follows:

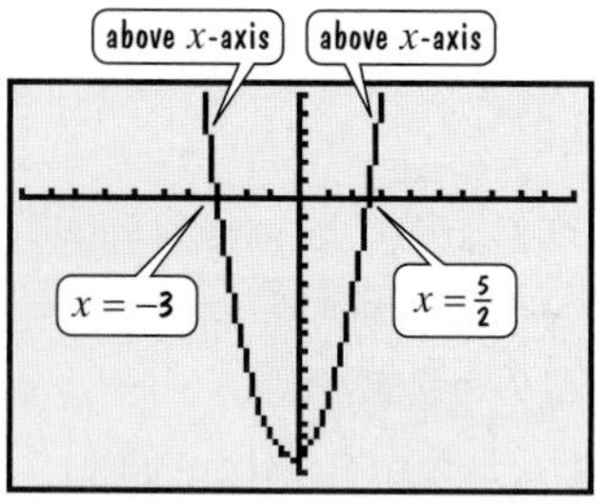

Check Point 1   Solve and graph the solution set: $x^2 - x > 20$.

EXAMPLE 2   Solving a Polynomial Inequality

Solve and graph the solution set on a real number line: $x^3 + x^2 \le 4x + 4$.

Solution

Step 1   Express the inequality in the form $f(x) \le 0$ or $f(x) \ge 0$. We begin by rewriting the inequality so that 0 is on the right side.

$$x^3 + x^2 \leq 4x + 4 \qquad \text{This is the given inequality.}$$

$$x^3 + x^2 - 4x - 4 \leq 4x + 4 - 4x - 4 \qquad \text{Subtract } 4x + 4 \text{ from both sides.}$$

$$x^3 + x^2 - 4x - 4 \leq 0 \qquad \text{Simplify.}$$

This inequality is equivalent to the one we wish to solve. It is in the form $f(x) \leq 0$, where $f(x) = x^3 + x^2 - 4x - 4$.

Step 2 Solve the equation $f(x) = 0$. We find the x-intercepts of $f(x) = x^3 + x^2 - 4x - 4$ by solving the equation $x^3 + x^2 - 4x - 4 = 0$.

<table>
<tr><td>$x^3 + x^2 - 4x - 4 = 0$</td><td>This polynomial equation is of degree 3.</td></tr>
<tr><td>$x^2(x + 1) - 4(x + 1) = 0$</td><td>Factor x^2 from the first two terms and -4 from the last two terms.</td></tr>
<tr><td>$(x + 1)(x^2 - 4) = 0$</td><td>A common factor of $x + 1$ is factored from the expression.</td></tr>
<tr><td>$x + 1 = 0 \quad \text{or} \quad x^2 - 4 = 0$</td><td>Set each factor equal to 0.</td></tr>
<tr><td>$x = -1 \qquad\qquad x^2 = 4$</td><td>Solve for x.</td></tr>
<tr><td>$x = \pm 2$</td><td>Use the square root property.</td></tr>
</table>

The x-intercepts of f are -2, -1, and 2. We will use these x-intercepts as boundary points on a number line.

Step 3 Locate the boundary points on a number line and separate the line into intervals. The number line with the boundary points is shown as follows:

$$\begin{array}{c}
-2 \quad -1 \qquad\quad 2 \\
\underset{-5\ \ -4\ \ -3\ \ -2\ \ -1\ \ \ 0\ \ \ 1\ \ \ 2\ \ \ 3\ \ \ 4\ \ \ 5}{\longleftarrow\!\!\!\!\!\!\!\!\!\!\!\bullet\ \ \bullet\ \ \ \ \ \ \ \bullet\ \ \ \ \ \ \longrightarrow} \ x
\end{array}$$

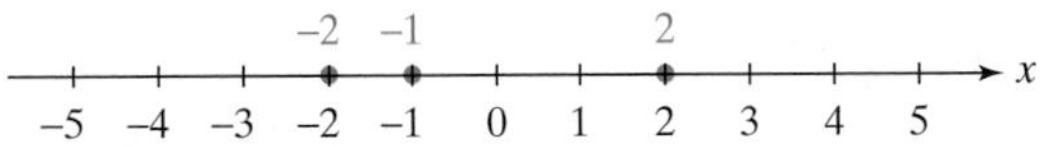

The boundary points divide the number line into four intervals:

$$(-\infty, -2) \quad (-2, -1) \quad (-1, 2) \quad (2, \infty).$$

Step 4 Choose one test value within each interval and evaluate f at that number.

Interval	Test Value	Substitute into $f(x) = x^3 + x^2 - 4x - 4$	Conclusion
$(-\infty, -2)$	-3	$f(-3) = (-3)^3 + (-3)^2 - 4(-3) - 4$ $= -10,\ \text{negative}$	$f(x) < 0$ for all x in $(-\infty, -2)$.
$(-2, -1)$	-1.5	$f(-1.5) = (-1.5)^3 + (-1.5)^2 - 4(-1.5) - 4$ $= 0.875,\ \text{positive}$	$f(x) > 0$ for all x in $(-2, -1)$.
$(-1, 2)$	0	$f(0) = 0^3 + 0^2 - 4 \cdot 0 - 4$ $= -4,\ \text{negative}$	$f(x) < 0$ for all x in $(-1, 2)$.
$(2, \infty)$	3	$f(3) = 3^3 + 3^2 - 4 \cdot 3 - 4$ $= 20,\ \text{positive}$	$f(x) > 0$ for all x in $(2, \infty)$.

Step 5 Write the solution set, selecting the interval or intervals that satisfy the given inequality. We are interested in solving $x^3 + x^2 - 4x - 4 \leq 0$, where $f(x) = x^3 + x^2 - 4x - 4$. Based on our work in step 4, we see that $f(x) < 0$ for all x in $(-\infty, -2)$ or $(-1, 2)$. However, because the inequality involves $\leq$ (less than or *equal to*), we must also include the solutions of $x^3 + x^2 - 4x - 4 = 0$, namely -2, -1, and 2, in the solution set. Thus, the solution set of the given inequality, $x^3 + x^2 \leq 4x + 4$, or, equivalently, $x^3 + x^2 - 4x - 4 \leq 0$, is

$$(-\infty, -2] \cup [-1, 2]$$

$$\text{or} \quad \{x | x \leq -2 \text{ or } -1 \leq x \leq 2\}.$$

The graph of the solution set on a number line is shown as follows:

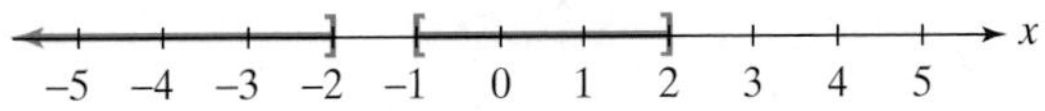

Technology

The solution set for

$$x^3 + x^2 \leq 4x + 4$$

or, equivalently,

$$x^3 + x^2 - 4x - 4 \leq 0$$

can be verified with a graphing utility. The graph of $f(x) = x^3 + x^2 - 4x - 4$ lies on or below the x-axis, representing $\leq$, for all x in $(-\infty, -2]$ or $[-1, 2]$.

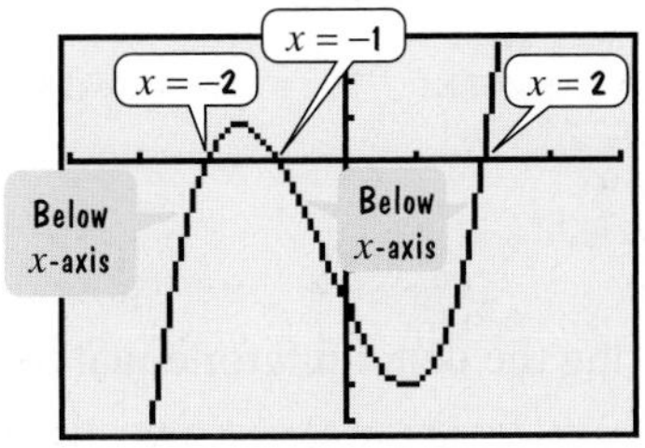

$[-4, 4, 1]$ by $[-7, 3, 1]$

Check Point 2 Solve and graph the solution set on a real number line: $x^3 + 3x^2 \le x + 3$.

2 Solve rational inequalities.

Solving Rational Inequalities

A **rational inequality** is any inequality that can be put into one of the forms

$$f(x) < 0, \quad f(x) > 0, \quad f(x) \le 0, \quad \text{or} \quad f(x) \ge 0,$$

where f is a rational function. An example of a rational inequality is

$$\frac{3x + 3}{2x + 4} > 0.$$

This inequality is in the form $f(x) > 0$, where f is the rational function given by

$$f(x) = \frac{3x + 3}{2x + 4}.$$

The graph of f is shown in Figure 3.41.

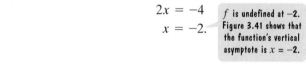

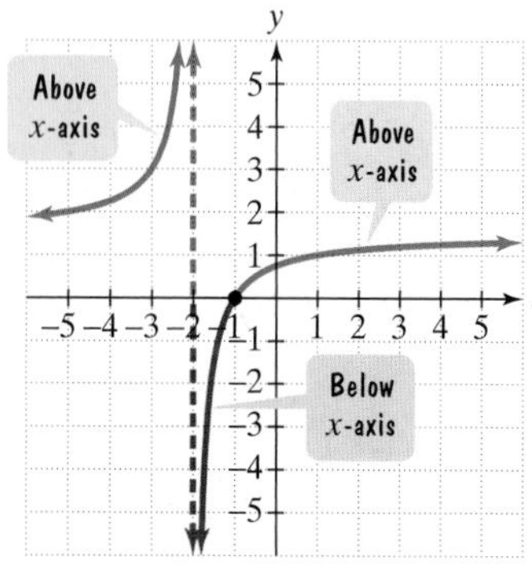

Figure 3.41 The graph of $f(x) = \frac{3x + 3}{2x + 4}$

We can find the x-intercept of f by setting the numerator equal to 0:

$$3x + 3 = 0$$
$$3x = -3$$
$$x = -1.$$

f has an x-intercept at -1 and passes through $(-1, 0)$.

We can determine where f is undefined by setting the denominator equal to 0:

$$2x + 4 = 0$$
$$2x = -4$$
$$x = -2.$$

f is undefined at -2. Figure 3.41 shows that the function's vertical asymptote is $x = -2$.

By setting both the numerator and the denominator of f equal to 0, we obtained -2 and -1. These numbers separate the x-axis into three intervals: $(-\infty, -2)$, $(-2, -1)$, and $(-1, \infty)$. On each interval, the graph of f is either above the x-axis $[f(x) > 0]$ or below the x-axis $[f(x) < 0]$.

Examine the graph in Figure 3.41 carefully. Can you see that it is above the x-axis for all x in $(-\infty, -2)$ or $(-1, \infty)$, shown in blue? Thus, the solution set of $\frac{3x + 3}{2x + 4} > 0$ is $(-\infty, -2) \cup (-1, \infty)$. By contrast, the graph of f lies below the x-axis for all x in $(-2, -1)$, shown in red. Thus, the solution set of $\frac{3x + 3}{2x + 4} < 0$ is $(-2, -1)$.

The first step in solving a rational inequality is to bring all terms to one side, obtaining zero on the other side. Then express the rational function on the nonzero side as a single quotient. The second step is to set the numerator and the denominator of f equal to zero. The solutions of these equations serve as boundary points that separate the real number line into intervals. At this point, the procedure is the same as the one we used for solving polynomial inequalities.

Study Tip

Do not begin solving

$$\frac{x + 1}{x + 3} \ge 2$$

by multiplying both sides by $x + 3$. We do not know if $x + 3$ is positive or negative. Thus, we do not know whether or not to change the sense of the inequality.

EXAMPLE 3 Solving a Rational Inequality

Solve and graph the solution set: $\frac{x + 1}{x + 3} \ge 2$.

Solution

Step 1 Express the inequality so that one side is zero and the other side is a single quotient. We subtract 2 from both sides to obtain zero on the right.

$\dfrac{x+1}{x+3} \geq 2$	This is the given inequality.
$\dfrac{x+1}{x+3} - 2 \geq 0$	Subtract 2 from both sides, obtaining 0 on the right.
$\dfrac{x+1}{x+3} - \dfrac{2(x+3)}{x+3} \geq 0$	The least common denominator is x + 3. Express 2 in terms of this denominator.
$\dfrac{x+1-2(x+3)}{x+3} \geq 0$	Subtract rational expressions.
$\dfrac{x+1-2x-6}{x+3} \geq 0$	Apply the distributive property.
$\dfrac{-x-5}{x+3} \geq 0$	Simplify.

This inequality is equivalent to the one we wish to solve. It is in the form $f(x) \geq 0$, where $f(x) = \dfrac{-x-5}{x+3}$.

Step 2 Set the numerator and the denominator of f equal to zero. The real solutions are the boundary points.

Study Tip

Never include the value that causes a rational function's denominator to equal zero in the solution set of a rational inequality. Division by zero is undefined.

$-x - 5 = 0$	$x + 3 = 0$	Set the numerator and denominator equal to 0. These are the values that make the previous quotient zero or undefined.
$x = -5$	$x = -3$	Solve for x.

We will use these solutions as boundary points on a number line.

Step 3 Locate the boundary points on a number line and separate the line into intervals. The number line with the boundary points is shown as follows:

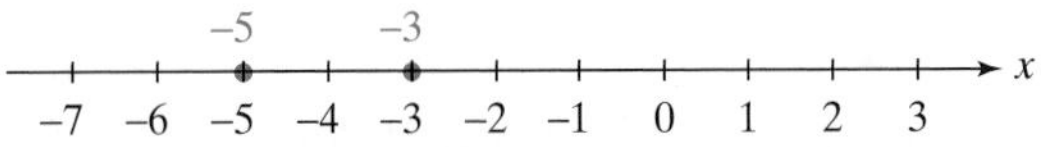

The boundary points divide the number line into three intervals:

$$(-\infty, -5) \quad (-5, -3) \quad (-3, \infty).$$

Step 4 Choose one test value within each interval and evaluate f at that number.

Interval	Test Value	Substitute into $f(x) = \dfrac{-x-5}{x+3}$	Conclusion
$(-\infty, -5)$	-6	$f(-6) = \dfrac{-(-6)-5}{-6+3} = -\frac{1}{3}$, negative	$f(x) < 0$ for all x in $(-\infty, -5)$.
$(-5, -3)$	-4	$f(-4) = \dfrac{-(-4)-5}{-4+3} = 1$, positive	$f(x) > 0$ for all x in $(-5, -3)$.
$(-3, \infty)$	0	$f(0) = \dfrac{-0-5}{0+3} = -\frac{5}{3}$, negative	$f(x) < 0$ for all x in $(-3, \infty)$.

Step 5 Write the solution set, selecting the interval or intervals that satisfy the given inequality. We are interested in solving $\dfrac{-x-5}{x+3} \geq 0$, where $f(x) = \dfrac{-x-5}{x+3}$. Based on our work in step 4, we see that $f(x) > 0$ for all x in $(-5, -3)$. However,

Discovery

Because $(x + 3)^2$ is positive, it is possible so solve

$$\frac{x + 1}{x + 3} \geq 2$$

by first multiplying both sides by $(x + 3)^2$ (where $x \neq -3$). This will not change the sense of the inequality and will clear the fraction. Try using this solution method and compare it to the solution on pages 374–376.

because the inequality involves $\geq$ (greater than or *equal to*), we must also include the solution of $f(x) = 0$, namely the value that we obtained when we set the numerator of f equal to zero. Thus, we must include -5 in the solution set. The solution set of the given inequality is

$$[-5, -3) \text{ or } \{x | -5 \leq x < -3\}.$$

The graph of the solution set on a number line is shown as follows:

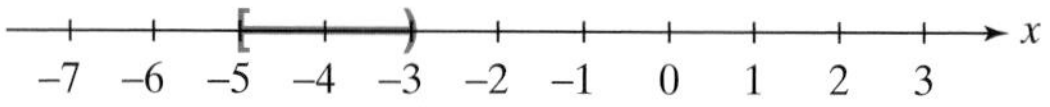

Technology

The solution set for

$$\frac{x + 1}{x + 3} \geq 2$$

or, equivalently,

$$\frac{-x - 5}{x + 3} \geq 0$$

can be verified with a graphing utility. The graph of $f(x) = \frac{-x - 5}{x + 3}$ lies on or above the x-axis, representing $\geq$, for all x in $[-5, -3)$.

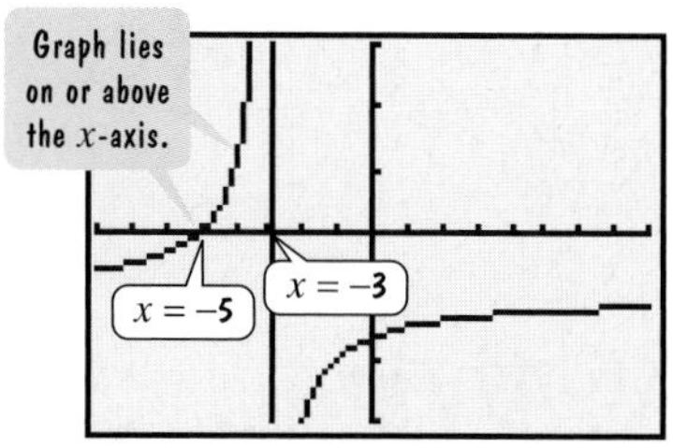

[−8, 8, 1] by [−3, 3, 1]

Check Point 3 Solve and graph the solution set: $\frac{2x}{x + 1} \geq 1$.

3 Solve problems modeled by polynomial or rational inequalities.

Applications

We are surrounded by evidence that the world is profoundly mathematical. For example, did you know that every time you throw an object vertically upward, its changing height above the ground can be described by a quadratic function? The same function can be used to describe objects that are falling, such as sky divers.

The Position Function for a Free-Falling Object Near Earth's Surface

An object that is falling or vertically projected into the air has its height above the ground, $s(t)$, in feet, given by

$$s(t) = -16t^2 + v_0 t + s_0,$$

where v_0 is the original velocity (initial velocity) of the object, in feet per second, t is the time that the object is in motion, in seconds, and s_0 is the original height (initial height) of the object, in feet.

In Example 4, we solve a polynomial inequality in a problem about the position of a free-falling object.

EXAMPLE 4 Using the Position Function

A ball is thrown vertically upward from the top of the Leaning Tower of Pisa (190 feet high) with an initial velocity of 96 feet per second (Figure 3.42). During which time period will the ball's height exceed that of the tower?

$t = 0$
$s_0 = 190$
$v_0 = 96$

190 feet

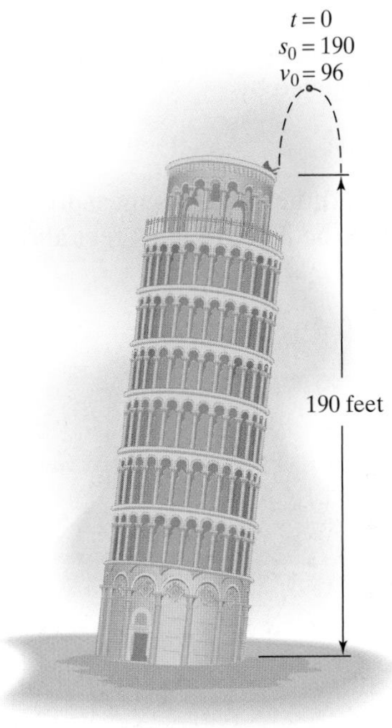

Figure 3.42 Throwing a ball from 190 feet with a velocity of 96 feet per second

Solution

$$s(t) = -16t^2 + v_0 t + s_0 \qquad \text{This is the position function for a free-falling object.}$$

$$s(t) = -16t^2 + 96t + 190 \qquad \text{Because } v_0 \text{ (initial velocity)} = 96 \text{ and } s_0 \text{ (initial position)} = 190, \text{ substitute these values into the formula.}$$

When will $s(t)$, the ball's height,   exceed that   of the tower?

$$-16t^2 + 96t + 190 > 190$$

$$-16t^2 + 96t + 190 > 190 \qquad \text{This is the inequality that models the problem's question. We must find } t.$$

$$-16t^2 + 96t > 0 \qquad \text{Subtract 190 from both sides. This inequality is in the form } f(t) > 0, \text{ where } f(t) = -16t^2 + 96t.$$

$$-16t^2 + 96t = 0 \qquad \text{Solve the equation } f(t) = 0.$$

$$-16t(t - 6) = 0 \qquad \text{Factor.}$$

$$-16t = 0 \quad \text{or} \quad t - 6 = 0 \qquad \text{Set each factor equal to 0.}$$

$$t = 0 \qquad\qquad t = 6 \qquad \text{Solve for } t. \text{ The boundary points are 0 and 6.}$$

(number line: $-2\ -1\ 0\ 1\ 2\ 3\ 4\ 5\ 6\ 7\ 8 \to t$ with points at 0 and 6)

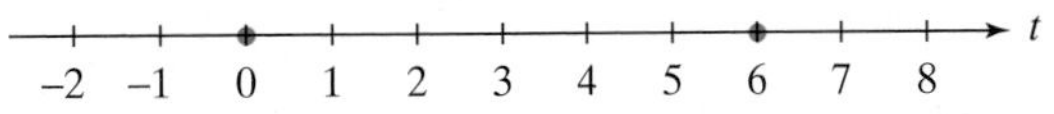

Locate these values on a number line, with $t \geq 0$.

The intervals are $(-\infty, 0)$, $(0, 6)$, and $(6, \infty)$. For our purposes, the mathematical model is useful only from $t = 0$ until the ball hits the ground. (By setting $-16t^2 + 96t + 190$ equal to zero, we find $t \approx 7.57$; the ball hits the ground after approximately 7.57 seconds.) Thus, we use $(0, 6)$ and $(6, 7.57)$ for our intervals.

Interval	Test Value	Substitute into $f(t) = -16t^2 + 96t$	Conclusion
$(0, 6)$	1	$f(1) = -16 \cdot 1^2 + 96 \cdot 1$ $= 80$, positive	$f(t) > 0$ for all t in $(0, 6)$.
$(6, 7.57)$	7	$f(7) = -16 \cdot 7^2 + 96 \cdot 7$ $= -112$, negative	$f(t) < 0$ for all t in $(6, 7.57)$.

We are interested in solving $-16t^2 + 96t > 0$, where $f(t) = -16t^2 + 96t$. We see that $f(t) > 0$ for all t in $(0, 6)$. This means that the ball's height exceeds that of the tower between 0 and 6 seconds.

Technology

The graphs of
$$y_1 = -16x^2 + 96x + 190$$
and
$$y_2 = 190$$
are shown in a
$$[0, 8, 1] \text{ by } [0, 360, 36]$$
seconds in motion   height, in feet

viewing rectangle. The graphs show that the ball's height exceeds that of the tower between 0 and 6 seconds.

Height of Ball:
$y_1 = -16x^2 + 96x + 190$

360

Height (feet)

Height of Tower:
$y_2 = 190$

Ball hits ground after 7.57 seconds.

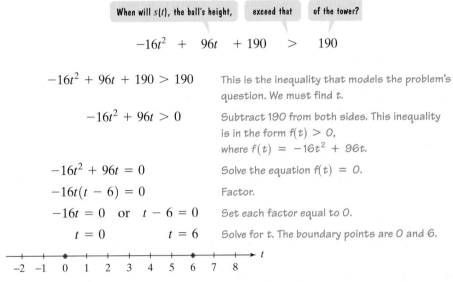

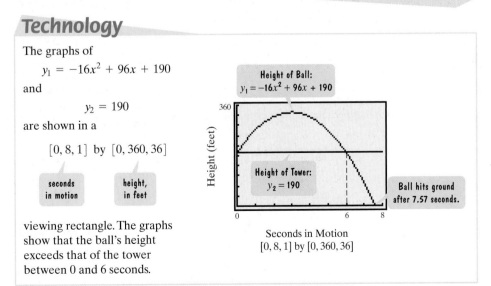

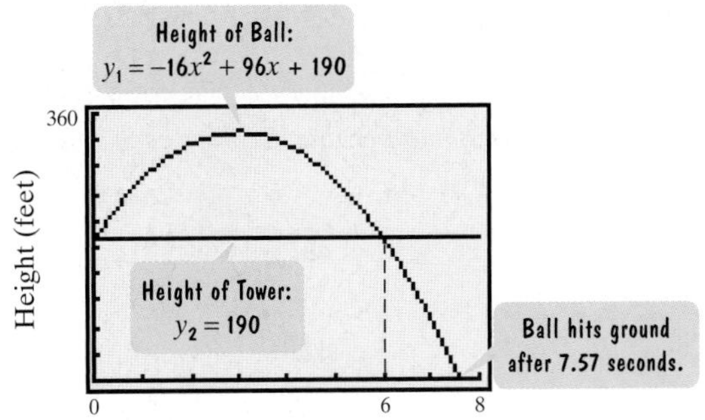

Seconds in Motion
$[0, 8, 1]$ by $[0, 360, 36]$

Check Point 4 An object is propelled straight up from ground level with an initial velocity of 80 feet per second. Its height at time t is modeled by

$$s(t) = -16t^2 + 80t,$$

where the height, $s(t)$, is measured in feet and the time, t, is measured in seconds. In which time interval will the object be more than 64 feet above the ground?

EXERCISE SET 3.6

Practice Exercises

Solve each polynomial inequality in Exercises 1–38 and graph the solution set on a real number line. Express each solution set in interval notation.

1. $(x - 4)(x + 2) > 0$
2. $(x + 3)(x - 5) > 0$
3. $(x - 7)(x + 3) \le 0$
4. $(x + 1)(x - 7) \le 0$
5. $x^2 - 5x + 4 > 0$
6. $x^2 - 4x + 3 < 0$
7. $x^2 + 5x + 4 > 0$
8. $x^2 + x - 6 > 0$
9. $x^2 - 6x + 9 < 0$
10. $x^2 - 2x + 1 > 0$
11. $3x^2 + 10x - 8 \le 0$
12. $9x^2 + 3x - 2 \ge 0$
13. $2x^2 + x < 15$
14. $6x^2 + x > 1$
15. $4x^2 + 7x < -3$
16. $3x^2 + 16x < -5$
17. $5x \le 2 - 3x^2$
18. $4x^2 + 1 \ge 4x$
19. $x^2 - 4x \ge 0$
20. $x^2 + 2x < 0$
21. $2x^2 + 3x > 0$
22. $3x^2 - 5x \le 0$
23. $-x^2 + x \ge 0$
24. $-x^2 + 2x \ge 0$
25. $x^2 \le 4x - 2$
26. $x^2 \le 2x + 2$
27. $x^2 - 6x + 9 < 0$
28. $4x^2 - 4x + 1 \ge 0$
29. $(x - 1)(x - 2)(x - 3) \ge 0$
30. $(x + 1)(x + 2)(x + 3) \ge 0$
31. $x^3 + 2x^2 - x - 2 \ge 0$
32. $x^3 + 2x^2 - 4x - 8 \ge 0$
33. $x^3 - 3x^2 - 9x + 27 < 0$
34. $x^3 + 7x^2 - x - 7 < 0$
35. $x^3 + x^2 + 4x + 4 > 0$
36. $x^3 - x^2 + 9x - 9 > 0$
37. $x^3 \ge 9x^2$
38. $x^3 \le 4x^2$

Solve each rational inequality in Exercises 39–56 and graph the solution set on a real number line. Express each solution set in interval notation.

39. $\frac{x - 4}{x + 3} > 0$
40. $\frac{x + 5}{x - 2} > 0$
41. $\frac{x + 3}{x + 4} < 0$
42. $\frac{x + 5}{x + 2} < 0$
43. $\frac{-x + 2}{x - 4} \ge 0$
44. $\frac{-x - 3}{x + 2} \le 0$
45. $\frac{4 - 2x}{3x + 4} \le 0$
46. $\frac{3x + 5}{6 - 2x} \ge 0$
47. $\frac{x}{x - 3} > 0$
48. $\frac{x + 4}{x} > 0$
49. $\frac{(x + 4)(x - 1)}{x + 2} \le 0$
50. $\frac{(x + 3)(x - 2)}{x + 1} \le 0$
51. $\frac{x + 1}{x + 3} < 2$
52. $\frac{x}{x - 1} > 2$
53. $\frac{x + 4}{2x - 1} \le 3$
54. $\frac{1}{x - 3} < 1$
55. $\frac{x - 2}{x + 2} \le 2$
56. $\frac{x}{x + 2} \ge 2$

Practice Plus

In Exercises 57–60, find the domain of each function.

57. $f(x) = \sqrt{2x^2 - 5x + 2}$
58. $f(x) = \frac{1}{\sqrt{4x^2 - 9x + 2}}$
59. $f(x) = \sqrt{\frac{2x}{x + 1} - 1}$
60. $f(x) = \sqrt{\frac{x}{2x - 1} - 1}$

Solve each inequality in Exercises 61–66 and graph the solution set on a real number line.

61. $|x^2 + 2x - 36| > 12$
62. $|x^2 + 6x + 1| > 8$
63. $\frac{3}{x + 3} > \frac{3}{x - 2}$
64. $\frac{1}{x + 1} > \frac{2}{x - 1}$
65. $\frac{x^2 - x - 2}{x^2 - 4x + 3} > 0$
66. $\frac{x^2 - 3x + 2}{x^2 - 2x - 3} > 0$

In Exercises 67–68, use the graph of the polynomial function to solve each inequality.

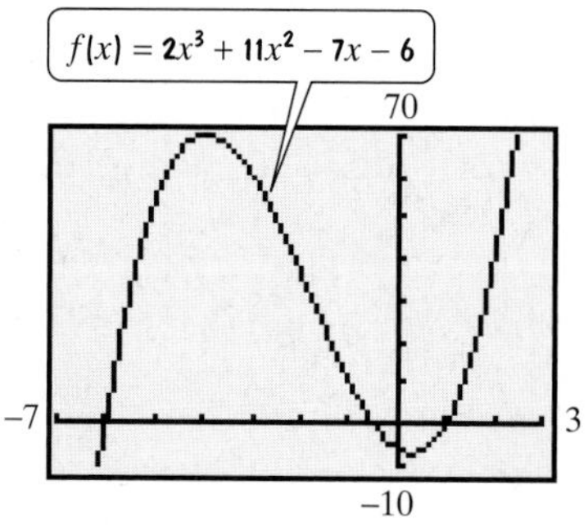

67. $2x^3 + 11x^2 \ge 7x + 6$
68. $2x^3 + 11x^2 < 7x + 6$

In Exercises 69–70, use the graph of the rational function to solve each inequality.

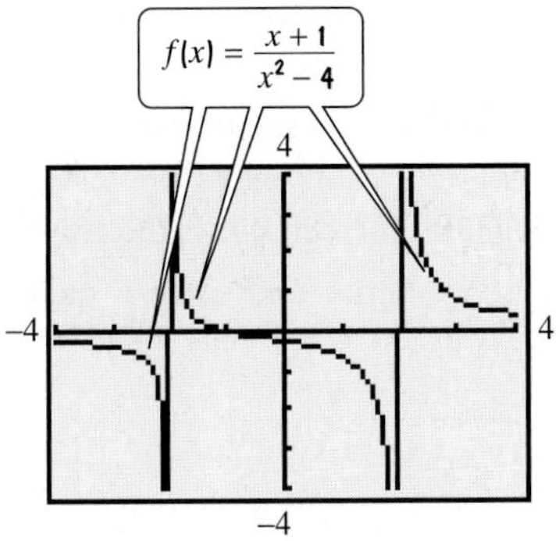

$$f(x) = \frac{x+1}{x^2-4}$$

69. $\dfrac{1}{4(x+2)} \leq -\dfrac{3}{4(x-2)}$

70. $\dfrac{1}{4(x+2)} > -\dfrac{3}{4(x-2)}$

 ## Application Exercises

Use the position function

$$s(t) = -16t^2 + v_0 t + s_0$$

(v_0 = initial velocity, s_0 = initial position, t = time)

to answer Exercises 71–72.

71. Divers in Acapulco, Mexico, dive headfirst at 8 feet per second from the top of a cliff 87 feet above the Pacific Ocean. During which time period will the diver's height exceed that of the cliff?

72. You throw a ball straight up from a rooftop 160 feet high with an initial velocity of 48 feet per second. During which time period will the ball's height exceed that of the rooftop?

The bar graph in Figure 3.39 on page 370 shows the cost of Medicare, in billions of dollars, through 2005. Using the regression feature of a graphing utility, these data can be modeled by

a linear function, $f(x) = 27x + 163$;

a quadratic function, $g(x) = 1.2x^2 + 15.2x + 181.4$.

In each function, x represents the number of years after 1995. Use these functions to solve Exercises 73–76.

73. The graph indicates that Medicare spending reached $379 billion in 2003. Find the amount predicted by each of the functions, f and g, for that year. How well do the functions model the value in the graph?

74. The graph indicates that Medicare spending reached $458 billion in 2005. Find the amount predicted by each of the functions, f and g, for that year. How well do the functions model the value in the graph? Which function serves as a better model for that year?

75. For which years does the quadratic model indicate that Medicare spending will exceed $536.6 billion?

76. For which years does the quadratic model indicate that Medicare spending will exceed $629.4 billion?

It's vacation time. You drive 90 miles along a scenic highway and then take a 5-mile run along a hiking trail. Your driving rate is nine times that of your running rate. The graph shows the total time you spend driving and running, $f(x)$, as a function of your running rate, x. Use the rational function and its graph to solve Exercises 77–81.

$$f(x) = \frac{90}{9x} + \frac{5}{x} = \frac{15}{x}$$

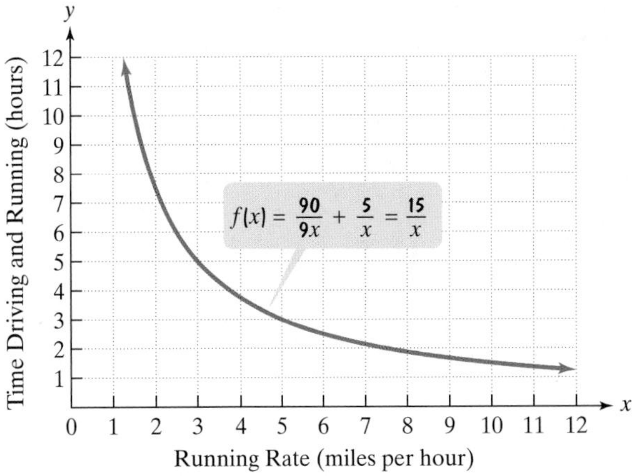

77. Describe your running rate if you have no more than a total of 3 hours for driving and running. Use a rational inequality to solve the problem. Then explain how your solution is shown on the graph.

78. Describe your running rate if you have no more than a total of 5 hours for driving and running. Use a rational inequality to solve the problem. Then explain how your solution is shown on the graph.

79. Describe the behavior of the graph as $x \to \infty$. What does this show about the time driving and running as a function of your running rate?

80. Describe the behavior of the graph as $x \to 0^+$. What does this show about the time driving and running as a function of your running rate?

81. Describe how to use the formula $t = \dfrac{d}{r}$ (time traveled equals distance traveled divided by the rate of travel) and the problem's verbal conditions to obtain the function's equation displayed in the voice balloon.

82. The perimeter of a rectangle is 50 feet. Describe the possible lengths of a side if the area of the rectangle is not to exceed 114 square feet.

83. The perimeter of a rectangle is 180 feet. Describe the possible lengths of a side if the area of the rectangle is not to exceed 800 square feet.

 ## Writing in Mathematics

84. What is a polynomial inequality?

85. What is a rational inequality?

86. If f is a polynomial or rational function, explain how the graph of f can be used to visualize the solution set of the inequality $f(x) < 0$.

Technology Exercises

87. Use a graphing utility to verify your solution sets to any three of the polynomial inequalities that you solved algebraically in Exercises 1–38.

88. Use a graphing utility to verify your solution sets to any three of the rational inequalities that you solved algebraically in Exercises 39–56.

Solve each inequality in Exercises 89–94 using a graphing utility.

89. $x^2 + 3x - 10 > 0$

90. $2x^2 + 5x - 3 \le 0$

91. $x^3 + x^2 - 4x - 4 > 0$

92. $\dfrac{x-4}{x-1} \le 0$

93. $\dfrac{x+2}{x-3} \le 2$

94. $\dfrac{1}{x+1} \le \dfrac{2}{x+4}$

Critical Thinking Exercises

95. Which one of the following is true?

a. The solution set of $x^2 > 25$ is $(5, \infty)$.

b. The inequality $\dfrac{x-2}{x+3} < 2$ can be solved by multiplying both sides by $x + 3$, resulting in the equivalent inequality $x - 2 < 2(x + 3)$.

c. $(x + 3)(x - 1) \ge 0$ and $\dfrac{x+3}{x-1} \ge 0$ have the same solution set.

d. None of these statements is true.

96. Write a polynomial inequality whose solution set is $[-3, 5]$.

97. Write a rational inequality whose solution set is $(-\infty, -4) \cup [3, \infty)$.

In Exercises 98–101, use inspection to describe each inequality's solution set. Do not solve any of the inequalities.

98. $(x - 2)^2 > 0$

99. $(x - 2)^2 \le 0$

100. $(x - 2)^2 < -1$

101. $\dfrac{1}{(x-2)^2} > 0$

102. The graphing utility screen shows the graph of $y = 4x^2 - 8x + 7$.

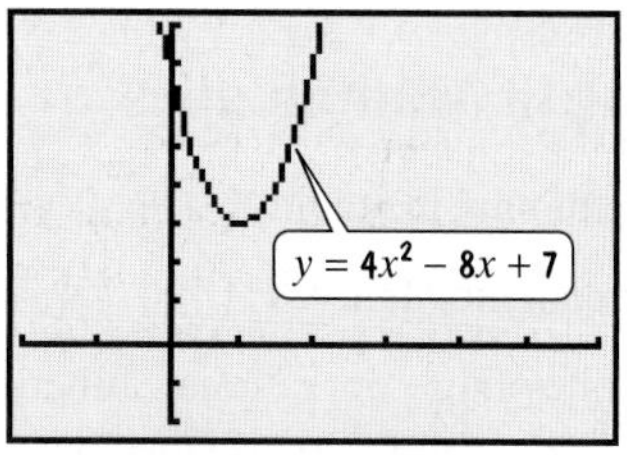

$[-2, 6, 1]$ by $[-2, 8, 1]$

a. Use the graph to describe the solution set of $4x^2 - 8x + 7 > 0$.

b. Use the graph to describe the solution set of $4x^2 - 8x + 7 < 0$.

c. Use an algebraic approach to verify each of your descriptions in parts (a) and (b).

103. The graphing utility screen shows the graph of $y = \sqrt{27 - 3x^2}$. Write and solve a quadratic inequality that explains why the graph only appears for $-3 \le x \le 3$.

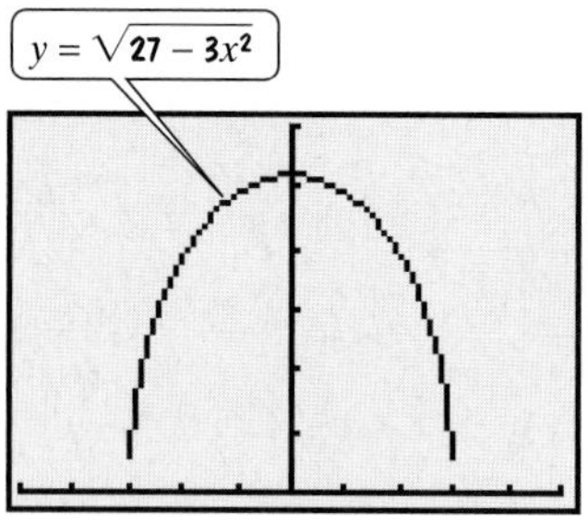

$[-5, 5, 1]$ by $[0, 6, 1]$

Group Exercise

104. This exercise is intended as a group learning experience and is appropriate for groups of three to five people. Before working on the various parts of the problem, reread the description of the position function on page 376.

a. Drop a ball from a height of 3 feet, 6 feet, and 12 feet. Record the number of seconds it takes for the ball to hit the ground.

b. For each of the three initial positions, use the position function to determine the time required for the ball to hit the ground.

c. What factors might result in differences between the times that you recorded and the times indicated by the function?

d. What appears to be happening to the time required for a free-falling object to hit the ground as its initial height is doubled? Verify this observation algebraically and with a graphing utility.

e. Repeat part (a) using a sheet of paper rather than a ball. What differences do you observe? What factor seems to be ignored in the position function?

f. What is meant by the acceleration of gravity and how does this number appear in the position function for a free-falling object?

SECTION 3.7 Modeling Using Variation

Objectives

1. Solve direct variation problems.
2. Solve inverse variation problems.
3. Solve combined variation problems.
4. Solve problems involving joint variation.

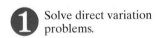

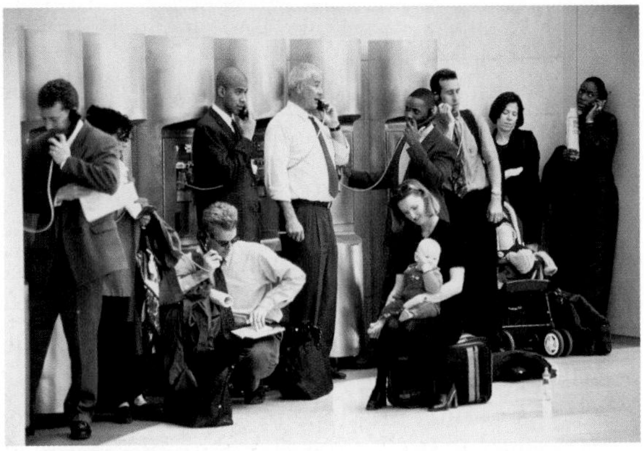

Have you ever wondered how telecommunication companies estimate the number of phone calls expected per day between two cities? The formula

$$C = \frac{0.02P_1P_2}{d^2}$$

shows that the daily number of phone calls, C, increases as the populations of the cities, P_1 and P_2, in thousands, increase and decreases as the distance, d, between the cities increases.

Certain formulas occur so frequently in applied situations that they are given special names. Variation formulas show how one quantity changes in relation to other quantities. Quantities can vary *directly, inversely*, or *jointly*. In this section, we look at situations that can be modeled by each of these kinds of variation. And think of this: The next time you get one of those "all-circuits-are-busy" messages, you will be able to use a variation formula to estimate how many other callers you're competing with for those precious 5-cent minutes.

1 Solve direct variation problems.

Direct Variation

When you swim underwater, the pressure in your ears depends on the depth at which you are swimming. The formula

$$p = 0.43d$$

describes the water pressure, p, in pounds per square inch, at a depth of d feet. We can use this linear function to determine the pressure in your ears at various depths:

If $d = 20, p = 0.43(20) = 8.6.$ At a depth of 20 feet, water pressure is 8.6 pounds per square inch.

Doubling the depth doubles the pressure.

If $d = 40, p = 0.43(40) = 17.2.$ At a depth of 40 feet, water pressure is 17.2 pounds per square inch.

Doubling the depth doubles the pressure.

If $d = 80, p = 0.43(80) = 34.4.$ At a depth of 80 feet, water pressure is 34.4 pounds per square inch.

The formula $p = 0.43d$ illustrates that water pressure is a constant multiple of your underwater depth. If your depth is doubled, the pressure is doubled; if your depth is tripled, the pressure is tripled; and so on. Because of this, the pressure in your ears is said to **vary directly** as your underwater depth. The **equation of variation** is

$$p = 0.43d.$$

Generalizing our discussion of pressure and depth on the previous page, we obtain the following statement:

Direct Variation

If a situation is described by an equation in the form

$$y = kx,$$

where k is a nonzero constant, we say that **y varies directly as x** or **y is directly proportional to x**. The number k is called the **constant of variation** or the **constant of proportionality**.

Can you see that **the direct variation equation, $y = kx$, is a special case of the linear function $y = mx + b$**? When $m = k$ and $b = 0$, $y = mx + b$ becomes $y = kx$. Thus, the slope of a direct variation equation is k, the constant of variation. Because b, the y-intercept, is 0, the graph of a direct variation equation is a line passing through the origin. This is illustrated in Figure 3.43, which shows the graph of $p = 0.43d$: Water pressure varies directly as depth.

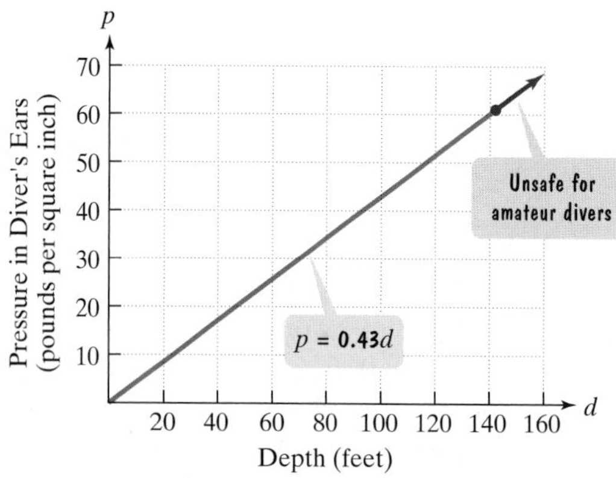

Figure 3.43 Water pressure at various depths

Problems involving direct variation can be solved using the following procedure. This procedure applies to direct variation problems, as well as to the other kinds of variation problems that we will discuss.

Solving Variation Problems

1. Write an equation that describes the given English statement.
2. Substitute the given pair of values into the equation in step 1 and solve for k, the constant of variation.
3. Substitute the value of k into the equation in step 1.
4. Use the equation from step 3 to answer the problem's question.

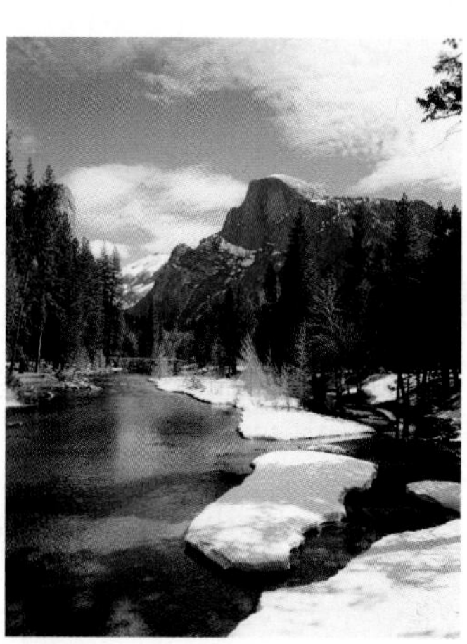

EXAMPLE 1 Solving a Direct Variation Problem

Many areas of Northern California depend on the snowpack of the Sierra Nevada mountain range for their water supply. The volume of water produced from melting snow varies directly as the volume of snow. Meteorologists have determined that 250 cubic centimeters of snow will melt to 28 cubic centimeters of water. How much water does 1200 cubic centimeters of melting snow produce?

Solution

Step 1 Write an equation. We know that *y varies directly as x* is expressed as

$$y = kx.$$

By changing letters, we can write an equation that describes the following English statement: Volume of water, W, varies directly as volume of snow, S.

$$W = kS$$

Step 2 Use the given values to find k. We are told that 250 cubic centimeters of snow will melt to 28 cubic centimeters of water. Substitute 250 for S and 28 for W in the direct variation equation. Then solve for k.

$W = kS$	Volume of water varies directly as volume of melting snow.
$28 = k(250)$	250 cubic centimeters of snow melt to 28 cubic centimeters of water.
$\frac{28}{250} = \frac{k(250)}{250}$	Divide both sides by 250.
$0.112 = k$	Simplify.

Step 3 Substitute the value of k into the equation.

$W = kS$	This is the equation from step 1.
$W = 0.112S$	Replace k, the constant of variation, with 0.112.

Step 4 Answer the problem's question. How much water does 1200 cubic centimeters of melting snow produce? Substitute 1200 for S in $W = 0.112S$ and solve for W.

$W = 0.112S$	Use the equation from step 3.
$W = 0.112(1200)$	Substitute 1200 for S.
$W = 134.4$	Multiply.

A snowpack measuring 1200 cubic centimeters will produce 134.4 cubic centimeters of water.

Check Point 1 The number of gallons of water, W, used when taking a shower varies directly as the time, t, in minutes, in the shower. A shower lasting 5 minutes uses 30 gallons of water. How much water is used in a shower lasting 11 minutes?

The direct variation equation $y = kx$ is a linear function. If $k > 0$, then the slope of the line is positive. Consequently, as x increases, y also increases.

A direct variation situation can involve variables to higher powers. For example, y can vary directly as $x^2 (y = kx^2)$ or as $x^3 (y = kx^3)$.

Direct Variation with Powers

y varies directly as the nth power of x if there exists some nonzero constant k such that

$$y = kx^n.$$

We also say that **y is directly proportional to the nth power of x.**

Direct variation with whole number powers is modeled by polynomial functions. In our next example, the graph of the variation equation is the familiar parabola.

EXAMPLE 2 Solving a Direct Variation Problem

The distance, s, that a body falls from rest varies directly as the square of the time, t, of the fall. If skydivers fall 64 feet in 2 seconds, how far will they fall in 4.5 seconds?

Solution

Step 1 Write an equation. We know that *y varies directly as the square of x* is expressed as

$$y = kx^2.$$

By changing letters, we can write an equation that describes the following English statement: Distance, s, varies directly as the square of time, t, of the fall.

$$s = kt^2$$

Step 2 Use the given values to find k. Skydivers fall 64 feet in 2 seconds. Substitute 64 for s and 2 for t in the direct variation equation. Then solve for k.

$$s = kt^2 \quad \text{Distance varies directly as the square of time.}$$
$$64 = k \cdot 2^2 \quad \text{Skydivers fall 64 feet in 2 seconds.}$$
$$64 = 4k \quad \text{Simplify: } 2^2 = 4.$$
$$\frac{64}{4} = \frac{4k}{4} \quad \text{Divide both sides by 4.}$$
$$16 = k \quad \text{Simplify.}$$

Step 3 Substitute the value of k into the equation.

$$s = kt^2 \quad \text{Use the equation from step 1.}$$
$$s = 16t^2 \quad \text{Replace } k\text{, the constant of variation, with 16.}$$

Step 4 Answer the problem's question. How far will the skydivers fall in 4.5 seconds? Substitute 4.5 for t in $s = 16t^2$ and solve for s.

$$s = 16(4.5)^2 = 16(20.25) = 324$$

Thus, in 4.5 seconds, the skydivers will fall 324 feet.

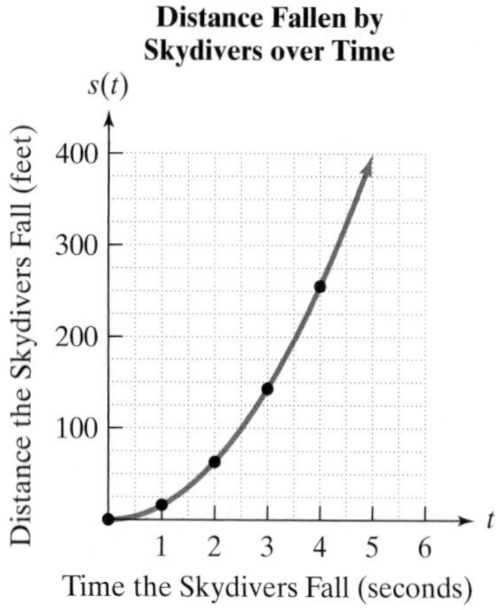

Figure 3.44 The graph of $s(t) = 16t^2$

We can express the variation equation from Example 2 in function notation, writing

$$s(t) = 16t^2.$$

The distance that a body falls from rest is a function of the time, t, of the fall. The parabola that is the graph of this quadratic function is shown in Figure 3.44. The graph increases rapidly from left to right, showing the effects of the acceleration of gravity.

Check Point 2 The distance required to stop a car varies directly as the square of its speed. If 200 feet are required to stop a car traveling 60 miles per hour, how many feet are required to stop a car traveling 100 miles per hour?

2 Solve inverse variation problems.

Inverse Variation

The distance from San Francisco to Los Angeles is 420 miles. The time that it takes to drive from San Francisco to Los Angeles depends on the rate at which one drives and is given by

$$\text{Time} = \frac{420}{\text{Rate}}.$$

For example, if you average 30 miles per hour, the time for the drive is

$$\text{Time} = \frac{420}{30} = 14,$$

or 14 hours. If you average 50 miles per hour, the time for the drive is

$$\text{Time} = \frac{420}{50} = 8.4,$$

or 8.4 hours. As your rate (or speed) increases, the time for the trip decreases and vice versa. This is illustrated by the graph in Figure 3.45.

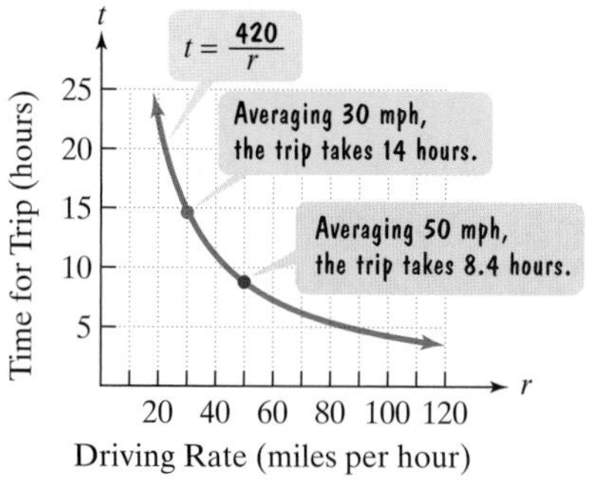

Figure 3.45

We can express the time for the San Francisco–Los Angeles trip using t for time and r for rate:

$$t = \frac{420}{r}.$$

This equation is an example of an **inverse variation** equation. Time, t, **varies inversely** as rate, r. When two quantities vary inversely, one quantity increases as the other decreases and vice versa.

Generalizing, we obtain the following statement:

Inverse Variation

If a situation is described by an equation in the form

$$y = \frac{k}{x},$$

where k is a nonzero constant, we say that **y varies inversely as x** or **y is inversely proportional to x**. The number k is called the **constant of variation**.

Notice that **the inverse variation equation**

$$y = \frac{k}{x}, \quad \text{or} \quad f(x) = \frac{k}{x},$$

is a rational function. For $k > 0$ and $x > 0$, the graph of the function takes on the shape shown in Figure 3.46.

We use the same procedure to solve inverse variation problems as we did to solve direct variation problems. Example 3 illustrates this procedure.

$y = \dfrac{k}{x}, k > 0 \text{ and } x > 0$

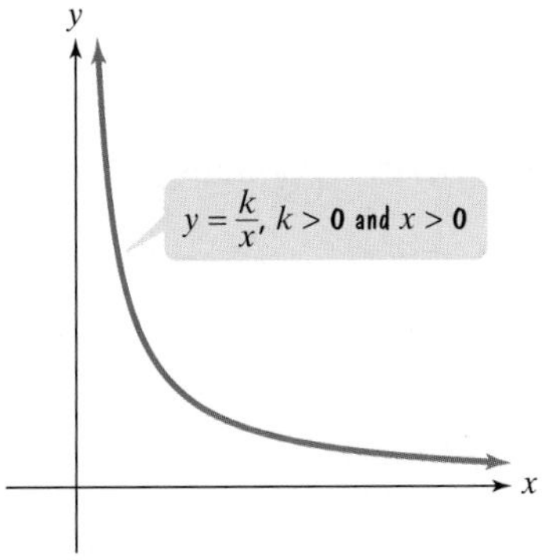

Figure 3.46 The graph of the inverse variation equation

EXAMPLE 3   Solving an Inverse Variation Problem

When you use a spray can and press the valve at the top, you decrease the pressure of the gas in the can. This decrease of pressure causes the volume of the gas in the can to increase. Because the gas needs more room than is provided in the can, it expands in spray form through the small hole near the valve. In general, if the temperature is constant, the pressure, P, of a gas in a container varies inversely as the volume, V, of the container. The pressure of a gas sample in a container whose volume is 8 cubic inches is 12 pounds per square inch. If the sample expands to a volume of 22 cubic inches, what is the new pressure of the gas?

$2P$

P

$2V$

V

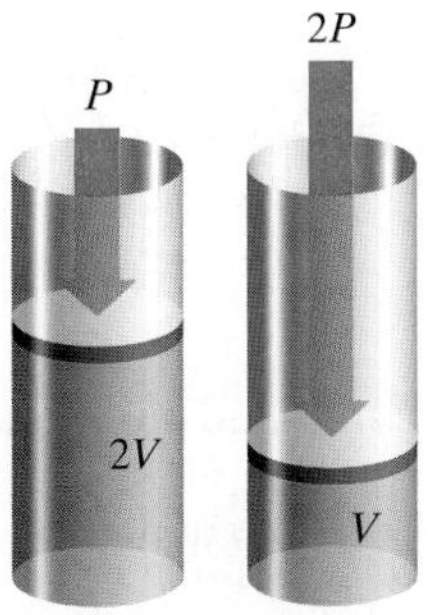

Doubling the pressure
halves the volume.

Solution

Step 1   Write an equation. We know that y *varies inversely as x* is expressed as

$$y = \frac{k}{x}.$$

By changing letters, we can write an equation that describes the following English statement: The pressure, P, of a gas in a container varies inversely as the volume, V.

$$P = \frac{k}{V}.$$

Step 2   Use the given values to find k. The pressure of a gas sample in a container whose volume is 8 cubic inches is 12 pounds per square inch. Substitute 12 for P and 8 for V in the inverse variation equation. Then solve for k.

$$P = \frac{k}{V} \qquad \text{Pressure varies inversely as volume.}$$

$$12 = \frac{k}{8} \qquad \text{The pressure in an 8 cubic-inch container is 12 pounds per square inch.}$$

$$12 \cdot 8 = \frac{k}{8} \cdot 8 \qquad \text{Multiply both sides by 8.}$$

$$96 = k \qquad \text{Simplify.}$$

Step 3 Substitute the value of k into the equation.

$$P = \frac{k}{V}$$ Use the equation from step 1.

$$P = \frac{96}{V}$$ Replace k, the constant of variation, with 96.

Step 4 Answer the problem's question. We need to find the pressure when the volume expands to 22 cubic inches. Substitute 22 for V and solve for P.

$$P = \frac{96}{V} = \frac{96}{22} = 4\frac{4}{11}$$

When the volume is 22 cubic inches, the pressure of the gas is $4\frac{4}{11}$ pounds per square inch.

Check Point 3 The length of a violin string varies inversely as the frequency of its vibrations. A violin string 8 inches long vibrates at a frequency of 640 cycles per second. What is the frequency of a 10-inch string?

3 Solve combined variation problems.

Combined Variation

In **combined variation**, direct and inverse variation occur at the same time. For example, as the advertising budget, A, of a company increases, its monthly sales, S, also increase. Monthly sales vary directly as the advertising budget:

$$S = kA.$$

By contrast, as the price of the company's product, P, increases, its monthly sales, S, decrease. Monthly sales vary inversely as the price of the product:

$$S = \frac{k}{P}.$$

We can combine these two variation equations into one combined equation:

$$S = \frac{kA}{P}.$$ Monthly sales, S, vary directly as the advertising budget, A, and inversely as the price of the product, P.

The following example illustrates an application of combined variation.

EXAMPLE 4 Solving a Combined Variation Problem

The owners of Rollerblades Plus determine that the monthly sales, S, of its skates vary directly as its advertising budget, A, and inversely as the price of the skates, P. When \$60,000 is spent on advertising and the price of the skates is \$40, the monthly sales are 12,000 pairs of rollerblades.

a. Write an equation of variation that describes this situation.

b. Determine monthly sales if the amount of the advertising budget is increased to \$70,000.

Solution

a. Write an equation.

$$S = \frac{kA}{P}.$$ Translate "sales vary directly as the advertising budget and inversely as the skates' price."

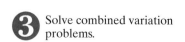

Use the given values to find k.

$$12{,}000 = \frac{k(60{,}000)}{40}$$ When \$60,000 is spent on advertising ($A = 60{,}000$) and the price is \$40 ($P = 40$), monthly sales are 12,000 units ($S = 12{,}000$).

$$12{,}000 = k \cdot 1500$$ Divide 60,000 by 40.

$$\frac{12{,}000}{1500} = \frac{k \cdot 1500}{1500}$$ Divide both sides of the equation by 1500.

$$8 = k$$ Simplify.

Therefore, the equation of variation that describes monthly sales is

$$S = \frac{8A}{P}.$$ Substitute 8 for k in $S = \frac{kA}{P}$.

b. The advertising budget is increased to \$70,000, so $A = 70{,}000$. The skates' price is still \$40, so $P = 40$.

$$S = \frac{8A}{P}$$ This is the equation from part (a).

$$S = \frac{8(70{,}000)}{40}$$ Substitute 70,000 for A and 40 for P.

$$S = 14{,}000$$ Simplify.

With a \$70,000 advertising budget and \$40 price, the company can expect to sell 14,000 pairs of rollerblades in a month (up from 12,000).

Check Point 4 The number of minutes needed to solve an exercise set of variation problems varies directly as the number of problems and inversely as the number of people working to solve the problems. It takes 4 people 32 minutes to solve 16 problems. How many minutes will it take 8 people to solve 24 problems?

4 Solve problems involving joint variation.

Joint Variation

Joint variation is a variation in which a variable varies directly as the product of two or more other variables. Thus, the equation $y = kxz$ is read "y varies jointly as x and z."

Joint variation plays a critical role in Isaac Newton's formula for gravitation:

$$F = G\frac{m_1m_2}{d^2}.$$

The formula states that the force of gravitation, F, between two bodies varies jointly as the product of their masses, m_1 and m_2, and inversely as the square of the distance between them, d. (G is the gravitational constant.) The formula indicates that gravitational force exists between any two objects in the universe, increasing as the distance between the bodies decreases. One practical result is that the pull of the moon on the oceans is greater on the side of Earth closer to the moon. This gravitational imbalance is what produces tides.

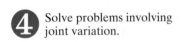

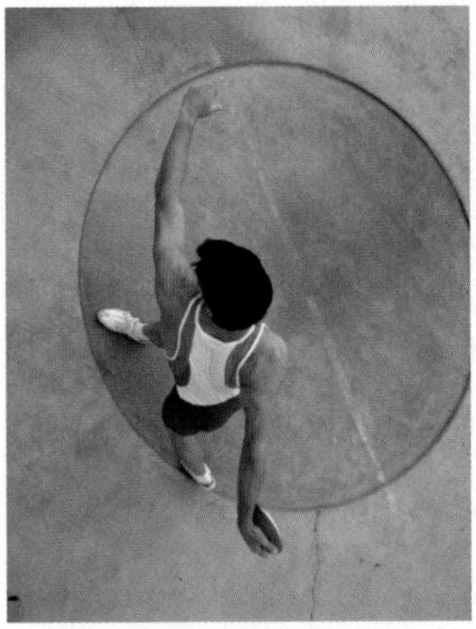

EXAMPLE 5 Modeling Centrifugal Force

The centrifugal force, C, of a body moving in a circle varies jointly with the radius of the circular path, r, and the body's mass, m, and inversely with the square of the time, t, it takes to move about one full circle. A 6-gram body moving in a circle with radius 100 centimeters at a rate of 1 revolution in 2 seconds has a centrifugal force of 6000 dynes. Find the centrifugal force of an 18-gram body moving in a circle with radius 100 centimeters at a rate of 1 revolution in 3 seconds.

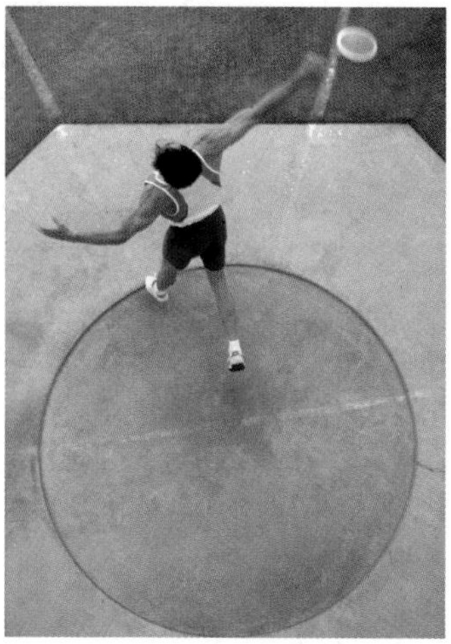

Solution

$$C = \frac{krm}{t^2}$$ Translate "Centrifugal force, C, varies jointly with radius, r, and mass, m, and inversely with the square of time, t."

$$6000 = \frac{k(100)(6)}{2^2}$$ A 6-gram body ($m = 6$) moving in a circle with radius 100 centimeters ($r = 100$) at 1 revolution in 2 seconds ($t = 2$) has a centifugal force of 6000 dynes ($C = 6000$).

$$6000 = 150k$$ Simplify.

$$40 = k$$ Divide both sides by 150 and solve for k.

$$C = \frac{40rm}{t^2}$$ Substitute 40 for k in the model for centrifugal force.

$$C = \frac{40(100)(18)}{3^2}$$ Find centifugal force, C, of an 18-gram body ($m = 18$) moving in a circle with radius 100 centimeters ($r = 100$) at 1 revolution in 3 seconds ($t = 3$).

$$= 8000$$ Simplify.

The centrifugal force is 8000 dynes.

Check Point 5 The volume of a cone, V, varies jointly as its height, h, and the square of its radius, r. A cone with a radius measuring 6 feet and a height measuring 10 feet has a volume of 120π cubic feet. Find the volume of a cone having a radius of 12 feet and a height of 2 feet.

EXERCISE SET 3.7

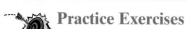

Practice Exercises

Use the four-step procedure for solving variation problems given on page 382 to solve Exercises 1–10.

1. y varies directly as x. $y = 65$ when $x = 5$. Find y when $x = 12$.
2. y varies directly as x. $y = 45$ when $x = 5$. Find y when $x = 13$.
3. y varies inversely as x. $y = 12$ when $x = 5$. Find y when $x = 2$.
4. y varies inversely as x. $y = 6$ when $x = 3$. Find y when $x = 9$.
5. y varies directly as x and inversely as the square of z. $y = 20$ when $x = 50$ and $z = 5$. Find y when $x = 3$ and $z = 6$.
6. a varies directly as b and inversely as the square of c. $a = 7$ when $b = 9$ and $c = 6$. Find a when $b = 4$ and $c = 8$.
7. y varies jointly as x and z. $y = 25$ when $x = 2$ and $z = 5$. Find y when $x = 8$ and $z = 12$.
8. C varies jointly as A and T. $C = 175$ when $A = 2100$ and $T = 4$. Find C when $A = 2400$ and $T = 6$.
9. y varies jointly as a and b and inversely as the square root of c. $y = 12$ when $a = 3, b = 2$, and $c = 25$. Find y when $a = 5, b = 3$, and $c = 9$.
10. y varies jointly as m and the square of n and inversely as p. $y = 15$ when $m = 2, n = 1$, and $p = 6$. Find y when $m = 3, n = 4$, and $p = 10$.

Practice Plus

In Exercises 11–20, write an equation that expresses each relationship. Then solve the equation for y.

11. x varies jointly as y and z.
12. x varies jointly as y and the square of z.
13. x varies directly as the cube of z and inversely as y.
14. x varies directly as the cube root of z and inversely as y.
15. x varies jointly as y and z and inversely as the square root of w.
16. x varies jointly as y and z and inversely as the square of w.
17. x varies jointly as z and the sum of y and w.
18. x varies jointly as z and the difference between y and w.
19. x varies directly as z and inversely as the difference between y and w.
20. x varies directly as z and inversely as the sum of y and w.

Application Exercises

Use the four-step procedure for solving variation problems given on page 382 to solve Exercises 21–36.

21. An alligator's tail length, T, varies directly as its body length, B. An alligator with a body length of 4 feet has a tail length of 3.6 feet. What is the tail length of an alligator whose body length is 6 feet?

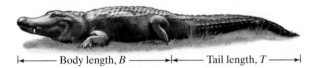

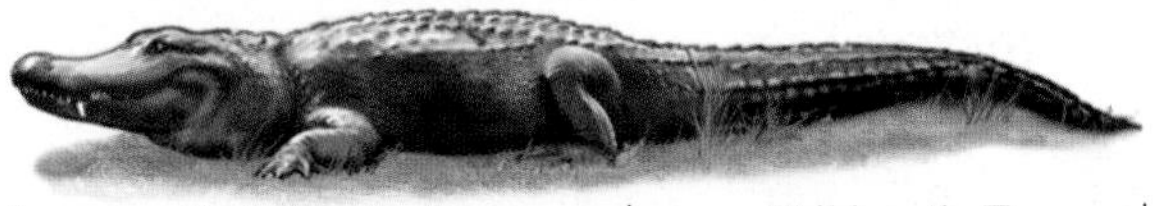

22. An object's weight on the moon, M, varies directly as its weight on Earth, E. Neil Armstrong, the first person to step on the moon on July 20, 1969, weighed 360 pounds on Earth (with all of his equipment on) and 60 pounds on the moon. What is the moon weight of a person who weighs 186 pounds on Earth?

23. The height that a ball bounces varies directly as the height from which it was dropped. A tennis ball dropped from 12 inches bounces 8.4 inches. From what height was the tennis ball dropped if it bounces 56 inches?

24. The distance that a spring will stretch varies directly as the force applied to the spring. A force of 12 pounds is needed to stretch a spring 9 inches. What force is required to stretch the spring 15 inches?

25. If all men had identical body types, their weight would vary directly as the cube of their height. Shown below is Robert Wadlow, who reached a record height of 8 feet 11 inches (107 inches) before his death at age 22. If a man who is 5 feet 10 inches tall (70 inches) with the same body type as Mr. Wadlow weighs 170 pounds, what was Robert Wadlow's weight shortly before his death?

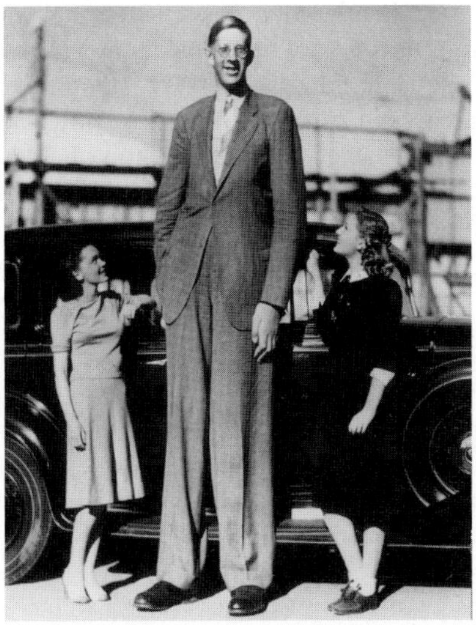

26. On a dry asphalt road, a car's stopping distance varies directly as the square of its speed. A car traveling at 45 miles per hour can stop in 67.5 feet. What is the stopping distance for a car traveling at 60 miles per hour?

27. The figure shows that a bicyclist tips the cycle when making a turn. The angle B, formed by the vertical direction and the bicycle, is called the banking angle. The banking angle varies inversely as the cycle's turning radius. When the turning radius is 4 feet, the banking angle is 28°. What is the banking angle when the turning radius is 3.5 feet?

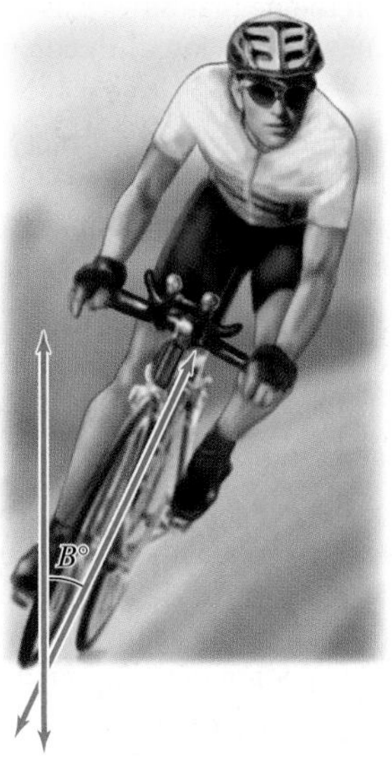

28. The water temperature of the Pacific Ocean varies inversely as the water's depth. At a depth of 1000 meters, the water temperature is 4.4° Celsius. What is the water temperature at a depth of 5000 meters?

29. Radiation machines, used to treat tumors, produce an intensity of radiation that varies inversely as the square of the distance from the machine. At 3 meters, the radiation intensity is 62.5 milliroentgens per hour. What is the intensity at a distance of 2.5 meters?

30. The illumination provided by a car's headlight varies inversely as the square of the distance from the headlight. A car's headlight produces an illumination of 3.75 footcandles at a distance of 40 feet. What is the illumination when the distance is 50 feet?

31. Body-mass index, or BMI, takes both weight and height into account when assessing whether an individual is underweight or overweight. BMI varies directly as one's weight, in pounds, and inversely as the square of one's height, in inches. In adults, normal values for the BMI are between 20 and 25, inclusive. Values below 20 indicate that an individual is underweight and values above 30 indicate that an individual is obese. A person who weighs 180 pounds and is 5 feet, or 60 inches, tall has a BMI of 35.15. What is the BMI, to the nearest tenth, for a 170 pound person who is 5 feet 10 inches tall. Is this person overweight?

32. One's intelligence quotient, or IQ, varies directly as a person's mental age and inversely as that person's chronological age. A person with a mental age of 25 and a chronological age of 20 has an IQ of 125. What is the chronological age of a person with a mental age of 40 and an IQ of 80?

33. The heat loss of a glass window varies jointly as the window's area and the difference between the outside and inside temperatures. A window 3 feet wide by 6 feet long loses 1200 Btu per hour when the temperature outside is 20° colder than the temperature inside. Find the heat loss through a glass window that is 6 feet wide by 9 feet long when the temperature outside is 10° colder than the temperature inside.

34. Kinetic energy varies jointly as the mass and the square of the velocity. A mass of 8 grams and velocity of 3 centimeters per second has a kinetic energy of 36 ergs. Find the kinetic energy for a mass of 4 grams and velocity of 6 centimeters per second.

35. Sound intensity varies inversely as the square of the distance from the sound source. If you are in a movie theater and you change your seat to one that is twice as far from the speakers, how does the new sound intensity compare to that of your original seat?

36. Many people claim that as they get older, time seems to pass more quickly. Suppose that the perceived length of a period of time is inversely proportional to your age. How long will a year seem to be when you are three times as old as you are now?

37. The average number of daily phone calls, C, between two cities varies jointly as the product of their populations, P_1 and P_2, and inversely as the square of the distance, d, between them.

a. Write an equation that expresses this relationship.

b. The distance between San Francisco (population: 777,000) and Los Angeles (population: 3,695,000) is 420 miles. If the average number of daily phone calls between the cities is 326,000, find the value of k to two decimal places and write the equation of variation.

c. Memphis (population: 650,000) is 400 miles from New Orleans (population: 490,000). Find the average number of daily phone calls, to the nearest whole number, between these cities.

38. The force of wind blowing on a window positioned at a right angle to the direction of the wind varies jointly as the area of the window and the square of the wind's speed. It is known that a wind of 30 miles per hour blowing on a window measuring 4 feet by 5 feet exerts a force of 150 pounds. During a storm with winds of 60 miles per hour, should hurricane shutters be placed on a window that measures 3 feet by 4 feet and is capable of withstanding 300 pounds of force?

39. The table shows the values for the current, I, in an electric circuit and the resistance, R, of the circuit.

***I* (amperes)**	0.5	1.0	1.5	2.0	2.5	3.0	4.0	5.0
***R* (ohms)**	12.0	6.0	4.0	3.0	2.4	2.0	1.5	1.2

a. Graph the ordered pairs in the table of values, with values of I along the x-axis and values of R along the y-axis. Connect the eight points with a smooth curve.

b. Does current vary directly or inversely as resistance? Use your graph and explain how you arrived at your answer.

c. Write an equation of variation for I and R, using one of the ordered pairs in the table to find the constant of variation. Then use your variation equation to verify the other seven ordered pairs in the table.

Writing in Mathematics

40. What does it mean if two quantities vary directly?

41. In your own words, explain how to solve a variation problem.

42. What does it mean if two quantities vary inversely?

43. Explain what is meant by combined variation. Give an example with your explanation.

44. Explain what is meant by joint variation. Give an example with your explanation.

In Exercises 45–46, describe in words the variation shown by the given equation.

45. $z = \dfrac{k\sqrt{x}}{y^2}$ **46.** $z = kx^2\sqrt{y}$

47. We have seen that the daily number of phone calls between two cities varies jointly as their populations and inversely as the square of the distance between them. This model, used by telecommunication companies to estimate the line capacities needed among various cities, is called the *gravity model*. Compare the model to Newton's formula for gravitation on page 387 and describe why the name *gravity model* is appropriate.

Technology Exercise

48. Use a graphing utility to graph any three of the variation equations in Exercises 21–30. Then TRACE along each curve and identify the point that corresponds to the problem's solution.

Critical Thinking Exercises

49. In a hurricane, the wind pressure varies directly as the square of the wind velocity. If wind pressure is a measure of a hurricane's destructive capacity, what happens to this destructive power when the wind speed doubles?

50. The illumination from a light source varies inversely as the square of the distance from the light source. If you raise a lamp from 15 inches to 30 inches over your desk, what happens to the illumination?

51. The heat generated by a stove element varies directly as the square of the voltage and inversely as the resistance. If the voltage remains constant, what needs to be done to triple the amount of heat generated?

52. Galileo's telescope brought about revolutionary changes in astronomy. A comparable leap in our ability to observe the universe took place as a result of the Hubble Space Telescope. The space telescope was able to see stars and galaxies whose brightness is $\frac{1}{50}$ of the faintest objects observable using ground-based telescopes. Use the fact that the brightness of a point source, such as a star, varies inversely as the square of its distance from an observer to show that the space telescope was able to see about seven times farther than a ground-based telescope.

Group Exercise

53. Begin by deciding on a product that interests the group because you are now in charge of advertising this product. Members were told that the demand for the product varies directly as the amount spent on advertising and inversely as the price of the product. However, as more money is spent on advertising, the price of your product rises. Under what conditions would members recommend an increased expense in advertising? Once you've determined what your product is, write formulas for the given conditions and experiment with hypothetical numbers. What other factors might you take into consideration in terms of your recommendation? How do these factor affect the demand for your product?

Chapter 3
Summary, Review, and Test

Summary

DEFINITIONS AND CONCEPTS	EXAMPLES
3.1 Quadratic Functions	
a. A quadratic function is of the form $f(x) = ax^2 + bx + c, a \neq 0$.	
b. The standard form of a quadratic function is $f(x) = a(x - h)^2 + k, a \neq 0$.	
c. The graph of a quadratic function is a parabola. The vertex is (h, k) or $\left(-\frac{b}{2a}, f\left(-\frac{b}{2a}\right)\right)$. A procedure for graphing a quadratic function is given in the box on page 300.	Ex. 1, p. 300; Ex. 2, p. 301; Ex. 3, p. 302
d. See the box on page 304 for minimum or maximum values of quadratic functions.	Ex. 4, p. 304; Ex. 5, p. 305
e. A strategy for solving problems involving maximizing or minimizing quadratic functions is given in the box on page 306.	Ex. 6, p. 306; Ex. 7, p. 308
3.2 Polynomial Functions and Their Graphs	
a. Polynomial Function of Degree n: $f(x) = a_n x^n + a_{n-1}x^{n-1} + \cdots + a_2 x^2 + a_1 x + a_0, a_n \neq 0$	
b. The graphs of polynomial functions are smooth and continuous.	Fig. 3.11, p. 314
c. The end behavior of the graph of a polynomial function depends on the leading term, given by the Leading Coefficient Test in the box on page 315.	Ex. 1, p. 315; Ex. 2, p. 315; Ex. 3, p. 316
d. The values of x for which $f(x)$ is equal to 0 are the zeros of the polynomial function f. These values are the roots, or solutions, of the polynomial equation $f(x) = 0$.	Ex. 4, p. 317; Ex. 5, p. 318
e. If $x - r$ occurs k times in a polynomial function's factorization, r is a repeated zero with multiplicity k. If k is even, the graph touches the x-axis at r. If k is odd, the graph crosses the x-axis at r.	Ex. 6, p. 319
f. The Intermediate Value Theorem: If f is a polynomial function and $f(a)$ and $f(b)$ have opposite signs, there is at least one value of c between a and b for which $f(c) = 0$.	Ex. 7, p. 319
g. If f is a polynomial of degree n, the graph of f has at most $n - 1$ turning points.	Fig. 3.21, p. 320
h. A strategy for graphing a polynomial function is given in the box on page 320.	Ex. 8, p. 321
3.3 Dividing Polynomials; Remainder and Factor Theorems	
a. Long division of polynomials is performed by dividing, multiplying, subtracting, bringing down the next term, and repeating this process until the degree of the remainder is less than the degree of the divisor. The details are given in the box on page 328.	Ex. 1, p. 327; Ex. 2, p. 328; Ex. 3, p. 330
b. The Division Algorithm: $f(x) = d(x)q(x) + r(x)$. The dividend is the product of the divisor and the quotient plus the remainder.	
c. Synthetic division is used to divide a polynomial by $x - c$. The details are given in the box on page 331.	Ex. 4, p. 331
d. The Remainder Theorem: If a polynomial $f(x)$ is divided by $x - c$, then the remainder is $f(c)$.	Ex. 5, p. 333
e. The Factor Theorem: If $x - c$ is a factor of a polynomial function $f(x)$, then c is a zero of f and a root of $f(x) = 0$. If c is a zero of f or a root of $f(x) = 0$, then $x - c$ is a factor of $f(x)$.	Ex. 6, p. 334
3.4 Zeros of Polynomial Functions	
a. The Rational Zero Theorem states that the possible rational zeros of a polynomial function $= \frac{\text{Factors of the constant term}}{\text{Factors of the leading coefficient}}$. The theorem is stated in the box on page 338.	Ex. 1, p. 338; Ex. 2, p. 339; Ex. 3, p. 339; Ex. 4, p. 340; Ex. 5, p. 341

DEFINITIONS AND CONCEPTS	EXAMPLES
b. Number of roots: If $f(x)$ is a polynomial of degree $n \geq 1$, then, counting multiple roots separately, the equation $f(x) = 0$ has n roots.	
c. If $a + bi$ is a root of $f(x) = 0$, then $a - bi$ is also a root.	
d. The Linear Factorization Theorem: An nth-degree polynomial can be expressed as the product of n linear factors. Thus, $f(x) = a_n(x - c_1)(x - c_2)\cdots(x - c_n)$.	Ex. 6, p. 344
e. Descartes's Rule of Signs: The number of positive real zeros of f equals the number of sign changes of $f(x)$ or is less than that number by an even integer. The number of negative real zeros of f applies a similar statement to $f(-x)$.	Table 3.1, p. 345; Ex. 7, p. 346

3.5 Rational Functions and Their Graphs

DEFINITIONS AND CONCEPTS	EXAMPLES
a. Rational function: $f(x) = \frac{p(x)}{q(x)}$; p and q are polynomial functions and $q(x) \neq 0$. The domain of f is the set of all real numbers excluding values of x that make $q(x)$ zero.	Ex. 1, p. 351
b. Arrow notation is summarized in the box on page 353.	
c. The line $x = a$ is a vertical asymptote of the graph of f if $f(x)$ increases or decreases without bound as x approaches a. Vertical asymptotes are identified using the location theorem in the box on page 354.	Ex. 2, p. 355
d. The line $y = b$ is a horizontal asymptote of the graph of f if $f(x)$ approaches b as x increases or decreases without bound. Horizontal asymptotes are identified using the location theorem in the box on page 357.	Ex. 3, p. 357
e. Table 3.2 on page 358 shows the graphs of $f(x) = \frac{1}{x}$ and $f(x) = \frac{1}{x^2}$. Some rational functions can be graphed using transformations of these common graphs.	Ex. 4, p. 359
f. A strategy for graphing rational functions is given in the box on page 359.	Ex. 5, p. 360; Ex. 6, p. 361; Ex. 7, p. 362
g. The graph of a rational function has a slant asymptote when the degree of the numerator is one more than the degree of the denominator. The equation of the slant asymptote is found using division and dropping the remainder term.	Ex. 8, p. 364

3.6 Polynomial and Rational Inequalities

DEFINITIONS AND CONCEPTS	EXAMPLES
a. A polynomial inequality can be expressed as $f(x) < 0, f(x) > 0, f(x) \leq 0$, or $f(x) \geq 0$, where f is a polynomial function. A procedure for solving polynomial inequalities is given in the box on page 371.	Ex. 1, p. 371; Ex. 2, p. 372
b. A rational inequality can be expressed as $f(x) < 0, f(x) > 0, f(x) \leq 0$, or $f(x) \geq 0$, where f is a rational function. The procedure for solving such inequalities begins with expressing them so that one side is zero and the other side is a single quotient. Find boundary points by setting the numerator and denominator equal to zero. Then follow a procedure similar to that for solving polynomial inequalities.	Ex. 3, p. 374

3.7 Modeling Using Variation

a. A procedure for solving variation problems is given in the box on page 382.

b.

English Statement	Equation	EXAMPLES
y varies directly as x. y is directly proportional to x.	$y = kx$	Ex. 1, p. 382
y varies directly as x^n. y is directly proportional to x^n.	$y = kx^n$	Ex. 2, p. 383
y varies inversely as x. y is inversely proportional to x.	$y = \frac{k}{x}$	Ex. 3, p. 385; Ex. 4, p. 386
y varies inversely as x^n. y is inversely proportional to x^n.	$y = \frac{k}{x^n}$	
y varies jointly as x and z.	$y = kxz$	Ex. 5, p. 387

Review Exercises

3.1

In Exercises 1–4, use the vertex and intercepts to sketch the graph of each quadratic function. Give the equation for the parabola's axis of symmetry. Use the graph to determine the function's domain and range.

1. $f(x) = -(x + 1)^2 + 4$ **2.** $f(x) = (x + 4)^2 - 2$

3. $f(x) = -x^2 + 2x + 3$ **4.** $f(x) = 2x^2 - 4x - 6$

In Exercises 5–6, use the function's equation, and not its graph, to find

a. *the minimum or maximum value and where it occurs.*

b. *the function's domain and its range.*

5. $f(x) = -x^2 + 14x - 106$

6. $f(x) = 2x^2 + 12x + 703$

7. The function

$$f(x) = -0.02x^2 + x + 1$$

models the yearly growth of a young redwood tree, $f(x)$, in inches, with x inches of rainfall per year. How many inches of rainfall per year result in maximum tree growth? What is the maximum yearly growth?

8. Suppose that a quadratic function is used to model the data shown in the graph using

(number of years after 1960, divorce rate per 1000 population).

U.S. Divorce Rate

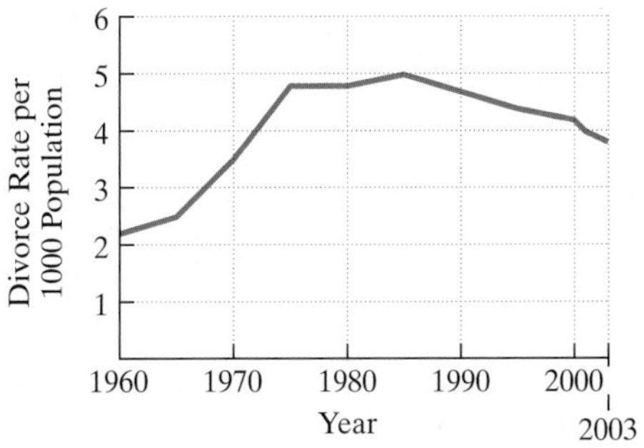

Source: National Center for Health Statistics

Determine, without obtaining an actual quadratic function that models the data, the approximate coordinates of the vertex for the function's graph. Describe what this means in practical terms.

9. A field bordering a straight stream is to be enclosed. The side bordering the stream is not to be fenced. If 1000 yards of fencing material is to be used, what are the dimensions of the largest rectangular field that can be fenced? What is the maximum area?

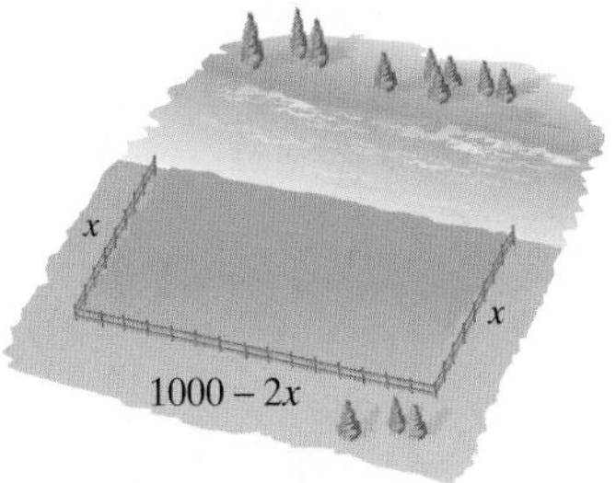

10. Among all pairs of numbers whose difference is 14, find a pair whose product is as small as possible. What is the minimum product?

3.2

In Exercises 11–14, use the Leading Coefficient Test to determine the end behavior of the graph of the given polynomial function. Then use this end behavior to match the polynomial function with its graph. [The graphs are labeled (a) through (d).]

11. $f(x) = -x^3 + x^2 + 2x$ **12.** $f(x) = x^6 - 6x^4 + 9x^2$

13. $f(x) = x^5 - 5x^3 + 4x$ **14.** $f(x) = -x^4 + 1$

a.

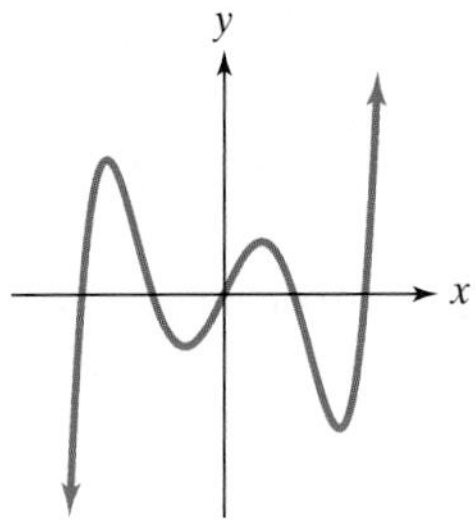

b.

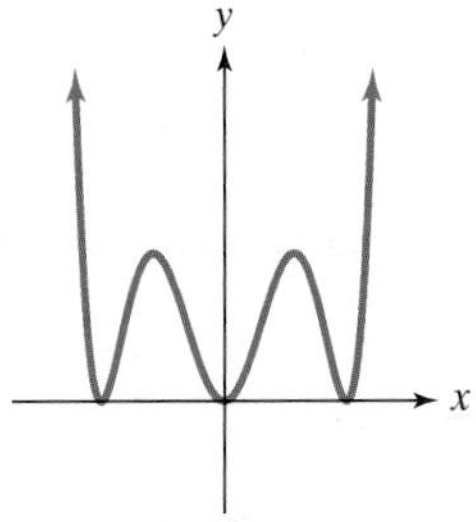

c.

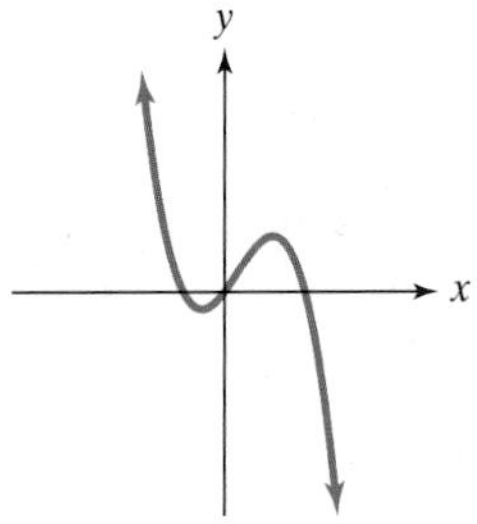

d.

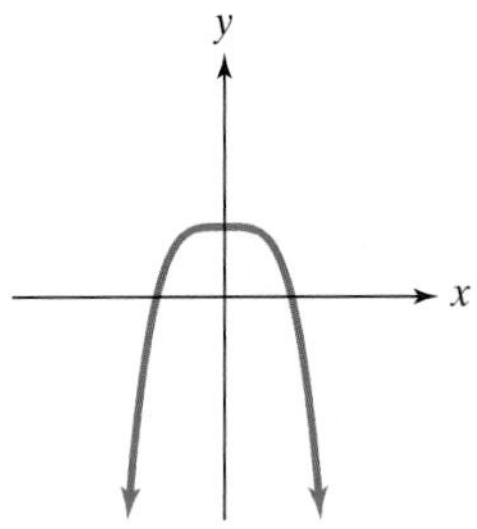

15. The polynomial function

$$f(x) = -0.87x^3 + 0.35x^2 + 81.62x + 7684.94$$

models the number of thefts, $f(x)$, in thousands, in the United States x years after 1987. Will this function be useful in modeling the number of thefts over an extended period of time? Explain your answer.

16. A herd of 100 elk is introduced to a small island. The number of elk, $f(x)$, after x years is modeled by the polynomial function

$$f(x) = -x^4 + 21x^2 + 100.$$

Use the Leading Coefficient Test to determine the graph's end behavior to the right. What does this mean about what will eventually happen to the elk population?

In Exercises 17–18, find the zeros for each polynomial function and give the multiplicity of each zero. State whether the graph crosses the x-axis, or touches the x-axis and turns around, at each zero.

17. $f(x) = -2(x - 1)(x + 2)^2(x + 5)^3$

18. $f(x) = x^3 - 5x^2 - 25x + 125$

19. Show that $f(x) = x^3 - 2x - 1$ has a real zero between 1 and 2.

In Exercises 20–25,

a. *Use the Leading Coefficient Test to determine the graph's end behavior.*

b. *Determine whether the graph has y-axis symmetry, origin symmetry, or neither.*

c. *Graph the function.*

20. $f(x) = x^3 - x^2 - 9x + 9$

21. $f(x) = 4x - x^3$

22. $f(x) = 2x^3 + 3x^2 - 8x - 12$

23. $f(x) = -x^4 + 25x^2$

24. $f(x) = -x^4 + 6x^3 - 9x^2$

25. $f(x) = 3x^4 - 15x^3$

In Exercises 26–27, graph each polynomial function.

26. $f(x) = 2x^2(x - 1)^3(x + 2)$

27. $f(x) = -x^3(x + 4)^2(x - 1)$

3.3

In Exercises 28–30, divide using long division.

28. $(4x^3 - 3x^2 - 2x + 1) \div (x + 1)$

29. $(10x^3 - 26x^2 + 17x - 13) \div (5x - 3)$

30. $(4x^4 + 6x^3 + 3x - 1) \div (2x^2 + 1)$

In Exercises 31–32, divide using synthetic division.

31. $(3x^4 + 11x^3 - 20x^2 + 7x + 35) \div (x + 5)$

32. $(3x^4 - 2x^2 - 10x) \div (x - 2)$

33. Given $f(x) = 2x^3 - 7x^2 + 9x - 3$, use the Remainder Theorem to find $f(-13)$.

34. Use synthetic division to divide $f(x) = 2x^3 + x^2 - 13x + 6$ by $x - 2$. Use the result to find all zeros of f.

35. Solve the equation $x^3 - 17x + 4 = 0$ given that 4 is a root.

3.4

In Exercises 36–37, use the Rational Zero Theorem to list all possible rational zeros for each given function.

36. $f(x) = x^4 - 6x^3 + 14x^2 - 14x + 5$

37. $f(x) = 3x^5 - 2x^4 - 15x^3 + 10x^2 + 12x - 8$

In Exercises 38–39, use Descartes's Rule of Signs to determine the possible number of positive and negative real zeros for each given function.

38. $f(x) = 3x^4 - 2x^3 - 8x + 5$

39. $f(x) = 2x^5 - 3x^3 - 5x^2 + 3x - 1$

40. Use Descartes's Rule of Signs to explain why $2x^4 + 6x^2 + 8 = 0$ has no real roots.

For Exercises 41–47,

a. *List all possible rational roots or rational zeros.*

b. *Use Descartes's Rule of Signs to determine the possible number of positive and negative real roots or real zeros.*

c. *Use synthetic division to test the possible rational roots or zeros and find an actual root or zero.*

d. *Use the quotient from part (c) to find all the remaining zeros or roots.*

41. $f(x) = x^3 + 3x^2 - 4$

42. $f(x) = 6x^3 + x^2 - 4x + 1$

43. $8x^3 - 36x^2 + 46x - 15 = 0$

44. $2x^3 + 9x^2 - 7x + 1 = 0$

45. $x^4 - x^3 - 7x^2 + x + 6 = 0$

46. $4x^4 + 7x^2 - 2 = 0$

47. $f(x) = 2x^4 + x^3 - 9x^2 - 4x + 4$

In Exercises 48–49, find an nth-degree polynomial function with real coefficients satisfying the given conditions. If you are using a graphing utility, graph the function and verify the real zeros and the given function value.

48. $n = 3$; 2 and $2 - 3i$ are zeros; $f(1) = -10$

49. $n = 4$; i is a zero; -3 is a zero of multiplicity 2; $f(-1) = 16$

In Exercises 50–51, find all the zeros of each polynomial function and write the polynomial as a product of linear factors.

50. $f(x) = 2x^4 + 3x^3 + 3x - 2$

51. $g(x) = x^4 - 6x^3 + x^2 + 24x + 16$

In Exercises 52–55, graphs of fifth-degree polynomial functions are shown. In each case, specify the number of real zeros and the number of imaginary zeros. Indicate whether there are any real zeros with multiplicity other than 1.

52.

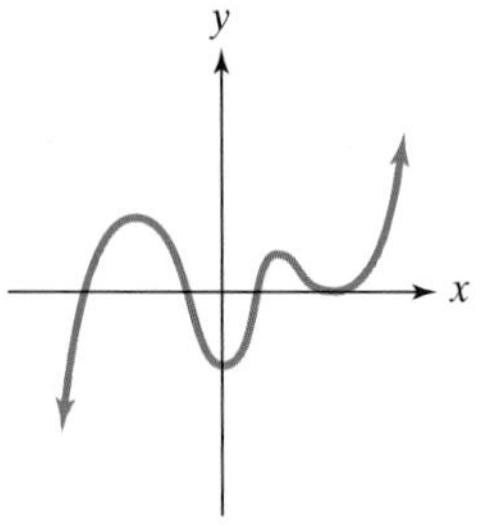

53.

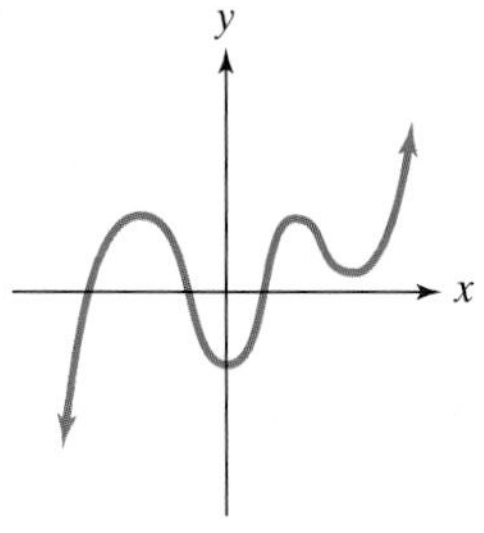

54.

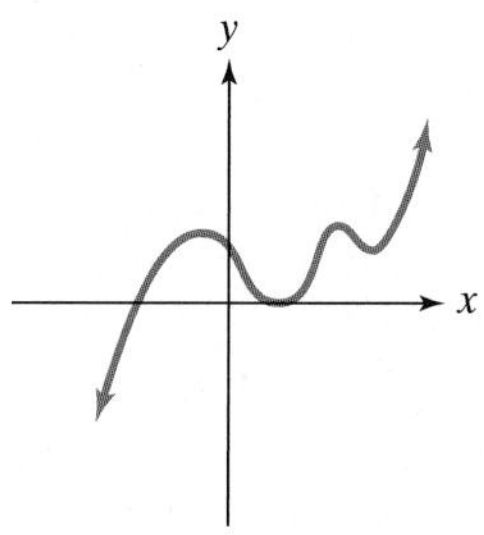

55.

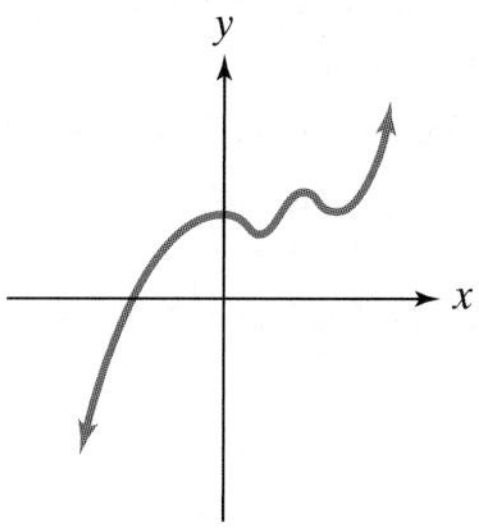

3.5

In Exercises 56–57, use transformations of $f(x) = \frac{1}{x}$ *or* $f(x) = \frac{1}{x^2}$ *to graph each rational function.*

56. $g(x) = \frac{1}{(x+2)^2} - 1$

57. $h(x) = \frac{1}{x-1} + 3$

In Exercises 58–65, find the vertical asymptotes, if any, the horizontal asymptote, if one exists, and the slant asymptote, if there is one, of the graph of each rational function. Then graph the rational function.

58. $f(x) = \frac{2x}{x^2 - 9}$

59. $g(x) = \frac{2x - 4}{x + 3}$

60. $h(x) = \frac{x^2 - 3x - 4}{x^2 - x - 6}$

61. $r(x) = \frac{x^2 + 4x + 3}{(x+2)^2}$

62. $y = \frac{x^2}{x + 1}$

63. $y = \frac{x^2 + 2x - 3}{x - 3}$

64. $f(x) = \frac{-2x^3}{x^2 + 1}$

65. $g(x) = \frac{4x^2 - 16x + 16}{2x - 3}$

66. A company is planning to manufacture affordable graphing calculators. The fixed monthly cost will be $50,000 and it will cost $25 to produce each calculator.

a. Write the cost function, C, of producing x graphing calculators.

b. Write the average cost function, $\overline{C}$, of producing x graphing calculators.

c. Find and interpret $\overline{C}(50)$, $\overline{C}(100)$, $\overline{C}(1000)$, and $\overline{C}(100{,}000)$.

d. What is the horizontal asymptote for the graph of this function and what does it represent?

67. In Palo Alto, California, a government agency ordered computer-related companies to contribute to a monetary pool to clean up underground water supplies. (The companies had stored toxic chemicals in leaking underground containers.) The rational function

$$C(x) = \frac{200x}{100 - x}$$

models the cost, $C(x)$, in tens of thousands of dollars, for removing x percent of the contaminants.

a. Find and interpret $C(90) - C(50)$.

b. What is the equation for the vertical asymptote? What does this mean in terms of the variables given by the function?

Exercises 68–69 involve rational functions that model the given situations. In each case, find the horizontal asymptote as $x \to \infty$ *and then describe what this means in practical terms.*

68. $f(x) = \frac{150x + 120}{0.05x + 1}$; the number of bass, $f(x)$, after x months in a lake that was stocked with 120 bass

69. $P(x) = \frac{72{,}900}{100x^2 + 729}$; the percentage, $P(x)$, of people in the United States with x years of education who are unemployed

70. The function $p(x) = 1.96x + 3.14$ models the number of nonviolent prisoners, $p(x)$, in thousands, in New York State prisons x years after 1980. The function $q(x) = 3.04x + 21.79$ models the total number of prisoners, $q(x)$, in thousands, in New York State prisons x years after 1980.

a. Write a function that models the fraction of nonviolent prisoners in New York State prisons x years after 1980.

b. What is the equation of the horizontal asymptote associated with the function in part (a)? Describe what this means about the percentage, to the nearest tenth of a percent, of nonviolent prisoners in New York State prisons over time.

c. Use your equation in part (b) to explain why, in 1998, New York State implemented a strategy where more nonviolent offenders are granted parole and more violent offenders are denied parole.

3.6

In Exercises 71–76, solve each inequality and graph the solution set on a real number line.

71. $2x^2 + 5x - 3 < 0$

72. $2x^2 + 9x + 4 \geq 0$

73. $x^3 + 2x^2 > 3x$

74. $\dfrac{x - 6}{x + 2} > 0$

75. $\dfrac{(x + 1)(x - 2)}{x - 1} \geq 0$

76. $\dfrac{x + 3}{x - 4} \leq 5$

77. Use the position function

$$s(t) = -16t^2 + v_0 t + s_0$$

to solve this problem. A projectile is fired vertically upward from ground level with an initial velocity of 48 feet per second. During which time period will the projectile's height exceed 32 feet?

3.7

Solve the variation problems in Exercises 78–83.

78. An electric bill varies directly as the amount of electricity used. The bill for 1400 kilowatts of electricity is $98. What is the bill for 2200 kilowatts of electricity?

79. The distance that a body falls from rest is directly proportional to the square of the time of the fall. If skydivers fall 144 feet in 3 seconds, how far will they fall in 10 seconds?

80. The time it takes to drive a certain distance is inversely proportional to the rate of travel. If it takes 4 hours at 50 miles per hour to drive the distance, how long will it take at 40 miles per hour?

81. The loudness of a stereo speaker, measured in decibels, varies inversely as the square of your distance from the speaker. When you are 8 feet from the speaker, the loudness is 28 decibels. What is the loudness when you are 4 feet from the speaker?

82. The time required to assemble computers varies directly as the number of computers assembled and inversely as the number of workers. If 30 computers can be assembled by 6 workers in 10 hours, how long would it take 5 workers to assemble 40 computers?

83. The volume of a pyramid varies jointly as its height and the area of its base. A pyramid with a height of 15 feet and a base with an area of 35 square feet has a volume of 175 cubic feet. Find the volume of a pyramid with a height of 20 feet and a base with an area of 120 square feet.

Chapter 3 Test

In Exercises 1–2, use the vertex and intercepts to sketch the graph of each quadratic function. Give the equation for the parabola's axis of symmetry. Use the graph to determine the function's domain and range.

1. $f(x) = (x + 1)^2 + 4$

2. $f(x) = x^2 - 2x - 3$

3. Determine, without graphing, whether the quadratic function $f(x) = -2x^2 + 12x - 16$ has a minimum value or a maximum value. Then find

   a. the minimum or maximum value and where it occurs.

   b. the function's domain and its range.

4. The function $f(x) = -x^2 + 46x - 360$ models the daily profit, $f(x)$, in hundreds of dollars, for a company that manufactures x computers daily. How many computers should be manufactured each day to maximize profit? What is the maximum daily profit?

5. Among all pairs of numbers whose sum is 14, find a pair whose product is as large as possible. What is the maximum product?

6. Consider the function $f(x) = x^3 - 5x^2 - 4x + 20$.

   a. Use factoring to find all zeros of f.

   b. Use the Leading Coefficient Test and the zeros of f to graph the function.

7. Use end behavior to explain why the graph at the top of the next column cannot be the graph of $f(x) = x^5 - x$. Then use intercepts to explain why the graph cannot represent $f(x) = x^5 - x$.

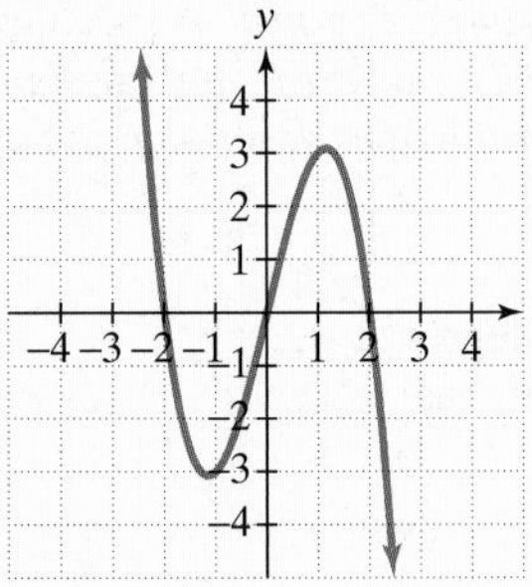

8. The graph of $f(x) = 6x^3 - 19x^2 + 16x - 4$ is shown in the figure.

   a. Based on the graph of f, find the root of the equation $6x^3 - 19x^2 + 16x - 4 = 0$ that is an integer.

   b. Use synthetic division to find the other two roots of $6x^3 - 19x^2 + 16x - 4 = 0$.

$f(x) = 6x^3 - 19x^2 + 16x - 4$

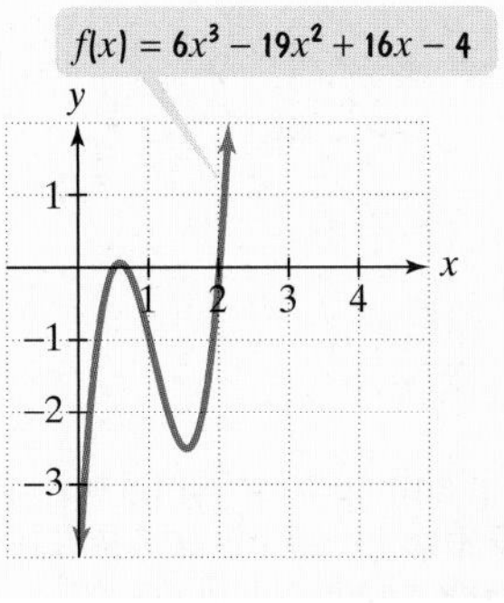

9. Use the Rational Zero Theorem to list all possible rational zeros of $f(x) = 2x^3 + 11x^2 - 7x - 6$.

10. Use Descartes's Rule of Signs to determine the possible number of positive and negative real zeros of
$$f(x) = 3x^5 - 2x^4 - 2x^2 + x - 1.$$

11. Solve: $x^3 + 9x^2 + 16x - 6 = 0$.

12. Consider the function whose equation is given by $f(x) = 2x^4 - x^3 - 13x^2 + 5x + 15$.

    a. List all possible rational zeros.

    b. Use the graph of f in the figure shown and synthetic division to find all zeros of the function.

$f(x) = 2x^4 - x^3 - 13x^2 + 5x + 15$

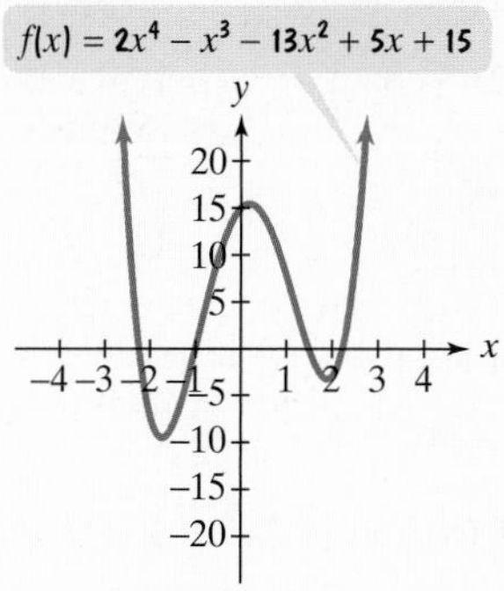

13. Use the graph of $f(x) = x^3 + 3x^2 - 4$ in the figure shown to factor $x^3 + 3x^2 - 4$.

$f(x) = x^3 + 3x^2 - 4$

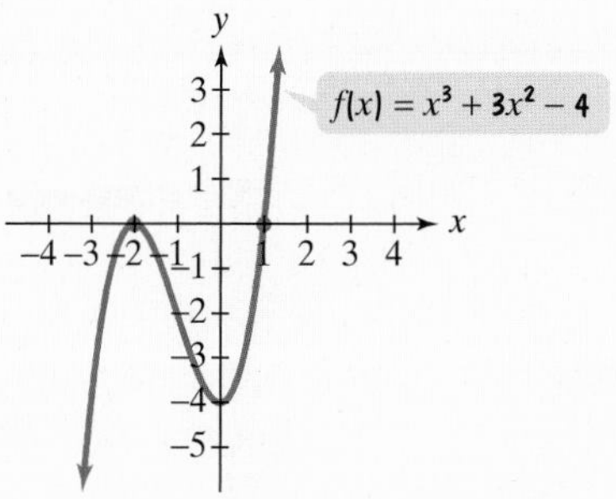

14. Find a fourth-degree polynomial function $f(x)$ with real coefficients that has -1, 1, and i as zeros and such that $f(3) = 160$.

15. The figure shows an incomplete graph of $f(x) = -3x^3 - 4x^2 + x + 2$. Find all the zeros of the function. Then draw a complete graph.

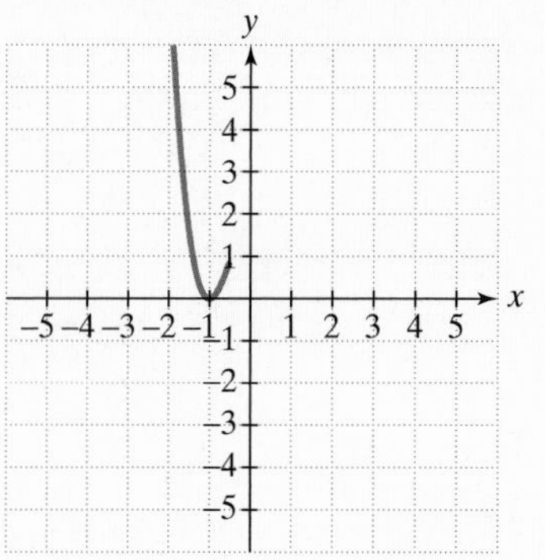

In Exercises 16–21, find the domain of each rational function and graph the function.

16. $f(x) = \dfrac{1}{(x+3)^2}$

17. $f(x) = \dfrac{1}{x-1} + 2$

18. $f(x) = \dfrac{x}{x^2 - 16}$

19. $f(x) = \dfrac{x^2 - 9}{x - 2}$

20. $f(x) = \dfrac{x+1}{x^2 + 2x - 3}$

21. $f(x) = \dfrac{4x^2}{x^2 + 3}$

22. A company is planning to manufacture pocket-sized televisions. The fixed monthly cost will be $300,000 and it will cost $10 to produce each television.

    a. Write the average cost function, $\overline{C}$, of producing x televisions.

    b. What is the horizontal asymptote for the graph of this function and what does it represent?

23. Rational functions can be used to model learning. Many of these functions model the proportion of correct responses as a function of the number of trials of a particular task. One such model, called a learning curve, is
$$f(x) = \frac{0.9x - 0.4}{0.9x + 0.1},$$
where $f(x)$ is the proportion of correct responses after x trials. If $f(x) = 0$, there are no correct responses. If $f(x) = 1$, all responses are correct. The graph of the rational function is shown.

A Learning Curve

$f(x) = \dfrac{0.9x - 0.4}{0.9x + 0.1}$

    a. According to the graph, what proportion of responses are correct after 5 learning trials?

    b. According to the graph, how many learning trials are necessary for 0.95 of the responses to be correct?

    c. Use the function's equation to write the equation of the horizontal asymptote. What does this mean in terms of the variables modeled by the learning curve?

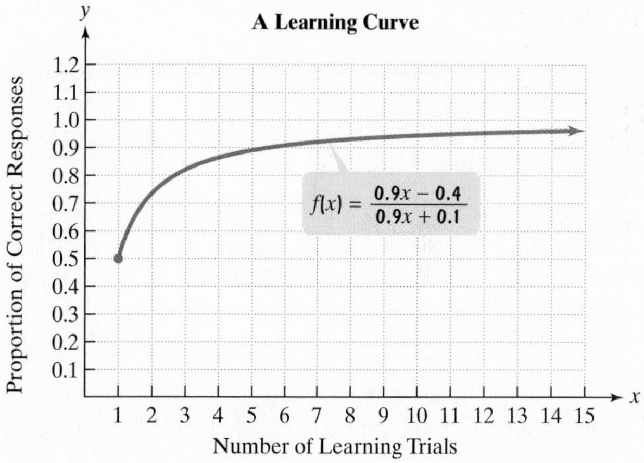

Solve each inequality in Exercises 24–25 and graph the solution set on a real number line. Express each solution set in interval notation.

24. $x^2 < x + 12$

25. $\dfrac{2x + 1}{x - 3} \le 3$

26. The intensity of light received at a source varies inversely as the square of the distance from the source. A particular light has an intensity of 20 foot-candles at 15 feet. What is the light's intensity at 10 feet?

Cumulative Review Exercises (Chapters 1–3)

Use the graph of $y = f(x)$ to solve Exercises 1–6.

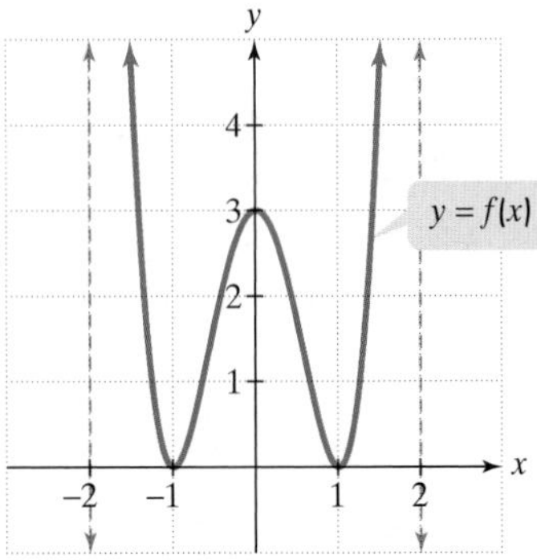

1. Find the domain and the range of f.

2. Find the zeros and the least possible multiplicity of each zero.

3. Where does the relative maximum occur?

4. Find $(f \circ f)(-1)$.

5. Use arrow notation to complete this statement: $f(x) \to \infty$ as _____ or as _____.

6. Graph $g(x) = f(x + 2) + 1$.

In Exercises 7–12, solve each equation or inequality.

7. $|2x - 1| = 3$

8. $3x^2 - 5x + 1 = 0$

9. $9 + \frac{3}{x} = \frac{2}{x^2}$

10. $x^3 + 2x^2 - 5x - 6 = 0$

11. $|2x - 5| > 3$

12. $3x^2 > 2x + 5$

In Exercises 13–18, graph each equation in a rectangular coordinate system. If two functions are given, graph both in the same system.

13. $f(x) = x^3 - 4x^2 - x + 4$

14. $f(x) = x^2 + 2x - 8$

15. $f(x) = x^2(x - 3)$

16. $f(x) = \frac{x - 1}{x - 2}$

17. $f(x) = |x|$ and $g(x) = -|x| - 1$

18. $x^2 + y^2 - 2x + 4y - 4 = 0$

In Exercises 19–20, let $f(x) = 2x^2 - x - 1$ and $g(x) = 4x - 1$.

19. Find $(f \circ g)(x)$.

20. Find $\frac{f(x + h) - f(x)}{h}$.

CHAPTER 4

Exponential and Logarithmic Functions

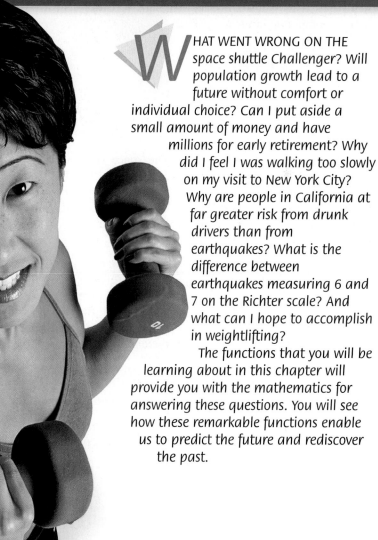

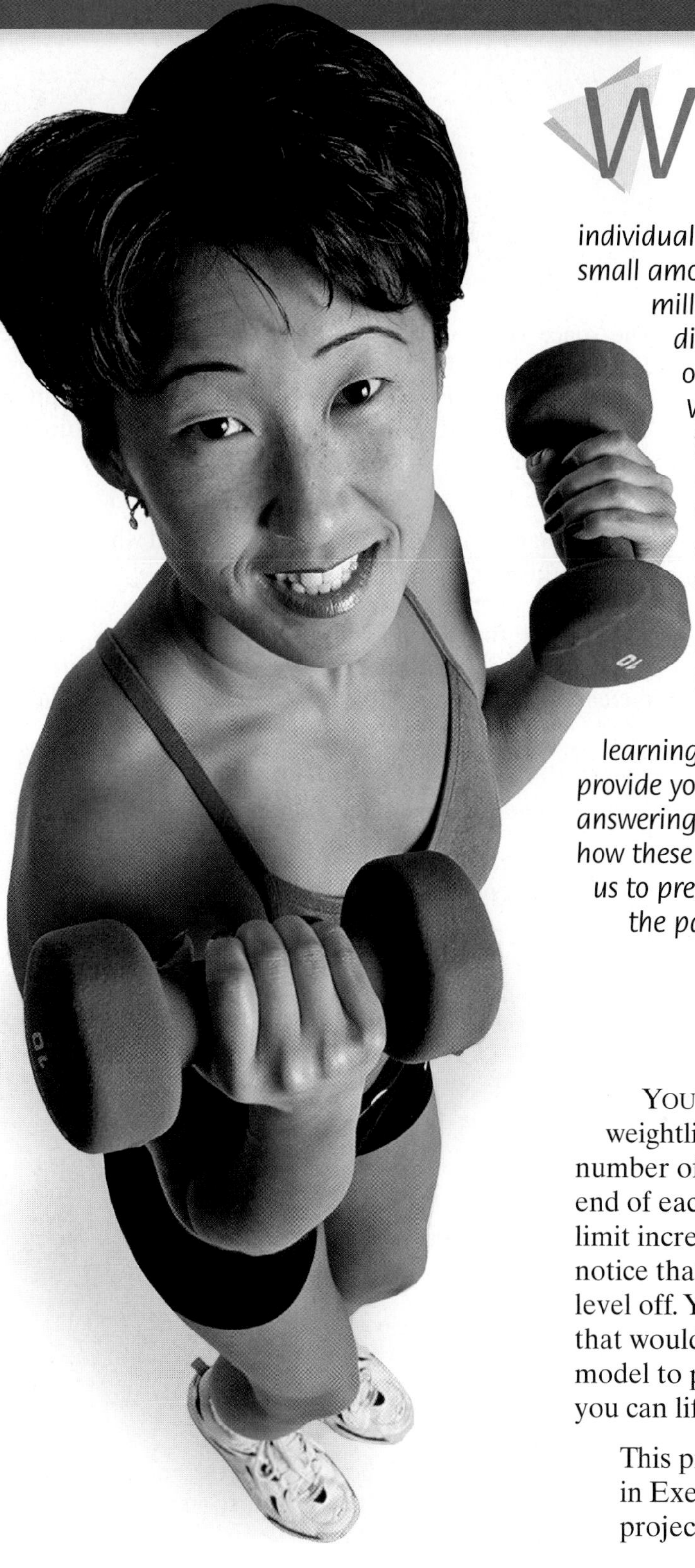

WHAT WENT WRONG ON THE *space shuttle Challenger? Will population growth lead to a future without comfort or individual choice? Can I put aside a small amount of money and have millions for early retirement? Why did I feel I was walking too slowly on my visit to New York City? Why are people in California at far greater risk from drunk drivers than from earthquakes? What is the difference between earthquakes measuring 6 and 7 on the Richter scale? And what can I hope to accomplish in weightlifting?*

The functions that you will be learning about in this chapter will provide you with the mathematics for answering these questions. You will see how these remarkable functions enable us to predict the future and rediscover the past.

YOU'VE RECENTLY TAKEN UP weightlifting, recording the maximum number of pounds you can lift at the end of each week. At first your weight limit increases rapidly, but now you notice that this growth is beginning to level off. You wonder about a function that would serve as a mathematical model to predict the number of pounds you can lift as you continue the sport.

This problem appears as Exercise 47 in Exercise Set 4.5 and as the group project (Exercise 60) on page 459.

SECTION 4.1   Exponential Functions

Objectives

① Evaluate exponential functions.

② Graph exponential functions.

③ Evaluate functions with base *e*.

④ Use compound interest formulas.

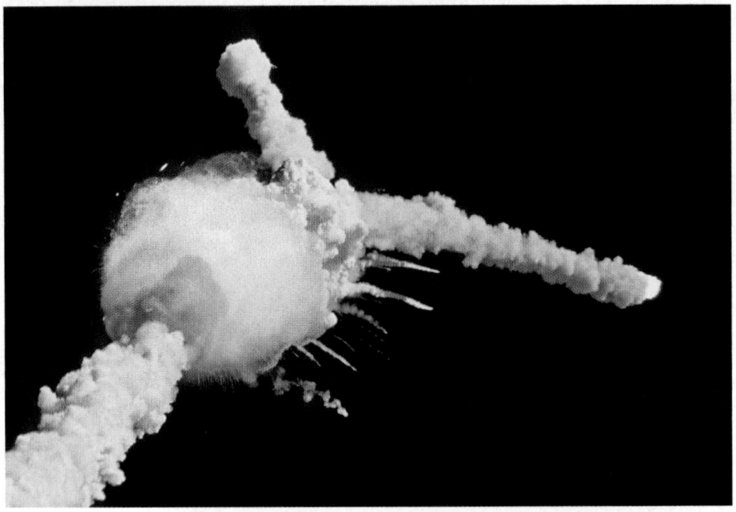

The space shuttle *Challenger* exploded approximately 73 seconds into flight on January 28, 1986. The tragedy involved damage to ○-rings, which were used to seal the connections between different sections of the shuttle engines. The number of ○-rings damaged increases dramatically as temperature falls.

The function

$$f(x) = 13.49(0.967)^x - 1$$

models the number of ○-rings expected to fail when the temperature is $x°$F. Can you see how this function is different from polynomial functions? The variable x is in the exponent. Functions whose equations contain a variable in the exponent are called **exponential functions**. Many real-life situations, including population growth, growth of epidemics, radioactive decay, and other changes that involve rapid increase or decrease, can be described using exponential functions.

Definition of the Exponential Function

The **exponential function** f **with base** b is defined by

$$f(x) = b^x \quad \text{or} \quad y = b^x,$$

where b is a positive constant other than 1 ($b > 0$ and $b \neq 1$) and x is any real number.

Here are some examples of exponential functions:

$$f(x) = 2^x \qquad g(x) = 10^x \qquad h(x) = 3^{x+1} \qquad j(x) = \left(\frac{1}{2}\right)^{x-1}.$$

| Base is 2. | Base is 10. | Base is 3. | Base is $\frac{1}{2}$. |

Each of these functions has a constant base and a variable exponent. By contrast, the following functions are not exponential functions:

$$F(x) = x^2 \qquad G(x) = 1^x \qquad H(x) = (-1)^x \qquad J(x) = x^x.$$

| Variable is the base and not the exponent. | The base of an exponential function must be a positive constant other than 1. | The base of an exponential function must be positive. | Variable is both the base and the exponent. |

Why is $G(x) = 1^x$ not classified as an exponential function? The number 1 raised to any power is 1. Thus, the function G can be written as $G(x) = 1$, which is a constant function.

Why is $H(x) = (-1)^x$ not an exponential function? The base of an exponential function must be positive to avoid having to exclude many values of x from the domain that result in nonreal numbers in the range:

$$H(x) = (-1)^x \qquad H\left(\frac{1}{2}\right) = (-1)^{\frac{1}{2}} = \sqrt{-1} = i.$$

Not an exponential function

All values of x resulting in even roots of negative numbers produce nonreal numbers.

1 Evaluate exponential functions.

You will need a calculator to evaluate exponential expressions. Most scientific calculators have a $\boxed{y^x}$ key. Graphing calculators have a $\boxed{\wedge}$ key. To evaluate expressions of the form b^x, enter the base b, press $\boxed{y^x}$ or $\boxed{\wedge}$, enter the exponent x, and finally press $\boxed{=}$ or $\boxed{\text{ENTER}}$.

EXAMPLE 1 Evaluating an Exponential Function

The exponential function $f(x) = 13.49(0.967)^x - 1$ describes the number of O-rings expected to fail, $f(x)$, when the temperature is x°F. On the morning the *Challenger* was launched, the temperature was 31°F, colder than any previous experience. Find the number of O-rings expected to fail at this temperature.

Solution Because the temperature was 31°F, substitute 31 for x and evaluate the function.

$$f(x) = 13.49(0.967)^x - 1 \quad \text{This is the given function.}$$
$$f(31) = 13.49(0.967)^{31} - 1 \quad \text{Substitute 31 for x.}$$

Use a scientific or graphing calculator to evaluate $f(31)$. Press the following keys on your calculator to do this:

Scientific calculator: 13.49 $\boxed{\times}$ .967 $\boxed{y^x}$ 31 $\boxed{-}$ 1 $\boxed{=}$

Graphing calculator: 13.49 $\boxed{\times}$ .967 $\boxed{\wedge}$ 31 $\boxed{-}$ 1 $\boxed{\text{ENTER}}$.

The display should be approximately 3.7668627.

$$f(31) = 13.49(0.967)^{31} - 1 \approx 3.8 \approx 4$$

Thus, four O-rings are expected to fail at a temperature of 31°F.

Check Point 1 Use the function in Example 1 to find the number of O-rings expected to fail at a temperature of 60°F. Round to the nearest whole number.

2 Graph exponential functions.

Graphing Exponential Functions

We are familiar with expressions involving b^x, where x is a rational number. For example,

$$b^{1.7} = b^{\frac{17}{10}} = \sqrt[10]{b^{17}} \quad \text{and} \quad b^{1.73} = b^{\frac{173}{100}} = \sqrt[100]{b^{173}}.$$

However, note that the definition of $f(x) = b^x$ includes all real numbers for the domain x. You may wonder what b^x means when x is an irrational number, such as $b^{\sqrt{3}}$ or b^{π}. Using closer and closer approximations for $\sqrt{3}$ ($\sqrt{3} \approx 1.73205$), we can think of $b^{\sqrt{3}}$ as the value that has the successively closer approximations

$$b^{1.7}, b^{1.73}, b^{1.732}, b^{1.73205}, \ldots.$$

In this way, we can graph exponential functions with no holes, or points of discontinuity, at the irrational domain values.

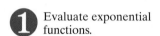

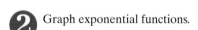

EXAMPLE 2 Graphing an Exponential Function

Graph: $f(x) = 2^x$.

Solution We begin by setting up a table of coordinates.

x	$f(x) = 2^x$
-3	$f(-3) = 2^{-3} = \frac{1}{8}$
-2	$f(-2) = 2^{-2} = \frac{1}{4}$
-1	$f(-1) = 2^{-1} = \frac{1}{2}$
0	$f(0) = 2^0 = 1$
1	$f(1) = 2^1 = 2$
2	$f(2) = 2^2 = 4$
3	$f(3) = 2^3 = 8$

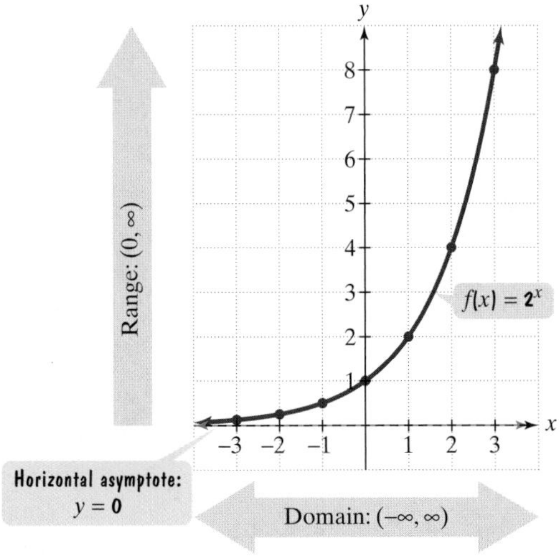

Figure 4.1 The graph of $f(x) = 2^x$

We plot these points, connecting them with a continuous curve. Figure 4.1 shows the graph of $f(x) = 2^x$. Observe that the graph approaches, but never touches, the negative portion of the x-axis. Thus, the x-axis, or $y = 0$, is a horizontal asymptote. The range is the set of all positive real numbers. Although we used integers for x in our table of coordinates, you can use a calculator to find additional points. For example, $f(0.3) = 2^{0.3} \approx 1.231$ and $f(0.95) = 2^{0.95} \approx 1.932$. The points $(0.3, 1.231)$ and $(0.95, 1.932)$ approximately fit the graph.

Check Point 2 Graph: $f(x) = 3^x$.

EXAMPLE 3 Graphing an Exponential Function

Graph: $g(x) = \left(\frac{1}{2}\right)^x$.

Solution We begin by setting up a table of coordinates. We compute the function values by noting that

$$g(x) = \left(\frac{1}{2}\right)^x = (2^{-1})^x = 2^{-x}.$$

x	$g(x) = \left(\frac{1}{2}\right)^x$ or 2^{-x}
-3	$g(-3) = 2^{-(-3)} = 2^3 = 8$
-2	$g(-2) = 2^{-(-2)} = 2^2 = 4$
-1	$g(-1) = 2^{-(-1)} = 2^1 = 2$
0	$g(0) = 2^{-0} = 1$
1	$g(1) = 2^{-1} = \frac{1}{2^1} = \frac{1}{2}$
2	$g(2) = 2^{-2} = \frac{1}{2^2} = \frac{1}{4}$
3	$g(3) = 2^{-3} = \frac{1}{2^3} = \frac{1}{8}$

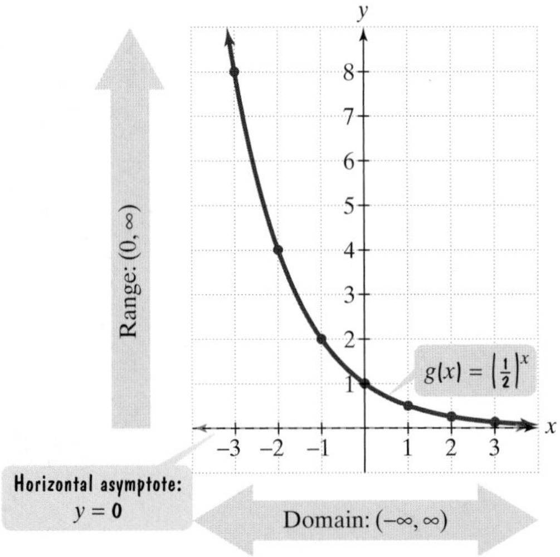

Figure 4.2 The graph of $g(x) = \left(\frac{1}{2}\right)^x$

We plot these points, connecting them with a continuous curve. Figure 4.2 shows the graph of $g(x) = \left(\frac{1}{2}\right)^x$. This time the graph approaches, but never touches, the *positive* portion of the x-axis. Once again, the x-axis, or $y = 0$, is a horizontal asymptote. The range consists of all positive real numbers.

Do you notice a relationship between the graphs of $f(x) = 2^x$ and $g(x) = \left(\frac{1}{2}\right)^x$ in Figures 4.1 and 4.2? The graph of $g(x) = \left(\frac{1}{2}\right)^x$ is the graph of $f(x) = 2^x$ reflected about the y-axis:

$$g(x) = \left(\frac{1}{2}\right)^x = 2^{-x} = f(-x)$$

Recall that the graph of $y = f(-x)$ is the graph of $y = f(x)$ reflected about the y-axis.

Check Point 3 Graph: $f(x) = \left(\frac{1}{3}\right)^x$. Note that $f(x) = \left(\frac{1}{3}\right)^x = (3^{-1})^x = 3^{-x}$.

Four exponential functions have been graphed in Figure 4.3. Compare the black and green graphs, where $b > 1$, to those in blue and red, where $b < 1$. When $b > 1$, the value of y increases as the value of x increases. When $b < 1$, the value of y decreases as the value of x increases. Notice that all four graphs pass through $(0, 1)$.

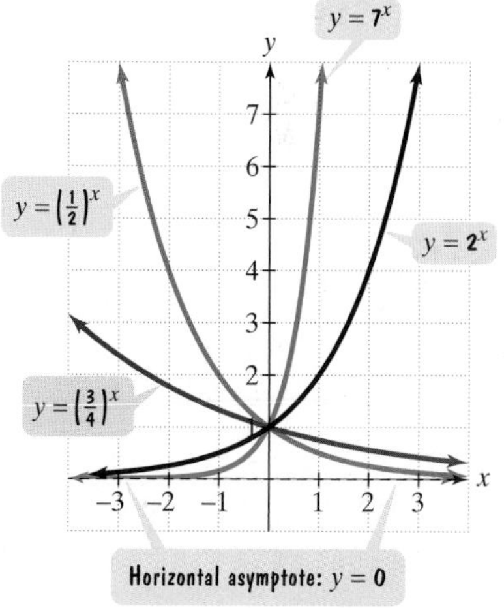

Figure 4.3 Graphs of four exponential functions

These graphs illustrate the following general characteristics of exponential functions:

Characteristics of Exponential Functions of the Form $f(x) = b^x$

1. The domain of $f(x) = b^x$ consists of all real numbers: $(-\infty, \infty)$. The range of $f(x) = b^x$ consists of all positive real numbers: $(0, \infty)$.
2. The graphs of all exponential functions of the form $f(x) = b^x$ pass through the point $(0, 1)$ because $f(0) = b^0 = 1 (b \neq 0)$. The y-intercept is 1.
3. If $b > 1$, $f(x) = b^x$ has a graph that goes up to the right and is an increasing function. The greater the value of b, the steeper the increase.
4. If $0 < b < 1$, $f(x) = b^x$ has a graph that goes down to the right and is a decreasing function. The smaller the value of b, the steeper the decrease.
5. $f(x) = b^x$ is one-to-one and has an inverse that is a function.
6. The graph of $f(x) = b^x$ approaches, but does not touch, the x-axis. The x-axis, or $y = 0$, is a horizontal asymptote.

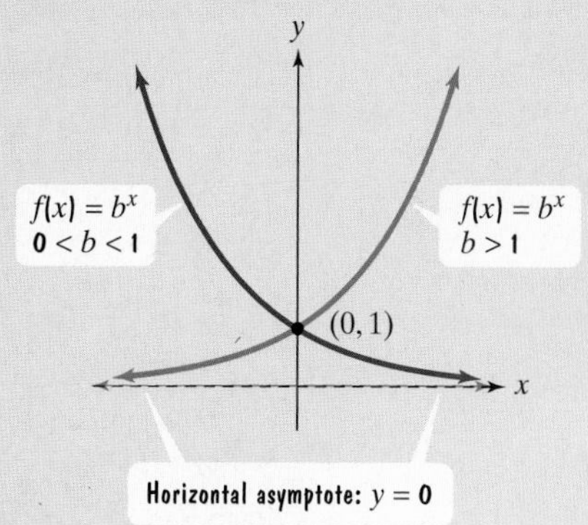

Transformations of Exponential Functions

The graphs of exponential functions can be translated vertically or horizontally, reflected, stretched, or shrunk. These transformations are summarized in Table 4.1.

Table 4.1 Transformations Involving Exponential Functions

In each case, c represents a positive real number.

Transformation	Equation	Description
Vertical translation	$g(x) = b^x + c$ $g(x) = b^x - c$	• Shifts the graph of $f(x) = b^x$ upward c units. • Shifts the graph of $f(x) = b^x$ downward c units.
Horizontal translation	$g(x) = b^{x+c}$ $g(x) = b^{x-c}$	• Shifts the graph of $f(x) = b^x$ to the left c units. • Shifts the graph of $f(x) = b^x$ to the right c units.
Reflection	$g(x) = -b^x$ $g(x) = b^{-x}$	• Reflects the graph of $f(x) = b^x$ about the x-axis. • Reflects the graph of $f(x) = b^x$ about the y-axis.
Vertical stretching or shrinking	$g(x) = cb^x$	• Vertically stretches the graph of $f(x) = b^x$ if $c > 1$. • Vertically shrinks the graph of $f(x) = b^x$ if $0 < c < 1$.
Horizontal stretching or shrinking	$g(x) = b^{cx}$	• Horizontally shrinks the graph of $f(x) = b^x$ if $c > 1$ • Horizontally stretches the graph of $f(x) = b^x$ if $0 < c < 1$.

EXAMPLE 4 Transformations Involving Exponential Functions

Use the graph of $f(x) = 3^x$ to obtain the graph of $g(x) = 3^{x+1}$.

Solution The graph of $g(x) = 3^{x+1}$ is the graph of $f(x) = 3^x$ shifted 1 unit to the left.

Begin with a table showing some of the coordinates for f.

x	$f(x) = 3^x$
−2	$f(-2) = 3^{-2} = \frac{1}{9}$
−1	$f(-1) = 3^{-1} = \frac{1}{3}$
0	$f(0) = 3^0 = 1$
1	$f(1) = 3^1 = 3$
2	$f(2) = 3^2 = 9$

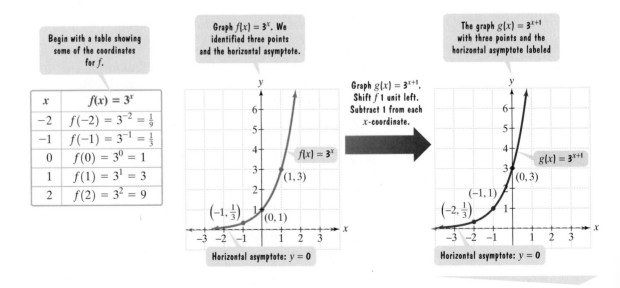

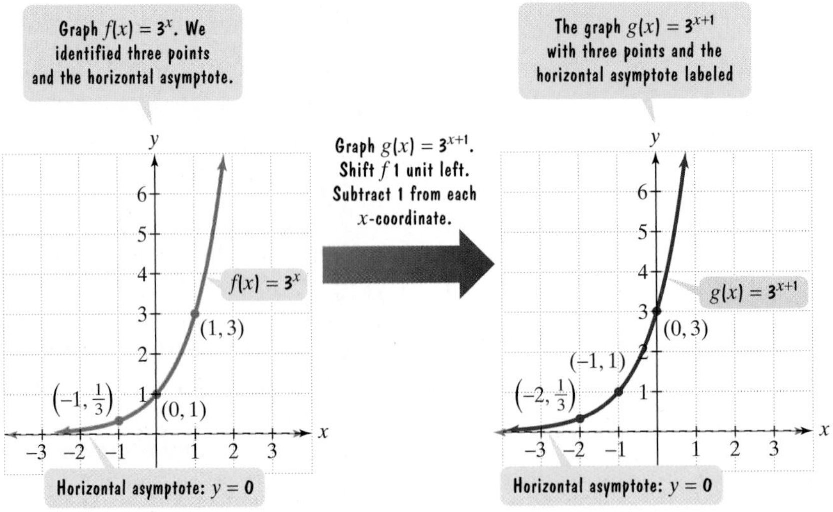
Graph f(x) = 3^x. We identified three points and the horizontal asymptote.
f(x) = 3^x
(−1, 1/3)
(0, 1)
(1, 3)
Horizontal asymptote: y = 0
Graph g(x) = 3^(x+1). Shift f 1 unit left. Subtract 1 from each x-coordinate.
The graph g(x) = 3^(x+1) with three points and the horizontal asymptote labeled
g(x) = 3^(x+1)
(−2, 1/3)
(−1, 1)
(0, 3)
Horizontal asymptote: y = 0

Check Point 4 Use the graph of $f(x) = 3^x$ to obtain the graph of $g(x) = 3^{x-1}$.

If an exponential function is translated upward or downward, the horizontal asymptote is shifted by the amount of the vertical shift.

EXAMPLE 5 Transformations Involving Exponential Functions

Use the graph of $f(x) = 2^x$ to obtain the graph of $g(x) = 2^x - 3$.

Solution The graph of $g(x) = 2^x - 3$ is the graph of $f(x) = 2^x$ shifted down 3 units.

Begin with a table showing some of the coordinates for *f*.

x	$f(x) = 2^x$
-2	$f(-2) = 2^{-2} = \frac{1}{4}$
-1	$f(-1) = 2^{-1} = \frac{1}{2}$
0	$f(0) = 2^0 = 1$
1	$f(1) = 2^1 = 2$
2	$f(2) = 2^2 = 4$

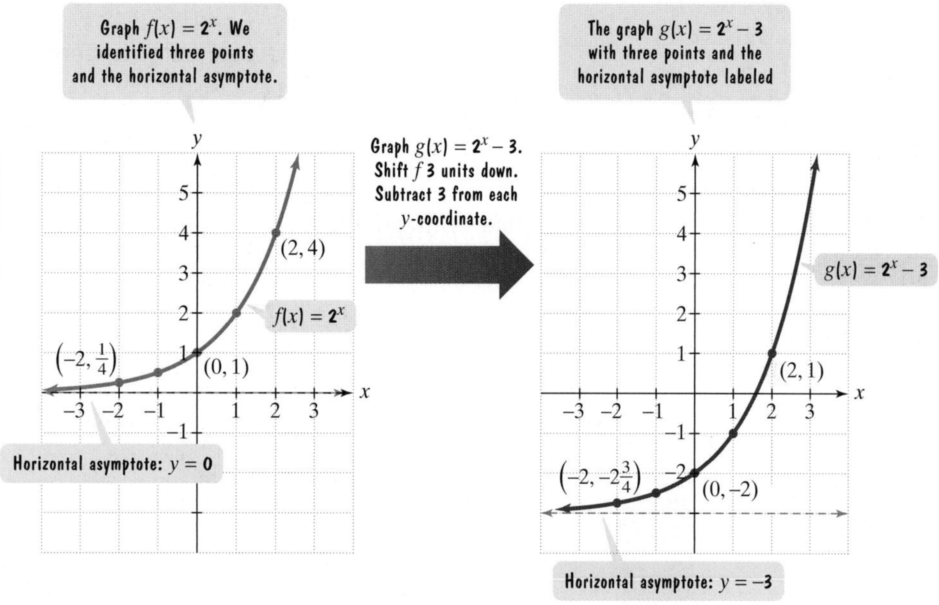

Graph f(x) = 2^x. We identified three points and the horizontal asymptote.
(2, 4)
f(x) = 2^x
(−2, 1/4)
(0, 1)
Horizontal asymptote: y = 0
Graph g(x) = 2^x − 3. Shift f 3 units down. Subtract 3 from each y-coordinate.
The graph g(x) = 2^x − 3 with three points and the horizontal asymptote labeled
g(x) = 2^x − 3
(2, 1)
(−2, −2 3/4)
(0, −2)
Horizontal asymptote: y = −3

Check Point 5 Use the graph of $f(x) = 2^x$ to obtain the graph of $g(x) = 2^x + 1$.

3 Evaluate functions with base *e*.

The Natural Base *e*

An irrational number, symbolized by the letter *e*, appears as the base in many applied exponential functions. The number *e* is defined as the value that $\left(1 + \frac{1}{n}\right)^n$ approaches as *n* gets larger and larger. Table 4.2 shows the values of $\left(1 + \frac{1}{n}\right)^n$ for increasingly large values of *n*. As $n \to \infty$, the approximate value of *e* to nine decimal places is

$$e \approx 2.718281827.$$

The irrational number *e*, approximately 2.72, is called the **natural base**. The function $f(x) = e^x$ is called the **natural exponential function**.

Use a scientific or graphing calculator with an e^x key to evaluate *e* to various powers. For example, to find e^2, press the following keys on most calculators:

Scientific calculator: 2 e^x

Graphing calculator: e^x 2 ENTER.

The display should be approximately 7.389.

$$e^2 \approx 7.389$$

Table 4.2

n	$\left(1 + \frac{1}{n}\right)^n$
1	2
2	2.25
5	2.48832
10	2.59374246
100	2.704813829
1000	2.716923932
10,000	2.718145927
100,000	2.718268237
1,000,000	2.718280469
1,000,000,000	2.718281827

As $n \to \infty$, $\left(1 + \frac{1}{n}\right)^n \to e$.

Technology

As $n \to \infty$, the graph of $y = \left(1 + \frac{1}{n}\right)^n$ approaches the graph of $y = e$.

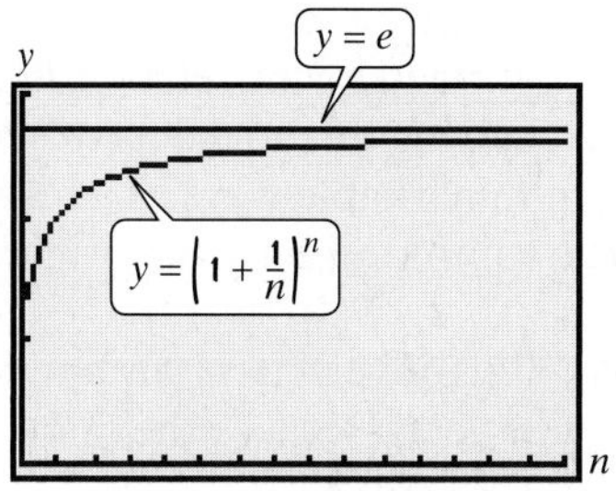

y = e
y = (1 + 1/n)^n

[0, 15, 1] by [0, 3, 1]

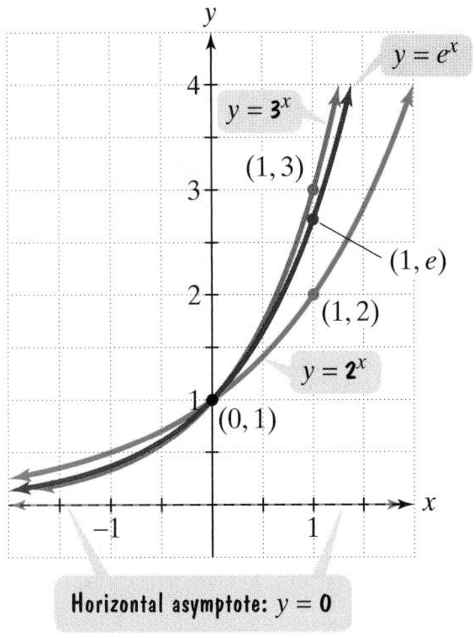

Figure 4.4 Graphs of three exponential functions

The number e lies between 2 and 3. Because $2^2 = 4$ and $3^2 = 9$, it makes sense that e^2, approximately 7.389, lies between 4 and 9.

Because $2 < e < 3$, the graph of $y = e^x$ is between the graphs of $y = 2^x$ and $y = 3^x$, shown in Figure 4.4.

EXAMPLE 6 World Population

In a report entitled *Resources and Man*, the U.S. National Academy of Sciences concluded that a world population of 10 billion "is close to (if not above) the maximum that an intensely managed world might hope to support with some degree of comfort and individual choice." At the time the report was issued in 1969, world population was approximately 3.6 billion, with a growth rate of 2% per year. The function

$$f(x) = 3.6e^{0.02x}$$

describes world population, $f(x)$, in billions, x years after 1969. Use the function to find world population in the year 2020. Is there cause for alarm?

Solution Because 2020 is 51 years after 1969, we substitute 51 for x in $f(x) = 3.6e^{0.02x}$:

$$f(51) = 3.6e^{0.02(51)}.$$

Perform this computation on your calculator.

Scientific calculator: 3.6 [×] [(] .02 [×] 51 [)] [e^x] [=]

Graphing calculator: 3.6 [×] [e^x] [(] .02 [×] 51 [)] [ENTER]

The display should be approximately 9.9835012. Thus,

$$f(51) = 3.6e^{0.02(51)} \approx 9.98.$$

This indicates that world population in the year 2020 will be approximately 9.98 billion. Because this number is quite close to 10 billion, the given function suggests that there may be cause for alarm.

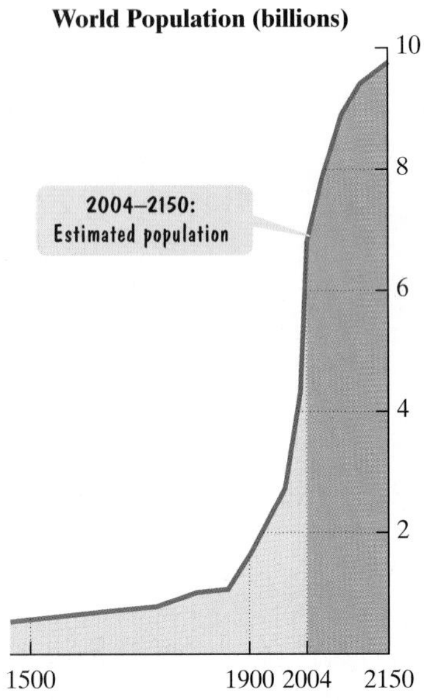

Source: U.N. Population Division

World population in 2004 was approximately 6.4 billion, but the growth rate was no longer 2%. It had slowed down to 1.23%. Using this current growth rate, exponential functions now predict a world population of 7.8 billion in the year 2020. Experts think the population may stabilize at 10 billion after 2200 if the growth rate continues to decline.

Check Point 6 The function $f(x) = 6.4e^{0.0123x}$ describes world population, $f(x)$, in billions, x years after 2004 subject to a growth rate of 1.23% annually. Use the function to predict world population in 2050.

4 Use compound interest formulas.

Compound Interest

We all want a wonderful life with fulfilling work, good health, and loving relationships. And let's be honest: Financial security wouldn't hurt! Achieving this goal depends on understanding how money in savings accounts grows in remarkable ways as a result of *compound interest*. **Compound interest** is interest computed on your original investment as well as on any accumulated interest.

Suppose a sum of money, called the **principal**, P, is invested at an annual percentage rate r, in decimal form, compounded once per year. Because the interest is added to the principal at year's end, the accumulated value, A, is

$$A = P + Pr = P(1 + r).$$

The accumulated amount of money follows this pattern of multiplying the previous principal by $(1 + r)$ for each successive year, as indicated in Table 4.3.

Table 4.3

Time in Years	Accumulated Value after Each Compounding
0	$A = P$
1	$A = P(1 + r)$
2	$A = P(1 + r)(1 + r) = P(1 + r)^2$
3	$A = P(1 + r)^2(1 + r) = P(1 + r)^3$
4	$A = P(1 + r)^3(1 + r) = P(1 + r)^4$
⋮	⋮
t	$A = P(1 + r)^t$

This formula gives the balance, A, that a principal, P, is worth after t years at interest rate r, compounded once a year.

Most savings institutions have plans in which interest is paid more than once a year. If compound interest is paid twice a year, the compounding period is six months. We say that the interest is **compounded semiannually**. When compound interest is paid four times a year, the compounding period is three months and the interest is said to be **compounded quarterly**. Some plans allow for monthly compounding or daily compounding.

In general, when compound interest is paid n times a year, we say that there are ***n* compounding periods per year**. The formula $A = P(1 + r)^t$ can be adjusted to take into account the number of compounding periods in a year. If there are n compounding periods per year, in each time period the interest rate is $i = \frac{r}{n}$ and there are nt time periods in t years. This results in the following formula for the balance, A, after t years:

$$A = P\left(1 + \frac{r}{n}\right)^{nt}.$$

Some banks use **continuous compounding**, where the number of compounding periods increases infinitely (compounding interest every trillionth of a second, every quadrillionth of a second, etc.). Let's see what happens to the balance, A, as $n \to \infty$.

$\frac{n}{r} \cdot rt = nt$

$$A = P\left(1 + \frac{r}{n}\right)^{nt} = P\left[\left(1 + \frac{1}{\frac{n}{r}}\right)^{\frac{n}{r}}\right]^{rt} = P\left[\left(1 + \frac{1}{h}\right)^{h}\right]^{rt} = Pe^{rt}$$

Let $h = \frac{n}{r}$. As $n \to \infty$, $h \to \infty$.

As $h \to \infty$, by definition $\left(1 + \frac{1}{h}\right)^h \to e$.

We see that the formula for continuous compounding is $A = Pe^{rt}$. Although continuous compounding sounds terrific, it yields only a fraction of a percent more interest over a year than daily compounding.

Formulas for Compound Interest

After t years, the balance, A, in an account with principal P and annual interest rate r (in decimal form) is given by the following formulas:

1. For n compoundings per year: $A = P\left(1 + \frac{r}{n}\right)^{nt}$
2. For continuous compounding: $A = Pe^{rt}$.

EXAMPLE 7 Choosing between Investments

You decide to invest \$8000 for 6 years and you have a choice between two accounts. The first pays 7% per year, compounded monthly. The second pays 6.85% per year, compounded continuously. Which is the better investment?

Solution The better investment is the one with the greater balance in the account after 6 years. Let's begin with the account with monthly compounding. We use the compound interest model with $P = 8000$, $r = 7\% = 0.07$, $n = 12$ (monthly compounding means 12 compoundings per year), and $t = 6$.

$$A = P\left(1 + \frac{r}{n}\right)^{nt} = 8000\left(1 + \frac{0.07}{12}\right)^{12 \cdot 6} \approx 12{,}160.84$$

The balance in this account after 6 years is \$12,160.84.

For the second investment option, we use the model for continuous compounding with $P = 8000$, $r = 6.85\% = 0.0685$, and $t = 6$.

$$A = Pe^{rt} = 8000e^{0.0685(6)} \approx 12{,}066.60$$

The balance in this account after 6 years is \$12,066.60, slightly less than the previous amount. Thus, the better investment is the 7% monthly compounding option.

Check Point 7 A sum of \$10,000 is invested at an annual rate of 8%. Find the balance in the account after 5 years subject to **a.** quarterly compounding and **b.** continuous compounding.

EXERCISE SET 4.1

Practice Exercises

In Exercises 1–10, approximate each number using a calculator. Round your answer to three decimal places.

1. $2^{3.4}$ **2.** $3^{2.4}$ **3.** $3^{\sqrt{5}}$ **4.** $5^{\sqrt{3}}$ **5.** $4^{-1.5}$

6. $6^{-1.2}$ **7.** $e^{2.3}$ **8.** $e^{3.4}$ **9.** $e^{-0.95}$ **10.** $e^{-0.75}$

In Exercises 11–18, graph each function by making a table of coordinates. If applicable, use a graphing utility to confirm your hand-drawn graph.

11. $f(x) = 4^x$ **12.** $f(x) = 5^x$

13. $g(x) = \left(\frac{3}{2}\right)^x$ **14.** $g(x) = \left(\frac{4}{3}\right)^x$

15. $h(x) = \left(\frac{1}{2}\right)^x$ **16.** $h(x) = \left(\frac{1}{3}\right)^x$

17. $f(x) = (0.6)^x$ **18.** $f(x) = (0.8)^x$

In Exercises 19–24, the graph of an exponential function is given. Select the function for each graph from the following options:

$$f(x) = 3^x,\ g(x) = 3^{x-1},\ h(x) = 3^x - 1,$$
$$F(x) = -3^x,\ G(x) = 3^{-x},\ H(x) = -3^{-x}.$$

19.

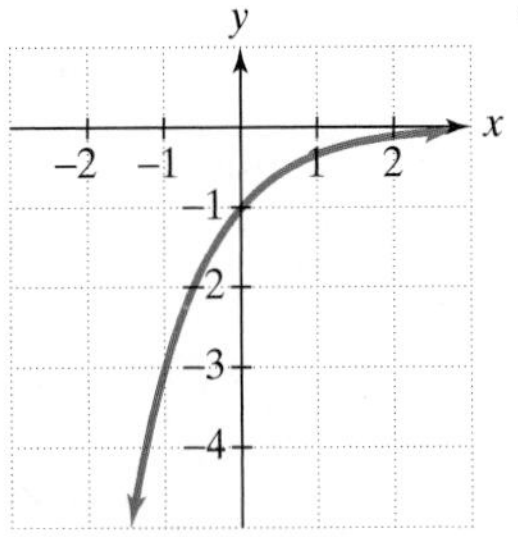

20.

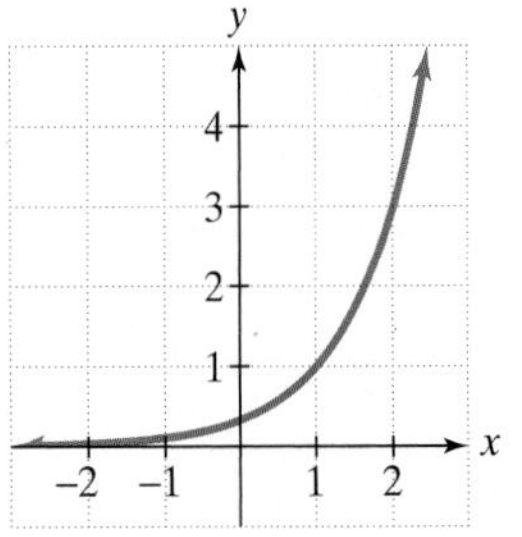

21.

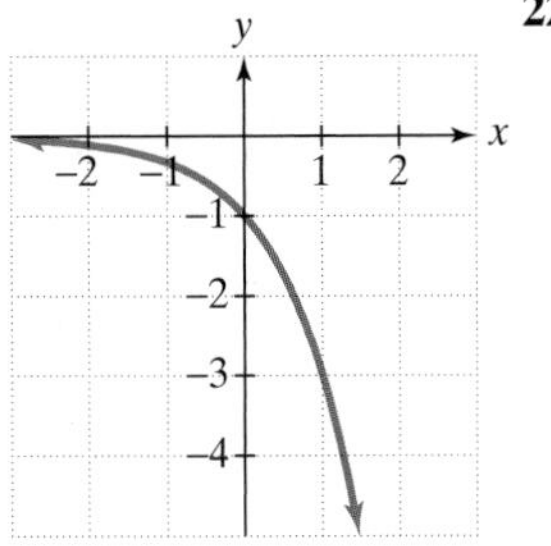

22.

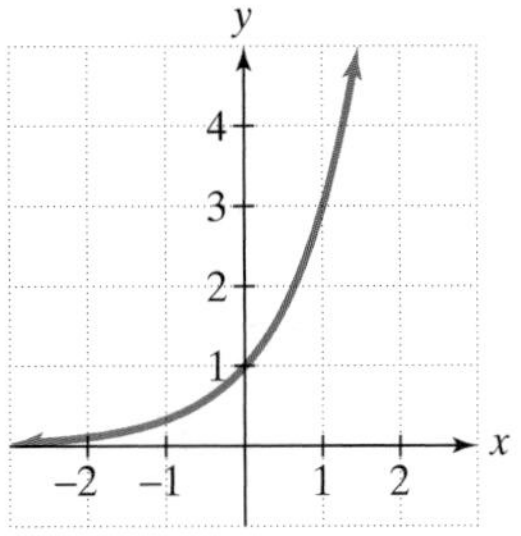

23.

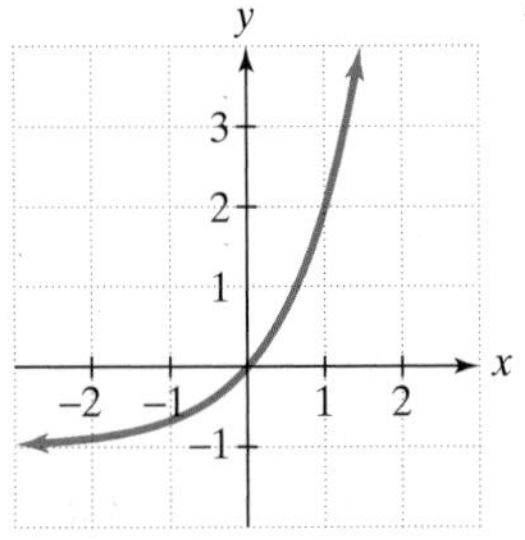

24.

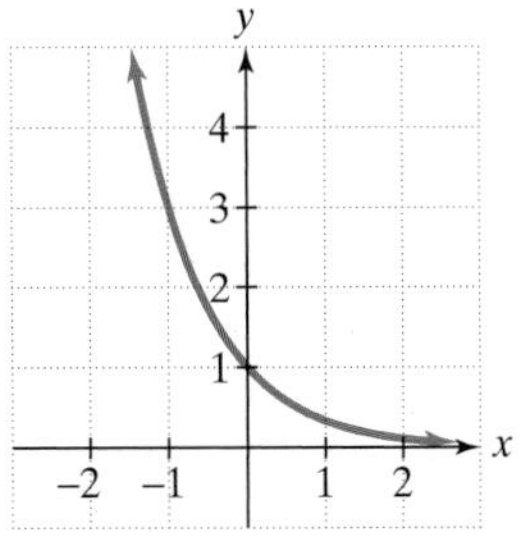

In Exercises 25–34, begin by graphing $f(x) = 2^x$. Then use transformations of this graph to graph the given function. Be sure to graph and give equations of the asymptotes. Use the graphs to determine each function's domain and range. If applicable, use a graphing utility to confirm your hand-drawn graphs.

25. $g(x) = 2^{x+1}$ **26.** $g(x) = 2^{x+2}$

27. $g(x) = 2^x - 1$ **28.** $g(x) = 2^x + 2$

29. $h(x) = 2^{x+1} - 1$ **30.** $h(x) = 2^{x+2} - 1$

31. $g(x) = -2^x$

32. $g(x) = 2^{-x}$

33. $g(x) = 2 \cdot 2^x$

34. $g(x) = \frac{1}{2} \cdot 2^x$

The figure shows the graph of $f(x) = e^x$. In Exercises 35–46, use transformations of this graph to graph each function. Be sure to give equations of the asymptotes. Use the graphs to determine each function's domain and range. If applicable, use a graphing utility to confirm your hand-drawn graphs.

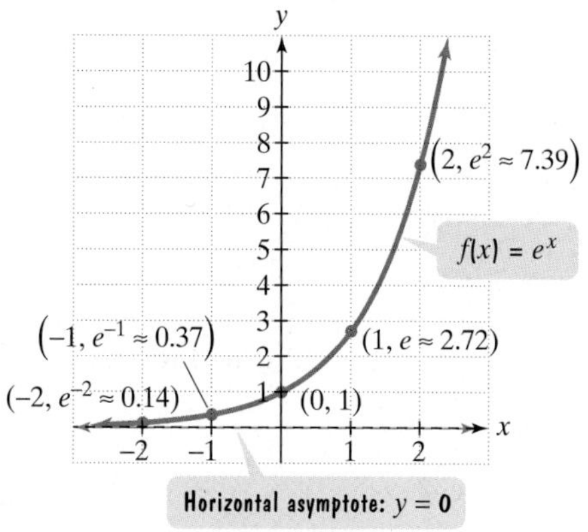

35. $g(x) = e^{x-1}$

36. $g(x) = e^{x+1}$

37. $g(x) = e^x + 2$

38. $g(x) = e^x - 1$

39. $h(x) = e^{x-1} + 2$

40. $h(x) = e^{x+1} - 1$

41. $h(x) = e^{-x}$

42. $h(x) = -e^x$

43. $g(x) = 2e^x$

44. $g(x) = \frac{1}{2}e^x$

45. $h(x) = e^{2x} + 1$

46. $h(x) = e^{\frac{x}{2}} + 2$

In Exercises 47–52, graph functions f and g in the same rectangular coordinate system. Graph and give equations of all asymptotes. If applicable, use a graphing utility to confirm your hand-drawn graphs.

47. $f(x) = 3^x$ and $g(x) = 3^{-x}$

48. $f(x) = 3^x$ and $g(x) = -3^x$

49. $f(x) = 3^x$ and $g(x) = \frac{1}{3} \cdot 3^x$

50. $f(x) = 3^x$ and $g(x) = 3 \cdot 3^x$

51. $f(x) = \left(\frac{1}{2}\right)^x$ and $g(x) = \left(\frac{1}{2}\right)^{x-1} + 1$

52. $f(x) = \left(\frac{1}{2}\right)^x$ and $g(x) = \left(\frac{1}{2}\right)^{x-1} + 2$

Use the compound interest formulas $A = P\left(1 + \frac{r}{n}\right)^{nt}$ and $A = Pe^{rt}$ to solve Exercises 53–56. Round answers to the nearest cent.

53. Find the accumulated value of an investment of \$10,000 for 5 years at an interest rate of 5.5% if the money is **a.** compounded semiannually; **b.** compounded quarterly; **c.** compounded monthly; **d.** compounded continuously.

54. Find the accumulated value of an investment of \$5000 for 10 years at an interest rate of 6.5% if the money is **a.** compounded semiannually; **b.** compounded quarterly; **c.** compounded monthly; **d.** compounded continuously.

55. Suppose that you have \$12,000 to invest. Which investment yields the greater return over 3 years: 7% compounded monthly or 6.85% compounded continuously?

56. Suppose that you have \$6000 to invest. Which investment yields the greater return over 4 years: 8.25% compounded quarterly or 8.3% compounded semiannually?

Practice Plus

In Exercises 57–58, graph f and g in the same rectangular coordinate system. Then find the point of intersection of the two graphs.

57. $f(x) = 2^x$, $g(x) = 2^{-x}$

58. $f(x) = 2^{x+1}$, $g(x) = 2^{-x+1}$

59. Graph $y = 2^x$ and $x = 2^y$ in the same rectangular coordinate system.

60. Graph $y = 3^x$ and $x = 3^y$ in the same rectangular coordinate system.

In Exercises 61–64, give the equation of each exponential function whose graph is shown.

61.

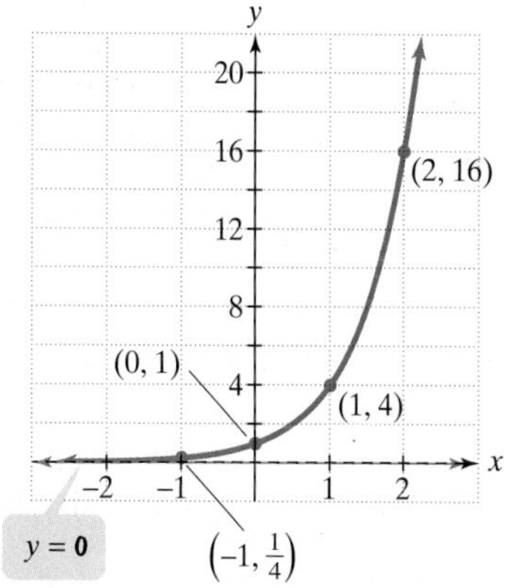

62.

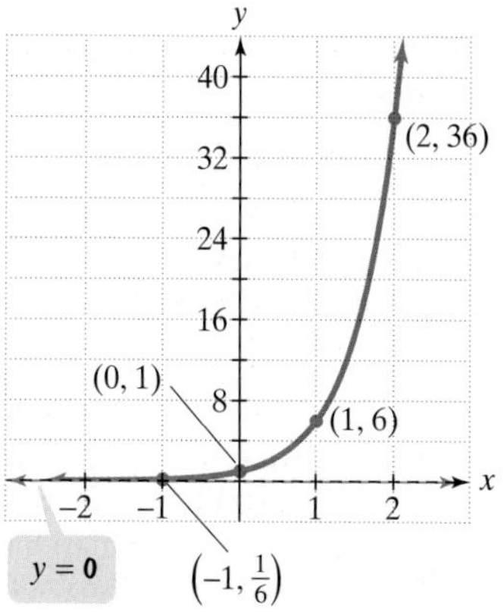

63.

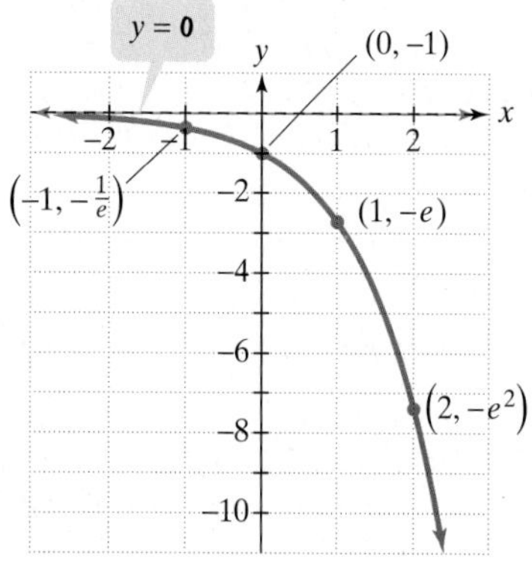

64.

70. A decimal approximation for π is 3.141593. Use a calculator to find $2^3, 2^{3.1}, 2^{3.14}, 2^{3.141}, 2^{3.1415}, 2^{3.14159}$, and $2^{3.141593}$. Now find 2^{π}. What do you observe?

Use a calculator with an $\boxed{e^x}$ key to solve Exercises 71–77.

The graph shows the number of words, in millions, in the U.S. federal tax code for selected years from 1955 through 2000. The data can be modeled by

$$f(x) = 0.16x + 1.43 \quad \text{and} \quad g(x) = 1.8e^{0.04x},$$

in which $f(x)$ and $g(x)$ represent the number of words, in millions, in the federal tax code x years after 1955. Use these functions to solve Exercises 71–72.

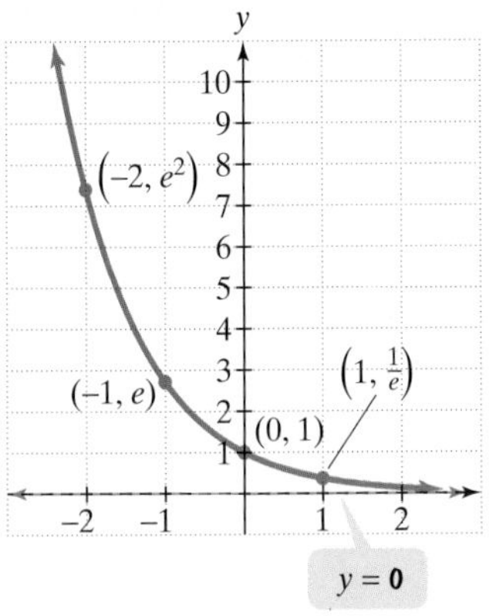

 Application Exercises

Use a calculator with a $\boxed{y^x}$ key or a $\boxed{\wedge}$ key to solve Exercises 65–70.

65. India is currently one of the world's fastest-growing countries. By 2040, the population of India will be larger than the population of China; by 2050, nearly one-third of the world's population will live in these two countries alone. The exponential function $f(x) = 574(1.026)^x$ models the population of India, $f(x)$, in millions, x years after 1974.

  a. Substitute 0 for x and, without using a calculator, find India's population in 1974.

  b. Substitute 27 for x and use your calculator to find India's population, to the nearest million, in the year 2001 as modeled by this function.

  c. Find India's population, to the nearest million, in the year 2028 as predicted by this function.

  d. Find India's population, to the nearest million, in the year 2055 as predicted by this function.

  e. What appears to be happening to India's population every 27 years?

66. The 1986 explosion at the Chernobyl nuclear power plant in the former Soviet Union sent about 1000 kilograms of radioactive cesium-137 into the atmosphere. The function $f(x) = 1000(0.5)^{\frac{x}{30}}$ describes the amount, $f(x)$, in kilograms, of cesium-137 remaining in Chernobyl x years after 1986. If even 100 kilograms of cesium-137 remain in Chernobyl's atmosphere, the area is considered unsafe for human habitation. Find $f(80)$ and determine if Chernobyl will be safe for human habitation by 2066.

The formula $S = C(1 + r)^t$ models inflation, where C = the value today, r = the annual inflation rate, and S = the inflated value t years from now. Use this formula to solve Exercises 67–68. Round answers to the nearest dollar.

67. If the inflation rate is 6%, how much will a house now worth $465,000 be worth in 10 years?

68. If the inflation rate is 3%, how much will a house now worth $510,000 be worth in 5 years?

69. A decimal approximation for $\sqrt{3}$ is 1.7320508. Use a calculator to find $2^{1.7}, 2^{1.73}, 2^{1.732}, 2^{1.73205}$, and $2^{1.7320508}$. Now find $2^{\sqrt{3}}$. What do you observe?

Number of Words, in Millions, in the Federal Tax Code

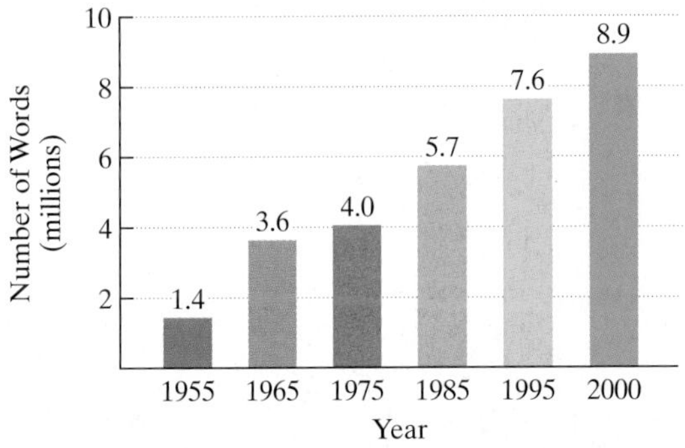

Source: The Tax Foundation

71. Which function, the linear or the exponential, is a better model for the data in 2000?

72. Which function, the linear or the exponential, is a better model for the data in 1985?

73. In college, we study large volumes of information—information that, unfortunately, we do not often retain for very long. The function

$$f(x) = 80e^{-0.5x} + 20$$

describes the percentage of information, $f(x)$, that a particular person remembers x weeks after learning the information.

  a. Substitute 0 for x and, without using a calculator, find the percentage of information remembered at the moment it is first learned.

  b. Substitute 1 for x and find the percentage of information that is remembered after 1 week.

  c. Find the percentage of information that is remembered after 4 weeks.

  d. Find the percentage of information that is remembered after one year (52 weeks).

74. In 1626, Peter Minuit convinced the Wappinger Indians to sell him Manhattan Island for $24. If the Native Americans had put the $24 into a bank account paying 5% interest, how much would the investment have been worth in the year 2005 if interest were compounded

  a. monthly?   **b.** continuously?

The bar graph shows the number of identity theft complaints to the Federal Trade Commission from 2000 through 2004. (The problem is much worse: The graph shows only the complaints. According to an FCC survey, 9.9 million Americans—about 1 in 30—were victims of identity theft from spring 2002 to spring 2003.)

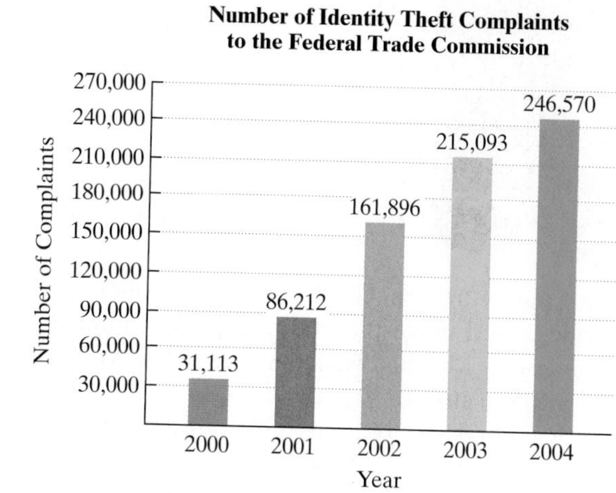

Source: Federal Trade Commission

The functions

$$f(x) = \frac{258{,}051}{1 + 6.78e^{-1.21x}} \quad \text{and} \quad g(x) = 55{,}979.5x + 36{,}217.8$$

model the number of identity theft complaints to the FCC, $f(x)$ or $g(x)$, x years after 2000. Use these functions to solve Exercises 75–76.

75. Which function is a better model for the number of complaints in 2004?

76. Which function is a better model for the number of complaints in 2003?

Writing in Mathematics

77. What is an exponential function?

78. What is the natural exponential function?

79. Use a calculator to evaluate $\left(1 + \frac{1}{x}\right)^x$ for $x = 10, 100, 1000,$ $10{,}000, 100{,}000,$ and $1{,}000{,}000$. Describe what happens to the expression as x increases.

80. Describe how you could use the graph of $f(x) = 2^x$ to obtain a decimal approximation for $\sqrt{2}$.

81. The exponential function $y = 2^x$ is one-to-one and has an inverse function. Try finding the inverse function by exchanging x and y and solving for y. Describe the difficulty that you encounter in this process. What is needed to overcome this problem?

82. In 2004, world population was approximately 6.4 billion with an annual growth rate of 1.23%. Discuss two factors that would cause this growth rate to slow down over the next ten years.

Technology Exercises

83. Graph $y = 13.49(0.967)^x - 1$, the function for the number of O-rings expected to fail at x°F, in a $[0, 90, 10]$ by $[0, 20, 5]$ viewing rectangle. If NASA engineers had used this function and its graph, is it likely they would have allowed the *Challenger* to be launched when the temperature was 31°F? Explain.

84. You have \$10,000 to invest. One bank pays 5% interest compounded quarterly and the other pays 4.5% interest compounded monthly.

a. Use the formula for compound interest to write a function for the balance in each account at any time t.

b. Use a graphing utility to graph both functions in an appropriate viewing rectangle. Based on the graphs, which bank offers the better return on your money?

85. a. Graph $y = e^x$ and $y = 1 + x + \frac{x^2}{2}$ in the same viewing rectangle.

b. Graph $y = e^x$ and $y = 1 + x + \frac{x^2}{2} + \frac{x^3}{6}$ in the same viewing rectangle.

c. Graph $y = e^x$ and $y = 1 + x + \frac{x^2}{2} + \frac{x^3}{6} + \frac{x^4}{24}$ in the same viewing rectangle.

d. Describe what you observe in parts (a)–(c). Try generalizing this observation.

Critical Thinking Exercises

86. Which one of the following is true?

a. As the number of compounding periods increases on a fixed investment, the amount of money in the account over a fixed interval of time will increase without bound.

b. The functions $f(x) = 3^{-x}$ and $g(x) = -3^x$ have the same graph.

c. If $f(x) = 2^x$, then $f(a + b) = f(a) + f(b)$.

d. The functions $f(x) = \left(\frac{1}{3}\right)^x$ and $g(x) = 3^{-x}$ have the same graph.

87. The graphs labeled (a)–(d) in the figure represent $y = 3^x$, $y = 5^x$, $y = \left(\frac{1}{3}\right)^x$, and $y = \left(\frac{1}{5}\right)^x$, but not necessarily in that order. Which is which? Describe the process that enables you to make this decision.

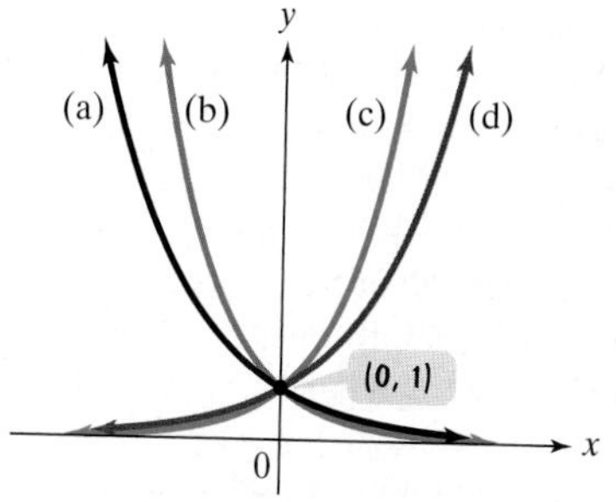

88. Graph $f(x) = 2^x$ and its inverse function in the same rectangular coordinate system.

89. The *hyperbolic cosine* and *hyperbolic sine* functions are defined by

$$\cosh x = \frac{e^x + e^{-x}}{2} \quad \text{and} \quad \sinh x = \frac{e^x - e^{-x}}{2}.$$

a. Show that $\cosh x$ is an even function.

b. Show that $\sinh x$ is an odd function.

c. Prove that $(\cosh x)^2 - (\sinh x)^2 = 1$.

SECTION 4.2 Logarithmic Functions

Objectives

1. Change from logarithmic to exponential form.
2. Change from exponential to logarithmic form.
3. Evaluate logarithms.
4. Use basic logarithmic properties.
5. Graph logarithmic functions.
6. Find the domain of a logarithmic function.
7. Use common logarithms.
8. Use natural logarithms.

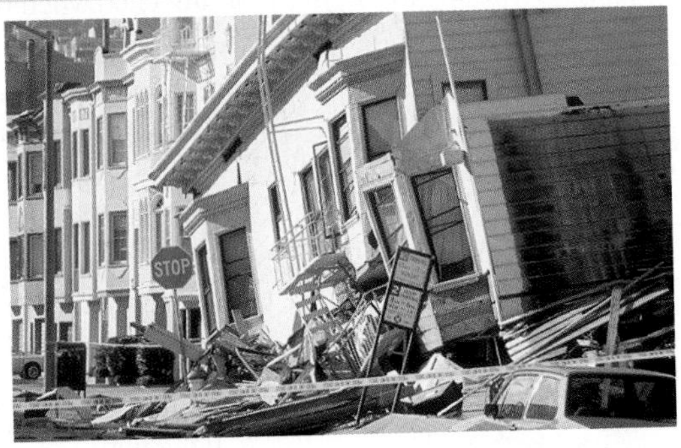

The earthquake that ripped through northern California on October 17, 1989, measured 7.1 on the Richter scale, killed more than 60 people, and injured more than 2400. Shown here is San Francisco's Marina district, where shock waves tossed houses off their foundations and into the street.

A higher measure on the Richter scale is more devastating than it seems because for each increase in one unit on the scale, there is a tenfold increase in the intensity of an earthquake. In this section, our focus is on the inverse of the exponential function, called the logarithmic function. The logarithmic function will help you to understand diverse phenomena, including earthquake intensity, human memory, and the pace of life in large cities.

The Definition of Logarithmic Functions

No horizontal line can be drawn that intersects the graph of an exponential function at more than one point. This means that the exponential function is one-to-one and has an inverse. The inverse function of the exponential function with base b is called the *logarithmic function with base b*.

Study Tip

The inverse of $y = b^x$ is $x = b^y$. Logarithms give us a way to express the inverse function $x = b^y$ for y in terms of x. Refer to the table of contents and the section titled *Inverse Functions* in case you need to review this topic. Here's a summary of what you should already know about functions and their inverses.

1. Only one-to-one functions have inverses that are functions. A function, f, has an inverse function, f^{-1}, if there is no horizontal line that intersects the graph of f at more than one point.
2. If a function is one-to-one, its inverse function can be found by interchanging x and y in the function's equation and solving for y.
3. If $f(a) = b$, then $f^{-1}(b) = a$. The domain of f is the range of f^{-1}. The range of f is the domain of f^{-1}.
4. $f(f^{-1}(x)) = x$ and $f^{-1}(f(x)) = x$.
5. The graph of f^{-1} is the reflection of the graph of f about the line $y = x$.

Definition of the Logarithmic Function

For $x > 0$ and $b > 0, b \neq 1$,

$$y = \log_b x \text{ is equivalent to } b^y = x.$$

The function $f(x) = \log_b x$ is the **logarithmic function with base *b*.**

The equations

$$y = \log_b x \quad \text{and} \quad b^y = x$$

are different ways of expressing the same thing. The first equation is in **logarithmic form** and the second equivalent equation is in **exponential form**.

Notice that a **logarithm, *y*, is an exponent**. You should learn the location of the base and exponent in each form.

Location of Base and Exponent in Exponential and Logarithmic Forms

Logarithmic Form: $y = \log_b x$ (Exponent: y; Base: b) Exponential Form: $b^y = x$ (Exponent: y; Base: b)

Study Tip

To change from logarithmic form to the more familiar exponential form, use this pattern:

$$y = \log_b x \text{ means } b^y = x.$$

1 Change from logarithmic to exponential form.

EXAMPLE 1 Changing from Logarithmic to Exponential Form

Write each equation in its equivalent exponential form:

a. $2 = \log_5 x$ **b.** $3 = \log_b 64$ **c.** $\log_3 7 = y$.

Solution We use the fact that $y = \log_b x$ means $b^y = x$.

a. $2 = \log_5 x$ means $5^2 = x$. **b.** $3 = \log_b 64$ means $b^3 = 64$.

Logarithms are exponents. Logarithms are exponents.

c. $\log_3 7 = y$ or $y = \log_3 7$ means $3^y = 7$.

Check Point 1 Write each equation in its equivalent exponential form:

a. $3 = \log_7 x$ **b.** $2 = \log_b 25$ **c.** $\log_4 26 = y$.

2 Change from exponential to logarithmic form.

EXAMPLE 2 Changing from Exponential to Logarithmic Form

Write each equation in its equivalent logarithmic form:

a. $12^2 = x$ **b.** $b^3 = 8$ **c.** $e^y = 9$.

Solution We use the fact that $b^y = x$ means $y = \log_b x$.

a. $12^2 = x$ means $2 = \log_{12} x$. **b.** $b^3 = 8$ means $3 = \log_b 8$.

Exponents are logarithms. Exponents are logarithms.

c. $e^y = 9$ means $y = \log_e 9$.

Check Point 2 Write each equation in its equivalent logarithmic form:

a. $2^5 = x$ **b.** $b^3 = 27$ **c.** $e^y = 33$.

3 Evaluate logarithms.

Remembering that logarithms are exponents makes it possible to evaluate some logarithms by inspection. The logarithm of x with base b, $\log_b x$, is the exponent to which b must be raised to get x. For example, suppose we want to evaluate $\log_2 32$. We ask, 2 to what power gives 32? Because $2^5 = 32$, $\log_2 32 = 5$.

EXAMPLE 3 Evaluating Logarithms

Evaluate:

a. $\log_2 16$ **b.** $\log_3 9$ **c.** $\log_{25} 5$.

Solution

Logarithmic Expression	Question Needed for Evaluation	Logarithmic Expression Evaluated
a. $\log_2 16$	2 to what power gives 16?	$\log_2 16 = 4$ because $2^4 = 16$.
b. $\log_3 9$	3 to what power gives 9?	$\log_3 9 = 2$ because $3^2 = 9$.
c. $\log_{25} 5$	25 to what power gives 5?	$\log_{25} 5 = \frac{1}{2}$ because $25^{\frac{1}{2}} = \sqrt{25} = 5$.

Check Point 3 Evaluate:

a. $\log_{10} 100$ **b.** $\log_3 3$ **c.** $\log_{36} 6$.

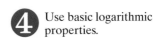 Use basic logarithmic properties.

Basic Logarithmic Properties

Because logarithms are exponents, they have properties that can be verified using properties of exponents.

> ### Basic Logarithmic Properties Involving One
>
> **1.** $\log_b b = 1$ because 1 is the exponent to which b must be raised to obtain b. $(b^1 = b)$
> **2.** $\log_b 1 = 0$ because 0 is the exponent to which b must be raised to obtain 1. $(b^0 = 1)$

EXAMPLE 4   Using Properties of Logarithms

Evaluate:

    a. $\log_7 7$ **b.** $\log_5 1$.

Solution

    a. Because $\log_b b = 1$, we conclude $\log_7 7 = 1$.   This means that $7^1 = 7$.

    b. Because $\log_b 1 = 0$, we conclude $\log_5 1 = 0$.   This means that $5^0 = 1$.

Check Point 4 Evaluate:

    a. $\log_9 9$ **b.** $\log_8 1$.

The inverse of the exponential function is the logarithmic function. Thus, if $f(x) = b^x$, then $f^{-1}(x) = \log_b x$. We have seen how inverse functions "undo" one another. In particular,

$$f(f^{-1}(x)) = x \text{ and } f^{-1}(f(x)) = x.$$

Applying these relationships to exponential and logarithmic functions, we obtain the following **inverse properties of logarithms**:

> ### Inverse Properties of Logarithms
> For $b > 0$ and $b \neq 1$,
>
> $$\log_b b^x = x$$ The logarithm with base b of b raised to a power equals that power.
>
> $$b^{\log_b x} = x$$ b raised to the logarithm with base b of a number equals that number.

Study Tip

The voice balloons should help you see the "undoing" that takes place between the exponential and logarithmic functions in the inverse properties.

Start with x.   End with x.   Start with x.   End with x.

$$\log_b \boxed{b^x} = x \qquad\qquad b^{\boxed{\log_b x}} = x$$

x is changed by the exponential function. x is changed by the logarithmic function.

The change is undone by the inverse logarithmic function.   The change is undone by the inverse exponential function.

EXAMPLE 5   Using Inverse Properties of Logarithms

Evaluate:

    a. $\log_4 4^5$ **b.** $6^{\log_6 9}$.

Solution

    a. Because $\log_b b^x = x$, we conclude $\log_4 4^5 = 5$.

    b. Because $b^{\log_b x} = x$, we conclude $6^{\log_6 9} = 9$.

> Check Point **5** Evaluate:
>
>   **a.** $\log_7 7^8$ **b.** $3^{\log_3 17}$.

 Graph logarithmic functions.

Graphs of Logarithmic Functions

How do we graph logarithmic functions? We use the fact that a logarithmic function is the inverse of an exponential function. This means that the logarithmic function reverses the coordinates of the exponential function. It also means that the graph of the logarithmic function is a reflection of the graph of the exponential function about the line $y = x$.

EXAMPLE 6   Graphs of Exponential and Logarithmic Functions

Graph $f(x) = 2^x$ and $g(x) = \log_2 x$ in the same rectangular coordinate system.

Solution   We first set up a table of coordinates for $f(x) = 2^x$. Reversing these coordinates gives the coordinates for the inverse function $g(x) = \log_2 x$.

Reverse coordinates.

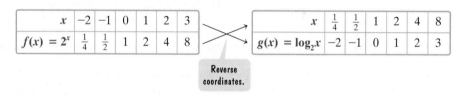

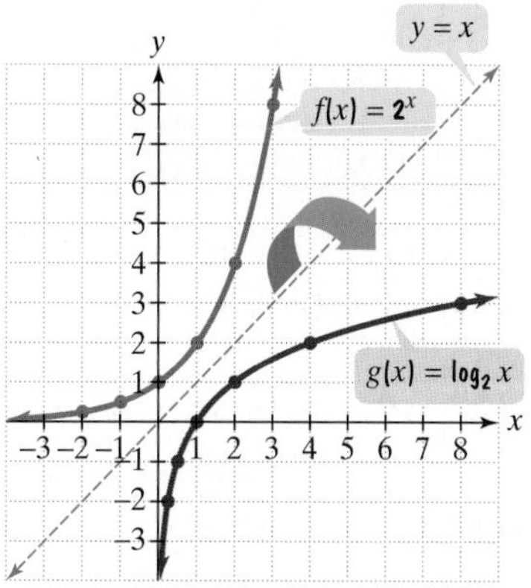

Figure 4.5 The graphs of $f(x) = 2^x$ and its inverse function

We now plot the ordered pairs from each table, connecting them with smooth curves. Figure 4.5 shows the graphs of $f(x) = 2^x$ and its inverse function $g(x) = \log_2 x$. The graph of the inverse can also be drawn by reflecting the graph of $f(x) = 2^x$ about the line $y = x$.

Study Tip

You can obtain a partial table of coordinates for $g(x) = \log_2 x$ without having to obtain and reverse coordinates for $f(x) = 2^x$. Because $g(x) = \log_2 x$ means $2^{g(x)} = x$, we begin with values for $g(x)$ and compute corresponding values for x:

Use $x = 2^{g(x)}$ to compute x. For example, if $g(x) = -2$, $x = 2^{-2} = \frac{1}{2^2} = \frac{1}{4}$.

Start with values for $g(x)$.

x	$\frac{1}{4}$	$\frac{1}{2}$	1	2	4	8
$g(x) = \log_2 x$	−2	−1	0	1	2	3

> Check Point **6** Graph $f(x) = 3^x$ and $g(x) = \log_3 x$ in the same rectangular coordinate system.

Figure 4.6 illustrates the relationship between the graph of an exponential function, shown in blue, and its inverse, a logarithmic function, shown in red, for bases greater than 1 and for bases between 0 and 1. Also shown and labeled are the exponential function's horizontal asymptote ($y = 0$) and the logarithmic function's vertical asymptote ($x = 0$).

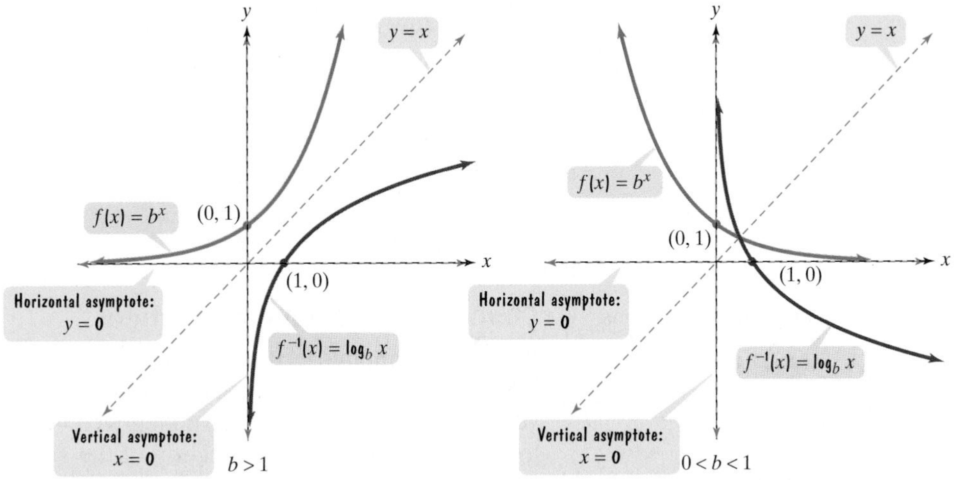

Figure 4.6 Graphs of exponential and logarithmic functions

Discovery

Verify each of the four characteristics in the box for the red graphs in Figure 4.6.

Characteristics of the Graphs of Logarithmic Functions of the Form $f(x) = \log_b x$

- The x-intercept is 1. There is no y-intercept.
- The y-axis, or $x = 0$, is a vertical asymptote. As $x \to 0^+$, $\log_b x \to -\infty$ or ∞.
- If $b > 1$, the function is increasing. If $0 < b < 1$, the function is decreasing.
- The graph is smooth and continuous. It has no sharp corners or gaps.

The graphs of logarithmic functions can be translated vertically or horizontally, reflected, stretched, or shrunk. These transformations are summarized in Table 4.4.

Table 4.4 Transformations Involving Logarithmic Functions

In each case, c represents a positive real number.

Transformation	Equation	Description
Vertical translation	$g(x) = \log_b x + c$ $g(x) = \log_b x - c$	• Shifts the graph of $f(x) = \log_b x$ upward c units. • Shifts the graph of $f(x) = \log_b x$ downward c units.
Horizontal translation	$g(x) = \log_b(x + c)$ $g(x) = \log_b(x - c)$	• Shifts the graph of $f(x) = \log_b x$ to the left c units. Vertical asymptote: $x = -c$ • Shifts the graph of $f(x) = \log_b x$ to the right c units. Vertical asymptote: $x = c$
Reflection	$g(x) = -\log_b x$ $g(x) = \log_b(-x)$	• Reflects the graph of $f(x) = \log_b x$ about the x-axis. • Reflects the graph of $f(x) = \log_b x$ about the y-axis.
Vertical stretching or shrinking	$g(x) = c \log_b x$	• Vertically stretches the graph of $f(x) = \log_b x$ if $c > 1$. • Vertically shrinks the graph of $f(x) = \log_b x$ if $0 < c < 1$.
Horizontal stretching or shrinking	$g(x) = \log_b(cx)$	• Horizontally shrinks the graph of $f(x) = \log_b x$ if $c > 1$. • Horizontally stretches the graph of $f(x) = \log_b x$ if $0 < c < 1$.

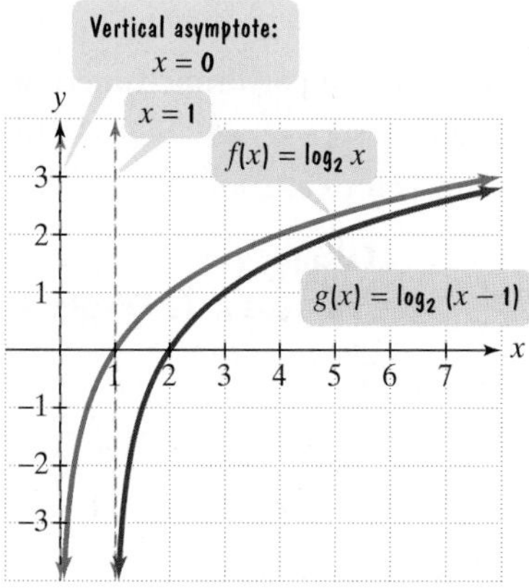

Figure 4.7 Shifting $f(x) = \log_2 x$ one unit to the right

For example, Figure 4.7 illustrates that the graph of $g(x) = \log_2(x - 1)$ is the graph of $f(x) = \log_2 x$ shifted one unit to the right. If a logarithmic function is translated to the left or to the right, both the x-intercept and the vertical asymptote are shifted by the amount of the horizontal shift. In Figure 4.7, the x-intercept of f is 1. Because g is shifted one unit to the right, its x-intercept is 2. Also observe that the vertical asymptote for f, the y-axis, or $x = 0$, is shifted one unit to the right for the vertical asymptote for g. Thus, $x = 1$ is the vertical asymptote for g.

Here are some other examples of transformations of graphs of logarithmic functions:

- The graph of $g(x) = 3 + \log_4 x$ is the graph of $f(x) = \log_4 x$ shifted up three units, shown in Figure 4.8.
- The graph of $h(x) = -\log_2 x$ is the graph of $f(x) = \log_2 x$ reflected about the x-axis, shown in Figure 4.9.
- The graph of $r(x) = \log_2(-x)$ is the graph of $f(x) = \log_2 x$ reflected about the y-axis, shown in Figure 4.10.

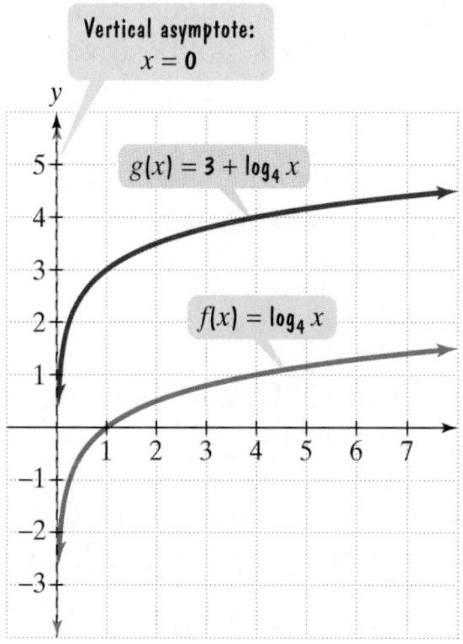

Figure 4.8 Shifting vertically up three units

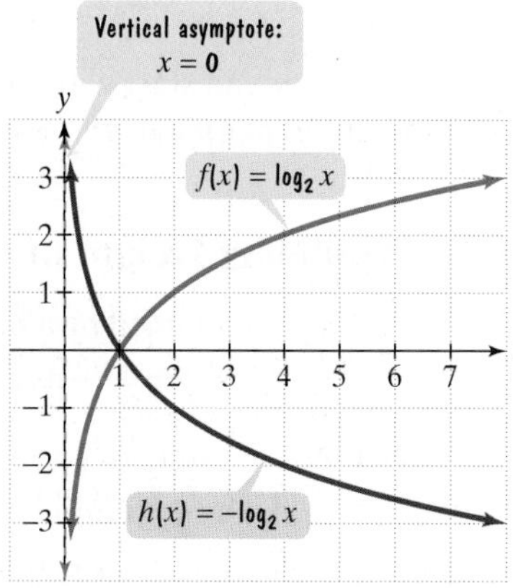

Figure 4.9 Reflection about the x-axis

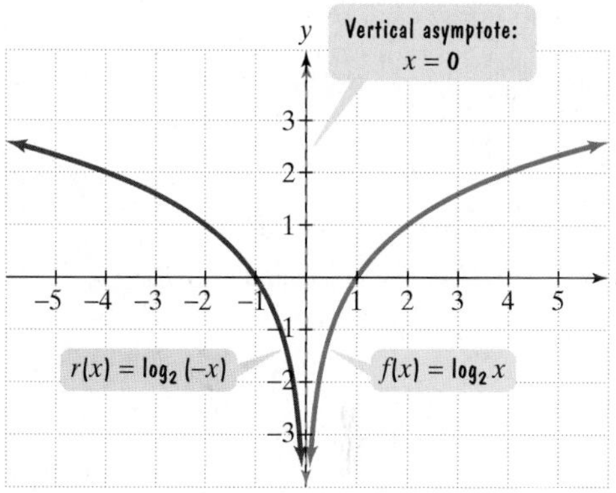

Figure 4.10 Reflection about the y-axis

6 Find the domain of a logarithmic function.

The Domain of a Logarithmic Function

In Section 4.1, we learned that the domain of an exponential function of the form $f(x) = b^x$ includes all real numbers and its range is the set of positive real numbers. Because the logarithmic function reverses the domain and the range of the exponential function, the **domain of a logarithmic function of the form $f(x) = \log_b x$ is the set of all positive real numbers.** Thus, $\log_2 8$ is defined because the value of x in the logarithmic expression, 8, is greater than zero and therefore is included in the domain of the logarithmic function $f(x) = \log_2 x$. However, $\log_2 0$ and $\log_2(-8)$ are not defined because 0 and -8 are not positive real numbers and therefore are excluded from the domain of the logarithmic function $f(x) = \log_2 x$. In general, **the domain of $f(x) = \log_b g(x)$ consists of all x for which $g(x) > 0$.**

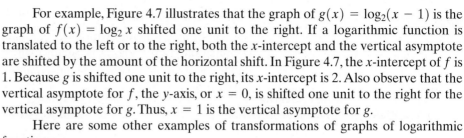

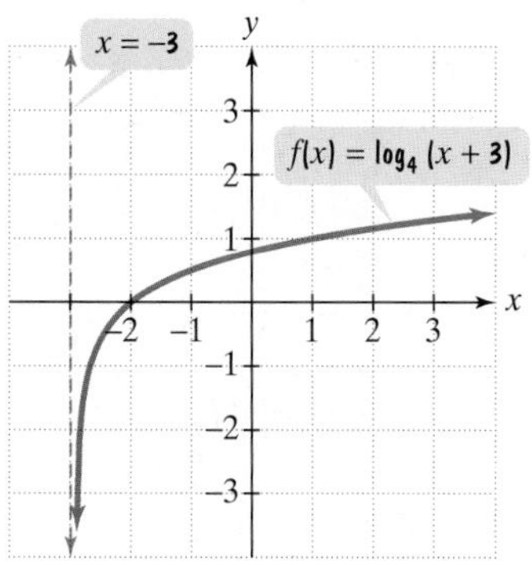

Figure 4.11 The domain of $g(x) = \log_4(x + 3)$ is $(-3, \infty)$.

EXAMPLE 7 Finding the Domain of a Logarithmic Function

Find the domain of $f(x) = \log_4(x + 3)$.

Solution The domain of f consists of all x for which $x + 3 > 0$. Solving this inequality for x, we obtain $x > -3$. Thus, the domain of f is $(-3, \infty)$. This is illustrated in Figure 4.11. The vertical asymptote is $x = -3$ and all points on the graph of f have x-coordinates that are greater than -3.

Check Point 7 Find the domain of $f(x) = \log_4(x - 5)$.

7 Use common logarithms.

Common Logarithms

The logarithmic function with base 10 is called the **common logarithmic function**. The function $f(x) = \log_{10} x$ is usually expressed as $f(x) = \log x$. A calculator with a LOG key can be used to evaluate common logarithms. Here are some examples:

Logarithm	Most Scientific Calculator Keystrokes	Most Graphing Calculator Keystrokes	Display (or Approximate Display)
$\log 1000$	1000 LOG	LOG 1000 ENTER	3
$\log\frac{5}{2}$	(5 ÷ 2) LOG	LOG (5 ÷ 2) ENTER	0.39794
$\frac{\log 5}{\log 2}$	5 LOG ÷ 2 LOG =	LOG 5 ÷ LOG 2 ENTER	2.32193
$\log(-3)$	3 +/− LOG	LOG (−) 3 ENTER	ERROR

Some graphing calculators display an open parenthesis when the LOG key is pressed. In this case, remember to close the set of parentheses after entering the function's domain value: LOG 5) ÷ LOG 2) ENTER.

The error message given by many calculators for $\log(-3)$ is a reminder that the domain of the common logarithmic function, $f(x) = \log x$, is the set of positive real numbers. In general, the domain of $f(x) = \log g(x)$ consists of all x for which $g(x) > 0$.

Many real-life phenomena start with rapid growth and then the growth begins to level off. This type of behavior can be modeled by logarithmic functions.

EXAMPLE 8 Modeling Height of Children

The percentage of adult height attained by a boy who is x years old can be modeled by

$$f(x) = 29 + 48.8\log(x + 1),$$

where x represents the boy's age and $f(x)$ represents the percentage of his adult height. Approximately what percentage of his adult height has a boy attained at age eight?

Solution We substitute the boy's age, 8, for x and evaluate the function.

$$f(x) = 29 + 48.8\log(x + 1)$$ This is the given function.

$$f(8) = 29 + 48.8\log(8 + 1)$$ Substitute 8 for x.

$$= 29 + 48.8\log 9$$ Graphing calculator keystrokes: 29 + 48.8 LOG 9 ENTER

$$\approx 76$$

Thus, an 8-year-old boy has attained approximately 76% of his adult height.

Check Point 8 Use the function in Example 8 to answer this question: Approximately what percentage of his adult height has a boy attained at age ten?

The basic properties of logarithms that were listed earlier in this section can be applied to common logarithms.

Properties of Common Logarithms

General Properties		Common Logarithm Properties
1. $\log_b 1 = 0$		**1.** $\log 1 = 0$
2. $\log_b b = 1$		**2.** $\log 10 = 1$
3. $\log_b b^x = x$	Inverse properties	**3.** $\log 10^x = x$
4. $b^{\log_b x} = x$		**4.** $10^{\log x} = x$

The property $\log 10^x = x$ can be used to evaluate common logarithms involving powers of 10. For example,

$$\log 100 = \log 10^2 = 2, \quad \log 1000 = \log 10^3 = 3, \quad \text{and} \quad \log 10^{7.1} = 7.1.$$

EXAMPLE 9 Earthquake Intensity

The magnitude, R, on the Richter scale of an earthquake of intensity I is given by

$$R = \log \frac{I}{I_0},$$

where I_0 is the intensity of a barely felt zero-level earthquake. The earthquake that destroyed San Francisco in 1906 was $10^{8.3}$ times as intense as a zero-level earthquake. What was its magnitude on the Richter scale?

Solution Because the earthquake was $10^{8.3}$ times as intense as a zero-level earthquake, the intensity, I, is $10^{8.3} I_0$.

$$R = \log \frac{I}{I_0}$$ This is the formula for magnitude on the Richter scale.

$$R = \log \frac{10^{8.3} I_0}{I_0}$$ Substitute $10^{8.3} I_0$ for I.

$$= \log 10^{8.3}$$ Simplify.

$$= 8.3$$ Use the property $\log 10^x = x$.

San Francisco's 1906 earthquake registered 8.3 on the Richter scale.

Check Point 9 Use the formula in Example 9 to solve this problem. If an earthquake is 10,000 times as intense as a zero-level quake ($I = 10{,}000 I_0$), what is its magnitude on the Richter scale?

8 Use natural logarithms.

Natural Logarithms

The logarithmic function with base e is called the **natural logarithmic function**. The function $f(x) = \log_e x$ is usually expressed as $f(x) = \ln x$, read "el en of x." A calculator with an LN key can be used to evaluate natural logarithms. Keystrokes are identical to those shown for common logarithmic evaluations on page 418.

Like the domain of all logarithmic functions, the domain of the natural logarithmic function $f(x) = \ln x$ is the set of all positive real numbers. Thus, the domain of $f(x) = \ln g(x)$ consists of all x for which $g(x) > 0$.

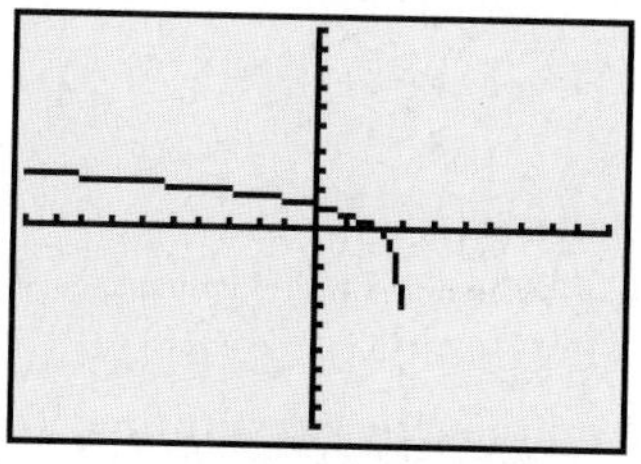

[−10, 10, 1] by [−10, 10, 1]

Figure 4.12 The domain of $f(x) = \ln(3 - x)$ is $(-\infty, 3)$.

EXAMPLE 10 Finding Domains of Natural Logarithmic Functions

Find the domain of each function:

a. $f(x) = \ln(3 - x)$ **b.** $h(x) = \ln(x - 3)^2$.

Solution

a. The domain of f consists of all x for which $3 - x > 0$. Solving this inequality for x, we obtain $x < 3$. Thus, the domain of f is $\{x \mid x < 3\}$ or $(-\infty, 3)$. This is verified by the graph in Figure 4.12.

b. The domain of h consists of all x for which $(x - 3)^2 > 0$. It follows that the domain of h is all real numbers except 3. Thus, the domain of h is $\{x \mid x \neq 3\}$ or $(-\infty, 3) \cup (3, \infty)$. This is shown by the graph in Figure 4.13. To make it more obvious that 3 is excluded from the domain, we used a DOT format.

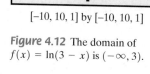

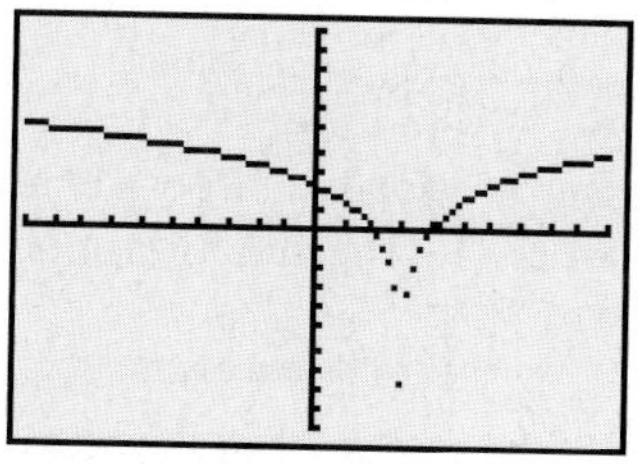

[−10, 10, 1] by [−10, 10, 1]

Figure 4.13 3 is excluded from the domain of $h(x) = \ln(x - 3)^2$.

Check Point 10 Find the domain of each function:

a. $f(x) = \ln(4 - x)$ **b.** $h(x) = \ln x^2$.

The basic properties of logarithms that were listed earlier in this section can be applied to natural logarithms.

Properties of Natural Logarithms

General Properties		Natural Logarithm Properties
1. $\log_b 1 = 0$		**1.** $\ln 1 = 0$
2. $\log_b b = 1$		**2.** $\ln e = 1$
3. $\log_b b^x = x$	Inverse properties	**3.** $\ln e^x = x$
4. $b^{\log_b x} = x$		**4.** $e^{\ln x} = x$

Examine the inverse properties, $\ln e^x = x$ and $e^{\ln x} = x$. Can you see how ln and e "undo" one another? For example,

$$\ln e^2 = 2,\ \ln e^{7x^2} = 7x^2,\ e^{\ln 2} = 2,\ \text{and}\ e^{\ln 7x^2} = 7x^2.$$

EXAMPLE 11 Dangerous Heat: Temperature in an Enclosed Vehicle

When the outside air temperature is anywhere from 72° to 96° Fahrenheit, the temperature in an enclosed vehicle climbs by 43° in the first hour. The bar graph in Figure 4.14 shows the temperature increase throughout the hour. The function

$$f(x) = 13.4 \ln x - 11.6$$

models the temperature increase, $f(x)$, in degrees Fahrenheit, after x minutes. Use the function to find the temperature increase, to the nearest degree, after 50 minutes. How well does the function model the actual increase shown in Figure 4.14?

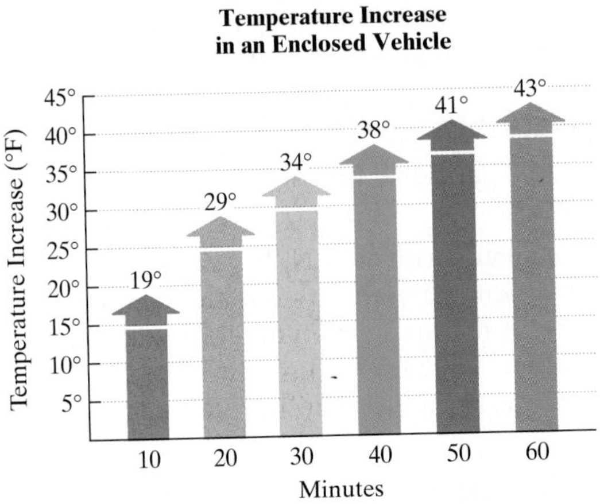

Figure 4.14

Source: Professor Jan Null, San Francisco State University

Solution We find the temperature increase after 50 minutes by substituting 50 for x and evaluating the function at 50.

$f(x) = 13.4 \ln x - 11.6$ This is the given function.

$f(50) = 13.4 \ln 50 - 11.6$ Substitute 50 for x.

≈ 41 Graphing calculator keystrokes: 13.4 [ln] 50 [−] 11.6 [ENTER]. On some calculators, a parenthesis is needed after 50.

According to the function, the temperature will increase by approximately 41° after 50 minutes. Because the increase shown in Figure 4.14 is 41°, the function models the actual increase extremely well.

Check Point 11 Use the function in Example 11 to find the temperature increase, to the nearest degree, after 30 minutes. How well does the function model the actual increase shown in Figure 4.14?

The Curious Number e

You will learn more about each curiosity mentioned below if you take calculus.

- The number e was named by the Swiss mathematician Leonhard Euler (1707–1783), who proved that it is the limit as $n \to \infty$ of $\left(1 + \frac{1}{n}\right)^n$.
- e features in Euler's remarkable relationship $e^{i\pi} = -1$, in which $i = \sqrt{-1}$.
- The first few decimal places of e are fairly easy to remember: $e = 2.7\,1828\,1828\,45\,90\,45\ldots.$
- The best approximation of e using numbers less than 1000 is also easy to remember: $e \approx \frac{878}{323} \approx 2.71826\ldots.$

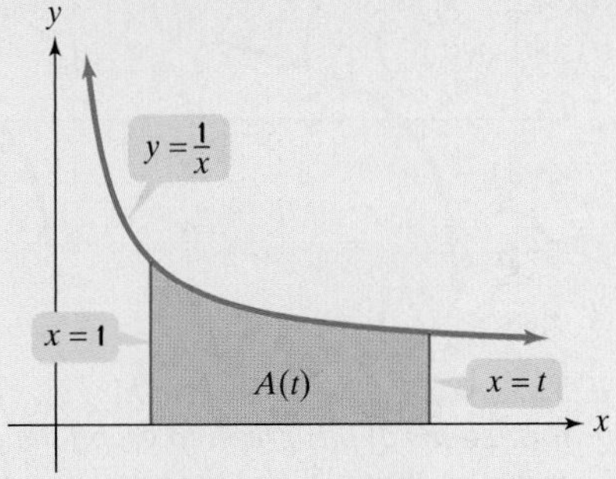

Figure 4.15

- Isaac Newton (1642–1727), one of the cofounders of calculus, showed that $e^x = 1 + x + \frac{x^2}{2!} + \frac{x^3}{3!} + \frac{x^4}{4!} + \cdots$, from which we obtain $e = 1 + 1 + \frac{1}{2!} + \frac{1}{3!} + \frac{1}{4!} + \cdots$, an infinite sum suitable for calculation because its terms decrease so rapidly. (*Note*: $n!$ (n factorial) is the product of all the consecutive integers from n down to 1: $n! = n(n-1)(n-2)(n-3)\cdot\cdots\cdot 3\cdot 2\cdot 1$.)
- The area of the region bounded by $y = \frac{1}{x}$, the x-axis, $x = 1$ and $x = t$ (shaded in Figure 4.15) is a function of t, designated by $A(t)$. Grégoire de Saint-Vincent, a Belgian Jesuit (1584–1667), spent his entire professional life attempting to find a formula for $A(t)$. With his student, he showed that $A(t) = \ln t$, becoming one of the first mathematicians to make use of the logarithmic function for something other than a computational device.

EXERCISE SET 4.2

Practice Exercises

In Exercises 1–8, write each equation in its equivalent exponential form.

1. $4 = \log_2 16$ **2.** $6 = \log_2 64$

3. $2 = \log_3 x$ **4.** $2 = \log_9 x$

5. $5 = \log_b 32$ **6.** $3 = \log_b 27$

7. $\log_6 216 = y$ **8.** $\log_5 125 = y$

In Exercises 9–20, write each equation in its equivalent logarithmic form.

9. $2^3 = 8$ **10.** $5^4 = 625$ **11.** $2^{-4} = \frac{1}{16}$ **12.** $5^{-3} = \frac{1}{125}$

13. $\sqrt[3]{8} = 2$ **14.** $\sqrt[3]{64} = 4$ **15.** $13^2 = x$ **16.** $15^2 = x$

17. $b^3 = 1000$ **18.** $b^3 = 343$ **19.** $7^y = 200$ **20.** $8^y = 300$

In Exercises 21–42, evaluate each expression without using a calculator.

21. $\log_4 16$ **22.** $\log_7 49$ **23.** $\log_2 64$ **24.** $\log_3 27$

25. $\log_5 \frac{1}{5}$ **26.** $\log_6 \frac{1}{6}$ **27.** $\log_2 \frac{1}{8}$ **28.** $\log_3 \frac{1}{9}$

29. $\log_7 \sqrt{7}$ **30.** $\log_6 \sqrt{6}$ **31.** $\log_2 \frac{1}{\sqrt{2}}$ **32.** $\log_3 \frac{1}{\sqrt{3}}$

33. $\log_{64} 8$ **34.** $\log_{81} 9$ **35.** $\log_5 5$ **36.** $\log_{11} 11$

37. $\log_4 1$ **38.** $\log_6 1$ **39.** $\log_5 5^7$ **40.** $\log_4 4^6$

41. $8^{\log_8 19}$ **42.** $7^{\log_7 23}$

43. Graph $f(x) = 4^x$ and $g(x) = \log_4 x$ in the same rectangular coordinate system.

44. Graph $f(x) = 5^x$ and $g(x) = \log_5 x$ in the same rectangular coordinate system.

45. Graph $f(x) = \left(\frac{1}{2}\right)^x$ and $g(x) = \log_{\frac{1}{2}} x$ in the same rectangular coordinate system.

46. Graph $f(x) = \left(\frac{1}{4}\right)^x$ and $g(x) = \log_{\frac{1}{4}} x$ in the same rectangular coordinate system.

In Exercises 47–52, the graph of a logarithmic function is given. Select the function for each graph from the following options:

$$f(x) = \log_3 x,\ g(x) = \log_3(x-1),\ h(x) = \log_3 x - 1,$$
$$F(x) = -\log_3 x,\ G(x) = \log_3(-x),\ H(x) = 1 - \log_3 x.$$

47.

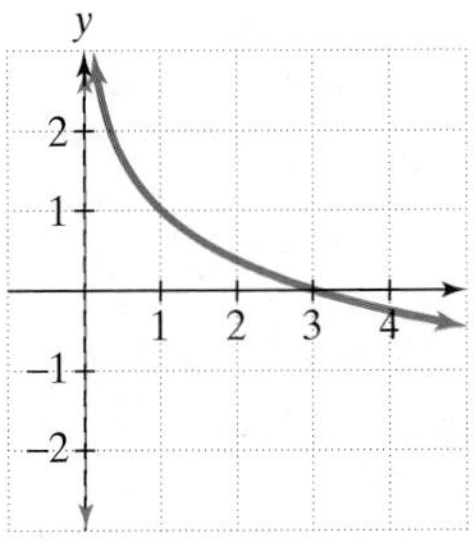

48.

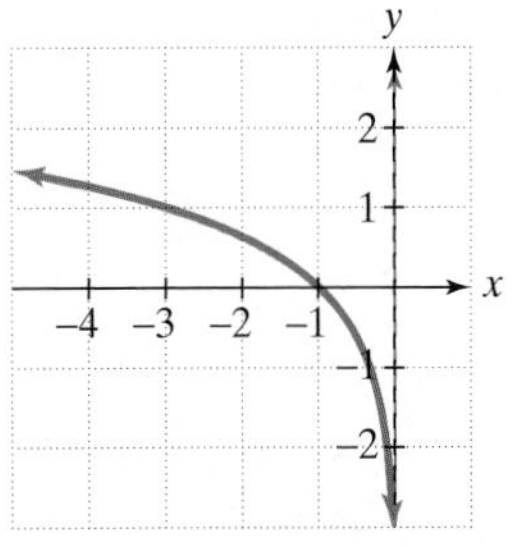

49.

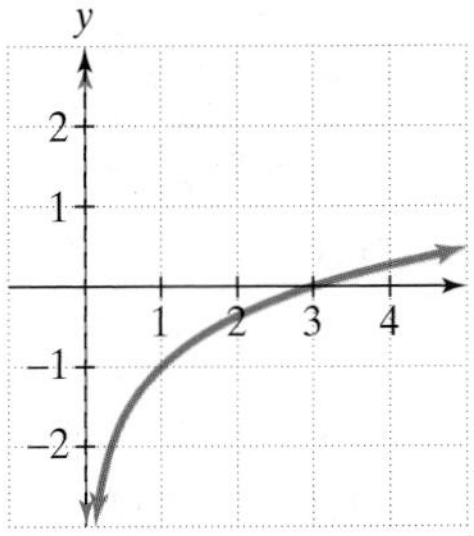

50.

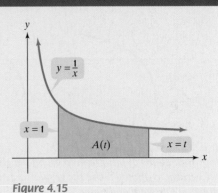

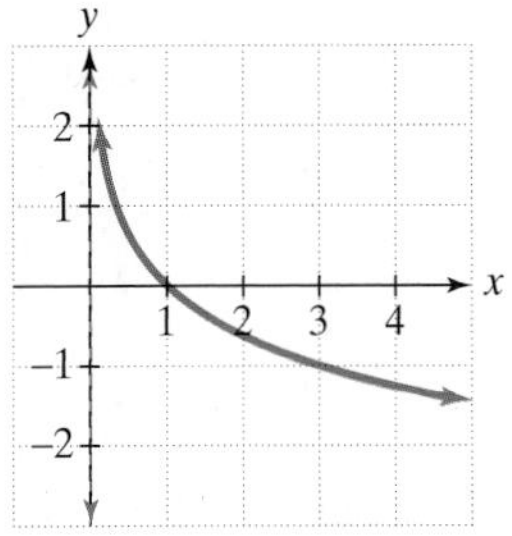

51.

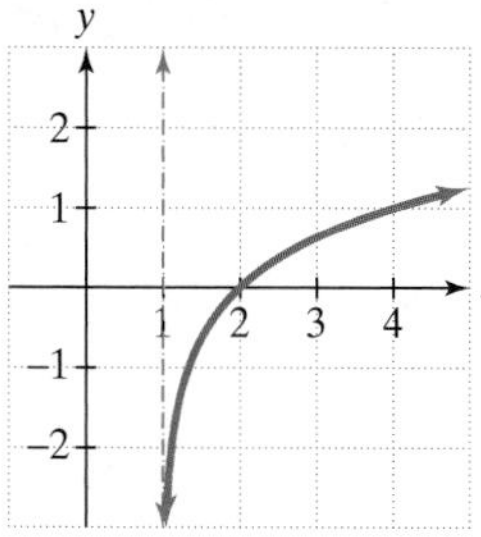

52.

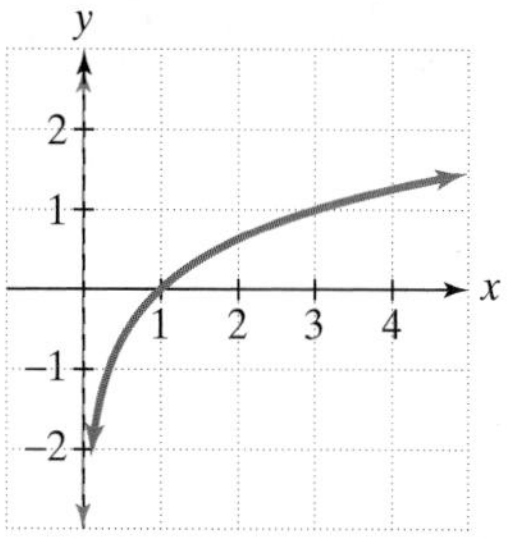

In Exercises 53–58, begin by graphing $f(x) = \log_2 x$. Then use transformations of this graph to graph the given function. What is the vertical asymptote? Use the graphs to determine each function's domain and range.

53. $g(x) = \log_2(x + 1)$ **54.** $g(x) = \log_2(x + 2)$

55. $h(x) = 1 + \log_2 x$ **56.** $h(x) = 2 + \log_2 x$

57. $g(x) = \frac{1}{2}\log_2 x$ **58.** $g(x) = -2\log_2 x$

The figure shows the graph of $f(x) = \log x$. In Exercises 59–64, use transformations of this graph to graph each function. Graph and give equations of the asymptotes. Use the graphs to determine each function's domain and range.

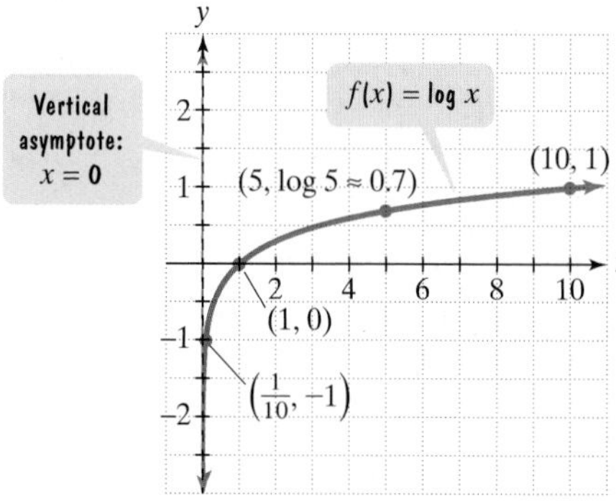

59. $g(x) = \log(x - 1)$ **60.** $g(x) = \log(x - 2)$

61. $h(x) = \log x - 1$ **62.** $h(x) = \log x - 2$

63. $g(x) = 1 - \log x$ **64.** $g(x) = 2 - \log x$

The figure shows the graph of $f(x) = \ln x$. In Exercises 65–74, use transformations of this graph to graph each function. Graph and give equations of the asymptotes. Use the graphs to determine each function's domain and range.

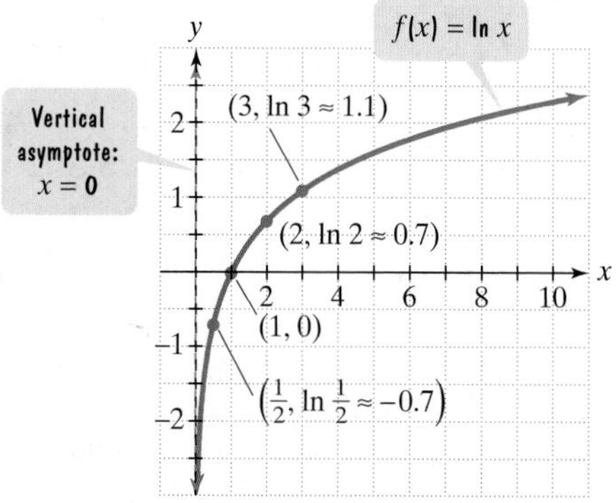

65. $g(x) = \ln(x + 2)$ **66.** $g(x) = \ln(x + 1)$

67. $h(x) = \ln(2x)$ **68.** $h(x) = \ln\left(\frac{1}{2}x\right)$

69. $g(x) = 2\ln x$ **70.** $g(x) = \frac{1}{2}\ln x$

71. $h(x) = -\ln x$ **72.** $h(x) = \ln(-x)$

73. $g(x) = 2 - \ln x$ **74.** $g(x) = 1 - \ln x$

In Exercises 75–80, find the domain of each logarithmic function.

75. $f(x) = \log_5(x + 4)$ **76.** $f(x) = \log_5(x + 6)$

77. $f(x) = \log(2 - x)$ **78.** $f(x) = \log(7 - x)$

79. $f(x) = \ln(x - 2)^2$ **80.** $f(x) = \ln(x - 7)^2$

In Exercises 81–100, evaluate or simplify each expression without using a calculator.

81. $\log 100$ **82.** $\log 1000$ **83.** $\log 10^7$ **84.** $\log 10^8$

85. $10^{\log 33}$ **86.** $10^{\log 53}$ **87.** $\ln 1$ **88.** $\ln e$

89. $\ln e^6$ **90.** $\ln e^7$ **91.** $\ln\dfrac{1}{e^6}$ **92.** $\ln\dfrac{1}{e^7}$

93. $e^{\ln 125}$ **94.** $e^{\ln 300}$ **95.** $\ln e^{9x}$ **96.** $\ln e^{13x}$

97. $e^{\ln 5x^2}$ **98.** $e^{\ln 7x^2}$ **99.** $10^{\log\sqrt{x}}$ **100.** $10^{\log\sqrt[3]{x}}$

Practice Plus

In Exercises 101–104, write each equation in its equivalent exponential form. Then solve for x.

101. $\log_3(x - 1) = 2$ **102.** $\log_5(x + 4) = 2$

103. $\log_4 x = -3$ **104.** $\log_{64} x = \dfrac{2}{3}$

In Exercises 105–108, evaluate each expression without using a calculator.

105. $\log_3(\log_7 7)$ **106.** $\log_5(\log_2 32)$

107. $\log_2(\log_3 81)$ **108.** $\log(\ln e)$

In Exercises 109–112, find the domain of each logarithmic function.

109. $f(x) = \ln(x^2 - x - 2)$ **110.** $f(x) = \ln(x^2 - 4x - 12)$

111. $f(x) = \log\left(\dfrac{x + 1}{x - 5}\right)$ **112.** $f(x) = \log\left(\dfrac{x - 2}{x + 5}\right)$

Application Exercises

The percentage of adult height attained by a girl who is x years old can be modeled by

$$f(x) = 62 + 35\log(x - 4),$$

where x represents the girl's age (from 5 to 15) and $f(x)$ represents the percentage of her adult height. Use the function to solve Exercises 113–114. Round answers to the nearest tenth of a percent.

113. Approximately what percentage of her adult height has a girl attained at age 13?

114. Approximately what percentage of her adult height has a girl attained at age ten?

The bar graph shows the percentage of U.S. companies that performed drug tests on employees or job applicants in five selected years from 1998 through 2003. The function

$$f(x) = -4.9 \ln x + 73.8$$

models the percentage of such companies x years after 1997. Use this function to solve Exercises 115–116. Round answers to the nearest percent.

Percentage of U.S. Companies Performing Drug Tests

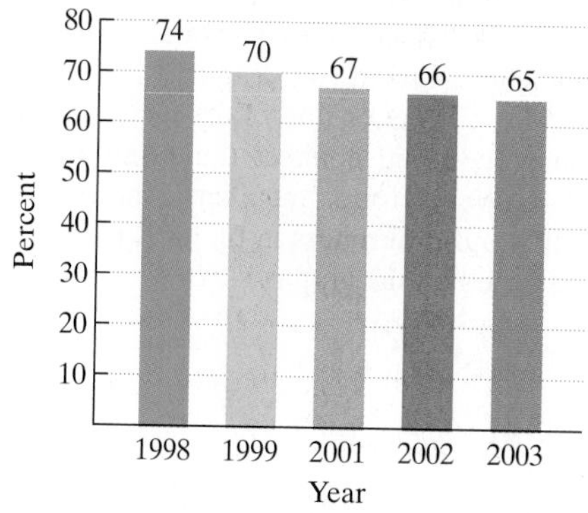

Source: American Management Association

115. Use the function to find the percentage of U.S. companies that performed drug tests in 2003. How well does this model the actual number shown for that year?

116. Use the function to predict the percentage of U.S. companies that will be performing drug tests in 2008.

The loudness level of a sound, D, in decibels, is given by the formula

$$D = 10 \log(10^{12} I),$$

where I is the intensity of the sound, in watts per meter^2. Decibel levels range from 0, a barely audible sound, to 160, a sound resulting in a ruptured eardrum. Use the formula to solve Exercises 117–118.

117. The sound of a blue whale can be heard 500 miles away, reaching an intensity of 6.3×10^6 watts per meter^2. Determine the decibel level of this sound. At close range, can the sound of a blue whale rupture the human eardrum?

118. What is the decibel level of a normal conversation, 3.2×10^{-6} watt per meter^2?

119. Students in a psychology class took a final examination. As part of an experiment to see how much of the course content they remembered over time, they took equivalent forms of the exam in monthly intervals thereafter. The average score for the group, $f(t)$, after t months was modeled by the function

$$f(t) = 88 - 15 \ln(t + 1), \qquad 0 \le t \le 12.$$

a. What was the average score on the original exam?

b. What was the average score after 2 months? 4 months? 6 months? 8 months? 10 months? one year?

c. Sketch the graph of f (either by hand or with a graphing utility). Describe what the graph indicates in terms of the material retained by the students.

Writing in Mathematics

120. Describe the relationship between an equation in logarithmic form and an equivalent equation in exponential form.

121. What question can be asked to help evaluate $\log_3 81$?

122. Explain why the logarithm of 1 with base b is 0.

123. Describe the following property using words: $\log_b b^x = x$.

124. Explain how to use the graph of $f(x) = 2^x$ to obtain the graph of $g(x) = \log_2 x$.

125. Explain how to find the domain of a logarithmic function.

126. Logarithmic models are well suited to phenomena in which growth is initially rapid but then begins to level off. Describe something that is changing over time that can be modeled using a logarithmic function.

127. Suppose that a girl is 4 feet 6 inches at age 10. Explain how to use the function in Exercises 113–114 to determine how tall she can expect to be as an adult.

Technology Exercises

In Exercises 128–131, graph f and g in the same viewing rectangle. Then describe the relationship of the graph of g to the graph of f.

128. $f(x) = \ln x,\ g(x) = \ln(x + 3)$

129. $f(x) = \ln x,\ g(x) = \ln x + 3$

130. $f(x) = \log x,\ g(x) = -\log x$

131. $f(x) = \log x,\ g(x) = \log(x - 2) + 1$

132. Students in a mathematics class took a final examination. They took equivalent forms of the exam in monthly intervals thereafter. The average score, $f(t)$, for the group after t months was modeled by the human memory function $f(t) = 75 - 10 \log(t + 1)$, where $0 \le t \le 12$. Use a graphing utility to graph the function. Then determine how many months will elapse before the average score falls below 65.

133. In parts (a)–(c), graph f and g in the same viewing rectangle.

a. $f(x) = \ln(3x),\ g(x) = \ln 3 + \ln x$

b. $f(x) = \log(5x^2),\ g(x) = \log 5 + \log x^2$

c. $f(x) = \ln(2x^3),\ g(x) = \ln 2 + \ln x^3$

d. Describe what you observe in parts (a)–(c). Generalize this observation by writing an equivalent expression for $\log_b(MN)$, where $M > 0$ and $N > 0$.

e. Complete this statement: The logarithm of a product is equal to ______________.

134. Graph each of the following functions in the same viewing rectangle and then place the functions in order from the one that increases most slowly to the one that increases most rapidly.

$$y = x,\ y = \sqrt{x},\ y = e^x,\ y = \ln x,\ y = x^x,\ y = x^2$$

Critical Thinking Exercises

135. Which one of the following is true?

a. $\dfrac{\log_2 8}{\log_2 4} = \dfrac{8}{4}$

b. $\log(-100) = -2$

c. The domain of $f(x) = \log_2 x$ is $(-\infty, \infty)$.

d. $\log_b x$ is the exponent to which b must be raised to obtain x.

136. Without using a calculator, find the exact value of

$$\frac{\log_3 81 - \log_\pi 1}{\log_{2\sqrt{2}} 8 - \log 0.001}.$$

137. Without using a calculator, find the exact value of $\log_4[\log_3(\log_2 8)]$.

138. Without using a calculator, determine which is the greater number: $\log_4 60$ or $\log_3 40$.

Group Exercise

139. This group exercise involves exploring the way we grow. Group members should create a graph for the function that models the percentage of adult height attained by a boy who is x years old, $f(x) = 29 + 48.8\log(x + 1)$. Let $x = 1, 2, 3, \ldots, 12$, find function values, and connect the resulting points with a smooth curve. Then create a graph for the function that models the percentage of adult height attained by a girl who is x years old, $g(x) = 62 + 35\log(x - 4)$. Let $x = 5, 6, 7, \ldots, 15$, find function values, and connect the resulting points with a smooth curve. Group members should then discuss similarities and differences in the growth patterns for boys and girls based on the graphs.

SECTION 4.3 Properties of Logarithms

Objectives

1. Use the product rule.
2. Use the quotient rule.
3. Use the power rule.
4. Expand logarithmic expressions.
5. Condense logarithmic expressions.
6. Use the change-of-base property.

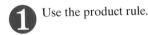

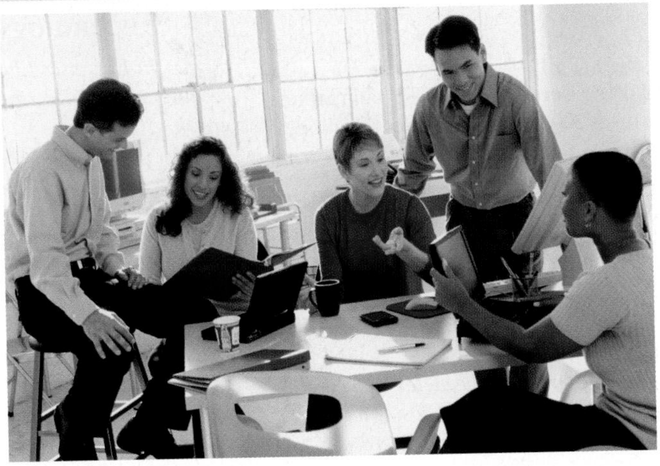

We all learn new things in different ways. In this section, we consider important properties of logarithms. What would be the most effective way for you to learn about these properties? Would it be helpful to use your graphing utility and discover one of these properties for yourself? To do so, work Exercise 133 in Exercise Set 4.2 before continuing. Would the properties become more meaningful if you could see exactly where they come from? If so, you will find details of the proofs of some of these properties in the appendix. The remainder of our work in this chapter will be based on the properties of logarithms that you learn in this section.

1 Use the product rule.

The Product Rule

Properties of exponents correspond to properties of logarithms. For example, when we multiply with the same base, we add exponents:

$$b^m \cdot b^n = b^{m+n}.$$

This property of exponents, coupled with an awareness that a logarithm is an exponent, suggests the following property, called the **product rule**:

Discovery

We know that $\log 100{,}000 = 5$. Show that you get the same result by writing 100,000 as $1000 \cdot 100$ and then using the product rule. Then verify the product rule by using other numbers whose logarithms are easy to find.

The Product Rule

Let b, M, and N be positive real numbers with $b \neq 1$.

$$\log_b(MN) = \log_b M + \log_b N$$

The logarithm of a product is the sum of the logarithms.

When we use the product rule to write a single logarithm as the sum of two logarithms, we say that we are **expanding a logarithmic expression**. For example, we can use the product rule to expand $\ln(7x)$:

$$\ln(7x) = \ln 7 + \ln x.$$

The logarithm of a product | is | the sum of the logarithms.

EXAMPLE 1 Using the Product Rule

Use the product rule to expand each logarithmic expression:

a. $\log_4(7 \cdot 5)$ **b.** $\log(10x)$.

Solution

a. $\log_4(7 \cdot 5) = \log_4 7 + \log_4 5$ The logarithm of a product is the sum of the logarithms.

b. $\log(10x) = \log 10 + \log x$ The logarithm of a product is the sum of the logarithms. These are common logarithms with base 10 understood.

$= 1 + \log x$ Because $\log_b b = 1$, then $\log 10 = 1$.

Check Point 1 Use the product rule to expand each logarithmic expression:

a. $\log_6(7 \cdot 11)$ **b.** $\log(100x)$.

2 Use the quotient rule.

The Quotient Rule

When we divide with the same base, we subtract exponents:

$$\frac{b^m}{b^n} = b^{m-n}.$$

This property suggests the following property of logarithms, called the **quotient rule**:

Discovery

We know that $\log_2 16 = 4$. Show that you get the same result by writing 16 as $\frac{32}{2}$ and then using the quotient rule. Then verify the quotient rule using other numbers whose logarithms are easy to find.

The Quotient Rule

Let b, M, and N be positive real numbers with $b \neq 1$.

$$\log_b\left(\frac{M}{N}\right) = \log_b M - \log_b N$$

The logarithm of a quotient is the difference of the logarithms.

When we use the quotient rule to write a single logarithm as the difference of two logarithms, we say that we are **expanding a logarithmic expression**. For example, we can use the quotient rule to expand $\log\frac{x}{2}$:

$$\log\left(\frac{x}{2}\right) = \log x - \log 2.$$

The logarithm of a quotient | is | the difference of the logarithms.

EXAMPLE 2 Using the Quotient Rule

Use the quotient rule to expand each logarithmic expression:

a. $\log_7\left(\frac{19}{x}\right)$ **b.** $\ln\left(\frac{e^3}{7}\right)$.

Solution

a. $\log_7\left(\frac{19}{x}\right) = \log_7 19 - \log_7 x$ The logarithm of a quotient is the difference of the logarithms.

b. $\ln\left(\frac{e^3}{7}\right) = \ln e^3 - \ln 7$ The logarithm of a quotient is the difference of the logarithms. These are natural logarithms with base e understood.

$= 3 - \ln 7$ Because $\ln e^x = x$, then $\ln e^3 = 3$.

Check Point 2 Use the quotient rule to expand each logarithmic expression:

a. $\log_8\left(\frac{23}{x}\right)$ **b.** $\ln\left(\frac{e^5}{11}\right)$.

3 Use the power rule.

The Power Rule

When an exponential expression is raised to a power, we multiply exponents:

$$(b^m)^n = b^{mn}.$$

This property suggests the following property of logarithms, called the **power rule**:

The Power Rule

Let b and M be positive real numbers with $b \neq 1$, and let p be any real number.

$$\log_b M^p = p \log_b M$$

The logarithm of a number with an exponent is the product of the exponent and the logarithm of that number.

When we use the power rule to "pull the exponent to the front," we say that we are **expanding a logarithmic expression**. For example, we can use the power rule to expand $\ln x^2$:

$$\ln x^2 = 2 \ln x.$$

The logarithm of a number with an exponent | is | the product of the exponent and the logarithm of that number.

Figure 4.16 shows the graphs of $y = \ln x^2$ and $y = 2 \ln x$ in $[-5, 5, 1]$ by $[-5, 5, 1]$ viewing rectangles. Are $\ln x^2$ and $2 \ln x$ the same? The graphs illustrate that $y = \ln x^2$ and $y = 2 \ln x$ have different domains. The graphs are only the same if $x > 0$. Thus, we should write

$$\ln x^2 = 2 \ln x \text{ for } x > 0.$$

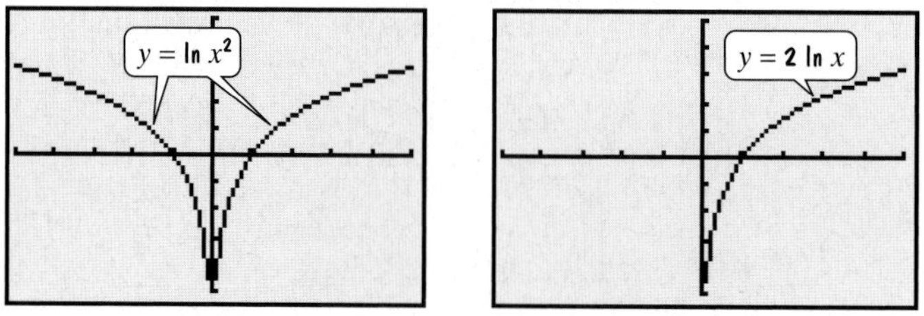

Domain: $(-\infty, 0) \cup (0, \infty)$ Domain: $(0, \infty)$

Figure 4.16 $\ln x^2$ and $2 \ln x$ have different domains.

When expanding a logarithmic expression, you might want to determine whether the rewriting has changed the domain of the expression. For the rest of this section, assume that all variables and variable expressions represent positive numbers.

EXAMPLE 3   Using the Power Rule

Use the power rule to expand each logarithmic expression:

    a. $\log_5 7^4$ **b.** $\ln \sqrt{x}$ **c.** $\log(4x)^5$.

Solution

    a. $\log_5 7^4 = 4 \log_5 7$ The logarithm of a number with an exponent is the exponent times the logarithm of the number.

    b. $\ln \sqrt{x} = \ln x^{\frac{1}{2}}$ Rewrite the radical using a rational exponent.

             $= \frac{1}{2} \ln x$ Use the power rule to bring the exponent to the front.

    c. $\log(4x)^5 = 5 \log(4x)$ We immediately apply the power rule because the entire variable expression, 4x, is raised to the 5th power.

Check Point 3   Use the power rule to expand each logarithmic expression:

    a. $\log_6 3^9$ **b.** $\ln \sqrt[3]{x}$ **c.** $\log(x + 4)^2$.

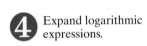

 ④ Expand logarithmic expressions.

Expanding Logarithmic Expressions

It is sometimes necessary to use more than one property of logarithms when you expand a logarithmic expression. Properties for expanding logarithmic expressions are as follows:

Properties for Expanding Logarithmic Expressions

For $M > 0$ and $N > 0$:

1. $\log_b(MN) = \log_b M + \log_b N$ Product rule

2. $\log_b\left(\dfrac{M}{N}\right) = \log_b M - \log_b N$ Quotient rule

3. $\log_b M^p = p \log_b M$ Power rule

Study Tip

The graphs show that

$$\ln(x + 3) \neq \ln x + \ln 3.$$

     $y = \ln x$ shifted 3 units left $y = \ln x$ shifted ln 3 units up

In general,

$$\log_b(M + N) \neq \log_b M + \log_b N.$$

$y_1 = \ln(x + 3)$

$y_2 = \ln x + \ln 3$

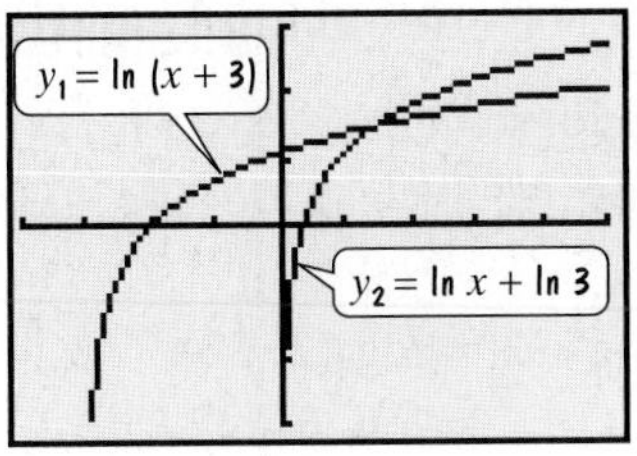

$[-4, 5, 1]$ by $[-3, 3, 1]$

Try to avoid the following errors:

Incorrect!

$\log_b(M + N) = \log_b M + \log_b N$

$\log_b(M - N) = \log_b M - \log_b N$

$\log_b(M \cdot N) = \log_b M \cdot \log_b N$

$\log_b\left(\dfrac{M}{N}\right) = \dfrac{\log_b M}{\log_b N}$

$\dfrac{\log_b M}{\log_b N} = \log_b M - \log_b N$

$\log_b(MN^p) = p \log_b(MN)$

EXAMPLE 4   Expanding Logarithmic Expressions

Use logarithmic properties to expand each expression as much as possible:

a. $\log_b(x^2\sqrt{y})$ **b.** $\log_6\left(\dfrac{\sqrt[3]{x}}{36y^4}\right)$.

Solution   We will have to use two or more of the properties for expanding logarithms in each part of this example.

a. $\log_b(x^2\sqrt{y}) = \log_b\left(x^2 y^{\frac{1}{2}}\right)$ Use exponential notation.

$\qquad\qquad\quad = \log_b x^2 + \log_b y^{\frac{1}{2}}$ Use the product rule.

$\qquad\qquad\quad = 2\log_b x + \dfrac{1}{2}\log_b y$ Use the power rule.

b. $\log_6\left(\dfrac{\sqrt[3]{x}}{36y^4}\right) = \log_6\left(\dfrac{x^{\frac{1}{3}}}{36y^4}\right)$ Use exponential notation.

$\qquad\qquad\quad = \log_6 x^{\frac{1}{3}} - \log_6(36y^4)$ Use the quotient rule.

$\qquad\qquad\quad = \log_6 x^{\frac{1}{3}} - (\log_6 36 + \log_6 y^4)$ Use the product rule on $\log_6(36y^4)$.

$\qquad\qquad\quad = \dfrac{1}{3}\log_6 x - (\log_6 36 + 4\log_6 y)$ Use the power rule.

$\qquad\qquad\quad = \dfrac{1}{3}\log_6 x - \log_6 36 - 4\log_6 y$ Apply the distributive property.

$\qquad\qquad\quad = \dfrac{1}{3}\log_6 x - 2 - 4\log_6 y$ $\log_6 36 = 2$ because 2 is the power to which we must raise 6 to get 36. $(6^2 = 36)$

 Check Point 4   Use logarithmic properties to expand each expression as much as possible:

a. $\log_b(x^4\sqrt[3]{y})$ **b.** $\log_5\left(\dfrac{\sqrt{x}}{25y^3}\right)$.

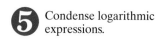 Condense logarithmic expressions.

Condensing Logarithmic Expressions

To **condense a logarithmic expression**, we write the sum or difference of two or more logarithmic expressions as a single logarithmic expression. We use the properties of logarithms to do so.

Properties for Condensing Logarithmic Expressions

For $M > 0$ and $N > 0$:

1. $\log_b M + \log_b N = \log_b(MN)$ Product rule

2. $\log_b M - \log_b N = \log_b\left(\dfrac{M}{N}\right)$ Quotient rule

3. $p\log_b M = \log_b M^p$ Power rule

EXAMPLE 5   Condensing Logarithmic Expressions

Write as a single logarithm:

a. $\log_4 2 + \log_4 32$ **b.** $\log(4x - 3) - \log x$.

Solution

a. $\log_4 2 + \log_4 32 = \log_4(2 \cdot 32)$ Use the product rule.

$\qquad\qquad\qquad\quad = \log_4 64$ We now have a single logarithm. However, we can simplify.

$\qquad\qquad\qquad\quad = 3$ $\log_4 64 = 3$ because $4^3 = 64$.

b. $\log(4x - 3) - \log x = \log\left(\frac{4x - 3}{x}\right)$ Use the quotient rule.

Check Point 5 Write as a single logarithm:

a. $\log 25 + \log 4$ **b.** $\log(7x + 6) - \log x$.

Coefficients of logarithms must be 1 before you can condense them using the product and quotient rules. For example, to condense

$$2 \ln x + \ln(x + 1),$$

the coefficient of the first term must be 1. We use the power rule to rewrite the coefficient as an exponent:

1. Use the power rule to make the number in front an exponent.

$$2 \ln x + \ln(x + 1) = \ln x^2 + \ln(x + 1) = \ln[x^2(x + 1)].$$

2. Use the product rule. The sum of logarithms with coefficients of 1 is the logarithm of the product.

EXAMPLE 6 Condensing Logarithmic Expressions

Write as a single logarithm:

a. $\frac{1}{2}\log x + 4\log(x - 1)$ **b.** $3\ln(x + 7) - \ln x$

c. $4\log_b x - 2\log_b 6 - \frac{1}{2}\log_b y$.

Solution

a. $\frac{1}{2}\log x + 4\log(x - 1)$

$= \log x^{\frac{1}{2}} + \log(x - 1)^4$ Use the power rule so that all coefficients are 1.

$= \log\left[x^{\frac{1}{2}}(x - 1)^4\right]$ Use the product rule. The condensed form can be expressed as $\log[\sqrt{x}(x - 1)^4]$.

b. $3\ln(x + 7) - \ln x$

$= \ln(x + 7)^3 - \ln x$ Use the power rule so that all coefficients are 1.

$= \ln\left[\frac{(x + 7)^3}{x}\right]$ Use the quotient rule.

c. $4\log_b x - 2\log_b 6 - \frac{1}{2}\log_b y$

$= \log_b x^4 - \log_b 6^2 - \log_b y^{\frac{1}{2}}$ Use the power rule so that all coefficients are 1.

$= \log_b x^4 - \left(\log_b 36 + \log_b y^{\frac{1}{2}}\right)$ Rewrite as a single subtraction.

$= \log_b x^4 - \log_b\left(36y^{\frac{1}{2}}\right)$ Use the product rule.

$= \log_b\left(\frac{x^4}{36y^{\frac{1}{2}}}\right)$ or $\log_b\left(\frac{x^4}{36\sqrt{y}}\right)$ Use the quotient rule.

Check Point 6 Write as a single logarithm:

a. $2\ln x + \frac{1}{3}\ln(x + 5)$ **b.** $2\log(x - 3) - \log x$

c. $\frac{1}{4}\log_b x - 2\log_b 5 - 10\log_b y$.

6 Use the change-of-base property.

The Change-of-Base Property

We have seen that calculators give the values of both common logarithms (base 10) and natural logarithms (base e). To find a logarithm with any other base, we can use the following change-of-base property:

The Change-of-Base Property

For any logarithmic bases a and b, and any positive number M,

$$\log_b M = \frac{\log_a M}{\log_a b}.$$

The logarithm of M with base b is equal to the logarithm of M with any new base divided by the logarithm of b with that new base.

In the change-of-base property, base b is the base of the original logarithm. Base a is a new base that we introduce. Thus, the change-of-base property allows us to change from base b to *any* new base a, as long as the newly introduced base is a positive number not equal to 1.

The change-of-base property is used to write a logarithm in terms of quantities that can be evaluated with a calculator. Because calculators contain keys for common (base 10) and natural (base e) logarithms, we will frequently introduce base 10 or base e.

Change-of-Base Property	Introducing Common Logarithms	Introducing Natural Logarithms
$\log_b M = \frac{\log_a M}{\log_a b}$	$\log_b M = \frac{\log_{10} M}{\log_{10} b}$	$\log_b M = \frac{\log_e M}{\log_e b}$
a is the new introduced base.	10 is the new introduced base.	e is the new introduced base.

Using the notations for common logarithms and natural logarithms, we have the following results:

The Change-of-Base Property: Introducing Common and Natural Logarithms

Introducing Common Logarithms

$$\log_b M = \frac{\log M}{\log b}$$

Introducing Natural Logarithms

$$\log_b M = \frac{\ln M}{\ln b}$$

EXAMPLE 7 Changing Base to Common Logarithms

Use common logarithms to evaluate $\log_5 140$.

Solution Because $\log_b M = \frac{\log M}{\log b}$,

$$\log_5 140 = \frac{\log 140}{\log 5} \approx 3.07.$$

Use a calculator: 140 LOG ÷ 5 LOG = or LOG 140 ÷ LOG 5 ENTER. On some calculators, parentheses are needed after 140 and 5.

This means that $\log_5 140 \approx 3.07$.

Discovery

Find a reasonable estimate of $\log_5 140$ to the nearest whole number. To what power can you raise 5 in order to get 140? Compare your estimate to the value obtained in Example 7.

Check Point 7 Use common logarithms to evaluate $\log_7 2506$.

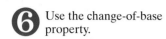

EXAMPLE 8 Changing Base to Natural Logarithms

Use natural logarithms to evaluate $\log_5 140$.

Solution Because $\log_b M = \dfrac{\ln M}{\ln b}$,

$$\log_5 140 = \frac{\ln 140}{\ln 5}$$

$$\approx 3.07.$$

Use a calculator: 140 LN ÷ 5 LN = or LN 140 ÷ LN 5 ENTER. On some calculators, parentheses are needed after 140 and 5.

We have again shown that $\log_5 140 \approx 3.07$.

Check Point 8 Use natural logarithms to evaluate $\log_7 2506$.

Technology

We can use the change-of-base property to graph logarithmic functions with bases other than 10 or e on a graphing utility. For example, Figure 4.17 shows the graphs of

$$y = \log_2 x \quad \text{and} \quad y = \log_{20} x$$

in a $[0, 10, 1]$ by $[-3, 3, 1]$ viewing rectangle. Because $\log_2 x = \dfrac{\ln x}{\ln 2}$ and $\log_{20} x = \dfrac{\ln x}{\ln 20}$, the functions are entered as

$y_1 =$ LN x ÷ LN 2

and $y_2 =$ LN x ÷ LN 20.

On some calculators, parentheses are needed after x, 2, and 20.

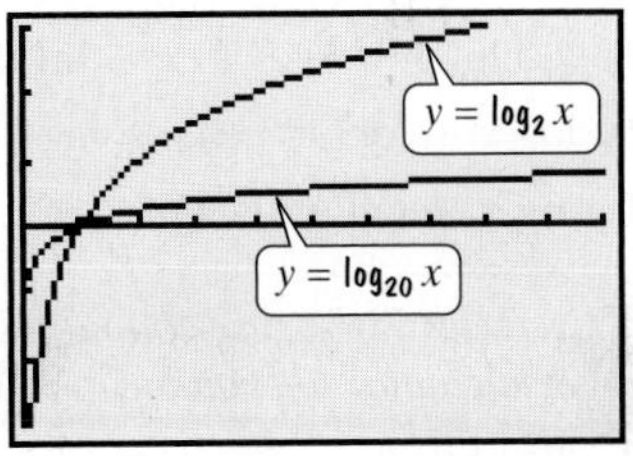

Figure 4.17 Using the change-of-base property to graph logarithmic functions

EXERCISE SET 4.3

Practice Exercises

In Exercises 1–40, use properties of logarithms to expand each logarithmic expression as much as possible. Where possible, evaluate logarithmic expressions without using a calculator.

1. $\log_5(7 \cdot 3)$
2. $\log_8(13 \cdot 7)$
3. $\log_7(7x)$
4. $\log_9(9x)$
5. $\log(1000x)$
6. $\log(10{,}000x)$
7. $\log_7\left(\frac{7}{x}\right)$
8. $\log_9\left(\frac{9}{x}\right)$
9. $\log\left(\frac{x}{100}\right)$
10. $\log\left(\frac{x}{1000}\right)$
11. $\log_4\left(\frac{64}{y}\right)$
12. $\log_5\left(\frac{125}{y}\right)$
13. $\ln\left(\frac{e^2}{5}\right)$
14. $\ln\left(\frac{e^4}{8}\right)$
15. $\log_b x^3$
16. $\log_b x^7$
17. $\log N^{-6}$
18. $\log M^{-8}$
19. $\ln \sqrt[5]{x}$
20. $\ln \sqrt[7]{x}$
21. $\log_b(x^2 y)$
22. $\log_b(xy^3)$
23. $\log_4\left(\frac{\sqrt{x}}{64}\right)$
24. $\log_5\left(\frac{\sqrt{x}}{25}\right)$
25. $\log_6\left(\frac{36}{\sqrt{x+1}}\right)$
26. $\log_8\left(\frac{64}{\sqrt{x+1}}\right)$
27. $\log_b\left(\frac{x^2 y}{z^2}\right)$
28. $\log_b\left(\frac{x^3 y}{z^2}\right)$
29. $\log \sqrt{100x}$
30. $\ln \sqrt{ex}$
31. $\log \sqrt[3]{\frac{x}{y}}$
32. $\log \sqrt[5]{\frac{x}{y}}$
33. $\log_b\left(\frac{\sqrt{x}y^3}{z^3}\right)$
34. $\log_b\left(\frac{\sqrt[3]{x}y^4}{z^5}\right)$
35. $\log_5 \sqrt[3]{\frac{x^2 y}{25}}$
36. $\log_2 \sqrt[5]{\frac{xy^4}{16}}$
37. $\ln\left[\frac{x^3\sqrt{x^2+1}}{(x+1)^4}\right]$
38. $\ln\left[\frac{x^4\sqrt{x^2+3}}{(x+3)^5}\right]$
39. $\log\left[\frac{10x^2\sqrt[3]{1-x}}{7(x+1)^2}\right]$
40. $\log\left[\frac{100x^3\sqrt[3]{5-x}}{3(x+7)^2}\right]$

In Exercises 41–70, use properties of logarithms to condense each logarithmic expression. Write the expression as a single logarithm whose coefficient is 1. Where possible, evaluate logarithmic expressions without using a calculator.

41. $\log 5 + \log 2$
42. $\log 250 + \log 4$
43. $\ln x + \ln 7$
44. $\ln x + \ln 3$
45. $\log_2 96 - \log_2 3$
46. $\log_3 405 - \log_3 5$
47. $\log(2x + 5) - \log x$
48. $\log(3x + 7) - \log x$
49. $\log x + 3 \log y$
50. $\log x + 7 \log y$
51. $\frac{1}{2}\ln x + \ln y$
52. $\frac{1}{3}\ln x + \ln y$
53. $2 \log_b x + 3 \log_b y$
54. $5 \log_b x + 6 \log_b y$
55. $5 \ln x - 2 \ln y$
56. $7 \ln x - 3 \ln y$
57. $3 \ln x - \frac{1}{3}\ln y$
58. $2 \ln x - \frac{1}{2}\ln y$
59. $4 \ln(x + 6) - 3 \ln x$
60. $8 \ln(x + 9) - 4 \ln x$
61. $3 \ln x + 5 \ln y - 6 \ln z$
62. $4 \ln x + 7 \ln y - 3 \ln z$
63. $\frac{1}{2}(\log x + \log y)$
64. $\frac{1}{3}(\log_4 x - \log_4 y)$
65. $\frac{1}{2}(\log_5 x + \log_5 y) - 2 \log_5(x + 1)$
66. $\frac{1}{3}(\log_4 x - \log_4 y) + 2 \log_4(x + 1)$
67. $\frac{1}{3}[2 \ln(x + 5) - \ln x - \ln(x^2 - 4)]$
68. $\frac{1}{3}[5 \ln(x + 6) - \ln x - \ln(x^2 - 25)]$
69. $\log x + \log(x^2 - 1) - \log 7 - \log(x + 1)$
70. $\log x + \log(x^2 - 4) - \log 15 - \log(x + 2)$

In Exercises 71–78, use common logarithms or natural logarithms and a calculator to evaluate to four decimal places.

71. $\log_5 13$
72. $\log_6 17$
73. $\log_{14} 87.5$
74. $\log_{16} 57.2$
75. $\log_{0.1} 17$
76. $\log_{0.3} 19$
77. $\log_\pi 63$
78. $\log_\pi 400$

In Exercises 79–82, use a graphing utility and the change-of-base property to graph each function.

79. $y = \log_3 x$
80. $y = \log_{15} x$
81. $y = \log_2(x + 2)$
82. $y = \log_3(x - 2)$

Practice Plus

In Exercises 83–88, let $\log_b 2 = A$ and $\log_b 3 = C$. Write each expression in terms of A and C.

83. $\log_b \frac{3}{2}$
84. $\log_b 6$
85. $\log_b 8$
86. $\log_b 81$
87. $\log_b \sqrt{\frac{2}{27}}$
88. $\log_b \sqrt{\frac{3}{16}}$

In Exercises 89–102, determine whether each equation is true or false. Where possible, show work to support your conclusion. If the statement is false, make the necessary change(s) to produce a true statement.

89. $\ln e = 0$
90. $\ln 0 = e$
91. $\log_4(2x^3) = 3 \log_4(2x)$
92. $\ln(8x^3) = 3 \ln(2x)$
93. $x \log 10^x = x^2$
94. $\ln(x + 1) = \ln x + \ln 1$
95. $\ln(5x) + \ln 1 = \ln(5x)$
96. $\ln x + \ln(2x) = \ln(3x)$
97. $\log(x + 3) - \log(2x) = \dfrac{\log(x + 3)}{\log(2x)}$
98. $\dfrac{\log(x + 2)}{\log(x - 1)} = \log(x + 2) - \log(x - 1)$
99. $\log_6\left(\dfrac{x - 1}{x^2 + 4}\right) = \log_6(x - 1) - \log_6(x^2 + 4)$
100. $\log_6[4(x + 1)] = \log_6 4 + \log_6(x + 1)$
101. $\log_3 7 = \dfrac{1}{\log_7 3}$
102. $e^x = \dfrac{1}{\ln x}$

Application Exercises

103. The loudness level of a sound can be expressed by comparing the sound's intensity to the intensity of a sound barely audible to the human ear. The formula

$$D = 10(\log I - \log I_0)$$

describes the loudness level of a sound, D, in decibels, where I is the intensity of the sound, in watts per meter^2, and I_0 is the intensity of a sound barely audible to the human ear.

a. Express the formula so that the expression in parentheses is written as a single logarithm.

b. Use the form of the formula from part (a) to answer this question: If a sound has an intensity 100 times the intensity of a softer sound, how much larger on the decibel scale is the loudness level of the more intense sound?

104. The formula

$$t = \frac{1}{c}[\ln A - \ln(A - N)]$$

describes the time, t, in weeks, that it takes to achieve mastery of a portion of a task, where A is the maximum learning possible, N is the portion of the learning that is to be achieved, and c is a constant used to measure an individual's learning style.

a. Express the formula so that the expression in brackets is written as a single logarithm.

b. The formula is also used to determine how long it will take chimpanzees and apes to master a task. For example, a typical chimpanzee learning sign language can master a maximum of 65 signs. Use the form of the formula from part (a) to answer this question: How many weeks will it take a chimpanzee to master 30 signs if c for that chimp is 0.03?

Writing in Mathematics

105. Describe the product rule for logarithms and give an example.

106. Describe the quotient rule for logarithms and give an example.

107. Describe the power rule for logarithms and give an example.

108. Without showing the details, explain how to condense $\ln x - 2 \ln(x + 1)$.

109. Describe the change-of-base property and give an example.

110. Explain how to use your calculator to find $\log_{14} 283$.

111. You overhear a student talking about a property of logarithms in which division becomes subtraction. Explain what the student means by this.

112. Find ln 2 using a calculator. Then calculate each of the following: $1 - \frac{1}{2}$; $1 - \frac{1}{2} + \frac{1}{3}$; $1 - \frac{1}{2} + \frac{1}{3} - \frac{1}{4}$; $1 - \frac{1}{2} + \frac{1}{3} - \frac{1}{4} + \frac{1}{5}$; .... Describe what you observe.

Technology Exercises

113. a. Use a graphing utility (and the change-of-base property) to graph $y = \log_3 x$.

b. Graph $y = 2 + \log_3 x$, $y = \log_3(x + 2)$, and $y = -\log_3 x$ in the same viewing rectangle as $y = \log_3 x$. Then describe the change or changes that need to be made to the graph of $y = \log_3 x$ to obtain each of these three graphs.

114. Graph $y = \log x$, $y = \log(10x)$, and $y = \log(0.1x)$ in the same viewing rectangle. Describe the relationship among the three graphs. What logarithmic property accounts for this relationship?

115. Use a graphing utility and the change-of-base property to graph $y = \log_3 x$, $y = \log_{25} x$, and $y = \log_{100} x$ in the same viewing rectangle.

a. Which graph is on the top in the interval $(0, 1)$? Which is on the bottom?

b. Which graph is on the top in the interval $(1, \infty)$? Which is on the bottom?

c. Generalize by writing a statement about which graph is on top, which is on the bottom, and in which intervals, using $y = \log_b x$ where $b > 1$.

Disprove each statement in Exercises 116–120 by first letting y equal a positive constant of your choice, and then using a graphing utility to graph the function on each side of the equal sign. The two functions should have different graphs, showing that the equation is not true in general.

116. $\log(x + y) = \log x + \log y$

117. $\log\left(\frac{x}{y}\right) = \frac{\log x}{\log y}$

118. $\ln(x - y) = \ln x - \ln y$

119. $\ln(xy) = (\ln x)(\ln y)$

120. $\frac{\ln x}{\ln y} = \ln x - \ln y$

Critical Thinking Exercises

121. Which one of the following is true?

a. $\frac{\log_7 49}{\log_7 7} = \log_7 49 - \log_7 7$

b. $\log_b(x^3 + y^3) = 3\log_b x + 3\log_b y$

c. $\log_b(xy)^5 = (\log_b x + \log_b y)^5$

d. $\ln\sqrt{2} = \frac{\ln 2}{2}$

122. Use the change-of-base property to prove that

$$\log e = \frac{1}{\ln 10}.$$

123. If $\log 3 = A$ and $\log 7 = B$, find $\log_7 9$ in terms of A and B.

124. Write as a single term that does not contain a logarithm:

$$e^{\ln 8x^5 - \ln 2x^2}.$$

125. If $f(x) = \log_b x$, show that

$$\frac{f(x + h) - f(x)}{h} = \log_b\left(1 + \frac{h}{x}\right)^{\frac{1}{h}}, h \neq 0.$$

CHAPTER 4 MID-CHAPTER CHECK POINT

What You Know: We evaluated and graphed exponential functions $[f(x) = b^x, b > 0 \text{ and } b \neq 1]$, including the natural exponential function $[f(x) = e^x, e \approx 2.718]$. A function has an inverse that is a function if there is no horizontal line that intersects the function's graph more than once. The exponential function passes this horizontal line test and we called the inverse of the exponential function with base b the logarithmic function with base b. We learned that $y = \log_b x$ is equivalent to $b^y = x$. We evaluated and graphed logarithmic functions, including the common logarithmic function $[f(x) = \log_{10} x \text{ or } f(x) = \log x]$ and the natural logarithmic function $[f(x) = \log_e x \text{ or } f(x) = \ln x]$. We learned to use transformations to graph exponential and logarithmic functions. Finally, we used properties of logarithms to expand and condense logarithmic expressions.

In Exercises 1–5, graph f and g in the same rectangular coordinate system. Graph and give equations of all asymptotes. Give each function's domain and range.

1. $f(x) = 2^x$ and $g(x) = 2^x - 3$
2. $f(x) = \left(\frac{1}{2}\right)^x$ and $g(x) = \left(\frac{1}{2}\right)^{x-1}$
3. $f(x) = e^x$ and $g(x) = \ln x$
4. $f(x) = \log_2 x$ and $g(x) = \log_2(x - 1) + 1$
5. $g(x) = \log_{\frac{1}{2}} x$ and $g(x) = -2\log_{\frac{1}{2}} x$

In Exercises 6–9, find the domain of each function.

6. $f(x) = \log_3(x + 6)$
7. $g(x) = \log_3 x + 6$
8. $h(x) = \log_3(x + 6)^2$
9. $f(x) = 3^{x+6}$

In Exercises 10–20, evaluate each expression without using a calculator. If evaluation is not possible, state the reason.

10. $\log_2 8 + \log_5 25$
11. $\log_3 \frac{1}{9}$
12. $\log_{100} 10$
13. $\log \sqrt[3]{10}$
14. $\log_2(\log_3 81)$
15. $\log_3\left(\log_2 \frac{1}{8}\right)$
16. $6^{\log_6 5}$
17. $\ln e^{\sqrt{7}}$
18. $10^{\log 13}$
19. $\log_{100} 0.1$
20. $\log_\pi \pi^{\sqrt{\pi}}$

In Exercises 21–22, expand and evaluate numerical terms.

21. $\log\left(\frac{\sqrt{xy}}{1000}\right)$
22. $\ln(e^{19}x^{20})$

In Exercises 23–25, write each expression as a single logarithm.

23. $8\log_7 x - \frac{1}{3}\log_7 y$
24. $7\log_5 x + 2\log_5 x$
25. $\frac{1}{2}\ln x - 3\ln y - \ln(z - 2)$
26. Use the formulas

$$A = P\left(1 + \frac{r}{n}\right)^{nt} \quad \text{and} \quad A = Pe^{rt}$$

to solve this exercise. You decide to invest \$8000 for 3 years at an annual rate of 8%. How much more is the return if the interest is compounded continuously than monthly? Round to the nearest dollar.

SECTION 4.4 Exponential and Logarithmic Equations

Objectives

1. Use like bases to solve exponential equations.
2. Use logarithms to solve exponential equations.
3. Use the definition of a logarithm to solve logarithmic equations.
4. Use the one-to-one property of logarithms to solve logarithmic equations.
5. Solve applied problems involving exponential and logarithmic equations.

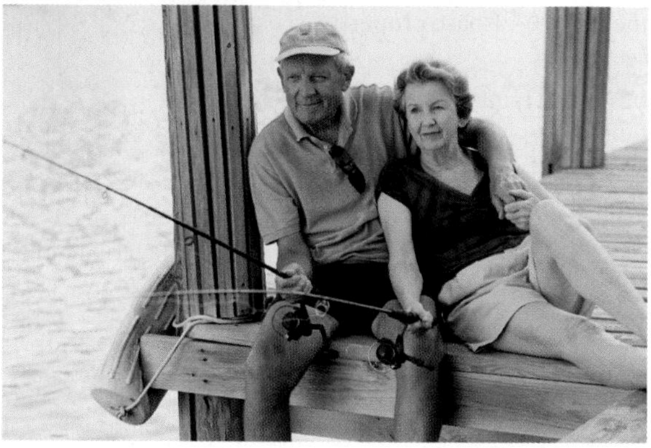

You inherited $30,000. You would like to put aside $25,000 and eventually have over half a million dollars for early retirement. Is this possible? In this section, you will see how techniques for solving equations with variable exponents provide an answer to the question.

1 Use like bases to solve exponential equations.

Exponential Equations

An **exponential equation** is an equation containing a variable in an exponent. Examples of exponential equations include

$$2^{3x-8} = 16, \quad 4^x = 15, \quad \text{and} \quad 40e^{0.6x} = 240.$$

Some exponential equations can be solved by expressing each side of the equation as a power of the same base. All exponential functions are one-to-one—that is, no two different ordered pairs have the same second component. Thus, if b is a positive number other than 1 and $b^M = b^N$, then $M = N$.

Solving Exponential Equations by Expressing Each Side as a Power of the Same Base

$$\text{If } b^M = b^N, \text{ then } M = N.$$

Express each side as a power of the same base. Set the exponents equal to each other.

1. Rewrite the equation in the form $b^M = b^N$.
2. Set $M = N$.
3. Solve for the variable.

Technology

The graphs of

$$y_1 = 2^{3x-8}$$
and $$y_2 = 16$$

have an intersection point whose x-coordinate is 4. This verifies that $\{4\}$ is the solution set of $2^{3x-8} = 16$.

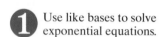

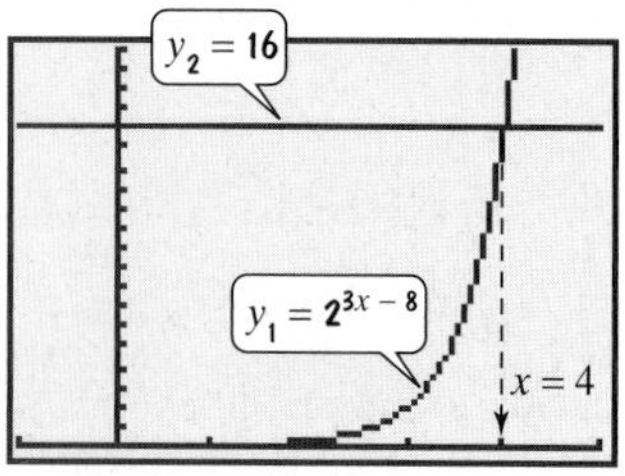

$[-1, 5, 1]$ by $[0, 20, 1]$

EXAMPLE 1 Solving Exponential Equations

Solve: **a.** $2^{3x-8} = 16$ **b.** $27^{x+3} = 9^{x-1}$.

Solution In each equation, express both sides as a power of the same base. Then set the exponents equal to each other and solve for the variable.

a. Because 16 is 2^4, we express each side of $2^{3x-8} = 16$ in terms of base 2.

$2^{3x-8} = 16$	This is the given equation.
$2^{3x-8} = 2^4$	Write each side as a power of the same base.
$3x - 8 = 4$	If $b^M = b^N$, $b > 0$ and $b \neq 1$, then $M = N$.
$3x = 12$	Add 8 to both sides.
$x = 4$	Divide both sides by 3.

Substituting 4 for x into the original equation produces the true statement $16 = 16$. The solution set is $\{4\}$.

b. Because $27 = 3^3$ and $9 = 3^2$, we can express both sides of $27^{x+3} = 9^{x-1}$ in terms of base 3.

$27^{x+3} = 9^{x-1}$	This is the given equation.
$(3^3)^{x+3} = (3^2)^{x-1}$	Write each side as a power of the same base.
$3^{3(x+3)} = 3^{2(x-1)}$	When an exponential expression is raised to a power, multiply exponents.
$3(x + 3) = 2(x - 1)$	If two powers of the same base are equal, then the exponents are equal.
$3x + 9 = 2x - 2$	Apply the distributive property.
$x + 9 = -2$	Subtract 2x from both sides.
$x = -11$	Subtract 9 from both sides.

Substituting -11 for x into the original equation produces $27^{-8} = 9^{-12}$, which simplifies to the true statement $3^{-24} = 3^{-24}$. The solution set is $\{-11\}$.

Check Point 1 Solve: **a.** $5^{3x-6} = 125$ **b.** $8^{x+2} = 4^{x-3}$.

② Use logarithms to solve exponential equations.

Most exponential equations cannot be rewritten so that each side has the same base. Logarithms are extremely useful in solving such equations. The solution begins with isolating the exponential expression and taking the natural logarithm on both sides. Why can we do this? All logarithmic relations are functions. Thus, if M and N are positive real numbers and $M = N$, then $\log_b M = \log_b N$.

Using Natural Logarithms to Solve Exponential Equations

1. Isolate the exponential expression.
2. Take the natural logarithm on both sides of the equation.
3. Simplify using one of the following properties:

$$\ln b^x = x \ln b \quad \text{or} \quad \ln e^x = x.$$

4. Solve for the variable.

EXAMPLE 2   Solving an Exponential Equation

Solve: $4^x = 15$.

Solution   Because the exponential expression, 4^x, is already isolated on the left, we begin by taking the natural logarithm on both sides of the equation.

$4^x = 15$	This is the given equation.
$\ln 4^x = \ln 15$	Take the natural logarithm on both sides.
$x \ln 4 = \ln 15$	Use the power rule and bring the variable exponent to the front: $\ln b^x = x \ln b$.
$x = \dfrac{\ln 15}{\ln 4}$	Solve for x by dividing both sides by $\ln 4$.

Discovery

The base that is used when taking the logarithm on both sides of an equation can be any base at all. Solve $4^x = 15$ by taking the common logarithm on both sides. Solve again, this time taking the logarithm with base 4 on both sides. Use the change-of-base property to show that the solutions are the same as the one obtained in Example 2.

We now have an exact value for x. We use the exact value for x in the equation's solution set. Thus, the equation's solution is $\dfrac{\ln 15}{\ln 4}$ and the solution set is $\left\{ \dfrac{\ln 15}{\ln 4} \right\}$.

We can obtain a decimal approximation by using a calculator: $x \approx 1.95$. Because $4^2 = 16$, it seems reasonable that the solution to $4^x = 15$ is approximately 1.95.

Check Point 2 Solve: $5^x = 134$. Find the solution set and then use a calculator to obtain a decimal approximation to two decimal places for the solution.

EXAMPLE 3 Solving an Exponential Equation

Solve: $40e^{0.6x} - 3 = 237$.

Solution We begin by adding 3 to both sides and dividing both sides by 40 to isolate the exponential expression, $e^{0.6x}$. Then we take the natural logarithm on both sides of the equation.

$$40e^{0.6x} - 3 = 237 \quad \text{This is the given equation.}$$

$$40e^{0.6x} = 240 \quad \text{Add 3 to both sides.}$$

$$e^{0.6x} = 6 \quad \text{Isolate the exponential factor by dividing both sides by 40.}$$

$$\ln e^{0.6x} = \ln 6 \quad \text{Take the natural logarithm on both sides.}$$

$$0.6x = \ln 6 \quad \text{Use the inverse property } \ln e^x = x \text{ on the left.}$$

$$x = \frac{\ln 6}{0.6} \approx 2.99 \quad \text{Divide both sides by 0.6 and solve for } x.$$

Thus, the solution of the equation is $\frac{\ln 6}{0.6} \approx 2.99$. Try checking this approximate solution in the original equation to verify that $\left\{\frac{\ln 6}{0.6}\right\}$ is the solution set.

Check Point 3 Solve: $7e^{2x} - 5 = 58$. Find the solution set and then use a calculator to obtain a decimal approximation to two decimal places for the solution.

EXAMPLE 4 Solving an Exponential Equation

Solve: $5^{x-2} = 4^{2x+3}$.

Solution Because each exponential expression is isolated on one side of the equation, we begin by taking the natural logarithm on both sides.

$$5^{x-2} = 4^{2x+3} \quad \text{This is the given equation.}$$

$$\ln 5^{x-2} = \ln 4^{2x+3} \quad \text{Take the natural logarithm on both sides.}$$

Be sure to insert parentheses around the binomials.

$$(x - 2)\ln 5 = (2x + 3)\ln 4 \quad \text{Use the power rule and bring the variable exponents to the front: } \ln b^x = x \ln b.$$

Remember that ln 5 and ln 4 are constants, not variables.

$$x\ln 5 - 2\ln 5 = 2x\ln 4 + 3\ln 4 \quad \text{Use the distributive property to distribute ln 5 and ln 4 to both terms in parentheses.}$$

$$x\ln 5 - 2x\ln 4 = 2\ln 5 + 3\ln 4 \quad \text{Collect variable terms involving } x \text{ on the left by subtracting } 2x\ln 4 \text{ and adding } 2\ln 5 \text{ on both sides.}$$

$$x(\ln 5 - 2\ln 4) = 2\ln 5 + 3\ln 4 \quad \text{Factor out } x \text{ from the two terms on the left.}$$

$$x = \frac{2\ln 5 + 3\ln 4}{\ln 5 - 2\ln 4} \quad \text{Isolate } x \text{ by dividing both sides by } \ln 5 - 2\ln 4.$$

The solution set is $\left\{\frac{2\ln 5 + 3\ln 4}{\ln 5 - 2\ln 4}\right\}$. The solution is approximately -6.34.

Discovery

Use properties of logarithms to show that the solution in Example 4 can be expressed as

$$\frac{\ln 1600}{\ln\left(\frac{5}{16}\right)}.$$

Check Point 4 Solve: $3^{2x-1} = 7^{x+1}$. Find the solution set and then use a calculator to obtain a decimal approximation to two decimal places for the solution.

EXAMPLE 5   Solving an Exponential Equation

Solve: $e^{2x} - 4e^x + 3 = 0$.

Solution   The given equation is quadratic in form. If $u = e^x$, the equation can be expressed as $u^2 - 4u + 3 = 0$. Because this equation can be solved by factoring, we factor to isolate the exponential term.

$e^{2x} - 4e^x + 3 = 0$	This is the given equation.
$(e^x - 3)(e^x - 1) = 0$	Factor on the left. Notice that if $u = e^x$, $u^2 - 4u + 3 = (u - 3)(u - 1)$.

$$e^x - 3 = 0 \quad \text{or} \quad e^x - 1 = 0 \qquad \text{Set each factor equal to 0.}$$
$$e^x = 3 \qquad\qquad e^x = 1 \qquad \text{Solve for } e^x.$$
$$\ln e^x = \ln 3 \qquad\qquad x = 0 \qquad \text{Take the natural logarithm on both sides of the first equation. The equation on the right can be solved by inspection.}$$
$$x = \ln 3 \qquad\qquad \text{ln } e^x = x$$

The solution set is $\{0, \ln 3\}$. The solutions are 0 and $\ln 3$, which is approximately 1.10.

Check Point 5   Solve: $e^{2x} - 8e^x + 7 = 0$. Find the solution set and then use a calculator to obtain a decimal approximation to two decimal places, if necessary, for the solutions.

❸ Use the definition of a logarithm to solve logarithmic equations.

Logarithmic Equations

A **logarithmic equation** is an equation containing a variable in a logarithmic expression. Examples of logarithmic equations include

$$\log_4(x + 3) = 2 \quad \text{and} \quad \ln(x + 2) - \ln(4x + 3) = \ln\left(\frac{1}{x}\right).$$

Some logarithmic equations can be expressed in the form $\log_b M = c$. We can solve such equations by rewriting them in exponential form.

Using the Definition of a Logarithm to Solve Logarithmic Equations

1. Express the equation in the form $\log_b M = c$.
2. Use the definition of a logarithm to rewrite the equation in exponential form:

$$\log_b M = c \quad \text{means} \quad b^c = M.$$

Logarithms are exponents.

3. Solve for the variable.
4. Check proposed solutions in the original equation. Include in the solution set only values for which $M > 0$.

EXAMPLE 6   Solving Logarithmic Equations

Solve: **a.** $\log_4(x + 3) = 2$ **b.** $3\ln(2x) = 12$.

Solution   The form $\log_b M = c$ involves a single logarithm whose coefficient is 1 on one side and a constant on the other side. Equation (a) is already in this form. We will need to divide both sides of equation (b) by 3 to obtain this form.

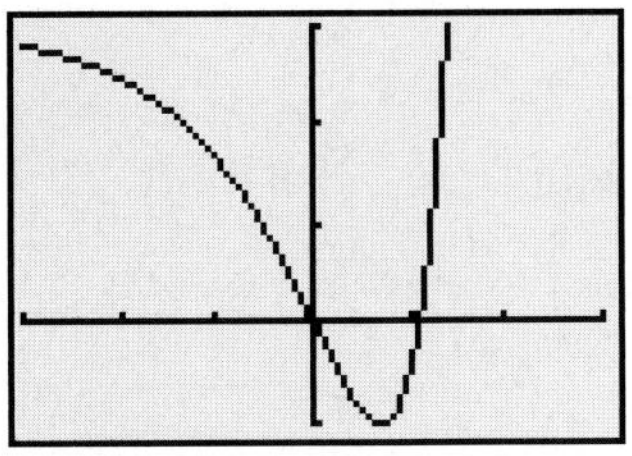

Technology

The graphs of

$y_1 = \log_4(x + 3)$ and $y_2 = 2$

have an intersection point whose x-coordinate is 13. This verifies that $\{13\}$ is the solution set for $\log_4(x + 3) = 2$.

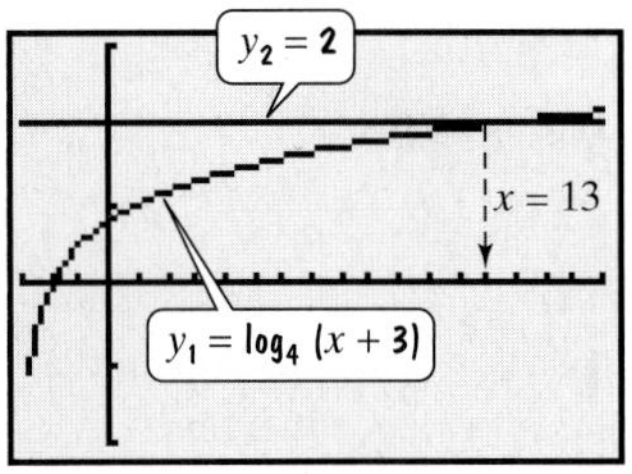

[−3, 17, 1] by [−2, 3, 1]

a. $\log_4(x + 3) = 2$ — This is the given equation.

$4^2 = x + 3$ — Rewrite in exponential form: $\log_b M = c$ means $b^c = M$.

$16 = x + 3$ — Square 4.

$13 = x$ — Subtract 3 from both sides.

Check 13:

$\log_4(x + 3) = 2$ — This is the given logarithmic equation.

$\log_4(13 + 3) \stackrel{?}{=} 2$ — Substitute 13 for x.

$\log_4 16 \stackrel{?}{=} 2$

$2 = 2$, true — $\log_4 16 = 2$ because $4^2 = 16$.

This true statement indicates that the solution set is $\{13\}$.

b. $3\ln(2x) = 12$ — This is the given equation.

$\ln(2x) = 4$ — Divide both sides by 3.

$\log_e(2x) = 4$ — Rewrite the natural logarithm showing base e. This step is optional.

$e^4 = 2x$ — Rewrite in exponential form: $\log_b M = c$ means $b^c = M$.

$\frac{e^4}{2} = x$ — Divide both sides by 2.

Check $\frac{e^4}{2}$:

$3\ln(2x) = 12$ — This is the given logarithmic equation.

$3\ln\left[2\left(\frac{e^4}{2}\right)\right] \stackrel{?}{=} 12$ — Substitute $\frac{e^4}{2}$ for x.

$3\ln e^4 \stackrel{?}{=} 12$ — Simplify: $\frac{\cancel{2}}{1}\cdot\frac{e^4}{\cancel{2}} = e^4$.

$3\cdot 4 \stackrel{?}{=} 12$ — Because $\ln e^x = x$, we conclude $\ln e^4 = 4$.

$12 = 12$, true

This true statement indicates that the solution set is $\left\{\frac{e^4}{2}\right\}$.

Check Point 6 Solve:

a. $\log_2(x - 4) = 3$ **b.** $4\ln(3x) = 8$.

Logarithmic expressions are defined only for logarithms of positive real numbers. **Always check proposed solutions of a logarithmic equation in the original equation. Exclude from the solution set any proposed solution that produces the logarithm of a negative number or the logarithm of 0.**

To rewrite the logarithmic equation $\log_b M = c$ in the equivalent exponential form $b^c = M$, we need a single logarithm whose coefficient is one. It is sometimes necessary to use properties of logarithms to condense logarithms into a single logarithm. In the next example, we use the product rule for logarithms to obtain a single logarithmic expression on the left side.

EXAMPLE 7 Solving a Logarithmic Equation

Solve: $\log_2 x + \log_2(x - 7) = 3$.

Solution

$\log_2 x + \log_2(x - 7) = 3$	This is the given equation.
$\log_2[x(x - 7)] = 3$	Use the product rule to obtain a single logarithm: $\log_b M + \log_b N = \log_b(MN)$.
$2^3 = x(x - 7)$	Rewrite in exponential form: $\log_b M = c$ means $b^c = M$.
$8 = x^2 - 7x$	Evaluate 2^3 on the left and apply the distributive property on the right.
$0 = x^2 - 7x - 8$	Set the equation equal to 0.
$0 = (x - 8)(x + 1)$	Factor.
$x - 8 = 0$ or $x + 1 = 0$	Set each factor equal to 0.
$x = 8$ $\quad$ $x = -1$	Solve for x.

Check 8:

$$\log_2 x + \log_2(x - 7) = 3$$
$$\log_2 8 + \log_2(8 - 7) \stackrel{?}{=} 3$$
$$\log_2 8 + \log_2 1 \stackrel{?}{=} 3$$
$$3 + 0 \stackrel{?}{=} 3$$
$$3 = 3, \text{ true}$$

Check −1:

$$\log_2 x + \log_2(x - 7) = 3$$
$$\log_2(-1) + \log_2(-1 - 7) \stackrel{?}{=} 3$$

The number −1 does not check. Negative numbers do not have logarithms.

The solution set is $\{8\}$.

Check Point 7 Solve: $\log x + \log(x - 3) = 1$.

4 Use the one-to-one property of logarithms to solve logarithmic equations.

Some logarithmic equations can be expressed in the form $\log_b M = \log_b N$. Because all logarithmic functions are one-to-one, we can conclude that $M = N$.

Using the One-to-One Property of Logarithms to Solve Logarithmic Equations

1. Express the equation in the form $\log_b M = \log_b N$. This form involves a single logarithm whose coefficient is 1 on each side of the equation.
2. Use the one-to-one property to rewrite the equation without logarithms: If $\log_b M = \log_b N$, then $M = N$.
3. Solve for the variable.
4. Check proposed solutions in the original equation. Include in the solution set only values for which $M > 0$ and $N > 0$.

EXAMPLE 8 Solving a Logarithmic Equation

Solve: $\ln(x + 2) - \ln(4x + 3) = \ln\left(\frac{1}{x}\right)$.

Solution In order to apply the one-to-one property of logarithms, we need a single logarithm whose coefficient is 1 on each side of the equation. The right side is already in this form. We can obtain a single logarithm on the left side by applying the quotient rule.

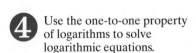

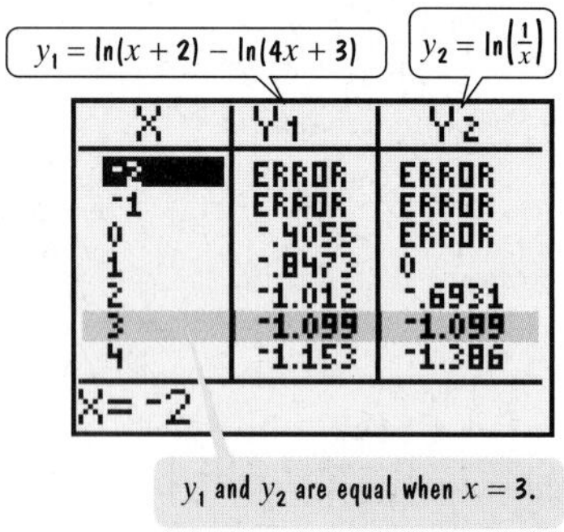

$$\ln(x + 2) - \ln(4x + 3) = \ln\left(\frac{1}{x}\right)$$ This is the given equation.

$$\ln\left(\frac{x + 2}{4x + 3}\right) = \ln\left(\frac{1}{x}\right)$$ Use the quotient rule to obtain a single logarithm on the left side:

$$\log_b M - \log_b N = \log_b\left(\frac{M}{N}\right).$$

$$\frac{x + 2}{4x + 3} = \frac{1}{x}$$ Use the one-to-one property: If $\log_b M = \log_b N$, then $M = N$.

$$x(4x + 3)\left(\frac{x + 2}{4x + 3}\right) = x(4x + 3)\left(\frac{1}{x}\right)$$ Multiply both sides by $x(4x + 3)$, the LCD.

$$x(x + 2) = 4x + 3$$ Simplify.

$$x^2 + 2x = 4x + 3$$ Apply the distributive property.

$$x^2 - 2x - 3 = 0$$ Subtract $4x + 3$ from both sides and set the equation equal to 0.

$$(x - 3)(x + 1) = 0$$ Factor.

$$x - 3 = 0 \quad \text{or} \quad x + 1 = 0$$ Set each factor equal to 0.

$$x = 3 \qquad\qquad x = -1$$ Solve for x.

Substituting 3 for x into the original equation produces the true statement $\ln\left(\frac{1}{3}\right) = \ln\left(\frac{1}{3}\right)$. However, substituting -1 produces logarithms of negative numbers. Thus, -1 is not a solution. The solution set is $\{3\}$.

Check Point 8 Solve: $\ln(x - 3) = \ln(7x - 23) - \ln(x + 1)$.

---

⑤ Solve applied problems involving exponential and logarithmic equations.

Applications

Our first applied example provides a mathematical perspective on the old slogan "Alcohol and driving don't mix." In California, where 38% of fatal traffic crashes involve drinking drivers, it is illegal to drive with a blood alcohol concentration of 0.08 or higher. At these levels, drivers may be arrested and charged with driving under the influence.

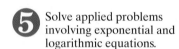

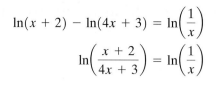

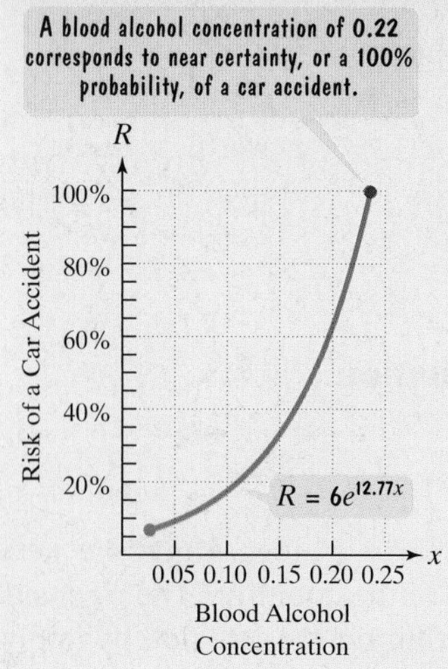

EXAMPLE 9 Alcohol and Risk of a Car Accident

Medical research indicates that the risk of having a car accident increases exponentially as the concentration of alcohol in the blood increases. The risk is modeled by
$$R = 6e^{12.77x},$$
where x is the blood alcohol concentration and R, given as a percent, is the risk of having a car accident. What blood alcohol concentration corresponds to a 20% risk of a car accident?

Solution For a risk of 20% we let $R = 20$ in the equation and solve for x, the blood alcohol concentration.

$$R = 6e^{12.77x}$$ This is the given equation.

$$6e^{12.77x} = 20$$ Substitute 20 for R and (optional) reverse the two sides of the equation.

$$e^{12.77x} = \frac{20}{6}$$ Isolate the exponential factor by dividing both sides by 6.

$$\ln e^{12.77x} = \ln\left(\frac{20}{6}\right)$$ Take the natural logarithm on both sides.

$$12.77x = \ln\left(\frac{20}{6}\right)$$ Use the inverse property $\ln e^x = x$ on the left.

$$x = \frac{\ln\left(\frac{20}{6}\right)}{12.77} \approx 0.09$$ Divide both sides by 12.77 and solve for x.

Visualizing the Relationship between Blood Alcohol Concentration and the Risk of a Car Accident

A blood alcohol concentration of 0.22 corresponds to near certainty, or a 100% probability, of a car accident.

Risk of a Car Accident

$R = 6e^{12.77x}$

Blood Alcohol Concentration

For a blood alcohol concentration of 0.09, the risk of a car accident is 20%. In many states, it is illegal to drive at 0.08, which is below this blood alcohol concentration.

Check Point 9 Use the formula in Example 9 to answer this question: What blood alcohol concentration corresponds to a 7% risk of a car accident? (In many states, drivers under the age of 21 can lose their licenses for driving at this level.)

Suppose that you inherit \$30,000. Is it possible to invest \$25,000 and have over half a million dollars for early retirement? Our next example illustrates the power of compound interest.

EXAMPLE 10 Revisiting the Formula for Compound Interest

The formula

$$A = P\left(1 + \frac{r}{n}\right)^{nt}$$

describes the accumulated value, A, of a sum of money, P, the principal, after t years at annual percentage rate r (in decimal form) compounded n times a year. How long will it take \$25,000 to grow to \$500,000 at 9% annual interest compounded monthly?

Solution

$$A = P\left(1 + \frac{r}{n}\right)^{nt}$$ This is the given formula.

$$500{,}000 = 25{,}000\left(1 + \frac{0.09}{12}\right)^{12t}$$ A(the desired accumulated value) = \$500,000, P(the principal) = \$25,000, r(the interest rate) = 9% = 0.09, and n = 12 (monthly compounding).

Our goal is to solve the equation for t. Let's reverse the two sides of the equation and then simplify within parentheses.

$$25{,}000\left(1 + \frac{0.09}{12}\right)^{12t} = 500{,}000$$ Reverse the two sides of the previous equation.

$$25{,}000(1 + 0.0075)^{12t} = 500{,}000$$ Divide within parentheses: $\frac{0.09}{12} = 0.0075$.

$$25{,}000(1.0075)^{12t} = 500{,}000$$ Add within parentheses.

$$(1.0075)^{12t} = 20$$ Divide both sides by 25,000.

$$\ln(1.0075)^{12t} = \ln 20$$ Take the natural logarithm on both sides.

$$12t \ln(1.0075) = \ln 20$$ Use the power rule to bring the exponent to the front: $\ln M^p = p \ln M$.

$$t = \frac{\ln 20}{12 \ln 1.0075}$$ Solve for t, dividing both sides by $12 \ln 1.0075$.

$$\approx 33.4$$ Use a calculator.

After approximately 33.4 years, the \$25,000 will grow to an accumulated value of \$500,000. If you set aside the money at age 20, you can begin enjoying a life of leisure at about age 53.

Check Point 10 How long, to the nearest tenth of a year, will it take \$1000 to grow to \$3600 at 8% annual interest compounded quarterly?

Playing Doubles: Interest Rates and Doubling Time

One way to calculate what your savings will be worth at some point in the future is to consider doubling time. The following table shows how long it takes for your money to double at different annual interest rates subject to continuous compounding.

Annual Interest Rate	Years to Double
5%	13.9 years
7%	9.9 years
9%	7.7 years
11%	6.3 years

Of course, the first problem is collecting some money to invest. The second problem is finding a reasonably safe investment with a return of 9% or more.

EXAMPLE 11 **The Growth in the Number of U.S. Internet Users**

The bar graph in Figure 4.18 shows the number, in millions, of Internet users in the United States from 2000 through 2003. The function

$$f(x) = 34.1 \ln x + 117.7$$

models the number of U.S. Internet users, $f(x)$, in millions, x years after 1999. By which year will there be 200 million Internet users in the United States?

Solution We substitute 200 for $f(x)$ and solve for x, the number of years after 1999.

$$f(x) = 34.1 \ln x + 117.7 \quad \text{This is the given function.}$$

$$200 = 34.1 \ln x + 117.7 \quad \text{Substitute 200 for } f(x).$$

Number of Internet Users in the U.S.

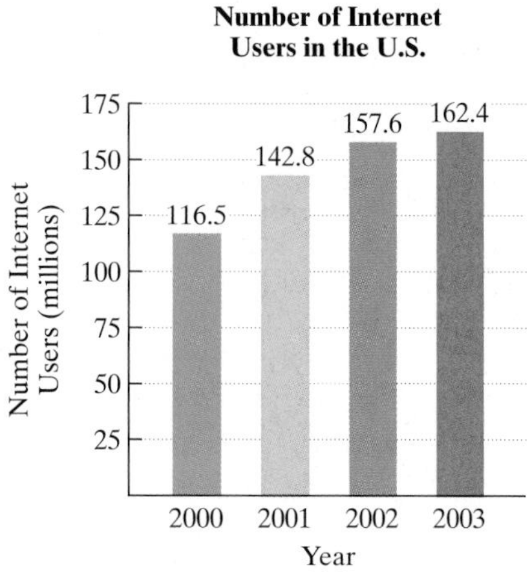

Figure 4.18

Source: Jupiter Media

Our goal is to isolate $\ln x$ in the equation $200 = 34.1 \ln x + 117.7$. We can then find x by using the definition of a logarithm to rewrite the equation in exponential form.

$$34.1 \ln x + 117.7 = 200 \quad \text{Reverse the two sides of the equation.}$$

$$34.1 \ln x = 82.3 \quad \text{Subtract 117.7 from both sides.}$$

$$\ln x = \frac{82.3}{34.1} \quad \text{Divide both sides by 34.1.}$$

$$\log_e x = \frac{82.3}{34.1} \quad \text{Rewrite the natural logarithm showing base } e. \text{ This step is optional.}$$

$$e^{\frac{82.3}{34.1}} = x \quad \text{Rewrite in exponential form: } \log_b M = c \text{ means } b^c = M.$$

$$11 \approx x \quad \text{Use a calculator.}$$

Approximately 11 years after 1999, in the year 2010, there will be 200 million Internet users in the United States.

Check Point 11 Use the function in Example 11 to find in which year there will be 210 million Internet users in the United States.

EXERCISE SET 4.4

 Practice Exercises

Solve each exponential equation in Exercises 1–22 by expressing each side as a power of the same base and then equating exponents.

1. $2^x = 64$

2. $3^x = 81$

3. $5^x = 125$

4. $5^x = 625$

5. $2^{2x-1} = 32$

6. $3^{2x+1} = 27$

7. $4^{2x-1} = 64$

8. $5^{3x-1} = 125$

9. $32^x = 8$

10. $4^x = 32$

11. $9^x = 27$

12. $125^x = 625$

13. $3^{1-x} = \frac{1}{27}$

14. $5^{2-x} = \frac{1}{125}$

15. $6^{\frac{x-3}{4}} = \sqrt{6}$

16. $7^{\frac{x-2}{6}} = \sqrt{7}$

17. $4^x = \frac{1}{\sqrt{2}}$

18. $9^x = \frac{1}{\sqrt[3]{3}}$

19. $8^{x+3} = 16^{x-1}$

20. $8^{1-x} = 4^{x+2}$

21. $e^{x+1} = \frac{1}{e}$

22. $e^{x+4} = \frac{1}{e^{2x}}$

Solve each exponential equation in Exercises 23–48. Express the solution set in terms of natural logarithms. Then use a calculator to obtain a decimal approximation, correct to two decimal places, for the solution.

23. $10^x = 3.91$

24. $10^x = 8.07$

25. $e^x = 5.7$ **26.** $e^x = 0.83$

27. $5^x = 17$ **28.** $19^x = 143$

29. $5e^x = 23$ **30.** $9e^x = 107$

31. $3e^{5x} = 1977$ **32.** $4e^{7x} = 10{,}273$

33. $e^{1-5x} = 793$ **34.** $e^{1-8x} = 7957$

35. $e^{5x-3} - 2 = 10{,}476$ **36.** $e^{4x-5} - 7 = 11{,}243$

37. $7^{x+2} = 410$ **38.** $5^{x-3} = 137$

39. $7^{0.3x} = 813$ **40.** $3^{\frac{x}{7}} = 0.2$

41. $5^{2x+3} = 3^{x-1}$ **42.** $7^{2x+1} = 3^{x+2}$

43. $e^{2x} - 3e^x + 2 = 0$ **44.** $e^{2x} - 2e^x - 3 = 0$

45. $e^{4x} + 5e^{2x} - 24 = 0$ **46.** $e^{4x} - 3e^{2x} - 18 = 0$

47. $3^{2x} + 3^x - 2 = 0$ **48.** $2^{2x} + 2^x - 12 = 0$

Solve each logarithmic equation in Exercises 49–90. Be sure to reject any value of x that is not in the domain of the original logarithmic expressions. Give the exact answer. Then, where necessary, use a calculator to obtain a decimal approximation, correct to two decimal places, for the solution.

49. $\log_3 x = 4$ **50.** $\log_5 x = 3$

51. $\ln x = 2$ **52.** $\ln x = 3$

53. $\log_4(x + 5) = 3$ **54.** $\log_5(x - 7) = 2$

55. $\log_3(x - 4) = -3$ **56.** $\log_7(x + 2) = -2$

57. $\log_4(3x + 2) = 3$ **58.** $\log_2(4x + 1) = 5$

59. $5\ln(2x) = 20$ **60.** $6\ln(2x) = 30$

61. $6 + 2\ln x = 5$ **62.** $7 + 3\ln x = 6$

63. $\ln\sqrt{x + 3} = 1$ **64.** $\ln\sqrt{x + 4} = 1$

65. $\log_5 x + \log_5(4x - 1) = 1$

66. $\log_6(x + 5) + \log_6 x = 2$

67. $\log_3(x - 5) + \log_3(x + 3) = 2$

68. $\log_2(x - 1) + \log_2(x + 1) = 3$

69. $\log_2(x + 2) - \log_2(x - 5) = 3$

70. $\log_4(x + 2) - \log_4(x - 1) = 1$

71. $2\log_3(x + 4) = \log_3 9 + 2$

72. $3\log_2(x - 1) = 5 - \log_2 4$

73. $\log_2(x - 6) + \log_2(x - 4) - \log_2 x = 2$

74. $\log_2(x - 3) + \log_2 x - \log_2(x + 2) = 2$

75. $\log(x + 4) = \log x + \log 4$

76. $\log(5x + 1) = \log(2x + 3) + \log 2$

77. $\log(3x - 3) = \log(x + 1) + \log 4$

78. $\log(2x - 1) = \log(x + 3) + \log 3$

79. $2\log x = \log 25$

80. $3\log x = \log 125$

81. $\log(x + 4) - \log 2 = \log(5x + 1)$

82. $\log(x + 7) - \log 3 = \log(7x + 1)$

83. $2\log x - \log 7 = \log 112$

84. $\log(x - 2) + \log 5 = \log 100$

85. $\log x + \log(x + 3) = \log 10$

86. $\log(x + 3) + \log(x - 2) = \log 14$

87. $\ln(x - 4) + \ln(x + 1) = \ln(x - 8)$

88. $\log_2(x - 1) - \log_2(x + 3) = \log_2\left(\frac{1}{x}\right)$

89. $\ln(x - 2) - \ln(x + 3) = \ln(x - 1) - \ln(x + 7)$

90. $\ln(x - 5) - \ln(x + 4) = \ln(x - 1) - \ln(x + 2)$

Practice Plus

In Exercises 91–100, solve each equation.

91. $5^{2x} \cdot 5^{4x} = 125$ **92.** $3^{x+2} \cdot 3^x = 81$

93. $2\,|\ln x| - 6 = 0$ **94.** $3\,|\log x| - 6 = 0$

95. $3^{x^2} = 45$ **96.** $5^{x^2} = 50$

97. $\ln(2x + 1) + \ln(x - 3) - 2\ln x = 0$

98. $\ln 3 - \ln(x + 5) - \ln x = 0$

99. $5^{x^2-12} = 25^{2x}$ **100.** $3^{x^2-12} = 9^{2x}$

Application Exercises

Use the formula $R = 6e^{12.77x}$, where x is the blood alcohol concentration and R, given as a percent, is the risk of having a car accident, to solve Exercises 101–102.

101. What blood alcohol concentration corresponds to a 25% risk of a car accident?

102. What blood alcohol concentration corresponds to a 50% risk of a car accident?

103. The formula $A = 18.9e^{0.0055t}$ models the population of New York State, A, in millions, t years after 2000.

a. What was the population of New York in 2000?

b. When will the population of New York reach 19.6 million?

104. The formula $A = 15.9e^{0.0235t}$ models the population of Florida, A, in millions, t years after 2000.

a. What was the population of Florida in 2000?

b. When will the population of Florida reach 19.2 million?

In Exercises 105–108, complete the table for a savings account subject to n compoundings yearly $\left[A = P\left(1 + \frac{r}{n}\right)^{nt}\right]$. *Round answers to one decimal place.*

Amount Invested	Number of Compounding Periods	Annual Interest Rate	Accumulated Amount	Time t in Years
105. \$12,500	4	5.75%	\$20,000	
106. \$7250	12	6.5%	\$15,000	
107. \$1000	360		\$1400	2
108. \$5000	360		\$9000	4

In Exercises 109–112, complete the table for a savings account subject to continuous compounding ($A = Pe^{rt}$). Round answers to one decimal place.

	Amount Invested	Annual Interest Rate	Accumulated Amount	Time t in Years
109.	\$8000	8%	Double the amount invested	
110.	\$8000		\$12,000	2
111.	\$2350		Triple the amount invested	7
112.	\$17,425	4.25%	\$25,000	

113. Fed up with junk mail clogging your computer? Despite high-profile legislation and lawsuits, the bar graph shows that spam has flourished.

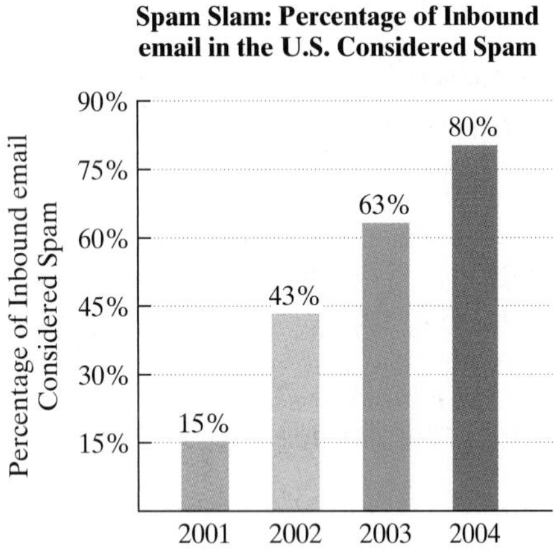

Source: Meta Group

The function $f(x) = 13.4 + 46.3 \ln x$ models the percentage of inbound email in the United States considered spam, $f(x)$, x years after 2000.

a. How well does the function model the data for 2003?

b. If law enforcement against spammers does not change and the model is projected into the future, when will 96% of inbound e-mail be spam? Round to the nearest year.

114. The bar graph shows the number of children under 18 as a percentage of the total U.S. population.

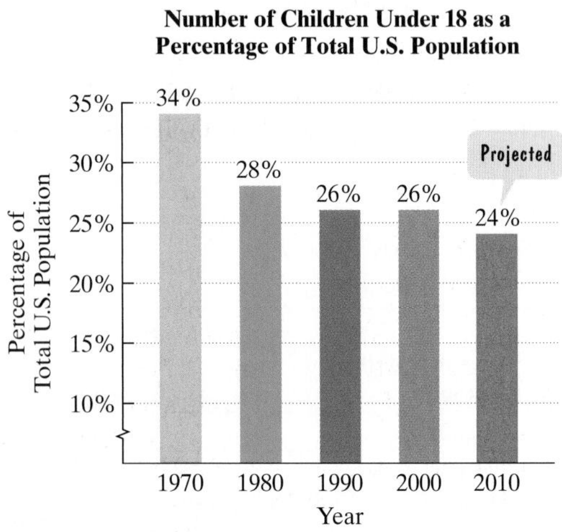

Source: www.childstats.gov

The function $f(x) = 34 - 2.6 \ln x$ models the number of children under 18 as a percentage of the total U.S. population, $f(x)$, x years after 1969.

a. How well does the function model the projected data for 2010?

b. According to the model, when will children under 18 decline to 23% of the total U.S. population? Round to the nearest year.

The function $P(x) = 95 - 30 \log_2 x$ models the percentage, $P(x)$, of students who could recall the important features of a classroom lecture as a function of time, where x represents the number of days that have elapsed since the lecture was given. The figure shows the graph of the function. Use this information to solve Exercises 115–116. Round answers to one decimal place.

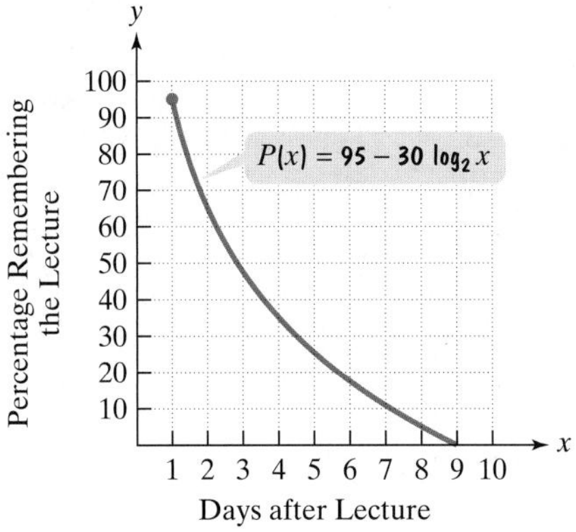

115. After how many days do only half the students recall the important features of the classroom lecture? (Let $P(x) = 50$ and solve for x.) Locate the point on the graph that conveys this information.

116. After how many days have all students forgotten the important features of the classroom lecture? (Let $P(x) = 0$ and solve for x.) Locate the point on the graph that conveys this information.

The pH of a solution ranges from 0 to 14. An acid solution has a pH less than 7. Pure water is neutral and has a pH of 7. Normal, unpolluted rain has a pH of about 5.6. The pH of a solution is given by

$$\text{pH} = -\log x,$$

where x represents the concentration of the hydrogen ions in the solution, in moles per liter. Use the formula to solve Exercises 117–118.

117. An environmental concern involves the destructive effects of acid rain. The most acidic rainfall ever had a pH of 2.4. What was the hydrogen ion concentration? Express the answer as a power of 10 and then round to the nearest thousandth.

118. The figure shows very acidic rain in the northeast United States. What is the hydrogen ion concentration of rainfall with a pH of 4.2? Express the answer as a power of 10 and then round to the nearest hundred-thousandth.

Acid Rain over Canada and the United States

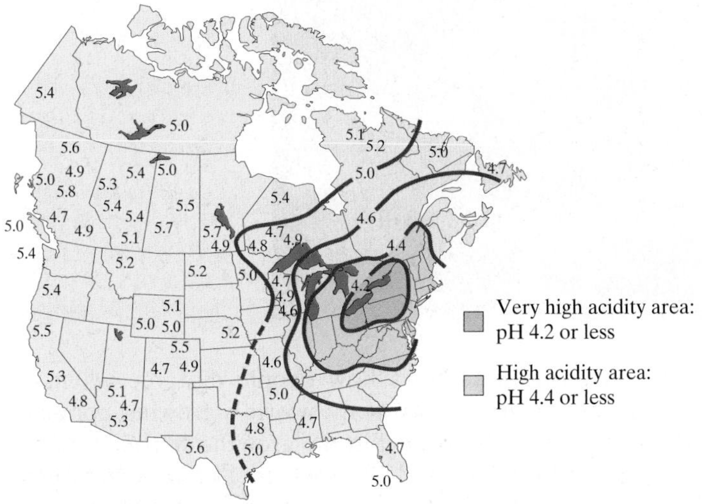

Source: National Atmospheric Program

Writing in Mathematics

119. Explain how to solve an exponential equation when both sides can be written as a power of the same base.

120. Explain how to solve an exponential equation when both sides cannot be written as a power of the same base. Use $3^x = 140$ in your explanation.

121. Explain the differences between solving $\log_3(x-1) = 4$ and $\log_3(x-1) = \log_3 4$.

122. In many states, a 17% risk of a car accident with a blood alcohol concentration of 0.08 is the lowest level for charging a motorist with driving under the influence. Do you agree with the 17% risk as a cutoff percentage, or do you feel that the percentage should be lower or higher? Explain your answer. What blood alcohol concentration corresponds to what you believe is an appropriate percentage?

Technology Exercises

In Exercises 123–130, use your graphing utility to graph each side of the equation in the same viewing rectangle. Then use the x-coordinate of the intersection point to find the equation's solution set. Verify this value by direct substitution into the equation.

123. $2^{x+1} = 8$

124. $3^{x+1} = 9$

125. $\log_3(4x - 7) = 2$

126. $\log_3(3x - 2) = 2$

127. $\log(x + 3) + \log x = 1$

128. $\log(x - 15) + \log x = 2$

129. $3^x = 2x + 3$

130. $5^x = 3x + 4$

Hurricanes are one of nature's most destructive forces. These low-pressure areas often have diameters of over 500 miles. The function $f(x) = 0.48\ln(x+1) + 27$ models the barometric air pressure, $f(x)$, in inches of mercury, at a distance of x miles from the eye of a hurricane. Use this function to solve Exercises 131–132.

131. Graph the function in a $[0, 500, 50]$ by $[27, 30, 1]$ viewing rectangle. What does the shape of the graph indicate about barometric air pressure as the distance from the eye increases?

132. Use an equation to answer this question: How far from the eye of a hurricane is the barometric air pressure 29 inches of mercury? Use the TRACE and ZOOM features or the intersect command of your graphing utility to verify your answer.

133. The function $P(t) = 145e^{-0.092t}$ models a runner's pulse, $P(t)$, in beats per minute, t minutes after a race, where $0 \le t \le 15$. Graph the function using a graphing utility. TRACE along the graph and determine after how many minutes the runner's pulse will be 70 beats per minute. Round to the nearest tenth of a minute. Verify your observation algebraically.

134. The function $W(t) = 2600(1 - 0.51e^{-0.075t})^3$ models the weight, $W(t)$, in kilograms, of a female African elephant at age t years. (1 kilogram $\approx$ 2.2 pounds) Use a graphing utility to graph the function. Then TRACE along the curve to estimate the age of an adult female elephant weighing 1800 kilograms.

Critical Thinking Exercises

135. Which one of the following is true?

a. If $\log(x + 3) = 2$, then $e^2 = x + 3$.

b. If $\log(7x + 3) - \log(2x + 5) = 4$, then the equation in exponential form is $10^4 = (7x + 3) - (2x + 5)$.

c. If $x = \frac{1}{k}\ln y$, then $y = e^{kx}$.

d. Examples of exponential equations include $10^x = 5.71$, $e^x = 0.72$, and $x^{10} = 5.71$.

136. If \$4000 is deposited into an account paying 3% interest compounded annually and at the same time \$2000 is deposited into an account paying 5% interest compounded annually, after how long will the two accounts have the same balance? Round to the nearest year.

Solve each equation in Exercises 137–139. Check each proposed solution by direct substitution or with a graphing utility.

137. $(\ln x)^2 = \ln x^2$

138. $(\log x)(2\log x + 1) = 6$

139. $\ln(\ln x) = 0$

Group Exercise

140. Research applications of logarithmic functions as mathematical models and plan a seminar based on your group's research. Each group member should research one of the following areas or any other area of interest: pH (acidity of solutions), intensity of sound (decibels), brightness of stars, human memory, progress over time in a sport, profit over time. For the area that you select, explain how logarithmic functions are used and provide examples.

SECTION 4.5 Exponential Growth and Decay; Modeling Data

Objectives

1. Model exponential growth and decay.
2. Use logistic growth models.
3. Model data with exponential and logarithmic functions.
4. Express an exponential model in base e.

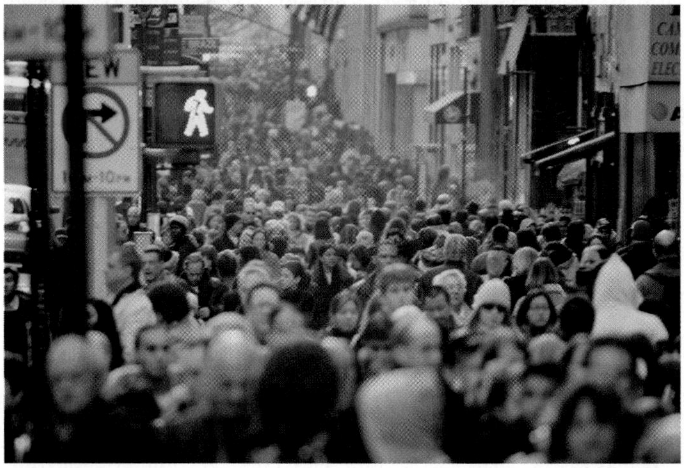

The most casual cruise on the Internet shows how people disagree when it comes to making predictions about the effects of the world's growing population. Some argue that there is a recent slowdown in the growth rate, economies remain robust, and famines in North Korea and Ethiopia are aberrations rather than signs of the future. Others say that the 6.3 billion people on Earth is twice as many as can be supported in middle-class comfort, and the world is running out of arable land and fresh water. Debates about entities that are growing exponentially can be approached mathematically: We can create functions that model data and use these functions to make predictions. In this section, we will show you how this is done.

1 Model exponential growth and decay.

Exponential Growth and Decay

One of algebra's many applications is to predict the behavior of variables. This can be done with *exponential growth* and *decay models*. With exponential growth or decay, quantities grow or decay at a rate directly proportional to their size. Populations that are growing exponentially grow extremely rapidly as they get larger because there are more adults to have offspring. For example, the **growth rate** for world population is 1.3%, or 0.013. This means that each year world population is 1.3% more than what it was in the previous year. In 2001, world population was approximately 6.2 billion. Thus, we compute the world population in 2002 as follows:

$$6.2 \text{ billion} + 13\% \text{ of } 6.2 \text{ billion} = 6.2 + (0.013)(6.2) = 6.2806.$$

This computation suggests that 6.2806 billion people populated the world in 2002. The 0.0806 billion represents an increase of 80.6 million people from 2001 to 2002, the equivalent of the population of Germany. Using 1.3% as the annual growth rate, world population for 2003 is found in a similar manner:

$$6.2806 + 1.3\% \text{ of } 6.2806 = 6.2806 + (0.013)(6.2806) \approx 6.3622.$$

This computation suggests that approximately 6.3622 billion people populated the world in 2003.

The explosive growth of world population may remind you of the growth of money in an account subject to compound interest. Just as the growth rate for world population is multiplied by the population plus any increase in the population, a compound interest rate is multiplied by your original investment plus any accumulated interest. The balance in an account subject to continuous compounding and world population are special cases of *exponential growth models*.

Study Tip

You have seen the formula for exponential growth before, but with different letters. It is the formula for compound interest with continous compounding.

$$A = Pe^{rt}$$

Amount at time t | Principal is the original amount. | Interest rate is the growth rate.

$$A = A_0e^{kt}$$

Exponential Growth and Decay Models

The mathematical model for **exponential growth** or **decay** is given by

$$f(t) = A_0e^{kt} \quad \text{or} \quad A = A_0e^{kt}.$$

- **If $k > 0$, the function models the amount, or size, of a *growing* entity.** A_0 is the original amount, or size, of the growing entity at time $t = 0$, A is the amount at time t, and k is a constant representing the growth rate.
- **If $k < 0$, the function models the amount, or size, of a *decaying* entity.** A_0 is the original amount, or size, of the decaying entity at time $t = 0$, A is the amount at time t, and k is a constant representing the decay rate.

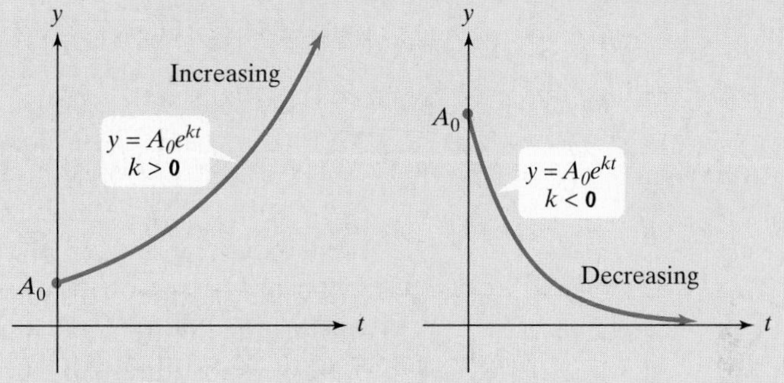

(a) Exponential growth (b) Exponential decay

Sometimes we need to use given data to determine k, the rate of growth or decay. After we compute the value of k, we can use the formula $A = A_0e^{kt}$ to make predictions. This idea is illustrated in our first two examples.

EXAMPLE 1 Modeling the Growth of the U.S. Population

The graph in Figure 4.19 shows the U.S. population, in millions, for five selected years from 1970 through 2003. In 1970, the U.S. population was 203.3 million. By 2003, it had grown to 294 million.

a. Find the exponential growth function that models the data for 1970 through 2003.

b. By which year will the U.S. population reach 315 million?

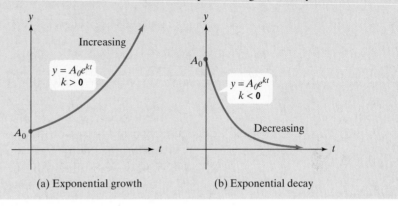

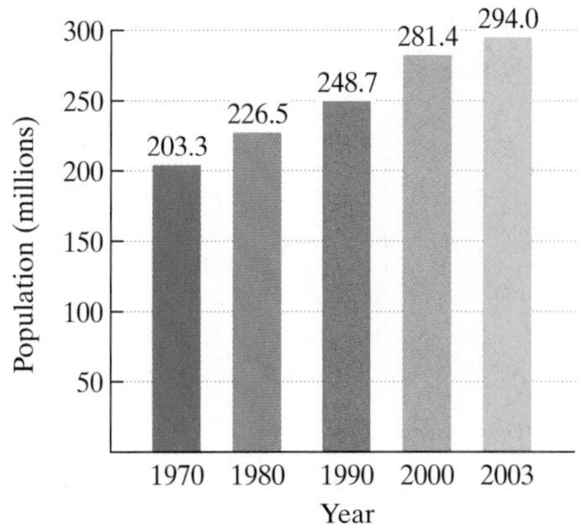

Figure 4.19

Source: Bureau of the Census

Solution

a. We use the exponential growth model

$$A = A_0e^{kt},$$

in which t is the number of years after 1970. This means that 1970 corresponds to $t = 0$. At that time the U.S. population was 203.3 million, so we substitute 203.3 for A_0 in the growth model:

$$A = 203.3e^{kt}.$$

We are given that 294 million is the population in 2003. Because 2003 is 33 years after 1970, when $t = 33$ the value of A is 294. Substituting these numbers into the growth model will enable us to find k, the growth rate. We know that $k > 0$ because the problem involves growth.

$A = 203.3e^{kt}$	Use the growth model $A = A_0e^{kt}$ with $A_0 = 203.3$.
$294 = 203.3e^{k \cdot 33}$	When $t = 33$, $A = 294$. Substitute these numbers into the model.
$e^{33k} = \frac{294}{203.3}$	Isolate the exponential factor by dividing both sides by 203.3. We also reversed the sides.
$\ln e^{33k} = \ln\left(\frac{294}{203.3}\right)$	Take the natural logarithm on both sides.
$33k = \ln\left(\frac{294}{203.3}\right)$	Simplify the left side using $\ln e^x = x$.
$k = \frac{\ln\left(\frac{294}{203.3}\right)}{33} \approx 0.011$	Divide both sides by 33 and solve for k. Then use a calculator.

The value of k, approximately 0.011, indicates a growth rate of about 1.1%. This means that the U.S. population is increasing by approximately 1.1% per year. We substitute 0.011 for k in the growth model, $A = 203.3e^{kt}$, to obtain the exponential growth function for the U.S. population. It is

$$A = 203.3e^{0.011t},$$

where t is measured in years after 1970.

b. To find the year in which the U.S. population will reach 315 million, substitute 315 for A in the model from part (a) and solve for t.

$A = 203.3e^{0.011t}$	This is the model from part (a).
$315 = 203.3e^{0.011t}$	Substitute 315 for A.
$e^{0.011t} = \frac{315}{203.3}$	Divide both sides by 203.3. We also reversed the sides.
$\ln e^{0.011t} = \ln\left(\frac{315}{203.3}\right)$	Take the natural logarithm on both sides.
$0.011t = \ln\left(\frac{315}{203.3}\right)$	Simplify on the left using $\ln e^x = x$.
$t = \frac{\ln\left(\frac{315}{203.3}\right)}{0.011} \approx 40$	Divide both sides by 33 and solve for t. Then use a calculator.

Because t represents the number of years after 1970, the model indicates that the U.S. population will reach 315 million by 1970 + 40, or in the year 2010.

Check Point 1 In 1990, the population of Africa was 643 million and by 2000 it had grown to 813 million.

a. Use the exponential growth model $A = A_0e^{kt}$, in which t is the number of years after 1990, to find the exponential growth function that models the data.

b. By which year will Africa's population reach 2000 million, or two billion?

Creating an Inaccurate Picture by Leaving Something Out

On Monday, October 19, 1987, the Dow Jones Industrial Average plunged 508 points, losing 22.6% of its value. The graph shown on the left, which appeared in a major newspaper following "Black Monday" (as it was instantly dubbed), creates the impression that the Dow average had been "bullish" from 1972 through 1987, increasing throughout this period. The graph creates this inaccurate picture by leaving something out. The graph on the right illustrates that the stock market rose and fell sharply over these years. The impressively smooth curve on the left was obtained by plotting only three of the data points. By ignoring most of the data, increases and decreases are not accounted for and the actual behavior of the market over the 15 years leading to "Black Monday" is inaccurately conveyed.

The Dow Jones Industrial Average: 1972–1987

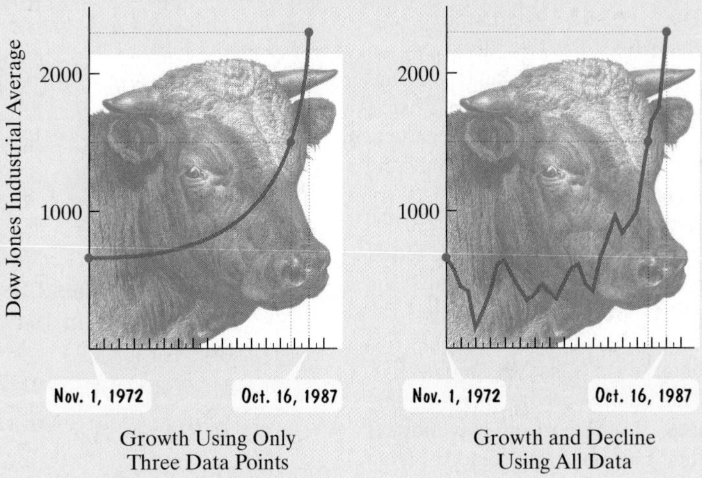

Source: A. K. Dewdney, *200% of Nothing*

In Example 1, we used only two data values, the population for 1970 and the population for 2003, to develop a model for U.S. population growth from 1970 through 2003. By not using data for any other years, have we created a model that inaccurately describes both the exisiting data and future population projections given by the U.S. Census Bureau? Something else to think about: Is an exponential model the best choice for describing U.S. population growth, or might a linear model provide a better description? We return to these issues in Exercises 50–54 in the exercise set.

Our next example involves exponential decay and its use in determining the age of fossils and artifacts. The method is based on considering the percentage of carbon-14 remaining in the fossil or artifact. Carbon-14 decays exponentially with a *half-life* of approximately 5715 years. The **half-life** of a substance is the time required for half of a given sample to disintegrate. Thus, after 5715 years a given amount of carbon-14 will have decayed to half the original amount. Carbon dating is useful for artifacts or fossils up to 80,000 years old. Older objects do not have enough carbon-14 left to determine age accurately.

EXAMPLE 2 Carbon-14 Dating: The Dead Sea Scrolls

a. Use the fact that after 5715 years a given amount of carbon-14 will have decayed to half the original amount to find the exponential decay model for carbon-14.

b. In 1947, earthenware jars containing what are known as the Dead Sea Scrolls were found by an Arab Bedouin herdsman. Analysis indicated that the scroll wrappings contained 76% of their original carbon-14. Estimate the age of the Dead Sea Scrolls.

Solution

a. We begin with the exponential decay model $A = A_0e^{kt}$. We know that $k < 0$ because the problem involves the decay of carbon-14. After 5715 years ($t = 5715$), the amount of carbon-14 present, A, is half the original amount, A_0. Thus, we can substitute $\frac{A_0}{2}$ for A in the exponential decay model. This will enable us to find k, the decay rate.

$$A = A_0e^{kt}$$ Begin with the exponential decay model.

$$\frac{A_0}{2} = A_0e^{k(5715)}$$ After 5715 years ($t = 5715$), $A = \frac{A_0}{2}$ (because the amount present, A, is half the original amount, A_0).

$$\frac{1}{2} = e^{5715k}$$ Divide both sides of the equation by A_0.

Carbon Dating and Artistic Development

The artistic community was electrified by the discovery in 1995 of spectacular cave paintings in a limestone cavern in France. Carbon dating of the charcoal from the site showed that the images, created by artists of remarkable talent, were 30,000 years old, making them the oldest cave paintings ever found. The artists seemed to have used the cavern's natural contours to heighten a sense of perspective. The quality of the painting suggests that the art of early humans did not mature steadily from primitive to sophisticated in any simple linear fashion.

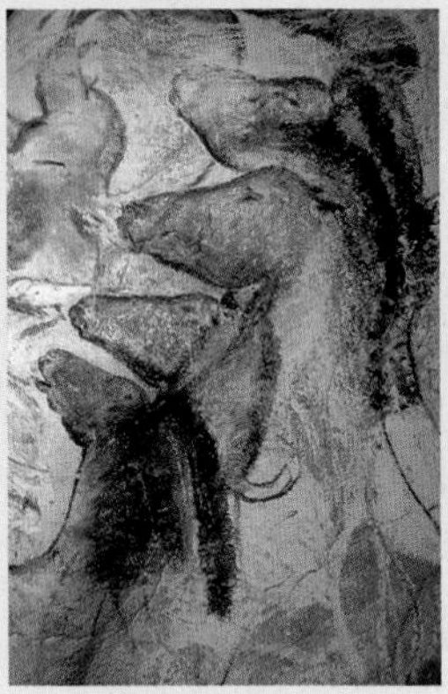

$$\ln\left(\frac{1}{2}\right) = \ln e^{5715k}$$ Take the natural logarithm on both sides.

$$\ln\left(\frac{1}{2}\right) = 5715k$$ Simplify the right side using $\ln e^x = x$.

$$k = \frac{\ln\left(\frac{1}{2}\right)}{5715} \approx -0.000121$$ Divide both sides by 5715 and solve for k.

Substituting for k in the decay model, $A = A_0e^{kt}$, the model for carbon-14 is

$$A = A_0e^{-0.000121t}.$$

b. In 1947, the Dead Sea Scrolls contained 76% of their original carbon-14. To find their age in 1947, substitute $0.76A_0$ for A in the model from part (a) and solve for t.

$$A = A_0e^{-0.000121t}$$ This is the decay model for carbon-14.

$$0.76A_0 = A_0e^{-0.000121t}$$ A, the amount present, is 76% of the original amount, so $A = 0.76A_0$.

$$0.76 = e^{-0.000121t}$$ Divide both sides of the equation by A_0.

$$\ln 0.76 = \ln e^{-0.000121t}$$ Take the natural logarithm on both sides.

$$\ln 0.76 = -0.000121t$$ Simplify the right side using $\ln e^x = x$.

$$t = \frac{\ln 0.76}{-0.000121} \approx 2268$$ Divide both sides by -0.000121 and solve for t.

The Dead Sea Scrolls are approximately 2268 years old plus the number of years between 1947 and the current year.

Check Point 2 Strontium-90 is a waste product from nuclear reactors. As a consequence of fallout from atmospheric nuclear tests, we all have a measurable amount of strontium-90 in our bones.

a. The half-life of strontium-90 is 28 years, meaning that after 28 years a given amount of the substance will have decayed to half the original amount. Find the exponential decay model for strontium-90.

b. Suppose that a nuclear accident occurs and releases 60 grams of strontium-90 into the atmosphere. How long will it take for strontium-90 to decay to a level of 10 grams?

2 Use logistic growth models.

Logistic Growth Models

From population growth to the spread of an epidemic, nothing on Earth can grow exponentially indefinitely. Growth is always limited. This is shown in Figure 4.20 by the horizontal asymptote. The *logistic growth model* is a function used to model situations of this type.

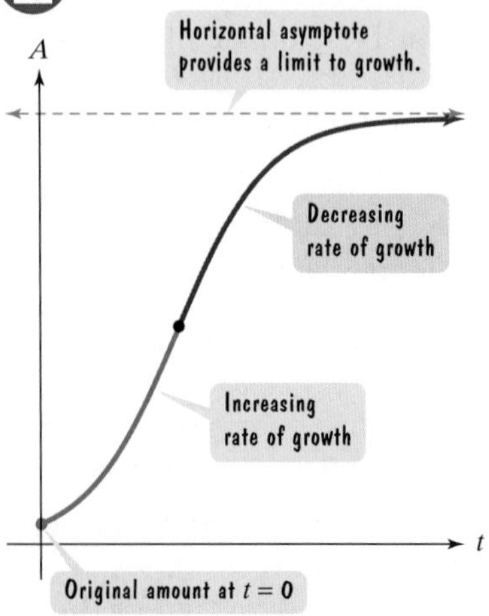

Figure 4.20 The logistic growth curve has a horizontal asymptote that identifies the limit of the growth of A over time.

Logistic Growth Model

The mathematical model for limited logistic growth is given by

$$f(t) = \frac{c}{1 + ae^{-bt}} \quad \text{or} \quad A = \frac{c}{1 + ae^{-bt}},$$

where a, b, and c are constants, with $c > 0$ and $b > 0$.

As time increases ($t \to \infty$), the expression ae^{-bt} in the model approaches 0, and A gets closer and closer to c. This means that $y = c$ is a horizontal asymptote for the graph of the function. Thus, the value of A can never exceed c and c represents the limiting size that A can attain.

EXAMPLE 3 Modeling the Spread of the Flu

The function

$$f(t) = \frac{30{,}000}{1 + 20e^{-1.5t}}$$

describes the number of people, $f(t)$, who have become ill with influenza t weeks after its initial outbreak in a town with 30,000 inhabitants.

a. How many people became ill with the flu when the epidemic began?

b. How many people were ill by the end of the fourth week?

c. What is the limiting size of $f(t)$, the population that becomes ill?

Solution

a. The time at the beginning of the flu epidemic is $t = 0$. Thus, we can find the number of people who were ill at the beginning of the epidemic by substituting 0 for t.

$$f(t) = \frac{30{,}000}{1 + 20e^{-1.5t}}$$ This is the given logistic growth function.

$$f(0) = \frac{30{,}000}{1 + 20e^{-1.5(0)}}$$ When the epidemic began, $t = 0$.

$$= \frac{30{,}000}{1 + 20}$$ $e^{-1.5(0)} = e^0 = 1$

$$\approx 1429$$

Approximately 1429 people were ill when the epidemic began.

b. We find the number of people who were ill at the end of the fourth week by substituting 4 for t in the logistic growth function.

$$f(t) = \frac{30{,}000}{1 + 20e^{-1.5t}}$$ Use the given logistic growth function.

$$f(4) = \frac{30{,}000}{1 + 20e^{-1.5(4)}}$$ To find the number of people ill by the end of week four, let $t = 4$.

$$\approx 28{,}583$$ Use a calculator.

Approximately 28,583 people were ill by the end of the fourth week. Compared with the number of people who were ill initially, 1429, this illustrates the virulence of the epidemic.

c. Recall that in the logistic growth model, $f(t) = \frac{c}{1 + ae^{-bt}}$, the constant c represents the limiting size that $f(t)$ can attain. Thus, the number in the numerator, 30,000, is the limiting size of the population that becomes ill.

Technology

The graph of the logistic growth function for the flu epidemic

$$y = \frac{30{,}000}{1 + 20e^{-1.5x}}$$

can be obtained using a graphing utility. We started x at 0 and ended at 10. This takes us to week 10. (In Example 3, we found that by week 4 approximately 28,583 people were ill.) We also know that 30,000 is the limiting size, so we took values of y up to 30,000. Using a [0, 10, 1] by [0, 30,000, 3000] viewing rectangle, the graph of the logistic growth function is shown below.

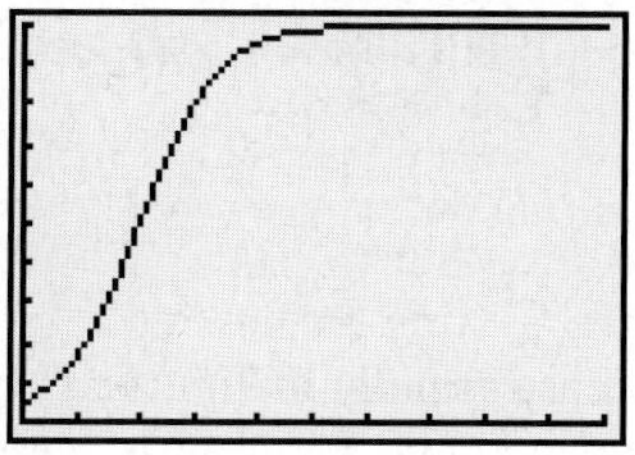

Check Point 3 In a learning theory project, psychologists discovered that

$$f(t) = \frac{0.8}{1 + e^{-0.2t}}$$

is a model for describing the proportion of correct responses, $f(t)$, after t learning trials.

a. Find the proportion of correct responses prior to learning trials taking place.

b. Find the proportion of correct responses after 10 learning trials.

c. What is the limiting size of $f(t)$, the proportion of correct responses, as continued learning trials take place?

Model data with exponential and logarithmic functions.

The Art of Modeling

Throughout this chapter, we have been working with models that were given. However, we can create functions that model data by observing patterns in scatter plots. Figure 4.21 shows scatter plots for data that are exponential or logarithmic.

| $y = ab^x, a > 0, b > 1$ | $y = ab^x, a > 0, 0 < b < 1$ | $y = a + b \ln x, a > 0, b > 0$ | $y = a + b \ln x, a > 0, b < 0$ |
| Exponential | Exponential | Logarithmic | Logarithmic |

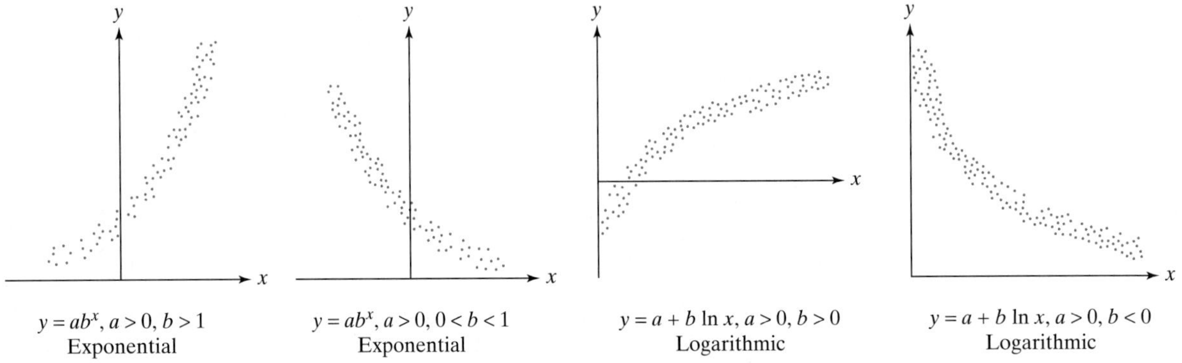

Figure 4.21 Scatter plots for exponential or logarithmic models

EXAMPLE 4   Choosing a Model for Data

Figure 4.22(a) shows the percentage of U.S. households with televisions that subscribe to cable television. The data are displayed for five selected years from 1980 through 2002. A scatter plot is shown in Figure 4.22(b). What function would be a good choice for modeling the data?

Percentage of U.S. Households with TVs with Cable Television

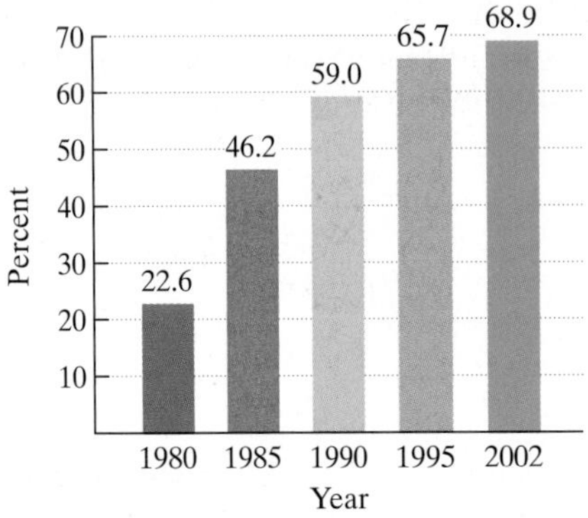

Figure 4.22(a)

Source: Nielsen Media Research

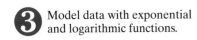

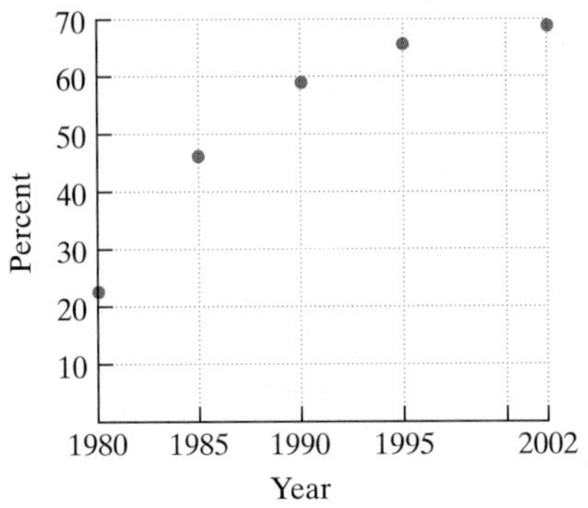

Figure 4.22(b)

Solution   Because the data in the scatter plot increase rapidly at first and then begin to level off a bit, the shape suggests that a logarithmic function is a good choice for modeling the data.

Check Point 4   Table 4.5 shows the populations of various cities, in thousands, and the average walking speed, in feet per second, of a person living in the city. Create a scatter plot for the data. Based on the scatter plot, what function would be a good choice for modeling the data?

Table 4.5   Population and Walking Speed

Population (thousands)	Walking Speed (feet per second)
5.5	0.6
14	1.0
71	1.6
138	1.9
342	2.2

Source: Mark and Helen Bornstein, "The Pace of Life"

How can we obtain a logarithmic function that models the data for the percentage of U.S. households with cable television shown in Figure 4.22(a)? A graphing utility can be used to obtain a logarithmic model of the form $y = a + b \ln x$. **Because the domain of the logarithmic function is the set of positive numbers, zero must not be a value for x.** What does this mean for our cable television data that begin in the year 1980? We must start values of x after 0. Thus, we'll assign x to represent the number of years after 1979. This gives us the data shown in Table 4.6. Using the Logarithmic REGression option, we obtain the equation in Figure 4.23.

Table 4.6

x, Number of Years after 1979	y, Percentage of U.S. Households with Cable TV
1 (1980)	22.6
6 (1985)	46.2
11 (1990)	59.0
16 (1995)	65.7
23 (2002)	68.9

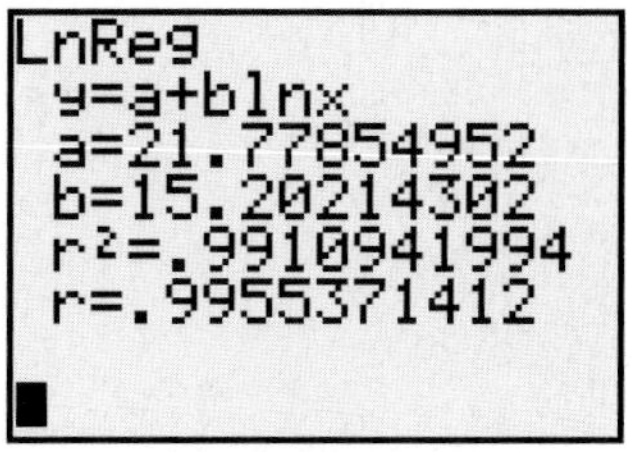

Figure 4.23 A logarithmic model for the data in Table 4.6

From Figure 4.23, we see that the logarithmic model of the data, with numbers rounded to three decimal places, is

$$y = 21.779 + 15.202 \ln x.$$

The number r that appears in Figure 4.23 is called the **correlation coefficient** and is a measure of how well the model fits the data. The value of r is such that $-1 \leq r \leq 1$. A positive r means that as the x-values increase, so do the y-values. A negative r means that as the x-values increase, the y-values decrease. **The closer that r is to -1 or 1, the better the model fits the data.** Because r is approximately 0.996, the model fits the data very well.

EXAMPLE 5 Choosing a Model for Data

Figure 4.24(a) shows world population, in billions, for seven selected years from 1950 through 2003. A scatter plot is shown in Figure 4.24(b). Suggest two functions that would be good choices for modeling the data.

World Population, 1950–2003

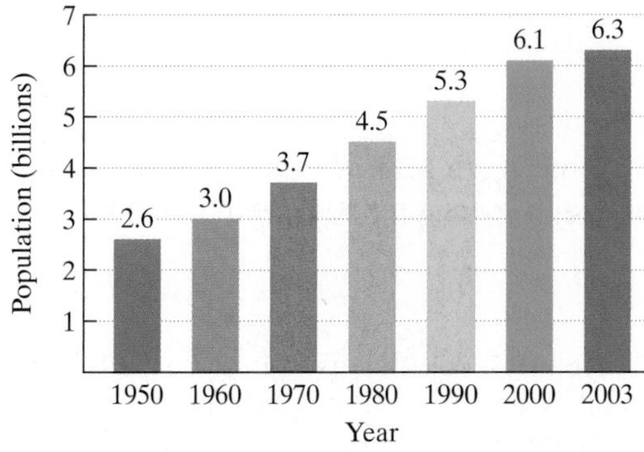

Figure 4.24(a)

Source: U.S. Census Bureau, International Database

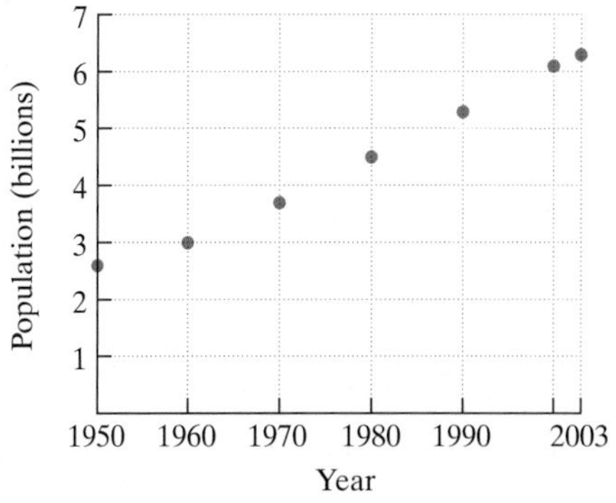

Figure 4.24(b)

Solution Because the data in the scatter plot appear to increase more and more rapidly, the shape suggests that an exponential model might be a good choice. Furthermore, we can probably draw a line that passes through or near the seven points. Thus, a linear function would also be a good choice for modeling the data.

Check Point 5 In 2003, 49.3 million tons of paper were recycled in the United States. Table 4.7 shows the percentage of all paper recycled for five selected years from 1970 through 2003. Create a scatter plot for the data. Based on the scatter plot, what function would be a good choice for modeling the data?

Table 4.7 Percentage of All Paper Recycled in the U.S.

Year	Percent
1970	22.4%
1980	26.2%
1990	33.5%
2000	46.0%
2003	50.3%

Source: The American Forest and Paper Association

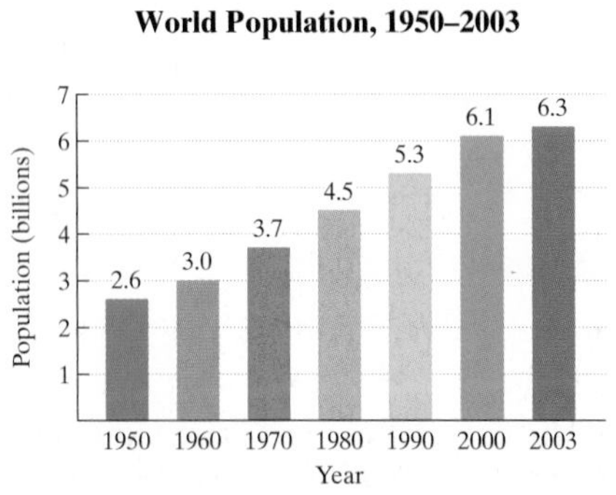

Figure 4.24(a) (repeated)

If we choose to model world population shown in Figure 4.24(a) with an exponential function, a graphing utility's Exponential REGression option can be used to obtain the function's equation. With this feature, a graphing utility fits the data to an exponential model of the form $y = ab^x$.

Although the domain of the exponential function $y = ab^x$ is the set of all real numbers, some graphing utilities only accept positive values for x. What does this mean for our data for world population that starts in the year 1950? We will start values of x after 0. Thus, we'll assign x to represent the number of years after 1949. This gives us the data shown in Table 4.8. Using the Exponential REGression option, we obtain the equation in Figure 4.25.

Table 4.8

x, Numbers of Years after 1949	y, World Population (billions)
1 (1950)	2.6
11 (1960)	3.0
21 (1970)	3.7
31 (1980)	4.5
41 (1990)	5.3
51 (2000)	6.1
54 (2003)	6.3

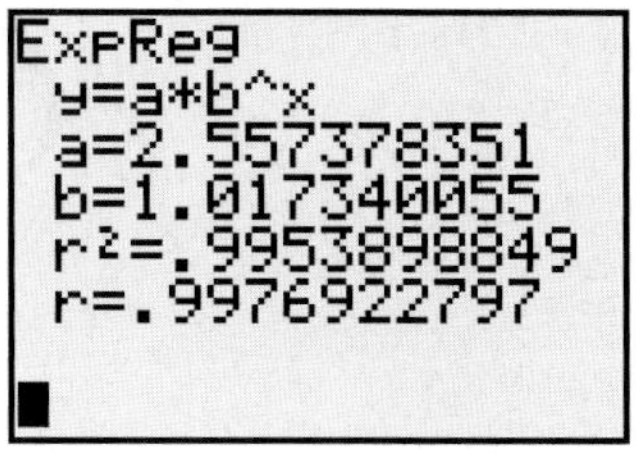

Figure 4.25 An exponential model for the data in Table 4.8

From Figure 4.25, we see that the exponential model of the data for world population x years after 1949, with numbers rounded to three decimal places, is

$$y = 2.557(1.017)^x.$$

The correlation coefficient, r, is close to 1, indicating that the model fits the data very well.

When using a graphing utility to model data, begin with a scatter plot, drawn either by hand or with the graphing utility, to obtain a general picture for the shape of the data. It might be difficult to determine which model best fits the data—linear, logarithmic, exponential, quadratic, or something else. If necessary, use your graphing utility to fit several models to the data. The best model is the one that yields the value of r, the correlation coefficient, closest to 1 or -1. Finding a proper fit for data can be almost as much art as it is mathematics. In this era of technology, the process of creating models that best fit data is one that involves more decision making than computation.

Study Tip

Once you have obtained one or more models for data, you can use a graphing utility's TABLE feature to numerically see how well each model describes the data. Enter the models as y_1, y_2, and so on. Create a table, scroll through the table, and compare the table values given by the models to the actual data.

4 Express an exponential model in base *e*.

Expressing $y = ab^x$ in Base *e*

Graphing utilities display exponential models in the form $y = ab^x$. However, our discussion of exponential growth involved base *e*. Because of the inverse property $b = e^{\ln b}$, we can rewrite any model in the form $y = ab^x$ in terms of base *e*.

Expressing an Exponential Model in Base *e*

$$y = ab^x \quad \text{is equivalent to} \quad y = ae^{(\ln b)\cdot x}$$

EXAMPLE 6 Rewriting the Model for World Population in Base *e*

We have seen that the function

$$y = 2.557(1.017)^x$$

models world population, *y*, in billions, *x* years after 1949. Rewrite the model in terms of base *e*.

Solution We use the two equivalent equations shown in the voice balloons to rewrite the model in terms of base *e*.

$y = ab^x$ $\qquad$ $y = ae^{(\ln b)\cdot x}$

$$y = 2.557(1.017)^x \quad \text{is equivalent to} \quad y = 2.557e^{(\ln 1.017)\cdot x}.$$

Using ln 1.017 ≈ 0.017, the exponential growth model for world population, *y*, in billions, *x* years after 1949 is

$$y = 2.557e^{0.017x}.$$

In Example 6, we can replace *y* with *A* and *x* with *t* so that the model has the same letters as those in the exponential growth model $A = A_0e^{kt}$.

$A = A_o \, e^{kt}$ This is the exponential growth model.

$A = 2.557e^{0.017t}$ This is the model for world population.

The value of *k*, 0.017, indicates a growth rate of 1.7%. Although this is an excellent model for the data, we must be careful about making projections about world population using this growth function. Why? World population growth rate is now 1.3%, not 1.7%, so our model will overestimate future populations.

Check Point 6 Rewrite $y = 4(7.8)^x$ in terms of base *e*. Express the answer in terms of a natural logarithm and then round to three decimal places.

EXERCISE SET 4.5

Practice Exercises and Application Exercises

The exponential models describe the population of the indicated country, A, in millions, t years after 2003. Use these models to solve Exercises 1–6.

India $A = 1049.7e^{0.015t}$

Iraq $A = 24.7e^{0.028t}$

Japan $A = 127.2e^{0.001t}$

Russia $A = 144.5e^{-0.004t}$

1. What was the population of Japan in 2003?
2. What was the population of Iraq in 2003?
3. Which country has the greatest growth rate? By what percentage is the population of that country increasing each year?
4. Which country has a decreasing population? By what percentage is the population of that country decreasing each year?
5. When will India's population be 1238 million?
6. When will India's population be 1416 million?

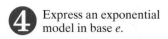

About the size of New Jersey, Israel has seen its population soar to more than 6 million since it was established. With the help of U.S. aid, the country now has a diversified economy rivaling those of other developed Western nations. By contrast, the Palestinians, living under Israeli occupation and a corrupt regime, endure bleak conditions. The graphs show that by 2050, Palestinians in the West Bank, Gaza Strip, and East Jerusalem will outnumber Israelis. Exercises 7–8, involve the projected growth of these two populations.

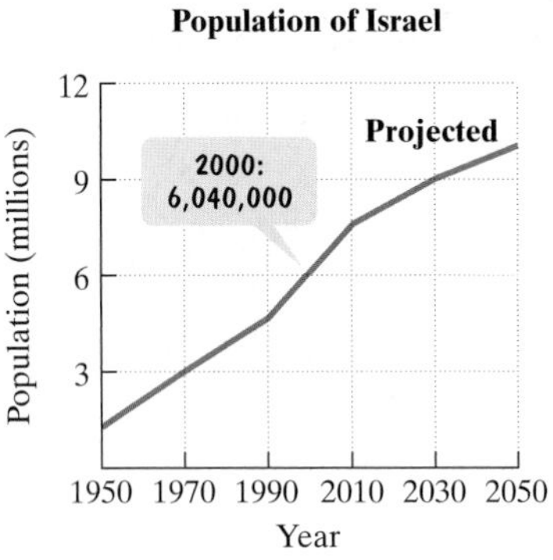

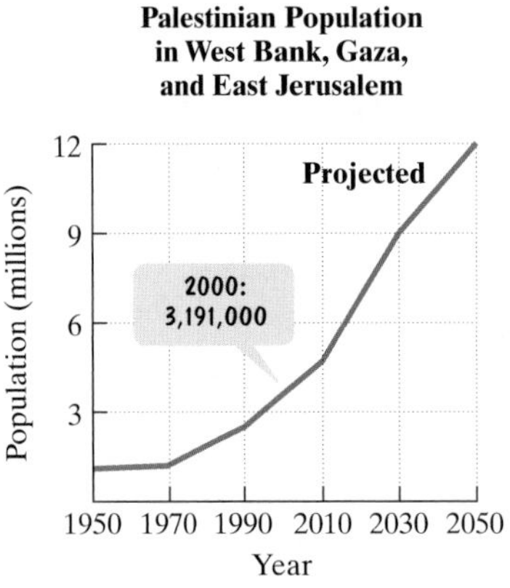

Source: Newsweek

7. a. In 2000, the population of Israel was approximately 6.04 million and by 2050 it is projected to grow to 10 million. Use the exponential growth model $A = A_0e^{kt}$, in which t is the number of years after 2000, to find an exponential growth function that models the data.

b. In which year will Israel's population be 9 million?

8. a. In 2000, the population of the Palestinians in the West Bank, Gaza Strip, and East Jerusalem was approximately 3.2 million and by 2050 it is projected to grow to 12 million. Use the exponential growth model $A = A_0e^{kt}$, in which t is the number of years after 2000, to find the exponential growth function that models the data.

b. In which year will the Palestinian population be 9 million?

An artifact originally had 16 grams of carbon-14 present. The decay model $A = 16e^{-0.000121t}$ describes the amount of carbon-14 present after t years. Use this model to solve Exercises 9–10.

9. How many grams of carbon-14 will be present in 5715 years?

10. How many grams of carbon-14 will be present in 11,430 years?

11. The half-life of the radioactive element krypton-91 is 10 seconds. If 16 grams of krypton-91 are initially present, how many grams are present after 10 seconds? 20 seconds? 30 seconds? 40 seconds? 50 seconds?

12. The half-life of the radioactive element plutonium-239 is 25,000 years. If 16 grams of plutonium-239 are initially present, how many grams are present after 25,000 years? 50,000 years? 75,000 years? 100,000 years? 125,000 years?

Use the exponential decay model for carbon-14, $A = A_0e^{-0.000121t}$, to solve Exercises 13–14.

13. Prehistoric cave paintings were discovered in a cave in France. The paint contained 15% of the original carbon-14. Estimate the age of the paintings.

14. Skeletons were found at a construction site in San Francisco in 1989. The skeletons contained 88% of the expected amount of carbon-14 found in a living person. In 1989, how old were the skeletons?

15. The August 1978 issue of *National Geographic* described the 1964 find of bones of a newly discovered dinosaur weighing 170 pounds, measuring 9 feet, with a 6-inch claw on one toe of each hind foot. The age of the dinosaur was estimated using potassium-40 dating of rocks surrounding the bones.

a. Potassium-40 decays exponentially with a half-life of approximately 1.31 billion years. Use the fact that after 1.31 billion years a given amount of potassium-40 will have decayed to half the original amount to show that the decay model for potassium-40 is given by $A = A_0e^{-0.52912t}$, where t is in billions of years.

b. Analysis of the rocks surrounding the dinosaur bones indicated that 94.5% of the original amount of potassium-40 was still present. Let $A = 0.945A_0$ in the model in part (a) and estimate the age of the bones of the dinosaur.

16. A bird species in danger of extinction has a population that is decreasing exponentially ($A = A_0e^{kt}$). Five years ago the population was at 1400 and today only 1000 of the birds are alive. Once the population drops below 100, the situation will be irreversible. When will this happen?

17. Use the exponential growth model, $A = A_0e^{kt}$, to show that the time it takes a population to double (to grow from A_0 to $2A_0$) is given by $t = \frac{\ln 2}{k}$.

18. Use the exponential growth model, $A = A_0e^{kt}$, to show that the time it takes a population to triple (to grow from A_0 to $3A_0$) is given by $t = \frac{\ln 3}{k}$.

Use the formula $t = \frac{\ln 2}{k}$ that gives the time for a population with a growth rate k to double to solve Exercises 19–20. Express each answer to the nearest whole year.

19. The growth model $A = 4e^{0.007t}$ describes New Zealand's population, A, in millions, t years after 2003.

a. What is New Zealand's growth rate?

b. How long will it take New Zealand to double its population?

20. The growth model $A = 104.9e^{0.017t}$ describes Mexico's population, A, in millions, t years after 2003.

a. What is Mexico's growth rate?

b. How long will it take Mexico to double its population?

21. The logistic growth function

$$f(t) = \frac{100{,}000}{1 + 5000e^{-t}}$$

describes the number of people, $f(t)$, who have become ill with influenza t weeks after its initial outbreak in a particular community.

a. How many people became ill with the flu when the epidemic began?

b. How many people were ill by the end of the fourth week?

c. What is the limiting size of the population that becomes ill?

Shown, again, is world population, in billions, for seven selected years from 1950 through 2003. Using a graphing utility's logistic REGression option, we obtain the equation shown on the screen.

x, Numbers of Years after 1949	y, World Population (billions)
1 (1950)	2.6
11 (1960)	3.0
21 (1970)	3.7
31 (1980)	4.5
41 (1990)	5.3
51 (2000)	6.1
54 (2003)	6.3

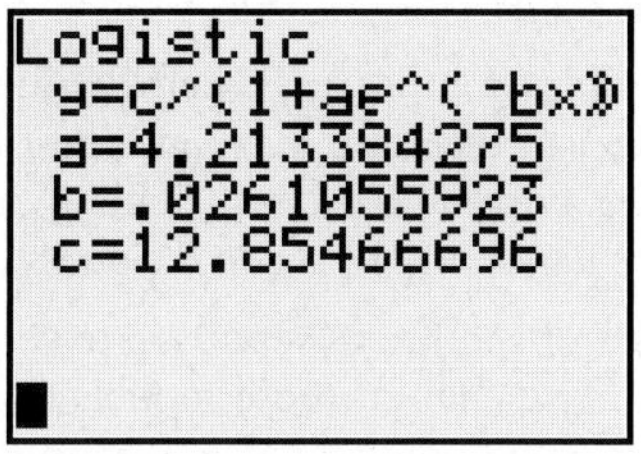

We see that a logistic growth model for world population, $f(x)$, in billions, x years after 1949 is

$$f(x) = \frac{12.85}{1 + 4.21e^{-0.026x}}.$$

Use this function to solve Exercises 22–26.

22. How well does the function model the data for 2000?

23. How well does the function model the data for 2003?

24. When will world population reach 7 billion?

25. When will world population reach 8 billion?

26. According to the model, what is the limiting size of the population that Earth will eventually sustain? What does this mean in terms of the statement made by the U.S. National Academy of Sciences that 10 billion is the maximum that the world can support with some degree of comfort and individual choice?

The logistic growth function

$$P(x) = \frac{90}{1 + 271e^{-0.122x}}$$

models the percentage, $P(x)$, of Americans who are x years old with some coronary heart disease. Use the function to solve Exercises 27–30.

27. What percentage of 20-year-olds have some coronary heart disease?

28. What percentage of 80-year-olds have some coronary heart disease?

29. At what age is the percentage of some coronary heart disease 50%?

30. At what age is the percentage of some coronary heart disease 70%?

Exercises 31–36 present data in the form of tables. For each data set shown by the table,

a. *Create a scatter plot for the data.*

b. *Use the scatter plot to determine whether an exponential function, a logarithmic function, or a linear function is the best choice for modeling the data. (If applicable, in Exercise 57, you will use your graphing utility to obtain these functions.)*

31. Percent of Miscarriages, by Age

Woman's Age	Percent of Miscarriages
22	9%
27	10%
32	13%
37	20%
42	38%
47	52%

Source: Time

32. Number of Countries Connected to the Internet

Year	Number of Countries Connected to the Internet
1985	11
1991	91
1994	146
1997	195
2002	220

Source: Medard Gabel, *Global Inc., 2003*

33. Number of Illegal Immigrants Living in the U.S.

Year	Number of Illegal Immigrants (millions)
1992	3.4
1996	5.0
2000	7.0
2004	8.0

Source: U.S. Department of Homeland Security

34. Number of U.S. Households with Pets

Year	Number with Pets (millions)
1998	54.0
1999	58.2
2000	61.1
2001	63.0
2002	64.2

Source: American Pet Products Manufacturers Association

35. Alcohol Use by U.S. High School Seniors

Year	Percentage Using Alcohol during 30 Days Preceding the Survey
1980	72.0%
1985	65.9%
1990	57.1%
1995	51.3%
2000	50.0%
2002	48.6%
2003	47.5%

Source: U.S. Department of Health and Human Services

36. U.S. Vehicle Fatality Rates

Year	Deaths per 100 Million Vehicle Miles Traveled
1965	5.50
1970	4.50
1975	3.40
1985	2.50
1990	2.00
1995	1.60
2000	1.50
2003	1.48

Source: National Highway Traffic Safety Administration

In Exercises 37–40, rewrite the equation in terms of base e. Express the answer in terms of a natural logarithm and then round to three decimal places.

37. $y = 100(4.6)^x$ **38.** $y = 1000(7.3)^x$

39. $y = 2.5(0.7)^x$ **40.** $y = 4.5(0.6)^x$

Writing in Mathematics

41. Nigeria has a growth rate of 0.025 or 2.5%. Describe what this means.

42. How can you tell whether an exponential model describes exponential growth or exponential decay?

43. Suppose that a population that is growing exponentially increases from 800,000 people in 2003 to 1,000,000 people in 2006. Without showing the details, describe how to obtain the exponential growth function that models the data.

44. What is the half-life of a substance?

45. Describe a difference between exponential growth and logistic growth.

46. Describe the shape of a scatter plot that suggests modeling the data with an exponential function.

47. You take up weightlifting and record the maximum number of pounds you can lift at the end of each week. You start off with rapid growth in terms of the weight you can lift from week to week, but then the growth begins to level off. Describe how to obtain a function that models the number of pounds you can lift at the end of each week. How can you use this function to predict what might happen if you continue the sport?

48. Would you prefer that your salary be modeled exponentially or logarithmically? Explain your answer.

49. One problem with all exponential growth models is that nothing can grow exponentially forever. Describe factors that might limit the size of a population.

Technology Exercises

In Example 1 on page 447, we used two data points and an exponential function to model the population of the United States from 1970 through 2003. The data are shown again in the table. Use all five data points to solve Exercises 50–54.

x, Number of Years after 1969	y, U.S. Population (millions)
1 (1970)	203.3
11 (1980)	226.5
21 (1990)	248.7
31 (2000)	281.4
34 (2003)	294.0

50. a. Use your graphing utility's Exponential REGression option to obtain a model of the form $y = ab^x$ that fits the data. How well does the correlation coefficient, r, indicate that the model fits the data?

b. Rewrite the model in terms of base e. By what percentage is the population of the United States increasing each year?

51. Use your graphing utility's Logarithmic REGression option to obtain a model of the form $y = a + b \ln x$ that fits the data. How well does the correlation coefficient, r, indicate that the model fits the data?

52. Use your graphing utility's Linear REGression option to obtain a model of the form $y = ax + b$ that fits the data. How well does the correlation coefficient, r, indicate that the model fits the data?

53. Use your graphing utility's Power REGression option to obtain a model of the form $y = ax^b$ that fits the data. How well does the correlation coefficient, r, indicate that the model fits the data?

54. Use the values of r in Exercises 50–53 to select the two models of best fit. Use each of these models to predict by which year the U.S. population will reach 315 million. How do these answers compare to the year we found in Example 1, namely 2010? If you obtained different years, how do you account for this difference?

55. In Exercises 27–30, you worked with the logistic growth function

$$P(x) = \frac{90}{1 + 271e^{-0.122x}}$$

which models the percentage, $P(x)$, of Americans who are x years old with some coronary heart disease. Use your graphing utility to graph the function in a $[0, 100, 10]$ by $[0, 100, 10]$ viewing rectangle. Describe as specifically as possible what the logistic curve indicates about aging and the percentage of Americans with coronary heart disease.

56. The figure shows the number of people in the United States age 65 and over, with projected figures for the year 2010 and beyond.

U.S. Population Age 65 and Over

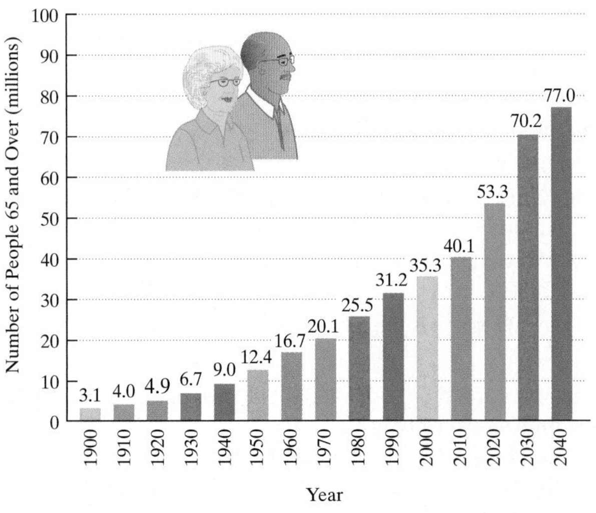

Source: U.S. Bureau of the Census

a. Let x represent the number of years after 1899 and let y represent the U.S. population age 65 and over, in millions. Use your graphing utility to find the model that best fits the data in the bar graph.

b. Rewrite the model in terms of base e. By what percentage is the 65 and over population increasing each year?

57. In Exercises 31–36, you determined the best choice for the kind of function that modeled the data in each table. For each of these exercises that you worked, use a graphing utility to find the actual function that best fits the data. Then use the model to make a reasonable prediction for a value that exceeds those shown in the table's first column.

Critical Thinking Exercises

58. The exponential growth models describe the population of the indicated country, A, in millions, t years after 2003.

Canada $A = 32.2e^{0.003t}$

Uganda $A = 25.6e^{0.03t}$

According to these models, which one of the following is true?

a. In 2003, Uganda's population was ten times that of Canada's.

b. In 2003, Canada's population exceeded Uganda's by 660,000.

c. In 2012, Uganda's population will exceed Canada's.

d. None of these statements is true.

59. Over a period of time, a hot object cools to the temperature of the surrounding air. This is described mathematically by Newton's Law of Cooling:

$$T = C + (T_0 - C)e^{-kt},$$

where t is the time it takes for an object to cool from temperature T_0 to temperature T, C is the surrounding air temperature, and k is a positive constant that is associated with the cooling object. A cake removed from the oven has a temperature of 210°F and is left to cool in a room that has a temperature of 70°F. After 30 minutes, the temperature of the cake is 140°F. What is the temperature of the cake after 40 minutes?

Group Exercises

60. This activity is intended for three or four people who would like to take up weightlifting. Each person in the group should record the maximum number of pounds that he or she can lift at the end of each week for the first 10 consecutive weeks. Use the Logarithmic REGression option of a graphing utility to obtain a model showing the amount of weight that group members can lift from week 1 through week 10. Graph each of the models in the same viewing rectangle to observe similarities and differences among weight-growth patterns of each member. Use the functions to predict the amount of weight that group members will be able to lift in the future. If the group continues to work out together, check the accuracy of these predictions.

61. Each group member should consult an almanac, newspaper, magazine, or the Internet to find data that can be modeled by exponential or logarithmic functions. Group members should select the two sets of data that are most interesting and relevant. For each set selected, find a model that best fits the data. Each group member should make one prediction based on the model and then discuss a consequence of this prediction. What factors might change the accuracy of the prediction?

Chapter 4 Summary, Review, and Test

Summary

DEFINITIONS AND CONCEPTS	EXAMPLES
4.1 Exponential Functions	
a. The exponential function with base b is defined by $f(x) = b^x$, where $b > 0$ and $b \neq 1$.	Ex. 1, p. 401
b. Characteristics of exponential functions and graphs for $0 < b < 1$ and $b > 1$ are shown in the box on page 403.	Ex. 2, p. 402; Ex. 3, p. 402
c. Transformations involving exponential functions are summarized in Table 4.1 on page 404.	Ex. 4, p. 404; Ex. 5, p. 405
d. The natural exponential function is $f(x) = e^x$. The irrational number e is called the natural base, where $e \approx 2.7183$. e is the value that $\left(1 + \frac{1}{n}\right)^n$ approaches as $n \to \infty$.	Ex. 6, p. 406
e. Formulas for compound interest: After t years, the balance, A, in an account with principal P and annual interest rate r (in decimal form) is given by one of the following formulas: **1.** For n compoundings per year: $A = P\left(1 + \frac{r}{n}\right)^{nt}$ **2.** For continuous compounding: $A = Pe^{rt}$	Ex. 7, p. 408

DEFINITIONS AND CONCEPTS	EXAMPLES
4.2 Logarithmic Functions	
a. Definition of the logarithmic function: For $x > 0$ and $b > 0, b \neq 1$, $y = \log_b x$ is equivalent to $b^y = x$. The function $f(x) = \log_b x$ is the logarithmic function with base b. This function is the inverse function of the exponential function with base b.	Ex. 1, p. 413; Ex. 2, p. 413; Ex. 3, p. 413
b. Graphs of logarithmic functions for $b > 1$ and $0 < b < 1$ are shown in Figure 4.6 on page 416. Characteristics of the graphs are summarized in the box that follows the figure.	Ex. 6, p. 415
c. Transformations involving logarithmic functions are summarized in Table 4.4 on page 416.	Figures 4.7–4.10, p. 417
d. The domain of a logarithmic function of the form $f(x) = \log_b x$ is the set of all positive real numbers. The domain of $f(x) = \log_b g(x)$ consists of all x for which $g(x) > 0$.	Ex. 7, p. 417; Ex. 10, p. 419
e. Common and natural logarithms: $f(x) = \log x$ means $f(x) = \log_{10} x$ and is the common logarithmic function. $f(x) = \ln x$ means $f(x) = \log_e x$ and is the natural logarithmic function.	Ex. 8, p. 418; Ex. 9, p. 419; Ex. 11, p. 420
f. Basic Logarithmic Properties	

Base b ($b > 0, b \neq 1$)	Base 10 (Common Logarithms)	Base e (Natural Logarithms)	
$\log_b 1 = 0$	$\log 1 = 0$	$\ln 1 = 0$	
$\log_b b = 1$	$\log 10 = 1$	$\ln e = 1$	Ex. 4, p. 414
$\log_b b^x = x$	$\log 10^x = x$	$\ln e^x = x$	
$b^{\log_b x} = x$	$10^{\log x} = x$	$e^{\ln x} = x$	Ex. 5, p. 415

DEFINITIONS AND CONCEPTS	EXAMPLES
4.3 Properties of Logarithms	
a. *The Product Rule:* $\log_b(MN) = \log_b M + \log_b N$	Ex. 1, p. 425
b. *The Quotient Rule:* $\log_b\left(\frac{M}{N}\right) = \log_b M - \log_b N$	Ex. 2, p. 426
c. *The Power Rule:* $\log_b M^p = p \log_b M$	Ex. 3, p. 427
d. *The Change-of-Base Property:*	

The General Property	Introducing Common Logarithms	Introducing Natural Logarithms	
$\log_b M = \frac{\log_a M}{\log_a b}$	$\log_b M = \frac{\log M}{\log b}$	$\log_b M = \frac{\ln M}{\ln b}$	Ex. 7, p. 430; Ex. 8, p. 431

DEFINITIONS AND CONCEPTS	EXAMPLES
e. Properties for expanding logarithmic expressions are given in the box on page 427.	Ex. 4, p. 428
f. Properties for condensing logarithmic expressions are given in the box on page 428.	Ex. 5, p. 428; Ex. 6, p. 429
4.4 Exponential and Logarithmic Equations	
a. An exponential equation is an equation containing a variable in an exponent. Some exponential equations can be solved by expressing each side as a power of the same base: If $b^M = b^N$, then $M = N$. Details are in the box on page 434.	Ex. 1, p. 434
b. The procedure for using natural logarithms to solve exponential equations is given in the box on page 435. The solution procedure involves isolating the exponential expression and taking the natural logarithm on both sides.	Ex. 2, p. 435; Ex. 3, p. 436; Ex. 4, p. 437; Ex. 5, p. 437
c. A logarithmic equation is an equation containing a variable in a logarithmic expression. Some logarithmic equations can be expressed in the form $\log_b M = c$. The definition of a logarithm is used to rewrite the equation in exponential form: $b^c = M$. See the box on page 437. When checking logarithmic equations, reject proposed solutions that produce the logarithm of a negative number or the logarithm of 0 in the original equation.	Ex. 6, p. 437; Ex. 7, p. 439

DEFINITIONS AND CONCEPTS | **EXAMPLES**

d. Some logarithmic equations can be expressed in the form $\log_b M = \log_b N$. Use the one-to-one property to rewrite the equation without logarithms: $M = N$. See the box on page 439. — Ex. 8, p. 439

4.5 Exponential Growth and Decay; Modeling Data

a. Exponential growth and decay models are given by $A = A_0e^{kt}$ in which t represents time, A_0 is the amount present at $t = 0$, and A is the amount present at time t. If $k > 0$, the model describes growth and k is the growth rate. If $k < 0$, the model describes decay and k is the decay rate. — Ex. 1, p. 447; Ex. 2, p. 449

b. The logistic growth model, given by $A = \dfrac{c}{1 + ae^{-bt}}$, describes situations in which growth is limited. $y = c$ is a horizontal asymptote for the graph, and growth, A, can never exceed c. — Ex. 3, p. 451

c. Scatter plots for exponential and logarithmic models are shown in Figure 4.21 on page 452. When using a graphing utility to model data, the closer that the correlation coefficient, r, is to -1 or 1, the better the model fits the data. — Ex. 4, p. 452; Ex. 5, p. 453

d. Expressing an Exponential Model in Base e: $y = ab^x$ is equivalent to $y = ae^{(\ln b)\cdot x}$. — Ex. 6, p. 455

Review Exercises

4.1

In Exercises 1–4, the graph of an exponential function is given. Select the function for each graph from the following options:

$$f(x) = 4^x,\ g(x) = 4^{-x},$$
$$h(x) = -4^{-x},\ r(x) = -4^{-x} + 3.$$

1.

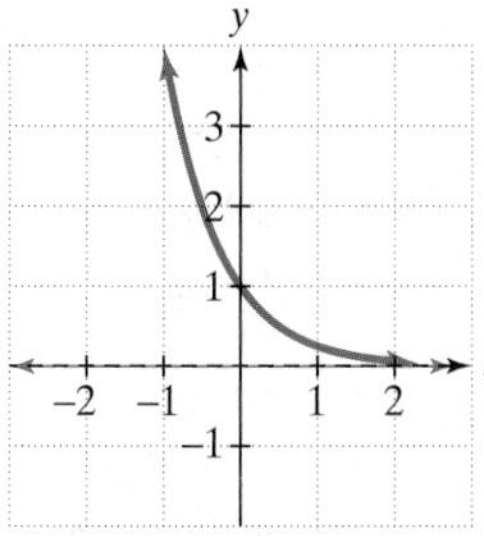

2.

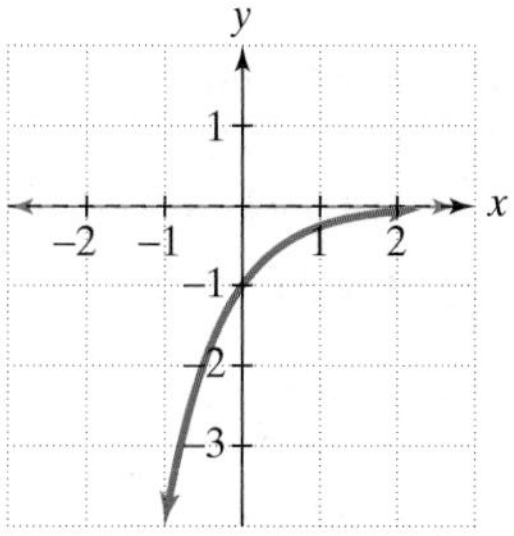

3.

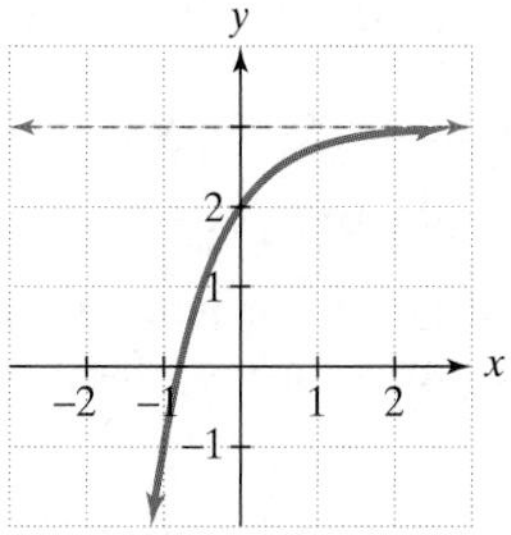

4.

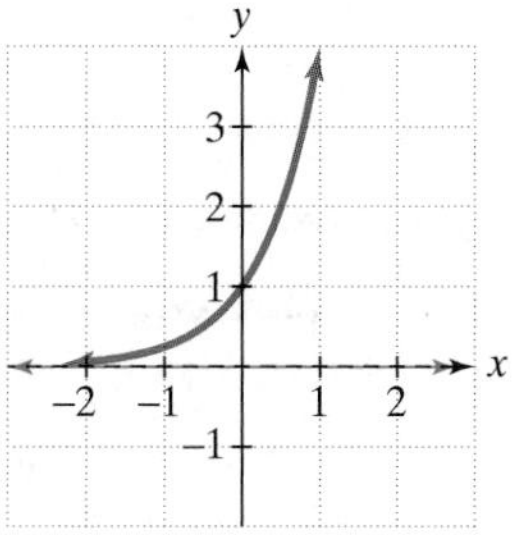

In Exercises 5–9, graph f and g in the same rectangular coordinate system. Use transformations of the graph of f to obtain the graph of g. Graph and give equations of all asymptotes. Use the graphs to determine each function's domain and range.

5. $f(x) = 2^x$ and $g(x) = 2^{x-1}$

6. $f(x) = 3^x$ and $g(x) = 3^x - 1$

7. $f(x) = 3^x$ and $g(x) = -3^x$

8. $f(x) = \left(\frac{1}{2}\right)^x$ and $g(x) = \left(\frac{1}{2}\right)^{-x}$

9. $f(x) = e^x$ and $g(x) = 2e^{\frac{x}{2}}$

Use the compound interest formulas to solve Exercises 10–11.

10. Suppose that you have \$5000 to invest. Which investment yields the greater return over 5 years: 5.5% compounded semiannually or 5.25% compounded monthly?

11. Suppose that you have \$14,000 to invest. Which investment yields the greater return over 10 years: 7% compounded monthly or 6.85% compounded continuously?

12. A cup of coffee is taken out of a microwave oven and placed in a room. The temperature, T, in degrees Fahrenheit, of the coffee after t minutes is modeled by the function $T = 70 + 130e^{-0.04855t}$. The graph of the function is shown in the figure.

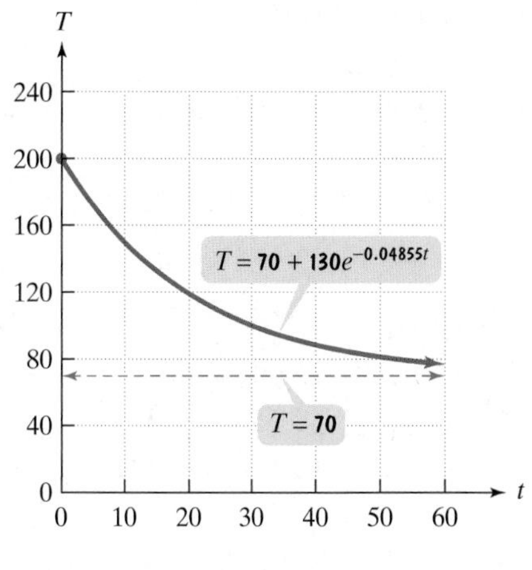

Use the graph shown at the bottom of the previous page to answer each of the following questions.

a. What was the temperature of the coffee when it was first taken out of the microwave?

b. What is a reasonable estimate of the temperature of the coffee after 20 minutes? Use your calculator to verify this estimate.

c. What is the limit of the temperature to which the coffee will cool? What does this tell you about the temperature of the room?

4.2

In Exercises 13–15, write each equation in its equivalent exponential form.

13. $\frac{1}{2} = \log_{49} 7$ **14.** $3 = \log_4 x$ **15.** $\log_3 81 = y$

In Exercises 16–18, write each equation in its equivalent logarithmic form.

16. $6^3 = 216$ **17.** $b^4 = 625$ **18.** $13^y = 874$

In Exercises 19–29, evaluate each expression without using a calculator. If evaluation is not possible, state the reason.

19. $\log_4 64$ **20.** $\log_5 \frac{1}{25}$ **21.** $\log_3(-9)$

22. $\log_{16} 4$ **23.** $\log_{17} 17$ **24.** $\log_3 3^8$

25. $\ln e^5$ **26.** $\log_3 \frac{1}{\sqrt{3}}$ **27.** $\ln \frac{1}{e^2}$

28. $\log \frac{1}{1000}$ **29.** $\log_3(\log_8 8)$

30. Graph $f(x) = 2^x$ and $g(x) = \log_2 x$ in the same rectangular coordinate system.

31. Graph $f(x) = \left(\frac{1}{3}\right)^x$ and $g(x) = \log_{\frac{1}{3}} x$ in the same rectangular coordinate system.

In Exercises 32–35, the graph of a logarithmic function is given. Select the function for each graph from the following options:

$$f(x) = \log x,\ g(x) = \log(-x),$$
$$h(x) = \log(2 - x),\ r(x) = 1 + \log(2 - x).$$

32.

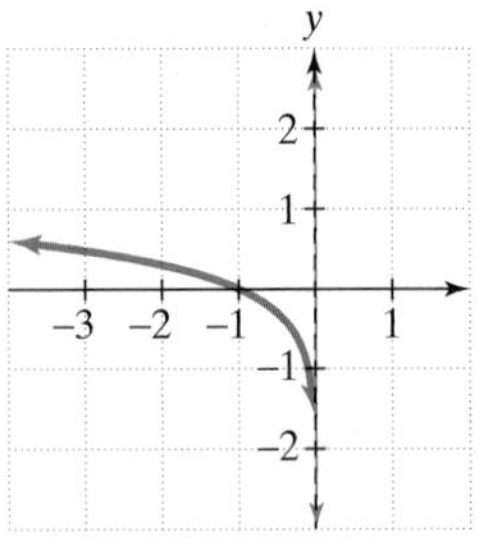

33.

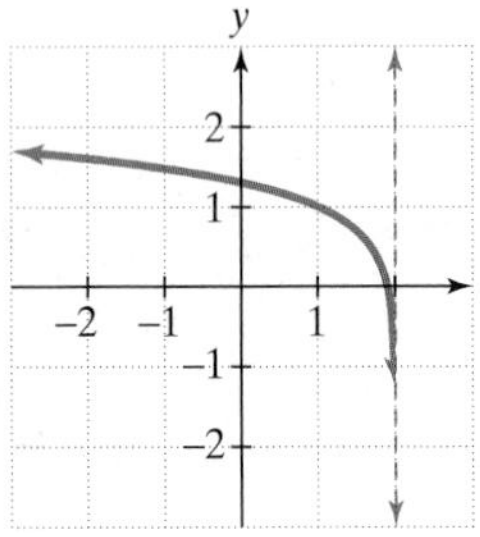

34.

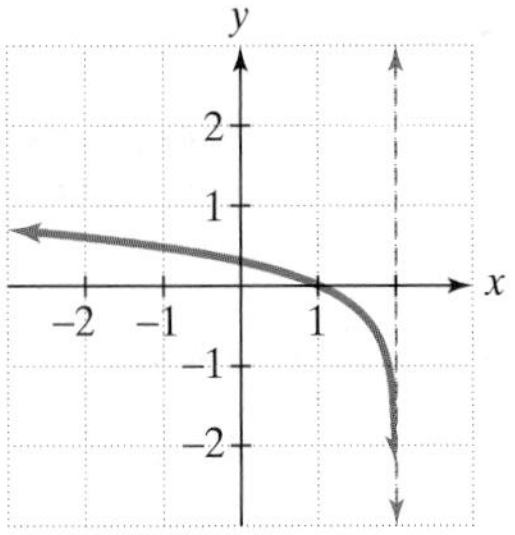

35.

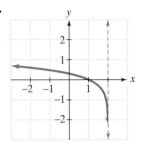

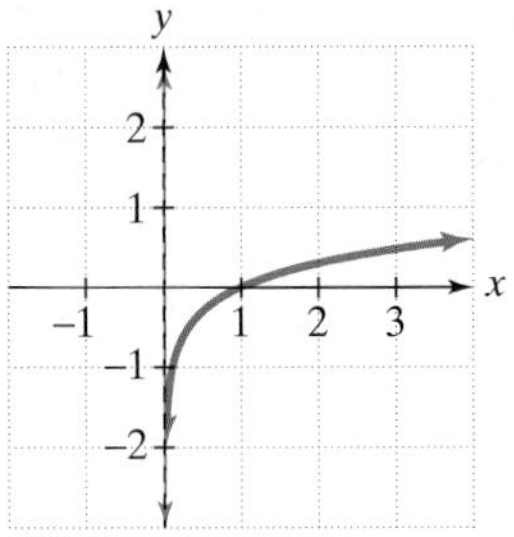

In Exercises 36–38, begin by graphing $f(x) = \log_2 x$. Then use transformations of this graph to graph the given function. What is the graph's x-intercept? What is the vertical asymptote? Use the graphs to determine each function's domain and range.

36. $g(x) = \log_2(x - 2)$ **37.** $h(x) = -1 + \log_2 x$

38. $r(x) = \log_2(-x)$

In Exercises 39–40, graph f and g in the same rectangular coordinate system. Use transformations of the graph of f to obtain the graph of g. Graph and give equations of all asymptotes. Use the graphs to determine each function's domain and range.

39. $f(x) = \log x$ and $g(x) = -\log(x + 3)$

40. $f(x) = \ln x$ and $g(x) = -\ln(2x)$

In Exercises 41–43, find the domain of each logarithmic function.

41. $f(x) = \log_8(x + 5)$ **42.** $f(x) = \log(3 - x)$

43. $f(x) = \ln(x - 1)^2$

In Exercises 44–46, use inverse properties of logarithms to simplify each expression.

44. $\ln e^{6x}$ **45.** $e^{\ln\sqrt{x}}$ **46.** $10^{\log 4x^2}$

47. On the Richter scale, the magnitude, R, of an earthquake of intensity I is given by $R = \log \frac{I}{I_0}$, where I_0 is the intensity of a barely felt zero-level earthquake. If the intensity of an earthquake is $1000I_0$, what is its magnitude on the Richter scale?

48. Students in a psychology class took a final examination. As part of an experiment to see how much of the course content they remembered over time, they took equivalent forms of the exam in monthly intervals thereafter. The average score, $f(t)$, for the group after t months is modeled by the function $f(t) = 76 - 18\log(t + 1)$, where $0 \le t \le 12$.

a. What was the average score when the exam was first given?

b. What was the average score after 2 months? 4 months? 6 months? 8 months? one year?

c. Use the results from parts (a) and (b) to graph f. Describe what the shape of the graph indicates in terms of the material retained by the students.

49. The formula

$$t = \frac{1}{c}\ln\left(\frac{A}{A - N}\right)$$

describes the time, t, in weeks, that it takes to achieve mastery of a portion of a task. In the formula, A represents maximum learning possible, N is the portion of the learning that is to be achieved, and c is a constant used to measure an individual's learning style. A 50-year-old man decides to start running as a way to maintain good health. He feels that the maximum rate he could ever hope to achieve is 12 miles per hour. How many weeks will it take before the man can run 5 miles per hour if $c = 0.06$ for this person?

4.3

In Exercises 50–53, use properties of logarithms to expand each logarithmic expression as much as possible. Where possible, evaluate logarithmic expressions without using a calculator.

50. $\log_6(36x^3)$

51. $\log_4\left(\frac{\sqrt{x}}{64}\right)$

52. $\log_2\left(\frac{xy^2}{64}\right)$

53. $\ln\sqrt[3]{\frac{x}{e}}$

In Exercises 54–57, use properties of logarithms to condense each logarithmic expression. Write the expression as a single logarithm whose coefficient is 1.

54. $\log_b 7 + \log_b 3$

55. $\log 3 - 3\log x$

56. $3\ln x + 4\ln y$

57. $\frac{1}{2}\ln x - \ln y$

In Exercises 58–59, use common logarithms or natural logarithms and a calculator to evaluate to four decimal places.

58. $\log_6 72{,}348$

59. $\log_4 0.863$

In Exercises 60–63, determine whether each equation is true or false. Where possible, show work to support your conclusion. If the statement is false, make the necessary change(s) to produce a true statement.

60. $(\ln x)(\ln 1) = 0$

61. $\log(x + 9) - \log(x + 1) = \frac{\log(x + 9)}{\log(x + 1)}$

62. $(\log_2 x)^4 = 4\log_2 x$

63. $\ln e^x = x\ln e$

4.4

In Exercises 64–72, solve each exponential equation. Where necessary, express the solution set in terms of natural logarithms and use a calculator to obtain a decimal approximation, correct to two decimal places, for the solution.

64. $2^{4x-2} = 64$

65. $125^x = 25$

66. $9^{x+2} = 27^{-x}$

67. $8^x = 12{,}143$

68. $9e^{5x} = 1269$

69. $e^{12-5x} - 7 = 123$

70. $5^{4x+2} = 37{,}500$

71. $3^{x+4} = 7^{2x-1}$

72. $e^{2x} - e^x - 6 = 0$

In Exercises 73–78, solve each logarithmic equation.

73. $\log_4(3x - 5) = 3$

74. $3 + 4\ln(2x) = 15$

75. $\log_2(x + 3) + \log_2(x - 3) = 4$

76. $\log_3(x - 1) - \log_3(x + 2) = 2$

77. $\ln(x + 4) - \ln(x + 1) = \ln x$

78. $\log_4(2x + 1) = \log_4(x - 3) + \log_4(x + 5)$

79. The function $P(x) = 14.7e^{-0.21x}$ models the average atmospheric pressure, $P(x)$, in pounds per square inch, at an altitude of x miles above sea level. The atmospheric pressure at the peak of Mt. Everest, the world's highest mountain, is 4.6 pounds per square inch. How many miles above sea level, to the nearest tenth of a mile, is the peak of Mt. Everest?

80. The amount of carbon dioxide in the atmosphere, measured in parts per million, has been increasing as a result of the burning of oil and coal. The buildup of gases and particles traps heat and raises the planet's temperature, a phenomenon called the greenhouse effect. Carbon dioxide accounts for about half of the warming. The function $f(t) = 364(1.005)^t$ projects carbon dioxide concentration, $f(t)$, in parts per million, t years after 2000. Using the projections given by the function, when will the carbon dioxide concentration be double the preindustrial level of 280 parts per million?

81. The function $W(x) = 0.37\ln x + 0.05$ models the average walking speed, $W(x)$, in feet per second, of residents in a city whose population is x thousand. Visitors to New York City frequently feel they are moving too slowly to keep pace with New Yorkers' average walking speed of 3.38 feet per second. What is the population of New York City? Round to the nearest thousand.

82. Use the formula for compound interest with n compoundings per year to solve this problem. How long, to the nearest tenth of a year, will it take \$12,500 to grow to \$20,000 at 6.5% annual interest compounded quarterly?

Use the formula for continuous compounding to solve Exercises 83–84.

83. How long, to the nearest tenth of a year, will it take \$50,000 to triple in value at 7.5% annual interest compounded continuously?

84. What interest rate, to the nearest percent, is required for an investment subject to continuous compounding to triple in 5 years?

4.5

85. According to the U.S. Bureau of the Census, in 1990 there were 22.4 million residents of Hispanic origin living in the United States. By 2000, the number had increased to 35.3 million. The exponential growth function $A = 22.4e^{kt}$ describes the U.S. Hispanic population, A, in millions, t years after 1990.

a. Find k, correct to three decimal places.

b. Use the resulting model to project the Hispanic resident population in 2010.

c. In which year will the Hispanic resident population reach 60 million?

86. Use the exponential decay model, $A = A_0e^{kt}$, to solve this exercise. The half-life of polonium-210 is 140 days. How long will it take for a sample of this substance to decay to 20% of its original amount?

87. The function

$$f(t) = \frac{500{,}000}{1 + 2499e^{-0.92t}}$$

models the number of people, $f(t)$, in a city who have become ill with influenza t weeks after its initial outbreak.

a. How many people became ill with the flu when the epidemic began?

b. How many people were ill by the end of the sixth week?

c. What is the limiting size of $f(t)$, the population that becomes ill?

Exercises 88–89 present data in the form of tables. For each data set shown by the table,

a. *Create a scatter plot for the data.*

b. *Use the scatter plot to determine whether an exponential function or a logarithmic function is the better choice for modeling the data.*

88. Percentage of the U.S. Population, Ages 25 or Older, with a College Degree

Year	Percent
1980	17.0%
1991	21.4%
1995	23.0%
2000	25.6%
2001	26.2%
2002	26.7%

Source: U.S. Census Bureau

89. Projection of U.S. Jobs Moving Overseas

Year	Number of Jobs Moving Overseas (millions)
2003	0.3
2008	1.0
2010	1.5
2012	2.5
2015	3.3

Source: Forrester Research, Inc.

In Exercises 90–91, rewrite the equation in terms of base e. Express the answer in terms of a natural logarithm and then round to three decimal places.

90. $y = 73(2.6)^x$

91. $y = 6.5(0.43)^x$

92. The figure shows world population projections through the year 2150. The data are from the United Nations Family Planning Program and are based on optimistic or pessimistic expectations for successful control of human population growth. Suppose that you are interested in modeling these data using exponential, logarithmic, linear, and quadratic functions. Which function would you use to model each of the projections? Explain your choices. For the choice corresponding to a quadratic model, would your formula involve one with a positive or negative leading coefficient? Explain.

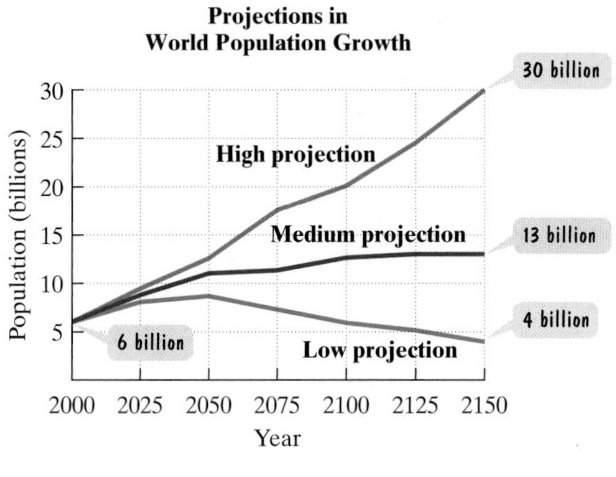

Source: U.N.

Chapter 4 Test

1. Graph $f(x) = 2^x$ and $g(x) = 2^{x+1}$ in the same rectangular coordinate system.

2. Graph $f(x) = \log_2 x$ and $g(x) = \log_2(x - 1)$ in the same rectangular coordinate system.

3. Write in exponential form: $\log_5 125 = 3$.

4. Write in logarithmic form: $\sqrt{36} = 6$.

5. Find the domain of $f(x) = \ln(3 - x)$.

In Exercises 6–7, use properties of logarithms to expand each logarithmic expression as much as possible. Where possible, evaluate logarithmic expressions without using a calculator.

6. $\log_4(64x^5)$
7. $\log_3\left(\frac{\sqrt[3]{x}}{81}\right)$

In Exercises 8–9, write each expression as a single logarithm.

8. $6 \log x + 2 \log y$
9. $\ln 7 - 3 \ln x$

10. Use a calculator to evaluate $\log_{15} 71$ to four decimal places.

In Exercises 11–18, solve each equation.

11. $3^{x-2} = 9^{x+4}$
12. $5^x = 1.4$
13. $400e^{0.005x} = 1600$
14. $e^{2x} - 6e^x + 5 = 0$
15. $\log_6(4x - 1) = 3$
16. $2 \ln(3x) = 8$
17. $\log x + \log(x + 15) = 2$
18. $\ln(x - 4) - \ln(x + 1) = \ln 6$

19. On the decibel scale, the loudness of a sound, D, in decibels, is given by $D = 10 \log \frac{I}{I_0}$, where I is the intensity of the sound, in watts per meter^2, and I_0 is the intensity of a sound barely audible to the human ear. If the intensity of a sound is $10^{12} I_0$, what is its loudness in decibels? (Such a sound is potentially damaging to the ear.)

In Exercises 20–22, simplify each expression.

20. $\ln e^{5x}$
21. $\log_b b$
22. $\log_6 1$

Use the compound interest formulas to solve Exercises 23–25.

23. Suppose you have $3000 to invest. Which investment yields the greater return over 10 years: 6.5% compounded semiannually or 6% compounded continuously? How much more (to the nearest dollar) is yielded by the better investment?

24. How long, to the nearest tenth of a year, will it take $4000 to grow to $8000 at 5% annual interest compounded quarterly?

25. What interest rate, to the nearest tenth of a percent, is required for an investment subject to continuous compounding to double in 10 years?

26. The function

$$A = 82.3e^{-0.002t}$$

models the population of Germany, A, in millions, t years after 2003.

   a. What was the population of Germany in 2003?
   b. Is the population of Germany increasing or decreasing? Explain.
   c. In which year will the population of Germany be 81.5 million?

27. The 1990 population of Europe was 509 million; in 2000, it was 729 million. Write the exponential growth function that describes the population of Europe, in millions, t years after 1990.

28. Use the exponential decay model for carbon-14, $A = A_0e^{-0.000121t}$, to solve this exercise. Bones of a prehistoric man were discovered and contained 5% of the original amount of carbon-14. How long ago did the man die?

29. The logistic growth function

$$f(t) = \frac{140}{1 + 9e^{-0.165t}}$$

describes the population, $f(t)$, of an endangered species of elk t years after they were introduced to a nonthreatening habitat.

   a. How many elk were initially introduced to the habitat?
   b. How many elk are expected in the habitat after 10 years?
   c. What is the limiting size of the elk population that the habitat will sustain?

In Exercises 30–33, determine whether the values in each table belong to an exponential function, a logarithmic function, a linear function, or a quadratic function.

30.

x	y
0	3
1	1
2	−1
3	−3
4	−5

31.

x	y
$\frac{1}{3}$	−1
1	0
3	1
9	2
27	3

32.

x	y
0	1
1	5
2	25
3	125
4	625

33.

x	y
0	12
1	3
2	0
3	3
4	12

34. Rewrite $y = 96(0.38)^x$ in terms of base e. Express the answer in terms of a natural logarithm and then round to three decimal places.

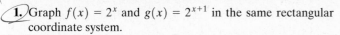

Cumulative Review Exercises (Chapters 1–4)

In Exercises 1–8, solve each equation or inequality.

1. $|3x - 4| = 2$
2. $\sqrt{2x - 5} - \sqrt{x - 3} = 1$
3. $x^4 + x^3 - 3x^2 - x + 2 = 0$
4. $e^{5x} - 32 = 96$
5. $\log_2(x + 5) + \log_2(x - 1) = 4$
6. $\ln(x + 4) + \ln(x + 1) = 2\ln(x + 3)$
7. $14 - 5x \geq -6$
8. $|2x - 4| \leq 2$

In Exercises 9–14, graph each equation in a rectangular coordinate system. If two functions are indicated, graph both in the same system.

9. $(x - 3)^2 + (y + 2)^2 = 4$
10. $f(x) = (x - 2)^2 - 1$
11. $f(x) = \dfrac{x^2 - 1}{x^2 - 4}$
12. $f(x) = (x - 2)^2(x + 1)$
13. $f(x) = 2x - 4$ and $f^{-1}(x)$
14. $f(x) = \ln x$ and $g(x) = \ln(x - 2) + 1$
15. Write the point-slope form and the slope-intercept form of the line passing through $(1, 3)$ and $(3, -3)$.
16. If $f(x) = x^2$ and $g(x) = x + 2$, find $(f \circ g)(x)$ and $(g \circ f)(x)$.
17. You discover that the number of hours you sleep each night varies inversely as the square of the number of cups of coffee consumed during the early evening. If 2 cups of coffee are consumed, you get 8 hours of sleep. If the number of cups of coffee is doubled, how many hours should you expect to sleep?

A baseball player hits a pop fly into the air. The function

$$s(t) = -16t^2 + 64t + 5$$

models the ball's height above the ground, $s(t)$, in feet, t seconds after it is hit. Use the function to solve Exercises 18–19.

18. When does the baseball reach its maximum height? What is that height?
19. After how many seconds does the baseball hit the ground? Round to the nearest tenth of a second.
20. You are paid time-and-a-half for each hour worked over 40 hours a week. Last week you worked 50 hours and earned \$660. What is your normal hourly salary?

CHAPTER 5

Systems of Equations and Inequalities

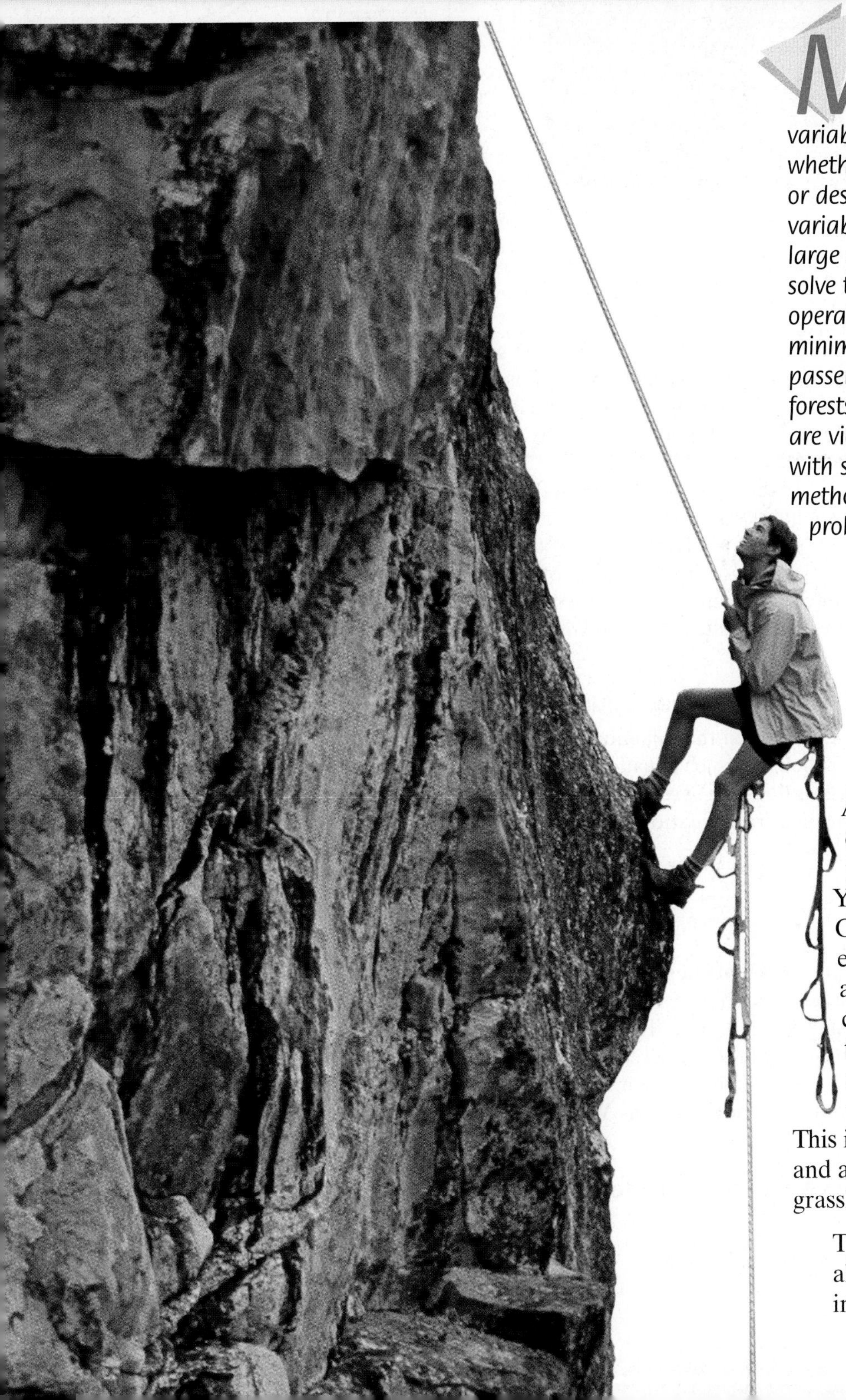

MOST THINGS IN LIFE DEPEND ON *many variables. Temperature and precipitation are two variables that have a critical effect on whether regions are forests, grasslands, or deserts. Airlines deal with numerous variables during weather disruptions at large connecting airports. They must solve the problem of putting their operations back together again to minimize the cost of the disruption and passenger inconvenience. In this chapter, forests, grasslands, and airline service are viewed in the same way—situations with several variables. You will learn methods for modeling and solving problems in these situations.*

A MAJOR WEATHER DISRUPTION delayed your flight for hours, but you finally made it. You are in Yosemite National Park in California, surrounded by evergreen forests, alpine meadows, and sheer walls of granite. Soaring cliffs, plunging waterfalls, gigantic trees, rugged canyons, mountains, and valleys stand in stark contrast to the angry chaos at the airport. This is so different from where you live and attend college, a region in which grasslands predominate.

This discussion is developed algebraically in Example 8 in Section 5.5.

SECTION 5.1 Systems of Linear Equations in Two Variables

Objectives

1. Decide whether an ordered pair is a solution of a linear system.
2. Solve linear systems by substitution.
3. Solve linear systems by addition.
4. Identify systems that do not have exactly one ordered-pair solution.
5. Solve problems using systems of linear equations.

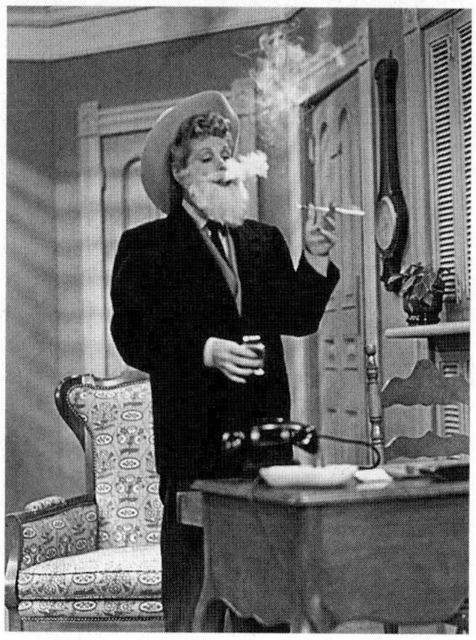

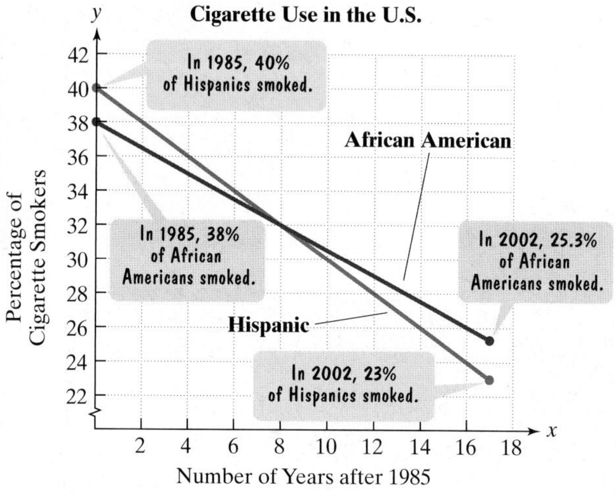

Figure 5.1

Source: Dept. of Health and Human Services

Although we still see celebrities smoking in movies, in music videos, and on television, there has been a remarkable decline in the percentage of cigarette smokers in the United States. The decline among African Americans and Hispanics, illustrated in Figure 5.1, can be analyzed using a pair of linear models in two variables.

In the first two sections of this chapter, you will learn to model your world with two equations in two variables and three equations in three variables. The methods you learn for solving these systems provide the foundation for solving complex problems involving thousands of equations containing thousands of variables. In the exercise set, you will apply these methods to analyze the linear decrease in cigarette use among whites, African Americans, and Hispanics.

1 Decide whether an ordered pair is a solution of a linear system.

Systems of Linear Equations and Their Solutions

We have seen that all equations in the form $Ax + By = C$ are straight lines when graphed. Two such equations are called a **system of linear equations** or a **linear system**. **A solution to a system of linear equations in two variables** is an ordered pair that satisfies both equations in the system. For example, (3, 4) satisfies the system

$$x + y = 7 \quad (3 + 4 \text{ is, indeed, } 7.)$$

$$x - y = -1. \quad (3 - 4 \text{ is, indeed, } -1.)$$

Thus, (3, 4) satisfies both equations and is a solution of the system. The solution can be described by saying that $x = 3$ and $y = 4$. The solution can also be described using set notation. The solution set to the system is $\{(3, 4)\}$—that is, the set consisting of the ordered pair (3, 4).

A system of linear equations can have exactly one solution, no solution, or infinitely many solutions. We begin with systems that have exactly one solution.

EXAMPLE 1 Determining Whether Ordered Pairs Are Solutions of a System

Consider the system:

$$x + 2y = 2$$

$$x - 2y = 6.$$

Determine if each ordered pair is a solution of the system:

a. $(4, -1)$ **b.** $(-4, 3)$.

Solution

a. We begin by determining whether $(4, -1)$ is a solution. Because 4 is the x-coordinate and -1 is the y-coordinate of $(4, -1)$, we replace x with 4 and y with -1.

$$\begin{aligned} x + 2y &= 2 \\ 4 + 2(-1) &\stackrel{?}{=} 2 \\ 4 + (-2) &\stackrel{?}{=} 2 \\ 2 &= 2, \quad \text{true} \end{aligned} \qquad \begin{aligned} x - 2y &= 6 \\ 4 - 2(-1) &\stackrel{?}{=} 6 \\ 4 - (-2) &\stackrel{?}{=} 6 \\ 4 + 2 &\stackrel{?}{=} 6 \\ 6 &= 6, \quad \text{true} \end{aligned}$$

The pair $(4, -1)$ satisfies both equations: It makes each equation true. Thus, the ordered pair is a solution of the system.

b. To determine whether $(-4, 3)$ is a solution, we replace x with -4 and y with 3.

$$\begin{aligned} x + 2y &= 2 \\ -4 + 2 \cdot 3 &\stackrel{?}{=} 2 \\ -4 + 6 &\stackrel{?}{=} 2 \\ 2 &= 2, \quad \text{true} \end{aligned} \qquad \begin{aligned} x - 2y &= 6 \\ -4 - 2 \cdot 3 &\stackrel{?}{=} 6 \\ -4 - 6 &\stackrel{?}{=} 6 \\ -10 &= 6, \quad \text{false} \end{aligned}$$

The pair $(-4, 3)$ fails to satisfy *both* equations: It does not make both equations true. Thus, the ordered pair is not a solution of the system.

Study Tip

When solving linear systems by graphing, neatly drawn graphs are essential for determining points of intersection.

- Use rectangular coordinate graph paper.
- Use a ruler or straightedge.
- Use a pencil with a sharp point.

The solution of a system of linear equations can sometimes be found by graphing both of the equations in the same rectangular coordinate system. For a system with one solution, the **coordinates of the point of intersection give the system's solution**. For example, the system in Example 1,

$$x + 2y = 2$$
$$x - 2y = 6$$

is graphed in Figure 5.2. The solution of the system, $(4, -1)$, corresponds to the point of intersection of the lines.

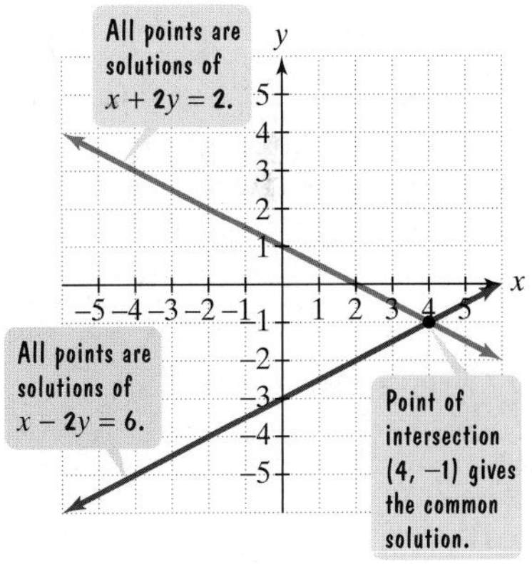

Figure 5.2 Visualizing a system's solution

Check Point 1 Consider the system:

$$2x - 3y = -4$$
$$2x + y = 4.$$

Determine if each ordered pair is a solution of the system:

a. $(1, 2)$ **b.** $(7, 6)$.

2 Solve linear systems by substitution.

Eliminating a Variable Using the Substitution Method

Finding the solution to a linear system by graphing equations may not be easy to do. For example, a solution of $\left(-\frac{2}{3}, \frac{157}{29}\right)$ would be difficult to "see" as an intersection point on a graph.

Let's consider a method that does not depend on finding a system's solution visually: the substitution method. This method involves converting the system to one equation in one variable by an appropriate substitution.

Study Tip

In step 1, you can choose which variable to isolate in which equation. If possible, solve for a variable whose coefficient is 1 or -1 to avoid working with fractions.

Solving Linear Systems by Substitution

1. Solve either of the equations for one variable in terms of the other. (If one of the equations is already in this form, you can skip this step.)
2. Substitute the expression found in step 1 into the *other* equation. This will result in an equation in one variable.
3. Solve the equation containing one variable.
4. Back-substitute the value found in step 3 into one of the original equations. Simplify and find the value of the remaining variable.
5. Check the proposed solution in both of the system's given equations.

EXAMPLE 2 Solving a System by Substitution

Solve by the substitution method:

$$5x - 4y = 9$$
$$x - 2y = -3.$$

Solution

Step 1 Solve either of the equations for one variable in terms of the other. We begin by isolating one of the variables in either of the equations. By solving for x in the second equation, which has a coefficient of 1, we can avoid fractions.

$$x - 2y = -3 \quad \text{This is the second equation in the given system.}$$
$$x = 2y - 3 \quad \text{Solve for } x \text{ by adding } 2y \text{ to both sides.}$$

Step 2 Substitute the expression from step 1 into the other equation. We substitute $2y - 3$ for x in the first equation.

$$x = \boxed{2y - 3} \qquad 5\boxed{x} - 4y = 9$$

This gives us an equation in one variable, namely

$$5(2y - 3) - 4y = 9.$$

The variable x has been eliminated.

Step 3 Solve the resulting equation containing one variable.

$$5(2y - 3) - 4y = 9 \quad \text{This is the equation containing one variable.}$$
$$10y - 15 - 4y = 9 \quad \text{Apply the distributive property.}$$
$$6y - 15 = 9 \quad \text{Combine like terms.}$$
$$6y = 24 \quad \text{Add 15 to both sides.}$$
$$y = 4 \quad \text{Divide both sides by 6.}$$

Step 4 Back-substitute the obtained value into one of the original equations. We back-substitute 4 for y into one of the original equations to find x. Let's use both equations to show that we obtain the same value for x in either case.

Using the first equation:

$$5x - 4y = 9$$
$$5x - 4(4) = 9$$
$$5x - 16 = 9$$
$$5x = 25$$
$$x = 5$$

Using the second equation:

$$x - 2y = -3$$
$$x - 2(4) = -3$$
$$x - 8 = -3$$
$$x = 5$$

With $x = 5$ and $y = 4$, the proposed solution is $(5, 4)$.

Step 5 Check. Take a moment to show that $(5, 4)$ satisfies both given equations. The solution set is $\{(5, 4)\}$.

Study Tip

The equation from step 1, in which one variable is expressed in terms of the other, is equivalent to one of the original equations. It is often easiest to back-substitute an obtained value into this equation to find the value of the other variable. After obtaining both values, get into the habit of checking the ordered-pair solution in *both* equations of the system.

Technology

A graphing utility can be used to solve the system in Example 2. Solve each equation for y, graph the equations, and use the intersection feature. The utility displays the solution $(5, 4)$ as $x = 5$, $y = 4$.

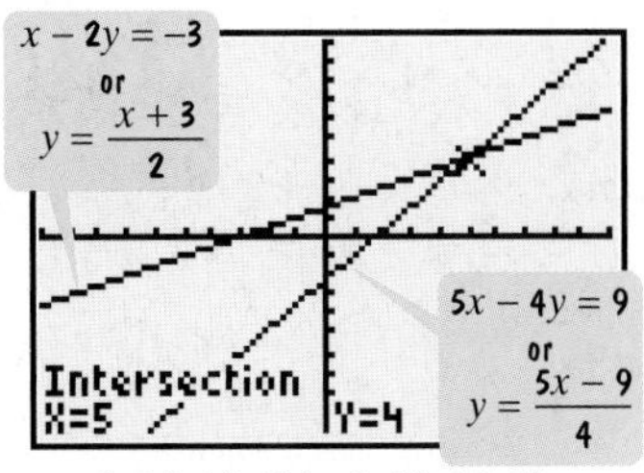

[−10, 10, 1] by [−10, 10, 1]

Check Point 2 Solve by the substitution method:

$$3x + 2y = 4$$
$$2x + \quad y = 1.$$

3 Solve linear systems by addition.

Eliminating a Variable Using the Addition Method

The substitution method is most useful if one of the given equations has an isolated variable. A second, and frequently the easiest, method for solving a linear system is the addition method. Like the substitution method, the addition method involves eliminating a variable and ultimately solving an equation containing only one variable. However, this time we eliminate a variable by adding the equations.

For example, consider the following system of linear equations:

$$3x - 4y = 11$$
$$-3x + 2y = -7.$$

When we add these two equations, the x-terms are eliminated. This occurs because the coefficients of the x-terms, 3 and -3, are opposites (additive inverses) of each other:

$$\begin{aligned} 3x - 4y &= 11 \\ \underline{-3x + 2y} &= \underline{-7} \end{aligned}$$

Add: $-2y = 4$ — The sum is an equation in one variable.

$y = -2$ — Solve for y by dividing both sides by -2.

Now we can back-substitute -2 for y into one of the original equations to find x. It does not matter which equation you use; you will obtain the same value for x in either case. If we use either equation, we can show that $x = 1$ and the solution $(1, -2)$ satisfies both equations in the system.

When we use the addition method, we want to obtain two equations whose sum is an equation containing only one variable. The key step is to **obtain, for one of the variables, coefficients that differ only in sign.** To do this, we may need to multiply one or both equations by some nonzero number so that the coefficients of one of the variables, x or y, become opposites. Then when the two equations are added, this variable is eliminated.

Study Tip

Although the addition method is also known as the elimination method, variables are eliminated when using both the substitution and addition methods. The name *addition method* specifically tells us that the elimination of a variable is accomplished by adding two equations.

Solving Linear Systems by Addition

1. If necessary, rewrite both equations in the form $Ax + By = C$.
2. If necessary, multiply either equation or both equations by appropriate nonzero numbers so that the sum of the x-coefficients or the sum of the y-coefficients is 0.
3. Add the equations in step 2. The sum is an equation in one variable.
4. Solve the equation in one variable.
5. Back-substitute the value obtained in step 4 into either of the given equations and solve for the other variable.
6. Check the solution in both of the original equations.

EXAMPLE 3 Solving a System by the Addition Method

Solve by the addition method:

$$3x + 2y = 48$$
$$9x - 8y = -24.$$

Solution

Step 1 Rewrite both equations in the form $Ax + By = C$. Both equations are already in this form. Variable terms appear on the left and constants appear on the right.

Step 2 If necessary, multiply either equation or both equations by appropriate numbers so that the sum of the x-coefficients or the sum of the y-coefficients is 0. We can eliminate x or y. Let's eliminate x. Consider the terms in x in each equation, that is, $3x$ and $9x$. To eliminate x, we can multiply each term of the first equation by -3 and then add the equations.

$$3x + 2y = 48 \xrightarrow{\text{Multiply by } -3.} -9x - 6y = -144$$
$$9x - 8y = -24 \xrightarrow{\text{No change}} \underline{9x - 8y = -24}$$

Step 3 Add the equations. Add: $-14y = -168$

Step 4 Solve the equation in one variable. We solve $-14y = -168$ by dividing both sides by -14.

$$\frac{-14y}{-14} = \frac{-168}{-14} \quad \text{Divide both sides by } -14.$$
$$y = 12 \quad \text{Simplify.}$$

Step 5 Back-substitute and find the value for the other variable. We can back-substitute 12 for y into either one of the given equations. We'll use the first one.

$$3x + 2y = 48 \quad \text{This is the first equation in the given system.}$$
$$3x + 2(12) = 48 \quad \text{Substitute 12 for } y.$$
$$3x + 24 = 48 \quad \text{Multiply.}$$
$$3x = 24 \quad \text{Subtract 24 from both sides.}$$
$$x = 8 \quad \text{Divide both sides by 3.}$$

We found that $y = 12$ and $x = 8$. The proposed solution is $(8, 12)$.

Step 6 Check. Take a few minutes to show that $(8, 12)$ satisfies both of the original equations in the system. The solution set is $\{(8, 12)\}$.

Check Point 3 Solve by the addition method:

$$4x + 5y = 3$$
$$2x - 3y = 7.$$

Some linear systems have solutions that are not integers. If the value of one variable turns out to be a "messy" fraction, back-substitution might lead to cumbersome arithmetic. If this happens, you can return to the original system and use the addition method to find the value of the other variable.

EXAMPLE 4 Solving a System by the Addition Method

Solve by the addition method:

$$2x = 7y - 17$$
$$5y = 17 - 3x.$$

Solution

Step 1 Rewrite both equations in the form $Ax + By = C$. We first arrange the system so that variable terms appear on the left and constants appear on the right. We obtain

$$2x - 7y = -17 \quad \text{Subtract 7y from both sides of the first equation.}$$
$$3x + 5y = 17. \quad \text{Add 3x to both sides of the second equation.}$$

Step 2 If necessary, multiply either equation or both equations by appropriate numbers so that the sum of the x-coefficients or the sum of the y-coefficients is 0. We can eliminate x or y. Let's eliminate x by multiplying the first equation by 3 and the second equation by -2.

$$2x - 7y = -17 \xrightarrow{\text{Multiply by 3.}} 6x - 21y = -51$$
$$3x + 5y = 17 \xrightarrow{\text{Multiply by } -2.} \underline{-6x - 10y = -34}$$

Step 3 Add the equations. Add: $-31y = -85$

Step 4 Solve the equation in one variable. We solve $-31y = -85$ by dividing both sides by -31.

$$\frac{-31y}{-31} = \frac{-85}{-31} \quad \text{Divide both sides by } -31.$$
$$y = \frac{85}{31} \quad \text{Simplify.}$$

Step 5 Back-substitute and find the value for the other variable. Back-substitution of $\frac{85}{31}$ for y into either of the given equations results in cumbersome arithmetic. Instead, let's use the addition method on the given system in the form $Ax + By = C$ to find the value for x. Thus, we eliminate y by multiplying the first equation by 5 and the second equation by 7.

$$2x - 7y = -17 \xrightarrow{\text{Multiply by 5.}} 10x - 35y = -85$$
$$3x + 5y = 17 \xrightarrow{\text{Multiply by 7.}} \underline{21x + 35y = 119}$$
$$\text{Add: } 31x = 34$$
$$x = \frac{34}{31} \quad \text{Divide both sides by 31.}$$

We found that $y = \frac{85}{31}$ and $x = \frac{34}{31}$. The proposed solution is $\left(\frac{34}{31}, \frac{85}{31}\right)$.

Step 6 Check. For this system, a calculator is helpful in showing that $\left(\frac{34}{31}, \frac{85}{31}\right)$ satisfies both of the original equations in the system. The solution set is $\left\{\left(\frac{34}{31}, \frac{85}{31}\right)\right\}$.

Check Point 4 Solve by the addition method:

$$2x = 9 + 3y$$
$$4y = 8 - 3x.$$

4 Identify systems that do not have exactly one ordered-pair solution.

Linear Systems Having No Solution or Infinitely Many Solutions

We have seen that a system of linear equations in two variables represents a pair of lines. The lines either intersect at one point, are parallel, or are identical. Thus, there are three possibilities for the number of solutions to a system of two linear equations.

The Number of Solutions to a System of Two Linear Equations

The number of solutions to a system of two linear equations in two variables is given by one of the following. (See Figure 5.3.)

Number of Solutions	What This Means Graphically
Exactly one ordered pair solution	The two lines intersect at one point.
No solution	The two lines are parallel.
Infinitely many solutions	The two lines are identical.

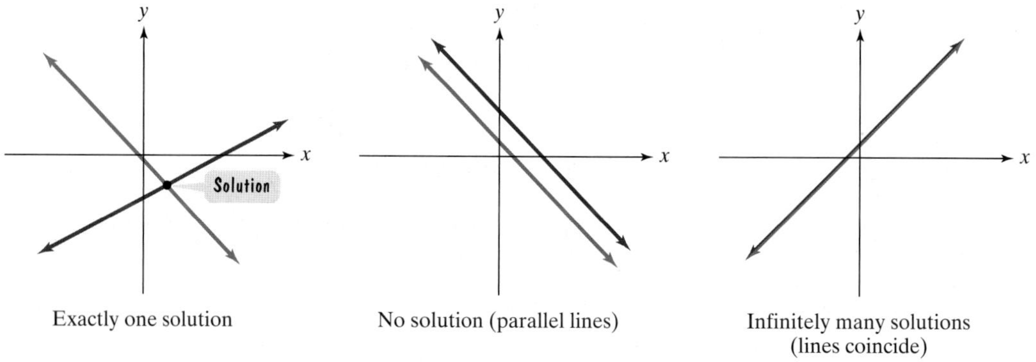

Figure 5.3 Possible graphs for a system of two linear equations in two variables

A linear system with no solution is called an **inconsistent system**. If you attempt to solve such a system by substitution or addition, you will eliminate both variables. A false statement, such as $0 = 12$, will be the result.

EXAMPLE 5 A System with No Solution

Solve the system:

$$4x + 6y = 12$$
$$6x + 9y = 12.$$

Solution Because no variable is isolated, we will use the addition method. To obtain coefficients of x that differ only in sign, we multiply the first equation by 3 and multiply the second equation by -2.

$4x + 6y = 12$ —Multiply by 3.→ $12x + 18y = 36$

$6x + 9y = 12$ —Multiply by -2.→ $-12x - 18y = -24$

Add: $0 = 12$

There are no values of x and y for which $0 = 12$. No values of x and y satisfy $0x + 0y = 12$.

The false statement $0 = 12$ indicates that the system is inconsistent and has no solution. The solution set is the empty set, $\varnothing$.

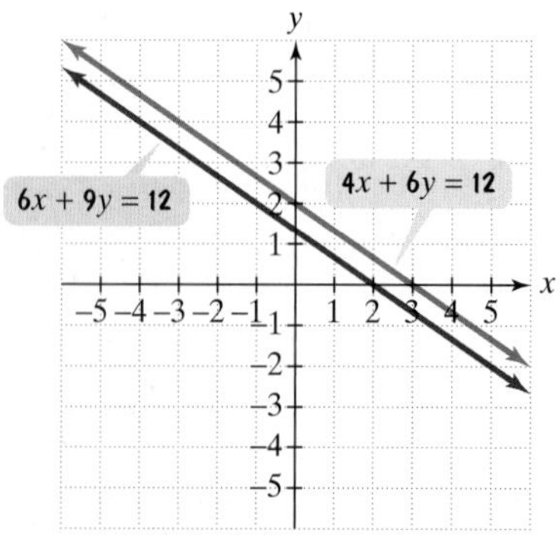

Figure 5.4 The graph of an inconsistent system

The lines corresponding to the two equations in Example 5 are shown in Figure 5.4. The lines are parallel and have no point of intersection.

Discovery

Show that the graphs of $4x + 6y = 12$ and $6x + 9y = 12$ must be parallel lines by solving each equation for y. What is the slope and y-intercept for each line? What does this mean? If a linear system is inconsistent, what must be true about the slopes and y-intercepts for the system's graphs?

Check Point 5 Solve the system:

$$5x - 2y = 4$$
$$-10x + 4y = 7.$$

A linear system that has at least one solution is called a **consistent system**. Lines that intersect and lines that coincide both represent consistent systems. If the lines coincide, then the consistent system has infinitely many solutions, represented by every point on either line.

The equations in a linear system with infinitely many solutions are called **dependent**. If you attempt to solve such a system by substitution or addition, you will eliminate both variables. However, a true statement, such as $10 = 10$, will be the result.

EXAMPLE 6 A System with Infinitely Many Solutions

Solve the system:

$$y = 3x - 2$$
$$15x - 5y = 10.$$

Solution Because the variable y is isolated in $y = 3x - 2$, the first equation, we can use the substitution method. We substitute the expression for y into the second equation.

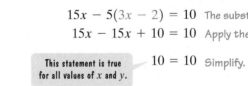

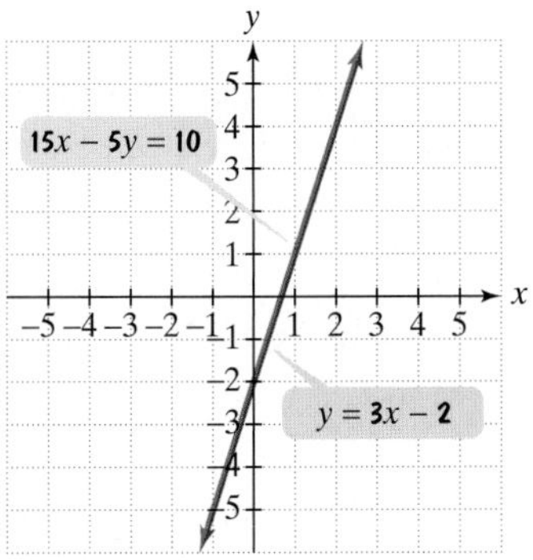

Figure 5.5 The graph of a system with infinitely many solutions

$y = \boxed{3x - 2}$ $\quad 15x - 5\boxed{y} = 10$ Substitute $3x - 2$ for y.

$15x - 5(3x - 2) = 10$ The substitution results in an equation in one variable.

$15x - 15x + 10 = 10$ Apply the distributive property.

$10 = 10$ Simplify. (This statement is true for all values of x and y.)

In our final step, both variables have been eliminated and the resulting statement, $10 = 10$, is true. This true statement indicates that the system has infinitely many solutions. The solution set consists of all points (x, y) lying on either of the coinciding lines, $y = 3x - 2$ or $15x - 5y = 10$, as shown in Figure 5.5.

We express the solution set for the system in one of two equivalent ways:

$$\{(x, y) \mid y = 3x - 2\} \quad \text{or} \quad \{(x, y) \mid 15x - 5y = 10\}.$$

The set of all ordered pairs (x, y) such that $y = 3x - 2$ — The set of all ordered pairs (x, y) such that $15x - 5y = 10$

Study Tip

Although the system in Example 6 has infinitely many solutions, this does not mean that any ordered pair of numbers you can form will be a solution. The ordered pair (x, y) must satisfy one of the system's equations, $y = 3 - 2x$ or $15x - 5y = 10$, and there are infinitely many such ordered pairs. Because the graphs are coinciding lines, the ordered pairs that are solutions of one of the equations are also solutions of the other equation.

Check Point 6 Solve the system:

$$\begin{aligned} x &= 4y - 8 \\ 5x - 20y &= -40. \end{aligned}$$

5 Solve problems using systems of linear equations.

Functions of Business: Break-Even Analysis

Suppose that a company produces and sells x units of a product. Its *revenue function* is the money generated by selling x units of the product. Its *cost function* is the cost of producing x units of the product.

Revenue and Cost Functions

A company produces and sells x units of a product.

Revenue Function

$$R(x) = (\text{price per unit sold})x$$

Cost Function

$$C(x) = \text{fixed cost} + (\text{cost per unit produced})x$$

The point of intersection of the graphs of the revenue and cost functions is called the **break-even point.** The x-coordinate of the point reveals the number of units that a company must produce and sell so that money coming in, the revenue, is equal to money going out, the cost. The y-coordinate of the break-even point gives the amount of money coming in and going out. Example 7 illustrates the use of the substitution method in determining a company's break-even point.

EXAMPLE 7 Finding a Break-Even Point

A company is planning to manufacture radically different wheelchairs. Fixed cost will be \$500,000 and it will cost \$400 to produce each wheelchair. Each wheelchair will be sold for \$600.

a. Write the cost function, C, of producing x wheelchairs.

b. Write the revenue function, R, from the sale of x wheelchairs.

c. Determine the break-even point. Describe what this means.

Solution

a. The cost function is the sum of the fixed cost and variable cost.

Fixed cost of \$500,000 | plus | Variable cost: \$400 for each chair produced

$$C(x) = 500{,}000 + 400x$$

b. The revenue function is the money generated from the sale of x wheelchairs.

Revenue per chair, \$600, times | the number of chairs sold

$$R(x) = 600x$$

c. The break-even point occurs where the graphs of C and R intersect. Thus, we find this point by solving the system

$$\begin{aligned} C(x) &= 500{,}000 + 400x \\ R(x) &= 600x \end{aligned} \quad \text{or} \quad \begin{aligned} y &= 500{,}000 + 400x \\ y &= 600x. \end{aligned}$$

Using substitution, we can substitute 600x for y in the first equation:

$$600x = 500{,}000 + 400x \quad \text{Substitute } 600x \text{ for } y \text{ in } y = 500{,}000 + 400x.$$

$$200x = 500{,}000 \quad \text{Subtract } 400x \text{ from both sides.}$$

$$x = 2500 \quad \text{Divide both sides by } 200.$$

Back-substituting 2500 for x in either of the system's equations (or functions), we obtain

$$R(2500) = 600(2500) = 1{,}500{,}000.$$

We used $R(x) = 600x$.

The break-even point is (2500, 1,500,000). This means that the company will break even if it produces and sells 2500 wheelchairs. At this level, the money coming in is equal to the money going out: $1,500,000.

Figure 5.6 shows the graphs of the revenue and cost functions for the wheelchair business. Similar graphs and models apply no matter how small or large a business venture may be.

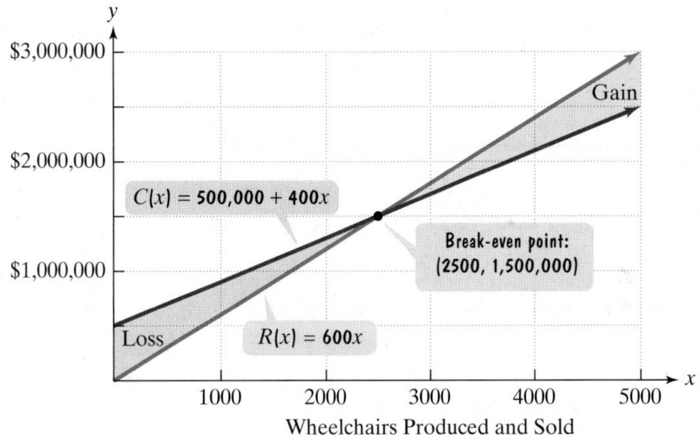

Figure 5.6

The intersection point confirms that the company breaks even by producing and selling 2500 wheelchairs. Can you see what happens for $x < 2500$? The red cost graph lies above the blue revenue graph. The cost is greater than the revenue and the business is losing money. Thus, if they sell fewer than 2500 wheelchairs, the result is a *loss*. By contrast, look at what happens for $x > 2500$. The blue revenue graph lies above the red cost graph. The revenue is greater than the cost and the business is making money. Thus, if they sell more than 2500 wheelchairs, the result is a *gain*.

Check Point 7 A company that manufactures running shoes has a fixed cost of $300,000. Additionally, it costs $30 to produce each pair of shoes. They are sold at $80 per pair.

a. Write the cost function, C, of producing x pairs of running shoes.

b. Write the revenue function, R, from the sale of x pairs of running shoes.

c. Determine the break-even point. Describe what this means.

What does every entrepreneur, from a kid selling lemonade to Donald Trump, want to do? Generate profit, of course. The *profit* made is the money taken in, or the revenue, minus the money spent, or the cost. This relationship between revenue and cost allows us to define the *profit function, P(x)*.

The Profit Function

The profit, $P(x)$, generated after producing and selling x units of a product is given by the **profit function**

$$P(x) = R(x) - C(x),$$

where R and C are the revenue and cost functions, respectively.

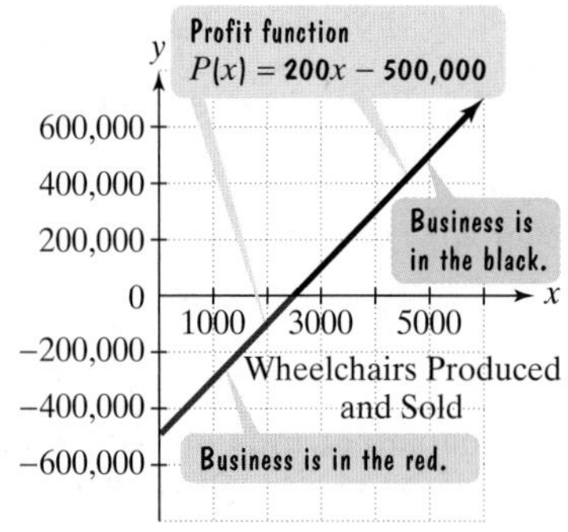

Figure 5.7

The profit function for the wheelchair business in Example 7 is

$$
\begin{aligned}
P(x) &= R(x) - C(x) \\
&= 600x - (500{,}000 + 400x) \\
&= 200x - 500{,}000.
\end{aligned}
$$

The graph of this profit function is shown in Figure 5.7. The red portion lies below the x-axis and shows a loss when fewer than 2500 wheelchairs are sold. The business is "in the red." The black portion lies above the x-axis and shows a gain when more than 2500 wheelchairs are sold. The wheelchair business is "in the black."

EXERCISE SET 5.1

Practice Exercises

In Exercises 1–4, determine whether the given ordered pair is a solution of the system.

1. $(2, 3)$
$x + 3y = 11$
$x - 5y = -13$

2. $(-3, 5)$
$9x + 7y = 8$
$8x - 9y = -69$

3. $(2, 5)$
$2x + 3y = 17$
$x + 4y = 16$

4. $(8, 5)$
$5x - 4y = 20$
$3y = 2x + 1$

In Exercises 5–18, solve each system by the substitution method.

5. $x + y = 4$
$y = 3x$

6. $x + y = 6$
$y = 2x$

7. $x + 3y = 8$
$y = 2x - 9$

8. $2x - 3y = -13$
$y = 2x + 7$

9. $x = 4y - 2$
$x = 6y + 8$

10. $x = 3y + 7$
$x = 2y - 1$

11. $5x + 2y = 0$
$x - 3y = 0$

12. $4x + 3y = 0$
$2x - y = 0$

13. $2x + 5y = -4$
$3x - y = 11$

14. $2x + 5y = 1$
$-x + 6y = 8$

15. $2x - 3y = 8 - 2x$
$3x + 4y = x + 3y + 14$

16. $3x - 4y = x - y + 4$
$2x + 6y = 5y - 4$

17. $y = \dfrac{1}{3}x + \dfrac{2}{3}$

$y = \dfrac{5}{7}x - 2$

18. $y = -\dfrac{1}{2}x + 2$

$y = \dfrac{3}{4}x + 7$

In Exercises 19–30, solve each system by the addition method.

19. $x + y = 1$
$x - y = 3$

20. $x + y = 6$
$x - y = -2$

21. $2x + 3y = 6$
$2x - 3y = 6$

22. $3x + 2y = 14$
$3x - 2y = 10$

23. $x + 2y = 2$
$-4x + 3y = 25$

24. $2x - 7y = 2$
$3x + y = -20$

25. $4x + 3y = 15$
$2x - 5y = 1$

26. $3x - 7y = 13$
$6x + 5y = 7$

27. $3x - 4y = 11$
$2x + 3y = -4$

28. $2x + 3y = -16$
$5x - 10y = 30$

29. $3x = 4y + 1$
$3y = 1 - 4x$

30. $5x = 6y + 40$
$2y = 8 - 3x$

In Exercises 31–42, solve by the method of your choice. Identify systems with no solution and systems with infinitely many solutions, using set notation to express their solution sets.

31. $x = 9 - 2y$
$x + 2y = 13$

32. $6x + 2y = 7$
$y = 2 - 3x$

33. $y = 3x - 5$
$21x - 35 = 7y$

34. $9x - 3y = 12$
$y = 3x - 4$

35. $3x - 2y = -5$
$4x + y = 8$

36. $2x + 5y = -4$
$3x - y = 11$

37. $x + 3y = 2$
$3x + 9y = 6$

38. $4x - 2y = 2$
$2x - y = 1$

39. $\dfrac{x}{4} - \dfrac{y}{4} = -1$

$x + 4y = -9$

40. $\dfrac{x}{6} - \dfrac{y}{2} = \dfrac{1}{3}$

$x + 2y = -3$

41. $2x = 3y + 4$
$4x = 3 - 5y$

42. $4x = 3y + 8$
$2x = -14 + 5y$

In Exercises 43–46, let x represent one number and let y represent the other number. Use the given conditions to write a system of equations. Solve the system and find the numbers.

43. The sum of two numbers is 7. If one number is subtracted from the other, their difference is -1. Find the numbers.

44. The sum of two numbers is 2. If one number is subtracted from the other, their difference is 8. Find the numbers.

45. Three times a first number decreased by a second number is 1. The first number increased by twice the second number is 12. Find the numbers.

46. The sum of three times a first number and twice a second number is 8. If the second number is subtracted from twice the first number, the result is 3. Find the numbers.

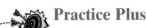

Practice Plus

In Exercises 47–48, solve each system by the method of your choice.

47. $\frac{x+2}{2} - \frac{y+4}{3} = 3$

$\frac{x+y}{5} = \frac{x-y}{2} - \frac{5}{2}$

48. $\frac{x-y}{3} = \frac{x+y}{2} - \frac{1}{2}$

$\frac{x+2}{2} - 4 = \frac{y+4}{3}$

In Exercises 49–50, solve each system for x and y, expressing either value in terms of a or b, if necessary. Assume that $a \neq 0$ and $b \neq 0$.

49. $5ax + 4y = 17$

$ax + 7y = 22$

50. $4ax + by = 3$

$6ax + 5by = 8$

51. For the linear function $f(x) = mx + b$, $f(-2) = 11$ and $f(3) = -9$. Find m and b.

52. For the linear function $f(x) = mx + b$, $f(-3) = 23$ and $f(2) = -7$. Find m and b.

Use the graphs of the linear functions to solve Exercises 53–54.

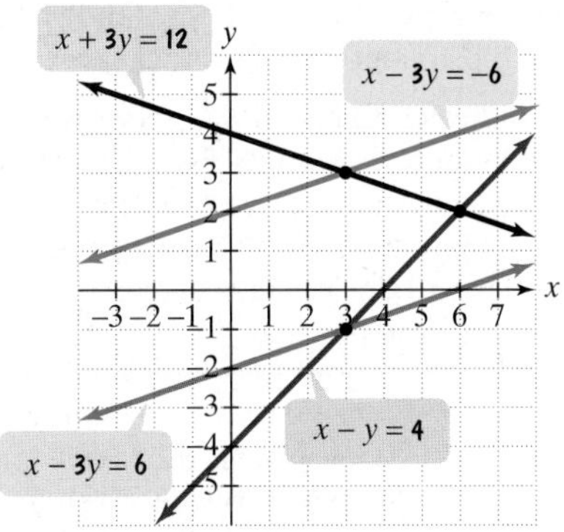

53. Write the linear system whose solution set is $\{(6, 2)\}$. Express each equation in the system in slope-intercept form.

54. Write the linear system whose solution set is $\varnothing$. Express each equation in the system in slope-intercept form.

Application Exercises

The figure shows the graphs of the cost and revenue functions for a company that manufactures and sells small radios. Use the information in the figure to solve Exercises 55–60.

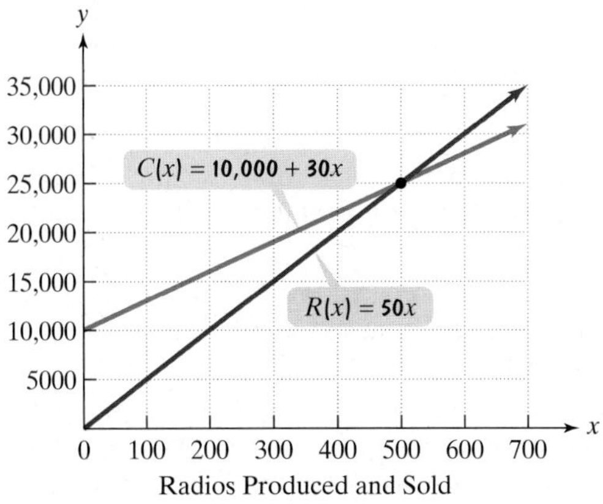

55. How many radios must be produced and sold for the company to break even?

56. More than how many radios must be produced and sold for the company to have a profit?

57. Use the formulas shown in the voice balloons to find $R(200) - C(200)$. Describe what this means for the company.

58. Use the formulas shown in the voice balloons to find $R(300) - C(300)$. Describe what this means for the company.

59. a. Use the formulas shown in the voice balloons to write the company's profit function, P, from producing and selling x radios.

b. Find the company's profit if 10,000 radios are produced and sold.

60. a. Use the formulas shown in the voice balloons to write the company's profit function, P, from producing and selling x radios.

b. Find the company's profit if 20,000 radios are produced and sold.

Exercises 61–64 describe a number of business ventures. For each exercise,

a. *Write the cost function, C.*

b. *Write the revenue function, R.*

c. *Determine the break-even point. Describe what this means.*

61. A company that manufactures small canoes has a fixed cost of \$18,000. It costs \$20 to produce each canoe. The selling price is \$80 per canoe. (In solving this exercise, let x represent the number of canoes produced and sold.)

62. A company that manufactures bicycles has a fixed cost of \$100,000. It costs \$100 to produce each bicycle. The selling price is \$300 per bike. (In solving this exercise, let x represent the number of bicycles produced and sold.)

63. You invest in a new play. The cost includes an overhead of \$30,000, plus production costs of \$2500 per performance. A sold-out performance brings in \$3125. (In solving this exercise, let x represent the number of sold-out performances.)

64. You invested \$30,000 and started a business writing greeting cards. Supplies cost 2¢ per card and you are selling each card for 50¢. (In solving this exercise, let x represent the number of cards produced and sold.)

An important application of systems of equations arises in connection with supply and demand. As the price of a product increases, the demand for that product decreases. However, at higher prices, suppliers are willing to produce greater quantities of the product. Exercises 65–66 involve supply and demand.

65. A chain of electronics stores sells hand-held color televisions. The weekly demand and supply models are given as follows:

Number sold per week — Demand model

$$N = -5p + 750$$

Price of television

Number supplied to the chain per week — Supply model

$$N = 2.5p.$$

a. How many hand-held color televisions can be sold and supplied at \$120 per television?

b. Find the price at which supply and demand are equal. At this price, how many televisions can be supplied and sold each week?

66. At a price of p dollars per ticket, the number of tickets to a rock concert that can be sold is given by the demand model $N = -25p + 7800$. At a price of p dollars per ticket, the number of tickets that the concert's promoters are willing to make available is given by the supply model $N = 5p + 6000$.

a. How many tickets can be sold and supplied for \$50 per ticket?

b. Find the ticket price at which supply and demand are equal. At this price, how many tickets will be supplied and sold?

67. The graphs shown below are based on 543 adults polled nationally by *Newsweek*.

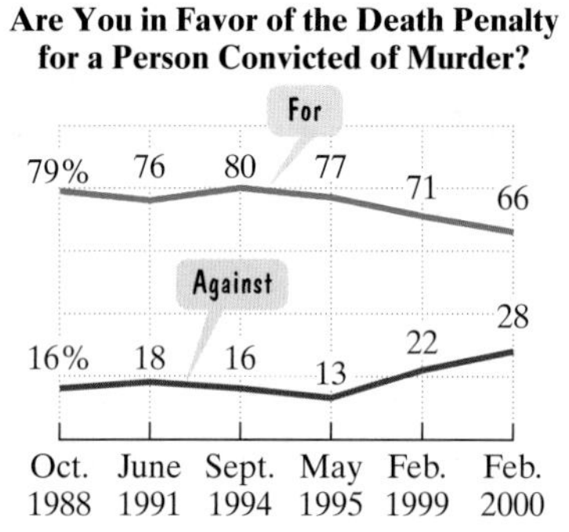

Source: Newsweek

The function $13x + 12y = 992$ models the percent, y, in favor of the death penalty x years after 1988. The function $-x + y = 16$ models the percent, y, against the death penalty x years after 1988. If the trends shown by the graphs continue, in which year will the percentage of Americans in favor of the death penalty be the same as the percentage of Americans who oppose it? For that year, what percent will be for the death penalty and what percent will be against it?

68. One of the most dramatic developments in the work force has been the increase in the number of women, at approximately $\frac{1}{2}\%$ per year. By contrast, the percentage of men is decreasing by $\frac{1}{4}\%$ per year. The graphs shown below illustrate these changes.

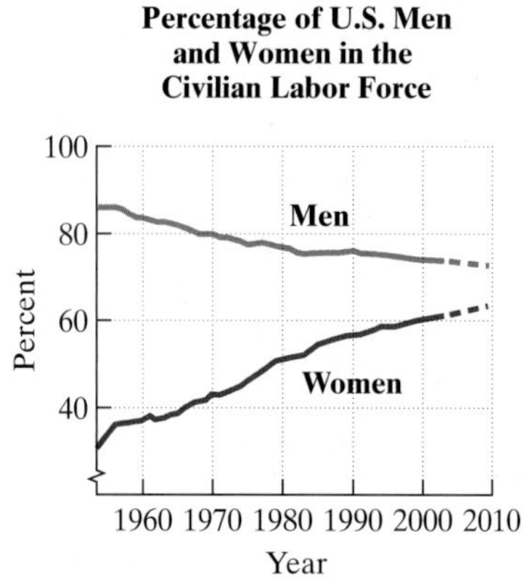

Source: U.S. Department of Labor

The function $y = 0.52x + 35.7$ models the percentage, y, of U.S. women in the work force x years after 1955. The function $0.25x + y = 85.4$ models the percentage, y, of U.S. men in the work force x years after 1955. Use these models to determine when the percentage of women in the work force will be the same as the percentage of men in the work force. Round to the nearest year. What percentage of women and what percentage of men will be in the work force at that time?

69. Although Social Security is a problem, some projections indicate that there's a much bigger time bomb ticking in the federal budget, and that's Medicare. In 2000, the cost of Social Security was 5.48% of the gross domestic product, increasing by 0.04% of the GDP per year. In 2000, the cost of Medicare was 1.84% of the gross domestic product, increasing by 0.17% of the GDP per year.

(*Source:* Congressional Budget Office)

a. Write a function that models the cost of Social Security as a percentage of the GDP x years after 2000.

b. Write a function that models the cost of Medicare as a percentage of the GDP x years after 2000.

c. In which year will the cost of Medicare and Social Security be the same? For that year, what will be the cost of each program as a percentage of the GDP? Which program will have the greater cost after that year?

70. The graph indicates that in 1984, there were 72 meals per person at take-out restaurants. For the period shown, this number increased by an average of 2.25 meals per person per year. In 1984, there were 94 meals per person at on-premise dining facilities and this number decreased by an average of 0.55 meals per person per year.

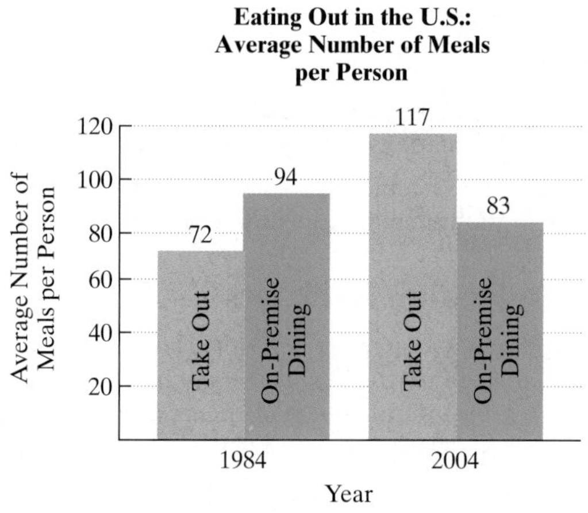

Source: The NPD Group

a. Write a function that models the average number of meals per person at take-out restaurants x years after 1984.

b. Write a function that models the average number of meals per person at on-premise dining facilities x years after 1984.

c. In which year, to the nearest whole year, was the average number of meals per person for take-out and on-premise restaurants the same? For that year, how many meals per person, to the nearest whole number, were there for each kind of restaurant? Which kind of restaurant had the greater number of meals per person after that year?

The bar graph shows the percentage of Americans who used cigarettes, by ethnicity, in 1985 and 2002. For each of the groups shown, cigarette use has been linearly decreasing. Use this information to solve Exercises 71–72.

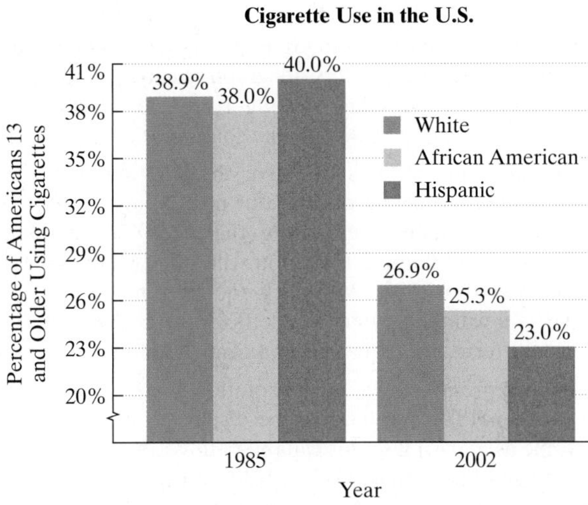

Source: Department of Health and Human Services

71. In this exercise, let x represent the number of years after 1985 and let y represent the percentage of Americans in one of the groups shown who used cigarettes.

a. Use the data points $(0, 38)$ and $(17, 25.3)$ to find the slope-intercept equation of the line that models the percentage of African Americans who used cigarettes, y, x years after 1985. Round the value of m to two decimal places.

b. Use the data points $(0, 40)$ and $(17, 23)$ to find the slope-intercept equation of the line that models the percentage of Hispanics who used cigarettes, y, x years after 1985.

c. Use the models from parts (a) and (b) to find the year during which cigarette use was the same for African Americans and Hispanics. What percentage of each group used cigarettes during that year?

72. In this exercise, let x represent the number of years after 1985 and let y represent the percentage of Americans in one of the groups shown who used cigarettes.

a. Use the data points $(0, 38.9)$ and $(17, 26.9)$ to find the slope-intercept equation of the line that models the percentage of whites who used cigarettes, y, x years after 1985. Round the value of m to two decimal places.

b. Use the data points $(0, 40)$ and $(17, 23)$ to find the slope-intercept equation of the line that models the percentage of Hispanics who used cigarettes, y, x years after 1985.

c. Use the models from parts (a) and (b) to find the year, to the nearest whole year, during which cigarette use was the same for whites and Hispanics. What percentage of each group, to the nearest percent, used cigarettes during that year?

Use a system of linear equations to solve Exercises 73–84.

The graph shows the calories in some favorite fast foods. Use the information in Exercises 73–74 to find the exact caloric content of the specified foods.

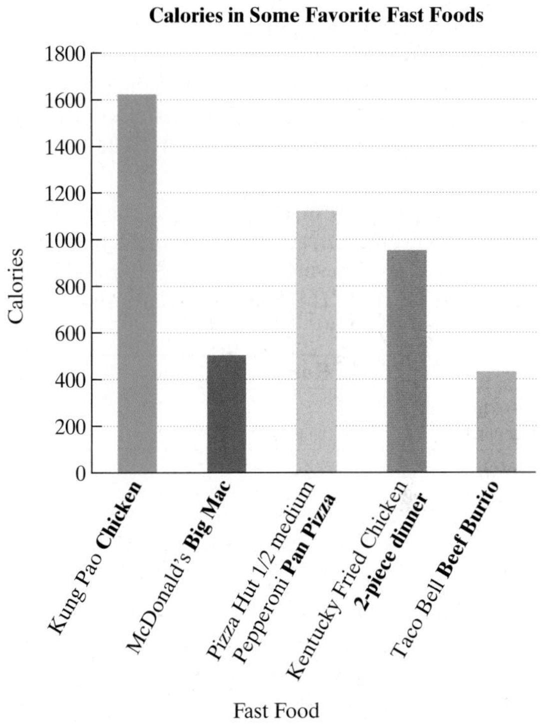

Source: Center for Science in the Public Interest

73. One pan pizza and two beef burritos provide 1980 calories. Two pan pizzas and one beef burrito provide 2670 calories. Find the caloric content of each item.

74. One Kung Pao chicken and two Big Macs provide 2620 calories. Two Kung Pao chickens and one Big Mac provide 3740 calories. Find the caloric content of each item.

75. Cholesterol intake should be limited to 300 mg or less each day. One serving of scrambled eggs from McDonalds and one Double Beef Whopper from Burger King exceed this intake by 241 mg. Two servings of scrambled eggs and three Double Beef Whoppers provide 1257 mg of cholesterol. Determine the cholesterol content in each item.

76. Two medium eggs and three cups of ice cream contain 701 milligrams of cholesterol. One medium egg and one cup of ice cream exceed the suggested daily cholesterol intake of 300 milligrams by 25 milligrams. Determine the cholesterol content in each item.

77. A hotel has 200 rooms. Those with kitchen facilities rent for \$100 per night and those without kitchen facilities rent for \$80 per night. On a night when the hotel was completely occupied, revenues were \$17,000. How many of each type of room does the hotel have?

78. A new restaurant is to contain two-seat tables and four-seat tables. Fire codes limit the restaurant's maximum occupancy to 56 customers. If the owners have hired enough servers to handle 17 tables of customers, how many of each kind of table should they purchase?

79. A rectangular lot whose perimeter is 360 feet is fenced along three sides. An expensive fencing along the lot's length costs \$20 per foot and an inexpensive fencing along the two side widths costs only \$8 per foot. The total cost of the fencing along the three sides comes to \$3280. What are the lot's dimensions?

80. A rectangular lot whose perimeter is 320 feet is fenced along three sides. An expensive fencing along the lot's length costs \$16 per foot and an inexpensive fencing along the two side widths costs only \$5 per foot. The total cost of the fencing along the three sides comes to \$2140. What are the lot's dimensions?

81. When a crew rows with the current, it travels 16 miles in 2 hours. Against the current, the crew rows 8 miles in 2 hours. Let x = the crew's rowing rate in still water and let y = the rate of the current. The following chart summarizes this information:

	Rate	×	Time	=	Distance
Rowing with current	$x + y$		2		16
Rowing against current	$x - y$		2		8

Find the rate of rowing in still water and the rate of the current.

82. When an airplane flies with the wind, it travels 800 miles in 4 hours. Against the wind, it takes 5 hours to cover the same distance. Find the plane's rate in still air and the rate of the wind.

In Exercises 83–84, an isosceles triangle containing two angles with equal measure is shown. The degree measure of each triangle's three interior angles and an exterior angle is represented with variables. Find the measure of the three interior angles.

83.

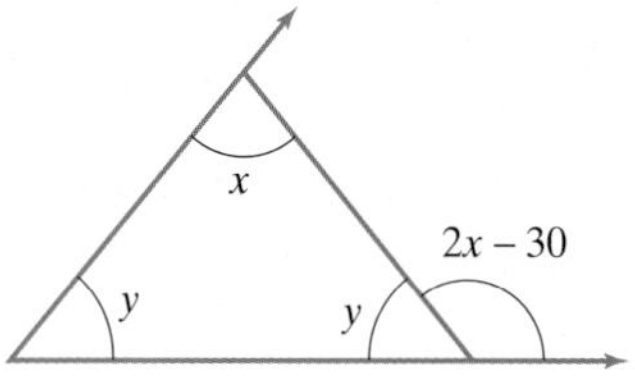

84.

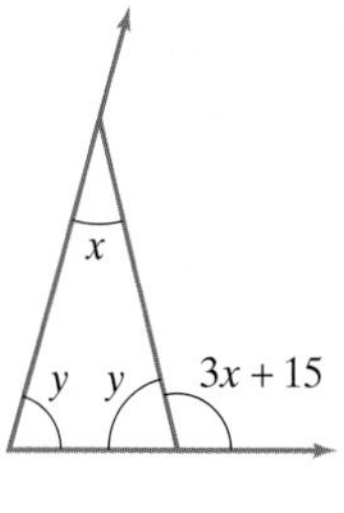

Writing in Mathematics

85. What is a system of linear equations? Provide an example with your description.

86. What is the solution of a system of linear equations?

87. Explain how to solve a system of equations using the substitution method. Use $y = 3 - 3x$ and $3x + 4y = 6$ to illustrate your explanation.

88. Explain how to solve a system of equations using the addition method. Use $3x + 5y = -2$ and $2x + 3y = 0$ to illustrate your explanation.

89. When is it easier to use the addition method rather than the substitution method to solve a system of equations?

90. When using the addition or substitution method, how can you tell if a system of linear equations has infinitely many solutions? What is the relationship between the graphs of the two equations?

91. When using the addition or substitution method, how can you tell if a system of linear equations has no solution? What is the relationship between the graphs of the two equations?

92. Describe the break-even point for a business.

Technology Exercises

93. Verify your solutions to any five exercises in Exercises 5–42 by using a graphing utility to graph the two equations in the system in the same viewing rectangle. Then use the intersection feature to display the solution.

94. Some graphing utilities can give the solution to a linear system of equations. (Consult your manual for details.) This capability is usually accessed with the SIMULT (simultaneous equations) feature. First, you will enter 2, for two equations in two variables. With each equation in $Ax + By = C$ form, you will then enter the coefficients for x and y and the constant term, one equation at a time. After entering all six numbers, press SOLVE. The solution will be displayed on the screen. (The x-value may be displayed as $x_1 =$ and the y-value as $x_2 =$.) Use this capability to verify the solution to any five of the exercises you solved in the practice exercises of this exercise set. Describe what happens when you use your graphing utility on a system with no solution or infinitely many solutions.

Critical Thinking Exercises

95. Write a system of equations having $\{(-2, 7)\}$ as a solution set. (More than one system is possible.)

96. Solve the system for x and y in terms of a_1, b_1, c_1, a_2, b_2, and c_2:

$$a_1x + b_1y = c_1$$
$$a_2x + b_2y = c_2.$$

97. Two identical twins can only be distinguished by the characteristic that one always tells the truth and the other always lies. One twin tells you of a lucky number pair: "When I multiply my first lucky number by 3 and my second lucky number by 6, the addition of the resulting numbers produces a sum of 12. When I add my first lucky number and twice my second lucky number, the sum is 5." Which twin is talking?

98. A marching band has 52 members and there are 24 in the pom-pom squad. They wish to form several hexagons and squares like those diagrammed below. Can it be done with no people left over?

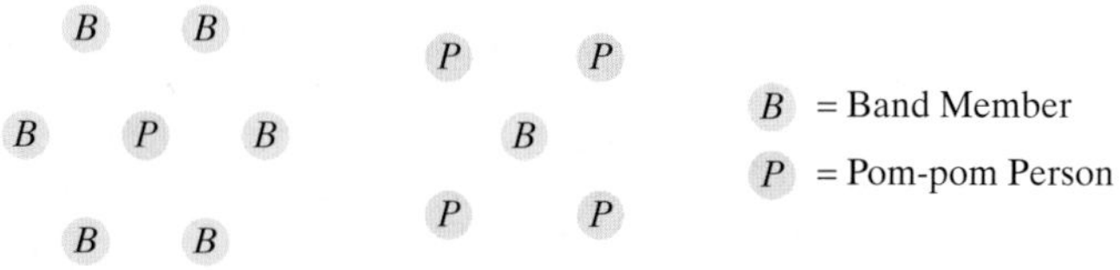

Group Exercise

99. The group should write four different word problems that can be solved using a system of linear equations in two variables. All of the problems should be on different topics. The group should turn in the four problems and their algebraic solutions.

SECTION 5.2 Systems of Linear Equations in Three Variables

Objectives

1. Verify the solution of a system of linear equations in three variables.
2. Solve systems of linear equations in three variables.
3. Solve problems using systems in three variables.

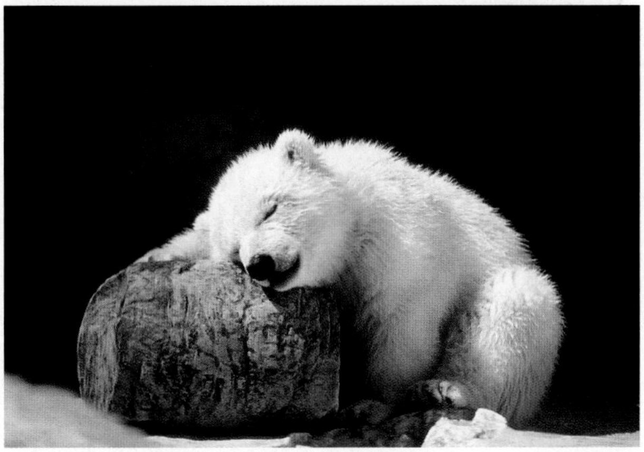

All animals sleep, but the length of time they sleep varies widely: Cattle sleep for only a few minutes at a time. We humans seem to need more sleep than other animals, up to eight hours a day. Without enough sleep, we have difficulty concentrating, make mistakes in routine tasks, lose energy, and feel bad-tempered. There is a relationship between hours of sleep and death rate per year per 100,000 people. How many hours of sleep will put you in the group with the minimum death rate? In this section, we will answer this question by solving a system of linear equations with more than two variables.

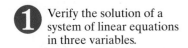

1 Verify the solution of a system of linear equations in three variables.

Systems of Linear Equations in Three Variables and Their Solutions

An equation such as $x + 2y - 3z = 9$ is called a *linear equation in three variables*. In general, any equation of the form

$$Ax + By + Cz = D,$$

where A, B, C, and D are real numbers such that A, B, and C are not all 0, is a **linear equation in three variables: x, y, and z.** The graph of this linear equation in three variables is a plane in three-dimensional space.

The process of solving a system of three linear equations in three variables is geometrically equivalent to finding the point of intersection (assuming that there is one) of three planes in space. (See Figure 5.8.) A **solution** of a system of linear equations in three variables is an ordered triple of real numbers that satisfies all equations of the system. The **solution set** of the system is the set of all its solutions.

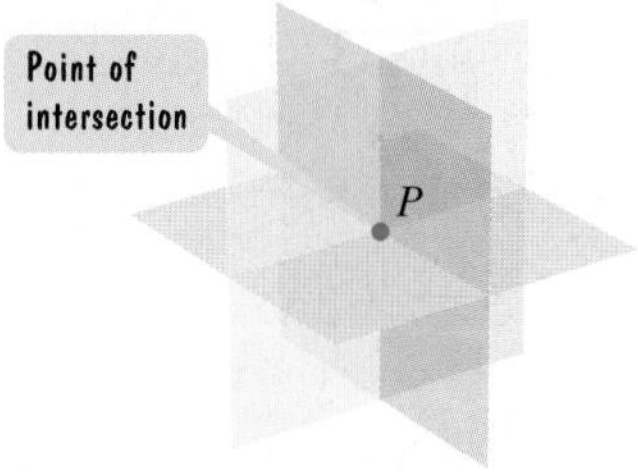

Figure 5.8

EXAMPLE 1 Determining Whether an Ordered Triple Satisfies a System

Show that the ordered triple $(-1, 2, -2)$ is a solution of the system:

$$\begin{aligned} x + 2y - 3z &= 9 \\ 2x - y + 2z &= -8 \\ -x + 3y - 4z &= 15. \end{aligned}$$

Solution Because -1 is the x-coordinate, 2 is the y-coordinate, and -2 is the z-coordinate of $(-1, 2, -2)$, we replace x with -1, y with 2, and z with -2 in each of the three equations.

$$\begin{aligned} x + 2y - 3z &= 9 \\ -1 + 2(2) - 3(-2) &\stackrel{?}{=} 9 \\ -1 + 4 + 6 &\stackrel{?}{=} 9 \\ 9 &= 9, \quad \text{true} \end{aligned}$$

$$\begin{aligned} 2x - y + 2z &= -8 \\ 2(-1) - 2 + 2(-2) &\stackrel{?}{=} -8 \\ -2 - 2 - 4 &\stackrel{?}{=} -8 \\ -8 &= -8, \quad \text{true} \end{aligned}$$

$$\begin{aligned} -x + 3y - 4z &= 15 \\ -(-1) + 3(2) - 4(-2) &\stackrel{?}{=} 15 \\ 1 + 6 + 8 &\stackrel{?}{=} 15 \\ 15 &= 15, \quad \text{true} \end{aligned}$$

The ordered triple $(-1, 2, -2)$ satisfies the three equations: It makes each equation true. Thus, the ordered triple is a solution of the system.

Check Point 1 Show that the ordered triple $(-1, -4, 5)$ is a solution of the system:

$$\begin{aligned} x - 2y + 3z &= 22 \\ 2x - 3y - z &= 5 \\ 3x + y - 5z &= -32. \end{aligned}$$

2 Solve systems of linear equations in three variables.

Solving Systems of Linear Equations in Three Variables by Eliminating Variables

The method for solving a system of linear equations in three variables is similar to that used on systems of linear equations in two variables. We use addition to eliminate any variable, reducing the system to two equations in two variables. Once we obtain a system of two equations in two variables, we use addition or substitution to eliminate a variable. The result is a single equation in one variable. We solve this equation to get the value of the remaining variable. Other variable values are found by back-substitution.

Study Tip

It does not matter which variable you eliminate, as long as you do it in two different pairs of equations.

Solving Linear Systems in Three Variables by Eliminating Variables

1. Reduce the system to two equations in two variables. This is usually accomplished by taking two different pairs of equations and using the addition method to eliminate the same variable from both pairs.
2. Solve the resulting system of two equations in two variables using addition or substitution. The result is an equation in one variable that gives the value of that variable.
3. Back-substitute the value of the variable found in step 2 into either of the equations in two variables to find the value of the second variable.
4. Use the values of the two variables from steps 2 and 3 to find the value of the third variable by back-substituting into one of the original equations.
5. Check the proposed solution in each of the original equations.

EXAMPLE 2 Solving a System in Three Variables

Solve the system:

$$\begin{aligned} 5x - 2y - 4z &= 3 && \text{Equation 1} \\ 3x + 3y + 2z &= -3 && \text{Equation 2} \\ -2x + 5y + 3z &= 3. && \text{Equation 3} \end{aligned}$$

Solution There are many ways to proceed. Because our initial goal is to reduce the system to two equations in two variables, **the central idea is to take two different pairs of equations and eliminate the same variable from both pairs.**

Step 1 Reduce the system to two equations in two variables. We choose any two equations and use the addition method to eliminate a variable. Let's eliminate z using Equations 1 and 2. We do so by multiplying Equation 2 by 2. Then we add equations.

$$\begin{array}{llcl} \text{(Equation 1)} & 5x - 2y - 4z = 3 & \xrightarrow{\text{No change}} & 5x - 2y - 4z = 3 \\ \text{(Equation 2)} & 3x + 3y + 2z = -3 & \xrightarrow{\text{Multiply by 2.}} & \underline{6x + 6y + 4z = -6} \\ & & \text{Add:} & 11x + 4y \quad = -3 \quad \text{Equation 4} \end{array}$$

Now we must eliminate the *same* variable using another pair of equations. We can eliminate z from Equations 2 and 3. First, we multiply Equation 2 by -3. Next, we multiply Equation 3 by 2. Finally, we add equations.

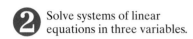

(Equation 2) $3x + 3y + 2z = -3$ $\underrightarrow{\text{Multiply by } -3.}$ $-9x - 9y - 6z = 9$

(Equation 3) $-2x + 5y + 3z = 3$ $\underrightarrow{\text{Multiply by 2.}}$ $-4x + 10y + 6z = \underline{6}$

Add: $-13x + y = 15$ Equation 5

Equations 4 and 5 give us a system of two equations in two variables:

$$11x + 4y = -3 \qquad \text{Equation 4}$$
$$-13x + y = 15. \qquad \text{Equation 5}$$

Step 2   Solve the resulting system of two equations in two variables. We will use the addition method to solve Equations 4 and 5 for x and y. To do so, we multiply Equation 5 on both sides by -4 and add this to Equation 4.

(Equation 4) $11x + 4y = -3$ $\underrightarrow{\text{No change}}$ $11x + 4y = -3$

(Equation 5) $-13x + y = 15$ $\underrightarrow{\text{Multiply by } -4.}$ $52x - 4y = \underline{-60}$

Add: $63x = -63$

$x = -1$ Divide both sides by 63.

Step 3   Use back-substitution in one of the equations in two variables to find the value of the second variable. We back-substitute -1 for x in either Equation 4 or 5 to find the value of y.

$$-13x + y = 15 \qquad \text{Equation 5}$$
$$-13(-1) + y = 15 \qquad \text{Substitute } -1 \text{ for } x.$$
$$13 + y = 15 \qquad \text{Multiply.}$$
$$y = 2 \qquad \text{Subtract 13 from both sides.}$$

Step 4   Back-substitute the values found for two variables into one of the original equations to find the value of the third variable. We can now use any one of the original equations and back-substitute the values of x and y to find the value for z. We will use Equation 2.

$$3x + 3y + 2z = -3 \qquad \text{Equation 2}$$
$$3(-1) + 3(2) + 2z = -3 \qquad \text{Substitute } -1 \text{ for } x \text{ and } 2 \text{ for } y.$$
$$3 + 2z = -3 \qquad \begin{array}{l}\text{Multiply and then add:}\\ 3(-1) + 3(2) = -3 + 6 = 3.\end{array}$$
$$2z = -6 \qquad \text{Subtract 3 from both sides.}$$
$$z = -3 \qquad \text{Divide both sides by 2.}$$

With $x = -1$, $y = 2$, and $z = -3$, the proposed solution is the ordered triple $(-1, 2, -3)$.

Step 5   Check. Check the proposed solution, $(-1, 2, -3)$, by substituting the values for x, y, and z into each of the three original equations. These substitutions yield three true statements. Thus, the solution set is $\{(-1, 2, -3)\}$.

Check Point **2**   Solve the system:

$$x + 4y - z = 20$$
$$3x + 2y + z = 8$$
$$2x - 3y + 2z = -16.$$

In some examples, one of the variables is already eliminated from a given equation. In this case, the same variable should be eliminated from the other two equations, thereby making it possible to omit one of the elimination steps. We illustrate this idea in Example 3.

EXAMPLE 3 Solving a System of Equations with a Missing Term

Solve the system:

$$\begin{aligned} x + \phantom{2y +{}} z &= 8 && \text{Equation 1}\\ x + y + 2z &= 17 && \text{Equation 2}\\ x + 2y + z &= 16. && \text{Equation 3}\end{aligned}$$

Solution

Step 1 Reduce the system to two equations in two variables. Because Equation 1 contains only x and z, we could omit one of the elimination steps by eliminating y using Equations 2 and 3. This will give us two equations in x and z. To eliminate y using Equations 2 and 3, we multiply Equation 2 by -2 and add Equation 3.

$$\begin{aligned} &\text{(Equation 2)}\quad x + y + 2z = 17 \quad \xrightarrow{\text{Multiply by } -2.} & -2x - 2y - 4z &= -34\\ &\text{(Equation 3)}\quad x + 2y + z = 16 \quad \xrightarrow{\text{No change}} & x + 2y + z &= 16\\ &\text{Add:} & -x \phantom{{}+2y} - 3z &= -18 \quad \text{Equation 4}\end{aligned}$$

Equation 4 and the given Equation 1 provide us with a system of two equations in two variables:

$$\begin{aligned} x + z &= 8 && \text{Equation 1}\\ -x - 3z &= -18. && \text{Equation 4}\end{aligned}$$

Step 2 Solve the resulting system of two equations in two variables. We will solve Equations 1 and 4 for x and z.

$$\begin{aligned} x + z &= 8 && \text{Equation 1}\\ -x - 3z &= -18 && \text{Equation 4}\\ \text{Add:}\quad -2z &= -10 &&\\ z &= 5 && \text{Divide both sides by } -2.\end{aligned}$$

Step 3 Use back-substitution in one of the equations in two variables to find the value of the second variable. To find x, we back-substitute 5 for z in either Equation 1 or 4. We will use Equation 1.

$$\begin{aligned} x + z &= 8 && \text{Equation 1}\\ x + 5 &= 8 && \text{Substitute 5 for } z.\\ x &= 3 && \text{Subtract 5 from both sides.}\end{aligned}$$

Step 4 Back-substitute the values found for two variables into one of the original equations to find the value of the third variable. To find y, we back-substitute 3 for x and 5 for z into Equation 2 or 3. We cannot use Equation 1 because y is missing in this equation. We will use Equation 2.

$$\begin{aligned} x + y + 2z &= 17 && \text{Equation 2}\\ 3 + y + 2(5) &= 17 && \text{Substitute 3 for } x \text{ and 5 for } z.\\ y + 13 &= 17 && \text{Multiply and add.}\\ y &= 4 && \text{Subtract 13 from both sides.}\end{aligned}$$

We found that $z = 5$, $x = 3$, and $y = 4$. Thus, the proposed solution is the ordered triple $(3, 4, 5)$.

Step 5 Check. Substituting 3 for x, 4 for y, and 5 for z into each of the three original equations yields three true statements. Consequently, the solution set is $\{(3, 4, 5)\}$.

Check Point 3 Solve the system:

$$\begin{aligned} 2y - z &= 7\\ x + 2y + z &= 17\\ 2x - 3y + 2z &= -1.\end{aligned}$$

A system of linear equations in three variables represents three planes. The three planes may not always intersect at one point. The planes may have no common point of intersection and represent an inconsistent system with no solution. By contrast, the planes may coincide or intersect along a line. In these cases, the planes have infinitely many points in common and represent systems with infinitely many solutions. Systems of linear equations in three variables that are inconsistent or that contain dependent equations will be discussed in Chapter 6.

③ Solve problems using systems in three variables.

Applications

Systems of equations may allow us to find models for data without using a graphing utility. Three data points that do not lie on or near a line determine the graph of a quadratic function of the form $y = ax^2 + bx + c, a \neq 0$. Quadratic functions often model situations in which values of y are decreasing and then increasing, suggesting the cuplike shape of a parabola.

EXAMPLE 4   Modeling Data Relating Sleep and Death Rate

In a study relating sleep and death rate, the following data were obtained. Use the function $y = ax^2 + bx + c$ to model the data.

x (Average Number of Hours of Sleep)	y (Death Rate per Year per 100,000 Males)
4	1682
7	626
9	967

Solution   We need to find values for a, b, and c in $y = ax^2 + bx + c$. We can do so by solving a system of three linear equations in a, b, and c. We obtain the three equations by using the values of x and y from the data as follows:

$$y = ax^2 + bx + c \qquad \textit{Use the quadratic function to model the data.}$$

When x = 4, y = 1682: $1682 = a \cdot 4^2 + b \cdot 4 + c$ or $16a + 4b + c = 1682$

When x = 7, y = 626: $626 = a \cdot 7^2 + b \cdot 7 + c$ or $49a + 7b + c = 626$

When x = 9, y = 967: $967 = a \cdot 9^2 + b \cdot 9 + c$ or $81a + 9b + c = 967.$

The easiest way to solve this system is to eliminate c from two pairs of equations, obtaining two equations in a and b. Solving this system gives $a = 104.5$, $b = -1501.5$, and $c = 6016$. We now substitute the values for a, b, and c into $y = ax^2 + bx + c$. The function that models the given data is

$$y = 104.5x^2 - 1501.5x + 6016.$$

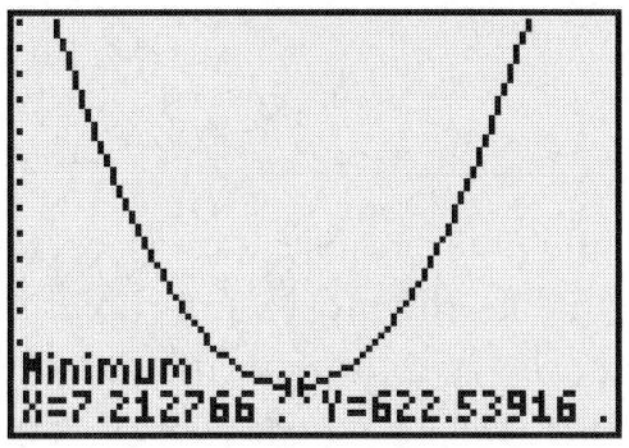

We can use the model that we obtained in Example 4 to find the death rate of males who average, say, 6 hours of sleep. First, write the model in function notation:

$$f(x) = 104.5x^2 - 1501.5x + 6016.$$

Substitute 6 for x:

$$f(6) = 104.5(6)^2 - 1501.5(6) + 6016 = 769.$$

According to the model, the death rate for males who average 6 hours of sleep is 769 deaths per 100,000 males.

Technology

The graph of
$$y = 104.5x^2 - 1501.5x + 6016$$
is displayed in a $[3, 12, 1]$ by $[500, 2000, 100]$ viewing rectangle. The minimum function feature shows that the lowest point on the graph, the vertex, is approximately $(7.2, 622.5)$. Men who average 7.2 hours of sleep are in the group with the lowest death rate, approximately 622.5 deaths per 100,000 males.

Check Point 4   Find the quadratic function $y = ax^2 + bx + c$ whose graph passes through the points $(1, 4)$, $(2, 1)$, and $(3, 4)$.

EXERCISE SET 5.2

Practice Exercises

In Exercises 1–4, determine if the given ordered triple is a solution of the system.

1. $(2, -1, 3)$

$$\begin{aligned} x + y + z &= 4 \\ x - 2y - z &= 1 \\ 2x - y - 2z &= -1 \end{aligned}$$

2. $(5, -3, -2)$

$$\begin{aligned} x + y + z &= 0 \\ x + 2y - 3z &= 5 \\ 3x + 4y + 2z &= -1 \end{aligned}$$

3. $(4, 1, 2)$

$$\begin{aligned} x - 2y \quad &= 2 \\ 2x + 3y \quad &= 11 \\ y - 4z &= -7 \end{aligned}$$

4. $(-1, 3, 2)$

$$\begin{aligned} x - 2z &= -5 \\ y - 3z &= -3 \\ 2x - z &= -4 \end{aligned}$$

Solve each system in Exercises 5–18.

5. $$\begin{aligned} x + y + 2z &= 11 \\ x + y + 3z &= 14 \\ x + 2y - z &= 5 \end{aligned}$$

6. $$\begin{aligned} 2x + y - 2z &= -1 \\ 3x - 3y - z &= 5 \\ x - 2y + 3z &= 6 \end{aligned}$$

7. $$\begin{aligned} 4x - y + 2z &= 11 \\ x + 2y - z &= -1 \\ 2x + 2y - 3z &= -1 \end{aligned}$$

8. $$\begin{aligned} x - y + 3z &= 8 \\ 3x + y - 2z &= -2 \\ 2x + 4y + z &= 0 \end{aligned}$$

9. $$\begin{aligned} 3x + 2y - 3z &= -2 \\ 2x - 5y + 2z &= -2 \\ 4x - 3y + 4z &= 10 \end{aligned}$$

10. $$\begin{aligned} 2x + 3y + 7z &= 13 \\ 3x + 2y - 5z &= -22 \\ 5x + 7y - 3z &= -28 \end{aligned}$$

11. $$\begin{aligned} 2x - 4y + 3z &= 17 \\ x + 2y - z &= 0 \\ 4x - y - z &= 6 \end{aligned}$$

12. $$\begin{aligned} x + z &= 3 \\ x + 2y - z &= 1 \\ 2x - y + z &= 3 \end{aligned}$$

13. $$\begin{aligned} 2x + y &= 2 \\ x + y - z &= 4 \\ 3x + 2y + z &= 0 \end{aligned}$$

14. $$\begin{aligned} x + 3y + 5z &= 20 \\ y - 4z &= -16 \\ 3x - 2y + 9z &= 36 \end{aligned}$$

15. $$\begin{aligned} x + y &= -4 \\ y - z &= 1 \\ 2x + y + 3z &= -21 \end{aligned}$$

16. $$\begin{aligned} x + y &= 4 \\ x + z &= 4 \\ y + z &= 4 \end{aligned}$$

17. $$\begin{aligned} 3(2x + y) + 5z &= -1 \\ 2(x - 3y + 4z) &= -9 \\ 4(1 + x) &= -3(z - 3y) \end{aligned}$$

18. $$\begin{aligned} 7z - 3 &= 2(x - 3y) \\ 5y + 3z - 7 &= 4x \\ 4 + 5z &= 3(2x - y) \end{aligned}$$

In Exercises 19–22, find the quadratic function $y = ax^2 + bx + c$ whose graph passes through the given points.

19. $(-1, 6), (1, 4), (2, 9)$

20. $(-2, 7), (1, -2), (2, 3)$

21. $(-1, -4), (1, -2), (2, 5)$

22. $(1, 3), (3, -1), (4, 0)$

In Exercises 23–24, let x represent the first number, y the second number, and z the third number. Use the given conditions to write a system of equations. Solve the system and find the numbers.

23. The sum of three numbers is 16. The sum of twice the first number, 3 times the second number, and 4 times the third number is 46. The difference between 5 times the first number and the second number is 31. Find the three numbers.

24. The following is known about three numbers: Three times the first number plus the second number plus twice the third number is 5. If 3 times the second number is subtracted from the sum of the first number and 3 times the third number, the result is 2. If the third number is subtracted from 2 times the first number and 3 times the second number, the result is 1. Find the numbers.

Practice Plus

Solve each system in Exercises 25–26.

25. $$\begin{aligned} \frac{x+2}{6} - \frac{y+4}{3} + \frac{z}{2} &= 0 \\ \frac{x+1}{2} + \frac{y-1}{2} - \frac{z}{4} &= \frac{9}{2} \\ \frac{x-5}{4} + \frac{y+1}{3} + \frac{z-2}{2} &= \frac{19}{4} \end{aligned}$$

26. $$\begin{aligned} \frac{x+3}{2} - \frac{y-1}{2} + \frac{z+2}{4} &= \frac{3}{2} \\ \frac{x-5}{2} + \frac{y+1}{3} - \frac{z}{4} &= -\frac{25}{6} \\ \frac{x-3}{4} - \frac{y+1}{2} + \frac{z-3}{2} &= -\frac{5}{2} \end{aligned}$$

In Exercises 27–28, find the equation of the quadratic function $y = ax^2 + bx + c$ whose graph is shown. Select three points whose coordinates appear to be integers.

27.

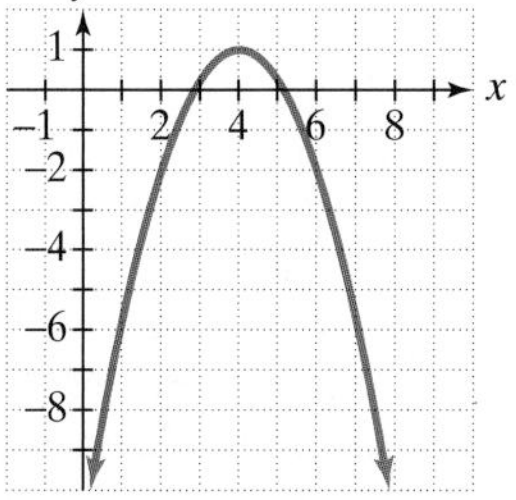

28.

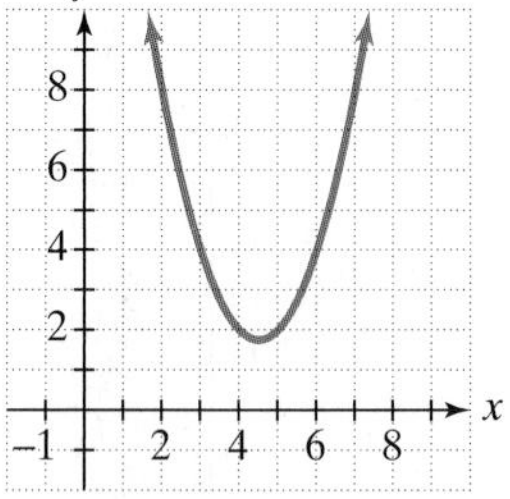

In Exercises 29–30, solve each system for (x, y, z) in terms of the nonzero constants a, b, and c.

29. $$\begin{aligned} ax - by - 2cz &= 21 \\ ax + by + cz &= 0 \\ 2ax - by + cz &= 14 \end{aligned}$$

30. $$\begin{aligned} ax - by + 2cz &= -4 \\ ax + 3by - cz &= 1 \\ 2ax + by + 3cz &= 2 \end{aligned}$$

Application Exercises

31. Although headlines about illegal steroids have focused on professional and Olympic athletes, the most vulnerable users may be high school students. The bar graph at the top of the next page shows the percentage of U.S. high school seniors who had taken steroids from 2000 through 2003.

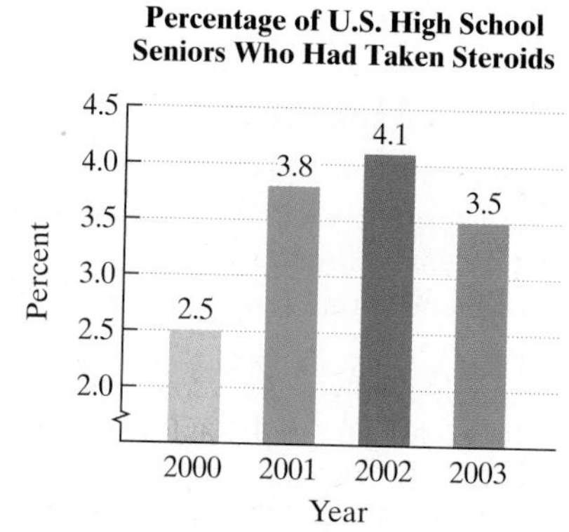

Source: University of Michigan

a. Write the data for 2000, 2002, and 2003 as ordered pairs (x, y), where x is the number of years after 2000 and y is the percentage of seniors who had taken steroids in that year.

b. The three data points in part (a) can be modeled by the quadratic function $y = ax^2 + bx + c$. Substitute each ordered pair into this function, one ordered pair at a time, and write a system of linear equations in three variables that can be used to find values for a, b, and c.

c. Solve the system in part (b) and write a quadratic function that models the percentage of U.S. high school seniors who had taken steroids x years after 2000.

32. The bar graph shows the percentage of people in the United States living below the poverty level for selected years from 1990 through 2003.

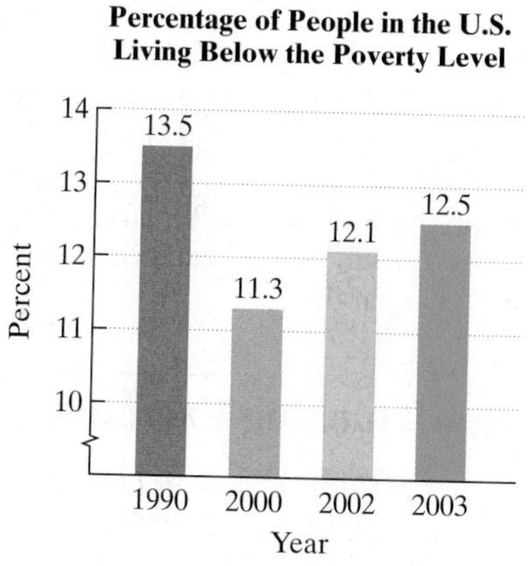

Source: Department of Health and Human Services

a. Write the data for 1990, 2002, and 2003 as ordered pairs (x, y), where x is the number of years after 1990 and y is the percentage of people living below the poverty level.

b. The three data points in part (a) can be modeled by the quadratic function $y = ax^2 + bx + c$. Substitute each ordered pair into this function, one ordered pair at a time, and write a system of linear equations in three variables that can be used to find values for a, b, and c. It is not necessary to solve the system.

33. You throw a ball straight up from a rooftop. The ball misses the rooftop on its way down and eventually strikes the ground. A mathematical model can be used to describe the relationship for the ball's height above the ground, y, after x seconds. Consider the following data:

x, seconds after the ball is thrown	y, ball's height, in feet, above the ground
1	224
3	176
4	104

a. Find the quadratic function $y = ax^2 + bx + c$ whose graph passes through the given points.

b. Use the function in part (a) to find the value for y when $x = 5$. Describe what this means.

34. A mathematical model can be used to describe the relationship between the number of feet a car travels once the brakes are applied, y, and the number of seconds the car is in motion after the brakes are applied, x. A research firm collects the following data:

x, seconds in motion after brakes are applied	y, feet car travels once the brakes are applied
1	46
2	84
3	114

a. Find the quadratic function $y = ax^2 + bx + c$ whose graph passes through the given points.

b. Use the function in part (a) to find the value for y when $x = 6$. Describe what this means.

Use a system of linear equations in three variables to solve Exercises 35–41.

35. In current U.S. dollars, John D. Rockefeller's 1913 fortune of \$900 million would be worth about \$189 billion. The bar graph shows that Rockefeller is the wealthiest among the world's five richest people of all time. The combined estimated wealth, in current billions of U.S. dollars, of Andrew Carnegie, Cornelius Vanderbilt, and Bill Gates is \$244 billion. The difference between Carnegie's estimated wealth and Vanderbilt's is \$4 billion. The difference between Vanderbilt's estimated wealth and Gates's is \$48 billion. Find the estimated wealth, in current billions of U.S. dollars, of Carnegie, Vanderbilt, and Gates.

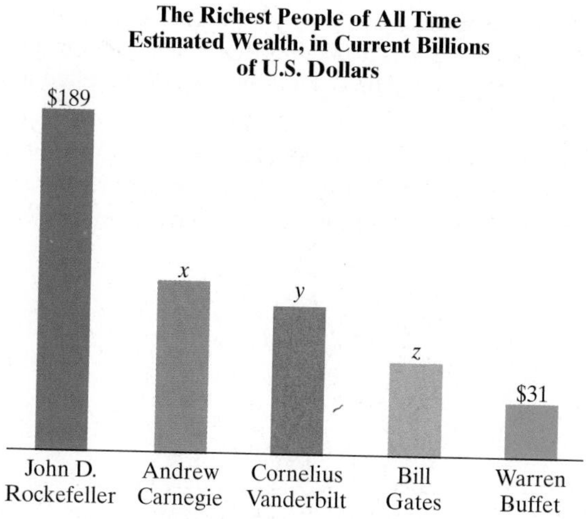

Source: Scholastic Book of World Records

36. The circle graph shows the percentage of Americans who drink caffeinated beverages on a daily basis and the number of cups consumed per day.

A Wired Nation: Percentage of American Adults Drinking Caffeinated Beverages

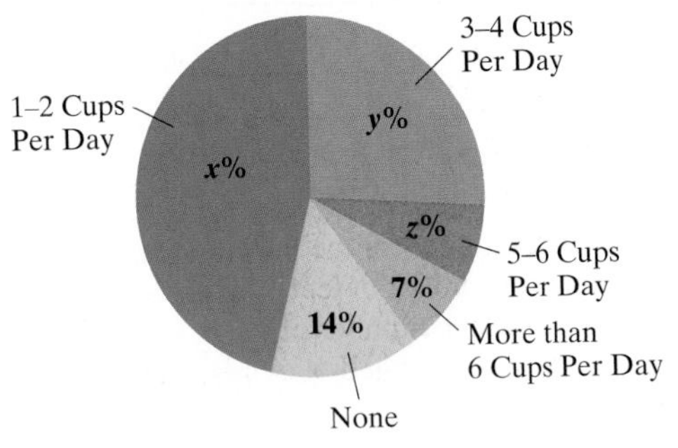

Source: Harris Interactive

72% of American adults drink from one to four cups of caffeinated beverages per day and 40% drink three or more cups per day. Find the percentage who drink from one to two cups, from three to four cups, and from five to six cups of caffeinated beverages per day.

37. At a college production of *Streetcar Named Desire*, 400 tickets were sold. The ticket prices were \$8, \$10, and \$12, and the total income from ticket sales was \$3700. How many tickets of each type were sold if the combined number of \$8 and \$10 tickets sold was 7 times the number of \$12 tickets sold?

38. A certain brand of razor blades comes in packages of 6, 12, and 24 blades, costing \$2, \$3, and \$4 per package, respectively. A store sold 12 packages containing a total of 162 razor blades and took in \$35. How many packages of each type were sold?

39. A person invested \$6700 for one year, part at 8%, part at 10%, and the remainder at 12%. The total annual income from these investments was \$716. The amount of money invested at 12% was \$300 more than the amount invested at 8% and 10% combined. Find the amount invested at each rate.

40. A person invested \$17,000 for one year, part at 10%, part at 12%, and the remainder at 15%. The total annual income from these investments was \$2110. The amount of money invested at 12% was \$1000 less than the amount invested at 10% and 15% combined. Find the amount invested at each rate.

41. In the following triangle, the degree measures of the three interior angles and two of the exterior angles are represented with variables. Find the measure of each interior angle.

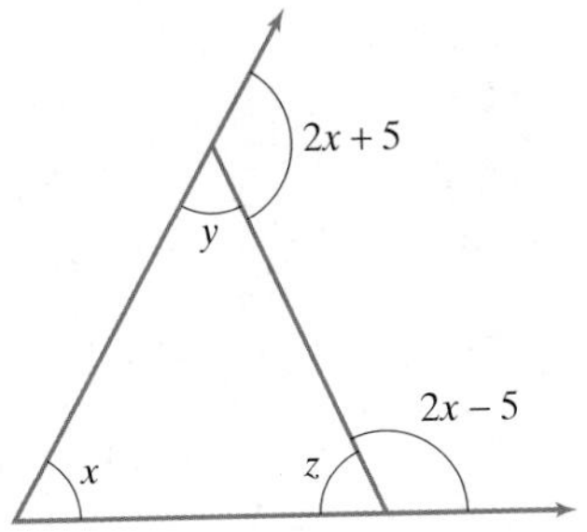

Writing in Mathematics

42. What is a system of linear equations in three variables?

43. How do you determine whether a given ordered triple is a solution of a system in three variables?

44. Describe in general terms how to solve a system in three variables.

45. AIDS is taking a deadly toll on southern Africa. Describe how to use the techniques that you learned in this section to obtain a model for African life span using projections with AIDS, shown by the red graph in the figure. Let x represent the number of years after 1985 and let y represent African life span in that year.

African Life Span

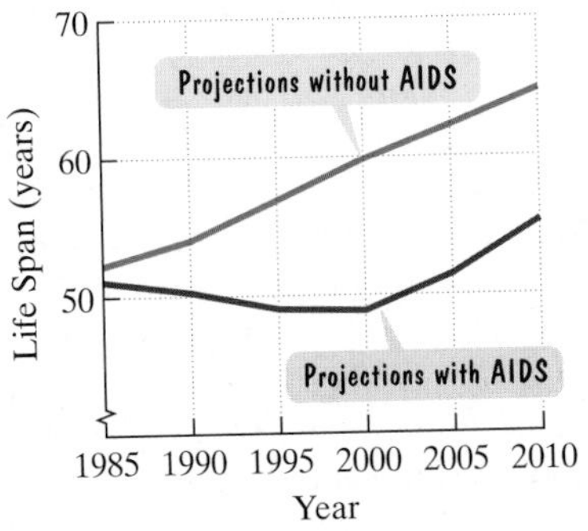

Source: United Nations

Technology Exercises

46. Does your graphing utility have a feature that allows you to solve linear systems by entering coefficients and constant terms? If so, use this feature to verify the solutions to any five exercises that you worked by hand from Exercises 5–16.

47. Verify your results in Exercises 19–22 by using a graphing utility to graph the resulting parabola. Trace along the curve and convince yourself that the three points given in the exercise lie on the parabola.

Critical Thinking Exercises

48. Describe how the system

$$\begin{aligned} x + y - z - 2w &= -8 \\ x - 2y + 3z + w &= 18 \\ 2x + 2y + 2z - 2w &= 10 \\ 2x + y - z + w &= 3 \end{aligned}$$

could be solved. Is it likely that in the near future a graphing utility will be available to provide a geometric solution (using intersecting graphs) to this system? Explain.

49. A modernistic painting consists of triangles, rectangles, and pentagons, all drawn so as to not overlap or share sides. Within each rectangle are drawn 2 red roses and each pentagon contains 5 carnations. How many triangles, rectangles, and pentagons appear in the painting if the painting contains a total of 40 geometric figures, 153 sides of geometric figures, and 72 flowers?

Group Exercise

50. Group members should develop appropriate functions that model each of the projections shown in Exercise 45.

SECTION 5.3 Partial Fractions

Objectives

1. Decompose $\frac{P}{Q}$, where Q has only distinct linear factors.
2. Decompose $\frac{P}{Q}$, where Q has repeated linear factors.
3. Decompose $\frac{P}{Q}$, where Q has a nonrepeated prime quadratic factor.
4. Decompose $\frac{P}{Q}$, where Q has a prime, repeated quadratic factor.

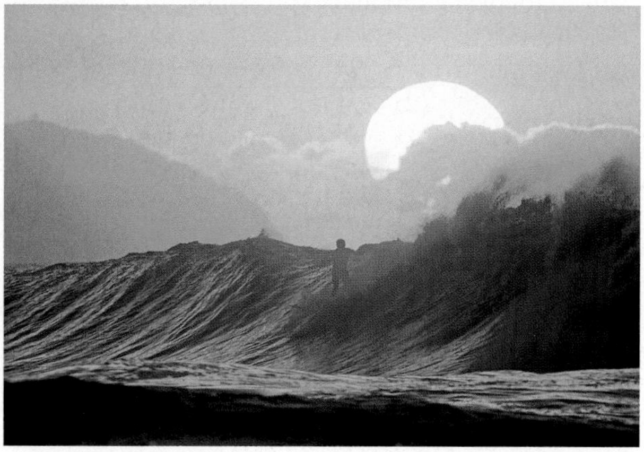

The rising and setting of the sun suggest the obvious: Things change over time. Calculus is the study of rates of change, allowing the motion of the rising sun to be measured by "freezing the frame" at one instant in time. If you are given a function, calculus reveals its rate of change at any "frozen" instant. In this section, you will learn an algebraic technique used in calculus to find a function if its rate of change is known. The technique involves expressing a given function in terms of simpler functions.

The Idea behind Partial Fraction Decomposition

We know how to use common denominators to write a sum or difference of rational expressions as a single rational expression. For example,

$$\frac{3}{x-4} - \frac{2}{x+2} = \frac{3(x+2) - 2(x-4)}{(x-4)(x+2)}$$

$$= \frac{3x+6-2x+8}{(x-4)(x+2)} = \frac{x+14}{(x-4)(x+2)}.$$

For solving the kind of calculus problem described in the section opener, we must reverse this process:

Partial fraction   Partial fraction

$\frac{x+14}{(x-4)(x+2)}$ is expressed as the sum of two simpler fractions.

$$\frac{x+14}{(x-4)(x+2)} = \frac{3}{x-4} + \frac{-2}{x+2}.$$

This is the partial fraction decomposition of $\frac{x+14}{(x-4)(x+2)}$.

Each of the two fractions on the right is called a **partial fraction**. The sum of these fractions is called the **partial fraction decomposition** of the rational expression on the left-hand side.

Partial fraction decompositions can be written for rational expressions of the form $\frac{P(x)}{Q(x)}$, where P and Q have no common factors and the highest power in the numerator is less than the highest power in the denominator. In this section, we will show you how to write the partial fraction decompositions for each of the following rational expressions:

$$\frac{9x^2 - 9x + 6}{(2x - 1)(x + 2)(x - 2)}$$

$P(x) = 9x^2 - 9x + 6$; highest power = 2

$Q(x) = (2x - 1)(x + 2)(x - 2)$; multiplying factors, highest power = 3.

$$\frac{5x^3 - 3x^2 + 7x - 3}{(x^2 + 1)^2}.$$

$P(x) = 5x^3 - 3x^2 + 7x - 3$; highest power = 3

$Q(x) = (x^2 + 1)^2$; squaring the expression, highest power = 4.

The partial fraction decomposition of a rational expression depends on the factors of the denominator. We consider four cases involving different kinds of factors in the denominator:

1. The denominator is a product of distinct linear factors.
2. The denominator is a product of linear factors, some of which are repeated.
3. The denominator has prime quadratic factors, none of which is repeated.
4. The denominator has a repeated prime quadratic factor.

1 Decompose $\frac{P}{Q}$, where Q has only distinct linear factors.

The Partial Fraction Decomposition of a Rational Expression with Distinct Linear Factors in the Denominator

If the denominator of a rational expression has a linear factor of the form $ax + b$, then the partial fraction decomposition will contain a term of the form

$$\frac{A}{ax + b}.$$

Constant

Linear factor

Each distinct linear factor in the denominator produces a partial fraction of the form *constant over linear factor*. For example,

$$\frac{9x^2 - 9x + 6}{(2x - 1)(x + 2)(x - 2)} = \frac{A}{2x - 1} + \frac{B}{x + 2} + \frac{C}{x - 2}.$$

We write a constant over each linear factor in the denominator.

The Partial Fraction Decomposition of $\frac{P(x)}{Q(x)}$: $Q(x)$ Has Distinct Linear Factors

The form of the partial fraction decomposition for a rational expression with distinct linear factors in the denominator is

$$\frac{P(x)}{(a_1x + b_1)(a_2x + b_2)(a_3x + b_3)\cdots(a_nx + b_n)}$$
$$= \frac{A_1}{a_1x + b_1} + \frac{A_2}{a_2x + b_2} + \frac{A_3}{a_3x + b_3} + \cdots + \frac{A_n}{a_nx + b_n}.$$

EXAMPLE 1 Partial Fraction Decomposition with Distinct Linear Factors

Find the partial fraction decomposition of

$$\frac{x + 14}{(x - 4)(x + 2)}.$$

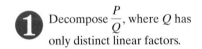

Solution We begin by setting up the partial fraction decomposition with the unknown constants. Write a constant over each of the two distinct linear factors in the denominator.

$$\frac{x+14}{(x-4)(x+2)} = \frac{A}{x-4} + \frac{B}{x+2}$$

Our goal is to find A and B. We do this by multiplying both sides of the equation by the least common denominator, $(x-4)(x+2)$.

$$(x-4)(x+2)\frac{x+14}{(x-4)(x+2)} = (x-4)(x+2)\left(\frac{A}{x-4} + \frac{B}{x+2}\right)$$

We use the distributive property on the right side.

$$\cancel{(x-4)}\cancel{(x+2)}\frac{x+14}{\cancel{(x-4)}\cancel{(x+2)}}$$

$$= \cancel{(x-4)}(x+2)\frac{A}{\cancel{(x-4)}} + (x-4)\cancel{(x+2)}\frac{B}{\cancel{(x+2)}}$$

Dividing out common factors in numerators and denominators, we obtain

$$x + 14 = A(x+2) + B(x-4).$$

To find values for A and B that make both sides equal, we'll express the sides in exactly the same form by writing the variable x-terms and then writing the constant terms. Apply the distributive property on the right side.

$x + 14 = Ax + 2A + Bx - 4B$	Distribute A and B over the parentheses.
$x + 14 = Ax + Bx + 2A - 4B$	Rearrange terms.
$1x + 14 = (A + B)x + (2A - 4B)$	Rewrite to identify the coefficient of x and the constant term.

As shown by the arrows, if two polynomials are equal, coefficients of like powers of x must be equal $(A + B = 1)$ and their constant terms must be equal $(2A - 4B = 14)$. Consequently, A and B satisfy the following two equations:

$$A + B = 1$$
$$2A - 4B = 14.$$

We can use the addition method to solve this linear system in two variables. By multiplying the first equation by -2 and adding equations, we obtain $A = 3$ and $B = -2$. Thus,

$$\frac{x+14}{(x-4)(x+2)} = \frac{A}{x-4} + \frac{B}{x+2} = \frac{3}{x-4} + \frac{-2}{x+2}\left(\text{or } \frac{3}{x-4} - \frac{2}{x+2}\right).$$

Study Tip

You will encounter some examples in which the denominator of the given rational expression is not already factored. If necessary, begin by factoring the denominator. Then apply the six steps needed to obtain the partial fraction decomposition.

Steps in Partial Fraction Decomposition

1. Set up the partial fraction decomposition with the unknown constants A, B, C, etc., in the numerator of the decomposition.
2. Multiply both sides of the resulting equation by the least common denominator.
3. Simplify the right-hand side of the equation.
4. Write both sides in descending powers, equate coefficients of like powers of x, and equate constant terms.
5. Solve the resulting linear system for A, B, C, etc.
6. Substitute the values for A, B, C, etc., into the equation in step 1 and write the partial fraction decomposition.

Check Point 1 Find the partial fraction decomposition of $\dfrac{5x - 1}{(x - 3)(x + 4)}$.

② Decompose $\dfrac{P}{Q}$, where Q has repeated linear factors.

The Partial Fraction Decomposition of a Rational Expression with Linear Factors in the Denominator, Some of Which Are Repeated

Suppose that $(ax + b)^n$ is a factor of the denominator. This means that the linear factor $ax + b$ is repeated n times. When this occurs, the partial fraction decomposition will contain a sum of n fractions for this factor of the denominator.

The Partial Fraction Decomposition of $\dfrac{P(x)}{Q(x)}$: $Q(x)$ Has Repeated Linear Factors

The form of the partial fraction decomposition for a rational expression containing the linear factor $ax + b$ occuring n times as its denominator is

$$\frac{P(x)}{(ax + b)^n} = \frac{A_1}{ax + b} + \frac{A_2}{(ax + b)^2} + \frac{A_3}{(ax + b)^3} + \cdots + \frac{A_n}{(ax + b)^n}.$$

> Include one fraction with a constant numerator for each power of $ax + b$.

EXAMPLE 2   Partial Fraction Decomposition with Repeated Linear Factors

Find the partial fraction decomposition of $\dfrac{x - 18}{x(x - 3)^2}$.

Solution

Step 1   Set up the partial fraction decomposition with the unknown constants. Because the linear factor $x - 3$ occurs twice, we must include one fraction with a constant numerator for each power of $x - 3$.

$$\frac{x - 18}{x(x - 3)^2} = \frac{A}{x} + \frac{B}{x - 3} + \frac{C}{(x - 3)^2}$$

Step 2   Multiply both sides of the resulting equation by the least common denominator. We clear fractions, multiplying both sides by $x(x - 3)^2$, the least common denominator.

$$x(x - 3)^2\left[\frac{x - 18}{x(x - 3)^2}\right] = x(x - 3)^2\left[\frac{A}{x} + \frac{B}{x - 3} + \frac{C}{(x - 3)^2}\right]$$

We use the distributive property on the right side.

$$\cancel{x}\,\cancel{(x - 3)^2} \cdot \frac{x - 18}{\cancel{x}\,\cancel{(x - 3)^2}} = \cancel{x}(x - 3)^2 \cdot \frac{A}{\cancel{x}} + x\cancel{(x - 3)^2} \cdot \frac{B}{\cancel{(x - 3)}} + x\cancel{(x - 3)^2} \cdot \frac{C}{\cancel{(x - 3)^2}}$$

Dividing out common factors in numerators and denominators, we obtain

$$x - 18 = A(x - 3)^2 + Bx(x - 3) + Cx.$$

Step 3   Simplify the right side of the equation. Square $x - 3$. Then apply the distributive property.

$$x - 18 = A(x^2 - 6x + 9) + Bx(x - 3) + Cx \qquad \text{Square } x - 3 \text{ using}$$
$$(A - B)^2 = A^2 - 2AB + B^2.$$

$$x - 18 = Ax^2 - 6Ax + 9A + Bx^2 - 3Bx + Cx \qquad \text{Apply the distributive property.}$$

Study Tip

Avoid this common error:

INCORRECT!

$$\cancel{\frac{x - 18}{x(x - 3)^2} = \frac{A}{x} + \frac{B}{x - 3} + \frac{C}{x - 3}}$$

> Listing $x - 3$ twice does not take into account $(x - 3)^2$.

Step 4 Write both sides in descending powers, equate coefficients of like powers of x, and equate constant terms. The left side, $x - 18$, is in descending powers of x: $x - 18x^0$. We will write the right side in descending powers of x.

$$x - 18 = Ax^2 + Bx^2 - 6Ax - 3Bx + Cx + 9A$$ Rearrange terms on the right side.

Express both sides in the same form.

$$0x^2 + 1x - 18 = (A + B)x^2 + (-6A - 3B + C)x + 9A$$ Rewrite to identify coefficients and the constant term.

Equating coefficients of like powers of x and equating constant terms results in the following system of linear equations:

$$\begin{aligned} A + B &= 0 \\ -6A - 3B + C &= 1 \\ 9A &= -18. \end{aligned}$$

Step 5 Solve the resulting system for A, B, and C. Dividing both sides of the last equation by 9, we obtain $A = -2$. Substituting -2 for A in the first equation, $A + B = 0$, gives $-2 + B = 0$, so $B = 2$. We find C by substituting -2 for A and 2 for B in the middle equation, $-6A - 3B + C = 1$. We obtain $C = -5$.

Step 6 Substitute the values of A, B, and C, and write the partial fraction decomposition. With $A = -2$, $B = 2$, and $C = -5$, the required partial fraction decomposition is

$$\frac{x - 18}{x(x - 3)^2} = \frac{A}{x} + \frac{B}{x - 3} + \frac{C}{(x - 3)^2} = -\frac{2}{x} + \frac{2}{x - 3} - \frac{5}{(x - 3)^2}.$$

Check Point 2 Find the partial fraction decomposition of $\frac{x + 2}{x(x - 1)^2}$.

3 Decompose $\frac{P}{Q}$, where Q has a nonrepeated prime quadratic factor.

The Partial Fraction Decomposition of a Rational Expression with Prime, Nonrepeated Quadratic Factors in the Denominator

Our final two cases of partial fraction decomposition involve prime quadratic factors of the form $ax^2 + bx + c$. Based on our work with the discriminant, we know that $ax^2 + bx + c$ is prime and cannot be factored over the integers if $b^2 - 4ac < 0$ or if $b^2 - 4ac$ is not a perfect square.

The Partial Fraction Decomposition of $\frac{P(x)}{Q(x)}$: $Q(x)$ Has a Nonrepeated, Prime Quadratic Factor

If $ax^2 + bx + c$ is a prime quadratic factor of $Q(x)$, the partial fraction decomposition will contain a term of the form

$$\frac{Ax + B}{ax^2 + bx + c}.$$

Linear numerator

Quadratic factor

The voice balloons in the box show that each distinct prime quadratic factor in the denominator produces a partial fraction of the form *linear numerator over quadratic factor*. For example,

$$\frac{3x^2 + 17x + 14}{(x - 2)(x^2 + 2x + 4)} = \frac{A}{x - 2} + \frac{Bx + C}{x^2 + 2x + 4}.$$

We write a constant over the linear factor in the denominator.

We write a linear numerator over the prime quadratic factor in the denominator.

Our next example illustrates how a linear system in three variables is used to determine values for A, B, and C.

EXAMPLE 3 Partial Fraction Decomposition

Find the partial fraction decomposition of

$$\frac{3x^2 + 17x + 14}{(x - 2)(x^2 + 2x + 4)}.$$

Solution

Step 1 Set up the partial fraction decomposition with the unknown constants. We put a constant (A) over the linear factor and a linear expression ($Bx + C$) over the prime quadratic factor.

$$\frac{3x^2 + 17x + 14}{(x - 2)(x^2 + 2x + 4)} = \frac{A}{x - 2} + \frac{Bx + C}{x^2 + 2x + 4}$$

Step 2 Multiply both sides of the resulting equation by the least common denominator. We clear fractions, multiplying both sides by $(x - 2)(x^2 + 2x + 4)$, the least common denominator.

$$(x - 2)(x^2 + 2x + 4)\left[\frac{3x^2 + 17x + 14}{(x - 2)(x^2 + 2x + 4)}\right] = (x - 2)(x^2 + 2x + 4)\left[\frac{A}{x - 2} + \frac{Bx + C}{x^2 + 2x + 4}\right]$$

We use the distributive property on the right side.

$$\cancel{(x - 2)}\,\cancel{(x^2 + 2x + 4)} \cdot \frac{3x^2 + 17x + 14}{\cancel{(x - 2)}\,\cancel{(x^2 + 2x + 4)}}$$

$$= \cancel{(x - 2)}(x^2 + 2x + 4) \cdot \frac{A}{\cancel{x - 2}} + (x - 2)\cancel{(x^2 + 2x + 4)} \cdot \frac{Bx + C}{\cancel{x^2 + 2x + 4}}$$

Dividing out common factors in numerators and denominators, we obtain

$$3x^2 + 17x + 14 = A(x^2 + 2x + 4) + (Bx + C)(x - 2).$$

Step 3 Simplify the right side of the equation. We multiply on the right side by distributing A over each term in parentheses and multiplying $(Bx + C)(x - 2)$ using the FOIL method.

$$3x^2 + 17x + 14 = Ax^2 + 2Ax + 4A + Bx^2 - 2Bx + Cx - 2C$$

Step 4 Write both sides in descending powers, equate coefficients of like powers of x, and equate constant terms. The left side, $3x^2 + 17x + 14$, is in descending powers of x. We write the right side in descending powers of x

$$3x^2 + 17x + 14 = Ax^2 + Bx^2 + 2Ax - 2Bx + Cx + 4A - 2C$$

and express both sides in the same form.

$$3x^2 + 17x + 14 = (A + B)x^2 + (2A - 2B + C)x + (4A - 2C)$$

Equating coefficients of like powers of x and equating constant terms results in the following system of linear equations:

$$\begin{aligned} A + B &= 3 \\ 2A - 2B + C &= 17 \\ 4A - 2C &= 14. \end{aligned}$$

Step 5 Solve the resulting system for A, B, and C. Because the first equation involves A and B, we can obtain another equation in A and B by eliminating C from the second and third equations. Multiply the second equation by 2 and add equations. Solving in this manner, we obtain $A = 5$, $B = -2$, and $C = 3$.

Step 6 Substitute the values of A, B, and C, and write the partial fraction decomposition. With $A = 5$, $B = -2$, and $C = 3$, the required partial fraction decomposition is

$$\frac{3x^2 + 17x + 14}{(x - 2)(x^2 + 2x + 4)} = \frac{A}{x - 2} + \frac{Bx + C}{x^2 + 2x + 4} = \frac{5}{x - 2} + \frac{-2x + 3}{x^2 + 2x + 4}.$$

Technology

You can use the TABLE feature of a graphing utility to check a partial fraction decomposition. To check the result of Example 3, enter the given rational function and its partial fraction decomposition:

$$y_1 = \frac{3x^2 + 17x + 14}{(x - 2)(x^2 + 2x + 4)}$$

$$y_2 = \frac{5}{x - 2} + \frac{-2x + 3}{x^2 + 2x + 4}.$$

X	Y1	Y2
-3	.28571	.28571
-2	.5	.5
-1	0	0
0	-1.75	-1.75
1	-4.857	-4.857
2	ERROR	ERROR
3	4.8421	4.8421

X= -3

No matter how far up or down we scroll, $y_1 = y_2$, so the decomposition appears to be correct.

Check Point 3 Find the partial fraction decomposition of

$$\frac{8x^2 + 12x - 20}{(x + 3)(x^2 + x + 2)}.$$

4 Decompose $\frac{P}{Q}$, where Q has a prime, repeated quadratic factor.

The Partial Fraction Decomposition of a Rational Expression with a Prime, Repeated Quadratic Factor in the Denominator

Suppose that $(ax^2 + bx + c)^n$ is a factor of the denominator and that $ax^2 + bx + c$ cannot be factored further. This means that the quadratic factor $ax^2 + bx + c$ occurs n times. When this occurs, the partial fraction decomposition will contain a linear numerator for each power of $ax^2 + bx + c$.

The Partial Fraction Decomposition of $\frac{P(x)}{Q(x)}$: $Q(x)$ Has a Prime, Repeated Quadratic Factor

The form of the partial fraction decomposition for a rational expression containing the prime factor $ax^2 + bx + c$ occuring n times as its denominator is

$$\frac{P(x)}{(ax^2 + bx + c)^n} = \frac{A_1x + B_1}{ax^2 + bx + c} + \frac{A_2x + B_2}{(ax^2 + bx + c)^2} + \frac{A_3x + B_3}{(ax^2 + bx + c)^3} + \cdots + \frac{A_nx + B_n}{(ax^2 + bx + c)^n}.$$

Include one fraction with a linear numerator for each power of $ax^2 + bx + c$.

Study Tip

When the denominator of a rational expression contains a **power of a linear factor**, set up the partial fraction decomposition with **constant numerators** (A, B, C, etc.). When the denominator of a rational expression contains a **power of a prime quadratic factor**, set up the partial fraction decomposition with **linear numerators** ($Ax + B, Cx + D$, etc.).

EXAMPLE 4 Partial Fraction Decomposition with a Repeated Quadratic Factor

Find the partial fraction decomposition of

$$\frac{5x^3 - 3x^2 + 7x - 3}{(x^2 + 1)^2}.$$

Solution

Step 1 Set up the partial fraction decomposition with the unknown constants. Because the quadratic factor $x^2 + 1$ occurs twice, we must include one fraction with a linear numerator for each power of $x^2 + 1$.

$$\frac{5x^3 - 3x^2 + 7x - 3}{(x^2 + 1)^2} = \frac{Ax + B}{x^2 + 1} + \frac{Cx + D}{(x^2 + 1)^2}$$

Step 2 Multiply both sides of the resulting equation by the least common denominator. We clear fractions, multiplying both sides by $(x^2 + 1)^2$, the least common denominator.

$$(x^2 + 1)^2\left[\frac{5x^3 - 3x^2 + 7x - 3}{(x^2 + 1)^2}\right] = (x^2 + 1)^2\left[\frac{Ax + B}{x^2 + 1} + \frac{Cx + D}{(x^2 + 1)^2}\right]$$

Now we multiply and simplify.

$$5x^3 - 3x^2 + 7x - 3 = (x^2 + 1)(Ax + B) + Cx + D$$

Step 3 Simplify the right side of the equation. We multiply $(x^2 + 1)(Ax + B)$ using the FOIL method.

$$5x^3 - 3x^2 + 7x - 3 = Ax^3 + Bx^2 + Ax + B + Cx + D$$

Step 4 Write both sides in descending powers, equate coefficients of like powers of x, and equate constant terms.

$$5x^3 - 3x^2 + 7x - 3 = Ax^3 + Bx^2 + Ax + Cx + B + D$$

$$5x^3 - 3x^2 + 7x - 3 = Ax^3 + Bx^2 + (A + C)x + (B + D)$$

Equating coefficients of like powers of x and equating constant terms results in the following system of linear equations:

$$A = 5$$
$$B = -3$$
$$A + C = 7 \quad \text{With } A = 5\text{, we immediately obtain } C = 2.$$
$$B + D = -3. \quad \text{With } B = -3\text{, we immediately obtain } D = 0.$$

Step 5 Solve the resulting system for A, B, C, and D. Based on our observations in step 4, $A = 5$, $B = -3$, $C = 2$, and $D = 0$.

Step 6 Substitute the values of A, B, C, and D, and write the partial fraction decomposition.

$$\frac{5x^3 - 3x^2 + 7x - 3}{(x^2 + 1)^2} = \frac{Ax + B}{x^2 + 1} + \frac{Cx + D}{(x^2 + 1)^2} = \frac{5x - 3}{x^2 + 1} + \frac{2x}{(x^2 + 1)^2}$$

Study Tip

When a rational expression contains a power of a factor in the denominator, be sure to set up the partial fraction decomposition to allow for every natural-number power of that factor less than or equal to the power. Example:

$$\frac{2x + 1}{(x - 5)^2 x^3}$$

$$= \frac{A}{x - 5} + \frac{B}{(x - 5)^2} + \frac{C}{x} + \frac{D}{x^2} + \frac{E}{x^3}$$

Although $(x - 5)^2$ and x^2 are quadratic, they are still expressed as powers of linear factors, $x - 5$ and x. Thus, the numerator is constant.

Check Point 4 Find the partial fraction decomposition of $\frac{2x^3 + x + 3}{(x^2 + 1)^2}$.

EXERCISE SET 5.3

Practice Exercises

In Exercises 1–8, write the form of the partial fraction decomposition of the rational expression. It is not necessary to solve for the constants.

1. $\dfrac{11x - 10}{(x - 2)(x + 1)}$

2. $\dfrac{5x + 7}{(x - 1)(x + 3)}$

3. $\dfrac{6x^2 - 14x - 27}{(x + 2)(x - 3)^2}$

4. $\dfrac{3x + 16}{(x + 1)(x - 2)^2}$

5. $\dfrac{5x^2 - 6x + 7}{(x - 1)(x^2 + 1)}$

6. $\dfrac{5x^2 - 9x + 19}{(x - 4)(x^2 + 5)}$

7. $\dfrac{x^3 + x^2}{(x^2 + 4)^2}$

8. $\dfrac{7x^2 - 9x + 3}{(x^2 + 7)^2}$

In Exercises 9–42, write the partial fraction decomposition of each rational expression.

9. $\dfrac{x}{(x - 3)(x - 2)}$

10. $\dfrac{1}{x(x - 1)}$

11. $\dfrac{3x + 50}{(x - 9)(x + 2)}$

12. $\dfrac{5x - 1}{(x - 2)(x + 1)}$

13. $\dfrac{7x - 4}{x^2 - x - 12}$

14. $\dfrac{9x + 21}{x^2 + 2x - 15}$

15. $\dfrac{4}{2x^2 - 5x - 3}$

16. $\dfrac{x}{x^2 + 2x - 3}$

17. $\dfrac{4x^2 + 13x - 9}{x(x - 1)(x + 3)}$

18. $\dfrac{4x^2 - 5x - 15}{x(x + 1)(x - 5)}$

19. $\dfrac{4x^2 - 7x - 3}{x^3 - x}$

20. $\dfrac{2x^2 - 18x - 12}{x^3 - 4x}$

21. $\dfrac{6x - 11}{(x - 1)^2}$

22. $\dfrac{x}{(x + 1)^2}$

23. $\dfrac{x^2 - 6x + 3}{(x - 2)^3}$

24. $\dfrac{2x^2 + 8x + 3}{(x + 1)^3}$

25. $\dfrac{x^2 + 2x + 7}{x(x - 1)^2}$

26. $\dfrac{3x^2 + 49}{x(x + 7)^2}$

27. $\dfrac{x^2}{(x - 1)^2(x + 1)}$

28. $\dfrac{x^2}{(x - 1)^2(x + 1)^2}$

29. $\dfrac{5x^2 - 6x + 7}{(x - 1)(x^2 + 1)}$

30. $\dfrac{5x^2 - 9x + 19}{(x - 4)(x^2 + 5)}$

31. $\dfrac{5x^2 + 6x + 3}{(x + 1)(x^2 + 2x + 2)}$

32. $\dfrac{9x + 2}{(x - 2)(x^2 + 2x + 2)}$

33. $\dfrac{x + 4}{x^2(x^2 + 4)}$

34. $\dfrac{10x^2 + 2x}{(x - 1)^2(x^2 + 2)}$

35. $\dfrac{6x^2 - x + 1}{x^3 + x^2 + x + 1}$

36. $\dfrac{3x^2 - 2x + 8}{x^3 + 2x^2 + 4x + 8}$

37. $\dfrac{x^3 + x^2 + 2}{(x^2 + 2)^2}$

38. $\dfrac{x^2 + 2x + 3}{(x^2 + 4)^2}$

39. $\dfrac{x^3 - 4x^2 + 9x - 5}{(x^2 - 2x + 3)^2}$

40. $\dfrac{3x^3 - 6x^2 + 7x - 2}{(x^2 - 2x + 2)^2}$

41. $\dfrac{4x^2 + 3x + 14}{x^3 - 8}$

42. $\dfrac{3x - 5}{x^3 - 1}$

Practice Plus

In Exercises 43–46, perform each long division and write the partial fraction decomposition of the remainder term.

43. $\dfrac{x^5 + 2}{x^2 - 1}$

44. $\dfrac{x^5}{x^2 - 4x + 4}$

45. $\dfrac{x^4 - x^2 + 2}{x^3 - x^2}$

46. $\dfrac{x^4 + 2x^3 - 4x^2 + x - 3}{x^2 - x - 2}$

In Exercises 47–50, write the partial fraction decomposition of each rational expression.

47. $\dfrac{1}{x^2 - c^2} \quad (c \neq 0)$

48. $\dfrac{ax + b}{x^2 - c^2} \quad (c \neq 0)$

49. $\dfrac{ax + b}{(x - c)^2} \quad (c \neq 0)$

50. $\dfrac{1}{x^2 - ax - bx + ab} \quad (a \neq b)$

Application Exercises

51. Find the partial fraction decomposition for $\dfrac{1}{x(x + 1)}$ and use the result to find the following sum:

$$\frac{1}{1 \cdot 2} + \frac{1}{2 \cdot 3} + \frac{1}{3 \cdot 4} + \cdots + \frac{1}{99 \cdot 100}.$$

52. Find the partial fraction decomposition for $\dfrac{2}{x(x + 2)}$ and use the result to find the following sum:

$$\frac{2}{1 \cdot 3} + \frac{2}{3 \cdot 5} + \frac{2}{5 \cdot 7} + \cdots + \frac{2}{99 \cdot 101}.$$

Writing in Mathematics

53. Explain what is meant by the partial fraction decomposition of a rational expression.

54. Explain how to find the partial fraction decomposition of a rational expression with distinct linear factors in the denominator.

55. Explain how to find the partial fraction decomposition of a rational expression with a repeated linear factor in the denominator.

56. Explain how to find the partial fraction decomposition of a rational expression with a prime quadratic factor in the denominator.

57. Explain how to find the partial fraction decomposition of a rational expression with a repeated, prime quadratic factor in the denominator.

58. How can you verify your result for the partial fraction decomposition for a given rational expression without using a graphing utility?

Technology Exercise

59. Use the TABLE feature of a graphing utility to verify any three of the decompositions that you obtained in Exercises 9–42.

Critical Thinking Exercises

60. Use an extension of the Study Tip at the top of page 498 to describe how to set up the partial fraction decomposition of a rational expression that contains powers of a prime cubic factor in the denominator. Give an example of such a decomposition.

61. Find the partial fraction decomposition of

$$\frac{4x^2 + 5x - 9}{x^3 - 6x - 9}.$$

SECTION 5.4 Systems of Nonlinear Equations in Two Variables

Objectives

1. Recognize systems of nonlinear equations in two variables.
2. Solve nonlinear systems by substitution.
3. Solve nonlinear systems by addition.
4. Solve problems using systems of nonlinear equations.

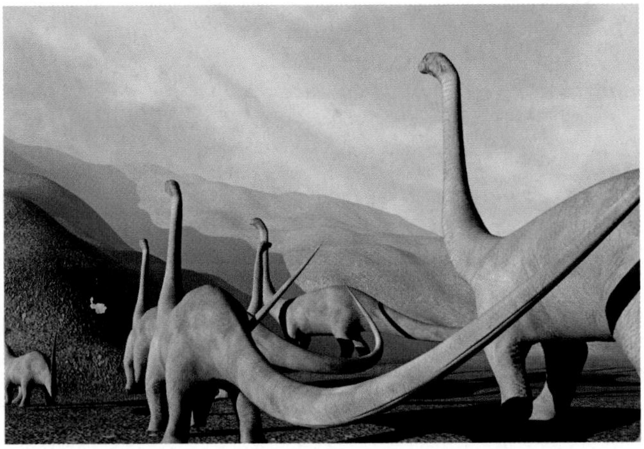

Scientists debate the probability that a "doomsday rock" will collide with Earth. It has been estimated that an asteroid, a tiny planet that revolves around the sun, crashes into Earth about once every 250,000 years, and that such a collision would have disastrous results. In 1908, a small fragment struck Siberia, leveling thousands of acres of trees. One theory about the extinction of dinosaurs 65 million years ago involves Earth's collision with a large asteroid and the resulting drastic changes in Earth's climate.

Understanding the path of Earth and the path of a comet is essential to detecting threatening space debris. Orbits about the sun are not described by linear equations in the form $Ax + By = C$. The ability to solve systems that contain more than just linear equations provides NASA scientists watching for troublesome asteroids with a way to locate possible collision points with Earth's orbit.

1 Recognize systems of nonlinear equations in two variables.

Systems of Nonlinear Equations and Their Solutions

A **system of** two **nonlinear equations** in two variables, also called a **nonlinear system**, contains at least one equation that cannot be expressed in the form $Ax + By = C$. Here are two examples:

$$x^2 = 2y + 10$$
$$3x - y = 9$$

Not in the form $Ax + By = C$. The term x^2 is not linear.

$$y = x^2 + 3$$
$$x^2 + y^2 = 9.$$

Neither equation is in the form $Ax + By = C$. The terms x^2 and y^2 are not linear.

A **solution** of a nonlinear system in two variables is an ordered pair of real numbers that satisfies both equations in the system. The **solution set** of the system is the set of all such ordered pairs. As with linear systems in two variables, the solution of a nonlinear system (if there is one) corresponds to the intersection point(s) of the graphs of the equations in the system. Unlike linear systems, the graphs can be circles, parabolas, or anything other than two lines. We will solve nonlinear systems using the substitution method and the addition method.

2 Solve nonlinear systems by substitution.

Eliminating a Variable Using the Substitution Method

The substitution method involves converting a nonlinear system to one equation in one variable by an appropriate substitution. The steps in the solution process are exactly the same as those used to solve a linear system by substitution. However, when you obtain an equation in one variable, this equation may not be linear. In our first example, this equation is quadratic.

EXAMPLE 1 Solving a Nonlinear System by the Substitution Method

Solve by the substitution method:

$$x^2 = 2y + 10 \quad \text{(The graph is a parabola.)}$$
$$3x - y = 9. \quad \text{(The graph is a line.)}$$

Solution

Step 1 Solve one of the equations for one variable in terms of the other. We begin by isolating one of the variables raised to the first power in either of the equations. By solving for y in the second equation, which has a coefficient of -1, we can avoid fractions.

$$3x - y = 9 \quad \text{This is the second equation in the given system.}$$
$$3x = y + 9 \quad \text{Add } y \text{ to both sides.}$$
$$3x - 9 = y \quad \text{Subtract 9 from both sides.}$$

Step 2 Substitute the expression from step 1 into the other equation. We substitute $3x - 9$ for y in the first equation.

$$y = \boxed{3x - 9} \qquad x^2 = 2\boxed{y} + 10$$

This gives us an equation in one variable, namely

$$x^2 = 2(3x - 9) + 10.$$

The variable y has been eliminated.

Step 3 Solve the resulting equation containing one variable.

$$x^2 = 2(3x - 9) + 10 \quad \text{This is the equation containing one variable.}$$
$$x^2 = 6x - 18 + 10 \quad \text{Use the distributive property.}$$
$$x^2 = 6x - 8 \quad \text{Combine numerical terms on the right.}$$
$$x^2 - 6x + 8 = 0 \quad \text{Move all terms to one side and set the quadratic equation equal to 0.}$$
$$(x - 4)(x - 2) = 0 \quad \text{Factor.}$$
$$x - 4 = 0 \quad \text{or} \quad x - 2 = 0 \quad \text{Set each factor equal to 0.}$$
$$x = 4 \qquad x = 2 \quad \text{Solve for } x.$$

Step 4 Back-substitute the obtained values into the equation from step 1. Now that we have the x-coordinates of the solutions, we back-substitute 4 for x and 2 for x into the equation $y = 3x - 9$.

If x is 4, $y = 3(4) - 9 = 3$, so $(4, 3)$ is a solution.
If x is 2, $y = 3(2) - 9 = -3$, so $(2, -3)$ is a solution.

Step 5 Check the proposed solutions in both of the system's given equations. We begin by checking $(4, 3)$. Replace x with 4 and y with 3.

$x^2 = 2y + 10$	$3x - y = 9$	These are the given equations.
$4^2 \stackrel{?}{=} 2(3) + 10$	$3(4) - 3 \stackrel{?}{=} 9$	Let $x = 4$ and $y = 3$.
$16 \stackrel{?}{=} 6 + 10$	$12 - 3 \stackrel{?}{=} 9$	Simplify.
$16 = 16$, true	$9 = 9$, true	True statements result.

The ordered pair $(4, 3)$ satisfies both equations. Thus, $(4, 3)$ is a solution of the system.

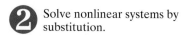

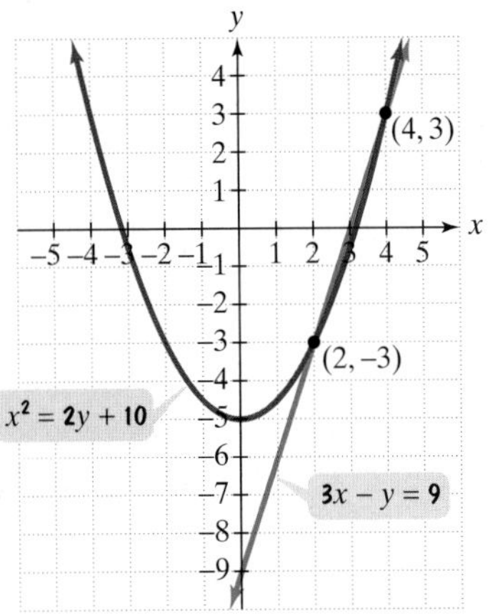

Figure 5.9 Points of intersection illustrate the nonlinear system's solutions.

Now let's check $(2, -3)$. Replace x with 2 and y with -3 in both given equations.

$x^2 = 2y + 10$	$3x - y = 9$	These are the given equations.
$2^2 \stackrel{?}{=} 2(-3) + 10$	$3(2) - (-3) \stackrel{?}{=} 9$	Let $x = 2$ and $y = -3$.
$4 \stackrel{?}{=} -6 + 10$	$6 + 3 \stackrel{?}{=} 9$	Simplify.
$4 = 4$, true	$9 = 9$, true	True statements result.

The ordered pair $(2, -3)$ also satisfies both equations and is a solution of the system. The solutions are $(4, 3)$ and $(2, -3)$, and the solution set is $\{(4, 3), (2, -3)\}$.

Figure 5.9 shows the graphs of the equations in the system and the solutions as intersection points.

Check Point 1 Solve by the substitution method:

$$x^2 = y - 1$$
$$4x - y = -1.$$

EXAMPLE 2 Solving a Nonlinear System by the Substitution Method

Solve by the substitution method:

$$x - y = 3 \quad \text{(The graph is a line.)}$$
$$(x - 2)^2 + (y + 3)^2 = 4. \quad \text{(The graph is a circle.)}$$

Study Tip

Recall that $(x - h)^2 + (y - k)^2 = r^2$ describes a circle with center (h, k) and radius r.

Solution Graphically, we are finding the intersection of a line and a circle with center $(2, -3)$ and radius 2.

Step 1 Solve one of the equations for one variable in terms of the other. We will solve for x in the linear equation—that is, the first equation. (We could also solve for y.)

$$x - y = 3 \quad \text{This is the first equation in the given system.}$$
$$x = y + 3 \quad \text{Add } y \text{ to both sides.}$$

Step 2 Substitute the expression from step 1 into the other equation. We substitute $y + 3$ for x in the second equation.

$$x = \boxed{y + 3} \qquad (\boxed{x} - 2)^2 + (y + 3)^2 = 4$$

This gives an equation in one variable, namely

$$(y + 3 - 2)^2 + (y + 3)^2 = 4.$$

The variable x has been eliminated.

Step 3 Solve the resulting equation containing one variable.

$(y + 3 - 2)^2 + (y + 3)^2 = 4$	This is the equation containing one variable.
$(y + 1)^2 + (y + 3)^2 = 4$	Combine numerical terms in the first parentheses.
$y^2 + 2y + 1 + y^2 + 6y + 9 = 4$	Use the formula $(A + B)^2 = A^2 + 2AB + B^2$ to square $y + 1$ and $y + 3$.
$2y^2 + 8y + 10 = 4$	Combine like terms on the left.
$2y^2 + 8y + 6 = 0$	Subtract 4 from both sides and set the quadratic equation equal to 0.
$2(y^2 + 4y + 3) = 0$	Factor out 2.
$2(y + 3)(y + 1) = 0$	Factor completely.
$y + 3 = 0$ or $y + 1 = 0$	Set each variable factor equal to 0.
$y = -3$ $\quad$ $y = -1$	Solve for y.

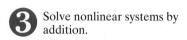

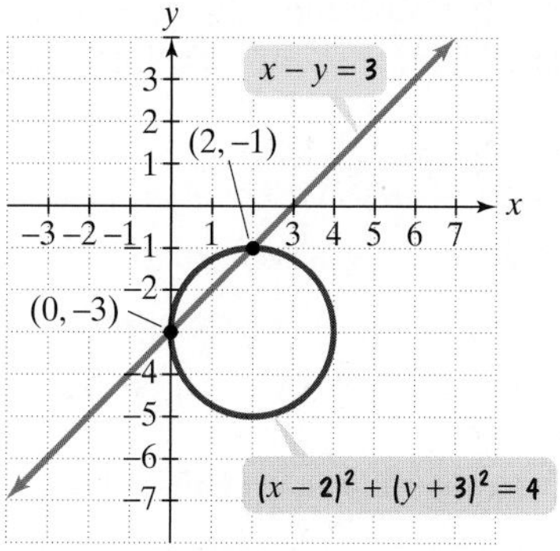

Figure 5.10 Points of intersection illustrate the nonlinear system's solutions.

Step 4 Back-substitute the obtained values into the equation from step 1. Now that we have the y-coordinates of the solutions, we back-substitute -3 for y and -1 for y in the equation $x = y + 3$.

If $y = -3$: $x = -3 + 3 = 0$, so $(0, -3)$ is a solution.
If $y = -1$: $x = -1 + 3 = 2$, so $(2, -1)$ is a solution.

Step 5 Check the proposed solutions in both of the system's given equations. Take a moment to show that each ordered pair satisfies both equations. The solutions are $(0, -3)$ and $(2, -1)$, and the solution set of the given system is $\{(0, -3), (2, -1)\}$.

Figure 5.10 shows the graphs of the equations in the system and the solutions as intersection points.

Check Point 2 Solve by the substitution method:

$$x + 2y = 0$$
$$(x - 1)^2 + (y - 1)^2 = 5.$$

3 Solve nonlinear systems by addition.

Eliminating a Variable Using the Addition Method

In solving linear systems with two variables, we learned that the addition method works well when each equation is in the form $Ax + By = C$. For nonlinear systems, the addition method can be used when each equation is in the form $Ax^2 + By^2 = C$. If necessary, we will multiply either equation or both equations by appropriate numbers so that the coefficients of x^2 or y^2 will have a sum of 0. We then add equations. The sum will be an equation in one variable.

EXAMPLE 3 Solving a Nonlinear System by the Addition Method

Solve the system:

$$4x^2 + y^2 = 13 \quad \text{Equation 1}$$
$$x^2 + y^2 = 10. \quad \text{Equation 2}$$

Solution We can use the same steps that we did when we solved linear systems by the addition method.

Step 1 Write both equations in the form $Ax^2 + By^2 = C$. Both equations are already in this form, so we can skip this step.

Step 2 If necessary, multiply either equation or both equations by appropriate numbers so that the sum of the x^2-coefficients or the sum of the y^2-coefficients is 0. We can eliminate y^2 by multiplying Equation 2 by -1.

$$4x^2 + y^2 = 13 \xrightarrow{\text{No change}} 4x^2 + y^2 = 13$$
$$x^2 + y^2 = 10 \xrightarrow{\text{Multiply by } -1.} -x^2 - y^2 = -10$$

Steps 3 and 4 Add equations and solve for the remaining variable.

$$4x^2 + y^2 = 13$$
$$\underline{-x^2 - y^2 = -10}$$
$$3x^2 = 3 \quad \text{Add equations.}$$
$$x^2 = 1 \quad \text{Divide both sides by 3.}$$
$$x = \pm 1 \quad \text{Use the square root property: If } x^2 = c, \text{ then } x = \pm\sqrt{c}.$$

Step 5 Back-substitute and find the values for the other variable. We must back-substitute each value of x into either one of the original equations. Let's use $x^2 + y^2 = 10$, Equation 2. If $x = 1$,

$$1^2 + y^2 = 10 \quad \text{Replace } x \text{ with 1 in Equation 2.}$$
$$y^2 = 9 \quad \text{Subtract 1 from both sides.}$$
$$y = \pm 3. \quad \text{Apply the square root property.}$$

$(1, 3)$ and $(1, -3)$ are solutions. If $x = -1$,

$$(-1)^2 + y^2 = 10 \quad \text{Replace } x \text{ with } -1 \text{ in Equation 2.}$$
$$y^2 = 9 \quad \text{The steps are the same as before.}$$
$$y = \pm 3.$$

$(-1, 3)$ and $(-1, -3)$ are solutions.

Step 6 Check. Take a moment to show that each of the four ordered pairs satisfies the given equations, $4x^2 + y^2 = 13$ and $x^2 + y^2 = 10$. The solution set of the given system is $\{(1, 3), (1, -3), (-1, 3), (-1, -3)\}$.

Figure 5.11 shows the graphs of the equations in the system and the solutions as intersection points.

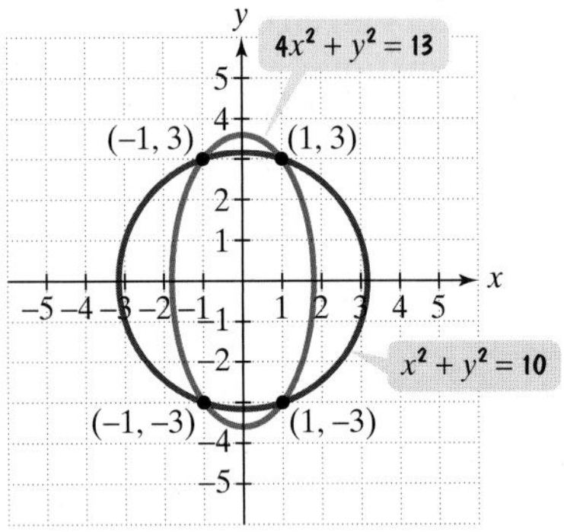

Figure 5.11 A system with four solutions

Study Tip

When solving nonlinear systems, extra solutions may be introduced that do not satisfy both equations in the system. Therefore, you should get into the habit of checking all proposed pairs in each of the system's two equations.

Check Point 3 Solve the system:

$$3x^2 + 2y^2 = 35$$
$$4x^2 + 3y^2 = 48.$$

In solving nonlinear systems, we include only ordered pairs with real numbers in the solution set. We have seen that each of these ordered pairs corresponds to a point of intersection of the system's graphs.

EXAMPLE 4 Solving a Nonlinear System by the Addition Method

Solve the system:

$$y = x^2 + 3 \quad \text{Equation 1 (The graph is a parabola.)}$$
$$x^2 + y^2 = 9. \quad \text{Equation 2 (The graph is a circle.)}$$

Solution We could use substitution because Equation 1, $y = x^2 + 3$, has y expressed in terms of x, but substituting $x^2 + 3$ for y in $x^2 + y^2 = 9$ would result in a fourth-degree equation. However, we can rewrite Equation 1 by subtracting x^2 from both sides and adding the equations to eliminate the x^2-terms.

Notice how like terms are arranged in columns.

$$-x^2 + y = 3 \quad \text{Subtract } x^2 \text{ from both sides of Equation 1.}$$
$$x^2 + y^2 = 9 \quad \text{This is Equation 2.}$$
$$y + y^2 = 12 \quad \text{Add the equations.}$$

We now solve this quadratic equation.

$$y + y^2 = 12 \quad \text{This is the equation containing one variable.}$$
$$y^2 + y - 12 = 0 \quad \text{Subtract 12 from both sides and set the quadratic equation equal to 0.}$$
$$(y + 4)(y - 3) = 0 \quad \text{Factor.}$$
$$y + 4 = 0 \quad \text{or} \quad y - 3 = 0 \quad \text{Set each factor equal to 0.}$$
$$y = -4 \qquad\qquad y = 3 \quad \text{Solve for } y.$$

To complete the solution, we must back-substitute each value of y into either one of the original equations. We will use $y = x^2 + 3$, Equation 1. First, we substitute -4 for y.

$$-4 = x^2 + 3$$
$$-7 = x^2 \quad \text{Subtract 3 from both sides.}$$

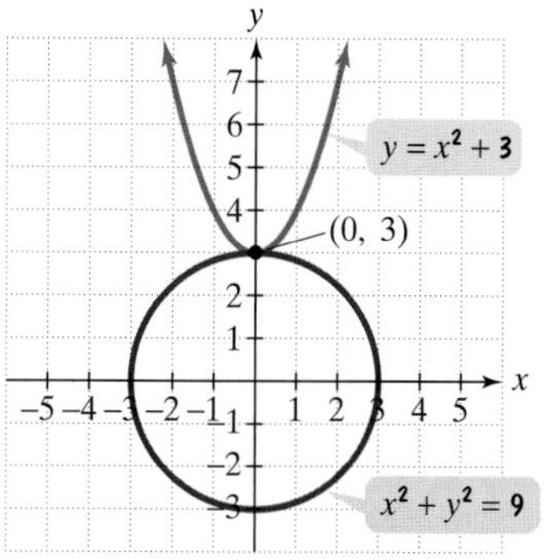

Figure 5.12 A system with one real solution

Because the square of a real number cannot be negative, the equation $x^2 = -7$ does not have real-number solutions. We will not include the imaginary solutions, $x = \pm\sqrt{-7}$, or $i\sqrt{7}$ and $-i\sqrt{7}$, in the ordered pairs that make up the solution set. Thus, we move on to our other value for y, 3, and substitute this value into Equation 1.

$$y = x^2 + 3 \quad \text{This is Equation 1.}$$
$$3 = x^2 + 3 \quad \text{Back-substitute 3 for } y.$$
$$0 = x^2 \quad \text{Subtract 3 from both sides.}$$
$$0 = x \quad \text{Solve for } x.$$

We see that if $y = 3$, then $x = 0$. Thus, $(0, 3)$ is the solution with a real ordered pair. Take a moment to show that $(0, 3)$ satisfies the given equations, $y = x^2 + 3$ and $x^2 + y^2 = 9$. The solution set of the system is $\{(0, 3)\}$. Figure 5.12 shows the system's graphs and the solution as an intersection point.

Check Point 4 Solve the system:

$$y = x^2 + 5$$
$$x^2 + y^2 = 25.$$

4 Solve problems using systems of nonlinear equations.

Applications

Many geometric problems can be modeled and solved by the use of systems of nonlinear equations. We will use our step-by-step strategy for solving problems using mathematical models that are created from verbal conditions.

EXAMPLE 5 An Application of a Nonlinear System

You have 36 yards of fencing to build the enclosure in Figure 5.13. Some of this fencing is to be used to build an internal divider. If you'd like to enclose 54 square yards, what are the dimensions of the enclosure?

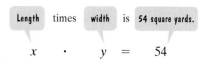

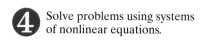

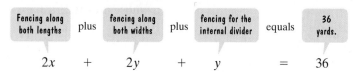

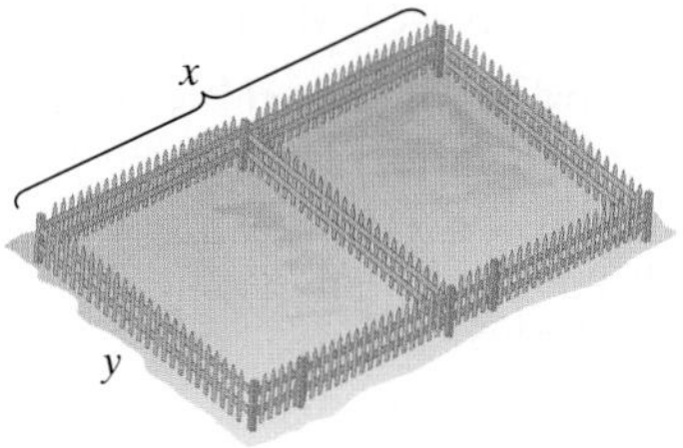

Figure 5.13 Building an enclosure

Solution

Step 1 Use variables to represent unknown quantities. Let x = the enclosure's length and y = the enclosure's width. These variables are shown in Figure 5.13.

Step 2 Write a system of equations describing the problem's conditions. The first condition is that you have 36 yards of fencing.

Fencing along both lengths plus fencing along both widths plus fencing for the internal divider equals 36 yards.

$$2x + 2y + y = 36$$

Adding like terms, we can express the equation that models the verbal conditions for the fencing as $2x + 3y = 36$.

The second condition is that you'd like to enclose 54 square yards. The rectangle's area, the product of its length and its width, must be 54 square yards.

Length times width is 54 square yards.

$$x \cdot y = 54$$

Step 3 Solve the system and answer the problem's question. We must solve the system

$$2x + 3y = 36 \quad \text{Equation 1}$$
$$xy = 54. \quad \text{Equation 2}$$

We will use substitution. Because Equation 1 has no coefficients of 1 or -1, we will work with Equation 2 and solve for y. Dividing both sides of $xy = 54$ by x, we obtain

$$y = \frac{54}{x}.$$

Now we substitute $\frac{54}{x}$ for y in Equation 1 and solve for x.

$$2x + 3y = 36 \quad \text{This is Equation 1.}$$
$$2x + 3 \cdot \frac{54}{x} = 36 \quad \text{Substitute } \frac{54}{x} \text{ for } y.$$
$$2x + \frac{162}{x} = 36 \quad \text{Multiply.}$$
$$x\left(2x + \frac{162}{x}\right) = 36 \cdot x \quad \text{Clear fractions by multiplying both sides by } x.$$
$$2x^2 + 162 = 36x \quad \text{Use the distributive property on the left side.}$$
$$2x^2 - 36x + 162 = 0 \quad \text{Subtract } 36x \text{ from both sides and set the quadratic equation equal to 0.}$$
$$2(x^2 - 18x + 81) = 0 \quad \text{Factor out 2.}$$
$$2(x - 9)^2 = 0 \quad \text{Factor completely using } A^2 - 2AB + B^2 = (A - B)^2.$$
$$x - 9 = 0 \quad \text{Set the repeated factor equal to zero.}$$
$$x = 9 \quad \text{Solve for x.}$$

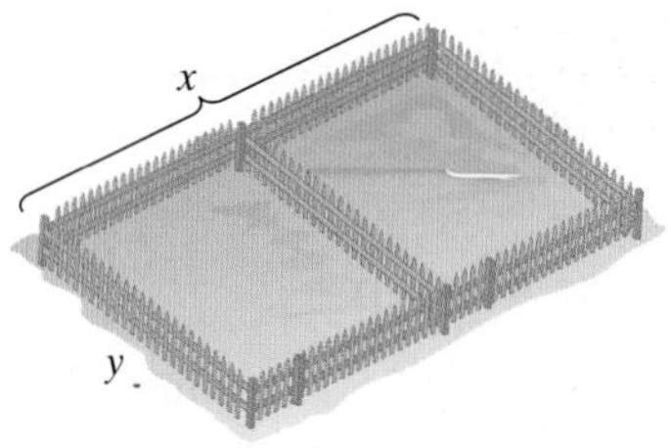

Figure 5.13 (repeated)

We back-substitute this value of x into $y = \frac{54}{x}$.

$$\text{If } x = 9, \quad y = \frac{54}{9} = 6.$$

This means that the dimensions of the enclosure in Figure 5.13 are 9 yards by 6 yards.

Step 4 Check the proposed solution in the original wording of the problem. Take a moment to check that a length of 9 yards and a width of 6 yards results in 36 yards of fencing and an area of 54 square yards.

Check Point 5 Find the length and width of a rectangle whose perimeter is 20 feet and whose area is 21 square feet.

EXERCISE SET 5.4

Practice Exercises

In Exercises 1–18, solve each system by the substitution method.

1. $x + y = 2$
$y = x^2 - 4$

2. $x - y = -1$
$y = x^2 + 1$

3. $x + y = 2$
$y = x^2 - 4x + 4$

4. $2x + y = -5$
$y = x^2 + 6x + 7$

5. $y = x^2 - 4x - 10$
$y = -x^2 - 2x + 14$

6. $y = x^2 + 4x + 5$
$y = x^2 + 2x - 1$

7. $x^2 + y^2 = 25$
$x - y = 1$

8. $x^2 + y^2 = 5$
$3x - y = 5$

9. $xy = 6$
$2x - y = 1$

10. $xy = -12$
$x - 2y + 14 = 0$

11. $y^2 = x^2 - 9$
$2y = x - 3$

12. $x^2 + y = 4$
$2x + y = 1$

13. $xy = 3$
$x^2 + y^2 = 10$

14. $xy = 4$
$x^2 + y^2 = 8$

15. $x + y = 1$
$x^2 + xy - y^2 = -5$

16. $x + y = -3$
$x^2 + 2y^2 = 12y + 18$

17. $x + y = 1$
$(x - 1)^2 + (y + 2)^2 = 10$

18. $2x + y = 4$
$(x + 1)^2 + (y - 2)^2 = 4$

In Exercises 19–28, solve each system by the addition method.

19. $x^2 + y^2 = 13$
$x^2 - y^2 = 5$

20. $4x^2 - y^2 = 4$
$4x^2 + y^2 = 4$

21. $x^2 - 4y^2 = -7$
$3x^2 + y^2 = 31$

22. $3x^2 - 2y^2 = -5$
$2x^2 - y^2 = -2$

23. $3x^2 + 4y^2 - 16 = 0$
$2x^2 - 3y^2 - 5 = 0$

24. $16x^2 - 4y^2 - 72 = 0$
$x^2 - y^2 - 3 = 0$

25. $x^2 + y^2 = 25$
$(x - 8)^2 + y^2 = 41$

26. $x^2 + y^2 = 5$
$x^2 + (y - 8)^2 = 41$

27. $y^2 - x = 4$
$x^2 + y^2 = 4$

28. $x^2 - 2y = 8$
$x^2 + y^2 = 16$

In Exercises 29–42, solve each system by the method of your choice.

29. $3x^2 + 4y^2 = 16$
$2x^2 - 3y^2 = 5$

30. $x + y^2 = 4$
$x^2 + y^2 = 16$

31. $2x^2 + y^2 = 18$
$xy = 4$

32. $x^2 + 4y^2 = 20$
$xy = 4$

33. $x^2 + 4y^2 = 20$
$x + 2y = 6$

34. $3x^2 - 2y^2 = 1$
$4x - y = 3$

35. $x^3 + y = 0$
$x^2 - y = 0$

36. $x^3 + y = 0$
$2x^2 - y = 0$

37. $x^2 + (y - 2)^2 = 4$
$x^2 - 2y = 0$

38. $x^2 - y^2 - 4x + 6y - 4 = 0$
$x^2 + y^2 - 4x - 6y + 12 = 0$

39. $y = (x + 3)^2$
$x + 2y = -2$

40. $(x - 1)^2 + (y + 1)^2 = 5$
$2x - y = 3$

41. $x^2 + y^2 + 3y = 22$
$2x + y = -1$

42. $x - 3y = -5$
$x^2 + y^2 - 25 = 0$

In Exercises 43–46, let x represent one number and let y represent the other number. Use the given conditions to write a system of nonlinear equations. Solve the system and find the numbers.

43. The sum of two numbers is 10 and their product is 24. Find the numbers.

44. The sum of two numbers is 20 and their product is 96. Find the numbers.

45. The difference between the squares of two numbers is 3. Twice the square of the first number increased by the square of the second number is 9. Find the numbers.

46. The difference between the squares of two numbers is 5. Twice the square of the second number subtracted from three times the square of the first number is 19. Find the numbers.

Practice Plus

In Exercises 47–52, solve each system by the method of your choice.

47. $2x^2 + xy = 6$
$x^2 + 2xy = 0$

48. $4x^2 + xy = 30$
$x^2 + 3xy = -9$

49. $-4x + y = 12$
$y = x^3 + 3x^2$

50. $-9x + y = 45$
$y = x^3 + 5x^2$

51. $\frac{3}{x^2} + \frac{1}{y^2} = 7$
$\frac{5}{x^2} - \frac{2}{y^2} = -3$

52. $\frac{2}{x^2} + \frac{1}{y^2} = 11$
$\frac{4}{x^2} - \frac{2}{y^2} = -14$

In Exercises 53–54, make a rough sketch in a rectangular coordinate system of the graphs representing the equations in each system.

53. The system, whose graphs are a line with positive slope and a parabola whose equation has a positive leading coefficient, has two solutions.

54. The system, whose graphs are a line with negative slope and a parabola whose equation has a negative leading coefficient, has one solution.

Application Exercises

55. A planet's orbit follows a path described by $16x^2 + 4y^2 = 64$. A comet follows the parabolic path $y = x^2 - 4$. Where might the comet intersect the orbiting planet?

56. A system for tracking ships indicates that a ship lies on a path described by $2y^2 - x^2 = 1$. The process is repeated and the ship is found to lie on a path described by $2x^2 - y^2 = 1$. If it is known that the ship is located in the first quadrant of the coordinate system, determine its exact location.

57. Find the length and width of a rectangle whose perimeter is 36 feet and whose area is 77 square feet.

58. Find the length and width of a rectangle whose perimeter is 40 feet and whose area is 96 square feet.

Use the formula for the area of a rectangle and the Pythagorean Theorem to solve Exercises 59–60.

59. A small television has a picture with a diagonal measure of 10 inches and a viewing area of 48 square inches. Find the length and width of the screen.

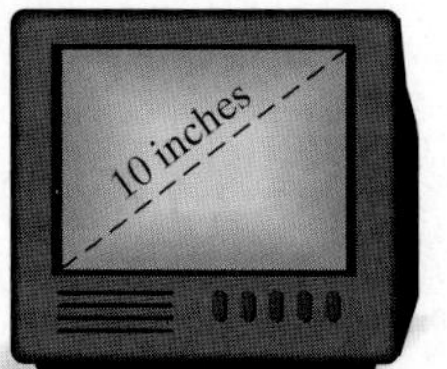

60. The area of a rug is 108 square feet and the length of its diagonal is 15 feet. Find the length and width of the rug.

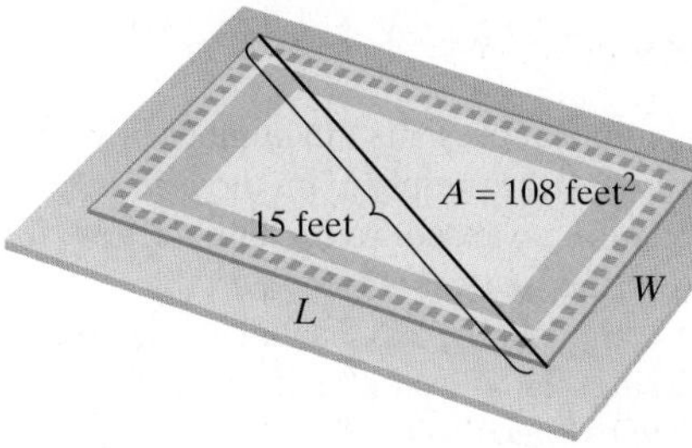

61. The figure shows a square floor plan with a smaller square area that will accommodate a combination fountain and pool. The floor with the fountain-pool area removed has an area of 21 square meters and a perimeter of 24 meters. Find the dimensions of the floor and the dimensions of the square that will accommodate the pool.

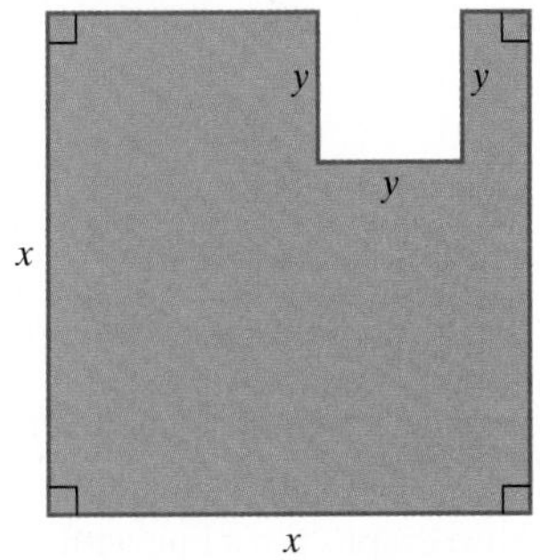

62. The area of the rectangular piece of cardboard shown on the left is 216 square inches. The cardboard is used to make an open box by cutting a 2-inch square from each corner and turning up the sides. If the box is to have a volume of 224 cubic inches, find the length and width of the cardboard that must be used.

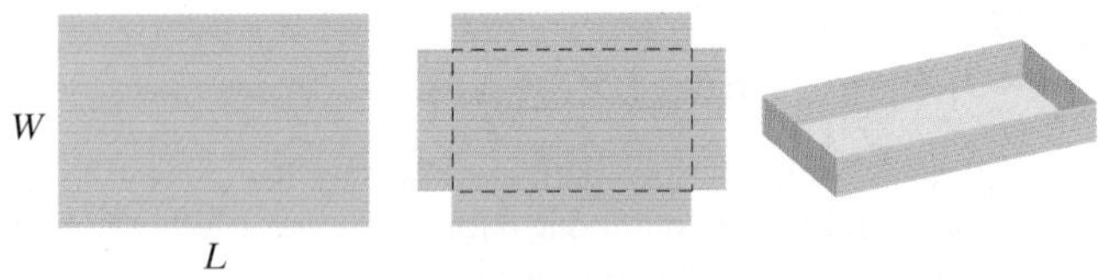

63. The graphs show the number, in thousands, of bachelor's degrees in science and engineering received by U.S. men and women from 1991 through 2001.

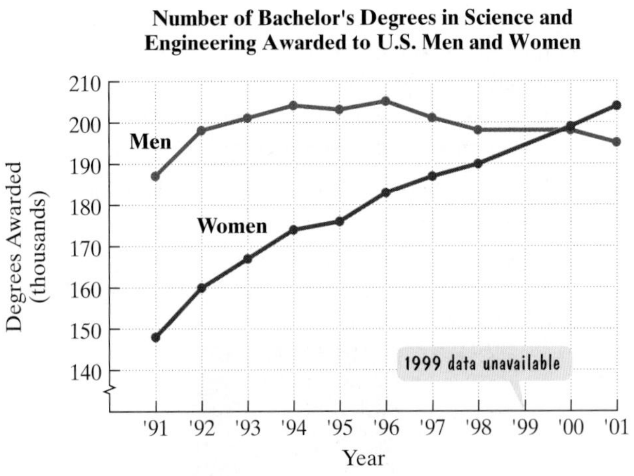

Source: National Science Foundation

The data can be modeled by quadratic and linear functions:

Men $y = -0.5x^2 + 5.8x + 185.8$

Women $5.4x - y + 146.5 = 0.$

In each function, x represents the number of years after 1990 and y represents the number, in thousands, of bachelor's degrees that were awarded. According to these functions, in which year, to the nearest whole year, did men and women receive the same number of bachelor's degrees in science and engineering? How well does this describe the information displayed by the graphs?

Writing in Mathematics

64. What is a system of nonlinear equations? Provide an example with your description.

65. Explain how to solve a nonlinear system using the substitution method. Use $x^2 + y^2 = 9$ and $2x - y = 3$ to illustrate your explanation.

66. Explain how to solve a nonlinear system using the addition method. Use $x^2 - y^2 = 5$ and $3x^2 - 2y^2 = 19$ to illustrate your explanation.

67. The daily demand and supply models for a carrot cake supplied by a bakery to a convenience store are given by the demand model $N = 40 - 3p$ and the supply model $N = \frac{p^2}{10}$, in which p is the price of the cake and N is the number of cakes sold or supplied each day to the convenience store. Explain how to determine the price at which supply and demand are equal. Then describe how to find how many carrot cakes can be supplied and sold each day at this price.

Technology Exercises

68. Verify your solutions to any five exercises from Exercises 1–42 by using a graphing utility to graph the two equations in the system in the same viewing rectangle. Then use the intersection feature to verify the solutions.

69. Write a system of equations, one equation whose graph is a line and the other whose graph is a parabola, that has no ordered pairs that are real numbers in its solution set. Graph the equations using a graphing utility and verify that you are correct.

Critical Thinking Exercises

70. Which one of the following is true?
    - **a.** A system of two equations in two variables whose graphs are a circle and a line can have four real ordered-pair solutions.
    - **b.** A system of two equations in two variables whose graphs are a parabola and a circle can have four real ordered-pair solutions.
    - **c.** A system of two equations in two variables whose graphs are two circles must have at least two real ordered-pair solutions.
    - **d.** A system of two equations in two variables whose graphs are a parabola and a circle cannot have only one real ordered-pair solution.

71. The points of intersection of the graphs of $xy = 20$ and $x^2 + y^2 = 41$ are joined to form a rectangle. Find the area of the rectangle.

72. Find a and b in this figure.

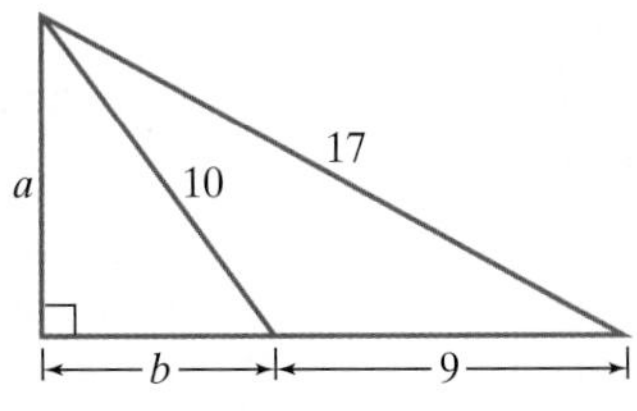

Solve the systems in Exercises 73–74.

73. $\log_y x = 3$
$\log_y(4x) = 5$

74. $\log x^2 = y + 3$
$\log x = y - 1$

CHAPTER 5
MID-CHAPTER CHECK POINT

What You Know: We learned to solve systems of equations. We solved linear and nonlinear systems in two variables by the substitution method and by the addition method. We solved linear systems in three variables by eliminating a variable, reducing the system to two equations in two variables. We saw that some linear systems, called inconsistent systems, have no solution, whereas other linear systems, called dependent systems, have infinitely many solutions. We applied systems to a variety of situations, including finding the break-even point for a business, finding a quadratic function from three points on its graph, and finding a rational function's partial fraction decomposition.

In Exercises 1–12, solve each system by the method of your choice.

1. $x = 3y - 7$
$4x + 3y = 2$

2. $3x + 4y = -5$
$2x - 3y = 8$

3. $\frac{2x}{3} + \frac{y}{5} = 6$
$\frac{x}{6} - \frac{y}{2} = -4$

4. $y = 4x - 5$
$8x - 2y = 10$

5. $2x + 5y = 3$
$3x - 2y = 1$

6. $\frac{x}{12} - y = \frac{1}{4}$
$4x - 48y = 16$

7. $2x - y + 2z = -8$
$x + 2y - 3z = 9$
$3x - y - 4z = 3$

8. $x - 3z = -5$
$2x - y + 2z = 16$
$7x - 3y - 5z = 19$

9. $x^2 + y^2 = 9$
$x + 2y - 3 = 0$

10. $3x^2 + 2y^2 = 14$
$2x^2 - y^2 = 7$

11. $y = x^2 - 6$
$x^2 + y^2 = 8$

12. $x - 2y = 4$
$2y^2 + xy = 8$

In Exercises 13–16, write the partial fraction decomposition of each rational expression.

13. $\frac{x^2 - 6x + 3}{(x - 2)^3}$

14. $\frac{10x^2 + 9x - 7}{(x + 2)(x^2 - 1)}$

15. $\frac{x^2 + 4x - 23}{(x + 3)(x^2 + 4)}$

16. $\frac{x^3}{(x^2 + 4)^2}$

17. A company is planning to manufacture PDAs (personal digital assistants). The fixed cost will be $400,000 and it will cost $20 to produce each PDA. Each PDA will be sold for $100.
a. Write the cost function, C, of producing x PDAs.
b. Write the revenue function, R, from the sale of x PDAs.
c. Write the profit function, P, from producing and selling x PDAs.
d. Determine the break-even point. Describe what this means.

18. Roses sell for $3 each and carnations for $1.50 each. If a mixed bouquet of 20 flowers consisting of roses and carnations costs $39, how many of each type of flower is in the bouquet?

19. Find the measure of each angle whose degree measure is represented with a variable.

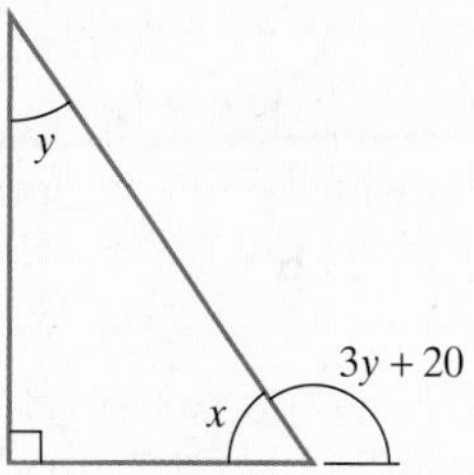

20. Find the quadratic function $y = ax^2 + bx + c$ whose graph passes through the points $(-1, 0)$, $(1, 4)$, and $(2, 3)$.

21. Find the length and width of a rectangle whose perimeter is 21 meters and whose area is 20 square meters.

SECTION 5.5 Systems of Inequalities

Objectives

1. Graph a linear inequality in two variables.
2. Graph a nonlinear inequality in two variables.
3. Graph a system of inequalities.
4. Solve applied problems involving systems of inequalities.

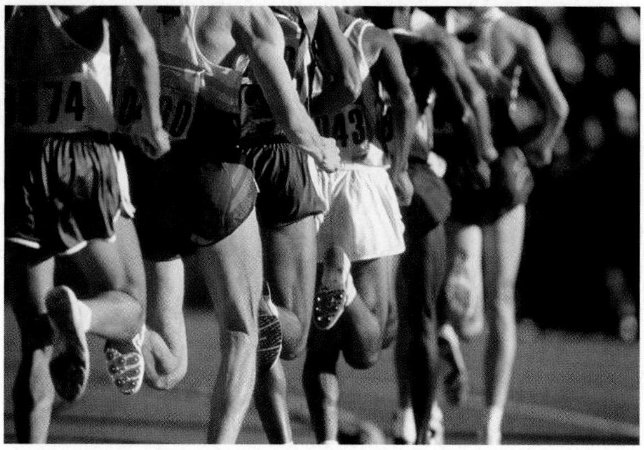

Had a good workout lately? If so, could you tell if you were overdoing it or not pushing yourself hard enough? In this section's exercise set, we will use systems of inequalities in two variables to help you establish a target zone for your workouts.

Linear Inequalities in Two Variables and Their Solutions

We have seen that equations in the form $Ax + By = C$ are straight lines when graphed. If we change the symbol $=$ to $>$, $<$, $\geq$, or $\leq$, we obtain a **linear inequality in two variables**. Some examples of linear inequalities in two variables are $x + y > 2$, $3x - 5y \leq 15$, and $2x - y < 4$.

A **solution of an inequality in two variables**, x and y, is an ordered pair of real numbers with the following property: When the x-coordinate is substituted for x and the y-coordinate is substituted for y in the inequality, we obtain a true statement. For example, $(3, 2)$ is a solution of the inequality $x + y > 1$. When 3 is substituted for x and 2 is substituted for y, we obtain the true statement $3 + 2 > 1$, or $5 > 1$. Because there are infinitely many pairs of numbers that have a sum greater than 1, the inequality $x + y > 1$ has infinitely many solutions. Each ordered-pair solution is said to **satisfy** the inequality. Thus, $(3, 2)$ satisfies the inequality $x + y > 1$.

1 Graph a linear inequality in two variables.

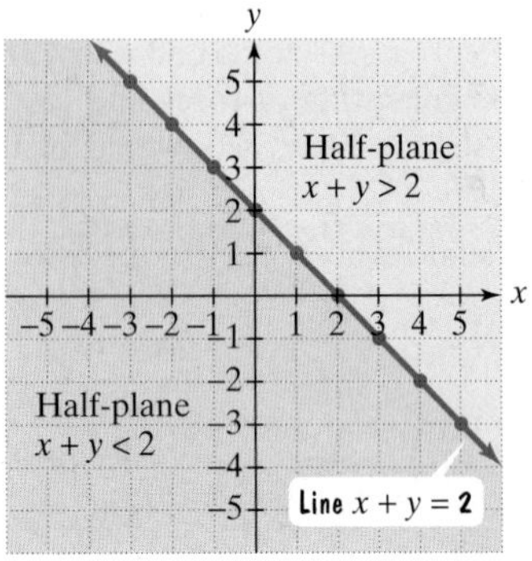

Figure 5.14

The Graph of a Linear Inequality in Two Variables

We know that the graph of an equation in two variables is the set of all points whose coordinates satisfy the equation. Similarly, the **graph of an inequality in two variables** is the set of all points whose coordinates satisfy the inequality.

Let's use Figure 5.14 to get an idea of what the graph of a linear inequality in two variables looks like. Part of the figure shows the graph of the linear equation $x + y = 2$. The line divides the points in the rectangular coordinate system into three sets. First, there is the set of points along the line, satisfying $x + y = 2$. Next, there is the set of points in the green region above the line. Points in the green region satisfy the linear inequality $x + y > 2$. Finally, there is the set of points in the purple region below the line. Points in the purple region satisfy the linear inequality $x + y < 2$.

A **half-plane** is the set of all the points on one side of a line. In Figure 5.14, the green region is a half-plane. The purple region is also a half-plane. A half-plane is the graph of a linear inequality that involves $>$ or $<$. The graph of an inequality that involves $\geq$ or $\leq$ is a half-plane and a line. A solid line is used to show that a line is part of a graph. A dashed line is used to show that a line is not part of a graph.

Graphing a Linear Inequality in Two Variables

1. Replace the inequality symbol with an equal sign and graph the corresponding linear equation. Draw a solid line if the original inequality contains a $\leq$ or $\geq$ symbol. Draw a dashed line if the original inequality contains a $<$ or $>$ symbol.
2. Choose a test point from one of the half-planes. (Do not choose a point on the line.) Substitute the coordinates of the test point into the inequality.
3. If a true statement results, shade the half-plane containing this test point. If a false statement results, shade the half-plane not containing this test point.

EXAMPLE 1 Graphing a Linear Inequality in Two Variables

Graph: $2x - 3y \geq 6$.

Solution

Step 1 Replace the inequality symbol by = and graph the linear equation. We need to graph $2x - 3y = 6$. We can use intercepts to graph this line.

We set $y = 0$ to find the x-intercept.	We set $x = 0$ to find the y-intercept.
$2x - 3y = 6$	$2x - 3y = 6$
$2x - 3 \cdot 0 = 6$	$2 \cdot 0 - 3y = 6$
$2x = 6$	$-3y = 6$
$x = 3$	$y = -2$

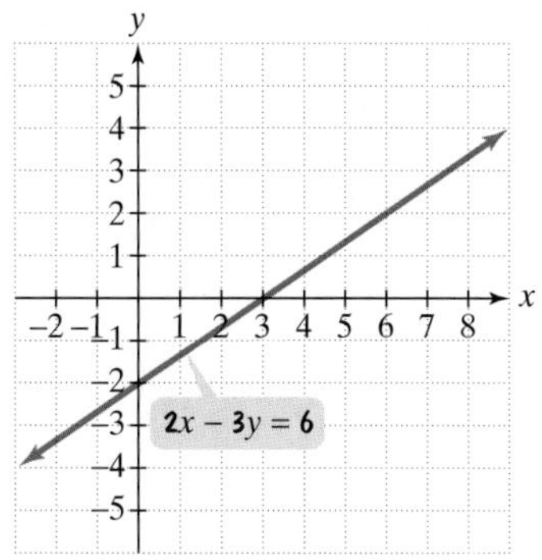

Figure 5.15 Preparing to graph $2x - 3y \geq 6$

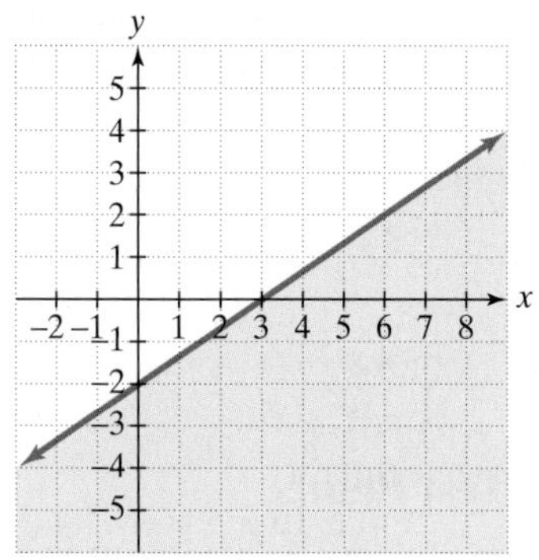

Figure 5.16 The graph of $2x - 3y \geq 6$

The x-intercept is 3, so the line passes through $(3, 0)$. The y-intercept is -2, so the line passes through $(0, -2)$. Using the intercepts, the line is shown in Figure 5.15 as a solid line. This is because the inequality $2x - 3y \geq 6$ contains a $\geq$ symbol, in which equality is included.

Step 2 Choose a test point from one of the half-planes and not from the line. Substitute its coordinates into the inequality. The line $2x - 3y = 6$ divides the plane into three parts—the line itself and two half-planes. The points in one half-plane satisfy $2x - 3y > 6$. The points in the other half-plane satisfy $2x - 3y < 6$. We need to find which half-plane belongs to the solution of $2x - 3y \geq 6$. To do so, we test a point from either half-plane. The origin, $(0, 0)$, is the easiest point to test.

$$2x - 3y \geq 6 \quad \text{This is the given inequality.}$$

$$2 \cdot 0 - 3 \cdot 0 \overset{?}{\geq} 6 \quad \text{Test } (0, 0) \text{ by substituting 0 for } x \text{ and 0 for } y.$$

$$0 - 0 \overset{?}{\geq} 6 \quad \text{Multiply.}$$

$$0 \geq 6 \quad \text{This statement is false.}$$

Step 3 If a false statement results, shade the half-plane not containing the test point. Because 0 is not greater than or equal to 6, the test point, $(0, 0)$, is not part of the solution set. Thus, the half-plane below the solid line $2x - 3y = 6$ is part of the solution set. The solution set is the line and the half-plane that does not contain the point $(0, 0)$, indicated by shading this half-plane. The graph is shown using green shading and a blue line in Figure 5.16.

Check Point 1 Graph: $4x - 2y \geq 8$.

When graphing a linear inequality, test a point that lies in one of the half-planes and *not on the line dividing the half-planes*. The test point $(0, 0)$ is convenient because it is easy to calculate when 0 is substituted for each variable. However, if $(0, 0)$ lies on the dividing line and not in a half-plane, a different test point must be selected.

EXAMPLE 2 Graphing a Linear Inequality in Two Variables

Graph: $y > -\frac{2}{3}x$.

Solution

Step 1 Replace the inequality symbol by = and graph the linear equation. Because we are interested in graphing $y > -\frac{2}{3}x$, we begin by graphing $y = -\frac{2}{3}x$. We can use the slope and the y-intercept to graph this linear function.

$$y = -\frac{2}{3}x + 0$$

Slope = $\frac{-2}{3} = \frac{\text{rise}}{\text{run}}$ y-intercept = 0

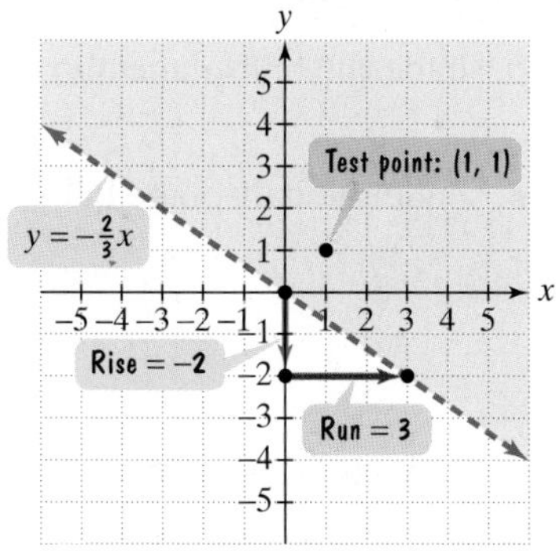

Figure 5.17 The graph of $y > -\frac{2}{3}x$

The y-intercept is 0, so the line passes through $(0, 0)$. Using the y-intercept and the slope, the line is shown in Figure 5.17 as a dashed line. This is because the inequality $y > -\frac{2}{3}x$ contains a $>$ symbol, in which equality is not included.

Step 2 Choose a test point from one of the half-planes and not from the line. Substitute its coordinates into the inequality. We cannot use $(0, 0)$ as a test point because it lies on the line and not in a half-plane. Let's use $(1, 1)$, which lies in the half-plane above the line.

$$y > -\frac{2}{3}x \quad \text{This is the given inequality.}$$

$$1 \overset{?}{>} -\frac{2}{3} \cdot 1 \quad \text{Test } (1, 1) \text{ by substituting 1 for } x \text{ and 1 for } y.$$

$$1 > -\frac{2}{3} \quad \text{This statement is true.}$$

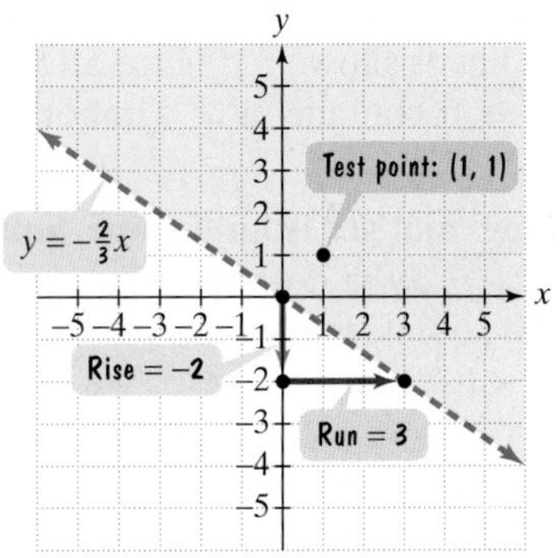

Figure 5.17 (repeated) The graph of $y > -\frac{2}{3}x$

Step 3 If a true statement results, shade the half-plane containing the test point. Because 1 is greater than $-\frac{2}{3}$, the test point $(1, 1)$ is part of the solution set. All the points on the same side of the line $y = -\frac{2}{3}x$ as the point $(1, 1)$ are members of the solution set. The solution set is the half-plane that contains the point $(1, 1)$, indicated by shading this half-plane. The graph is shown using green shading and a dashed blue line in Figure 5.17.

Technology

Most graphing utilities can graph inequalities in two variables with the SHADE feature. The procedure varies by model, so consult your manual. For most graphing utilities, you must first solve for y if it is not already isolated. The figure shows the graph of $y > -\frac{2}{3}x$. Most displays do not distinguish between dashed and solid boundary lines.

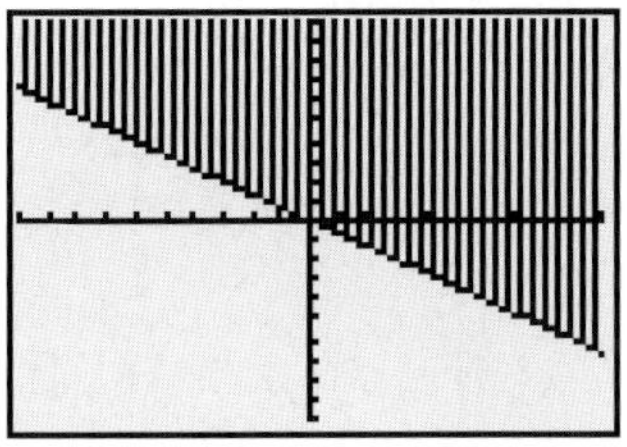

Check Point 2 Graph: $y > -\frac{3}{4}x$.

Graphing Linear Inequalities without Using Test Points

You can graph inequalities in the form $y > mx + b$ or $y < mx + b$ without using test points. The inequality symbol indicates which half-plane to shade.

- If $y > mx + b$, shade the half-plane above the line $y = mx + b$.
- If $y < mx + b$, shade the half-plane below the line $y = mx + b$.

It is also not necessary to use test points when graphing inequalities involving half-planes on one side of a vertical or a horizontal line.

For the Vertical Line $x = a$:

- If $x > a$, shade the half-plane to the right of $x = a$.
- If $x < a$, shade the half-plane to the left of $x = a$.

For the Horizontal Line $y = b$:

- If $y > b$, shade the half-plane above $y = b$.
- If $y < b$, shade the half-plane below $y = b$.

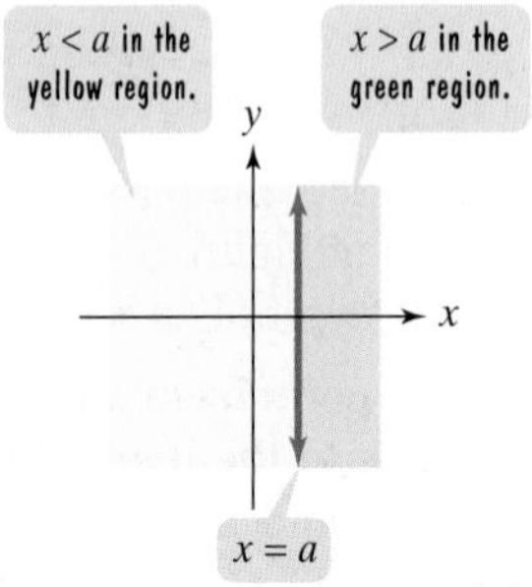

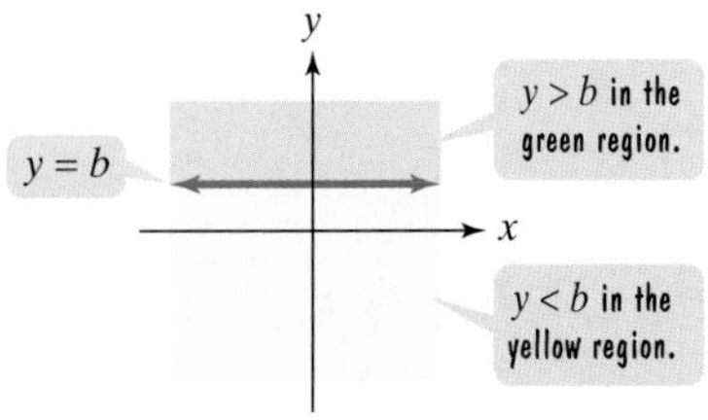

Study Tip

Continue using test points to graph inequalities in the form $Ax + By > C$ or $Ax + By < C$. The graph of $Ax + By > C$ can lie above or below the line of $Ax + By = C$, depending on the value of B. The same comment applies to the graph of $Ax + By < C$.

EXAMPLE 3 Graphing Inequalities without Using Test Points

Graph each inequality in a rectangular coordinate system:

a. $y \leq -3$ **b.** $x > 2$.

Solution

a. $y \leq -3$

Graph $y = -3$, a horizontal line with y-intercept -3. The line is solid because equality is included in $y \leq -3$. Because of the less than part of $\leq$, shade the half-plane below the horizontal line.

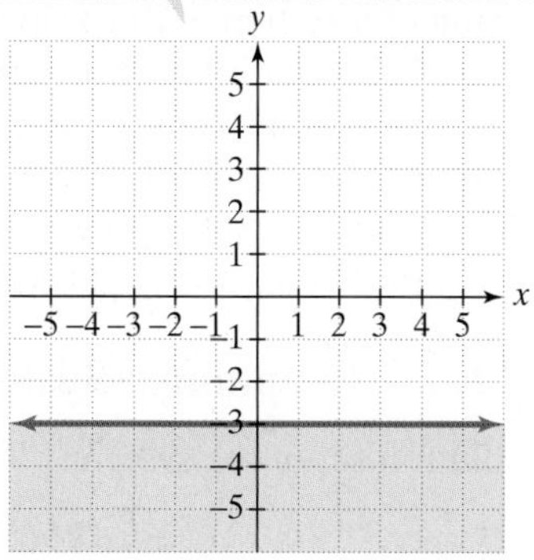

b. $x > 2$

Graph $x = 2$, a vertical line with x-intercept 2. The line is dashed because equality is not included in $x > 2$. Because of $>$, the greater than symbol, shade the half-plane to the right of the vertical line.

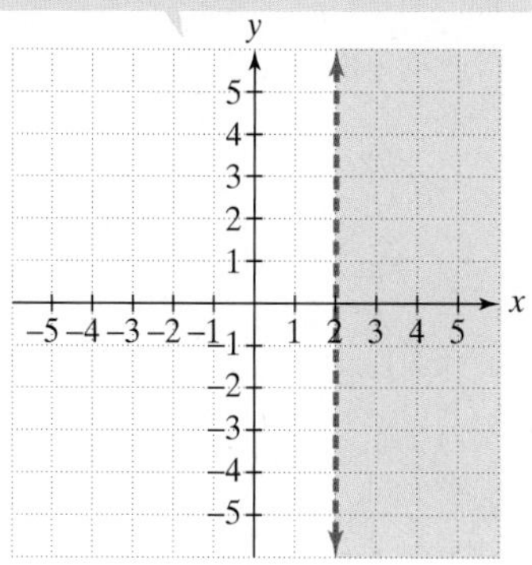

Check Point 3 Graph each inequality in a rectangular coordinate system:

a. $y > 1$ **b.** $x \leq -2$.

2 Graph a nonlinear inequality in two variables.

Graphing a Nonlinear Inequality in Two Variables

Example 4 illustrates that a nonlinear inequality in two variables is graphed in the same way that we graph a linear inequality.

EXAMPLE 4 Graphing a Nonlinear Inequality in Two Variables

Graph: $x^2 + y^2 \leq 9$.

Solution

Step 1 Replace the inequality symbol with = and graph the nonlinear equation. We need to graph $x^2 + y^2 = 9$. The graph is a circle of radius 3 with its center at the origin. The graph is shown in Figure 5.18 as a solid circle because equality is included in the $\leq$ symbol.

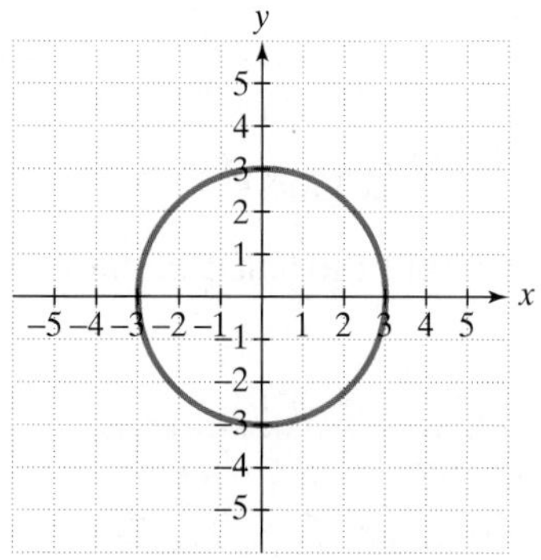

Figure 5.18 Preparing to graph $x^2 + y^2 \leq 9$

Step 2 Choose a test point from one of the regions and not from the circle. Substitute its coordinates into the inequality. The circle divides the plane into three parts—the circle itself, the region inside the circle, and the region outside the circle. We need to determine whether the region inside or outside the circle is included in the solution. To do so, we will use the test point (0, 0) from inside the circle.

$x^2 + y^2 \leq 9$	This is the given inequality.
$0^2 + 0^2 \overset{?}{\leq} 9$	Test (0, 0) by substituting 0 for x and 0 for y.
$0 + 0 \overset{?}{\leq} 9$	Square 0: $0^2 = 0$.
$0 \leq 9$	Add. This statement is true.

Step 3 If a true statement results, shade the region containing the test point. The true statement tells us that all the points inside the circle satisfy $x^2 + y^2 \leq 9$. The graph is shown using green shading and a solid blue circle in Figure 5.19.

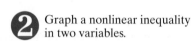

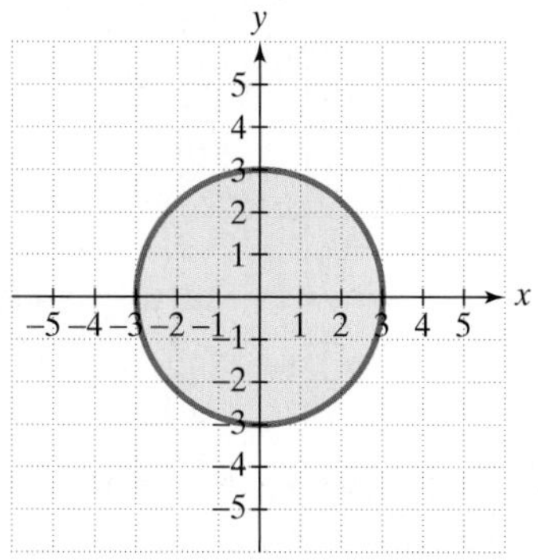

Figure 5.19 The graph of $x^2 + y^2 \leq 9$

Check Point 4 Graph: $x^2 + y^2 \geq 16$.

3 Graph a system of inequalities.

Systems of Inequalities in Two Variables

Just as two linear equations make up a system of linear equations, two or more linear inequalities make up a **system of linear inequalities**. Here is an example of a system of linear inequalities:

$$x - y < 1$$
$$2x + 3y \geq 12.$$

A **solution of a system of linear inequalities** in two variables is an ordered pair that satisfies each inequality in the system. The set of all such ordered pairs is the **solution set** of the system. Thus, to graph a system of inequalities in two variables, begin by graphing each individual inequality in the same rectangular coordinate system. Then find the region, if there is one, that is common to every graph in the system. This region of intersection gives a picture of the system's solution set.

EXAMPLE 5 Graphing a System of Linear Inequalities

Graph the solution set of the system:

$$x - y < 1$$
$$2x + 3y \geq 12.$$

Solution Replacing each inequality symbol with an equal sign indicates that we need to graph $x - y = 1$ and $2x + 3y = 12$. We can use intercepts to graph these lines.

$x - y = 1$		$2x + 3y = 12$

x-intercept: $x - 0 = 1$ [Set $y = 0$ in each equation.] x-intercept: $2x + 3 \cdot 0 = 12$
$x = 1$ $2x = 12$
The line passes through $(1, 0)$. $x = 6$
The line passes through $(6, 0)$.

y-intercept: $0 - y = 1$ [Set $x = 0$ in each equation.] y-intercept: $2 \cdot 0 + 3y = 12$
$-y = 1$ $3y = 12$
$y = -1$ $y = 4$
The line passes through $(0, -1)$.   The line passes through $(0, 4)$.

Now we are ready to graph the solution set of the system of linear inequalities.

Graph $x - y < 1$. The blue line, $x - y = 1$, is dashed: Equality is not included in $x - y < 1$. Because $(0, 0)$ makes the inequality true $(0 - 0 < 1$, or $0 < 1$, is true), shade the half-plane containing $(0, 0)$ in yellow.

Add the graph of $2x + 3y \geq 12$. The red line, $2x + 3y = 12$, is solid: Equality is included in $2x + 3y \geq 12$. Because $(0, 0)$ makes the inequality false $(2 \cdot 0 + 3 \cdot 0 \geq 12$, or $0 \geq 12$, is false), shade the half-plane not containing $(0, 0)$ using green vertical shading.

The solution set of the system is graphed as the intersection (the overlap) of the two half-planes. This is the region in which the yellow shading and the green vertical shading overlap.

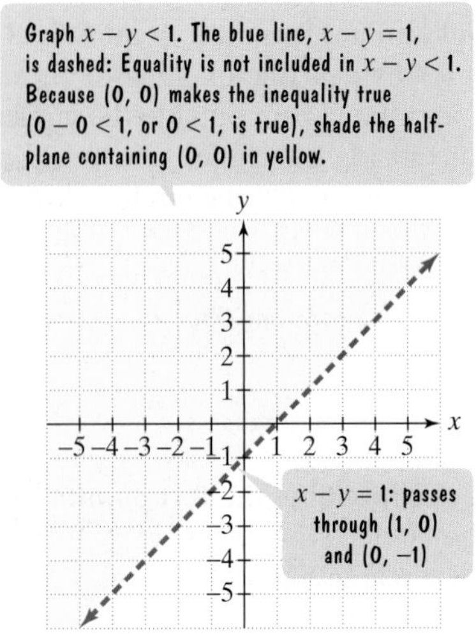

The graph of $x - y < 1$

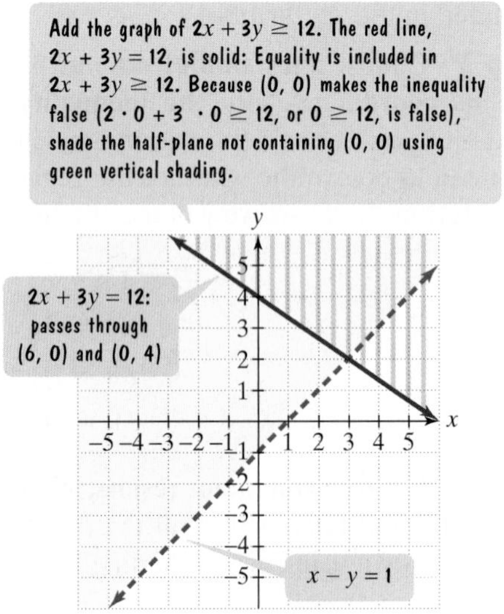

Adding the graph of $2x + 3y \geq 12$

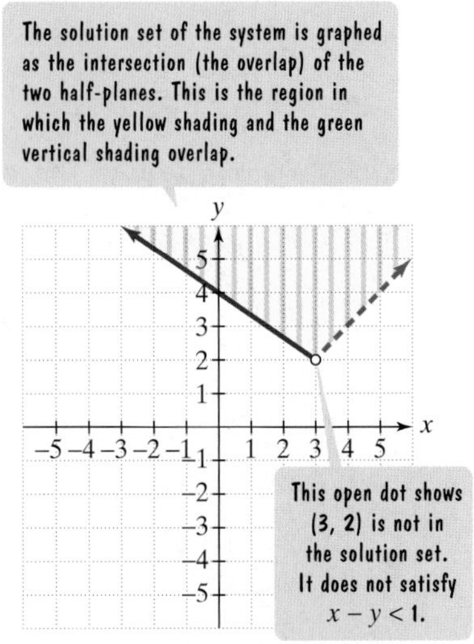

The graph of $x - y < 1$ and $2x + 3y \geq 12$

Check Point 5 Graph the solution set of the system:

$$x - 3y < 6$$
$$2x + 3y \geq -6.$$

EXAMPLE 6 Graphing a System of Inequalities

Graph the solution set of the system:

$$y \geq x^2 - 4$$
$$x - y \geq 2.$$

Solution We begin by graphing $y \geq x^2 - 4$. Because equality is included in $\geq$, we graph $y = x^2 - 4$ as a solid parabola. Because $(0, 0)$ makes the inequality $y \geq x^2 - 4$ true (we obtain $0 \geq -4$), we shade the interior portion of the parabola containing $(0, 0)$, shown in yellow in Figure 5.20.

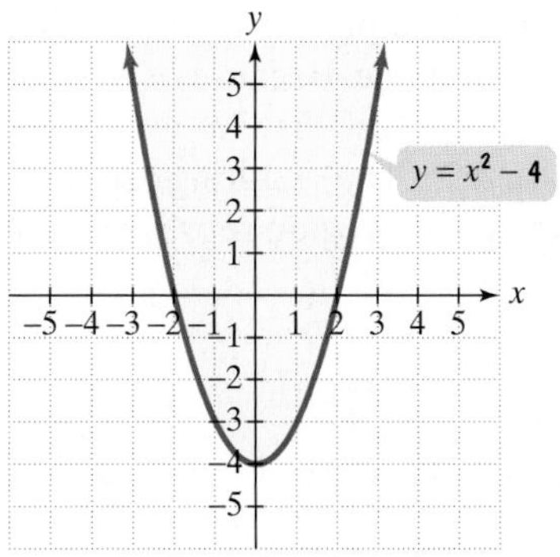

Figure 5.20 The graph of $y \geq x^2 - 4$

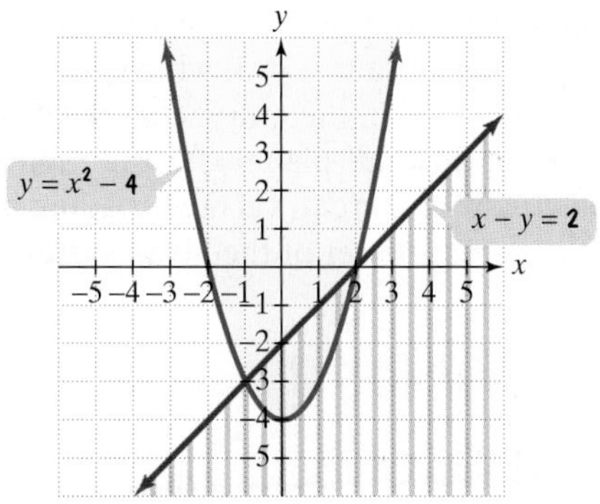

Figure 5.21 Adding the graph of $x - y \geq 2$

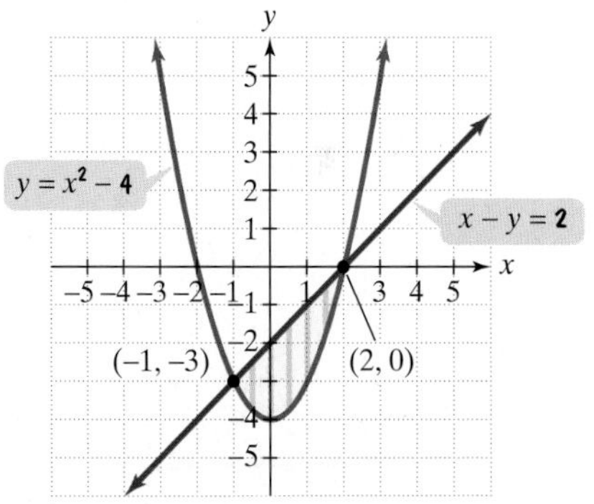

Figure 5.22 The graph of $y \geq x^2 - 4$ and $x - y \geq 2$

Now we graph $x - y \geq 2$ in the same rectangular coordinate system. First we graph the line $x - y = 2$ using its x-intercept, 2, and its y-intercept, -2. Because $(0, 0)$ makes the inequality $x - y \geq 2$ false (we obtain $0 \geq 2$), we shade the half-plane below the line. This is shown in Figure 5.21 using green vertical shading.

The solution of the system is shown in Figure 5.22 by the intersection (the overlap) of the solid yellow and green vertical shadings. The graph of the system's solution set consists of the region enclosed by the parabola and the line. To find the points of intersection of the parabola and the line, use the substitution method to solve the nonlinear system

$$y = x^2 - 4$$
$$x - y = 2.$$

Take a moment to show that the solutions are $(-1, -3)$ and $(2, 0)$, as shown in Figure 5.22.

Check Point 6 Graph the solution set of the system:

$$y \geq x^2 - 4$$
$$x + y \leq 2.$$

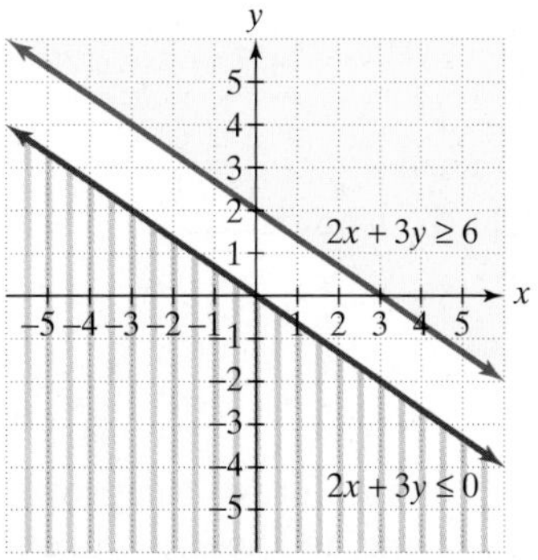

Figure 5.23 A system of inequalities with no solution

A system of inequalities has no solution if there are no points in the rectangular coordinate system that simultaneously satisfy each inequality in the system. For example, the system

$$2x + 3y \geq 6$$
$$2x + 3y \leq 0,$$

whose separate graphs are shown in Figure 5.23, has no overlapping region. Thus, the system has no solution. The solution set is $\varnothing$, the empty set.

EXAMPLE 7 Graphing a System of Inequalities

Graph the solution set of the system:

$$\begin{aligned} x - y &< 2 \\ -2 \leq x &< 4 \\ y &< 3. \end{aligned}$$

Solution We begin by graphing $x - y < 2$, the first given inequality. The line $x - y = 2$ has an x-intercept of 2 and a y-intercept of -2. The test point $(0, 0)$ makes the inequality $x - y < 2$ true, and its graph is shown in Figure 5.24.

Now, let's consider the second given inequality, $-2 \leq x < 4$. Replacing the inequality symbols by $=$, we obtain $x = -2$ and $x = 4$, graphed as red vertical lines in Figure 5.25. The line of $x = 4$ is not included. Using $(0, 0)$ as a test point and substituting the x-coordinate, 0, into $-2 \leq x < 4$, we obtain the true statement $-2 \leq 0 < 4$. We therefore shade the region between the vertical lines. We must intersect this region with the yellow region in Figure 5.24. The resulting region is shown in yellow and green vertical shading in Figure 5.25.

Finally, let's consider the third given inequality, $y < 3$. Replacing the inequality symbol by $=$, we obtain $y = 3$, which graphs as a horizontal line. Because of the less than symbol in $y < 3$, the graph consists of the half-plane below the line $y = 3$. We must intersect this half-plane with the region in Figure 5.25. The resulting region is shown in yellow and green vertical shading in Figure 5.26. This region represents the graph of the solution set of the given system.

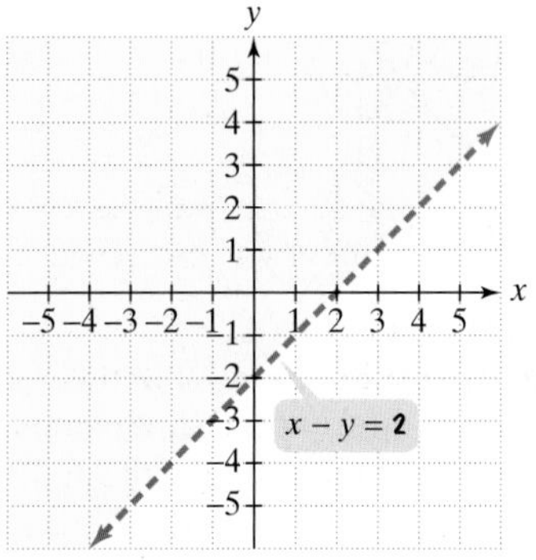

Figure 5.24 The graph of $x - y < 2$

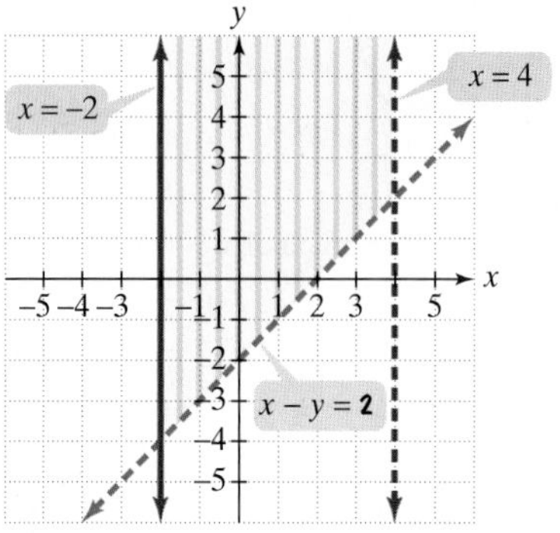

Figure 5.25 The graph of $x - y < 2$ and $-2 \leq x < 4$

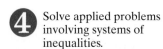

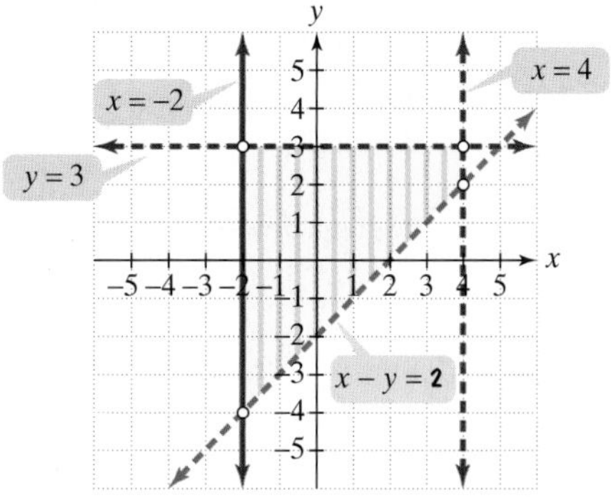

Figure 5.26 The graph of $x - y < 2$ and $-2 \leq x < 4$ and $y < 3$

In Figure 5.26, it may be difficult to tell where the graph of $x - y = 2$ intersects the vertical line $x = 4$. Using the substitution method, it can be determined that this intersection point is $(4, 2)$. Take a moment to verify that the four intersection points in Figure 5.26 are, clockwise from upper left, $(-2, 3)$, $(4, 3)$, $(4, 2)$, and $(-2, -4)$. These points are shown as open dots because none satisfies all three of the system's inequalities.

Check Point 7 Graph the solution set of the system:

$$\begin{aligned} x + y &< 2 \\ -2 \leq x &< 1 \\ y &> -3. \end{aligned}$$

4 Solve applied problems involving systems of inequalities.

Applications

Temperature and precipitation affect whether or not trees and forests can grow. At certain levels of precipitation and temperature, only grasslands and deserts will exist. Figure 5.27 shows three kinds of regions—deserts, grasslands, and forests—that result from various ranges of temperature, T, and precipitation, P.

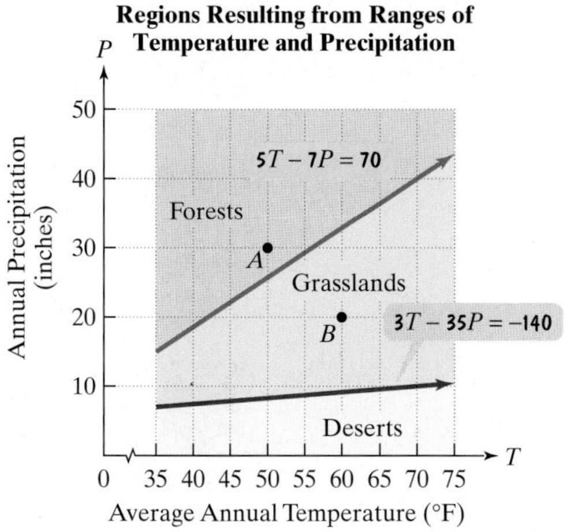

Figure 5.27

Source: A. Miller and J. Thompson, *Elements of Meteorology*

Systems of inequalities can be used to describe where forests, grasslands, and deserts occur. Because these regions occur when the average annual temperature, T, is 35°F or greater, each system contains the inequality $T \geq 35$.

Forests occur if	**Grasslands occur if**	**Deserts occur if**
$T \geq 35$	$T \geq 35$	$T \geq 35$
$5T - 7P < 70.$	$5T - 7P \geq 70$	$3T - 35P > -140.$
	$3T - 35P \leq -140.$	

EXAMPLE 8 Forests and Systems of Inequalities

Show that point A in Figure 5.27 is a solution of the system of inequalities that describes where forests occur.

Solution Point A has coordinates (50, 30). This means that if a region has an average annual temperature of 50°F and an average annual precipitation of 30 inches, a forest occurs. We can show that (50, 30) satisfies the system of inequalities for forests by substituting 50 for T and 30 for P in each inequality in the system.

$$T \geq 35 \qquad 5T - 7P < 70$$

$$50 \geq 35, \text{ true} \qquad 5 \cdot 50 - 7 \cdot 30 \stackrel{?}{<} 70$$

$$250 - 210 \stackrel{?}{<} 70$$

$$40 < 70, \text{ true}$$

The coordinates (50, 30) make each inequality true. Thus, (50, 30) satisfies the system for forests.

Check Point 8 Show that point B in Figure 5.27 is a solution of the system of inequalities that describes where grasslands occur.

EXERCISE SET 5.5

Practice Exercises

In Exercises 1–26, graph each inequality.

1. $x + 2y \leq 8$

2. $3x - 6y \leq 12$

3. $x - 2y > 10$

4. $2x - y > 4$

5. $y \leq \frac{1}{3}x$

6. $y \leq \frac{1}{4}x$

7. $y > 2x - 1$

8. $y > 3x + 2$

9. $x \leq 1$

10. $x \leq -3$

11. $y > 1$

12. $y > -3$

13. $x^2 + y^2 \le 1$

14. $x^2 + y^2 \le 4$

15. $x^2 + y^2 > 25$

16. $x^2 + y^2 > 36$

17. $(x - 2)^2 + (y + 1)^2 < 9$

18. $(x + 2)^2 + (y - 1)^2 < 16$

19. $y < x^2 - 1$

20. $y < x^2 - 9$

21. $y \ge x^2 - 9$

22. $y \ge x^2 - 1$

23. $y > 2^x$

24. $y \le 3^x$

25. $y \ge \log_2(x + 1)$

26. $y \ge \log_3(x - 1)$

In Exercises 27–62, graph the solution set of each system of inequalities or indicate that the system has no solution.

27. $3x + 6y \le 6$, $2x + y \le 8$

28. $x - y \ge 4$, $x + y \le 6$

29. $2x - 5y \le 10$, $3x - 2y > 6$

30. $2x - y \le 4$, $3x + 2y > -6$

31. $y > 2x - 3$, $y < -x + 6$

32. $y < -2x + 4$, $y < x - 4$

33. $x + 2y \le 4$, $y \ge x - 3$

34. $x + y \le 4$, $y \ge 2x - 4$

35. $x \le 2$, $y \ge -1$

36. $x \le 3$, $y \le -1$

37. $-2 \le x < 5$

38. $-2 < y \le 5$

39. $x - y \le 1$, $x \ge 2$

40. $4x - 5y \ge -20$, $x \ge -3$

41. $x + y > 4$, $x + y < -1$

42. $x + y > 3$, $x + y < -2$

43. $x + y > 4$, $x + y > -1$

44. $x + y > 3$, $x + y > -2$

45. $y \ge x^2 - 1$, $x - y \ge -1$

46. $y \ge x^2 - 4$, $x - y \ge 2$

47. $x^2 + y^2 \le 16$, $x + y > 2$

48. $x^2 + y^2 \le 4$, $x + y > 1$

49. $x^2 + y^2 > 1$, $x^2 + y^2 < 4$

50. $x^2 + y^2 > 1$, $x^2 + y^2 < 9$

51. $(x - 1)^2 + (y + 1)^2 < 25$, $(x - 1)^2 + (y + 1)^2 \ge 16$

52. $(x + 1)^2 + (y - 1)^2 < 16$, $(x + 1)^2 + (y - 1)^2 \ge 4$

53. $x^2 + y^2 \le 1$, $y - x^2 > 0$

54. $x^2 + y^2 < 4$, $y - x^2 \ge 0$

55. $x^2 + y^2 < 16$, $y \ge 2^x$

56. $x^2 + y^2 \le 16$, $y < 2^x$

57. $x - y \le 2$, $x > -2$, $y \le 3$

58. $3x + y \le 6$, $x > -2$, $y \le 4$

59. $x \ge 0$, $y \ge 0$, $2x + 5y < 10$, $3x + 4y \le 12$

60. $x \ge 0$, $y \ge 0$, $2x + y < 4$, $2x - 3y \le 6$

61. $3x + y \le 6$, $2x - y \le -1$, $x > -2$, $y < 4$

62. $2x + y \le 6$, $x + y > 2$, $1 \le x \le 2$, $y < 3$

Practice Plus

In Exercises 63–64, write each sentence as an inequality in two variables. Then graph the inequality.

63. The y-variable is at least 4 more than the product of -2 and the x-variable.

64. The y-variable is at least 2 more than the product of -3 and the x-variable.

In Exercises 65–68, write the given sentences as a system of inequalities in two variables. Then graph the system.

65. The sum of the x-variable and the y-variable is at most 4. The y-variable added to the product of 3 and the x-variable does not exceed 6.

66. The sum of the x-variable and the y-variable is at most 3. The y-variable added to the product of 4 and the x-variable does not exceed 6.

67. The sum of the x-variable and the y-variable is no more than 2. The y-variable is no less than the difference between the square of the x-variable and 4.

68. The sum of the squares of the x-variable and the y-variable is no more than 25. The sum of twice the y-variable and the x-variable is no less than 5.

In Exercises 69–70, rewrite each inequality in the system without absolute value bars. Then graph the rewritten system in rectangular coordinates.

69. $|x| \le 2$, $|y| \le 3$

70. $|x| \le 1$, $|y| \le 2$

The graphs of solution sets of systems of inequalities involve finding the intersection of the solution sets of two or more inequalities. By contrast, in Exercises 71–72, you will be graphing the ***union*** *of the solution sets of two inequalities.*

71. Graph the union of $y > \frac{3}{2}x - 2$ and $y < 4$.

72. Graph the union of $x - y \ge -1$ and $5x - 2y \le 10$.

Without graphing, in Exercises 73–76, determine if each system has no solution or infinitely many solutions.

73. $3x + y < 9$, $3x + y > 9$

74. $6x - y \le 24$, $6x - y > 24$

75. $(x + 4)^2 + (y - 3)^2 \le 9$, $(x + 4)^2 + (y - 3)^2 \ge 9$

76. $(x - 4)^2 + (y + 3)^2 \le 24$, $(x - 4)^2 + (y + 3)^2 \ge 24$

Application Exercises

Maximum heart rate, H, in beats per minute is a function of age, a, modeled by the formula

$$H = 220 - a,$$

where $10 \le a \le 70$. The bar graph shows the target heart rate ranges as a percentage of the maximum heart rate for four types of exercise goals.

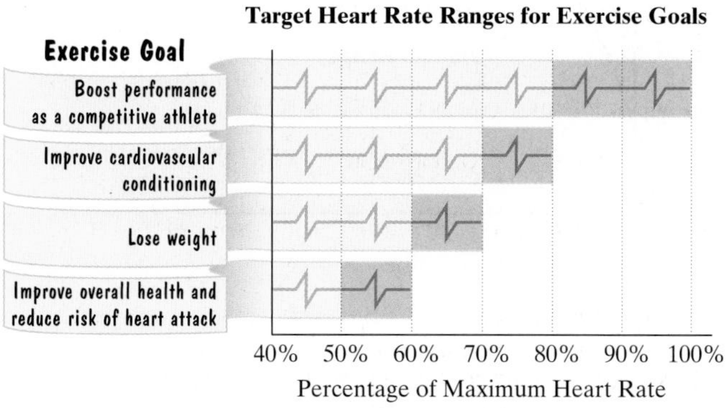

Source: Vitality

In Exercises 77–80, systems of inequalities will be used to model three of the target heart rate ranges shown in the bar graph. We begin with the target heart rate range for cardiovascular conditioning, modeled by the following system of inequalities:

$10 \le a \le 70$ — Heart rate ranges apply to ages 10 through 70, inclusive.

$H \ge 0.7(220 - a)$ — Target heart rate range is greater than or equal to 70% of maximum heart rate

$H \le 0.8(220 - a)$ — and less than or equal to 80% of maximum heart rate.

The graph of this system is shown in the figure. Use the graph to solve Exercises 77–78.

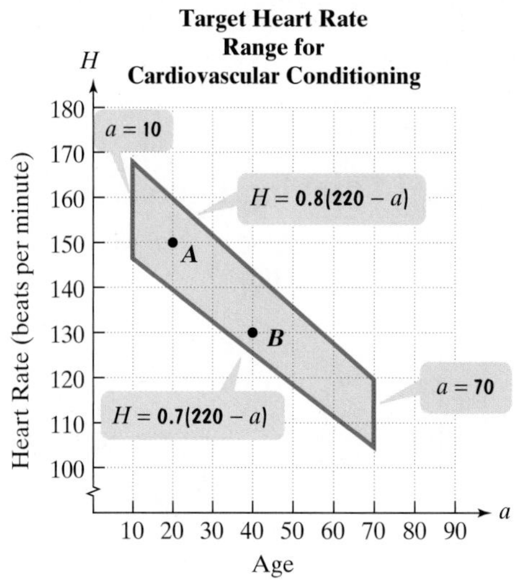

77. **a.** What are the coordinates of point A and what does this mean in terms of age and heart rate?

b. Show that point A is a solution of the system of inequalities.

78. **a.** What are the coordinates of point B and what does this mean in terms of age and heart rate?

b. Show that point B is a solution of the system of inequalities.

79. Write a system of inequalities that models the target heart rate range for the goal of losing weight.

80. Write a system of inequalities that models the target heart rate range for improving overall health.

81. Many elevators have a capacity of 2000 pounds.

a. If a child averages 50 pounds and an adult 150 pounds, write an inequality that describes when x children and y adults will cause the elevator to be overloaded.

b. Graph the inequality. Because x and y must be positive, limit the graph to quadrant I only.

c. Select an ordered pair satisfying the inequality. What are its coordinates and what do they represent in this situation?

82. A patient is not allowed to have more than 330 milligrams of cholesterol per day from a diet of eggs and meat. Each egg provides 165 milligrams of cholesterol. Each ounce of meat provides 110 milligrams.

a. Write an inequality that describes the patient's dietary restrictions for x eggs and y ounces of meat.

b. Graph the inequality. Because x and y must be positive, limit the graph to quadrant I only.

c. Select an ordered pair satisfying the inequality. What are its coordinates and what do they represent in this situation?

83. On your next vacation, you will divide lodging between large resorts and small inns. Let x represent the number of nights spent in large resorts. Let y represent the number of nights spent in small inns.

a. Write a system of inequalities that models the following conditions:

> You want to stay at least 5 nights. At least one night should be spent at a large resort. Large resorts average $200 per night and small inns average $100 per night. Your budget permits no more than $700 for lodging.

b. Graph the solution set of the system of inequalities in part (a).

c. Based on your graph in part (b), what is the greatest number of nights you could spend at a large resort and still stay within your budget?

84. A person with no more than $15,000 to invest plans to place the money in two investments. One investment is high risk, high yield; the other is low risk, low yield. At least $2000 is to be placed in the high-risk investment. Furthermore, the amount invested at low risk should be at least three times the amount invested at high risk. Find and graph a system of inequalities that describes all possibilities for placing the money in the high- and low-risk investments.

The graph of an inequality in two variables is a region in the rectangular coordinate system. Regions in coordinate systems have numerous applications. For example, the regions in the following two graphs indicate whether a person is obese, overweight, borderline overweight, normal weight, or underweight.

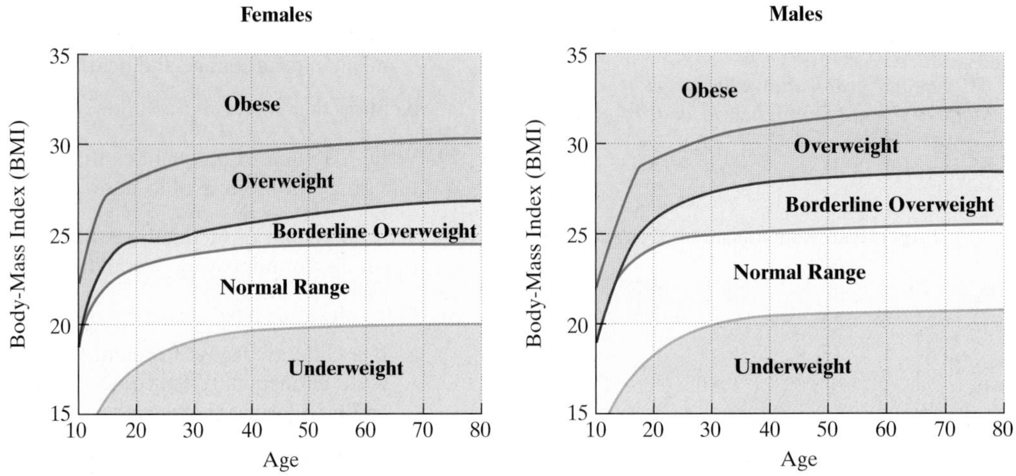

Source: Centers for Disease Control and Prevention

The horizontal axis shows a person's age. The vertical axis shows that person's body-mass index (BMI), computed using the following formula:

$$\text{BMI} = \frac{703W}{H^2}.$$

The variable W represents weight, in pounds. The variable H represents height, in inches. Use this information to solve Exercises 85–86.

85. A man is 20 years old, 72 inches (6 feet) tall, and weighs 200 pounds.

a. Compute the man's BMI. Round to the nearest tenth.

b. Use the man's age and his BMI to locate this information as a point in the coordinate system for males. Is this person obese, overweight, borderline overweight, normal weight, or underweight?

86. A woman is 25 years old, 66 inches (5 feet, 6 inches) tall, and weighs 105 pounds.

a. Compute the woman's BMI. Round to the nearest tenth.

b. Use the woman's age and her BMI to locate this information as a point in the coordinate system for females. Is this person obese, overweight, borderline overweight, normal weight, or underweight?

Writing in Mathematics

87. What is a linear inequality in two variables? Provide an example with your description.

88. How do you determine if an ordered pair is a solution of an inequality in two variables, x and y?

89. What is a half-plane?

90. What does a solid line mean in the graph of an inequality?

91. What does a dashed line mean in the graph of an inequality?

92. Compare the graphs of $3x - 2y > 6$ and $3x - 2y \le 6$. Discuss similarities and differences between the graphs.

93. What is a system of linear inequalities?

94. What is a solution of a system of linear inequalities?

95. Explain how to graph the solution set of a system of inequalities.

96. What does it mean if a system of linear inequalities has no solution?

Technology Exercises

Graphing utilities can be used to shade regions in the rectangular coordinate system, thereby graphing an inequality in two variables. Read the section of the user's manual for your graphing utility that describes how to shade a region. Then use your graphing utility to graph the inequalities in Exercises 97–102.

97. $y \le 4x + 4$

98. $y \ge \frac{2}{3}x - 2$

99. $y \ge x^2 - 4$

100. $y \ge \frac{1}{2}x^2 - 2$

101. $2x + y \le 6$

102. $3x - 2y \ge 6$

103. Does your graphing utility have any limitations in terms of graphing inequalities? If so, what are they?

104. Use a graphing utility with a SHADE feature to verify any five of the graphs that you drew by hand in Exercises 1–26.

105. Use a graphing utility with a SHADE feature to verify any five of the graphs that you drew by hand for the systems in Exercises 27–62.

Critical Thinking Exercises

In Exercises 106–109, write a system of inequalities for each graph.

106.

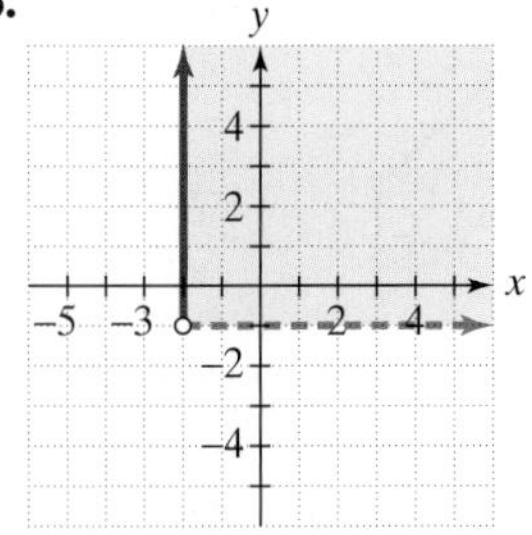

107.

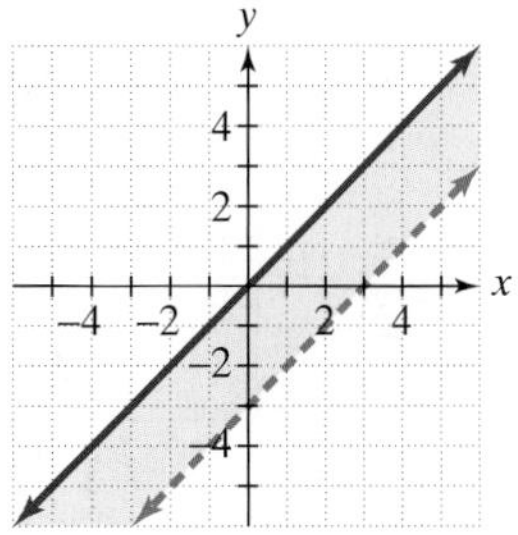

108.

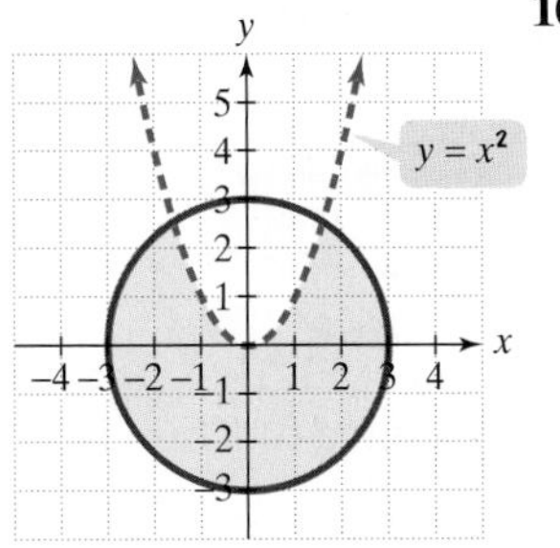

109.

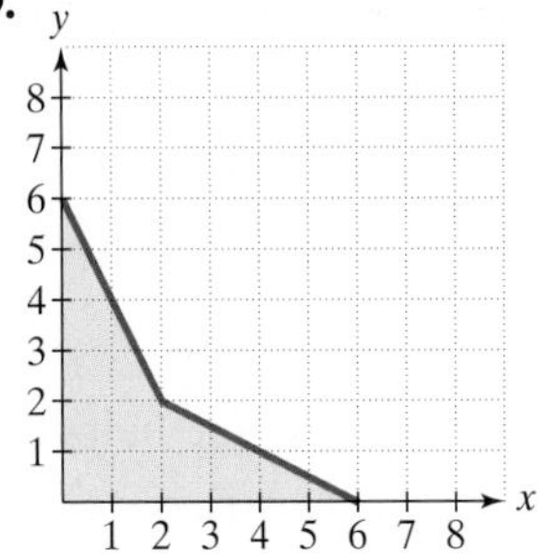

110. Write a system of inequalities whose solution set includes every point in the rectangular coordinate system.

111. Sketch the graph of the solution set for the following system of inequalities:

$$y \geq nx + b\ (n < 0, b > 0)$$
$$y \leq mx + b\ (m > 0, b > 0).$$

SECTION 5.6 Linear Programming

Objectives

1. Write an objective function describing a quantity that must be maximized or minimized.
2. Use inequalities to describe limitations in a situation.
3. Use linear programming to solve problems.

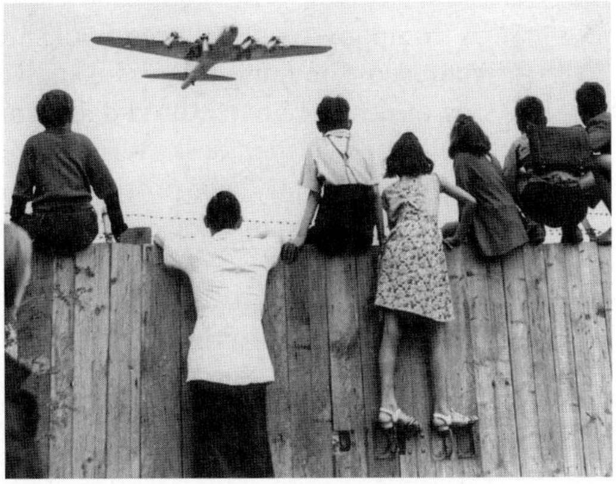

West Berlin children at Tempelhof airport watch fleets of U.S. airplanes bringing in supplies to circumvent the Soviet blockade. The airlift began June 28, 1948 and continued for 15 months.

The Berlin Airlift (1948–1949) was an operation by the United States and Great Britain in response to military action by the former Soviet Union: Soviet troops closed all roads and rail lines between West Germany and Berlin, cutting off supply routes to the city. The Allies used a mathematical technique developed during World War II to maximize the amount of supplies transported. During the 15-month airlift, 278,228 flights provided basic necessities to blockaded Berlin, saving one of the world's great cities.

In this section, we will look at an important application of systems of linear inequalities. Such systems arise in **linear programming**, a method for solving problems in which a particular quantity that must be maximized or minimized is limited by other factors. Linear programming is one of the most widely used tools in management science. It helps businesses allocate resources to manufacture products in a way that will maximize profit. Linear programming accounts for more than 50% and perhaps as much as 90% of all computing time used for management decisions in business. The Allies used linear programming to save Berlin.

1 Write an objective function describing a quantity that must be maximized or minimized.

Objective Functions in Linear Programming

Many problems involve quantities that must be maximized or minimized. Businesses are interested in maximizing profit. An operation in which bottled water and medical kits are shipped to earthquake victims needs to maximize the number of victims helped by this shipment. An **objective function** is an algebraic expression in two or more variables describing a quantity that must be maximized or minimized.

EXAMPLE 1 Writing an Objective Function

Bottled water and medical supplies are to be shipped to victims of an earthquake by plane. Each container of bottled water will serve 10 people and each medical kit will aid 6 people. If x represents the number of bottles of water to be shipped and y represents the number of medical kits, write the objective function that describes the number of people who can be helped.

Solution Because each bottle of water serves 10 people and each medical kit aids 6 people, we have

The number of people helped	is	10 times the number of bottles of water	plus	6 times the number of medical kits.
	$=$	$10x$	$+$	$6y$.

Using z to represent the number of people helped, the objective function is

$$z = 10x + 6y.$$

Unlike the functions that we have seen so far, the objective function is an equation in three variables. For a value of x and a value of y, there is one and only one value of z. Thus, z is a function of x and y.

Check Point 1 A company manufactures bookshelves and desks for computers. Let x represent the number of bookshelves manufactured daily and y the number of desks manufactured daily. The company's profits are \$25 per bookshelf and \$55 per desk. Write the objective function that describes the company's total daily profit, z, from x bookshelves and y desks. (Check Points 2 through 4 are related to this situation, so keep track of your answers.)

2 Use inequalities to describe limitations in a situation.

Constraints in Linear Programming

Ideally, the number of earthquake victims helped in Example 1 should increase without restriction so that every victim receives water and medical kits. However, the planes that ship these supplies are subject to weight and volume restrictions. In linear programming problems, such restrictions are called **constraints**. Each constraint is expressed as a linear inequality. The list of constraints forms a system of linear inequalities.

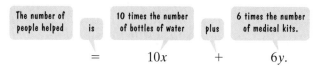

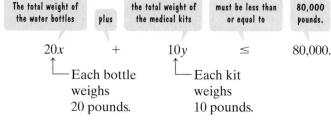

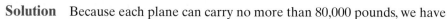

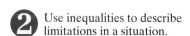

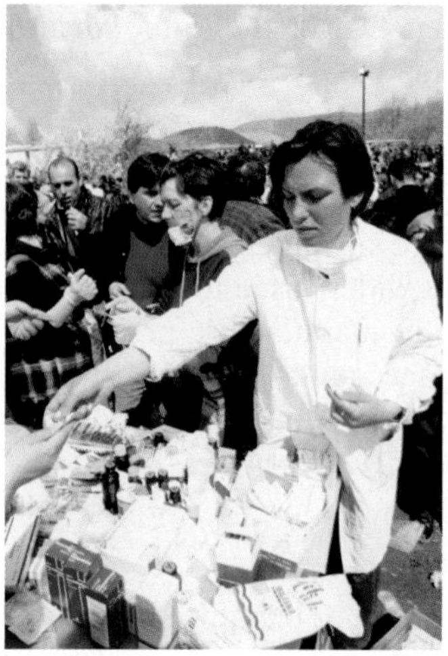

EXAMPLE 2 Writing a Constraint

Each plane can carry no more than 80,000 pounds. The bottled water weighs 20 pounds per container and each medical kit weighs 10 pounds. Let x represent the number of bottles of water to be shipped and y the number of medical kits. Write an inequality that describes this constraint.

Solution Because each plane can carry no more than 80,000 pounds, we have

The total weight of the water bottles	plus	the total weight of the medical kits	must be less than or equal to	80,000 pounds.
$20x$	$+$	$10y$	$\leq$	80,000.
↑ Each bottle weighs 20 pounds.		↑ Each kit weighs 10 pounds.		

The plane's weight constraint is described by the inequality

$$20x + 10y \leq 80{,}000.$$

Check Point 2 To maintain high quality, the company in Check Point 1 should not manufacture more than a total of 80 bookshelves and desks per day. Write an inequality that describes this constraint.

In addition to a weight constraint on its cargo, each plane has a limited amount of space in which to carry supplies. Example 3 demonstrates how to express this constraint.

EXAMPLE 3   Writing a Constraint

Each plane can carry a total volume of supplies that does not exceed 6000 cubic feet. Each water bottle is 1 cubic foot and each medical kit also has a volume of 1 cubic foot. With x still representing the number of water bottles and y the number of medical kits, write an inequality that describes this second constraint.

Solution   Because each plane can carry a volume of supplies that does not exceed 6000 cubic feet, we have

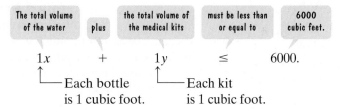

$$1x \quad + \quad 1y \quad \leq \quad 6000.$$

Each bottle is 1 cubic foot.   Each kit is 1 cubic foot.

The plane's volume constraint is described by the inequality $x + y \leq 6000$.

In summary, here's what we have described so far in this aid-to-earthquake-victims situation:

$$z = 10x + 6y$$ *This is the objective function describing the number of people helped with x bottles of water and y medical kits.*

$$20x + 10y \leq 80{,}000$$ *These are the constraints based on each plane's weight and*
$$x + y \leq 6000.$$ *volume limitations.*

Check Point **3**   To meet customer demand, the company in Check Point 1 must manufacture between 30 and 80 bookshelves per day, inclusive. Furthermore, the company must manufacture at least 10 and no more than 30 desks per day. Write an inequality that describes each of these sentences. Then summarize what you have described about this company by writing the objective function for its profits and the three constraints.

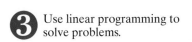

3 Use linear programming to solve problems.

Solving Problems with Linear Programming

The problem in the earthquake situation described previously is to maximize the number of victims who can be helped, subject to each plane's weight and volume constraints. The process of solving this problem is called *linear programming*, based on a theorem that was proven during World War II.

> **Solving a Linear Programming Problem**
>
> Let $z = ax + by$ be an objective function that depends on x and y. Furthermore, z is subject to a number of constraints on x and y. If a maximum or minimum value of z exists, it can be determined as follows:
>
> **1.** Graph the system of inequalities representing the constraints.
> **2.** Find the value of the objective function at each corner, or **vertex**, of the graphed region. The maximum and minimum of the objective function occur at one or more of the corner points.

EXAMPLE 4   Solving a Linear Programming Problem

Determine how many bottles of water and how many medical kits should be sent on each plane to maximize the number of earthquake victims who can be helped.

Solution   We must maximize $z = 10x + 6y$ subject to the following constraints:

$$20x + 10y \leq 80{,}000$$
$$x + y \leq 6000.$$

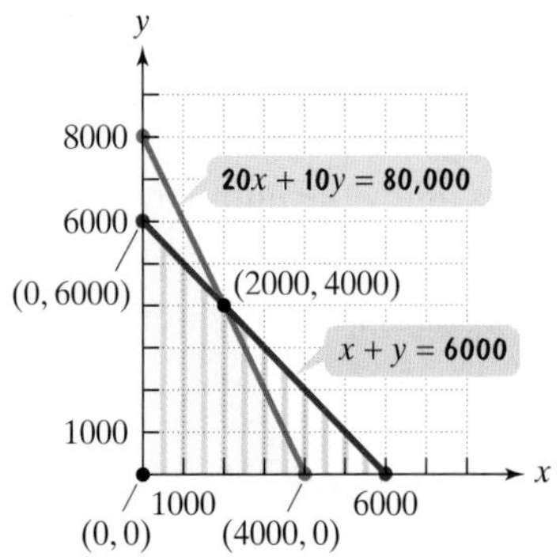

Figure 5.28 The region in quadrant I representing the constraints $20x + 10y \le 80{,}000$ and $x + y \le 6000$

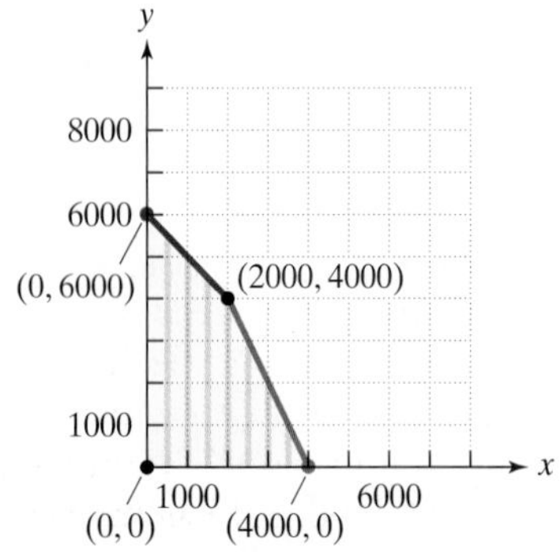

Figure 5.29

Step 1 Graph the system of inequalities representing the constraints. Because x (the number of bottles of water per plane) and y (the number of medical kits per plane) must be nonnegative, we need to graph the system of inequalities in quadrant I and its boundary only.

To graph the inequality $20x + 10y \le 80{,}000$, we graph the equation $20x + 10y = 80{,}000$ as a solid blue line (Figure 5.28). Setting $y = 0$, the x-intercept is 4000 and setting $x = 0$, the y-intercept is 8000. Using $(0, 0)$ as a test point, the inequality is satisfied, so we shade below the blue line, as shown in yellow in Figure 5.28.

Now we graph $x + y \le 6000$ by first graphing $x + y = 6000$ as a solid red line. Setting $y = 0$, the x-intercept is 6000. Setting $x = 0$, the y-intercept is 6000. Using $(0, 0)$ as a test point, the inequality is satisfied, so we shade below the red line, as shown using green vertical shading in Figure 5.28.

We use the addition method to find where the lines $20x + 10y = 80{,}000$ and $x + y = 6000$ intersect.

$$\begin{aligned} 20x + 10y &= 80{,}000 \quad \xrightarrow{\text{No change}} & 20x + 10y &= 80{,}000 \\ x + y &= 6000 \quad \xrightarrow{\text{Multiply by } -10.} & -10x - 10y &= -60{,}000 \\ & \qquad\qquad\qquad\text{Add:} & 10x &= 20{,}000 \\ & & x &= 2000 \end{aligned}$$

Back-substituting 2000 for x in $x + y = 6000$, we find $y = 4000$, so the intersection point is $(2000, 4000)$.

The system of inequalities representing the constraints is shown by the region in which the yellow shading and the green vertical shading overlap in Figure 5.28. The graph of the system of inequalities is shown again in Figure 5.29. The red and blue line segments are included in the graph.

Step 2 Find the value of the objective function at each corner of the graphed region. The maximum and minimum of the objective function occur at one or more of the corner points. We must evaluate the objective function, $z = 10x + 6y$, at the four corners, or vertices, of the region in Figure 5.29.

Corner (x, y)	**Objective Function** $z = 10x + 6y$
$(0, 0)$	$z = 10(0) + 6(0) = 0$
$(4000, 0)$	$z = 10(4000) + 6(0) = 40{,}000$
$(2000, 4000)$	$z = 10(2000) + 6(4000) = 44{,}000$ ← maximum
$(0, 6000)$	$z = 10(0) + 6(6000) = 36{,}000$

Thus, the maximum value of z is 44,000, and this occurs when $x = 2000$ and $y = 4000$. In practical terms, this means that the maximum number of earthquake victims who can be helped with each plane shipment is 44,000. This can be accomplished by sending 2000 water bottles and 4000 medical kits per plane.

Check Point 4 For the company in Check Points 1–3, how many bookshelves and how many desks should be manufactured per day to obtain maximum profit? What is the maximum daily profit?

EXAMPLE 5 Solving a Linear Programming Problem

Find the maximum value of the objective function

$$z = 2x + y$$

subject to the following constraints:

$$\begin{aligned} x \ge 0, y &\ge 0 \\ x + 2y &\le 5 \\ x - y &\le 2. \end{aligned}$$

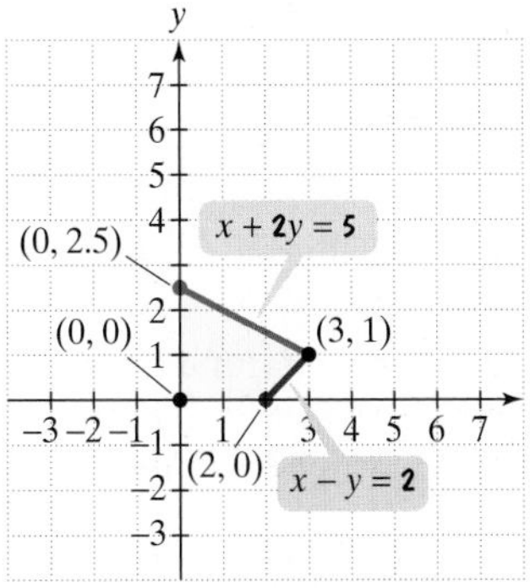

Figure 5.30 The graph of $x + 2y \leq 5$ and $x - y \leq 2$ in quadrant I

Solution We begin by graphing the region in quadrant I ($x \geq 0, y \geq 0$) formed by the constraints. The graph is shown in Figure 5.30.

Now we evaluate the objective function at the four vertices of this region.

Objective function: $z = 2x + y$

At (0, 0): $z = 2 \cdot 0 + 0 = 0$

At (2, 0): $z = 2 \cdot 2 + 0 = 4$

At (3, 1): $z = 2 \cdot 3 + 1 = 7$ — Maximum value of z

At (0, 2.5): $z = 2 \cdot 0 + 2.5 = 2.5$

Thus, the maximum value of z is 7, and this occurs when $x = 3$ and $y = 1$.

We can see why the objective function in Example 5 has a maximum value that occurs at a vertex by solving the equation for y.

$z = 2x + y$ — This is the objective function of Example 5.

$y = -2x + z$ — Solve for y. Recall that the slope-intercept form of a line is $y = mx + b$.

Slope = −2 y-intercept = z

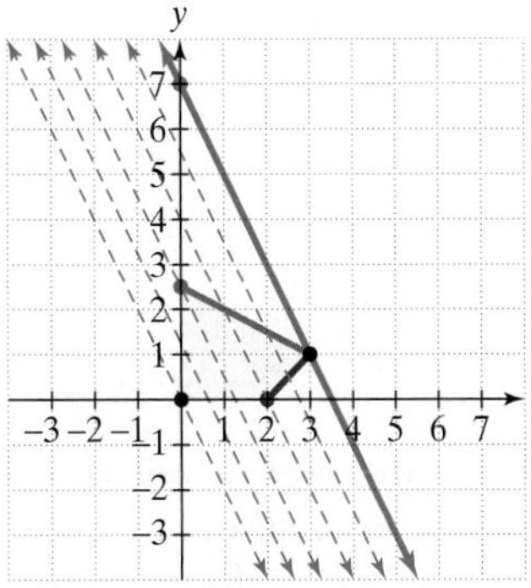

Figure 5.31 The line with slope −2 with the greatest y-intercept that intersects the shaded region passes through one of its vertices.

In this form, z represents the y-intercept of the objective function. The equation describes infinitely many parallel lines, each with slope −2. The process in linear programming involves finding the maximum z-value for all lines that intersect the region determined by the constraints. Of all the lines whose slope is −2, we're looking for the one with the greatest y-intercept that intersects the given region. As we see in Figure 5.31, such a line will pass through one (or possibly more) of the vertices of the region.

Check Point 5 Find the maximum value of the objective function $z = 3x + 5y$ subject to the constraints $x \geq 0, y \geq 0, x + y \geq 1, x + y \leq 6$.

EXERCISE SET 5.6

Practice Exercises

In Exercises 1–4, find the value of the objective function at each corner of the graphed region. What is the maximum value of the objective function? What is the minimum value of the objective function?

1. Objective Function
$z = 5x + 6y$

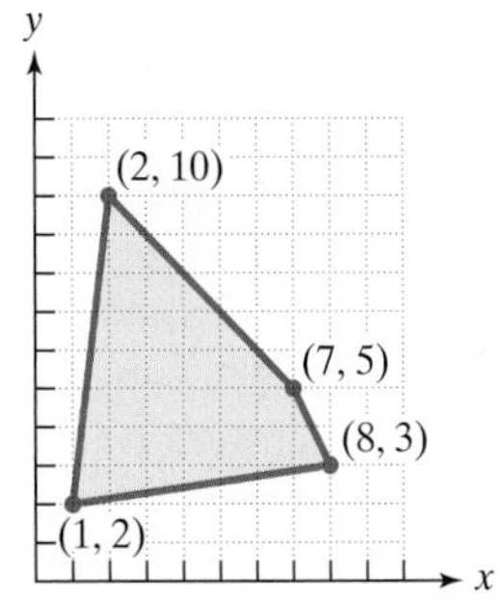

2. Objective Function
$z = 3x + 2y$

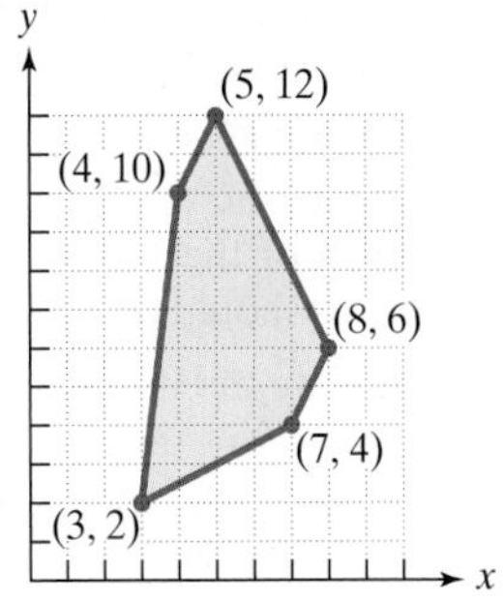

3. Objective Function
$z = 40x + 50y$

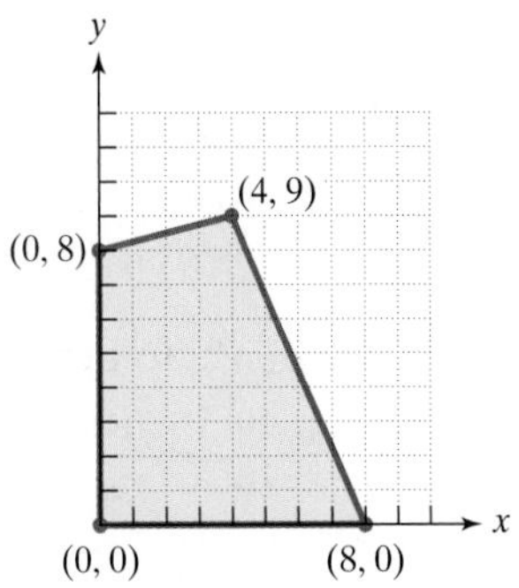

4. Objective Function
$z = 30x + 45y$

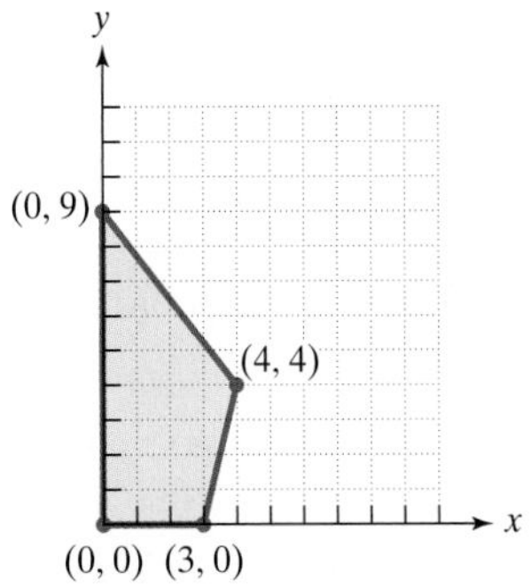

In Exercises 5–14, an objective function and a system of linear inequalities representing constraints are given.

a. *Graph the system of inequalities representing the constraints.*

b. *Find the value of the objective function at each corner of the graphed region.*

c. *Use the values in part (b) to determine the maximum value of the objective function and the values of x and y for which the maximum occurs.*

5. Objective Function $z = 3x + 2y$
Constraints
$$x \geq 0,\ y \geq 0$$
$$2x + y \leq 8$$
$$x + y \geq 4$$

6. Objective Function $z = 2x + 3y$
Constraints
$$x \geq 0,\ y \geq 0$$
$$2x + y \leq 8$$
$$2x + 3y \leq 12$$

7. Objective Function $z = 4x + y$
Constraints
$$x \geq 0,\ y \geq 0$$
$$2x + 3y \leq 12$$
$$x + y \geq 3$$

8. Objective Function $z = x + 6y$
Constraints
$$x \geq 0,\ y \geq 0$$
$$2x + y \leq 10$$
$$x - 2y \geq -10$$

9. Objective Function $z = 3x - 2y$
Constraints
$$1 \leq x \leq 5$$
$$y \geq 2$$
$$x - y \geq -3$$

10. Objective Function $z = 5x - 2y$
Constraints
$$0 \leq x \leq 5$$
$$0 \leq y \leq 3$$
$$x + y \geq 2$$

11. Objective Function $z = 4x + 2y$
Constraints
$$x \geq 0,\ y \geq 0$$
$$2x + 3y \leq 12$$
$$3x + 2y \leq 12$$
$$x + y \geq 2$$

12. Objective Function $z = 2x + 4y$
Constraints
$$x \geq 0,\ y \geq 0$$
$$x + 3y \geq 6$$
$$x + y \geq 3$$
$$x + y \leq 9$$

13. Objective Function $z = 10x + 12y$
Constraints
$$x \geq 0,\ y \geq 0$$
$$x + y \leq 7$$
$$2x + y \leq 10$$
$$2x + 3y \leq 18$$

14. Objective Function $z = 5x + 6y$
Constraints
$$x \geq 0,\ y \geq 0$$
$$2x + y \geq 10$$
$$x + 2y \geq 10$$
$$x + y \leq 10$$

Application Exercises

15. A television manufacturer makes console and wide-screen televisions. The profit per unit is \$125 for the console televisions and \$200 for the wide-screen televisions.

a. Let x = the number of consoles manufactured in a month and let y = the number of wide-screens manufactured in a month. Write the objective function that describes the total monthly profit.

b. The manufacturer is bound by the following constraints:

- Equipment in the factory allows for making at most 450 console televisions in one month.
- Equipment in the factory allows for making at most 200 wide-screen televisions in one month.
- The cost to the manufacturer per unit is \$600 for the console televisions and \$900 for the wide-screen televisions. Total monthly costs cannot exceed \$360,000.

Write a system of three inequalities that describes these constraints.

c. Graph the system of inequalities in part (b). Use only the first quadrant and its boundary, because x and y must both be nonnegative.

d. Evaluate the objective function for total monthly profit at each of the five vertices of the graphed region. [The vertices should occur at $(0, 0)$, $(0, 200)$, $(300, 200)$, $(450, 100)$, and $(450, 0)$.]

e. Complete the missing portions of this statement: The television manufacturer will make the greatest profit by manufacturing _____ console televisions each month and _____ wide-screen televisions each month. The maximum monthly profit is \$ _____.

16. a. A student earns \$10 per hour for tutoring and \$7 per hour as a teacher's aid. Let x = the number of hours each week spent tutoring and let y = the number of hours each week spent as a teacher's aid. Write the objective function that describes total weekly earnings.

b. The student is bound by the following constraints:

- To have enough time for studies, the student can work no more than 20 hours per week.
- The tutoring center requires that each tutor spend at least three hours per week tutoring.
- The tutoring center requires that each tutor spend no more than eight hours per week tutoring.

Write a system of three inequalities that describes these constraints.

c. Graph the system of inequalities in part (b). Use only the first quadrant and its boundary, because x and y are nonnegative.

d. Evaluate the objective function for total weekly earnings at each of the four vertices of the graphed region. [The vertices should occur at $(3, 0)$, $(8, 0)$, $(3, 17)$, and $(8, 12)$.]

e. Complete the missing portions of this statement: The student can earn the maximum amount per week by tutoring for _____ hours per week and working as a teacher's aid for _____ hours per week. The maximum amount that the student can earn each week is \$ _____.

Use the two steps for solving a linear programming problem, given in the box on page 523, to solve the problems in Exercises 17–23.

17. A manufacturer produces two models of mountain bicycles. The times (in hours) required for assembling and painting each model are given in the following table:

	Model A	Model B
Assembling	5	4
Painting	2	3

The maximum total weekly hours available in the assembly department and the paint department are 200 hours and 108 hours, respectively. The profits per unit are $25 for model A and $15 for model B. How many of each type should be produced to maximize profit?

18. A large institution is preparing lunch menus containing foods A and B. The specifications for the two foods are given in the following table:

Food	Units of Fat per Ounce	Units of Carbohydrates per Ounce	Units of Protein per Ounce
A	1	2	1
B	1	1	1

Each lunch must provide at least 6 units of fat per serving, no more than 7 units of protein, and at least 10 units of carbohydrates. The institution can purchase food A for $0.12 per ounce and food B for $0.08 per ounce. How many ounces of each food should a serving contain to meet the dietary requirements at the least cost?

19. Food and clothing are shipped to victims of a natural disaster. Each carton of food will feed 12 people, while each carton of clothing will help 5 people. Each 20-cubic-foot box of food weighs 50 pounds and each 10-cubic-foot box of clothing weighs 20 pounds. The commercial carriers transporting food and clothing are bound by the following constraints:

- The total weight per carrier cannot exceed 19,000 pounds.
- The total volume must be less than 8000 cubic feet.

How many cartons of food and clothing should be sent with each plane shipment to maximize the number of people who can be helped?

20. On June 24, 1948, the former Soviet Union blocked all land and water routes through East Germany to Berlin. A gigantic airlift was organized using American and British planes to supply food, clothing, and other supplies to the more than 2 million people in West Berlin. The cargo capacity was 30,000 cubic feet for an American plane and 20,000 cubic feet for a British plane. To break the Soviet blockade, the Western Allies had to maximize cargo capacity, but were subject to the following restrictions:

- No more than 44 planes could be used.
- The larger American planes required 16 personnel per flight, double that of the requirement for the British planes. The total number of personnel available could not exceed 512.
- The cost of an American flight was $9000 and the cost of a British flight was $5000. Total weekly costs could not exceed $300,000.

Find the number of American and British planes that were used to maximize cargo capacity.

21. A theater is presenting a program on drinking and driving for students and their parents. The proceeds will be donated to a local alcohol information center. Admission is $2.00 for parents and $1.00 for students. However, the situation has two constraints: The theater can hold no more than 150 people and every two parents must bring at least one student. How many parents and students should attend to raise the maximum amount of money?

22. You are about to take a test that contains computation problems worth 6 points each and word problems worth 10 points each. You can do a computation problem in 2 minutes and a word problem in 4 minutes. You have 40 minutes to take the test and may answer no more than 12 problems. Assuming you answer all the problems attempted correctly, how many of each type of problem must you answer to maximize your score? What is the maximum score?

23. In 1978, a ruling by the Civil Aeronautics Board allowed Federal Express to purchase larger aircraft. Federal Express's options included 20 Boeing 727s that United Airlines was retiring and/or the French-built Dassault Fanjet Falcon 20. To aid in their decision, executives at Federal Express analyzed the following data:

	Boeing 727	Falcon 20
Direct Operating Cost	$1400 per hour	$500 per hour
Payload	42,000 pounds	6000 pounds

Federal Express was faced with the following constraints:

- Hourly operating cost was limited to $35,000.
- Total payload had to be at least 672,000 pounds.
- Only twenty 727s were available.

Given the constraints, how many of each kind of aircraft should Federal Express have purchased to maximize the number of aircraft?

Writing in Mathematics

24. What kinds of problems are solved using the linear programming method?

25. What is an objective function in a linear programming problem?

26. What is a constraint in a linear programming problem? How is a constraint represented?

27. In your own words, describe how to solve a linear programming problem.

28. Describe a situation in your life in which you would really like to maximize something, but you are limited by at least two constraints. Can linear programming be used in this situation? Explain your answer.

Critical Thinking Exercises

29. Suppose that you inherit \$10,000. The will states how you must invest the money. Some (or all) of the money must be invested in stocks and bonds. The requirements are that at least \$3000 be invested in bonds, with expected returns of \$0.08 per dollar, and at least \$2000 be invested in stocks, with expected returns of \$0.12 per dollar. Because the stocks are medium risk, the final stipulation requires that the investment in bonds should never be less than the investment in stocks. How should the money be invested so as to maximize your expected returns?

30. Consider the objective function $z = Ax + By$ ($A > 0$ and $B > 0$) subject to the following constraints: $2x + 3y \leq 9$, $x - y \leq 2$, $x \geq 0$, and $y \geq 0$. Prove that the objective function will have the same maximum value at the vertices $(3, 1)$ and $(0, 3)$ if $A = \frac{2}{3}B$.

Group Exercises

31. Group members should choose a particular field of interest. Research how linear programming is used to solve problems in that field. If possible, investigate the solution of a specific practical problem. Present a report on your findings, including the contributions of George Dantzig, Narendra Karmarkar, and L. G. Khachion to linear programming.

32. Members of the group should interview a business executive who is in charge of deciding the product mix for a business. How are production policy decisions made? Are other methods used in conjunction with linear programming? What are these methods? What sort of academic background, particularly in mathematics, does this executive have? Present a group report addressing these questions, emphasizing the role of linear programming for the business.

Chapter 5 Summary, Review, and Test

Summary

DEFINITIONS AND CONCEPTS — **EXAMPLES**

5.1 Systems of Linear Equations in Two Variables

a. Two equations in the form $Ax + By = C$ are called a system of linear equations. A solution of the system is an ordered pair that satisfies both equations in the system.
Ex. 1, p. 468

b. Systems of linear equations in two variables can be solved by eliminating a variable, using the substitution method (see the box on page 470) or the addition method (see the box on page 471).
Ex. 2, p. 470; Ex. 3, p. 472; Ex. 4, p. 473

c. Some linear systems have no solution and are called inconsistent systems; others have infinitely many solutions. The equations in a linear system with infinitely many solutions are called dependent. For details, see the box on page 474.
Ex. 5, p. 474; Ex. 6, p. 475

d. Functions of Business
Ex. 7, p. 476; Figure 5.7, p. 478

Revenue Function

$$R(x) = (\text{price per unit sold})x$$

Cost Function

$$C(x) = \text{fixed cost} + (\text{cost per unit produced})x$$

Profit Function

$$P(x) = R(x) - C(x)$$

The point of intersection of the graphs of R and C is the break-even point. The x-coordinate of the point reveals the number of units that a company must produce and sell so that the money coming in, the revenue, is equal to the money going out, the cost. The y-coordinate gives the amount of money coming in and going out.

DEFINITIONS AND CONCEPTS	EXAMPLES
5.2 Systems of Linear Equations in Three Variables	
a. Three equations in the form $Ax + By + Cz = D$ are called a system of linear equations in three variables. A solution of the system is an ordered triple that satisfies all three equations in the system.	Ex. 1, p. 483
b. A system of linear equations in three variables can be solved by eliminating variables. Use the addition method to eliminate any variable, reducing the system to two equations in two variables. Use substitution or the addition method to solve the resulting system in two variables. Details are found in the box on page 484.	Ex. 2, p. 484; Ex. 3, p. 486
c. Three points that do not lie on a line determine the graph of a quadratic function $y = ax^2 + bx + c$. Use the three given points to create a system of three equations. Solve the system to find a, b, and c.	Ex 4, p. 487
5.3 Partial Fractions	
a. Partial fraction decomposition is used on rational expressions in which the numerator and denominator have no common factors and the highest power in the numerator is less than the highest power in the denominator. The steps in partial fraction decomposition are given in the box on page 493.	
b. Include one partial fraction with a constant numerator for each distinct linear factor in the denominator. Include one partial fraction with a constant numerator for each power of a repeated linear factor in the denominator.	Ex. 1, p. 492; Ex. 2, p. 494
c. Include one partial fraction with a linear numerator for each distinct prime quadratic factor in the denominator. Include one partial fraction with a linear numerator for each power of a prime, repeated quadratic factor in the denominator.	Ex. 3, p. 496; Ex. 4, p. 498
5.4 Systems of Nonlinear Equations in Two Variables	
a. A system of two nonlinear equations in two variables contains at least one equation that cannot be expressed as $Ax + By = C$.	
b. Systems of nonlinear equations in two variables can be solved algebraically by eliminating all occurrences of one of the variables by the substitution or addition methods.	Ex. 1, p. 501; Ex. 2, p. 502; Ex. 3, p. 503; Ex. 4, p. 504
5.5 Systems of Inequalities	
a. A linear inequality in two variables can be written in the form $Ax + By > C$, $Ax + By \geq C$, $Ax + By < C$, or $Ax + By \leq C$.	
b. The procedure for graphing a linear inequality in two variables is given in the box on page 510. A nonlinear inequality in two variables is graphed using the same procedure.	Ex. 1, p. 510; Ex. 2, p. 511; Ex. 3, p. 512; Ex. 4, p. 513
c. To graph the solution set of a system of inequalities, graph each inequality in the system in the same rectangular coordinate system. Then find the region, if there is one, that is common to every graph in the system.	Ex. 5, p. 514; Ex. 6, p. 515; Ex. 7, p. 516
5.6 Linear Programming	
a. An objective function is an algebraic expression in three variables describing a quantity that must be maximized or minimized.	Ex. 1, p. 521
b. Constraints are restrictions, expressed as linear inequalities.	Ex. 2, p. 522; Ex. 3, p. 523
c. Steps for solving a linear programming problem are given in the box on page 523.	Ex. 4, p. 523; Ex. 5, p. 524

Review Exercises

5.1

In Exercises 1–5, solve by the method of your choice. Identify systems with no solution and systems with infinitely many solutions, using set notation to express their solution sets.

1. $y = 4x + 1$
$3x + 2y = 13$

2. $x + 4y = 14$
$2x - y = 1$

3. $5x + 3y = 1$
$3x + 4y = -6$

4. $2y - 6x = 7$
$3x - y = 9$

5. $4x - 8y = 16$
$3x - 6y = 12$

6. A company is planning to manufacture computer desks. The fixed cost will be $60,000 and it will cost $200 to produce each desk. Each desk will be sold for $450.

a. Write the cost function, C, of producing x desks.

b. Write the revenue function, R, from the sale of x desks.

c. Determine the break-even point. Describe what this means.

7. The bar graph shows the five countries with the longest healthy life expectancy at birth. Combined, people in Japan and Switzerland can expect to spend 146.4 years in good health. The difference between healthy life expectancy between these two countries is 0.8 years. Find the healthy life expectancy at birth in Japan and Switzerland.

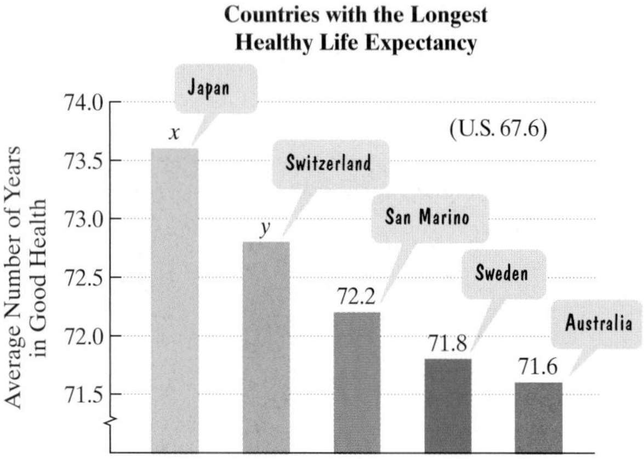

Source: World Health Organization

8. The perimeter of a rectangular table top is 34 feet. The difference between 4 times the length and 3 times the width is 33 feet. Find the dimensions.

9. A travel agent offers two package vacation plans. The first plan costs $360 and includes 3 days at a hotel and a rental car for 2 days. The second plan costs $500 and includes 4 days at a hotel and a rental car for 3 days. The daily charge for the hotel is the same under each plan, as is the daily charge for the car. Find the cost per day for the hotel and for the car.

10. The calorie-nutrient information for an apple and an avocado is given in the table. How many of each should be eaten to get exactly 1000 calories and 100 grams of carbohydrates?

	One Apple	One Avocado
Calories	100	350
Carbohydrates (grams)	24	14

5.2

Solve each system in Exercises 11–12.

11. $2x - y + z = 1$
$3x - 3y + 4z = 5$
$4x - 2y + 3z = 4$

12. $x + 2y - z = 5$
$2x - y + 3z = 0$
$2y + z = 1$

13. Find the quadratic function $y = ax^2 + bx + c$ whose graph passes through the points $(1, 4)$, $(3, 20)$, and $(-2, 25)$.

14. The bar graph shows that the U.S. divorce rate increased from 1970 to 1985 and then decreased from 1985 to 2003.

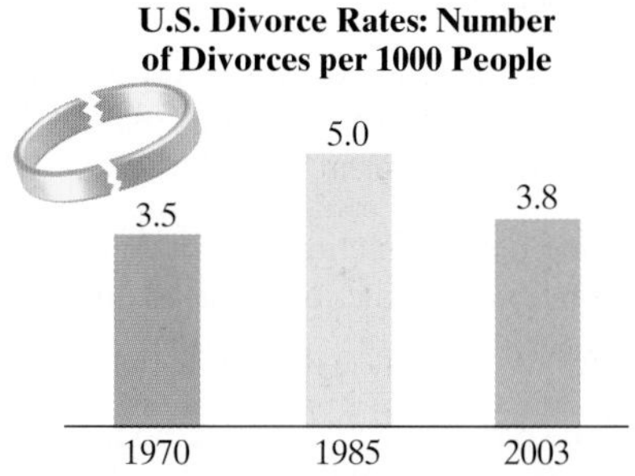

Source: U.S. Census Bureau

a. Write the data for 1970, 1985, and 2003 as ordered pairs (x, y), where x is the number of years after 1970 and y is that year's divorce rate.

b. The three data points in part (a) can be modeled by the quadratic function $y = ax^2 + bx + c$. Write a system of linear equations in three variables that can be used to find values for a, b, and c. It is not necessary to solve the system.

15. The bar graph shows the top five purebred dogs in the United States in 2004 compared with how many there were a decade earlier.

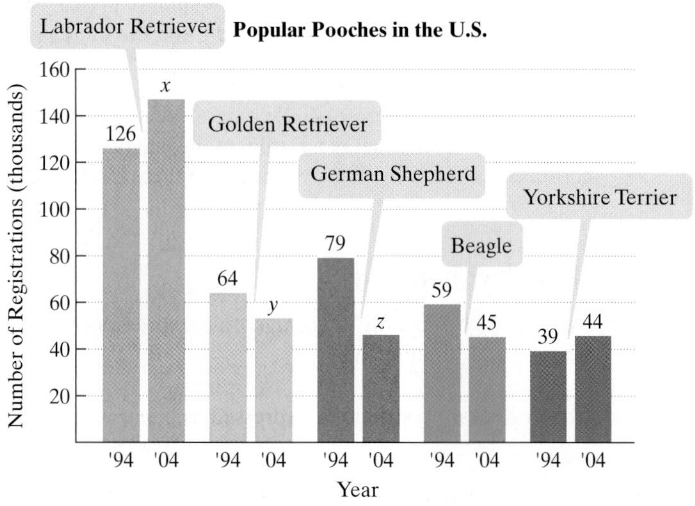

Source: American Kennel Club

In 2004, for the 15th year in a row, the Labrador retriever was the most popular purebred dog. The number of Labs exceeded the number of Golden retrievers and German shepherds combined by 48 thousand. Furthermore, the number of Labs was 9 thousand more than three times the number of German shepherds. If there were 246 thousand registrations for Labrador retrievers, Golden retrievers, and German shepherds, how many registrations, in thousands, were there for each of these breeds?

5.3

In Exercises 16–24, write the partial fraction decomposition of each rational expression.

16. $\dfrac{x}{(x-3)(x+2)}$

17. $\dfrac{11x-2}{x^2-x-12}$

18. $\dfrac{4x^2-3x-4}{x(x+2)(x-1)}$

19. $\dfrac{2x+1}{(x-2)^2}$

20. $\dfrac{2x-6}{(x-1)(x-2)^2}$

21. $\dfrac{3x}{(x-2)(x^2+1)}$

22. $\dfrac{7x^2-7x+23}{(x-3)(x^2+4)}$

23. $\dfrac{x^3}{(x^2+4)^2}$

24. $\dfrac{4x^3+5x^2+7x-1}{(x^2+x+1)^2}$

5.4

In Exercises 25–35, solve each system by the method of your choice.

25. $5y = x^2 - 1$
$x - y = 1$

26. $y = x^2 + 2x + 1$
$x + y = 1$

27. $x^2 + y^2 = 2$
$x + y = 0$

28. $2x^2 + y^2 = 24$
$x^2 + y^2 = 15$

29. $xy - 4 = 0$
$y - x = 0$

30. $y^2 = 4x$
$x - 2y + 3 = 0$

31. $x^2 + y^2 = 10$
$y = x + 2$

32. $xy = 1$
$y = 2x + 1$

33. $x + y + 1 = 0$
$x^2 + y^2 + 6y - x = -5$

34. $x^2 + y^2 = 13$
$x^2 - y = 7$

35. $2x^2 + 3y^2 = 21$
$3x^2 - 4y^2 = 23$

36. The perimeter of a rectangle is 26 meters and its area is 40 square meters. Find its dimensions.

37. Find the coordinates of all points (x, y) that lie on the line whose equation is $2x + y = 8$, so that the area of the rectangle shown in the figure is 6 square units.

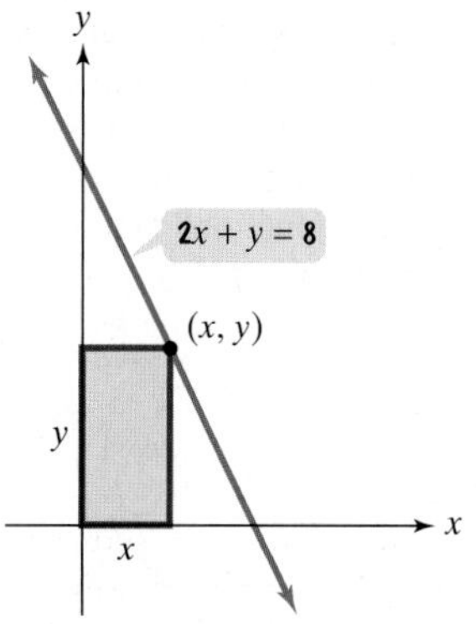

38. Two adjoining square fields with an area of 2900 square feet are to be enclosed with 240 feet of fencing. The situation is represented in the figure. Find the length of each side where a variable appears.

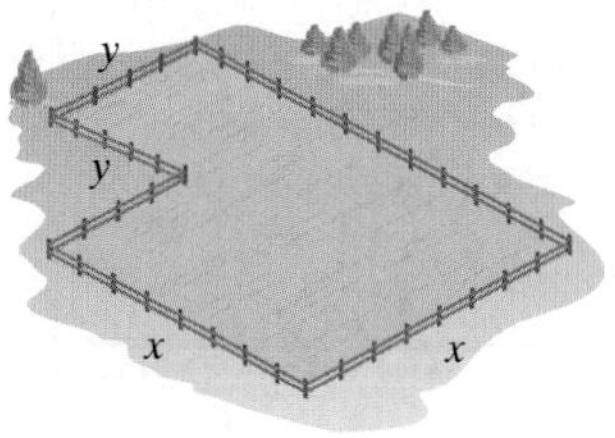

5.5

In Exercises 39–45, graph each inequality.

39. $3x - 4y > 12$

40. $y \le -\frac{1}{2}x + 2$

41. $x < -2$

42. $y \ge 3$

43. $x^2 + y^2 > 4$

44. $y \le x^2 - 1$

45. $y \le 2^x$

In Exercises 46–55, graph the solution set of each system of inequalities or indicate that the system has no solution.

46. $3x + 2y \ge 6$
$2x + y \ge 6$

47. $2x - y \ge 4$
$x + 2y < 2$

48. $y < x$
$y \le 2$

49. $x + y \le 6$
$y \ge 2x - 3$

50. $0 \le x \le 3$
$y > 2$

51. $2x + y < 4$
$2x + y > 6$

52. $x^2 + y^2 \le 16$
$x + y < 2$

53. $x^2 + y^2 \le 9$
$y < -3x + 1$

54. $y > x^2$
$x + y < 6$
$y < x + 6$

55. $y \ge 0$
$3x + 2y \ge 4$
$x - y \le 3$

5.6

56. Find the value of the objective function $z = 2x + 3y$ at each corner of the graphed region shown. What is the maximum value of the objective function? What is the minimum value of the objective function?

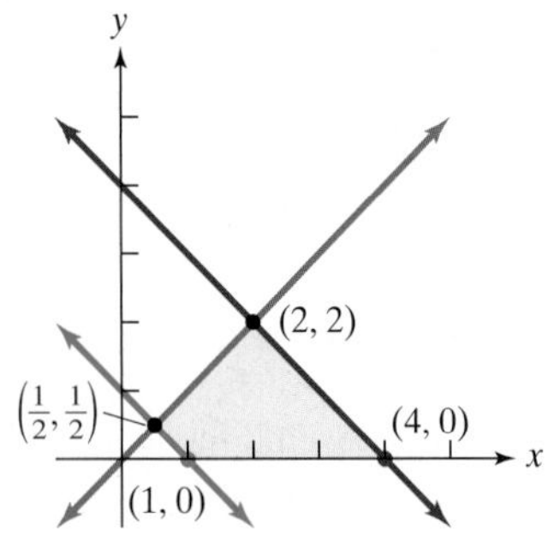

In Exercises 57–59, graph the region determined by the constraints. Then find the maximum value of the given objective function, subject to the constraints.

57. Objective Function $z = 2x + 3y$
Constraints $x \ge 0,\ y \ge 0$
$x + y \le 8$
$3x + 2y \ge 6$

58. Objective Function $z = x + 4y$
Constraints $0 \le x \le 5,\ 0 \le y \le 7$
$x + y \ge 3$

59. Objective Function $z = 5x + 6y$
Constraints $x \ge 0,\ y \ge 0$
$y \le x$
$2x + y \le 12$
$2x + 3y \ge 6$

60. A paper manufacturing company converts wood pulp to writing paper and newsprint. The profit on a unit of writing paper is \$500 and the profit on a unit of newsprint is \$350.

a. Let x represent the number of units of writing paper produced daily. Let y represent the number of units of newsprint produced daily. Write the objective function that models total daily profit.

b. The manufacturer is bound by the following constraints:
- Equipment in the factory allows for making at most 200 units of paper (writing paper and newsprint) in a day.
- Regular customers require at least 10 units of writing paper and at least 80 units of newsprint daily.

Write a system of inequalities that models these constraints.

c. Graph the inequalities in part (b). Use only the first quadrant, because x and y must both be positive. (*Suggestion:* Let each unit along the x- and y-axes represent 20.)

d. Evaluate the objective function at each of the three vertices of the graphed region.

e. Complete the missing portions of this statement: The company will make the greatest profit by producing _____ units of writing paper and _____ units of newsprint each day. The maximum daily profit is \$ _____.

61. A manufacturer of lightweight tents makes two models whose specifications are given in the following table:

	Cutting Time per Tent	Assembly Time per Tent
Model A	0.9 hour	0.8 hour
Model B	1.8 hours	1.2 hours

On a monthly basis, the manufacturer has no more than 864 hours of labor available in the cutting department and at most 672 hours in the assembly division. The profits come to \$25 per tent for model A and \$40 per tent for model B. How many of each should be manufactured monthly to maximize the profit?

Chapter 5 Test

In Exercises 1–5, solve the system.

1. $x = y + 4$
$3x + 7y = -18$

2. $2x + 5y = -2$
$3x - 4y = 20$

3. $x + y + z = 6$
$3x + 4y - 7z = 1$
$2x - y + 3z = 5$

4. $x^2 + y^2 = 25$
$x + y = 1$

5. $2x^2 - 5y^2 = -2$
$3x^2 + 2y^2 = 35$

6. Find the partial fraction decomposition for
$$\frac{x}{(x+1)(x^2+9)}.$$

In Exercises 7–10, graph the solution set of each inequality or system of inequalities.

7. $x - 2y < 8$

8. $x \geq 0, y \geq 0$
$3x + \quad y \leq 9$
$2x + \quad 3y \geq 6$

9. $x^2 + y^2 > 1$
$x^2 + y^2 < 4$

10. $y \leq 1 - x^2$
$x^2 + y^2 \leq 9$

11. Find the maximum value of the objective function $z = 3x + 5y$ subject to the following constraints: $x \geq 0, y \geq 0, x + y \leq 6, x \geq 2$.

12. Health experts agree that cholesterol intake should be limited to 300 mg or less each day. Three ounces of shrimp and 2 ounces of scallops contain 156 mg of cholesterol. Five ounces of shrimp and 3 ounces of scallops contain 45 mg of cholesterol less than the suggested maximum daily intake. Determine the cholesterol content in an ounce of each item.

13. A company is planning to produce and sell a new line of computers. The fixed cost will be \$360,000 and it will cost \$850 to produce each computer. Each computer will be sold for \$1150.

a. Write the cost function, C, of producing x computers.

b. Write the revenue function, R, from the sale of x computers.

c. Determine the break-even point. Describe what this means.

14. Find the quadratic function whose graph passes through the points $(-1, -2)$, $(2, 1)$, and $(-2, 1)$.

15. The rectangular plot of land shown in the figure is to be fenced along three sides using 39 feet of fencing. No fencing is to be placed along the river's edge. The area of the plot is 180 square feet. What are its dimensions?

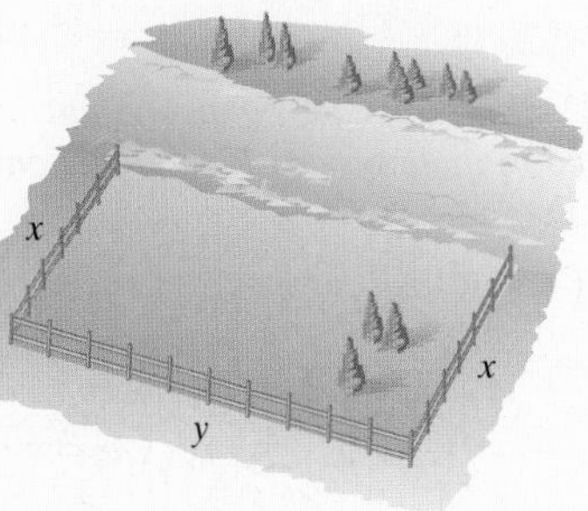

16. A manufacturer makes two types of jet skis, regular and deluxe. The profit on a regular jet ski is \$200 and the profit on the deluxe model is \$250. To meet customer demand, the company must manufacture at least 50 regular jet skis per week and at least 75 deluxe models. To maintain high quality, the total number of both models of jet skis manufactured by the company should not exceed 150 per week. How many jet skis of each type should be manufactured per week to obtain maximum profit? What is the maximum weekly profit?

Cumulative Review Exercises (Chapters 1–5)

The figure shows the graph of $y = f(x)$ and its two vertical asymptotes. Use the graph to solve Exercises 1–10.

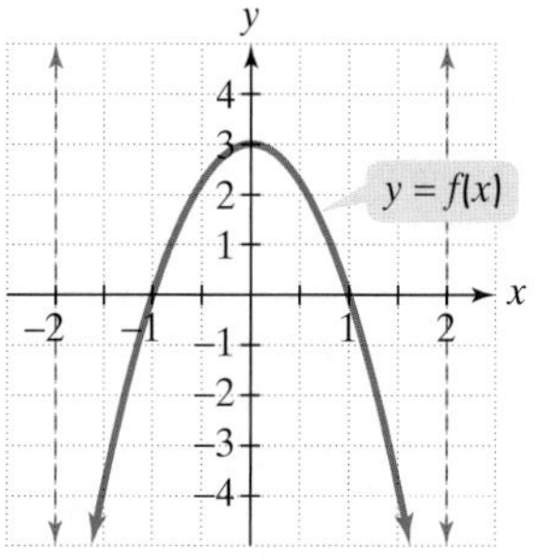

1. Find the domain and the range of f.

2. Find the zeros and the least possible multiplicity of each zero.

3. What is the relative maximum and where does it occur?

4. Find the interval(s) on which f is decreasing.

5. Is $f(-0.7)$ positive or negative?

6. Find $(f \circ f)(-1)$.

7. Use arrow notation to complete this statement:

$f(x) \to -\infty$ as __________ or as __________.

8. Does f appear to be even, odd, or neither?

9. Graph $g(x) = f(x + 2) - 1$.

10. Graph $h(x) = \frac{1}{2}f\left(\frac{1}{2}x\right)$.

In Exercises 11–21, solve each equation, inequality, or system of equations.

11. $\sqrt{x^2 - 3x} = 2x - 6$

12. $4x^2 = 8x - 7$

13. $\left|\frac{x}{3} + 2\right| < 4$

14. $\frac{x + 5}{x - 1} > 2$

15. $2x^3 + x^2 - 13x + 6 = 0$

16. $6x - 3(5x + 2) = 4(1 - x)$

17. $\log(x + 3) + \log x = 1$

18. $3^{x+2} = 11$

19. $x^{\frac{1}{2}} - 2x^{\frac{1}{4}} - 15 = 0$

20. $3x - y = -2$
$2x^2 - y = 0$

21. $x + 2y + 3z = -2$
$3x + 3y + 10z = -2$
$2y - 5z = 6$

In Exercises 22–28, graph each equation, function, or inequality in a rectangular coordinate system. If two functions are indicated, graph both in the same system.

22. $f(x) = (x + 2)^2 - 4$

23. $2x - 3y \leq 6$

24. $y = 3^{x-2}$

25. $f(x) = \dfrac{x^2 - x - 6}{x + 1}$

26. $f(x) = 2x - 4$ and f^{-1}

27. $(x - 2)^2 + (y - 4)^2 > 9$

28. $f(x) = |x|$ and $g(x) = -|x - 2|$

In Exercises 29–30, let $f(x) = 2x^2 - x - 1$ and $g(x) = 1 - x$.

29. Find $(f \circ g)(x)$ and $(g \circ f)(x)$.

30. Find $\dfrac{f(x + h) - f(x)}{h}$ and simplify.

In Exercises 31–32, write the linear function in slope-intercept form satisfying the given conditions.

31. Graph of f passes through $(2, 4)$ and $(4, -2)$.

32. Graph of g passes through $(-1, 0)$ and is perpendicular to the line whose equation is $x + 3y - 6 = 0$.

33. You invested \$4000 in two stocks paying 12% and 14% annual interest, respectively. At the end of the year, the total interest from these investments was \$508. How much was invested at each rate?

34. The length of a rectangle is 1 meter more than twice the width. If the rectangle's area is 36 square meters, find its dimensions.

35. What interest rate is required for an investment of \$6000 subject to continuous compounding to grow to \$18,000 in 10 years?

Appendix
Where Did That Come From? Selected Proofs

SECTION 4.3 Properties of Logarithms

The Product Rule

Let b, M, and N be positive real numbers with $b \neq 1$.

$$\log_b(MN) = \log_b M + \log_b N$$

Proof

We begin by letting $\log_b M = R$ and $\log_b N = S$.
Now we write each logarithm in exponential form.

$$\log_b M = R \quad \text{means} \quad b^R = M.$$

$$\log_b N = S \quad \text{means} \quad b^S = N.$$

By substituting and using a property of exponents, we see that

$$MN = b^R b^S = b^{R+S}.$$

Now we change $MN = b^{R+S}$ to logarithmic form.

$$MN = b^{R+S} \quad \text{means} \quad \log_b(MN) = R + S.$$

Finally, substituting $\log_b M$ for R and $\log_b N$ for S gives us

$$\log_b(MN) = \log_b M + \log_b N,$$

the property that we wanted to prove.

The quotient and power rules for logarithms are proved using similar procedures.

The Change-of-Base Property

For any logarithmic bases a and b, and any positive number M,

$$\log_b M = \frac{\log_a M}{\log_a b}.$$

Proof

To prove the change-of-base property, we let x equal the logarithm on the left side:

$$\log_b M = x.$$

Now we rewrite this logarithm in exponential form.

$$\log_b M = x \quad \text{means} \quad b^x = M.$$

Because b^x and M are equal, the logarithms with base a for each of these expressions must be equal. This means that

$$\log_a b^x = \log_a M$$

$$x \log_a b = \log_a M \qquad \text{Apply the power rule for logarithms on the left side.}$$

$$x = \frac{\log_a M}{\log_a b} \qquad \text{Solve for } x \text{ by dividing both sides by } \log_a b.$$

In our first step we let x equal $\log_b M$. Replacing x on the left side by $\log_b M$ gives us

$$\log_b M = \frac{\log_a M}{\log_a b},$$

which is the change-of-base property.

Answers to Selected Exercises

CHAPTER P

Section P.1

Check Point Exercises

1. 608 **2.** 2311; The formula models the data quite well. **3.** $\{3, 7\}$ **4.** $\{3, 4, 5, 6, 7, 8, 9\}$ **5. a.** $\sqrt{9}$ **b.** $0, \sqrt{9}$ **c.** $-9, 0, \sqrt{9}$ **d.** $-9, -1.3, 0, 0.\bar{3}, \sqrt{9}$ **e.** $\frac{\pi}{2}, \sqrt{10}$ **f.** $-9, -1.3, 0, 0.\bar{3}, \frac{\pi}{2}, \sqrt{9}, \sqrt{10}$ **6. a.** $\sqrt{2} - 1$ **b.** $\pi - 3$ **c.** 1 **7.** 9 **8.** $38x^2 + 23x$ **9.** $42 - 4x$

Exercise Set P.1

1. 57 **3.** 10 **5.** 88 **7.** 10 **9.** 44 **11.** 46 **13.** 10 **15.** −8 **17.** 10°C **19.** 60 ft **21.** $\{2, 4\}$ **23.** $\{s, e, t\}$ **25.** $\varnothing$ **27.** $\varnothing$ **29.** $\{1, 2, 3, 4, 5\}$ **31.** $\{1, 2, 3, 4, 5, 6, 7, 8, 10\}$ **33.** $\{a, e, i, o, u\}$ **35. a.** $\sqrt{100}$ **b.** $0, \sqrt{100}$ **c.** $-9, 0, \sqrt{100}$ **d.** $-9, -\frac{4}{5}, 0, 0.25, 9.2, \sqrt{100}$ **e.** $\sqrt{3}$ **f.** $-9, -\frac{4}{5}, 0, 0.25, \sqrt{3}, 9.2, \sqrt{100}$ **37. a.** $\sqrt{64}$ **b.** $0, \sqrt{64}$ **c.** $-11, 0, \sqrt{64}$ **d.** $-11, -\frac{5}{6}, 0, 0.75, \sqrt{64}$ **e.** $\sqrt{5}, \pi$ **f.** $-11, -\frac{5}{6}, 0, 0.75, \sqrt{5}, \pi, \sqrt{64}$ **39.** 0 **41.** Answers may vary. **43.** true **45.** true **47.** true **49.** true **51.** 300 **53.** $12 - \pi$ **55.** $5 - \sqrt{2}$ **57.** −1 **59.** 4 **61.** 3 **63.** 7 **65.** −1 **67.** $|17 - 2|; 15$ **69.** $|5 - (-2)|; 7$ **71.** $|-4 - (-19)|; 15$ **73.** $|-1.4 - (-3.6)|; 2.2$ **75.** commutative property of addition **77.** associative property of addition **79.** commutative property of addition **81.** distributive property of multiplication over addition **83.** inverse property of multiplication **85.** $15x + 16$ **87.** $27x - 10$ **89.** $29y - 29$ **91.** $8y - 12$ **93.** $16y - 25$ **95.** $12x^2 + 11$ **97.** $14x$ **99.** $-2x + 3y + 6$ **101.** x **103.** $>$ **105.** $=$ **107.** $<$ **109.** $=$ **111.** $x - (x + 4); -4$ **113.** $6(-5x); -30x$ **115.** $5x - 2x; 3x$ **117.** $8x - (3x + 6); 5x - 6$ **119.** 313; very well **121.** 522 **123.** Model 3 **125.** Model 3 **127. a.** $1200 - 0.07x$ **b.** \$780 **141.** d **143.** $>$ **145. a.** \$50.50, \$5.50, \$1.00 **b.** no

Section P.2

Check Point Exercises

1. a. 3^5 or 243 **b.** $40x^5y^{10}$ **2. a.** $(-3)^3$ or −27 **b.** $9x^{11}y^3$ **3. a.** $\frac{1}{25}$ **b.** $-\frac{1}{27}$ **c.** 16 **d.** $\frac{3y^4}{x^6}$ **4. a.** 3^6 or 729 **b.** $\frac{1}{y^{14}}$ **c.** b^{12} **5.** $-64x^3$ **6. a.** $-\frac{32}{y^5}$ **b.** $\frac{x^{15}}{27}$ **7. a.** $16x^{12}y^{24}$ **b.** $-18x^3y^8$ **c.** $\frac{5y^6}{x^4}$ **d.** $\frac{y^8}{25x^2}$ **8. a.** −2,600,000,000 **b.** 0.000003017 **9. a.** 5.21×10^9 **b.** -6.893×10^{-8} **10.** 4.1×10^9 **11.** a. 3.55×10^{-1} **b.** 4×10^8 **12.** \$7,014

Exercise Set P.2

1. 50 **3.** 64 **5.** −64 **7.** 1 **9.** −1 **11.** $\frac{1}{64}$ **13.** 32 **15.** 64 **17.** 16 **19.** $\frac{1}{9}$ **21.** $\frac{1}{16}$ **23.** $\frac{y}{x^2}$ **25.** y^5 **27.** x^{10} **29.** x^5 **31.** x^{21} **33.** $\frac{1}{x^{15}}$ **35.** x^7 **37.** x^{21} **39.** $64x^6$ **41.** $-\frac{64}{x^3}$ **43.** $9x^4y^{10}$ **45.** $6x^{11}$ **47.** $18x^9y^5$ **49.** $4x^{16}$ **51.** $-5a^{11}b$ **53.** $\frac{2}{b^7}$ **55.** $\frac{1}{16x^6}$ **57.** $\frac{3y^{14}}{4x^4}$ **59.** $\frac{y^2}{25x^6}$ **61.** $-\frac{27b^{15}}{a^{18}}$ **63.** 1 **65.** 380 **67.** 0.0006 **69.** −7,160,000 **71.** 0.79 **73.** −0.00415 **75.** −60,000,100,000 **77.** 3.2×10^4 **79.** 6.38×10^{17} **81.** -5.716×10^3 **83.** 2.7×10^{-3} **85.** -5.04×10^{-9} **87.** 6.3×10^7 **89.** 6.4×10^4 **91.** 1.22×10^{-11} **93.** 2.67×10^{13} **95.** 2.1×10^3 **97.** 4×10^5 **99.** 2×10^{-8} **101.** 5×10^3 **103.** 4×10^{15} **105.** 9×10^{-3} **107.** 1 **109.** $\frac{y}{16x^8z^6}$ **111.** $\frac{1}{x^{12}y^{16}z^{20}}$ **113.** $\frac{x^{18}y^6}{4}$ **115.** 6.26×10^7 people **117.** 9.63×10^7 people **119.** approximately 68 hot dogs per person **121.** $2.5 \times 10^2 = 250$ chickens **123.** approximately \$23,448 **133.** b **135.** $A = C + D$

Section P.3

Check Point Exercises

1. a. 9 **b.** −3 **c.** $\frac{1}{5}$ **d.** 10 **e.** 14 **2. a.** $5\sqrt{3}$ **b.** $5x\sqrt{2}$ **3. a.** $\frac{5}{4}$ **b.** $5x\sqrt{3}$ **4. a.** $17\sqrt{13}$ **b.** $-19\sqrt{17x}$ **5. a.** $17\sqrt{3}$ **b.** $10\sqrt{2x}$ **6. a.** $\frac{5\sqrt{3}}{3}$ **b.** $\sqrt{3}$ **7.** $\frac{32 - 8\sqrt{5}}{11}$ **8. a.** $2\sqrt[3]{5}$ **b.** $2\sqrt[5]{2}$ **c.** $\frac{5}{3}$ **9.** $5\sqrt[3]{3}$ **10. a.** 5 **b.** 2 **c.** −3 **d.** −2 **e.** $\frac{1}{3}$ **11. a.** 81 **b.** 8 **c.** $\frac{1}{4}$ **12. a.** $10x^4$ **b.** $4x^{5/2}$ **13.** $\sqrt{x}$

Exercise Set P.3

1. 6 **3.** −6 **5.** not a real number **7.** 3 **9.** 1 **11.** 13 **13.** $5\sqrt{2}$ **15.** $3|x|\sqrt{5}$ **17.** $2x\sqrt{3}$ **19.** $x\sqrt{x}$ **21.** $2x\sqrt{3x}$ **23.** $\frac{1}{9}$ **25.** $\frac{7}{4}$ **27.** $4x$ **29.** $5x\sqrt{2x}$ **31.** $2x^2\sqrt{5}$ **33.** $13\sqrt{3}$ **35.** $-2\sqrt{17x}$ **37.** $5\sqrt{2}$ **39.** $3\sqrt{2x}$ **41.** $34\sqrt{2}$ **43.** $20\sqrt{2} - 5\sqrt{3}$ **45.** $\frac{\sqrt{7}}{7}$ **47.** $\frac{\sqrt{10}}{5}$ **49.** $\frac{13(3 - \sqrt{11})}{-2}$ **51.** $7(\sqrt{5} + 2)$ **53.** $3(\sqrt{5} - \sqrt{3})$ **55.** 5 **57.** −2 **59.** not a real number **61.** 3

63. -3 **65.** $-\frac{1}{2}$ **67.** $2\sqrt[3]{4}$ **69.** $x\sqrt[3]{x}$ **71.** $3\sqrt[3]{2}$ **73.** $2x$ **75.** $7\sqrt[5]{2}$ **77.** $13\sqrt[3]{2}$ **79.** $-y\sqrt[3]{2x}$ **81.** $\sqrt{2}+2$ **83.** 6
85. 2 **87.** 25 **89.** $\frac{1}{16}$ **91.** $14x^{7/12}$ **93.** $4x^{1/4}$ **95.** x^2 **97.** $5x^2|y|^3$ **99.** $27y^{2/3}$ **101.** $\sqrt{5}$ **103.** x^2 **105.** $\sqrt[3]{x^2}$
107. $\sqrt[5]{x^2y}$ **109.** 3 **111.** $\frac{x^2}{7y^{3/2}}$ **113.** $\frac{x^3}{y^2}$ **115.** $6\sqrt{3}$ miles; 10.4 miles **117.** 70 mph; He was speeding.
119. $\frac{7\sqrt{2\cdot 2\cdot 3}}{6}=\frac{7\cdot 2\sqrt{3}}{6}=\frac{14\sqrt{3}}{6}=\frac{7\sqrt{3}}{3}=\frac{7}{3}\sqrt{3}$ **121. a.** $C = 35.74 + 0.6215t - 35.74v^{4/25} + 0.4275tv^{4/25}$ **b.** 8°F
123. $P = 18\sqrt{5}$ ft ; $A = 100$ sq ft **133.** d **135.** Let □ = 25 and □ = 14. **137. a.** > **b.** >

Section P.4

Check Point Exercises

1. a. $-x^3 + x^2 - 8x - 20$ **b.** $20x^3 - 11x^2 - 2x - 8$ **2.** $15x^3 - 31x^2 + 30x - 8$ **3.** $28x^2 - 41x + 15$ **4. a.** $49x^2 - 64$ **b.** $4y^6 - 25$
5. a. $x^2 + 20x + 100$ **b.** $25x^2 + 40x + 16$ **6. a.** $x^2 - 18x + 81$ **b.** $49x^2 - 42x + 9$ **7.** $2x^2y + 5xy^2 - 2y^3$ **8. a.** $21x^2 - 25xy + 6y^2$
b. $4x^2 + 16xy + 16y^2$

Exercise Set P.4

1. yes; $3x^2 + 2x - 5$ **3.** no **5.** 2 **7.** 4 **9.** $11x^3 + 7x^2 - 12x - 4$; 3 **11.** $12x^3 + 4x^2 + 12x - 14$; 3 **13.** $6x^2 - 6x + 2$; 2
15. $x^3 + 1$ **17.** $2x^3 - 9x^2 + 19x - 15$ **19.** $x^2 + 10x + 21$ **21.** $x^2 - 2x - 15$ **23.** $6x^2 + 13x + 5$ **25.** $10x^2 - 9x - 9$
27. $15x^4 - 47x^2 + 28$ **29.** $8x^5 - 40x^3 + 3x^2 - 15$ **31.** $x^2 - 9$ **33.** $9x^2 - 4$ **35.** $25 - 49x^2$ **37.** $16x^4 - 25x^2$ **39.** $1 - y^{10}$
41. $x^2 + 4x + 4$ **43.** $4x^2 + 12x + 9$ **45.** $x^2 - 6x + 9$ **47.** $16x^4 - 8x^2 + 1$ **49.** $4x^2 - 28x + 49$ **51.** $x^3 + 3x^2 + 3x + 1$
53. $8x^3 + 36x^2 + 54x + 27$ **55.** $x^3 - 9x^2 + 27x - 27$ **57.** $27x^3 - 108x^2 + 144x - 64$ **59.** $7x^2y - 4xy$ is of degree 3
61. $2x^2y + 13xy + 13$ is of degree 3 **63.** $-5x^3 + 8xy - 9y^2$ is of degree 3 **65.** $x^4y^2 + 8x^3y + y - 6x$ is of degree 6
67. $7x^2 + 38xy + 15y^2$ **69.** $2x^2 + xy - 21y^2$ **71.** $15x^2y^2 + xy - 2$ **73.** $49x^2 + 70xy + 25y^2$ **75.** $x^4y^4 - 6x^2y^2 + 9$
77. $x^3 - y^3$ **79.** $9x^2 - 25y^2$ **81.** $49x^2y^4 - 100y^2$ **83.** $48xy$ **85.** $-9x^2 + 3x + 9$ **87.** $16x^4 - 625$ **89.** $4x^2 - 28x + 49$
91. Model 4; $0.01x^3 + 0.09x^2 + 1.1x + 5.64$ **93.** Model 1 **95.** very well **97.** $4x^3 - 36x^2 + 80x$ **99.** $6x + 22$
109. $49x^2 + 70x + 25 - 16y^2$ **111.** $6x^n - 13$

Mid-Chapter P Check Point

1. $12x^2 - x - 35$ **2.** $-x + 12$ **3.** $10\sqrt{6}$ **4.** $3\sqrt{3}$ **5.** $x + 45$ **6.** $6x^2 - 48x + 9$ **7.** $\frac{x^2}{y^3}$ **8.** $\frac{3}{4}$ **9.** $-x^2 + 5x - 6$
10. $2x^3 - 11x^2 + 17x - 5$ **11.** $-x^6 + 2x^3$ **12.** $18a^2 - 11ab - 10b^2$ **13.** $\{a, c, d, e, f, h\}$ **14.** $\{c, d\}$ **15.** $5x^2y^3 + 2xy - y^2$
16. $-\frac{12y^{15}}{x^3}$ **17.** $\frac{6y^3}{x^7}$ **18.** $|\sqrt[3]{x}|$ **19.** 1.2×10^{-2} **20.** $2\sqrt[3]{2}$ **21.** $x^6 - 4$ **22.** $x^4 + 4x^2 + 4$ **23.** $10\sqrt{3}$
24. $\frac{77 + 11\sqrt{3}}{46}$ **25.** $\frac{11\sqrt{3}}{3}$ **26.** $-11, -\frac{3}{7}, 0, 0.45, \sqrt{25}$ **27.** $\sqrt{13} - 2$ **28.** $-x^3$ **29.** $\$3.48 \times 10^{10}$ **30.** 4 times
31. a. Model 3 **b.** \$1001 million, or \$1,001,000,000

Section P.5

Check Point Exercises

1. a. $2x^2(5x - 2)$ **b.** $(x - 7)(2x + 3)$ **2.** $(x + 5)(x^2 - 2)$ **3. a.** $(x + 8)(x + 5)$ **b.** $(x - 7)(x + 2)$ **4.** $(3x - 1)(2x + 7)$
5. $(3x - y)(x - 4y)$ **6. a.** $(x + 9)(x - 9)$ **b.** $(6x + 5)(6x - 5)$ **7.** $(9x^2 + 4)(3x + 2)(3x - 2)$ **8. a.** $(x + 7)^2$ **b.** $(4x - 7)^2$
9. a. $(x + 1)(x^2 - x + 1)$ **b.** $(5x - 2)(25x^2 + 10x + 4)$ **10.** $3x(x - 5)^2$ **11.** $(x + 10 + 6a)(x + 10 - 6a)$ **12.** $\frac{2x - 1}{(x - 1)^{1/2}}$

Exercise Set P.5

1. $9(2x + 3)$ **3.** $3x(x + 2)$ **5.** $9x^2(x^2 - 2x + 3)$ **7.** $(x + 5)(x + 3)$ **9.** $(x - 3)(x^2 + 12)$ **11.** $(x - 2)(x^2 + 5)$
13. $(x - 1)(x^2 + 2)$ **15.** $(3x - 2)(x^2 - 2)$ **17.** $(x + 2)(x + 3)$ **19.** $(x - 5)(x + 3)$ **21.** $(x - 5)(x - 3)$ **23.** $(3x + 2)(x - 1)$
25. $(3x - 28)(x + 1)$ **27.** $(2x - 1)(3x - 4)$ **29.** $(2x + 3)(2x + 5)$ **31.** $(3x - 2)(3x - 1)$ **33.** $(5x + 8)(4x - 1)$
35. $(2x + y)(x + y)$ **37.** $(3x + 2y)(2x - 3y)$ **39.** $(x + 10)(x - 10)$ **41.** $(6x + 7)(6x - 7)$ **43.** $(3x + 5y)(3x - 5y)$
45. $(x^2 + 4)(x + 2)(x - 2)$ **47.** $(4x^2 + 9)(2x + 3)(2x - 3)$ **49.** $(x + 1)^2$ **51.** $(x - 7)^2$ **53.** $(2x + 1)^2$ **55.** $(3x - 1)^2$
57. $(x + 3)(x^2 - 3x + 9)$ **59.** $(x - 4)(x^2 + 4x + 16)$ **61.** $(2x - 1)(4x^2 + 2x + 1)$ **63.** $(4x + 3)(16x^2 - 12x + 9)$
65. $3x(x + 1)(x - 1)$ **67.** $4(x + 2)(x - 3)$ **69.** $2(x^2 + 9)(x + 3)(x - 3)$ **71.** $(x - 3)(x + 3)(x + 2)$ **73.** $2(x - 8)(x + 7)$
75. $x(x - 2)(x + 2)$ **77.** prime **79.** $(x - 2)(x + 2)^2$ **81.** $y(y^2 + 9)(y + 3)(y - 3)$ **83.** $5y^2(2y + 3)(2y - 3)$
85. $(x - 6 + 7y)(x - 6 - 7y)$ **87.** $(x + y)(3b + 4)(3b - 4)$ **89.** $(y - 2)(x + 4)(x - 4)$ **91.** $2x(x + 6 + 2a)(x + 6 - 2a)$
93. $x^{1/2}(x - 1)$ **95.** $\frac{4(1 + 2x)}{x^{2/3}}$ **97.** $-(x + 3)^{1/2}(x + 2)$ **99.** $\frac{x + 4}{(x + 5)^{3/2}}$ **101.** $-\frac{4(4x - 1)^{1/2}(x - 1)}{3}$ **103.** $(x + 1)(5x - 6)(2x + 1)$
105. $(x^2 + 6)(6x^2 - 1)$ **107.** $y(y^2 + 1)(y^4 - y^2 + 1)$ **109.** $(x + 2y)(x - 2y)(x + y)(x - y)$ **111.** $(x - y)^2(x - y + 2)(x - y - 2)$
113. $(2x - y^2)(x - 3y^2)$ **115. a.** $(x - 0.4x)(1 - 0.4) = (0.6x)(0.6) = 0.36x$ **b.** no; 36% **117. a.** $9x^2 - 16$ **b.** $(3x + 4)(3x - 4)$
119. a. $x(x + y) - y(x + y)$ **b.** $(x + y)(x - y)$ **121.** $4a^3 - 4ab^2 = 4a(a + b)(a - b)$ **131.** $(x^n + 4)(x^n + 2)$ **133.** $(x - y)^3(x + y)$
135. $b = 8, -8, 16, -16$

Section P.6

Check Point Exercises

1. a. -5 **b.** $6, -6$ **2. a.** $x^2, x \neq -3$ **b.** $\frac{x - 1}{x + 1}, x \neq -1$ **3.** $\frac{x - 3}{(x - 2)(x + 3)}, x \neq 2, x \neq -2, x \neq -3$

4. $\frac{3(x-1)}{x(x+2)}, x \neq 1, x \neq 0, x \neq -2$ **5.** $-2, x \neq -1$ **6.** $\frac{2(4x+1)}{(x+1)(x-1)}, x \neq 1, x \neq -1$ **7.** $(x-3)(x-3)(x+3)$ **8.** $\frac{-x^2+11x-20}{2(x-5)^2}, x \neq 5$ **9.** $\frac{2(2-3x)}{4+3x}, x \neq 0, x \neq -\frac{4}{3}$ **10.** $-\frac{1}{x(x+7)}, x \neq 0, x \neq -7$

Exercise Set P.6

1. 3 **3.** 5, −5 **5.** −1, −10 **7.** $\frac{3}{x-3}, x \neq 3$ **9.** $\frac{x-6}{4}, x \neq 6$ **11.** $\frac{y+9}{y-1}, y \neq 1, 2$ **13.** $\frac{x+6}{x-6}, x \neq 6, -6$ **15.** $\frac{1}{3}, x \neq 2, -3$ **17.** $\frac{(x-3)(x+3)}{x(x+4)}, x \neq 0, -4, 3$ **19.** $\frac{x-1}{x+2}, x \neq -2, -1, 2, 3$ **21.** $\frac{x^2+2x+4}{3x}, x \neq -2, 0, 2$ **23.** $\frac{7}{9}, x \neq -1$ **25.** $\frac{(x-2)^2}{x}, x \neq 0, -2, 2$ **27.** $\frac{2(x+3)}{3}, x \neq 3, -3$ **29.** $\frac{x-5}{2}, x \neq 1, -5$ **31.** $\frac{(x+2)(x+4)}{x-5}, x \neq -6, -3, -1, 3, 5$ **33.** $2, x \neq -\frac{5}{6}$ **35.** $\frac{2x-1}{x+3}, x \neq 0, -3$ **37.** $3, x \neq 2$ **39.** $\frac{3}{x-3}, x \neq 3, -4$ **41.** $\frac{9x+39}{(x+4)(x+5)}, x \neq -4, -5$ **43.** $-\frac{3}{x(x+1)}, x \neq -1, 0$ **45.** $\frac{3x^2+4}{(x+2)(x-2)}, x \neq -2, 2$ **47.** $\frac{2x^2+50}{(x-5)(x+5)}, x \neq -5, 5$ **49.** $\frac{4x+16}{(x+3)^2}, x \neq -3$ **51.** $\frac{x^2-x}{(x+5)(x-2)(x+3)}, x \neq -5, 2, -3$ **53.** $\frac{x-1}{x+2}, x \neq -2, -1$ **55.** $\frac{1}{3}, x \neq 3$ **57.** $\frac{x+1}{3x-1}, x \neq 0, \frac{1}{3}$ **59.** $\frac{1}{xy}, x \neq 0, y \neq 0, x \neq -y$ **61.** $\frac{x}{x+3}, x \neq -2, -3$ **63.** $-\frac{x-14}{7}, x \neq -2, 2$ **65.** $\frac{x-3}{x+2}, x \neq -2, -1, 3$ **67.** $-\frac{2x+h}{x^2(x+h)^2}, x \neq 0, h \neq 0, x \neq -h$ **69.** $\frac{x^2+5x+8}{(x+2)(x+1)}$ **71.** 2 **73.** $\frac{1}{y(y+5)}$ **75.** $\frac{2d}{a^2+ab+b^2}$

77. a. 86.67, 520, 1170; It costs \$86,670,000 to inoculate 40% of the population against this strain of flu, \$520,000,000 to inoculate 80% of the population, and \$1,170,000,000 to inoculate 90% of the population. **b.** $x = 100$ **c.** increases rapidly; impossible to inoculate 100% of the population. **79. a.** $\frac{-0.3t+14}{3.6t+260}$ **b.** 0.04; 4000 **c.** fairly well **81.** $\frac{4x^2+14x}{(x+3)(x+4)}$ **97.** $\frac{1}{x^{2n}-1}$ **99.** $\frac{x-y+1}{(x-y)(x-y)}$

Chapter P Review Exercises

1. 51 **2.** 16 **3.** 124 ft **4.** $\{a, c\}$ **5.** $\{a, b, c, d, e\}$ **6.** $\{a, b, c, d, f, g\}$ **7.** $\{a\}$ **8. a.** $\sqrt{81}$ **b.** $0, \sqrt{81}$ **c.** $-17, 0, \sqrt{81}$ **d.** $-17, -\frac{9}{13}, 0, 0.75, \sqrt{81}$ **e.** $\sqrt{2}, \pi$ **f.** $-17, -\frac{9}{13}, 0, 0.75, \sqrt{2}, \pi, \sqrt{81}$ **9.** 103 **10.** $\sqrt{2}-1$ **11.** $\sqrt{17}-3$ **12.** $|4-(-17)|$; 21 **13.** commutative property of addition **14.** associative property of multiplication **15.** distributive property of multiplication over addition **16.** commutative property of multiplication **17.** commutative property of multiplication **18.** commutative property of addition **19.** $17x - 15$ **20.** $2x$ **21.** $5y - 17$ **22.** $10x$ **23.** $E = 0.04x^2 + 9.2x + 169$ **24.** −108 **25.** $\frac{5}{16}$ **26.** $\frac{1}{25}$ **27.** $\frac{1}{27}$ **28.** $-8x^{12}y^9$ **29.** $\frac{10}{x^8}$ **30.** $\frac{1}{16x^{12}}$ **31.** $\frac{y^8}{4x^{10}}$ **32.** 37,400 **33.** 0.0000745 **34.** 3.59×10^6 **35.** 7.25×10^{-3} **36.** 390,000 **37.** 0.023 **38.** 10^3 or 1000 yr **39.** $\$4.35 \times 10^{10}$ **40.** $10\sqrt{3}$ **41.** $2|x|\sqrt{3}$ **42.** $2x\sqrt{5}$ **43.** $r\sqrt{r}$ **44.** $\frac{11}{2}$ **45.** $4x\sqrt{3}$ **46.** $20\sqrt{5}$ **47.** $16\sqrt{2}$ **48.** $24\sqrt{2} - 8\sqrt{3}$ **49.** $6\sqrt{5}$ **50.** $\frac{\sqrt{6}}{3}$ **51.** $\frac{5(6-\sqrt{3})}{33}$ **52.** $7(\sqrt{7}+\sqrt{5})$ **53.** 5 **54.** −2 **55.** not a real number **56.** 5 **57.** $3\sqrt[3]{3}$ **58.** $y\sqrt[3]{y^2}$ **59.** $2\sqrt[4]{5}$ **60.** $13\sqrt[3]{2}$ **61.** $x\sqrt[4]{2}$ **62.** 4 **63.** $\frac{1}{5}$ **64.** 5 **65.** $\frac{1}{3}$ **66.** 16 **67.** $\frac{1}{81}$ **68.** $20x^{11/12}$ **69.** $3x^{1/4}$ **70.** $25x^4$ **71.** $\sqrt{y}$ **72.** $8x^3 + 10x^2 - 20x - 4$; degree 3 **73.** $8x^4 - 5x^3 + 6$; degree 4 **74.** $12x^3 + x^2 - 21x + 10$ **75.** $6x^2 - 7x - 5$ **76.** $16x^2 - 25$ **77.** $4x^2 + 20x + 25$ **78.** $9x^2 - 24x + 16$ **79.** $8x^3 + 12x^2 + 6x + 1$ **80.** $125x^3 - 150x^2 + 60x - 8$ **81.** $-x^2 - 17xy - 3y^2$; degree 2 **82.** $24x^3y^2 + x^2y - 12x^2 + 4$; degree 5 **83.** $3x^2 + 16xy - 35y^2$ **84.** $9x^2 - 30xy + 25y^2$ **85.** $9x^4 + 12x^2y + 4y^2$ **86.** $49x^2 - 16y^2$ **87.** $a^3 - b^3$ **88.** $3x^2(5x+1)$ **89.** $(x-4)(x-7)$ **90.** $(3x+1)(5x-2)$ **91.** $(8-x)(8+x)$ **92.** prime **93.** $3x^2(x-5)(x+2)$ **94.** $4x^3(5x^4-9)$ **95.** $(x+3)(x-3)^2$ **96.** $(4x-5)^2$ **97.** $(x^2+4)(x+2)(x-2)$ **98.** $(y-2)(y^2+2y+4)$ **99.** $(x+4)(x^2-4x+16)$ **100.** $3x^2(x-2)(x+2)$ **101.** $(3x-5)(9x^2+15x+25)$ **102.** $x(x-1)(x+1)(x^2+1)$ **103.** $(x^2-2)(x+5)$ **104.** $(x+9+y)(x+9-y)$ **105.** $\frac{16(1+2x)}{x^{3/4}}$ **106.** $(x+2)(x-2)(x^2+3)^{1/2}(-x^4+x^2+13)$ **107.** $\frac{6(2x+1)}{x^{3/2}}$ **108.** $x^2, x \neq -2$ **109.** $\frac{x-3}{x-6}, x \neq -6, 6$ **110.** $\frac{x}{x+2}, x \neq -2$ **111.** $\frac{(x+3)^3}{(x-2)^2(x+2)}, x \neq 2, -2$ **112.** $\frac{2}{x(x+1)}, x \neq 0, 1, -1, -\frac{1}{3}$ **113.** $\frac{x+3}{x-4}, x \neq -3, 4, 2, 8$ **114.** $\frac{1}{x-3}, x \neq 3, -3$ **115.** $\frac{4x(x-1)}{(x+2)(x-2)}, x \neq 2, -2$ **116.** $\frac{2x^2-3}{(x-3)(x+3)(x-2)}, x \neq 3, -3, 2$ **117.** $\frac{11x^2-x-11}{(2x-1)(x+3)(3x+2)}, x \neq \frac{1}{2}, -3, -\frac{2}{3}$ **118.** $\frac{3}{x}, x \neq 0, 2$ **119.** $\frac{3x}{x-4}, x \neq 0, 4, -4$ **120.** $\frac{3x+8}{3x+10}, x \neq -3, -\frac{10}{3}$

Chapter P Test

1. $6x^2 - 27x$ **2.** $-6x + 17$ **3.** $\{5\}$ **4.** $\{1, 2, 5, a\}$ **5.** $6x^2y^3 + 4xy + 2y^2$ **6.** $\frac{5y^8}{x^6}$ **7.** $3r\sqrt{2}$ **8.** $11\sqrt{2}$ **9.** $\frac{3(5-\sqrt{2})}{23}$ **10.** $2x\sqrt[3]{2x}$ **11.** $\frac{x+3}{x-2}, x \neq 2, 1$ **12.** 2.5×10^1 **13.** $2x^3 - 13x^2 + 26x - 15$ **14.** $25x^2 + 30xy + 9y^2$ **15.** $\frac{2(x+3)}{x+1}, x \neq 3, -1, -4, -3$ **16.** $\frac{x^2+2x+15}{(x+3)(x-3)}, x \neq 3, -3$ **17.** $\frac{11}{(x-3)(x-4)}, x \neq 3, 4$ **18.** $\frac{3-x}{3}, x \neq 0$ **19.** $(x-3)(x-6)$ **20.** $(x^2+3)(x+2)$ **21.** $(5x-3)(5x+3)$ **22.** $(6x-7)^2$ **23.** $(y-5)(y^2+5y+25)$ **24.** $(x+5+3y)(x+5-3y)$ **25.** $\frac{2x+3}{(x+3)^{3/5}}$ **26.** $-7, -\frac{4}{5}, 0, 0.25, \sqrt{4}, \frac{22}{7}$ **27.** commutative property of addition **28.** distributive property of multiplication over addition **29.** 7.6×10^{-4} **30.** $\frac{1}{243}$ **31.** 1.26×10^{10} **32. a.** men: Model 2; women: Model 1 **b.** 74; fairly well

CHAPTER 1

Section 1.1

Check Point Exercises

1.

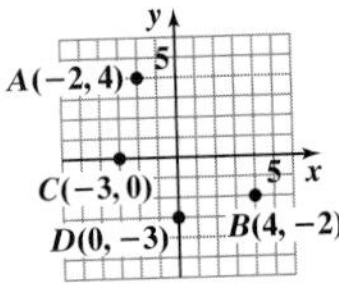

2.

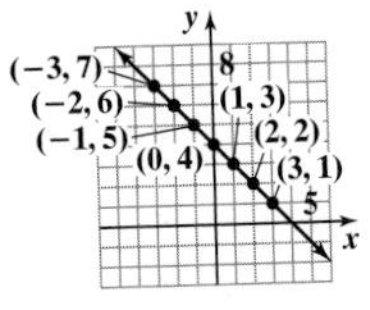

$y = 4 - x$

3.

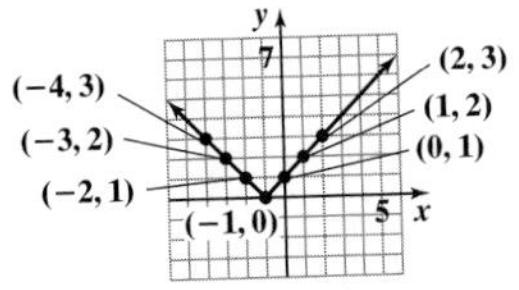

$y = |x + 1|$

4. minimum x-value: −100; maximum x-value: 100; distance between tick marks on x-axis: 50; minimum y-value: −100; maximum y-value: 100; distance between tick marks on y-axis: 10

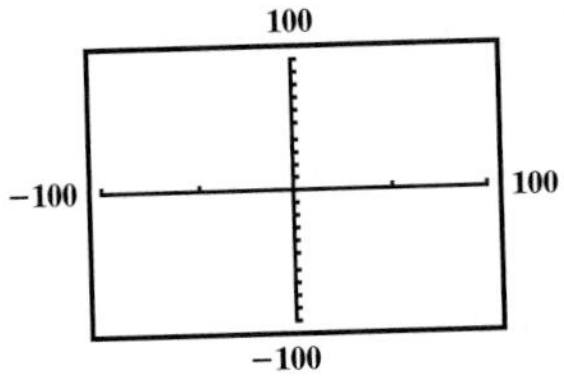

5. a. x-intercept: −3; y-intercept: 5 **b.** no x-intercept; y-intercept: 4 **c.** x-intercept: 0; y-intercept: 0

6. approximately 96,000

Exercise Set 1.1

1.

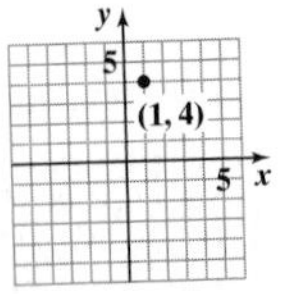

3.

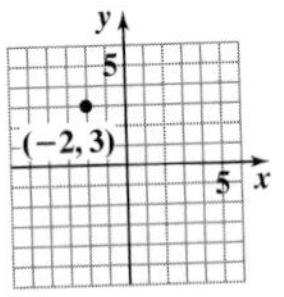

5.

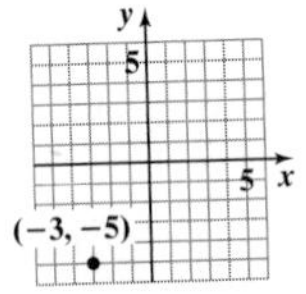

7.

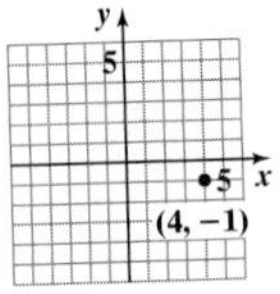

9.

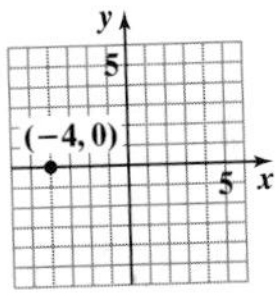

11.

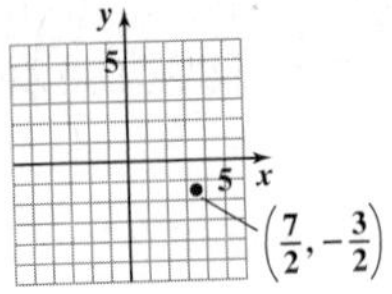

13.

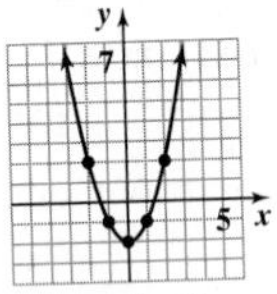

$y = x^2 - 2$

15.

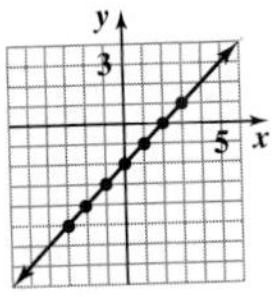

$y = x - 2$

17.

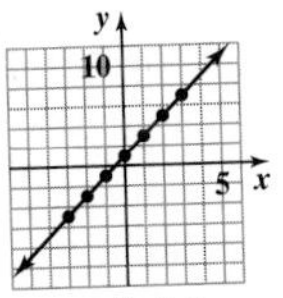

$y = 2x + 1$

19.

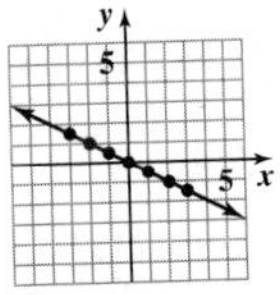

$y = -\frac{1}{2}x$

21.

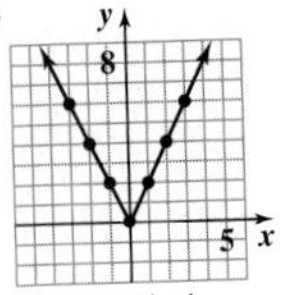

$y = 2|x|$

23.

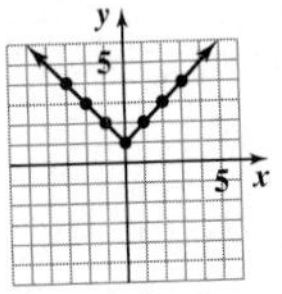

$y = |x| + 1$

25.

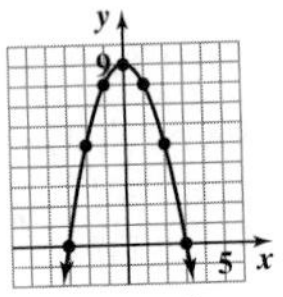

$y = 9 - x^2$

27.

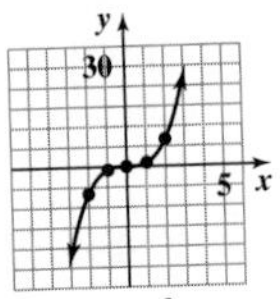

$y = x^3$

29. (c) **31.** (b) **33.** c **35.** no **37.** (2, 0) **39.** (−2, 4) and (1, 1) **41. a.** 2 **b.** −4 **43. a.** 1, −2 **b.** 2 **45. a.** −1 **b.** none

47. $y = 2x + 4$

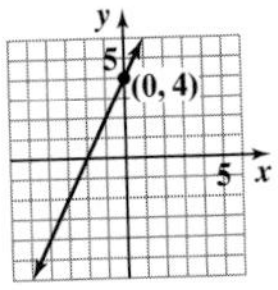

49. $y = 3 - x^2$

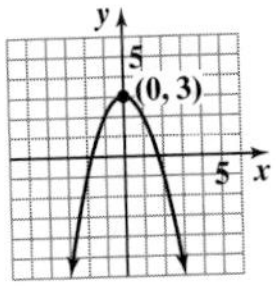

51.

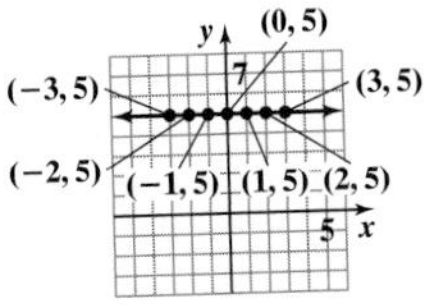

53.

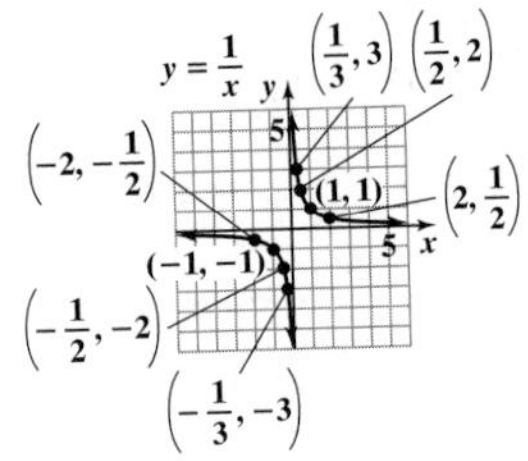

55. 65 **57.** 1989–1993 **59.** 1977 **61.** 135 beats/min; (40, 135) on blue graph **63. a.** 36 cm **b.** 44.7 cm **c.** 46.9 cm **d.** describes healthy children

71.

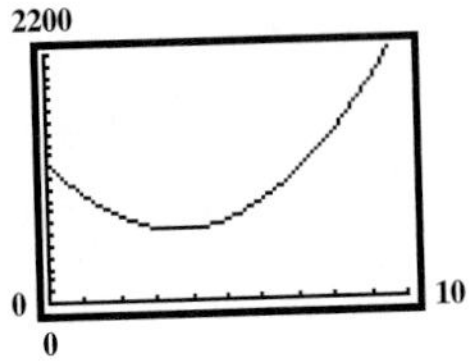

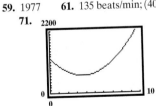

73. a **75.** b **77.** b

Section 1.2

Check Point Exercises

1. {6} **2.** {5} **3.** {1} **4.** {3} **5.** ∅ **6.** 11 **7.** ∅; inconsistent equation

Exercise Set 1.2

1. {11} **3.** {7} **5.** {13} **7.** {2} **9.** {9} **11.** {−5} **13.** {6} **15.** {−2} **17.** {12} **19.** {24} **21.** {−15} **23.** {5} **25.** $\left\{\frac{33}{2}\right\}$ **27.** {−12} **29.** $\left\{\frac{46}{5}\right\}$ **31. a.** 0 **b.** $\left\{\frac{1}{2}\right\}$ **33. a.** 0 **b.** {−2} **35. a.** 0 **b.** {2} **37. a.** 0 **b.** {4}

39. a. 1 **b.** {3} **41. a.** −1 **b.** ∅ **43. a.** 1 **b.** {2} **45. a.** −2, 2 **b.** ∅ **47. a.** −1, 1 **b.** {−3} **49. a.** −2, 4 **b.** ∅ **51.** 6 **53.** −7 **55.** 2 **57.** 19 **59.** −1 **61.** identity **63.** inconsistent equation **65.** conditional equation **67.** inconsistent equation **69.** {−7}; conditional equation **71.** ∅; inconsistent equation **73.** {−4}; conditional equation **75.** {8}; conditional equation **77.** {−1}; conditional equation **79.** ∅; inconsistent equation **81.** $3(x-4)=3(2-2x)$; {2} **83.** $-3(x-3)=5(2-x)$; {0.5} **85.** 2 **87.** −7 **89.** {−2} **91.** ∅ or no solution **93.** {10} **95.** {−2} **97.** 2006 **99.** 5.5; point (5.5, 3.5) on high-humor graph **101.** 12 years old **103.** (12, 500) on blue graph **105.** no **107.** 11 learning trials; (11, 0.95) **109.** 125 liters

121. {5}

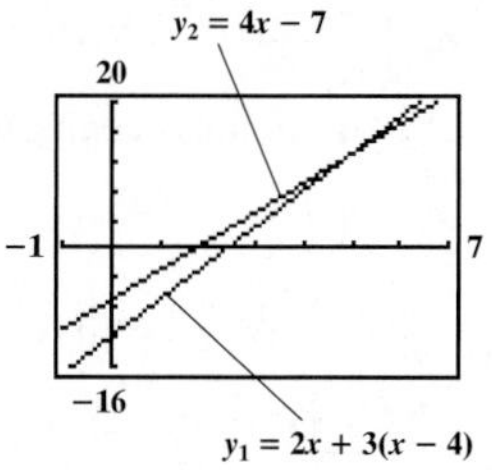

123. {−5}

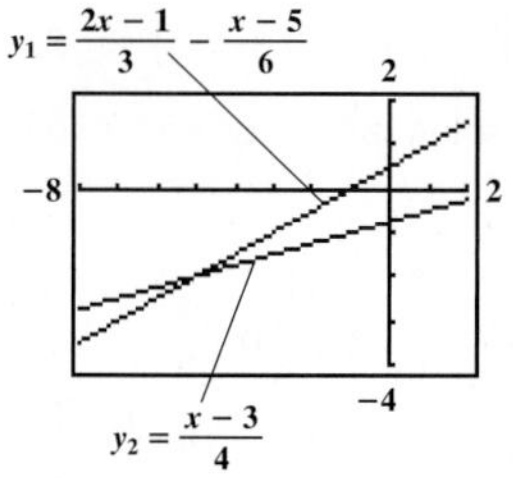

127. 20

Section 1.3

Check Point Exercises

1. basketball: 1.6 million; bicycle riding: 1.3 million; football: 1 million **2.** 2011 **3.** 300 min **4.** $1200 **5.** $3150 at 9%; $1850 at 11% **6.** 50 ft by 94 ft **7.** $w=\dfrac{P-2l}{2}$ **8.** $C=\dfrac{P}{1+M}$

Exercise Set 1.3

1. 6 **3.** 25 **5.** 120 **7.** 320 **9.** 19 and 45 **11.** 2 **13.** 8 **15.** all real numbers **17.** 5 **19.** births: 375 thousand; deaths: 146 thousand **21.** U.S.: 169 million; Japan: 56 million; China: 46 million **23.** liberals: 39.6%; conservatives: 17.6% **25.** 2050 **27. a.** $y=1.7x+39.8$ **b.** 2008 **29.** after 7 years **31.** after 5 months; $165 **33.** 30 times

c.

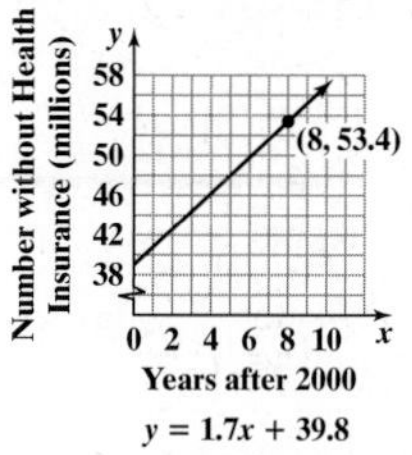

35. a. 2014; 22,300 students **b.** $y_1=13{,}300+1000x$; $y_2=26{,}800-500x$ **37.** $420 **39.** $150 **41.** $36,000 **43.** $467.20 **45.** $2000 at 6%; $5000 at 8% **47.** $6000 at 12%; $2000 at a 5% loss **49.** 50 yd by 100 yd **51.** 36 ft by 78 ft **53.** 2 in. **55.** 11 hr **57.** 5 ft 7 in. **59.** 7 oz **61.** $\omega=\dfrac{A}{l}$ **63.** $b=\dfrac{2A}{h}$ **65.** $p=\dfrac{I}{rt}$ **67.** $m=\dfrac{E}{c^2}$ **69.** $p=\dfrac{T-D}{m}$ **71.** $a=\dfrac{2A}{h}-b$ **73.** $r=\dfrac{S-P}{Pt}$ **75.** $S=\dfrac{F}{B}+V$ **77.** $I=\dfrac{E}{R+r}$ **79.** $f=\dfrac{pq}{p+q}$

89. a. $F=30+5x$; $F=7.5x$

b.

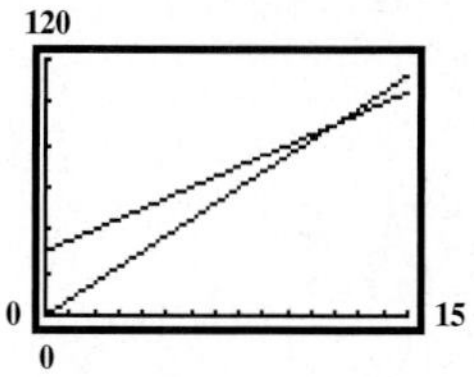

c. (12, 90); For 12 hours, both options cost the same, $90.

91. $200
93. 10 correct answers
95. 36 plants

Section 1.4

Check Point Exercises

1. a. $8+i$ **b.** $-10+7i$ **2. a.** $63+14i$ **b.** $58-11i$ **3.** $\dfrac{16}{17}+\dfrac{21}{17}i$ **4. a.** $7i\sqrt{3}$ **b.** $1-4i\sqrt{3}$ **c.** $-7+i\sqrt{3}$

Exercise Set 1.4

1. $8-2i$ **3.** $-2+9i$ **5.** $24-3i$ **7.** $-14+17i$ **9.** $21+15i$ **11.** $-19+7i$ **13.** $-29-11i$ **15.** 34 **17.** 26 **19.** $-5+12i$ **21.** $\dfrac{3}{5}+\dfrac{1}{5}i$ **23.** $1+i$ **25.** $-\dfrac{24}{25}+\dfrac{32}{25}i$ **27.** $\dfrac{7}{5}+\dfrac{4}{5}i$ **29.** $3i$ **31.** $47i$ **33.** $-8i$ **35.** $2+6i\sqrt{7}$ **37.** $-\dfrac{1}{3}+\dfrac{\sqrt{2}}{6}i$ **39.** $-\dfrac{1}{8}-\dfrac{\sqrt{3}}{24}i$ **41.** $-2\sqrt{6}-2i\sqrt{10}$ **43.** $24\sqrt{15}$ **45.** $-11-5i$ **47.** $-5+10i$ **49.** $0+47i$ or $47i$ **51.** 0 **53.** $\dfrac{20}{13}+\dfrac{30}{13}i$ **55.** $(47+13i)$ volts **57.** $(5+i\sqrt{15})+(5-i\sqrt{15})=10$; $(5+i\sqrt{15})(5-i\sqrt{15})=25-15i^2=25+15=40$

67. d **69.** $\frac{6}{5} + 0i$ or $\frac{6}{5}$

Section 1.5

Check Point Exercises

1. a. $\{0, 3\}$ **b.** $\left\{-1, \frac{1}{2}\right\}$ **2. a.** $\{-\sqrt{7}, \sqrt{7}\}$ **b.** $\pm 3i$ **c.** $\{-5 + \sqrt{11}, -5 - \sqrt{11}\}$ **3. a.** $9; x^2 + 6x + 9 = (x + 3)^2$
b. $\frac{25}{4}; x^2 - 5x + \frac{25}{4} = \left(x - \frac{5}{2}\right)^2$ **c.** $\frac{1}{9}; x^2 + \frac{2}{3}x + \frac{1}{9} = \left(x + \frac{1}{3}\right)^2$ **4.** $\{-2 \pm \sqrt{5}\}$ **5.** $\left\{\frac{-3 \pm \sqrt{41}}{4}\right\}$ **6.** $\left\{\frac{-1 + \sqrt{3}}{2}, \frac{-1 - \sqrt{3}}{2}\right\}$
7. $\{1 + i, 1 - i\}$ **8. a.** 0; one real solution **b.** 81; two rational solutions **c.** -44; two imaginary solutions that are complex conjugates
9. approximately 26 years old **10.** 12 in.

Exercise Set 1.5

1. $\{-2, 5\}$ **3.** $\{3, 5\}$ **5.** $\left\{-\frac{5}{2}, \frac{2}{3}\right\}$ **7.** $\left\{-\frac{4}{3}, 2\right\}$ **9.** $\{-4, 0\}$ **11.** $\left\{0, \frac{1}{3}\right\}$ **13.** $\{-3, 1\}$ **15.** $\{-3, 3\}$ **17.** $\{-\sqrt{10}, \sqrt{10}\}$
19. $\{\pm 5i\}$ **21.** $\{-7, 3\}$ **23.** $\{4 \pm \sqrt{5}\}$ **25.** $\{-3 \pm 4i\}$ **27.** $\{3 \pm i\sqrt{5}\}$ **29.** $\left\{-\frac{5}{3}, \frac{1}{3}\right\}$ **31.** $\left\{\frac{1 - \sqrt{7}}{5}, \frac{1 + \sqrt{7}}{5}\right\}$
33. $\left\{\frac{4 - 2\sqrt{2}}{3}, \frac{4 + 2\sqrt{2}}{3}\right\}$ **35.** $36; x^2 + 12x + 36 = (x + 6)^2$ **37.** $25; x^2 - 10x + 25 = (x - 5)^2$ **39.** $\frac{9}{4}; x^2 + 3x + \frac{9}{4} = \left(x + \frac{3}{2}\right)^2$
41. $\frac{49}{4}; x^2 - 7x + \frac{49}{4} = \left(x - \frac{7}{2}\right)^2$ **43.** $\frac{1}{9}; x^2 - \frac{2}{3}x + \frac{1}{9} = \left(x - \frac{1}{3}\right)^2$ **45.** $\frac{1}{36}; x^2 - \frac{1}{3}x + \frac{1}{36} = \left(x - \frac{1}{6}\right)^2$ **47.** $\{-7, 1\}$
49. $\{1 + \sqrt{3}, 1 - \sqrt{3}\}$ **51.** $\{3 + 2\sqrt{5}, 3 - 2\sqrt{5}\}$ **53.** $\{-2 + \sqrt{3}, -2 - \sqrt{3}\}$ **55.** $\{2, 3\}$ **57.** $\left\{\frac{-3 + \sqrt{13}}{2}, \frac{-3 - \sqrt{13}}{2}\right\}$
59. $\left\{\frac{1}{2}, 3\right\}$ **61.** $\left\{\frac{1 + \sqrt{2}}{2}, \frac{1 - \sqrt{2}}{2}\right\}$ **63.** $\left\{\frac{1 + \sqrt{7}}{3}, \frac{1 - \sqrt{7}}{3}\right\}$ **65.** $\{-5, -3\}$ **67.** $\left\{\frac{-5 + \sqrt{13}}{2}, \frac{-5 - \sqrt{13}}{2}\right\}$
69. $\left\{\frac{3 + \sqrt{57}}{6}, \frac{3 - \sqrt{57}}{6}\right\}$ **71.** $\left\{\frac{1 + \sqrt{29}}{4}, \frac{1 - \sqrt{29}}{4}\right\}$ **73.** $\{3 + i, 3 - i\}$ **75.** 36; 2 unequal real solutions **77.** 97; 2 unequal real solutions
79. 0; 1 real solution **81.** 37; 2 unequal real solutions **83.** $\left\{-\frac{1}{2}, 1\right\}$ **85.** $\left\{\frac{1}{5}, 2\right\}$ **87.** $\{-2\sqrt{5}, 2\sqrt{5}\}$ **89.** $\{1 + \sqrt{2}, 1 - \sqrt{2}\}$
91. $\left\{\frac{-11 + \sqrt{33}}{4}, \frac{-11 - \sqrt{33}}{4}\right\}$ **93.** $\left\{0, \frac{8}{3}\right\}$ **95.** $\{2\}$ **97.** $\{-2, 2\}$ **99.** $\{3 + 2i, 3 - 2i\}$ **101.** $\{2 + i\sqrt{3}, 2 - i\sqrt{3}\}$
103. $\left\{0, \frac{7}{2}\right\}$ **105.** $\{2 + \sqrt{10}, 2 - \sqrt{10}\}$ **107.** $\{-5, -1\}$ **109.** -1 and 5; d **111.** -3 and 1; f **113.** no x-intercepts; b
115. $-\frac{1}{2}$ and 2 **117.** -6 and 3 **119.** $\frac{-5 - \sqrt{33}}{2}$ and $\frac{-5 + \sqrt{33}}{2}$ **121.** $\frac{5 - \sqrt{7}}{3}$ and $\frac{5 + \sqrt{7}}{3}$ **123.** $\frac{-2 - \sqrt{22}}{2}$ and $\frac{-2 + \sqrt{22}}{2}$
125. $1 + \sqrt{7}$ **127.** $\left\{\frac{-1 \pm \sqrt{21}}{2}\right\}$ **129.** $\left\{-2\sqrt{2}, \frac{\sqrt{2}}{2}\right\}$ **131.** 33-year-olds and 58-year-olds; The formula models the actual data well.
133. 77.8 ft; (b) **135.** $4\sqrt{5}$ ft or 8.9 ft **137.** 13.23 ft **139.** length: 9 ft; width: 6 ft **141.** 5 in. **143.** 5 m **145.** 10 in.
147. 9.3 in. and 0.7 in. **161.** $x^2 - 2x - 15 = 0$ **163.** 2.4 m; yes

Mid-Chapter 1 Check Point

1. $\{6\}$ **2.** $\left\{-1, \frac{7}{5}\right\}$ **3.** $\{-7\}$ **4.** $\left\{\frac{3 \pm \sqrt{15}}{3}\right\}$ **5.** all real numbers **6.** $\left\{\pm\frac{6\sqrt{5}}{5}\right\}$ **7.** $\left\{\frac{3}{4} \pm \frac{\sqrt{23}}{4}i\right\}$ **8.** $\{3\}$
9. $\{-3 \pm 2\sqrt{6}\}$ **10.** $\{2 \pm \sqrt{3}\}$ **11.** $\varnothing$ **12.** $\{4\}$ **13.** $-3 - \sqrt{7}$ and $-3 + \sqrt{7}$ **14.** 1 **15.** no x-intercepts
16. $\frac{-3 - \sqrt{41}}{4}$ and $\frac{-3 + \sqrt{41}}{4}$ **17.** no x-intercepts **18.** 0 **19.** -4 and $\frac{1}{2}$ **20.** $\{-5 \pm 2\sqrt{7}\}$
21. two imaginary solutions **22.** two rational solutions

23.

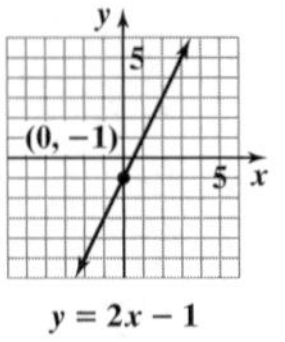

$y = 2x - 1$

24.

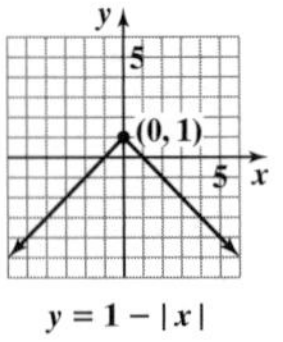

$y = 1 - |x|$

25.

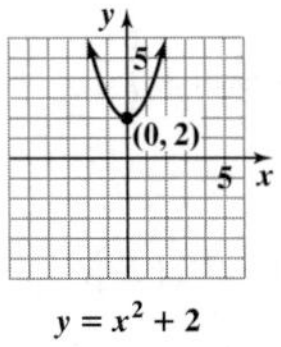

$y = x^2 + 2$

26. $n = \frac{L - a}{d} + 1$ **27.** $l = \frac{A - 2wh}{2w + 2h}$
28. $f_1 = -\frac{ff_2}{f - f_2}$ or $f_1 = -\frac{ff_2}{f_2 - f}$
29. U.S.: \$291 billion; Russia: \$44 billion; Japan: \$40 billion
30. 6 months **31.** \$11,500 at 8%; \$13,500 at 9%
32. 20 prints; \$3.80 **33.** 129 pounds **34.** \$2500 at 4%; \$1500 at 3% loss **35.** length: 17 ft; width: 6 ft **36.** length: 7 ft; width: 4 ft
37. 12 yd **38.** 1995; quite well **39.** 2011 **40.** $-1 - i$ **41.** $-3 + 6i$ **42.** $7 + i$ **43.** i **44.** $3i\sqrt{3}$ **45.** $1 - 4i\sqrt{3}$

Section 1.6

Check Point Exercises

1. $\{-\sqrt{3}, 0, \sqrt{3}\}$ **2.** $\left\{-2, -\frac{3}{2}, 2\right\}$ **3.** $\{6\}$ **4.** $\{4\}$ **5. a.** $\{\sqrt[3]{25}\}$ or $\{5^{2/3}\}$ **b.** $\{-8, 8\}$ **6.** $\{-\sqrt{3}, -\sqrt{2}, \sqrt{2}, \sqrt{3}\}$
7. $\left\{-\frac{1}{27}, 64\right\}$ **8.** $\{-2, 3\}$ **9.** $\{-2, 3\}$

Exercise Set 1.6

1. $\{-4, 0, 4\}$ **3.** $\left\{-2, -\frac{2}{3}, 2\right\}$ **5.** $\left\{-\frac{1}{2}, \frac{1}{2}, \frac{3}{2}\right\}$ **7.** $\left\{-2, -\frac{1}{2}, \frac{1}{2}\right\}$ **9.** $\{0, 2, -1 + i\sqrt{3}, -1 - i\sqrt{3}\}$ **11.** $\{6\}$ **13.** $\{6\}$

15. $\{-6\}$ **17.** $\{10\}$ **19.** $\{-5\}$ **21.** $\{12\}$ **23.** $\{8\}$ **25.** $\varnothing$ **27.** $\varnothing$ **29.** $\left\{\frac{13+\sqrt{105}}{6}\right\}$ **31.** $\{4\}$ **33.** $\{13\}$ **35.** $\{\sqrt[5]{4}\}$
37. $\{-60, 68\}$ **39.** $\{-4, 5\}$ **41.** $\{-2, -1, 1, 2\}$ **43.** $\left\{-\frac{4}{3}, -1, 1, \frac{4}{3}\right\}$ **45.** $\{25, 64\}$ **47.** $\left\{-\frac{1}{4}, \frac{1}{5}\right\}$ **49.** $\{-8, 27\}$ **51.** $\{1\}$
53. $\left\{\frac{1}{4}, 1\right\}$ **55.** $\{2, 12\}$ **57.** $\{-3, -1, 2, 4\}$ **59.** $\{-8, -2, 1, 4\}$ **61.** $\{-8, 8\}$ **63.** $\{-5, 9\}$ **65.** $\{-2, 3\}$ **67.** $\left\{-\frac{5}{3}, 3\right\}$
69. $\left\{-\frac{2}{5}, \frac{2}{5}\right\}$ **71.** $\left\{-\frac{4}{5}, 4\right\}$ **73.** $\varnothing$ **75.** $\left\{\frac{1}{2}\right\}$ **77.** $\{-1, 3\}$ **79.** 2; c **81.** 1; e **83.** 2 and 3; f **85.** $\left\{-\frac{3}{2}, 4\right\}$ **87.** $\{4\}$
89. $\left\{-2, -\frac{1}{2}, 2\right\}$ **91.** $\{0\}$ **93.** $\{-2, 0, 2\}$ **95.** $\{-8, -6, 4, 6\}$ **97.** $\{-7, -1, 6\}$ **99.** 8 **101.** $V = \frac{\pi r^2 h}{3}$ or $\frac{1}{3}\pi r^2 h$ **103.** -16 and 12
105. 16 years after 1996, or 2012 **107.** High: 66.0°; Low: 61.5° **109.** Using H: 2014; Using L: 2010 **111.** 36 years old; (36, 40,000)
113. 149 million km **115.** either 1.2 feet or 7.5 feet from the base of the 6 foot pole **125.** $\{-3, -1, 1\}$ **127.** $\{-2\}$
129. d **131.** $\left\{\frac{2}{5}, \frac{1}{2}\right\}$ **133.** $\{0, 1\}$

Section 1.7

Check Point Exercises

1. a. $\{x \mid -2 \le x < 5\}$

b. $\{x \mid 1 \le x \le 3.5\}$

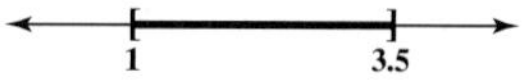

c. $\{x \mid x < -1\}$

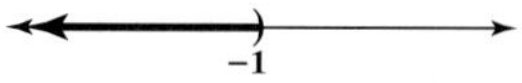

2. a. $(2, 3]$ **b.** $[1, 6)$

3. $[-1, \infty)$ or $\{x \mid x \ge -1\}$

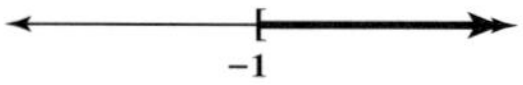

4. $\{x \mid x < 4\}$ or $(-\infty, 4)$

5. a. $\{x \mid x$ is a real number$\}$ or $\mathbb{R}$ or $(-\infty, \infty)$ **b.** $\varnothing$

6. $[-1, 4)$ or $\{x \mid -1 \le x < 4\}$

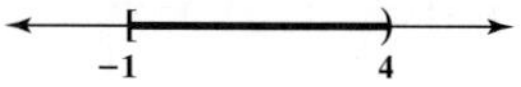

7. $(-3, 7)$ or $\{x \mid -3 < x < 7\}$

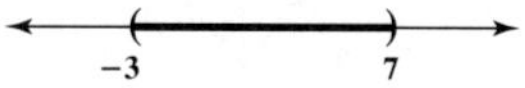

8. $\left\{x \mid -\frac{11}{5} \le x \le 3\right\}$ or $\left[-\frac{11}{5}, 3\right]$

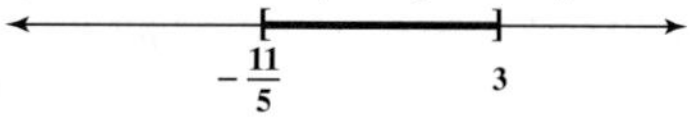

9. $\{x \mid x < -4$ or $x > 8\}$ or $(-\infty, -4) \cup (8, \infty)$

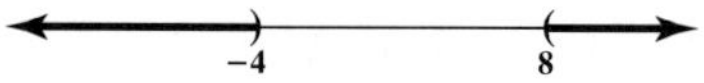

10. more than 720 mi per week

Exercise Set 1.7

1. $1 < x \le 6$

3. $-5 \le x < 2$

5. $-3 \le x \le 1$

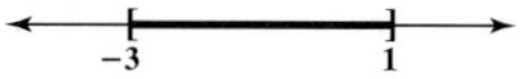

7. $x > 2$

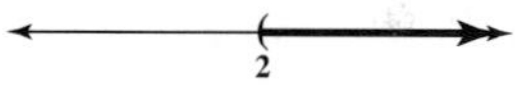

9. $x \ge -3$

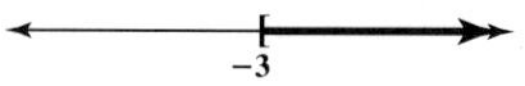

11. $x < 3$

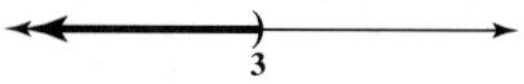

13. $x < 5.5$

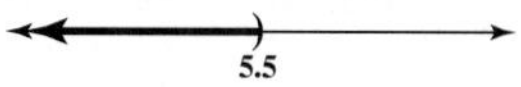

15. $[-1, 0)$ **17.** $(-3, 2]$
19. $[1, 5)$ **21.** $(-\infty, 8)$
23. $(6, \infty)$ **25.** $[3, \infty)$

27. $(-\infty, 3)$

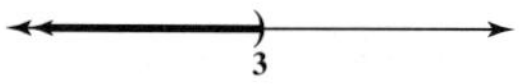

29. $\left[\frac{20}{3}, \infty\right)$

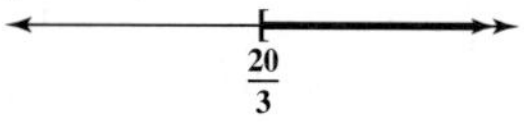

31. $(-\infty, -4]$

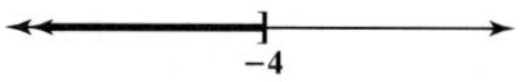

33. $\left(-\infty, -\frac{2}{5}\right]$

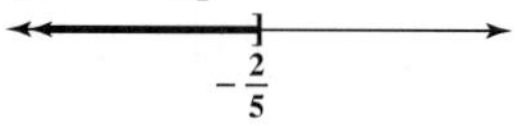

35. $[0, \infty)$

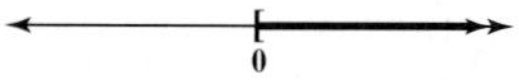

37. $(-\infty, 1)$

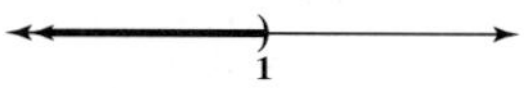

39. $[6, \infty)$

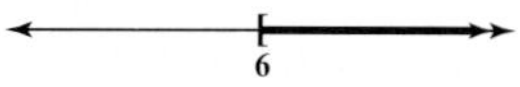

41. $[-10, \infty)$

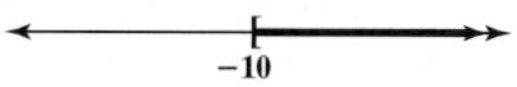

43. $(-\infty, -6)$

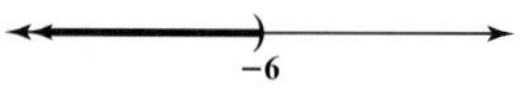

45. $[13, \infty)$

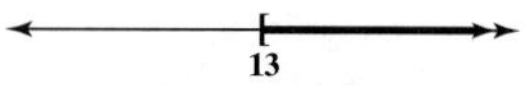

47. $(-\infty, \infty)$

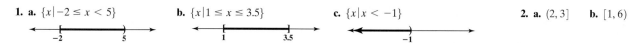

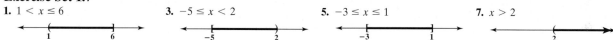

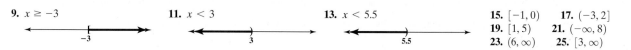

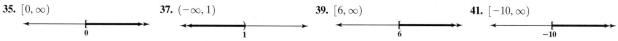

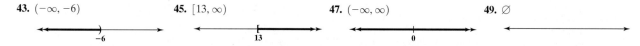

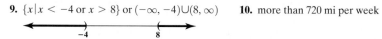

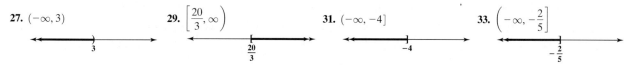

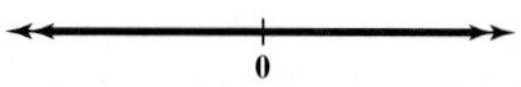

49. $\varnothing$

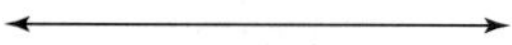

51. $(3, 5)$ **53.** $[-1, 3)$ **55.** $(-5, -2]$ **57.** $[3, 6)$ **59.** $(-3, 3)$ **61.** $[-1, 3]$ **63.** $(-1, 7)$ **65.** $[-5, 3]$ **67.** $(-6, 0)$
69. $(-\infty, -3)$ or $(3, \infty)$ **71.** $(-\infty, -1]$ or $[3, \infty)$ **73.** $\left(-\infty, \frac{1}{3}\right)$ or $(5, \infty)$ **75.** $(-\infty, -5]$ or $[3, \infty)$ **77.** $(-\infty, -3)$ or $(12, \infty)$
79. $(-\infty, -1]$ or $[3, \infty)$ **81.** $[2, 6]$ **83.** $(-\infty, -3) \cup (5, \infty)$ **85.** $(-\infty, 1] \cup [2, \infty)$ **87.** $(-1, 9)$ **89.** $\left(-\infty, \frac{1}{3}\right) \cup (1, \infty)$
91. $\left(-\infty, -\frac{75}{14}\right) \cup \left(\frac{87}{14}, \infty\right)$ **93.** $(-\infty, -6]$ or $[24, \infty)$ **95.** $(-\infty, -3]$ **97.** $[6, \infty)$ **99.** $\left(-\frac{2}{3}, \frac{10}{3}\right)$ **101.** $(-\infty, -10] \cup [2, \infty)$
103. $(-1, 9)$ **105.** $[-1, 2)$ **107.** $\left(-\infty, -\frac{1}{3}\right] \cup [3, \infty)$ **109.** $(0, 4)$ **111.** intimacy $\ge$ passion or passion $\le$ intimacy
113. commitment $>$ passion or passion $<$ commitment **115.** 9; after 3 years **117.** voting years after 2006 **119.** between 80 and 110 minutes, inclusive **121.** $h \le 41$ or $h \ge 59$ **123.** $15 + 0.08x < 3 + 0.12x$; more than 300 min **125.** $2 + 0.08x < 8 + 0.05x$; 199 checks or less **127.** $5.50x > 3000 + 3x$; more than 1200 packages **129.** $245 + 95x \le 3000$; at most 29 bags

131. **a.** $\dfrac{86 + 88 + x}{3} \ge 90$; at least a 96 **b.** $\dfrac{86 + 88 + x}{3} < 80$; a grade less than 66

133. more than 3 and less than 15 crossings per three-month period

143.

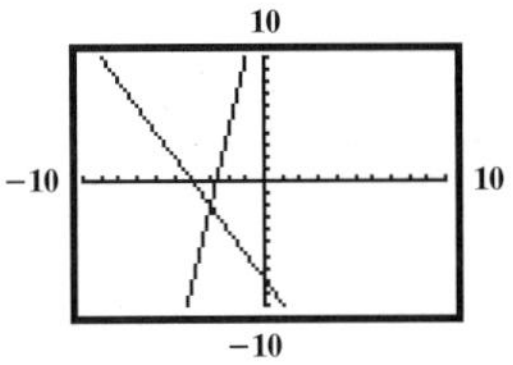

$x < -3$

145. **a.** $C = 4 + 0.10x$; $C = 2 + 0.15x$

b.

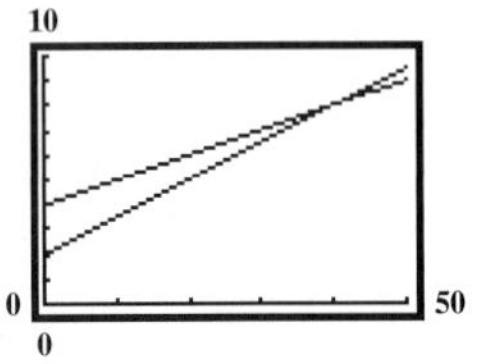

c. 41 or more checks

d. $x > 40$

147. Because $x > y$, $y - x$ represents a negative number, so when both sides are multiplied by $(y - x)$, the inequality must be reversed.

149. Albany: Model 2; San Francisco: Model 1

Chapter 1 Review Exercises

1.

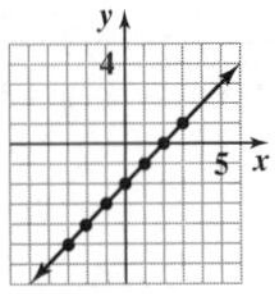

$y = 2x - 2$

2.

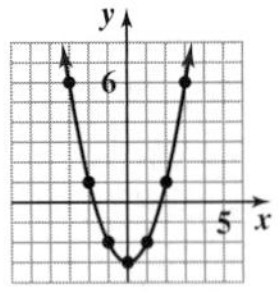

$y = x^2 - 3$

3.

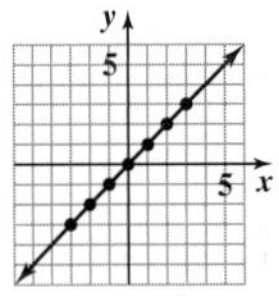

$y = x$

4.

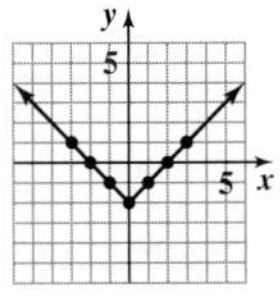

$y = |x| - 2$

5.

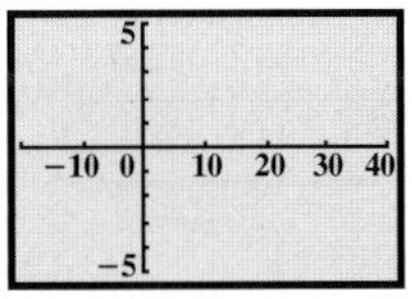

6. x-intercept: -2; y-intercept: 2 **7.** x-intercepts: $-2, 2$; y-intercept: -4 **8.** x-intercept: 5; no y-intercept **9.** (91, 125); In 1991, 125 thousand acres were used for cultivation. **10.** 1997 **11.** 2001; 25 thousand acres **12.** 2004; 300 thousand acres **13.** 1991 and 1992 **14.** 2001 and 2002; 155 thousand acres **15.** {6}; conditional equation **16.** $\{-10\}$; conditional equation **17.** {5}; conditional equation **18.** $\{-13\}$; conditional equation **19.** $\{-3\}$; conditional equation **20.** $\{-1\}$; conditional equation **21.** $\varnothing$; inconsistent equation **22.** all real numbers; identity **23.** {2}; conditional equation **24.** {2}; conditional equation **25.** $\left\{\frac{72}{11}\right\}$; conditional equation **26.** $\left\{-\frac{36}{7}\right\}$; conditional equation **27.** $\left\{\frac{77}{15}\right\}$; conditional equation **28.** {2}; conditional equation **29.** $\varnothing$; inconsistent equation **30.** all real numbers except -1 and 1; conditional equation **31.** $\left\{\frac{5}{2}\right\}$; conditional equation **32.** $\varnothing$; inconsistent equation **33.** $\left\{\frac{4}{7}\right\}$; conditional equation **34.** $\left\{\frac{1}{2}\right\}$; conditional equation **35.** {2}; conditional equation **36.** Chicken Caesar: 495; Express Taco: 620; Mandarin Chicken: 590 **37.** 2008 **38.** 500 min **39.** \$60 **40.** \$10,000 in sales **41.** \$2500 at 4%; \$6500 at 7% **42.** \$4500 at 2%; \$3500 at 5% **43.** 44 yd by 126 yd **44.** **a.** $14{,}100 + 1500x = 41{,}700 - 800x$ **b.** 2019; 32,100 **45.** $g = \dfrac{s - vt}{t^2}$ **46.** $g = \dfrac{T}{r + vt}$ **47.** $P = \dfrac{A}{1 + rt}$ **48.** $-9 + 4i$ **49.** $-12 - 8i$ **50.** $17 + 19i$ **51.** $-7 - 24i$ **52.** 113 **53.** $\frac{15}{13} - \frac{3}{13}i$ **54.** $\frac{1}{5} + \frac{11}{10}i$ **55.** $i\sqrt{2}$ **56.** $-96 - 40i$ **57.** $2 + i\sqrt{2}$ **58.** $\left\{-8, \frac{1}{2}\right\}$ **59.** $\{-4, 0\}$ **60.** $\{-8, 8\}$ **61.** $\{-4i, 4i\}$ **62.** $\{-3 - i\sqrt{10}, -3 + i\sqrt{10}\}$ **63.** $\left\{\frac{4 - 3\sqrt{2}}{3}, \frac{4 + 3\sqrt{2}}{3}\right\}$ **64.** 100; $(x + 10)^2$ **65.** $\frac{9}{4}$; $\left(x - \frac{3}{2}\right)^2$ **66.** {3, 9} **67.** $\left\{2 + \frac{\sqrt{3}}{3}, 2 - \frac{\sqrt{3}}{3}\right\}$ **68.** $\{1 + \sqrt{5}, 1 - \sqrt{5}\}$ **69.** $\{1 + 3i\sqrt{2}, 1 - 3i\sqrt{2}\}$ **70.** $\left\{\frac{-2 + \sqrt{10}}{2}, \frac{-2 - \sqrt{10}}{2}\right\}$ **71.** -36; 2 complex imaginary solutions **72.** 81; 2 unequal real solutions **73.** $\left\{\frac{1}{2}, 5\right\}$ **74.** $\left\{-2, \frac{10}{3}\right\}$ **75.** $\left\{\frac{7 + \sqrt{37}}{6}, \frac{7 - \sqrt{37}}{6}\right\}$ **76.** $\{-3, 3\}$ **77.** $\{-2, 8\}$ **78.** $\left\{\frac{1}{6} + i\frac{\sqrt{23}}{6}, \frac{1}{6} - i\frac{\sqrt{23}}{6}\right\}$ **79.** $\left\{-\frac{2}{3}, 4\right\}$ **80.** $\{-2 - 2i, -2 + 2i\}$ **81.** $\{4 + \sqrt{5}, 4 - \sqrt{5}\}$ **82.** 14 weeks **83.** 2012 **84.** length = 5 yd; width = 3 yd **85.** approximately 134 m **86.** $\{-5, 0, 5\}$ **87.** $\left\{-3, \frac{1}{2}, 3\right\}$ **88.** {2} **89.** {8} **90.** {16} **91.** {132} **92.** $\{-2, -1, 1, 2\}$ **93.** {16} **94.** $\{-4, 3\}$ **95.** $\{-5, 11\}$ **96.** $\left\{-1, -\frac{2\sqrt{6}}{9}, \frac{2\sqrt{6}}{9}, 1\right\}$ **97.** {2} **98.** {1, 4} **99.** $\{-3, -2, 3\}$ **100.** {2} **101.** $\{-3, -\sqrt{2}, \sqrt{2}\}$ **102.** $\{-4, 2\}$ **103.** 2007

104. $\{x \mid -3 \le x < 5\}$

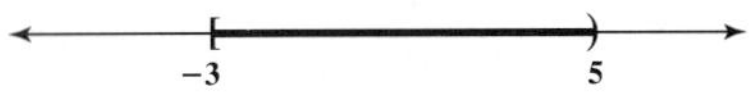

105. $\{x \mid x > -2\}$

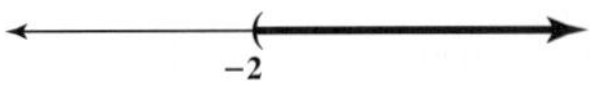

106. $\{x \mid x \le 0\}$

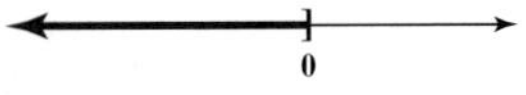

107. $[-1, 1]$ **108.** $(-2, 3)$ **109.** $[1, 3)$ **110.** $(0, 4)$

111. $[-2, \infty)$
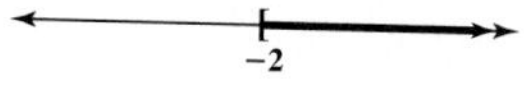

112. $\left[\frac{3}{5}, \infty\right)$
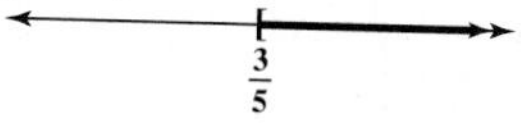

113. $\left(-\infty, -\frac{21}{2}\right)$
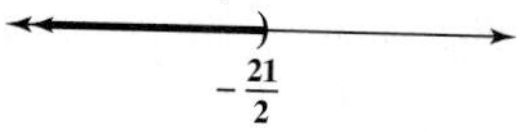

114. $(-3, \infty)$
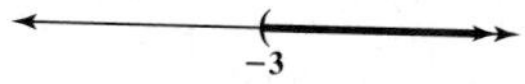

115. $(-\infty, -2]$
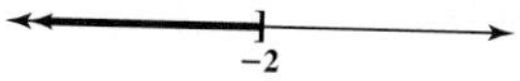

116. $\varnothing$

117. $(2, 3]$
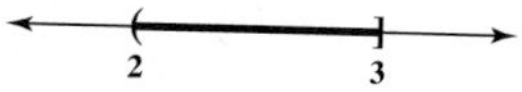

118. $[-9, 6]$
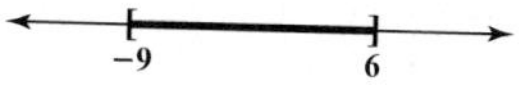

119. $(-\infty, -6)$ or $(0, \infty)$
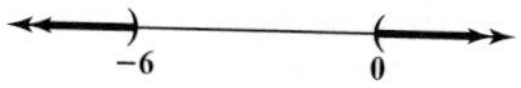

120. $(-\infty, -3]$ or $[-2, \infty)$
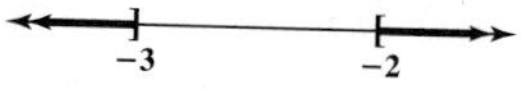

121. $(-\infty, -5]\cup[1, \infty)$;
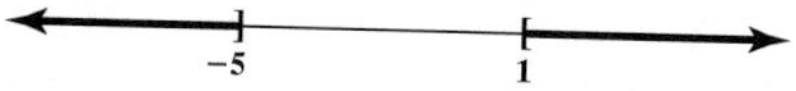

122. $(-\infty, -1)$ **123.** $[-2, 7]$ **124.** no more than 80 miles per day **125.** $[49\%, 99\%)$ **126.** at least \$120,000

Chapter 1 Test

1. $\{-1\}$ **2.** $\{-1\}$ **3.** $\{-6\}$ **4.** $\{5\}$ **5.** $\left\{-\frac{1}{2}, 2\right\}$ **6.** $\left\{\frac{1-5\sqrt{3}}{3}, \frac{1+5\sqrt{3}}{3}\right\}$ **7.** $\{-3-5i, -3+5i\}$
8. $\{1-\sqrt{5}, 1+\sqrt{5}\}$ **9.** $\left\{1+\frac{1}{2}i, 1-\frac{1}{2}i\right\}$ **10.** $\{-1, 1, 4\}$ **11.** $\{7\}$ **12.** $\{2\}$ **13.** $\{5\}$ **14.** $\{\sqrt[3]{4}\}$ **15.** $\{1, 512\}$
16. $\{6, 12\}$ **17.** $\left\{\frac{1}{2}, 3\right\}$ **18.** $\{2-\sqrt{3}, 2+\sqrt{3}\}$ **19.** $\{4\}$

20. $(-\infty, 12]$
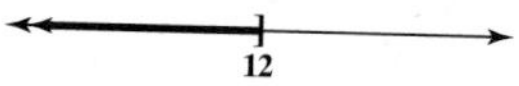

21. $\left[\frac{21}{8}, \infty\right)$
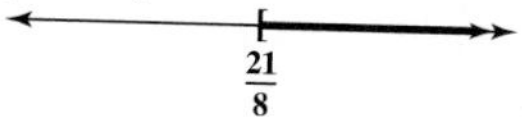

22. $\left[-7, \frac{13}{2}\right)$
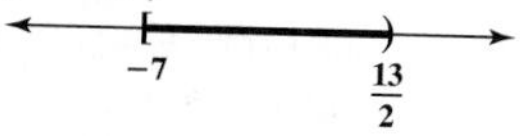

23. $\left(-\infty, -\frac{5}{3}\right]$ or $\left[\frac{1}{3}, \infty\right)$
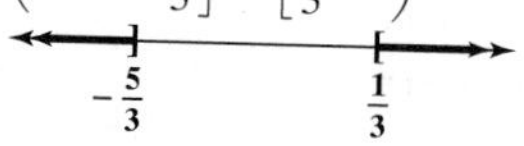

24. $[1, 6]$ **25.** $(-\infty, -2]\cup[6, \infty)$ **26.** $[-1, 5]$ **27.** $(0, 2)$ **28.** $h = \frac{3V}{lw}$ **29.** $x = \frac{y - y_1}{m} + x_1$

30.
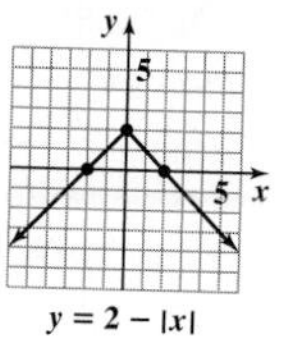

31.
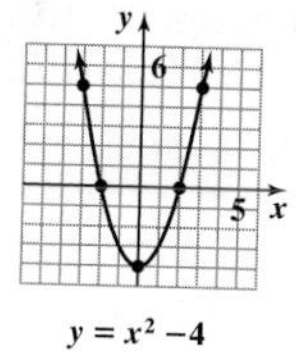

32. $47 + 16i$ **33.** $2 + i$ **34.** $38i$ **35.** 2018 **36.** 2018
37. quite well **38.** 2002: 782 books; 2003: 844 books; 2003: 972 books
39. 26 yr; \$33,600 **40.** \$3000 at 8%; \$7000 at 10%
41. length = 12 ft; width = 4 ft **42.** 10 ft **43.** \$50
44. more than 200 calls

CHAPTER 2

Section 2.1

Check Point Exercises

1. domain: $\{5, 10, 15, 20, 25\}$; range: $\{12.8, 16.2, 18.9, 20.7, 21.8\}$ **2. a.** not a function **b.** function **3. a.** $y = 6 - 2x$; function **b.** $y = \pm\sqrt{1 - x^2}$; not a function **4. a.** 42 **b.** $x^2 + 6x + 15$ **c.** $x^2 + 2x + 7$

5.
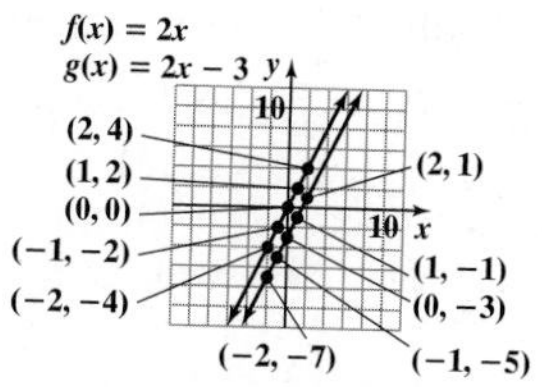

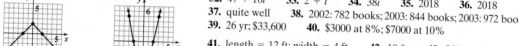

; The graph of g is the graph of f shifted down by 3 units.

6. a. function **b.** function **c.** not a function **7. a.** $f(10) \approx 16$ **b.** $x \approx 8$
8. a. Domain = $\{x|-2 \le x \le 1\}$; Range = $\{y|0 \le y \le 3\}$
b. Domain = $\{x|-2 < x \le 1\}$; Range = $\{y|-1 \le y < 2\}$
c. Domain = $\{x|-3 \le x < 0\}$; Range = $\{-3, -2, -1\}$

Exercise Set 2.1

1. function; $\{1, 3, 5\}$; $\{2, 4, 5\}$ **3.** not a function; $\{3, 4\}$; $\{4, 5\}$ **5.** function; $\{3, 4, 5, 7\}$; $\{-2, 1, 9\}$ **7.** function; $\{-3, -2, -1, 0\}$; $\{-3, -2, -1, 0\}$ **9.** not a function; $\{1\}$; $\{4, 5, 6\}$ **11.** y is a function of x. **13.** y is a function of x. **15.** y is not a function of x. **17.** y is not a function of x. **19.** y is a function of x. **21.** y is a function of x. **23.** y is a function of x. **25.** y is a function of x. **27. a.** 29 **b.** $4x + 9$ **c.** $-4x + 5$ **29. a.** 2 **b.** $x^2 + 12x + 38$ **c.** $x^2 - 2x + 3$ **31. a.** 13 **b.** 1 **c.** $x^4 - x^2 + 1$ **d.** $81a^4 - 9a^2 + 1$ **33. a.** 3 **b.** 7 **c.** $\sqrt{x} + 3$ **35. a.** $\frac{15}{4}$ **b.** $\frac{15}{4}$ **c.** $\frac{4x^2 - 1}{x^2}$ **37. a.** 1 **b.** −1 **c.** 1

39.
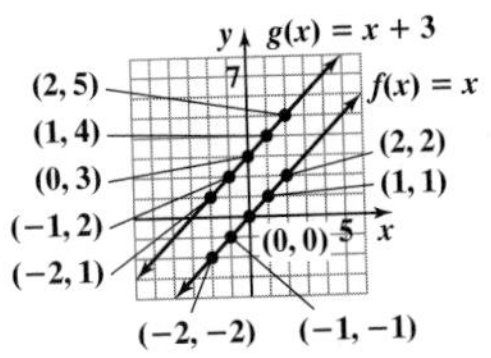

The graph of g is the graph of f shifted up by 3 units.

41.
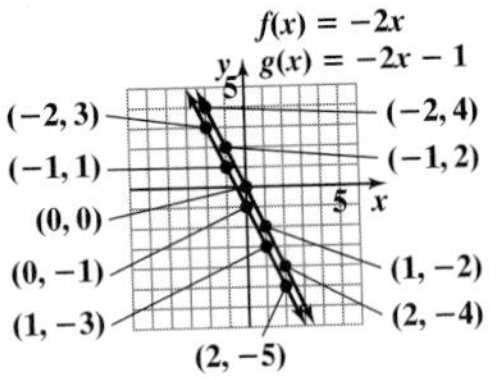

The graph of g is the graph of f shifted down by 1 unit.

43.
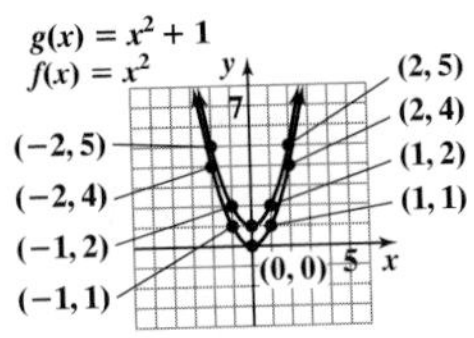

The graph of g is the graph of f shifted up by 1 unit.

45.
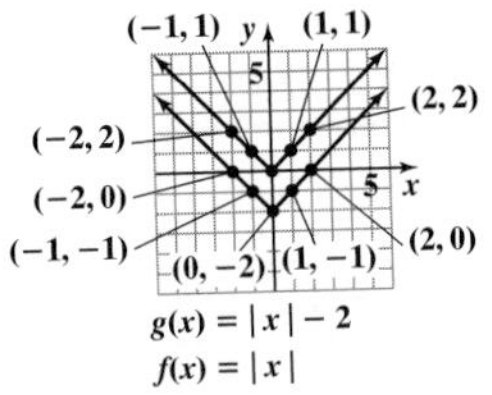

The graph of g is the graph of f shifted down by 2 units.

47.
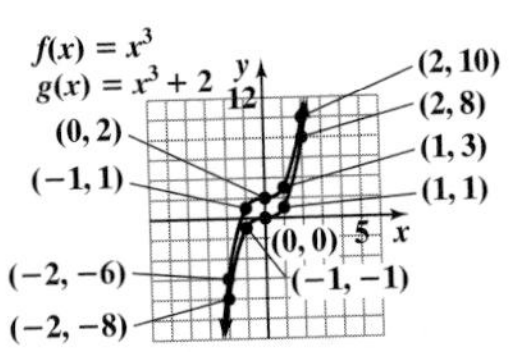

The graph of g is the graph of f shifted up by 2 units.

49.
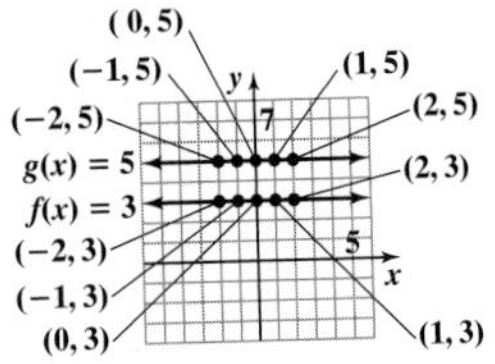

The graph of g is the graph of f shifted up by 2 units.

51.
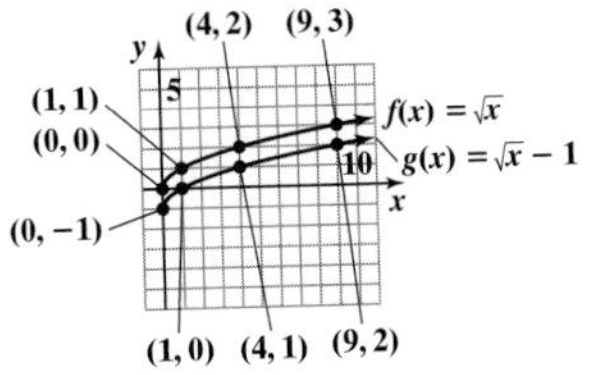

The graph of g is the graph of f shifted down by 1 unit.

53.
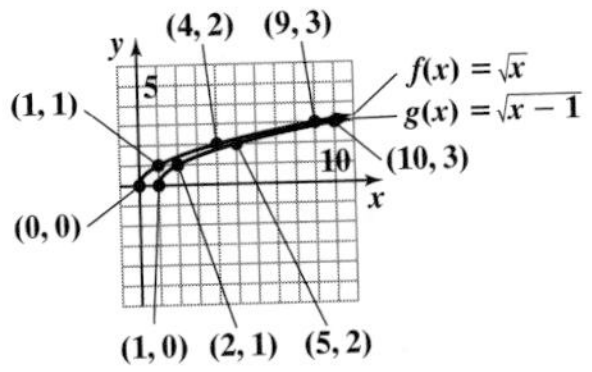

The graph of g is the graph of f shifted to the right by 1 unit.

55. function **57.** function **59.** not a function **61.** function **63.** function **65.** -4 **67.** 4 **69.** 0 **71.** 2 **73.** 2 **75.** -2 **77. a.** $(-\infty, \infty)$ **b.** $[-4, \infty)$ **c.** -3 and 1 **d.** -3 **e.** $f(-2) = -3$ and $f(2) = 5$ **79. a.** $(-\infty, \infty)$ **b.** $[1, \infty)$ **c.** none **d.** 1 **e.** $f(-1) = 2$ and $f(3) = 4$ **81. a.** $[0, 5)$ **b.** $[-1, 5)$ **c.** 2 **d.** -1 **e.** $f(3) = 1$ **83. a.** $[0, \infty)$ **b.** $[1, \infty)$ **c.** none **d.** 1 **e.** $f(4) = 3$ **85. a.** $[-2, 6]$ **b.** $[-2, 6]$ **c.** 4 **d.** 4 **e.** $f(-1) = 5$ **87. a.** $(-\infty, \infty)$ **b.** $(-\infty, -2]$ **c.** none **d.** -2 **e.** $f(-4) = -5$ and $f(4) = -2$ **89. a.** $(-\infty, \infty)$ **b.** $(0, \infty)$ **c.** none **d.** 1.5 **e.** $f(4) = 6$ **91. a.** $\{-5, -2, 0, 1, 3\}$ **b.** $\{2\}$ **c.** none **d.** 2 **e.** $f(-5) + f(3) = 4$ **93.** -2; 10 **95.** -38 **97.** $-2x^3 - 2x$ **99. a.** {(U.S., 80%), (Japan, 64%), (France, 64%), (Germany, 61%), (England, 59%), (China, 47%)} **b.** Yes; Each country corresponds to a unique percent. **c.** {(80%, U.S.), (64%, Japan), (64%, France), (61%, Germany), (59% England), (47%, China)} **d.** No; 64% in the domain corresponds to two members of the range, Japan and France. **101.** 5.22; In 2000, there were 5.22 million women enrolled in U.S. colleges; (2000, 5.22). **103.** 1.4; In 2004, there were 1.4 million more women than men enrolled in U.S. colleges. **105. a.** 73% **b.** 72.8% **c.** 73.3% **107.** $C = 100{,}000 + 100x$, where x is the number of bicycles produced; $C(90) = 109{,}000$; It cost $109,000 to produce 90 bicycles. **109.** $T = \frac{40}{x} + \frac{40}{x + 30}$, where x is the rate on the outgoing trip; $T(30) = 2$; It takes 2 hours, traveling 30 mph outgoing and 60 mph returning. **121.** c **123.** Answers will vary; an example is $\{(1, 1), (2, 1)\}$.

Section 2.2

Check Point Exercises

1. a. $-2x^2 - 4xh - 2h^2 + x + h + 5$ **b.** $-4x - 2h + 1$ **2. a.** 20; With 40 calling minutes, the cost is $20; (40, 20). **b.** 28; With 80 calling minutes, the cost is $28; (80, 28). **3.** increasing on $(-\infty, -1)$, decreasing on $(-1, 1)$, increasing on $(1, \infty)$ **4. a.** even **b.** odd **c.** neither

Exercise Set 2.2

1. $4, h \neq 0$ **3.** $3, h \neq 0$ **5.** $2x + h, h \neq 0$ **7.** $2x + h - 4, h \neq 0$ **9.** $4x + 2h + 1, h \neq 0$ **11.** $-2x - h + 2, h \neq 0$ **13.** $-4x - 2h + 5, h \neq 0$ **15.** $-4x - 2h - 1, h \neq 0$ **17.** $0, h \neq 0$ **19.** $-\frac{1}{x(x + h)}, h \neq 0$ **21.** $\frac{1}{\sqrt{x + h} + \sqrt{x}}, h \neq 0$ **23. a.** -1 **b.** 7 **c.** 19 **25. a.** 3 **b.** 3 **c.** 0 **27. a.** 8 **b.** 3 **c.** 6 **29. a.** increasing: $(-1, \infty)$ **b.** decreasing: $(-\infty, -1)$ **c.** constant: none **31. a.** increasing: $(0, \infty)$ **b.** decreasing: none **c.** constant: none **33. a.** increasing: none **b.** decreasing: $(-2, 6)$ **c.** constant: none **35. a.** increasing: $(-\infty, -1)$ **b.** decreasing: none **c.** constant: $(-1, \infty)$ **37. a.** increasing: $(-\infty, 0)$ or $(1.5, 3)$ **b.** decreasing: $(0, 1.5)$ or $(3, \infty)$ **c.** constant: none **39. a.** increasing: $(-2, 4)$ **b.** decreasing: none **c.** constant: $(-\infty, -2)$ or $(4, \infty)$ **41. a.** 0; $f(0) = 4$ **b.** -3, 3; $f(-3) = f(3) = 0$ **43. a.** -2; $f(-2) = 21$ **b.** 1; $f(1) = -6$ **45.** odd **47.** neither **49.** even **51.** even **53.** even **55.** odd **57.** even **59.** odd **61. a.** $(-\infty, \infty)$ **b.** $[-4, \infty)$ **c.** 1 and 7 **d.** 4 **e.** $(4, \infty)$ **f.** $(0, 4)$ **g.** $(-\infty, 0)$ **h.** 4 **i.** -4 **j.** 4 **k.** 2 and 6 **l.** neither **63. a.** $(-\infty, 3]$ **b.** $(-\infty, 4]$ **c.** -3 and 3 **d.** 3 **e.** $(-\infty, 1)$ **f.** $(1, 3)$ **g.** $(-\infty, -3]$ **h.** A relative maximum of 4 occurs at 1. **i.** 1 **j.** positive

65. $f(1.06) = 1$ **67.** $f\left(\frac{1}{3}\right) = 0$ **69.** $f(-2.3) = -3$ **71.** -18

73. $0.30t - 6$ **75.** $C(t) = \begin{cases} 50 & \text{if } 0 \leq t \leq 400 \\ 50 + 0.30(t - 400) & \text{if } t > 400 \end{cases}$

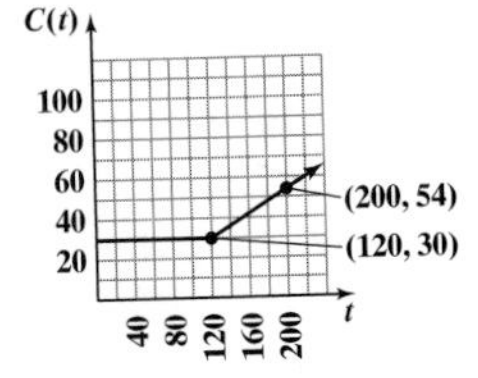

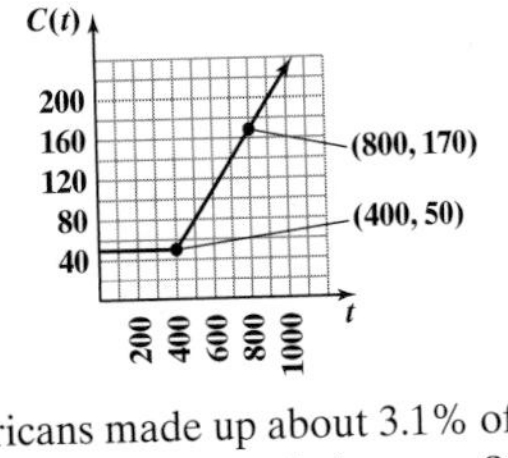

77. $f(60) \approx 3.1$; In 1960, Jewish Americans made up about 3.1% of the U.S. population. **79.** $x \approx 19$ and $x \approx 64$; In 1919 and 1964, Jewish Americans made up about 3% of the U.S. population. **81.** 1940; 3.7% **83.** Each year corresponds to only one percentage. **85.** increasing: (45, 74); decreasing: (16, 45); The number of accidents occurring per 50,000 miles driven increases with age starting at age 45, while it decreases with age starting at age 16. **87.** Answers will vary; an example is 16 and 74 years old. For those ages, the number of accidents is 526.4 per 50 million miles. **89.** 1989 cigarettes per adult; quite well **91.** 1960; 4100 cigarettes per adult; The function estimates 3797 cigarettes per adult, which models the graph reasonably well.

93.

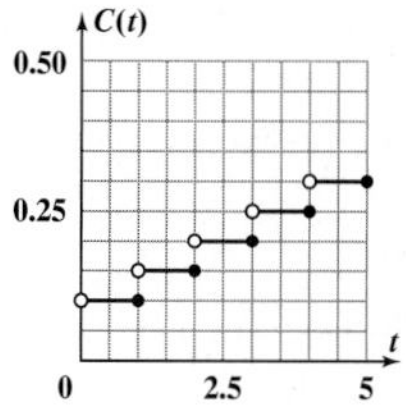

103. a.

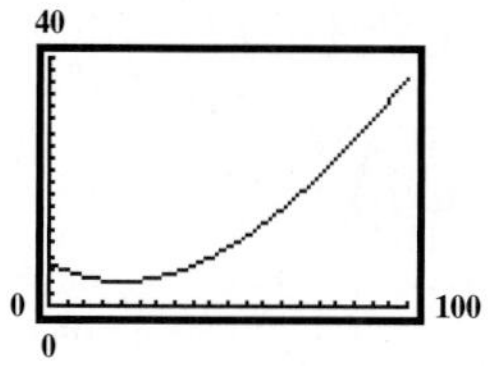

The number of doctor visits decreases during childhood and then increases as you get older. The minimum is (20.29, 3.99), which means that the minimum number of annual doctor visits, about 4, occurs at around age 20.

105.

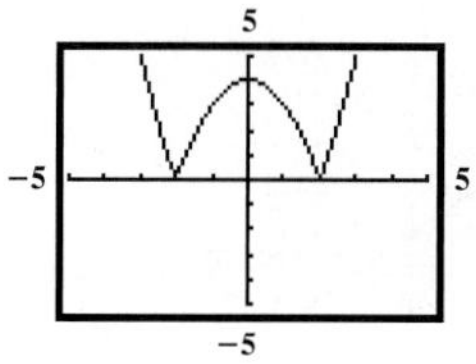

Increasing: $(-2, 0)$ or $(2, \infty)$
Decreasing: $(-\infty, -2)$ or $(0, 2)$

107.

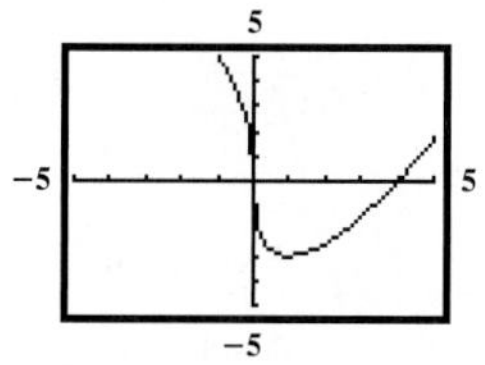

Increasing: $(1, \infty)$
Decreasing: $(-\infty, 1)$

109.

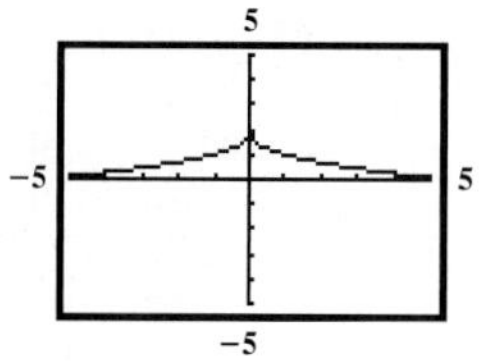

Increasing: $(-\infty, 0)$
Decreasing: $(0, \infty)$

113. a. h is even if both f and g are even or if both f and g are odd.
b. h is odd if f is odd and g is even or if f is even and g is odd.

Section 2.3

Check Point Exercises

1. a. 6 **b.** $-\frac{7}{5}$ **2.** $y + 5 = 6(x - 2)$; $y = 6x - 17$ **3.** $y + 1 = -5(x + 2)$; $y = -5x - 11$

4.

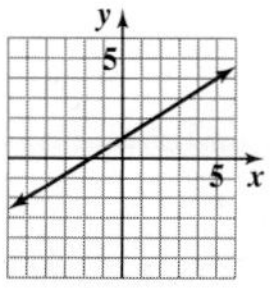

$f(x) = \frac{3}{5}x + 1$

5. $x = -3$

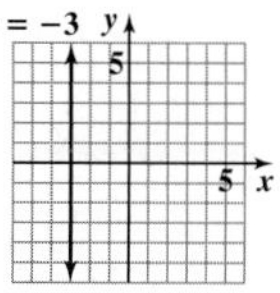

6. slope: $-\frac{1}{2}$; y-intercept: 2

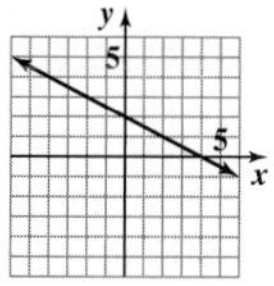

$3x + 6y - 12 = 0$

7.

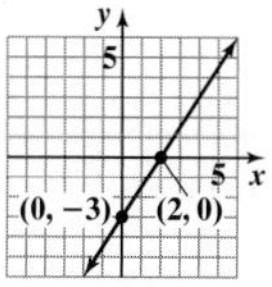

$3x - 2y = 6$

8. $y = 0.28x + 27.2$; 41.2

Exercise Set 2.3

1. $\frac{3}{4}$; rises **3.** $\frac{1}{4}$; rises **5.** 0; horizontal **7.** -5; falls **9.** undefined; vertical **11.** $y - 5 = 2(x - 3)$; $y = 2x - 1$
13. $y - 5 = 6(x + 2)$; $y = 6x + 17$ **15.** $y + 3 = -3(x + 2)$; $y = -3x - 9$ **17.** $y - 0 = -4(x + 4)$; $y = -4x - 16$
19. $y + 2 = -1\left(x + \frac{1}{2}\right)$; $y = -x - \frac{5}{2}$ **21.** $y - 0 = \frac{1}{2}(x - 0)$; $y = \frac{1}{2}x$ **23.** $y + 2 = -\frac{2}{3}(x - 6)$; $y = -\frac{2}{3}x + 2$
25. using $(1, 2)$, $y - 2 = 2(x - 1)$; $y = 2x$ **27.** using $(-3, 0)$, $y - 0 = 1(x + 3)$; $y = x + 3$ **29.** using $(-3, -1)$, $y + 1 = 1(x + 3)$; $y = x + 2$
31. using $(-3, -2)$, $y + 2 = \frac{4}{3}(x + 3)$; $y = \frac{4}{3}x + 2$ **33.** using $(-3, -1)$, $y + 1 = 0(x + 3)$; $y = -1$ **35.** using $(2, 4)$, $y - 4 = 1(x - 2)$; $y = x + 2$
37. using $(0, 4)$, $y - 4 = 8(x - 0)$; $y = 8x + 4$

39. $m = 2$; $b = 1$

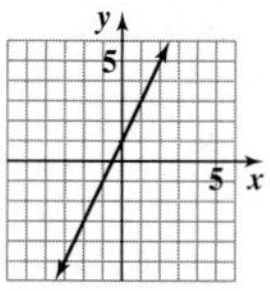

$y = 2x + 1$

41. $m = -2$; $b = 1$

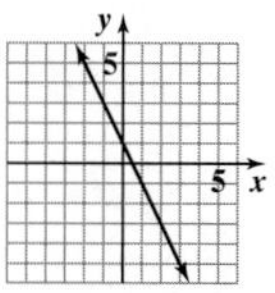

$f(x) = -2x + 1$

43. $m = \frac{3}{4}$; $b = -2$

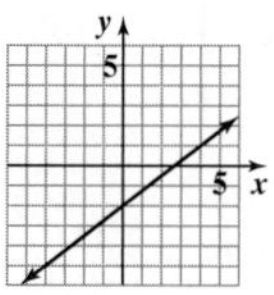

$f(x) = \frac{3}{4}x - 2$

45. $m = -\frac{3}{5}$; $b = 7$

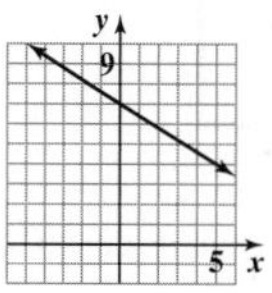

$y = -\frac{3}{5}x + 7$

47. $m = -\frac{1}{2}$; $b = 0$

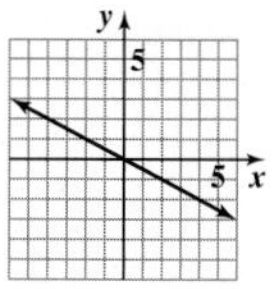

$g(x) = -\frac{1}{2}x$

49. $y = -2$

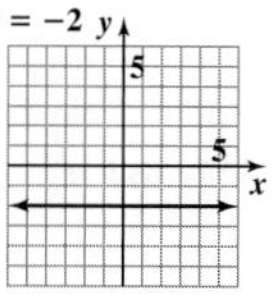

51. $x = -3$

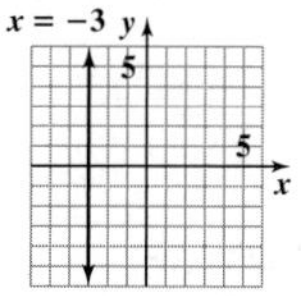

53. $y = 0$

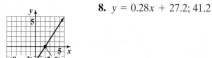

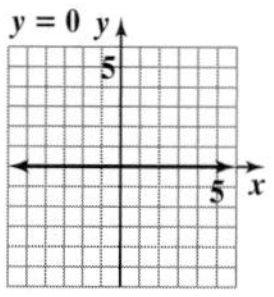

55.

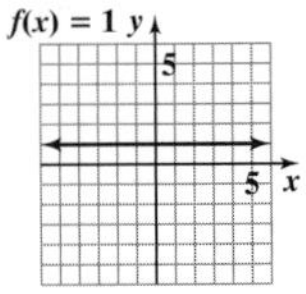

57.

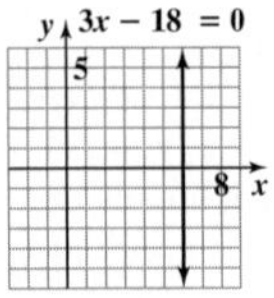

59. a. $y = -3x + 5$

b. $m = -3; b = 5$

c.

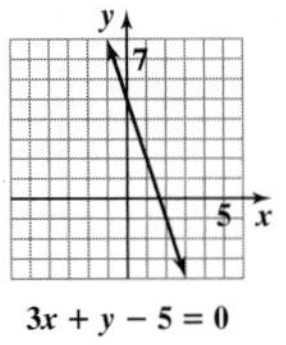

61. a. $y = -\frac{2}{3}x + 6$

b. $m = -\frac{2}{3}; b = 6$

c.

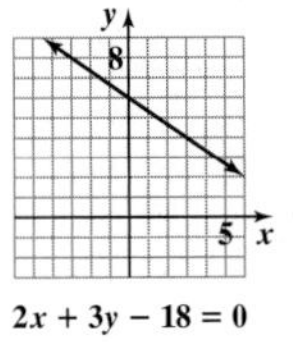

63. a. $y = 2x - 3$

b. $m = 2; b = -3$

c.

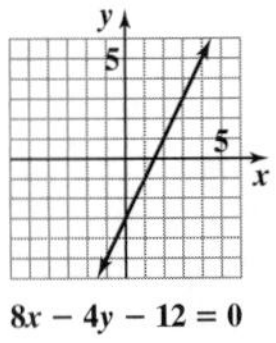

65. a. $y = 3$

b. $m = 0; b = 3$

c.

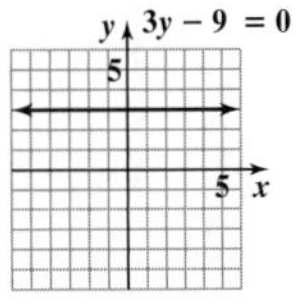

67.

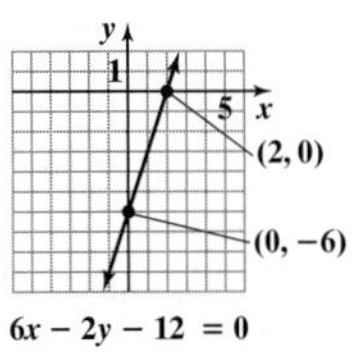

69.

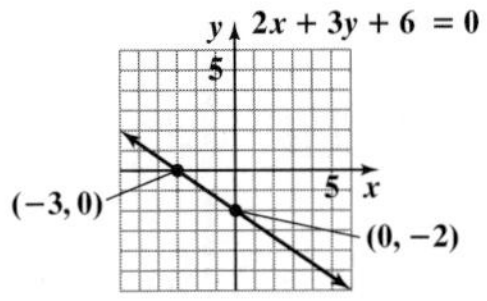

71.

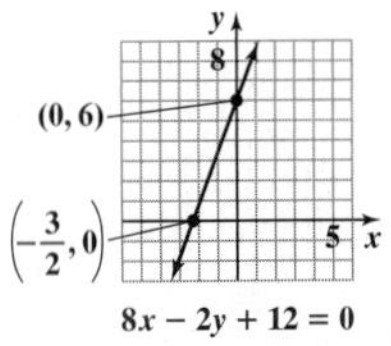

73. $m = -\frac{a}{b}$; falls **75.** undefined slope; vertical **77.** $m = -\frac{A}{B}; b = \frac{C}{B}$ **79.** -2

81.

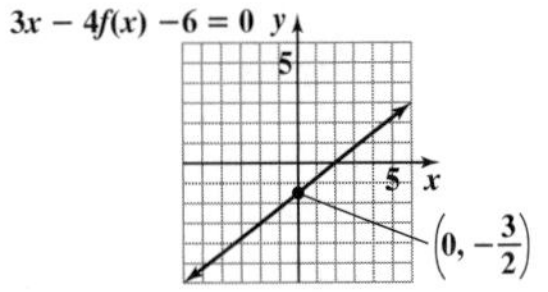

83. 5 **85.** m_1, m_3, m_2, m_4

87. a. $y - 16 = -0.55(x - 10)$ or $y - 12.7 = -0.55(x - 16)$ **b.** $f(x) = -0.55x + 21.5$ **c.** 10.5%

89. a.

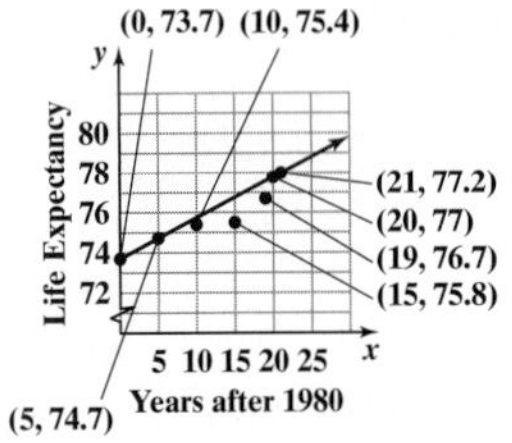

b. $y - 74.7 = 0.15(x - 5)$ or $y - 77.0 = 0.15(x - 20)$;
$y = 0.15x + 73.95$ or $y = 0.15x + 74$

c. $E(x) = 0.15x + 73.95$ or $E(x) = 0.15x + 74$; 80 years

91. $y = -2.3x + 255$, where x is the percentage of adult females who are literate and y is under-five mortality per thousand; For each percent increase in adult female literacy, under-five mortality decreases by 2.3 per thousand.

101. $m = -3$

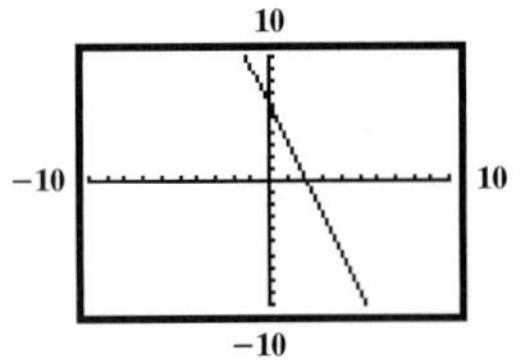

103. $m = \frac{3}{4}$

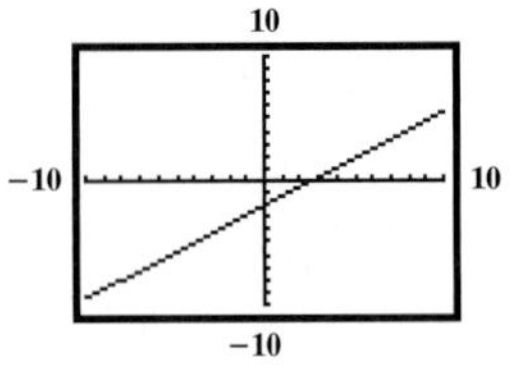

105. c **107.** coefficient of x: 1; coefficient of y: -2 **109.** $E = 2.4M - 20$

Section 2.4

Check Point Exercises

1. $y - 5 = 3(x + 2)$; $y = 3x + 11$ or $f(x) = 3x + 11$ **2. a.** 3 **b.** $3x - y = 0$ **3.** $\frac{2}{15} \approx 0.13$; The number of U.S. men living alone is projected to increase by 0.13 million each year. **4. a.** 1 **b.** 7 **c.** 4 **5.** 0.01 mg per 100 ml per hr

Exercise Set 2.4

1. $y - 2 = 2(x - 4)$; $y = 2x - 6$ or $f(x) = 2x - 6$ **3.** $y - 4 = -\frac{1}{2}(x - 2)$; $y = -\frac{1}{2}x + 5$ or $f(x) = -\frac{1}{2}x + 5$ **5.** $y + 10 = -4(x + 8)$; $y = -4x - 42$ **7.** $y + 3 = -5(x - 2)$; $y = -5x + 7$ **9.** $y - 2 = \frac{2}{3}(x + 2)$; $2x - 3y + 10 = 0$ **11.** $y + 7 = -2(x - 4)$; $2x + y - 1 = 0$ **13.** 3 **15.** 10 **17.** $\frac{1}{5}$ **19.** $f(x) = 5$ **21.** $f(x) = -\frac{1}{2}x + 1$ **23.** $f(x) = -\frac{2}{3}x - 2$

25. $m = 0.01$; The temperature of Earth is increasing by 0.01°F per year. **27.** $m = -0.52$; The percentage of U.S. adults who smoke cigarettes is decreasing by 0.52% each year. **29.** $f(x) = 13x + 222$ **31.** $f(x) = -2.40x + 52.40$ **33.** $m = -1.22$; average decrease of 1.22% of total sales per year

41. a. The product of their slopes is -1.

b.

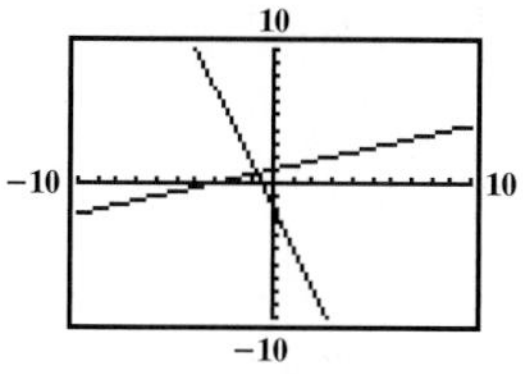

c.

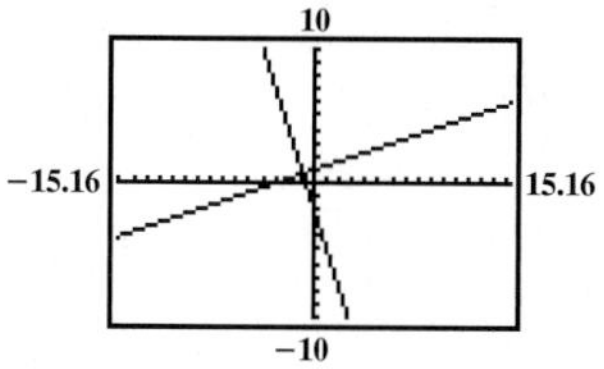

43.

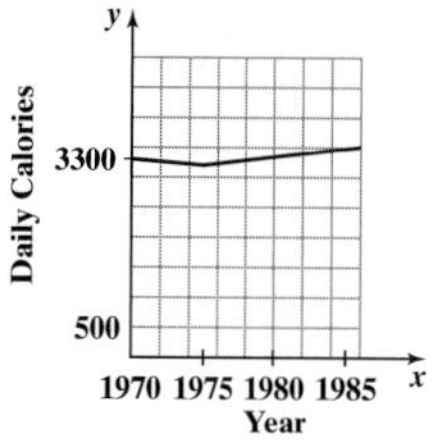

45. $-\frac{3}{7}$

Mid-Chapter 2 Check Point

1. not a function; Domain: $\{1, 2\}$; Range: $\{-6, 4, 6\}$ **2.** function; Domain: $\{0, 2, 3\}$; Range: $\{1, 4\}$ **3.** function; Domain: $\{x|-2 \le x < 2\}$; Range: $\{y|0 \le y \le 3\}$ **4.** not a function; Domain: $\{x|-3 < x \le 4\}$; Range: $\{y|-1 \le y \le 2\}$ **5.** not a function; Domain: $\{-2, -1, 0, 1, 2\}$; Range: $\{-2, -1, 1, 3\}$ **6.** function; Domain: $\{x|x \le 1\}$; Range: $\{y|y \ge -1\}$ **7.** y is a function of x **8.** y is not a function of x **9.** No vertical line intersects the graph in more than one point. **10.** $(-\infty, \infty)$ **11.** $(-\infty, 4]$ **12.** -6 and 2 **13.** 3 **14.** $(-\infty, -2)$ **15.** $(-2, \infty)$ **16.** -2 **17.** 4 **18.** 3 **19.** -7 and 3 **20.** -6 and 2 **21.** $(-6, 2)$ **22.** negative **23.** neither **24.** -1

25. $y = -2x$

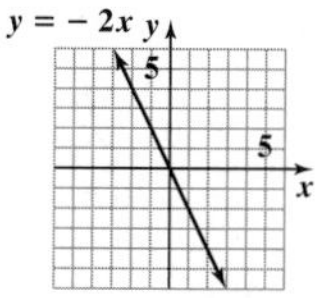

26. $y = -2$

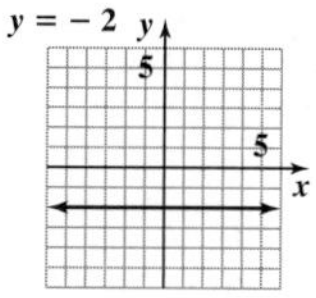

27. $x + y = -2$

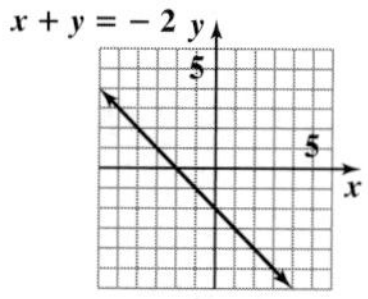

28. $y = \frac{1}{3}x - 2$

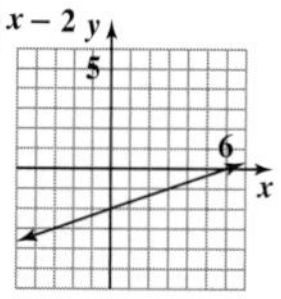

29. $x = 3.5$

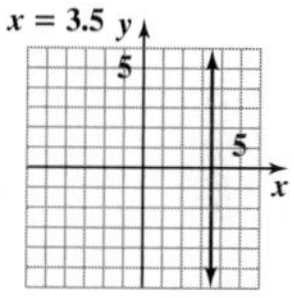

30.

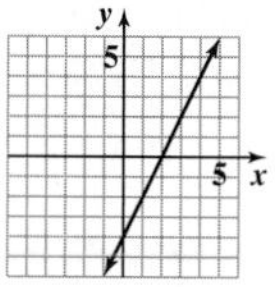

$4x - 2y = 8$

31.

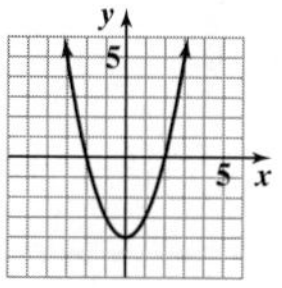

$f(x) = x^2 - 4$

32.

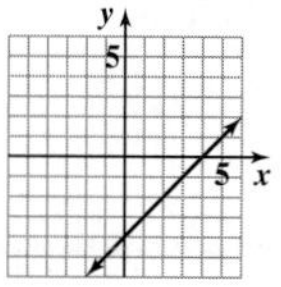

$f(x) = x - 4$

33.

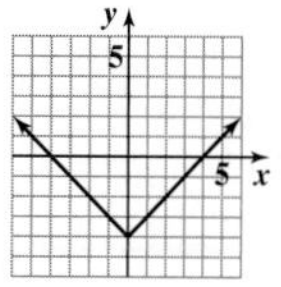

$f(x) = |x| - 4$

34.

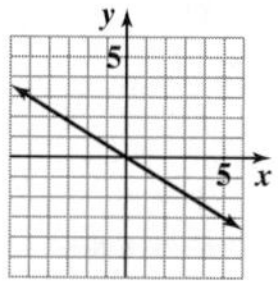

$5y = -3x$

35.

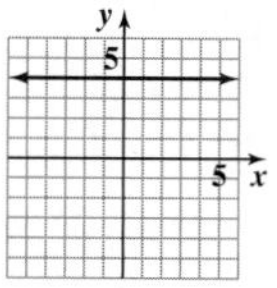

$5y = 20$

36.

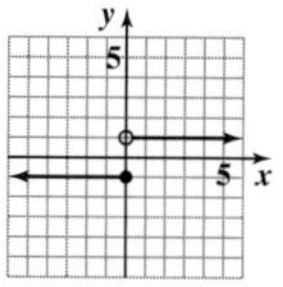

$f(x) = \begin{cases} -1 \text{ if } x \le 0 \\ 1 \text{ if } x > 0 \end{cases}$

37. a. $f(x) = -2x^2 - x - 5$; neither **b.** $-4x - 2h + 1$ **38. a.** 30 **b.** 50 **39.** $f(x) = -2x - 5$ **40.** $f(x) = 2x - 3$ **41.** $f(x) = 3x - 13$ **42.** $f(x) = -\frac{5}{2}x - 13$ **43.** The lines are parallel. **44.** $m = 7.8$; The percentage of U.S. colleges offering distance learning is increasing by 7.8% each year. **45.** 2

Section 2.5

Check Point Exercises

1.

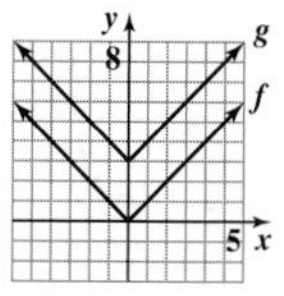

2.

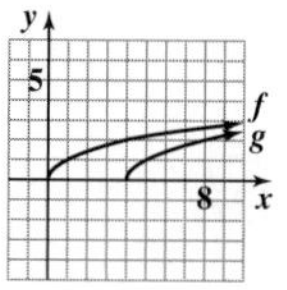

3.

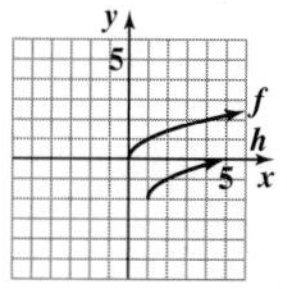

4.

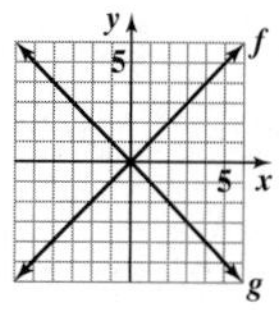

5.

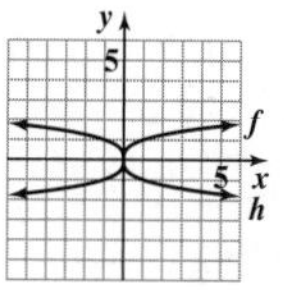

6.

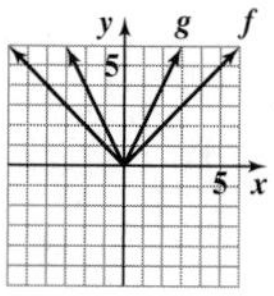

7. a.

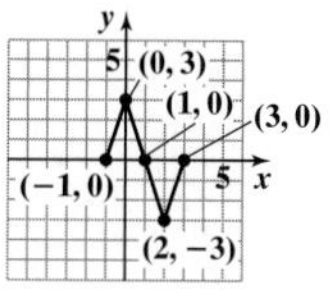

$g(x) = f(2x)$

b.

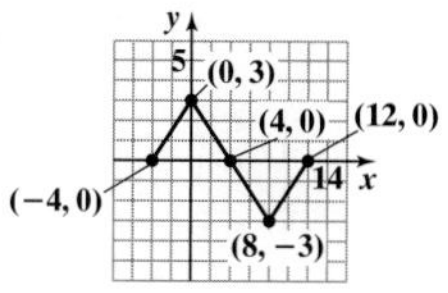

$h(x) = f\left(\frac{1}{2}x\right)$

8.

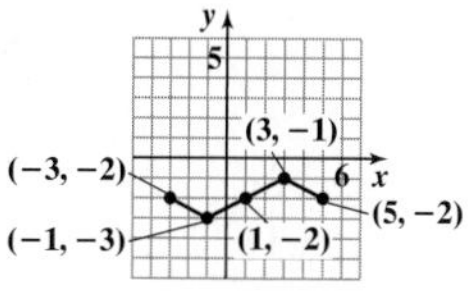

$y = -\frac{1}{3}f(x + 1) - 2$

9.

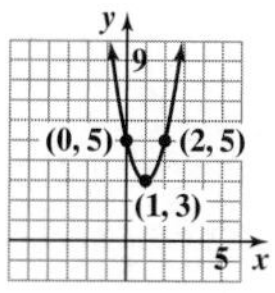

$g(x) = 2(x - 1)^2 + 3$

Exercise Set 2.5

1.

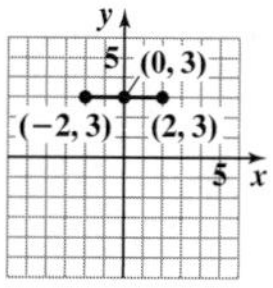

$g(x) = f(x) + 1$

3.

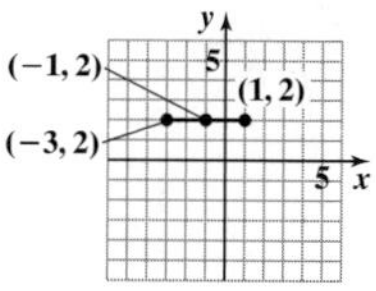

$g(x) = f(x + 1)$

5.

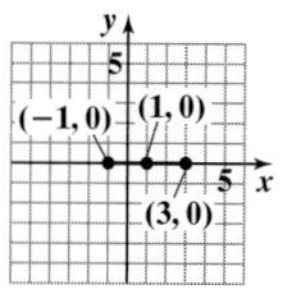

$g(x) = f(x - 1) - 2$

7.

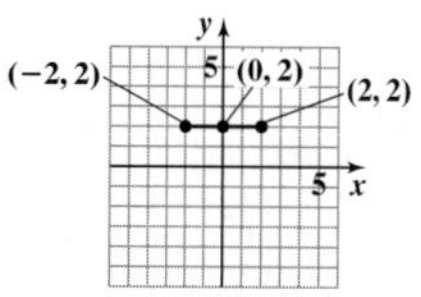

$g(x) = f(-x)$

9.

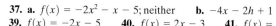

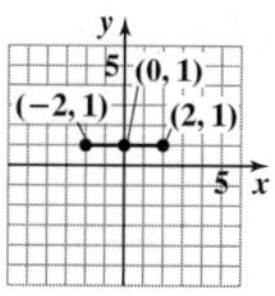

$g(x) = -f(x) + 3$

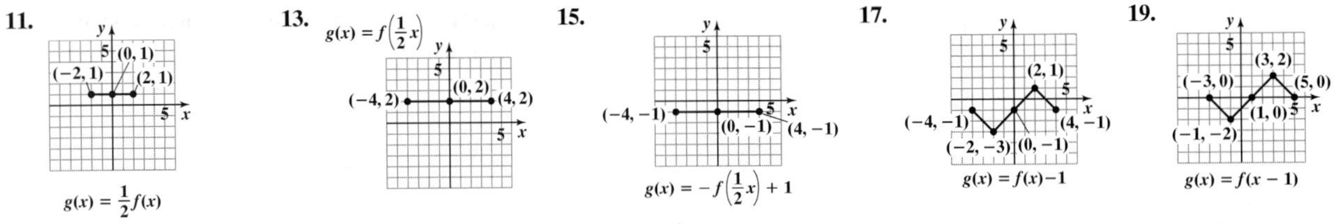

11. $g(x) = \frac{1}{2}f(x)$ 15. $g(x) = -f\left(\frac{1}{2}x\right) + 1$ 17. $g(x) = f(x) - 1$ 19. $g(x) = f(x - 1)$

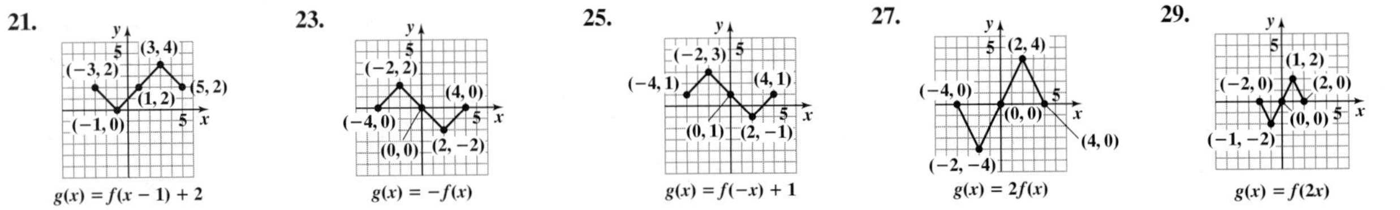

21. $g(x) = f(x - 1) + 2$ 23. $g(x) = -f(x)$ 25. $g(x) = f(-x) + 1$ 27. $g(x) = 2f(x)$ 29. $g(x) = f(2x)$

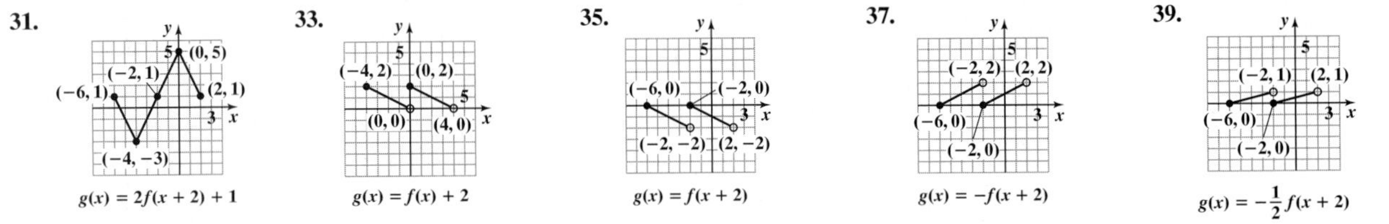

31. $g(x) = 2f(x + 2) + 1$ 33. $g(x) = f(x) + 2$ 35. $g(x) = f(x + 2)$ 37. $g(x) = -f(x + 2)$ 39. $g(x) = -\frac{1}{2}f(x + 2)$

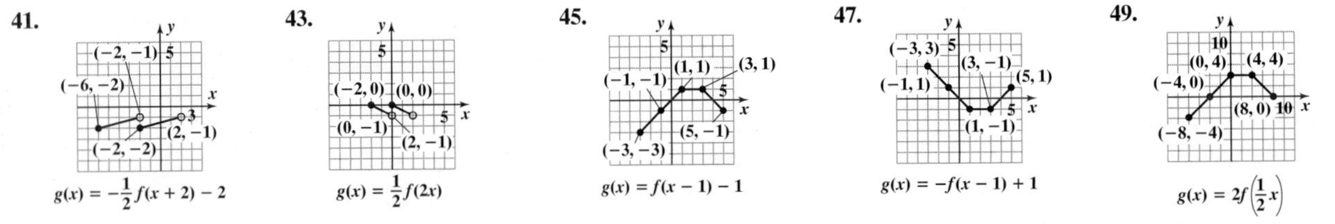

41. $g(x) = -\frac{1}{2}f(x + 2) - 2$ 43. $g(x) = \frac{1}{2}f(2x)$ 45. $g(x) = f(x - 1) - 1$ 47. $g(x) = -f(x - 1) + 1$ 49. $g(x) = 2f\left(\frac{1}{2}x\right)$

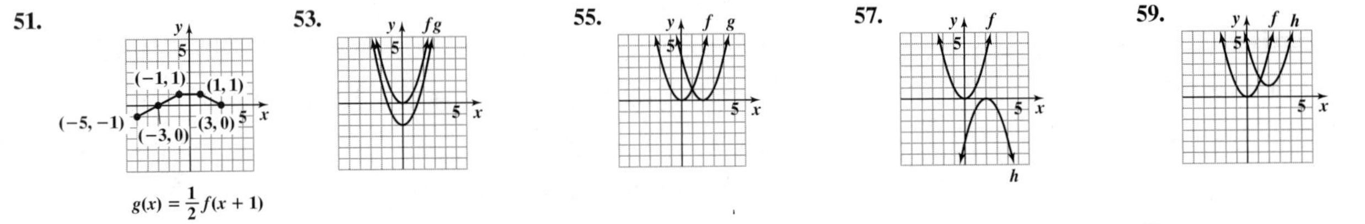

51. $g(x) = \frac{1}{2}f(x + 1)$

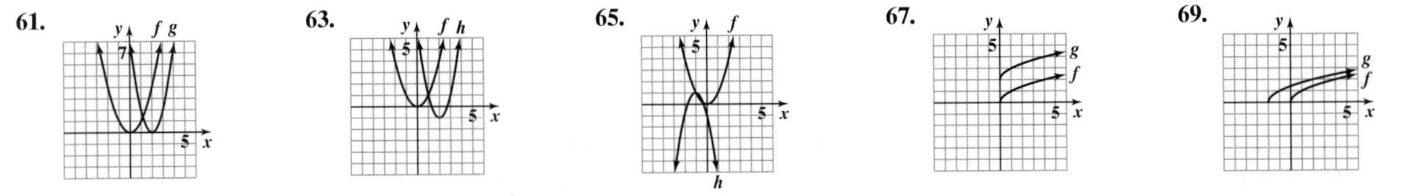

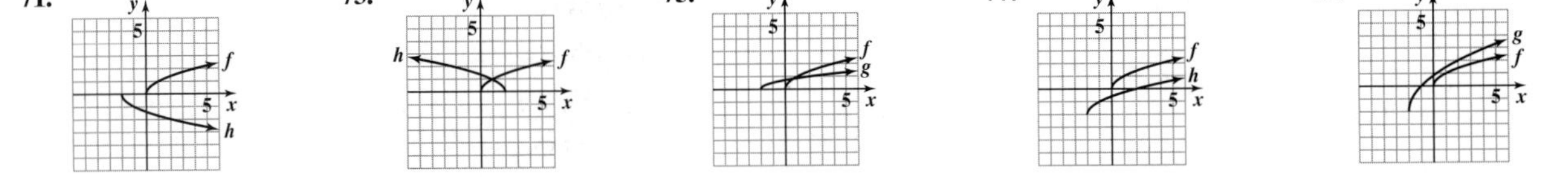

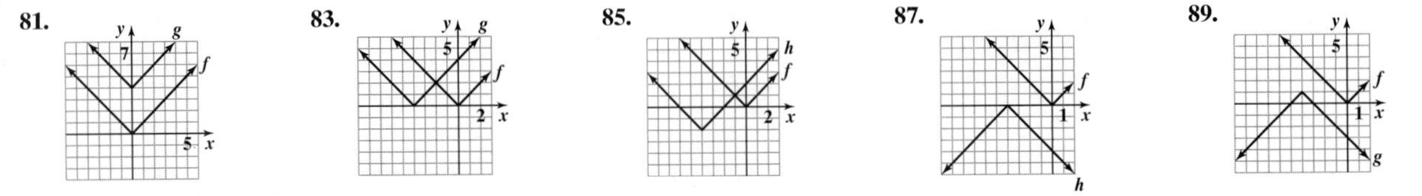

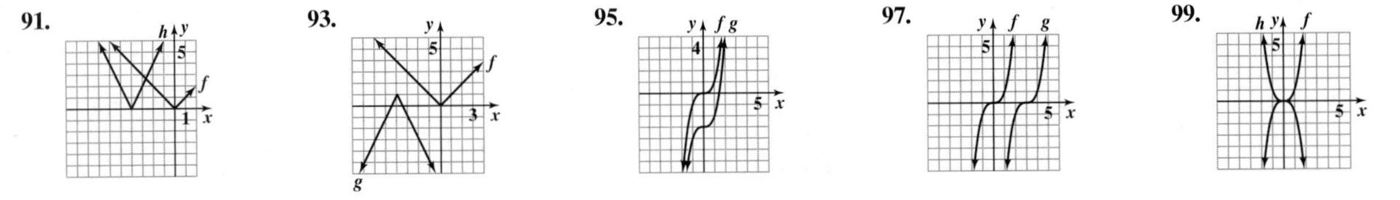

101. 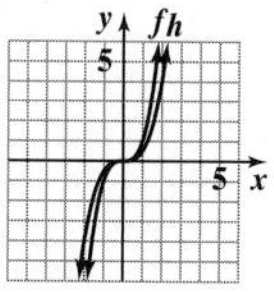
103. 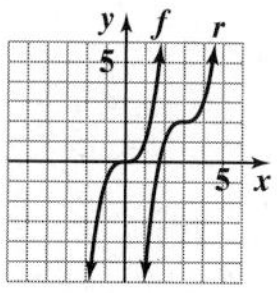
105. 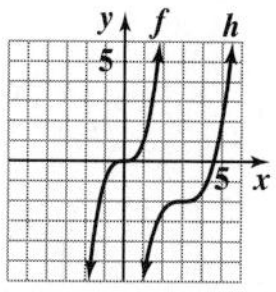
107. 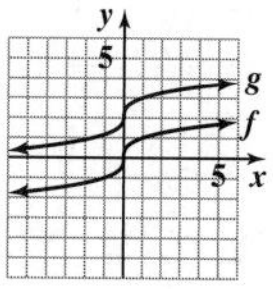
109. 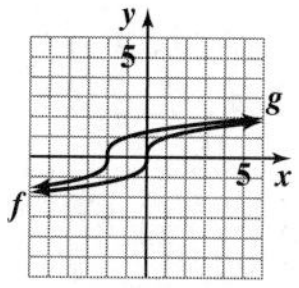

111. 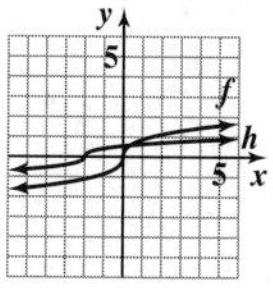
113. 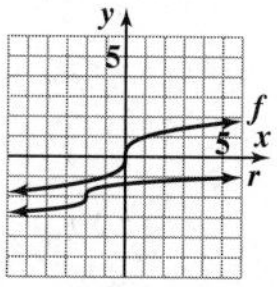
115. 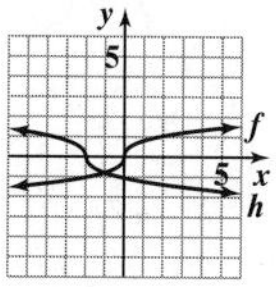
117. 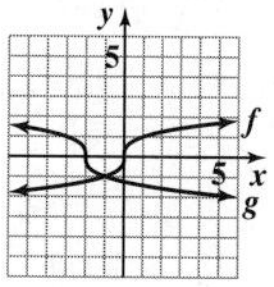
119. 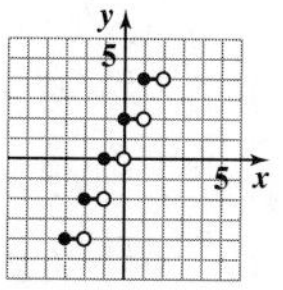

$g(x) = 2 \text{ int } (x + 1)$

121. 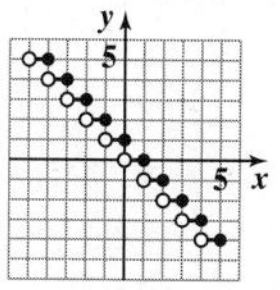

$h(x) = \text{int } (-x) + 1$

123. $y = \sqrt{x - 2}$ **125.** $y = (x + 1)^2 - 4$

127. a. First, vertically stretch the graph of $f(x) = \sqrt{x}$ by the factor 2.9; then, shift the result up 20.1 units.
b. 40.2 in.; very well
c. 0.9 in. per month
d. 0.2 in. per month; This is a much smaller rate of change; The graph is not as steep between 50 and 60 as it is between 0 and 10.

135. a. 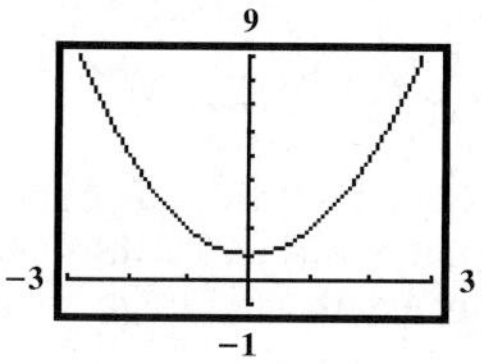

b. 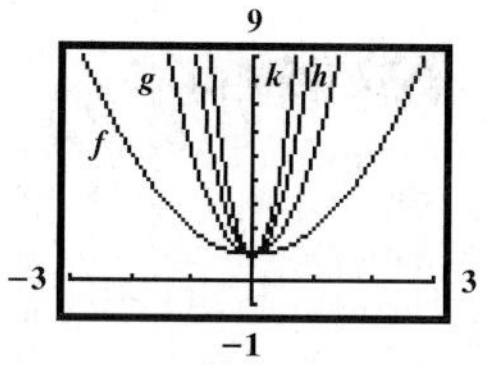

137. d **139.** $g(x) = -|x - 5| + 1$ **141.** $g(x) = -\frac{1}{4}\sqrt{16 - x^2} - 1$ **143.** $(a, 2b)$ **145.** $(a, b - 3)$

Section 2.6

Check Point Exercises

1. a. $(-\infty, \infty)$ **b.** $(-\infty, -7)\cup(-7, 7)\cup(7, \infty)$ **c.** $[3, \infty)$ **2. a.** $(f + g)(x) = x^2 + x - 6$ **b.** $(f - g)(x) = -x^2 + x - 4$
c. $(fg)(x) = x^3 - 5x^2 - x + 5$ **d.** $\left(\frac{f}{g}\right)(x) = \frac{x - 5}{x^2 - 1}, x \neq \pm 1$ **3. a.** $(f + g)(x) = \sqrt{x - 3} + \sqrt{x + 1}$ **b.** $[3, \infty)$
4. a. $(f \circ g)(x) = 10x^2 - 5x + 1$ **b.** $(g \circ f)(x) = 50x^2 + 115x + 65$ **5. a.** $(f \circ g)(x) = \frac{4x}{1 + 2x}$ **b.** $\left(-\infty, -\frac{1}{2}\right)\cup\left(-\frac{1}{2}, 0\right)\cup(0, \infty)$
6. if $f(x) = \sqrt{x}$ and $g(x) = x^2 + 5$, then $h(x) = (f \circ g)(x)$

Exercise Set 2.6

1. $(-\infty, \infty)$ **3.** $(-\infty, 4)\cup(4, \infty)$ **5.** $(-\infty, \infty)$ **7.** $(-\infty, -3)\cup(-3, 5)\cup(5, \infty)$ **9.** $(-\infty, -7)\cup(-7, 9)\cup(9, \infty)$
11. $(-\infty, -1)\cup(-1, 1)\cup(1, \infty)$ **13.** $(-\infty, 0)\cup(0, 3)\cup(3, \infty)$ **15.** $(-\infty, 1)\cup(1, 3)\cup(3, \infty)$ **17.** $[3, \infty)$ **19.** $(3, \infty)$ **21.** $[-7, \infty)$
23. $(-\infty, 12]$ **25.** $[2, \infty)$ **27.** $[2, 5)\cup(5, \infty)$ **29.** $(-\infty, -2)\cup(-2, 2)\cup(2, 5)\cup(5, \infty)$
31. $(f + g)(x) = 3x + 2$; Domain: $(-\infty, \infty)$; $(f - g)(x) = x + 4$; Domain: $(-\infty, \infty)$; $(fg)(x) = 2x^2 + x - 3$; Domain: $(-\infty, \infty)$; $\left(\frac{f}{g}\right)(x) = \frac{2x + 3}{x - 1}$; Domain: $(-\infty, 1)\cup(1, \infty)$ **33.** $(f + g)(x) = 3x^2 + x - 5$; Domain: $(-\infty, \infty)$; $(f - g)(x) = -3x^2 + x - 5$; Domain: $(-\infty, \infty)$; $(fg)(x) = 3x^3 - 15x^2$; Domain: $(-\infty, \infty)$; $\left(\frac{f}{g}\right)(x) = \frac{x - 5}{3x^2}$; Domain: $(-\infty, 0)\cup(0, \infty)$ **35.** $(f + g)(x) = 2x^2 - 2$; Domain: $(-\infty, \infty)$; $(f - g)(x) = 2x^2 - 2x - 4$; Domain: $(-\infty, \infty)$; $(fg)(x) = 2x^3 + x^2 - 4x - 3$; Domain: $(-\infty, \infty)$; $\left(\frac{f}{g}\right)(x) = 2x - 3$; Domain: $(-\infty, -1)\cup(-1, \infty)$
37. $(f + g)(x) = 2x - 12$; Domain: $(-\infty, \infty)$; $(f - g)(x) = -2x^2 - 2x + 18$; Domain: $(-\infty, \infty)$; $(fg)(x) = -x^4 - 2x^3 + 18x^2 + 6x - 45$; Domain: $(-\infty, \infty)$; $\left(\frac{f}{g}\right)(x) = \frac{3 - x^2}{x^2 + 2x - 15}$; Domain: $(-\infty, -5) \cup (-5, 3)\cup(3, \infty)$ **39.** $(f + g)(x) = \sqrt{x} + x - 4$; Domain: $[0, \infty)$; $(f - g)(x) = \sqrt{x} - x + 4$; Domain: $[0, \infty)$; $(fg)(x) = \sqrt{x}(x - 4)$; Domain: $[0, \infty)$; $\left(\frac{f}{g}\right)(x) = \frac{\sqrt{x}}{x - 4}$; Domain: $[0, 4)\cup(4, \infty)$
41. $(f + g)(x) = \frac{2x + 2}{x}$; Domain: $(-\infty, 0)\cup(0, \infty)$; $(f - g)(x) = 2$; Domain: $(-\infty, 0)\cup(0, \infty)$; $(fg)(x) = \frac{2x + 1}{x^2}$; Domain: $(-\infty, 0)\cup(0, \infty)$; $\left(\frac{f}{g}\right)(x) = 2x + 1$; Domain: $(-\infty, 0)\cup(0, \infty)$ **43.** $(f + g)(x) = \frac{9x - 1}{x^2 - 9}$; Domain: $(-\infty, -3)\cup(-3, 3)\cup(3, \infty)$; $(f - g)(x) = \frac{x + 3}{x^2 - 9} = \frac{1}{x - 3}$; Domain: $(-\infty, -3)\cup(-3, 3)\cup(3, \infty)$; $(fg)(x) = \frac{20x^2 - 6x - 2}{(x^2 - 9)^2}$;

Domain: $(-\infty, -3)\cup(-3, 3)\cup(3, \infty)$; $\left(\frac{f}{g}\right)(x) = \frac{5x + 1}{4x - 2}$; Domain: $(-\infty, -3)\cup\left(-3, \frac{1}{2}\right)\cup\left(\frac{1}{2}, 3\right)\cup(3, \infty)$
45. $(f + g)(x) = \sqrt{x + 4} + \sqrt{x - 1}$; Domain: $[1, \infty)$; $(f - g)(x) = \sqrt{x + 4} - \sqrt{x - 1}$; Domain: $[1, \infty)$; $(fg)(x) = \sqrt{x^2 + 3x - 4}$; Domain: $[1, \infty)$; $\left(\frac{f}{g}\right)(x) = \frac{\sqrt{x + 4}}{\sqrt{x - 1}}$; Domain: $(1, \infty)$ **47.** $(f + g)(x) = \sqrt{x - 2} + \sqrt{2 - x}$; Domain: $\{2\}$; $(f - g)(x) = \sqrt{x - 2} - \sqrt{2 - x}$; Domain: $\{2\}$; $(fg)(x) = \sqrt{x - 2} \cdot \sqrt{2 - x}$; Domain: $\{2\}$; $\left(\frac{f}{g}\right)(x) = \frac{\sqrt{x - 2}}{\sqrt{2 - x}}$; Domain: $\varnothing$ **49. a.** $(f \circ g)(x) = 2x + 14$ **b.** $(g \circ f)(x) = 2x + 7$ **c.** $(f \circ g)(2) = 18$ **51. a.** $(f \circ g)(x) = 2x + 5$ **b.** $(g \circ f)(x) = 2x + 9$ **c.** $(f \circ g)(2) = 9$ **53. a.** $(f \circ g)(x) = 20x^2 - 11$ **b.** $(g \circ f)(x) = 80x^2 - 120x + 43$ **c.** $(f \circ g)(2) = 69$ **55. a.** $(f \circ g)(x) = x^4 - 4x^2 + 6$ **b.** $(g \circ f)(x) = x^4 + 4x^2 + 2$ **c.** $(f \circ g)(2) = 6$ **57. a.** $(f \circ g)(x) = -2x^2 - x - 1$ **b.** $(g \circ f)(x) = 2x^2 - 17x + 41$ **c.** -11 **59. a.** $(f \circ g)(x) = \sqrt{x - 1}$ **b.** $(g \circ f)(x) = \sqrt{x} - 1$ **c.** $(f \circ g)(2) = 1$ **61. a.** $(f \circ g)(x) = x$ **b.** $(g \circ f)(x) = x$ **c.** $(f \circ g)(2) = 2$ **63. a.** $(f \circ g)(x) = x$ **b.** $(g \circ f)(x) = x$ **c.** 2 **65. a.** $(f \circ g)(x) = \frac{2x}{1 + 3x}$ **b.** $\left(-\infty, -\frac{1}{3}\right)\cup\left(-\frac{1}{3}, 0\right)\cup(0, \infty)$ **67. a.** $(f \circ g)(x) = \frac{4}{4 + x}$ **b.** $(-\infty, -4)\cup(-4, 0)\cup(0, \infty)$ **69. a.** $(f \circ g)(x) = \sqrt{x - 2}$ **b.** $[2, \infty)$ **71. a.** $(f \circ g)(x) = 5 - x$ **b.** $(-\infty, 1]$ **73.** $f(x) = x^4, g(x) = 3x - 1$ **75.** $f(x) = \sqrt[3]{x}, g(x) = x^2 - 9$ **77.** $f(x) = |x|, g(x) = 2x - 5$ **79.** $f(x) = \frac{1}{x}, g(x) = 2x - 3$ **81.** 5 **83.** -1 **85.** $\{x|-4 \le x \le 3\}$

87.

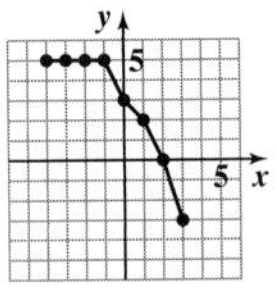

89. 1 **91.** -6 **93.** 1 and 2 **95.** $\{x|x = 0, 1, 2, \ldots, 8\}$
97. a. $(B - D)(x) = 8244x + 1{,}569{,}712$; change in U.S. population **b.** 1,635,664; In 2003, the U.S. population increased by 1,635,664. **c.** 1,670,000; fairly well
99. $f + g$ represents the total world population in year x.
101. $(f + g)(2000) \approx 6$ billion people **103.** $(R - C)(20{,}000) = -200{,}000$; The company lost \$200,000 since costs exceeded revenues; $(R - C)(30{,}000) = 0$; The company broke even since revenues equaled cost; $(R - C)(40{,}000) = 200{,}000$; The company made a profit of \$200,000. **105. a.** f gives the price of the computer after a \$400 discount. g gives the price of the computer after a 25% discount. **b.** $(f \circ g)(x) = 0.75x - 400$; This models the price of a computer after first a 25% discount and then a \$400 discount. **c.** $(g \circ f)(x) = 0.75(x - 400)$; This models the price of a computer after first a \$400 discount and then a 25% discount. **d.** The function $f \circ g$ models the greater discount, since the 25% discount is taken on the regular price first.

113.

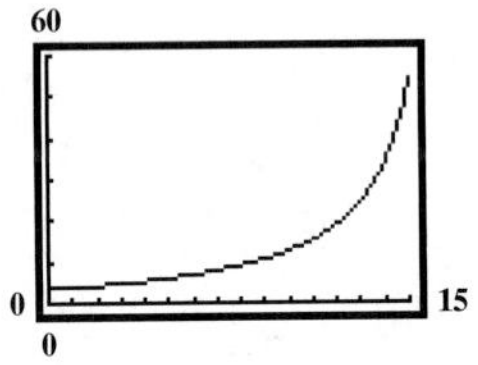

The per capita costs are increasing over time.

115.

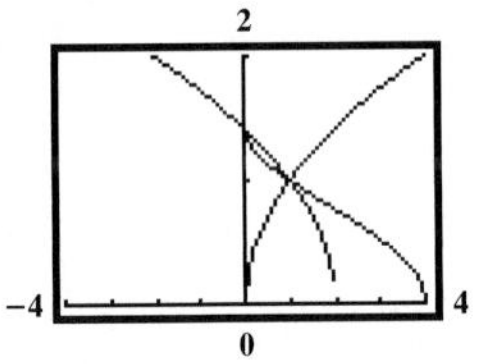

Domain: $[0, 4]$

117. Assume f and g are even; then $f(-x) = f(x)$ and $g(-x) = g(x)$. $(fg)(-x) = f(-x)g(-x) = f(x)g(x) = (fg)(x)$, so fg is even.

119.

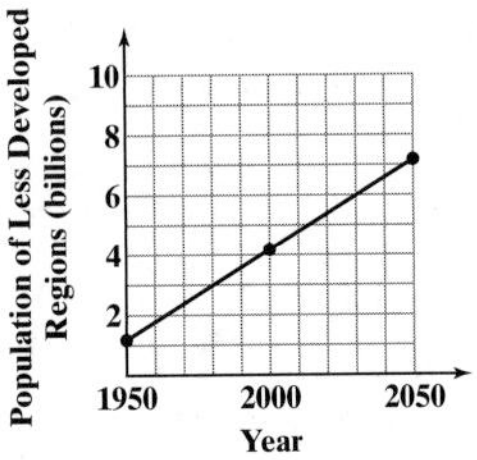

Section 2.7

Check Point Exercises

1. $f(g(x)) = 4\left(\frac{x + 7}{4}\right) - 7 = x + 7 - 7 = x$; $g(f(x)) = \frac{(4x - 7) + 7}{4} = \frac{4x}{4} = x$ **2.** $f^{-1}(x) = \frac{x - 7}{2}$ **3.** $f^{-1}(x) = \sqrt[3]{\frac{x + 1}{4}}$
4. $f^{-1}(x) = \frac{3}{x + 1}$ **5.** (b) and (c)

6.

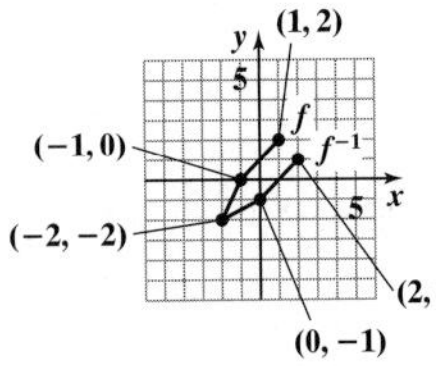

7. $f^{-1}(x) = \sqrt{x - 1}$

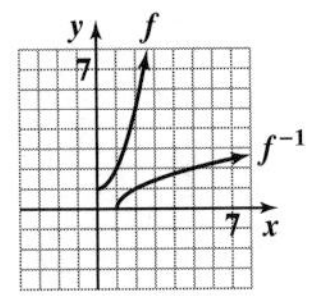

Exercise Set 2.7

1. $f(g(x)) = x$; $g(f(x)) = x$; f and g are inverses. **3.** $f(g(x)) = x$; $g(f(x)) = x$; f and g are inverses.
5. $f(g(x)) = \frac{5x - 56}{9}$; $g(f(x)) = \frac{5x - 4}{9}$; f and g are not inverses. **7.** $f(g(x)) = x$; $g(f(x)) = x$; f and g are inverses.

9. $f(g(x)) = x$; $g(f(x)) = x$; f and g are inverses. **11.** $f^{-1}(x) = x - 3$ **13.** $f^{-1}(x) = \frac{x}{2}$ **15.** $f^{-1}(x) = \frac{x-3}{2}$ **17.** $f^{-1}(x) = \sqrt[3]{x-2}$ **19.** $f^{-1}(x) = \sqrt[3]{x} - 2$ **21.** $f^{-1}(x) = \frac{1}{x}$ **23.** $f^{-1}(x) = x^2, x \geq 0$ **25.** $f^{-1}(x) = \frac{7}{x+3}$ **27.** $f^{-1}(x) = \frac{3x+1}{x-2}; x \neq 2$ **29.** The function is not one-to-one, so it does not have an inverse function. **31.** The function is not one-to-one, so it does not have an inverse function. **33.** The function is one-to-one, so it does have an inverse function.

35.

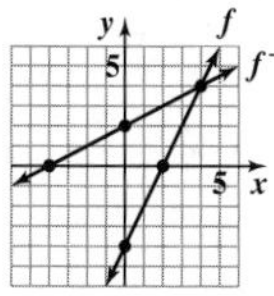

37.

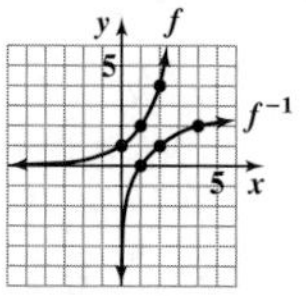

39. a. $f^{-1}(x) = \frac{x+1}{2}$
b.

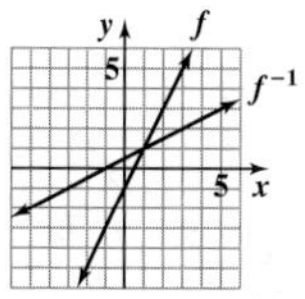

c. Domain of f = Range of $f^{-1} = (-\infty, \infty)$; Range of f = Domain of $f^{-1} = (-\infty, \infty)$

41. a. $f^{-1}(x) = \sqrt{x+4}$
b.

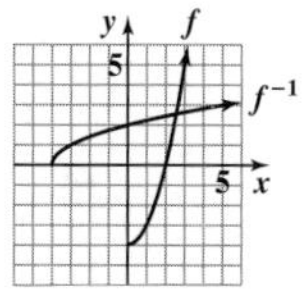

c. Domain of f = Range of $f^{-1} = [0, \infty)$; Range of f = Domain of $f^{-1} = [-4, \infty)$

43. a. $f^{-1}(x) = -\sqrt{x} + 1$
b.

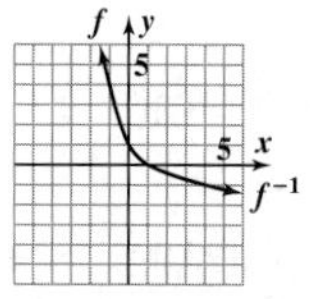

c. Domain of f = Range of $f^{-1} = (-\infty, 1]$; Range of f = Domain of $f^{-1} = [0, \infty)$

45. a. $f^{-1}(x) = \sqrt[3]{x+1}$
b.

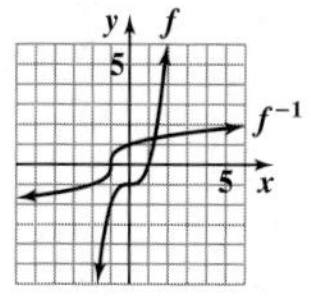

c. Domain of f = Range of $f^{-1} = (-\infty, \infty)$; Range of f = Domain of $f^{-1} = (-\infty, \infty)$

47. a. $f^{-1}(x) = \sqrt[3]{x} - 2$
b.

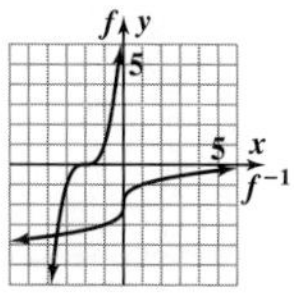

c. Domain of f = Range of $f^{-1} = (-\infty, \infty)$; Range of f = Domain of $f^{-1} = (-\infty, \infty)$

49. a. $f^{-1}(x) = x^2 + 1, x \geq 0$
b.

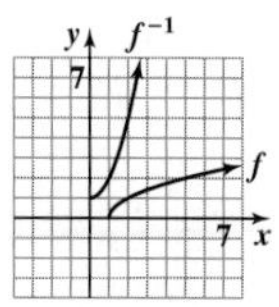

c. Domain of f = Range of $f^{-1} = [1, \infty)$; Range of f = Domain of $f^{-1} = [0, \infty)$

51. a. $f^{-1}(x) = (x-1)^3$
b.

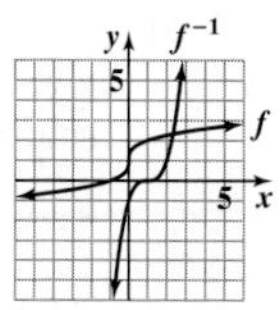

c. Domain of f = Range of $f^{-1} = (-\infty, \infty)$; Range of f = Domain of $f^{-1} = (-\infty, \infty)$

53. 5 **55.** 1 **57.** 2 **59.** −7 **61.** 3 **63.** 11 **65. a.** f = {(Zambia, −7.3), (Columbia, −4.5), (Poland, −2.8), (Italy, −2.8), (United States, −1.9)} **b.** f^{-1} = {(−7.3, Zambia), (−4.5, Columbia), (−2.8, Poland), (−2.8, Italy), (−1.9, United States)}; No; One member of the domain, −2.8, corresponds to more than one member of the range, Poland and Italy. **67. a.** f is a one-to-one function. **b.** $f^{-1}(0.25)$ is the number of people in a room for a 25% probability of two people sharing a birthday. $f^{-1}(0.5)$ is the number of people in a room for a 50% probability of two people sharing a birthday. $f^{-1}(0.7)$ is the number of people in a room for a 70% probability of two people sharing a birthday.

69. $f(g(x)) = \frac{9}{5}\left[\frac{5}{9}(x-32)\right] + 32 = x$ and $g(f(x)) = \frac{5}{9}\left[\left(\frac{9}{5}x + 32\right) - 32\right] = x$

77.

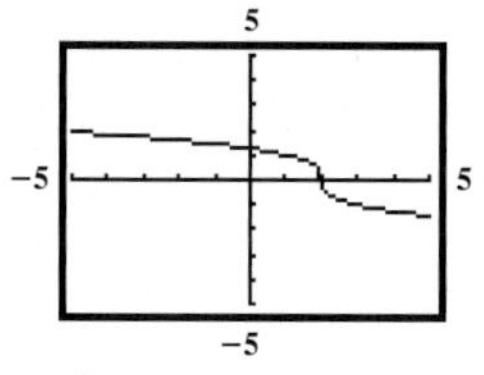

one-to-one

79.

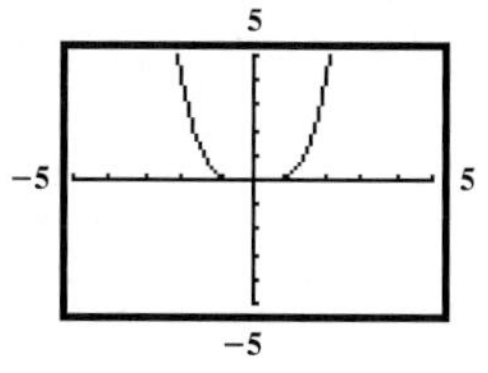

not one-to-one

81.

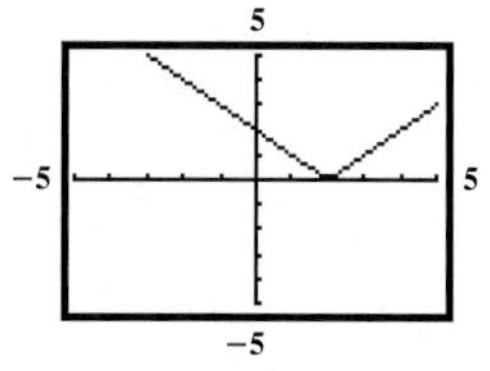

not one-to-one

83.

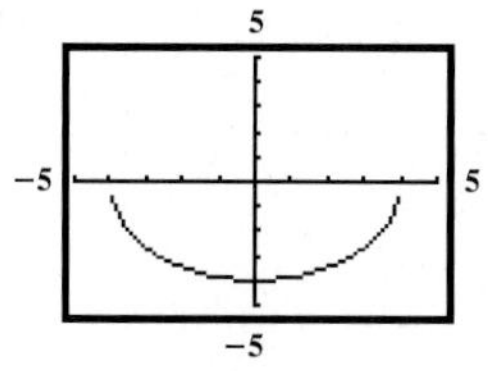

not one-to-one

85.

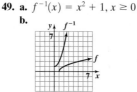

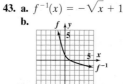

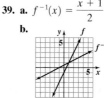

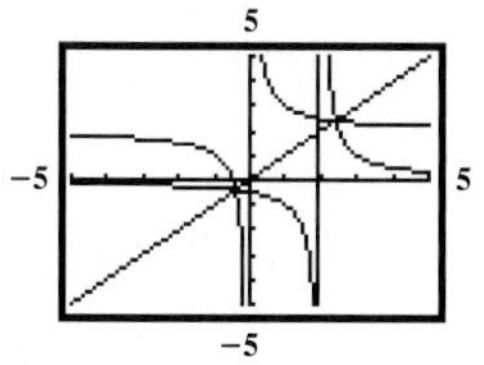

f and g are inverses.

87. d
89. $(f \circ f)(x) = x$, so f is its own inverse.
91. 7

Section 2.8

Check Point Exercises

1. 13 **2.** $\left(4, -\frac{1}{2}\right)$ **3.** $x^2 + y^2 = 16$ **4.** $(x-5)^2 + (y+6)^2 = 100$

5. a. center: $(-3, 1)$; radius: 2   **6.** $(x + 2)^2 + (y - 2)^2 = 9$

b.

$(x + 3)^2 + (y - 1)^2 = 4$

$x^2 + y^2 + 4x - 4y - 1 = 0$

c. Domain: $[-5, -1]$;
   Range: $[-1, 3]$

Exercise Set 2.8

1. 13   **3.** $2\sqrt{29} \approx 10.77$ **5.** 5   **7.** $\sqrt{29} \approx 5.39$ **9.** $4\sqrt{2} \approx 5.66$ **11.** $2\sqrt{5} \approx 4.47$ **13.** $2\sqrt{2} \approx 2.83$ **15.** $\sqrt{93} \approx 9.64$

17. $\sqrt{5} \approx 2.24$ **19.** $(4, 6)$ **21.** $(-4, -5)$ **23.** $\left(\dfrac{3}{2}, -6\right)$ **25.** $(-3, -2)$ **27.** $(1, 5\sqrt{5})$ **29.** $(2\sqrt{2}, 0)$ **31.** $x^2 + y^2 = 49$

33. $(x - 3)^2 + (y - 2)^2 = 25$ **35.** $(x + 1)^2 + (y - 4)^2 = 4$ **37.** $(x + 3)^2 + (y + 1)^2 = 3$ **39.** $(x + 4)^2 + (y - 0)^2 = 100$

41. center: $(0, 0)$
   radius: 4
   Domain: $[-4, 4]$;
   Range: $[-4, 4]$

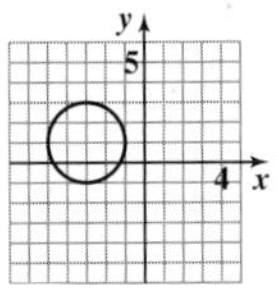

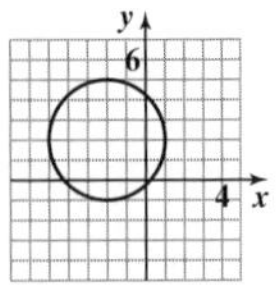

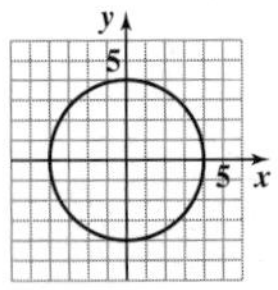

43. center: $(3, 1)$
   radius: 6
   Domain: $[-3, 9]$;
   Range: $[-5, 7]$

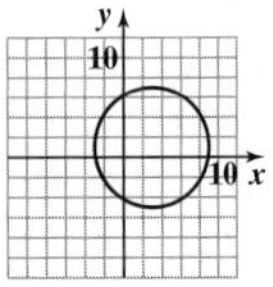

45. center: $(-3, 2)$
   radius: 2
   Domain: $[-5, -1]$;
   Range: $[0, 4]$

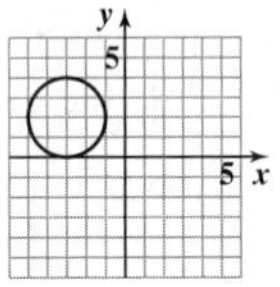

47. center: $(-2, -2)$
   radius: 2
   Domain: $[-4, 0]$;
   Range: $[-4, 0]$

$x^2 + y^2 = 16$

$(x - 3)^2 + (y - 1)^2 = 36$

$(x + 3)^2 + (y - 2)^2 = 4$

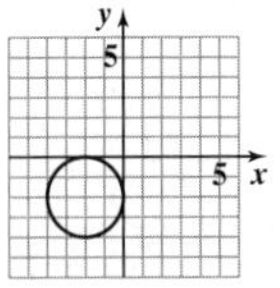

$(x + 2)^2 + (y + 2)^2 = 4$

49. $(x + 3)^2 + (y + 1)^2 = 4$
   center: $(-3, -1)$
   radius: 2

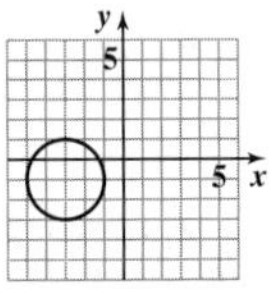

51. $(x - 5)^2 + (y - 3)^2 = 64$
   center: $(5, 3)$
   radius: 8

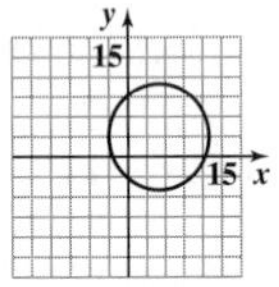

53. $(x + 4)^2 + (y - 1)^2 = 25$
   center: $(-4, 1)$
   radius: 5

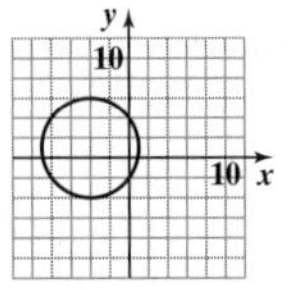

55. $(x - 1)^2 + (y - 0)^2 = 16$
   center: $(1, 0)$
   radius: 4

$x^2 + y^2 + 6x + 2y + 6 = 0$

$x^2 + y^2 - 10x - 6y - 30 = 0$

$x^2 + y^2 + 8x - 2y - 8 = 0$

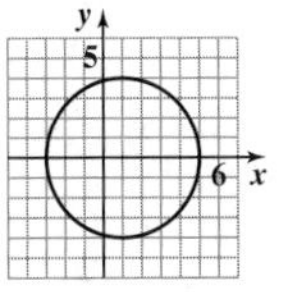

$x^2 - 2x + y^2 - 15 = 0$

57. $\left(x - \dfrac{1}{2}\right)^2 + (y + 1)^2 = \dfrac{1}{4}$ **59.** $\left(x + \dfrac{3}{2}\right)^2 + (y - 1)^2 = \dfrac{17}{4}$

   center: $\left(\dfrac{1}{2}, -1\right)$

   radius: $\dfrac{1}{2}$

   center: $\left(-\dfrac{1}{2}, -\dfrac{1}{2}\right)$

   radius: $\dfrac{\sqrt{17}}{2}$

$x^2 + y^2 - x + 2y + 1 = 0$

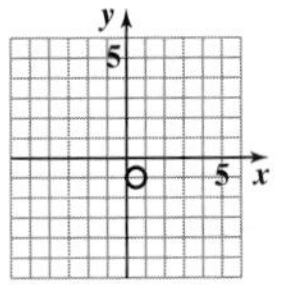

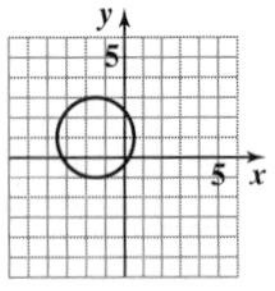

$x^2 + y^2 + 3x - 2y - 1 = 0$

61. a. $(5, 10)$ **b.** $\sqrt{5}$ **c.** $(x - 5)^2 + (y - 10)^2 = 5$

63. $\{(0, -4), (4, 0)\}$ **65.** $\{(0, -3), (2, -1)\}$

$x^2 + y^2 = 16$
$x - y = 4$

$(x - 2)^2 + (y + 3)^2 = 4$
$y = x - 3$

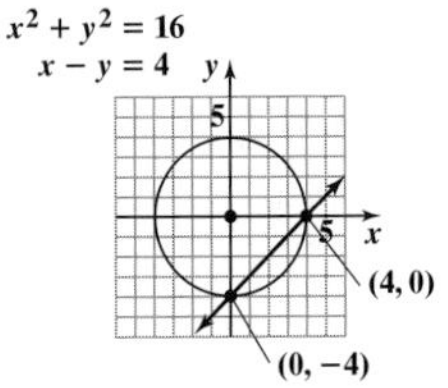

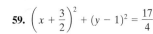

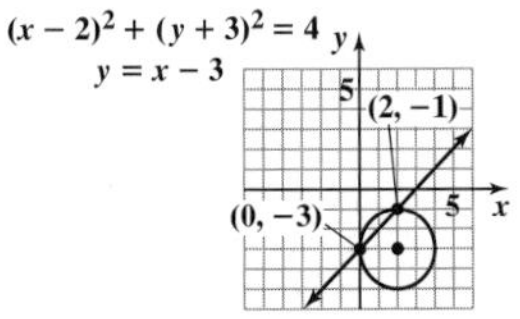

67. 0.5 hr; 30 min   **69.** $(x + 2.4)^2 + (y + 2.7)^2 = 900$

77.   **79.** **81.** $2\sqrt{2} + 3\sqrt{2} = 5\sqrt{2}$
83. 11π

Chapter 2 Review Exercises

1. Function; Domain: $\{2, 3, 5\}$; Range: $\{7\}$ **2.** Function; Domain: $\{1, 2, 13\}$; Range: $\{10, 500, \pi\}$ **3.** Not a function; Domain: $\{12, 14\}$; Range: $\{13, 15, 19\}$ **4.** y is a function of x.   **5.** y is a function of x.   **6.** y is not a function of x.   **7. a.** $f(4) = -23$ **b.** $f(x + 3) = -7x - 16$
c. $f(-x) = 5 + 7x$ **8. a.** $g(0) = 2$ **b.** $g(-2) = 24$ **c.** $g(x - 1) = 3x^2 - 11x + 10$ **d.** $g(-x) = 3x^2 + 5x + 2$
9. a. $g(13) = 3$ **b.** $g(0) = 4$ **c.** $g(-3) = 7$ **10. a.** -1 **b.** 12 **c.** 3 **11.** not a function   **12.** function   **13.** function
14. not a function   **15.** not a function   **16.** function   **17.** 8 **18.** $-4x - 2h + 1$ **19. a.** Domain: $[-3, 5)$
b. Range: $[-5, 0]$ **c.** x-intercept: -3 **d.** y-intercept: -2 **e.** increasing: $(-2, 0)$; or $(3, 5)$ decreasing: $(-3, -2)$ or $(0, 3)$
f. $f(-2) = -3$ and $f(3) = -5$ **20. a.** Domain: $(-\infty, \infty)$ **b.** Range: $(-\infty, \infty)$ **c.** x-intercepts: -2 and 3 **d.** y-intercept: 3
e. increasing: $(-5, 0)$; decreasing: $(-\infty, -5)$ or $(0, \infty)$ **f.** $f(-2) = 0$ and $f(6) = -3$ **21. a.** Domain: $(-\infty, \infty)$ **b.** Range: $[-2, 2]$
c. x-intercept: 0 **d.** y-intercept: 0 **e.** increasing: $(-2, 2)$; constant: $(-\infty, -2)$ or $(2, \infty)$ **f.** $f(-9) = -2$ and $f(14) = 2$
22. a. $0; f(0) = -2$ **b.** $-2, 3; f(-2) = -3, f(3) = -5$ **23. a.** $0; f(0) = 3$ **b.** $-5; f(-5) = -6$ **24.** odd; symmetric with respect to
the origin   **25.** even; symmetric with respect to the y-axis   **26.** odd; symmetric with respect to the origin   **27. a.** yes; The graph passes the
vertical line test.   **b.** Decreasing: $(3, 12)$; The eagle descended.   **c.** Constant: $(0, 3)$ and $(12, 17)$; The eagle's height held steady during the
first 3 seconds and the eagle was on the ground for 5 seconds.   **d.** Increasing: $(17, 30)$; The eagle was ascending.

28.   **29.** $m = -\dfrac{1}{2}$; falls   **30.** $m = 1$; rises   **31.** $m = 0$; horizontal   **32.** $m = $ undefined; vertical
33. $y - 2 = -6(x + 3)$; $y = -6x - 16$ **34.** using $(1, 6)$, $y - 6 = 2(x - 1)$; $y = 2x + 4$
35. $y + 7 = -3(x - 4)$; $y = -3x + 5$ **36.** $y - 6 = -3(x + 3)$; $y = -3x - 3$ **37.** $x + 6y + 18 = 0$

38. Slope: $\dfrac{2}{5}$; y-intercept: -1 **39.** Slope: -4; y-intercept: 5 **40.** Slope: $-\dfrac{2}{3}$; y-intercept: -2 **41.** Slope: 0; y-intercept: 4

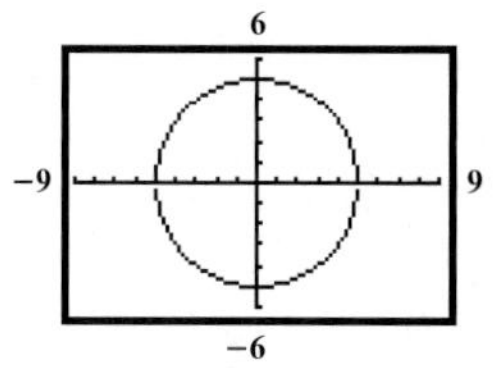

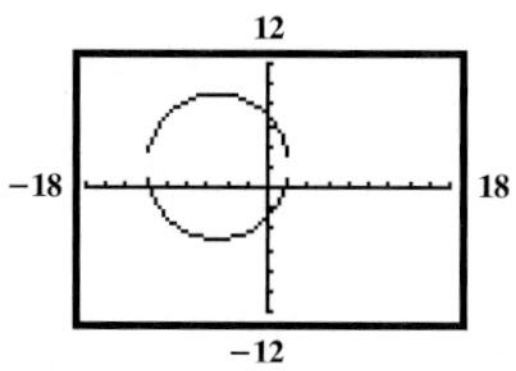

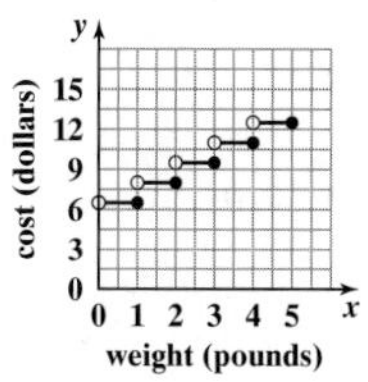

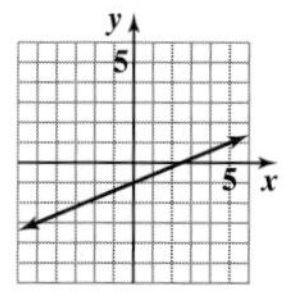

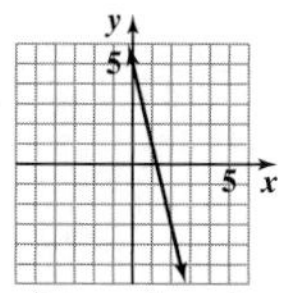

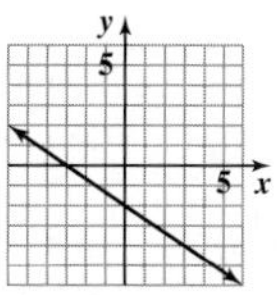

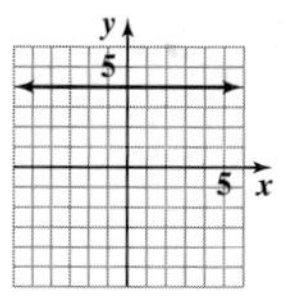

$y = \dfrac{2}{5}x - 1$ $f(x) = -4x + 5$ $2x + 3y + 6 = 0$ $2y - 8 = 0$

42. $2x - 5y - 10 = 0$ **43.** $2x - 10 = 0$ **44. a.** $y - 1.5 = 0.95(x - 1)$, or $y - 3.4 = 0.95(x - 3)$ **b.** $y = 0.95x + 0.55$
c. \$10.05 billion   **45. a.** $m = -44$; The number of new AIDS diagnoses decreased
at a rate of 44 each year from 1999 to 2001.   **b.** $m = 909$; The number of new AIDS
diagnoses increased at a rate of 909 each year from 2001 to 2003.   **c.** $m = 432.5$; yes;
Answers will vary.   **46.** 10

47. $y = g(x)$ **48.** $y = g(x)$ **49.**   **50.** $y = g(x)$ **51.**

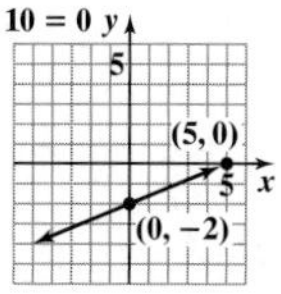

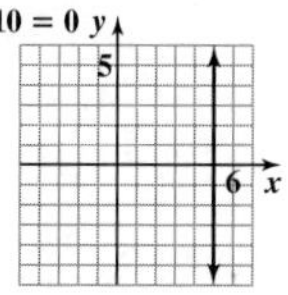

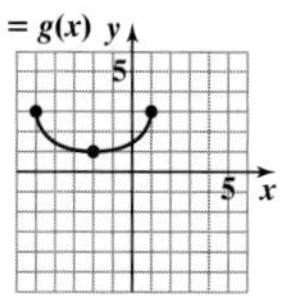

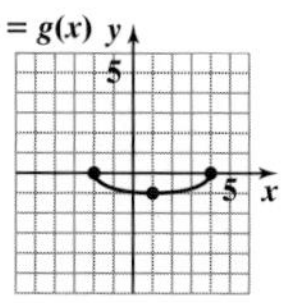

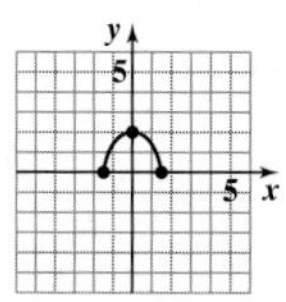

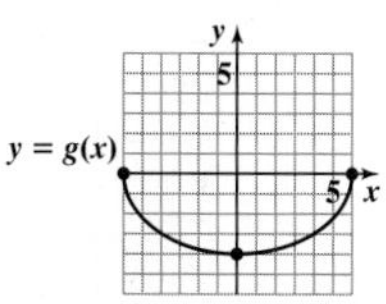

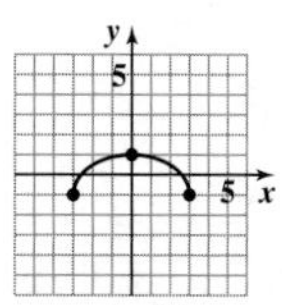

52.   **53.**   **54.**   **55.**   **56.**

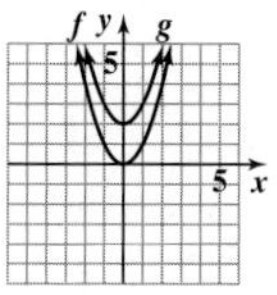

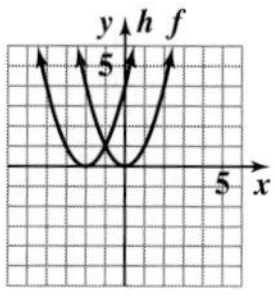

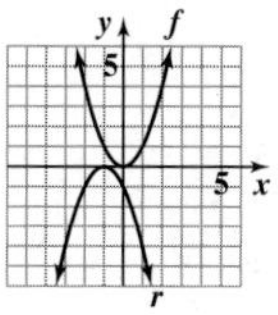

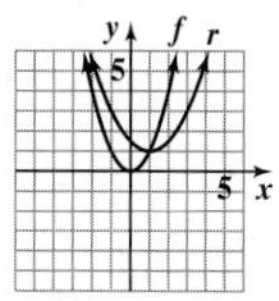

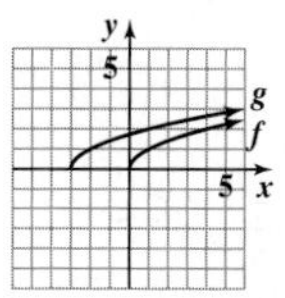

57.   **58.**   **59.**   **60.**   **61.**

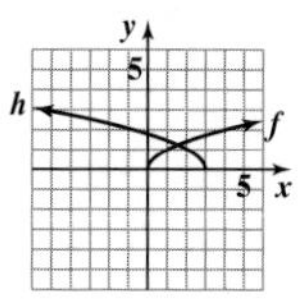

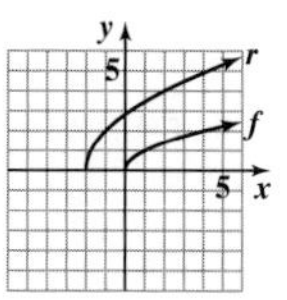

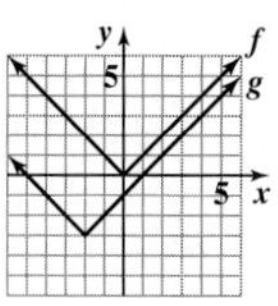

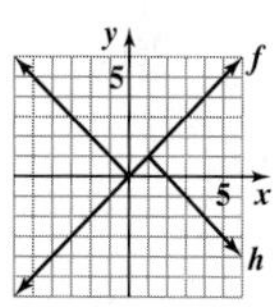

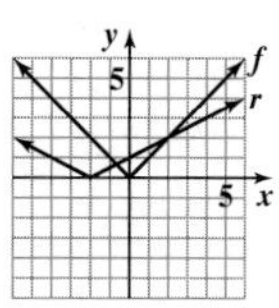

62. **63.** **64.** **65.** **66.**

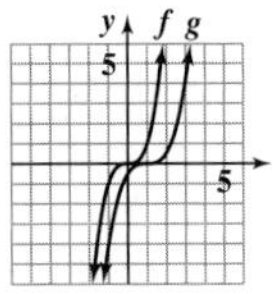

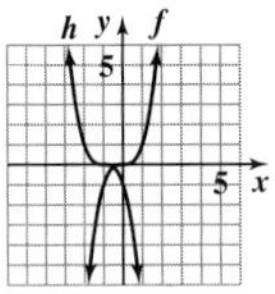

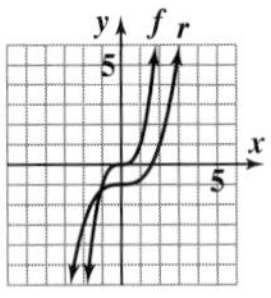

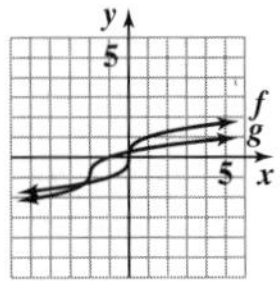

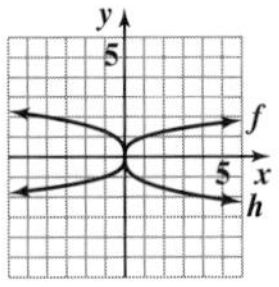

67. **68.** $(-\infty, \infty)$ **69.** $(-\infty, 7)\cup(7, \infty)$ **70.** $(-\infty, 4]$ **71.** $(-\infty, -7)\cup(-7, 3)\cup(3, \infty)$

72. $[2, 5)\cup(5, \infty)$ **73.** $[1, \infty)$ **74.** $(f + g)(x) = 4x - 6$; Domain: $(-\infty, \infty)$; $(f - g)(x) = 2x + 4$;

Domain: $(-\infty, \infty)$; $(fg)(x) = 3x^2 - 16x + 5$; Domain: $(-\infty, \infty)$; $\left(\dfrac{f}{g}\right)(x) = \dfrac{3x - 1}{x - 5}$; Domain: $(-\infty, 5)\cup(5, \infty)$

75. $(f + g)(x) = 2x^2 + x$; Domain: $(-\infty, \infty)$; $(f - g)(x) = x + 2$; Domain: $(-\infty, \infty)$; $(fg)(x) = x^4 + x^3 - x - 1$;

Domain: $(-\infty, \infty)$; $\left(\dfrac{f}{g}\right)(x) = \dfrac{x^2 + x + 1}{x^2 - 1}$; Domain: $(-\infty, -1)\cup(-1, 1)\cup(1, \infty)$

76. $(f + g)(x) = \sqrt{x + 7} + \sqrt{x - 2}$; Domain: $[2, \infty)$; $(f - g)(x) = \sqrt{x + 7} - \sqrt{x - 2}$; Domain: $[2, \infty)$; $(fg)(x) = \sqrt{x^2 + 5x - 14}$; Domain:

$[2, \infty)$; $\left(\dfrac{f}{g}\right)(x) = \dfrac{\sqrt{x + 7}}{\sqrt{x - 2}}$; Domain: $(2, \infty)$ **77. a.** $(f \circ g)(x) = 16x^2 - 8x + 4$ **b.** $(g \circ f)(x) = 4x^2 + 11$ **c.** $(f \circ g)(3) = 124$

78. a. $(f \circ g)(x) = \sqrt{x + 1}$ **b.** $(g \circ f)(x) = \sqrt{x} + 1$ **c.** $(f \circ g)(3) = 2$ **79. a.** $(f \circ g)(x) = \dfrac{1 + x}{1 - 2x}$ **b.** $\left\{x \,|\, x \neq 0 \text{ and } x \neq \dfrac{1}{2}\right\}$

80. a. $(f \circ g)(x) = \sqrt{x + 2}$ **b.** $\{x \,|\, x \geq -2\}$ **81.** $f(x) = x^4, g(x) = x^2 + 2x - 1$ **82.** $f(x) = \sqrt[3]{x}, g(x) = 7x + 4$ **83.** $f(g(x)) = x - \dfrac{7}{10}$;

$g(f(x)) = x - \dfrac{7}{6}$; f and g are not inverses of each other. **84.** $f(g(x)) = x$; $g(f(x)) = x$; f and g are inverses of each other. **85.** $f^{-1}(x) = \dfrac{x + 3}{4}$

86. $f^{-1}(x) = \sqrt[3]{\dfrac{x - 1}{8}}$ or $\dfrac{\sqrt[3]{x - 1}}{2}$ **87.** $f^{-1}(x) = \dfrac{2}{x - 5}$ **88.** Inverse function exists. **89.** Inverse function does not exist.

90. Inverse function exists. **91.** Inverse function does not exist.

92. **93.** $f^{-1}(x) = \sqrt{1 - x}$ **94.** $f^{-1}(x) = (x - 1)^2, x \geq 1$ **95.** 13 **96.** $2\sqrt{2} \approx 2.83$ **97.** $(-5, 5)$

98. $\left(-\dfrac{11}{2}, -2\right)$ **99.** $x^2 + y^2 = 9$

100. $(x + 2)^2 + (y - 4)^2 = 36$

101. Center: $(0, 0)$; radius: 1 **102.** Center: $(-2, 3)$; radius: 3 **103.** Center: $(2, -1)$; radius: 3
Domain: $[-1, 1]$; Range: $[-1, 1]$ Domain: $[-5, 1]$; Range: $[0, 6]$ Domain: $[-1, 5]$; Range: $[-4, 2]$

$x^2 + y^2 = 1$ $(x + 2)^2 + (y - 3)^2 = 9$ $x^2 + y^2 - 4x + 2y - 4 = 0$

Chapter 2 Test

1. b, c, d **2. a.** $f(4) - f(-3) = 5$ **b.** Domain: $(-5, 6]$ **c.** Range: $[-4, 5]$ **d.** Increasing: $(-1, 2)$ **e.** Decreasing: $(-5, -1)$ or $(2, 6)$
f. 2; $f(2) = 5$ **g.** $-1; f(-1) = -4$ **h.** $-4, 1,$ and 5 **i.** -3 **3. a.** -2 and 2 **b.** -1 and 1 **c.** 0 **d.** even **e.** no
f. relative minimum

g. **h.** **i.** **j.** $-\dfrac{1}{3}$

4. **5.** **6.** **7.** **8.**

$f(x) = -\dfrac{1}{3}x + 2$ $(x + 2)^2 + (y - 1)^2 = 9$

Domain: $(-\infty, \infty)$; Domain: $[-2, 2]$; Domain: $(-\infty, \infty)$; Domain: $(-\infty, \infty)$; Domain: $[-5, 1]$;
Range: $(-\infty, \infty)$ Range: $[-2, 2]$ Range: $\{4\}$ Range: $(-\infty, \infty)$ Range: $[-2, 4]$

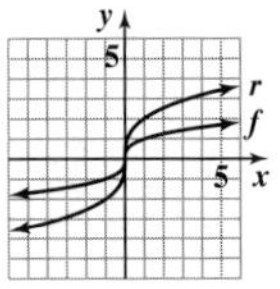

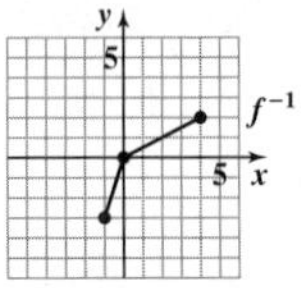

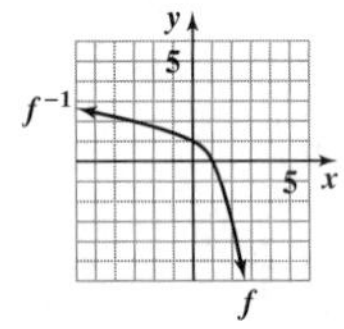

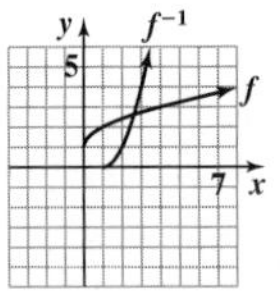

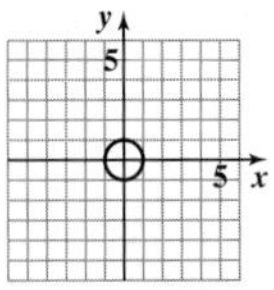

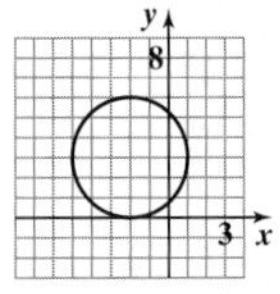

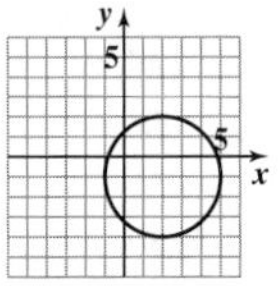

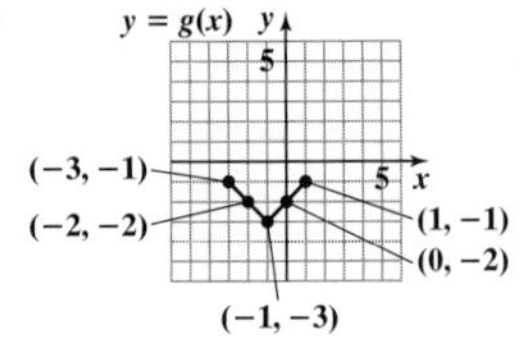

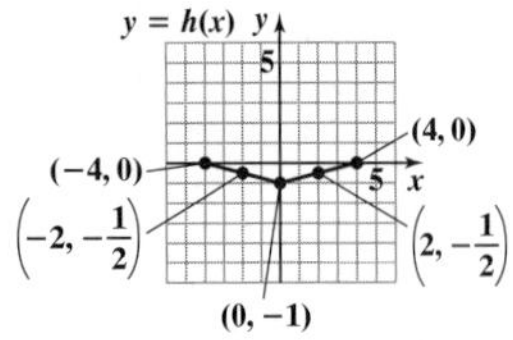

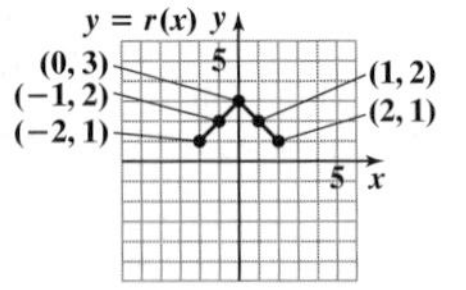

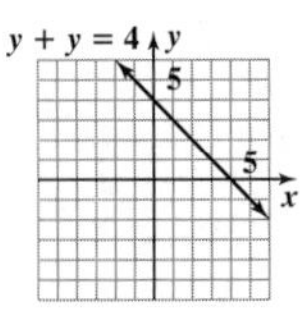

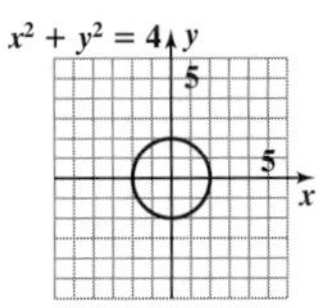

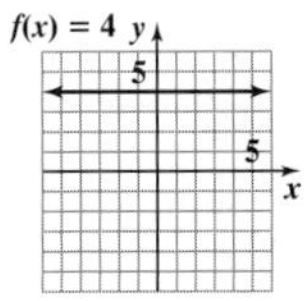

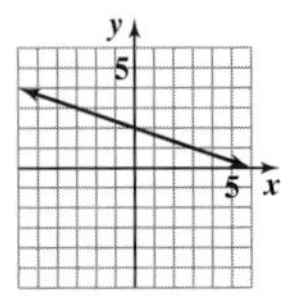

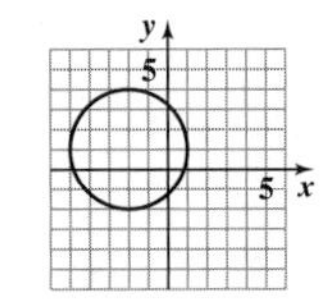

9.

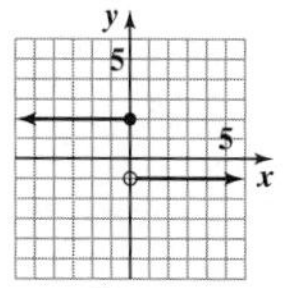

$f(x) = \begin{cases} 2 \text{ if } x \le 0 \\ -1 \text{ if } x > 0 \end{cases}$

Domain: $(-\infty, \infty)$;
Range: $\{-1, 2\}$

10.

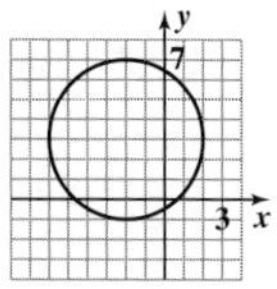

$x^2 + y^2 + 4x - 6y - 3 = 0$

Domain: $[-6, 2]$;
Range: $[-1, 7]$

11.

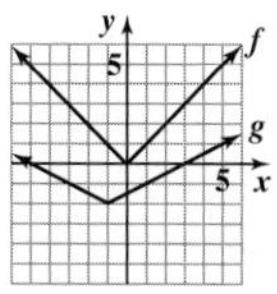

Domain of f = Domain of $g = (-\infty, \infty)$;
Range of $f = [0, \infty)$; Range of $g = [-2, \infty)$

12.

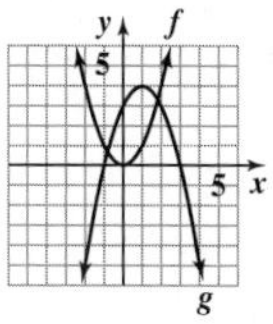

Domain of f = Domain of $g = (-\infty, \infty)$;
Range of $f = [0, \infty)$; Range of $g = (-\infty, 4]$

13.

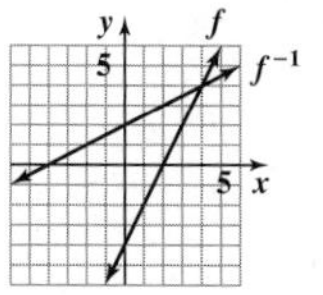

Domain of f = Range of $f^{-1} = (-\infty, \infty)$;
Range of f = Domain of $f^{-1} = (-\infty, \infty)$

14.

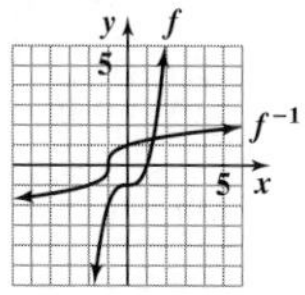

Domain of f = Range of $f^{-1} = (-\infty, \infty)$;
Range of f = Domain of $f^{-1} = (-\infty, \infty)$

15.

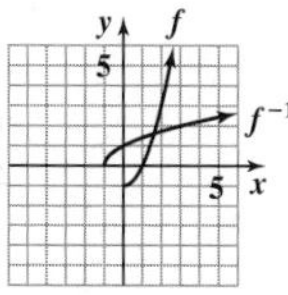

Domain of f = Range of $f^{-1} = [0, \infty)$;
Range of f = Domain of $f^{-1} = [-1, \infty)$

16. $f(x - 1) = x^2 - 3x - 2$ **17.** $2x + h - 1$ **18.** $(g - f)(x) = -x^2 + 3x - 2$
19. $\left(\frac{f}{g}\right)(x) = \frac{x^2 - x - 4}{2x - 6}$; $(-\infty, 3)\cup(3, \infty)$ **20.** $(f \circ g)(x) = 4x^2 - 26x + 38$ **21.** $(g \circ f)(x) = 2x^2 - 2x - 14$
22. -10 **23.** $f(-x) = x^2 + x - 4$; neither **24.** using $(2, 1)$, $y - 1 = 3(x - 2)$; $y = 3x - 5$
25. $y - 6 = 4(x + 4)$; $y = 4x + 22$ **26.** $2x + y + 24 = 0$ **27.** **a.** $y - 4.85 = -0.12(x - 2)$ or $y - 4.49 = -0.12(x - 5)$
b. $f(x) = -0.12x + 5.09$ **c.** \$3.89 **28.** 48 **29.** $g(-1) = 4$; $g(7) = 2$ **30.** $(-\infty, -5)\cup(-5, 1)\cup(1, \infty)$

31. $[1, \infty)$ **32.** $\frac{7x}{2 - 4x}$; $\left\{x \mid x \neq 0 \text{ or } x \neq \frac{1}{2}\right\}$ **33.** $f(x) = x^7$, $g(x) = 2x + 13$ **34.** 5; $(3.5, 0)$

Cumulative Review Exercises (Chapters 1–2)

1. Domain: $[0, 2)$; Range: $[0, 2]$ **2.** $\frac{1}{2}$ and $\frac{3}{2}$ **3.** 2

4.

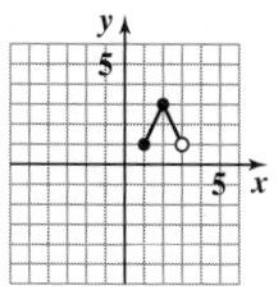

$g(x) = f(x - 1) + 1$

5.

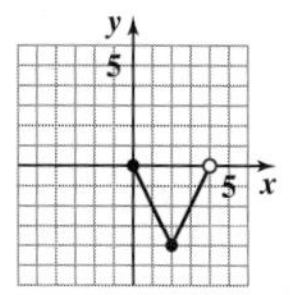

6. $\{-4, 5\}$
7. $\left\{\frac{25}{18}\right\}$
8. $\{4\}$
9. $\{-8, 27\}$
10. $(-\infty, 20)$

11.

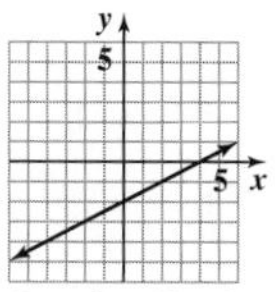

$3x - 6y - 12 = 0$

Domain: $(-\infty, \infty)$;
Range: $(-\infty, \infty)$

12.

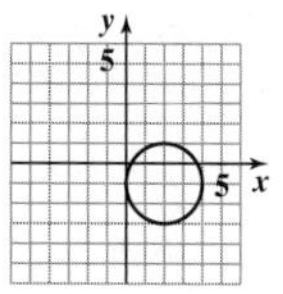

$(x - 2)^2 + (y + 1)^2 = 4$

Domain: $[0, 4]$;
Range: $[-3, 1]$

13.

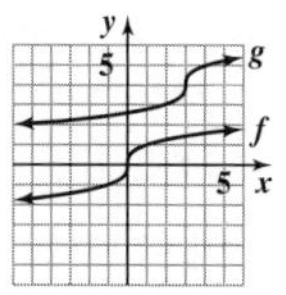

Domain of f = Domain of $g = (-\infty, \infty)$;
Range of f = Range of $g = (-\infty, \infty)$

14.

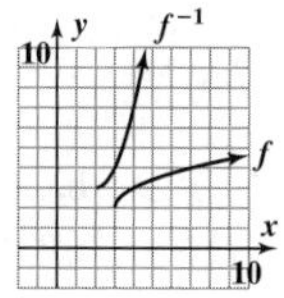

Domain of f = Range of $f^{-1} = [3, \infty)$;
Range of f = Domain of $f^{-1} = [2, \infty)$

15. $-2x - h$ **16.** -7 and -3 **17.** $y - 5 = 4(x + 2)$; $y = 4x + 13$; $4x - y = -13$ **18.** \$1500 at 7%; \$4500 at 9% **19.** \$2000
20. 3 ft by 8 ft

CHAPTER 3

Section 3.1

Check Point Exercises

1.

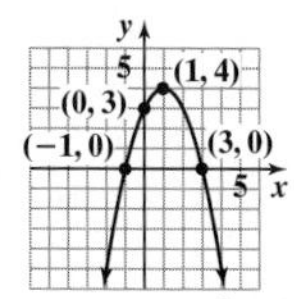

$f(x) = -(x - 1)^2 + 4$

2.

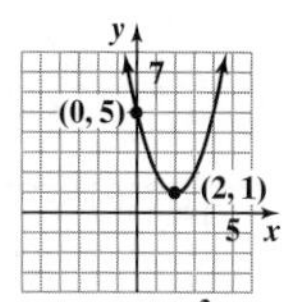

$f(x) = (x - 2)^2 + 1$

3.

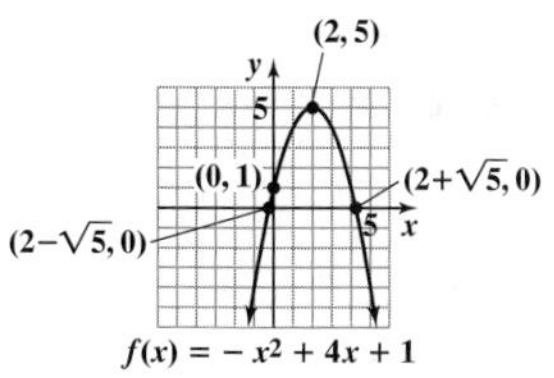

$f(x) = -x^2 + 4x + 1$

Domain: $(-\infty, \infty)$; Range: $(-\infty, 5]$

4. **a.** minimum **b.** Minimum is 984 at $x = 2$.
c. Domain: $(-\infty, \infty)$; Range: $[984, \infty)$
5. 45; 190 **6.** 4, -4; -16
7. 30 ft by 30 ft; 900 sq ft

Exercise Set 3.1

1. $h(x) = (x-1)^2 + 1$ **3.** $j(x) = (x-1)^2 - 1$ **5.** $h(x) = x^2 - 1$ **7.** $g(x) = x^2 - 2x + 1$ **9.** $(3, 1)$ **11.** $(-1, 5)$
13. $(2, -5)$ **15.** $(-1, 9)$

17. Domain: $(-\infty, \infty)$
Range: $[-1, \infty)$
axis of symmetry: $x = 4$

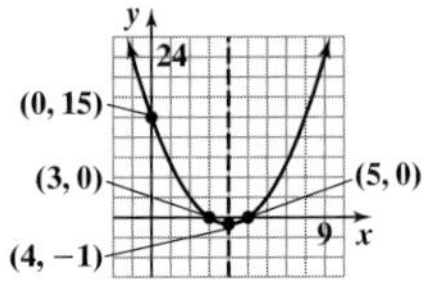

$f(x) = (x-4)^2 - 1$

19. Domain: $(-\infty, \infty)$
Range: $[2, \infty)$
axis of symmetry: $x = 1$

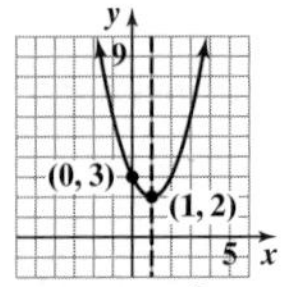

$f(x) = (x-1)^2 + 2$

21. Domain: $(-\infty, \infty)$
Range: $[1, \infty)$
axis of symmetry: $x = 3$

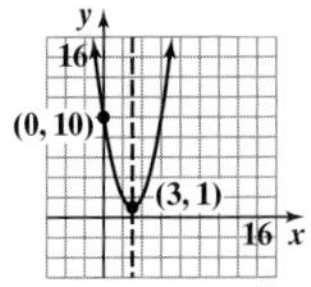

$y - 1 = (x-3)^2$

23. Domain: $(-\infty, \infty)$
Range: $[-1, \infty)$
axis of symmetry: $x = -2$

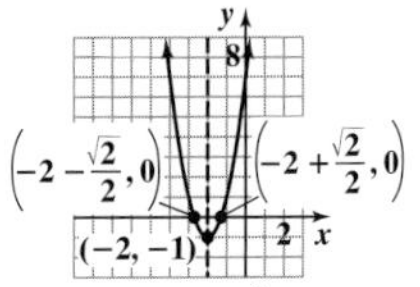

$f(x) = 2(x+2)^2 - 1$

25. Domain: $(-\infty, \infty)$
Range: $(-\infty, 4]$
axis of symmetry: $x = 1$

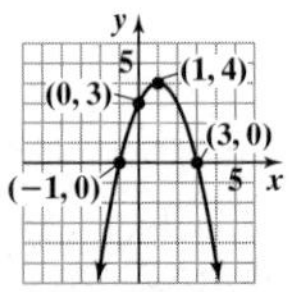

$f(x) = 4 - (x-1)^2$

27. Domain: $(-\infty, \infty)$
Range: $[-4, \infty)$
axis of symmetry: $x = 1$

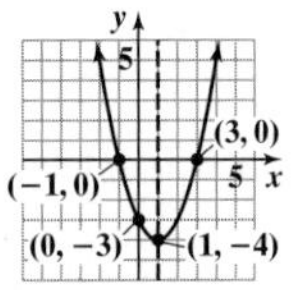

$f(x) = x^2 - 2x - 3$

29. Domain: $(-\infty, \infty)$
Range: $\left[-\frac{49}{4}, \infty\right)$
axis of symmetry: $x = -\frac{3}{2}$

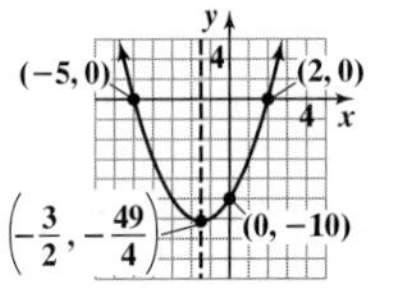

$f(x) = x^2 + 3x - 10$

31. Domain: $(-\infty, \infty)$
Range: $(-\infty, 4]$
axis of symmetry: $x = 1$

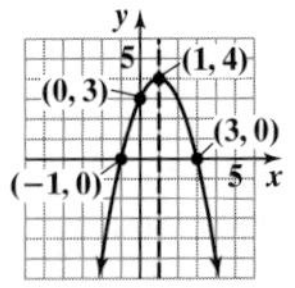

$f(x) = 2x - x^2 + 3$

33. Domain: $(-\infty, \infty)$
Range: $[-6, \infty)$
axis of symmetry: $x = -3$

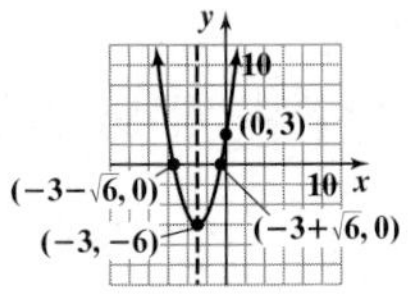

$f(x) = x^2 + 6x + 3$

35. Domain: $(-\infty, \infty)$
Range: $[-5, \infty)$
axis of symmetry: $x = -1$

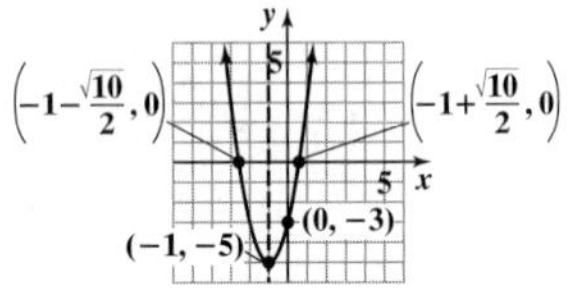

$f(x) = 2x^2 + 4x - 3$

37. Domain: $(-\infty, \infty)$
Range: $(-\infty, -1]$
axis of symmetry: $x = 1$

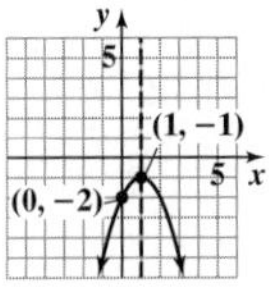

$f(x) = 2x - x^2 - 2$

39. a. minimum **b.** Minimum is -13 at $x = 2$. **c.** Domain: $(-\infty, \infty)$; Range: $[-13, \infty)$
41. a. maximum **b.** Maximum is 1 at $x = 1$. **c.** Domain: $(-\infty, \infty)$; Range: $(-\infty, 1]$
43. a. minimum **b.** Minimum is $-\frac{5}{4}$ at $x = \frac{1}{2}$. **c.** Domain: $(-\infty, \infty)$; Range: $\left[-\frac{5}{4}, \infty\right)$
45. Domain: $(-\infty, \infty)$; Range: $[-2, \infty)$ **47.** Domain: $(-\infty, \infty)$; Range: $(-\infty, -6]$ **49.** $f(x) = 2(x-5)^2 + 3$
51. $f(x) = 2(x+10)^2 - 5$ **53.** $f(x) = -3(x+2)^2 + 4$ **55.** $f(x) = 3(x-11)^2$ **57.** 1990; 2.1 gal; quite well

59. a. 2 sec; 264 ft
b. 6.1 sec
c. 200; 200 feet is the height of the building.
d.

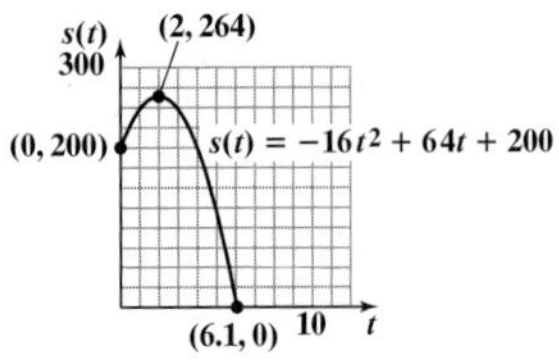

61. 8 and 8; 64
63. 8, −8; −64
65. length: 300 ft; width: 150 ft; maximum area: 45,000 sq ft
67. 12.5 yd by 12.5 yd; 156.25 sq yd
69. 150 ft by 100 ft; 15,000 sq ft
71. 5 in.; 50 sq in.
73. a. $C(x) = 0.55x + 525$
b. $P(x) = -0.001x^2 + 2.45x - 525$
c. 1225 roast beef sandwiches; $\approx$ \$975.63

81. a.

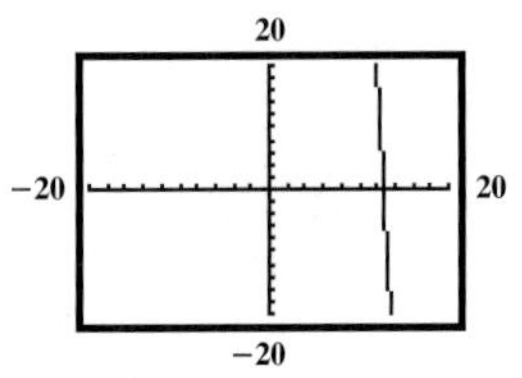

You can only see a little of the parabola.

b. $(20.5, -120.5)$
c. Ymax = 750
d. You can choose Xmin and Xmax so the x-value of the vertex is in the center of the graph. Choose Ymin to include the y-value of the vertex.

83. $(2.5, 185)$

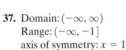

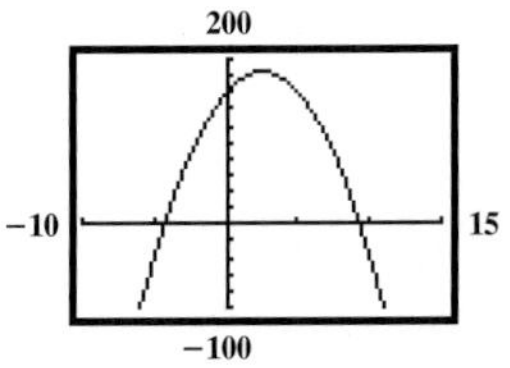

85. $(-30, 91)$

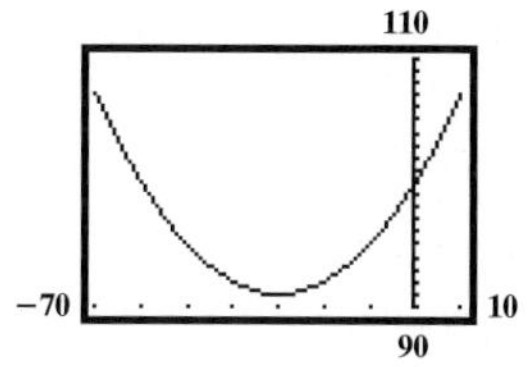

87. a
89. $x = 3; (0, 11)$
91. $f(x) = -2(x + 3)^2 - 1$
93. 65 trees; 16,900 lb

Section 3.2

Check Point Exercises

1. The graph rises to the left and to the right. **2.** Since n is odd and the leading coefficient is negative, the function falls to the right. Since the ratio cannot be negative, the model won't be appropriate. **3.** No; the graph should fall to the left, but doesn't appear to. **4.** $\{-2, 2\}$ **5.** $\{-2, 0, 2\}$ **6.** $-\frac{1}{2}$ with multiplicity 2 and 5 with multiplicity 3; touches and turns at $-\frac{1}{2}$ and crosses at 5 **7.** $f(-3) = -42; f(-2) = 5$
8.

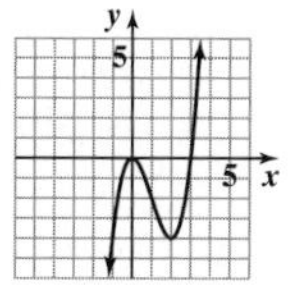

$f(x) = x^3 - 3x^2$

Exercise Set 3.2

1. polynomial function; degree: 3 **3.** polynomial function; degree: 5 **5.** not a polynomial function **7.** not a polynomial function **9.** not a polynomial function **11.** polynomial function **13.** not a polynomial function **15.** b **17.** a **19.** falls to the left and rises to the right **21.** rises to the left and to the right **23.** falls to the left and to the right **25.** $x = 5$ has multiplicity 1; The graph crosses the x-axis; $x = -4$ has multiplicity 2; The graph touches the x-axis and turns around. **27.** $x = 3$ has multiplicity 1; The graph crosses the x-axis; $x = -6$ has multiplicity 3; The graph crosses the x-axis. **29.** $x = 0$ has multiplicity 1; The graph crosses the x-axis; $x = 1$ has multiplicity 2; The graph touches the x-axis and turns around. **31.** $x = 2, x = -2$ and $x = -7$ have multiplicity 1; The graph crosses the x-axis. **33.** $f(1) = -1; f(2) = 5; 1.3$ **35.** $f(-1) = -1; f(0) = 1; -0.5$ **37.** $f(-3) = -11; f(-2) = 1; -2.1$ **39.** $f(-3) = -42; f(-2) = 5; -2.2$ **40.** $f(2) = -4; f(3) = 14; 2.4$

41. **a.** $f(x)$ rises to the right and falls to the left.
b. $x = -2, x = 1, x = -1$;
$f(x)$ crosses the x-axis at each.
c. The y-intercept is -2.
d. neither
e.

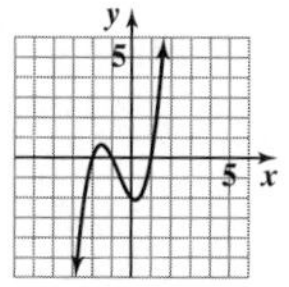

$f(x) = x^3 + 2x^2 - x - 2$

43. **a.** $f(x)$ rises to the left and the right.
b. $x = 0, x = 3, x = -3$;
$f(x)$ crosses the x-axis at -3 and 3;
$f(x)$ touches the x-axis at 0.
c. The y-intercept is 0.
d. y-axis symmetry
e.

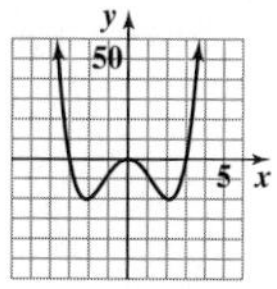

$f(x) = x^4 - 9x^2$

45. **a.** $f(x)$ falls to the left and the right.
b. $x = 0, x = 4, x = -4$;
$f(x)$ crosses the x-axis at -4 and 4;
$f(x)$ touches the x-axis at 0.
c. The y-intercept is 0.
d. y-axis symmetry
e.

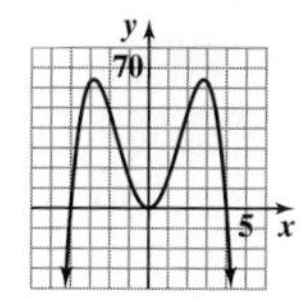

$f(x) = -x^4 + 16x^2$

47. **a.** $f(x)$ rises to the left and the right.
b. $x = 0, x = 1$;
$f(x)$ touches the x-axis at 0 and 1.
c. The y-intercept is 0.
d. neither
e.

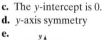

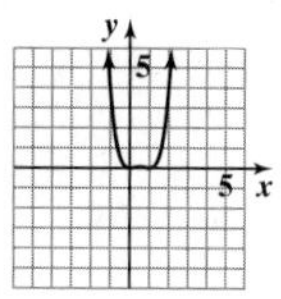

$f(x) = x^4 - 2x^3 + x^2$

49. **a.** $f(x)$ falls to the left and the right.
b. $x = 0, x = 2;$
$f(x)$ crosses the x-axis at 0 and 2.
c. The y-intercept is 0.
d. neither
e.

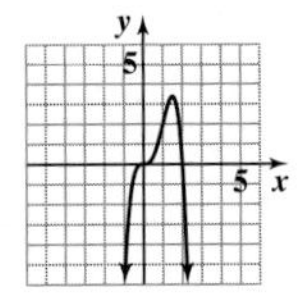

$f(x) = -2x^4 + 4x^3$

51. **a.** $f(x)$ rises to the left and falls to the right.
b. $x = 0, x = \pm\sqrt{3};$
$f(x)$ crosses the x-axis at 0;
$f(x)$ touches the x-axis at $\sqrt{3}$ and $-\sqrt{3}$.
c. The y-intercept is 0.
d. origin symmetry
e.

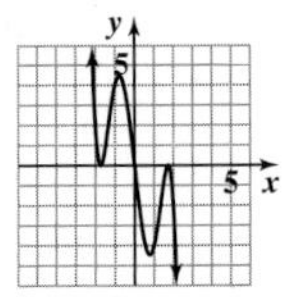

$f(x) = 6x^3 - 9x - x^5$

53. **a.** $f(x)$ rises to the left and falls to the right.
b. $x = 0, x = 3;$
$f(x)$ crosses the x-axis at 3;
$f(x)$ touches the x-axis at 0.
c. The y-intercept is 0.
d. neither
e.

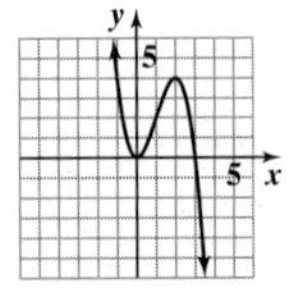

$f(x) = 3x^2 - x^3$

55. **a.** $f(x)$ falls to the left and the right.
b. $x = 1, x = -2, x = 2;$
$f(x)$ crosses the x-axis at -2 and 2;
$f(x)$ touches the x-axis at 1.
c. The y-intercept is 12.
d. neither
e.

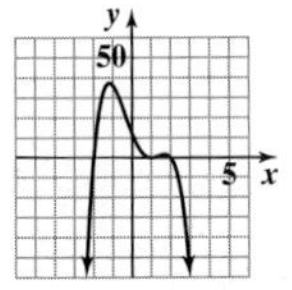

$f(x) = -3(x - 1)^2 (x^2 - 4)$

57. **a.** $f(x)$ rises to the left and the right.
b. $x = -2, x = 0, x = 1;$
$f(x)$ crosses the x-axis at -2 and 1;
$f(x)$ touches the x-axis at 0.
c. The y-intercept is 0.
d. neither
e.

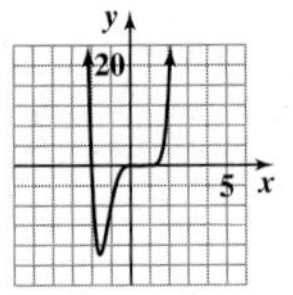

$f(x) = x^2 (x - 1)^3 (x + 2)$

59. **a.** $f(x)$ falls to the left and the right.
b. $x = -3, x = 0, x = 1;$
$f(x)$ crosses the x-axis at -3 and 1;
$f(x)$ touches the x-axis at 0.
c. The y-intercept is 0.
d. neither
e.

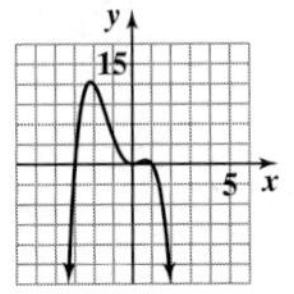

$f(x) = -x^2 (x - 1)(x + 3)$

61. **a.** $f(x)$ falls to the left and the right.
b. $x = -5, x = 0, x = 1;$
$f(x)$ crosses the x-axis at -5 and 0;
$f(x)$ touches the x-axis at 1.
c. The y-intercept is 0.
d. neither
e.

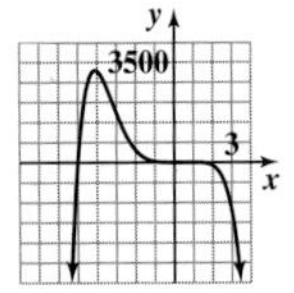

$f(x) = -2x^3 (x - 1)^2 (x + 5)$

63. **a.** $f(x)$ rises to the left and the right.
b. $x = -4, x = 1, x = 2;$
$f(x)$ crosses the x-axis at -4 and 1;
$f(x)$ touches the x-axis at 2.
c. The y-intercept is -16.
d. neither
e.

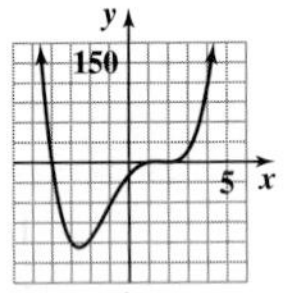

$f(x) = (x - 2)^2 (x + 4)(x - 1)$

65. **a.** -2, odd; 1, odd; 4, odd
b. $f(x) = (x + 2)(x - 1)(x - 4)$
c. 8

67. **a.** -1, odd; 3, even
b. $f(x) = (x + 1)(x - 3)^2$
c. 9

69. a. -3, even; 2, even **b.** $f(x) = -(x + 3)^2(x - 2)^2$ **c.** -36 **71. a.** -2, even; -1, odd; 1, odd **b.** $f(x) = (x + 2)^2(x + 1)(x - 1)^3$ **c.** -4 **73.** $f(10) = 462{,}446$; $g(10) = 465{,}325$; f **75.** Falls to the right: no; cumulative number of deaths cannot decrease.
77. a. 1970 through 1980 and 1985 through 1990 **b.** 1980 through 1985 and 1990 through 2002 **c.** 3 **d.** 4 **e.** negative; The graph falls to the left and falls to the right.

97.

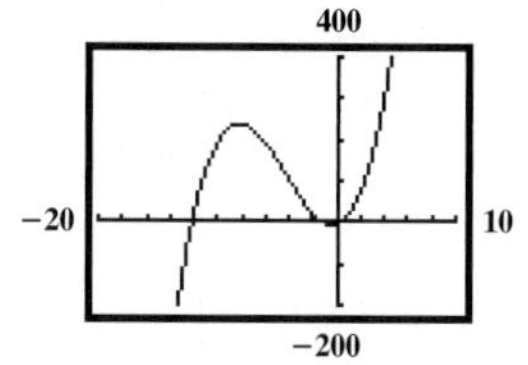

99.

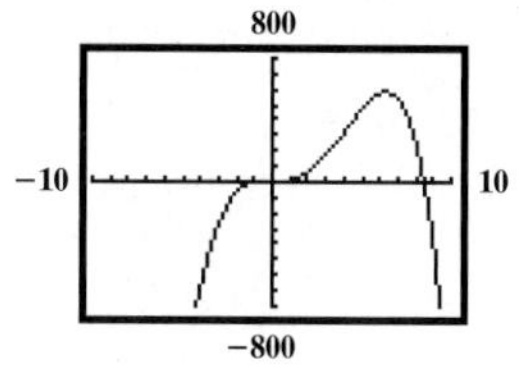

101.

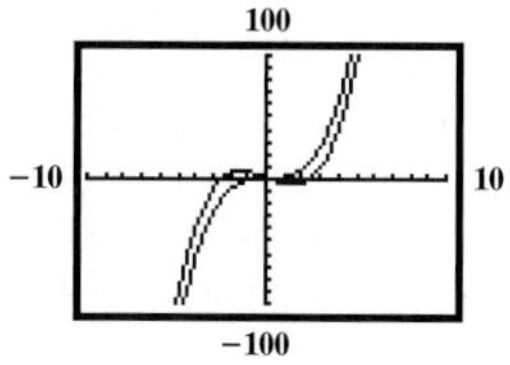

103. c **105.** $f(x) = x^3 - 2x^2$

Section 3.3

Check Point Exercises

1. $x + 5$ **2.** $2x^2 + 3x - 2 + \frac{1}{x - 3}$ **3.** $2x^2 + 7x + 14 + \frac{21x - 10}{x^2 - 2x}$ **4.** $x^2 - 2x - 3$ **5.** -105 **6.** $\left\{-1, -\frac{1}{3}, \frac{2}{5}\right\}$

Exercise Set 3.3

1. $x + 3$ **3.** $x^2 + 3x + 1$ **5.** $2x^2 + 3x + 5$ **7.** $4x + 3 + \frac{2}{3x - 2}$ **9.** $2x^2 + x + 6 - \frac{38}{x + 3}$ **11.** $4x^3 + 16x^2 + 60x + 246 + \frac{984}{x - 4}$
13. $2x + 5$ **15.** $6x^2 + 3x - 1 - \frac{3x - 1}{3x^2 + 1}$ **17.** $2x + 5$ **19.** $3x - 8 + \frac{20}{x + 5}$ **21.** $4x^2 + x + 4 + \frac{3}{x - 1}$
23. $6x^4 + 12x^3 + 22x^2 + 48x + 93 + \frac{187}{x - 2}$ **25.** $x^3 - 10x^2 + 51x - 260 + \frac{1300}{x + 5}$ **27.** $x^4 + x^3 + 2x^2 + 2x + 2$
29. $x^3 + 4x^2 + 16x + 64$ **31.** $2x^4 - 7x^3 + 15x^2 - 31x + 64 - \frac{129}{x + 2}$ **33.** -25 **35.** -133 **37.** 240 **39.** 1
41. $x^2 - 5x + 6$; $x = -1, x = 2, x = 3$ **43.** $\left\{-\frac{1}{2}, 1, 2\right\}$ **45.** $\left\{-\frac{3}{2}, -\frac{1}{3}, \frac{1}{2}\right\}$ **47.** 2; The remainder is zero; $\{-3, -1, 2\}$
49. 1; The remainder is zero; $\left\{\frac{1}{3}, \frac{1}{2}, 1\right\}$ **51. a.** The remainder is 0. **b.** 3 mm **53.** $0.5x^2 - 0.4x + 0.3$ **55. a.** 70
b. $80 + \frac{800}{x - 110}$; $f(30) = 70$; yes **c.** No, f is a rational function because it is a quotient of two polynomials.

65. For Exercise 43:

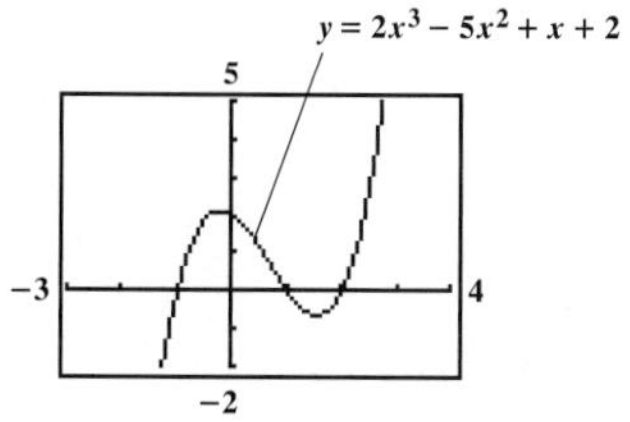

For Exercise 44:

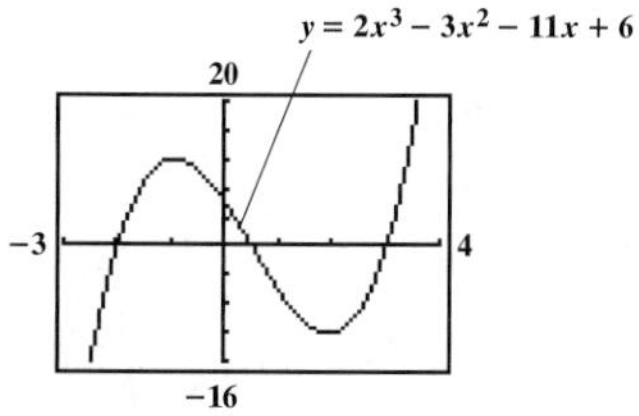

For Exercise 45:

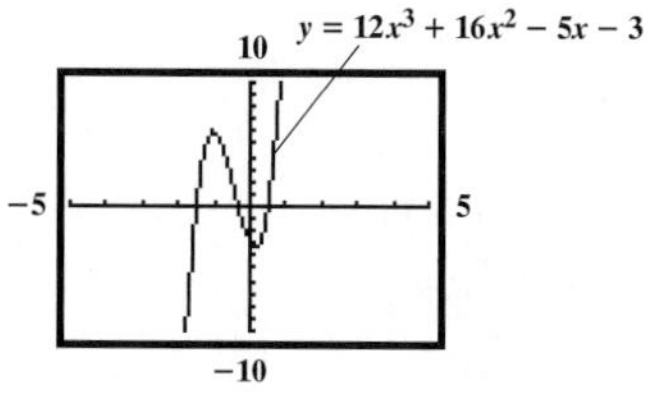

For Exercise 46:

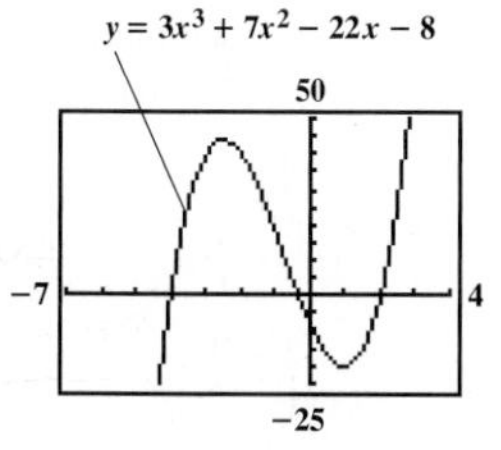

67. $k = -12$ **69.** $x^{2n} - x^n + 1$ **71.** The remainder is 0.; $\{-2, -1, 2, 5\}$

Section 3.4

Check Point Exercises

1. $\pm 1, \pm 2, \pm 3, \pm 6$ **2.** $\pm 1, \pm 3, \pm\frac{1}{2}, \pm\frac{1}{4}, \pm\frac{3}{2}, \pm\frac{3}{4}$ **3.** $\{-5, -4, 1\}$ **4.** $\left\{2, \frac{-3 - \sqrt{5}}{2}, \frac{-3 + \sqrt{5}}{2}\right\}$ **5.** $\{1, 2 - 3i, 2 + 3i\}$
6. $f(x) = x^3 + 3x^2 + x + 3$ **7.** 4, 2, or 0 positive zeros, no possible negative zeros

Exercise Set 3.4

1. $\pm 1, \pm 2, \pm 4$ **3.** $\pm 1, \pm 2, \pm 3, \pm 6, \pm\frac{1}{3}, \pm\frac{2}{3}$ **5.** $\pm 1, \pm 2, \pm 3, \pm 6, \pm\frac{1}{2}, \pm\frac{1}{4}, \pm\frac{3}{2}, \pm\frac{3}{4}$ **7.** $\pm 1, \pm 2, \pm 3, \pm 4, \pm 6, \pm 12$ **9. a.** $\pm 1, \pm 2, \pm 4$
b. 2 is a zero **c.** $\{2, -2, -1\}$ **11. a.** $\pm 1, \pm 2, \pm 3, \pm 6, \pm\frac{1}{2}, \pm\frac{3}{2}$ **b.** 3 is a zero **c.** $\left\{3, \frac{1}{2}, -2\right\}$

13. a. $\pm1, \pm2, \pm3, \pm6$ **b.** -1 is a zero **c.** $\left\{-1, \frac{-3-\sqrt{33}}{2}, \frac{-3+\sqrt{33}}{2}\right\}$ **15. a.** $\pm1, \pm\frac{1}{2}, \pm2$ **b.** -2 is a zero **c.** $\left\{-2, \frac{-1+i}{2}, \frac{-1-i}{2}\right\}$ **17. a.** $\pm1, \pm2, \pm3, \pm4, \pm6, \pm12$ **b.** 4 is a root **c.** $\{-3, 1, 4\}$ **19. a.** $\pm1, \pm2, \pm3, \pm4, \pm6, \pm12$ **b.** -2 is a root **c.** $\{-2, 1+\sqrt{7}, 1-\sqrt{7}\}$ **21. a.** $\pm1, \pm5, \pm\frac{1}{2}, \pm\frac{5}{2}, \pm\frac{1}{3}, \pm\frac{5}{3}, \pm\frac{1}{6}, \pm\frac{5}{6}$ **b.** -5 is a root **c.** $\left\{-5, \frac{1}{2}, \frac{1}{3}\right\}$
23. a. $\pm1, \pm2, \pm4$ **b.** 2 is a root **c.** $\{-2, 2, 1+\sqrt{2}, 1-\sqrt{2}\}$ **25.** $f(x) = 2x^3 - 2x^2 + 50x - 50$ **27.** $f(x) = x^3 - 3x^2 - 15x + 125$
29. $f(x) = x^4 + 10x^2 + 9$ **31.** $f(x) = x^4 - 9x^3 + 21x^2 + 21x - 130$ **33.** no positive real roots; 3 or 1 negative real roots
35. 3 or 1 positive real roots; no negative real roots **37.** 2 or 0 positive real roots; 2 or 0 negative real roots
39. $x = -2, x = 5, x = 1$ **41.** $\left\{-\frac{1}{2}, \frac{1+\sqrt{17}}{2}, \frac{1-\sqrt{17}}{2}\right\}$ **43.** $\{-1, -2+2i, -2-2i\}$ **45.** $\{-1, -2, 3+\sqrt{13}, 3-\sqrt{13}\}$
47. $x = -1, x = 2, x = -\frac{1}{3}, x = 3$ **49.** $\left\{1, -\frac{3}{4}, i\sqrt{2}, -i\sqrt{2}\right\}$ **51.** $\left\{-2, \frac{1}{2}, \sqrt{2}, -\sqrt{2}\right\}$

53. a. $-4, 1$, and 4 **b.**

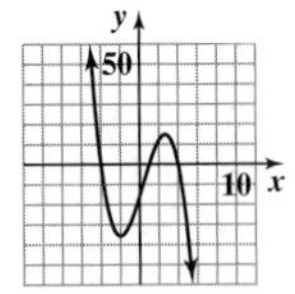

$f(x) = -x^3 + x^2 + 16x - 16$

55. a. -1 and $\frac{3}{2}$ **b.**

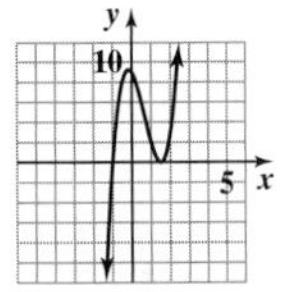

$f(x) = 4x^3 - 8x^2 - 3x + 9$

57. a. $\frac{1}{2}, 3, -1 \pm i$ **b.**

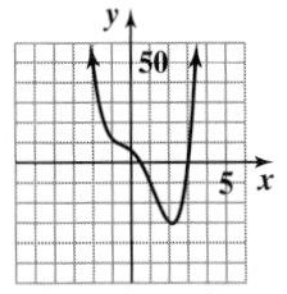

$f(x) = 2x^4 - 3x^3 - 7x^2 - 8x + 6$

59. a. $-2, -1, -\frac{2}{3}, 1$, and 2 **b.**

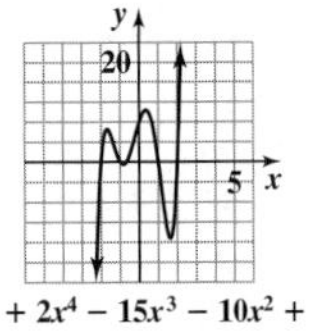

$f(x) = 3x^5 + 2x^4 - 15x^3 - 10x^2 + 12x + 8$

61. a. $x = 40$; at age 40, about 27% of art productivity occurs **b.** degree 2; leading coefficient: negative
63. ≈ 3 yr **67.** $x = 1$ in. or $x = 2$ in. **77.** $\frac{1}{2}, \frac{2}{3}, 2$ **79.** $\pm\frac{1}{2}$ **81.** 5, 3, or 1 positive real roots; no negative real roots

83.

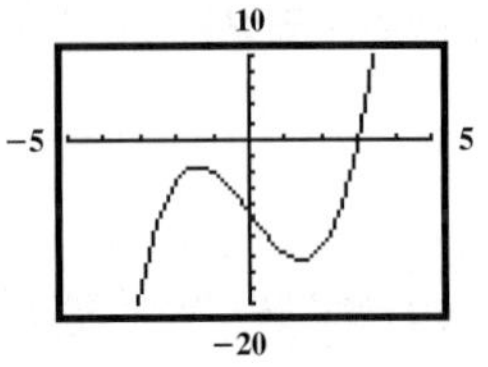

1 real zero, 2 nonreal complex zeros

85.

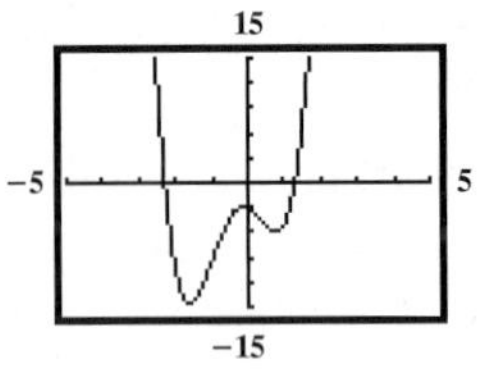

2 real zeros, 2 nonreal complex zeros

87. d **89.** 3 in. **91.** 3 **93.** 5

Mid-Chapter 3 Check Point

1.

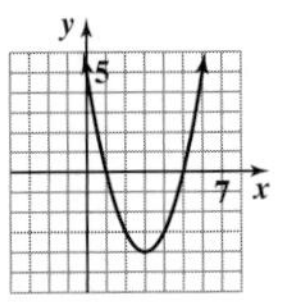

$f(x) = (x-3)^2 - 4$;
Domain: $(-\infty, \infty)$; Range: $[-4, \infty)$

2.

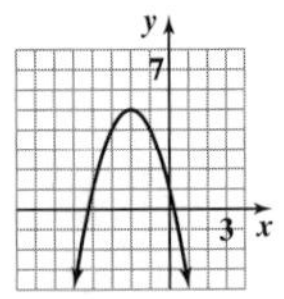

$f(x) = 5 - (x+2)^2$;
Domain: $(-\infty, \infty)$; Range: $(-\infty, 5]$

3.

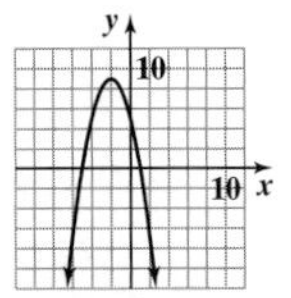

$f(x) = -x^2 - 4x + 5$;
Domain: $(-\infty, \infty)$; Range: $(-\infty, 9]$

4.

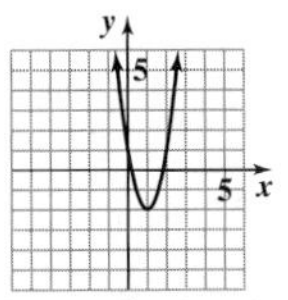

$f(x) = 3x^2 - 6x + 1$;
Domain: $(-\infty, \infty)$; Range: $[-2, \infty)$

5. -1 and 2

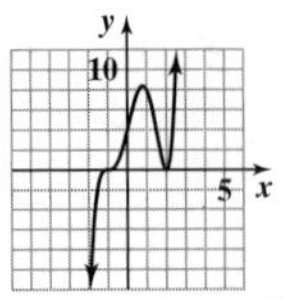

$f(x) = (x-2)^2(x+1)^3$

6. -1 and 2

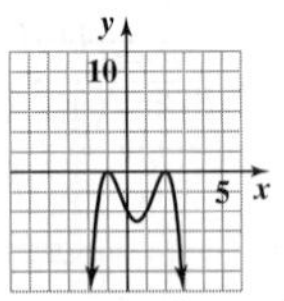

$f(x) = -(x-2)^2(x+1)^2$

7. $-2, 1$, and 2

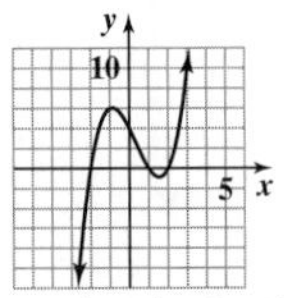

$f(x) = x^3 - x^2 - 4x + 4$

8. $-2, -1, 1$, and 2

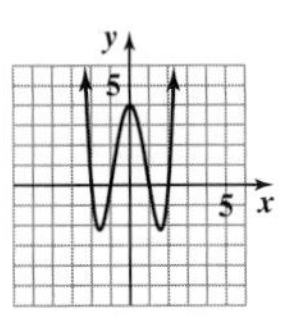

$f(x) = x^4 - 5x^2 + 4$

9. -1

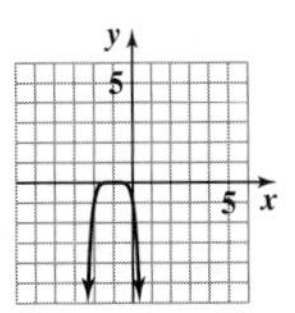

$f(x) = -(x+1)^6$

10. $-\frac{1}{3}, \frac{1}{2}$, and 1

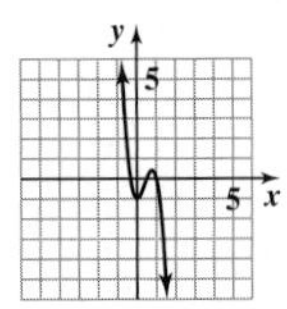

$f(x) = -6x^3 + 7x^2 - 1$

11. $-1, 0$, and 1

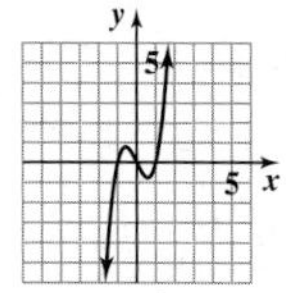

$f(x) = 2x^3 - 2x$

12. $0, 1 \pm 5i$

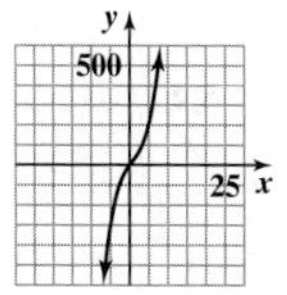

$f(x) = x^3 - 2x^2 + 26x$

13. $3, 1 \pm \sqrt{2}$

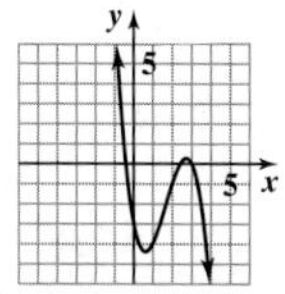

$f(x) = -x^3 + 5x^2 - 5x - 3$

14. $\{-2, 1\}$ **15.** $\left\{\frac{1}{3}, \frac{1}{2}, 1\right\}$ **16.** $\left\{-\frac{1}{2}, \frac{2}{3}, \frac{7}{2}\right\}$ **17.** $\left\{-10, -\frac{5}{2}, 10\right\}$ **18.** $\{-3, 4, \pm i\}$ **19.** $\left\{-3, \frac{1}{2}, 1 \pm \sqrt{3}\right\}$

20. 75 cabinets per day; \$1200 **21.** $-9, -9; 81$ **22.** 10 in.; 100 sq in. **23.** $2x^2 - x - 3 + \dfrac{x + 1}{3x^2 - 1}$ **24.** $2x^3 - 5x^2 - 3x + 6$

25. $f(x) = -2x^3 + 2x^2 - 2x + 2$ **26.** $f(x) = x^4 - 4x^3 + 13x^2 - 36x + 36$ **27.** yes

Section 3.5

Check Point Exercises

1. a. $\{x|x \neq 5\}$ **b.** $\{x|x \neq -5, x \neq 5\}$ **c.** all real numbers **2. a.** $x = 1, x = -1$ **b.** $x = -1$ **c.** none
3. a. $y = 3$ **b.** $y = 0$ **c.** none

4.

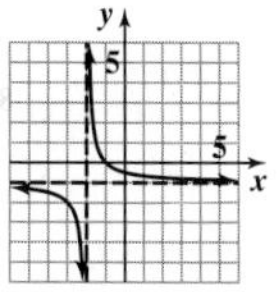

$g(x) = \dfrac{1}{x + 2} - 1$

5.

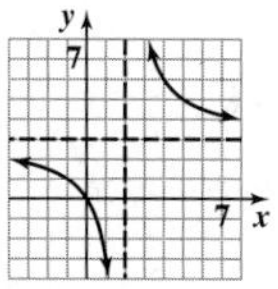

$f(x) = \dfrac{3x}{x - 2}$

6.

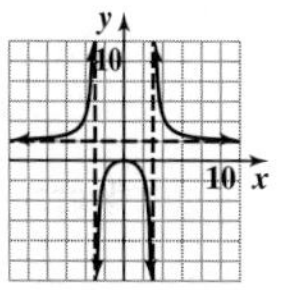

$f(x) = \dfrac{2x^2}{x^2 - 9}$

7.

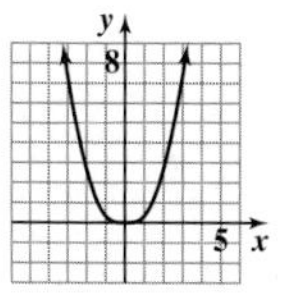

$f(x) = \dfrac{x^4}{x^2 + 2}$

8. $y = 2x - 1$ **9. a.** $C(x) = 500x + 600{,}000$ **b.** $\overline{C}(x) = \dfrac{500x + 600{,}000}{x}$

c. $\overline{C}(1000) = 1100$, when 1000 new systems are produced, it costs \$1100 to produce each system; $\overline{C}(10{,}000) = 560$, when 10,000 new systems are produced, it costs \$560 to produce each system, $\overline{C}(100{,}000) = 506$, when 100,000 new systems are produced, it costs \$506 to produce each system.
d. $y = 500$; The cost per system approaches \$500 as more systems are produced.

Exercise Set 3.5

1. $\{x|x \neq 4\}$ **3.** $\{x|x \neq 5, x \neq -4\}$ **5.** $\{x|x \neq 7, x \neq -7\}$ **7.** All real numbers **9.** $-\infty$ **11.** $-\infty$ **13.** 0 **15.** $+\infty$
17. $-\infty$ **19.** 1 **21.** $x = -4$ **23.** $x = 0, x = -4$ **25.** $x = -4$ **27.** no vertical asymptotes **29.** $y = 0$ **31.** $y = 4$
33. no horizontal asymptote **35.** $y = -\dfrac{2}{3}$

37.

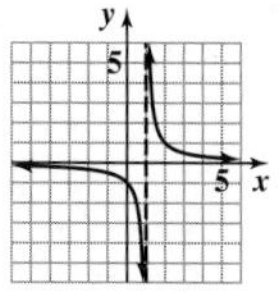

$g(x) = \dfrac{1}{x - 1}$

39.

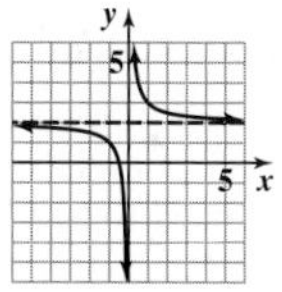

$h(x) = \dfrac{1}{x} + 2$

41.

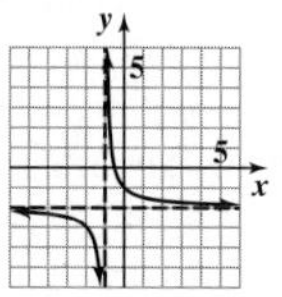

$g(x) = \dfrac{1}{x + 1} - 2$

43.

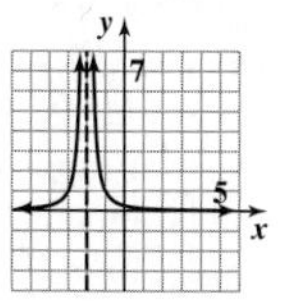

$g(x) = \dfrac{1}{(x + 2)^2}$

45.

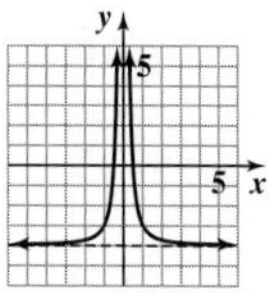

$h(x) = \dfrac{1}{x^2} - 4$

47.

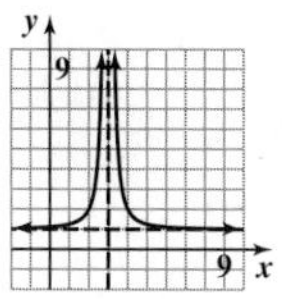

$h(x) = \dfrac{1}{(x - 3)^2} + 1$

49.

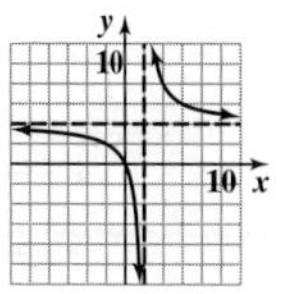

$f(x) = \dfrac{4x}{x - 2}$

51.

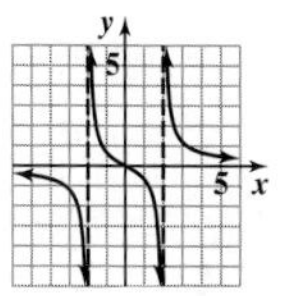

$f(x) = \dfrac{2x}{x^2 - 4}$

53.

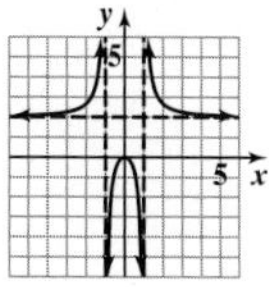

$f(x) = \dfrac{2x^2}{x^2 - 1}$

55.

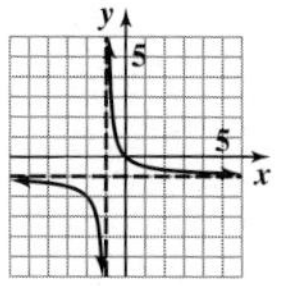

$f(x) = \dfrac{-x}{x + 1}$

57.

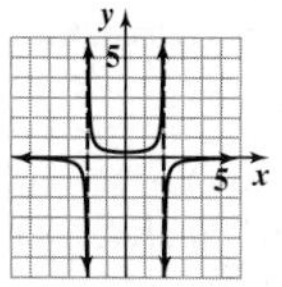

$f(x) = -\dfrac{1}{x^2 - 4}$

59.

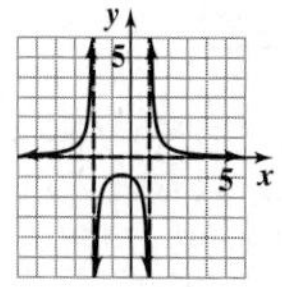

$f(x) = \dfrac{2}{x^2 + x - 2}$

61. 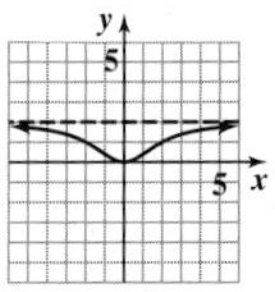

$f(x) = \dfrac{2x^2}{x^2 + 4}$

63. 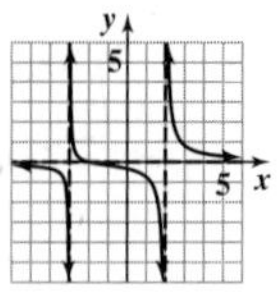

$f(x) = \dfrac{x + 2}{x^2 + x - 6}$

65. 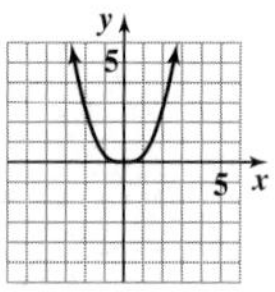

$f(x) = \dfrac{x^4}{x^2 + 2}$

67. 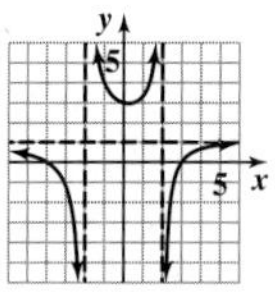

$f(x) = \dfrac{x^2 + x - 12}{x^2 - 4}$

69. 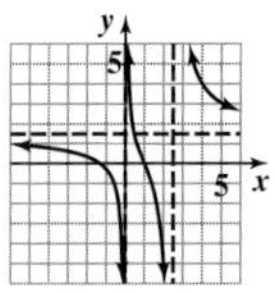

$f(x) = \dfrac{3x^2 + x - 4}{2x^2 - 5x}$

71. a. Slant asymptote: $y = x$ **b.** 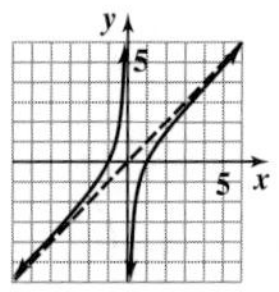

$f(x) = \dfrac{x^2 - 1}{x}$

73. a. Slant asymptote: $y = x$ **b.** 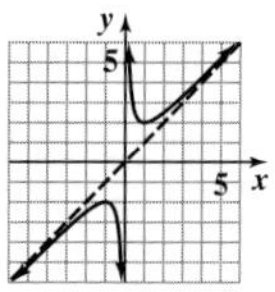

$f(x) = \dfrac{x^2 + 1}{x}$

75. a. Slant asymptote: $y = x + 4$ **b.** 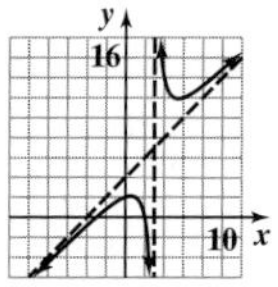

$f(x) = \dfrac{x^2 + x - 6}{x - 3}$

77. a. Slant asymptote: $y = x - 2$ **b.** 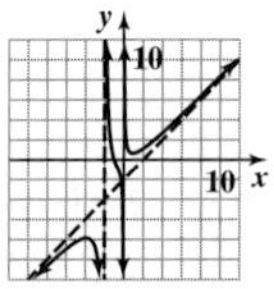

$f(x) = \dfrac{x^3 + 1}{x^2 + 2x}$

79. 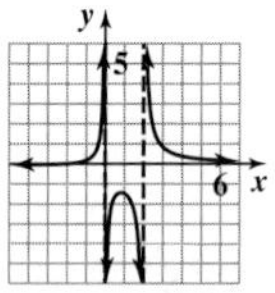

$f(x) = \dfrac{x + 2}{2x(x - 2)}$

81. 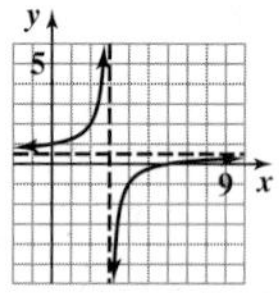

$f(x) = \dfrac{x - 6}{2(x - 3)}$

83. 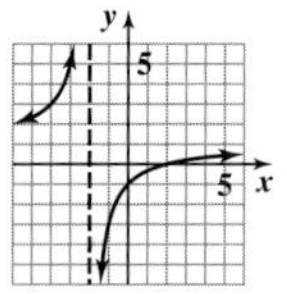

$f(x) = \dfrac{x - 2}{x + 2}$

85. $f(x) = \dfrac{1}{x + 3} + 2$

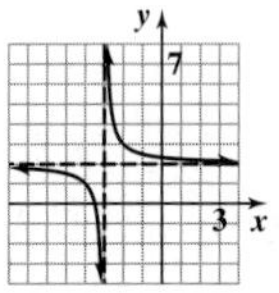

$g(x) = \dfrac{1}{x + 3} + 2$

87. $f(x) = \dfrac{-1}{x - 2} + 3$

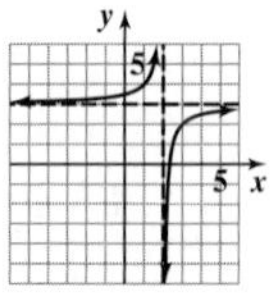

$g(x) = \dfrac{-1}{x - 2} + 3$

89. a. $C(x) = 100x + 100{,}000$ **b.** $\overline{C}(x) = \dfrac{100x + 100{,}000}{x}$
c. $\overline{C}(500) = 300$, when 500 bicycles are produced, it costs \$300 to produce each bicycle; $\overline{C}(1000) = 200$, when 1000 bicycles are produced, it costs \$200 to produce each bicycle; $\overline{C}(2000) = 150$, when 2000 bicycles are produced, it costs \$150 to produce each bicycle; $\overline{C}(4000) = 125$, when 4000 bicycles are produced, it costs \$125 to produce each bicycle.
d. $y = 100$; The cost per bicycle approaches \$100 as more bicycles are produced.
91. a. 6.0 **b.** after 6 minutes; about 4.8 **c.** 6.5 **d.** $y = 6.5$; Over time, the pH level rises back to normal. **e.** It quickly drops below normal and then slowly begins to approach the normal level.
93. 90; An incidence ratio of 10 means 90% of the deaths are smoking related. **95.** $y = 100$; The percentage of deaths cannot exceed 100% as the incidence ratios increase. **97. a.** $\dfrac{128.3}{134.5} \approx 0.954$; 954 males per 1000 females **b.** approximately 966 males per 1000 females
c. $f(x) = \dfrac{1.256x + 74.2}{1.324x + 76.71}$ **d.** approximately 959 males per 1000 females; Answers will vary. **e.** approximately 958 males per 1000 females; Answers will vary. **f.** $y \approx 0.949$; Over time, the number of males per 1000 females will approach 949.

109. 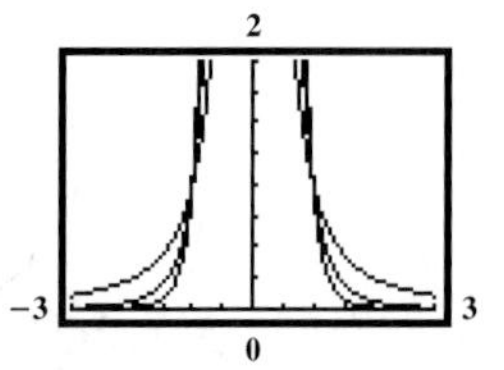

The graph approaches the horizontal asymptote faster and the vertical asymptote slower as n increases.

111. a. 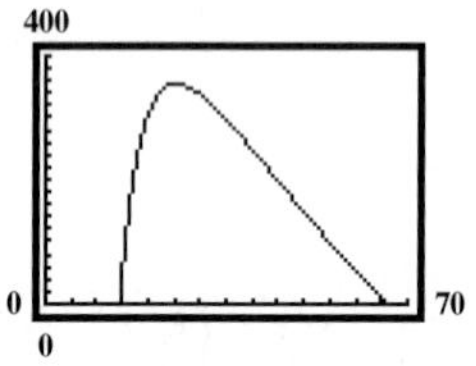

b. The graph increases and reaches a maximum of about 356 arrests per 100,000 drivers at age 25.
c. at age 25, about 356 arrests

113. d

Section 3.6

Check Point Exercises

1. $\{x \mid x < -4 \text{ or } x > 5\}$ or $(-\infty, -4) \cup (5, \infty)$ **2.** $\{x \mid x \le -3 \text{ or } -1 \le x \le 1\}$ or $(-\infty, -3] \cup [-1, 1]$ **3.** $\{x \mid x < -1 \text{ or } x \ge 1\}$ or $(-\infty, -1) \cup [1, \infty)$ **4.** between 1 and 4 seconds, excluding $t = 1$ and $t = 4$

Exercise Set 3.6

1. $(-\infty, -2)$ or $(4, \infty)$ **3.** $[-3, 7]$ **5.** $(-\infty, 1)$ or $(4, \infty)$ **7.** $(-\infty, -4)$ or $(-1, \infty)$ **9.** $\varnothing$ **11.** $\left[-4, \frac{2}{3}\right]$ **13.** $\left(-3, \frac{5}{2}\right)$ **15.** $\left(-1, -\frac{3}{4}\right)$ **17.** $\left[-2, \frac{1}{3}\right]$ **19.** $(-\infty, 0]$ or $[4, \infty)$ **21.** $\left(-\infty, -\frac{3}{2}\right)$ or $(0, \infty)$ **23.** $[0, 1]$ **25.** $[2 - \sqrt{2}, 2 + \sqrt{2}]$ **27.** $\varnothing$ **29.** $[1, 2]$ or $[3, \infty)$ **31.** $[-2, -1]$ or $[1, \infty)$ **33.** $(-\infty, -3)$ **35.** $(-1, \infty)$ **37.** $\{0\}$ or $[9, \infty)$ **39.** $(-\infty, -3)$ or $(4, \infty)$ **41.** $(-4, -3)$ **43.** $[2, 4)$ **45.** $\left(-\infty, -\frac{4}{3}\right)$ or $[2, \infty)$ **47.** $(-\infty, 0)$ or $(3, \infty)$ **49.** $(-\infty, -4] \cup (-2, 1]$ **51.** $(-\infty, -5)$ or $(-3, \infty)$ **53.** $\left(-\infty, \frac{1}{2}\right)$ or $\left[\frac{7}{5}, \infty\right)$ **55.** $(-\infty, -6]$ or $(-2, \infty)$ **57.** $\left(-\infty, \frac{1}{2}\right] \cup [2, \infty)$ **59.** $(-\infty, -1) \cup [1, \infty)$ **61.** $(-\infty, -8) \cup (-6, 4) \cup (6, \infty)$ **63.** $(-3, 2)$ **65.** $(-\infty, -1) \cup (1, 2) \cup (3, \infty)$ **67.** $\left[-6, -\frac{1}{2}\right] \cup [1, \infty)$ **69.** $(-\infty, -2) \cup [-1, 2)$ **71.** between 0 and $\frac{1}{2}$ second **73.** f: \$379 billion; g: \$379.8 billion; The functions model the value in the graph well. **75.** years after 2007 **77.** The running rate is at least 5 miles per hour.; For running rates of at least 5 miles per hour, the graph of f is on or below the line $y = 3$. **79.** The graph approaches the x-axis.; As the running rate increases, the total time decreases. **83.** The sides (in feet) are in $(0, 10]$ or $[80, 90)$.

89. $(-\infty, -5)$ or $(2, \infty)$ **91.** $(-2, -1)$ or $(2, \infty)$ **93.** $(-\infty, 3) \cup [8, \infty)$ **95.** d **97.** Answers may vary. One possible solution is $\frac{x-3}{x+4} \ge 0$.

99. $\{2\}$ **101.** $(-\infty, 2)$ or $(2, \infty)$

103.
$$\begin{aligned} 27 - 3x^2 &\ge 0 \\ 3x^2 &\le 27 \\ x^2 &\le 9 \\ -3 \le x &\le 3 \end{aligned}$$

Section 3.7

Check Point Exercises

1. 66 gal **2.** about 556 ft **3.** 512 cycles per second **4.** 24 min **5.** 96π cubic feet

Exercise Set 3.7

1. 156 **3.** 30 **5.** $\frac{5}{6}$ **7.** 240 **9.** 50 **11.** $x = kyz$; $y = \frac{x}{kz}$ **13.** $x = \frac{kz^3}{y}$; $y = \frac{kz^3}{x}$ **15.** $x = \frac{kyz}{\sqrt{w}}$; $y = \frac{x\sqrt{w}}{kz}$ **17.** $x = kz(y + w)$; $y = \frac{x - kzw}{kz}$ **19.** $x = \frac{kz}{y - w}$; $y = \frac{xw + kz}{x}$ **21.** 5.4 ft **23.** 80 in. **25.** about 607 lb **27.** 32°

29. 90 milliroentgens per hour **31.** This person has a BMI of 24.4 and is not overweight. **33.** 1800 Btu per hour
35. $\frac{1}{4}$ of what it was originally **37. a.** $C = \frac{kP_1P_2}{d^2}$ **b.** $k \approx 0.02$; $C = \frac{0.02\,P_1P_2}{d^2}$ **c.** approximately 39,813 daily phone calls

39. a.

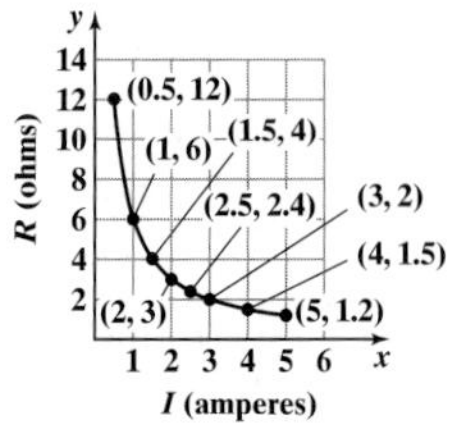

b. Current varies inversely as resistance. **c.** $R = \frac{6}{I}$

45. z varies directly as the square root of x and inversely as the square of y. **49.** The destructive power is four times as much.
51. Reduce the resistance by a factor of $\frac{1}{3}$.

Chapter 3 Review Exercises

1.

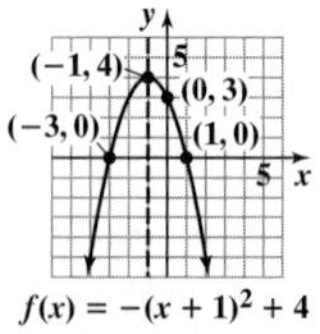

$f(x) = -(x + 1)^2 + 4$

axis of symmetry; $x = -1$
Domain: $(-\infty, \infty)$; Range: $(-\infty, 4]$

2.

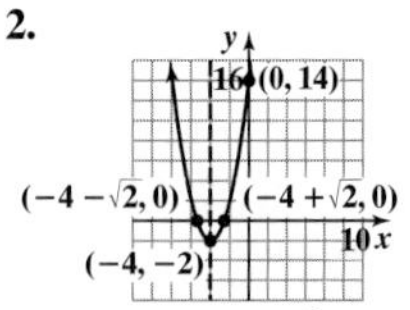

$f(x) = (x + 4)^2 - 2$

axis of symmetry: $x = -4$
Domain: $(-\infty, \infty)$; Range: $[-2, \infty)$

3.

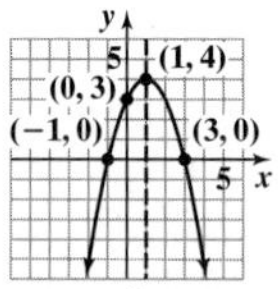

$f(x) = -x^2 + 2x + 3$

axis of symmetry: $x = 1$
Domain: $(-\infty, \infty)$; Range: $(-\infty, 4]$

4.

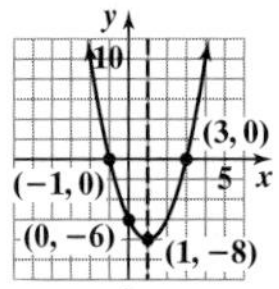

$f(x) = 2x^2 - 4x - 6$

axis of symmetry: $x = 1$
Domain: $(-\infty, \infty)$; Range: $[-8, \infty)$

5. a. maximum is -57 at $x = 7$ **b.** Domain: $(-\infty, \infty)$; Range: $(-\infty, -57]$
6. a. minimum is 685 at $x = -3$ **b.** Domain: $(-\infty, \infty)$; Range: $[685, \infty)$ **7.** 25 in. of rainfall per year; 13.5 in. of growth
8. (25, 5); The maximum divorce rate of about 5 per 1000 population was in 1985. **9.** 250 yd by 500 yd; 125,000 sq yard **10.** -7 and 7; -49
11. c **12.** b **13.** a **14.** d **15.** No; the graph falls to the right, so eventually there would be a negative number of thefts, which is not possible.
16. The graph falls to the right; eventually the elk population will be extinct.
17. $x = 1$, multiplicity 1, crosses; $x = -2$, multiplicity 2, touches; $x = -5$, multiplicity 3, crosses
18. $x = -5$, multiplicity 1, crosses; $x = 5$, multiplicity 2, touches
19. $f(1)$ is negative and $f(2)$ is positive, so by the Intermediate Value Theorem, f has a real zero between 1 and 2.

20. a. The graph falls to the left and rises to the right.
b. no symmetry
c.

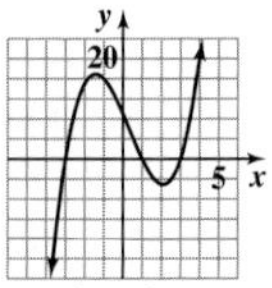

$f(x) = x^3 - x^2 - 9x + 9$

21. a. The graph rises to the left and falls to the right.
b. origin symmetry
c.

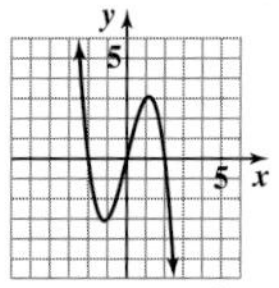

$f(x) = 4x - x^3$

22. a. The graph falls to the left and rises to the right.
b. no symmetry
c.

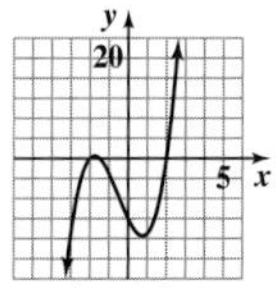

$f(x) = 2x^3 + 3x^2 - 8x - 12$

23. a. The graph falls to the left and to the right.
b. y-axis symmetry
c.

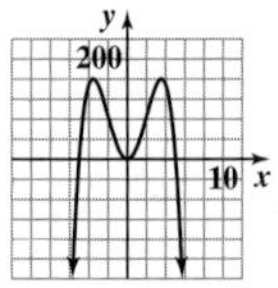

$f(x) = -x^4 + 25x^2$

24. a. The graph falls to the left and to the right.
b. no symmetry
c.

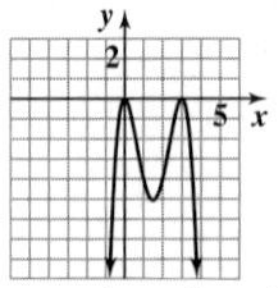

$f(x) = -x^4 + 6x^3 - 9x^2$

25. a. The graph rises to the left and to the right.
b. no symmetry
c.

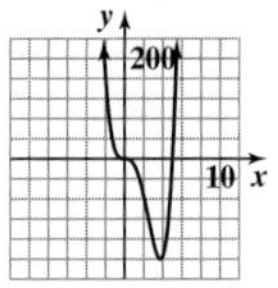

$f(x) = 3x^4 - 15x^3$

26.

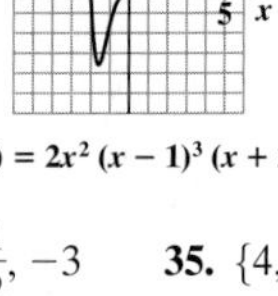

$f(x) = 2x^2(x - 1)^3(x + 2)$

27.

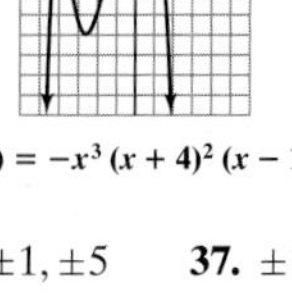

$f(x) = -x^3(x + 4)^2(x - 1)$

28. $4x^2 - 7x + 5 - \frac{4}{x + 1}$
29. $2x^2 - 4x + 1 - \frac{10}{5x - 3}$
30. $2x^2 + 3x - 1$ **31.** $3x^3 - 4x^2 + 7$
32. $3x^3 + 6x^2 + 10x + 10 + \frac{20}{x - 2}$ **33.** -5697

34. $2, \frac{1}{2}, -3$ **35.** $\{4, -2 \pm \sqrt{5}\}$ **36.** $\pm 1, \pm 5$ **37.** $\pm 1, \pm 2, \pm 4, \pm 8, \pm\frac{8}{3}, \pm\frac{4}{3}, \pm\frac{2}{3}, \pm\frac{1}{3}$ **38.** 2 or 0 positive solutions; no negative solutions
39. 3 or 1 positive real roots; 2 or 0 negative solutions **40.** No sign variations exist for either $f(x)$ or $f(-x)$, so no real roots exist.
41. a. $\pm 1, \pm 2, \pm 4$ **b.** 1 positive real zero; 2 or no negative real zeros **c.** 1 is a zero **d.** $\{1, -2\}$
42. a. $\pm 1, \pm\frac{1}{2}, \pm\frac{1}{3}, \pm\frac{1}{6}$ **b.** 2 or 0 positive real zeros; 1 negative real zero **c.** -1 is a zero **d.** $\left\{-1, \frac{1}{3}, \frac{1}{2}\right\}$

43. **a.** $\pm 1, \pm 3, \pm 5, \pm 15, \pm\frac{1}{2}, \pm\frac{1}{4}, \pm\frac{1}{8}, \pm\frac{3}{2}, \pm\frac{3}{4}, \pm\frac{3}{8}, \pm\frac{5}{2}, \pm\frac{5}{4}, \pm\frac{5}{8}, \pm\frac{15}{2}, \pm\frac{15}{4}, \pm\frac{15}{8}$ **b.** 3 or 1 positive real solutions; no negative real solutions **c.** $\frac{1}{2}$ is a zero **d.** $\left\{\frac{1}{2}, \frac{3}{2}, \frac{5}{2}\right\}$

44. **a.** $\pm 1, \pm\frac{1}{2}$ **b.** 2 or 0 positive real solutions; 1 negative solution **c.** $\frac{1}{2}$ is a zero **d.** $\left\{\frac{1}{2}, \frac{-5-\sqrt{29}}{2}, \frac{-5+\sqrt{29}}{2}\right\}$

45. **a.** $\pm 1, \pm 2, \pm 3, \pm 6$ **b.** 2 or zero positive real solutions; 2 or zero negative real solutions **c.** -2 is a zero **d.** $\{-2, -1, 1, 3\}$

46. **a.** $\pm 1, \pm 2, \pm\frac{1}{2}, \pm\frac{1}{4}$ **b.** 1 positive real root; 1 negative real root **c.** $\frac{1}{2}$ is a zero **d.** $\left\{\frac{1}{2}, -\frac{1}{2}, i\sqrt{2}, -i\sqrt{2}\right\}$

47. **a.** $\pm 1, \pm 2, \pm 4, \pm\frac{1}{2}$ **b.** 2 or no positive zeros; 2 or no negative zeros **c.** 2 is a zero **d.** $\left\{2, -2, \frac{1}{2}, -1\right\}$

48. $f(x) = x^3 - 6x^2 + 21x - 26$ **49.** $f(x) = 2x^4 + 12x^3 + 20x^2 + 12x + 18$ **50.** $-2, \frac{1}{2}, \pm i$; $f(x) = (x - i)(x + i)(x + 2)(2x - 1)$

51. $-1, 4$; $g(x) = (x + 1)^2(x - 4)^2$ **52.** 4 real zeros, one with multiplicity two **53.** 3 real zeros; 2 nonreal complex zeros

54. 2 real zeros, one with multiplicity two; 2 nonreal complex zeros **55.** 1 real zero; 4 nonreal complex zeros

56.

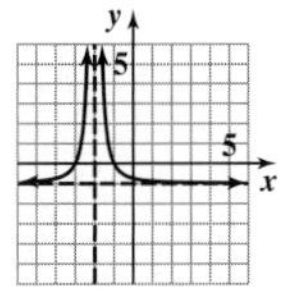

$g(x) = \frac{1}{(x+2)^2} - 1$

57.

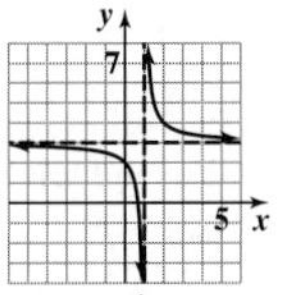

$h(x) = \frac{1}{x-1} + 3$

58. Vertical asymptote: $x = 3$ and $x = -3$
horizontal asymptote: $y = 0$

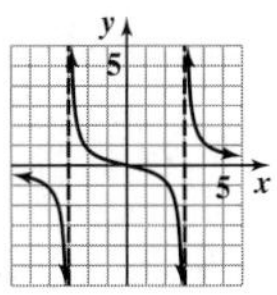

$f(x) = \frac{2x}{x^2 - 9}$

59. Vertical asymptote: $x = -3$
horizontal asymptote: $y = 2$

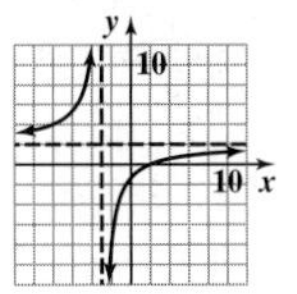

$f(x) = \frac{2x - 4}{x + 3}$

60. Vertical asymptotes: $x = 3, -2$
horizontal asymptote: $y = 1$

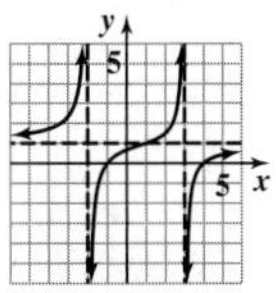

$h(x) = \frac{x^2 - 3x - 4}{x^2 - x - 6}$

61. Vertical asymptote: $x = -2$
horizontal asymptote: $y = 1$

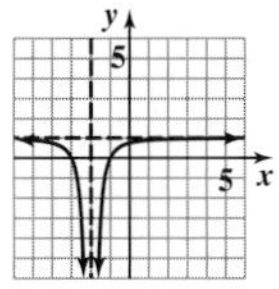

$r(x) = \frac{x^2 + 4x + 3}{(x + 2)^2}$

62. Vertical asymptote: $x = -1$
no horizontal asymptote
slant asymptote: $y = x - 1$

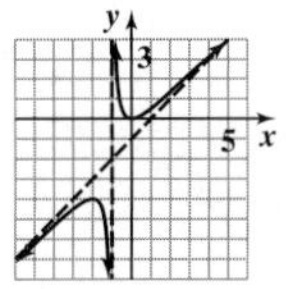

$y = \frac{x^2}{x + 1}$

63. Vertical asymptote: $x = 3$
no horizontal asymptote
slant asymptote: $y = x + 5$

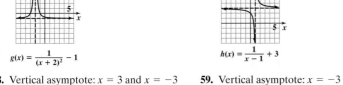

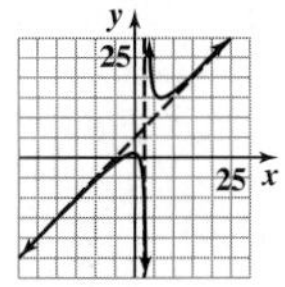

$y = \frac{x^2 + 2x - 3}{x - 3}$

64. No vertical asymptote

no horizontal asymptote

slant asymptote: $y = -2x$

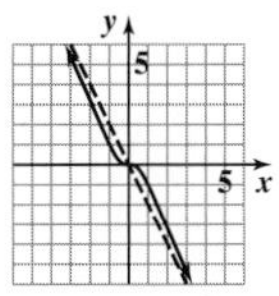

$f(x) = \frac{-2x^3}{x^2 + 1}$

65. Vertical asymptote: $x = \frac{3}{2}$

no horizontal asymptote

slant asymptote: $y = 2x - 5$

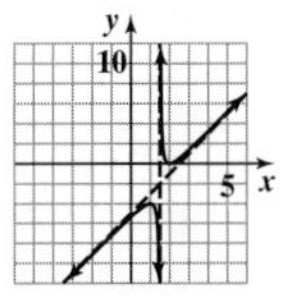

$g(x) = \frac{4x^2 - 16x + 16}{2x - 3}$

66. **a.** $C(x) = 25x + 50{,}000$
b. $\overline{C}(x) = \frac{25x + 50{,}000}{x}$
c. $\overline{C}(50) = 1025$, when 50 calculators are manufactured, it costs \$1025 to manufacture each; $\overline{C}(100) = 525$, when 100 calculators are manufactured, it costs \$525 to manufacture each; $\overline{C}(1000) = 75$, when 1000 calculators are manufactured, it costs \$75 to manufacture each; $\overline{C}(100{,}000) = 25.5$, when 100,000 calculators are manufactured, it costs \$25.50 to manufacture each.
d. $y = 25$; costs will approach \$25.

67. **a.** 1600; The difference in cost of removing 90% versus 50% of the contaminants is 16 million dollars.
b. $x = 100$; No amount of money can remove 100% of the contaminants, since $C(x)$ increases without bound as x approaches 100.

68. $y = 3000$; The number of fish in the pond approaches 3000.

69. $y = 0$; As the number of years of education increases the percentage rate of unemployment approaches zero.

70. **a.** $f(x) = \frac{1.96x + 3.14}{3.04x + 21.79}$ **b.** $y = 0.6$; As the years increase, the fraction of nonviolent prisoners approaches 0.6. **c.** Answers may vary.

71. $\left(-3, \frac{1}{2}\right)$;

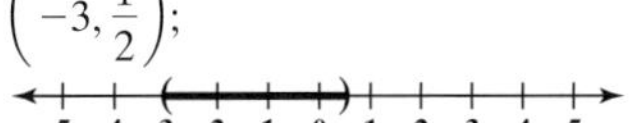

72. $(-\infty, -4] \cup \left[-\frac{1}{2}, \infty\right)$;

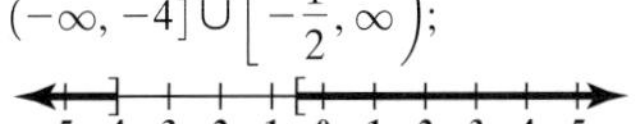

73. $(-3, 0) \cup (1, \infty)$;

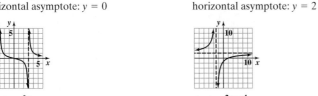

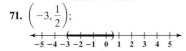

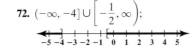

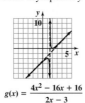

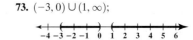

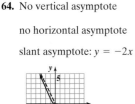

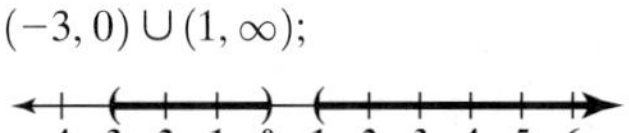

74. $(-\infty, -2) \cup (6, \infty)$;

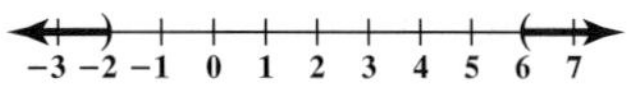

75. $[-1, 1) \cup [2, \infty)$;

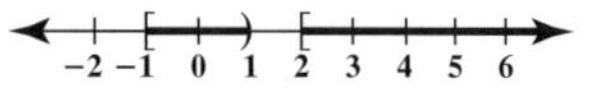

76. $(-\infty, 4) \cup \left[\frac{23}{4}, \infty\right)$;

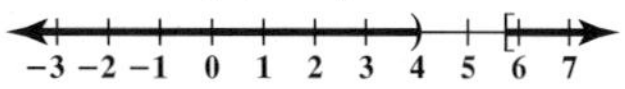

77. from 1 to 2 sec **78.** \$154 **79.** 1600 ft **80.** 5 hr **81.** 112 decibels **82.** 16 hr **83.** 800 ft^3

Chapter 3 Test

1.

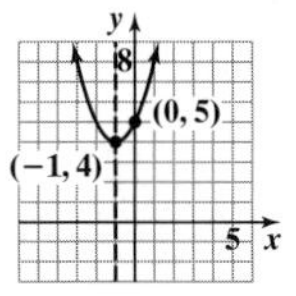

$f(x) = (x + 1)^2 + 4$

axis of symmetry: $x = -1$
Domain: $(-\infty, \infty)$; Range: $[4, \infty)$

2.

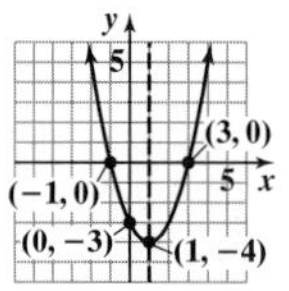

$f(x) = x^2 - 2x - 3$

axis of symmetry: $x = 1$
Domain: $(-\infty, \infty)$; Range: $[-4, \infty)$

3. maximum of 2 at $x = 3$;
Domain: $(-\infty, \infty)$; Range: $(-\infty, 2]$
4. 23 computers;
maximum daily profit = \$16,900
5. 7 and 7; 49

6. a. 5, 2, −2
b.

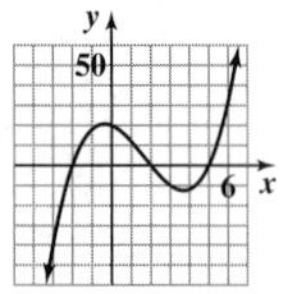

$f(x) = x^3 - 5x^2 - 4x + 20$

7. Since the degree of the polynomial is odd and the leading coefficient is positive, the graph of f should fall to the left and rise to the right. The x-intercepts should be −1 and 1.

8. a. 2 **b.** $\frac{1}{2}, \frac{2}{3}$ **9.** $\pm1, \pm2, \pm3, \pm6, \pm\frac{1}{2}, \pm\frac{3}{2}$
10. 3 or 1 positive real zeros; no negative real zeros.

11. $\{-3, -3 - \sqrt{11}, -3 + \sqrt{11}\}$ **12. a.** $\pm1, \pm3, \pm5, \pm15, \pm\frac{1}{2}, \pm\frac{3}{2}, \pm\frac{5}{2}, \pm\frac{15}{2}$ **b.** $-\sqrt{5}, -1, \frac{3}{2}$, and $\sqrt{5}$ **13.** $(x - 1)(x + 2)^2$

14. $f(x) = 2x^4 - 2$

15. −1 and $\frac{2}{3}$

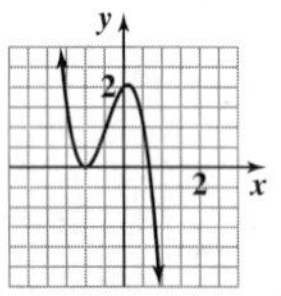

$f(x) = -3x^3 - 4x^2 + x + 2$

16. $(-\infty, -3) \cup (-3, \infty)$

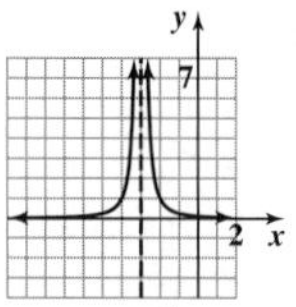

$f(x) = \frac{1}{(x + 3)^2}$

17. $(-\infty, 1) \cup (1, \infty)$

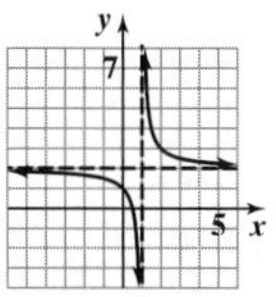

$f(x) = \frac{1}{x - 1} + 2$

18. domain: $\{x | x \neq 4, x \neq -4\}$

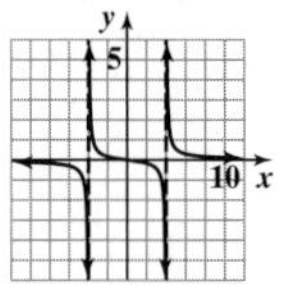

$f(x) = \frac{x}{x^2 - 16}$

19. domain: $\{x | x \neq 2\}$

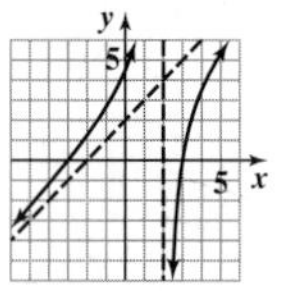

$f(x) = \frac{x^2 - 9}{x - 2}$

20. domain: $\{x | x \neq -3, x \neq 1\}$

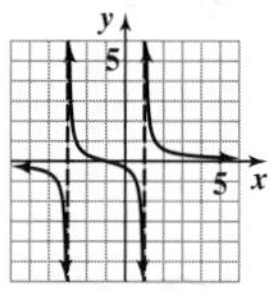

$f(x) = \frac{x + 1}{x^2 + 2x - 3}$

21. domain: all real numbers

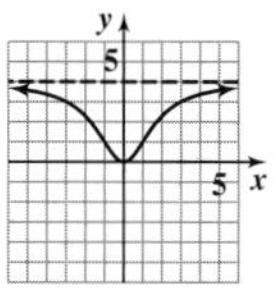

$f(x) = \frac{4x^2}{x^2 + 3}$

22. a. $\overline{C}(x) = \frac{300{,}000 + 10x}{x}$ **b.** $y = 10$; As the number of televisions increases, the average cost approaches \$10.
23. a. 0.89 **b.** 11 **c.** $y = 1$; as the number of learning trials increases, the proportion of correct responses approaches 1.

24. $(-3, 4)$

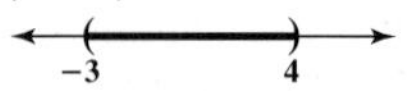

25. $(-\infty, 3) \cup [10, \infty)$

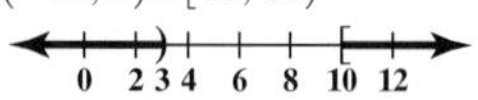

26. 45 foot-candles

Cumulative Review Exercises (Chapters 1–3)

1. Domain: $(-2, 2)$; Range: $[0, \infty)$ **2.** −1 and 1, both of multiplicity 2 **3.** 0 **4.** 3 **5.** $x \to -2^+; x \to 2^-$

6.

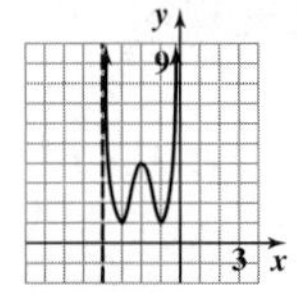

$g(x) = f(x + 2) + 1$

7. $\{2, -1\}$ **8.** $\left\{\frac{5 + \sqrt{13}}{6}, \frac{5 - \sqrt{13}}{6}\right\}$ **9.** $\left\{\frac{1}{3}, -\frac{2}{3}\right\}$ **10.** $\{-3, -1, 2\}$ **11.** $(-\infty, 1)$ or $(4, \infty)$

12. $(-\infty, -1)$ or $\left(\frac{5}{3}, \infty\right)$

13.

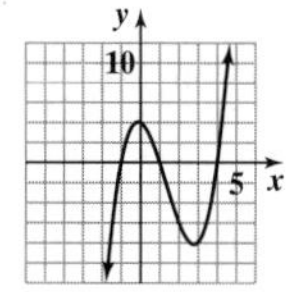

$f(x) = x^3 - 4x^2 - x + 4$

14.

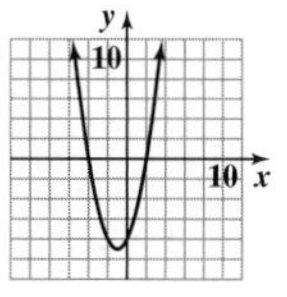

$f(x) = x^2 + 2x - 8$

15.

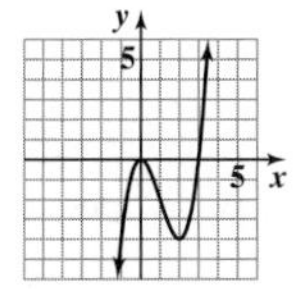

$f(x) = x^2(x - 3)$

16.

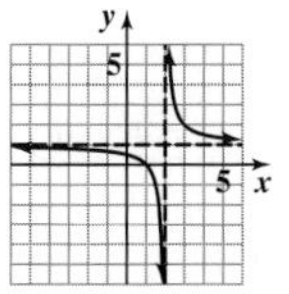

$f(x) = \frac{x - 1}{x - 2}$

17.

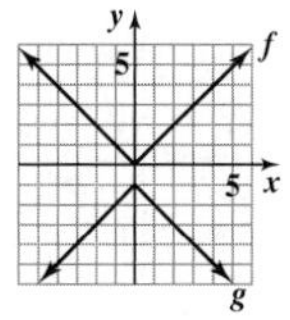

18.

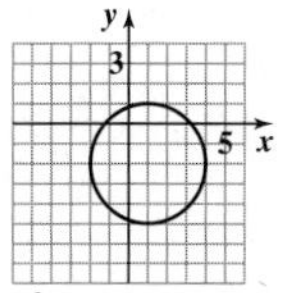

$x^2 + y^2 - 2x + 4y - 4 = 0$

19. $(f \circ g)(x) = 32x^2 - 20x + 2$ **20.** $4x + 2h - 1$

CHAPTER 4

Section 4.1

Check Point Exercises

1. 1 O-ring

2.

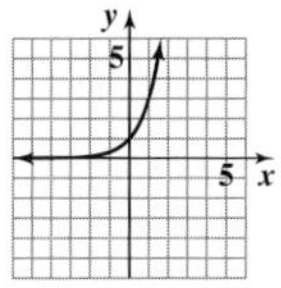

$f(x) = 3^x$

3.

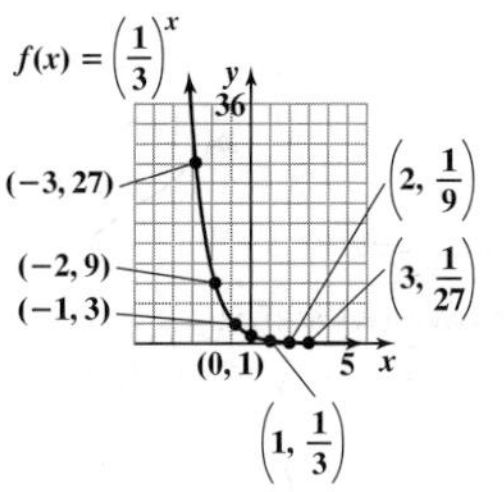

4.

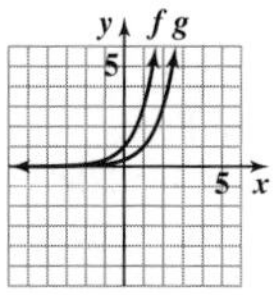

$f(x) = 3^x$
$g(x) = 3^{x-1}$

5.

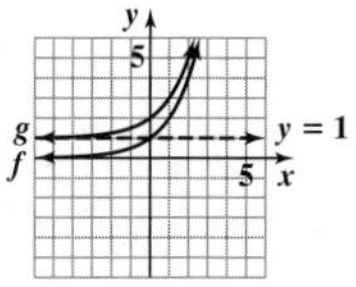

$f(x) = 2^x$
$g(x) = 2^x + 1$

6. approximately 11.27 billion
7. a. \$14,859.47 **b.** \$14,918.25

Exercise Set 4.1

1. 10.556 **3.** 11.665 **5.** 0.125 **7.** 9.974 **9.** 0.387

11.

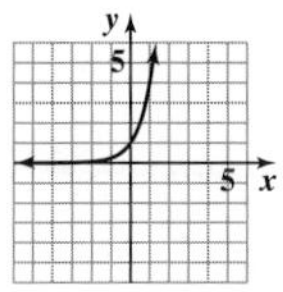

$f(x) = 4^x$

13.

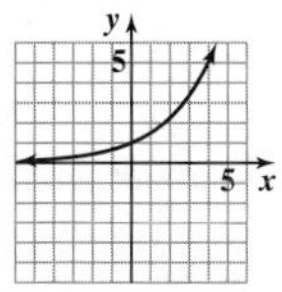

$g(x) = \left(\frac{3}{2}\right)^x$

15.

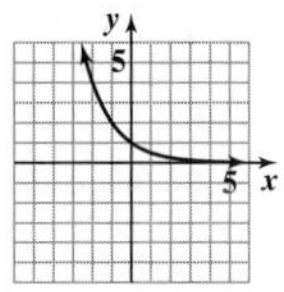

$h(x) = \left(\frac{1}{2}\right)^x$

17.

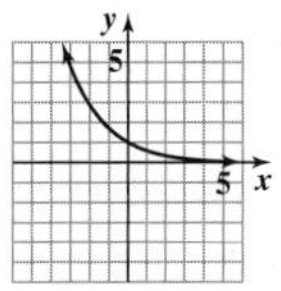

$f(x) = (0.6)^x$

19. $H(x) = -3^{-x}$
21. $F(x) = -3^x$
23. $h(x) = 3^x - 1$

25.

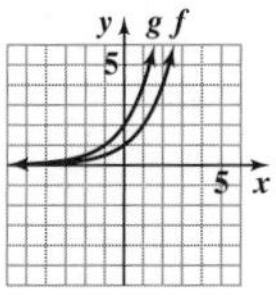

$f(x) = 2^x$
$g(x) = 2^{x+1}$

asymptote: $y = 0$; Domain: $(-\infty, \infty)$; Range: $(0, \infty)$

27.

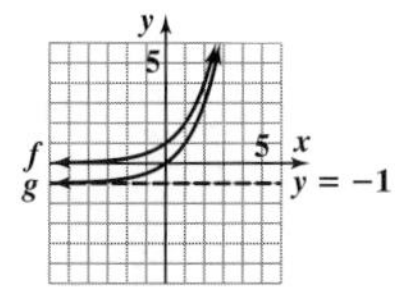

$f(x) = 2^x$
$g(x) = 2^x - 1$

asymptote: $y = -1$; Domain: $(-\infty, \infty)$; Range: $(-1, \infty)$

29.

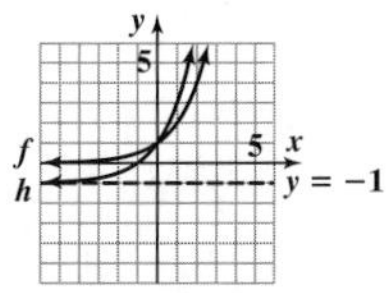

$f(x) = 2^x$
$h(x) = 2^{x+1} - 1$

asymptote: $y = -1$; Domain: $(-\infty, \infty)$; Range: $(-1, \infty)$

31.

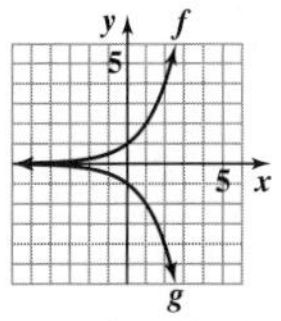

$f(x) = 2^x$
$g(x) = -2^x$

asymptote: $y = 0$; Domain: $(-\infty, \infty)$; Range: $(-\infty, 0)$

33.

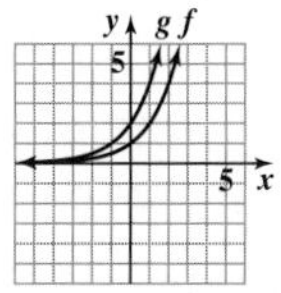

$f(x) = 2^x$
$g(x) = 2 \cdot 2^x$

asymptote: $y = 0$; Domain: $(-\infty, \infty)$; Range: $(0, \infty)$

35.

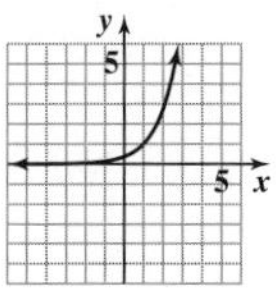

$g(x) = e^{x-1}$

asymptote: $y = 0$; Domain: $(-\infty, \infty)$; Range: $(0, \infty)$

37.

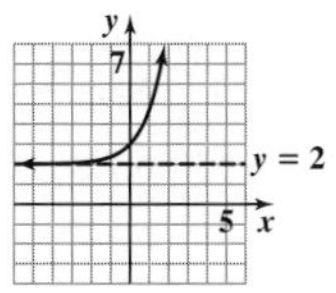

$g(x) = e^x + 2$

asymptote: $y = 2$; Domain: $(-\infty, \infty)$; Range: $(2, \infty)$

39.

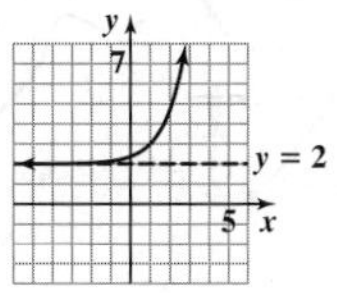

$h(x) = e^{x-1} + 2$

asymptote: $y = 2$; Domain: $(-\infty, \infty)$; Range: $(2, \infty)$

41.

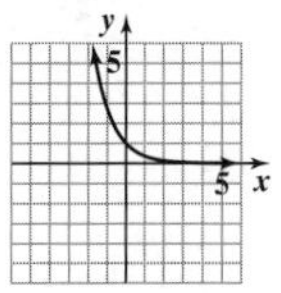

$h(x) = e^{-x}$

asymptote: $y = 0$; Domain: $(-\infty, \infty)$; Range: $(0, \infty)$

43.

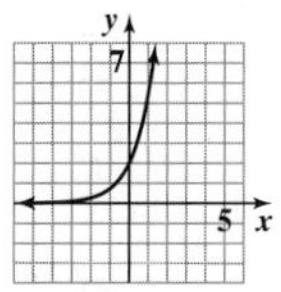

$g(x) = 2e^x$

asymptote: $y = 0$; Domain: $(-\infty, \infty)$; Range: $(0, \infty)$

45.

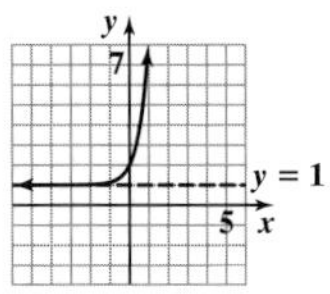

$h(x) = e^{2x} + 1$

asymptote: $y = 1$; Domain: $(-\infty, \infty)$; Range: $(1, \infty)$

47.

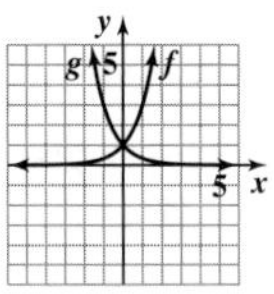

$f(x) = 3^x$
$g(x) = 3^{-x}$

asymptote of f: $y = 0$; asymptote of g: $y = 0$

49.

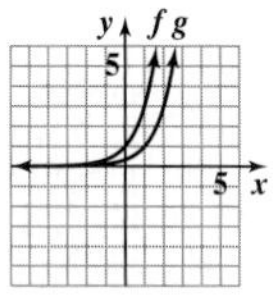

$f(x) = 3^x$
$g(x) = \frac{1}{3} \cdot 3^x$

asymptote of f: $y = 0$; asymptote of g: $y = 0$

51.

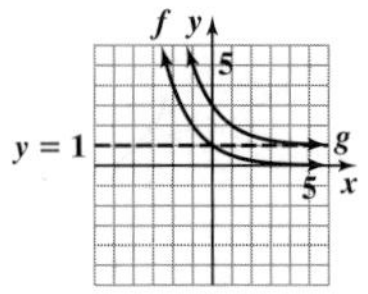

$f(x) = \left(\frac{1}{2}\right)^x$

$g(x) = \left(\frac{1}{2}\right)^{x-1} + 1$

asymptote of f: $y = 0$; asymptote of g: $y = 1$

53. a. \$13,116.51 **b.** \$13,140.67 **c.** \$13,157.04 **d.** \$13,165.31

55. 7% compounded monthly

57. (0, 1)

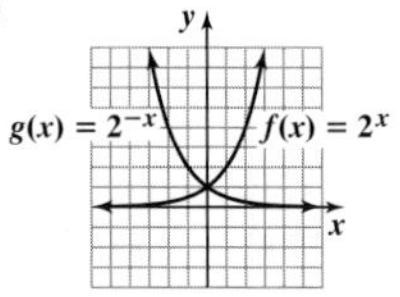

59.

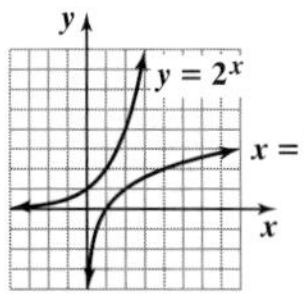

61. $y = 4^x$ **63.** $y = -e^x$ **65. a.** 574 million **b.** 1148 million **c.** 2295 million **d.** 4590 million **e.** It appears to double. **67.** \$832,744 **69.** 3.249009585; 3.317278183; 3.321880096; 3.321995226; 3.321997068; $2^{\sqrt{3}} \approx 3.321997085$; The closer the exponent is to $\sqrt{3}$, the closer the value is to $2^{\sqrt{3}}$. **71.** linear **73. a.** 100% **b.** $\approx 68.5\%$ **c.** $\approx 30.8\%$ **d.** $\approx 20\%$ **75.** f

83.

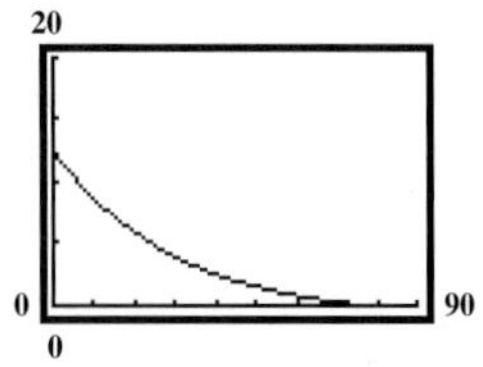

no; Nearly 4 O-rings are expected to fail.

85. a.

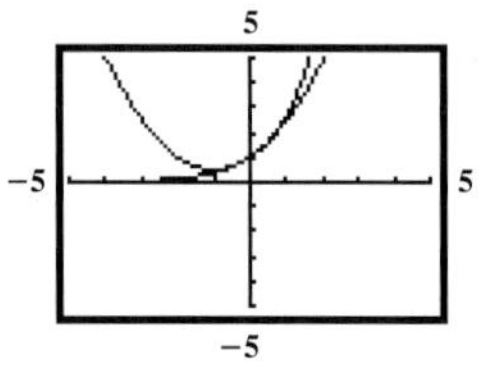

b.

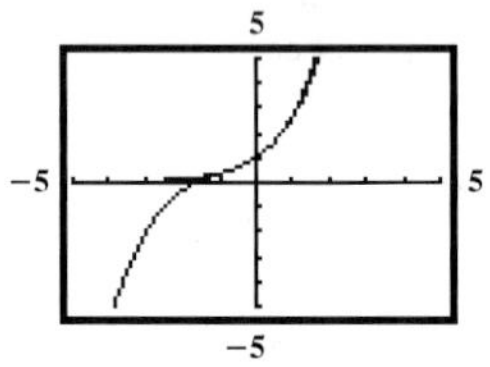

c.

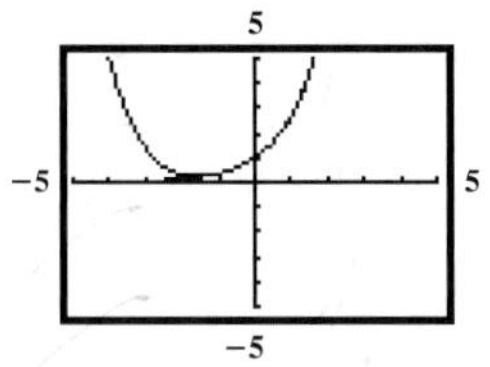

87. $y = 3^x$ is (d); $y = 5^x$ is (c); $y = \left(\frac{1}{3}\right)^x$ is (a); $y = \left(\frac{1}{5}\right)^x$ is (b).

89. a. $\cosh(-x) = \dfrac{e^{-x} + e^{-(-x)}}{2} = \dfrac{e^{-x} + e^x}{2} = \dfrac{e^x + e^{-x}}{2} = \cosh x$

b. $\sinh(-x) = \dfrac{e^{-x} - e^{-(-x)}}{2} = \dfrac{e^{-x} - e^x}{2} = -\dfrac{e^x - e^{-x}}{2} = -\sinh x$

c. $\left(\dfrac{e^x + e^{-x}}{2}\right)^2 - \left(\dfrac{e^x - e^{-x}}{2}\right)^2 \stackrel{?}{=} 1$

$\dfrac{e^{2x} + 2 + e^{-2x}}{4} - \dfrac{e^{2x} - 2 + e^{-2x}}{4} \stackrel{?}{=} 1$

$\dfrac{e^{2x} + 2 + e^{-2x} - e^{2x} + 2 - e^{-2x}}{4} \stackrel{?}{=} 1$

$\dfrac{4}{4} \stackrel{?}{=} 1$

$1 = 1$

Section 4.2

Check Point Exercises

1. a. $7^3 = x$ **b.** $b^2 = 25$ **c.** $4^y = 26$ **2. a.** $5 = \log_2 x$ **b.** $3 = \log_b 27$ **c.** $y = \log_e 33$ **3. a.** 2 **b.** 1 **c.** $\frac{1}{2}$ **4. a.** 1 **b.** 0 **5. a.** 8 **b.** 17 **6.**

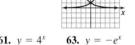

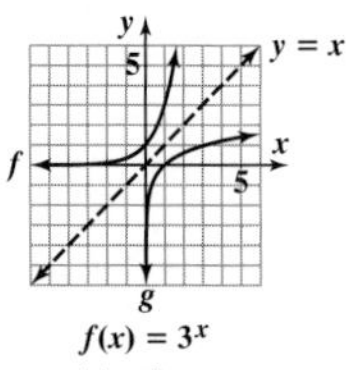

$f(x) = 3^x$
$g(x) = \log_3 x$

7. $(5, \infty)$ **8.** 80% **9.** 4.0 **10. a.** $(-\infty, 4)$ **b.** $(-\infty, 0)$ or $(0, \infty)$ **11.** 34°; quite well

Exercise Set 4.2

1. $2^4 = 16$ **3.** $3^2 = x$ **5.** $b^5 = 32$ **7.** $6^y = 216$ **9.** $\log_2 8 = 3$ **11.** $\log_2 \frac{1}{16} = -4$ **13.** $\log_8 2 = \frac{1}{3}$ **15.** $\log_{13} x = 2$ **17.** $\log_b 1000 = 3$ **19.** $\log_7 200 = y$ **21.** 2 **23.** 6 **25.** −1 **27.** −3 **29.** $\frac{1}{2}$ **31.** $-\frac{1}{2}$ **33.** $\frac{1}{2}$ **35.** 1 **37.** 0 **39.** 7 **41.** 19

43.

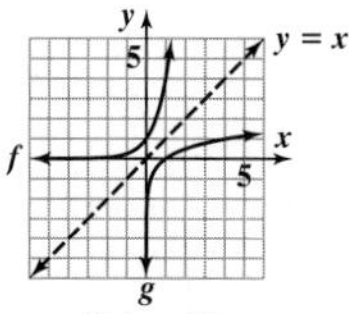

$f(x) = 4^x$
$g(x) = \log_4 x$

45.

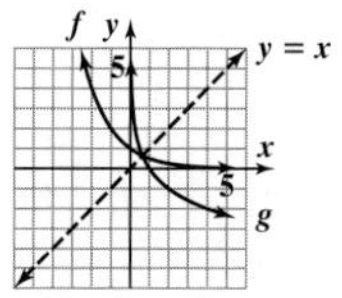

$f(x) = \left(\frac{1}{2}\right)^x$
$g(x) = \log_{1/2} x$

47. $H(x) = 1 - \log_3 x$ **49.** $h(x) = \log_3 x - 1$ **51.** $g(x) = \log_3 (x - 1)$

53.

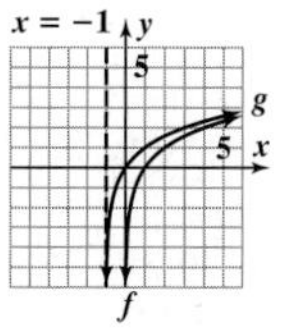

$f(x) = \log_2 x$
$g(x) = \log_2(x+1)$

vertical asymptote: $x = -1$
Domain: $(-1, \infty)$;
Range: $(-\infty, \infty)$

55.

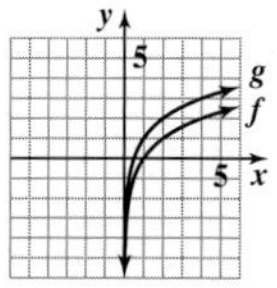

$f(x) = \log_2 x$
$g(x) = 1 + \log_2 x$

vertical asymptote: $x = 0$
Domain: $(0, \infty)$;
Range: $(-\infty, \infty)$

57.

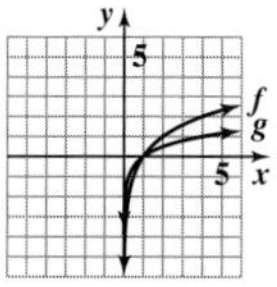

$f(x) = \log_2 x$
$g(x) = \frac{1}{2} \log_2 x$

vertical asymptote: $x = 0$
Domain: $(0, \infty)$;
Range: $(-\infty, \infty)$

59.

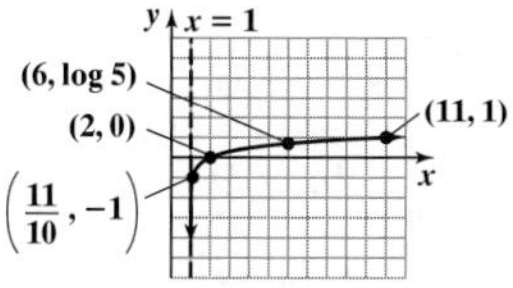

$g(x) = \log(x - 1)$

asymptote: $x = 1$;
Domain: $(1, \infty)$;
Range: $(-\infty, \infty)$

61.

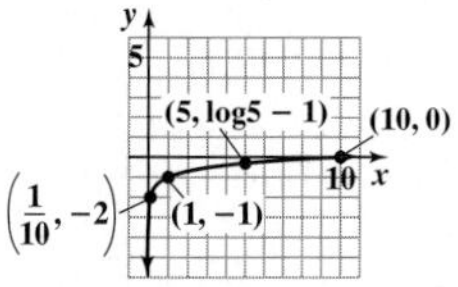

$h(x) = \log x - 1$

asymptote: $x = 0$;
Domain: $(0, \infty)$;
Range: $(-\infty, \infty)$

63.

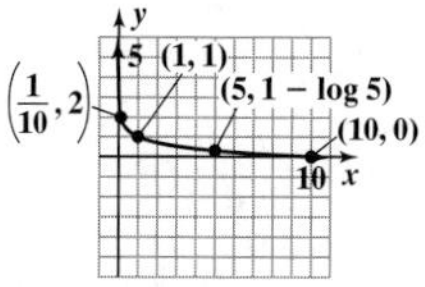

$g(x) = 1 - \log x$

asymptote: $x = 0$;
Domain: $(0, \infty)$;
Range: $(-\infty, \infty)$

65.

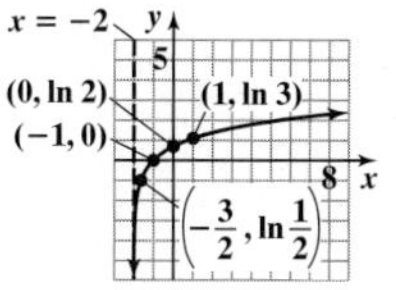

$g(x) = \ln(x + 2)$

asymptote: $x = -2$;
Domain: $(-2, \infty)$;
Range: $(-\infty, \infty)$

67.

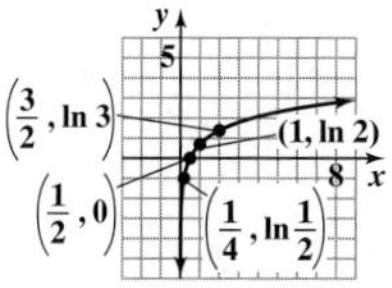

$h(x) = \ln(2x)$

asymptote: $x = 0$;
Domain: $(0, \infty)$;
Range: $(-\infty, \infty)$

69.

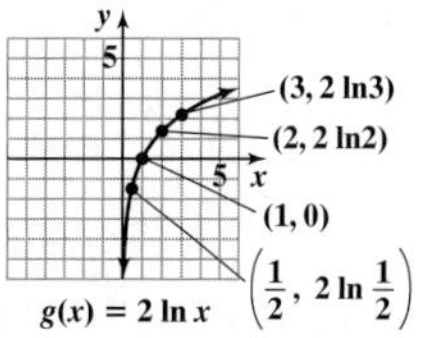

$g(x) = 2 \ln x$

asymptote: $x = 0$;
Domain: $(0, \infty)$;
Range: $(-\infty, \infty)$

71.

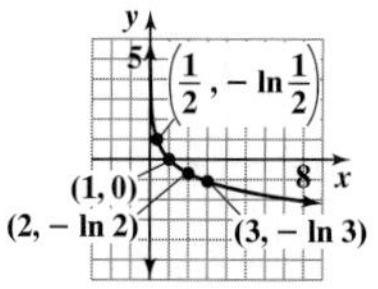

$h(x) = -\ln x$

asymptote: $x = 0$;
Domain: $(0, \infty)$;
Range: $(-\infty, \infty)$

73.

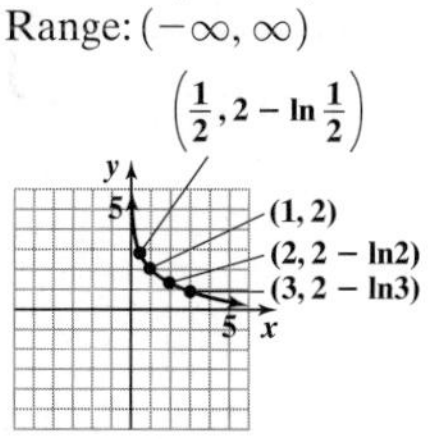

$g(x) = 2 - \ln x$

asymptote: $x = 0$;
Domain: $(0, \infty)$;
Range: $(-\infty, \infty)$

75. $(-4, \infty)$ **77.** $(-\infty, 2)$ **79.** $(-\infty, 2)$ or $(2, \infty)$ **81.** 2 **83.** 7
85. 33 **87.** 0 **89.** 6 **91.** −6 **93.** 125 **95.** $9x$ **97.** $5x^2$
99. $\sqrt{x}$ **101.** $3^2 = x - 1$; $\{10\}$ **103.** $4^{-3} = x$; $\left\{\frac{1}{64}\right\}$ **105.** 0
107. 2 **109.** $(-\infty, -1) \cup (2, \infty)$ **111.** $(-\infty, -1) \cup (5, \infty)$ **113.** 95.4%
115. 65%; extremely well **117.** ≈ 188 db; yes

119. a. 88
b. 71.5; 63.9; 58.8; 55; 52; 49.5
c.

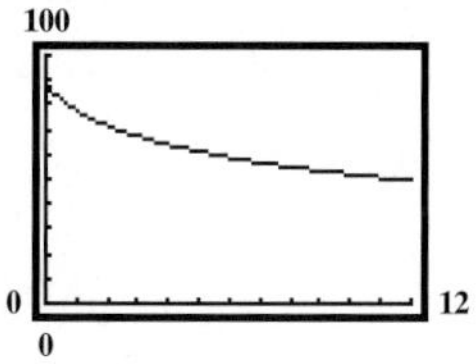

Material retention decreases as time passes.

129.

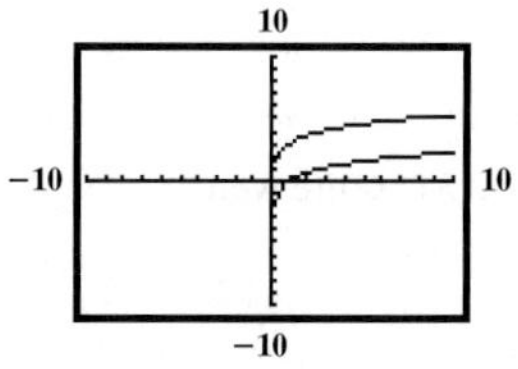

$g(x)$ is $f(x)$ shifted upward 3 units.

131.

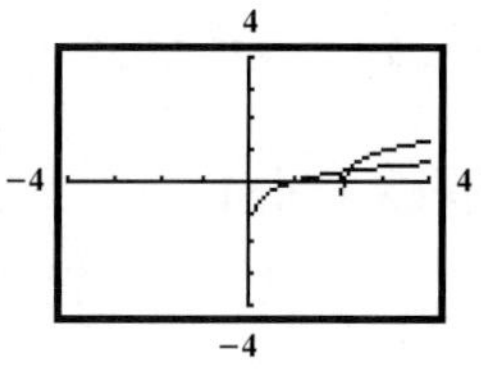

$g(x)$ is $f(x)$ shifted right 2 units and upward 1 unit.

133. a.

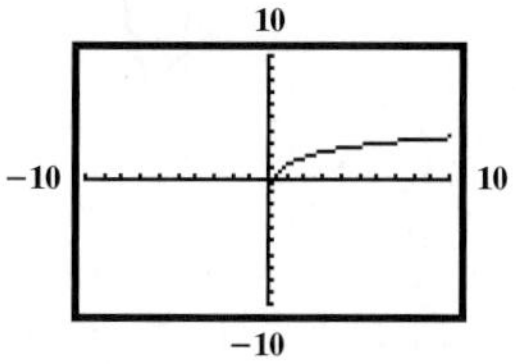

b.

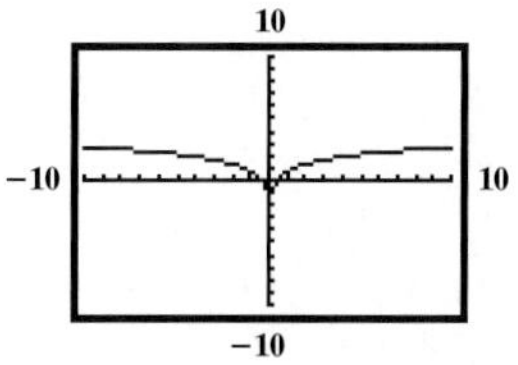

c.

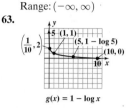

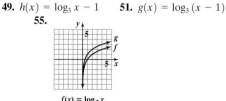

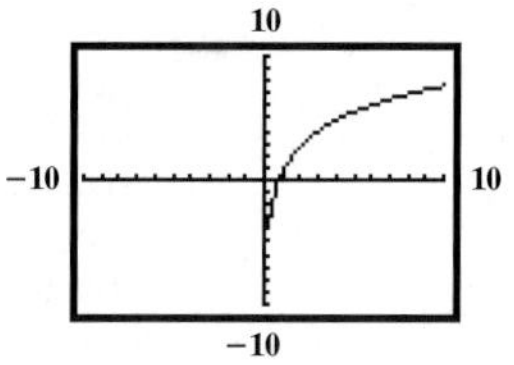

d. They are the same. $\log_b M + \log_b N$ **e.** the sum of the logarithms of its factors
135. d **137.** 0

Section 4.3

Check Point Exercises

1. a. $\log_6 7 + \log_6 11$ **b.** $2 + \log x$ **2. a.** $\log_8 23 - \log_8 x$ **b.** $5 - \ln 11$ **3. a.** $9 \log_6 3$ **b.** $\frac{1}{3} \ln x$ **c.** $2 \log(x + 4)$

4. a. $4 \log_b x + \frac{1}{3} \log_b y$ **b.** $\frac{1}{2} \log_5 x - 2 - 3 \log_5 y$ **5. a.** $\log 100 = 2$ **b.** $\log \frac{7x + 6}{x}$ **6. a.** $\ln x^2 \sqrt[3]{x + 5}$ **b.** $\log \frac{(x - 3)^2}{x}$

c. $\log_b \frac{\sqrt[4]{x}}{25y^{10}}$ **7.** 4.02 **8.** 4.02

Exercise Set 4.3

1. $\log_5 7 + \log_5 3$ **3.** $1 + \log_7 x$ **5.** $3 + \log x$ **7.** $1 - \log_7 x$ **9.** $\log x - 2$ **11.** $3 - \log_4 y$ **13.** $2 - \ln 5$ **15.** $3 \log_b x$

17. $-6 \log N$ **19.** $\frac{1}{5} \ln x$ **21.** $2 \log_b x + \log_b y$ **23.** $\frac{1}{2} \log_4 x - 3$ **25.** $2 - \frac{1}{2} \log_6(x + 1)$ **27.** $2 \log_b x + \log_b y - 2 \log_b z$

29. $1 + \frac{1}{2} \log x$ **31.** $\frac{1}{3} \log x - \frac{1}{3} \log y$ **33.** $\frac{1}{2} \log_b x + 3 \log_b y - 3 \log_b z$ **35.** $\frac{2}{3} \log_5 x + \frac{1}{3} \log_5 y - \frac{2}{3}$ **37.** $3 \ln x + \frac{1}{2} \ln(x^2 + 1) - 4 \ln(x + 1)$

39. $1 + 2 \log x + \frac{1}{3} \log(1 - x) - \log 7 - 2 \log(x + 1)$ **41.** 1 **43.** $\ln(7x)$ **45.** 5 **47.** $\log\left(\frac{2x + 5}{x}\right)$ **49.** $\log(xy^3)$

51. $\ln(x^{1/2}y)$ or $\ln(y\sqrt{x})$ **53.** $\log_b(x^2y^3)$ **55.** $\ln\left(\frac{x^5}{y^2}\right)$ **57.** $\ln\left(\frac{x^3}{y^{1/3}}\right)$ or $\ln\left(\frac{x^3}{\sqrt[3]{y}}\right)$ **59.** $\ln \frac{(x + 6)^4}{x^3}$ **61.** $\ln\left(\frac{x^3y^5}{z^6}\right)$ **63.** $\log \sqrt{xy}$

65. $\log_5\left(\frac{\sqrt{xy}}{(x + 1)^2}\right)$ **67.** $\ln \sqrt[3]{\frac{(x + 5)^2}{x(x^2 - 4)}}$ **69.** $\log \frac{x(x^2 - 1)}{7(x + 1)} = \log \frac{x(x - 1)}{7}$ **71.** 1.5937 **73.** 1.6944 **75.** −1.2304 **77.** 3.6193

79.

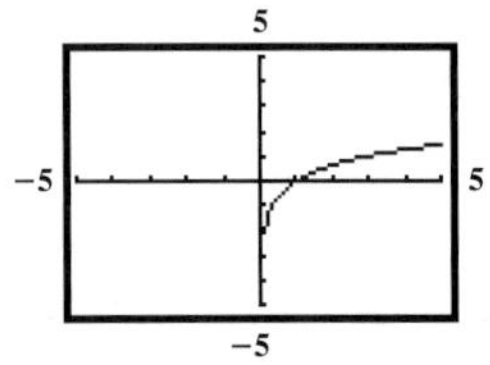

81.

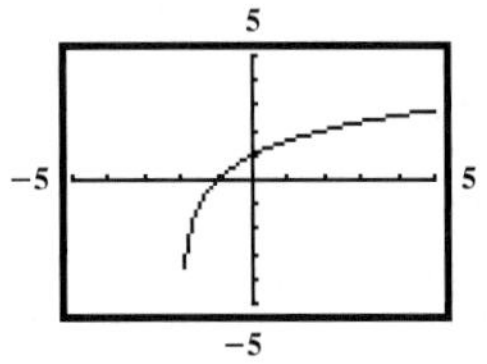

83. $C - A$ **85.** $3A$ **87.** $\frac{1}{2}A - \frac{3}{2}C$ **89.** false; $\ln e = 1$ **91.** false; $\log_4 (2x)^3 = 3 \log_4 (2x)$ **93.** true **95.** true

97. false; $\log(x + 3) - \log(2x) = \log\left(\frac{x + 3}{2x}\right)$ **99.** true **101.** true **103. a.** $D = 10 \log \frac{I}{I_0}$ **b.** 20 decibels louder

113. a.

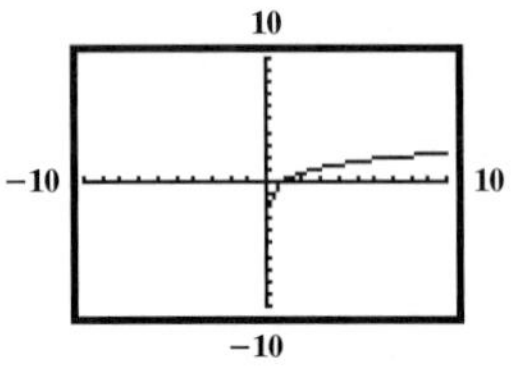

b.

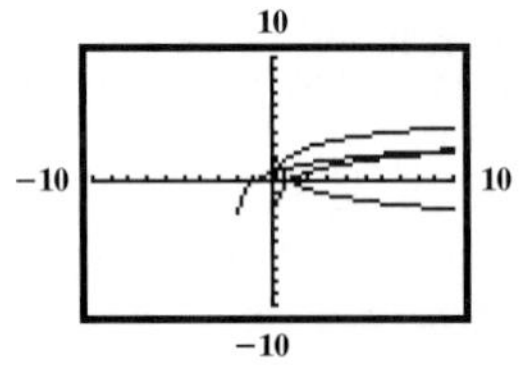

$y = 2 + \log_3 x$ shifts the graph of $y = \log_3 x$ two units upward; $y = \log_3 (x + 2)$ shifts the graph of $y = \log_3 x$ two units left; $y = -\log_3 x$ reflects the graph of $y = \log_3 x$ about the x-axis.

115.

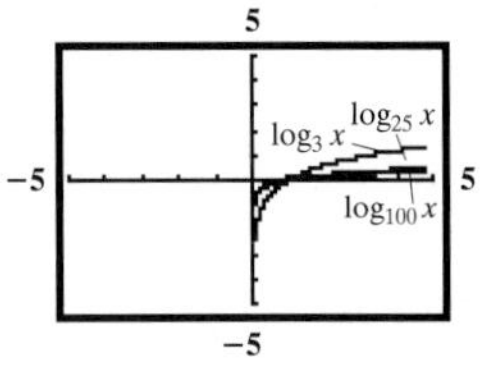

a. top graph: $y = \log_{100} x$; bottom graph: $y = \log_3 x$
b. top graph: $y = \log_3 x$; bottom graph: $y = \log_{100} x$
c. The graph of the equation with the largest b will be on the top in the interval $(0, 1)$ and on the bottom in the interval $(1, \infty)$.

121. d **123.** $\frac{2A}{B}$

Mid-Chapter 4 Check Point

1.

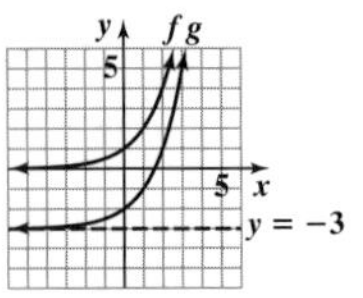

$f(x) = 2^x$
$g(x) = 2^x - 3$

asymptote of f: $y = 0$;
asymptote of g: $y = -3$;
Domain of f = Domain of g = $(-\infty, \infty)$;
Range of f = $(0, \infty)$; Range of g = $(-3, \infty)$

2.

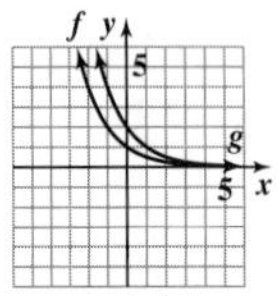

$f(x) = \left(\frac{1}{2}\right)^x$
$g(x) = \left(\frac{1}{2}\right)^{x-1}$

asymptote of f: $y = 0$;
asymptote of g: $y = 0$;
Domain of f = Domain of g = $(-\infty, \infty)$;
Range of f = Range of g = $(0, \infty)$

3.

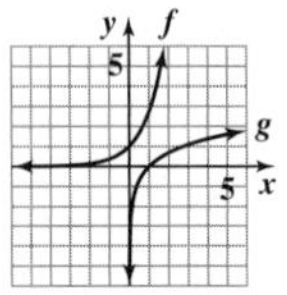

$f(x) = e^x$
$g(x) = \ln x$

asymptote of f: $y = 0$;
asymptote of g: $x = 0$;
Domain of f = Range of g = $(-\infty, \infty)$;
Range of f = Domain of g = $(0, \infty)$

4.

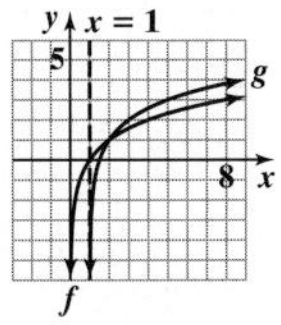

$f(x) = \log_2 x$
$g(x) = \log_2 (x - 1) + 1$

asymptote of f: $x = 0$;
asymptote of g: $x = 1$;
Domain of $f = (0, \infty)$; Domain of $g = (1, \infty)$;
Range of f = Range of $g = (-\infty, \infty)$

5.

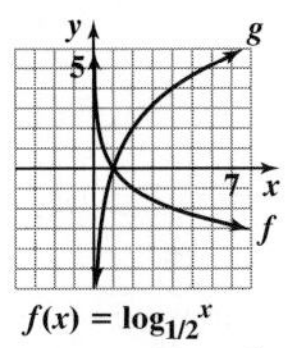

$f(x) = \log_{1/2} x$
$g(x) = -2 \log_{1/2} x$

asymptote of f: $x = 0$;
asymptote of g: $x = 0$;
Domain of f = Domain of $g = (0, \infty)$;
Range of f = Range of $g = (-\infty, \infty)$

6. $(-6, \infty)$ **7.** $(0, \infty)$ **8.** $(-\infty, -6) \cup (-6, \infty)$ **9.** $(-\infty, \infty)$ **10.** 5 **11.** -2 **12.** $\frac{1}{2}$ **13.** $\frac{1}{3}$ **14.** 2

15. Evaluation not possible; $\log_2 \frac{1}{8} = -3$ and $\log_3(-3)$ is undefined. **16.** 5 **17.** $\sqrt{7}$ **18.** 13 **19.** $-\frac{1}{2}$ **20.** $\sqrt{\pi}$

21. $\frac{1}{2}\log x + \frac{1}{2}\log y - 3$ **22.** $19 + 20 \ln x$ **23.** $\log_7\left(\frac{x^8}{\sqrt[3]{y}}\right)$ **24.** $\log_5 x^9$ **25.** $\ln\left[\frac{\sqrt{x}}{y^3(z-2)}\right]$ **26.** \$8

Section 4.4

Check Point Exercises

1. a. $\{3\}$ **b.** $\{-12\}$ **2.** $\left\{\frac{\ln 134}{\ln 5}\right\}$; ≈ 3.04 **3.** $\left\{\frac{\ln 9}{2}\right\}$; ≈ 1.10 **4.** $\left\{\frac{\ln 3 + \ln 7}{2\ln 3 - \ln 7}\right\}$; ≈ 12.11 **5.** $\{0, \ln 7\}$; $\ln 7 \approx 1.95$

6. a. $\{12\}$ **b.** $\left\{\frac{e^2}{3}\right\}$ **7.** $\{5\}$ **8.** $\{4, 5\}$ **9.** 0.01 **10.** 16.2 yr **11.** 2014

Exercise Set 4.4

1. $\{6\}$ **3.** $\{3\}$ **5.** $\{3\}$ **7.** $\{2\}$ **9.** $\left\{\frac{3}{5}\right\}$ **11.** $\left\{\frac{3}{2}\right\}$ **13.** $\{4\}$ **15.** $\{5\}$ **17.** $\left\{-\frac{1}{4}\right\}$ **19.** $\{13\}$ **21.** $\{-2\}$

23. $\left\{\frac{\ln 3.91}{\ln 10}\right\}$; ≈ 0.59 **25.** $\{\ln 5.7\}$; ≈ 1.74 **27.** $\left\{\frac{\ln 17}{\ln 5}\right\}$; ≈ 1.76 **29.** $\left\{\ln \frac{23}{5}\right\}$; ≈ 1.53 **31.** $\left\{\frac{\ln 659}{5}\right\}$; ≈ 1.30

33. $\left\{\frac{\ln 793 - 1}{-5}\right\}$; ≈ -1.14 **35.** $\left\{\frac{\ln 10{,}478 + 3}{5}\right\}$; ≈ 2.45 **37.** $\left\{\frac{\ln 410}{\ln 7} - 2\right\}$; ≈ 1.09 **39.** $\left\{\frac{\ln 813}{0.3 \ln 7}\right\}$; ≈ 11.48

41. $\left\{\frac{3\ln 5 + \ln 3}{\ln 3 - 2\ln 5}\right\}$; ≈ -2.80 **43.** $\{0, \ln 2\}$; $\ln 2 \approx 0.69$ **45.** $\left\{\frac{\ln 3}{2}\right\}$; ≈ 0.55 **47.** $\{0\}$ **49.** $\{81\}$ **51.** $\{e^2\}$; ≈ 7.39 **53.** $\{59\}$

55. $\left\{\frac{109}{27}\right\}$ **57.** $\left\{\frac{62}{3}\right\}$ **59.** $\left\{\frac{e^4}{2}\right\}$; ≈ 27.30 **61.** $\{e^{-1/2}\}$; ≈ 0.61 **63.** $\{e^2 - 3\}$; ≈ 4.39 **65.** $\left\{\frac{5}{4}\right\}$ **67.** $\{6\}$ **69.** $\{6\}$

71. $\{5\}$ **73.** $\{12\}$ **75.** $\left\{\frac{4}{3}\right\}$ **77.** $\varnothing$ **79.** $\{5\}$ **81.** $\left\{\frac{2}{9}\right\}$ **83.** $\{28\}$ **85.** $\{2\}$ **87.** $\varnothing$ **89.** $\left\{\frac{11}{3}\right\}$ **91.** $\left\{\frac{1}{2}\right\}$

93. $\{e^3, e^{-3}\}$ **95.** $\left\{\pm\sqrt{\frac{\ln 45}{\ln 3}}\right\}$ **97.** $\left\{\frac{5 + \sqrt{37}}{2}\right\}$ **99.** $\{-2, 6\}$ **101.** about 0.11 **103. a.** 18.9 million **b.** ≈ 2007

105. 8.2 yr **107.** 16.8% **109.** 8.7 yr **111.** 15.7% **113. a.** quite well **b.** 2006

115. 2.8 days; Yes, the point (2.8, 50) appears to lie on the graph of P. **117.** $10^{-2.4}$; 0.004 moles per liter

123. $\{2\}$ **125.** $\{4\}$ **127.** $\{2\}$ **129.** $\{-1.391606, 1.6855579\}$

131.

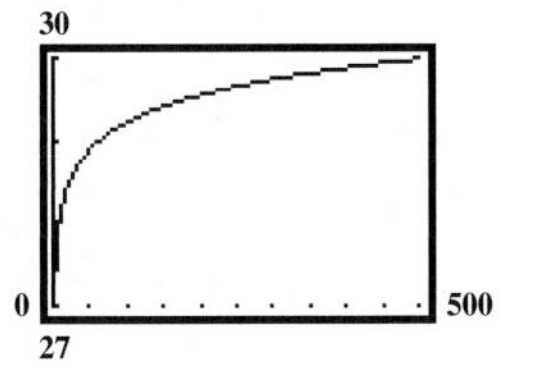

As distance from eye increases, barometric air pressure increases.

133.

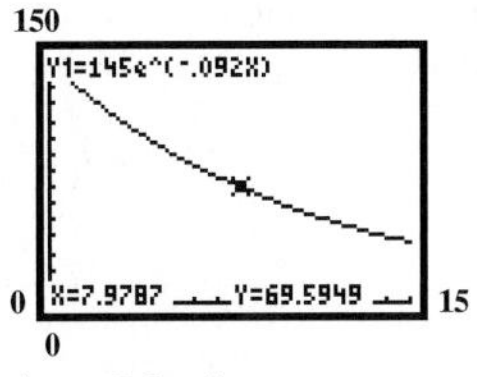

about 7.9 min

135. c **137.** $\{1, e^2\}$, $e^2 \approx 7.389$ **139.** $\{e\}$, $e \approx 2.718$

Section 4.5

Check Point Exercises

1. a. $A = 643\,e^{0.023t}$ **b.** 2039 **2. a.** $A = A_0 e^{-0.0248t}$ **b.** about 72 yr **3. a.** 0.4 correct responses **b.** 0.7 correct responses
c. 0.8 correct responses

4. 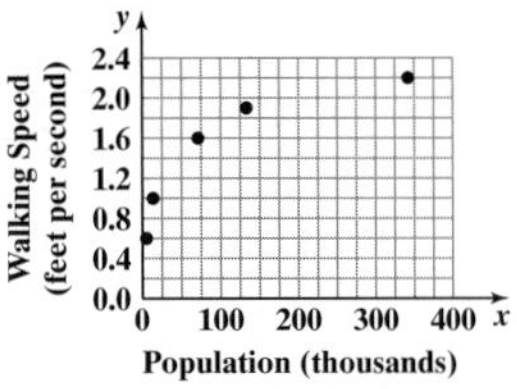

; logarithmic function

5. 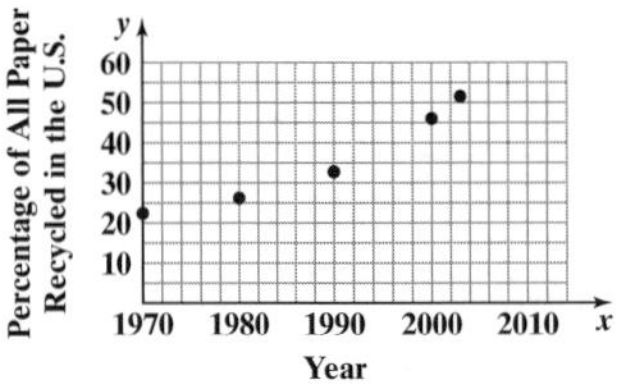

; exponential function

6. $y = 4e^{(\ln 7.8)x}$; $y = 4e^{2.054x}$

Exercise Set 4.5

1. 127.2 million **3.** Iraq; 2.8% **5.** 2014 **7. a.** $A = 6.04e^{0.01t}$ **b.** 2040 **9.** approximately 8 grams **11.** 8 grams after 10 seconds; 4 grams after 20 seconds; 2 grams after 30 seconds; 1 gram after 40 seconds; 0.5 gram after 50 seconds **13.** approximately 15,679 years old

15. a. $\frac{1}{2} = e^{1.31k}$ yields $k = \frac{\ln\left(\frac{1}{2}\right)}{1.31} \approx -0.52912$. **b.** about 0.1069 billion or 106,900,000 years old

17. $2A_0 = A_0e^{kt}$; $2 = e^{kt}$; $\ln 2 = \ln e^{kt}$; $\ln 2 = kt$; $\frac{\ln 2}{k} = t$ **19. a.** 0.7% **b.** about 99 years **21. a.** about 20 people **b.** about 1080 people **c.** 100,000 people **23.** quite well **25.** 2024 **27.** about 3.7% **29.** about 48 years old

31. a. 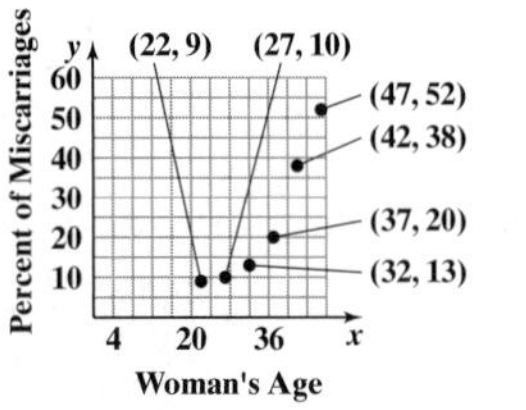

b. exponential function

33. a. 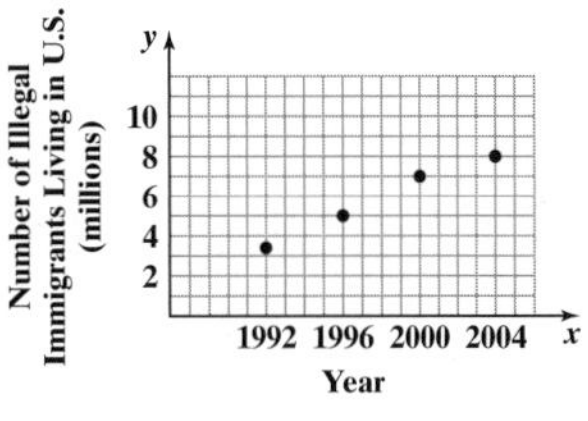

b. linear function

35. a. 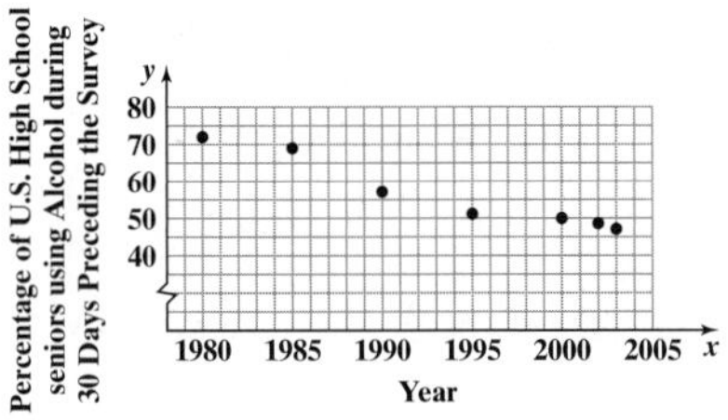

b. exponential function

37. $y = 100e^{(\ln 4.6)x}$; $y = 100e^{1.526x}$ **39.** $y = 2.5e^{(\ln 0.7)x}$; $y = 2.5e^{-0.357x}$ **51.** $y = 194.328 + 22.758 \ln x$; $r \approx 0.881$; Fit is ok, but not great.

53. $y = 196.619x^{0.094}$; $r \approx 0.903$; Fits data fairly well.

55. 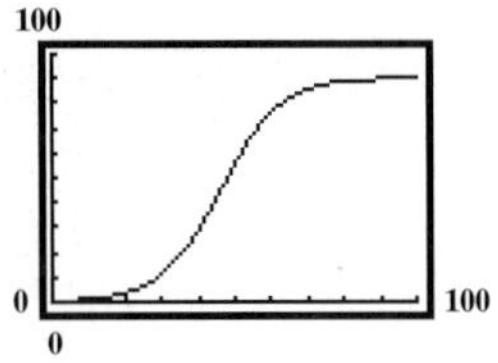

The probability of coronary heart disease starts increasing at a more rapid rate at about age 20. At about age 60, the rate of increase starts to slow down.

57. Models will vary. Examples are given. Predictions will vary. For Exercise 31: $y = 1.402(1.078)^x$;
For Exercise 32: $y = 12.93x + 18.78$ (x = years since 1985);
For Exercise 33: $y = 0.395x + 3.48$ (x = years since 1992);
For Exercise 34: $y = 53.936 + 6.438 \ln x$ (x = years since 1997);
For Exercise 35: $y = 70.64\,(0.9823)^x$ (x = years since 1980);
For Exercise 36: $y = 5.193\,(0.9646)^x$ (x = years since 1965);

59. about 126°F

Chapter 4 Review Exercises

1. $g(x) = 4^{-x}$ **2.** $h(x) = -4^{-x}$ **3.** $r(x) = -4^{-x} + 3$ **4.** $f(x) = 4^x$

5. 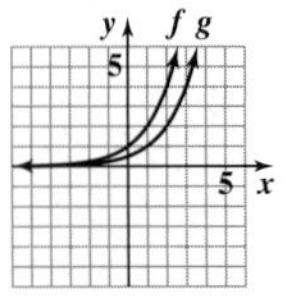

$f(x) = 2^x$
$g(x) = 2^x - 1$

asymptote of f: $y = 0$;
asymptote of g: $y = 0$;
Domain of f = Domain of $g = (-\infty, \infty)$;
Range of f = Range of $g = (0, \infty)$

6. 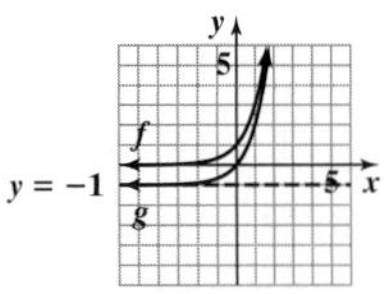

$f(x) = 3^x$
$g(x) = 3^x - 1$

asymptote of f: $y = 0$;
asymptote of g: $y = -1$;
Domain of f = Domain of $g = (-\infty, \infty)$;
Range of $f = (0, \infty)$; Range of $g = (-1, \infty)$

7. 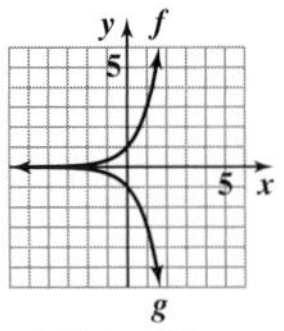

$f(x) = 3^x$
$g(x) = -3^x$

asymptote of f: $y = 0$;
asymptote of g: $y = 0$;
Domain of f = Domain of $g = (-\infty, \infty)$;
Range of $f = (0, \infty)$; Range of $g = (-\infty, 0)$

8.

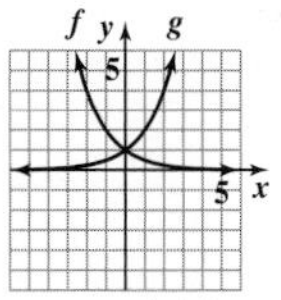

$f(x) = \left(\dfrac{1}{2}\right)^x$

$g(x) = \left(\dfrac{1}{2}\right)^{-x}$

asymptote of f: $y = 0$;
asymptote of g: $y = 0$;
Domain of f = Domain of $g = (-\infty, \infty)$;
Range of f = Range of $g = (0, \infty)$

9.

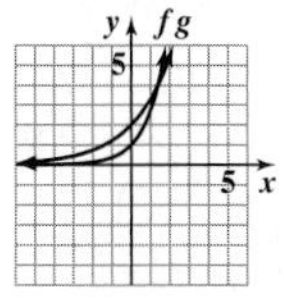

$f(x) = e^x$
$g(x) = 2e^{x/2}$

asymptote of f: $y = 0$;
asymptote of g: $y = 0$;
Domain of f = Domain of $g = (-\infty, \infty)$;
Range of f = Range of $g = (0, \infty)$

10. 5.5% compounded semiannually   **11.** 7% compounded monthly   **12. a.** $200°$ **b.** $120°; 119°$ **c.** $70°$; The temperature in the room is $70°$.
13. $49^{1/2} = 7$ **14.** $4^3 = x$ **15.** $3^y = 81$ **16.** $\log_6 216 = 3$ **17.** $\log_b 625 = 4$ **18.** $\log_{13} 874 = y$ **19.** 3   **20.** -2 **21.** undefined;
$\log_b x$ is defined only for $x > 0$.   **22.** $\dfrac{1}{2}$ **23.** 1   **24.** 8   **25.** 5   **26.** $-\dfrac{1}{2}$ **27.** -2 **28.** -3 **29.** 0

30.

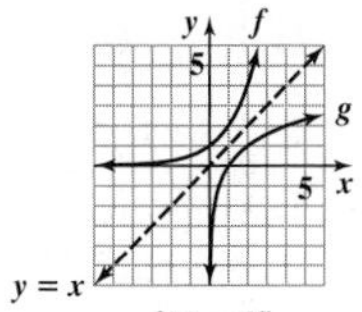

$f(x) = 2^x$
$g(x) = \log_2 x$

31.

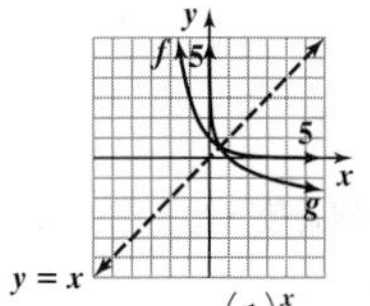

$f(x) = \left(\dfrac{1}{3}\right)^x$

$g(x) = \log_{1/3} x$

32. $g(x) = \log(-x)$
33. $r(x) = 1 + \log(2 - x)$
34. $h(x) = \log(2 - x)$
35. $f(x) = \log x$

36.

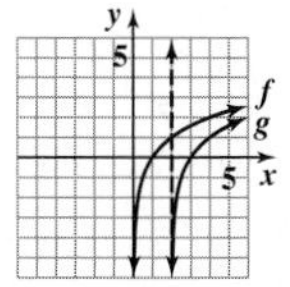

$x = 2$

$f(x) = \log_2 x$

$g(x) = \log_2(x - 2)$

x-intercept: $(3, 0)$
vertical asymptote: $x = 2$
Domain: $(2, \infty)$; Range: $(-\infty, \infty)$

37.

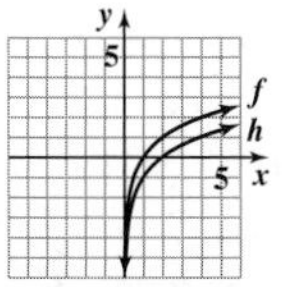

$f(x) = \log_2 x$
$h(x) = -1 + \log_2 x$

x-intercept: $(2, 0)$
vertical asymptote: $x = 0$
Domain: $(0, \infty)$; Range: $(-\infty, \infty)$

38.

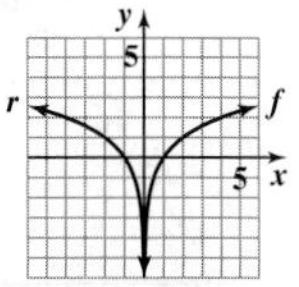

$f(x) = \log_2 x$
$r(x) = \log_2(-x)$

x-intercept: $(-1, 0)$
vertical asymptote: $x = 0$
Domain: $(-\infty, 0)$; Range: $(-\infty, \infty)$

39.

$x = -3$

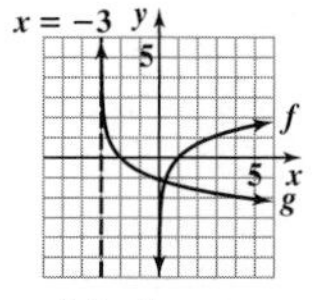

$f(x) = \log x$
$g(x) = -\log(x + 3)$

asymptote of f: $x = 0$;
asymptote of g: $x = -3$;
Domain of $f = (0, \infty)$; Domain of $g = (-3, \infty)$;
Range of f = Range of $g = (-\infty, \infty)$

40.

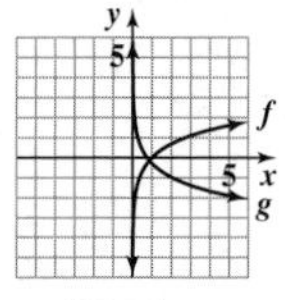

$f(x) = \ln x$
$g(x) = -\ln(2x)$

asymptote of f: $x = 0$;
asymptote of g: $x = 0$;
Domain of f = Domain of $g = (0, \infty)$;
Range of f = Range of $g = (-\infty, \infty)$

41. $(-5, \infty)$ **42.** $(-\infty, 3)$ **43.** $(-\infty, 1) \cup (1, \infty)$ **44.** $6x$ **45.** $\sqrt{x}$ **46.** $4x^2$ **47.** 3.0

48. a. 76   **b.** $\approx 67, \approx 63, \approx 61, \approx 59, \approx 56$ **49.** about 9 weeks   **50.** $2 + 3\log_6 x$ **51.** $\dfrac{1}{2}\log_4 x - 3$

c.

$f(t) = 76 - 18\log(t + 1)$

Time (months)

52. $\log_2 x + 2\log_2 y - 6$ **53.** $\dfrac{1}{3}\ln x - \dfrac{1}{3}$ **54.** $\log_b 21$ **55.** $\log \dfrac{3}{x^3}$

56. $\ln(x^3 y^4)$ **57.** $\ln \dfrac{\sqrt{x}}{y}$ **58.** 6.2448   **59.** -0.1063

60. true   **61.** false; $\log(x + 9) - \log(x + 1) = \log\left(\dfrac{x + 9}{x + 1}\right)$

62. false; $4\log_2 x = \log_2 x^4$ **63.** true   **64.** $\{2\}$ **65.** $\left\{\dfrac{2}{3}\right\}$

66. $\left\{-\dfrac{4}{5}\right\}$ **67.** $\left\{\dfrac{\ln 12{,}143}{\ln 8}\right\}; \approx 4.52$ **68.** $\left\{\dfrac{1}{5}\ln 141\right\}; \approx 0.99$ **69.** $\left\{\dfrac{12 - \ln 130}{5}\right\}; \approx 1.43$ **70.** $\left\{\dfrac{\ln 37{,}500 - 2\ln 5}{4\ln 5}\right\}; \approx 1.14$

71. $\left\{\dfrac{\ln 7 + 4\ln 3}{2\ln 7 - \ln 3}\right\}; \approx 2.27$ **72.** $\{\ln 3\}; \approx 1.10$ **73.** $\{23\}$ **74.** $\left\{\dfrac{e^3}{2}\right\}; \approx 10.04$ **75.** $\{5\}$ **76.** $\varnothing$ **77.** $\{2\}$ **78.** $\{4\}$
79. 5.5 mi   **80.** approximately 2086   **81.** approximately 8103 thousand or 8,103,000   **82.** 7.3 yr   **83.** 14.6 yr   **84.** about 22%

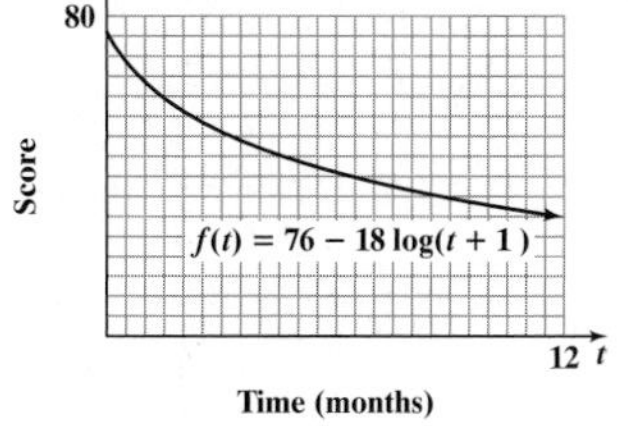

85. a. 0.045   **b.** 55.1 million   **c.** 2012   **86.** 325 days   **87. a.** 200 people   **b.** about 45,411 people   **c.** 500,000 people
88. a.   **89. a.**   **90.** $y = 73e^{(\ln 2.6)x}$; $y = 73e^{0.956x}$
91. $y = 6.5e^{(\ln 0.43)x}$; $y = 6.5e^{-0.844x}$

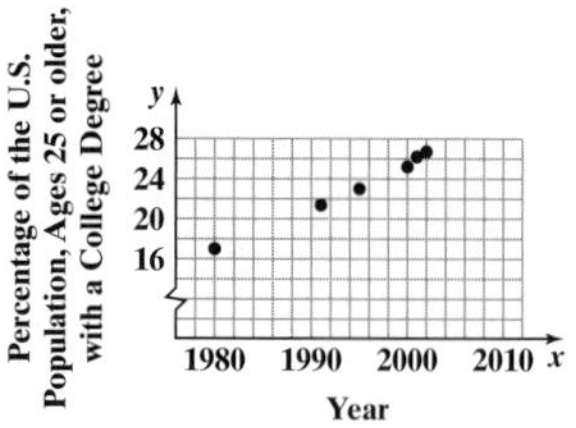

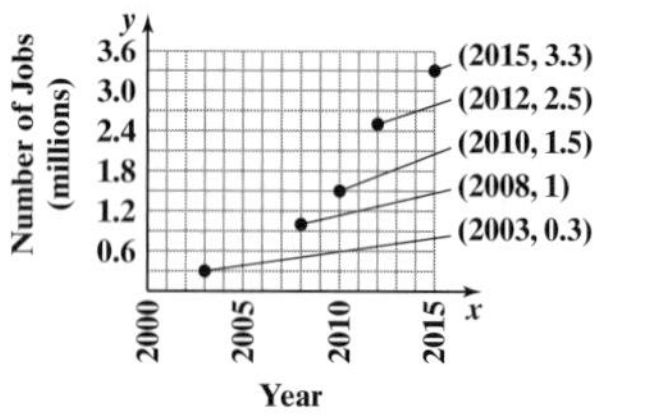

b. exponential function   **b.** exponential function

Chapter 4 Test

1.   **2.**   **3.** $5^3 = 125$ **4.** $\log_{36} 6 = \dfrac{1}{2}$ **5.** $(-\infty, 3)$ **6.** $3 + 5\log_4 x$

7. $\dfrac{1}{3}\log_3 x - 4$ **8.** $\log(x^6 y^2)$ **9.** $\ln \dfrac{7}{x^3}$ **10.** 1.5741   **11.** $\{-10\}$

12. $\left\{\dfrac{\ln 1.4}{\ln 5}\right\}$ **13.** $\left\{\dfrac{\ln 4}{0.005}\right\}$ **14.** $\{0, \ln 5\}$ **15.** $\{54.25\}$

16. $\left\{\dfrac{e^4}{3}\right\}$ **17.** $\{5\}$ **18.** $\varnothing$ **19.** 120 db   **20.** $5x$ **21.** 1   **22.** 0

$f(x) = 2^x$
$g(x) = 2^x + 1$

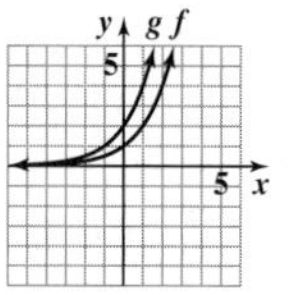

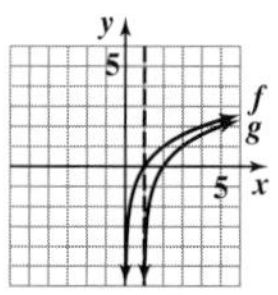

$x = 1$
$f(x) = \log_2 x$
$g(x) = \log_2(x - 1)$

23. 6.5% compounded semiannually; $221 more   **24.** 13.9 years   **25.** about 6.9%   **26. a.** 82.3 million   **b.** decreasing; The growth rate, -0.002, is negative.   **c.** 2008   **27.** 20. $A = 509e^{0.036t}$ **28.** about 24,758 years ago   **29. a.** 14 elk   **b.** about 51 elk   **c.** 140 elk
30. linear   **31.** logarithmic   **32.** exponential   **33.** quadratic   **34.** $y = 96e^{(\ln 0.38)x}$; $y = 96e^{-0.968x}$

Cumulative Review Exercises (Chapters 1–4)

1. $\left\{\dfrac{2}{3}, 2\right\}$ **2.** $\{3, 7\}$ **3.** $\{-2, -1, 1\}$ **4.** $\left\{\dfrac{\ln 128}{5}\right\}$ **5.** $\{3\}$ **6.** $\varnothing$ **7.** $(-\infty, 4]$ **8.** $[1, 3]$
9.   **10.**   **11.**   **12.**

$(x - 3)^2 + (y + 2)^2 = 4$ $f(x) = (x - 2)^2 - 1$ $f(x) = \dfrac{x^2 - 1}{x^2 - 4}$ $f(x) = (x - 2)^2(x + 1)$

13.   **14.**   **15.** using $(1, 3)$, $y - 3 = -3(x - 1)$; $y = -3x + 6$
16. $(f \circ g)(x) = (x + 2)^2$; $(g \circ f)(x) = x^2 + 2$
17. You can expect to sleep 2 hours.
18. after 2 sec; 69 ft
19. 4.1 sec
20. $12 per hr

$f(x) = 2x - 4$
$f^{-1}(x) = \dfrac{x + 4}{2}$

$x = 2$
$f(x) = \ln x$
$g(x) = \ln(x - 2) + 1$

CHAPTER 5

Section 5.1

Check Point Exercises

1. a. solution   **b.** not a solution   **2.** $\{(-2, 5)\}$ **3.** $\{(2, -1)\}$ **4.** $\left\{\left(\dfrac{60}{17}, -\dfrac{11}{17}\right)\right\}$ **5.** no solution or $\varnothing$
6. $\{(x, y)|x = 4y - 8\}$ or $\{(x, y)|5x - 20y = -40\}$ **7. a.** $C(x) = 300,000 + 30x$ **b.** $R(x) = 80x$ **c.** $(6000, 480,000)$; The company will break even if it produces and sells 6000 pairs of shoes.

Exercise Set 5.1

1. solution   **3.** not a solution   **5.** $\{(1, 3)\}$ **7.** $\{(5, 1)\}$ **9.** $\{(-22, -5)\}$ **11.** $\{(0, 0)\}$ **13.** $\{(3, -2)\}$ **15.** $\{(5, 4)\}$ **17.** $\{(7, 3)\}$

19. $\{(2, -1)\}$ **21.** $\{(3, 0)\}$ **23.** $\{(-4, 3)\}$ **25.** $\{(3, 1)\}$ **27.** $\{(1, -2)\}$ **29.** $\left\{\left(\dfrac{7}{25}, -\dfrac{1}{25}\right)\right\}$ **31.** $\varnothing$ **33.** $\{(x, y)|y = 3x - 5\}$

35. $\{(1, 4)\}$ **37.** $\{(x, y)|x + 3y = 2\}$ **39.** $\{(-5, -1)\}$ **41.** $\left\{\left(\dfrac{29}{22}, -\dfrac{5}{11}\right)\right\}$ **43.** $x + y = 7$; $x - y = -1$; 3 and 4

45. $3x - y = 1$; $x + 2y = 12$; 2 and 5   **47.** $(6, -1)$ **49.** $\left\{\left(\dfrac{1}{a}, 3\right)\right\}$ **51.** $m = -4, b = 3$ **53.** $y = x - 4$; $y = -\dfrac{1}{3}x + 4$

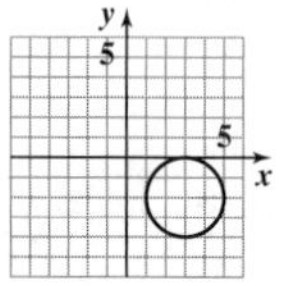

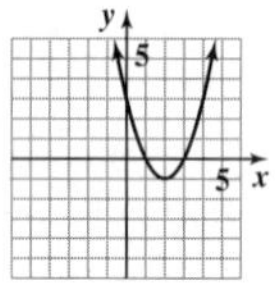

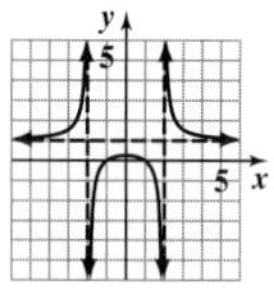

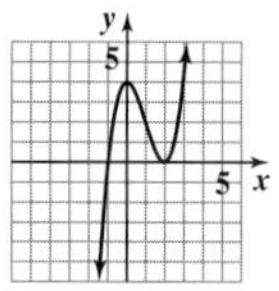

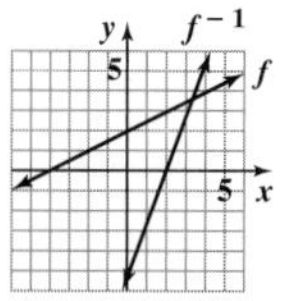

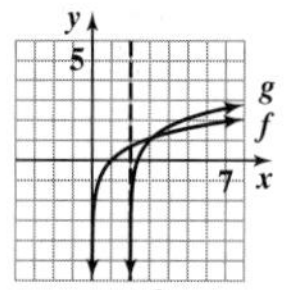

55. 500 radios **57.** −6000; When the company produces and sells 200 radios, the loss is \$6000. **59. a.** $P(x) = 20x - 10{,}000$ **b.** \$190,000 **61. a.** $C(x) = 18{,}000 + 20x$ **b.** $R(x) = 80x$ **c.** (300, 24,000); When 300 canoes are produced and sold, both revenue and cost are \$24,000. **63. a.** $C(x) = 30{,}000 + 2500x$ **b.** $R(x) = 3125x$ **c.** (48, 150,000); For 48 sold-out performances, both cost and revenue are \$150,000. **65. a.** 150 sold; 300 supplied **b.** \$100; 250 **67.** 2020; 48% **69. a.** $y = 5.48 + 0.04x$ **b.** $y = 1.84 + 0.17x$ **c.** 2028; 6.6%; Medicare **71. a.** $y = -0.75x + 38$ **b.** $y = -x + 40$ **c.** 1993; 32% **73.** Pan pizza: 1120 calories; beef burrito: 430 calories **75.** Scrambled eggs: 366 mg cholesterol; Double Beef Whopper: 175 mg cholesterol **77.** 50 rooms with kitchen facilities, 150 rooms without kitchen facilities **79.** 100 ft long by 80 ft wide **81.** Rate rowing in still water: 6 mph; rate of the current: 2 mph **83.** 80°, 50°, 50° **97.** the twin who always lies

Section 5.2

Check Point Exercises

1. $(-1) - 2(-4) + 3(5) = 22;\ 2(-1) - 3(-4) - 5 = 5;\ 3(-1) + (-4) - 5(5) = -32$ **2.** $\{(1, 4, -3)\}$ **3.** $\{(4, 5, 3)\}$ **4.** $y = 3x^2 - 12x + 13$

Exercise Set 5.2

1. solution **3.** solution **5.** $\{(2, 3, 3)\}$ **7.** $\{(2, -1, 1)\}$ **9.** $\{(1, 2, 3)\}$ **11.** $\{(3, 1, 5)\}$ **13.** $\{(1, 0, -3)\}$ **15.** $\{(1, -5, -6)\}$ **17.** $\left\{\left(\frac{1}{2}, \frac{1}{3}, -1\right)\right\}$ **19.** $y = 2x^2 - x + 3$ **21.** $y = 2x^2 + x - 5$ **23.** 7, 4, and 5 **25.** $\{(4, 8, 6)\}$ **27.** $y = -\frac{3}{4}x^2 + 6x - 11$ **29.** $\left\{\left(\frac{8}{a}, -\frac{3}{b}, -\frac{5}{c}\right)\right\}$ **31. a.** (0, 2.5), (2, 4.1), (3, 3.5) **b.** $c = 2.5;\ 4a + 2b + c = 4.1;\ 9a + 3b + c = 3.5$ **c.** $y = -\frac{7}{15}x^2 + \frac{26}{15}x + \frac{5}{2}$ **33. a.** $y = -16x^2 + 40x + 200$ **b.** $y = 0$ when $x = 5$; The ball hits the ground after 5 seconds **35.** Carnegie: \$100 billion; Vanderbilt: \$96 billion; Gates: \$48 billion **37.** 200 \$8 tickets; 150 \$10 tickets; 50 \$12 tickets **39.** \$1200 at 8%, \$2000 at 10%, and \$3500 at 12% **41.** $x = 60, y = 55, z = 65$ **49.** 13 triangles, 21 rectangles, and 6 pentagons

Section 5.3

Check Point Exercises

1. $\frac{2}{x-3} + \frac{3}{x+4}$ **2.** $\frac{2}{x} - \frac{2}{x-1} + \frac{3}{(x-1)^2}$ **3.** $\frac{2}{x+3} + \frac{6x-8}{x^2+x+2}$ **4.** $\frac{2x}{x^2+1} + \frac{-x+3}{(x^2+1)^2}$

Exercise Set 5.3

1. $\frac{A}{x-2} + \frac{B}{x+1}$ **3.** $\frac{A}{x+2} + \frac{B}{x-3} + \frac{C}{(x-3)^2}$ **5.** $\frac{A}{x-1} + \frac{Bx+C}{x^2+1}$ **7.** $\frac{Ax+B}{x^2+4} + \frac{Cx+D}{(x^2+4)^2}$ **9.** $\frac{3}{x-3} - \frac{2}{x-2}$ **11.** $\frac{7}{x-9} - \frac{4}{x+2}$ **13.** $\frac{24}{7(x-4)} + \frac{25}{7(x+3)}$ **15.** $\frac{4}{7(x-3)} - \frac{8}{7(2x+1)}$ **17.** $\frac{3}{x} + \frac{2}{x-1} - \frac{1}{x+3}$ **19.** $\frac{3}{x} + \frac{4}{x+1} - \frac{3}{x-1}$ **21.** $\frac{6}{x-1} - \frac{5}{(x-1)^2}$ **23.** $\frac{1}{x-2} - \frac{2}{(x-2)^2} - \frac{5}{(x-2)^3}$ **25.** $\frac{7}{x} - \frac{6}{x-1} + \frac{10}{(x-1)^2}$ **27.** $\frac{1}{4(x+1)} + \frac{3}{4(x-1)} + \frac{1}{2(x-1)^2}$ **29.** $\frac{3}{x-1} + \frac{2x-4}{x^2+1}$ **31.** $\frac{2}{x+1} + \frac{3x-1}{x^2+2x+2}$ **33.** $\frac{1}{4x} + \frac{1}{x^2} - \frac{x+4}{4(x^2+4)}$ **35.** $\frac{4}{x+1} + \frac{2x-3}{x^2+1}$ **37.** $\frac{x+1}{x^2+2} - \frac{2x}{(x^2+2)^2}$ **39.** $\frac{x-2}{x^2-2x+3} + \frac{2x+1}{(x^2-2x+3)^2}$ **41.** $\frac{3}{x-2} + \frac{x-1}{x^2+2x+4}$ **43.** $x^3 + x - \frac{1}{2(x+1)} + \frac{3}{2(x-1)}$ **45.** $x + 1 - \frac{2}{x} - \frac{2}{x^2} + \frac{2}{x-1}$ **47.** $\frac{\frac{1}{2c}}{x-c} - \frac{\frac{1}{2c}}{x+c}$ **49.** $\frac{a}{x-c} + \frac{ac+b}{(x-c)^2}$ **51.** $\frac{1}{x} - \frac{1}{x+1};\ \frac{99}{100}$ **61.** $\frac{2}{x-3} + \frac{2x+5}{x^2+3x+3}$

Section 5.4

Check Point Exercises

1. $\{(0, 1), (4, 17)\}$ **2.** $\left\{\left(-\frac{6}{5}, \frac{3}{5}\right), (2, -1)\right\}$ **3.** $\{(3, 2), (3, -2), (-3, 2), (-3, -2)\}$ **4.** $\{(0, 5)\}$ **5.** length: 7 ft; width: 3 ft or length: 3 ft; width: 7 ft

Exercise Set 5.4

1. $\{(-3, 5), (2, 0)\}$ **3.** $\{(1, 1), (2, 0)\}$ **5.** $\{(4, -10), (-3, 11)\}$ **7.** $\{(4, 3), (-3, -4)\}$ **9.** $\left\{\left(-\frac{3}{2}, -4\right), (2, 3)\right\}$ **11.** $\{(-5, -4), (3, 0)\}$ **13.** $\{(3, 1), (-3, -1), (1, 3), (-1, -3)\}$ **15.** $\{(4, -3), (-1, 2)\}$ **17.** $\{(0, 1), (4, -3)\}$ **19.** $\{(3, 2), (3, -2), (-3, 2), (-3, -2)\}$ **21.** $\{(3, 2), (3, -2), (-3, 2), (-3, -2)\}$ **23.** $\{(2, 1), (2, -1), (-2, 1), (-2, -1)\}$ **25.** $\{(3, 4), (3, -4)\}$ **27.** $\{(0, 2), (0, -2), (-1, \sqrt{3}), (-1, -\sqrt{3})\}$ **29.** $\{(2, 1), (2, -1), (-2, 1), (-2, -1)\}$ **31.** $\{(-2\sqrt{2}, -\sqrt{2}), (-1, -4), (1, 4), (2\sqrt{2}, \sqrt{2})\}$ **33.** $\{(2, 2), (4, 1)\}$ **35.** $\{(0, 0), (-1, 1)\}$ **37.** $\{(0, 0), (-2, 2), (2, 2)\}$ **39.** $\left\{(-4, 1), \left(-\frac{5}{2}, \frac{1}{4}\right)\right\}$ **41.** $\left\{\left(\frac{12}{5}, -\frac{29}{5}\right), (-2, 3)\right\}$ **43.** 4 and 6 **45.** 2 and 1, 2 and −1, −2 and 1, or −2 and −1 **47.** $\{(2, -1), (-2, 1)\}$ **49.** $\{(2, 20), (-2, 4), (-3, 0)\}$ **51.** $\left\{\left(-1, -\frac{1}{2}\right), \left(-1, \frac{1}{2}\right), \left(1, -\frac{1}{2}\right), \left(1, \frac{1}{2}\right)\right\}$

53.

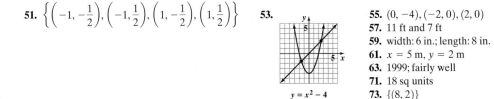

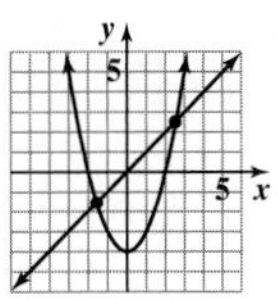

$y = x^2 - 4$
$y = x$

55. (0, −4), (−2, 0), (2, 0)
57. 11 ft and 7 ft
59. width: 6 in.; length: 8 in.
61. $x = 5$ m, $y = 2$ m
63. 1999; fairly well
71. 18 sq units
73. $\{(8, 2)\}$

Mid-Chapter 5 Check Point

1. $\{(-1, 2)\}$ **2.** $\{(1, -2)\}$ **3.** $\{(6, 10)\}$ **4.** $\{(x, y)|y = 4x - 5\}$ or $\{(x, y)|8x - 2y = 10\}$ **5.** $\left\{\left(\frac{11}{19}, \frac{7}{19}\right)\right\}$ **6.** $\varnothing$ **7.** $\{(-1, 2, -2)\}$ **8.** $\{(4, -2, 3)\}$ **9.** $\left\{\left(-\frac{9}{5}, \frac{12}{5}\right), (3, 0)\right\}$ **10.** $\{(-2, -1), (-2, 1), (2, -1), (2, 1)\}$ **11.** $\{(-\sqrt{7}, 1), (-2, -2), (2, -2), (\sqrt{7}, 1)\}$ **12.** $\{(0, -2), (6, 1)\}$ **13.** $\frac{1}{x-2} - \frac{2}{(x-2)^2} - \frac{5}{(x-2)^3}$ **14.** $\frac{5}{x+2} + \frac{3}{x+1} + \frac{2}{x-1}$ **15.** $-\frac{2}{x+3} + \frac{3x-5}{x^2+4}$ **16.** $\frac{x}{x^2+4} - \frac{4x}{(x^2+4)^2}$ **17. a.** $C(x) = 400{,}000 + 20x$ **b.** $R(x) = 100x$ **c.** $P(x) = 80x - 400{,}000$ **d.** (5000, 500,000); The company will break even when it produces and sells 5000 PDAs. At this level, both revenue and cost are \$500,000. **18.** 6 roses and 14 carnations **19.** $x = 55°$, $y = 35°$ **20.** $y = -x^2 + 2x + 3$ **21.** length: 8 m; width: 2.5 m

Section 5.5

Check Point Exercises

1. $4x - 2y \geq 8$

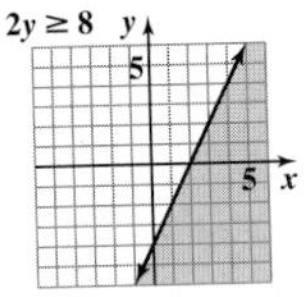

2. $y > -\frac{3}{4}x$

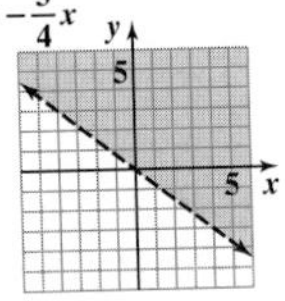

3. a. $y > 1$

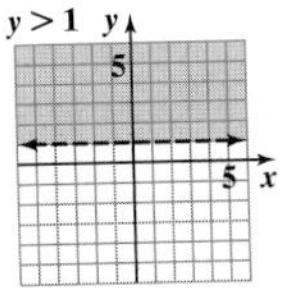

b. $x \leq -2$

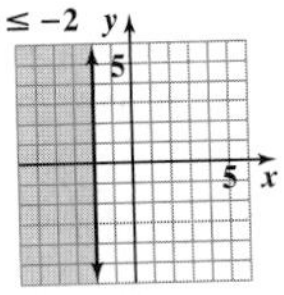

4.

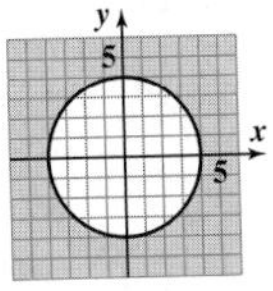

$x^2 + y^2 \geq 16$

5. $x - 3y < 6$, $2x + 3y \geq -6$

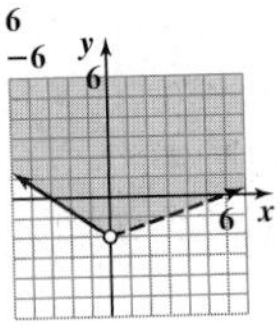

6.

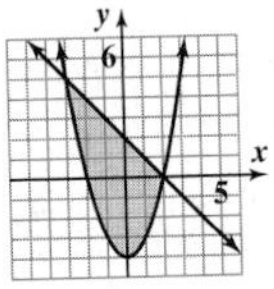

$y \geq x^2 - 4$
$x + y \leq 2$

7. $x + y < 2$, $-2 \leq x < 1$, $y > -3$

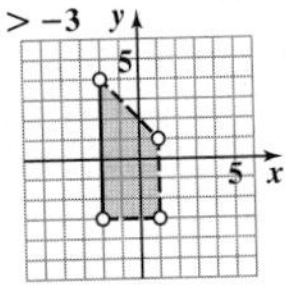

8. $B = (60, 20)$; Using $T = 60$ and $P = 20$, each of the three inequalities for grasslands is true: $60 \geq 35$, true; $5(60) - 7(20) \geq 70$, true; $3(60) - 35(20) \leq -140$, true.

Exercise Set 5.5

1.

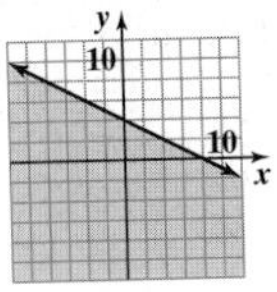

$x + 2y \leq 8$

3.

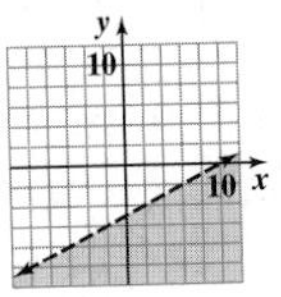

$x - 2y > 10$

5.

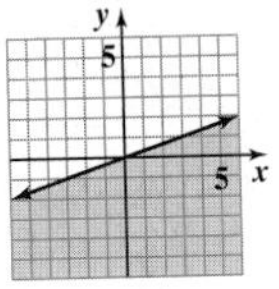

$y \leq \frac{1}{3}x$

7.

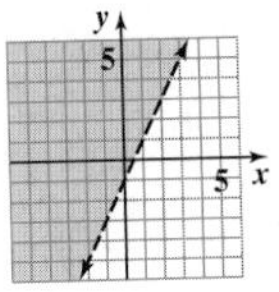

$y > 2x - 1$

9.

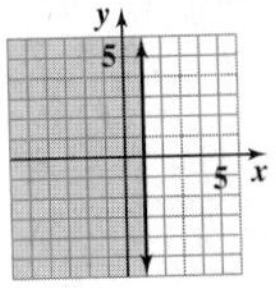

$x \leq 1$

11.

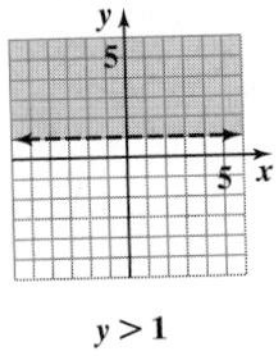

$y > 1$

13.

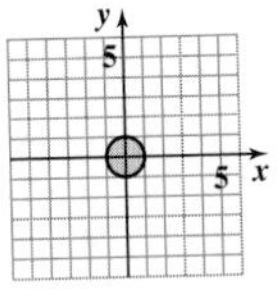

$x^2 + y^2 \leq 1$

15.

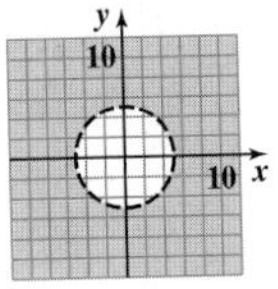

$x^2 + y^2 > 25$

17.

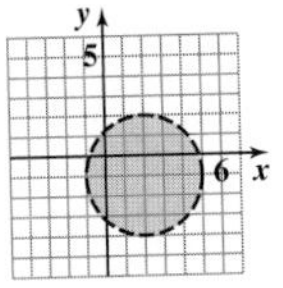

$(x - 2)^2 + (y + 1)^2 < 9$

19.

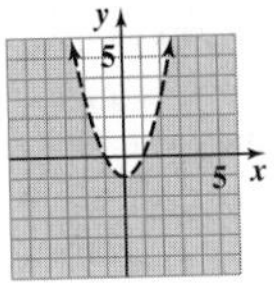

$y < x^2 - 1$

21.

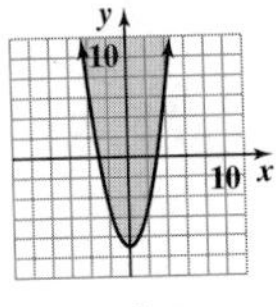

$y \geq x^2 - 9$

23.

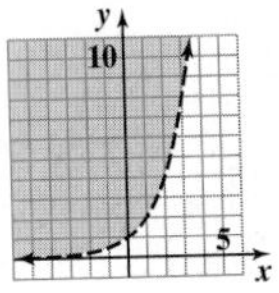

$y > 2^x$

25. $x = -1$

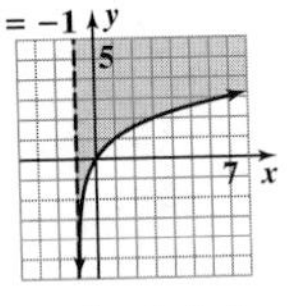

$y \geq \log_2 (x + 1)$

27. $3x + 6y \leq 6$, $2x + y \leq 8$

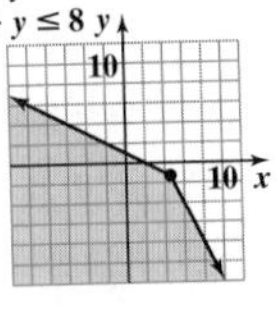

29. $2x - 5y \leq 10$, $3x - 2y > 6$

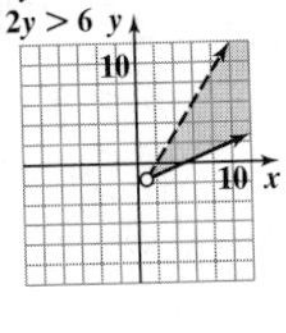

31. $y > 2x - 3$, $y < -x + 6$

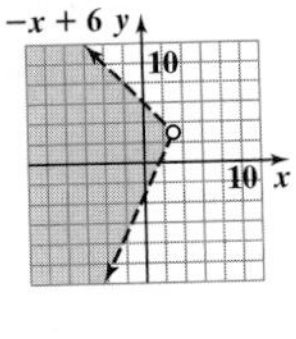

33. $x + 2y \leq 4$, $y \geq x - 3$

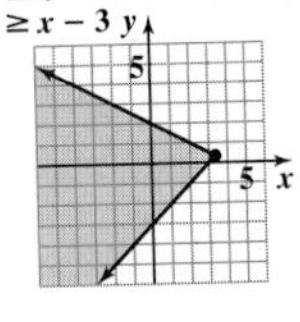

35. $x \leq 2$, $y \geq -1$

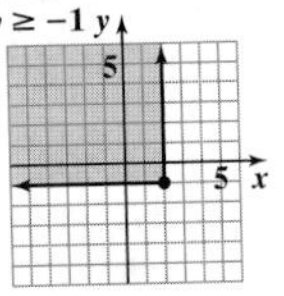

37. $-2 \leq x < 5$

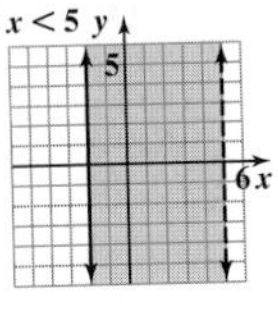

39. $x - y \leq 1$, $x \geq 2$

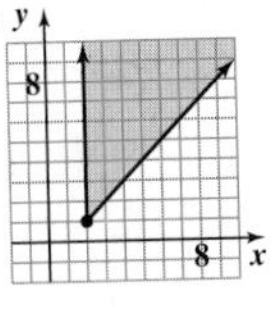

41. ∅

43. $x + y > 4$
$x + y > -1$

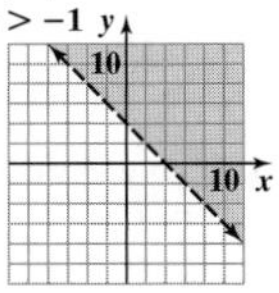

45.

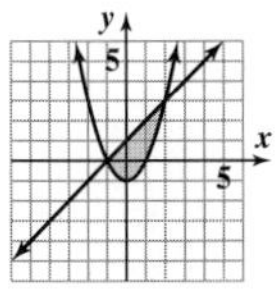

$y \geq x^2 - 1$
$x - y \geq -1$

47.

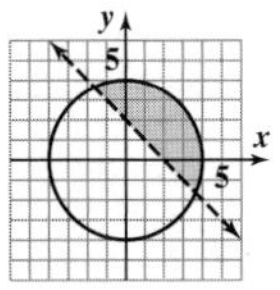

$x^2 + y^2 \leq 16$
$x + y > 2$

49.

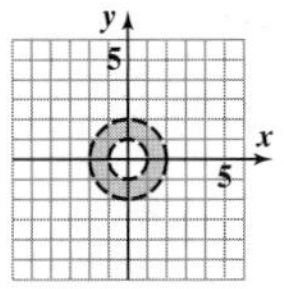

$x^2 + y^2 > 1$
$x^2 + y^2 < 4$

51.

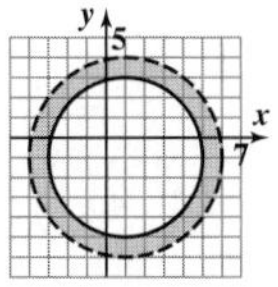

$(x - 1)^2 + (y + 1)^2 < 25$
$(x - 1) + (y + 1)^2 \geq 16$

53.

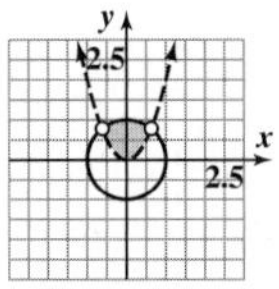

$x^2 + y^2 \leq 1$
$y - x^2 > 0$

55.

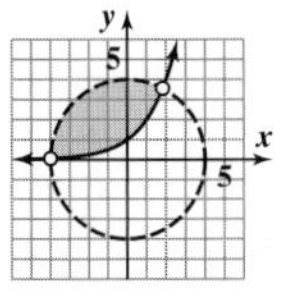

$x^2 + y^2 < 16$
$y \geq 2^x$

57. $x - y \leq 2$
$x > -2$
$y \leq 3$

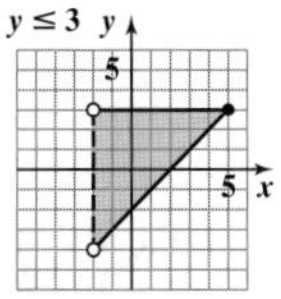

59. $x \geq 0$
$y \geq 0$
$2x + 5y < 10$
$3x + 4y \leq 12$

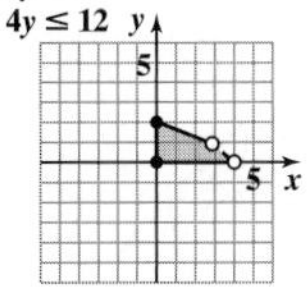

61. $3x + y \leq 6$
$2x - y \leq -1$
$x > -2$
$y < 4$

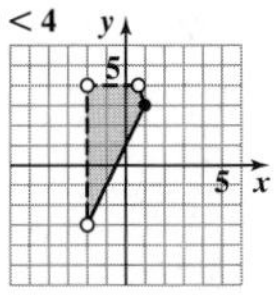

63. $y \geq -2x + 4$

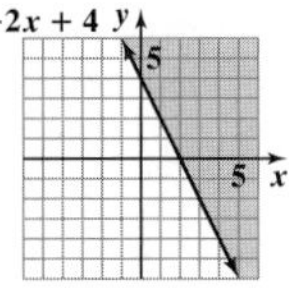

65. $x + y \leq 4$
$3x + y \leq 6$

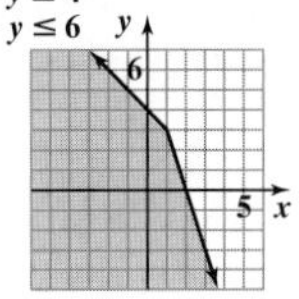

67.

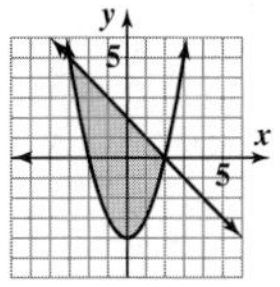

$x + y \leq 2$
$y \geq x^2 - 4$

69. $-2 \leq x \leq 2$
$-3 \leq y \leq 3$

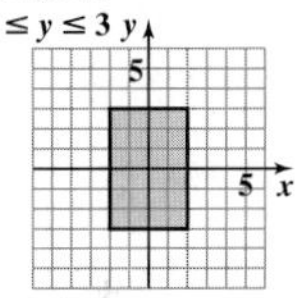

71. $y > \frac{3}{2}x - 2$ or $y < 4$

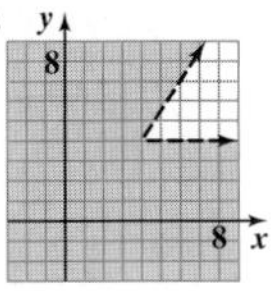

73. no solution
75. infinitely many solutions

77. a. $A = (20, 150)$; A 20-year-old with a heart rate of 150 beats per minute is within the target range. **b.** $10 \leq 20 \leq 70$, true; $150 \geq 0.7(220 - 20)$, true; $150 \leq 0.8(220 - 20)$, true **79.** $10 \leq a \leq 70$; $H \geq 0.6(220 - a)$; $H \leq 0.7(220 - a)$

81. a. $50x + 150y > 2000$

b.

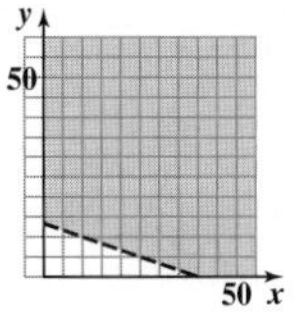

$50x + 150y > 2000$

c. Answers may vary. Example: (20, 20): 20 children and 20 adults will cause the elevator to be overloaded.

83. a. $y \geq 0$; $x + y \geq 5$; $x \geq 1$; $200x + 100y \leq 700$

b. $y \geq 0$
$x + y \geq 5$
$x \geq 1$
$200x + 100y \leq 700$

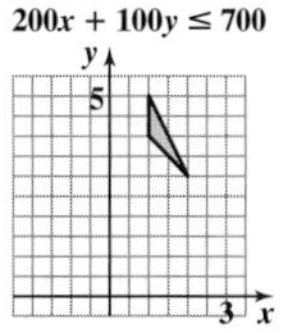

c. 2 nights

85. a. 27.1 **b.** overweight

97.

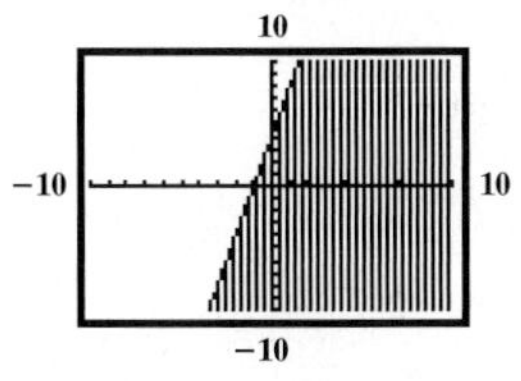

99.

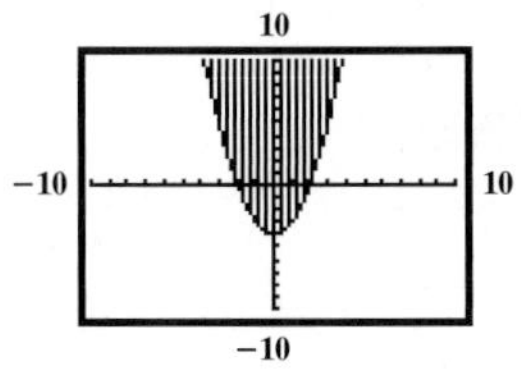

101.

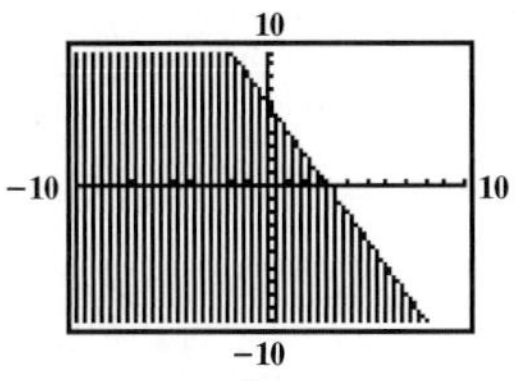

107. $y > x - 3$; $y \leq x$ **109.** $x + 2y \leq 6$ or $2x + y \leq 6$ **111.** $y \geq nx + b$
$y \leq mx + b$

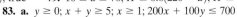

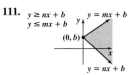

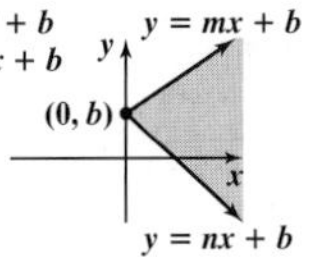

Section 5.6

Check Point Exercises

1. $z = 25x + 55y$ **2.** $x + y \le 80$ **3.** $30 \le x \le 80$; $10 \le y \le 30$; objective function: $z = 25x + 55y$; constraints: $x + y \le 80$; $30 \le x \le 80$; $10 \le y \le 30$ **4.** 50 bookshelves and 30 desks; \$2900 **5.** 30

Exercise Set 5.6

1. (1, 2): 17; (2, 10): 70; (7, 5): 65; (8, 3): 58; maximum: $z = 70$; minimum: $z = 17$
3. (0, 0): 0; (0, 8): 400; (4, 9): 610; (8, 0): 320; maximum: $z = 610$; minimum: $z = 0$

5. a.

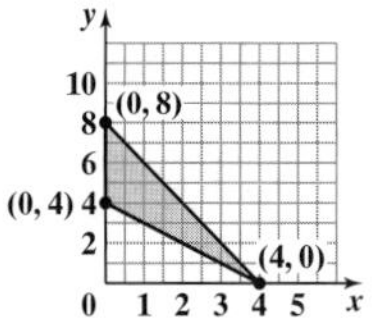

b. (0, 8): 16; (0, 4): 8; (4, 0): 12
c. maximum value: 16 at $x = 0$ and $y = 8$

7. a.

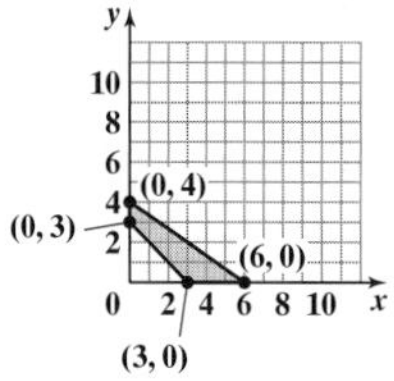

b. (0, 4): 4; (0, 3): 3; (3, 0): 12; (6, 0): 24
c. maximum value: 24 at $x = 6$ and $y = 0$

9. a.

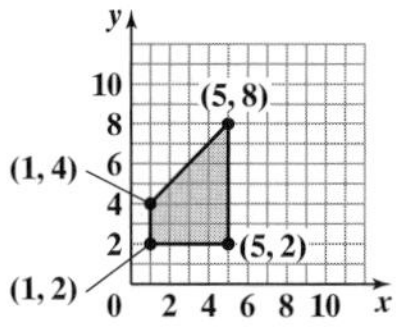

b. (1, 2): −1; (1, 4): −5; (5, 8): −1; (5, 2): 11
c. maximum value: 11 at $x = 5$ and $y = 2$

11. a.

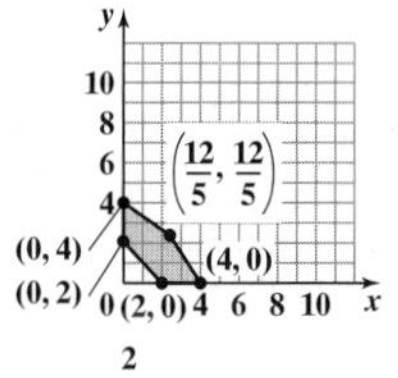

b. (0, 4): 8; (0, 2): 4; (2, 0): 8; (4, 0): 16; $\left(\frac{12}{5}, \frac{12}{5}\right)$: $\frac{72}{5}$
c. maximum value: 16 at $x = 4$ and $y = 0$

13. a.

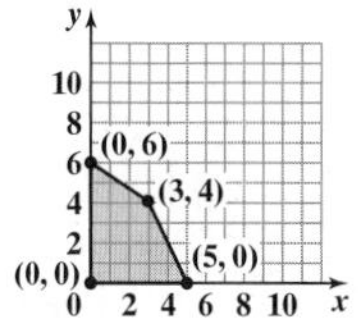

b. (0, 6): 72, (0, 0): 0; (5, 0): 50; (3, 4): 78
c. maximum value: 78 at $x = 3$ and $y = 4$

15. a. $z = 125x + 200y$
b. $x \le 450$; $y \le 200$; $600x + 900y \le 360{,}000$
c.

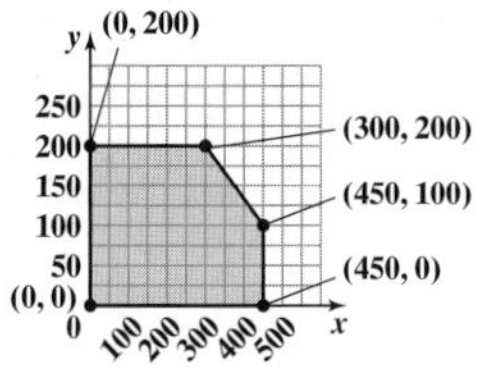

d. (0, 0): 0; (0, 200): 40,000; (300, 200): 77,500; (450, 100): 76,250; (450, 0): 56,250
e. 300; 200; \$77,500

17. 40 model A bicycles and no model B bicycles **19.** 300 cartons of food and 200 cartons of clothing **21.** 50 students and 100 parents **23.** 10 Boeing 727s and 42 Falcon 20s **29.** \$5000 in stocks and \$5000 in bonds

Chapter 5 Review Exercises

1. $\{(1, 5)\}$ **2.** $\{(2, 3)\}$ **3.** $\{(2, -3)\}$ **4.** $\varnothing$ **5.** $\{(x, y) \mid 3x - 6y = 12\}$ **6. a.** $C(x) = 60{,}000 + 200x$ **b.** $R(x) = 450x$ **c.** (240, 108,000); This means the company will break even if it produces and sells 240 desks. **7.** Japan: 73.6 yr; Switzerland: 72.8 yr **8.** 12 ft by 5 ft **9.** \$80 per day for the room, \$60 per day for the car **10.** 3 apples and 2 avocados **11.** $\{(0, 1, 2)\}$ **12.** $\{(2, 1, -1)\}$ **13.** $y = 3x^2 - 4x + 5$ **14. a.** (0, 3.5), (15, 5.0), (33, 3.8) **b.** $c = 3.5$; $225a + 15b + c = 5.0$; $1089a + 33b + c = 3.8$ **15.** Labrador retrievers: 147; Golden retrievers: 53; German shepherds: 46 **16.** $\frac{3}{5(x-3)} + \frac{2}{5(x+2)}$ **17.** $\frac{6}{x-4} + \frac{5}{x+3}$ **18.** $\frac{2}{x} + \frac{3}{x+2} - \frac{1}{x-1}$ **19.** $\frac{2}{x-2} + \frac{5}{(x-2)^2}$ **20.** $-\frac{4}{x-1} + \frac{4}{x-2} - \frac{2}{(x-2)^2}$ **21.** $\frac{6}{5(x-2)} + \frac{-6x+3}{5(x^2+1)}$ **22.** $\frac{5}{x-3} + \frac{2x-1}{x^2+4}$ **23.** $\frac{x}{x^2+4} - \frac{4x}{(x^2+4)^2}$ **24.** $\frac{4x+1}{x^2+x+1} + \frac{2x-2}{(x^2+x+1)^2}$ **25.** $\{(4, 3), (1, 0)\}$ **26.** $\{(0, 1), (-3, 4)\}$ **27.** $\{(1, -1), (-1, 1)\}$ **28.** $\{(3, \sqrt{6}), (3, -\sqrt{6}), (-3, \sqrt{6}), (-3, -\sqrt{6})\}$ **29.** $\{(2, 2), (-2, -2)\}$ **30.** $\{(9, 6), (1, 2)\}$ **31.** $\{(-3, -1), (1, 3)\}$ **32.** $\left\{\left(\frac{1}{2}, 2\right), (-1, -1)\right\}$ **33.** $\left\{\left(\frac{5}{2}, -\frac{7}{2}\right), (0, -1)\right\}$ **34.** $\{(2, -3), (-2, -3), (3, 2), (-3, 2)\}$ **35.** $\{(3, 1), (3, -1), (-3, 1), (-3, -1)\}$ **36.** 8 m and 5 m **37.** (1, 6), (3, 2) **38.** $x = 46$ and $y = 28$ or $x = 50$ and $y = 20$

39.
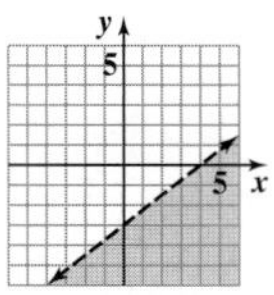
$3x - 4y > 12$

40.
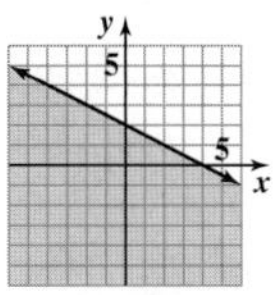
$y \le -\frac{1}{2}x + 2$

41.
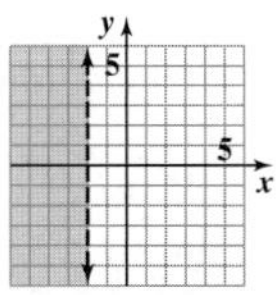
$x < -2$

42.
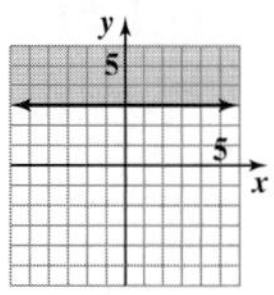
$y \ge 3$

43.
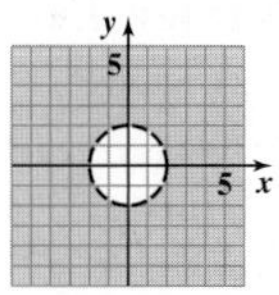
$x^2 + y^2 > 4$

44.
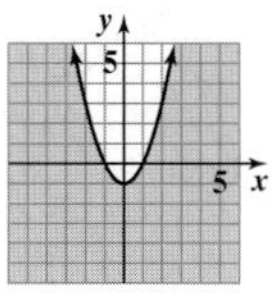
$y \le x^2 - 1$

45.
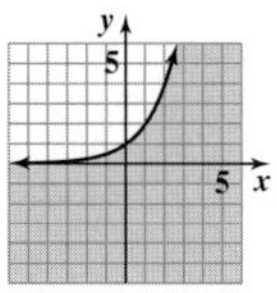
$y \le 2^x$

46.
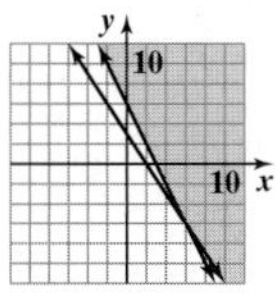
$3x + 2y \ge 6$
$2x + y \ge 6$

47.
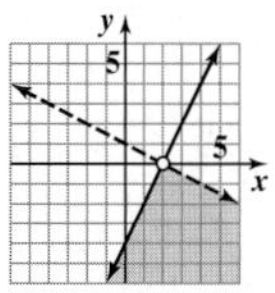
$2x - y \ge 4$
$x + 2y < 2$

48.
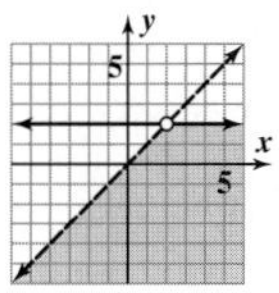
$y < x$
$y \le 2$

49.
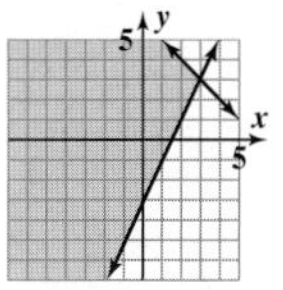
$x + y \le 6$
$y \ge 2x - 3$

50.
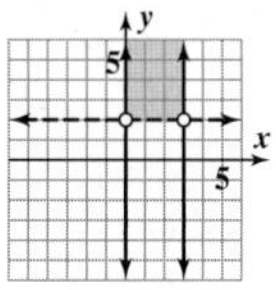
$0 \le x \le 3$
$y > 2$

51. no solution

52.
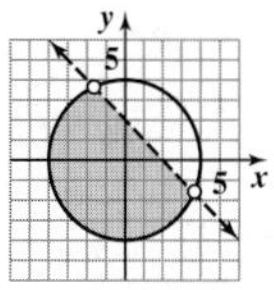
$x^2 + y^2 \le 16$
$x + y < 2$

53.
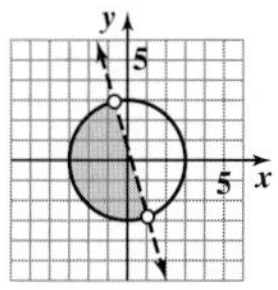
$x^2 + y^2 \le 9$
$y < -3x + 1$

54.
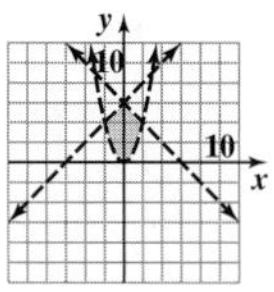
$y > x^2$
$x + y < 6$
$y < x + 6$

55.
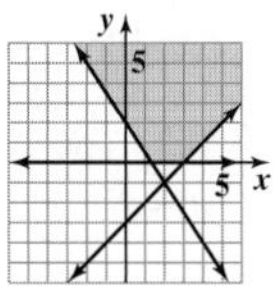
$y \ge 0$
$3x + 2y \ge 4$
$x - y \le 3$

56. (2, 2): 10; (4, 0): 8; $\left(\frac{1}{2}, \frac{1}{2}\right)$: $\frac{5}{2}$; (1, 0): 2; maximum value: 10; minimum value: 2

57.
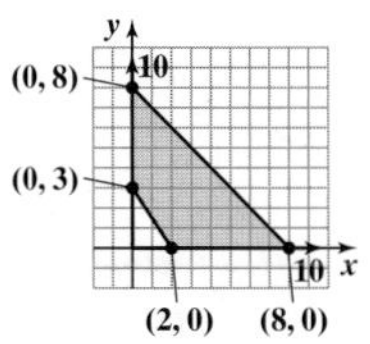

$x \ge 0,\ y \ge 0$
$x + y \le 8$
$3x + 2y \ge 6$

Maximum is 24 at $x = 0$, $y = 8$.

58.
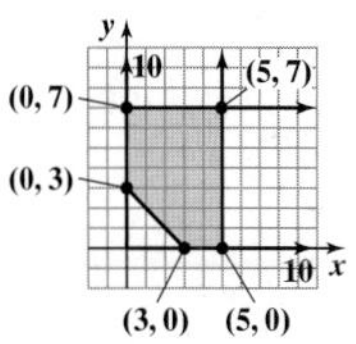

$0 \le x \le 5$
$0 \le y \le 7$
$x + y \ge 3$

Maximum is 33 at $x = 5$, $y = 7$.

59.
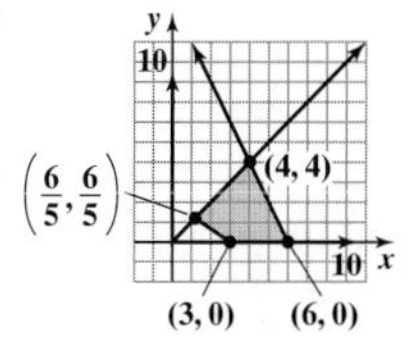

$x \ge 0, y \ge 0$
$y \le x$
$2x + y \le 12$
$2x + 3y \ge 6$

Maximum is 44 at $x = y = 4$.

60. a. $z = 500x + 350y$
b. $x + y \le 200$; $x \ge 10$; $y \ge 80$
c.
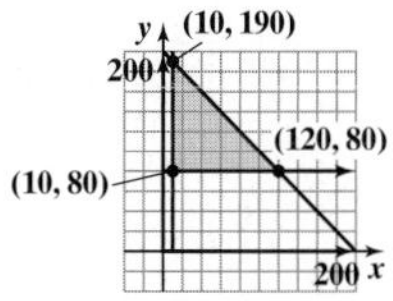

$x + y \le 200$
$x \ge 10, y \ge 80$

d. (10, 80): 33,000; (10, 190): 71,500; (120, 80): 88,000
e. 120; 80; 88,000

61. 480 of model A and 240 of model B

Chapter 5 Test

1. $\{(1, -3)\}$ **2.** $\{(4, -2)\}$ **3.** $\{(1, 3, 2)\}$ **4.** $\{(4, -3), (-3, 4)\}$ **5.** $\{(3, 2), (3, -2), (-3, 2), (-3, -2)\}$ **6.** $\frac{-1}{10(x + 1)} + \frac{x + 9}{10(x^2 + 9)}$

7.
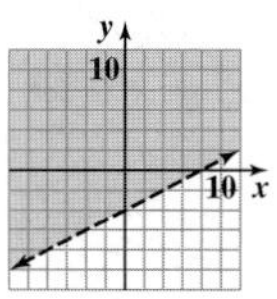
$x - 2y < 8$

8.
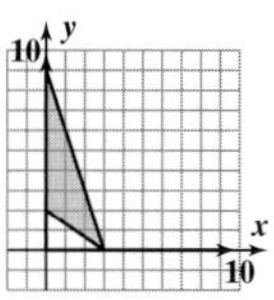
$x \ge 0,\ y \ge 0$
$3x + y \le 9$
$2x + 3y \ge 6$

9.
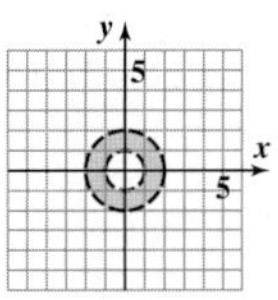
$x^2 + y^2 > 1$
$x^2 + y^2 < 4$

10.
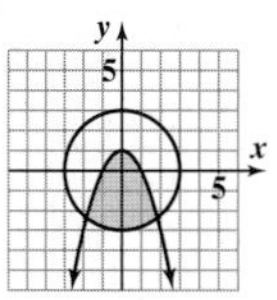
$y \le 1 - x^2$
$x^2 + y^2 \le 9$

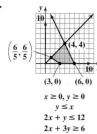

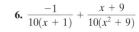

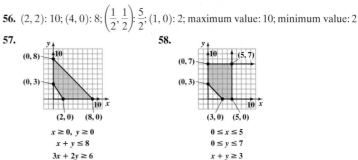

11. 26 **12.** Shrimp: 42 mg; scallops: 15mg **13. a.** $C(x) = 360{,}000 + 850x$ **b.** $R(x) = 1150x$ **c.** (1200, 1,380,000); The company will break even if it produces and sells 1200 computers. **14.** $y = x^2 - 3$ **15.** $x = 7.5$ ft and $y = 24$ ft or $x = 12$ ft and $y = 15$ ft
16. 50 regular and 100 deluxe jet skis; $35,000

Cumulative Review Exercises (Chapters 1–5)

1. Domain: $(-2, 2)$; Range: $(-\infty, 3]$ **2.** -1 and 1, each of multiplicity 1 **3.** maximum of 3 at $x = 0$ **4.** $(0, 2)$
5. positive **6.** 3 **7.** $x \to -2^+$; $x \to 2^-$ **8.** even

9.

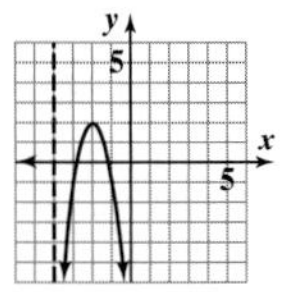

$g(x) = f(x + 2) - 1$

10.

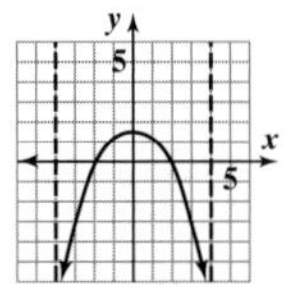

$h(x) = \frac{1}{2} f(\frac{1}{2} x)$

11. $\{3, 4\}$ **12.** $\left\{\frac{2 + i\sqrt{3}}{2}, \frac{2 - i\sqrt{3}}{2}\right\}$
13. $\{(-18, 6)\}$ **14.** $\{(1, 7)\}$
15. $\left\{-3, \frac{1}{2}, 2\right\}$ **16.** $\{-2\}$
17. $\{2\}$ **18.** $\{-2 + \log_3 11\}$
19. $\{625\}$ **20.** $\left\{\left(-\frac{1}{2}, \frac{1}{2}\right), (2, 8)\right\}$ **21.** $\{(8, -2, -2)\}$

22.

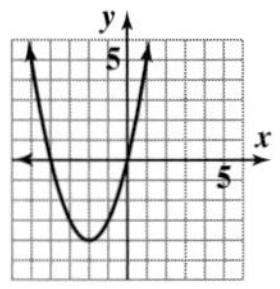

$f(x) = (x + 2)^2 - 4$

23.

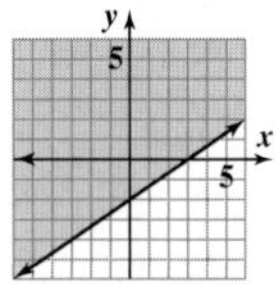

$2x - 3y \leq 6$

24.

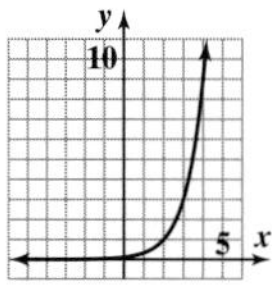

$y = 3^{x-2}$

25.

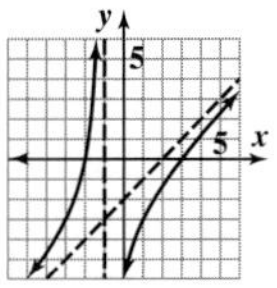

$f(x) = \frac{x^2 - x - 6}{x + 1}$

26.

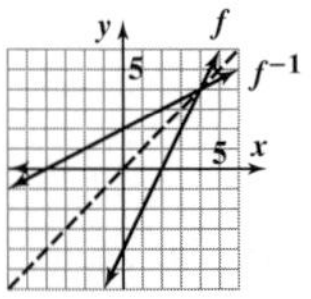

$f(x) = 2x - 4$

$f^{-1}(x) = \frac{x + 4}{2}$

27.

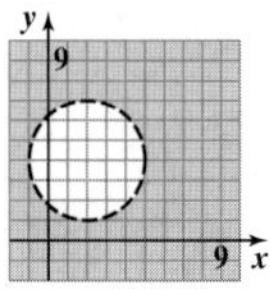

$(x - 2)^2 + (y - 4)^2 > 9$

28.

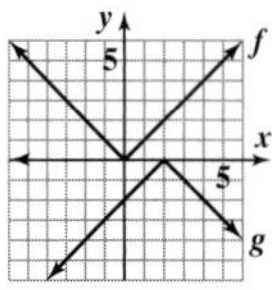

$f(x) = |x|$

$g(x) = -|x - 2|$

29. $(f \circ g)(x) = 2x^2 - 3x$;
$(g \circ f)(x) = -2x^2 + x + 2$
30. $4x + 2h - 1$
31. $y = -3x + 10$
32. $y = 3x + 3$
33. \$2600 at 12%; \$1400 at 14%
34. 4 m by 9 m
35. 10.99%

Subject Index

H

I

J

L

Photo Credits

Table of Contents Martin Barraud/Getty Images, © Chip Simons/Taxi/Getty Images, Corbis Royalty Free, Stockbyte, RubberBall/SuperStock, Inc., Bill Hatcher/Getty-National Geographic Society

About the Author Robert F. Blitzer

CHAPTER P Martin Barraud/Getty Images **p2** Tim Davis/Lynn Image **p17** SuperStock Inc. **p31** William Sallaz/Duomo Photography Incorporated **p35** Photofest **p45** Corbis Royalty Free **p57** Gabe Palmer/Kane Studio **p68** Peter Arnold, Inc.

CHAPTER 1 © Chip Simons/Taxi/Getty Images **p84** Globe Photos, Inc. **p94** DreamWorks/Apatow Prod./Frank Masi/Picture Desk, Inc./Kobal Collection **p108** Michael Agliolo/Grant Heilman Photography, Inc. **p111** © 318/Gallo Images/CORBIS **p127** Stamp from the private collection of Professor C. M. Lang, photography by Gary J. Shulfer, University of Wisconsin, Stevens Point. "Germany: #5"; Scott Standard Postage Stamp Catalogue, Scott Pub. Co., Sidney, Ohio. **p129** Michael Keller/Corbis/Bettmann **p138** Corbis/Bettmann **p150** © Lester Lefkowitz/CORBIS **p164** Ron Kimball Photography

CHAPTER 2 Corbis Royalty Free **p186** Hans Neleman/Neleman Inc. **p202** Howard Grey/Getty Images Inc.—Stone Allstock **p216** David Schmidt/Masterfile Corporation **p217** Carol Simowitz/San Francisco Convention and Visitors Bureau **p231** (a) Spantrans/AGE Fotostock America, Inc. (b) LWA-Stephen/Corbis/Bettmann **p241** Douglas Kirkland/Corbis/Sygma **p246** SuperStock, Inc. **p256** Howard Koby Photography **p269** Fred Prouser/Reuters/Landov LLC **p280** Skip Moody/Dembinsky Photo Associates

CHAPTER 3 Stockbyte **p298** Getty Images, Inc—Liaison **p299** (Left) Simon Bruty/Getty Images, Inc.—Allsport Photography (Right) Copyright © 1981 Scott Kim, scottkim.com. All rights reserved. **p312** National Institute for Biological Standards and Control (U.K.)/Science Photo Library/Photo Researchers, Inc. **p326** Treat Davidson/Photo Researchers, Inc. **p337** From Hieraymi Candi, "Ni Mediolanensis Medio (Basel, 1554)," Cardano portrait, frontispiece. Smithsonian Institution Libraries. © 2003 Smithsonian Institution. **p345** Courtesy of the Library of Congress **p351** AP Wide World Photos **p370** Mark Gamba/Corbis/Bettmann **p381** Stephen Simpson/Getty Images, Inc.—Taxi **p382** Gerald French/Corbis/Bettmann **p387** David Madison/Duomo Photography Incorporated **p388** Duomo Photography Incorporated **p389** UPI/Corbis/Bettmann

CHAPTER 4 RubberBall/SuperStock, Inc. **p400** AP Wide World Photos **p412** David Weintraub/Photo Researchers, Inc. **p424** Paul Barton/Corbis/Bettmann **p434** George Shelley/Masterfile Corporation **p446** Mark Leibowitz/Masterfile Corporation **p450** AP Wide World Photos

CHAPTER 5 Bill Hatcher/Getty-National Geographic Society **p468** IPOL/Globe Photos, Inc. **p483** David W. Hamilton/Getty Images Inc.—Image Bank **p491** © LOOK GmbH/eStock Photo **p500** Christian Darkin/SPL/Photo Researchers, Inc. **p509** Simon Burty/Getty Images Inc.—Stone Allstock **p521** AP Wide World Photos **p522** Pascal Parrot/Corbis/Sygma